Student Solutions Manual

Elementary and Intermediate Algebra

FIFTH EDITION

Alan Tussy
Citrus College

R. David Gustafson
Rock Valley College

Prepared by

Alexander Lee
Hinds Community College

Kristy Hill
Hinds Community College

BROOKS/COLE
CENGAGE Learning

Australia • Brazil • Japan • Korea • Mexico • Singapore • Spain • United Kingdom • United States

BROOKS/COLE
CENGAGE Learning·

For product information and technology assistance, contact us at **Cengage Learning Customer & Sales Support, 1-800-354-9706**

For permission to use material from this text or product, submit all requests online at **www.cengage.com/permissions** Further permissions questions can be emailed to **permissionrequest@cengage.com**

ISBN-13: 978-1-111-57847-3
ISBN-10: 1-111-57847-8

Brooks/Cole
20 Davis Drive
Belmont, CA 94002-3098
USA

Cengage Learning is a leading provider of customized learning solutions with office locations around the globe, including Singapore, the United Kingdom, Australia, Mexico, Brazil, and Japan. Locate your local office at: **www.cengage.com/global**

Cengage Learning products are represented in Canada by Nelson Education, Ltd.

To learn more about Brooks/Cole, visit **www.cengage.com/brookscole**

Purchase any of our products at your local college store or at our preferred online store **www.cengagebrain.com**

Printed in the United States of America
1 2 3 4 5 6 7 15 14 13 12 11

TABLE OF CONTENTS

SECTION 1.1
VOCABULARY
Fill in the blanks.

1. A **sum** is the result of an addition.
   A **difference** is the result of a subtraction.
   A **product** is the result of a multiplication.
   A **quotient** is the result of a division.

3. A number, such as 8, is called a **constant** because it does not change.

5. An **equation** is a mathematical sentence that contains an = symbol. An algebraic **expression** does not.

7. The **horizontal** axis of a graph extends left and right and the vertical axis extends up and down.

CONCEPTS
Classify each item as an algebraic expression or an equation.

9. a. **Equation**
    b. **Algebraic expression**

11. a. **Algebraic expression**
    b. **Equation**

13. **Addition, multiplication, division**
    t

15. **Construct a line graph using the data in the table.**

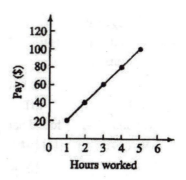

NOTATION
Fill in the blanks.

17. The symbol ≠ means **is not equal to**.

19. Write the multiplication 5 × 6 using a raised dot and then using parentheses. **5·6 , 5(6)**

Write each expression without using a multiplication symbol or parentheses.

21. **4x**   23. **2w**

Write each division using a fraction bar.

25. $\dfrac{32}{x}$ 27. $\dfrac{55}{5}$

GUIDED PRACTICE
29. ACCOUNTING

   15-year-old machinery is worth $35,000.

31. BUSINESS

   30 customers, $250

Express each statement using one of the words sum, product, difference, or quotient.

33. The product of 8 and 2 equals 16.

35. The difference of 11 and 9 equals 2.

37. The sum of *x* and 2 equals 10.

39. The quotient of 66 and 11 equals 6.

Translate each verbal model into an equation. (Answers may vary, depending on the variables chosen.)

41. $p = 100 - d$

43. $7d = h$

45. $s = 3c$

47. $w = e + 1,200$

49. $p = r - 600$

51. $\dfrac{l}{4} = m$

Use the formula to complete each table.

53. $d = 360 + L$

Lunch Time (minutes) L	School Day (minutes) d
30	$d = 360 + L$ $d = 360 + 30$ **$d = 390$**
40	$d = 360 + L$ $d = 360 + 40$ **$d = 400$**
45	$d = 360 + L$ $d = 360 + 45$ **$d = 405$**

55. $t = 1,500 - d$

Deductions d	Take-home pay t
200	$t = 1,500 - d$ $t = 1,500 - 200$ $t = \mathbf{1,300}$
300	$t = 1,500 - d$ $t = 1,500 - 300$ $t = \mathbf{1,200}$
400	$t = 1,500 - d$ $t = 1,500 - 400$ $t = \mathbf{1,100}$

57. $d = \dfrac{e}{\boxed{12}}$

Eggs e	Dozens d
24	2
36	3
48	4

59. $I = \boxed{2}\, c$

Couples c	Individuals I
20	40
100	200
200	400

APPLICATIONS

61. EXERCISE

a. The number of calories burned is the product of 3 and the number of minutes cleaning.

b. $c = 3m$

c.

m	10	20	30	40	50	60
c	**30**	**60**	**90**	**120**	**150**	**180**

d.

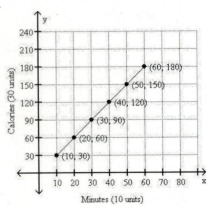

WRITING

63-65. Answers will vary.

CHALLENGE PROBLEMS

67. Complete formula.

$$t = \boxed{3}\, s + \boxed{1}$$

s	t
18	**55**
33	100
47	142

Fill in the blanks.

1. A factor is a number being **multiplied**.

3. When we write 60 as $2 \cdot 2 \cdot 3 \cdot 5$, we say that we have written 60 in **prime-factored** form.

5. Two fractions that represent the same number, such as $\frac{1}{2}$ and $\frac{2}{4}$, are called **equivalent** fractions.

7. The **least or lowest** common denominator for a set of fractions is the smallest number each denominator will divide exactly.

CONCEPTS

9. a. **1** b. **a**

    c. $\dfrac{a \cdot c}{b \cdot d}$ d. $\dfrac{a \cdot d}{b \cdot c}$

    e. $\dfrac{a + b}{d}$ f. $\dfrac{a - b}{d}$

11. Complete each statement.

    a. To simplify a fraction, we remove factors equal to **1** in the form of $\frac{2}{2}$, $\frac{3}{3}$, or $\frac{4}{4}$, and so on.

    b. To build a fraction, we multiply it by **1** in the form of $\frac{2}{2}$, $\frac{3}{3}$, or $\frac{4}{4}$, and so on.

NOTATION

13. a. $\dfrac{5}{6} \cdot \dfrac{\mathbf{5}}{\mathbf{5}} = \dfrac{\mathbf{25}}{\mathbf{30}}$

    b. $\dfrac{12}{42} = \dfrac{\cancel{2} \cdot 2 \cdot \cancel{3}}{\cancel{2} \cdot \cancel{3} \cdot 7}$

    $= \dfrac{\mathbf{2}}{\mathbf{7}}$

GUIDED PRACTICE

Find the prime factorization of each number.

15. $75 = 3 \cdot 5 \cdot 5$

17. $28 = 2 \cdot 2 \cdot 7$

19. $81 = 3 \cdot 3 \cdot 3 \cdot 3$

21. $117 = 3 \cdot 3 \cdot 13$

23. $220 = 2 \cdot 2 \cdot 5 \cdot 11$

25. $1{,}254 = 2 \cdot 3 \cdot 11 \cdot 19$

Perform each operation.

27. $\dfrac{5}{6} \cdot \dfrac{1}{8} = \dfrac{5}{48}$

29. $\dfrac{7}{11} \cdot \dfrac{3}{5} = \dfrac{21}{55}$

31. $\dfrac{3}{4} \div \dfrac{2}{5} = \dfrac{3}{4} \cdot \dfrac{5}{2}$

    $= \dfrac{15}{8}$

33. $\dfrac{6}{5} \div \dfrac{5}{7} = \dfrac{6}{5} \cdot \dfrac{7}{5}$

    $= \dfrac{42}{25}$

Build each fraction or whole number to an equivalent fraction with the indicated denominator.

35. $\dfrac{1}{3} \cdot \dfrac{\mathbf{3}}{\mathbf{3}} = \dfrac{3}{9}$

37. $\dfrac{4}{9} \cdot \dfrac{\mathbf{6}}{\mathbf{6}} = \dfrac{24}{54}$

39. $\dfrac{7}{1} \cdot \dfrac{\mathbf{5}}{\mathbf{5}} = \dfrac{35}{5}$

41. $\dfrac{5}{1} \cdot \dfrac{\mathbf{7}}{\mathbf{7}} = \dfrac{35}{7}$

Simplify each fraction, if possible.

43. $\dfrac{6}{18} = \dfrac{\cancel{2} \cdot \cancel{3}}{\cancel{2} \cdot \cancel{3} \cdot 3}$

    $= \dfrac{1}{3}$

45. $\dfrac{24}{28} = \dfrac{\cancel{2} \cdot \cancel{2} \cdot 6}{\cancel{2} \cdot \cancel{2} \cdot 7}$

    $= \dfrac{6}{7}$

47. $\dfrac{15}{40} = \dfrac{3 \cdot \cancel{5}}{2 \cdot 2 \cdot 2 \cdot \cancel{5}}$

    $= \dfrac{3}{8}$

49. $\dfrac{33}{56} = \dfrac{3 \cdot 11}{2 \cdot 2 \cdot 2 \cdot 7}$

    Simplest form

51. $\dfrac{26}{39} = \dfrac{2 \cdot \cancel{13}}{3 \cdot \cancel{13}}$

    $= \dfrac{2}{3}$

53. $\dfrac{36}{225} = \dfrac{2 \cdot 2 \cdot \cancel{3} \cdot \cancel{3}}{5 \cdot 5 \cdot \cancel{3} \cdot \cancel{3}}$

    $= \dfrac{4}{25}$

Perform the operations and, if possible, simplify.

55. $\dfrac{3}{5} + \dfrac{3}{5} = \dfrac{3 + 3}{5}$

    $= \dfrac{6}{5}$

57. $\dfrac{6}{7} - \dfrac{2}{7} = \dfrac{6 - 2}{7}$

    $= \dfrac{4}{7}$

59.
$$\frac{1}{6}+\frac{1}{24}=\frac{1}{6}\cdot\frac{\mathbf{4}}{\mathbf{4}}+\frac{1}{24}$$
$$=\frac{4+1}{24}$$
$$=\frac{5}{24}$$

61.
$$\frac{7}{10}-\frac{1}{14}=\frac{7}{10}\cdot\frac{\mathbf{7}}{\mathbf{7}}-\frac{1}{14}\cdot\frac{\mathbf{5}}{\mathbf{5}}$$
$$=\frac{49-5}{70}$$
$$=\frac{44}{70}$$
$$=\frac{\overset{1}{\cancel{2}}\cdot22}{\underset{1}{\cancel{2}}\cdot35}$$
$$=\frac{22}{35}$$

63.
$$\frac{2}{15}+\frac{7}{9}=\frac{2}{15}\cdot\frac{\mathbf{3}}{\mathbf{3}}+\frac{7}{9}\cdot\frac{\mathbf{5}}{\mathbf{5}}$$
$$=\frac{6+35}{45}$$
$$=\frac{41}{45}$$

65.
$$\frac{13}{28}-\frac{1}{21}=\frac{13}{28}\cdot\frac{\mathbf{3}}{\mathbf{3}}-\frac{1}{21}\cdot\frac{\mathbf{4}}{\mathbf{4}}$$
$$=\frac{39-4}{84}$$
$$=\frac{35}{84}$$
$$=\frac{5\cdot\overset{1}{\cancel{7}}}{\underset{1}{\cancel{7}}\cdot12}$$
$$=\frac{5}{12}$$

Perform the operations and, if possible, simplify.

67.
$$16\left(\frac{3}{2}\right)=\frac{16}{1}\cdot\frac{3}{2}$$
$$=\frac{\overset{1}{\cancel{2}}\cdot8\cdot3}{1\cdot\underset{1}{\cancel{2}}}$$
$$=24$$

69.
$$18\left(\frac{2}{9}\right)=\frac{18}{1}\cdot\frac{2}{9}$$
$$=\frac{\overset{1}{\cancel{9}}\cdot2\cdot2}{1\cdot\underset{1}{\cancel{9}}}$$
$$=4$$

71.
$$\frac{2}{3}+\frac{5}{18}-\frac{1}{6}=\frac{2}{3}\cdot\frac{\mathbf{6}}{\mathbf{6}}+\frac{5}{18}-\frac{1}{6}\cdot\frac{\mathbf{3}}{\mathbf{3}}$$
$$=\frac{12+5-3}{18}$$
$$=\frac{14}{18}$$
$$=\frac{\overset{1}{\cancel{2}}\cdot7}{\underset{1}{\cancel{2}}\cdot9}$$
$$=\frac{7}{9}$$

73.
$$\frac{5}{12}+\frac{1}{3}-\frac{2}{5}=\frac{5}{12}\cdot\frac{\mathbf{5}}{\mathbf{5}}+\frac{1}{3}\cdot\frac{\mathbf{20}}{\mathbf{20}}-\frac{2}{5}\cdot\frac{\mathbf{12}}{\mathbf{12}}$$
$$=\frac{25+20-24}{60}$$
$$=\frac{21}{60}$$
$$=\frac{\overset{1}{\cancel{3}}\cdot7}{\underset{1}{\cancel{3}}\cdot20}$$
$$=\frac{7}{20}$$

Perform the operations and, if possible, simplify.

75.
$$4\frac{2}{3}\cdot7=\frac{14}{3}\cdot\frac{7}{1}$$
$$=\frac{14\cdot7}{3\cdot1}$$
$$=\frac{98}{3}$$
$$=32\frac{2}{3}$$

77.
$$8\div3\frac{1}{5}=\frac{8}{1}\div\frac{16}{5}$$
$$=\frac{8}{1}\cdot\frac{5}{16}$$
$$=\frac{5\cdot\overset{1}{\cancel{8}}}{1\cdot2\cdot\underset{1}{\cancel{8}}}$$
$$=\frac{5}{2}$$
$$=2\frac{1}{2}$$

79. $8\dfrac{2}{9} - 7\dfrac{2}{3} = \dfrac{74}{9} - \dfrac{23}{3}$

$\qquad = \dfrac{74}{9} - \dfrac{23}{3} \cdot \dfrac{3}{3}$

$\qquad = \dfrac{74-69}{9}$

$\qquad = \dfrac{5}{9}$

81. $3\dfrac{3}{16} + 2\dfrac{5}{24} = \dfrac{51}{16} + \dfrac{53}{24}$

$\qquad = \dfrac{51}{16} \cdot \dfrac{3}{3} + \dfrac{53}{24} \cdot \dfrac{2}{2}$

$\qquad = \dfrac{153+106}{48}$

$\qquad = \dfrac{259}{48}$

$\qquad = 5\dfrac{19}{48}$

TRY IT YOURSELF

Perform the operations and, if possible, simplify.

83. $\dfrac{3}{5} + \dfrac{2}{3} = \dfrac{3}{5} \cdot \dfrac{3}{3} + \dfrac{2}{3} \cdot \dfrac{5}{5}$

$\qquad = \dfrac{9+10}{15}$

$\qquad = \dfrac{19}{15}$

$\qquad = 1\dfrac{4}{15}$

85. $21\left(\dfrac{10}{3}\right) = \dfrac{21}{1} \cdot \dfrac{10}{3}$

$\qquad = \dfrac{\cancel{3}^{\,} \cdot 7 \cdot 10}{1 \cdot \cancel{3}_{\,1}}$

$\qquad = 70$

87. $6 \cdot 2\dfrac{7}{24} = \dfrac{6}{1} \cdot \dfrac{55}{24}$

$\qquad = \dfrac{\cancel{6}^{\,1} \cdot 55}{1 \cdot \cancel{6} \cdot 4}$

$\qquad = \dfrac{55}{4}$

$\qquad = 13\dfrac{3}{4}$

89. $\dfrac{2}{3} - \dfrac{1}{4} + \dfrac{1}{12} = \dfrac{2}{3} \cdot \dfrac{4}{4} - \dfrac{1}{4} \cdot \dfrac{3}{3} + \dfrac{1}{12}$

$\qquad = \dfrac{8-3+1}{12}$

$\qquad = \dfrac{\cancel{6}^{\,1}}{2 \cdot \cancel{6}_{\,1}}$

$\qquad = \dfrac{1}{2}$

91. $\dfrac{21}{35} \div \dfrac{3}{14} = \dfrac{21}{35} \cdot \dfrac{14}{3}$

$\qquad = \dfrac{\cancel{7}^{\,1} \cdot \cancel{3}^{\,1} \cdot 14}{5 \cdot \cancel{7}_{\,1} \cdot \cancel{3}_{\,1}}$

$\qquad = \dfrac{14}{5}$

$\qquad = 2\dfrac{4}{5}$

93. $\dfrac{4}{3}\left(\dfrac{6}{5}\right) = \dfrac{4 \cdot 6}{3 \cdot 5}$

$\qquad = \dfrac{4 \cdot 2 \cdot \cancel{3}^{\,1}}{\cancel{3}_{\,1} \cdot 5}$

$\qquad = \dfrac{8}{5}$

$\qquad = 1\dfrac{3}{5}$

95. $\dfrac{4}{63} + \dfrac{1}{45} = \dfrac{4}{63} \cdot \dfrac{5}{5} + \dfrac{1}{45} \cdot \dfrac{7}{7}$

$\qquad = \dfrac{20+7}{315}$

$\qquad = \dfrac{27}{315}$

$\qquad = \dfrac{3 \cdot \cancel{9}^{\,1}}{\cancel{9}_{\,1} \cdot 35}$

$\qquad = \dfrac{3}{35}$

Section 1.2

97. $3 - \dfrac{3}{4} = \dfrac{3}{1} \cdot \dfrac{\mathbf{4}}{\mathbf{4}} - \dfrac{3}{4}$

$= \dfrac{12 - 3}{4}$

$= \dfrac{9}{4}$

$= 2\dfrac{1}{4}$

99. $\dfrac{1}{5} \cdot \dfrac{3}{5} = \dfrac{1 \cdot 3}{5 \cdot 5}$

$= \dfrac{3}{25}$

101. $3\dfrac{1}{3} \div 1\dfrac{5}{6} = \dfrac{10}{3} \div \dfrac{11}{6}$

$= \dfrac{10}{3} \cdot \dfrac{6}{11}$

$= \dfrac{10 \cdot 2 \cdot \cancel{3}^{1}}{\cancel{3}_{1} \cdot 11}$

$= \dfrac{20}{11}$

$= 1\dfrac{9}{11}$

103. $\dfrac{11}{21} - \dfrac{8}{21} = \dfrac{11 - 8}{21}$

$= \dfrac{\cancel{3}^{1}}{7 \cdot \cancel{3}_{1}}$

$= \dfrac{1}{7}$

105. $\dfrac{7}{30} + \dfrac{1}{50} - \dfrac{19}{75} = \dfrac{7}{30} \cdot \dfrac{\mathbf{5}}{\mathbf{5}} + \dfrac{1}{50} \cdot \dfrac{\mathbf{3}}{\mathbf{3}} - \dfrac{19}{75} \cdot \dfrac{\mathbf{2}}{\mathbf{2}}$

$= \dfrac{35 + 3 - 38}{150}$

$= \dfrac{38 - 38}{150}$

$= \dfrac{0}{150}$

$= 0$

107. $1\dfrac{31}{32} \cdot 7\dfrac{1}{9} = \dfrac{63}{32} \cdot \dfrac{64}{9}$

$= \dfrac{7 \cdot \cancel{9}^{1}}{\cancel{32}_{1}} \cdot \dfrac{2 \cdot \cancel{32}^{1}}{\cancel{9}_{1}}$

$= \dfrac{14}{1}$

$= 14$

LOOK ALIKES

109. a. $\dfrac{4}{9} + \dfrac{3}{7} = \dfrac{4}{9} \cdot \dfrac{\mathbf{7}}{\mathbf{7}} + \dfrac{3}{7} \cdot \dfrac{\mathbf{9}}{\mathbf{9}}$

$= \dfrac{28 + 27}{63}$

$= \dfrac{55}{63}$

b. $\dfrac{4}{9} - \dfrac{3}{7} = \dfrac{4}{9} \cdot \dfrac{\mathbf{7}}{\mathbf{7}} - \dfrac{3}{7} \cdot \dfrac{\mathbf{9}}{\mathbf{9}}$

$= \dfrac{28 - 27}{63}$

$= \dfrac{1}{63}$

c. $\dfrac{4}{9} \cdot \dfrac{3}{7} = \dfrac{4 \cdot \cancel{3}^{1}}{\cancel{3}_{1} \cdot 3 \cdot 7}$

$= \dfrac{4}{27}$

d. $\dfrac{4}{9} \div \dfrac{3}{7} = \dfrac{4}{9} \cdot \dfrac{7}{3}$

$= \dfrac{28}{27}$

APPLICATION

111. FORESTRY

a. $\dfrac{5}{32} + \dfrac{1}{16} = \dfrac{5}{32} + \dfrac{1}{16} \cdot \dfrac{\mathbf{2}}{\mathbf{2}}$

$= \dfrac{5 + 2}{32}$

$= \dfrac{7}{32}$

The growth was $\dfrac{7}{32}$ in.

b. $\dfrac{5}{32} - \dfrac{1}{16} = \dfrac{5}{32} - \dfrac{1}{16} \cdot \dfrac{\mathbf{2}}{\mathbf{2}}$

$= \dfrac{5 - 2}{32}$

$= \dfrac{3}{32}$

The difference is $\dfrac{3}{32}$ in.

113. CAMPUS TO CAREERS

$$22\frac{3}{4} + 11\frac{1}{2} + 28\frac{7}{8} = \frac{91}{4} + \frac{23}{2} + \frac{231}{8}$$

$$= \frac{91}{4} \cdot \frac{2}{2} + \frac{23}{2} \cdot \frac{4}{4} + \frac{231}{8}$$

$$= \frac{182 + 92 + 231}{8}$$

$$= \frac{505}{8}$$

$$= 63\frac{1}{8}$$

The total dimensions is $63\frac{1}{8}$ inches.

115. FRAMES

$$4 \cdot 10\frac{1}{8} = \frac{4}{1} \cdot \frac{81}{8}$$

$$= \frac{\overset{1}{\cancel{4}} \cdot 81}{1 \cdot 2 \cdot \underset{1}{\cancel{4}}}$$

$$= \frac{81}{2}$$

$$= 40\frac{1}{2}$$

$40\frac{1}{2}$ inches of molding will be needed.

WRITING

117-119. Answers will vary.

REVIEW

121. **Variables** are letters (or symbols) that stand for numbers.

CHALLENGE PROBLEMS

123. $\quad \dfrac{11}{12} \cdot \dfrac{9}{9} = \dfrac{99}{108} \qquad \dfrac{8}{9} \cdot \dfrac{12}{12} = \dfrac{96}{108}$

$\dfrac{11}{12}$ is larger than $\dfrac{8}{9}$.

SECTION 1.3
VOCABULARY

Fill in the blanks.

1. The set of **whole** numbers is
   {0, 1, 2, 3, 4, 5, …}.

3. The figure is called a
   number line .

5. Positive and negative numbers are called
   signed numbers.

7. The symbols < and > are **inequality** symbols.

9. 0.25 is called a **terminating** decimal
   and 0.333 … is called a **repeating**
   decimal.

11. An irrational number is a nonterminating,
    nonrepeating **decimal** .

CONCEPTS

13. Represent each situation using a signed
    number.

   a. **–$15 million**   b. $\frac{5}{16}$ **in. or** $+\frac{5}{16}$ **in.**

15. a. **–20**   b. $\frac{2}{3}$

17. **–14** and **–4**

NOTATION
Fill in the blanks.

19. $\sqrt{2}$ is read "the **square root** of 2."

21. The symbol $\approx$ means **is approximately
    equal to** .

23. The symbol π is a letter from the **Greek**
    alphabet.

25. $-\dfrac{4}{5} = \dfrac{\boxed{-4}}{5} = \dfrac{4}{\boxed{-5}}$.

GUIDED PRACTICE
27.

	5	0	-3	$\frac{7}{8}$	0.17	$-9\frac{1}{4}$	$\sqrt{2}$	π
Real	✔	✔	✔	✔	✔	✔	✔	✔
Irrational							✔	✔
Rational	✔	✔	✔	✔	✔	✔		
Integer	✔	✔	✔					
Whole	✔	✔						
Natural	✔							

Determine whether each statement is true or false.

29. **True**

31. **False**

33. **True**

35. **True**

**Use one of the symbols < or > to make each
statement true.**

37. $0 > -4$

39. $917 < 971$

41. $-2 > -3$

43. $-\frac{5}{8} < -\frac{3}{8}$

45. $\frac{2}{3} > \frac{3}{5}$

47. $-6.19 < -5.8$

**Write each fraction as a decimal. If the result is a
repeating decimal, use an over bar.**

49. $\dfrac{5}{8} = 8\overline{)5.000}^{\,0.625}$

   $= 0.625$

51. $\dfrac{1}{30} = 30\overline{)1.0000}^{\,0.0333}$

   $= 0.0\overline{3}$

53. $\dfrac{1}{60} = 60\overline{)1.0000}^{\,0.0166}$

   $= 0.01\overline{6}$

55. $\dfrac{21}{50} = 50\overline{)21.00}^{\,0.42}$

   $= 0.42$

Graph each set of numbers on a number line.

57.

59.

Find each absolute value.

61. $|83| = 83$

63. $\left|\dfrac{4}{3}\right| = \dfrac{4}{3}$

65. $|-11| = 11$

67. $|-6.1| = 6.1$

Insert one of the symbols <, >, or = in the blank to make each statement true.

69. $|3.4|\;\boxed{?}\;-3$

    $3.4\;\boxed{?}\;-3$

    $3.4 > -3$

Thus, $|3.4| > -3$

71. $|-1.1|\;\boxed{?}\;1.2$

    $1.1\;\boxed{?}\;1.2$

    $1.1 < 1.2$

Thus, $|-1.1| < 1.2$

73. $\left|-\dfrac{15}{2}\right|\;\boxed{?}\;7.5$

    $\dfrac{15}{2}\;\boxed{?}\;7.5$

    $7.5 = 7.5$

    Thus, $\left|-\dfrac{15}{2}\right| = 7.5$

75. $\dfrac{99}{100} = 0.99$

77. $0.3 < 0.333...$

79. $1\;\boxed{?}\;\left|-\dfrac{15}{16}\right|$

    $1\;\boxed{?}\;\dfrac{15}{16}$

    $1 > \dfrac{15}{16}$

    Thus, $1 > \left|-\dfrac{15}{16}\right|$

APPLICATIONS

81. DRAFTING

natural: 9   whole: 9

integers: 9   rational: $9, \dfrac{15}{16}, 3\dfrac{1}{8}, 1.765$

irrational: $2\pi, 3\pi, \sqrt{89}$ real: all

83. iPHONES

iii is the iPhone with the strongest strength of -49.

85. TRADE

  a. 2006 had a deficit of $-\$90$ billion.

  b. The smallest deficit was in 2009 and it was $-\$45$ billion.

WRITING

87-91. Answers will vary.

REVIEW

93. $\dfrac{24}{54} = \dfrac{2 \cdot 2 \cdot 2 \cdot 3}{2 \cdot 3 \cdot 3 \cdot 3}$

    $= \dfrac{\overset{1}{\cancel{2}} \cdot 2 \cdot 2 \cdot \overset{1}{\cancel{3}}}{\underset{1}{\cancel{2}} \cdot 3 \cdot 3 \cdot \underset{1}{\cancel{3}}}$

    $= \dfrac{4}{9}$

95. $5\dfrac{2}{3} \div 2\dfrac{5}{9} = \dfrac{17}{3} \div \dfrac{23}{9}$

    $= \dfrac{17}{3} \cdot \dfrac{9}{23}$

    $= \dfrac{17 \cdot \overset{1}{\cancel{3}} \cdot 3}{\underset{1}{\cancel{3}} \cdot 23}$

    $= \dfrac{51}{23}$

    $= 2\dfrac{5}{23}$

CHALLENGE PROBLEMS

97. 1,999 integers

99. $\dfrac{1}{8} \cdot \dfrac{9}{9}$ $\dfrac{1}{9} \cdot \dfrac{8}{8}$

  $\dfrac{9}{72}$ $\dfrac{8}{72}$

  $\dfrac{9}{72} \cdot \dfrac{2}{2}$ $\dfrac{8}{72} \cdot \dfrac{2}{2}$

  $\dfrac{18}{144}$ $\dfrac{16}{144}$

$\dfrac{17}{144}$ is between the 2 fractions.

SECTION 1.4
VOCABULARY

Fill in the blanks.

1. In the addition statement $-2 + 5 = 3$, the result, 3, is called the **sum**.

3. The **commutative** property of addition states that changing the order when adding does not affect the answer. The **associative** property of addition states that changing the grouping when adding does not affect the answer.

CONCEPTS

5. a. **6**   b. -9.2

7. a. **Negative**   b. **Positive**

9. a. $1+(-5)$ b. $-80.5+15$
   c. $20+4$ d. $3+2.1$

11. **Commutative property of addition**
    Associative property of addition

NOTATION

13. a. $x + y = y + x$
    b. $(x + y) + z = x + (y + z)$

GUIDED PRACTICE

Add.

15. $-8 + (-1) = -9$

17. $-5 + (-12) = -17$

19. $-29 + (-45) = -74$

21. $-4.2 + (-6.1) = -10.3$

23. $-\dfrac{3}{4} + \left(-\dfrac{2}{3}\right) = -\dfrac{3}{4} \cdot \dfrac{3}{3} + \left(-\dfrac{2}{3}\right)\left(\dfrac{4}{4}\right)$

$$= \dfrac{-9 + (-8)}{12}$$

$$= -\dfrac{17}{12}$$

25. $-\dfrac{1}{4} + \left(-\dfrac{1}{10}\right) = -\dfrac{1}{4} \cdot \dfrac{5}{5} + \left(-\dfrac{1}{10}\right)\left(\dfrac{2}{2}\right)$

$$= \dfrac{-5 + (-2)}{20}$$

$$= -\dfrac{7}{20}$$

Add.

27. $-7 + 4 = -3$ 29. $50 + (-11) = 39$

33. $-6.25 + 8.5 = 2.25$ 31. $15.84 + (-15.84) = 0$

35. $-\dfrac{7}{15} + \dfrac{3}{15} = \dfrac{-7 + 3}{15}$

$$= -\dfrac{4}{15}$$

37. $\dfrac{1}{2} + \left(-\dfrac{1}{8}\right) = \dfrac{1}{2} \cdot \dfrac{4}{4} + \left(-\dfrac{1}{8}\right)$

$$= \dfrac{4 + (-1)}{8}$$

$$= \dfrac{3}{8}$$

Add.

39. $8 + (-5) + 13 = 8 + 13 + (-5)$
$$= 21 + (-5)$$
$$= 16$$

41. $21 + (-27) + (-9) = 21 + (-36)$
$$= -15$$

43. $-27 + (-3) + (-13) + 22 = -30 + (-13) + 22$
$$= -43 + 22$$
$$= -21$$

45. $-60 + 70 + (-10) + (-10) + 205$
$$= -60 + (-10) + (-10) + 70 + 205$
$$= -80 + 275$$
$$= 195$$

Apply the associative property of addition to find the sum.

47. $-99 + (99 + 215) = (-99 + 99) + 215$
$$= 0 + 215$$
$$= 215$$

49. $(-112 + 56) + (-56) = -112 + [(56 + (-56)]$
$$= -112 + 0$$
$$= -112$$

51. $\dfrac{1}{8} + \left(\dfrac{7}{8} + \dfrac{2}{3}\right) = \left(\dfrac{1}{8} + \dfrac{7}{8}\right) + \dfrac{2}{3}$

$$= \dfrac{8}{8} + \dfrac{2}{3}$$

$$= 1 + \dfrac{2}{3}$$

$$= 1\dfrac{2}{3}$$

53. $(12.4+1.9)+1.1=12.4+(1.9+1.1)$
$=12.4+3$
$=15.4$

Add.

55. $-1+9+1=-1+1+9$
$=0+9$
$=9$

57. $-8+11+(-11)+8+1=-8+0+8+1$
$=-8+8+1$
$=0+1$
$=1$

TRY IT YOURSELF
Add.

59. $-9+81+(-2)=-9+(-2)+81$
$=-11+81$
$=70$

61. $0+(-6.6)=-6.6$

63. $-\dfrac{9}{16}+\dfrac{7}{16}=\dfrac{-9+7}{16}$
$=\dfrac{-2}{16}$
$=-\dfrac{\overset{1}{\cancel{2}}}{\underset{1}{\cancel{2}}\cdot 8}$
$=-\dfrac{1}{8}$

65. $-6+(-8)=-14$ 67. $-167+167=0$

69. $-20+(-16)+10=-36+10$
$=-26$

71. $19.35+(-20.21)+1.53$
$=19.35+1.53+(-20.21)$
$=20.88+(-20.21)$
$=0.67$

73. $-7+5+(-10)+7=-7+7+5+(-10)$
$=0+5+(-10)$
$=5+(-10)$
$=-5$

75. $19.2+(-41.3)=-22.1$

77. $2{,}345+(-178)=2{,}167$

79. $-2.1+6.5+(-8.2)+2.1$
$=-2.1+2.1+6.5+(-8.2)$
$=0+6.5+(-8.2)$
$=-1.7$

81. $3+(-6)+(-3)+74=3+(-3)+(-6)+74$
$=0+(-6)+74$
$=68$

83. $-\dfrac{1}{4}+\left(-\dfrac{2}{7}\right)=-\dfrac{1}{4}\cdot\dfrac{\mathbf{7}}{\mathbf{7}}+\left(-\dfrac{2}{7}\right)\left(\dfrac{\mathbf{4}}{\mathbf{4}}\right)$
$=\dfrac{-7+(-8)}{28}$
$=-\dfrac{15}{28}$

85. $-0.2+(-0.3)+(-0.4)=-0.5+(-0.4)$
$=-0.9$

LOOK ALIKES

87. a. $12+15=27$ b. $-12+15=3$
c. $-12+(-15)=-27$ d. $12+(-15)=-3$

89. a. $\dfrac{1}{2}+\dfrac{2}{9}=\dfrac{1}{2}\left(\dfrac{\mathbf{9}}{\mathbf{9}}\right)+\dfrac{2}{9}\left(\dfrac{\mathbf{2}}{\mathbf{2}}\right)$
$=\dfrac{9}{18}+\dfrac{4}{18}$
$=\dfrac{13}{18}$

   b. $-\dfrac{1}{2}+\dfrac{2}{9}=-\dfrac{1}{2}\left(\dfrac{\mathbf{9}}{\mathbf{9}}\right)+\dfrac{2}{9}\left(\dfrac{\mathbf{2}}{\mathbf{2}}\right)$
$=-\dfrac{9}{18}+\dfrac{4}{18}$
$=-\dfrac{5}{18}$

   c. $-\dfrac{1}{2}+\left(-\dfrac{2}{9}\right)=-\dfrac{1}{2}\left(\dfrac{\mathbf{9}}{\mathbf{9}}\right)+\left(-\dfrac{2}{9}\right)\left(\dfrac{\mathbf{2}}{\mathbf{2}}\right)$
$=-\dfrac{9}{18}+\left(-\dfrac{4}{18}\right)$
$=-\dfrac{13}{18}$

   d. $\dfrac{1}{2}+\left(-\dfrac{2}{9}\right)=\dfrac{1}{2}\left(\dfrac{\mathbf{9}}{\mathbf{9}}\right)+\left(-\dfrac{2}{9}\right)\left(\dfrac{\mathbf{2}}{\mathbf{2}}\right)$
$=\dfrac{9}{18}+\left(-\dfrac{4}{18}\right)$
$=\dfrac{5}{18}$

APPLICATIONS

91. MILITARY SCIENCE

$$-1,500 + 2,400 + 1,250 = -1,500 + 3,650$$
$$= 2,150$$

The net gain is 2,150 meters.

93. GOLF

$$+1 + (-2) + (-5) + (-2) = -1 + (-5) + (-2)$$
$$= -6 + (-2)$$
$$= -8$$

Y. E. Yang is 8 strokes under par.

$$-5 + (-2) + (-1) + (+3) = -7 + (-1) + (+3)$$
$$= -8 + (+3)$$
$$= -5$$

Tiger Woods is 5 strokes under par.

$$-2 + 0 + (+1) + (-2) = -1 + (-2)$$
$$= -3$$

Lee Westwood is 3 strokes under par.

$$-1 + (+1) + (-1) + (-2) = 0 + (-1) + (-2)$$
$$= -1 + (-2)$$
$$= -3$$

Rory McIlroy is 3 strokes under par.

95. CREDIT CARDS

a. Previous Balance: $3,660.66

New Purchases: $1,408.78

b. $-3,660.66 + (-1,408.78) + 3,826.58$

$$= -5,069.44 + 3,826.58$$
$$= -1,242.86$$

The new balance is $-\$1,242.86$.

97. MOVIE LOSSES

$$-176,350,000 + 76,700,000$$
$$= -99,650,000$$

The loss is $99,650,000.

99. CHEMISTRY

a. $-8 + 7 = -1$

The net charge is -1.

b. $-10 + 13 = 3$

The net charge is 3.

101. THE BIG EASY

$$-6 + 85 = 79$$

Top of the building is 79 feet above sea level.

103. ACCOUNTING

$$4,247 + 3,224 + (-1,001) + (-194)$$
$$= 7,471 + (-1,195)$$
$$= 6,276$$

Net income in 2009 was $6,276 million.

WRITING

105. Answers will vary.

REVIEW

107. True

109. $-3 + (-6) = -9$ & $-3 + 6 = 3$

CHALLENGE PROBLEMS

111. No, $1 + 1 = 2$, and 2 is not a member of the set.

SECTION 1.5

VOCABULARY

Fill in the blanks.

1. **Subtraction** finds the difference between two numbers.

3. The difference between the maximum and the minimum value of a collection of measurements is called the **range** of the values.

CONCEPTS

5. a. -12 b. $\dfrac{1}{5}$ c. -2.71 d. 0

7. $1-(-9)=1\boxed{+9}=\boxed{10}$

9. $-10+(-8)+(-23)+5+34$

11. a. $-6\boxed{-}(-4)$ b. $7+(-3)\boxed{-}5\boxed{-}(-2)$

NOTATION

13. a. $1-(-7)=8$ b. $-(-2)=2$

  c. $-|-3|=-3$ d. $2-6=-4$

GUIDED PRACTICE

Simplify each expression. See Example 1.

15. $-(-55)=55$

17. $-(-x)=x$

19. $-|-25|=-25$

21. $-\left|-\dfrac{3}{16}\right|=-\dfrac{3}{16}$

Subtract. See Example 2.

23. $4-7=4+(-7)$
$=-3$

25. $-6-4=-6+(-4)$
$=-10$

27. $8-(-3)=8+3$
$=11$

29. $0-6=0+(-6)$
$=-6$

31. $-1-(-3)=-1+3$
$=2$

33. $20-(-20)=20+20$
$=40$

35. $-2-(-7)=-2+7$
$=5$

37. $0-(-12)=0+12$
$=12$

39. $-1.4-5.5=-1.4+(-5.5)$
$=-6.9$

41. $-1.5-0.81=-1.5+(-0.81)$
$=-2.31$

43. $-\dfrac{1}{8}-\dfrac{3}{8}=-\dfrac{1}{8}+\left(-\dfrac{3}{8}\right)$
$\phantom{-\dfrac{1}{8}-\dfrac{3}{8}}=\dfrac{-1+(-3)}{8}$
$\phantom{-\dfrac{1}{8}-\dfrac{3}{8}}=\dfrac{-4}{8}$
$\phantom{-\dfrac{1}{8}-\dfrac{3}{8}}=-\dfrac{\cancel{4}}{2\cdot\cancel{4}}$
$\phantom{-\dfrac{1}{8}-\dfrac{3}{8}}=-\dfrac{1}{2}$

45. $\dfrac{1}{3}-\dfrac{3}{4}=\dfrac{1}{3}+\left(-\dfrac{3}{4}\right)$
$\phantom{\dfrac{1}{3}-\dfrac{3}{4}}=\dfrac{1}{3}\cdot\dfrac{4}{4}+\left(-\dfrac{3}{4}\right)\cdot\dfrac{3}{3}$
$\phantom{\dfrac{1}{3}-\dfrac{3}{4}}=\dfrac{4+(-9)}{12}$
$\phantom{\dfrac{1}{3}-\dfrac{3}{4}}=-\dfrac{5}{12}$

Perform the indicated operation. See Example 3.

47. $17-(-5)=17+5$
$=22$

49. $-13-12=-13+(-12)$
$=-25$

Perform the operations. See Example 4.

51. $-6+8-(-1)-10=-6+8+1+(-10)$
$=-16+9$
$=-7$

53. $61-(-62)+(-64)-60$
$=61+62+(-64)+(-60)$
$=123+(-124)$
$=-1$

Section 1.5

TRY IT YOURSELF

Perform the operations.

55. $244 - (-12) = 244 + 12$
$$= 256$$

57. $-20 - (-30) - 50 + 40$
$$= -20 + 30 + (-50) + 40$$
$$= -70 + 70$$
$$= 0$$

59. $-1.2 - 0.9 = -1.2 + (-0.9)$
$$= -2.1$$

61. $\dfrac{1}{8} - \left(-\dfrac{5}{7}\right) = \dfrac{1}{8} + \dfrac{5}{7}$
$$= \dfrac{1}{8} \cdot \dfrac{7}{7} + \dfrac{5}{7} \cdot \dfrac{8}{8}$$
$$= \dfrac{7 + 40}{56}$$
$$= \dfrac{47}{56}$$

63. $-62 - 71 - (-37) + 99$
$$= -62 + (-71) + 37 + 99$$
$$= -133 + 136$$
$$= 3$$

65. $0 - 47.5 = 0 + (-47.5)$
$$= -47.5$$

67. $12 - (-137) = 12 + 137$
$$= 149$$

69. $-1{,}903 - (-1{,}732) = -1{,}903 + 1{,}732$
$$= -171$$

71. $2.83 - (-1.8) = 2.83 + 1.8$
$$= 4.63$$

73. $-\dfrac{5}{6} - \dfrac{3}{4} = -\dfrac{5}{6} + \left(-\dfrac{3}{4}\right)$
$$= -\dfrac{5}{6} \cdot \dfrac{2}{2} + \left(-\dfrac{3}{4}\right) \cdot \dfrac{3}{3}$$
$$= \dfrac{-10 + (-9)}{12}$$
$$= -\dfrac{19}{12}$$

75. $8 - 9 - 10 = 8 + (-9) + (-10)$
$$= 8 + (-19)$$
$$= -11$$

77. $-44 - 44 = -44 + (-44)$
$$= -88$$

79. $-0.9 - 0.2 = -0.9 + (-0.2)$
$$= -1.1$$

81. $-25 - (-50) - 75 = -25 + 50 + (-75)$
$$= -100 + 50$$
$$= -50$$

83. $6.3 - 9.8 = 6.3 + (-9.8)$
$$= -3.5$$

85. $-\dfrac{9}{16} - \left(-\dfrac{1}{4}\right) = -\dfrac{9}{16} + \dfrac{1}{4}$
$$= -\dfrac{9}{16} + \dfrac{1}{4} \cdot \dfrac{4}{4}$$
$$= \dfrac{-9 + 4}{16}$$
$$= -\dfrac{5}{16}$$

87. $0 - (-1) = 0 + 1$
$$= 1$$

89. $2 - 15 = 2 + (-15)$
$$= -13$$

LOOK ALIKES

91. a. $-50 + (-3) = -53$

 b. $-50 - (-3) = -50 + 3$
$$= -47$$

93. a. $-\dfrac{5}{9} + \left(-\dfrac{1}{6}\right) = -\dfrac{5}{9}\left(\dfrac{2}{2}\right) + \left(-\dfrac{1}{6}\right)\left(\dfrac{3}{3}\right)$
$$= -\dfrac{10}{18} + \left(-\dfrac{3}{18}\right)$$
$$= -\dfrac{13}{18}$$

 b. $-\dfrac{5}{9} - \left(-\dfrac{1}{6}\right) = -\dfrac{5}{9}\left(\dfrac{2}{2}\right) + \left(\dfrac{1}{6}\right)\left(\dfrac{3}{3}\right)$
$$= -\dfrac{10}{18} + \dfrac{3}{18}$$
$$= -\dfrac{7}{18}$$

APPLICATIONS

95. THE EMPIRE STATE
$$108 - (-52) = 108 + 52$$
$$= 160$$
The range is $160°$.

97. LAW ENFORCEMENT
$$3 - (-18) = 3 + 18$$
$$= 21$$
The change is 21.

FROM CAMPUS TO CAREERS

99. LEAD TRANS. SECURITY OFFICER
$$1,296,000 - 1,411,000$$
$$= 1,296,000 + (-1,411,000)$$
$$= -115,000$$
Orlando's change was $-115,000$.

$$1,148,000 - 1,198,000$$
$$= 1,148,000 + (-1,198,000)$$
$$= -50,000$$
Fort Lauderdale's change was $-50,000$.

101. GEOGRAPHY
$$-282 - (-1,312) = -282 + 1,312$$
$$= 1,030$$
The difference in elevations is 1,030 ft.

103. WORLD'S COLDEST ICE CREAM
$$-40 - (-355) = -40 + 355$$
$$= 315$$
There is an increase of $315°$ F.

105. NASCAR
Mark Martin
$$6,511 - 6,652 = 6,511 + (-6,652)$$
$$= -141$$

Jeff Gordan
$$6,473 - 6,652 = 6,473 + (-6,652)$$
$$= -179$$

Kurt Busch
$$6,446 - 6,652 = 6,446 + (-6,652)$$
$$= -206$$

WRITING

107-109. Answers will vary.

REVIEW

111. $30 = 2 \cdot 3 \cdot 5$

113. True

CHALLENGE PROBLEMS

115. a. True b. True c. False d. False

Section 1.5

SECTION 1.6
VOCABULARY
Fill in the blanks.

1. The answer to a multiplication problem is called a **product**. The answer to a division problem is called a **quotient** .

3. The **associative** property of multiplication states that changing the grouping when multiplying does not affect the answer.

CONCEPTS

5. a. The product or quotient of two numbers with like signs is **positive**.

   b. The product or quotient of two numbers with unlike signs is **negative**.

7. a. $\dfrac{-9}{3} = -3$ because $-3 \cdot 3 = -9$.

   b. $\dfrac{0}{8} = 0$ because $0 \cdot 8 = 0$.

9. a. $\dfrac{a}{1} = a$ b. $\dfrac{a}{a} = 1$

   c. $\dfrac{0}{a} = 0$ d. $\dfrac{a}{0}$ is **undefined**

11. a. positive   b. negative   c. negative

13. a. $8 \cdot 5$ b. $(-2 \cdot 6)9$

   c. $\dfrac{1}{5}$ d. 1

Let POS stand for a positive number and NEG stand for a negative number. Determine the sign of each result, if possible.

15. a. NEG   b. Not possible to tell

   c. POS   d. NEG

NOTATION
Write each sentence using symbols.

17. $-4 \cdot (-5) = 20$

GUIDED PRACTICE
Multiply. See Example 1.

19. $4(-1) = -4$

21. $-2 \cdot 8 = -16$

23. $12(-5) = -60$

25. $3(-22) = -66$

27. $1.2(-0.4) = -0.48$

29. $\dfrac{1}{3}\left(-\dfrac{3}{4}\right) = -\dfrac{1 \cdot \cancel{3}}{\cancel{3} \cdot 4}$

$\qquad = -\dfrac{1}{4}$

Multiply. See Example 2.

31. $(-1)(-7) = 7$

33. $(-6)(-9) = 54$

35. $-3(-3) = 9$

37. $63(-7) = -441$

39. $-0.6(-4) = 2.4$

41. $\left(-\dfrac{7}{8}\right)\left(-\dfrac{2}{21}\right) = \dfrac{\cancel{7} \cdot \cancel{2}}{\cancel{2} \cdot 4 \cdot 3 \cdot \cancel{7}}$

$\qquad = \dfrac{1}{12}$

Multiply. See Example 3.

43. $3.3(-4)(-5) = 3.3(20)$

$\qquad = 66$

45. $-2(-3)(-4)(-5)(-6) = 6(20)(-6)$

$\qquad = 120(-6)$

$\qquad = -720$

47. $(-41)(3)(-7)(-1) = -123(7)$

$\qquad = -861$

49. $(-6)(-6)(-6) = 36(-6)$

$\qquad = -216$

Find the reciprocal of each number. Then find the product of the given number and its reciprocal. See Example 4.

51. $\dfrac{9}{7}, \quad \dfrac{7}{9} \cdot \dfrac{9}{7} = \dfrac{\cancel{7} \cdot \cancel{9}}{\cancel{9} \cdot \cancel{7}} \cdot$

$\qquad = 1$

53. $-\dfrac{1}{13}, \quad \dfrac{-13}{1} \cdot \dfrac{1}{-13} = \dfrac{\cancel{13} \cdot 1}{1 \cdot \cancel{13}}$

$\qquad = 1$

Divide. See Example 5.

55. $-30 \div (-3) = 10$

57. $-6 \div (-2) = 3$

59. $\dfrac{85}{-5} = -17$

61. $\dfrac{-110}{-110} = 1$

63. $\dfrac{-10.8}{1.2} = -9$

65. $\dfrac{0.5}{-100} = -0.005$

67. $-\dfrac{1}{3} \div \dfrac{4}{5} = -\dfrac{1}{3} \cdot \dfrac{5}{4}$

$= -\dfrac{1 \cdot 5}{3 \cdot 4}$

$= -\dfrac{5}{12}$

69. $-\dfrac{9}{16} \div \left(-\dfrac{3}{20}\right) = -\dfrac{9}{16} \cdot \left(-\dfrac{20}{3}\right)$

$= \dfrac{3 \cdot \overset{1}{\cancel{3}} \cdot 5 \cdot \overset{1}{\cancel{4}}}{4 \cdot \underset{1}{\cancel{4}} \cdot \underset{1}{\cancel{3}}}$

$= \dfrac{15}{4}$

TRY IT YOURSELF
Perform the operations.

71. $\dfrac{0}{150} = 0$

73. $\dfrac{-17}{0}$ is undefined.

75. $\dfrac{24}{-6} = -4$

77. $\dfrac{17}{-17} = -1$

79. $(-2)(-2)(-2)(-2) = 4(4)$

$= 16$

81. $-3(-4)(0) = 0$

83. $\dfrac{-23.5}{5} = -4.7$

85. $-5.2 \cdot 100 = -520$

87. $\dfrac{1}{2}\left(-\dfrac{1}{3}\right)\left(-\dfrac{1}{4}\right) = \dfrac{1 \cdot 1 \cdot 1}{2 \cdot 3 \cdot 4}$

$= \dfrac{1}{24}$

89. $\dfrac{550}{-50} = -11$

91. $-3\dfrac{3}{8} \div \left(-2\dfrac{1}{4}\right) = -\dfrac{27}{8} \div \left(-\dfrac{9}{4}\right)$

$= -\dfrac{27}{8} \cdot \left(-\dfrac{4}{9}\right)$

$= \dfrac{3 \cdot \overset{1}{\cancel{9}} \cdot \overset{1}{\cancel{4}}}{2 \cdot \underset{1}{\cancel{4}} \cdot \underset{1}{\cancel{9}}}$

$= \dfrac{3}{2}$

$= 1\dfrac{1}{2}$

93. $7.2(-2.1)(-2) = 7.2(4.2)$

$= 30.24$

95. $\dfrac{1}{2}\left(-\dfrac{3}{4}\right) = -\dfrac{1 \cdot 3}{2 \cdot 4}$

$= -\dfrac{3}{8}$

97. $-\dfrac{16}{25} \div \dfrac{64}{15} = -\dfrac{16}{25} \cdot \dfrac{15}{64}$

$= -\dfrac{\overset{1}{\cancel{16}} \cdot 3 \cdot \overset{1}{\cancel{5}}}{5 \cdot \underset{1}{\cancel{5}} \cdot 4 \cdot \underset{1}{\cancel{16}}}$

$= -\dfrac{3}{20}$

99. $\dfrac{-24.24}{-0.8} = 30.3$

101. $-1\dfrac{1}{4}\left(-\dfrac{3}{4}\right) = -\dfrac{5}{4}\left(-\dfrac{3}{4}\right)$

$= \dfrac{5 \cdot 3}{4 \cdot 4}$

$= \dfrac{15}{16}$

LOOK ALIKES

103. a. $2.7 + (-0.9) = 1.8$

   b. $2.7 - (-0.9) = 3.6$

   c. $2.7(-0.9) = -2.43$

   d. $\dfrac{2.7}{-0.9} = -3$

Use the associative property of multiplication to find each product.

105. $-\dfrac{1}{2}(2 \cdot 67) = \left(-\dfrac{1}{2} \cdot 2\right) \cdot 67$

$$= -1 \cdot 67$$

$$= -67$$

107. $-0.2(-10 \cdot 3) = (-0.2 \cdot -10) \cdot 3$

$$= 2 \cdot 3$$

$$= 6$$

APPLICATIONS

109. REAL ESTATE

$\$200,000 - \$160,000 = \$40,000$

$\$40,000 \div 5 = \$8,000$

Depreciation per year was $8,000.

111. FLUID FLOW

$-6 \cdot 12 = -72$

The solution $-72°$ indicates that the temperature dropped $72°$.

113. WEIGHT LOSS

   a. *ii*

   b. $-4\dfrac{1}{2} \cdot (-8) = \dfrac{-9}{2} \cdot \dfrac{-8}{1}$

$$= \dfrac{9 \cdot 4 \cdot \cancel{2}}{\cancel{2} \cdot 1}$$

$$= 36$$

Tom was 36 pounds heavier.

115. CAR RADIATORS

$1\dfrac{1}{2}(-34) = \dfrac{3}{2} \cdot \dfrac{-34}{1}$

$$= -\dfrac{3 \cdot 17 \cdot \cancel{2}}{\cancel{2} \cdot 1}$$

$$= -51$$

60/40 mixture protects to $-51°$F.

117. AIRLINES

$-25 \cdot 6.4 = -160$

2nd qrt losses was \$160 million or (160).

$-800 \cdot \dfrac{5}{16} = -250$

3rd qrt losses was \$250 million or (250).

$-160 \cdot 5 = -800$

4th qrt losses was \$800 million or (800).

119. PHYSICS

   a. $5, -10$

   b. $5 \cdot 0.5 = 2.5$, $-10 \cdot 0.5 = -5$

   c. $5 \cdot 1.5 = 7.5$, $-10 \cdot 1.5 = -15$

   d. $5 \cdot 2 = 10$, $-10 \cdot 2 = -20$

WRITING

121-123. Answers will vary.

REVIEW

125. $-3 + (-4) + (-5) + 4 + 3 = -5$

127. $\dfrac{1}{2} + \dfrac{1}{4} + \dfrac{1}{3} = \dfrac{1}{2} \cdot \dfrac{6}{6} + \dfrac{1}{4} \cdot \dfrac{3}{3} + \dfrac{1}{3} \cdot \dfrac{4}{4}$

$$= \dfrac{6 + 3 + 4}{12}$$

$$= \dfrac{13}{12}$$

$$\begin{array}{r} 1.0833 \\ 12\overline{)13.0000} \end{array}$$

$$\dfrac{13}{12} = 1.08\overline{3}$$

CHALLENGE PROBLEMS

129. An odd number of factors must be negative.

SECTION 1.7

VOCABULARY

Fill in the blanks.

1. In the exponential expression 7^5, 7 is the **base**, and 5 is the **exponent**. 7^5 is the fifth **power** of seven.

3. An **exponent** is used to represent repeated multiplication.

5. The rules for the **order** of operations guarantee that an evaluation of a numerical expression will result in a single answer.

CONCEPTS

7. a. **Subtraction**   b. **Division**
   c. **Addition**   d. **Power**

NOTATION

9. a. **Parentheses, brackets, braces, absolute value symbols, fraction bar**

   b. **Innermost : parentheses; outermost : brackets**

11. a. -5 b. 5

13. $-19 - 2[(1+2)^2 \cdot 3] = -19 - 2[\mathbf{3}^2 \cdot 3]$
$$= -19 - 2[\mathbf{9} \cdot 3]$$
$$= -19 - 2[\mathbf{27}]$$
$$= -19 - \mathbf{54}$$
$$= \mathbf{-73}$$

GUIDED PRACTICE

Write each product using exponents. See Example 1.

15. $8 \cdot 8 \cdot 8 = 8^3$

17. $7 \cdot 7 \cdot 7 \cdot 12 \cdot 12 = 7^3 12^2$

19. $x \cdot x \cdot x = x^3$

21. $r \cdot r \cdot r \cdot r \cdot s \cdot s = r^4 s^2$

Evaluate each expression. See Example 2.

23. $7^2 = 7 \cdot 7$
$$= 49$$

25. $6^3 = 6 \cdot 6 \cdot 6$
$$= 36 \cdot 6$$
$$= 216$$

27. $(-5)^4 = (-5) \cdot (-5) \cdot (-5) \cdot (-5)$
$$= 25 \cdot (-5) \cdot (-5)$$
$$= -125 \cdot (-5)$$
$$= 625$$

29. $(-0.1)^2 = (-0.1) \cdot (-0.1)$
$$= 0.01$$

31. $\left(-\dfrac{1}{4}\right)^3 = \left(-\dfrac{1}{4}\right) \cdot \left(-\dfrac{1}{4}\right) \cdot \left(-\dfrac{1}{4}\right)$
$$= -\dfrac{1 \cdot 1 \cdot 1}{4 \cdot 4 \cdot 4}$$
$$= -\dfrac{1}{64}$$

33. $\left(\dfrac{2}{3}\right)^3 = \left(\dfrac{2}{3}\right) \cdot \left(\dfrac{2}{3}\right) \cdot \left(\dfrac{2}{3}\right)$
$$= \dfrac{2 \cdot 2 \cdot 2}{3 \cdot 3 \cdot 3}$$
$$= \dfrac{8}{27}$$

Evaluate each expression. See Example 3.

35. $(-6)^2 = (-6) \cdot (-6)$
$$= 36$$

$$-6^2 = -(6 \cdot 6)$$
$$= -36$$

37. $(-8)^2 = (-8) \cdot (-8)$
$$= 64$$

$$-8^2 = -(8 \cdot 8)$$
$$= -64$$

Evaluate each expression. See Example 4.

39. $3 - 5 \cdot 4 = 3 - 20$
$$= 3 + (-20)$$
$$= -17$$

41. $32 - 16 \div 4 + 2 = 32 - 4 + 2$
$$= 28 + 2$$
$$= 30$$

43. $9 \cdot 5 - 6 \div 3 = 45 - 2$
$$= 45 + (-2)$$
$$= 43$$

45. $12 \div 3 \cdot 2 = 4 \cdot 2$
$$= 8$$

47. $-22-15+3 = -22+(-15)+3$
$= -37+3$
$= -34$

49. $-2(9)-2(5)(10) = -18-100$
$= -18+(-100)$
$= -118$

Evaluate each expression. See Example 5.

51. $-4(6+5) = -4(11)$
$= -44$

53. $(9-3)(9-9)^2 = [9+(-3)][9+(-9)]^2$
$= (6)(0)^2$
$= (6)(0)$
$= 0$

55. $(-1-3^2 \cdot 4)2^2 = [-1-(3\cdot3)\cdot4](2\cdot2)$
$= [-1-9\cdot4]4$
$= [-1-36]4$
$= [-1+(-36)]4$
$= (-37)4$
$= -148$

57. $1+5(10+2\cdot5)-1 = 1+5(10+10)-1$
$= 1+5(20)-1$
$= 1+100-1$
$= 0+100$
$= 100$

Evaluate each expression. See Example 6.

59. $(-1)^9[-7^2-(-2)^2] = (-1)[-49-4]$
$= (-1)[-49+(-4)]$
$= (-1)(-53)$
$= 53$

61. $64-6[15+2(-3+8)] = 64-6[15+2(5)]$
$= 64-6[15+10]$
$= 64-6[25]$
$= 64-150$
$= 64+(-150)$
$= -86$

63. $-2[2+4^2(8-9)]^2 = -2\{2+16[8+(-9)]\}^2$
$= -2[2+16(-1)]^2$
$= -2[2+(-16)]^2$
$= -2[-14]^2$
$= -2(196)$
$= -392$

65. $3+2[-1-(4-5)] = 3+2[-1-(-1)]$
$= 3+2[-1+1]$
$= 3+2[0]$
$= 3+0$
$= 3$

Evaluate each expression. See Example 7.

67. $\dfrac{-2-5}{-7+(-7)} = \dfrac{-2+(-5)}{-7+(-7)}$
$= \dfrac{-7}{-14}$
$= \dfrac{1}{2}$

69. $\dfrac{2\cdot2^5-60+(-4)}{5^4-(-4)(-5)} = \dfrac{2\cdot32-60+(-4)}{625-20}$
$= \dfrac{64-60+(-4)}{625+(-20)}$
$= \dfrac{4+(-4)}{605}$
$= \dfrac{0}{605}$
$= 0$

71. $\dfrac{2(-4-2\cdot2)}{3(-3)(-2)} = \dfrac{2(-4-4)}{3(6)}$
$= \dfrac{2[-4+(-4)]}{18}$
$= \dfrac{2(-8)}{18}$
$= \dfrac{-16}{18}$
$= -\dfrac{\overset{1}{\cancel{2}}\cdot8}{\underset{1}{\cancel{2}}\cdot9}$
$= -\dfrac{8}{9}$

73. $\dfrac{72-(2-2\cdot4)}{10^2-(9\cdot10+2^2)} = \dfrac{72-(2-8)}{100-(90+4)}$

$= \dfrac{72-[2+(-8)]}{100-94}$

$= \dfrac{72-(-6)}{100+(-94)}$

$= \dfrac{72+6}{6}$

$= \dfrac{78}{6}$

$= \dfrac{\overset{1}{\cancel{6}}\cdot13}{\underset{1}{\cancel{6}}}$

$= 13$

Evaluate each expression.

75. $10-2|4-8| = 10-2|4+(-8)|$

$= 10-2|-4|$

$= 10-2(4)$

$= 10-8$

$= 2$

77. $-|7-2^3(4-7)| = -|7-8[4+(-7)]|$

$= -|7-8(-3)|$

$= -|7-(-24)|$

$= -|7+24|$

$= -|31|$

$= -31$

79. $\dfrac{(3+5)^2+|-2|}{-2(5-8)} = \dfrac{(8)^2+2}{-2[5+(-8)]}$

$= \dfrac{64+2}{-2(-3)}$

$= \dfrac{66}{6}$

$= \dfrac{\overset{1}{\cancel{6}}\cdot11}{\underset{1}{\cancel{6}}}$

$= 11$

81. $\dfrac{|6-4|+2|-4|}{226-6^3} = \dfrac{|6+(-4)|+2(4)}{226-216}$

$= \dfrac{|2|+8}{10}$

$= \dfrac{2+8}{10}$

$= \dfrac{10}{10}$

$= \dfrac{\overset{1}{\cancel{10}}}{\underset{1}{\cancel{10}}}$

$= 1$

83. $-(2\cdot3-2^2)^5 = -(6-4)^5$

$= -(2)^5$

$= -(2\cdot2\cdot2\cdot2\cdot2)$

$= -32$

85. $2\cdot5^2+4\cdot3^2 = 2\cdot(5\cdot5)+4\cdot(3\cdot3)$

$= 2\cdot25+4\cdot9$

$= 50+36$

$= 86$

87. $-2(-1)^2+3(-1)-3 = -2(1)+(-3)-3$

$= -2+(-3)+(-3)$

$= -8$

89. $8-3[5^2-(7-3)^2] = 8-3[5^2-(4)^2]$

$= 8-3(25-16)$

$= 8-3[25+(-16)]$

$= 8-3(9)$

$= 8-27$

$= 8+(-27)$

$= -19$

TRY IT YOURSELF
Evaluate each expression.

91. $[6(5)-5(5)]^3(-4) = [30-25]^3(-4)$

$= [30+(-25)]^3(-4)$

$= [5]^3(-4)$

$= 125(-4)$

$= -500$

93. $8-6[(130-4^3)-2]$
$\qquad = 8-6[(130-64)-2]$
$\qquad = 8-6[130+(-64)+(-2)]$
$\qquad = 8-6[64]$
$\qquad = 8-384$
$\qquad = 8+(-384)$
$\qquad = -376$

95. $-2\left(\dfrac{15}{-5}\right)-\dfrac{6}{2}+9 = -2(-3)-3+9$
$\qquad = 6+(-3)+9$
$\qquad = 3+9$
$\qquad = 12$

97. $-5(-2)^3-|-2+1| = -5(-8)-|-1|$
$\qquad = 40-1$
$\qquad = 39$

99. $\dfrac{18-[2+(1-6)]}{16-(-4)^2} = \dfrac{18-[2+(1+(-6))]}{16-(16)}$
$\qquad = \dfrac{18-[2+(-5)]}{16+(-16)}$
$\qquad = \dfrac{18-[-3]}{0}$
$\qquad$ Undefined

101. $-|-5\cdot 7^2|-30 = -|-5\cdot 49|+(-30)$
$\qquad = -|-245|+(-30)$
$\qquad = -245+(-30)$
$\qquad = -275$

103. $(-3)^3\left(\dfrac{-4}{2}\right)(-1) = -27(-2)(-1)$
$\qquad = 54(-1)$
$\qquad = -54$

105. $\dfrac{1}{2}\left(\dfrac{1}{8}\right)+\left(-\dfrac{1}{4}\right)^2 = \dfrac{1\cdot 1}{2\cdot 8}+\dfrac{1\cdot 1}{4\cdot 4}$
$\qquad = \dfrac{1}{16}+\dfrac{1}{16}$
$\qquad = \dfrac{1+1}{16}$
$\qquad = \dfrac{2}{16}$
$\qquad = \dfrac{\overset{1}{\cancel{2}}}{\underset{1}{\cancel{2}}\cdot 8}$
$\qquad = \dfrac{1}{8}$

107. $\dfrac{-5^2\cdot 10+5\cdot 2^5}{-5-3-1} = \dfrac{-25\cdot 10+5\cdot 32}{-5+(-3)+(-1)}$
$\qquad = \dfrac{-250+160}{-9}$
$\qquad = \dfrac{-90}{-9}$
$\qquad = \dfrac{\overset{1}{\cancel{9}}\cdot 10}{\underset{1}{\cancel{9}}}$
$\qquad = 10$

109. $-\left(\dfrac{40-1^3-2^4}{3(2+5)+2}\right) = -\left(\dfrac{40-1-16}{3(7)+2}\right)$
$\qquad = -\left(\dfrac{40+(-1)+(-16)}{21+2}\right)$
$\qquad = -\left(\dfrac{23}{23}\right)$
$\qquad = -\dfrac{\overset{1}{\cancel{23}}}{\underset{1}{\cancel{23}}}$
$\qquad = -1$

LOOK ALIKES

111. a. $(-7-4)(-2) = (-11)(-2)$
$\qquad = 22$

   b. $(-7-4)-2 = (-11)+(-2)$
$\qquad = -13$

113. a. $-100 \div 5 \cdot 2 = -20 \cdot 2$
$= -40$

b. $-100 \div (5 \cdot 2) = -100 \div 10$
$= -10$

APPLICATIONS

115. LIGHT

2 yards $= 2^2$ square units

3 yards $= 3^2$ square units

4 yards $= 4^2$ square units

117. CAMPUS TO CAREERS

Step 1: Find the total minutes.

Gate A: $3:21 - 3:05 = 16$

Gate B: $3:13 - 3:03 = 10$

Gate C: $3:09 - 3:01 = 8$

Gate D: $3:16 - 3:02 = 14$

Step 2: Find the average.

$\dfrac{16+10+8+14}{4} = \dfrac{48}{4}$
$= 12$

The average wait time is about 12 minutes.

119. CASH AWARDS

a. $2,500 + 4(500) + 35(150) + 85(25)$
$= 2,500 + 2,000 + 5,250 + 2,125$
$= 4,500 + 5,250 + 2,125$
$= 9,750 + 2,125$
$= 11,875$

The total amount of money is $11,875.

b. $1 + 4 + 35 + 85 = 125$

125 is the total number of cash prizes.

$\dfrac{11,875}{125} = 95$

The average cash prize is $95.

121. WRAPPING GIFTS

$15 + 4(4) + 2(16) + 2(9)$
$= 15 + 16 + 32 + 18$
$= 31 + 32 + 18$
$= 63 + 18$
$= 81$

The total amount of ribbon is 81 in.

WRITING

123-125. Answers will vary.

REVIEW

127. $-11 + (-6) = -17$
$-11 + 6 = -5$

CHALLENGE PROBLEMS

129. $\left(3^2\right)^4 = 9^4$
$= 6,561$

Translate the set of instructions to an expression and then evaluate it.

131. $(-3)^3(-4) - (-9+8) = -27(-4) - (-1)$
$= 108 + 1$
$= 109$

Section 1.7

SECTION 1.8
VOCABULARY

Fill in the blanks.

1. Variables and/or numbers can be combined with the operations of arithmetic to create algebraic **expressions**.

3. Addition symbols separate algebraic expressions into parts called **terms**.

5. The **coefficient** of the term $10x$ is 10.

CONCEPTS

7. Number of days in w weeks

$$7, \quad 14, \quad 21, \quad 7w$$

9. $(12-h)$ inches

11. $(x+20)$ ounces

13. a. $b-15$ b. $p+15$

NOTATION

15. $9a - a^2 = 9(5) - (5)^2$
$$= 9(5) - 25$$
$$= 45 - 25$$
$$= 20$$

17. a. $8y$ b. $2cd$ c. **Commutative**

GUIDED PRACTICE . See Example 1.

19. a. $3x^3 + 11x^2 - x + 9$ has 4 terms.

    b. Coefficient of $3x^3$ is 3.

       Coefficient of $11x^2$ is 11.

       Coefficient of $-x$ is -1.

       Coefficient of constant 9 is 9.

Determine whether the variable c is used as a factor or as a term. See Example 2.

21. $c+32$, Term

23. $5c$, Factor

Translate each phrase to an algebraic expression. If no variable is given, use x as the variable. See Example 3.

25. $l+15$ 27. $50x$

29. $\dfrac{w}{l}$ 31. $P+\dfrac{2}{3}p$

33. $k^2 - 2{,}005$ 35. $2a-1$

37. $\dfrac{1{,}000}{n}$

39. $2p+90$

41. $3(35+h+300)$

43. $p-680$

45. $4d-15$

47. $2(200+t)$

49. $|a-2|$

51. $0.1d$ or $\dfrac{1}{10}d$

Translate each algebraic expression into an English phrase. (Answers may vary.) See Example 3.

53. Three-fourths of r

55. Fifty less than t

57. The product of x, y, and z.

59. Twice m, increased by 5

Answer with an algebraic expression. See Example 4.

61. $(x+2)$ inches

63. $(36-x)$ inches

Answer with an algebraic expression. See Example 8.

65. $60h$ minutes

67. $\dfrac{i}{12}$ feet

Answer with an algebraic expression. See Example 9.

69. $\$8x$

71. $49x¢$

73. $\$2t$

75. $\$25(x+2)$

Evaluate each expression, for $x=3$, $y=-2$, and $z=-4$. See Example 10.

77. $-y = -(-2)$
$$= 2$$

79. $-z+3x = -(-4)+3(3)$
$$= 4+9$$
$$= 13$$

81. $3y^2 - 6y - 4 = 3(-2)^2 - 6(-2) - 4$
$$= 3(4) + 12 - 4$$
$$= 12 + 12 + (-4)$$
$$= 24 + (-4)$$
$$= 20$$

83. $(3+x)y = (3+3)(-2)$
$$= 6(-2)$$
$$= -12$$

85. $(x+y)^2 - |z+y| = [3+(-2)]^2 - |-4+(-2)|$
$$= (1)^2 - |-6|$$
$$= 1 - 6$$
$$= 1 + (-6)$$
$$= -5$$

87. $-\dfrac{2x+y^3}{y+2z} = -\dfrac{2(3)+(-2)^3}{(-2)+2(-4)}$
$$= -\dfrac{6+(-8)}{(-2)+(-8)}$$
$$= -\dfrac{-2}{-10}$$
$$= -\dfrac{\overset{1}{\cancel{2}}}{\underset{1}{\cancel{2}}\cdot 5}$$
$$= -\dfrac{1}{5}$$

Evaluate each expression. See Example 10.

89. $b^2 - 4ac = (5)^2 - 4(-1)(-2)$
$$= 25 - 8$$
$$= 25 + (-8)$$
$$= 17$$

91. $a^2 + 2ab + b^2 = (-5)^2 + 2(-5)(-1) + (-1)^2$
$$= 25 + 10 + 1$$
$$= 35 + 1$$
$$= 36$$

93. $\dfrac{n}{2}[2a + (n-1)d]$
$$= \dfrac{10}{2}[2(-4.2) + (10-1)6.6]$$
$$= 5[-8.4 + 9(6.6)]$$
$$= 5[-8.4 + 59.4]$$
$$= 5(51)$$
$$= 255$$

95. $(27c^2 - 4d^2)^3 = \left[27\left(\dfrac{1}{3}\right)^2 - 4\left(\dfrac{1}{2}\right)^2\right]^3$
$$= \left[27\left(\dfrac{1}{9}\right) - 4\left(\dfrac{1}{4}\right)\right]^3$$
$$= \left(\dfrac{27\cdot 1}{9} - \dfrac{4\cdot 1}{4}\right)^3$$
$$= \left(\dfrac{\overset{1}{\cancel{9}}\cdot 3}{\underset{1}{\cancel{9}}} - \dfrac{\overset{1}{\cancel{4}}}{\underset{1}{\cancel{4}}}\right)^3$$
$$= (3 + (-1))^3$$
$$= (2)^3$$
$$= 8$$

Complete each table. See Example 11.

97.

x	$x^3 - 1$
0	$(0)^3 - 1 = 0 - 1$ $= -1$
-1	$(-1)^3 - 1 = -1 - 1$ $= -2$
-3	$(-3)^3 - 1 = -27 - 1$ $= -28$

99.

s	$\dfrac{5s + 36}{s}$
1	$\dfrac{5(1)+36}{1} = \dfrac{5+36}{1}$ $= \dfrac{41}{1}$ $= 41$
6	$\dfrac{5(6)+36}{6} = \dfrac{30+36}{6}$ $= \dfrac{66}{6}$ $= 11$
-12	$\dfrac{5(-12)+36}{-12} = \dfrac{-60+36}{-12}$ $= \dfrac{-24}{-12}$ $= 2$

101.

Input x	Output $2x - \dfrac{x}{2}$
100	$2(\mathbf{100}) - \dfrac{\mathbf{100}}{2} = 200 - 50$ $= 150$
−300	$2(\mathbf{-300}) - \dfrac{\mathbf{-300}}{2} = -600 - (-150)$ $= -600 + 150$ $= -450$

103.

x	$(x + 1)(x + 5)$
−1	$(\mathbf{-1} + 1)(\mathbf{-1} + 5) = (0)(4)$ $= 0$
−5	$(\mathbf{-5} + 1)(\mathbf{-5} + 5) = (-4)(0)$ $= 0$
−6	$(\mathbf{-6} + 1)(\mathbf{-6} + 5) = (-5)(-1)$ $= 5$

TRY IT YOURSELF

105. a. $x + 7$ b. $(x + 7)^2$

107. a. $4(x + 2)$ b. $4x + 2$

APPLICATIONS

109. VEHICLE WEIGHTS

    a. Let x = weight of the Element

        $2x - 340$ = weight of the Hummer

    b. Elements weights 3,370 pounds

        Hummer $= 2(3,370) + (-340)$

                $= 6,740 + (-340)$

                $= 6,400$

    The Hummer weights 6,400 pounds.

111. COMPUTER COMPANIES

    a. Let x = age of Apple

        $x + 80$ = age of IBM

        $x - 9$ = age of Dell

    b. Let $x = 32$ (Apple's age in 2008)

        IBM's age $= x + 80$

                $= 32 + 80$

                $= 112$

        IBM is 112 years old.

        Dell's age $= x - 9$

                $= 32 - 9$

                $= 32 + (-9)$

                $= 23$

    Dell is 23 years old.

WRITING

113-115. Answers will vary.

REVIEW

117. $12 = 2 \cdot 2 \cdot 3$

    $15 = 3 \cdot 5$

    $\text{LCD} = 2 \cdot 2 \cdot 3 \cdot 5 = 60$

119. $\left(\dfrac{2}{3}\right)^3 = \dfrac{2 \cdot 2 \cdot 2}{3 \cdot 3 \cdot 3}$

          $= \dfrac{8}{27}$

CHALLENGE PROBLEMS

121. The answer would be 0, because $(8 - 8) = 0$ is a factor of the expression.

SECTION 1.9

VOCABULARY

Fill in the blanks.

1. To **simplify** the expression $5(6x)$ means to write it in simpler form: $5(6x) = 30x$.

3. To perform the multiplication $2(x + 8)$, we use the **distributive** property.

5. Terms such as $7x^2$ and $5x^2$, which have the same variables raised to exactly the same power, are called **like** terms.

CONCEPTS

7. a. $4(9t) = (4 \cdot 9)t = 36y$

   b. Associative Property of Multiplication

9. a. $2(x+4) = 2x+8$

   b. $2(x-4) = 2x-8$

   c. $-2(x+4) = -2x-8$

   d. $-2(-x-4) = 2x+8$

11. a. $5(2x) = 10x$

   b. $5+2x$; Can't be simplified

   c. $6(-7x) = -42x$

   d. $6-7x$; Can't be simplified

   e. $2(3x)(3) = 18x$

   f. $2+3x+3 = 3x+5$

NOTATION

13. a. $6(h-4)$ b. $-(z+16)$

GUIDED PRACTICE

Use the given property to complete each statement. See Example 1.

15. $8+(7+a) = \underline{(8+7)+a}$

17. $y \cdot 11 = \underline{11y}$

19. $(8d \cdot 2)6 = \underline{8d(2 \cdot 6)}$

21. $9t+(4+t) = 9t+\underline{(t+4)}$

Simplify each expression. See Example 2.

23. $3 \cdot 4t = (3 \cdot 4)t$

   $= 12t$

25. $5(-7q) = (5 \cdot -7)q$

   $= -35q$

27. $(-5.6x)(-2) = (-5.6 \cdot -2)x$

   $= 11.2x$

29. $5(4c)(3) = [5(4)(3)]c$

   $= 60c$

31. $\dfrac{5}{3} \cdot \dfrac{3}{5} g = \dfrac{5 \cdot 3}{3 \cdot 5} g$

   $= \dfrac{\cancel{5} \cdot \cancel{3}}{\cancel{3} \cdot \cancel{5}} g$

   $= g$

33. $12\left(\dfrac{5}{12}x\right) = \dfrac{12}{1} \cdot \dfrac{5}{12}x$

   $= \dfrac{\cancel{12} \cdot 5}{1 \cdot \cancel{12}}x$

   $= 5x$

Multiply. See Example 3.

35. $5(x+3) = 5 \cdot x + 5 \cdot 3$

   $= 5x+15$

37. $-3(4x+9) = -3(4x)-3(9)$

   $= -12x-27$

39. $45\left(\dfrac{x}{5}+\dfrac{2}{9}\right) = 45\left(\dfrac{x}{5}\right)+45\left(\dfrac{2}{9}\right)$

   $= \dfrac{45 \cdot 1}{5}x+\left(\dfrac{45 \cdot 2}{9}\right)$

   $= \dfrac{\cancel{5} \cdot 9}{\cancel{5}}x+\left(\dfrac{\cancel{9} \cdot 5 \cdot 2}{\cancel{9}}\right)$

   $= 9x+10$

41. $0.4(x+4) = 0.4(x)+(0.4)(4)$

   $= 0.4x+1.6$

Multiply. See Example 4.

43. $6(6c-7) = 6(6c)-6(7)$

   $= 36c-42$

45. $-6(13c-3) = -6(13c)-(-6)(3)$

   $= -78c-(-18)$

   $= -78c+18$

47. $-15(-2t-6) = -15(-2t) - (-15)(6)$
$$= 30t - (-90)$$
$$= 30t + 90$$

49. $-1(-4a+1) = -1(-4a) + (-1)(1)$
$$= 4a + (-1)$$
$$= 4a - 1$$

Multiply. See Example 5.

51. $(3t+2)8 = (3t)8 + (2)8$
$$= 24t + 16$$

53. $(3w-6)\dfrac{2}{3} = 3w\left(\dfrac{2}{3}\right) - 6\left(\dfrac{2}{3}\right)$
$$= \dfrac{2\cdot 3}{3}w - \left(\dfrac{2\cdot 6}{3}\right)$$
$$= \dfrac{2\cdot \cancel{3}^{1}}{\cancel{3}_{1}}w - \left(\dfrac{2\cdot 2\cdot \cancel{3}^{1}}{\cancel{3}_{1}}\right)$$
$$= 2w - 4$$

55. $4(7y+4)2 = 4\cdot 2(7y+4)$
$$= 8(7y+4)$$
$$= 8(7y) + 8(4)$$
$$= 56y + 32$$

57. $2.5(2a-3b+1)$
$$= 2.5(2a) - (2.5)(3b) + (2.5)(1)$$
$$= 5a - 7.5b + 2.5$$

Multiply. See Example 6.

59. $-(x-7) = -1(x) - (-1)(7)$
$$= -x - (-7)$$
$$= -x + 7$$

61. $-(-5.6y+7) = -1(-5.6y) + (-1)(7)$
$$= 5.6y + (-7)$$
$$= 5.6y - 7$$

List the like terms in each expression, if any. See Example 7.

63. $3x+2-2x$, $3x$ and $-2x$

65. $-12m^4 - 3m^3 + 2m^2 - m^3$
$-3m^3$ and $-m^3$

Simplify by combining like terms. See Example 8.

67. $3x+7x = (3+7)x$
$$= 10x$$

69. $-7b^2 + 27b^2 = (-7+27)b^2$
$$= 20b^2$$

Simplify by combining like terms. See Example 9.

71. $36y + y - 9y = [36+1+(-9)]y$
$$= 28y$$

73. $\dfrac{3}{5}t + \dfrac{1}{5}t = \left(\dfrac{3}{5} + \dfrac{1}{5}\right)t$
$$= \left(\dfrac{3+1}{5}\right)t$$
$$= \dfrac{4}{5}t$$

75. $13r - 12r = (13-12)r$
$$= 1r$$
$$= r$$

77. $43s^3 - 44s^3 = [43+(-44)]s^3$
$$= -1s^3$$
$$= -s^3$$

Simplify by combining like terms. See Example 10.

79. $15y - 10 - y - 20y$
$$= [15+(-1)+(-20)]y - 10$$
$$= -6y - 10$$

81. $3x+4-5x+1 = (3-5)x + (4+1)$
$$= -2x + 5$$

83. $9m^2 - 6m + 12m - 4 = 9m^2 + (-6+12)m - 4$
$$= 9m^2 + 6m - 4$$

85. $4x^2 + 5x - 8x + 9 = 4x^2 + (5-8)x + 9$
$$= 4x^2 - 3x + 9$$

Simplify. See Example 11.

87. $2z + 5(z-3) - 10 = 2z + 5(z) - (5)(3) - 10$
$$= 2z + 5z + (-15-10)$$
$$= (2+5)z + (-25)$$
$$= 7z - 25$$

89. $2(s^2 - 7) - (s^2 - 2)$
$$= 2(s^2) - (2)(7) - 1(s^2) - (-1)(2)$$
$$= 2s^2 - 14 - s^2 + 2$$
$$= (2-1)s^2 + (-14+2)$$
$$= 1s^2 + (-12)$$
$$= s^2 - 12$$

TRY IT YOURSELF
Simplify each expression, if possible.

91. $-\dfrac{7}{16}x - \dfrac{3}{16}x = \left(-\dfrac{7}{16} - \dfrac{3}{16}\right)x$
$$= \left(\dfrac{-7 + (-3)}{16}\right)x$$
$$= -\dfrac{10}{16}x$$
$$= -\dfrac{\overset{1}{\cancel{2}} \cdot 5}{\underset{1}{\cancel{2}} \cdot 8}x$$
$$= -\dfrac{5}{8}x$$

93. $-9.8c + 6.2c = (-9.8 + 6.2)c$
$$= -3.6c$$

95. $-4(-6)(-4m) = \left[-4(-6)(-4)\right]m$
$$= -96m$$

97. $-4x + 4x = (-4 + 4)x$
$$= 0x$$
$$= 0$$

99. $-0.2r - (-0.6r) = \left[-0.2 - (-0.6)\right]r$
$$= (-0.2 + 0.6)r$$
$$= 0.4r$$

101. $8\left(\dfrac{3}{4}y\right) = \dfrac{8}{1} \cdot \dfrac{3}{4}y$
$$= \dfrac{\overset{1}{\cancel{4}} \cdot 2 \cdot 3}{1 \cdot \underset{1}{\cancel{4}}}y$$
$$= 6y$$

103. $-9(3r - 9) - 7(2r - 7)$
$$= -9(3r) - 9(-9) - 7(2r) - (-7)(7)$$
$$= -27r + 81 - 14r - (-49)$$
$$= -27r + 81 - 14r + 49$$
$$= (-27 - 14)r + (81 + 49)$$
$$= -41r + 130$$

105. $9(7m) = (9 \cdot 7)m$
$$= 63m$$

107. $6 - 4(-3c - 7) = 6 - 4(-3c) - (-4)(7)$
$$= 6 + 12c + 28$$
$$= 12c + (6 + 28)$$
$$= 12c + 34$$

109. $5t \cdot 60 = (5 \cdot 60)t$
$$= 300t$$

111. $36\left(\dfrac{2}{9}x - \dfrac{3}{4}\right) + 36\left(\dfrac{1}{2}\right)$
$$= \dfrac{36}{1}\left(\dfrac{2}{9}\right)x - \left(\dfrac{36}{1}\right)\left(\dfrac{3}{4}\right) + \left(\dfrac{36}{1}\right)\left(\dfrac{1}{2}\right)$$
$$= \dfrac{36 \cdot 2}{1 \cdot 9}x - \left(\dfrac{36 \cdot 3}{1 \cdot 4}\right) + \left(\dfrac{36 \cdot 1}{1 \cdot 2}\right)$$
$$= \dfrac{\overset{1}{\cancel{9}} \cdot 4 \cdot 2}{\underset{1}{\cancel{9}}}x - \left(\dfrac{\overset{1}{\cancel{4}} \cdot 9 \cdot 3}{\underset{1}{\cancel{4}}}\right) + \left(\dfrac{\overset{1}{\cancel{2}} \cdot 18}{\underset{1}{\cancel{2}}}\right)$$
$$= 8x - (27) + 18$$
$$= 8x + (-27 + 18)$$
$$= 8x - 9$$

113. $-4r - 7r + 2r - r = \left[-4 + (-7) + 2 + (-1)\right]r$
$$= (-10)r$$
$$= -10r$$

Section 1.9

115. $24\left(-\dfrac{5}{6}r\right) = \dfrac{24}{1}\left(-\dfrac{5}{6}\right)r$

$\qquad = \left(-\dfrac{24\cdot 5}{1\cdot 6}\right)r$

$\qquad = \left(-\dfrac{\overset{1}{\cancel{6}}\cdot 4\cdot 5}{\underset{1}{\cancel{6}}}\right)r$

$\qquad = -\dfrac{20}{1}r$

$\qquad = -20r$

117. $a+a+a = (1+1+1)a$

$\qquad = 3a$

119. $60\left(\dfrac{3}{20}r - \dfrac{4}{15}\right) = \left(\dfrac{60}{1}\cdot\dfrac{3}{20}\right)r - \left(\dfrac{60}{1}\cdot\dfrac{4}{15}\right)$

$\qquad = \left(\dfrac{60\cdot 3}{1\cdot 20}\right)r - \left(\dfrac{60\cdot 4}{1\cdot 15}\right)$

$\qquad = \left(\dfrac{\overset{1}{\cancel{20}}\cdot 3\cdot 3}{\underset{1}{\cancel{20}}}\right)r - \left(\dfrac{\overset{1}{\cancel{15}}\cdot 4\cdot 4}{\underset{1}{\cancel{15}}}\right)$

$\qquad = 9r - 16$

121. $4a+4b+4c$, Doesn't simplify.

123. $-(c+7)+2(c-3)$

$\qquad = -1(c)+(-1)(7)+2(c)+2(-3)$

$\qquad = -c+(-7)+2c+(-6)$

$\qquad = (-1+2)c+(-7-6)$

$\qquad = c-13$

125. $a^3+2a^2+4a-2a^2-4a-8$

$\qquad = a^3+(2-2)a^2+(4-4)a-8$

$\qquad = a^3+(0)a^2+(0)a-8$

$\qquad = a^3-8$

LOOK ALIKES

127. a. $2(7x)5 = (2\cdot 7\cdot 5)x$

$\qquad = 70x$

b. $2(7x+5) = 2\cdot 7x+2\cdot 5$

$\qquad = 14x+10$

APPLICATIONS

129. RED CROSS

$x+x+x+x+x+x$
$\qquad +x+x+x+x+x+x = 12x$

Perimeter of the cross is $12x$ units.

WRITING

131. Answers will vary.

REVIEW

Evaluate each expression for $x = -3$, $y = -5$, and $z = 0$.

133. $\dfrac{x-y^2}{2y-1+x} = \dfrac{-3-(-5)^2}{2(-5)-1+(-3)}$

$\qquad = \dfrac{-3-(25)}{-10-1+(-3)}$

$\qquad = \dfrac{-3+(-25)}{-10+(-1)+(-3)}$

$\qquad = \dfrac{-28}{-14}$

$\qquad = \dfrac{\overset{1}{\cancel{14}}\cdot 2}{\underset{1}{\cancel{14}}}$

$\qquad = 2$

CHALLENGE PROBLEMS

135. $\boxed{-17}(\boxed{11x}+\boxed{7}) = -187x-119$

CHAPTER 1 REVIEW

SECTION 1.1
Introducing the Language of Algebra

1. 1 hour; 100 cars
2. 100
3. 7 P.M.
4. 12 A.M. (midnight)
5. The difference of 15 and 3 equals 12.
6. The sum of 15 and 3 equals 18.
7. The quotient of 15 and 3 equals 5.
8. The product of 15 and 3 equals 45.

9. a. $4 \cdot 9$ and $4(9)$ b. $\dfrac{9}{3}$

10. a. $8b$ b. Prt

11. a. Equation   b. Expression

12. $n = b + 5$

Brackets (b)	Nails (n)
5	**10**
10	**15**
20	**25**

SECTION 1.2
Fractions

13. a. $2 \cdot 12$, $3 \cdot 8$ (Answers may vary)

    b. $2 \cdot 2 \cdot 6$ (Answers may vary)

    c. $1, 2, 3, 4, 6, 8, 12, 24$

14. Equivalent

15. $54 = 2 \cdot 3^3$

16. $147 = 3 \cdot 7^2$

17. $385 = 5 \cdot 7 \cdot 11$

18. 41 is prime.

19. $\dfrac{20}{35} = \dfrac{2 \cdot 2 \cdot \overset{1}{\cancel{5}}}{\cancel{5} \cdot 7}$

      $= \dfrac{4}{7}$

20. $\dfrac{24}{18} = \dfrac{2 \cdot 2 \cdot \overset{1}{\cancel{2}} \cdot \overset{1}{\cancel{3}}}{\cancel{2} \cdot \cancel{3} \cdot 3}$

      $= \dfrac{4}{3}$

21. $\dfrac{5}{8} \cdot \dfrac{\mathbf{8}}{\mathbf{8}} = \dfrac{40}{64}$

22. $\dfrac{12}{1} \cdot \dfrac{\mathbf{3}}{\mathbf{3}} = \dfrac{36}{3}$

23. $\quad 10 = 2 \cdot 5$

      $18 = 2 \cdot 3 \cdot 3$

   $LCD = 2 \cdot 3 \cdot 3 \cdot 5$

        $= 90$

24. $\quad 21 = 3 \cdot 7$

      $70 = 2 \cdot 5 \cdot 7$

   $LCD = 2 \cdot 3 \cdot 5 \cdot 7$

        $= 210$

25. $\dfrac{1}{8} \cdot \dfrac{7}{8} = \dfrac{1 \cdot 7}{8 \cdot 8}$

         $= \dfrac{7}{64}$

26. $\dfrac{16}{35} \cdot \dfrac{25}{48} = \dfrac{\overset{1}{\cancel{16}} \cdot \overset{1}{\cancel{5}} \cdot 5}{\cancel{5} \cdot 7 \cdot \cancel{16} \cdot 3}$

           $= \dfrac{5}{21}$

27. $\dfrac{1}{3} \div \dfrac{15}{16} = \dfrac{1}{3} \cdot \dfrac{16}{15}$

         $= \dfrac{1 \cdot 16}{3 \cdot 15}$

         $= \dfrac{16}{45}$

28. $16\dfrac{1}{4} \div 5 = \dfrac{65}{4} \div \dfrac{5}{1}$

           $= \dfrac{65}{4} \cdot \dfrac{1}{5}$

           $= \dfrac{\overset{1}{\cancel{5}} \cdot 13}{4 \cdot \cancel{5}}$

           $= \dfrac{13}{4}$

           $= 3\dfrac{1}{4}$

29.
$$\frac{17}{25} - \frac{7}{25} = \frac{17-7}{25}$$
$$= \frac{10}{25}$$
$$= \frac{2 \cdot \cancel{5}^{1}}{5 \cdot \cancel{5}_{1}}$$
$$= \frac{2}{5}$$

30.
$$\frac{8}{11} - \frac{1}{2} = \frac{8}{11} \cdot \frac{\mathbf{2}}{\mathbf{2}} - \frac{1}{2} \cdot \frac{\mathbf{11}}{\mathbf{11}}$$
$$= \frac{16-11}{22}$$
$$= \frac{5}{22}$$

31.
$$\frac{17}{24} + \frac{11}{40} = \frac{17}{24} \cdot \frac{\mathbf{5}}{\mathbf{5}} + \frac{11}{40} \cdot \frac{\mathbf{3}}{\mathbf{3}}$$
$$= \frac{85+33}{120}$$
$$= \frac{118}{120}$$
$$= \frac{59 \cdot \cancel{2}^{1}}{60 \cdot \cancel{2}_{1}}$$
$$= \frac{59}{60}$$

32.
$$4\frac{1}{9} = 4\frac{2}{18} = 3\frac{18}{18} + \frac{2}{18} = 3\frac{20}{18}$$
$$-3\frac{5}{6} = -3\frac{15}{18} = -3\frac{15}{18} \qquad = -3\frac{15}{18}$$
$$\overline{}$$
$$\frac{5}{18}$$

33. THE INTERNET
$$30 \cdot 1\frac{3}{4} = \frac{30}{1} \cdot \frac{7}{4}$$
$$= \frac{\cancel{2}^{1} \cdot 15 \cdot 7}{\cancel{2}_{1} \cdot 2}$$
$$= \frac{105}{2}$$
$$= 52\frac{1}{2}$$

It received $52\frac{1}{2}$ million hits.

34. MACHINE SHOPS
$$\frac{17}{24} - \frac{17}{32} = \frac{17}{24} \cdot \frac{\mathbf{4}}{\mathbf{4}} - \frac{17}{32} \cdot \frac{\mathbf{3}}{\mathbf{3}}$$
$$= \frac{68-51}{96}$$
$$= \frac{17}{96}$$

$\frac{17}{96}$ of an inch needs to be milled.

SECTION 1.3
The Real Numbers

35. a. 0   b. $\{\ldots, -2, -1, 0, 1, 2, \ldots\}$

36. -206 feet

37. a. $<$ b. $>$

38. a. $\frac{7}{10}$ b. $\frac{14}{3}$

39.
$$\frac{1}{250} = 250\overline{)1.000}^{\,0.004}$$
$$\frac{1}{250} = 0.004$$

40.
$$\frac{17}{22} = 22\overline{)17.00000}^{\,0.77272}$$
$$\frac{17}{22} = 0.7\overline{72}$$

41.

42. Natural: 8

Whole: 0, 8

Integer: $0, -12, 8$

Rational: $-\dfrac{4}{5}, 99.99, 0, -12, 4\dfrac{1}{2}, 0.666\ldots, 8$

Irrational: $\sqrt{2}$

Real: All

43. False

44. False

45. True

46. True

47. $\left|-6\right|\boxed{?}\left|-5\right|$

$\quad 6 > 5$

$\quad \left|-6\right| > \left|-5\right|$

48. $-9\boxed{?}\left|-10\right|$

$\quad -9 < 10$

$\quad -9\boxed{<}\left|-10\right|$

SECTION 1.4
Adding Real Numbers; Properties of Addition

49. $-45 + (-37) = -82$

50. $25 + (-13) = 12$

51. $0 + (-7) = -7$

52. $-7 + 7 = 0$

53. $12 + (-8) + (-15) = 4 + (-15)$
$$= -11$$

54. $-9.9 + (-2.4) = -12.3$

55. $\dfrac{5}{16} + \left(-\dfrac{1}{2}\right) = \dfrac{5}{16} + \left(-\dfrac{1}{2} \cdot \dfrac{\mathbf{8}}{\mathbf{8}}\right)$
$$= \dfrac{5 + (-8)}{16}$$
$$= -\dfrac{3}{16}$$

56. $35 + (-13) + (-17) + 6 = 41 + (-30)$
$$= 11$$

57. a. Commutative Property of Addition

b. Associative Property of Addition

c. Addition Property of Opposites
(Inverse Property of Addition)

d. Addition Property of 0
(Identity Property of Addition)

58. TEMPERATURES

$\quad -48 + 166 = 118$

The record high temperature is 118°F.

SECTION 1.5
Subtracting Real Numbers

59. a. -10 b. 3

60. a. $\dfrac{9}{16}$ b. -4

61. $45 - 64 = 45 + (-64)$
$$= -19$$

62. $-\dfrac{3}{5} - \dfrac{1}{3} = -\dfrac{3}{5} + \left(-\dfrac{1}{3}\right)$
$$= -\dfrac{3}{5} \cdot \dfrac{\mathbf{3}}{\mathbf{3}} + \left(-\dfrac{1}{3} \cdot \dfrac{\mathbf{5}}{\mathbf{5}}\right)$$
$$= \dfrac{-9 + (-5)}{15}$$
$$= -\dfrac{14}{15}$$

63. $-7 - (-12) = -7 + 12$
$$= 5$$

64. $3.6 - (-2.1) = 3.6 + 2.1$
$$= 5.7$$

65. $0 - 10 = 0 + (-10)$
$$= -10$$

66. $-33 + 7 - 5 - (-2) = -33 + 7 + (-5) + 2$
$$= -26 + (-5) + 2$$
$$= -31 + 2$$
$$= -29$$

Chapter 1 Review and Test

67. GEOGRAPHY

$$29,028-(-36,205)=29,028+36,205$$
$$=65,233$$

The difference in elevations is 65,233 feet.

Check: $65,233+(-36,205)=29,028$

68. HISTORY

$$-212+(-75)=-287$$

He was born in 287 B.C.

Check: $-287+75=-212$

SECTION 1.6
Multiplying and Dividing Real Numbers
Properties of Multiplication and Division

69. $-8\cdot7=-56$

70. $-9\left(-\dfrac{1}{9}\right)=-\dfrac{9}{1}\left(-\dfrac{1}{9}\right)$

$$=\frac{\cancel{9}^{1}}{\cancel{9}_{1}}$$

$$=1$$

71. $2(-3)(-2)=-6(-2)$

$$=12$$

72. $(-4)(-1)(-3)=4(-3)$

$$=-12$$

73. $-1.2(-5.3)=6.36$

74. $0.002(-1,000)=-2$

75. $-\dfrac{2}{3}\left(\dfrac{1}{5}\right)=\dfrac{-2\cdot1}{3\cdot5}$

$$=-\frac{2}{15}$$

76. $-6(-3)(0)(-1)=0$

77. ELECTRONICS

$$2(1.5)=3$$

The new high is 3.

$$-3(1.5)=-4.5$$

The new low is -4.5.

78. a. Associative Property of Multiplication

　　b. Commutative Property of Multiplication

　　c. Multiplication Property of 1

　　　(Identity Property of Multiplication)

　　d. Inverse Property of Multiplication

79. $\dfrac{44}{-44}=-1$

80. $\dfrac{-272}{16}=-17$

81. $\dfrac{-81}{-27}=3$

82. $-\dfrac{3}{5}\div\dfrac{1}{2}=-\dfrac{3}{5}\cdot\dfrac{2}{1}$

$$=-\frac{3\cdot2}{5\cdot1}$$

$$=-\frac{6}{5}$$

83. $\dfrac{-60}{0}$ is undefined

84. $\dfrac{-4.5}{1}=-4.5$

85. $\dfrac{0}{18}=0$ because $0\cdot18=0$.

86. GEMSTONES

$$3,000-1,200=1,800$$
$$1,800\div5=360$$

The average annual depreciation is $-\$360$.

SECTION 1.7
Exponents and Order of Operations

87. a. $8\cdot8\cdot8\cdot8\cdot8=8^{5}$　　b. $9\cdot\pi\cdot r\cdot r=9\pi r^{2}$

88. a. $9^2 = 9 \cdot 9$
$$= 81$$

   b. $\left(-\dfrac{2}{3}\right)^3 = \left(-\dfrac{2}{3}\right)\left(-\dfrac{2}{3}\right)\left(-\dfrac{2}{3}\right)$
$$= -\dfrac{2 \cdot 2 \cdot 2}{3 \cdot 3 \cdot 3}$$
$$= -\dfrac{8}{27}$$

   c. $2^5 = 2 \cdot 2 \cdot 2 \cdot 2 \cdot 2$

   d. $50^1 = 50$
$$= 32$$

89. $2 + 5 \cdot 3 = 2 + 15$
$$= 17$$

90. $-24 \div 2 \cdot 3 = -12 \cdot 3$
$$= -36$$

91. $-(16 - 3)^2 = -(13)^2$
$$= -(13)(13)$$
$$= -169$$

92. $43 + 2(-6 - 2 \cdot 2) = 43 + 2(-6 - 4)$
$$= 43 + 2[-6 + (-4)]$$
$$= 43 + 2(-10)$$
$$= 43 + (-20)$$
$$= 23$$

93. $10 - 5[-3 - 2(5 - 7^2)] - 5$
$$= 10 - 5[-3 - 2(5 - 49)] - 5$$
$$= 10 - 5[-3 - 2(-44)] - 5$$
$$= 10 - 5[-3 + 88] - 5$$
$$= 10 - 5[85] - 5$$
$$= 10 - 425 - 5$$
$$= -415 - 5$$
$$= -420$$

94. $\dfrac{-4(4 + 2) - 4}{2|-18 - 4(5)|} = \dfrac{-4(6) - 4}{2|-18 - 20|}$
$$= \dfrac{-24 - 4}{2|-38|}$$
$$= \dfrac{-28}{2(38)}$$
$$= -\dfrac{\cancel{2} \cdot \cancel{2} \cdot 7}{\cancel{2} \cdot \cancel{2} \cdot 19}$$
$$= -\dfrac{7}{19}$$

95. $(-3)^3\left(\dfrac{-8}{2}\right) + 5 = -27(-4) + 5$
$$= 108 + 5$$
$$= 113$$

96. $\dfrac{2^4 - (4 - 6)(3 - 6)}{12 + 4\left[(-1)^8 - 2^2\right]} = \dfrac{16 - (-2)(-3)}{12 + 4[1 - 4]}$
$$= \dfrac{16 - 6}{12 + 4[-3]}$$
$$= \dfrac{10}{12 + (-12)}$$
$$= \dfrac{10}{0}$$
Undefined

97. a. $(-9)^2 = (-9)(-9)$
$$= 81$$

   b. $-9^2 = -(9 \cdot 9)$
$$= -81$$

98. WALK-A-THONS
$$\dfrac{20 \cdot 5 + 65 \cdot 10 + 25 \cdot 20 + 5 \cdot 50 + 10 \cdot 100}{20 + 65 + 25 + 5 + 10}$$
$$= \dfrac{100 + 650 + 500 + 250 + 1,000}{125}$$
$$= \dfrac{2,500}{125}$$
$$= 20$$
The average donation was $20.

Chapter 1 Review and Test

SECTION 1.8
Algebraic Expressions

99. a. 3 terms   b. 1 term

100. a. $16x^2 - 5x + 25$; $16, -5, 25$

    b. $\dfrac{x}{2} + y$; $\dfrac{1}{2}, 1$

101. $h + 25$

102. $3s - 15$

103. $\dfrac{1}{2}t - 6$

104. $\left| 2 - a^2 \right|$

105. HARDWARE

    $(n + 4)$ inch

106. HARDWARE

    $(b - 4)$ inch

107. $10d$

108. $(x - 5)$ years

109.

Type of coin	Number	· Value (in cents)	= Total value (in cents)
Nickel	6	5	**30**
Dime	d	10	**10d**

110.

x	$20x - x^3$
0	$20(0) - (0)^3 = 0 - 0$ $= \mathbf{0}$
1	$20(1) - (1)^3 = 20 - 1$ $= \mathbf{19}$
-4	$20(-4) - (-4)^3 = -80 - (-64)$ $= \mathbf{-16}$

111. $b^2 - 4ac = (-10)^2 - 4(3)(5)$

$\qquad\qquad = 100 - 60$

$\qquad\qquad = 40$

112. $\dfrac{x + y}{-x - z} = \dfrac{19 + 17}{-(19) - (-18)}$

$\qquad\quad = \dfrac{36}{-19 + 18}$

$\qquad\quad = \dfrac{36}{-1}$

$\qquad\quad = -36$

SECTION 1.9
Simplify Algebraic Expressions Using Properties of Real Numbers

113. $a \cdot 150 = \underline{150a}$

114. $9 + (1 + 7y) = \underline{(9 + 1) + 7y}$

115. $2.7(10b) = \underline{(2.7 \cdot 10)b}$

116. $x + 2x^2 = \underline{2x^2 + x}$

117. $-4(7w) = -4(7)w$

$\qquad\qquad = -28w$

118. $3(-2x)(-4) = 3(-2)(-4)x$

$\qquad\qquad\quad = 24x$

119. $0.4(5.2f) = 0.4(5.2)f$

$\qquad\qquad = 2.08f$

120. $\dfrac{7}{2} \cdot \dfrac{2}{7} r = \dfrac{\overset{1}{\cancel{7}} \cdot \overset{1}{\cancel{2}}}{\underset{1}{\cancel{2}} \cdot \underset{1}{\cancel{7}}} r$

$\qquad\quad = r$

121. $5(x + 3) = 5x + 5(3)$

$\qquad\qquad = 5x + 15$

122. $-(2x + 3 - y) = -(2x) + (-1)(3) - (-1)y$

$\qquad\qquad\qquad = -2x - 3 + y$

123. $\dfrac{3}{4}(4c - 8) = \dfrac{3}{4}\left(\dfrac{4}{1}\right)c - \dfrac{3}{4}\left(\dfrac{8}{1}\right)$

$\qquad\qquad = \dfrac{3 \cdot \overset{1}{\cancel{4}}}{\underset{1}{\cancel{4}} \cdot 1} c - \dfrac{3 \cdot \overset{1}{\cancel{4}} \cdot 2}{\underset{1}{\cancel{4}} \cdot 1}$

$\qquad\qquad = 3c - 6$

124. $-2(-3c - 7)(2.1) = -2(2.1)(-3c - 7)$

$\qquad\qquad\qquad = -4.2(-3c - 7)$

$\qquad\qquad\qquad = -4.2(-3c) - (-4.2)(7)$

$\qquad\qquad\qquad = 12.6c + 29.4$

125. $8p + 5p - 4p = (8 + 5 - 4)p$

$\qquad\qquad\qquad = 9p$

126. $-5m + 2 - 2m - 2 = -5m - 2m + 2 - 2$

$\qquad\qquad\qquad = (-5 - 2)m + (2 - 2)$

$\qquad\qquad\qquad = -7m + 0$

$\qquad\qquad\qquad = -7m$

127. $n+n+n+n = (1+1+1+1)n$
$$= 4n$$

128. $5(p-2)-2(3p+4)$
$$= 5(p)-5(2)-2(3p)+(-2)(4)$$
$$= 5p-10-6p-8$$
$$= 5p-6p-10-8$$
$$= (5-6)p+(-10-8)$$
$$= -p-18$$

129. $55.7k^2 - 55.6k^2 = (55.7-55.6)k^2$
$$= 0.1k^2$$

130. $8a^3 + 4a^3 + 2a - 4a^3 - 2a - 1$
$$= (8+4-4)a^3 + (2-2)a - 1$$
$$= 8a^3 + 0a - 1$$
$$= 8a^3 - 1$$

131. $\dfrac{3}{5}w - \left(-\dfrac{2}{5}w\right) = \dfrac{3}{5}w + \dfrac{2}{5}w$
$$= \left(\dfrac{3+2}{5}\right)w$$
$$= \dfrac{\cancel{5}^{1}}{\cancel{5}_{1}}w$$
$$= w$$

132. $36\left(\dfrac{1}{9}h - \dfrac{3}{4}\right) + 36\left(\dfrac{1}{3}\right)$
$$= \dfrac{36}{1}\left(\dfrac{1}{9}\right)h - \dfrac{36}{1}\left(\dfrac{3}{4}\right) + \dfrac{36}{1}\left(\dfrac{1}{3}\right)$$
$$= \dfrac{\cancel{9}\cdot 4}{\cancel{9}}h - \dfrac{\cancel{4}\cdot 9 \cdot 3}{\cancel{4}} + \dfrac{\cancel{3}\cdot 12}{\cancel{3}}$$
$$= 4h - 27 + 12$$
$$= 4h - 15$$

133. $-(7.6t-1.9)+(1.4t-1.2)8+t$
$$= -1(7.6t)-(-1)(1.9)+(1.4t)8-1.2(8)+t$$
$$= -7.6t+1.9+11.2t-9.6+t$$
$$= (-7.6t+11.2t+t)+(1.9-9.6)$$
$$= 4.6t-7.7$$

134. a. $1x = x$

    b. $-1x = -x$

    c. $4x-(-1) = 4x+1$

    d. $4x+(-1) = 4x-1$

CHAPTER 1 TEST

1. a. Two fractions. such as $\dfrac{1}{2}$ and $\dfrac{5}{10}$, that represent the same number are called **equivalent** fractions.

    b. The results of a multiplication is called a **product** .

    c. $\dfrac{8}{7}$ is the **reciprocal** of $\dfrac{7}{8}$ because $\dfrac{8}{7}\cdot\dfrac{7}{8} = 1$.

    d. $9x^2$ and $7x^2$ are **like** **terms** because they have the same variable raised to exactly the same power.

    e. For any nonzero raeal number a, $\dfrac{a}{0}$ is **undefined** .

2. SECURITY GUARDS

    a. $24   b. 5 hours

3. $f = \dfrac{a}{5}$

a	$f = \dfrac{a}{5}$
15	$f = \dfrac{a}{5}$ $= \dfrac{15}{5}$ $= 3$
100	$f = \dfrac{a}{5}$ $= \dfrac{100}{5}$ $= 20$
350	$f = \dfrac{a}{5}$ $= \dfrac{350}{5}$ $= 70$

4. $180 = 2 \cdot 2 \cdot 3 \cdot 3 \cdot 5$

$\quad\quad = 2^2 \cdot 3^2 \cdot 5$

5. $\dfrac{42}{105} = \dfrac{2 \cdot 3 \cdot 7}{3 \cdot 5 \cdot 7}$

$\quad = \dfrac{2 \cdot \cancel{3} \cdot \cancel{7}}{\cancel{3} \cdot 5 \cdot \cancel{7}}$

$\quad = \dfrac{2}{5}$

6. $\dfrac{15}{16} \div \dfrac{5}{8} = \dfrac{15}{16} \cdot \dfrac{8}{5}$

$\quad\quad = \dfrac{\cancel{5} \cdot 3 \cdot \cancel{8}}{\cancel{8} \cdot 2 \cdot \cancel{5}}$

$\quad\quad = \dfrac{3}{2}$

$\quad\quad = 1\dfrac{1}{2}$

7. $\dfrac{7}{10} + \dfrac{1}{14} = \dfrac{7}{10} \cdot \dfrac{7}{7} + \dfrac{1}{14} \cdot \dfrac{5}{5}$

$\quad\quad = \dfrac{49 + 5}{70}$

$\quad\quad = \dfrac{54}{70}$

$\quad\quad = \dfrac{\cancel{2} \cdot 27}{\cancel{2} \cdot 35}$

$\quad\quad = \dfrac{27}{35}$

8. $8\dfrac{2}{5} = 8\dfrac{2}{5} \cdot \dfrac{3}{3} = 8\dfrac{6}{15} = 7\dfrac{15}{15} + \dfrac{6}{15} = 7\dfrac{21}{15}$

$-1\dfrac{2}{3} = -1\dfrac{2}{3} \cdot \dfrac{5}{5} = -1\dfrac{10}{15} = -1\dfrac{10}{15} \quad\quad = -1\dfrac{10}{15}$

$\quad\quad\quad\quad\quad\quad\quad\quad\quad\quad\quad\quad\quad\quad\quad\quad\quad 6\dfrac{11}{15}$

9. SHOPPING

a. $4\dfrac{1}{4}$ pounds

b. $4\dfrac{1}{4} \cdot 84 = \dfrac{17}{4} \cdot \dfrac{84}{1}$

$\quad\quad\quad\quad\quad = \dfrac{17 \cdot 84}{4}$

$\quad\quad\quad\quad\quad = \dfrac{17 \cdot \cancel{4} \cdot 21}{\cancel{4}}$

$\quad\quad\quad\quad\quad = 357$

The fruit cost $3.57.

10. $6\overline{)5.000}$ quotient 0.833

$\dfrac{5}{6} = 0.8\bar{3}$

11a.

11b. Natural: 2

Whole: 0, 2

Integer: $0, 2, -3$

Rational: $-1\dfrac{1}{4}, 0, -3.75, 2, \dfrac{7}{2}, 0.5, -3$

Irrational: $\sqrt{2}$

Real: All

12. a. True b. False c. True
    d. True e. True

13. a. $-2 \boxed{>} -3$

b. $-|-9| \; ? \; 8$

$\quad -9 \; ? \; 8$

$\quad -9 \; < \; 8$

$\quad -|-9| \; \boxed{<} \; 8$

c. $|-4| \; ? \; -(-5)$

$\quad 4 \; ? \; 5$

$\quad 4 \; < \; 5$

$\quad |-4| \boxed{<} -(-5)$

d. $\left|-\dfrac{7}{8}\right| \; ? \; 0.5$

$\quad \dfrac{7}{8} \; ? \; 0.5$

$\quad 0.875 > 0.5$

$\quad \left|-\dfrac{7}{8}\right| \boxed{>} 0.5$

14. TELEVISION

$$\frac{0.6+(-0.3)+1.7+1.5+(-0.2)+1.1+(-0.2)}{7}$$

$$=\frac{4.9+(-0.7)}{7}$$

$$=\frac{4.2}{7}$$

$$=0.6$$

A gain of 0.6 of a rating point.

15. $-5.6+(-2)=-7.6$

16. $(-6)+8+(-4)=2+(-4)$
$$=-2$$

17. $-\dfrac{1}{2}+\dfrac{7}{8}=-\dfrac{1}{2}\cdot\dfrac{4}{4}+\dfrac{7}{8}$
$$=\frac{-4+7}{8}$$
$$=\frac{3}{8}$$

18. a. $-10-(-4)=-10+4$
$$=-6$$
   b. $-6+(-4)=-10$

19. a. $\dfrac{-12.6}{-0.9}=14$
   b. $14(-0.9)=-12.6$

20. $(-2)(-3)(-5)=6(-5)$
$$=-30$$

21. $-6.1(0.4)=-2.44$

22. $\dfrac{0}{-3}=0$

23. $\left(-\dfrac{3}{5}\right)^3=\left(-\dfrac{3}{5}\right)\left(-\dfrac{3}{5}\right)\left(-\dfrac{3}{5}\right)$
$$=\frac{-3\cdot-3\cdot-3}{5\cdot5\cdot5}$$
$$=-\frac{27}{125}$$

24. $3+(-3)=0$

25. $0-3=0+(-3)$
$$=-3$$

26. $-30+50-10-(-40)$
$$=-30+50+(-10)+40$$
$$=-40+90$$
$$=50$$

27. ASTRONOMY
$-12.5-(-26.5)=-12.5+26.5$
$$=14$$

They differ by a magnitude of 14.

28. GLACIERS
$87-165=87+(-165)$
$$=-78$$

There was a net lost of 78 inches.

29. INVENTORY
$85\cdot(-15)=-1,275$

There was a financial lost of $1,275.

30. a. $[-12+(97+3)]$
   b. $2x+14$
   c. $-2(5)m$
   d. 1
   e. $15x$

31. a. $9(9)(9)(9)(9)=9^5$
   b. $3\cdot x\cdot x\cdot z\cdot z\cdot z=3x^2z^3$

32. $2x-\dfrac{30}{x}$

x	$2x-\dfrac{30}{x}$
5	$2x-\dfrac{30}{x}=2(\mathbf{5})-\dfrac{30}{\mathbf{5}}$ $=10-6$ $=\mathbf{4}$
10	$2x-\dfrac{30}{x}=2(\mathbf{10})-\dfrac{30}{\mathbf{10}}$ $=20-3$ $=\mathbf{17}$
-30	$2x-\dfrac{30}{x}=2(\mathbf{-30})-\dfrac{30}{\mathbf{-30}}$ $=-60+1$ $=\mathbf{-59}$

Chapter 1 Review and Test

33. $8 + 2 \cdot 3^4 = 8 + 2 \cdot 81$

$$= 8 + 162$$
$$= 170$$

34. $\dfrac{3(40 - 2^3)}{-2(6-4)^2} = \dfrac{3(40-8)}{-2(2)^2}$

$$= \dfrac{3[(40 + (-8)]}{-2(4)}$$
$$= \dfrac{3[32]}{-8}$$
$$= \dfrac{96}{-8}$$
$$= -12$$

35. $-10^2 - 5 + 6 = -(10)(10) - 5 + 6$

$$= -100 - 5 + 6$$
$$= -105 + 6$$
$$= -99$$

36. $9 - 3\left[45 - 5^2(1^5 - 4)\right]$

$$= 9 - 3\left[45 - 25(1 - 4)\right]$$
$$= 9 - 3\left[45 - 25[1 + (-4)]\right]$$
$$= 9 - 3\left[45 - 25(-3)\right]$$
$$= 9 - 3\left[45 + 75\right]$$
$$= 9 - 3\left[120\right]$$
$$= 9 - 360$$
$$= 9 + (-360)$$
$$= -351$$

37. $|-50 \div 5 \cdot 2| = |-10 \cdot 2|$

$$= |-20|$$
$$= 20$$

38. $3(10x - y) - 5(x + y^2)$

$$= 3[10(2) - (-5)] - 5[2 + (-5)^2]$$
$$= 3[20 + 5] - 5[2 + 25]$$
$$= 3[25] - 5[27]$$
$$= 75 - 135$$
$$= -60$$

39. $2w - 7$

40. a. Music   b. Money

    $x - 2$ $25q¢$

41. 3 terms

42. $1,\ -6,\ -1,\ 10$

43. $5(-4x) = 5(-4)x$

$$= -20x$$

44. $-8(-7t)(4) = -8(-7)(4)t$

$$= 224t$$

45. $\dfrac{4}{5}(15a + 5) - 16a = \dfrac{4}{5} \cdot \dfrac{15}{1}a + \dfrac{4}{5} \cdot \dfrac{5}{1} - 16a$

$$= \dfrac{4 \cdot 15}{5 \cdot 1}a + \dfrac{4 \cdot 5}{5 \cdot 1} - 16a$$
$$= \dfrac{4 \cdot \cancel{5} \cdot 3}{\cancel{5}}a + \dfrac{4 \cdot \cancel{5}}{\cancel{5}} - 16a$$
$$= 12a - 16a + 4$$
$$= 12a + (-16a) + 4$$
$$= -4a + 4$$

46. $-1.1d^3 - 3.8d^3 - d^3$

$$= [-1.1 + (-3.8) + (-1)]d^3$$
$$= -5.9d^3$$

47. $9x + 2(7x - 3) - 9(x - 1)$

$$= 9x + 2(7)x + 2(-3) - 9x - 9(-1)$$
$$= 9x + 14x - 6 - 9x + 9$$
$$= (9x - 9x + 14x) + (-6 + 9)$$
$$= 14x + 3$$

48. a. True   b. False   c. False
   d. False   e. True   f. False

49. $m^2 + 4m^2 + 5m - 2m^2 - 3m - 4$

$$= (1 + 4 - 2)m^2 + (5 - 3)m - 4$$
$$= 3m^2 + 2m - 4$$

SECTION 2.1
VOCABULARY

Fill in the blanks.
1. A statement indicating that two expressions are equal, such as $x + 1 = 7$, is called an **equation**.
3. To **solve** an equation means to find all values of the variable that make the equation true.
5. Equations with the same solutions are called **equivalent** equations.

CONCEPTS
7. a. $x+6$ b. Neither   c. No   d. Yes

9. a. $a+c = b + \boxed{c}$ and $a - c = b - \boxed{c}$.

 b. $ca = \boxed{c}b$ and $\dfrac{a}{c} = \dfrac{b}{\boxed{c}} (c \neq 0)$.

11. a. x b. y c. t d. h

NOTATION
Complete each solution to solve the equation.
13. $x - 5 = 45$ Check: $x - 5 = 45$

$x - 5 + \boxed{5} = 45 + \boxed{5}$ $\boxed{50} - 5 \overset{?}{=} 45$

$x = \boxed{50}$ $\boxed{45} = 45$ True

$\boxed{50}$ is the solution.

15. a. Is possibly equal to   b. Yes

GUIDED PRACTICE
Check to determine whether the number in red is a solution of the equation. See Example 1.

17. $6 + 12 \overset{?}{=} 28$

$18 = 28$ False

6 is not a solution.

19. $2(-8) + 3 \overset{?}{=} -15$

$-16 + 3 \overset{?}{=} -15$

$-13 = -15$ False

−8 is not a solution.

21. $0.5(5) \overset{?}{=} 2.9$

$2.5 = 2.9$ False

5 is not a solution.

23. $33 - \dfrac{-6}{2} \overset{?}{=} 30$

$33 + 3 \overset{?}{=} 30$

$36 = 30$ False

−6 is not a solution.

25. $|-2 - 8| \overset{?}{=} 10$

$|-10| \overset{?}{=} 10$

$10 = 10$ True

−2 is a solution.

27. $3(12) - 2 \overset{?}{=} 4(12) - 5$

$36 - 2 \overset{?}{=} 48 - 5$

$34 = 43$ False

12 is not a solution.

29. $(-3)^2 - (-3) - 6 \overset{?}{=} 0$

$9 + 3 - 6 \overset{?}{=} 0$

$12 - 6 \overset{?}{=} 0$

$6 = 0$ False

−3 is not a solution.

31. $\dfrac{2}{1+1} + 5 \overset{?}{=} \dfrac{12}{1+1}$

$\dfrac{2}{2} + 5 \overset{?}{=} \dfrac{12}{2}$

$1 + 5 \overset{?}{=} 6$

$6 = 6$ True

1 is a solution.

33. $\dfrac{3}{4} - \dfrac{1}{8} \overset{?}{=} \dfrac{5}{8}$

$\dfrac{6}{8} - \dfrac{1}{8} \overset{?}{=} \dfrac{5}{8}$

$\dfrac{5}{8} = \dfrac{5}{8}$ True

3/4 is a solution.

35. $(-3 - 4)(-3 + 3) \overset{?}{=} 0$

$(-7)(0) \overset{?}{=} 0$

$0 = 0$ True

− 3 is a solution.

Use a property of equality to solve each equation. Then check the result. See Example 2.

37.
$$a - 5 = 66$$
$$a - 5 + \mathbf{5} = 66 + \mathbf{5}$$
$$a = 71$$
Check: $a - 5 = 66$
$$71 - 5 \overset{?}{=} 66$$
$$66 = 66 \quad \text{True}$$
71 is the solution.

39.
$$9 = p - 9$$
$$9 + \mathbf{9} = p - 9 + \mathbf{9}$$
$$18 = p$$
Check: $9 = p - 9$
$$9 \overset{?}{=} 18 - 9$$
$$9 = 9 \quad \text{True}$$
18 is the solution.

Use a property of equality to solve each equation. Then check the result. See Example 3.

41.
$$-16 = y - 4$$
$$-16 + \mathbf{4} = y - 4 + \mathbf{4}$$
$$-12 = y$$
Check: $-16 = y - 4$
$$-16 \overset{?}{=} -12 - 4$$
$$-16 = -16 \quad \text{True}$$
-12 is the solution.

43.
$$-3 + a = 0$$
$$-3 + a + \mathbf{3} = 0 + \mathbf{3}$$
$$a = 3$$
Check: $-3 + a = 0$
$$-3 + 3 \overset{?}{=} 0$$
$$0 = 0 \quad \text{True}$$
3 is the solution.

Use a property of equality to solve each equation. Then check the result. See Example 4.

45.
$$x + \frac{1}{10} = \frac{6}{5}$$
$$x + \frac{1}{10} - \frac{\mathbf{1}}{\mathbf{10}} = \frac{6}{5} - \frac{\mathbf{1}}{\mathbf{10}}$$
$$x = \frac{6}{5} \cdot \frac{2}{2} - \frac{1}{10}$$
$$x = \frac{12 - 1}{10}$$
$$x = \frac{11}{10}$$

Check:
$$x + \frac{1}{10} = \frac{6}{5}$$
$$\frac{11}{10} + \frac{1}{10} \overset{?}{=} \frac{6}{5}$$
$$\frac{12}{10} \overset{?}{=} \frac{6}{5}$$
$$\frac{\overset{1}{\cancel{2}} \cdot 6}{\cancel{2} \cdot 5} \overset{?}{=} \frac{6}{5}$$
$$\frac{6}{5} = \frac{6}{5} \quad \text{True}$$

$\frac{11}{10}$ is the solution.

47.
$$3.5 + f = 1.2$$
$$3.5 + f - \mathbf{3.5} = 1.2 - \mathbf{3.5}$$
$$f = -2.3$$
Check:
$$3.5 + f = 1.2$$
$$3.5 + (-2.3) \overset{?}{=} 1.2$$
$$1.2 = 1.2 \quad \text{True}$$
-2.3 is the solution.

Use a property of equality to solve each equation. Then check the result. See Example 5.

49.
$$\frac{x}{15} = 3$$
$$\mathbf{15} \cdot \frac{x}{15} = \mathbf{15} \cdot 3$$
$$x = 45$$
Check:
$$\frac{x}{15} = 3$$
$$\frac{45}{15} \overset{?}{=} 3$$
$$3 = 3 \quad \text{True}$$
45 is the solution.

51.
$$\frac{d}{8} = -6$$
$$\mathbf{8} \cdot \frac{d}{8} = \mathbf{8} \cdot -6$$
$$d = -48$$
Check:
$$\frac{d}{8} = -6$$
$$\frac{-48}{8} \overset{?}{=} -6$$
$$-6 = -6 \quad \text{True}$$
-48 is the solution.

Use a property of equality to solve each equation. Then check the result. See Example 6.

53. $\dfrac{4}{5}t = 16$

$$\left(\dfrac{5}{4}\right)\cdot\dfrac{4}{5}t = \left(\dfrac{5}{4}\right)\cdot 16$$

$$\left(\dfrac{5}{4}\cdot\dfrac{4}{5}\right)t = \dfrac{5\cdot 4\cdot \overset{1}{\cancel{4}}}{\cancel{4}_{1}}$$

$$t = 20$$

Check: $\dfrac{4}{5}t = 16$

$$\dfrac{4}{5}\cdot 20 \overset{?}{=} 16$$

$$\dfrac{4\cdot 4\cdot \overset{1}{\cancel{5}}}{\cancel{5}_{1}} \overset{?}{=} 16$$

$$16 = 16 \quad \text{True}$$

20 is the solution.

55. $-\dfrac{7r}{2} = \dfrac{5}{12}$

$$\left(-\dfrac{2}{7}\right)\cdot -\dfrac{7r}{2} = \left(-\dfrac{2}{7}\right)\cdot\dfrac{5}{12}$$

$$\left(-\dfrac{2}{7}\cdot -\dfrac{7}{2}\right)r = -\dfrac{\overset{1}{\cancel{2}}\cdot 5}{7\cdot\cancel{2}\cdot 6}_{1}$$

$$r = -\dfrac{5}{42}$$

Check: $-\dfrac{7r}{2} = \dfrac{5}{12}$

$$-\dfrac{7}{2}\cdot\left(-\dfrac{5}{42}\right) \overset{?}{=} \dfrac{5}{12}$$

$$\dfrac{\overset{1}{\cancel{7}}\cdot 5}{2\cdot\cancel{7}\cdot 6}_{1} \overset{?}{=} \dfrac{5}{12}$$

$$\dfrac{5}{12} = \dfrac{5}{12} \quad \text{True}$$

$-\dfrac{5}{42}$ is the solution.

Use a property of equality to solve each equation. Then check the result. See Example 7.

57. $4x = 16$

$$\dfrac{4x}{4} = \dfrac{16}{4}$$

$$x = 4$$

Check: $4x = 16$

$$4\cdot 4 \overset{?}{=} 16$$

$$16 = 16 \quad \text{True}$$

4 is the solution.

59. $-1.7 = -3.4y$

$$\dfrac{-1.7}{-3.4} = \dfrac{-3.4y}{-3.4}$$

$$0.5 = y$$

Check: $-1.7 = -3.4y$

$$-1.7 \overset{?}{=} -3.4\cdot 0.5$$

$$-1.7 = -1.7 \quad \text{True}$$

0.5 is the solution.

Use a property of equality to solve each equation. Then check the result. See Example 8.

61. $-x = 18$

$$(-1)-x = (-1)18$$

$$x = -18$$

Check: $-x = 18$

$$-(-18) \overset{?}{=} 18$$

$$18 = 18 \quad \text{True}$$

-18 is the solution.

63. $-n = \dfrac{4}{21}$

$$(-1)-n = (-1)\dfrac{4}{21}$$

$$n = -\dfrac{4}{21}$$

Check: $-n = \dfrac{4}{21}$

$$-\left(-\dfrac{4}{21}\right) \overset{?}{=} \dfrac{4}{21}$$

$$\dfrac{4}{21} = \dfrac{4}{21} \quad \text{True}$$

$-\dfrac{4}{21}$ is the solution.

Section 2.1

TRY IT YOURSELF

Solve each equation. Then check the results.

65. $63 = 9c$ Check: $63 = 9c$

$$\frac{63}{\mathbf{9}} = \frac{9c}{\mathbf{9}}$$

$$63 \overset{?}{=} 9 \cdot 7$$

$$7 = c$$

$$63 = 63 \quad \text{True}$$

7 is the solution.

67. $d - \dfrac{1}{9} = \dfrac{7}{9}$

$$d - \frac{1}{9} + \frac{\mathbf{1}}{\mathbf{9}} = \frac{7}{9} + \frac{\mathbf{1}}{\mathbf{9}}$$

$$d = \frac{8}{9}$$

Check: $d - \dfrac{1}{9} = \dfrac{7}{9}$

$$\frac{8}{9} - \frac{1}{9} \overset{?}{=} \frac{7}{9}$$

$$\frac{7}{9} = \frac{7}{9} \quad \text{True}$$

$\dfrac{8}{9}$ is the solution.

69. $0 = \dfrac{v}{11}$

$$\mathbf{11} \cdot 0 = \mathbf{11} \cdot \frac{v}{11}$$

$$0 = v$$

Check: $0 = \dfrac{v}{11}$

$$0 \overset{?}{=} \frac{0}{11}$$

$$0 = 0 \quad \text{True}$$

0 is the solution.

71. $x - 1.6 = -2.5$

$$x - 1.6 + \mathbf{1.6} = -2.5 + \mathbf{1.6}$$

$$x = -0.9$$

Check: $x - 1.6 = -2.5$

$$-0.9 - 1.6 \overset{?}{=} -2.5$$

$$-2.5 = -2.5$$

-0.9 is the solution.

73. $\dfrac{2}{3}c = 10$

$$\left(\frac{\mathbf{3}}{\mathbf{2}}\right) \cdot \frac{2}{3}c = \left(\frac{\mathbf{3}}{\mathbf{2}}\right) \cdot 10$$

$$\left(\frac{3}{2} \cdot \frac{2}{3}c\right) = \frac{3 \cdot 5 \cdot \cancel{2}}{\cancel{2}}$$

$$c = 15$$

Check: $\dfrac{2}{3}c = 10$

$$\frac{2}{3} \cdot 15 \overset{?}{=} 10$$

$$\frac{2 \cdot 5 \cdot \cancel{3}}{\cancel{3}} \overset{?}{=} 10$$

$$10 = 10 \quad \text{True}$$

15 is the solution.

75. $-100 = -5g$

$$\frac{-100}{\mathbf{-5}} = \frac{-5g}{\mathbf{-5}}$$

$$20 = g$$

Check: $-100 = -5g$

$$-100 \overset{?}{=} -5 \cdot 20$$

$$-100 = -100 \quad \text{True}$$

20 is the solution.

77. $s + \dfrac{1}{5} = \dfrac{4}{25}$

$$s + \frac{1}{5} - \frac{\mathbf{1}}{\mathbf{5}} = \frac{4}{25} - \frac{\mathbf{1}}{\mathbf{5}}$$

$$s = \frac{4}{25} - \frac{1}{5} \cdot \frac{\mathbf{5}}{\mathbf{5}}$$

$$s = \frac{4}{25} - \frac{5}{25}$$

$$s = -\frac{1}{25}$$

Check: $s + \dfrac{1}{5} = \dfrac{4}{25}$

$-\dfrac{1}{25} + \dfrac{1}{5} \cdot \dfrac{5}{5} \overset{?}{=} \dfrac{4}{25}$

$-\dfrac{1}{25} + \dfrac{5}{25} \overset{?}{=} \dfrac{4}{25}$

$\dfrac{4}{25} = \dfrac{4}{25}$ True

$-\dfrac{1}{25}$ is the solution.

79.　　$\dfrac{d}{-7} = -3$

$(-7)\dfrac{d}{-7} = (-7) \cdot -3$

$d = 21$

Check: $\dfrac{d}{-7} = -3$

$\dfrac{21}{-7} \overset{?}{=} -3$

$-3 = -3$ True

21 is the solution.

81.　$8h = 0$

$\dfrac{8h}{8} = \dfrac{0}{8}$

$h = 0$

Check: $8h = 0$

$8 \cdot 0 \overset{?}{=} 0$

$0 = 0$ True

0 is the solution.

83.　　$\dfrac{y}{0.6} = -4.4$

$(0.6) \cdot \dfrac{y}{0.6} = (0.6) \cdot -4.4$

$y = -2.64$

Check: $\dfrac{y}{0.6} = -4.4$

$\dfrac{-2.64}{0.6} \overset{?}{=} -4.4$

$-4.4 = -4.4$ True

-2.64 is the solution.

85.　$23b = 23$

$\dfrac{23b}{23} = \dfrac{23}{23}$

$b = 1$

Check:　$23b = 23$

$23 \cdot 1 \overset{?}{=} 23$

$23 = 23$ True

1 is the solution.

87.　　　$-\dfrac{5}{4}h = -5$

$\left(-\dfrac{4}{5}\right) \cdot -\dfrac{5}{4}h = \left(-\dfrac{4}{5}\right) \cdot -5$

$\left(-\dfrac{4}{5} \cdot -\dfrac{5}{4}\right)h = \dfrac{-4 \cdot -1 \cdot \cancel{5}^{\,1}}{\cancel{5}_{\,1}}$

$h = 4$

Check:　$-\dfrac{5}{4}h = -5$

$-\dfrac{5}{4} \cdot 4 \overset{?}{=} -5$

$\dfrac{-5 \cdot \cancel{4}^{\,1}}{\cancel{4}_{\,1}} \overset{?}{=} -5$

$-5 = -5$ True

4 is the solution.

89.　　　$8.9 = -4.1 + t$

$8.9 + \mathbf{4.1} = -4.1 + t + \mathbf{4.1}$

$13 = t$

Check:　$8.9 = -4.1 + t$

$8.9 \overset{?}{=} -4.1 + 13$

$8.9 = 8.9$ True

13 is the solution.

91.　　　$-2.5 = -m$

$(-1) - 2.5 = (-1)(-m)$

$2.5 = m$

Check:　$-2.5 = -m$

$-2.5 \overset{?}{=} (-1)2.5$

$-2.5 = -2.5$ True

2.5 is the solution.

93.

$$-\frac{9}{8}x = 3$$

$$\left(-\frac{8}{9}\right) \cdot -\frac{9}{8}x = \left(-\frac{8}{9}\right) \cdot 3$$

$$\left(-\frac{8}{9} \cdot -\frac{9}{8}\right)x = \frac{-8 \cdot \overset{1}{\cancel{3}}}{3 \cdot \underset{1}{\cancel{3}}}$$

$$x = -\frac{8}{3}$$

Check: $\quad -\frac{9}{8}x = 3$

$$-\frac{9}{8} \cdot -\frac{8}{3} \overset{?}{=} 3$$

$$\frac{3 \cdot \cancel{3} \cdot \cancel{8}}{\cancel{8} \cdot \cancel{3}} \overset{?}{=} 3$$

$$3 = 3 \quad \text{True}$$

$-\frac{8}{3}$ is the solution.

95.

$$\frac{2}{3}n = -\frac{7}{8}$$

$$\left(\frac{3}{2}\right) \cdot \frac{2}{3}n = \left(\frac{3}{2}\right) \cdot -\frac{7}{8}$$

$$\left(\frac{3}{2} \cdot \frac{2}{3}\right)n = \frac{3 \cdot -7}{2 \cdot 8}$$

$$n = -\frac{21}{16}$$

Check: $\quad \frac{2}{3}n = -\frac{7}{8}$

$$\frac{2}{3} \cdot \left(-\frac{21}{16}\right) \overset{?}{=} -\frac{7}{8}$$

$$\frac{\cancel{2} \cdot \cancel{3} \cdot -7}{\cancel{3} \cdot \cancel{2} \cdot 8} \overset{?}{=} -\frac{7}{8}$$

$$-\frac{7}{8} = -\frac{7}{8} \quad \text{True}$$

$-\frac{21}{16}$ is the solution.

97.

$$-10 = n - 5$$

$$-10 + 5 = n - 5 + 5$$

$$-5 = n$$

Check: $\quad -10 = n - 5$

$$-10 \overset{?}{=} -5 - 5$$

$$-10 = -10 \quad \text{True}$$

-5 is the solution.

99.

$$\frac{h}{-40} = 5$$

$$(-40) \cdot \frac{h}{-40} = (-40) \cdot 5$$

$$h = -200$$

Check: $\quad \frac{h}{-40} = 5$

$$\frac{-200}{-40} \overset{?}{=} 5$$

$$5 = 5 \quad \text{True}$$

-200 is the solution.

101.

$$-\frac{15}{16}a = -\frac{5}{4}$$

$$\left(-\frac{16}{15}\right) \cdot -\frac{15}{16}a = \left(-\frac{16}{15}\right) \cdot -\frac{5}{4}$$

$$\left(-\frac{16}{15} \cdot -\frac{15}{16}\right)a = \frac{-4 \cdot \cancel{4} \cdot -1 \cdot \cancel{5}}{3 \cdot \cancel{5} \cdot \cancel{4}}$$

$$a = \frac{4}{3}$$

Check: $\quad -\frac{15}{16}a = -\frac{5}{4}$

$$-\frac{15}{16} \cdot \left(\frac{4}{3}\right) \overset{?}{=} -\frac{5}{4}$$

$$\frac{\cancel{3} \cdot -5 \cdot \cancel{4}}{4 \cdot \cancel{4} \cdot \cancel{3}} \overset{?}{=} -\frac{5}{4}$$

$$-\frac{5}{4} = -\frac{5}{4} \quad \text{True}$$

$\frac{4}{3}$ is the solution.

103. $-15x = -60$

$$\frac{-15x}{\mathbf{-15}} = \frac{-60}{\mathbf{-15}}$$

$$x = 4$$

Check: $-15x = -60$

$$-15 \cdot 4 \overset{?}{=} -60$$

$$-60 = -60 \quad \text{True}$$

4 is the solution.

LOOK ALIKES

105. a. $d + \dfrac{1}{10} = \dfrac{3}{4}$

$$d + \frac{1}{10} - \frac{\mathbf{1}}{\mathbf{10}} = \frac{3}{4} - \frac{\mathbf{1}}{\mathbf{10}}$$

$$d = \frac{3}{4}\left(\frac{\mathbf{5}}{\mathbf{5}}\right) - \frac{1}{10}\left(\frac{\mathbf{2}}{\mathbf{2}}\right)$$

$$d = \frac{15 - 2}{20}$$

$$d = \frac{13}{20}$$

Check: $d + \dfrac{1}{10} = \dfrac{3}{4}$

$$\frac{13}{20} + \frac{1}{10} \cdot \frac{\mathbf{2}}{\mathbf{2}} \overset{?}{=} \frac{3}{4}$$

$$\frac{13}{20} + \frac{2}{20} \overset{?}{=} \frac{3}{4}$$

$$\frac{15}{20} \overset{?}{=} \frac{3}{4}$$

$$\frac{\overset{1}{\cancel{5}} \cdot 3}{\cancel{5} \cdot 4} \overset{?}{=} \frac{3}{4}$$

$$\frac{3}{4} = \frac{3}{4} \quad \text{True}$$

$\dfrac{13}{20}$ is the solution.

105. b. $d - \dfrac{1}{10} = \dfrac{3}{4}$

$$d - \frac{1}{10} + \frac{\mathbf{1}}{\mathbf{10}} = \frac{3}{4} + \frac{\mathbf{1}}{\mathbf{10}}$$

$$d = \frac{3}{4}\left(\frac{\mathbf{5}}{\mathbf{5}}\right) + \frac{1}{10}\left(\frac{\mathbf{2}}{\mathbf{2}}\right)$$

$$d = \frac{15 + 2}{20}$$

$$d = \frac{17}{20}$$

Check: $d - \dfrac{1}{10} = \dfrac{3}{4}$

$$\frac{17}{20} - \frac{1}{10} \cdot \frac{\mathbf{2}}{\mathbf{2}} \overset{?}{=} \frac{3}{4}$$

$$\frac{17}{20} - \frac{2}{20} \overset{?}{=} \frac{3}{4}$$

$$\frac{15}{20} \overset{?}{=} \frac{3}{4}$$

$$\frac{\overset{1}{\cancel{5}} \cdot 3}{\cancel{5} \cdot 4} \overset{?}{=} \frac{3}{4}$$

$$\frac{3}{4} = \frac{3}{4} \quad \text{True}$$

$\dfrac{17}{20}$ is the solution.

105. c. $\dfrac{1}{10}d = \dfrac{3}{4}$

$$\frac{\mathbf{10}}{\mathbf{1}}\left(\frac{1}{10}d\right) = \frac{\mathbf{10}}{\mathbf{1}} \cdot \frac{3}{4}$$

$$d = \frac{\overset{1}{\cancel{2}} \cdot 5 \cdot 3}{1 \cdot \cancel{2} \cdot 2}$$

$$d = \frac{15}{2}$$

Section 2.1

Check: $\dfrac{1}{10}d = \dfrac{3}{4}$

$$\dfrac{1}{10}\cdot\dfrac{15}{2} \overset{?}{=} \dfrac{3}{4}$$

$$\dfrac{15}{20} \overset{?}{=} \dfrac{3}{4}$$

$$\dfrac{\overset{1}{\cancel{5}}\cdot 3}{\underset{1}{\cancel{5}}\cdot 4} \overset{?}{=} \dfrac{3}{4}$$

$$\dfrac{3}{4} = \dfrac{3}{4} \quad \text{True}$$

$\dfrac{15}{2}$ is the solution.

105. d. $\qquad 10d = \dfrac{3}{4}$

$$\dfrac{1}{10}\left(\dfrac{10}{1}d\right) = \dfrac{1}{10}\cdot\dfrac{3}{4}$$

$$d = \dfrac{3}{40}$$

Check: $\qquad 10d = \dfrac{3}{4}$

$$\dfrac{10}{1}\cdot\dfrac{3}{40} \overset{?}{=} \dfrac{3}{4}$$

$$\dfrac{30}{40} \overset{?}{=} \dfrac{3}{4}$$

$$\dfrac{\overset{1}{\cancel{10}}\cdot 3}{\underset{1}{\cancel{10}}\cdot 4} \overset{?}{=} \dfrac{3}{4}$$

$$\dfrac{3}{4} = \dfrac{3}{4} \quad \text{True}$$

$\dfrac{3}{40}$ is the solution.

APPLICATIONS

107. SYNTHESIZERS

$$x + 115 = 180$$

$$x + 115 - \mathbf{115} = 180 - \mathbf{115}$$

$$x = 65$$

Check: $x + 115 = 180$

$$65 + 115 \overset{?}{=} 180$$

$$180 = 180 \quad \text{True}$$

$65°$ is the measure of the angle.

109. SHARING THE WINNING TICKET

$$\dfrac{x}{16} = 375{,}000$$

$$\mathbf{16}\cdot\dfrac{x}{16} = \mathbf{16}\cdot 375{,}000$$

$$x = 6{,}000{,}000$$

Check: $\qquad \dfrac{x}{16} = 375{,}000$

$$\dfrac{6{,}000{,}000}{16} \overset{?}{=} 375{,}000$$

$$375{,}000 = 375{,}000 \quad \text{True}$$

$\$6{,}000{,}000$ is the amount of the jackpot.

WRITING
111-113. Answers will vary.

REVIEW
115. Evaluate:

$$-9 - 3x = -9 - 3(-3)$$

$$= -9 + 9$$

$$= 0$$

117. Translate to symbols:
Subtract x from 45.

$$45 - x$$

CHALLENGE PROBLEMS
119. If $a + 81 = 49$, what is $a - 81$?

$$a + 81 = 49$$

$$a + 81 - \mathbf{81} = 49 - \mathbf{81}$$

$$a = -32$$

$$a - 81 = -32 - 81$$

$$= -113$$

SECTION 2.2

VOCABULARY
Fill in the blanks.

1. $3x + 8 = 10$ is an example of a linear **equation** in one variable.

3. A linear equation that is true for any permissible replacement value for the variable is called an **identity**.

CONCEPTS
Fill in the blanks.

5. a. To solve $3x - 5 = 1$, we first undo the **subtraction** of 5 by adding 5 to both sides. Then we undo the **multiplication** by 3 by dividing both sides by 3.

   b. To solve $\frac{x}{2} + 3 = 5$, we can undo the **addition** of 3 by subtracting 3 from both sides. Then we can undo the **division** by 2 by multiplying both sides by 2.

7. a.
$$6x + 5 = 7$$
$$6(\mathbf{-2}) + 5 \overset{?}{=} 7$$
$$-12 + 5 \overset{?}{=} 7$$
$$-7 = 7$$
$$\text{No}$$

9. a. 6   b. 10

LOOK ALIKES ...

10. a. $2x + 5$ b. $\dfrac{4}{3}$ c. 23   d. No

NOTATION
Complete the solution.

11. Solve: $2x - 7 = 21$
$$2x - 7\boxed{+7} = 21\boxed{+7}$$
$$2x = 28$$
$$\frac{2x}{\boxed{2}} = \frac{28}{\boxed{2}}$$
$$x = 14$$

Check: $2x - 7 = 21$
$$2(\boxed{14}) - 7 \overset{?}{\boxed{=}} 21$$
$$\boxed{28} - 7 \overset{?}{=} 21$$
$$\boxed{21} = 21$$
$$\boxed{14} \text{ is the solution.}$$

GUIDED PRACTICE
Solve each equation and check the result.

13.
$$-8x + 1 = 73$$
$$-8x + 1 - \mathbf{1} = 73 - \mathbf{1}$$
$$-8x = 72$$
$$\frac{-8x}{\mathbf{-8}} = \frac{72}{\mathbf{-8}}$$
$$x = -9$$
$$-8 \text{ is the solution.}$$

Check: $-8x + 1 = 73$
$$-8(-9) + 1 \overset{?}{=} 73$$
$$72 + 1 \overset{?}{=} 73$$
$$73 = 73$$
$$-9 \text{ is the solution.}$$

15.
$$-5q - 2 = 23$$
$$-5q - 2 + \mathbf{2} = 23 + \mathbf{2}$$
$$-5q = 25$$
$$\frac{-5q}{\mathbf{-5}} = \frac{25}{\mathbf{-5}}$$
$$q = -5$$
$$-10 \text{ is the solution.}$$

Check: $-5q - 2 = 23$
$$-5(-5) - 2 \overset{?}{=} 23$$
$$25 - 2 \overset{?}{=} 23$$
$$23 = 23$$
$$-5 \text{ is the solution.}$$

17.
$$\frac{5}{6}k - 5 = 10$$
$$\frac{5}{6}k - 5 + \mathbf{5} = 10 + \mathbf{5}$$
$$\frac{\mathbf{6}}{\mathbf{5}}\left(\frac{5}{6}k\right) = \frac{\mathbf{6}}{\mathbf{5}}(15)$$
$$k = \frac{6 \cdot 3 \cdot \cancel{5}}{\cancel{5}}$$
$$k = 18$$

Check: $\frac{5}{6}k - 5 = 10$
$$\frac{5}{6}(18) - 5 \overset{?}{=} 10$$
$$\frac{5 \cdot 3 \cdot \cancel{6}}{\cancel{6}} - 5 \overset{?}{=} 10$$
$$15 - 5 \overset{?}{=} 10$$
$$10 = 10$$
$$18 \text{ is the solution.}$$

19.
$$-\frac{7}{16}h + 28 = 21$$
$$-\frac{7}{16}h + 28 - \mathbf{28} = 21 - \mathbf{28}$$
$$-\frac{\mathbf{16}}{\mathbf{7}}\left(-\frac{7}{16}h\right) = -\frac{\mathbf{16}}{\mathbf{7}}(-7)$$
$$h = \frac{-16 \cdot \cancel{7}}{\cancel{7}}$$
$$h = 16$$

Check: $-\frac{7}{16}h + 28 = 21$
$$-\frac{7}{16}(16) + 28 \overset{?}{=} 21$$
$$\frac{-7 \cdot \cancel{16}}{\cancel{16}} + 28 \overset{?}{=} 21$$
$$-7 + 28 \overset{?}{=} 21$$
$$21 = 21$$
$$16 \text{ is the solution.}$$

21.
$$-6 - y = -2$$
$$-6 - y + \mathbf{6} = -2 + \mathbf{6}$$
$$-y = 4$$
$$(-1)(-y) = (-1)4$$
$$y = -4$$

Check: $-6 - y = -2$
$$-6 - (-4) \overset{?}{=} -2$$
$$-6 + 4 \overset{?}{=} -2$$
$$-2 = -2$$
$$-4 \text{ is the solution.}$$

23.
$$-1.7 = 1.2 - x$$
$$-1.7 - \mathbf{1.2} = 1.2 - x - \mathbf{1.2}$$
$$-2.9 = -x$$
$$(-1)(-2.9) = (-1)(-x)$$
$$2.9 = x$$

Check: $-1.7 = 1.2 - x$
$$-1.7 \overset{?}{=} 1.2 - (2.9)$$
$$-1.7 \overset{?}{=} 1.2 + (-2.9)$$
$$-1.7 = -1.7$$
$$2.9 \text{ is the solution.}$$

25.
$$3(2y - 2) - y = 5$$
$$6y - 6 - y = 5$$
$$5y - 6 = 5$$
$$5y - 6 + \mathbf{6} = 5 + \mathbf{6}$$
$$5y = 11$$
$$\frac{5y}{\mathbf{5}} = \frac{11}{\mathbf{5}}$$
$$y = \frac{11}{5}$$

Check: $3(2y - 2) - y = 5$
$$6y - 6 - y = 5$$
$$5y - 6 = 5$$
$$5 \cdot \frac{11}{5} - 6 \overset{?}{=} 5$$
$$11 - 6 \overset{?}{=} 5$$
$$5 = 5 \quad \text{True}$$
$$\frac{11}{5} \text{ is the solution.}$$

Section 2.2

27.
$$6a - 3(3a - 4) = 30$$
$$6a - 9a + 12 = 30$$
$$-3a + 12 = 30$$
$$-3a + 12 - \mathbf{12} = 30 - \mathbf{12}$$
$$-3a = 18$$
$$\frac{-3a}{-3} = \frac{18}{-3}$$
$$a = -6$$

Check: $6a - 3(3a - 4) = 30$
$$6(-6) - 3(3 \cdot -6 - 4) \overset{?}{=} 30$$
$$-36 - 3(-22) \overset{?}{=} 30$$
$$-36 + 66 \overset{?}{=} 30$$
$$30 = 30 \quad \text{True}$$
-6 is the solution.

29.
$$7a - 12 = 8a + 9$$
$$7a - 12 - \mathbf{7a} = 8a + 9 - \mathbf{7a}$$
$$-12 = a + 9$$
$$-12 - \mathbf{9} = a + 9 - \mathbf{9}$$
$$-21 = a$$

Check: $7a - 12 = 8a + 9$
$$7(-21) - 12 \overset{?}{=} 8(-21) + 9$$
$$-147 - 12 \overset{?}{=} -168 + 9$$
$$-159 = -159$$
-21 is the solution.

31.
$$60r - 50 = 15r - 5$$
$$60r - 50 - \mathbf{15r} = 15r - 5 - \mathbf{15r}$$
$$45r - 50 = -5$$
$$45r - 50 + \mathbf{50} = -5 + \mathbf{50}$$
$$45r = 45$$
$$\frac{45r}{\mathbf{45}} = \frac{45}{\mathbf{45}}$$
$$r = 1$$

Check: $60r - 50 = 15r - 5$
$$60 \cdot 1 - 50 \overset{?}{=} 15 \cdot 1 - 5$$
$$60 - 50 \overset{?}{=} 15 - 5$$
$$10 = 10 \quad \text{True}$$
1 is the solution.

33. LCD $= 18$
$$\frac{5}{6}x + \frac{2}{9} = \frac{1}{3}$$
$$\mathbf{18}\left(\frac{5}{6}x\right) + \mathbf{18}\left(\frac{2}{9}\right) = \mathbf{18}\left(\frac{1}{3}\right)$$
$$15x + 4 = 6$$
$$15x + 4 - \mathbf{4} = 6 - \mathbf{4}$$
$$15x = 2$$
$$\frac{15x}{\mathbf{15}} = \frac{2}{\mathbf{15}}$$
$$x = \frac{2}{15}$$

Check: $\dfrac{1}{3} = \dfrac{5}{6}x + \dfrac{2}{9}$
$$18\left(\frac{1}{3}\right) = 18\left(\frac{5}{6}x\right) + 18\left(\frac{2}{9}\right)$$
$$6 = 15x + 4$$
$$6 \overset{?}{=} 15 \cdot \left(\frac{2}{15}\right) + 4$$
$$6 \overset{?}{=} 2 + 4$$
$$6 = 6 \quad \text{True}$$
$\dfrac{2}{15}$ is the solution.

35. LCD $= 8$
$$\frac{1}{8}y - \frac{1}{2} = \frac{1}{4}$$
$$\mathbf{8}\left(\frac{1}{8}y\right) - \mathbf{8}\left(\frac{1}{2}\right) = \mathbf{8}\left(\frac{1}{4}\right)$$
$$y - 4 = 2$$
$$y - 4 + \mathbf{4} = 2 + \mathbf{4}$$
$$y = 6$$

Check: $\dfrac{1}{8}y - \dfrac{1}{2} = \dfrac{1}{4}$
$$\frac{1}{8}(6) - \frac{1}{2} \overset{?}{=} \frac{1}{4}$$
$$\frac{6}{8} - \frac{1}{2} \overset{?}{=} \frac{1}{4}$$
$$\frac{3}{4} - \frac{2}{4} \overset{?}{=} \frac{1}{4}$$
$$\frac{1}{4} = \frac{1}{4}$$
6 is the solution.

Solve each equation and check the result.

37.
$$0.02(62) - 0.08s = 0.06(s + 9)$$
$$\mathbf{100}[0.02(62) - 0.08s] = \mathbf{100}[0.06(s + 9)]$$
$$100[0.02(62)] - 100[0.08s] = 100[0.06(s + 9)]$$
$$2(62) - 8s = 6(s + 9)$$
$$124 - 8s = 6s + 54$$
$$124 - 8s + \mathbf{8s} = 6s + \mathbf{8s} + 54$$
$$124 = 14s + 54$$
$$124 - \mathbf{54} = 14s + 54 - \mathbf{54}$$
$$70 = 14s$$
$$\frac{70}{\mathbf{14}} = \frac{14s}{\mathbf{14}}$$
$$5 = s$$

Check: $0.02(62) - 0.08s = 0.06(s + 9)$
$$0.02(62) - 0.08(5) \overset{?}{=} 0.06(5 + 9)$$
$$1.24 - 0.4 \overset{?}{=} 0.06(14)$$
$$0.84 = 0.84$$
5 is the solution.

39.
$$0.09(t + 50) + 0.15t = 52.5$$
$$\mathbf{100}[0.09(t + 50)] + \mathbf{100}[0.15t] = \mathbf{100}(52.5)$$
$$19(t + 50) + 15t = 5,250$$
$$9t + 450 + 15t = 5,250$$
$$(9 + 15)t + 450 = 5,250$$
$$24t + 450 = 5,250$$
$$24t + 450 - \mathbf{450} = 5,250 - \mathbf{450}$$
$$24t = 4,800$$
$$\frac{24t}{\mathbf{24}} = \frac{4,800}{\mathbf{24}}$$
$$t = 200$$

Check: $0.09(t + 50) + 0.15t = 52.5$
$$0.09(200 + 50) + 0.15(200) \overset{?}{=} 52.5$$
$$0.09(250) + 30 \overset{?}{=} 52.5$$
$$22.5 + 30 \overset{?}{=} 52.5$$
$$52.5 = 52.5 \quad \text{True}$$
200 is the solution.

Solve each equation and check the result.

41. LCD $= 3$
$$\frac{10 - 5s}{3} = -s + 6$$
$$\mathbf{3}\left(\frac{10 - 5s}{3}\right) = \mathbf{3}(-s + 6)$$
$$10 - 5s = -3s + 18$$
$$10 - 5s + \mathbf{5s} = -3s + 18 + \mathbf{5s}$$
$$10 = 2s + 18$$
$$10 - \mathbf{18} = 2s + 18 - \mathbf{18}$$
$$-8 = 2s$$
$$\frac{-8}{\mathbf{2}} = \frac{2s}{\mathbf{2}}$$
$$-4 = s$$

Check:
$$\frac{10 - 5s}{3} = -s + 6$$
$$\frac{10 - 5(-4)}{3} \overset{?}{=} -(-4) + 6$$
$$\frac{10 + 20}{3} \overset{?}{=} 4 + 6$$
$$\frac{30}{3} \overset{?}{=} 10$$
$$10 = 10$$
-4 is the solution.

43. $LCD = 16$

$$\frac{7t-9}{16} = t$$

$$16\left(\frac{7t-9}{16}\right) = 16t$$

$$7t-9 = 16t$$

$$7t-9-\mathbf{7t} = 16t-\mathbf{7t}$$

$$-9 = 9t$$

$$\frac{-9}{\mathbf{9}} = \frac{9t}{\mathbf{9}}$$

$$-1 = t$$

Check: $\quad = \dfrac{7t-9}{16} = t$

$$\frac{7(-1)-9}{16} \overset{?}{=} -1$$

$$\frac{-7-9}{16} \overset{?}{=} -1$$

$$\frac{-16}{16} \overset{?}{=} -1$$

$$-1 = -1 \quad \text{True}$$

-1 is the solution.

45. $\quad 8x + 3(2-x) = 5x+6$

$$8x + 6 - 3x = 5x+6$$

$$5x+6 = 5x+6$$

$$5x+6-\mathbf{5x} = 5x+6-\mathbf{5x}$$

$$6 = 6$$

The terms involving x drop out and the result is true. This means that *all real numbers* are solutions and this equation is an identity.

47. $\quad -3(s+2) = 5(s-4)-s$

$$-3s+10 = 5s-8-s$$

$$5x+10 \quad 3s-5s = 3x-8-5x$$

$$-3s-6+3s = -3s-8+3s$$

$$10 = -2$$

$$-6 = -8$$

The terms involving s drop out and the result is false. This means the equation has *no solution* and it is a contradiction.

TRY IT YOURSELF
Solve each equation, if possible.

49. $\quad 3x-8-4x-7x = -2-8$

$$-8x-8 = -10$$

$$-8x-8+\mathbf{8} = -10+\mathbf{8}$$

$$-8x = -2$$

$$\frac{-8x}{-8} = \frac{-2}{-8}$$

$$x = \frac{\overset{1}{\cancel{2}}}{4 \cdot \underset{1}{\cancel{2}}}$$

$$x = \frac{1}{4}$$

Check: $\quad 3x-8-4x-7x = -2-8$

$$\text{let } \frac{1}{4} = 0.25$$

$$3(0.25)-8-4(0.25)-7(0.25) \overset{?}{=} -2-8$$

$$0.75-8-1-1.75 \overset{?}{=} -10$$

$$-8-1-1 \overset{?}{=} -10$$

$$-10 = -10 \quad \text{True}$$

$\dfrac{1}{4}$ or 0.25 is the solution.

51. $\quad \dfrac{t}{3}+2 = 6$

$$\frac{t}{3}+2-\mathbf{2} = 6-\mathbf{2}$$

$$\frac{3}{1}\left(\frac{t}{3}\right) = \frac{3}{1}(4)$$

$$t = 12$$

Check: $\quad \dfrac{t}{3}+2 = 6$

$$\frac{12}{3}+2 \overset{?}{=} 6$$

$$4+2 \overset{?}{=} 6$$

$$6 = 6$$

12 is the solution.

53. $\quad 4(5b)+2(6b-1) = -34$

$$20b+12b-2 = -34$$

$$32b-2 = -34$$

$$32b-2-\mathbf{2} = -34-\mathbf{2}$$

$$32b = -32$$

$$\frac{32b}{\mathbf{32}} = \frac{-32}{\mathbf{32}}$$

$$b = -1$$

Check: $\quad 4(5b)+2(6b-1) = -34$

$$4(5 \cdot -1)+2(6 \cdot -1-1) \overset{?}{=} -34$$

$$4(-5)+2(-7) \overset{?}{=} -34$$

$$-20+(-14) \overset{?}{=} -34$$

$$-34 = -34 \quad \text{True}$$

-1 is the solution.

55. $\quad 2x+5 = 17$

$$2x+5-\mathbf{5} = 17-\mathbf{5}$$

$$2x = 12$$

$$\frac{2x}{\mathbf{2}} = \frac{12}{\mathbf{2}}$$

$$x = 6$$

Check: $2x+5 = 17$

$$2(6)+5 \overset{?}{=} 17$$

$$12+5 \overset{?}{=} 17$$

$$17 = 17$$

6 is the solution.

57. $LCD = 6$

$$\frac{5}{6}(1-x) = -x+1$$

$$6\left(\frac{5}{6}\right)(1-x) = 6(-x+1)$$

$$5(1-x) = 6(-x+1)$$

$$5-5x = -6x+6$$

$$5-5x+\mathbf{6x} = -6x+6+\mathbf{6x}$$

$$x+5 = 6$$

$$x+5-\mathbf{5} = 6-\mathbf{5}$$

$$x = 1$$

Check: $\quad \dfrac{5}{6}(1-x) = -x+1$

$$\frac{5}{6}(1-1) \overset{?}{=} -1+1$$

$$\frac{5}{6}(0) \overset{?}{=} 0$$

$$0 = 0$$

1 is the solution.

59. $\quad 0.05a+0.01(90) = 0.02(a+90)$

$$\mathbf{100}[0.05a+0.01(90)] = \mathbf{100}[0.02(a+90)]$$

$$100(0.05a)+100[0.01(90)] = 100[0.02(a+90)]$$

$$5a+1(90) = 2(a+90)$$

$$5a+90 = 2a+180$$

$$5a+90-\mathbf{2a} = 2a+180-\mathbf{2a}$$

$$3a+90 = 180$$

$$3a+90-\mathbf{90} = 180-\mathbf{90}$$

$$3a = 90$$

$$\frac{3a}{\mathbf{3}} = \frac{90}{\mathbf{3}}$$

$$a = 30$$

Section 2.2

Check: $0.05a + 0.01(90) = 0.02(a+90)$

$0.05(30) + 0.01(90) \overset{?}{=} 0.02(30+90)$

$1.5 + 0.9 \overset{?}{=} 0.02(120)$

$2.4 = 2.4$ True

30 is the solution.

61. $\dfrac{7}{2} + \dfrac{3}{2}d = -9 + 1.5d$

$3.5 + 1.5d = -9 + 1.5d$

$3.5 + 1.5d - \mathbf{1.5d} = -9 + 1.5d - \mathbf{1.5d}$

$3.5 = -9$

The terms involving d drop out and the result is false. This means the equation has *no solution* and is a contration.

63. $-(19 - 3s) - (8s + 1) = 35$

$-19 + 3s - 8s - 1 = 35$

$-5s - 20 = 35$

$-5s - 20 + \mathbf{20} = 35 + \mathbf{20}$

$-5s = 55$

$\dfrac{-5s}{\mathbf{-5}} = \dfrac{55}{\mathbf{-5}}$

$s = -11$

Check: $-(19 - 3s) - (8s + 1) = 35$

$-(19 - 3 \cdot -11) - (8 \cdot -11 + 1) \overset{?}{=} 35$

$-(19 + 33) - (-88 + 1) \overset{?}{=} 35$

$-(52) - (-87) \overset{?}{=} 35$

$-52 + 87 \overset{?}{=} 35$

$35 = 35$ True

-11 is the solution.

65. $5x = 4x + 7$

$5x - \mathbf{4x} = 4x + 7 - \mathbf{4x}$

$x = 7$

Check: $5x = 4x + 7$

$5(7) \overset{?}{=} 4(7) + 7$

$35 \overset{?}{=} 28 + 7$

$35 = 35$

7 is the solution.

67. LCD $= 4$

$\dfrac{3(b+2)}{2} = \dfrac{4b - 10}{4}$

$4\left(\dfrac{3(b+2)}{2}\right) = 4\left(\dfrac{4b-10}{4}\right)$

$2[3(b+2)] = 4b - 10$

$2(3b + 6) = 4b - 10$

$6b + 12 = 4b - 10$

$6b + 12 - \mathbf{4b} = 4b - 10 - \mathbf{4b}$

$2b + 12 = -10$

$2b + 12 - \mathbf{12} = -10 - \mathbf{12}$

$2b = -22$

$\dfrac{2b}{\mathbf{2}} = \dfrac{-22}{\mathbf{2}}$

$b = -11$

Check: $\dfrac{3(b+2)}{2} = \dfrac{4b-10}{4}$

$\dfrac{3(-11+2)}{2} \overset{?}{=} \dfrac{4(-11)-10}{4}$

$\dfrac{3(-9)}{2} \overset{?}{=} \dfrac{-44-10}{4}$

$\dfrac{-27}{2} \overset{?}{=} \dfrac{-54}{4}$

$\dfrac{-27}{2} \overset{?}{=} \dfrac{\overset{1}{\cancel{2}} \cdot -27}{\underset{1}{\cancel{2}} \cdot 2}$

$-\dfrac{27}{2} = -\dfrac{27}{2}$ True

-11 is the solution.

69. $8y - 2 = 4y + 16$

$8y - 2 - \mathbf{4y} = 4y + 16 - \mathbf{4y}$

$4y - 2 = 16$

$4y - 2 + \mathbf{2} = 16 + \mathbf{2}$

$4y = 18$

$\dfrac{4y}{\mathbf{4}} = \dfrac{18}{\mathbf{4}}$

$y = \dfrac{9 \cdot \overset{1}{\cancel{2}}}{2 \cdot \underset{1}{\cancel{2}}}$

$y = \dfrac{9}{2}$

Check: $8y - 2 = 4y + 16$

$8 \cdot \dfrac{9}{2} - 2 \overset{?}{=} 4 \cdot \dfrac{9}{2} + 16$

$36 - 2 \overset{?}{=} 18 + 16$

$34 = 34$ True

$\dfrac{9}{2}$ is the solution.

71. LCD $= 12$

$\dfrac{1}{6}y + \dfrac{1}{4}y = -1$

$\mathbf{12}\left(\dfrac{1}{6}y\right) + \mathbf{12}\left(\dfrac{1}{4}y\right) = \mathbf{12}(-1)$

$2y + 3y = -12$

$5y = -12$

$\dfrac{5y}{\mathbf{5}} = \dfrac{-12}{\mathbf{5}}$

$y = -\dfrac{12}{5}$

Check:
$$\frac{1}{6}y + \frac{1}{4}y = -1$$

$$\left(\frac{1}{6} \cdot -\frac{12}{5}\right) + \left(\frac{1}{4} \cdot -\frac{12}{5}\right) \overset{?}{=} -1$$

$$-\frac{2}{5} + \left(-\frac{3}{5}\right) \overset{?}{=} -1$$

$$-\frac{5}{5} \overset{?}{=} -1$$

$$-1 = -1$$

$-\dfrac{12}{5}$ is the solution.

73.
$$0.7 - 4y = 1.7$$
$$0.7 - 4y - \mathbf{0.7} = 1.7 - \mathbf{0.7}$$
$$-4y = 1$$
$$\frac{-4y}{-4} = \frac{1}{-4}$$
$$y = -0.25$$

Check:
$$0.7 - 4y = 1.7$$
$$0.7 - 4(-0.25) \overset{?}{=} 1.7$$
$$0.7 + 1 \overset{?}{=} 1.7$$
$$1.7 = 1.7$$
-0.25 is the solution.

75.
$$-33 = 5t + 2$$
$$-33 - \mathbf{2} = 5t + 2 - \mathbf{2}$$
$$-35 = 5t$$
$$\frac{-35}{5} = \frac{5t}{5}$$
$$-7 = t$$

Check: $-33 = 5t + 2$
$$-33 \overset{?}{=} 5(-7) + 2$$
$$-33 \overset{?}{=} -35 + 2$$
$$-33 = -33$$
-7 is the solution.

77.
$$-3p + 7 = -3$$
$$-3p + 7 - \mathbf{7} = -3 - \mathbf{7}$$
$$-3p = -10$$
$$\frac{-3p}{-3} = \frac{-10}{-3}$$
$$p = \frac{10}{3}$$

Check: $-3p + 7 = -3$
$$-3\left(\frac{10}{3}\right) + 7 \overset{?}{=} -3$$
$$-10 + 7 \overset{?}{=} -3$$
$$-3 = -3$$
$\dfrac{10}{3}$ is the solution.

79.
$$2(-3) + 4y = 14$$
$$-6 + 4y = 14$$
$$-6 + 4y + \mathbf{6} = 14 + \mathbf{6}$$
$$4y = 20$$
$$\frac{4y}{4} = \frac{20}{4}$$
$$y = 5$$

Check: $2(-3) + 4y = 14$
$$-6 + 4(5) \overset{?}{=} 14$$
$$-6 + 20 \overset{?}{=} 14$$
$$14 = 14 \quad \text{True}$$
5 is the solution.

81.
$$0.06(a + 200) + 0.1a = 172$$
$$\mathbf{100}[0.06(a + 200) + 0.1a] = \mathbf{100}(172)$$
$$\mathbf{100}[0.06(a + 200)] + \mathbf{100}[0.1a] = \mathbf{100}(172)$$
$$6(a + 200) + 100(0.1a) = 100(172)$$
$$6a + 1,200 + 10a = 17,200$$
$$16a + 1,200 = 17,200$$
$$16a + 1,200 - \mathbf{1,200} = 17,200 - \mathbf{1,200}$$
$$16a = 16,000$$
$$\frac{16a}{16} = \frac{16,000}{16}$$
$$a = 1,000$$

Check:
$$0.06(a + 200) + 0.1a = 172$$
$$0.06(1,000 + 200) + (0.1)(1,000) \overset{?}{=} 172$$
$$0.06(1,200) + 100 \overset{?}{=} 172$$
$$72 + 100 \overset{?}{=} 172$$
$$172 = 172 \quad \text{True}$$
$1,000$ is the solution.

83.
$$8.6y + 3.4 = 4.2y - 9.8$$
$$8.6y + 3.4 - \mathbf{4.2y} = 4.2y - 9.8 - \mathbf{4.2y}$$
$$4.4y + 3.4 = -9.8$$
$$4.4y + 3.4 - \mathbf{3.4} = -9.8 - \mathbf{3.4}$$
$$4.4y = -13.2$$
$$\frac{4.4y}{\mathbf{4.4}} = \frac{-13.2}{\mathbf{4.4}}$$
$$y = -3$$

Check: $8.6y + 3.4 = 4.2y - 9.8$
$$8.6(-3) + 3.4 \overset{?}{=} 4.2(-3) - 9.8$$
$$-25.8 + 3.4 \overset{?}{=} -12.6 - 9.8$$
$$-22.4 = -22.4$$
-3 is the solution.

85. LCD $= 15$
$$\frac{2}{3}y + 2 = \frac{1}{5} + y$$
$$\mathbf{15}\left(\frac{2}{3}y\right) + \mathbf{15}(2) = \mathbf{15}\left(\frac{1}{5}\right) + \mathbf{15}(y)$$
$$10y + 30 = 3 + 15y$$
$$10y + 30 - \mathbf{10y} = 3 + 15y - \mathbf{10y}$$
$$30 = 3 + 5y$$
$$30 - \mathbf{3} = 3 + 5y - \mathbf{3}$$
$$27 = 5y$$
$$\frac{27}{5} = \frac{5y}{5}$$
$$\frac{27}{5} = y$$

Check:
$$\frac{2}{3}y + 2 = \frac{1}{5} + y$$
$$15\left(\frac{2}{3}y\right) + 15(2) = 15\left(\frac{1}{5}\right) + 15(y)$$
$$10y + 30 = 3 + 15y$$
$$10 \cdot \left(\frac{27}{5}\right) + 30 \overset{?}{=} 3 + 15 \cdot \left(\frac{27}{5}\right)$$
$$54 + 30 \overset{?}{=} 3 + 81$$
$$84 = 84 \quad \text{True}$$
$\dfrac{27}{5}$ is the solution.

Section 2.2

87.
$$0.4b - 0.1(b - 100) = 70$$
$$\mathbf{10}(0.4b) - \mathbf{10}[0.1(b - 100)] = \mathbf{10}(70)$$
$$4b - 1(b - 100) = 700$$
$$4b - b + 100 = 700$$
$$3b + 100 = 700$$
$$3b + 100 - \mathbf{100} = 700 - \mathbf{100}$$
$$3b = 600$$
$$\frac{3b}{\mathbf{3}} = \frac{600}{\mathbf{3}}$$
$$b = 200$$

Check: $0.4b - 0.1(b - 100) = 70$
$$0.4(200) - 0.1(200 - 100) \overset{?}{=} 70$$
$$80 - 0.1(100) \overset{?}{=} 70$$
$$80 - 10 \overset{?}{=} 70$$
$$70 = 70 \quad \text{True}$$

200 is the solution.

89. LCD $= 4$
$$\frac{1}{4}(10 - 2y) = 8$$
$$\mathbf{4}\left[\frac{1}{4}(10 - 2y)\right] = \mathbf{4}(8)$$
$$10 - 2y = 32$$
$$10 - 2y - \mathbf{10} = 32 - \mathbf{10}$$
$$-2y = 22$$
$$\frac{-2y}{\mathbf{-2}} = \frac{22}{\mathbf{-2}}$$
$$y = -11$$

Check: $\frac{1}{4}(10 - 2y) = 8$
$$\frac{1}{4}[10 - 2(-11)] \overset{?}{=} 8$$
$$\frac{1}{4}[10 + 22] \overset{?}{=} 8$$
$$\frac{1}{4}[32] \overset{?}{=} 8$$
$$8 = 8$$

-11 is the solution.

91.
$$2 - 3(x - 5) = 4(x - 1)$$
$$2 - 3x + 15 = 4x - 4$$
$$-3x + 17 = 4x - 4$$
$$-3x + 17 + \mathbf{3x} = 4x - 4 + \mathbf{3x}$$
$$17 = 7x - 4$$
$$17 + 4 = 7x - 4 + \mathbf{4}$$
$$21 = 7x$$
$$\frac{21}{\mathbf{7}} = \frac{7x}{\mathbf{7}}$$
$$3 = x$$

Check: $2 - 3(x - 5) = 4(x - 1)$
$$2 - 3(3 - 5) \overset{?}{=} 4(3 - 1)$$
$$2 - 3(-2) \overset{?}{=} 4(2)$$
$$2 + 6 \overset{?}{=} 8$$
$$8 = 8 \quad \text{True}$$

3 is the solution.

93. LCD $= 12$
$$2n - \frac{3}{4}n = \frac{1}{2}n + \frac{13}{3}$$
$$\mathbf{12}\left(2n - \frac{3}{4}n\right) = \mathbf{12}\left(\frac{1}{2}n + \frac{13}{3}\right)$$
$$12(2n) - 12\left(\frac{3}{4}n\right) = 12\left(\frac{1}{2}n\right) + 12\left(\frac{13}{3}\right)$$
$$24n - 3(3n) = 6n + 4(13)$$
$$24n - 9n = 6n + 52$$
$$15n - \mathbf{6n} = 6n + 52 - \mathbf{6n}$$
$$9n = 52$$
$$\frac{9n}{9} = \frac{52}{9}$$
$$n = \frac{52}{9}$$

Check: $2n - \frac{3}{4}n = \frac{1}{2}n + \frac{13}{3}$
$$2\left(\frac{52}{9}\right) - \frac{3}{4}\left(\frac{52}{9}\right) \overset{?}{=} \frac{1}{2}\left(\frac{52}{9}\right) + \frac{13}{3}$$
$$\frac{104}{9} - \frac{39}{9} \overset{?}{=} \frac{26}{9} + \frac{39}{9}$$
$$\frac{65}{9} = \frac{65}{9} \quad \text{True}$$

$\frac{52}{9}$ is the solution.

95.
$$10.08 = 4(0.5x + 2.5)$$
$$10.08 = 2x + 10$$
$$10.08 - \mathbf{10} = 2x + 10 - \mathbf{10}$$
$$0.08 = 2x$$
$$\frac{0.08}{\mathbf{2}} = \frac{2x}{\mathbf{2}}$$
$$0.04 = x$$

Check:
$$10.08 = 4(0.5x + 2.5)$$
$$10.08 \overset{?}{=} 4(0.5 \cdot 0.04 + 2.5)$$
$$10.08 \overset{?}{=} 4(0.02 + 2.5)$$
$$10.08 \overset{?}{=} 4(2.52)$$
$$10.08 = 10.08 \quad \text{True}$$

0.04 is the solution.

97. LCD $=12$

$$\frac{3}{4}(d-8)=\frac{2}{3}(d+1)$$

$$\mathbf{12}\left(\frac{3}{4}\right)(d-8)=\mathbf{12}\left(\frac{2}{3}\right)(d+1)$$

$$3[3(d-8)]=\mathbf{4}[2(d+1)]$$

$$9(d-8)=8(d+1)$$

$$9d-72=8d+8$$

$$9d-72-\mathbf{8d}=8d+8-\mathbf{8d}$$

$$d-72=8$$

$$d-72+\mathbf{72}=8+\mathbf{72}$$

$$d=80$$

Check: $\dfrac{3(d-8)}{4}=\dfrac{2(d+1)}{3}$

$$\dfrac{3(80-8)}{4}\overset{?}{=}\dfrac{2(80+1)}{3}$$

$$\dfrac{3(72)}{4}\overset{?}{=}\dfrac{2(81)}{3}$$

$$54=54 \quad \text{True}$$

80 is the solution.

99. $\quad 2d+5=0$

$$2d+5-\mathbf{5}=0-\mathbf{5}$$

$$2d=-5$$

$$\frac{2d}{\mathbf{2}}=\frac{-5}{\mathbf{2}}$$

$$d=-\frac{5}{2}$$

$-\dfrac{5}{2}$ is the solution.

Check: $2d+5=0$

$$2\left(-\frac{5}{2}\right)+5\overset{?}{=}0$$

$$-5+5\overset{?}{=}0$$

$$0=0$$

101. $\quad 3(A+2)=2(A-7)$

$$3A+6=2A-14$$

$$3A+6-\mathbf{2A}=2A-14-\mathbf{2A}$$

$$A+6=-14$$

$$A+6-\mathbf{6}=-14-\mathbf{6}$$

$$A=-20$$

Check: $\quad 3(A+2)=2(A-7)$

$$3(-20+2)\overset{?}{=}2(-20-7)$$

$$3(-18)\overset{?}{=}2(-27)$$

$$-54=-54 \quad \text{True}$$

-20 is the solution.

103. $\quad 4(a-3)=-2(a-6)+6a$

$$4a-12=-2a+12+6a$$

$$4a-12=4a+12$$

$$4a-12-\mathbf{4a}=4a+12-\mathbf{4a}$$

$$-12=+12$$

The terms involving a drop out and the result is false. This means the equation has *no solution* and it is a contradiction.

105. $\quad 4(y-3)-y=3(y-4)$

$$4y-12-y=3y-12$$

$$3y-12=3y-12$$

$$3y-12-\mathbf{3y}=3y-12-\mathbf{3y}$$

$$-12=-12$$

The terms involving y drop out and the result is true. This means that *all real numbers* are solutions and this equation is an identity.

107. $\quad -(4-m)=-10$

$$-4+m=-10$$

$$-4+m+\mathbf{4}=-10+\mathbf{4}$$

$$m=-6$$

Check: $-(4-m)=-10$

$$-[4-(-6)]\overset{?}{=}-10$$

$$-(4+6)\overset{?}{=}-10$$

$$-10=-10$$

-6 is the solution.

109. LCD $=3$

$$-\frac{2}{3}z+4=8$$

$$\mathbf{3}\left(-\frac{2}{3}z+4\right)=\mathbf{3}(8)$$

$$3\left(-\frac{2}{3}z\right)+3(4)=3(8)$$

$$-2z+12=24$$

$$-2z+12-\mathbf{12}=24-\mathbf{12}$$

$$-2z=12$$

$$\frac{-2z}{\mathbf{-2}}=\frac{12}{\mathbf{-2}}$$

$$z=-6$$

Check: $\quad -\dfrac{2}{3}z+4=8$

$$-\frac{2}{3}(-6)+4\overset{?}{=}8$$

$$4+4\overset{?}{=}8$$

$$8=8 \quad \text{True}$$

-6 is the solution.

Look Alikes . . .
Simplify each expression and solve each equation.

111. a. $-2(9-3x)-(5x+2)$

$$=\mathbf{-2}(9)-\mathbf{2}(-3x)-\mathbf{1}(5x)-\mathbf{1}(2)$$

$$=-18+6x-5x-2$$

$$=(6x-5x)+(-18-2)$$

$$=x-20$$

Section 2.2

111. b.
$$-2(9-3x)-(5x+2)=-25$$
$$\mathbf{-2}(9)-\mathbf{2}(-3x)-\mathbf{1}(5x)-\mathbf{1}(2)=-25$$
$$-18+6x-5x-2=-25$$
$$(6x-5x)+(-18-2)=-25$$
$$x-20=-25$$
$$x-20+\mathbf{20}=-25+\mathbf{20}$$
$$x=-5$$

Check:
$$-2(9-3x)-(5x+2)=-25$$
$$-2[9-3(-5)]-[5(-5)+2]\overset{?}{=}-25$$
$$-2(9+15)-(-25+2)\overset{?}{=}-25$$
$$-2(24)-(-23)\overset{?}{=}-25$$
$$-48+23\overset{?}{=}-25$$
$$-25=-25$$

−5 is the solution.

113. a.
$$0.6-0.2(x+1)=0.6-\mathbf{0.2}(x)-\mathbf{0.2}(1)$$
$$=0.6-0.2x-0.2$$
$$=(0.6-0.2)-0.2x$$
$$=0.4-0.2x$$

113. b.
$$0.6-0.2(x+1)=0.4$$
$$0.6-\mathbf{0.2}(x)-\mathbf{0.2}(1)=0.4$$
$$0.6-0.2x-0.2=0.4$$
$$(0.6-0.2)-0.2x=0.4$$
$$0.4-0.2x=0.4$$
$$0.4-0.2x-\mathbf{0.4}=0.4-\mathbf{0.4}$$
$$-0.2x=0$$
$$\frac{-0.2x}{\mathbf{-0.2}}=\frac{0}{\mathbf{-0.2}}$$
$$x=0$$

Check:
$$0.6-0.2(x+1)=0.4$$
$$0.6-0.2(0+1)\overset{?}{=}0.4$$
$$0.6-0.2(1)\overset{?}{=}0.4$$
$$0.6-0.2\overset{?}{=}0.4$$
$$0.4=0.4$$

0 is the solution.

WRITING
115.−117. Answers will vary.

REVIEW
Name the property that is used.

119. $x \cdot 9 = 9x$; **Com. Prop. Mult.**

121. $(x+1)+2 = x+(1+2)$; **Assoc. Prop. Add.**

CHALLENGE PROBLEMS
123. Solve: LCD $=6$
$$\frac{5}{6}\left(-\frac{3}{4}m+1\right)=-\frac{2}{3}\left(\frac{1}{2}m-1\right)$$
$$\mathbf{6}\left[\frac{5}{6}\left(-\frac{3}{4}m+1\right)\right]=\mathbf{6}\left[-\frac{2}{3}\left(\frac{1}{2}m-1\right)\right]$$
$$5\left(-\frac{3}{4}m+1\right)=-4\left(\frac{1}{2}m-1\right)$$
$$\mathbf{5}\left(-\frac{3}{4}m\right)+\mathbf{5}(1)=-4\left(\frac{1}{2}m\right)-4(-1)$$
$$-\frac{15}{4}m+5=-2m+4$$
$$\text{LCD}=4$$
$$\mathbf{4}\left(-\frac{15}{4}m+5\right)=\mathbf{4}(-2m+4)$$
$$\mathbf{4}\left(-\frac{15}{4}m\right)+\mathbf{4}(5)=\mathbf{4}(-2m)+\mathbf{4}(4)$$
$$-15m+20=-8m+16$$
$$-15m+20+\mathbf{15m}=-8m+16+\mathbf{15m}$$
$$20=7m+16$$
$$20-\mathbf{16}=7m+16-\mathbf{16}$$
$$4=7m$$
$$\frac{4}{7}=\frac{7m}{7}$$
$$\frac{4}{7}=m$$

Check:
$$\frac{5}{6}\left(-\frac{3}{4}m+1\right)=-\frac{2}{3}\left(\frac{1}{2}m-1\right)$$
$$\frac{5}{6}\left(-\frac{3}{4}\cdot\frac{4}{7}+1\right)\overset{?}{=}-\frac{2}{3}\left(\frac{1}{2}\cdot\frac{4}{7}-1\right)$$
$$\frac{5}{6}\left(-\frac{3}{7}+1\right)\overset{?}{=}-\frac{2}{3}\left(\frac{2}{7}-1\right)$$
$$\frac{5}{6}\left(-\frac{3}{7}+\frac{7}{7}\right)\overset{?}{=}-\frac{2}{3}\left(\frac{2}{7}-\frac{7}{7}\right)$$
$$\frac{5}{6}\left(\frac{4}{7}\right)\overset{?}{=}-\frac{2}{3}\left(-\frac{5}{7}\right)$$
$$\frac{10}{21}=\frac{10}{21}$$

$\dfrac{4}{7}$ is the solution.

SECTION 2.3
VOCABULARY

Fill in the blanks.

1. **Percent** means parts per one hundred.

3. In percent questions, the word *of* means
   multiplication , and the word **is** means equals.

CONCEPTS

5. $\dfrac{51}{100}$, 0.51 , 51%

7. Fill in the blanks using the words percent,
   amount, and base.

   Amount = **percent** · **base**

9. a. Find the amount of increase in the number
      of earthquakes.

   $$4,257 - 3,618 = 639$$

   b. Fill in blanks to find the percent of
      increase in earthquakes:

   639 is **what** % of **3,618** ?

NOTATION

11. Change each percent to a decimal.

   a. 35% **0.35**   b. 8.5% **0.085**

   c. 150% **1.5**   d. $2\frac{3}{4}\%$ **0.0275**

   e. 9.25% **0.0925**   f. $1\frac{1}{2}\%$ **0.015**

GUIDED PRACTICE

13. What number is 48% of 650?

   $$x = 0.48 \cdot 650$$
   $$x = 312$$
   312 is the solution.

15. What number is 92.4% of 50?

   $$x = 0.924 \cdot 50$$
   $$x = 46.2$$
   46.2 is the solution.

17. 75 is 25% of what number?

   $$75 = 0.25 \cdot x$$
   $$\frac{75}{\mathbf{0.25}} = \frac{0.25x}{\mathbf{0.25}}$$
   $$300 = x$$
   300 is the solution.

19. 128.1 is 8.75% of what number?

   $$128.1 = 0.0875 \cdot x$$
   $$\frac{128.1}{\mathbf{0.0875}} = \frac{0.0875x}{\mathbf{0.0875}}$$
   $$1,464 = x$$
   1,464 is the solution.

21. 78 is what percent of 300?

   $$78 = x \cdot 300$$
   $$\frac{78}{\mathbf{300}} = \frac{300x}{\mathbf{300}}$$
   $$0.26 = x$$
   $$26\% = x$$
   26% is the solution.

23. 0.42 is what percent of 16.8?

   $$0.42 = x \cdot 16.8$$
   $$\frac{0.42}{\mathbf{16.8}} = \frac{16.8x}{\mathbf{16.8}}$$
   $$0.025 = x$$
   $$2.5\% = x$$
   2.5% is the solution.

APPLICATIONS

25. ANTISEPTICS

   Let x = amt of pure hydrogen peroxide
   What number is 3% of 16?

   $$x = 0.03 \cdot 16$$
   $$x = 0.48$$
   0.48 oz is pure hydrogen peroxide.

27. U.S. FEDERAL BUDGET

   a. Let x = amt spent on SS, Medicare
      What number is 37% of 2,980 billion?

   $$x = 0.37 \cdot 2,980 \text{ billion}$$
   $$x = 1,102.6 \text{ billion}$$
   $1,102.6 billion was spent on SS, Medicare.
   ----

   b. Let y = amt spent on Nat Def, vet
      What number is 24% of 2,980 billion?

   $$y = 0.24 \cdot 2,980 \text{ billion}$$
   $$y = 715.2 \text{ billion}$$
   $715.2 billion was spent on Nat Def, Vet.

29. PAYPAL

   Step 1: Let x = fee charged by PayPal
           What number is 2.9% of 350?

   $$x = 0.029 \cdot 350$$
   $$x = 10.15$$
   $10.15 is the fee charged by PayPal.

   Step 2: $10.15 + $0.30 = $10.45
           $10.45 is the final fee paid to PayPal.

31. PRICE GUARANTEES

Step 1: Let x = amt of difference saved

$$\$120 - \$98 = \$22$$

What amount is 10% of 22?

$$x = 0.1 \cdot 22$$
$$x = 2.2$$

$2.20 is the amount reimbursed on $22.

Step 2: $\$120 - \$98 = \$22$

$22 is the difference in prices.

Step 3: $\$22.00 + \$2.20 = \$24.20$

$24.20 is the total reimbursement.

33. COMPUTER MEMORY

Step 1: Let x = percent of used disk space

44.7 is what percent of 74.5 GB?

$$44.7 = x \cdot 74.5 \text{ GB}$$
$$\frac{44.7}{74.5} = \frac{74.5x}{74.5}$$
$$0.6 \approx x$$
$$60\% \approx x$$

60% of the disk is being used.

Step 2: Let y = amt of free disk space

29.8 is what percent of 74.5 GB?

$$29.8 = y \cdot 74.5 \text{ GB}$$
$$\frac{29.8}{74.5} = \frac{74.5y}{74.5}$$
$$0.4 \approx y$$
$$40\% \approx y$$

40% of the disk space is free.

35. DENTISTRY

Let x = percent of teeth have filling

6 is what percent of 32?

$$6 = x \cdot 32$$
$$\frac{6}{32} = \frac{32x}{32}$$
$$0.1875 = x$$
$$19\% = x$$

19% of the teeth have filling.

37. DMV WRITTEN TEST

Let x = test score of teenager

33 is what percent of 50?

$$33 = x \cdot 50$$
$$\frac{33}{50} = \frac{50x}{50}$$
$$0.66 = x$$
$$66\% = x$$

66% is the test score.

No, the teenager did not pass.

39. CHILD CARE

Let x = maximum number of children

84 is 70% of what number?

$$84 = 0.70 \cdot x$$
$$\frac{84}{0.70} = \frac{0.70x}{0.70}$$
$$120 = x$$

120 is the maximum number of children.

41. NUTRITION

a. 5 grams and 25%

b. Let x = total amount of fat in grams

5 is 25% of what number?

$$5 = 0.25 \cdot x$$
$$\frac{5}{0.25} = \frac{0.25x}{0.25}$$
$$20 = x$$

20 grams is the total amount of fat.

43. RAINFOREST

Part a:

Step 1: Let x = amt of increase

$$4,984 - 4,490 = 494$$

494 is the amount of increase.

Step 2: 494 is what percent of 4,490?

$$494 = x \cdot 4,490$$
$$\frac{494}{4,490} = \frac{4,490x}{4,490}$$
$$0.11 \approx x$$

There was a 11% increase from year 2007 to 2008.

Part b:

Step 1: Let x = amt of decrease

$$4,984 - 2,705 = 2,279$$

2,279 is the amount of decrease.

Step 2: 2,279 is what percent of 4,984?

$$2,279 = x \cdot 4,984$$
$$\frac{2,279}{4,984} = \frac{4,984x}{4,984}$$
$$0.457 \approx x$$

There was a 46% decrease from year 2008 to 2009.

45. INSURANCE COSTS

Step 1: $\$1,050 - \$925 = \$125$

This is the amount of decrease.

Step 2: 125 is what percent of 1,050?

$$125 = x \cdot 1,050$$

$$\frac{125}{1,050} = \frac{1,050x}{1,050}$$

$$0.119 \approx x$$

12% is the percent of decrease.

from Campus To Careers

47. AUTO SERVICE TECHNICIAN

a. 4 tires times 1.5% equals what number?

$$4 \cdot 0.015 = x$$

$$0.06 = x$$

6% is the percent that the mileage decreases.

b. 6% of 25 miles equals what number?

$$0.06 \cdot 25 = x$$

$$1.5 = x$$

$25 - 1.5 = 23.5$ the mileage per gallon.

23.5 is the mileage per gallon of gas.

49. TV SHOPPING

Let $x =$ the original price of the toy

15 is 20% of what number?

$$15 = 0.20 \cdot x$$

$$\frac{15}{0.20} = \frac{0.20x}{0.20}$$

$$75 = x$$

$75 is the original price of the toy.

51. SALES

Step 1: $\$210 \div 2 = \$105$

$105 is the savings for one set.

Step 2: Let $x =$ the original price of one set

105 is 35% of what number?

$$105 = 0.35 \cdot x$$

$$\frac{105}{0.35} = \frac{0.35x}{0.35}$$

$$300 = x$$

$300 is the original price of one set.

53. REAL ESTATE

Let $x =$ the selling price of the condo

3,325 is $3\frac{1}{2}$% of what number?

$$3,325 = 0.035 \cdot x$$

$$\frac{3,325}{0.035} = \frac{0.035x}{0.035}$$

$$95,000 = x$$

$95,000 is the selling price of the condo.

54. CONSIGNMENT

Let $x =$ the selling price of the sculpture

13,500 is 45% of what number?

$$13,500 = 0.45 \cdot x$$

$$\frac{13,500}{0.45} = \frac{0.45x}{0.45}$$

$$30,000 = x$$

$30,000 is the selling price of the sculpture.

56. AGENTS

Let $x =$ the amount of the contract

1,000,000 is 12.5% of what number?

$$1.000,000 = 0.125 \cdot x$$

$$\frac{1,000,000}{0.125} = \frac{0.125x}{0.125}$$

$$8,000,000 = x$$

$8 million is the amount of the contract.

WRITING

57 - 59. Answers will vary.

REVIEW

61. $-\dfrac{16}{25} \div \left(-\dfrac{4}{15}\right) = -\dfrac{16}{25} \cdot \left(-\dfrac{15}{4}\right)$

$$= \frac{16 \cdot 15}{25 \cdot 4}$$

$$= \frac{\overset{1}{\cancel{2}} \cdot \overset{1}{\cancel{2}} \cdot 3 \cdot 4 \cdot \overset{1}{\cancel{5}}}{\underset{1}{\cancel{2}} \cdot \underset{1}{\cancel{2}} \cdot \underset{1}{\cancel{5}} \cdot 5}$$

$$= \frac{12}{5}$$

$$= 2\frac{2}{5}$$

63. $x + 15 = -49$

$$-34 + 15 \overset{?}{=} -49$$

$$-19 \overset{?}{=} -49 \quad \text{False}$$

-34 is not a solution.

CHALLENGE PROBLEMS

65. SOAPS

Step 1: $100\% - 99\dfrac{44}{100}\% = \dfrac{56}{100}\%$

$$= \frac{14}{25}\%$$

This is the amount of impurities.

Step 2: $\dfrac{56}{100}\% = 0.56\%$

$$= 0.0056$$

Section 2.3

SECTION 2.4
VOCABULARY

Fill in the blanks.

1. A **formula** is an equation that states a mathematical relationship between two or more variables.

3. The **volume** of a three-dimensional geometric solid is the amount of space it encloses.

CONCEPTS

5. a. Time, distance, rate $\quad d = rt$

   b. Markup, retail price, cost $\quad r = c + m$

   c. Costs, revenue, profit $\quad p = r - c$

   d. Interest rate, time, interest, principal
$$I = Prt$$

7. Complete the table.

rate • time = distance

Light	186,282 mi/sec	60 sec	**11,176,920 mi**
Sound	1,088 ft/sec	60 sec	**65,280 ft**

NOTATION

9. Solve $\quad Ax + By = C$ for y.

$$Ax + By = C$$
$$Ax + By - \boxed{Ax} = C - \boxed{Ax}$$
$$By = C - Ax$$
$$\frac{By}{\boxed{B}} = \frac{C - Ax}{\boxed{B}}$$
$$y = \frac{C - Ax}{\boxed{B}}$$

11. a. Write $\pi \cdot r^2 \cdot h$ in simpler form. $\quad \boldsymbol{\pi r^2 h}$

   b. In the formula $V = \pi r^2 h$, what does r represent?

   The r represents the radius of the cylinder.

   What does h represent?

   The h represents the height of the cylinder.

GUIDED PRACTICE

Use a formula to solve each problem.

13. HOLLYWOOD

Let $c =$ the cost of the movie in millions

$$\text{profit} = \text{revenue} - \text{cost}$$
$$p = r - c$$
$$1,602 = 1,842 - c$$
$$1,602 - \mathbf{1,842} = 1,842 - c - \mathbf{1,842}$$
$$-240 = -c$$
$$\frac{-240}{-1} = \frac{-c}{-1}$$
$$240 = c$$

$240 million is the cost of the movie.

15. SERVICE CLUBS

Let $r =$ the gross profit

$$\text{profit} = \text{revenue} - \text{cost}$$
$$p = r - c$$
$$875.85 = r - 55.15$$
$$875.85 - \mathbf{55.15} = r - 55.15 - \mathbf{55.15}$$
$$931 = r$$

$931 is the gross profit made.

17. ENTREPRENEURS

Let $r =$ the simple interest rate

Principal •	rate •	time =	interest
2,500	r	2	175

$$2,500 \cdot r \cdot 2 = 175$$
$$5,000r = 175$$
$$\frac{5,000r}{\mathbf{5,000}} = \frac{175}{\mathbf{5,000}}$$
$$r = 0.035$$
$$r = 3.5\%$$

3.5% is the simple interest rate.

19. LOANS

Let $P =$ the amount borrowed

Principal •	rate •	time =	interest
P	0.02	3	360

$$p \cdot 0.02 \cdot 3 = 360$$
$$0.06p = 360$$
$$\frac{0.06p}{\mathbf{0.06}} = \frac{360}{\mathbf{0.06}}$$
$$p = 6,000$$

He borrowed $6,000.

21. SWIMMING

Let r = the average rate in mph

rate	•	time	=	distance
r		742		1,826

$$r \cdot 742 = 1,826$$
$$\frac{742r}{742} = \frac{1,826}{742}$$
$$r \approx 2.46$$
$$r \approx 2.5$$

2.5 mph was his average rate.

23. HOT AIR BALLOONS

Let t = the number of hours

rate	•	time	=	distance
37		t		166.5

$$37t = 166.5$$
$$\frac{37t}{37} = \frac{166.5}{37}$$
$$t = 4.5$$

It will take 4.5 hours.

25. FRYING FOODS

$$C = \frac{5}{9}(F - 32)$$
$$C = \frac{5}{9}(365 - 32)$$
$$C = \frac{5}{9}(333)$$
$$C = \frac{\overset{1}{\cancel{3}} \cdot \overset{1}{\cancel{3}} \cdot 5 \cdot 37}{\underset{1}{\cancel{3}} \cdot \underset{1}{\cancel{3}}}$$
$$C = 185$$

The temperature is 185° Celsius.

27. BIOLOGY

$$C = \frac{5}{9}(F - 32)$$
$$-270 = \frac{5}{9}(F - 32)$$
$$\frac{9}{5} \cdot (-270) = \frac{9}{5} \cdot \frac{5}{9}(F - 32)$$
$$-486 = F - 32$$
$$-486 + 32 = F - 32 + 32$$
$$-454 = F$$

The temperature is $-454°$ Fahrenheit.

29. ENERGY SAVINGS

$$P = 2l + 2w$$
$$100 = 2 \cdot 30 + 2w$$
$$100 = 60 + 2w$$
$$100 - 60 = 60 + 2w - 60$$
$$40 = 2w$$
$$\frac{40}{2} = \frac{2w}{2}$$
$$20 = w$$

The width is 20 inches.

31. STRAWS

$$V = \pi r^2 h \quad, \pi = 3.141592654$$
$$= 3.141592654 \cdot 2^2 \cdot 150$$
$$= 3.141592654 \cdot 4 \cdot 150$$
$$\approx 1,884.9$$
$$\approx 1,885$$

The volume is about 1,885 mm³.

GUIDED PRACTICE

Solve for the specified variable.

33. $r = c + m$ for c
$$r = c + m$$
$$r - m = c + m - m$$
$$r - m = c$$
$$c = r - m$$

35. $P = a + b + c$ for b
$$P = a + b + c$$
$$P - a = a + b + c - a$$
$$P - a = b + c$$
$$P - a - c = b + c - c$$
$$P - a - c = b$$
$$b = P - a - c$$

37. $V = \frac{1}{3}Bh$ for h
$$V = \frac{1}{3}Bh$$
$$3 \cdot V = 3 \cdot \frac{1}{3}Bh$$
$$3V = Bh$$
$$\frac{3V}{B} = \frac{Bh}{B}$$
$$\frac{3V}{B} = h$$
$$h = \frac{3V}{B}$$

39. $E = IR$ for R
$$E = IR$$
$$\frac{E}{I} = \frac{IR}{I}$$
$$\frac{E}{I} = R$$
$$R = \frac{E}{I}$$

Section 2.4

41. $T = 2r + 2t$ for r

$$T = 2r + 2t$$
$$T - \mathbf{2t} = 2r + 2t - \mathbf{2t}$$
$$T - 2t = 2r$$
$$\frac{T - 2t}{\mathbf{2}} = \frac{2r}{\mathbf{2}}$$
$$\frac{T - 2t}{2} = r$$
$$r = \frac{T - 2t}{2}$$

43. $Ax + By = C$ for x

$$Ax + By = C$$
$$Ax + By - \mathbf{By} = C - \mathbf{By}$$
$$Ax = C - By$$
$$\frac{Ax}{\mathbf{A}} = \frac{C - By}{\mathbf{A}}$$
$$x = \frac{C - By}{A}$$

45. $2x + 7y = 21$ for y

$$2x + 7y = 21$$
$$2x + 7y - \mathbf{2x} = 21 - \mathbf{2x}$$
$$7y = -2x + 21$$
$$\frac{7y}{\mathbf{7}} = \frac{-2x}{\mathbf{7}} + \frac{21}{\mathbf{7}}$$
$$y = -\frac{2}{7}x + 3$$

47. $9x - 2y = -8$ for y

$$9x - 2y = -8$$
$$9x - 2y - \mathbf{9x} = -8 - \mathbf{9x}$$
$$-2y = -9x - 8$$
$$\frac{-2y}{\mathbf{-2}} = \frac{-9x}{\mathbf{-2}} - \frac{8}{\mathbf{-2}}$$
$$y = \frac{9}{2}x + 4$$

49. $T = 4b(a + am)$ for m

$$T = 4b(a + am)$$
$$T = \mathbf{4b}(a) + \mathbf{4b}(am)$$
$$T = 4ab + 4abm$$
$$T - \mathbf{4ab} = 4ab + 4abm - \mathbf{4ab}$$
$$T - 4ab = 4abm$$
$$\frac{T - 4ab}{\mathbf{4ab}} = \frac{4abm}{\mathbf{4ab}}$$
$$\frac{T - 4ab}{4ab} = m$$
$$m = \frac{T - 4ab}{4ab}$$

51. $G = g(4r - 1)$ for r

$$G = g(4r - 1)$$
$$G = \mathbf{g}(4r) - \mathbf{g}(1)$$
$$G = 4gr - g$$
$$G + \mathbf{g} = 4gr - gz - \mathbf{g}$$
$$G + g = 4gr$$
$$\frac{G + g}{\mathbf{4g}} = \frac{4gr}{\mathbf{4g}}$$
$$\frac{G + g}{4g} = r$$
$$r = \frac{G + g}{4g}$$

TRY IT YOURSELF
Solve for the specified variable or expression.

53. $A = \dfrac{a + b + c}{3}$ for c

$$A = \frac{a + b + c}{3}$$
$$\mathbf{3} \cdot A = \mathbf{3} \cdot \left(\frac{a + b + c}{3} \right)$$
$$3A = a + b + c$$
$$3A - \mathbf{a} = a + b + c - \mathbf{a}$$
$$3A - a = b + c$$
$$3A - a - \mathbf{b} = b + c - \mathbf{b}$$
$$3A - a - b = c$$
$$c = 3A - a - b$$

55. $3x + y = 9$ for y

$$3x + y = 9$$
$$3x + y - \mathbf{3x} = 9 - \mathbf{3x}$$
$$y = -3x + 9$$

57. $K = \dfrac{1}{2}mv^2$ for m

$$K = \frac{1}{2}mv^2$$
$$\mathbf{2} \cdot K = \mathbf{2} \cdot \left(\frac{1}{2}mv^2 \right)$$
$$2K = mv^2$$
$$\frac{2K}{\mathbf{v^2}} = \frac{mv^2}{\mathbf{v^2}}$$
$$\frac{2K}{v^2} = m$$
$$m = \frac{2K}{v^2}$$

59. $C = 2\pi r$ for r

$$C = 2\pi r$$
$$\frac{C}{\mathbf{2\pi}} = \frac{2\pi r}{\mathbf{2\pi}}$$
$$\frac{C}{2\pi} = r$$
$$r = \frac{C}{2\pi}$$

61. $\dfrac{M}{2} - 9.9 = 2.1B$ for M

$$\dfrac{M}{2} - 9.9 = 2.1B$$

$$2 \cdot \left(\dfrac{M}{2} - 9.9 \right) = 2 \cdot (2.1B)$$

$$2 \cdot \dfrac{M}{2} - 2 \cdot 9.9 = 2(2.1B)$$

$$M - 19.8 = 4.2B$$

$$M - 19.8 + \mathbf{19.8} = 4.2B + \mathbf{19.8}$$

$$M = 4.2B + 19.8$$

63. $w = \dfrac{s}{f}$ for f

$$w = \dfrac{s}{f}$$

$$\mathbf{f} \cdot w = \mathbf{f} \cdot \dfrac{s}{f}$$

$$fw = s$$

$$\dfrac{fw}{\mathbf{w}} = \dfrac{s}{\mathbf{w}}$$

$$f = \dfrac{s}{w}$$

65. $-x + 3y = 9$ for y

$$-x + 3y = 9$$

$$-x + 3y + \mathbf{x} = 9 + \mathbf{x}$$

$$3y = x + 9$$

$$\dfrac{3y}{\mathbf{3}} = \dfrac{x + 9}{\mathbf{3}}$$

$$\dfrac{3y}{3} = \dfrac{x}{3} + \dfrac{9}{3}$$

$$y = \dfrac{1}{3}x + 3$$

67. $A = \dfrac{1}{2}h(b + d)$ for b

$$A = \dfrac{1}{2}h(b + d)$$

$$\mathbf{2} \cdot A = \mathbf{2} \cdot \left(\dfrac{1}{2}h(b + d) \right)$$

$$2A = h(b + d)$$

$$2A = bh + dh$$

$$2A - \mathbf{dh} = bh + dh - \mathbf{dh}$$

$$2A - dh = bh$$

$$\dfrac{2A - dh}{\mathbf{h}} = \dfrac{bh}{\mathbf{h}}$$

$$\dfrac{2A - dh}{h} = b$$

$$b = \dfrac{2A - dh}{h} \ \text{ or } \ \dfrac{2A}{h} - d$$

69. $c^2 = a^2 + b^2$ for a^2

$$c^2 = a^2 + b^2$$

$$c^2 - \mathbf{b^2} = a^2 + b^2 - \mathbf{b^2}$$

$$c^2 - b^2 = a^2$$

$$a^2 = c^2 - b^2$$

71. $\dfrac{7}{8}c + w = 9$ for c

$$\dfrac{7}{8}c + w = 9$$

$$\mathbf{8} \cdot \left(\dfrac{7}{8}c + w \right) = \mathbf{8} \cdot 9$$

$$8 \cdot \left(\dfrac{7}{8}c \right) + 8 \cdot w = 8 \cdot 9$$

$$7c + 8w = 72$$

$$7c + 8w - \mathbf{8w} = 72 - \mathbf{8w}$$

$$7c = 72 - 8w$$

$$\dfrac{7c}{\mathbf{7}} = \dfrac{72 - 8w}{\mathbf{7}}$$

$$c = \dfrac{72 - 8w}{7}$$

73. $m = 70 + t(a + b)$ for b

$$m = 70 + t(a + b)$$

$$m = 70 + \mathbf{t}(a) + \mathbf{t}(b)$$

$$m = 70 + at + bt$$

$$m - \mathbf{70} - \mathbf{at} = 70 + at + bt - \mathbf{70} - \mathbf{at}$$

$$m - 70 - at = bt$$

$$\dfrac{m - 70 - at}{\mathbf{t}} = \dfrac{bt}{\mathbf{t}}$$

$$\dfrac{m - 70 - at}{t} = t$$

Section 2.4

75. $V = lwh$ for l

$$V = lwh$$

$$\frac{V}{wh} = \frac{lwh}{wh}$$

$$\frac{V}{wh} = l$$

$$l = \frac{V}{wh}$$

77. $2E = \dfrac{T-t}{9}$ for t

$$2E = \frac{T-t}{9}$$

$$9 \cdot 2E = 9 \cdot \left(\frac{T-t}{9} \right)$$

$$18E = T - t$$

$$18E + t = T - t + t$$

$$18E + t = T$$

$$18E + t - 18E = T - 18E$$

$$t = T - 18E$$

79. $s = 4\pi r^2$ for r^2

$$s = 4\pi r^2$$

$$\frac{s}{4\pi} = \frac{4\pi r^2}{4\pi}$$

$$\frac{s}{4\pi} = r^2$$

$$r^2 = \frac{s}{4\pi}$$

Look Alikes . . .

81. Solve $A = R + ab$

a. for R

$$A = R + ab$$
$$A - ab = R + ab - ab$$
$$A - ab = R$$
$$R = A - ab$$

b. for a

$$A = R + ab$$
$$A - R = R + ab - R$$
$$A - R = ab$$
$$\frac{A - R}{b} = \frac{ab}{b}$$
$$\frac{A - R}{b} = a$$
$$a = \frac{A - R}{b}$$

83. Solve $S = 2(2lw + wh)$

a. for h

$$S = 2(2lw + wh)$$
$$S = 2(2lw) + 2(wh)$$
$$S = 4lw + 2wh$$
$$S - 4lw = 4lw + 2wh - 4lw$$
$$S - 4lw = 2wh$$
$$\frac{S - 4lw}{2w} = \frac{2wh}{2w}$$
$$\frac{S - 4lw}{2w} = h$$
$$h = \frac{S - 4lw}{2w}$$

b. for l

$$S = 2(2lw + wh)$$
$$S = 2(2lw) + 2(wh)$$
$$S = 4lw + 2wh$$
$$S - 2wh = 4lw + 2wh - 2wh$$
$$S - 2wh = 4lw$$
$$\frac{S - 2wh}{4w} = \frac{4lw}{4w}$$
$$\frac{S - 2wh}{4w} = l$$
$$l = \frac{S - 2wh}{4w}$$

APPLICATIONS
from Campus To Careers

85. Solve $\text{Torque} = \dfrac{5,252 \ \cdot \ \text{Horsepower}}{\text{RPM}}$

    for Horsepower

$$\text{Torque} = \frac{5,252 \ \cdot \ \text{Horsepower}}{\text{RPM}}$$

$$\mathbf{RPM} \cdot \text{Torque} = \mathbf{RPM} \left(\frac{5,252 \ \cdot \ \text{Horsepower}}{\text{RPM}} \right)$$

$$\text{RPM} \cdot \text{Torque} = 5,252 \ \cdot \ \text{Horsepower}$$

$$\frac{\text{RPM} \cdot \text{Torque}}{5,252} = \frac{5,252 \ \cdot \ \text{Horsepower}}{5,252}$$

$$\frac{\text{RPM} \cdot \text{Torque}}{5,252} = \text{Horsepower}$$

$$\text{Horsepower} = \frac{\text{RPM} \cdot \text{Torque}}{5,252}$$

87. AVON PRODUCTS

Let p = the operating profit ending Mar. '09

operating profit = revenue – cost

$$p = r - c$$
$$= 2,186.9 - 2,018.5$$
$$= 168.4$$

$168.4 million is the operating profit for the Quarter ending Mar. '09.

Let p = the operating profit ending Mar. '10

operating profit = revenue – cost

$$p = r - c$$
$$= 2,490.4 - 2,297.6$$
$$= 192.8$$

$192.8 million is the operating profit for the Quarter ending Mar. '10.

89. CAMPERS

Let w = the length of the shorter side in inches

$$P = 2l + 2w$$
$$140 = 2 \cdot 56 + 2w$$
$$140 = 112 + 2w$$
$$140 - \mathbf{112} = 112 + 2w - \mathbf{112}$$
$$28 = 2w$$
$$\frac{28}{\mathbf{2}} = \frac{2w}{\mathbf{2}}$$
$$14 = w$$

The length of the shorter side is 14 inches.

91. KITES

Let b = the wing span of the kite in inches

$$A = \frac{1}{2}bh$$

$A = 650 \text{ in}^2, \ h = 26 \text{ inches}$

$$A = \frac{1}{2}bh$$
$$650 = \frac{1}{2}(26b)$$
$$650 = 13b$$
$$\frac{650}{\mathbf{13}} = \frac{13b}{\mathbf{13}}$$
$$50 = b$$

The wing span of the kite is 50 in.

93. WHEELCHAIRS

Find the diameter of the rear wheel.

$$12.5 \cdot 2 = 25$$

The diameter of the wheel is 25 inches.

Find the radius of the front wheel.

$$5 \div 2 = 2.5$$

The radius of the wheel is 2.5 inches.

95. BULLS-EYE

Step 1: Find the radius of the bulls-eye.

$$4.8 \div 2 = 2.4$$

Step 2: Let A = the area of bulls-eye

$$A = \pi r^2 \quad , \quad \pi = 3.141592654$$
$$A = 3.141592654 \cdot (2.4)^2$$
$$A = 3.141592654 \cdot 5.76$$
$$A \approx 18.09$$
$$A \approx 18.1$$

The area of the bulls-eye is about 18.1 in^2.

97. HORSES

Let A = the area of the circle

$$A = \pi r^2 \quad , \quad \pi = 3.141592654$$
$$A = 3.141592654 \cdot (28)^2$$
$$A = 3.141592654 \cdot 784$$
$$A \approx 2,463.0$$
$$A \approx 2,463$$

The area of the circle is about $2,463 \text{ ft}^2$.

99. WORLD HISTORY

Let A = the area of the window

$$A = \frac{1}{2}h(B + b)$$
$$A = \frac{1}{2} \cdot 70 \cdot (50 + 40)$$
$$A = \frac{1}{2} \cdot 70 \cdot 90$$
$$A = 35 \cdot 90$$
$$A = 3,150$$

The area of the window is $3,150 \text{ cm}^2$.

101. TIRES

Let w = the width of the tire

$$A = lw$$
$$45 = 7\frac{1}{2} \cdot w$$
$$45 = 7.5 \cdot w$$
$$\frac{45}{\mathbf{7.5}} = \frac{7.5w}{\mathbf{7.5}}$$
$$6 = w$$

The width of the tire is 6 inches.

103. FIREWOOD

Let l = the length of the cord of wood

$$V = lwh$$
$$128 = l \cdot 4 \cdot 4$$
$$128 = 16l$$
$$\frac{128}{\mathbf{16}} = \frac{16l}{\mathbf{16}}$$
$$8 = l$$

The length of the cord of wood is 8 feet.

Section 2.4

105. IGLOOS

Let V = the volume of a sphere in ft^3

Step 1: $V = \dfrac{4}{3}\pi r^3$, $\pi = 3.141592654$

$$V = \dfrac{4}{3}\cdot 3.141592654 \cdot (5.5)^3$$

$$V = \dfrac{4}{3}\cdot 3.141592654 \cdot (166.375)$$

$$V \approx 696.9099703$$

Step 2: Divide the answer from Step 1 by 2.

$$\dfrac{696.9099703}{2} = 348.4$$

The volume of the igloo is 348 ft^3.

107. COOKING

Let A = the area of the grill

Find the radius: $r = \dfrac{18}{2}$

$$= 9$$

$$A = \pi r^2 \quad , \quad \pi = 3.141592654$$
$$A = 3.141592654 \cdot (9)^2$$
$$A = 3.141592654 \cdot 81$$
$$A \approx 254.4$$
$$A = 254$$

The area of the grill is about 254 in^2.

109. PULLEYS

$L = 2D + 3.25(r + R)$ Solve for R.

$$L = 2D + 3.25(r + R)$$
$$L = 2D + 3.25r + 3.25R$$
$$L - \mathbf{3.25r} = 2D + 3.25r + 3.25R - \mathbf{3.25r}$$
$$L - 3.25r = 2D + 3.25R$$
$$L - 3.25r - \mathbf{2D} = 2D + 3.25R - \mathbf{2D}$$
$$L - 2D - 3.25r = 3.25R$$
$$\dfrac{L - 2D - 3.25r}{\mathbf{3.25}} = \dfrac{3.25R}{\mathbf{3.25}}$$
$$\dfrac{L - 2D - 3.25r}{3.25} = R$$
$$R = \dfrac{L - 2D - 3.25r}{3.25}$$

or

$$R = \dfrac{L - 2D}{3.25} - \dfrac{3.25r}{3.25}$$
$$R = \dfrac{L - 2D}{3.25} - r$$

WRITING

111 - 113. Answers will vary.

REVIEW

115. What number is 82% of 168?

$$x = 0.82 \cdot 168$$
$$x = 137.76$$

137.76 is the solution.

117. What percent of 200 is 30?

$$x \cdot 200 = 30$$
$$\dfrac{200x}{\mathbf{200}} = \dfrac{30}{\mathbf{200}}$$
$$x = 0.15$$
$$x = 15\%$$

15% is the solution.

CHALLENGE PROBLEMS

119. Solve $-7(\alpha - \beta) - (4\alpha - \theta) = \dfrac{\alpha}{2}$ for α

$$-7(\alpha - \beta) - (4\alpha - \theta) = \dfrac{\alpha}{2}$$
$$\mathbf{2}[-7(\alpha - \beta) - (4\alpha - \theta)] = \mathbf{2}\left(\dfrac{\alpha}{2}\right)$$
$$2[-7(\alpha - \beta)] - 2[(4\alpha - \theta)] = 2\left(\dfrac{\alpha}{2}\right)$$
$$-14(\alpha - \beta) - 2(4\alpha - \theta) = \alpha$$
$$-14\alpha + 14\beta - 8\alpha + 2\theta = \alpha$$
$$-22\alpha + 14\beta + 2\theta = \alpha$$
$$-22\alpha + 14\beta + 2\theta + \mathbf{22\alpha} = \alpha + \mathbf{22\alpha}$$
$$14\beta + 2\theta = 23\alpha$$
$$\dfrac{14\beta + 2\theta}{\mathbf{23}} = \dfrac{23\alpha}{\mathbf{23}}$$
$$\dfrac{14\beta + 2\theta}{23} = \alpha$$
$$\alpha = \dfrac{14\beta + 2\theta}{23}$$

SECTION 2.5
VOCABULARY

Fill in the blanks.

1. Integers that follow one another, such as 7 and 8, are called **consecutive** integers.

3. The equal sides of an isosceles triangle meet to form the **vertex** angle. The angles opposite the equal sides are called **base** angles, and they have equal measures.

CONCEPTS

5. $\boxed{\text{17 feet}}$ = total length

   $x = $ length of shortest section

   $\boxed{x+2}$ = length of middle-sized section

   $\boxed{3x}$ = length of longest section

7. **$\$0.03x$**

9. **180^0**

NOTATION

11. a. $x+1$ b. $x+2$ c. $x+2,\ x+4$

GUIDED PRACTICE

13. **Analyze**
- 12 ft board is cut into 2 pieces.
- Longer piece is twice as long as shorter piece.
- Find the length of each piece.

Assign

Let x = length of shorter piece in feet

$2x$ = length of longer piece in feet

Form

The length of the shorter board	plus	the length of the longer board	equals	the total length.
x	$+$	$2x$	$=$	12

Solve

$$x + 2x = 12$$
$$3x = 12$$
$$\frac{3x}{3} = \frac{12}{3}$$
$$x = 4$$

Longer length board
$$2x = 2(4)$$
$$= 8$$

State

Length of shorter piece is 4 ft.
Length of longer piece is 8 ft.

Check

$4\ \text{ft} + 8\ \text{ft} = 12\ \text{ft}$
The results check.

APPLICATIONS

15. NATIONAL PARKS

Analyze
- Total route is 444 miles.
- It took 4 days to drive.
- Each day they drove 6 miles further than the previous day.
- How many miles did they drive each day?

Assign

Let x = miles driven the first day

$x + 6$ = miles driven the second day

$(x+6)+6$ = miles driven the third day

$(x+12)+6$ = miles driven the fourth day

Form

The number of miles driven on the 1ˢᵗ day	plus	the number of miles driven on the 2ⁿᵈ day	plus	the number of miles driven on the 3ʳᵈ day	plus	the number of miles driven on the 4ᵗʰ day	equals	the total distance.
x	$+$	$x+6$	$+$	$x+12$	$+$	$x+18$	$=$	444

Solve

$$x + x + 6 + x + 12 + x + 18 = 444$$
$$4x + 36 = 444$$
$$4x + 36 - 36 = 444 - 36$$
$$4x = 408$$
$$\frac{4x}{4} = \frac{408}{4}$$
$$x = 102$$

2nd day
$$x + 6 = 102 + 6$$
$$= 108$$

3rd day
$$x + 12 = 102 + 12$$
$$= 114$$

4th day
$$x + 18 = 102 + 18$$
$$= 120$$

State

They drove 102 miles the 1st day.
They drove 108 miles the 2nd day.
They drove 114 miles the 3rd day.
They drove 120 miles the 4th day.

Check

$102 + 108 + 114 + 120 = 444$

17. SOLAR HEATING

Analyze
- The width of both is 18 feet.
- One panel is 3.4 feet wider than the other.
- What is the width of each panel?

Assign
Let x = width of smaller panel in feet
$x + 3.4$ = width of larger panel in feet

Form

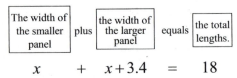

The width of the smaller panel	plus	the width of the larger panel	equals	the total lengths.
x	+	$x + 3.4$	=	18

Solve
$$x + x + 3.4 = 18$$
$$2x + 3.4 = 18$$
$$2x + 3.4 - \mathbf{3.4} = 18 - \mathbf{3.4}$$
$$2x = 14.6$$
$$\frac{2x}{\mathbf{2}} = \frac{14.6}{\mathbf{2}}$$
$$x = 7.3$$

Width of
larger panel
$$x + 3.4 = 7.3 + 3.4$$
$$= 10.7$$

State
The width of the smaller panel is 7.3 feet.
The width of the larger panel is 10.7 feet.

Check
$7.3 + 10.7 = 18$

19. iPHONE APPS

Analyze
- Total of $52.97 spent on 3 applications.
- Call of Duty: Zombies II cost $7 more than Guitar Hero.
- Tom Tom cost $4.11 more than twelve times Guitar Hero.
- Find the cost of each application.

Assign
Let x = cost of Guitar Hero
$x + 7$ = cost of Call of Duty: Zombies
$12x + 4.11$ = cost of Tom Tom

Form

Guitar Hero cost	plus	Call of Duty: Zombies cost	plus	Tom Tom cost	equals	the total cost.
x	+	$(x + 7)$	+	$(12x + 4.11)$	=	52.97

Solve
$$x + (x + 7) + (12x + 4.11) = 52.97$$
$$14x + 11.11 = 52.97$$
$$14x + 11.1 - \mathbf{11.11} = 52.97 - \mathbf{11.11}$$
$$14x = 41.86$$
$$\frac{14x}{\mathbf{14}} = \frac{41.86}{\mathbf{14}}$$
$$x = 2.99$$

WW: Zombie
$$x + 7 = 2.99 + 7$$
$$= 9.99$$

Tom Tom
$$12x + 4.11 = 12(2.99) + 4.11$$
$$= 35.88 + 4.11$$
$$= 39.99$$

State
The cost of Guitar Hero is $2.99
The cost of Call of Duty: Zombies is $9.99
The cost of Tom Tom is $39.99.

Check
$2.99 + $9.99 + $39.99 = $52.97

21. COUNTING CALORIES

Analyze
- Both have a total of 850 calories.
- The pie has 100 more than twice the calories in ice cream.
- How many calories are in each food?

Assign

Let x = number of calories in the ice cream
$2x + 100$ = number of calories in the pie

Form

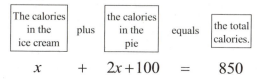

$$x \quad + \quad 2x + 100 \quad = \quad 850$$

Solve
$$x + 2x + 100 = 850$$
$$3x + 100 = 850$$
$$3x + 100 - \mathbf{100} = 850 - \mathbf{100}$$
$$3x = 750$$
$$\frac{3x}{\mathbf{3}} = \frac{750}{\mathbf{3}}$$
$$x = 250$$

Calories in the pie
$$2x + 100 = 2(250) + 100$$
$$= 500 + 100$$
$$= 600$$

State
The ice cream has 250 calories.
The pie has 600 calories.

Check
$$250 + 600 = 850$$

23. CONCERTS

Analyze
- Fee to rent the concert hall is $2,250.
- $150 an hour to pay support staff.
- Budget is $3,300.
- How many hours can the hall be rented?

Assign

Let x = number of hours

Form

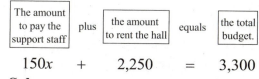

$$150x \quad + \quad 2,250 \quad = \quad 3,300$$

Solve
$$150x + 2,250 = 3,300$$
$$150x + 2,250 - \mathbf{2,250} = 3,300 - \mathbf{2,250}$$
$$150x = 1,050$$
$$\frac{150x}{\mathbf{150}} = \frac{1,050}{\mathbf{150}}$$
$$x = 7$$

State
The concert hall can be rented for 7 hours.

Check
$$\$150(7) + \$2,250 = \$3,300$$
$$\$1050 + \$2,250 = \$3,300$$

25. FIELD TRIPS

Analyze
- Budget is $275.
- Charged $0.25 per mile to rent van.
- Cost $65 a day to rent van (2 days).
- How many miles can be driven?

Assign

Let x = number of miles

Form

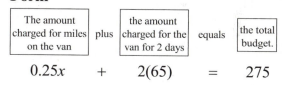

$$0.25x \quad + \quad 2(65) \quad = \quad 275$$

Solve
$$0.25x + 2(65) = 275$$
$$0.25x + 130 = 275$$
$$0.25x + 130 - \mathbf{130} = 275 - \mathbf{130}$$
$$0.25x = 145$$
$$\frac{0.25x}{\mathbf{0.25}} = \frac{145}{\mathbf{0.25}}$$
$$x = 580$$

State
The van can be driven for 580 miles.

Check
$$\$0.25(580) + 2(\$65) = \$275$$
$$\$145 + \$130 = \$275$$

Section 2.5

27. TUTORING

Analyze
- Money set aside for tutoring is $400.
- Placement test cost $25.
- Tutoring cost $18.75 per hour.
- How many hours of tutoring can be had?

Assign
Let x = number of hours

Form

The cost of the test	plus	the amount charged for tutoring	equals	the total money.
25	+	18.75x	=	400

Solve
$$25 + 18.75x = 400$$
$$25 + 18.75x - 25 = 400 - 25$$
$$18.75x = 375$$
$$\frac{18.75x}{18.75} = \frac{375}{18.75}$$
$$x = 20$$

State
They can afford 20 hours of tutoring.

Check
$$\$25 + \$18.75(20) = \$400$$
$$\$25 + \$375 = \$400$$

29. CATTLE AUCTIONS

Analyze
- Wants to make $45,500 off the auction.
- Charged 9% commision of the sale.
- What is the selling price?

Assign
Let x = the selling price of the bull

Form

The selling price of the bull	minus	the amount of the commission	equals	the amount made on the auction.
x	−	0.09x	=	45,500

Solve
$$x - 0.09x = 45,500$$
$$0.91x = 45,500$$
$$\frac{0.91x}{0.91} = \frac{45,500}{0.91}$$
$$x = 50,000$$

State
The selling price was $50,000.

Check
$$\$50,000 - \$50,000(0.09) = \$45,500$$
$$\$50,000 - \$4,500 = \$45,500$$

31. SELLING USED CLOTHING

Analyze
- Wants to make $210 on the sale of the coat.
- Consignment charge is $12\frac{1}{2}\%$ of the cost.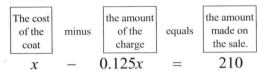
- What is the price of the coat?

Assign
Let x = the price of the coat

Form

The cost of the coat	minus	the amount of the charge	equals	the amount made on the sale.
x	−	0.125x	=	210

Solve
$$x - 0.125x = 210$$
$$0.875x = 210$$
$$\frac{0.875x}{0.875} = \frac{210}{0.875}$$
$$x = 240$$

State
The price of the coat should be $240.

Check
$$\$240 - \$240(0.125) = \$210$$
$$\$240 - \$30 = \$210$$

33. SAVINGS ACCOUNTS

Analyze
- Balance after one year was $5,512.50.
- Interest rate was 5%.
- What was the beginning balance?

Assign
Let x = the beginning balance

Form

The beginning balance	plus	the amount of the interest	equals	the ending balance.
x	+	0.05x	=	5,512.50

Solve
$$x + 0.05x = 5,512.50$$
$$1.05x = 5,512.50$$
$$\frac{1.05x}{1.05} = \frac{5,512.50}{1.05}$$
$$x = 5,250$$

State
The beginning balance was $5,250.

Check
$$\$5,250 + \$5,250(0.05) = \$5,512.50$$
$$\$5,250 + \$262.50 = \$5,512.50$$

CONSECUTIVE INTEGER PROBLEMS

35. SOCCER

Analyze

- Total number of goals is 29.
- Ronaldo scored one more goal than Mueller.
- The respective number of goals made by each are consecutive integers.
- How many goals did each make?

Assign

Let x = number of goals made by Mueller

$x+1$ = number of goals made by Ronaldo

Form

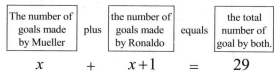

$$x \quad + \quad x+1 \quad = \quad 29$$

Solve

$$x + x + 1 = 29$$
$$2x + 1 = 29$$
$$2x + 1 - 1 = 29 - 1$$
$$2x = 28$$
$$\frac{2x}{2} = \frac{28}{2}$$
$$x = 14$$

Ronaldo
$$x + 1 = 14 + 1$$
$$= 15$$

State

Mueller scored 14 goals.
Ronaldo scored 15 goals.

Check

$$14 + 15 = 29$$

37. TV HISTORY

Analyze

- Total number of episodes is 470.
- *Friends* has two more episodes than *Leave it to Beaver*.
- The respective number of eposides each are consecutive even integers.
- How many episodes of each are there?

Assign

Let x = number of *Leave It to Beaver* episodes

$x+2$ = number of *Friends* episodes

Form

The number of *Leave it to Beaver* episodes made	plus	the number of *Friends* episodes made	equals	the total number of episodes.

$$x \quad + \quad x+2 \quad = \quad 470$$

Solve

$$x + x + 2 = 470$$
$$2x + 2 = 470$$
$$2x + 2 - 2 = 470 - 2$$
$$2x = 468$$
$$\frac{2x}{2} = \frac{468}{2}$$
$$x = 234$$

Friends
$$x + 2 = 234 + 2$$
$$= 236$$

State

Number of *Leave It to Beaver* eposides is 234.
Number *Friends* eposides is 236.

Check

$$234 + 236 = 470$$

39. CELEBRITY BIRTHDAYS

Analyze

- Total of the three calendar dates is 72.
- The respective value for each birthday is a consecutive even integers.
- What is the birth date of each person?

Assign

Let x = birth date of Selena

$x + 2$ = birth date of Jennifer

$x + 4$ = birth date of Sandra

Form

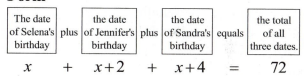

$$x \quad + \quad x+2 \quad + \quad x+4 \quad = \quad 72$$

Solve

$$x + x + 2 + x + 4 = 72$$
$$3x + 6 = 72$$
$$3x + 6 - 6 = 72 - 6$$
$$3x = 66$$
$$\frac{3x}{3} = \frac{66}{3}$$
$$x = 22$$

Jennifer $\qquad$ Sandra
$$x + 2 = 22 + 2 \qquad x + 4 = 22 + 4$$
$$= 24 \qquad\qquad = 26$$

State

The birthdate of Selena is July 22th.
The birthdate of Jennifer is July 24th.
The birthdate of Sandra is July 26th.

Check

$$22 + 24 + 26 = 72$$

Section 2.5

GEOMETRY PROBLEMS

41. TENNIS

Analyze
- Perimeter of the rectangular court is 210 feet.
- The length is 3 feet less than three times the width.
- What are the dimensions of the court?

Assign

Let x = width of the court in feet

$3x - 3$ = length of the court in feet

Form

Two widths of the court	plus	two lengths of the court	equals	the perimeter of the court.
$2x$	$+$	$2(3x-3)$	$=$	210

Solve

$$2x + 2(3x-3) = 210$$
$$2x + 6x - 6 = 210$$
$$8x - 6 = 210$$
$$8x - 6 + \mathbf{6} = 210 + \mathbf{6}$$
$$8x = 216$$
$$\frac{8x}{\mathbf{8}} = \frac{216}{\mathbf{8}}$$
$$x = 27$$

length
$$3x - 3 = 3(27) - 3$$
$$= 81 - 3$$
$$= 78$$

State

The width of the court is 27 feet.
The length of the court is 78 feet.

Check

$$27 + 78 + 27 + 78 = 210$$

43. ART

Analyze
- Perimeter of the rectangular picture is 102.5 inches.
- The length is 11.75 inches shorter than twice the width.
- What are the dimensions of the picture?

Assign

Let x = width of the picture in inches

$2x - 11.75$ = length of the picture in inches

Form

Two widths of the picture	plus	two lengths of the picture	equals	the perimeter of the picture.
$2x$	$+$	$2(2x-11.75)$	$=$	102.5

Solve

$$2x + 2(2x-11.75) = 102.5$$
$$2x + 4x - 23.5 = 102.5$$
$$6x - 23.5 = 102.5$$
$$6x - 23.5 + \mathbf{23.5} = 102.5 + \mathbf{23.5}$$
$$6x = 126$$
$$\frac{6x}{\mathbf{6}} = \frac{126}{\mathbf{6}}$$
$$x = 21$$

length
$$2x - 11.75 = 2(21) - 11.75$$
$$= 42 - 11.75$$
$$= 30.25$$

State

The width of the picture is 21 inches.
The length of the picture is 30.25 inches.

Check

$$21 + 30.25 + 21 + 30.25 = 102.5$$

45. ENGINEERING

Analyze
- Perimeter of the isosceles triangular truss is 25 feet.
- Each of the two equal sides is 4 feet shorter than the third side
- What are the lengths of the sides?

Assign

Let x = length of third side in feet

$x - 4$ = length of 1 of the 2 equal sides in feet

Form

The length of the third side	plus	the 2 equal lengths	equals	the perimeter of the truss.
x	$+$	$2(x-4)$	$=$	25

Solve

$$x + 2(x - 4) = 25$$
$$x + 2x - 8 = 25$$
$$3x - 8 + \mathbf{8} = 25 + \mathbf{8}$$
$$3x = 33$$
$$\frac{3x}{\mathbf{3}} = \frac{33}{\mathbf{3}}$$
$$x = 11$$

1 equal side
$$x - 4 = 11 - 4$$
$$= 7$$

State

Length of the third side is 11 feet.
Length of each of the 2 equal sides is 7 feet.

Check

$11 + 7 + 7 = 25$

47. TV TOWERS

Analyze
- Internal measure of the 3 angles of a triangle is 180°.
- Each base angle is 4 times the vertex angle.
- What is the measure of the vertex angle?

Assign

Let x = measure of the vertex angle

$4x$ = measure of each base angle

Form

The measure of the vertex angle	plus	the measure of the 2 base angles	equals	the measure of the angles of a triangle.
x	$+$	$2(4x)$	$=$	180

Solve

$$x + 2(4x) = 180$$
$$x + 8x = 180$$
$$9x = 180$$
$$\frac{9x}{9} = \frac{180}{9}$$
$$x = 20$$

State

Measure of the vertex angle is 20°.

Check

$$20° + 4(20°) + 4(20°) = 180°$$
$$20° + 80° + 80° = 180°$$

Section 2.5

49. MOUNTAIN BICYCLE

Analyze

- Internal measure of the 3 angles of a triangle is $180°$.
- The angle the crossbar makes with the seat is $15°$ less than twice the angle at the steering.
- The angle at the pedal is $25°$ more than the angle at the steering column.
- What is the measures of the three angle?

Assign

Let x = measure of the steering column angle

$2x - 15$ = measure of crossbar angle

$x + 25$ = measure of pedal gear angle

Form

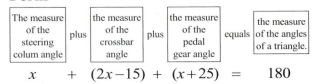

$$x \quad + \quad (2x-15) \quad + \quad (x+25) \quad = \quad 180$$

Solve

$$x + (2x - 15) + (x + 25) = 180$$
$$4x + 10 = 180$$
$$4x + 10 - \mathbf{10} = 180 - \mathbf{10}$$
$$4x = 170$$
$$\frac{4x}{4} = \frac{170}{4}$$
$$x = 42.5$$

$$\begin{array}{cc}
\text{crossbar angle} & \text{pedal gear angle} \\
2x - 15 = 2(42.5) - 15 & x + 25 = 42.5 + 25 \\
\qquad = 85 - 15 & \qquad = 67.5 \\
\qquad = 70 &
\end{array}$$

State

Measure of the steering colum angle is $42.5°$.

Measure of the crossbar angle is $70°$.

Measure of the pedal gear angle is $67.5°$.

Check

$$42.5° + 70° + 67.5° = 180°$$

51. ANGLES

Analyze

- The measure of complementary angles is $90°$.
- The measure of the first angle is $2x$.
- The measure of the second angle is $(6x + 2)°$.
- What are the measure of each angles?

Assign

Let $2x$ = measure of $\angle 1$

$6x + 2$ = measure of $\angle 2$

Form

The measure of $\angle 1$	plus	the measure of $\angle 2$	equals	the measure of the complementary angles.
$2x$	$+$	$6x + 2$	$=$	90

Solve

$$2x + 6x + 2 = 90$$
$$8x + 2 = 90$$
$$8x + 2 - \mathbf{2} = 90 - \mathbf{2}$$
$$8x = 88$$
$$\frac{8x}{\mathbf{8}} = \frac{88}{\mathbf{8}}$$
$$x = 11$$

$$\begin{array}{cc}
\angle 1 & \angle 2 \\
2x = 2(11) & 6x + 2 = 6(11) + 2 \\
\quad = 22 & \qquad = 66 + 2 \\
& \qquad = 68
\end{array}$$

State

$x = 11$

Measure of $\angle 1$ is $22°$.

Measure of $\angle 2$ is $68°$.

Check

$$22° + 68° = 90°$$

53. "LIGHTNING BOLT"

Analyze
- The measure of supplementary angles is $180°$.
- The measure of the maximum stride angle is $18°$ less than twice its supplement.
- What is the angle of the maximum stride?

Assign

Let $x =$ measure of the supplement angle
$2x - 18 =$ measure of the maximum stide angle

Form

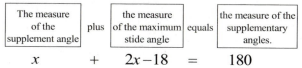

The measure of the supplement angle	plus	the measure of the maximum stide angle	equals	the measure of the supplementary angles.
x	$+$	$2x - 18$	$=$	180

Solve

$$x + 2x - 18 = 180$$
$$3x - 18 = 180$$
$$3x - 18 + 18 = 180 + 18$$
$$3x = 198$$
$$\frac{3x}{3} = \frac{198}{3}$$
$$x = 66$$

stide angle
$$2x - 18 = 2(66) - 18$$
$$= 132 - 18$$
$$= 114$$

State

Measure of supplemental angle is $66°$.
Measure of stride angle is $114°$.

Check

$$66° + 114° = 180°$$

WRITING

55 - 57. Answers will vary.

REVIEW

Solve.

59.
$$\frac{5}{8}x = -15$$
$$\frac{8}{5}\left(\frac{5}{8}x\right) = \frac{8}{5} \cdot -15$$
$$x = -\frac{8 \cdot 3 \cdot \cancel{5}^{1}}{\cancel{5}_{1}}$$
$$x = -24$$

-24 is the solution.
The solution checks.

61.
$$\frac{3}{4}y = \frac{2}{5}y - \frac{3}{2}y - 2$$
$$20 \cdot \left(\frac{3}{4}y\right) = 20 \cdot \left(\frac{2}{5}y - \frac{3}{2}y - 2\right)$$
$$20 \cdot \left(\frac{3}{4}y\right) = 20 \cdot \left(\frac{2}{5}y\right) - 20 \cdot \left(\frac{3}{2}y\right) - 20(2)$$
$$15y = 8y - 30y - 40$$
$$15y = -22y - 40$$
$$15y + 22y = -22y - 40 + 22y$$
$$37y = -40$$
$$\frac{37y}{37} = \frac{-40}{37}$$
$$y = -\frac{40}{37}$$

$-\frac{40}{37}$ is the solution.

The solution checks.

CHALLENGE PROBLEMS

63. Consecutive integers
01:02:03 04/05/06

Section 2.5

SECTION 2.6
VOCABULARY

Fill in the blanks.

1. Problems that involve depositing money are called __investment__ problems, and problems that involve moving vehicles are called uniform __motion__ problems.

CONCEPTS

3. Complete the table.

	Principal	• rate	• time	= interest
Stocks	x			
Art	$30,000 - x$			

5. Complete the table.

	rate	• time	= distance
West	r		
East	$r - 150$		

7. Complete the table.

	rate	• time	= distance
Husband	35	t	$35t$
Wife	45	t	$45t$
		Total:	**80**

Form

$$35t + 45t = 80$$

9a. Complete the table.

	Amount	• Strength	= Amount of pure antifreeze
Strong	6	0.50	**0.50(6)**
Weak	x	0.25	**0.25x**
Mixture	**6 + x**	0.30	**0.30(6 + x)**

Form

$$0.50(6) + 0.25x = 0.30(6 + x)$$

9b. Complete the table.

	Amount	• Strength	= Amount of pure vinegar
Strong	x	0.06	**0.06x**
Weak	**10 − x**	0.03	**0.03(10 − x)**
Mixture	10	0.05	**0.05(10)**

Form

$$0.06x + 0.03(10 - x) = 0.05(10)$$

NOTATION

11. $6\% = \mathbf{0.06}$, $15.2\% = \mathbf{0.152}$

GUIDED PRACTICE

Solve each equation.

13.
$$0.18x + 0.45(12 - x) = 0.36(12)$$
$$\mathbf{100}[0.18x + 0.45(12 - x)] = \mathbf{100}[0.36(12)]$$
$$100(0.18x) + 100[0.45(12 - x)] = 100[0.36(12)]$$
$$18x + 45(12 - x) = 36(12)$$
$$18x + 540 - 45x = 432$$
$$-27x + 540 = 432$$
$$-27x + 540 - \mathbf{540} = 432 - \mathbf{540}$$
$$-27x = -108$$
$$\frac{-27x}{-27} = \frac{-108}{-27}$$
$$x = 4$$
$$0.18x + 0.45(12 - x) = 0.36(12)$$
$$0.18(4) + 0.45(12 - 4) \overset{?}{=} 0.36(12)$$
$$0.72 + 0.45(8) \overset{?}{=} 4.32$$
$$0.72 + 3.6 \overset{?}{=} 4.32$$
$$4.32 = 4.32$$

4 is the solution.

15.
$$0.08x + 0.07(15,000 - x) = 1,110$$
$$\mathbf{100}[0.08x + 0.07(15,000 - x)] = \mathbf{100}(1,110)$$
$$100(0.08x) + 100[0.07(15,000 - x)] = 100(1,110)$$
$$8x + 7(15,000 - x) = 111,000$$
$$8x + 105,000 - 7x = 111,000$$
$$x + 105,000 = 111,000$$
$$x + 105,000 - \mathbf{105,000} = 111,000 - \mathbf{105,000}$$
$$x = 6,000$$

$$0.08x + 0.07(15,000 - x) = 1,110$$
$$0.08(6,000) + 0.07(15,000 - 6,000) \overset{?}{=} 1,110$$
$$480 + 0.07(9,000) \overset{?}{=} 1,110$$
$$480 + 630 \overset{?}{=} 1,110$$
$$1,110 = 1,110$$

6,000 is the solution.

APPLICATIONS
Investment problems

17. CORPORATE INVESTMENTS
Analyze
- $25,000 is the total investment.
- 1st part at 4% annual interest.
- 2nd part at 7% annual interest.
- $1,300 is the first-year combined income.
- How much is invested at each rate?

Assign
Let x = amount invested at 4%

$25,000 - x$ = amount invested at 7%

Form

	Principal	• rate	• time =	interest
1st part	x	0.04	1	**0.04x**
2nd part	$25,000 - x$	0.07	1	**0.07(25,000 − x)**
			Total:	**1,300**

The interest earned at 4%	plus	the interest earned at 7%	equals	the total interest.
$0.04x$	+	$0.07(25,000 - x)$	=	$1,300$

Solve
$$0.04x + 0.07(25,000 - x) = 1,300$$
$$\mathbf{100}[0.04x + 0.07(25,000 - x)] = \mathbf{100}(1,300)$$
$$100(0.04x) + 100[0.07(25,000 - x)] = 100(1,300)$$
$$4x + 7(25,000 - x) = 130,000$$
$$4x + 175,000 - 7x = 130,000$$
$$-3x + 175,000 = 130,000$$
$$-3x + 175,000 - \mathbf{175,000} = 130,000 - \mathbf{175,000}$$
$$-3x = -45,000$$
$$\frac{-3x}{-3} = \frac{-45,000}{-3}$$
$$x = 15,000$$

$$7\%$$
$$25,000 - x = 25,000 - 15,000$$
$$= 10,000$$

State
$15,000 was invested at 4%.
$10,000 was invested at 7%.

Check
The first investment earned 0.04($15,000), or $600.
The second earned 0.07($10,000), or $700. Since the total return was $600 + $700 = $1,300, the results check.

19. OLD COINS
Analyze
- $3,500 was used to buy coins.
- Gold coins earning 15% annual interest.
- Silver coins earning 12% annual interest.
- $480 was the return after 1 year.
- How much was invested in each?

Assign
Let x = amount invested in gold at 15%

$3,500 - x$ = amount invested in silver at 12%

Form

	Principal	• rate	• time =	interest
Gold	x	0.15	1	**0.15x**
Silver	$3,500 - x$	0.12	1	**0.12(3,500 − x)**
			Total:	**480**

The return on gold coin at 15%	plus	the return on silver coins at 12%	equals	the total return.
$0.15x$	+	$0.12(3,500 - x)$	=	480

Solve
$$0.15x + 0.12(3,500 - x) = 480$$
$$\mathbf{100}[0.15x + 0.12(3,500 - x)] = \mathbf{100}(480)$$
$$100(0.15x) + 100[0.12(3,500 - x)] = 100(480)$$
$$15x + 12(3,500 - x) = 48,000$$
$$15x + 42,000 - 12x = 48,000$$
$$3x + 42,000 = 48,000$$
$$3x + 42,000 - \mathbf{42,000} = 48,000 - \mathbf{42,000}$$
$$3x = 6,000$$
$$\frac{3x}{3} = \frac{6,000}{3}$$
$$x = 2,000$$

$$12\%$$
$$3,500 - x = 3,500 - 2,000$$
$$= 1,500$$

State
$2,000 was invested in gold coins at 15%.
$1,500 was invested in silver coins at 12%.

Check
The gold coins earned 0.15($2,000), or $300.
The silver coins earned 0.12($1,500), or $180.
Since the total interest was $300 + $180 = $480, the results check.

21. RETIREMENT

Analyze

- $28,000 invested at 6% annual interest.
- $3,500 is the return goal.
- What amount is invested at 7% annual interest.

Assign

Let $x =$ amount invested at 7%

Form

	Principal	• rate	• time =	interest
1st part	28,000	0.06	1	**1,680**
2nd part	x	0.07	1	**0.07x**
			Total:	**3,500**

The interest on 6% investment	plus	the interest on 7% investment	equals	the total return goal.
1,680	+	0.07x	=	3,500

Solve

$$1,680 + 0.07x = 3,500$$
$$\mathbf{100}(1,680 + 0.07x) = \mathbf{100}(3,500)$$
$$100(1,680) + 100(0.07x) = 100(3,500)$$
$$168,000 + 7x = 350,000$$
$$168,000 + 7x - \mathbf{168,000} = 350,000 - \mathbf{168,000}$$
$$7x = 182,000$$
$$\frac{7x}{7} = \frac{182,000}{7}$$
$$x = 26,000$$

State

$26,000 is needed to be invested at 7%.

Check

The 1st investment earned 0.06($28,000), or $1,680.
The 2nd investment earned 0.07($26,000), or $1,820.
Since the total interest was $1,680 + $1,820 = $3,500, the results check.

23. 1099 FORMS

Analyze

- $15,000 is the total deposit.
- 822 part at 5% annual interest.
- 721 part at 4.5% annual interest.
- $720 is the total interest income.
- How much is invested at each rate?

Assign

Let $\quad x =$ amount invested at 5%
$15,000 - x =$ amount invested at 4.5%

Form

	Principal	• rate	• time =	interest
822	x	0.05	1	**0.05x**
721	$15,000 - x$	0.045	1	**0.045(15,000 − x)**
			Total:	**720**

The interest earned at 5%	plus	the interest earned at 4.5%	equals	the total interest.
0.05x	+	0.045(15,000 − x)	=	720

Solve

$$0.05x + 0.045(15,000 - x) = 720$$
$$\mathbf{1,000}\left[0.05x + 0.045(15,000 - x)\right] = \mathbf{1,000}(720)$$
$$1,000(0.05x) + 1,000[0.045(15,000 - x)] = 1,000(720)$$
$$50x + 45(15,000 - x) = 720,000$$
$$50x + 675,000 - 45x = 720,000$$
$$5x + 675,000 = 720,000$$
$$5x + 675,000 - \mathbf{675,000} = 720,000 - \mathbf{675,000}$$
$$5x = 45,000$$
$$\frac{5x}{5} = \frac{45,000}{5}$$
$$x = 9,000$$

$$4.5\%$$
$$15,000 - x = 15,000 - 9,000$$
$$= 6,000$$

State

$9,000 was invested in the 822 at 5%.
$6,000 was invested in the 721 at 4.5%.

Check

The 1st investment earned 0.05($9,000), or $450.
The 2nd investment earned 0.045($6,000), or $270.
Since the total interest was $450 + $270 = $720, the results check.

25. INVESTMENTS

Analyze

- 1st equal amount invested at 7%.
- 2nd equal amount invested at 8%.
- 3rd equal amount invested at 10.5%.
- $1,249.50 is the total interest income.
- How much is invested at each rate?

Assign

Let x = equal amount invested at all rates

Form

	Principal	rate	time =	interest
1st part	x	0.07	1	**0.07x**
2nd part	x	0.08	1	**0.08x**
3rd part	x	0.105	1	**0.105x**
			Total:	**1,249.50**

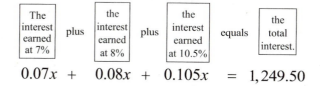

$$0.07x + 0.08x + 0.105x = 1,249.50$$

Solve

$$0.07x + 0.08x + 0.105x = 1,249.50$$
$$0.255x = 1,249.50$$
$$\mathbf{1,000}(0.255x) = \mathbf{1,000}(1,249.50)$$
$$255x = 1,249,500$$
$$\frac{255x}{\mathbf{255}} = \frac{1,249,500}{\mathbf{255}}$$
$$x = 4,900$$

State

$4,900 was invested at each of the rates.

Check

The 1st investment earned 0.07($4,900), or $343.

The 2nd investment earned 0.08($4,900), or $392.

The 3rd investment earned 0.105($4,900), or $514.50.

Since the total interest was
$343 + $392 + $514.50 = $1,249.50,
the results check.

27. BAD INVESTMENTS

Analyze

- $18,000 is the total investment.
- Credit union earns 3% annual interest.
- Utility stocks earns 7% annual interest, but suffered a loss.
- $90 was the net income after first year..
- How much was invested in each investment?

Assign

Let x = amount invested at 3%

$18,000 - x$ = amount invested at 7%

Form

	Principal	rate	time =	interest
CU	x	0.03	1	**0.03x**
US	$18,000 - x$	0.07	1	**0.07(18,000 − x)**
			Total:	**90**

$$0.03x - 0.07(18,000 - x) = 90$$

Solve

$$0.03x - 0.07(18,000 - x) = 90$$
$$\mathbf{100}\left[0.03x - 0.07(18,000 - x)\right] = \mathbf{100}(90)$$
$$100(0.03x) - 100[0.07(18,000 - x)] = 100(90)$$
$$3x - 7(18,000 - x) = 9,000$$
$$3x - 126,000 + 7x = 9,000$$
$$10x - 126,000 = 9,000$$
$$10x - 126,000 + \mathbf{126,000} = 9,000 + \mathbf{126,000}$$
$$10x = 135,000$$
$$\frac{10x}{\mathbf{10}} = \frac{135,000}{\mathbf{10}}$$
$$x = 13,500$$

$$7\%$$
$$18,000 - x = 18,000 - 13,500$$
$$= 4,500$$

State

$13,500 was invested in the credit union.
$4,500 was invested in utility stock.

Check

The CU investment earned 0.03($13,500), or $405.
The US investment lost value 0.07($4,500), or $315.
Since the net interest was $405 − $315 = $90,
the results check.

Section 2.6

Uniform Motion Problems

29. TORNADOES

Analyze

- One team travels east at 20 mph.
- One team travels west at 25 mph.
- Both leave from same place at the same time.
- Radios have a range of 90 miles.
- How long before they lose radio contact?

Assign

Let t = time in hours before they lose contact

Form

	rate	• time	= distance
East	20	t	$20t$
West	25	t	$25t$
		Total:	**90**

The distance traveled east	plus	the distance traveled west	equals	the total distance.
$20t$	$+$	$25t$	$=$	90

Solve

$$20t + 25t = 90$$
$$45t = 90$$
$$\frac{45t}{45} = \frac{90}{45}$$
$$t = 2$$

State

After 2 hours they will be out of contact with each other.

Check

The distance traveled east is 20(2), or 40 mi.
The distance traveled west is 25(2), or 50 mi.
Since the total was 40 mi + 50 mi = 90 mi, the result checks.

31. HELLO/GOOD-BYE

Analyze

- Husband travels towards home at 45 mph.
- Wife travels towards work at 35 mph.
- Both leave at the same time.
- Combined distance is 20 miles.
- How long before they wave at each other?

Assign

Let t = time in hours they wave to each other

Form

	rate	• time	= distance
Husband	45	t	$45t$
Wife	35	t	$35t$
		Total:	**20**

The distance traveled by the husband	plus	the distance traveled by the wife	equals	the total distance.
$45t$	$+$	$35t$	$=$	20

Solve

$$45t + 35t = 20$$
$$80t = 20$$
$$\frac{80t}{80} = \frac{20}{80}$$
$$t = 0.25$$
$$t = \frac{1}{4}$$

State

After $\frac{1}{4}$ hour or 15 minutes they will wave.

Check

The distance traveled by husband is 45(0.25), or 11.25 mi.
The distance traveled by wife is 35(0.25), or 8.75 mi.
Since the total was 11.25 mi + 8.75 mi = 20 mi, the result checks.

33. CYCLING

Analyze

- Cyclist leaves base at 18 mph.
- Staff leaves 1.5 hours later in the same direction at 45 mph.
- Both leave from the same place.
- Both travel the same distances.
- How long before the staff catches the cylist?

Assign

Let t = time in hrs before car catches cyclist

Form

	rate	• time	= distance
Cyclist	18	$t + 1.5$	$18(t + 1.5)$
Staff	45	t	$45t$

Distance: same

The distance traveled by the cyclist	equals	the distance traveled by the staff.

$$18(t+1.5) \quad = \quad 45t$$

Solve

$$18(t+1.5) = 45t$$
$$18t + 27 = 45t$$
$$18t + 27 - \mathbf{18t} = 45t - \mathbf{18t}$$
$$27 = 27t$$
$$\frac{27}{\mathbf{27}} = \frac{27t}{\mathbf{27}}$$
$$1 = t$$

State

After 1 hour, the staff car will catch the cyclist.

Check

The distance traveled by cyclist is 18(2.5), or 45 mi.
The distance traveled by car is 45(1), or 45 mi.
Since the each distance is the same 45 mi = 45 mi, the result checks.

35. ROAD TRIPS

Analyze

- 1st part of the trip averaged 40 mph.
- 2nd part of the trip averaged 50 mph.
- Total time of the trip was 5 hours.
- Total distance covered is 210 miles.
- How long did the car average 40 mph?

Assign

Let t = time in hours at 40 mph

Form

	rate	• time	= distance
1st part	40	t	$40t$
2nd part	50	$5 - t$	$50(5 - t)$

Total: 210

The distance traveled at 40 mph	plus	the distance traveled at 50 mph	equals	the total distance.

$$40t \quad + \quad 50(5-t) \quad = \quad 210$$

Solve

$$40t + 50(5 - t) = 210$$
$$40t + 250 - 50t = 210$$
$$-10t + 250 = 210$$
$$-10t + 250 - \mathbf{250} = 210 - \mathbf{250}$$
$$-10t = -40$$
$$\frac{-10t}{\mathbf{-10}} = \frac{-40}{\mathbf{-10}}$$
$$t = 4$$

State

The 40 mph part of the trip averaged 4 hours.

Check

The distance traveled at 40 mph is 40(4), or 160 mi.
The distance traveled at 50 mph is 50(1), or 50 mi.
Since the total of the distances is 160 mi + 50 mi, or 210 miles, the result checks.

Section 2.6

37. WINTER DRIVING

Analyze

- 1st part of the trip was for 4 hours.
- 2nd part of the trip was for 3 hours at 20 mph less.
- Total distance of the trip was 325 miles.
- What was the average speed for the 1st part of the trip?

Assign

Let r = rate of 1st part of the trip in mph

Form

	rate •	time =	distance
1st part	r	4	$4r$
2nd part	$r-20$	3	$3(t-20)$
		Total:	325

The distance traveled by the 1st part	plus	the distance traveled by the 2nd part	equals	the total distance.
$4r$	$+$	$3(r-20)$	$=$	325

Solve

$$4r + 3(r-20) = 325$$
$$4r + 3r - 60 = 325$$
$$7r - 60 = 325$$
$$7r - 60 + \mathbf{60} = 325 + \mathbf{60}$$
$$7r = 385$$
$$\frac{7r}{7} = \frac{385}{7}$$
$$r = 55$$

$$2^{nd} \ part$$
$$r - 20 = 55 - 20$$
$$= 35$$

State

1st part was averaged at 55 mph.

2nd part was averaged at 35 mph.

Check

The distance traveled at 55 mph is 55(4), or 220 mi.

The distance traveled at 33 mph is 35(3), or 105 mi.

Since the total of the distances is 220 mi + 105 mi, or 325 miles, the results check.

Liquid Mixture Problems

39. SALT SOLUTION

Analyze

- Weak salt solution is 3%.
- Strong salt solution is 7%.
- A mixture of the two is to be 5%.
- Start with 50 gallons of 7% solution.
- How many gallons of 3% solution are needed?

Assign

Let x = the amount of 3% solution in gallons

Form

	Amount •	Strength =	Amount of pure salt
Weak	x	0.03	$0.03x$
Strong	50	0.07	3.5
Mixture	$x+50$	0.05	$0.05(x+50)$

The salt in the 3% solution	plus	the salt in the 7% solution	equals	the salt in the 5% solution.
$0.03x$	$+$	3.5	$=$	$0.05(x+50)$

Solve

$$0.03x + 3.5 = 0.05(x+50)$$
$$\mathbf{100}[0.03x + 3.5] = \mathbf{100}[0.05(x+50)]$$
$$100(0.03x) + 100(3.5) = 100[0.05(x+50)]$$
$$3x + 350 = 5(x+50)$$
$$3x + 350 = 5x + 250$$
$$3x + 350 - \mathbf{3x} = 5x + 250 - \mathbf{3x}$$
$$350 = 2x + 250$$
$$350 - \mathbf{250} = 2x + 250 - \mathbf{250}$$
$$100 = 2x$$
$$\frac{100}{2} = \frac{2x}{2}$$
$$50 = x$$

State

50 gallons of the 3% salt solution will be needed.

Check

The salt in the 3% solution is 50(0.03), or 1.5
The salt in the 7% solution is 50(0.07), or 3.5
The salt in the 5% solution is 100(0.05), or 5.0
Since the total amount of salt in the ingredients is 1.5 + 3.5, or 5.0, and that equals the amount of salt in the mixture, 5.0, the result checks.

41. MAKING CHEESE

Analyze
- Milk contains 4% butterfat.
- Milk contains 1% butterfat.
- A mixture of 15 gallons is to be 2%.
- How many gallons of each are needed?

Assign

Let x = the amount of 4% milk in gal

$15 - x$ = the amount of 1% milk in gal

Form

	Amount •	Strength =	Amount of pure butterfat
4% Milk	x	0.04	$0.04x$
1% Milk	$15 - x$	0.01	$0.01(15 - x)$
Mixture	15	0.02	0.3

The butterfat in the 4% solution	plus	the butterfat in the 1% solution	equals	the butterfat in the 2% solution.
$0.04x$	$+$	$0.01(15 - x)$	$=$	0.3

Solve

$$0.04x + 0.01(15 - x) = 0.3$$
$$\mathbf{100}[0.04x + 0.01(15 - x)] = \mathbf{100}(0.3)$$
$$100(0.04x) + 100[0.01(15 - x)] = 100(0.3)$$
$$4x + 1(15 - x) = 30$$
$$4x + 15 - x = 30$$
$$3x + 15 = 30$$
$$3x + 15 - \mathbf{15} = 30 - \mathbf{15}$$
$$3x = 15$$
$$\frac{3x}{3} = \frac{15}{3}$$
$$x = 5$$

State

5 gallons of the 4% milk is needed.

$15 - 5 = 10$ gallons of 1% milk is needed.

Check

The butterfat in the 4% solution is 5(0.04), or 0.2
The butterfat in the 1% solution is 10(0.01), or 0.1
The butterfat in the 2% solution is 15(0.02), or 0.3
Since the total amount of butterfat in the ingredients is $0.2 + 0.1$, or 0.3 and that equals the amount of butterfat in the mixture, 0.3, the result checks.

43. PRINTING

Analyze
- 8% cobalt blue color ink.
- 22% cobalt blue color ink.
- A mixture of 64 ounces is to be 15%.
- How many ounces of each are needed?

Assign

Let x = the amount of 8% ink in ounces

$64 - x$ = the amount of 22% ink in ounces

Form

	Amount •	Strength =	Amount of pure cobalt blue
Weaker	x	0.08	$0.08x$
Stronger	$64 - x$	0.22	$0.22(64 - x)$
Mixture	64	0.15	9.6

The cobalt blue in the 8% solution	plus	the cobalt blue in the 22% solution	equals	the cobalt blue in the 15% solution.
$0.08x$	$+$	$0.22(64 - x)$	$=$	9.6

Solve

$$0.08x + 0.22(64 - x) = 9.6$$
$$\mathbf{100}[0.08x + 0.22(64 - x)] = \mathbf{100}(9.6)$$
$$100(0.08x) + 100[0.22(64 - x)] = 100(9.6)$$
$$8x + 22(64 - x) = 960$$
$$8x + 1,408 - 22x = 960$$
$$-14x + 1,408 = 960$$
$$-14x + 1,408 - \mathbf{1,408} = 960 - \mathbf{1,408}$$
$$-14x = -448$$
$$\frac{-14x}{-14} = \frac{-448}{-14}$$
$$x = 32$$
$$22\%$$
$$64 - x = 64 - 32$$
$$= 32$$

State

32 ounces of the 8% cobalt blue is needed.
32 ounces of the 22% cobalt blue is needed.

Check

The coblt blue in the 8% solution is 32(0.08), or 2.56 ounces.
The cobalt blue in the 22% solution is 32(0.22), or 7.04 ounces.
The cobalt blue in the 15% solution is 64(0.15), or 9.6 ounces.
Since the total amount of C B in the ingredients is 2.56 oz + 7.04 oz, or 9.6 oz and that equals the amount of C B in the mixture, 9.6 oz, the result checks.

45. INTERIOR DECORATING

Analyze
- 7% Desert Sunrise tint.
- 1 gal of 18.2% Bright Pumpkin tint.
- Mixture is Cool Cantaloupe of 8.6% tint.
- How many gal of Desert tint is needed?

Assign

Let x = the gal of 3% Desert tint

Form

	Amount •	Strength =	Amount of pure tint
Desert	x	0.07	**0.07x**
Bright	1	0.182	**0.182(1)**
Cool	$x + 1$	0.086	**0.086(x + 1)**

The tint in the 7% Desert	plus	the tint in the 18.2% Bright	equals	the tint in the 8.6% Cool.
0.07x	+	0.182(1)	=	0.086(x+1)

Solve

$$0.07x + 0.182 = 0.086(x+1)$$
$$\mathbf{1,000}[0.07x + 0.182] = \mathbf{1,000}[0.086(x+1)]$$
$$1,000(0.07x) + 1,000(0.182) = 1,000[0.086(x+1)]$$
$$70x + 182 = 86(x+1)$$
$$70x + 182 = 86x + 86$$
$$70x + 182 - \mathbf{70x} = 86x + 86 - \mathbf{70x}$$
$$182 = 16x + 86$$
$$182 - \mathbf{86} = 16x + 86 - \mathbf{86}$$
$$96 = 16x$$
$$\frac{96}{\mathbf{16}} = \frac{16x}{\mathbf{16}}$$
$$6 = x$$

State

6 gal of 7% Desert Sunrise is needed.

Check

The tint in the 7% paint is 6(0.07), or 0.42 gal.
The tint in the 18.2% paint is 1(0.182), or 0.182 gal.
The tint in the 8.6% paint is 7(0.086), or 0.602 gal.
Since the total amount of tint in the ingredients is 0.42 gal + 0.182 gal, or 0.602 gal and that equals the amount of tint in the mixture, 0.602 gal, the result checks.

Dry Mixture Problems

47. LAWN SEED

Analyze
- Bluegrass seed sells for $6 per lb.
- Ryegrass seed sells for $3 per lb.
- A blend of the two is to sell for $5 per lb.
- Start with 100 lb of bluegrass.
- How many pounds of ryegrass are needed?

Assign

Let x = the lb of ryegrass needed

Form

	Number •	Value =	Total value
Bluegrass	100	6	**600**
Ryegrass	x	3	**3x**
Blend	$x + 100$	5	**5(x + 100)**

The value of the bluegrass	plus	the value of the ryegrass	equals	the total value of the blend.
600	+	3x	=	5(x+100)

Solve

$$600 + 3x = 5(x+100)$$
$$600 + 3x = 5x + 500$$
$$600 + 3x - \mathbf{3x} = 5x + 500 - \mathbf{3x}$$
$$600 = 2x + 500$$
$$600 - \mathbf{500} = 2x + 500 - \mathbf{500}$$
$$100 = 2x$$
$$\frac{100}{\mathbf{2}} = \frac{2x}{\mathbf{2}}$$
$$50 = x$$

State

50 lb of the ryegrass seed will be needed.

Check

The value of the bluegrass is 100($6), or $600.
The value of the ryegrass is 50($3), or $150.
The value of the blend is 150($5), or $750.
Since the total was $600 + $150 = $750, the result checks.

49. RAISINS

Analyze
- Natural raisins sells for $3.45 per scoop.
- Golden raisins sells for $2.55 per scoop.
- A blend of the two is to sell for $3 per scoop.
- Start with 20 scoops of golden raisins.
- How many scoops of natural raisins are needed?

Assign

Let x = the scoops of natural raisins needed

Form

	Number •	Value =	Total value
Golden	20	2.55	**51**
Natural	x	3.45	**3.45x**
Blend	$x + 20$	3	**3(x + 20)**

The value of the golden	plus	the value of the natural	equals	the total value of the blend.
51	+	3.45x	=	3(x + 20)

Solve

$$51 + 3.45x = 3(x + 20)$$
$$51 + 3.45x = 3x + 60$$
$$51 + 3.45x - \mathbf{3x} = 3x + 60 - \mathbf{3x}$$
$$51 + 0.45x = 60$$
$$51 + 0.45x - \mathbf{51} = 60 - \mathbf{51}$$
$$0.45x = 9$$
$$\frac{0.45x}{\mathbf{0.45}} = \frac{9}{\mathbf{0.45}}$$
$$x = 20$$

State

20 scoops of the golden raisins will be needed.

Check

The value of the golden is 20($2.55), or $51.
The value of the natural is 20($3.45), or $69.
The value of the blend is 40($3), or $120.
Since the total was $51 + $69 = $120, the

result checks.

51. PACKAGED SALAD

Analyze
- Iceberg sells for $2.20 per 10 oz bag.
- Romaine sells for $3.50 per 10 oz bag.
- Start with 50 bags of Iceberg.
- How many bags of Romaine are needed a blend sells for $2.50 per 10 oz bag.?

Assign

Let x = the bags of Roamine needed

Form

	Number •	Value =	Total value
Romaine	x	3.50	**3.50x**
Iceberg	50	2.20	**50(2.20)**
Blend	$x + 50$	2.50	**2.50(x + 50)**

The value of the Romaine	plus	the value of the Iceberg	equals	the total value of the blend.
3.50x	+	50(2.20)	=	2.50(x + 50)

Solve

$$3.50x + 50(2.20) = 2.50(x + 50)$$
$$3.50x + 110 = 2.50x + 125$$
$$3.50x + 110 - \mathbf{2.50x} = 2.50x + 125 - \mathbf{2.50x}$$
$$x + 110 = 125$$
$$x + 110 - \mathbf{110} = 125 - \mathbf{110}$$
$$x = 15$$

State

15 bags of Romaine will be needed.
Optional: Needed for checking
50 bags of Iceberg will be needed.

Check

The value of Romaine is 15($3.50), or $52.50.
The value of Iceberg is 50($2.20), or $110.
The value of blend is 65($2.50), or $162.50.
Since the total was $52.50 + $110 = $162.50, the result checks.

Section 2.6

53. BRONZE

Analyze
- 4 pounds of tin.
- 6 pounds of copper.
- Tin cost $1 more than copper.
- The bronze is 10 pounds.
- How much is a pound of tin?

Assign
Let x = the cost of tin

$x - 1$ = the cost of copper

Form

	Number $\bullet$	Value =	Total value
Tin	4	x	$4x$
Copper	6	$x - 1$	$6(x - 1)$
Bronze	10	3.65	36.5

The value of the tin	plus	the value of the copper	equals	the value of the bronze.
$4x$	$+$	$6(x-1)$	$=$	36.5

Solve

$$4x + 6(x - 1) = 36.5$$
$$4x + 6x - 6 = 36.5$$
$$10x - 6 = 36.5$$
$$10x - 6 + 6 = 36.5 + 6$$
$$10x = 42.5$$
$$\frac{10x}{10} = \frac{42.5}{10}$$
$$x = 4.25$$

copper
$$x - 1 = 4.25 - 1$$
$$= 3.25$$

State
Tin cost $4.25 per pound.

Optional: Needed for checking
Copper cost $3.25 per pound.

Check
The value of the tin is 4($4.25), or $17.
The value of the copper is 6($3.25), or $19.50.
The value of the blend is 10($3.65), or $36.50.
Since the total was $17 + $19.50 = $36.50, the results check.

Number - Value Problems
55. RENTALS

Analyze
- 1-room rents for $550.
- 2-room rents for $700.
- 3-room rents for $900.
- An equal number is rented.
- Total monthly rent is $36,550.
- How many of each apartment are rented?

Assign
Let x = the number of each apartment rented

Form

	Number $\bullet$	Value =	Total value
1-room	x	550	$550x$
2-room	x	700	$700x$
3-room	x	900	$900x$
		Total:	36,550

The value of the 1-room	plus	the value of the 2-room	plus	the value of the 3-room	equals	the total rent.
$550x$	$+$	$700x$	$+$	$900x$	$=$	$36,550$

Solve

$$550x + 700x + 900x = 36,550$$
$$2,150x = 36,550$$
$$\frac{2,150x}{2,150} = \frac{36,550}{2,150}$$
$$x = 17$$

State
17 of each type of bedroom was rented.

Check
The value of the 1-room is 17($550), or $9,350.
The value of the 2-room is 17($700), or $11,900.
The value of the 3-room is 17($900), or $15,300.
Since the total was $9,350 + $11,000 + $15,300 = $36,550, the result checks.

57. SOFTWARE

Analyze
- Spreadsheet cost $150. Database cost $195.
- Word processing cost $210.
- An equal number of SS and data were sold.
- 15 more Word than the other two combined were sold.
- $72,000 in sales were generated by all 3.
- How many spreadsheets were sold?

Assign

Let x = the number of spreadsheets
    and database sold

Form

	Number	• Value =	Total value
Spreadsheet	x	150	$150x$
Database	x	195	$195x$
Word	$2x + 15$	210	$210(2x + 15)$
		Total:	**72,000**

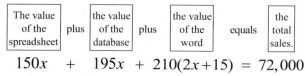

The value of the spreadsheet	plus	the value of the database	plus	the value of the word	equals	the total sales.

$$150x \;+\; 195x \;+\; 210(2x+15) = 72{,}000$$

Solve

$$150x + 195x + 210(2x+15) = 72{,}000$$
$$150x + 195x + 420x + 3{,}150 = 72{,}000$$
$$765x + 3{,}150 = 72{,}000$$
$$765x + 3{,}150 - \mathbf{3{,}150} = 72{,}000 - \mathbf{3{,}150}$$
$$765x = 68{,}850$$
$$\frac{765x}{\mathbf{765}} = \frac{68{,}850}{\mathbf{765}}$$
$$x = 90$$

word processing
$$2x + 15 = 2(90) + 15$$
$$= 180 + 15$$
$$= 195$$

State

90 spreadsheets were sold.
Optional: Needed for checking
90 database were sold.
195 word processing were sold.

Check

The value of the SS is 90($150), or $13,500.
The value of the data is 90($195), or $17,550.
The value of the word is 195($210), or $40,950.
Since the total was $13,500 + $17,550 + $40,950
= $72,000, the result checks.

59. PIGGY BANKS

Analyze
- Pennies are worth $0.01.
- Nickels are worth $0.05.
- Dimes are worth $0.10.
- 20 more pennies than dimes.
- Nickels triple the number of dimes.
- $5.40 is the values of all the coins.
- How many of each type coins are there?

Assign

Let x = the number of dimes
  $x + 20$ = the number of pennies
    $3x$ = the number of nickels

Form

	Number	• Value =	Total value
Dimes	x	0.10	$0.10x$
Pennies	$x + 20$	0.01	$0.01(x + 20)$
Nickels	$3x$	0.05	$0.15x$
		Total:	**5.40**

The value of the dimes	plus	the value of pennies	plus	the value of nickels	equals	the value of all coins.

$$0.10x \;+\; 0.01(x+20) + \; 0.15x \;=\; 5.40$$

Solve

$$0.10x + 0.01(x + 20) + 0.15x = 5.40$$
$$0.10x + 0.01x + 0.20 + 0.15x = 5.40$$
$$0.26x + 0.20 = 5.40$$
$$0.26x + 0.20 - \mathbf{0.20} = 5.40 - \mathbf{0.20}$$
$$0.26x = 5.20$$
$$\frac{0.26x}{\mathbf{0.26}} = \frac{5.20}{\mathbf{0.26}}$$
$$x = 20$$

pennies   nickels
$x + 20 = 20 + 20$ $3x = 3(20)$
  $= 40$ $= 60$

State

There are 20 dimes.
There are 40 pennies.
There are 60 nickels.

Check

The value of the dimes is 20($0.10), or $2.
The value of the pennies is 40($0.01), or $0.40.
The value of the nickels is 60($0.05), or $3.
Since the total was $2 + $0.40 + $3 = $5.40,
the results check.

Section 2.6

61. BASKETBALL

Analyze
- Made 46 more 2-point baskets than 3-point.
- Made only 1 free throw.
- 113 total points were scored.
- How many 2-point and 3-points baskets were made?

Assign

Let x = the number of 3-point baskets

$x + 46$ = the number of 2-point baskets

Form

	Number	• Value =	Total value
3-point	x	3	$3x$
2-point	$x + 46$	2	$2(x + 46)$
Free throw	1	1	1
		Total:	113

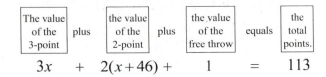

The value of the 3-point	plus	the value of the 2-point	plus	the value of the free throw	equals	the total points.
$3x$	+	$2(x+46)$ +		1	=	113

Solve

$$3x + 2(x + 46) + 1 = 113$$
$$3x + 2x + 92 + 1 = 113$$
$$5x + 93 = 113$$
$$5x + 93 - \mathbf{93} = 113 - \mathbf{93}$$
$$5x = 20$$
$$\frac{5x}{5} = \frac{20}{5}$$
$$x = 4$$

2-point
$$x + 46 = 4 + 46$$
$$= 50$$

State

4 3-point baskets were made.

50 2-point baskets were made.

1 free throw.

Check

The value of the 3-point is 4(3), or 12.
The value of the 2-point is 50(2), or 100.
The value of the free throw is 1(1), or 1.
Since the total was $12 + 100 + 1 = 113$,
the results check.

63-65. Answers will vary.

67. $-12(3a + 4b - 32)$
$$= -12(3a) - 12(4b) - 12(-32)$$
$$= -36a - 48b + 384$$

69. $3(5t + 1)2 = [3(5t) + 3(1)]2$
$$= (15t + 3)2$$
$$= 2(15t) + 2(3)$$
$$= 30t + 6$$

CHALLENGE

71. EVAPORATION

Analyze
- 300 ml of a 2% salt solution.
- Reduce to a 3% salt solution.
- How much water needs to be boiled away?

Assign

Let x = the amount of water in ml

Form

	Amount	• Strength =	Amount of pure salt
Weak	300	0.02	**6**
Strong	**300 − x**	0.03	**30.0(300 − x)**

The salt in the 2% solution	equals	the salt in the 3% solution.
6	=	0.03(300 − x)

Solve

$$6 = 0.03(100 - x)$$
$$\mathbf{100}(6) = \mathbf{100}[0.03(100 - x)]$$
$$100(6) = 100[0.03(100 - x)]$$
$$600 = 3(100 - x)$$
$$600 = 300 + 3x$$
$$600 - \mathbf{300} = 300 + 3x - \mathbf{300}$$
$$300 = 3x$$
$$\frac{300}{3} = \frac{3x}{3}$$
$$100 = x$$

State

100 milliliters of water needs to be boiled off.

Check

The salt in the 2% solution is 300(0.02), or 6 ml.
The salt in the 3% solution is 200(0.03), or 6 ml.
Since the salt in the weaker is 6 ml and the salt in the stronger is 6 ml, the result checks.

73. FINANCIAL PLANNING

Analyze

- Insured fund pays 11% interest.
- Risky investment pays 13% interest.
- Same amount invested in both.
- Higher rate generates $150 extra.
- How much is invested at each rate?

Assign

Let x = amount invested at 11% and 13%

Form

	Principal •	rate •	time =	interest
Insured	x	0.11	1	**0.11x**
Risky	x	0.13	1	**0.13x**

In order for the two amounts of interest to be equal, one must add $150 to the Insured interest before solving.

The interest earned at 11%	+	$150	equals	the interest earned at 13%
0.11x	+	150	=	0.13x

Solve

$$0.11x + 150 = 0.13x$$
$$\mathbf{100}[(0.11x) + 150] = \mathbf{100}(0.13x)$$
$$100(0.11x) + 100(150) = 100(0.13x)$$
$$11x + 15,000 = 13x$$
$$11x + 15,000 - \mathbf{11x} = 13x - \mathbf{11x}$$
$$15,000 = 2x$$
$$\frac{15,000}{\mathbf{2}} = \frac{2x}{\mathbf{2}}$$
$$7,500 = x$$

State

$7,500 was invested in each.

Check

The 1st investment earned 0.11($7,500) + 150, or $975.
The 2nd investment earned 0.13($7,500), or $975.
Since the 2 interests are equal, $975 = $975, the result checks.

Section 2.6

SECTION 2.7
VOCABULARY

Fill in the blanks.

1. An **inequality** is a statement that contains one of the symbols: $>$, $\geq$, $<$, or $\leq$. An equation is a statement that contains an $=$ symbol.

3. The solution set of $x > 2$ can be expressed in **interval** notation as $(2, \infty)$.

CONCEPTS

5. a. Adding the same number to **both** sides of an inequality does not change the solutions.

   b. Multiplying or dividing both sides of an inequality by the same **positive** number does not change the solutions.

   c. If we multiply or divide both sides of an inequality by a **negative** number, the direction of the inequality symbol must be reversed for the inequalities to have the same solutions.

7. Rewrite the inequality $32 < x$ in an equivalent form with the variable on the left side. $\mathbf{x > 32}$

9. a. $x < \mathbf{-1}$, $(-\infty, -1)$ b. $x \geq \mathbf{2}$, $[2, \infty)$

NOTATION

11. a. $\leq$ b. ∞ c. $[\ \text{or}\ ]$ d. $>$

13. $\quad 4x - 5 \geq 7$

$$4x - 5 + \boxed{5} \geq 7 + \boxed{5}$$
$$4x \geq \boxed{12}$$
$$\frac{4x}{\boxed{4}} \geq \frac{12}{\boxed{4}}$$
$$x \geq 3$$

Solution set: $[\boxed{3}, \infty)$

GUIDED PRACTICE

15. Determine whether each number is a solution of $3x - 2 > 5$.

   a. $\quad 3x - 2 > 5$
   $$3(\mathbf{5}) - 2 \overset{?}{>} 5$$
   $$15 - 2 \overset{?}{>} 5$$
   $$13 > 5 \ \text{True}$$
   5 is a solution.

   b. $\quad 3x - 2 > 5$
   $$3(\mathbf{-4}) - 2 \overset{?}{>} 5$$
   $$-12 - 2 \overset{?}{>} 5$$
   $$-14 > 5 \ \text{False}$$
   -4 is not a solution.

17. Determine whether each number is a solution of $-5(x-1) \geq 2x + 12$.

   a. $\quad -5(x-1) \geq 2x + 12$
   $$-5(\mathbf{1} - 1) \overset{?}{\geq} 2(\mathbf{1}) + 12$$
   $$-5(0) \overset{?}{\geq} 2 + 12$$
   $$0 \geq 14 \ \text{False}$$
   1 is not a solution.

   b. $\quad -5(x-1) \geq 2x + 12$
   $$-5(\mathbf{-1} - 1) \overset{?}{\geq} 2(\mathbf{-1}) + 12$$
   $$-5(-2) \overset{?}{\geq} -2 + 12$$
   $$10 \geq 10 \ \text{True}$$
   -1 is a solution.

Graph each inequality and describe the graph using interval notation.

19. $x < 5$

$(-\infty, 5)$

21. $-3 < x \leq 1$

$(-3, 1]$

Solve each inequality. Write the solution set in interval notation and graph it.

23. $\quad x + 2 > 5$
$$x + 2 - \mathbf{2} > 5 - \mathbf{2}$$
$$x > 3$$
$$(3, \infty)$$

25. $\quad g - 30 \geq -20$
$$g - 30 + \mathbf{30} \geq -20 + \mathbf{30}$$
$$g \geq 10$$
$$[10, \infty)$$

27. $\quad -\dfrac{3}{16}x \geq -9$
$$-\frac{\mathbf{16}}{\mathbf{3}}\left(-\frac{3}{16}x\right) \leq -\frac{\mathbf{16}}{\mathbf{3}}(-9)$$
$$x \leq 48$$
$$(-\infty, 48]$$

29.
$$\frac{2}{3}x \ge 2$$
$$\frac{3}{2}\left(\frac{2}{3}x\right) \ge \frac{3}{2}(2)$$
$$x \ge 3$$
$$[3, \infty)$$

31.
$$-3y \le -6$$
$$\frac{-3y}{-3} \ge \frac{-6}{-3}$$
$$y \ge 2$$
$$[2, \infty)$$

33.
$$8h < 48$$
$$\frac{8h}{8} < \frac{48}{8}$$
$$h < 6$$
$$(-\infty, 6)$$

35.
$$64 < 9x + 1$$
$$64 - \mathbf{1} < 9x + 1 - \mathbf{1}$$
$$63 < 9x$$
$$\frac{63}{9} < \frac{9x}{9}$$
$$7 < x$$
$$x > 7$$
$$(7, \infty)$$

37.
$$-20 \ge 3m - 5$$
$$-20 + \mathbf{5} \ge 3m - 5 + \mathbf{5}$$
$$-15 \ge 3m$$
$$\frac{-15}{3} \ge \frac{3m}{3}$$
$$-5 \ge m$$
$$m \le -5$$
$$(-\infty, -5]$$

39.
$$1.3 - 2x \ge 0.5$$
$$1.3 - 2x - \mathbf{1.3} \ge 0.5 - \mathbf{1.3}$$
$$-2x \ge -0.8$$
$$\frac{-2x}{-2} \le \frac{-0.8}{-2}$$
$$x \le 0.4$$
$$(-\infty, 0.4]$$

41.
$$24.9 - 12a < -3.9$$
$$24.9 - 12a - \mathbf{24.9} < -3.9 - \mathbf{24.9}$$
$$-12a < -28.8$$
$$\frac{-12a}{-12} > \frac{-28.8}{-12}$$
$$a > 2.4$$
$$(2.4, \infty)$$

43.
$$9a + 4 > 5a - 16$$
$$9a + 4 - \mathbf{5a} > 5a - 16 - \mathbf{5a}$$
$$4a + 4 > -16$$
$$4a + 4 - \mathbf{4} > -16 - \mathbf{4}$$
$$4a > -20$$
$$\frac{4a}{4} > \frac{-20}{4}$$
$$x > -5$$
$$(-5, \infty)$$

45.
$$8(2n + 1) \le 4(6n + 7) + 4n$$
$$\mathbf{8}(2n) + \mathbf{8}(1) \le \mathbf{4}(6n) + \mathbf{4}(7) + 4n$$
$$16n + 8 \le 24n + 28 + 4n$$
$$16n + 8 \le 28n + 28$$
$$16n + 8 - \mathbf{28n} \le 28n + 28 - \mathbf{28n}$$
$$-12n + 8 \le 28$$
$$-12n + 8 - \mathbf{8} \le 28 - \mathbf{8}$$
$$-12n \le 20$$
$$\frac{-12n}{-12} \ge \frac{20}{-12}$$
$$n \ge -\frac{5}{3}$$
$$\left[-\frac{5}{3}, \infty\right)$$

Section 2.7

47.
$$\frac{1}{2}+\frac{n}{5}>\frac{3}{4} \quad , \text{ LCD}=20$$
$$\mathbf{20}\left(\frac{1}{2}+\frac{n}{5}\right)>\mathbf{20}\left(\frac{3}{4}\right)$$
$$20\left(\frac{1}{2}\right)+20\left(\frac{n}{5}\right)>20\left(\frac{3}{4}\right)$$
$$10+4n>15$$
$$10+4n-\mathbf{10}>15-\mathbf{10}$$
$$4n>5$$
$$\frac{4n}{\mathbf{4}}>\frac{5}{\mathbf{4}}$$
$$n>\frac{5}{4}$$
$$\left(\frac{5}{4},\infty\right)$$

49.
$$\frac{1}{2}-\frac{x}{24}\geq-\frac{1}{8} \quad , \text{ LCD}=24$$
$$\mathbf{24}\left(\frac{1}{2}-\frac{x}{24}\right)\geq\mathbf{24}\left(-\frac{1}{8}\right)$$
$$24\left(\frac{1}{2}\right)-24\left(\frac{x}{24}\right)\geq24\left(-\frac{1}{8}\right)$$
$$12-x\geq-3$$
$$12-x-\mathbf{12}\geq-3-\mathbf{12}$$
$$-x\geq-15$$
$$\frac{-x}{\mathbf{-1}}\leq\frac{-15}{\mathbf{-1}}$$
$$x\leq15$$
$$(-\infty,15]$$

Graph each inequality and describe the graph using interval notation.

51. $\quad -2\leq x<3$
$$[-2,3)$$

53. $\quad -\dfrac{7}{4}<x<2$
$$\left(-\frac{7}{4},2\right)$$

Solve each compound inequality. Write the solution set in interval notation and graph it.

55. $\qquad 2<x-5<5$
$$2+5<x-5+5<5+5$$
$$7<x<10$$
$$(7,10)$$

57. $\qquad 0\leq x+10\leq10$
$$0-\mathbf{10}\leq x+10-\mathbf{10}\leq10-\mathbf{10}$$
$$-10\leq x\leq0$$
$$[-10,0]$$

59. $\qquad 3\leq 2x-1<5$
$$3+\mathbf{1}\leq 2x-1+\mathbf{1}<5+\mathbf{1}$$
$$4\leq 2x<6$$
$$\frac{4}{\mathbf{2}}\leq\frac{2x}{\mathbf{2}}<\frac{6}{\mathbf{2}}$$
$$2\leq x<3$$
$$[2,3)$$

61. $\qquad -9<6x+9\leq45$
$$-9-\mathbf{9}<6x+9-\mathbf{9}\leq45-\mathbf{9}$$
$$-18<6x\leq36$$
$$\frac{-18}{\mathbf{6}}<\frac{6x}{\mathbf{6}}\leq\frac{36}{\mathbf{6}}$$
$$-3<x\leq6$$
$$(-3,6]$$

TRY IT YOURSELF

Solve each inequality or compound inequality.
Write the solution set in interval notation and
graph it.

63.
$$\frac{6x+1}{4} \le x+1 \ , \ \text{LCD} = 4$$
$$4\left(\frac{6x+1}{4}\right) \le 4(x+1)$$
$$6x+1 \le 4x+4$$
$$6x+1-\mathbf{4x} \le 4x+4-\mathbf{4x}$$
$$2x+1 \le 4$$
$$2x+1-\mathbf{1} \le 4-\mathbf{1}$$
$$2x \le 3$$
$$\frac{2x}{\mathbf{2}} \le \frac{3}{\mathbf{2}}$$
$$x \le \frac{3}{2}$$
$$\left(-\infty, \frac{3}{2}\right]$$

65.
$$17(3-x) \ge 3-13x$$
$$51-17x \ge 3-13x$$
$$51-17x+\mathbf{13x} \ge 3-13x+\mathbf{13x}$$
$$51-4x \ge 3$$
$$51-4x-\mathbf{51} \ge 3-\mathbf{51}$$
$$-4x \ge -48$$
$$\frac{-4x}{\mathbf{-4}} \le \frac{-48}{\mathbf{-4}}$$
$$x \le 12$$
$$(-\infty, 12]$$

67.
$$0 < 5(x+2) \le 15$$
$$0 < 5x+10 \le 15$$
$$0-\mathbf{10} < 5x+10-\mathbf{10} \le 15-\mathbf{10}$$
$$-10 < 5x \le 5$$
$$\frac{-10}{\mathbf{5}} < \frac{5x}{\mathbf{5}} \le \frac{5}{\mathbf{5}}$$
$$-2 < x \le 1$$
$$(-2, 1]$$

69.
$$0.4x \le 0.1x + 0.45$$
$$0.4x - \mathbf{0.1x} \le 0.1x + 0.45 - \mathbf{0.1x}$$
$$0.3x \le 0.45$$
$$\frac{0.3x}{\mathbf{0.3}} \le \frac{0.45}{\mathbf{0.3}}$$
$$x \le 1.5$$
$$(-\infty, 1.5]$$

71.
$$-\frac{2}{3} \ge \frac{2y}{3} - \frac{3}{4} \ , \ \text{LCD} = 12$$
$$\mathbf{12}\left(-\frac{2}{3}\right) \ge \mathbf{12}\left(\frac{2y}{3} - \frac{3}{4}\right)$$
$$12\left(-\frac{2}{3}\right) \ge 12\left(\frac{2y}{3}\right) - 12\left(\frac{3}{4}\right)$$
$$-8 \ge 8y - 9$$
$$-8 + \mathbf{9} \ge 8y - 9 + \mathbf{9}$$
$$1 \ge 8y$$
$$\frac{1}{\mathbf{8}} \ge \frac{8y}{\mathbf{8}}$$
$$\frac{1}{8} \ge y$$
$$y \le \frac{1}{8}$$
$$\left(-\infty, \frac{1}{8}\right]$$

73.
$$\frac{m}{-42} - 1 > -1$$
$$\frac{m}{-42} - 1 + \mathbf{1} > -1 + \mathbf{1}$$
$$\frac{m}{-42} > 0$$
$$\mathbf{-42}\left(\frac{m}{-42}\right) < \mathbf{-42}(0)$$
$$m < 0$$
$$(-\infty, 0)$$

- 93 -

Section 2.7

75.

$$6 - x \le 3(x-1)$$
$$6 - x \le 3(x) - 3(1)$$
$$6 - x \le 3x - 3$$
$$6 - x - 3x \le 3x - 3 - 3x$$
$$6 - 4x \le -3$$
$$6 - 4x - 6 \le -3 - 6$$
$$-4x \le -9$$
$$\frac{-4x}{-4} \ge \frac{-9}{-4}$$
$$x \ge \frac{9}{4}$$
$$\left[\frac{9}{4}, \infty\right)$$

77.

$$6 < -2(x-1) < 12$$
$$6 < -2x + 2 < 12$$
$$6 - 2 < -2x + 2 - 2 < 12 - 2$$
$$4 < -2x < 10$$
$$\frac{4}{-2} > \frac{-2x}{-2} > \frac{10}{-2}$$
$$-2 > x > -5$$
$$-5 < x < -2$$
$$(-5, -2)$$

79.

$$-1 \le -\frac{1}{2}n \ , \ \text{LCD} = -2$$
$$-2(-1) \ge -2\left(-\frac{1}{2}n\right)$$
$$2 \ge n$$
$$n \le 2$$
$$(-\infty, 2]$$

81.

$$-m - 12 > 15$$
$$-m - 12 + 12 > 15 + 12$$
$$-m > 27$$
$$\frac{-m}{-1} < \frac{27}{-1}$$
$$m < -27$$
$$(-\infty, -27)$$

83.

$$y - \frac{1}{7} \le \frac{2}{3} \ , \ \text{LCD} = 21$$
$$21\left(y - \frac{1}{7}\right) \le 21\left(\frac{2}{3}\right)$$
$$21(y) - 21\left(\frac{1}{7}\right) \le 21\left(\frac{2}{3}\right)$$
$$21y - 3 \le 14$$
$$21y - 3 + 3 \le 14 + 3$$
$$21y \le 17$$
$$\frac{21y}{21} \le \frac{17}{21}$$
$$y \le \frac{17}{21}$$
$$\left(-\infty, \frac{17}{21}\right]$$

85.

$$9x + 13 \ge 2x + 6x$$
$$9x + 13 \ge 8x$$
$$9x + 13 - 8x \ge 8x - 8x$$
$$x + 13 \ge 0$$
$$x + 13 - 13 \ge 0 - 13$$
$$x \ge -13$$
$$[-13, \infty)$$

87.

$$7 < \frac{5}{3}a + (-3)$$

$$7 + \mathbf{3} < \frac{5}{3}a - 3 + \mathbf{3}$$

$$10 < \frac{5}{3}a$$

$$\frac{\mathbf{3}}{\mathbf{5}} \cdot 10 < \frac{\mathbf{3}}{\mathbf{5}} \cdot \left(\frac{5}{3}a\right)$$

$$6 < a$$

$$a > 6$$

$$(6, \infty)$$

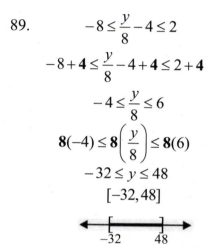

89.

$$-8 \le \frac{y}{8} - 4 \le 2$$

$$-8 + \mathbf{4} \le \frac{y}{8} - 4 + \mathbf{4} \le 2 + \mathbf{4}$$

$$-4 \le \frac{y}{8} \le 6$$

$$\mathbf{8}(-4) \le \mathbf{8}\left(\frac{y}{8}\right) \le \mathbf{8}(6)$$

$$-32 \le y \le 48$$

$$[-32, 48]$$

91.

$$0.04x + 1.04 \le 0.01x + 1.085$$

$$0.04x + 1.04 - \mathbf{0.01x} \le 0.01x + 1.085 - \mathbf{0.01x}$$

$$0.03x + 1.04 \le 1.085$$

$$0.03x + 1.04 - \mathbf{1.04} \le 1.085 - \mathbf{1.04}$$

$$0.03x \le 0.045$$

$$\frac{0.03x}{\mathbf{0.03}} \le \frac{0.045}{\mathbf{0.03}}$$

$$x \le 1.5$$

$$(-\infty, 1.5]$$

93.

$$\frac{5}{3}(x+1) \ge -x + \frac{2}{3} \quad , \ \text{LCD} = 3$$

$$\mathbf{3}\left(\frac{5}{3}(x+1)\right) \ge \mathbf{3}\left(-x + \frac{2}{3}\right)$$

$$3\left(\frac{5}{3}(x+1)\right) \ge 3(-x) + 3\left(\frac{2}{3}\right)$$

$$5(x+1) \ge -3x + 2$$

$$5x + 5 \ge -3x + 2$$

$$5x + 5 + \mathbf{3x} \ge -3x + 2 + \mathbf{3x}$$

$$8x + 5 \ge 2$$

$$8x + 5 - \mathbf{5} \ge 2 - \mathbf{5}$$

$$8x \ge -3$$

$$\frac{8x}{\mathbf{8}} \ge \frac{-3}{\mathbf{8}}$$

$$x \ge -\frac{3}{8}$$

$$\left[-\frac{3}{8}, \infty\right)$$

95.

$$\frac{4}{5}x < \frac{2}{5} \quad , \ \text{LCD} = 5$$

$$\mathbf{5}\left(\frac{4}{5}x\right) < \mathbf{5}\left(\frac{2}{5}\right)$$

$$4x < 2$$

$$\frac{4x}{\mathbf{4}} < \frac{2}{\mathbf{4}}$$

$$x < \frac{1}{2}$$

$$\left(-\infty, \frac{1}{2}\right)$$

Section 2.7

97.
$$2x + 3(2x + 3) \le 7(x + 1) + 1$$
$$2x + 3(2x) + 3(3) \le 7(x) + 7(1) + 1$$
$$2x + 6x + 9 \le 7x + 7 + 1$$
$$8x + 9 \le 7x + 8$$
$$8x + 9 - 7x \le 7x + 8 - 7x$$
$$x + 9 \le 8$$
$$x + 9 - 9 \le 8 - 9$$
$$x \le -1$$
$$(-\infty, -1]$$

Look Alikes . . .

Solve each equation and inequality. Write the solution set of each inequality in interval notation and graph it.

99. a.
$$\frac{3}{8} + \frac{b}{3} > \frac{5}{12} \quad , \text{LCD} = 24$$
$$24\left(\frac{3}{8}\right) + 24\left(\frac{b}{3}\right) > 24\left(\frac{5}{12}\right)$$
$$3(3) + 8b > 2(5)$$
$$9 + 8b > 10$$
$$9 + 8b - 9 > 10 - 9$$
$$8b > 1$$
$$\frac{8b}{8} > \frac{1}{8}$$
$$b > \frac{1}{8}$$
$$\left(\frac{1}{8}, \infty\right)$$

99. b.
$$\frac{3}{8} + \frac{b}{3} = \frac{5}{12} \quad , \text{LCD} = 24$$
$$24\left(\frac{3}{8}\right) + 24\left(\frac{b}{3}\right) = 24\left(\frac{5}{12}\right)$$
$$3(3) + 8b = 2(5)$$
$$9 + 8b = 10$$
$$9 + 8b - 9 = 10 - 9$$
$$8b = 1$$
$$\frac{8b}{8} = \frac{1}{8}$$
$$b = \frac{1}{8}$$

101. a.
$$4 \le 2x - 6$$
$$2x - 6 + 6 \ge 4 + 6$$
$$2x \ge 10$$
$$\frac{2x}{2} \ge \frac{10}{2}$$
$$x \ge 5$$
$$[5, \infty)$$

101. b.
$$4 \le 2x - 6 < 18$$
$$4 + 6 \le 2x - 6 + 6 < 18 + 6$$
$$10 \le 2x < 24$$
$$\frac{10}{2} \le \frac{2x}{2} < \frac{24}{2}$$
$$5 \le x < 12$$
$$[5, 12)$$

APPLICATIONS

103. GRADES

Analyze
- A student has test scores of 68%, 75% and 79%.
- What should be the score of the next test to have a B (80%) or better?

Assign

Let x = the 4th test score

Form

The average of the four test grades	must be no less than	80.
$\dfrac{68 + 75 + 79 + x}{4}$	$\ge$	80

Solve
$$\frac{68 + 75 + 79 + x}{4} \ge 80$$
$$4\left(\frac{68 + 75 + 79 + x}{4}\right) \ge 4(80)$$
$$x + 222 \ge 320$$
$$x + 228 - 222 \ge 320 - 222$$
$$x \ge 98$$

State

The 4th test has to be a 98% or better.

Check
$$\frac{68 + 75 + 79 + 98}{4} \ge 80$$
$$\frac{320}{4} \ge 80$$
$$80 \ge 80$$

105. GAS MILEAGE

Analyze
- Two cars average 17 and 19 mpg.
- What should be the mpg of the 3rd car to have an average of 21 mpg or better?

Assign

Let x = the mpg of the 3rd car

Form

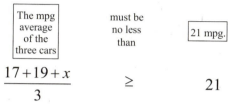

$$\frac{17+19+x}{3} \geq 21$$

Solve

$$\frac{17+19+x}{3} \geq 21$$

$$3\left(\frac{17+19+x}{3}\right) \geq 3(21)$$

$$x+36 \geq 63$$

$$x+36-\mathbf{36} \geq 63-\mathbf{36}$$

$$x \geq 27$$

State

The 3rd car will need to average 27 mpg or better.

Check

$$\frac{17+19+27}{3} \geq 21$$

$$\frac{63}{3} \geq 21$$

$$21 \geq 21$$

107. GEOMETRY

Analyze
- The perimeter of an equilateral triangle is at most 57 feet.
- What could be the length of a side?

Assign

Let x = the length of one side in feet

Form

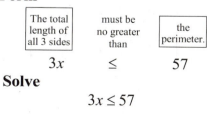

$$3x \qquad \leq \qquad 57$$

Solve

$$3x \leq 57$$

$$\frac{3x}{\mathbf{3}} \leq \frac{57}{\mathbf{3}}$$

$$x \leq 19$$

State

The length of one side is 19 feet or less.

Check

$$3x \leq 57$$

$$3(19) \leq 57$$

$$57 \leq 57$$

109. COUNTER SPACE

Analyze
- The width is x feet.
- The length is $x+5$ feet.
- The outside perimeter needs to exceed 30 feet.
- What is the value of x?

Assign

Let x = the value

Form

Two widths of the counter	plus	two lengths of the counter	must be no less than	the perimeter of the counter.
$2x$	$+$	$2(x+5)$	$\geq$	30

Solve

$$2x+2(x+5) \geq 30$$

$$2x+2x+10 \geq 30$$

$$4x+10 \geq 30$$

$$4x+10-\mathbf{10} \geq 30-\mathbf{10}$$

$$4x \geq 20$$

$$\frac{4x}{\mathbf{4}} \geq \frac{20}{\mathbf{4}}$$

$$x \geq 5$$

State

The value of x is 5 feet or more.

Check

$$2x+2(x+5) \geq 30$$

$$2(5)+2(5+5) \geq 30$$

$$10+20 \geq 30$$

$$30 \geq 30$$

Section 2.7

111. GRADUATIONS

Analyze
- Cost $18 to rent a cap and gown.
- $0.80 per announcement.
- Does not want to spend over $50.
- How many announcements can she order?

Assign
Let x = the number of announcements

Form

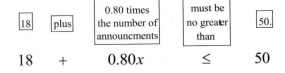

$$18 \quad + \quad 0.80x \quad \leq \quad 50$$

Solve
$$18 + 0.80x \leq 50$$
$$18 + 0.80x - \mathbf{18} \leq 50 - \mathbf{18}$$
$$0.80x \leq 32$$
$$\frac{0.80x}{\mathbf{0.80}} \leq \frac{32}{\mathbf{0.80}}$$
$$x \leq 40$$

State
She can buy 40 or less announcements.

Check
$$18 + 0.80x \leq 50$$
$$18 + 0.80(40) \leq 50$$
$$18 + 32 \leq 50$$
$$50 \leq 50$$

113. WINDOWS

Analyze
- Triangular window area is no greater than 100 in^2.
- Base must be 16 inches.
- Formula: $A = \frac{1}{2}bh$ or $A = 0.5bh$.
- What is the height of the window?

Assign
Let h = the height of the window in inches

Form
$$A \geq 0.5bh$$

Solve
$$A \geq 0.5bh$$
$$100 \geq 0.5(16)h$$
$$100 \geq 8h$$
$$\frac{100}{\mathbf{8}} \geq \frac{8h}{\mathbf{8}}$$
$$12.5 \geq h$$
$$h \leq 12.5$$

State
The height can be 12.5 inches or less.

Check
$$100 \geq 0.5(16)h$$
$$100 \geq 0.5(16)(12.5)$$
$$100 \geq 100$$

115. NUMBER PUZZLES

Analyze
- What *whole* numbers satisfy the condition: Twice the number decreased by 1 is between 50 and 60?

Assign
Let x = whole # between the 2 numbers

Form

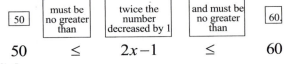

$$50 \quad \leq \quad 2x - 1 \quad \leq \quad 60$$

Solve
$$50 \leq 2x - 1 \leq 60$$
$$50 + \mathbf{1} \leq 2x - 1 + \mathbf{1} \leq 60 + \mathbf{1}$$
$$51 \leq 2x \leq 61$$
$$\frac{51}{\mathbf{2}} \leq \frac{2x}{\mathbf{2}} \leq \frac{61}{\mathbf{2}}$$
$$25.5 \leq x \leq 30.5$$

State
The whole numbers are 26, 27, 28, 29, 30.

Check
$$50 \leq 2x - 1 \leq 60$$
$$50 \leq 2(\mathbf{27}) - 1 \leq 60$$
$$50 \leq 53 \leq 60$$

WRITING

117. Answers will vary.

REVIEW

119.

x	$x^2 - 3$
-2	$(-2)^2 - 3$ $= 4 - 3$ $= \mathbf{1}$
0	$0^2 - 3$ $= 0 - 3$ $= \mathbf{-3}$
3	$3^2 - 3$ $= 9 - 3$ $= \mathbf{6}$

CHALLENGE PROBLEMS

121. Solve the inequality. Write the solution set in interval notation and graph it.

$$3 - x < 5 < 7 - x$$
$$3 - x + \boldsymbol{x} < 5 + \boldsymbol{x} < 7 - x + \boldsymbol{x}$$
$$3 < 5 + x < 7$$
$$3 - \mathbf{5} < 5 + x - \mathbf{5} < 7 - \mathbf{5}$$
$$-2 < x < 2$$

$$(-2, 2)$$

Section 2.7

CHAPTER 2 REVIEW
SECTION 2.1
Solving Equations Using Properties of Equality

Determine whether the given number is a solution of the equation.

1. 84, $x - 34 = 50$

$$84 - 34 \overset{?}{=} 50$$
$$50 = 50 \quad \text{True}$$

Since the resulting statement is true, 84 is a solution.

2. 3, $5y + 2 = 12$

$$5(3) + 2 \overset{?}{=} 12$$
$$15 + 2 \overset{?}{=} 12$$
$$17 = 12 \quad \text{False}$$

Since the resulting statement is false, 3 is not a solution.

3. -30, $\dfrac{x}{5} = 6$

$$\dfrac{-30}{5} \overset{?}{=} 6$$
$$-6 = 6 \quad \text{False}$$

Since the resulting statement is false, -30 is not a solution.

4. 2, $\left| a^2 - a - 1 \right| = 0$

$$\left| 2^2 - 2 - 1 \right| \overset{?}{=} 0$$
$$\left| 4 - 2 - 1 \right| \overset{?}{=} 0$$
$$\left| 2 - 1 \right| \overset{?}{=} 0$$
$$\left| 1 \right| \overset{?}{=} 0$$
$$1 = 0 \quad \text{False}$$

Since the resulting statement is false, 2 is not a solution.

5. -3, $5b - 2 = 3b - 8$

$$5(-3) - 2 \overset{?}{=} 3(-3) - 8$$
$$-15 - 2 \overset{?}{=} -9 - 8$$
$$-17 = -17 \quad \text{True}$$

Since the resulting statement is true, -3 is a solution.

6. 1, $\dfrac{2}{y+1} = \dfrac{12}{y+1} - 5$

$$\dfrac{2}{1+1} \overset{?}{=} \dfrac{12}{1+1} - 5$$
$$\dfrac{2}{2} \overset{?}{=} \dfrac{12}{2} - 5$$
$$1 \overset{?}{=} 6 - 5$$
$$1 = 1 \quad \text{True}$$

Since the resulting statement is true, 1 is a solution.

Fill in the blanks.

7. An **equation** is a statement indicating that two expressions are equal.

8. To solve $x - 8 = 10$ means to find all the values of the variable that make the equation a **true** statement.

Solve each equation and check the result.

9. $$x - 9 = 12$$
$$x - 9 + 9 = 12 + 9$$
$$x = 21$$

$$x - 9 = 12$$
$$21 - 9 \overset{?}{=} 12$$
$$12 = 12$$

The solution is 21. The result checks.

10. $$-y = -32 \qquad\qquad -y = -32$$
$$\dfrac{-y}{-1} = \dfrac{-32}{-1} \qquad -(32) \overset{?}{=} -32$$
$$y = 32 \qquad\qquad -32 = -32$$

The solution is 32. The result checks.

11. $$a + 3.7 = -16.9$$
$$a + 3.7 - \mathbf{3.7} = -16.9 - \mathbf{3.7}$$
$$a = -20.6$$

$$a + 3.7 = -16.9$$
$$\mathbf{-20.6} + 3.7 \overset{?}{=} -16.9$$
$$-16.9 = -16.9$$

The solution is -2.6. The result checks.

12. $$100 = -7 + r \qquad\qquad 100 = -7 + r$$
$$100 + \mathbf{7} = -7 + r + \mathbf{7} \qquad 100 \overset{?}{=} -7 + \mathbf{107}$$
$$107 = r \qquad\qquad 100 = 100$$

The solution is 107. The result checks.

13. $120 = 5c$ $120 = 5c$
$$\frac{120}{5} = \frac{5c}{5}$$
$$120 \overset{?}{=} 5(24)$$
$$24 = c$$
$$120 = 120$$

The solution is 24. The result checks.

14. $$t - \frac{1}{3} = \frac{3}{7}$$
$$t - \frac{1}{3} + \frac{1}{3} = \frac{3}{7} + \frac{1}{3}$$
$$t = \frac{3}{7} \cdot \frac{3}{3} + \frac{1}{3} \cdot \frac{7}{7}$$
$$t = \frac{9+7}{21}$$
$$t = \frac{16}{21}$$

The solution is $\frac{16}{21}$. The result checks.

15. $$\frac{4}{3}t = -12$$
$$\frac{3}{4} \cdot \frac{4}{3}t = \frac{3}{4} \cdot (-12)$$
$$t = -\frac{3 \cdot 3 \cdot \overset{1}{\cancel{4}}}{\cancel{4}}$$
$$t = -9$$
$$\frac{4}{3}t = -12$$
$$\frac{4}{3}(-9) \overset{?}{=} -12$$
$$\frac{4}{\cancel{3}} \cdot \frac{-3 \cdot \overset{1}{\cancel{3}}}{1} \overset{?}{=} -12$$
$$-12 = -12$$

The solution is -9. The result checks.

16. $$3 = \frac{q}{-2.6}$$
$$\frac{-2.6}{1} \cdot 3 = \frac{-2.6}{1} \cdot \frac{q}{-2.6}$$
$$-7.8 = q$$

The solution is -7.8. The result checks.

17. $6b = 0$ $6b = 0$
$$\frac{6b}{6} = \frac{0}{6}$$
$$6 \cdot 0 \overset{?}{=} 0$$
$$b = 0$$
$$0 = 0$$

The solution is 0. The result checks.

18. $$\frac{15}{16}s = -3$$
$$\frac{16}{15} \left(\frac{15}{16}s \right) = \frac{16}{15} \cdot (-3)$$
$$s = -\frac{\overset{1}{\cancel{3}} \cdot 16}{\cancel{3} \cdot 5}$$
$$s = -\frac{16}{5}$$

The solution is $\frac{16}{5}$. The result checks.

SECTION 2.2
More about Solving Equations

Solve each equation and check the answer.

19. $5x + 4 = 14$
$$5x + 4 - 4 = 14 - 4$$
$$5x = 10$$
$$\frac{5x}{5} = \frac{10}{5}$$
$$x = 2$$

$$5x + 4 = 14$$
$$5(2) + 4 \overset{?}{=} 14$$
$$10 + 4 \overset{?}{=} 14$$
$$14 = 14$$

The solution is 2. The result checks.

20. $98.6 - t = 129.2$
$$98.6 - t - 98.6 = 129.2 - 98.6$$
$$-t = 30.6$$
$$\frac{-t}{-1} = \frac{30.6}{-1}$$
$$t = -30.6$$

The solution is -30.6. The result checks.

21. $$\frac{n}{5} + (-2) = 4$$
$$\frac{n}{5} + (-2) + 2 = 4 + 2$$
$$\frac{n}{5} = 6$$
$$\frac{5}{1} \cdot \frac{n}{5} = \frac{5}{1} \cdot (6)$$
$$n = 30$$

$$\frac{n}{5} + (-2) = 4$$
$$\frac{30}{5} + (-2) \overset{?}{=} 4$$
$$6 + (-2) \overset{?}{=} 4$$
$$4 = 4$$

The solution is 30. The result checks.

22. $\dfrac{b-5}{4} = -6$

$\left(\dfrac{4}{1}\right) \cdot \dfrac{b-5}{4} = \left(\dfrac{4}{1}\right) \cdot (-6)$

$b - 5 = -24$

$b - 5 + 5 = -24 + 5$

$b = -19$

The solution is -19. The solution checks.

23. $5(2x - 4) - 5x = 0$

$10x - 20 - 5x = 0$

$5x - 20 = 0$

$5x - 20 + 20 = 0 + 20$

$5x = 20$

$\dfrac{5x}{5} = \dfrac{20}{5}$

$x = 4$

$5(2x - 4) - 5x = 0$

$5(2 \cdot 4 - 4) - 5 \cdot 4 \overset{?}{=} 0$

$5(4) - 20 \overset{?}{=} 0$

$20 - 20 \overset{?}{=} 0$

$0 = 0$

The solution is 4. The solution checks.

24. $-2(x - 5) = 5(-3x + 4) + 3$

$-2x + 10 = -15x + 20 + 3$

$-2x + 10 = -15x + 23$

$-2x + 10 - 10 = -15x + 23 - 10$

$-2x = -15x + 13$

$-2x + 15x = -15x + 13 + 15x$

$13x = 13$

$\dfrac{13x}{13} = \dfrac{13}{13}$

$x = 1$

The solution is 1. The solution checks.

25. $\dfrac{3}{4} = \dfrac{1}{2} + \dfrac{d}{5}$ $\qquad$ LCD $= 20$

$20 \cdot \dfrac{3}{4} = 20 \cdot \left(\dfrac{1}{2} + \dfrac{d}{5}\right)$

$20 \cdot \dfrac{3}{4} = 20 \cdot \left(\dfrac{1}{2}\right) + 20 \cdot \left(\dfrac{d}{5}\right)$

$15 = 10 + 4d$

$15 - 10 = 10 + 4d - 10$

$5 = 4d$

$\dfrac{5}{4} = \dfrac{4d}{4}$

$\dfrac{5}{4} = d$

$\dfrac{3}{4} = \dfrac{1}{2} + \dfrac{d}{5}$

$\dfrac{3}{4} \overset{?}{=} \dfrac{1}{2} + \dfrac{\frac{5}{4}}{5}$

$\dfrac{3}{4} \overset{?}{=} \dfrac{1}{2} + \dfrac{5}{4} \cdot \dfrac{1}{5}$

$\dfrac{3}{4} \overset{?}{=} \dfrac{1}{2} + \dfrac{1}{4}$

$\dfrac{3}{4} = \dfrac{3}{4}$

The solution is $\dfrac{5}{4}$. The solution checks.

26.

$$\frac{5(7-x)}{4} = 2x - 3 \quad \text{LCD} = 4$$

$$4 \cdot \left(\frac{5(7-x)}{4}\right) = 4 \cdot (2x - 3)$$

$$4 \cdot \left(\frac{5(7-x)}{4}\right) = 4 \cdot (2x) - 4 \cdot (3)$$

$$5(7-x) = 8x - 12$$

$$35 - 5x = 8x - 12$$

$$35 - 5x + \mathbf{12} = 8x - 12 + \mathbf{12}$$

$$47 - 5x = 8x$$

$$47 - 5x + 5\boldsymbol{x} = 8x + 5\boldsymbol{x}$$

$$47 = 13x$$

$$\frac{47}{13} = \frac{47x}{13}$$

$$\frac{47}{13} = x$$

The solution is $\dfrac{47}{13}$. The solution checks.

27.

$$\frac{3(2-c)}{2} = \frac{-2(2c+3)}{5} \quad , \quad \text{LCD} = 10$$

$$\mathbf{10} \cdot \left(\frac{3(2-c)}{2}\right) = \mathbf{10} \cdot \left(\frac{-2(2c+3)}{5}\right)$$

$$5[3(2-c)] = 2[-2(2c+3)]$$

$$15(2-c) = -4(2c+3)$$

$$30 - 15c = -8c - 12$$

$$30 - 15c + \mathbf{12} = -8c - 12 + \mathbf{12}$$

$$42 - 15c = -8c$$

$$42 - 15c + \mathbf{15c} = -8c + \mathbf{15c}$$

$$42 = 7c$$

$$\frac{42}{7} = \frac{7c}{7}$$

$$6 = c$$

$$\frac{3(2-c)}{2} = \frac{-2(2c+3)}{5}$$

$$\frac{3(2-6)}{2} \overset{?}{=} \frac{-2(2\cdot 6 + 3)}{5}$$

$$\frac{3(-4)}{2} \overset{?}{=} \frac{-2(15)}{5}$$

$$\frac{-12}{2} \overset{?}{=} \frac{-30}{5}$$

$$-6 = -6$$

The solution is 6. The solution checks.

28.

$$\frac{b}{3} + \frac{11}{9} + 3b = -\frac{5}{6}b \quad , \quad \text{LCD} = 18$$

$$18 \cdot \left(\frac{b}{3} + \frac{11}{9} + 3b\right) = 18 \cdot \left(-\frac{5}{6}b\right)$$

$$18 \cdot \left(\frac{b}{3}\right) + 18 \cdot \left(\frac{11}{9}\right) + 18 \cdot (3b) = 18 \cdot \left(-\frac{5}{6}b\right)$$

$$6b + 22 + 54b = -15b$$

$$60b + 22 = -15b$$

$$60b + 22 - \mathbf{60b} = -15b - \mathbf{60b}$$

$$22 = -75b$$

$$\frac{22}{-75} = \frac{-75b}{-75}$$

$$-\frac{22}{75} = b$$

The solution is $-\dfrac{22}{75}$. The solution checks.

29.

$$0.15(x+2) + 0.3 = 0.35x - 0.4$$

$$\mathbf{100}[0.15(x+2) + 0.3] = \mathbf{100}[0.35x - 0.4]$$

$$100[0.15(x+2)] + 100(0.3) = 100(0.35x) - 100(0.4)$$

$$15(x+2) + 30 = 35x - 40$$

$$15x + 30 + 30 = 35x - 40$$

$$15x + 60 = 35x - 40$$

$$15x + 60 - \mathbf{15x} = 35x - 40 - \mathbf{15x}$$

$$60 = 20x - 40$$

$$60 + \mathbf{40} = 20x - 40 + \mathbf{40}$$

$$100 = 20x$$

$$\frac{100}{20} = \frac{20x}{20}$$

$$5 = x$$

$$0.15(x+2) + 0.3 = 0.35x - 0.4$$

$$0.15(5+2) + 0.3 \overset{?}{=} 0.35(5) - 0.4$$

$$0.15(7) + 0.3 \overset{?}{=} 1.75 - 0.4$$

$$1.05 + 0.3 \overset{?}{=} 1.35$$

$$1.35 = 1.35$$

The solution is 5. The solution checks.

30.
$$0.5 - 0.02(y-2) = 0.16 + 0.36y$$
$$\mathbf{100}[0.5 - 0.02(y-2)] = \mathbf{100}[0.16 + 0.36y]$$
$$100(0.5) - 100[0.02(y-2)] = 100(0.16) + 100(0.36y)$$
$$50 - 2(y-2) = 16 + 36y$$
$$50 - 2y + 4 = 16 + 36y$$
$$54 - 2y = 16 + 36y$$
$$54 - 2y + \mathbf{2y} = 16 + 36y + \mathbf{2y}$$
$$54 = 16 + 38y$$
$$54 - \mathbf{16} = 16 + 38y - \mathbf{16}$$
$$38 = 38y$$
$$\frac{38}{\mathbf{38}} = \frac{38y}{\mathbf{38}}$$
$$1 = y$$

The solution is 1. The solution checks.

31.
$$3(a+8) = 6(a+4) - 3a$$
$$3a + 24 = 6a + 24 - 3a$$
$$3a + 24 = 3a + 24$$
$$3a + 24 - \mathbf{3a} = 3a + 24 - \mathbf{3a}$$
$$24 = 24$$

The terms involving a drop out and the result is true. This means that *all real numbers* are solutions and this equation is called an identity.

32.
$$2(y+10) + y = 3(y+8)$$
$$2y + 20 + y = 3y + 24$$
$$3y + 20 = 3y + 24$$
$$3y + 20 - \mathbf{3y} = 3y + 24 - \mathbf{3y}$$
$$20 = 24$$

The terms involving x drop out and the result is false. This means the equation has *no solution* and it is a contradiction.

SECTION 2.3
Application of Percent

33. Fill in the blanks
   a. **Percent** means parts per one hundred.
   b. When the price of an item is reduced, we call the amount of the reduction a **discount**.
   c. An employee who is paid a **commission** is paid a percent of the goods or services that he or she sells.

34. 4.81 is 2.5% of what number?
$$4.81 = 0.025 \cdot x$$
$$\frac{4.81}{\mathbf{0.025}} = \frac{0.025x}{\mathbf{0.025}}$$
$$192.4 = x$$

192.4 is the solution.

35. What number is 15% of 950?
$$x = 0.15 \cdot 950$$
$$x = 142.5$$

142.5 is the solution.

36. What percent of 410 is 49.2?
$$x \cdot 410 = 49.2$$
$$\frac{410x}{\mathbf{410}} = \frac{49.2}{\mathbf{410}}$$
$$x = 0.12$$
$$x = 12\%$$

12% is the solution.

37. INTERNET USERS
   a. What percent did not use the Internet?
$$100\% - 71.2\% = 28.8\%$$
   The percent of non-users is 28.8%.
   b. How many people used the Internet?
   Let x = number of people using the Internet
   What number is 71.2% of 310 million?
$$x = 0.712 \cdot 310$$
$$x = 220.72$$
$$x \approx 221$$
   About 221 million people used the Internet.

38. COST OF LIVING
   Let x = amt of cost of living
   What number is 3.5% of $764?
$$x = 0.035 \cdot 764$$
$$x = 26.74$$
   The increase is $26.74.

39. FAMILY BUDGETS

Let x = percent of housing cost

625 is what percent of 1,890?

$$625 = x \cdot 1,890$$

$$\frac{625}{1,890} = \frac{1,890x}{1,890}$$

$$0.33068 \approx x$$

$$33\% \approx x$$

About 33% of the income is spent on housing. No, they are not within the range.

40. DISCOUNTS

Let x = the original cost of food processor

148.50 is 33% of what number?

$$148.50 = 0.33 \cdot x$$

$$\frac{148.50}{0.33} = \frac{0.33x}{0.33}$$

$$450 = x$$

$450 was the original cost of the food processor.

41. TUPPERWARE

Let x = amt of the commission

What number is 25% of $600?

$$x = 0.25 \cdot 600$$

$$x = 150$$

The commission is $150.

42. COLLECTIBLES

Step 1: $100 - \$6 = \$94$

This is the amount of increase.

Step 2: 94 is what percent of 6?

$$94 = x \cdot 6$$

$$\frac{94}{6} = \frac{6x}{6}$$

$$15.6\overline{6} \approx x$$

$$15.67 \approx x$$

The percent of increase is 1,567%.

SECTION 2.4
Formulas

43. SHOPPING

Let m = the amount of markup

$$r = c + m$$

$$395 = 219 + m$$

$$395 - 219 = 219 + m - 219$$

$$176 = m$$

$176 is the amount of the markup.

44. RESTAURANTS

Let r = the revenue

$$r = p + e$$

$$13,500 = p + e$$

$$13,500 - 1,700 = 1,700 + e - 1,700$$

$$11,800 = e$$

$11,800 is the amount of expenses.

45. SNAILS

Let t = the number of minutes

Rate	•	time	=	distance
2.5		t		20

$$2.5 \cdot t = 20$$

$$\frac{2.5t}{2.5} = \frac{20}{2.5}$$

$$t = 8$$

It will take 8 minutes.

46. CERTIFICATES OF DEPOSIT

Let r = the annual interest rate

Principal	•	rate	• time	=	interest
26,000		r	1		1,170

$$26,000 \cdot r \cdot 1 = 1,170$$

$$26,000r = 1,170$$

$$\frac{26,000r}{26,000} = \frac{1,170}{26,000}$$

$$r = 0.045$$

$$r\% = 4.5\%$$

The rate is 4.5% annually.

Chapter 2 Review and Test

© 2013 Cengage Learning. All Rights Reserved. May not be scanned, copied or duplicated, or posted to a publicly accessible website, in whole or in part.

47. JEWELRY

$$C = \frac{5}{9}(F - 32)$$

$$1{,}065 = \frac{5}{9}(F - 32)$$

$$\mathbf{\frac{9}{5}} \cdot (1{,}065) = \mathbf{\frac{9}{5}} \cdot \frac{5}{9}(F - 32)$$

$$1{,}917 = F - 32$$

$$1{,}917 + \mathbf{32} = F - 32 + \mathbf{32}$$

$$1{,}949 = F$$

The temperature is $1{,}949°$ Fahrenheit.

48. CAMPING

a. $P = 2l + 2w$

$$= 2 \cdot 60 + 2 \cdot 24$$

$$= 120 + 48$$

$$= 168$$

The perimeter is 168 inches.

b. $A = lw$

$$= 60 \cdot 24$$

$$= 1{,}440$$

The area is 1,440 square inches.

c. $V = lwh$

$$= 60 \cdot 24 \cdot 3$$

$$= 4{,}320$$

The volume is 4,320 cubic inches.

49. $A = \frac{1}{2}bh$

$$= \frac{1}{2} \cdot 17 \cdot 9$$

$$= \frac{1}{2} \cdot 153$$

$$= 76.5$$

The area is 76.5 square meters.

50. $A = \frac{1}{2}h(b + d)$

$$= \frac{1}{2} \cdot 12 \cdot (11 + 13)$$

$$= \frac{1}{2} \cdot 12 \cdot 24$$

$$= 6 \cdot 24$$

$$= 144$$

The area is 144 square inches.

51. a. $C = 2\pi r$, $\pi = 3.141592654$

$$= 2 \cdot 3.141592654 \cdot 8$$

$$\approx 50.265$$

$$\approx 50.27$$

The circumference is about 50.27 cm.

b. $A = \pi r^2$, $\pi = 3.141592654$

$$= 3.141592654 \cdot 8 \cdot 8$$

$$\approx 201.061$$

$$\approx 201$$

The area is about 201 sq cm.

52. $V = \pi r^2 h$, $\pi = 3.141592654$

$$= 3.141592654 \cdot (0.5)^2 \cdot 12$$

$$= 3.141592654 \cdot 0.25 \cdot 12$$

$$\approx 9.42$$

$$\approx 9.4$$

The volume is about 9.4 cubic feet.

53. HALLOWEEN

$$V = \frac{4}{3}\pi r^3 \ , \ \pi = 3.1415926542$$

$$= \frac{4}{3} \cdot 3.1415926542 \cdot (4.5)^3$$

$$= \frac{4}{3} \cdot 3.1415926542 \cdot (91.125)$$

$$\approx 381.70$$

$$\approx 381.7$$

The volume is about 381.7 cubic inches.

54. $V = \frac{1}{3}Bh$

$$= \frac{1}{3} \cdot 6 \cdot 6 \cdot 10$$

$$= 120$$

The volume is 120 cubic feet.

Solve each formula for the specified variable.

55. $A = 2\pi rh$ for h

$$A = 2\pi rh$$

$$\frac{A}{\mathbf{2\pi r}} = \frac{2\pi rh}{\mathbf{2\pi r}}$$

$$\frac{A}{2\pi r} = h$$

$$h = \frac{A}{2\pi r}$$

56. $A - BC = \dfrac{G-K}{3}$ for G

$$3(A-BC) = 3\left(\dfrac{G-K}{3}\right)$$
$$3(A-BC) = G-K$$
$$3(A-BC) + \mathbf{K} = G-K+\mathbf{K}$$
$$3(A-BC) + K = G$$
$$G = 3(A-BC) + K$$
$$or$$
$$G = 3A - 3BC + K$$

57. $C = \dfrac{1}{4}s(t-d)$ for t

$$\mathbf{4} \cdot C = \mathbf{4} \cdot \left[\dfrac{1}{4}s(t-d)\right]$$
$$4C = s(t-d)$$
$$\dfrac{4C}{\mathbf{s}} = \dfrac{s(t-d)}{\mathbf{s}}$$
$$\dfrac{4C}{s} = t-d$$
$$\dfrac{4C}{s} + \mathbf{d} = t-d+\mathbf{d}$$
$$\dfrac{4C}{s} + d = t$$
$$t = \dfrac{4C}{s} + d \ \text{ or } \ \dfrac{4C+ds}{s}$$

58. $4y - 3x = 16$ for y

$$4y - 3x + \mathbf{3x} = 16 + \mathbf{3x}$$
$$4y = 3x + 16$$
$$\dfrac{4y}{\mathbf{4}} = \dfrac{3x}{\mathbf{4}} + \dfrac{16}{\mathbf{4}}$$
$$y = \dfrac{3}{4}x + 4$$

SECTION 2.5
Problem Solving
59. SOUND SYSTEMS

Analyze
- A 45-foot wire is cut into 3 pieces.
- One piece is 15 feet long.
- 2nd piece is 2 feet less than 3 times the 3rd
- Find the shorter piece.

Assign
Let x = length of 3rd piece in feet
$3x - 2$ = length of 2nd piece in feet

Form

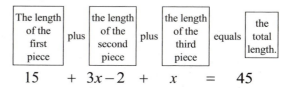

The length of the first piece	plus	the length of the second piece	plus	the length of the third piece	equals	the total length.
15	+	$3x-2$	+	x	=	45

Solve
$$15 + 3x - 2 + x = 45$$
$$4x + 13 = 45$$
$$4x + 13 - \mathbf{13} = 45 - \mathbf{13}$$
$$4x = 32$$
$$\dfrac{4x}{\mathbf{4}} = \dfrac{32}{\mathbf{4}}$$
$$x = 8$$
Second length
$$3x - 2 = 3(8) - 2$$
$$= 24 - 2$$
$$= 22$$

State
The length of the shorter piece is 8 ft.

Optional answers for checking.
Length of second piece is 22 ft.
Length of first piece is 15 ft.

Check
15 ft + 22 ft + 8 ft = 45 ft
The results check.

Chapter 2 Review and Test

60. SIGNING PETITIONS

Analyze
- Collector makes $50 a day.
- Collector makes $2.25 per verified signature.
- How many signature are needed to earn $500 a day?

Assign

Let x = number of signatures

Form

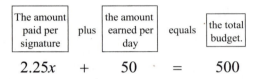

The amount paid per signature	plus	the amount earned per day	equals	the total budget.
$2.25x$	+	50	=	500

Solve
$$2.25x + 50 = 500$$
$$2.25x + 50 - \mathbf{50} = 500 - \mathbf{50}$$
$$2.25x = 450$$
$$\frac{2.25x}{\mathbf{2.25}} = \frac{450}{\mathbf{2.25}}$$
$$x = 200$$

State

200 signatures are needed.

Check
$$2.25x + 50 = 500$$
$$2.25(200) + 50 = 500$$
$$450 + 50 = 500$$
$$500 = 500$$

The result checks.

61. LOTTERY WINNINGS

Analyze
- $1,800,000 was the lump sum after taxes.
- 28% of the original prize was federal income tax.
- What was the original cash prize?

Assign

Let x = the original amount

Form

The original amount	minus	the amount of federal taxes	equals	the lum sum.
x	−	$0.28x$	=	1,800,000

Solve
$$x - 0.28x = 1,800,000$$
$$0.72x = 1,800,000$$
$$\frac{0.72x}{\mathbf{0.72}} = \frac{1,800,000}{\mathbf{0.72}}$$
$$x = 2,500,000$$

State

The original price was $2,500,000.

Check
$$\$2,500,000 - 0.28(\$2,500,000)$$
$$= \$2,500,000 - \$700,000$$
$$= \$1,800,000$$

The result checks.

62. NASCAR

Analyze
- The sum of the two consecutive odd numbers is 88.
- The smaller number belongs to Labonte.
- Find the two numbers.

Assign

Let $x = 1^{st}$ consecutive odd integer

$x + 2 = 2^{nd}$ consecutive odd integer

Form

The first consecutive odd integer	plus	the second consecutive odd integer	equals	the total of the two odd integers.
x	+	$x+2$	=	88

Solve
$$x + x + 2 = 88$$
$$2x + 2 = 88$$
$$2x + 2 - \mathbf{2} = 88 - \mathbf{2}$$
$$2x = 86$$
$$\frac{2x}{\mathbf{2}} = \frac{86}{\mathbf{2}}$$
$$x = 43$$

State

43 is Labonte's number.

$43 + 2 = 45$ is Petty's number.

Check
$$x + x + 2 = 88$$
$$43 + 43 + 2 = 88$$
$$88 = 88$$

The results check.

63. ART HISTORY

Analyze
- Perimeter of the rectangular painting is 109.5 $(109\frac{1}{2})$ inches.
- The length is 5 inches more than its width.
- What are the dimensions of the painting?

Assign

Let x = width of the painting in inches

$x + 5$ = length of the painting in inches

Form

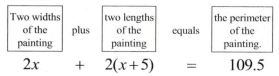

Two widths of the painting	plus	two lengths of the painting	equals	the perimeter of the painting.
$2x$	+	$2(x+5)$	=	109.5

Solve

$$2x + 2(x + 5) = 109.5$$
$$2x + 2x + 10 = 109.5$$
$$4x + 10 = 109.5$$
$$4x + 10 - \mathbf{10} = 109.5 - \mathbf{10}$$
$$4x = 99.5$$
$$\frac{4x}{\mathbf{4}} = \frac{99.5}{\mathbf{4}}$$
$$x = 24.875$$
$$or$$
$$x = 24\tfrac{7}{8}$$

$$\text{length}$$
$$x + 5 = 24.875 + 5$$
$$= 29.875$$
$$or$$
$$= 29\tfrac{7}{8}$$

State

The width of the painting is 24.875 inches.
The length of the painting is 29.875 inches.

Check

$$2x + 2(x + 5) = 109.5$$
$$2(24.875) + 2(29.875) = 109.5$$
$$49.75 + 59.75 = 109.5$$
$$109.5 = 109.5$$

The results check.

64. GEOMETRY

Analyze
- Internal measure of the 3 angles of a triangle is 180°.
- The measure of the vertex angle is 27°?
- What is the measure of each base angle which are equal to each other.

Assign

Let x = measure of each base angle

Form

The measure of the vertex angle	plus	the measure of the 2 base angles	equals	the measure of the angles of a triangle.
27	+	$2x$	=	180

Solve

$$27 + 2x = 180$$
$$27 + 2x - \mathbf{27} = 180 - \mathbf{27}$$
$$2x = 153$$
$$\frac{2x}{\mathbf{2}} = \frac{153}{\mathbf{2}}$$
$$x = 76.5$$

State

Measure of each base angle is 76.5°.

Check

$$27 + 2x = 180$$
$$27 + 2(76.5) = 180$$
$$27 + 153 = 180$$
$$180 = 180$$

The result checks.

Chapter 2 Review and Test

SECTION 2.6
More on Problem Solving

65. INVESTMENT INCOME
Analyze
- $27,000 is the total investment.
- CD investment at 7% annual interest.
- CMF investment at 9% annual interest.
- $2,110 is the first-year combined interest.
- How much is invested at each rate?

Assign

Let x = amount invested at 7%

$27,000 - x$ = amount invested at 9%

Form

	Principal	• rate	• time =	interest
CD	x	0.07	1	**0.07x**
CMF	$27,000 - x$	0.09	1	**0.09(27,000 − x)**
			Total:	**2,110**

The interest earned at 7%	plus	the interest earned at 9%	equals	the total interest.
0.07x	+	0.09(27,000 − x)	=	2,110

Solve

$$0.07x + 0.09(27,000 - x) = 2,110$$
$$\mathbf{100}[0.07x + 0.09(27,000 - x)] = \mathbf{100}(2,110)$$
$$100(0.07x) + 100[0.09(27,000 - x)] = 100(2,110)$$
$$7x + 9(27,000 - x) = 211,000$$
$$7x + 243,000 - 9x = 211,000$$
$$-2x + 243,000 = 211,000$$
$$-2x + 243,000 - \mathbf{243,000} = 211,000 - \mathbf{243,000}$$
$$-2x = -32,000$$
$$\frac{-2x}{\mathbf{-2}} = \frac{-32,000}{\mathbf{-2}}$$
$$x = 16,000$$

State

$16,000 was invested at 7%.

$27,000 - \$16,000 = \$11,000$ was the amount invested at 9%.

Check

The CD investment earned 0.07($16,000), or $1,120. The CMF earned 0.09($11,000), or $990. Since the total return was $1,120 + \$990 = \$2,110$, the results check.

66. WALKING AND BICYCLING
Analyze
- Man walks toward biker at 3 mph.
- Friend bikes toward walker at 12 mph.
- Both leave at the same time.
- Combined distance is 5 miles.
- How long before they meet at each other?

Assign

Let t = time in hours they meet to each other

Form

	rate	• time	= distance
Walker	3	t	**3t**
Biker	12	t	**12t**
		Total:	**5**

The distance traveled by the walker	plus	the distance traveled by the biker	equals	the total distance.
3t	+	12t	=	5

Solve

$$3t + 12t = 5$$
$$15t = 5$$
$$\frac{15t}{15} = \frac{5}{15}$$
$$t = \frac{1}{3}$$

State

After $\frac{1}{3}$ hour or 20 minutes they will meet.

Check

The distance traveled by walker is $3\left(\dfrac{1}{3}\right)$, or 1 mi.

The distance traveled by biker is $12\left(\dfrac{1}{3}\right)$, or 4 mi.

Since the total was $1 \text{ mi} + 4 \text{ mi} = 5 \text{ mi}$, the result checks.

67. AIRPLANES

Analyze
- Propeller plane's rate is 180 mph.
- Propeller plane leaves base 2.5 hours earlier.
- Jet's rate is 450 mph.
- Both leave from the same place.
- Both travel in the same direction.
- How long before the jet catches the prop plane?

Assign
Let t = time in hours before jet catches propeller plane

Form

	rate	time	distance
Prop	180	$t + 2.5$	$180(t + 2.5)$
Jet	450	t	$450t$

rate $\cdot$ time $=$ distance

Distance: **same**

The distance traveled by the prop	equals	the distance traveled by the jet.

$$180(t+2.5) \quad = \quad 450t$$

Solve

$$180(t + 2.5) = 450t$$
$$180t + 450 = 450t$$
$$180t + 450 - \mathbf{180t} = 450t - \mathbf{180t}$$
$$450 = 270t$$
$$\frac{450}{\mathbf{270}} = \frac{270t}{\mathbf{270}}$$
$$1\tfrac{2}{3} = t$$

or

$$1 \text{ hr } 40 \text{ minutes} = t$$

State
After $1\tfrac{2}{3}$ hours, the jet will catch the prop plane.

Check
Distance traveled by prop is $180(2\tfrac{1}{2} + 1\tfrac{2}{3} = 4\tfrac{1}{6})$, or 750 mi.
Distance traveled by jet is $450(1\tfrac{2}{3})$, or 750 mi.
Since the each distance is the same 750 mi = 750 mi, the result checks.

68. AUTOGRAPHS

Analyze
- Kesha has 8 more TV autographs than movie star autographs.
- Each TV autograph is worth $75.
- Each movie star autograph is worth $250.
- Total collection is worth $1,900.
- How many of each does she have?

Assign
Let x = the # of movie star autographs
$x + 8$ = the # of TV autographs

Form

	Number	$\cdot$ Value	= Total value
Movie Star	x	250	$250x$
TV	$x + 8$	75	$75(x + 8)$
		Total:	**1,900**

The value of the movie star autographs	plus	the value of the TV autographs	equals	the total value of the autographs.

$$250x \quad + \quad 75(x+8) \quad = \quad 1,900$$

Solve

$$250x + 75(x + 8) = 1,900$$
$$250x + 75x + 600 = 1,900$$
$$325x + 600 = 1,900$$
$$325x + 600 - \mathbf{600} = 1,900 - \mathbf{600}$$
$$325x = 1,300$$
$$\frac{325x}{\mathbf{325}} = \frac{1,300}{\mathbf{325}}$$
$$x = 4$$

$$\text{TV}$$
$$x + 8 = 4 + 8$$
$$= 12$$

State
She has 4 movie star autographs.
She has 12 TV celebrity autographs.

Check
The value of the movie is 4($250), or $1,000.
The value of the TV is 12($75), or $900.
Since the total was $1,000 + $900 = $1,900, the results check.

Chapter 2 Review and Test

69. MIXTURES

Analyze
- Candy sells for $0.90 per lb.
- Gumdrops sells for $1.50 per lb.
- A mixture of the two sells for $1.20 per lb.
- Blend of 20 lb is needed.
- How many pounds of each is needed?

Assign

Let x = the pounds of candy

$20 - x$ = the pounds of gumdrops

Form

	Number •	Value =	Total value
Candy	x	0.90	**0.90x**
Gumdrops	**20 − x**	1.50	**1.50(20 − x)**
Mixture	20	1.20	**24**

The value of the candy	plus	the value of the gumdrops	equals	the total value of the mixture.
$0.90x$	+	$1.50(20-x)$	=	24

Solve

$$0.90x + 1.50(20 - x) = 24$$
$$0.90x + 30 - 1.50x = 24$$
$$-0.60x + 30 = 24$$
$$-0.60x + 30 - \mathbf{30} = 24 - \mathbf{30}$$
$$-0.60x = -6$$
$$\frac{-0.60x}{\mathbf{-0.60}} = \frac{-6}{\mathbf{-0.60}}$$
$$x = 10$$

gumdrops
$$20 - x = 20 - 10$$
$$= 10$$

State

10 lb of the candy will be needed.

10 lb of gumdrops will be needed.

Check

The value of the candy is 10($0.90), or $9.
The value of the gumdrops is 10($1.50), or $15.
The value of the blend is 20($1.20), or $24.
Since the total was $9 + $15 = $24,
the results check.

70. ELIMINATING MILDEW

Analyze
- Weak solution is 2% fungicide.
- Start with 4 gallons of 5% fungicide, the strong solution.
- How many gallons 2% fungicide is needed to make a solution of 4% fungicide?

Assign

Let x = Number of gallons of 2% fungicide

Form

	Amount •	Strength =	Amount of pure fungicide
Weak	x	0.02	**0.02x**
Strong	4	0.05	**0.20**
4% solution	**$x + 4$**	0.04	**0.04(x + 4)**

The pure fungicide in the 2% solution	plus	the pure fungicide in the 5% solution	equals	the pure fungicide in the 4% solution.
$0.02x$	+	0.20	=	$0.04(x+4)$

Solve

$$0.02x + 0.20 = 0.04(x + 4)$$
$$\mathbf{100}(0.02x + 0.20) = \mathbf{100}[0.04(x + 4)]$$
$$100(0.02x) + 100(0.20) = 100[(0.04(x + 4)]$$
$$2x + 20 = 4(x + 4)$$
$$2x + 20 = 4x + 16$$
$$2x + 20 - \mathbf{20} = 4x + 16 - \mathbf{20}$$
$$2x = 4x - 4$$
$$2x - \mathbf{4x} = 4x - 4 - \mathbf{4x}$$
$$-2x = -4$$
$$\frac{-2x}{\mathbf{-2}} = \frac{-4}{\mathbf{-2}}$$
$$x = 2$$

State

2 gallons of the 2% fungicide is needed.

Check

The fungicide in the 2% solution is 2(0.02), or 0.04.
The fungicide in the 5% solution is 4(0.05), or 0.20.
The fungicide in the 4% solution is 6(0.04), or 0.24.
Since the total was 0.04 + 0.20 = 0.24, the results check.

SECTION 2.7
Solving Inequalities

Solve each inequality. Write the solution set in interval notation and graph it.

71. $3x + 2 < 5$

$$3x + 2 - \mathbf{2} < 5 - \mathbf{2}$$

$$3x < 3$$

$$\frac{3x}{\mathbf{3}} < \frac{3}{\mathbf{3}}$$

$$x < 1$$

$$(-\infty, 1)$$

72. $-\frac{3}{4}x \geq -9$

$$-\frac{\mathbf{4}}{\mathbf{3}}\left(-\frac{3}{4}x\right) \leq -\frac{\mathbf{4}}{\mathbf{3}}(-9)$$

$$x \leq 12$$

$$(-\infty, 12]$$

73. $\frac{3}{4} < \frac{d}{5} + \frac{1}{2}$, LCD $= 20$

$$\mathbf{20} \cdot \left(\frac{3}{4}\right) < \mathbf{20} \cdot \left(\frac{d}{5} + \frac{1}{2}\right)$$

$$20 \cdot \left(\frac{3}{4}\right) < 20 \cdot \left(\frac{d}{5}\right) + 20 \cdot \left(\frac{1}{2}\right)$$

$$15 < 4d + 10$$

$$15 - \mathbf{10} < 4d + 10 - \mathbf{10}$$

$$5 < 4d$$

$$\frac{5}{\mathbf{4}} < \frac{4d}{\mathbf{4}}$$

$$\frac{5}{4} < d$$

$$d > \frac{5}{4}$$

$$\left(\frac{5}{4}, \infty\right)$$

74. $5(3 - x) \leq 3(x - 3)$

$$15 - 5x \leq 3x - 9$$

$$15 - 5x - \mathbf{3x} \leq 3x - 9 - \mathbf{3x}$$

$$15 - 8x \leq -9$$

$$15 - 8x - \mathbf{15} \leq -9 - \mathbf{15}$$

$$-8x \leq -24$$

$$\frac{-8x}{-\mathbf{8}} \geq \frac{-24}{-\mathbf{8}}$$

$$x \geq 3$$

$$[3, \infty)$$

75. $\frac{t}{-5} - (-1.8) \geq -6.2$

$$\frac{t}{-5} + 1.8 \geq -6.2$$

$$\frac{t}{-5} + 1.8 - \mathbf{1.8} \geq -6.2 - \mathbf{1.8}$$

$$\frac{t}{-5} \geq -8$$

$$-\mathbf{5} \cdot \frac{t}{-5} \leq -\mathbf{5} \cdot (-8)$$

$$t \leq 40$$

$$(-\infty, 40]$$

76. $a + 5 - 2(10 - a) > 6$

$$a + 5 - \mathbf{2}(10) - \mathbf{2}(-a) > 6$$

$$a + 5 - 20 + 2a > 6$$

$$3a - 15 > 6$$

$$3a - 15 + \mathbf{15} > 6 + \mathbf{15}$$

$$3a > 21$$

$$\frac{3a}{\mathbf{3}} > \frac{21}{\mathbf{3}}$$

$$a > 7$$

$$(7, \infty)$$

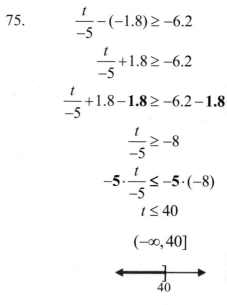

Chapter 2 Review and Test

77.

$$24 < 3(x+2) < 39$$
$$24 < \mathbf{3}(x) + \mathbf{3}(2) < 39$$
$$24 < 3x + 6 < 39$$
$$24 - \mathbf{6} < 3x + 6 - \mathbf{6} < 39 - \mathbf{6}$$
$$18 < 3x < 33$$
$$\frac{18}{\mathbf{3}} < \frac{3x}{\mathbf{3}} < \frac{33}{\mathbf{3}}$$
$$6 < x < 11$$
$$(6, 11)$$

78.

$$0 \le 3 - 2x < 10$$
$$0 - \mathbf{3} \le 3 - 2x - \mathbf{3} < 10 - \mathbf{3}$$
$$-3 \le -2x < 7$$
$$\frac{-3}{\mathbf{-2}} \ge \frac{-2x}{\mathbf{-2}} > \frac{7}{\mathbf{-2}}$$
$$\frac{3}{2} \ge x > -\frac{7}{2}$$
$$-\frac{7}{2} < x \le \frac{3}{2}$$
$$\left(-\frac{7}{2}, \frac{3}{2}\right]$$

79. SPORTS EQUIPMENT

$$2.40 \text{ g} \le w \le 2.53 \text{ g}$$

80. SIGNS

Analyze

• The width must be 18 inches.
• The perimeter must not exceed 132 inches.
• What are the possible lengths?

Assign

Let x = the length in inches

Form

Two 18 inch widths	plus	two lengths	must be no more than	the perimeter of the sign.
2(18)	+	2x	≤	132

Solve

$$2(18) + 2x \le 132$$
$$36 + 2x \le 132$$
$$36 + 2x - \mathbf{36} \le 132 - \mathbf{36}$$
$$2x \le 96$$
$$\frac{2x}{\mathbf{2}} \le \frac{96}{\mathbf{2}}$$
$$x \le 48$$

State

The lengths must be 48 inches or less or
0 inches $< l \le 48$ inches.

Check

$$2(18) + 2x \le 132$$
$$2(18) + 2(48) \le 132$$
$$36 + 96 \le 132$$
$$132 \le 132$$

The results check.

CHAPTER 2 TEST

1. Fill in the blanks.
   a. To **solve** an equation means to find all of the values of the variable that make the equation true.
   b. **Percent** means parts per one hundred.
   c. The distance around a circle is called its **circumference** .
   d. An **inequality** is a statement that contains one of the symbols $>, \ge, <,$ or $\le$.
   e. The **multiplication** property of **equality** says that multiplying both sides of an equation by the same nonzero number does not change its solution.

2. $\quad 5y + 2 = 12$
   $$5(3) + 2 \overset{?}{=} 12$$
   $$15 + 2 \overset{?}{=} 12$$
   $$17 = 12 \quad \text{False}$$
   3 is not a solution.

Solve each equation.

3. $\quad 3h + 2 = 8 \qquad\qquad$ Checking
   $$3h + 2 - \mathbf{2} = 8 - \mathbf{2} \qquad 3h + 2 = 8$$
   $$3h = 6 \qquad\qquad 3(2) + 2 \overset{?}{=} 8$$
   $$\frac{3h}{\mathbf{3}} = \frac{6}{\mathbf{3}} \qquad\qquad 6 + 2 \overset{?}{=} 8$$
   $$h = 2 \qquad\qquad\quad 8 = 8$$

   The solution is 2. The result checks.

4. $-22 = -x$

$\dfrac{-22}{-1} = \dfrac{-x}{-1}$

$22 = x$

Checking

$-22 = -x$

$-22 \overset{?}{=} -(22)$

$-22 = -22$

The solution is 22. The result checks.

5. $\dfrac{4}{5}t = -4$

$\dfrac{5}{4} \cdot \dfrac{4}{5}t = \dfrac{5}{4} \cdot (-4)$

$t = -\dfrac{5 \cdot \overset{1}{\cancel{4}}}{\underset{1}{\cancel{4}}}$

$t = -5$

Checking

$\dfrac{4}{5}t = -4$

$\dfrac{4}{5}(-5) \overset{?}{=} -4$

$-4 = -4$

The solution is -5. The result checks.

6. $\dfrac{11b - 11}{5} = \dfrac{3b - 2}{2}$

$\left(\dfrac{10}{1}\right)\left(\dfrac{11b - 11}{5}\right) = \left(\dfrac{10}{1}\right)\left(\dfrac{3b - 2}{2}\right)$

$2(11b - 11) = 5(3b - 2)$

$22b - 22 = 15b - 10$

$22b - 22 - \mathbf{15b} = 15b - 10 - \mathbf{15b}$

$7b - 22 = -10$

$7b - 22 + \mathbf{22} = -10 + \mathbf{22}$

$7b = 12$

$\dfrac{7b}{7} = \dfrac{12}{7}$

$b = \dfrac{12}{7}$

This problem was not checked due to its complexity.

The solution is $\dfrac{12}{7}$. The result checks.

7. $0.8(x - 1,000) + 1.3 = 2.9 + 0.2x$

$10[0.8(x - 1,000) + 1.3] = 10[2.9 + 0.2x]$

$10[0.8(x - 1,000)] + 10(1.3) = 10(2.9) + 10(0.2x)$

$8(x - 1,000) + 13 = 29 + 2x$

$8x - 8,000 + 13 = 29 + 2x$

$8x - 7,987 = 29 + 2x$

$8x - 7,987 - \mathbf{2x} = 29 + 2x - \mathbf{2x}$

$6x - 7,987 = 29$

$6x - 7,987 + \mathbf{7,987} = 29 + \mathbf{7,987}$

$6x = 8,016$

$\dfrac{6x}{6} = \dfrac{8,016}{6}$

$x = 1,336$

$0.8(x - 1,000) + 1.3 = 2.9 + 0.2x$

$0.8(\mathbf{1,336} - 1,000) + 1.3 \overset{?}{=} 2.9 + 0.2(\mathbf{1,336})$

$0.8(336) + 1.3 \overset{?}{=} 2.9 + 267.2$

$268.8 + 1.3 \overset{?}{=} 270.1$

$270.1 = 270.1$

The solution is $1,336$. The result checks.

8. $2(y - 7) - 3y = -(y - 3) - 17$

$2y - 14 - 3y = -y + 3 - 17$

$-y - 14 = -y - 14$

$-y - 14 + \mathbf{y} = -y - 14 + \mathbf{y}$

$-14 = -14$ True

The terms involving y drop out and the result is true. This means that *all real numbers* are solutions and this equation is an identity.

Chapter 2 Review and Test

9.
$$\frac{m}{2} - \frac{1}{3} = \frac{1}{4} + \frac{m}{6}, \quad LCD = 12$$

$$\mathbf{12} \cdot \left(\frac{m}{2} - \frac{1}{3}\right) = \mathbf{12} \cdot \left(\frac{1}{4} + \frac{m}{6}\right)$$

$$12 \cdot \frac{m}{2} - 12 \cdot \frac{1}{3} = 12 \cdot \left(\frac{1}{4}\right) + 12 \cdot \left(\frac{m}{6}\right)$$

$$6m - 4 = 3 + 2m$$

$$6m - 4 - \mathbf{2m} = 3 + 2m - \mathbf{2m}$$

$$4m - 4 = 3$$

$$4m - 4 + \mathbf{4} = 3 + \mathbf{4}$$

$$4m = 7$$

$$\frac{4m}{\mathbf{4}} = \frac{7}{\mathbf{4}}$$

$$m = \frac{7}{4}$$

This problem was not checked due to its complexity.

The solution is $\frac{7}{4}$. The solution checks.

10.
$$\frac{3}{4}(6n - 2) = 246 \qquad\qquad \frac{3}{4}(6n - 2) = 246$$

$$\mathbf{4}\left[\frac{3}{4}(6n - 2)\right] = \mathbf{4}(246) \qquad \frac{3}{4}(6 \cdot \mathbf{55} - 2) \overset{?}{=} 246$$

$$3(6n - 2) = 984 \qquad\qquad \frac{3}{4}(330 - 2) \overset{?}{=} 246$$

$$\mathbf{3}(6n) - \mathbf{3}(2) = 984 \qquad\qquad \frac{3}{4}(328) \overset{?}{=} 246$$

$$18n - 6 = 984$$

$$18n - 6 + \mathbf{6} = 984 + \mathbf{6} \qquad\qquad 3 \cdot \frac{\overset{82}{\cancel{328}}}{\underset{1}{\cancel{4}}} \overset{?}{=} 246$$

$$18n = 990$$

$$\frac{18n}{\mathbf{18}} = \frac{990}{\mathbf{18}} \qquad\qquad 246 = 246$$

$$n = 55$$

The solution is 55. The solution checks.

11.
$$5x = 0 \qquad\qquad \text{Checking}$$

$$\frac{5x}{\mathbf{5}} = \frac{0}{\mathbf{5}} \qquad\qquad 5x = 0$$

$$x = 0 \qquad\qquad 5(0) \overset{?}{=} 0$$

$$\qquad\qquad\qquad 0 = 0$$

The solution is 0. The solution checks.

12.
$$6a + (-7) = 3a - 7 + 2a$$

$$6a - 7 = 5a - 7$$

$$6a - 7 + \mathbf{7} = 5a - 7 + \mathbf{7}$$

$$6a = 5a$$

$$6a - \mathbf{5a} = 5a - \mathbf{5a}$$

$$a = 0$$

$$6a + (-7) = 3a - 7 + 2a$$

$$6(0) + (-7) \overset{?}{=} 3(0) - 7 + 2(0)$$

$$-7 = -7$$

The solution is 0. The solution checks.

13.
$$9 - 5(2x + 10) = -1$$

$$9 - 10x - 50 = -1$$

$$-10x - 41 = -1$$

$$-10x - 41 + \mathbf{41} = -1 + \mathbf{41}$$

$$-10x = 40$$

$$\frac{-10x}{\mathbf{-10}} = \frac{40}{\mathbf{-10}}$$

$$x = -4$$

$$9 - 5(2x + 10) = -1$$

$$9 - 5[2(-4) + 10] \overset{?}{=} -1$$

$$9 - 5(-8 + 10) \overset{?}{=} -1$$

$$9 - 5(2) \overset{?}{=} -1$$

$$9 - 10 \overset{?}{=} -1$$

$$-1 = -1$$

The solution is -4. The solution checks.

14.
$$24t = -6(8 - 4t)$$

$$24t = -48 + 24t$$

$$24t - \mathbf{24t} = -48 + 24t - \mathbf{24t}$$

$$0 = -48 \quad \text{False}$$

The terms involving t drop out and the result is false. This means the equation has *no solution* and it is a contradiction.

15. What number is 15.2% of 80?

$$x = 0.152 \cdot 80$$

$$x = 12.16$$

12.16 is the solution.

16. DOWN PAYMENTS

Let $x =$ the selling price of the house

11,400 is 15% of what number?

$11,400 = 0.15 \cdot x$

$\dfrac{11,400}{\textbf{0.15}} = \dfrac{0.15x}{\textbf{0.15}}$

$76,000 = x$

$76,000 is the selling price.

17. BODY TEMPERATURES

Step 1: $105 - 98.6 = 6.4$

This is the amount of increase.

Step 2: 6.4 is what percent of 98.6?

$6.4 = x \cdot 98.6$

$\dfrac{6.4}{\textbf{98.6}} = \dfrac{98.6x}{\textbf{98.6}}$

$0.064 \approx x$

$6\% \approx x$

The percent of increase is about 6%.

18. COMMISSIONS

Let $x =$ amt of the commission

What number is 5% of $599.99?

$x = 0.05 \cdot 599.99$

$x = 30$

$30 is amount of the commission.

19. GRAND OPENINGS

Let $c =$ the cost

$r = p + c$

$445 = 150 + c$

$445 - \textbf{150} = 150 + c - \textbf{150}$

$295 = c$

$295 is the amount of the cost.

20. Find the Celsius temperature.

$C = \dfrac{5}{9}(F - 32)$

$= \dfrac{5}{9}(14 - 32)$

$= \dfrac{5}{9}(-18)$

$= -10$

The temperature is $-10°$ Celsius.

21. SOUND

$d = rt$

$= 1,108 \text{ft/second } (60 \text{ seconds})$

$= 66,480$

In one minute it will travel 66,480 ft.

22. PETS

$V = \dfrac{4}{3}\pi r^3$, Use this formula to find the volume of water in the bowl that is $\dfrac{3}{4}$ full.

$V = \dfrac{4}{3}\pi r^3$, $\pi = 3.141592654$

$= \dfrac{3}{4} \cdot \dfrac{4}{3} \cdot 3.1415926542 \cdot (5)^3$

$= 1 \cdot 3.1415926542 \cdot (125)$

≈ 392.6

≈ 393

The volume of water is about 393 in^3.

Solve for the specified variable.

23. $V = \pi r^2 h$ for h.

$V = \pi r^2 h$

$\dfrac{V}{\pi r^2} = \dfrac{\pi r^2 h}{\pi r^2}$

$\dfrac{V}{\pi r^2} = h$

$h = \dfrac{V}{\pi r^2}$

Chapter 2 Review and Test

24. $A = P + Prt$ for r.

$$A = P + Prt$$
$$A - \textbf{P} = P + Prt - \textbf{P}$$
$$A - P = Prt$$
$$\frac{A - P}{\textbf{Pt}} = \frac{Prt}{\textbf{Pt}}$$
$$\frac{A - P}{Pt} = r$$
$$r = \frac{A - P}{Pt}$$

N25. $A = \dfrac{a + b + c + d}{4}$ for c.

$$A = \frac{a + b + c + d}{4}$$
$$\textbf{4} \cdot A = \textbf{4} \cdot \frac{a + b + c + d}{4}$$
$$4A = a + b + c + d$$
$$4A - \textbf{a} - \textbf{b} - \textbf{d} = a + b + c + d - \textbf{a} - \textbf{b} - \textbf{d}$$
$$4A - a - b - d = c$$
$$c = 4A - a - b - d$$

N26. $2x - 3y = 9$ for y.

$$2x - 3y = 9$$
$$2x - 3y - \textbf{2x} = 9 - \textbf{2x}$$
$$-3y = 9 - 2x$$
$$\frac{-3y}{\textbf{-3}} = -\frac{2x}{\textbf{-3}} + \frac{9}{\textbf{-3}}$$
$$y = \frac{2x}{3} - 3$$

27. IRONS

$$A = \frac{1}{2}bh$$
$$= \frac{1}{2} \cdot 8 \cdot 5$$
$$= 4 \cdot 5$$
$$= 20$$

The area is 20 square inches.

28. TELEVISION

Analyze
- 30-minute block of TV time
- Programming minutes are 2 less than three times the number of commercials minutes.
- Find the number of minutes for both.

Assign

Let x = number of commercial minutes
$3x - 2$ = number of programming minutes

Form

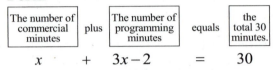

The number of commercial minutes	plus	The number of programming minutes	equals	the total 30 minutes.
x	$+$	$3x - 2$	$=$	30

Solve

$$x + 3x - 2 = 30$$
$$4x - 2 = 30$$
$$4x - 2 + \textbf{2} = 30 + \textbf{2}$$
$$4x = 32$$
$$\frac{4x}{4} = \frac{32}{4}$$
$$x = 8$$

Programming
$$3x - 2 = 3(8) - 2$$
$$= 24 - 2$$
$$= 22$$

State
8 minutes of commercials.
22 minutes of programming.

Check
8 minutes + 22 minutes = 30 minutes
The results check.

29. PLUMBING BILLS

Analyze
- Standard service charge is $25.75
- Parts cost $38.75.
- 4 hours of labor.
- Total charges are $226.70.
- Find the cost per hour for the labor.

Assign

Let x = cost per hour for labor

Form

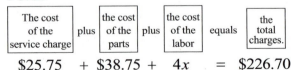

The cost of the service charge	plus	the cost of the parts	plus	the cost of the labor	equals	the total charges.
$25.75	$+$	$38.75	$+$	$4x$	$=$	$226.70

Solve

$$25.75 + 38.75 + 4x = 226.70$$
$$4x + 64.50 = 226.70$$
$$4x + 64.50 - \mathbf{64.50} = 226.70 - \mathbf{64.50}$$
$$4x = 162.20$$
$$\frac{4x}{\mathbf{4}} = \frac{162.20}{\mathbf{4}}$$
$$x = 40.55$$

State

The cost per hour for the labor is $40.55.

Check

$$\$25.75 + \$38.75 + 4x = \$226.70$$
$$\$25.75 + \$38.75 + 4(\$40.55) = \$226.70$$
$$\$25.75 + \$38.75 + \$162.55 = \$226.70$$
$$\$226.70 = \$226.70$$

The result checks.

30. CONCERT SEATING

Analyze

- Floor seats cost $12.50 each.
- Balcony seats cost $20.50 each.
- Ten times as many floor seats were sold as balcony seats (smaller quanity of the two).
- Total receipts is $11,640.
- How many of each type of tickets were sold?

Assign

Let x = the number of balcony seats sold
$10x$ = the number of floor seats sold

Form

	Number	•	Value	=	Total value
Balcony	x		20.50		**20.50x**
Floor	$10x$		12.50		**12.50(10x)**
				Total:	**11,640**

The value of the balcony seats	plus	the value of the floor seats	equals	the total value of the receipts.
20.50x	+	12.50(10x)	=	11,640

Solve

$$20.50x + 12.50(10x) = 11,640$$
$$20.50x + 125x = 11,640$$
$$145.50x = 11,640$$
$$\frac{145.50x}{\mathbf{145.50}} = \frac{11,640}{\mathbf{145.50}}$$
$$x = 80$$

Floor seats
$$10x = 10(80)$$
$$= 800$$

State

80 balcony seats tickets were sold.
800 floor seats tickets were sold.

Check

The value of the balcony tickets is 80($20.50), or $1,640.
The value of the floor seat tickets is 800($12.50), or $10,000.
Since the total was $1,640 + 10,000 = \$11,640$, the results check.

31. HOME SALES

Analyze

- Cleared $114,600 on the sale of a house.
- Paid agent 4.5% of the selling price.
- What was the selling price of the house?

Assign

Let x = the selling price of the house

Form

The selling price of the house	minus	the amount paid to the agent	equals	cleared amount.
x	−	0.045x	=	114,600

Solve

$$x - 0.045x = 114,600$$
$$0.955x = 114,600$$
$$\frac{0.955x}{\mathbf{0.955}} = \frac{114,600}{\mathbf{0.955}}$$
$$x = 120,000$$

State

The selling price was $120,000.

Check

$$x - 0.045x = 120,000 - 0.045(120,000)$$
$$= 120,000 - 5,400$$
$$= 114,600$$

The result checks.

Chapter 2 Review and Test

32. COLORADO

Analyze
- Perimeter of the state is 1,320 miles.
- Length is 100 miles longer than the width.
- What are the dimensions of the state?

Assign

Let x = width of the state in miles

$x + 100$ = length of the state in miles

Form

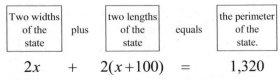

Two widths of the state	plus	two lengths of the state	equals	the perimeter of the state.
$2x$	$+$	$2(x+100)$	$=$	$1{,}320$

Solve

$$2x + 2(x+100) = 1{,}320$$
$$2x + 2x + 200 = 1{,}320$$
$$4x + 200 = 1{,}320$$
$$4x + 200 - \mathbf{200} = 1{,}320 - \mathbf{200}$$
$$4x = 1{,}120$$
$$\frac{4x}{\mathbf{4}} = \frac{1{,}120}{\mathbf{4}}$$
$$x = 280$$

length
$$x + 100 = 280 + 100$$
$$= 380$$

State

Width is 280 miles.
Length is 380 miles.

Check

$$2(280) + 2(380) = 560 + 760$$
$$= 1{,}320$$

The results check.

33. TEA

Analyze
- Green tea is worth $40 a lb.
- Herbal tea is worth $50 a lb.
- A 20 pound blend is worth $42 a lb.
- How many pounds of each is needed?

Assign

Let x = the pounds of green tea

$20 - x$ = the pounds of herbal tea

Form

	Number $\cdot$	Value $=$	Total value
Green	x	40	$40x$
Herbal	$20 - x$	50	$50(20 - x)$
Blend	20	42	840

The value of the green tea	plus	the value of the herbal tea	equals	the total value of the blend.
$40x$	$+$	$50(20 - x)$	$=$	840

Solve

$$40x + 50(20 - x) = 840$$
$$40x + 1{,}000 - 50x = 840$$
$$-10x + 1{,}000 = 840$$
$$-10x + 1{,}000 - \mathbf{1{,}000} = 840 - \mathbf{1{,}000}$$
$$-10x = -160$$
$$\frac{-10x}{\mathbf{-10}} = \frac{-160}{\mathbf{-10}}$$
$$x = 16$$

herbal
$$20 - x = 20 - 16$$
$$= 4$$

State

16 lb of green tea will be needed.

4 lb of herbal tea will be needed

Check

$$40x + 50(20 - x) = 840$$
$$40(16) + 50(20 - 16) = 840$$
$$640 + 200 = 840$$
$$840 = 840$$

The results check.

34. READING

Analyze
- Total of the two page numbers is 825.
- The respective number for each page is a consecutive integers.
- What are the two page numbers?

Assign

Let $x = 1^{st}$ page number

$x + 1 = 2^{nd}$ page number

Form

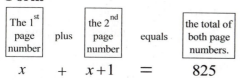

The 1st page number	plus	the 2nd page number	equals	the total of both page numbers.
x	$+$	$x+1$	$=$	825

Solve

$$x + x + 1 = 825$$
$$2x + 1 = 825$$
$$2x + 1 - 1 = 825 - 1$$
$$2x = 824$$
$$\frac{2x}{2} = \frac{824}{2}$$
$$x = 412$$
$$2^{nd} \text{ page}$$
$$x + 1 = 412 + 1$$
$$= 413$$

State

The 1st page number is 412.
The 2nd page number is 413.

Check

$$x + x + 1 = 825$$
$$412 + 412 + 1 = 825$$
$$825 = 825$$

The results check.

35. TRAVEL TIMES

Analyze
- A car drives towards Madison at 65 mph.
- A truck drives towards Rockford at 55 mph.
- Both leave at the same time using the same road.
- The distance between the 2 cities is 72 miles.
- How long before they meet at each other?

Assign

Let t = time in hours they meet to each other

	rate	• time	= distance
Car	65	t	$65t$
Truck	55	t	$55t$
		Total:	**72**

Form

The distance traveled by the car	plus	the distance traveled by the truck	equals	the total distance.
$65t$	$+$	$55t$	$=$	72

Solve

$$65t + 55t = 72$$
$$120t = 72$$
$$\frac{120t}{120} = \frac{72}{120}$$
$$t = \frac{3}{5}$$
$$or$$
$$t = 0.6$$

State

After $\frac{3}{5}$ hour or 36 minutes they will meet.

Check

$$65t + 55t = 72$$
$$65\left(\frac{3}{5}\right) + 55\left(\frac{3}{5}\right) = 72$$
$$13(3) + 11(3) = 72$$
$$39 + 33 = 72$$
$$72 = 72$$

The result checks.

Chapter 2 Review and Test

36. PICKLES

Analyze
- Start with 30 liters of 10% brine
- Want to dilute 10% brine to 8%
- How many liters of 2% is needed?

Assign

Let x = the number of liters of 2% brine

	Amount	Strength	=	Amount of pure salt
2%	x	0.02		**0.02x**
10%	30	0.10		**3**
8%	**x + 30**	0.08		**0.08(x + 30)**

Form

The salt in the 2% brine	plus	the salt in the 10% brine	equals	the salt in the 8% brine.

$$0.02x + 3 = 0.08(x+30)$$

Solve

$$0.02x + 3 = 0.08(x+30)$$
$$\mathbf{100}(0.02x+3) = \mathbf{100}[0.08(x+30)]$$
$$100(0.02x) + 100(3) = 100[(0.08(x+30))]$$
$$2x + 300 = 8(x+30)$$
$$2x + 300 = 8x + 240$$
$$2x + 300 - \mathbf{300} = 8x + 240 - \mathbf{300}$$
$$2x = 8x - 60$$
$$2x - \mathbf{8x} = 8x - 60 - \mathbf{8x}$$
$$-6x = -60$$
$$\frac{-6x}{\mathbf{-6}} = \frac{-60}{\mathbf{-6}}$$
$$x = 10$$

State

10 liters of 2% brine is needed.

Check

$$0.02x + 3 = 0.08(x+30)$$
$$0.02(10) + 3 = 0.08(10+30)$$
$$0.2 + 3 = 0.08(40)$$
$$3.2 = 3.2$$

The result checks.

37. EXERCISE

Analyze
- Jogger jogs at 8 mph with a half-hour head start.
- Bicyclist travels at 20 mph.
- How long before the biker catches the jogger?

Assign

Let t = time in hours before the biker catches the jogger

Form

	rate	•	time	= distance
Jogger	8		$t + 0.5$	**8(t + 0.5)**
Biker	20		t	**20t**

Distance same

The distance traveled by the biker	equals	the distance traveled by the jogger.

$$20t = 8(t+0.5)$$

Solve

$$20t = 8(t+0.5)$$
$$20t = \mathbf{8}(t) + \mathbf{8}(0.5)$$
$$20t = 8t + 4$$
$$20t - \mathbf{8t} = 8t + 4 - \mathbf{8t}$$
$$12t = 4$$
$$\frac{12t}{\mathbf{12}} = \frac{4}{\mathbf{12}}$$
$$t = \frac{1}{3}$$

State

The biker catches the jogger after $\frac{1}{3}$ hour or 20 minutes.

Check

$$20t = 8(t+0.5)$$
$$20\left(\frac{1}{3}\right) = 8\left(\frac{1}{3} + \frac{1}{2}\right),$$
$$\frac{20}{3} = 8\left(\frac{5}{6}\right)$$
$$\frac{20}{3} = \frac{20}{3}$$

The distances are equal, the result checks.

38. GEOMETRY

Analyze

- Internal measure of the 3 angles of a triangle is $180°$.
- The measure of the vertex angle is $44°$?
- What is the measure of each base angle which are equal to each other.

Assign

Let x = measure of each base angle

Form

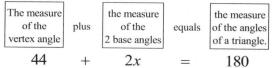

The measure of the vertex angle	plus	the measure of the 2 base angles	equals	the measure of the angles of a triangle.
44	+	$2x$	=	180

Solve

$$44 + 2x = 180$$
$$44 + 2x - \mathbf{44} = 180 - \mathbf{44}$$
$$2x = 136$$
$$\frac{2x}{\mathbf{2}} = \frac{136}{\mathbf{2}}$$
$$x = 68$$

State

The measure of each base angle is $68°$.

Check

$$44 + 2x = 180$$
$$44 + 2(68) = 180$$
$$44 + 136 = 180$$
$$180 = 180$$

The result checks.

39. INVESTMENTS

Analyze

- $13,750 is the total investment.
- 1st part invested at 9% annual interest.
- 2nd part invested at 8% annual interest.
- $1,185 is the first-year combined interest.
- How much is invested at the lower rate?

Assign

Let x = amount invested at 8%
$13,750 - x$ = amount invested at 9%

Form

	Principal	• rate	• time =	interest
1st	x	0.08	1	$0.08x$
2nd	$13,750 - x$	0.09	1	$0.09(13,750 - x)$
			Total:	1,185

The interest earned at 8%	plus	the interest earned at 9%	equals	the total interest.
$0.08x$	+	$0.09(13,750 - x)$	=	1,185

Solve

$$0.08x + 0.09(13,750 - x) = 1,185$$
$$\mathbf{100}[0.08x + 0.09(13,750 - x)] = \mathbf{100}(1,185)$$
$$100(0.08x) + 100[0.09(13,750 - x)] = 100(1,185)$$
$$8x + 9(13,750 - x) = 118,500$$
$$8x + 123,750 - 9x = 118,500$$
$$-x + 123,750 = 118,500$$
$$-x + 123,750 - \mathbf{123,750} = 118,500 - \mathbf{123,750}$$
$$-x = -5,250$$
$$\frac{-x}{\mathbf{-1}} = \frac{-5,250}{\mathbf{-1}}$$
$$x = 5,250$$

State

$5,250 was invested at 8%.
For checking purposes the other amount is needed. $13,750 - \$5,250 = \$8,500$.

Check

$$0.08(5,250) + 0.09(8,509) = 1,185$$
$$420 + 765.81 = 1,185$$

The results check.

Determine if -3 is a solution.

40. $4 - 9w < -4w + 19$, Is -3 a solution?

$$4 - 9(-3) \overset{?}{<} -4(-3) + 19$$
$$4 + 27 \overset{?}{<} 12 + 19$$
$$31 \overset{?}{<} 31$$
$$\text{False}$$

Solve each inequality. Write the solution set in interval notation and graph it.

41. $\qquad -8x - 20 \le 4$

$$-8x - 20 + \mathbf{20} \le 4 + \mathbf{20}$$
$$-8x \le 24$$
$$\frac{-8x}{\mathbf{-8}} \ge \frac{24}{\mathbf{-8}}$$
$$x \ge -3$$
$$[-3, \infty)$$

42.
$$-8.1 > \frac{t}{2} + (-11.3)$$
$$-8.1 > \frac{t}{2} - 11.3$$
$$-8.1 + \mathbf{11.3} > \frac{t}{2} - 11.3 + \mathbf{11.3}$$
$$3.2 > \frac{t}{2}$$
$$\mathbf{2} \cdot 3.2 > \mathbf{2} \cdot \left(\frac{t}{2}\right)$$
$$6.4 > t$$
$$t < 6.4$$
$$(-\infty, 6.4)$$

43.
$$-12 \le 2(x+1) < 10$$
$$-12 \le 2x + 2 < 10$$
$$-12 - \mathbf{2} \le 2x + 2 - \mathbf{2} < 10 - \mathbf{2}$$
$$-14 \le 2x < 8$$
$$\frac{-14}{\mathbf{2}} \le \frac{2x}{\mathbf{2}} < \frac{8}{\mathbf{2}}$$
$$-7 \le x < 4$$
$$[-7, 4)$$

44.
$$\frac{1}{3}(a-5) > \frac{1}{2}(a+1)$$
$$\mathbf{6}\left[\frac{1}{3}(a-5)\right] > \mathbf{6}\left[\frac{1}{2}(a+1)\right]$$
$$2(a-5) > 3(a+1)$$
$$\mathbf{2}(a) - \mathbf{2}(5) > \mathbf{3}(a) + \mathbf{3}(1)$$
$$2a - 10 > 3a + 3$$
$$2a - 10 - \mathbf{3a} > 3a + 3 - \mathbf{3a}$$
$$-a - 10 > 3$$
$$-a - 10 + \mathbf{10} > 3 + \mathbf{10}$$
$$-a > 13$$
$$\frac{-a}{\mathbf{-1}} < \frac{13}{\mathbf{-1}}$$
$$a < -13$$
$$(-\infty, -13)$$

45.
$$-9(h-3) + 2h \le 8(4-h)$$
$$\mathbf{-9}(h) - \mathbf{9}(-3) + 2h \le \mathbf{8}(4) - \mathbf{8}(h)$$
$$-9h + 27 + 2h \le 32 - 8h$$
$$-7h + 27 \le 32 - 8h$$
$$-7h + 27 + \mathbf{8h} \le 32 - 8h + \mathbf{8h}$$
$$h + 27 \le 32$$
$$h + 27 - \mathbf{27} \le 32 - \mathbf{27}$$
$$h \le 5$$
$$(-\infty, 5]$$

46. AWARDS

Analyze
- The frame cost $15.
- Each word cost $0.75.
- Total cost cannot exceed $150.
- What is the maximum number of words?

Assign

Let x = the number of words

Form

The cost of the frame	plus	the cost of the words	must be no more than	the maximum amount.
15	+	0.75x	≤	150

Solve
$$15 + 0.75x \le 150$$
$$15 + 0.75x \le 150$$
$$15 + 0.75x - \mathbf{15} \le 150 - \mathbf{15}$$
$$0.75x \le 135$$
$$\frac{0.75x}{\mathbf{0.75}} \le \frac{135}{\mathbf{0.75}}$$
$$x \le 180$$

State

The maximum numer of words is 180.

Check
$$15 + 0.75(180) \le 150$$
$$15 + 135 \le 150$$
$$150 \le 150$$

The result checks.

SECTION 3.1

VOCABULARY

Fill in the blanks.

1. (7, 1) is called an **ordered** pair.

3. A rectangular coordinate system is formed by two perpendicular number lines called the *x*-**axis** and the *y*-**axis**. The point where the axes cross is called the **origin**.

5. The point with coordinates (4, 2) can be graphed on a **rectangular** coordinate system.

CONCEPTS

7. **Fill in the blanks.**

   a. To plot (–5, 4), we start at the **origin** and move 5 units to the **left** and then move 4 units **up**.

   b. To plot $\left(6, -\dfrac{3}{2}\right)$, we start at the **origin** and move 6 units to the **right** and then move $\dfrac{3}{2}$ units **down**.

9. a. I and II   b. II and III

   c. IV   d. The *y*-axis

NOTATION

11. Explain the difference between (3, 5) and 3(5).

   (3, 5) is an ordered pair, 3(5) = 3 • 5.

13. Do these ordered pairs name the same point?

   $\left(2.5, -\dfrac{7}{2}\right), \left(2\frac{1}{2}, -3.5\right), \left(2.5, -3\frac{1}{2}\right)$

   Yes

15. In the ordered pair (4, 5), is the number 4 associated with the horizontal or the vertical axis?

   Horizontal

GUIDED PRACTICE

17. Graph each point: $(-3, 4), (4, 3.5), \left(-2, -\dfrac{5}{2}\right)$

   $(0, -4), \left(\dfrac{3}{2}, 0\right), (2.7, -4.1)$.

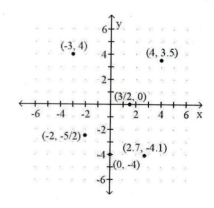

19. Complete the coordinates for each point.

   (4, 3), (0, 4), (–5, 0), (–4, –5), (3, –3)

21. a. **60 beats/min**   b. **10 min**

23. a. **5 min and 50 min after starting**

   b. **20 min**

25. a **2 hrs**   b. **–1,000 ft**

27. a **It ascends (rises) 500 ft**   b. **–500 ft**

APPLICATIONS

29. BRIDGE CONSTRUCTION

   Rivets: $(-60, 0), (-20, 0), (20, 0), (60, 0)$

   Welds: $(-40, 30), (0, 30), (40, 30)$

   Anchors: $(-60, -30), (60, -30)$

31. GAMES

   (G, 2), (G, 3), (G, 4)

from Campus to Careers

33. DENTAL ASSISTANT

   a. 8

   b. It represents the patient's left side.

35. WATER PRESSURE

   a. 60°; 4 ft   b. 30°; 4 ft

37. AREA

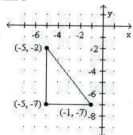

   Area: $A = 0.5bh$

   $= 0.5(4)(5)$

   $= 10$

   The solution is 10 sq. units.

39. TRUCKS

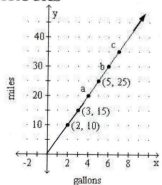

   a. 20 miles   b. 6 gallons   c. 35 miles

Section 3.1

41. DEPRECIATION

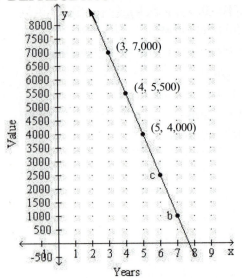

a. A 3-year old copier is worth $7,000.

b. $1,000

c. 6 years

WRITING

43- 45. Answers will vary.

REVIEW

47. Solve $AC = \dfrac{2}{3}h - T$ for h.

$$AC = \frac{2}{3}h - T$$

$$AC + \boldsymbol{T} = \frac{2}{3}h - T + \boldsymbol{T}$$

$$AC + T = \frac{2}{3}h$$

$$AC + T = \frac{2}{3}h$$

$$\frac{\boldsymbol{3}}{\boldsymbol{2}}(AC + T) = \frac{\boldsymbol{3}}{\boldsymbol{2}}\left(\frac{2}{3}h\right)$$

$$\frac{3(AC + T)}{2} = h$$

$$h = \frac{3(AC + T)}{2}$$

49.
$$\frac{-4(4+2) - 2^3}{|-12 - 4(5)|} = \frac{-4(6) - 8}{|-12 - 20|}$$

$$= \frac{-24 - 8}{|-32|}$$

$$= \frac{-32}{32}$$

$$= -1$$

CHALLENGE PROBLEMS

51. Quadrant III is the quadrant.

Example point. $(-3, -5)$

Sum: $-3 + (-5) = -8$

Product: $-3(-5) = 15$

SECTION 3.2
VOCABULARY

Fill in the blanks.

1. $y = 9x + 5$ is an equation in **two** variables, x and y.

3. Solutions of equations in two variables are often listed in a **table** of solutions.

5. $y = 3x + 8$ is called a **linear** equation because its graph is a line.

CONCEPTS

7. Consider $y = -3x + 6$.
   a. How many variables does the equation contain? **2**
   b. Does $(4, -6)$ satisfy the equation? **Yes**
   c. Is $(-2, 0)$ a solution? **No**
   d. How many solutions does this equation have? **Infinitely many**

9. Every point on the graph represents an ordered-pair **solution** of $y = -2x - 3$ and every ordered-pair solution is a **point** on the graph.

11. a. One could choose $-5, 0, 5$. These numbers are multiples of the denominator. There are many others to choose from.
    b. One could choose $-10, 0, 10$. These numbers would simplify the decimal. There are many others to choose from.

NOTATION

13. Verify that $(-2, 6)$ is a solution of $y = -x + 4$.
$$y = -x + 4$$
$$\boxed{6} \overset{?}{=} -(\boxed{-2}) + 4$$
$$6 \overset{?}{=} \boxed{2} + 4$$
$$6 = \boxed{6}$$

15. a. In the linear equation $y = \frac{1}{2}x + 7$ what are the understood exponents on the variables.
    They are 1's.

    b. Explain why $y = x^2 + 2$ and $y = x^3 - 4$ are not linear equations.
    The exponent on x is not 1.

GUIDED PRACTICE

Determine whether each equation has the given ordered pair as a solution.

17. $y = 5x - 4$; $(1, 1)$
$$1 \overset{?}{=} 5(1) - 4$$
$$1 \overset{?}{=} 5 - 4$$
$$1 = 1 \quad \text{True}$$
$(1, 1)$ is a solution.

19. $7x - 2y = 3$; $(2, 6)$
$$7(2) - 2(6) \overset{?}{=} 3$$
$$14 - 12 \overset{?}{=} 3$$
$$2 = 3 \quad \text{False}$$
$(2, 6)$ is not a solution.

21. $x + 12y = -12$; $(0, -1)$
$$0 + 12(-1) \overset{?}{=} -12$$
$$0 - 12 \overset{?}{=} -12$$
$$-12 = -12 \quad \text{True}$$
$(0, -1)$ is a solution.

23. $3x - 6y = 12$; $(-3.6, -3.8)$
$$3(-3.6) - 6(-3.8) \overset{?}{=} 12$$
$$-10.8 + 22.8 \overset{?}{=} 12$$
$$12 = 12 \quad \text{True}$$
$(-3.6, -3.8)$ is a solution.

25. $y - 6x = 12$; $\left(\frac{5}{6}, 7\right)$
$$7 - 6\left(\frac{5}{6}\right) \overset{?}{=} 12$$
$$7 - 5 \overset{?}{=} 12$$
$$2 = 12 \quad \text{False}$$
$\left(\frac{5}{6}, 7\right)$ is not a solution.

27. $y = -\frac{3}{4}x + 8$; $(-8, 12)$
$$12 \overset{?}{=} -\frac{3}{4}(-8) + 8$$
$$12 \overset{?}{=} 6 + 8$$
$$12 = 14 \quad \text{False}$$
$(-8, 12)$ is not a solution.

For each equation, complete the solution.

29. $y = -5x - 4$; $(-3, ?)$

$y = -5(-3) - 4$

$y = 15 - 4$

$y = 11$

$(-3, \boxed{11})$ is the completed ordered pair.

31. $4x - 5y = -4$; $(?, 4)$

$4x - 5(4) = -4$

$4x - 20 = -4$

$4x - 20 + 20 = -4 + 20$

$4x = 16$

$\dfrac{4x}{4} = \dfrac{16}{4}$

$x = 4$

$(\boxed{4}, 4)$ is the completed ordered pair.

33. $y = \frac{x}{4} + 9$; $(16, ?)$

$y = \frac{16}{4} + 9$

$y = 4 + 9$

$y = 13$

$(16, \boxed{13})$ is the completed ordered pair.

35. $7x = 4y$; $(?, -2)$

$7x = 4(-2)$

$7x = -8$

$\dfrac{7x}{7} = \dfrac{-8}{7}$

$x = -\dfrac{8}{7}$

$\left(\boxed{-\dfrac{8}{7}}, -2\right)$ is the completed ordered pair.

Complete each table of solutions.

37. $y = 2x - 4$ $y = 2x - 4$

    $(8, ?)$ $(?, 8)$

$y = 2(8) - 4$ $8 = 2x - 4$

$y = 16 - 4$ $8 + 4 = 2x - 4 + 4$

$y = 12$ $12 = 2x$

$\dfrac{12}{2} = \dfrac{2x}{2}$

$6 = x$

x	y	(x, y)
8	12	(8,12)
6	8	(6,8)

39. $3x - y = -2$ $3x - y = -2$

    $(-5, ?)$ $(?, -1)$

$3(-5) - y = -2$ $3x - (-1) = -2$

$-15 - y = -2$ $3x + 1 - 1 = -2 - 1$

$-15 - y + 15 = -2 + 15$ $3x = -3$

$-y = 13$ $\dfrac{3x}{3} = \dfrac{-3}{3}$

$\dfrac{-y}{-1} = \dfrac{13}{-1}$ $x = -1$

$y = -13$

x	y	(x, y)
-5	-13	(-5,-13)
-1	-1	(-1,-1)

Complete a table of solutions and then graph each equation.

41. $y = 2x - 3$

x	y	(x, y)
-2	$y = 2x - 3$ $= 2(-2) - 3$ $= -4 - 3$ $= -7$	(-2, -7)
0	$y = 2x - 3$ $= 2(0) - 3$ $= 0 - 3$ $= -3$	(0, -3)
2	$y = 2x - 3$ $= 2(2) - 3$ $= 4 - 3$ $= 1$	(2, 1)

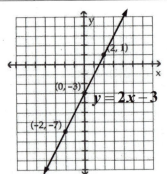

43. $y = 5x - 4$

x	y	(x, y)
-1	$\begin{aligned} y &= 5x - 4 \\ &= 5(-1) - 4 \\ &= -5 - 4 \\ &= -9 \end{aligned}$	$(-1, -9)$
0	$\begin{aligned} y &= 5x - 4 \\ &= 5(0) - 4 \\ &= 0 - 4 \\ &= -4 \end{aligned}$	$(0, -4)$
1	$\begin{aligned} y &= 5x - 4 \\ &= 5(1) - 4 \\ &= 5 - 4 \\ &= 1 \end{aligned}$	$(1, 1)$

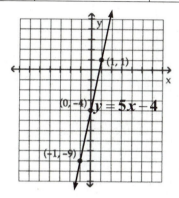

45. $y = -6x$

x	y	(x, y)
-1	$\begin{aligned} y &= -6x \\ &= -6(-1) \\ &= 6 \end{aligned}$	$(-1, 6)$
0	$\begin{aligned} y &= -6x \\ &= -6(0) \\ &= 0 \end{aligned}$	$(0, 0)$
1	$\begin{aligned} y &= -6x \\ &= -6(1) \\ &= -6 \end{aligned}$	$(1, -6)$

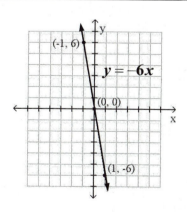

47. $y = -7x$

x	y	(x, y)
-1	$\begin{aligned} y &= -7x \\ &= -7(-1) \\ &= 7 \end{aligned}$	$(-1, 7)$
0	$\begin{aligned} y &= -7x \\ &= -7(0) \\ &= 0 \end{aligned}$	$(0, 0)$
1	$\begin{aligned} y &= -7x \\ &= -7(1) \\ &= -7 \end{aligned}$	$(1, -7)$

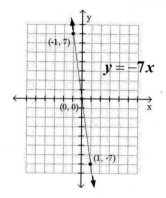

49.
$$\begin{aligned} 2x + 3y &= -3 \\ 2x + 3y - 2x &= -3 - 2x \\ 3y &= -2x - 3 \\ \frac{3y}{3} &= \frac{-2x}{3} - \frac{3}{3} \\ y &= -\frac{2}{3}x - 1 \end{aligned}$$

x	y	(x, y)
-3	$\begin{aligned} y &= -\frac{2}{3}x - 1 \\ &= -\frac{2}{3}(-3) - 1 \\ &= 2 - 1 \\ &= 1 \end{aligned}$	$(-3, 1)$
0	$\begin{aligned} y &= -\frac{2}{3}x - 1 \\ &= -\frac{2}{3}(0) - 1 \\ &= -1 \end{aligned}$	$(0, -1)$
3	$\begin{aligned} y &= -\frac{2}{3}x - 1 \\ &= -\frac{2}{3}(3) - 1 \\ &= -2 - 1 \\ &= -3 \end{aligned}$	$(3, -3)$

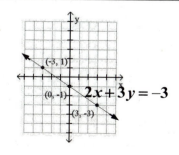

- 129 -

Section 3.2

51.
$$5y - x = 20$$
$$5y - x + x = 20 + x$$
$$5y = x + 20$$
$$\frac{5y}{5} = \frac{x}{5} + \frac{20}{5}$$
$$y = \tfrac{1}{5}x + 4$$

x	y	(x, y)
-5	$y = \tfrac{1}{5}x + 4$ $= \tfrac{1}{5}(-5) + 4$ $= -1 + 4$ $= 3$	$(-5, 3)$
0	$y = \tfrac{1}{5}x + 4$ $= \tfrac{1}{5}(0) + 4$ $= 4$	$(0, 4)$
5	$y = \tfrac{1}{5}x + 4$ $= \tfrac{1}{5}(5) + 4$ $= 1 + 4$ $= 5$	$(5, 5)$

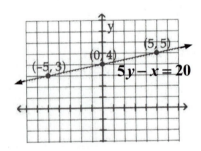

53. $y = x$

x	y	(x, y)
-3	-3	$(-3, -3)$
0	0	$(0, 0)$
4	4	$(4, 4)$

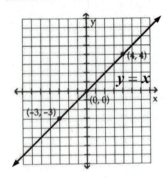

55. $y = -x - 1$

x	y	(x, y)
-2	$y = -x - 1$ $= -(-2) - 1$ $= 2 - 1$ $= 1$	$(-2, 1)$
0	$y = -x - 1$ $= -(0) - 1$ $= 0 - 1$ $= -1$	$(0, -1)$
2	$y = -x - 1$ $= -(2) - 1$ $= -2 - 1$ $= -3$	$(2, -3)$

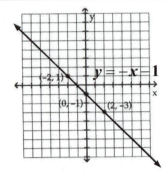

57.
$$3y = 12x + 15$$
$$\frac{3y}{3} = \frac{12x}{3} + \frac{15}{3}$$
$$y = 4x + 5$$

x	y	(x, y)
-1	$y = 4x + 5$ $= 4(-1) + 5$ $= -4 + 5$ $= 1$	$(-1, 1)$
0	$y = 4x + 5$ $= 4(0) + 5$ $= 0 + 5$ $= 5$	$(0, 5)$
1	$y = 4x + 5$ $= 4(1) + 5$ $= 4 + 5$ $= 9$	$(1, 9)$

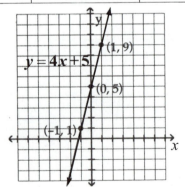

59. $y = \dfrac{3}{8}x - 6$

x	y	(x, y)
-8	$y = \dfrac{3}{8}x - 6$ $= \dfrac{3}{8}(-8) - 6$ $= -3 - 6$ $= -9$	$(-8, -9)$
0	$y = \dfrac{3}{8}x - 6$ $= \dfrac{3}{8}(0) - 6$ $= 0 - 6$ $= -6$	$(0, -6)$
8	$y = \dfrac{3}{8}x - 6$ $= \dfrac{3}{8}(8) - 6$ $= 3 - 6$ $= -3$	$(8, -3)$

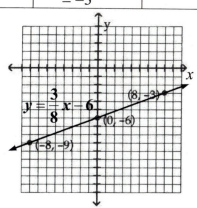

61. $y = 1.5x - 4$

x	y	(x, y)
-2	$y = 1.5x - 4$ $= 1.5(-2) - 4$ $= -3 - 4$ $= -7$	$(-2, -7)$
0	$y = 1.5x - 4$ $= 1.5(0) - 4$ $= 0 - 4$ $= -4$	$(0, -4)$
2	$y = 1.5x - 4$ $= 1.5(2) - 4$ $= 3 - 4$ $= -1$	$(2, -1)$

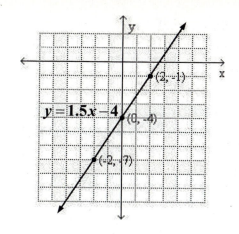

63.
$$8x + 4y = 16$$
$$8x + 4y - \mathbf{8x} = -\mathbf{8x} + 16$$
$$4y = -8x + 16$$
$$\frac{4y}{\mathbf{4}} = \frac{-8x}{\mathbf{4}} + \frac{16}{\mathbf{4}}$$
$$y = -2x + 4$$

x	y	(x, y)
-1	$y = -2x + 4$ $= -2(-1) + 4$ $= 2 + 4$ $= 6$	$(-1, 6)$
0	$y = -2x + 4$ $= -2(0) + 4$ $= 0 + 4$ $= 4$	$(0, 4)$
1	$y = -2x + 4$ $= -2(1) + 4$ $= -2 + 4$ $= 2$	$(1, 2)$

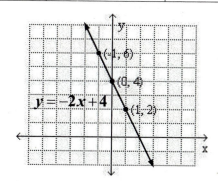

65. $y = -\dfrac{1}{2}x$

x	y	(x, y)
-2	$y = -\dfrac{1}{2}x$ $= -\dfrac{1}{2}(-2)$ $= 1$	$(-2, 1)$
0	$y = -\dfrac{1}{2}x$ $= -\dfrac{1}{2}(0)$ $= 0$	$(0, 0)$
2	$y = -\dfrac{1}{2}x$ $= -\dfrac{1}{2}(2)$ $= -1$	$(2, -1)$

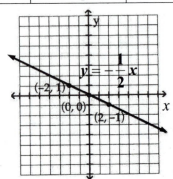

67. $y = \dfrac{5}{6}x - 5$

x	y	(x, y)
-6	$y = \dfrac{5}{6}x - 5$ $= \dfrac{5}{6}(-6) - 5$ $= -5 - 5$ $= -10$	$(-6, -10)$
0	$y = \dfrac{5}{6}x - 5$ $= \dfrac{5}{6}(0) - 5$ $= 0 - 5$ $= -5$	$(0, -5)$
6	$y = \dfrac{5}{6}x - 5$ $= \dfrac{5}{6}(6) - 5$ $= 5 - 5$ $= 0$	$(6, 0)$

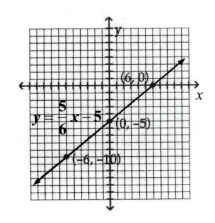

69. $-6y = 30x + 12$

$$\dfrac{-6y}{-6} = \dfrac{30x}{-6} + \dfrac{12}{-6}$$

$$y = -5x - 2$$

x	y	(x, y)
-1	$y = -5x - 2$ $= -5(-1) - 2$ $= 5 - 2$ $= 3$	$(-1, 3)$
0	$y = -5x - 2$ $= -5(0) - 2$ $= 0 - 2$ $= -2$	$(0, -2)$
1	$y = -5x - 2$ $= -5(1) - 2$ $= -5 - 2$ $= -7$	$(1, -7)$

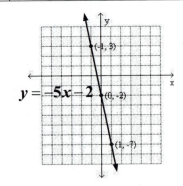

71. $y = \dfrac{x}{3}$

x	y	(x, y)
-3	$y = \dfrac{x}{3}$ $= \dfrac{-3}{3}$ $= -1$	$(-3, -1)$
0	$y = \dfrac{x}{3}$ $= \dfrac{0}{3}$ $= 0$	$(0, 0)$
3	$y = \dfrac{x}{3}$ $= \dfrac{3}{3}$ $= 1$	$(3, 1)$

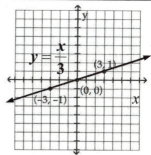

73. $y = -2x + 1$

x	y	(x, y)
-1	$y = -2x + 1$ $= -2(-1) + 1$ $= 2 + 1$ $= 3$	$(-1, 3)$
0	$y = -2x + 1$ $= -2(0) + 1$ $= 0 + 1$ $= 1$	$(0, 1)$
1	$y = -2x + 1$ $= -2(1) + 1$ $= -2 + 1$ $= -1$	$(1, -1)$

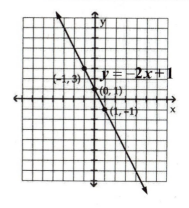

75.
$$7x - y = 1$$
$$7x - y + y = y + 1$$
$$7x = y + 1$$
$$7x - 1 = y + 1 - 1$$
$$y = 7x - 1$$

x	y	(x, y)
-1	$y = 7x - 1$ $= 7(-1) - 1$ $= -7 - 1$ $= -8$	$(-1, -8)$
0	$y = 7x - 1$ $= 7(0) - 1$ $= 0 - 1$ $= -1$	$(0, -1)$
1	$y = 7x - 1$ $= 7(1) - 1$ $= 7 - 1$ $= 6$	$(1, 6)$

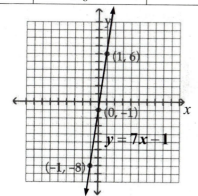

77.
$$7y = -2x$$
$$\dfrac{7y}{7} = \dfrac{-2x}{7}$$
$$y = -\dfrac{2}{7}x$$

x	y	(x, y)
-7	$y = -\dfrac{2}{7}x$ $= -\dfrac{2}{7}(-7)$ $= 2$	$(-7, 2)$
0	$y = -\dfrac{2}{7}x$ $= -\dfrac{2}{7}(0)$ $= 0$	$(0, 0)$
7	$y = -\dfrac{2}{7}x$ $= -\dfrac{2}{7}(7)$ $= -2$	$(7, -2)$

Section 3.2

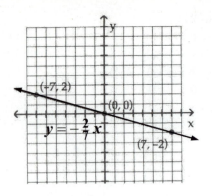

81. BILLARDS

$y = 2x - 4$			$y = -2x + 12$		
x	y	(x, y)	x	y	(x, y)
1	–2	(1, –2)	4	4	(4, 4)
2	0	(2, 0)	6	0	(6, 0)
4	4	(4, 4)	8	–4	(8, –4)

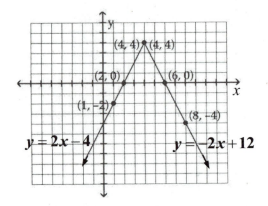

79. $y = -2.5x + 5$

x	y	(x, y)
–2	$\begin{aligned} y &= -2.5x + 5 \\ &= -2.5(-2) + 5 \\ &= 5 + 5 \\ &= 10 \end{aligned}$	(–2, 10)
0	$\begin{aligned} y &= -2.5x + 5 \\ &= -2.5(0) + 5 \\ &= 5 \end{aligned}$	(0, 5)
2	$\begin{aligned} y &= -2.5x + 5 \\ &= -2.5(2) + 5 \\ &= -5 + 5 \\ &= 0 \end{aligned}$	(2, 0)

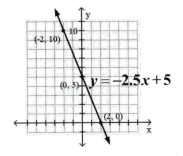

83. DEFROSTING POULTRY

$$h = 5p$$

p	h	(p, h)
5	25	(5, 25)
15	75	(15, 75)
20	100	(20, 100)
25	?	(25, ?)

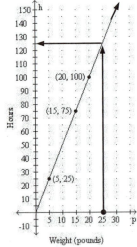

It will take a 25-pound turkey
125 hours to defrost in the refrigerator.

85. HOUSEKEEPING

The problem wants you to estimate (not calculate) the amount of polish that is left in the bottle after 650 sprays. Use a graph to find the answer.

n = number of sprays (x-axis) scale of 50
A = ounces remaining (y-axis) scale of 2

$$A = -0.02n + 16$$

n	A	(n, A)
100	$A = -0.02(100) + 16$ $= -2 + 16$ $= 14$	$(100, 14)$
200	12	$(200, 12)$
300	$A = -0.02(300) + 16$ $= -6 + 16$ $= 10$	$(300, 10)$
500	6	$(500, 6)$
650	?	$(650, \ ?)$

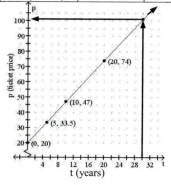

The amount of polish left in the bottle after 650 sprays is 3 ounces.

87. NFL TICKETS

The problem wants you to estimate (not calculate) the price of a ticket in 2020. Use a graph to find the answer.

t = the year (x-axis) scale of 2
p = price of the ticket (y-axis) scale of 5

$$p = 2.7t + 20$$

t	p	(t, p)
0 (1990)	20	$(0, 20)$
5 (1995)	33.50	$(5, 33.50)$
10 (2000)	47	$(10, 47)$
20 (2010)	74	$(20, 74)$
30 (2020)	?	$(30, \ ?)$

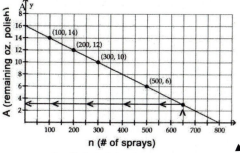

89. RAFFLES

The problem wants you to predict (not calculate) the number of raffle tickets that will be sold at $6. Use a graph to find the answer.

p = the price (x-axis) scale of 1
n = the number of tickets (y-axis) scale of 10

$$n = -20p + 300$$

p	n	(p, n)
1	280	$(1, 280)$
3	240	$(3, 240)$
5	200	$(5, 200)$
6	?	$(6, \ ?)$

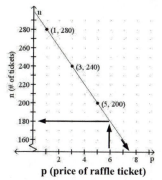

The predicted number of $6 raffle tickets to be sold is around 180 tickets.

91. U.S. SPACE PROGRAM

The problem wants you to predict (not calculate) when 70% will say "yes". Use a graph to find the answer.

n = the year (x-axis) scale of 2
p = the percent of "yes" responses (y-axis) scale of 5

$$p = \tfrac{3}{5}n + 40$$

n	p	(n, p)
0 (1980)	40	$(0, 40)$
5 (1985)	43	$(5, 43)$
10 (1990)	46	$(10, 46)$
20 (2000)	52	$(20, 52)$
?	70	$(? \ , 70)$

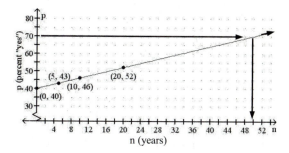

The predicted year that 70% will respond with a "yes" is the year 2030 (1980 + 50).

Section 3.2

WRITING

93–99. Answers will vary.

REVIEW

101. Simplify: $-(-5-4c) = -(-5)-(-4c)$
$$= 5+4c$$

103. $V = \frac{4}{3}\pi r^3$, $\pi = 3.141592654$
$$= \frac{4}{3}(3.141592654)(6)(6)(6)$$
$$\approx 904.77$$
$$\approx 904.8$$

The volume of the sphere is approximately 904.8 ft^3.

CHALLENGE PROBLEMS

105. Graph $y = x^2 + 1$.

x	$y = x^2 + 1$	(x, y)
-3	$y = (-3)^2 + 1$ $= 9+1$ $= 10$	$(-3, 10)$
-2	$y = (-2)^2 + 1$ $= 4+1$ $= 5$	$(-2, 5)$
-1	$y = (-1)^2 + 1$ $= 1+1$ $= 2$	$(-1, 2)$
0	$y = (0)^2 + 1$ $= 0+1$ $= 1$	$(0, 1)$
1	$y = (1)^2 + 1$ $= 1+1$ $= 2$	$(1, 2)$
2	$y = (2)^2 + 1$ $= 4+1$ $= 5$	$(2, 5)$
3	$y = (3)^2 + 1$ $= 9+1$ $= 10$	$(3, 10)$

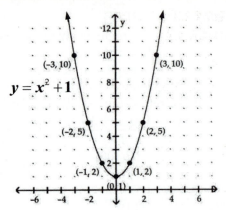

This type of graph is called a **Parabola**.

107. Graph $y = |x| - 2$.

x	$y = |x| - 2$	(x, y)
-3	$y = |-3| - 2$ $= 3-2$ $= 1$	$(-3, 1)$
-2	$y = |-2| - 2$ $= 2-2$ $= 0$	$(-2, 0)$
-1	$y = |-1| - 2$ $= 1-2$ $= -1$	$(-1, -1)$
0	$y = |0| - 2$ $= 0-2$ $= -2$	$(0, -2)$
1	$y = |1| - 2$ $= 1-2$ $= -1$	$(1, -1)$
2	$y = |2| - 2$ $= 2-2$ $= 0$	$(2, 0)$
3	$y = |3| - 2$ $= 3-2$ $= 1$	$(3, 1)$

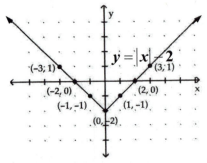

This kind of graph is called the **Absolute Value** graph.

SECTION 3.3

VOCABULARY

Fill in the blanks.

1. The **x-intercept** of a line is the point where the line intersects/crosses the x-axis.

3. The graph of $y = 4$ is a **horizontal** line and the graph of $x = 6$ is a **vertical** line.

CONCEPTS

5. **Fill in the blanks.**
   a. To find the y-intercept of the graph of a line, substitute **0** for x in the equation and solve for **y** .
   b. To find the x-intercept of the graph of a line, substitute **0** for y in the equation and solve for **x** .

7. a. Which intercept tells the purchase price of the machinery? **y-intercept: (0, 80,000)**
      What was that price? **$80,000**
   b. Which intercept indicates when the machinery will have lost all of its value?
      x-intercept: (30, 0)
      When is that? **30 years after purchase**

NOTATION

9. What is the equation of the x-axis? **$y = 0$**
   What is the equation of the y-axis? **$x = 0$**

GUIDED PRACTICE

Give the coordinates of the intercepts of each graph.

11. The x-intercept is (4, 0) and the y-intercept is (0, 3).

13. The x-intercept is (–5, 0) and the y-intercept is (0, –4).

15. The y-intercept is (0, 2). There is no x-intercept because the line does not cross the x-axis.

Estimate the coordinates of the intercepts of each graph.

17. The x-intercept is $\left(-2\frac{1}{2}, 0\right)$ and the y-intercept is $\left(0, \frac{2}{3}\right)$.

Find the x- and y-intercepts of the graph of each equation. *Do not graph the line.*

19. x–intercept: $y = 0$
$$8x + 3y = 24$$
$$8x + 3(0) = 24$$
$$8x = 24$$
$$x = 3$$
The x–intercept is $(3, 0)$.

   y–intercept: $x = 0$
$$8x + 3y = 24$$
$$8(0) + 3y = 24$$
$$3y = 24$$
$$y = 8$$
The y–intercept is $(0, 8)$.

21. x–intercept: $y = 0$
$$7x - 2y = 28$$
$$7x - 2(0) = 28$$
$$7x = 28$$
$$x = 4$$
The x–intercept is $(4, 0)$.

   y–intercept: $x = 0$
$$7x - 2y = 28$$
$$7(0) - 2y = 28$$
$$-2y = 28$$
$$y = -14$$
The y–intercept is $(0, -14)$.

23. x–intercept: $y = 0$
$$-5x - 3y = 10$$
$$-5x - 3(0) = 10$$
$$-5x = 10$$
$$x = -2$$
The x–intercept is $(-2, 0)$.

   y–intercept: $x = 0$
$$-5x - 3y = 10$$
$$-5(0) - 3y = 10$$
$$-3y = 10$$
$$y = -\frac{10}{3}$$
The y–intercept is $\left(0, -\frac{10}{3}\right)$.

Section 3.3

25. x-intercept: $y = 0$

$$6x + y = 9$$
$$6x + (0) = 9$$
$$6x = 9$$
$$x = \frac{9}{6}$$
$$x = \frac{3}{2}$$

The x-intercept is $\left(\frac{3}{2}, 0\right)$.

y-intercept: $x = 0$

$$6x + y = 9$$
$$6(0) + y = 9$$
$$y = 9$$

The y-intercept is $(0, 9)$.

Use the intercept method to graph each equation.

27. $4x + 5y = 20$

y-intercept:	x-intercept:
If $x = 0$,	If $y = 0$
$4(0) + 5y = 20$	$4x + 5(0) = 20$
$5y = 20$	$4x = 20$
$y = 4$	$x = 5$

The y-intercept is $(0, 4)$, and the x-intercept is $(5, 0)$.

Check Point
$$4x + 5y = 20$$
$$4(2) + 5y = 20$$
$$5y = 12$$
$$y = \frac{12}{5}$$
$$\left(2, \frac{12}{5}\right)$$

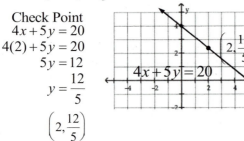

29. $5x + 15y = -15$

y-intercept:	x-intercept:
If $x = 0$,	If $y = 0$
$5(0) + 15y = -15$	$5x + 15(0) = -15$
$15y = -15$	$5x = -15$
$y = -1$	$x = -3$

The y-intercept is $(0, -1)$, and the x-intercept is $(-3, 0)$.

Check Point
$$5x + 15y = -15$$
$$5(3) + 15y = -15$$
$$15y = -30$$
$$y = -2$$
$$(3, -2)$$

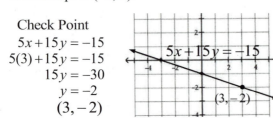

31. $x - y = -3$

y-intercept:	x-intercept:
If $x = 0$,	If $y = 0$
$(0) - y = -3$	$x - (0) = -3$
$-y = -3$	$x = -3$
$y = 3$	

The y-intercept is $(0, 3)$, and the x-intercept is $(-3, 0)$.

Check Point
$$x - y = -3$$
$$-2 - y = -3$$
$$-y = -1$$
$$y = 1$$
$$(-2, 1)$$

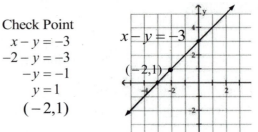

33. $x + 2y = -2$

y-intercept:	x-intercept:
If $x = 0$,	If $y = 0$
$(0) + 2y = -2$	$x + 2(0) = -2$
$2y = -2$	$x = -2$
$y = -1$	

The y-intercept is $(0, -1)$, and the x-intercept is $(-2, 0)$.

Check Point
$$x + 2y = -2$$
$$2 + 2y = -2$$
$$2y = -4$$
$$y = -2$$
$$(2, -2)$$

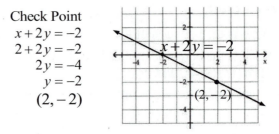

Use the intercept method to graph each equation.

35. $30x + y = -30$

y-intercept:	x-intercept:
If $x = 0$,	If $y = 0$
$30(0) + y = -30$	$30x + (0) = -30$
$y = -30$	$30x = -30$
	$x = -1$

The y-intercept is $(0, -30)$, and the x-intercept is $(-1, 0)$.

Check Point
$$30x + y = -30$$
$$30(-0.5) + y = -30$$
$$-15 + y = -30$$
$$y = -15$$
$$(-0.5, -15)$$

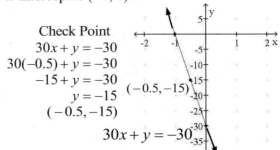

37. $4x - 20y = 60$

y-intercept:	x-intercept:
If $x = 0$,	If $y = 0$
$4(0) - 20y = 60$	$4x - 20(0) = 60$
$-20y = 60$	$4x = 60$
$y = -3$	$x = 15$

The y-intercept is $(0, -3)$, and the
x-intercept is $(15, 0)$.

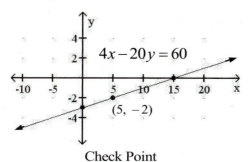

Check Point
$4x - 20y = 60$
$4(5) - 20y = 60$
$20 - 20y = 60$
$-20y = 40$
$y = -2$
$(5, -2)$

Use the intercept method to graph each equation.

39. $3x + 4y = 8$

y-intercept:	x-intercept:
If $x = 0$,	If $y = 0$
$3(0) + 4y = 8$	$3x + 4(0) = 8$
$4y = 8$	$3x = 8$
$y = 2$	$x = \dfrac{8}{3}$

The y-intercept is $(0, 2)$, and the

x-intercept is $(\dfrac{8}{3}, 0)$.

Check Point
$3x + 4y = 8$
$3(1) + 4y = 8$
$3 + 4y = 8$
$4y = 5$
$y = \dfrac{5}{4}$

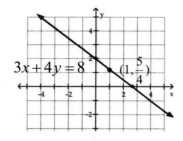

41. $-9x + 4y = 9$

y-intercept:	x-intercept:
If $x = 0$,	If $y = 0$
$-9(0) + 4y = 9$	$-9x + 4(0) = 9$
$4y = 9$	$-9x = 9$
$y = \dfrac{9}{4}$	$x = -1$

The y-intercept is $(0, \dfrac{9}{4})$, and the

x-intercept is $(-1, 0)$.

Check Point
$-9x + 4y = 9$
$-9(-2) + 4y = 9$
$18 + 4y = 9$
$4y = -9$
$y = -\dfrac{9}{4}$

$\left(-2, -\dfrac{9}{4}\right)$

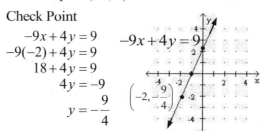

43. $3x - 4y = 11$

y-intercept:	x-intercept:
If $x = 0$,	If $y = 0$
$3(0) - 4y = 11$	$3x - 4(0) = 11$
$-4y = 11$	$3x = 11$
$y = -\dfrac{11}{4}$	$x = \dfrac{11}{3}$

The y-intercept is $(0, -\dfrac{11}{4})$, and the

x-intercept is $(\dfrac{11}{3}, 0)$.

Check Point
$3x - 4y = 11$
$3(1) - 4y = 11$
$3 - 4y = 11$
$-4y = 8$
$y = -2$
$(1, -2)$

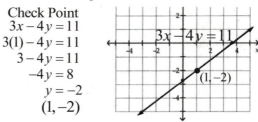

45. $9x + 3y = 10$

y-intercept:	x-intercept:
If $x = 0$,	If $y = 0$
$9(0) + 3y = 10$	$9x + 3(0) = 10$
$3y = 10$	$9x = 10$
$y = \dfrac{10}{3}$	$x = \dfrac{10}{9}$

The y-intercept is $(0, \dfrac{10}{3})$, and the

x-intercept is $(\dfrac{10}{9}, 0)$.

Check Point
$9x + 3y = 10$
$9(2) + 3y = 10$
$18 + 3y = 10$
$3y = -8$
$y = -\dfrac{8}{3}$

$\left(2, -\dfrac{8}{3}\right)$

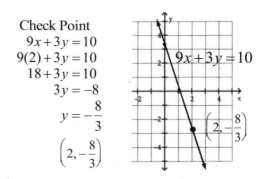

Use the intercept method to graph each equation.

47. $3x + 5y = 0$

y-intercept:	x-intercept:
If $x = 0$,	If $y = 0$
$3(0) + 5y = 0$	$3x + 5(0) = 0$
$5y = 0$	$3x = 0$
$y = 0$	$x = 0$

The y-intercept is $(0, 0)$, and the x-intercept is $(0, 0)$.

Find two other points.

If $x = 5$,	If $y = 3$
$3(5) + 5y = 0$	$3x + 5(3) = 0$
$15 + 5y = 0$	$3x + 15 = 0$
$5y = -15$	$3x = -15$
$y = -3$	$x = -5$
$(5, -3)$	$(-5, 3)$

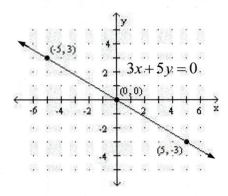

49. $2x - 7y = 0$

y-intercept:	x-intercept:
If $x = 0$,	If $y = 0$
$2(0) - 7y = 0$	$2x - 7(0) = 0$
$-7y = 0$	$2x = 0$
$y = 0$	$x = 0$

The y-intercept is $(0, 0)$, and the x-intercept is $(0, 0)$.

Find two other points.

If $x = 7$,	If $y = -2$
$2(7) - 7y = 0$	$2x - 7(-2) = 0$
$14 - 7y = 0$	$2x + 14 = 0$
$-7y = -14$	$2x = -14$
$y = 2$	$x = -7$
$(7, 2)$	$(-7, -2)$

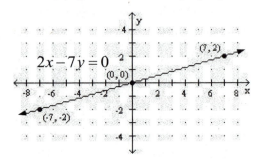

Graph each equation.

51. $y = 5$

x	y	(x, y)
-2	5	$(-2, 5)$
0	5	$(0, 5)$
3	5	$(3, 5)$

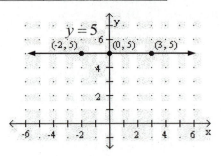

53. $y = 0$

x	y	(x, y)
-2	0	$(-2, 0)$
0	0	$(0, 0)$
3	0	$(3, 0)$

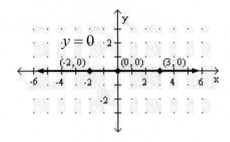

55. $x = -2$

x	y	(x, y)
-2	-3	$(-2, -3)$
-2	0	$(-2, 0)$
-2	2	$(-2, 2)$

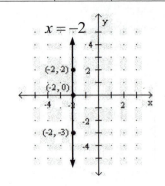

57. $x = \dfrac{4}{3}$

x	y	(x, y)
$\dfrac{4}{3}$	-3	$\left(\dfrac{4}{3}, -3\right)$
$\dfrac{4}{3}$	0	$\left(\dfrac{4}{3}, 0\right)$
$\dfrac{4}{3}$	2	$\left(\dfrac{4}{3}, 2\right)$

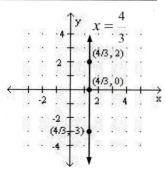

59. $y - 2 = 0$
$\qquad y = 2$

x	y	(x, y)
-2	2	$(-2, 2)$
0	2	$(0, 2)$
3	2	$(3, 2)$

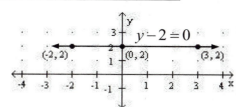

61. $5x = 7.5$
$\qquad \dfrac{5x}{5} = \dfrac{7.5}{5}$
$\qquad x = 1.5$

x	y	(x, y)
1.5	-3	$(1.5, -3)$
1.5	0	$(1.5, 0)$
1.5	2	$(1.5, 2)$

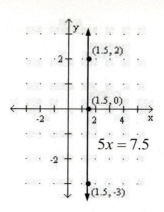

$5x = 7.5$

TRY IT YOURSELF
Graph each equation.

63. $7x = 4y - 12$

y-intercept:	x-intercept:
If $x = 0$,	If $y = 0$
$7(\mathbf{0}) = 4y - 12$	$7x = 4(\mathbf{0}) - 12$
$0 = 4y - 12$	$7x = -12$
$-4y = -12$	$x = -\dfrac{12}{7}$
$y = 3$	

The y-intercept is $(0, 3)$, and the
x-intercept is $\left(-\frac{12}{7}, 0\right)$.

Check Point
$7x = 4y - 12$
$7(-4) = 4y - 12$
$-28 = 4y - 12$
$-16 = 4y$
$-4 = y$

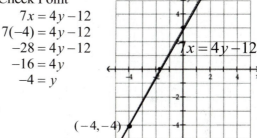

$(-4, -4)$

65. $4x - 3y = 12$

y-intercept:	x-intercept:
If $x = 0$,	If $y = 0$
$4(\mathbf{0}) - 3y = 12$	$4x - 3(\mathbf{0}) = 12$
$-3y = 12$	$4x = 12$
$y = -4$	$x = 3$

The y-intercept is $(0, -4)$, and the
x-intercept is $(3, 0)$.

Check Point
$4x - 3y = 12$
$4(1) - 3y = 12$
$-3y = 8$
$\dfrac{-3y}{-3} = \dfrac{8}{-3}$
$y = -\dfrac{8}{3}$
$\left(1, -\dfrac{8}{3}\right)$

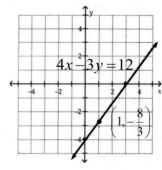

$4x - 3y = 12$
$\left(1, -\dfrac{8}{3}\right)$

Section 3.3

67. $x = -\dfrac{5}{3}$

x	y	(x, y)
$-\dfrac{5}{3}$	-3	$\left(-\dfrac{5}{3}, -3\right)$
$-\dfrac{5}{3}$	0	$\left(-\dfrac{5}{3}, 0\right)$
$-\dfrac{5}{3}$	2	$\left(-\dfrac{5}{3}, 2\right)$

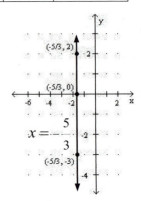

69. $y - 3x = -\dfrac{4}{3}$

y-intercept:
If $x = 0$,
$$y - 3(0) = -\dfrac{4}{3}$$
$$y = -\dfrac{4}{3}$$

x-intercept:
If $y = 0$
$$0 - 3x = -\dfrac{4}{3}$$
$$-3x = -\dfrac{4}{3}$$
$$x = -\dfrac{4}{3} \cdot -\dfrac{1}{3}$$
$$x = \dfrac{4}{9}$$

The y-intercept is $\left(0, -\dfrac{4}{3}\right)$, and the

x-intercept is $\left(\dfrac{4}{9}, 0\right)$.

Check Point
$$y - 3x = -\dfrac{4}{3}$$
$$y - 3\left(\dfrac{3}{3}\right) = -\dfrac{4}{3}$$
$$y - \dfrac{9}{3} = -\dfrac{4}{3}$$
$$y = \dfrac{5}{3}$$
$$\left(1, \dfrac{5}{3}\right)$$

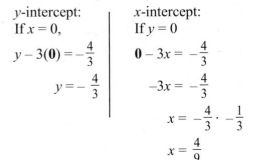

71. $7x + 3y = 0$

y-intercept:
If $x = 0$,
$$7(0) + 3y = 0$$
$$3y = 0$$
$$y = 0$$

x-intercept:
If $y = 0$
$$7x + 3(0) = 0$$
$$7x = 0$$
$$x = 0$$

The y-intercept is $(0, 0)$, and the
x-intercept is $(0, 0)$.

Find two other points.
If $x = 3$,
$$7(3) + 3y = 0$$
$$21 + 3y = 0$$
$$3y = -21$$
$$y = -7$$
$$(3, -7)$$

If $y = 7$
$$7x + 3(7) = 0$$
$$7x + 21 = 0$$
$$7x = -21$$
$$x = -3$$
$$(-3, 7)$$

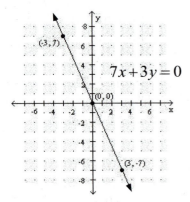

73. $-4x = 8 - 2y$

y-intercept:
If $x = 0$,
$$-4(0) = 8 - 2y$$
$$0 = 8 - 2y$$
$$2y = 8$$
$$y = 4$$

x-intercept:
If $y = 0$
$$-4x = 8 - 2(0)$$
$$-4x = 8$$
$$x = -2$$

The y-intercept is $(0, 4)$, and the
x-intercept is $(-2, 0)$.

Check Point
$$-4x = 8 - 2y$$
$$-4(-1) = 8 - 2y$$
$$4 = 8 - 2y$$
$$-4 = -2y$$
$$2 = y$$
$$(-1, 2)$$

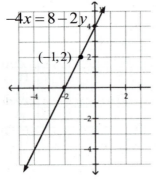

75. $3x = -150 - 5y$

y-intercept:	x-intercept:
If $x = 0$,	If $y = 0$
$3(0) = -150 - 5y$	$3x = -150 - 5(0)$
$0 = -150 - 5y$	$3x = -150$
$5y = -150$	$x = -50$
$y = -30$	

The y-intercept is $(0, -30)$, and the x-intercept is $(-50, 0)$.

Check Point
$$3x = -150 - 5y$$
$$3(-25) = -150 - 5y$$
$$-75 = -150 - 5y$$
$$75 = -5y$$
$$-15 = y$$

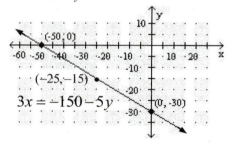

77. $-3y = 3$

y-intercept:	x-intercept:
If $x = 0$,	none
$-3y = 3$	
$y = -1$	

The y-intercept is $(0, -1)$, and the x-intercept does not exist.

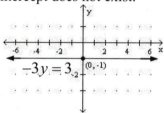

APPLICATIONS

79. CHEMISTRY
   a. Estimate absolute zero. **about −270ºC**
   b. What is the volume of the gas when the temperature is absolute zero? **0 milliliters**

81. BOTTLED WATER DISPENSER
   The g-intercept is $(0, 5)$: Before any cups of water have been served from the bottle, it contains 5 gallons of water. The c-intercept is $(106\frac{2}{3}, 0)$: The bottle will be empty after $106\frac{2}{3}$ 6-ounce cups of water have been served from it.

83. LANDSCAPING
   x – axis represents the number of trees
   y – axis represents the number of shrubs
   $$50x + 25y = 5,000$$

y-intercept:	x-intercept:
If $x = 0$,	If $y = 0$
$50(0) + 25y = 5,000$	$50x + 25(0) = 5,000$
$25y = 5,000$	$50x = 5,000$
$y = 200$	$x = 100$

The y-intercept is $(0, 200)$, and the x-intercept is $(100, 0)$.

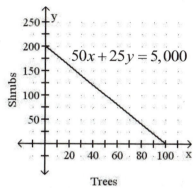

   a. What information is given by the y-intercept?
      If only shrubs are purchased, he can buy 200.
   b. What information is given by the x-intercept?
      If only trees are purchased, he can buy 100.

WRITING
85-87. Answers will vary.

REVIEW

89. $$\frac{3 \cdot 5 \cdot 5}{3 \cdot 5 \cdot 5 \cdot 5} = \frac{\overset{1}{\cancel{3}} \cdot \overset{1}{\cancel{5}} \cdot \overset{1}{\cancel{5}}}{\underset{1}{\cancel{3}} \cdot \underset{1}{\cancel{5}} \cdot \underset{1}{\cancel{5}} \cdot 5}$$
$$= \frac{1}{5}$$

91. Translate: Six less than twice x.
   $$2x - 6$$

CHALLENGE PROBLEMS
93. Where will the line $y = b$ intersect the line $x = a$? **Intersect at (a, b).**
95. What is the least number of intercepts a line can have? **1;** What is the greatest number a line can have? **Infinitely many**

Section 3.3

SECTION 3.4

VOCABULARY

Fill in the blanks.

1. The **slope** of a line is a measure of the line's steepness. It is the **ratio** of the vertical change to the horizontal change.

3. The rate of **change** of a linear relationship can be found by finding the slope of the graph of the line and attaching the proper units.

CONCEPTS

5. Which line graphed has
   a. a positive slope? **Line 2**
   b. a negative slope? **Line 1**
   c. zero slope? **Line 4**
   d. undefined slope? **Line 3**

7. For each graph, determine which line has the greater slope.
   a. **Line 1** b. **Line 1** c. **Line 2** d. **Line 2**

9. a. Rise of 1 and a run of 3, slope is $\frac{1}{3}$

   b. Rise of 4 and a run of 12, slope is $\frac{4}{12} = \frac{1}{3}$

   c. When calculating the slope of a line, the **same** value will be obtained no matter which two points on the line are used.

11. Write each slope in a better way.

   a. $m = \frac{0}{6}$, $m = 0$ b. $m = \frac{8}{0}$, **Undefined**

   c. $m = \frac{3}{12}$, $m = \frac{1}{4}$ d. $m = \frac{-10}{-5}$, $m = 2$

13. The grade of an incline is its slope expressed as a percent. Express the slope $\frac{2}{5}$ as a grade.
   40%

15. Find the negative reciprocal of each number.

   a. $6, -\frac{1}{6}$ b. $-\frac{7}{8}, \frac{8}{7}$ c. $-1, 1$

NOTATION

17. a. What is the formula used to find the slope of a line passing through (x_1, y_1) and (x_2, y_2)?

$$m = \frac{y_2 - y_1}{x_2 - x_1}$$

   b. Fill in the blanks to state the slope formula in words: m equals y **sub** two minus

y **sub** one **over (divided by)** x sub **two** minus x sub **one**.

19. Consider the points $(7, 2)$ and $(-4, 1)$. If we let $x_1 = 7$, then what is y_2? **1**

GUIDED PRACTICE

21. $m = \frac{2}{3}$ 23. $m = \frac{4}{3}$

25. $m = -2$ 27. $m = 0$

29. $m = -\frac{1}{5}$ 31. $m = \frac{1}{2}$

Find the slope of the line passing through the given points, when possible.

33. Let $(x_1, y_1) = (1, 3)$
    Let $(x_2, y_2) = (2, 4)$

$$m = \frac{y_2 - y_1}{x_2 - x_1}$$
$$= \frac{4-3}{2-1}$$
$$= \frac{1}{1}$$
$$= 1$$

35. Let $(x_1, y_1) = (3, 4)$
    Let $(x_2, y_2) = (2, 7)$

$$m = \frac{y_2 - y_1}{x_2 - x_1}$$
$$= \frac{7-4}{2-3}$$
$$= \frac{3}{-1}$$
$$= -3$$

37. Let $(x_1, y_1) = (0, 0)$
    Let $(x_2, y_2) = (4, 5)$

$$m = \frac{y_2 - y_1}{x_2 - x_1}$$
$$= \frac{5-0}{4-0}$$
$$= \frac{5}{4}$$

39. Let $(x_1, y_1) = (-3, 5)$
    Let $(x_2, y_2) = (-5, 6)$

$$m = \frac{y_2 - y_1}{x_2 - x_1}$$
$$= \frac{6-5}{-5-(-3)}$$
$$= \frac{1}{-5+3}$$
$$= -\frac{1}{2}$$

41. Let $(x_1, y_1) = (-2, -2)$
    Let $(x_2, y_2) = (-12, -8)$

$$m = \frac{y_2 - y_1}{x_2 - x_1}$$
$$= \frac{-8-(-2)}{-12-(-2)}$$
$$= \frac{-8+2}{-12+2}$$
$$= \frac{-6}{-10}$$
$$= \frac{3}{5}$$

43. Let $(x_1, y_1) = (5, 7)$
    Let $(x_2, y_2) = (-4, 7)$

$$m = \frac{y_2 - y_1}{x_2 - x_1}$$
$$= \frac{7-7}{-4-5}$$
$$= \frac{0}{-9}$$
$$= 0$$

45. Let $(x_1, y_1) = (8, -4)$
Let $(x_2, y_2) = (8, -3)$

$m = \dfrac{y_2 - y_1}{x_2 - x_1}$

$= \dfrac{-3 - (-4)}{8 - 8}$

$= \dfrac{-3 + 4}{8 - 8}$

$= \dfrac{1}{0}$

Undefined

47. Let $(x_1, y_1) = (-6, 0)$
Let $(x_2, y_2) = (0, -4)$

$m = \dfrac{y_2 - y_1}{x_2 - x_1}$

$= \dfrac{-4 - 0}{0 - (-6)}$

$= \dfrac{-4}{6}$

$= -\dfrac{2}{3}$

49. Let $(x_1, y_1) = (-2.5, 1.75)$
Let $(x_2, y_2) = (-0.5, -7.75)$

$m = \dfrac{y_2 - y_1}{x_2 - x_1}$

$= \dfrac{-7.75 - 1.75}{-0.5 - (-2.5)}$

$= \dfrac{-9.5}{-0.5 + 2.5}$

$= \dfrac{-9.5}{2}$

$= -4.75$

51. Let $(x_1, y_1) = (-2.2,\ 18.6)$
Let $(x_2, y_2) = (-1.7,\ 18.6)$

$m = \dfrac{y_2 - y_1}{x_2 - x_1}$

$= \dfrac{18.6 - 18.6}{-1.7 - (-2.2)}$

$= \dfrac{0}{-1.7 + 2.2}$

$= 0$

53. Let $(x_1, y_1) = \left(-\dfrac{4}{7}, -\dfrac{1}{5}\right)$

Let $(x_2, y_2) = \left(\dfrac{3}{7}, \dfrac{6}{5}\right)$

$m = \dfrac{y_2 - y_1}{x_2 - x_1}$

$= \dfrac{\dfrac{6}{5} - \left(-\dfrac{1}{5}\right)}{\dfrac{3}{7} - \left(-\dfrac{4}{7}\right)}$

$= \dfrac{\dfrac{6}{5} + \dfrac{1}{5}}{\dfrac{3}{7} + \dfrac{4}{7}}$

$= \dfrac{\dfrac{7}{5}}{\dfrac{7}{7}}$

$= \dfrac{7}{5}$

55. Let $(x_1, y_1) = \left(-\dfrac{3}{4}, \dfrac{2}{3}\right)$

Let $(x_2, y_2) = \left(\dfrac{4}{3}, -\dfrac{1}{6}\right)$

$m = \dfrac{y_2 - y_1}{x_2 - x_1}$

$= \dfrac{-\dfrac{1}{6} - \dfrac{2}{3}}{\dfrac{4}{3} - \left(-\dfrac{3}{4}\right)}$

$= \dfrac{-\dfrac{1}{6} - \dfrac{4}{6}}{\dfrac{16}{12} + \dfrac{9}{12}}$

$= \dfrac{-\dfrac{5}{6}}{\dfrac{25}{12}}$

$= -\dfrac{5}{6} \cdot \dfrac{12}{25}$

$= -\dfrac{5 \cdot 12}{6 \cdot 25}$

$= -\dfrac{\overset{1}{\cancel{5}} \cdot 2 \cdot \overset{1}{\cancel{6}}}{\underset{1}{\cancel{6}} \cdot \underset{1}{\cancel{5}} \cdot 5}$

$= -\dfrac{2}{5}$

Determine the slope of the graph of the line that has the given table of solutions.

57. Let $(x_1, y_1) = (-3, -1)$
Let $(x_2, y_2) = (1, 2)$

$m = \dfrac{y_2 - y_1}{x_2 - x_1}$

$= \dfrac{2 - (-1)}{1 - (-3)}$

$= \dfrac{2 + 1}{1 + 3}$

$= \dfrac{3}{4}$

59. Let $(x_1, y_1) = (-3, 6)$
Let $(x_2, y_2) = (0, 6)$

$m = \dfrac{y_2 - y_1}{x_2 - x_1}$

$= \dfrac{6 - 6}{0 - (-3)}$

$= \dfrac{0}{3}$

$= 0$

Find the slope of each line, if possible.

61. $y = -11$ Select any two points that lie on the horizontal line.

Let $(x_1, y_1) = (3, -11)$
Let $(x_2, y_2) = (5, -11)$

$m = \dfrac{y_2 - y_1}{x_2 - x_1}$

$= \dfrac{-11 - (-11)}{5 - 3}$

$= \dfrac{-11 + 11}{2}$

$= \dfrac{0}{2}$

$= 0$

63. $y = 0$ Select any two points that lie on the horizontal line.

Let $(x_1, y_1) = (3, 0)$
Let $(x_2, y_2) = (5, 0)$

$m = \dfrac{y_2 - y_1}{x_2 - x_1}$

$= \dfrac{0 - 0}{5 - 3}$

$= \dfrac{0}{2}$

$= 0$

Section 3.4

65. $x = 6$ Select any two points that lie
on the vertical line.
Let $(x_1, y_1) = (6, 4)$
Let $(x_2, y_2) = (6, 5)$

$$m = \frac{y_2 - y_1}{x_2 - x_1}$$

$$= \frac{5 - 4}{6 - 6}$$

$$= \frac{1}{0}$$

Undefined

67. $x = -10$ Select any two points that lie
on the vertical line.
Let $(x_1, y_1) = (-10, 4)$
Let $(x_2, y_2) = (-10, 5)$

$$m = \frac{y_2 - y_1}{x_2 - x_1}$$

$$= \frac{5 - 4}{-10 - (-10)}$$

$$= \frac{1}{-10 + 10}$$

$$= \frac{1}{0}$$

Undefined

69. $y - 9 = 0$ Solve for y.
   $y = 9$ Select any two points that lie
on the horizontal line.
Let $(x_1, y_1) = (3, 9)$
Let $(x_2, y_2) = (5, 9)$

$$m = \frac{y_2 - y_1}{x_2 - x_1}$$

$$= \frac{9 - 9}{5 - 3}$$

$$= \frac{0}{2}$$

$$= 0$$

71. $3x = -12$ Solve for x.
   $x = -4$ Select any two points that lie
on the vertical line.
Let $(x_1, y_1) = (-4, 4)$
Let $(x_2, y_2) = (-4, 5)$

$$m = \frac{y_2 - y_1}{x_2 - x_1}$$

$$= \frac{5 - 4}{-4 - (-4)}$$

$$= \frac{1}{-4 + 4}$$

$$= \frac{1}{0}$$

Undefined

Determine whether the lines through each pair of points are parallel, perpendicular, or neither.

73. Step 1: Determine the slopes.

Let $(x_1, y_1) = (5, 3)$ | Let $(x_1, y_1) = (-3, -4)$
Let $(x_2, y_2) = (1, 4)$ | Let $(x_2, y_2) = (1, -5)$

$$m_1 = \frac{y_2 - y_1}{x_2 - x_1} \qquad m_2 = \frac{y_2 - y_1}{x_2 - x_1}$$

$$= \frac{4 - 3}{1 - 5} \qquad = \frac{-5 - (-4)}{1 - (-3)}$$

$$= -\frac{1}{4} \qquad = \frac{-5 + 4}{1 + 3}$$

$$\qquad = -\frac{1}{4}$$

Step 2: Compare the slopes.
The slopes are equal.
The lines are parallel.

75. Step 1: Determine the slopes.

Let $(x_1, y_1) = (-4, -2)$ | Let $(x_1, y_1) = (7, 1)$
Let $(x_2, y_2) = (2, -3)$ | Let $(x_2, y_2) = (8, 7)$

$$m_1 = \frac{y_2 - y_1}{x_2 - x_1} \qquad m_2 = \frac{y_2 - y_1}{x_2 - x_1}$$

$$= \frac{-3 - (-2)}{2 - (-4)} \qquad = \frac{7 - 1}{8 - 7}$$

$$= \frac{-3 + 2}{2 + 4} \qquad = \frac{6}{1}$$

$$= -\frac{1}{6} \qquad = 6$$

Step 2: Compare the slopes.
The slopes are negative reciprocals.
The lines are perpendicular.

77. Step 1: Determine the slopes.

Let $(x_1, y_1) = (2,2)$ | Let $(x_1, y_1) = (-3,4)$
Let $(x_2, y_2) = (4,-3)$ | Let $(x_2, y_2) = (-1,9)$

$$m_1 = \frac{y_2 - y_1}{x_2 - x_1}$$ | $$m_2 = \frac{y_2 - y_1}{x_2 - x_1}$$

$$= \frac{-3-2}{4-2}$$ | $$= \frac{9-4}{-1-(-3)}$$

$$= -\frac{5}{2}$$ | $$= \frac{5}{-1+3}$$

| $$= \frac{5}{2}$$

Step 2: Compare the slopes.

The slopes are not equal.
The slopes are not negative reciprocals.
The lines are neither parallel nor perpendicular.

79. Step 1: Determine the slopes.

Let $(x_1, y_1) = (-1,8)$ | Let $(x_1, y_1) = (3,3)$
Let $(x_2, y_2) = (-6,8)$ | Let $(x_2, y_2) = (3,7)$

$$m_1 = \frac{y_2 - y_1}{x_2 - x_1}$$ | $$m_2 = \frac{y_2 - y_1}{x_2 - x_1}$$

$$= \frac{8-8}{-6-(-1)}$$ | $$= \frac{7-3}{3-3}$$

$$= \frac{0}{-6+1}$$ | $$= \frac{4}{0}$$

$$= 0$$ | Undefined

Step 2: Compare the slopes.

The slopes are of a horizontal and a vertical line. The lines are perpendicular.

81. Step 1: Determine the slopes.

Let $(x_1, y_1) = (6,4)$ | Let $(x_1, y_1) = (-2,-3)$
Let $(x_2, y_2) = (2,5)$ | Let $(x_2, y_2) = (2,-4)$

$$m_1 = \frac{y_2 - y_1}{x_2 - x_1}$$ | $$m_2 = \frac{y_2 - y_1}{x_2 - x_1}$$

$$= \frac{5-4}{2-6}$$ | $$= \frac{-4-(-3)}{2-(-2)}$$

$$= -\frac{1}{4}$$ | $$= \frac{-4+3}{2+2}$$

| $$= -\frac{1}{4}$$

Step 2: Compare the slopes.
The slopes are equal.
The lines are parallel.

83. Step 1: Determine the slopes.

Let $(x_1, y_1) = (4,2)$ | Let $(x_1, y_1) = (-5,3)$
Let $(x_2, y_2) = (5,-3)$ | Let $(x_2, y_2) = (-2,9)$

$$m_1 = \frac{y_2 - y_1}{x_2 - x_1}$$ | $$m_2 = \frac{y_2 - y_1}{x_2 - x_1}$$

$$= \frac{-3-2}{5-4}$$ | $$= \frac{9-3}{-2-(-5)}$$

$$= -5$$ | $$= \frac{6}{-2+5}$$

| $$= \frac{6}{3}$$

| $$= 2$$

Step 2: Compare the slopes.
The slopes are not equal.
The slopes are not negative reciprocals.
The lines are neither parallel nor perpendicular.

Find the slope of a line perpendicular to the line passing through the given two points.

85. Step 1: Determine the slope.

Let $(x_1, y_1) = (0,0)$
Let $(x_2, y_2) = (5,-9)$

$$m = \frac{y_2 - y_1}{x_2 - x_1}$$

$$= \frac{-9-0}{5-0}$$

$$= -\frac{9}{5}$$

Step 2: Determine the negative reciprocal.

The negative reciprocal is $\frac{5}{9}$.

Section 3.4

87. Step 1: Determine the slope.

Let $(x_1, y_1) = (-1, 7)$
Let $(x_2, y_2) = (1, 10)$

$$m = \frac{y_2 - y_1}{x_2 - x_1}$$

$$= \frac{10 - 7}{1 - (-1)}$$

$$= \frac{3}{1 + 1}$$

$$= \frac{3}{2}$$

Step 2: Determine the negative reciprocal.

The negative reciprocal is $-\dfrac{2}{3}$.

89. Step 1: Determine the slope.

Let $(x_1, y_1) = \left(-2, \dfrac{1}{2}\right)$

Let $(x_2, y_2) = \left(-1, \dfrac{3}{2}\right)$

$$m = \frac{y_2 - y_1}{x_2 - x_1}$$

$$= \frac{\dfrac{3}{2} - \dfrac{1}{2}}{-1 - (-2)}$$

$$= \frac{\dfrac{2}{2}}{-1 + 2}$$

$$= \frac{1}{1}$$

$$= 1$$

Step 2: Determine the negative reciprocal.
The negative reciprocal is -1.

91. Step 1: Determine the slope.

Let $(x_1, y_1) = (-1, 2)$
Let $(x_2, y_2) = (-3, 6)$

$$m = \frac{y_2 - y_1}{x_2 - x_1}$$

$$= \frac{6 - 2}{-3 - (-1)}$$

$$= \frac{4}{-3 + 1}$$

$$= \frac{4}{-2}$$

$$= -2$$

Step 2: Determine the negative reciprocal.

The negative reciprocal is $\dfrac{1}{2}$.

APPLICATIONS

93. POOL DESIGN

Step 1: Determine the rise of the slope by finding the difference of the two given vertical heights. **9 ft – 3 ft = 6 ft**

Step 2: Determine the run of the slope by looking at the distance between the two depths. **15 ft**

Compute the ratio.

$$m = \frac{\text{rise}}{\text{run}}$$

$$= \frac{6}{15}$$

$$= \frac{2}{5}$$

Step 3: The slope falls from left to right thus indicating it is negative.

$$m = -\frac{2}{5}$$

95. GRADE OF A ROAD

$$m = \frac{\text{rise}}{\text{run}}$$

$$= \frac{264}{5,280}$$

$$= \frac{1}{20}$$

Change the fraction to a percent by dividing:

$$\frac{1}{20} = 0.05$$

$$= 5\%$$

97. TREADMILLS

The height setting is the change in y and the length of the treadmill is the change in x.

For 6 inches:

$$m = \frac{\text{rise}}{\text{run}}$$

$$= \frac{6}{50}$$

$$= \frac{3}{25}$$

$$= 0.12$$

$$= 12\%$$

99. CARPENTRY

The pitch of the front of the roof is

$$\frac{9}{6} = \frac{3}{2}$$

The pitch of the side of the roof is

$$\frac{9}{15} = \frac{3}{5}$$

101. IRRIGATION

Find an order pair for each of the endpoints: (0, 8,000) and (8, 1,000). Now apply the slope formula:

Let $(x_1, y_1) = (0, 8,000)$ Let $(x_2, y_2) = (8, 1,000)$

$$m = \frac{y_2 - y_1}{x_2 - x_1}$$

$$= \frac{1,000 - 8,000}{8 - 0}$$

$$= \frac{-7,000}{8}$$

$$= -875$$

The rate of change would be −875 gal per hour.

103. MILK PRODUCTION

Find an order pair for each of the endpoints: (1996, 16,433) and (2009, 20,580). Now apply the slope formula:

Let $(x_1, y_1) = (1996, \ 16,433)$
Let $(x_2, y_2) = (2009, \ 20,580)$

$$m = \frac{y_2 - y_1}{x_2 - x_1}$$

$$= \frac{20,580 - 16,433}{2009 - 1996}$$

$$= \frac{4,147}{13}$$

$$= 319$$

The rate of change would be 319 lb per year.

From Campus to Careers
105. DENTAL ASSISTANT

Find an order pair for each of the endpoints: (2000, 6,600) and (2008, 9,200). Now apply the slope formula:

Let $(x_1, y_1) = (2000, \ 6,600)$
Let $(x_2, y_2) = (2008, \ 9,200)$

$$m = \frac{y_2 - y_1}{x_2 - x_1}$$

$$= \frac{9,200 - 6,600}{2008 - 2000}$$

$$= \frac{2,600}{8}$$

$$= 325$$

There is an increase of 325 students per year.

WRITING

107 − 109. Answers will vary.

REVIEW

111. HALLOWEEN CANDY

Analyze

- Black licorice sells for $1.90 per lb.
- Orange gumdrops sells for $2.20 per lb.
- A mixture of the two is to sell for $2 per lb.
- A mixture of 60 pounds is needed.
- How many pounds of each is needed?

Assign

Let x = the # of lbs of black licorice

$60 - x$ = the # of lbs of orange gumdrops

Form

	Number	• Value =	Total value
Licorice	x	1.90	**1.90x**
Gumdrops	60 - x	2.20	**2.20(60 - x)**
Blend	60	2.00	**60(2)**

The value of the black licorice	plus	the value of the orange gumdrops	equals	the total value of the blend.
$1.90x$	+	$2.20(60 - x)$	=	$60(2)$

Solve

$$1.90x + 2.20(60 - x) = 60(2)$$
$$1.90x + 132 - 2.20x = 120$$
$$-0.30x + 132 = 120$$
$$-0.30x + 132 - \mathbf{132} = 120 - \mathbf{132}$$
$$-0.30x = -12$$
$$\frac{-0.30x}{\mathbf{-0.30}} = \frac{-12}{\mathbf{-0.30}}$$
$$x = 40$$

gumdrops
$$60 - x = 60 - 40$$
$$= 20$$

State

40 lb of black licorice will be needed.
20 lb of orange gumdrops will be needed.

Check

The value of the licorice is 40($1.90), or $76.
The value of the gumdrops is 20($2.20), or $44.
The value of the blend is 60($2), or $120.
Since the total was $76 + $44 = $120, the results check.

CHALLENGE

113. Find the slope from A to B.

Let $(x_1, y_1) = A(-50, -10)$
Let $(x_2, y_2) = B(20, 0)$

$$m_{\overline{AB}} = \frac{y_2 - y_1}{x_2 - x_1}$$
$$= \frac{0 - (-10)}{20 - (-50)}$$
$$= \frac{10}{70}$$
$$= \frac{1}{7}$$

Find the slope from A to C.

Let $(x_1, y_1) = A(-50, -10)$
Let $(x_2, y_2) = C(34, 2)$

$$m_{\overline{AC}} = \frac{y_2 - y_1}{x_2 - x_1}$$
$$= \frac{2 - (-10)}{34 - (-50)}$$
$$= \frac{12}{84}$$
$$= \frac{1}{7}$$

Find the slope from C to B.

Let $(x_1, y_1) = C(34, 2)$
Let $(x_2, y_2) = B(20, 0)$

$$m_{\overline{CB}} = \frac{y_2 - y_1}{x_2 - x_1}$$
$$= \frac{0 - 2}{20 - 34}$$
$$= \frac{-2}{-14}$$
$$= \frac{1}{7}$$

The slopes are the same; therefore all three points do lie on the same line.

SECTION 3.5
VOCABULARY

Fill in the blanks.

1. The equation $y = mx + b$ is called the <u>**slope–**</u> <u>**intercept**</u> form of the equation of a line.

CONCEPTS

3. a. $7x + 4y = 2$ **No**   b. $5y = 2x - 3$ **No**

   c. $y = 6x + 1$ **Yes**   d. $x = 4y - 8$ **No**

5. **Simplify the right side of each equation.**

   a. $y = \dfrac{4x}{2} + \dfrac{16}{2}$ b. $y = \dfrac{15x}{-3} + \dfrac{9}{-3}$

     $y = 2x + 8$ $y = -5x - 3$

   c. $y = \dfrac{2x}{6} - \dfrac{6}{6}$ d. $y = \dfrac{-9x}{-5} - \dfrac{20}{-5}$

     $y = \dfrac{1}{3}x - 1$ $y = \dfrac{9}{5}x + 4$

NOTATION

7.
$$2x + 5y = 15$$
$$2x + 5y - 2x = \boxed{-2x} + 15$$
$$\boxed{5y} = -2x + 15$$
$$\dfrac{5y}{\boxed{5}} = \dfrac{-2x}{\boxed{5}} + \dfrac{15}{\boxed{5}}$$
$$y = \boxed{-\dfrac{2}{5}}x + \boxed{3}$$

The slope is $\boxed{-\frac{2}{5}}$ and the y-intercept is $\boxed{(\boxed{0}, \boxed{3})}$.

9. $-\dfrac{3}{2} = \dfrac{3}{\boxed{-2}} = \dfrac{\boxed{-3}}{2}$

GUIDED PRACTICE

Find the slope and the y - intercept of the line with the given equation.

11. The slope is 4. The y-intercept is $(0,2)$.

13. The slope is -5. The y-intercept is $(0,-8)$.

15. The slope is 25. The y-intercept is $(0,-9)$.

17. The slope is -1. The y-intercept is $(0,11)$.

19. The slope is $\dfrac{1}{2}$. The y-intercept is $(0,6)$.

21. The slope is $\dfrac{1}{4}$. The y-intercept is $\left(0, -\dfrac{1}{2}\right)$.

23. The slope is -5. The y-intercept is $(0,0)$.

25. The slope is 1. The y-intercept is $(0,0)$.

27. The slope is 0. The y-intercept is $(0,-2)$.

29.
$$-5y - 2 = 0$$
$$-5y - 2 + 2 = 0 + 2$$
$$-5y = 2$$
$$\dfrac{-5y}{-5} = \dfrac{2}{-5}$$
$$y = -\dfrac{2}{5}$$

The slope is 0. The y-intercept is $\left(0, -\dfrac{2}{5}\right)$.

31.
$$x + y = 8$$
$$x + y - x = 8 - x$$
$$y = -x + 8$$

The slope is -1. The y-intercept is $(0,8)$.

33.
$$6y = x - 6$$
$$\dfrac{6y}{6} = \dfrac{x}{6} - \dfrac{6}{6}$$
$$y = \dfrac{x}{6} - 1$$

The slope is $\dfrac{1}{6}$. The y-intercept is $(0,-1)$.

35.
$$-4y = 6x - 4$$
$$\dfrac{-4y}{-4} = \dfrac{6x}{-4} - \dfrac{4}{-4}$$
$$y = -\dfrac{3}{2}x + 1$$

The slope is $-\dfrac{3}{2}$. The y-intercept is $(0,1)$.

37.
$$2x + 3y = 6$$
$$2x + 3y - 2x = 6 - 2x$$
$$3y = -2x + 6$$
$$\dfrac{3y}{3} = \dfrac{-2x}{3} + \dfrac{6}{3}$$
$$y = -\dfrac{2}{3}x + 2$$

The slope is $-\dfrac{2}{3}$. The y-intercept is $(0,2)$.

39.
$$3x - 5y = 15$$
$$3x - 5y - 3x = 15 - 3x$$
$$-5y = -3x + 15$$
$$\dfrac{-5y}{-5} = \dfrac{-3x}{-5} + \dfrac{15}{-5}$$
$$y = \dfrac{3}{5}x - 3$$

The slope is $\dfrac{3}{5}$. The y-intercept is $(0,-3)$.

41.
$$-6x + 6y = -11$$
$$-6x + 6y + 6x = -11 + 6x$$
$$6y = 6x - 11$$
$$\frac{6y}{6} = \frac{6x}{6} - \frac{11}{6}$$
$$y = x - \frac{11}{6}$$

The slope is 1. The y-intercept is $\left(0, -\frac{11}{6}\right)$.

Write an equation of the line with the given slope and y-intercept and graph it.

43. Slope 5, y-intercept $(0, -3)$

The equation of the line is $y = 5x - 3$.

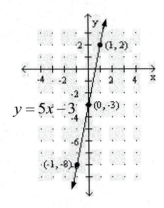

45. Slope -3, y-intercept $(0, 6)$

The equation of the line is $y = -3x + 6$.

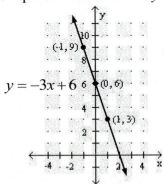

47. Slope $\frac{1}{4}$, y-intercept $(0, -2)$

The equation of the line is $y = \frac{1}{4}x - 2$.

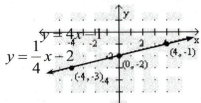

49. Slope $-\frac{8}{3}$, y-intercept $(0, 5)$

The equation of the line is $y = -\frac{8}{3}x + 5$.

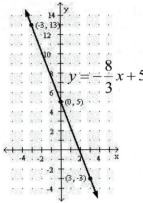

51. Slope $\frac{6}{5}$, y-intercept $(0, 0)$

The equation of the line is $y = \frac{6}{5}x$.

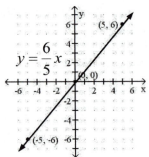

53. Slope -2, y-intercept $\left(0, \frac{1}{2}\right)$

The equation of the line is $y = -2x + \frac{1}{2}$.

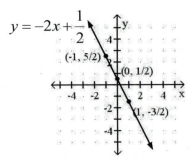

Write an equation for each line shown.

55. Slope 5, y-intercept $(0, -1)$

The equation of the line is $y = 5x - 1$.

57. Slope -2, y-intercept $(0, 3)$

The equation of the line is $y = -2x + 3$.

59. Slope $\dfrac{4}{5}$, y-intercept $(0, -2)$

The equation of the line is $y = \dfrac{4}{5}x - 2$.

61. Slope $-\dfrac{5}{3}$, y-intercept $(0, 2)$

The equation of the line is $y = -\dfrac{5}{3}x + 2$.

Find the slope and the y-intercept of the graph of each equation and graph it.

63. $y = 3x + 3$

$m = 3$, y-intercept $= (0, 3)$

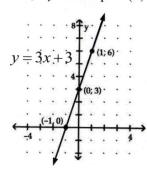

65. $y = \dfrac{1}{2}x + 2$

$m = \dfrac{1}{2}$, y-intercept $= (0, 2)$

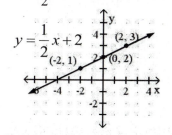

67. $y = -3x$

$m = -3$, y-intercept $= (0, 0)$

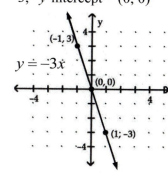

69.
$$4x + y = -4$$
$$4x + y - \mathbf{4x} = -4 - \mathbf{4x}$$
$$y = -4x - 4$$
$$m = -4, \quad y\text{-intercept} = (0, -4)$$

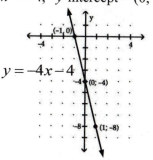

71.
$$3x + 4y = 16$$
$$3x + 4y - \mathbf{3x} = 16 - \mathbf{3x}$$
$$4y = -3x + 16$$
$$\dfrac{4y}{4} = \dfrac{-3x}{4} + \dfrac{16}{4}$$
$$y = -\dfrac{3}{4}x + 4$$
$$m = -\dfrac{3}{4}, \quad y\text{-intercept} = (0, 4)$$

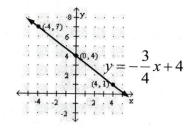

73.
$$10x - 5y = 5$$
$$10x - 5y - \mathbf{10x} = 5 - \mathbf{10x}$$
$$-5y = -10x + 5$$
$$\dfrac{-5y}{-5} = \dfrac{-10x}{-5} + \dfrac{5}{-5}$$
$$y = 2x - 1$$
$$m = 2, \quad y\text{-intercept} = (0, -1)$$

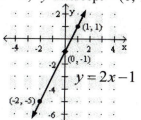

For each pair of equations, determine whether their graphs are parallel, perpendicular, or neither.

75. $y = 6x + 8$ | $y = 6x$

$m = 6$ | $m = 6$

The slopes are equal.
The lines are parallel.

77. $y = x$ | $y = -x$

$m = 1$ | $m = -1$

The slopes are negative reciprocals.
The lines are perpendicular.

79. $y = \dfrac{1}{2}x - \dfrac{4}{5}$ | $y = 0.5x + 3$

$m = \dfrac{1}{2}$ | $m = 0.5$

The slopes are equal.
The lines are parallel.

81. $y = -2x - 9$ | $2x - y = 9$

$m = -2$ | $2x - y + y = 9 + y$

$2x = y + 9$

$2x - 9 = y + 9 - 9$

$y = 2x - 9$

$m = 2$

The slopes are not equal.
The slopes are not negative reciprocals.
The lines are neither parallel nor perpendicular.

83. $3x = 5y - 10$ | $5x = 1 - 3y$

$3x + 10 = 5y - 10 + 10$ | $5x - 1 = 1 - 3y - 1$

$3x + 10 = 5y$ | $5x - 1 = -3y$

$\dfrac{3x}{5} + \dfrac{10}{5} = \dfrac{5y}{5}$ | $\dfrac{5x}{-3} - \dfrac{1}{-3} = \dfrac{-3y}{-3}$

$y = \dfrac{3}{5}x + 2$ | $y = -\dfrac{5}{3}x + \dfrac{1}{3}$

$m = \dfrac{3}{5}$ | $m = -\dfrac{5}{3}$

The slopes are negative reciprocals.
The lines are perpendicular.

85. $x - y = 12$ | $-2x + 2y = -23$

$x - y - x = 12 - x$ | $-2x + 2y + 2x = -23 + 2x$

$-y = -x + 12$ | $2y = 2x - 23$

$\dfrac{-y}{-1} = \dfrac{-x}{-1} + \dfrac{12}{-1}$ | $\dfrac{2y}{2} = \dfrac{2x}{2} - \dfrac{23}{2}$

$y = x - 12$ | $y = x - \dfrac{23}{2}$

$m = 1$ | $m = 1$

The slopes are equal.
The lines are parallel.

87. $x = 9$ | $y = 8$

$m =$ undefined | $m = 0$

The slopes are of a horizontal and a vertical line. The lines are perpendicular.

89. $-4x + 3y = -12$ | $8x + 6y = 54$

$-4x + 3y + 4x = -12 + 4x$ | $8x + 6y - 8x = 54 - 8x$

$3y = 4x - 12$ | $6y = -8x + 54$

$\dfrac{3y}{3} = \dfrac{4x}{3} - \dfrac{12}{3}$ | $\dfrac{6y}{6} = \dfrac{-8x}{6} + \dfrac{54}{6}$

$y = \dfrac{4}{3}x - 4$ | $y = -\dfrac{4}{3}x + 9$

$m = \dfrac{4}{3}$ | $m = -\dfrac{4}{3}$

The slopes are not equal.
The slopes are not negative reciprocals.
The lines are neither parallel nor perpendicular.

APPLICATIONS

91. PRODUCTION COSTS

a. basic fee is $5,000 (constant)
extra cost is $2,000/hr (not constant)
total hours used in not known (h)
total cost is not known (c)

Total product cost	equals	extra cost of $2,000/hr	plus	basic fee of $5,000.
c	$=$	$2,000h$	$+$	$5,000$

b. The cost for 8 hours of filming is found by replacing h with 8 and then calculating the results.

$c = 2,000h + 5,000$

$c = 2,000(8) + 5,000$

$c = 16,000 + 5,000$

$c = 21,000$

The total cost for the 8 hours is $21,000.

93. CHEMISTRY
$F = 5t - 10$

95. EMPLOYMENT SERVICE
$c = -20m + 500$

97. SEWING COSTS
$c = 5x + 20$

99. iPADS

a. $c = 14.95m + 629.99$

b. $c = 14.95m + 629.99$

$m = 24$ months (2 years)

$c = 14.95(24) + 629.99$

$c = 358.80 + 629.99$

$c = 988.79$

The total cost for the iPad after 2 years
is \$988.79.

101. NAVIGATION

a. When there are no head waves, the
ship can travel at 18 knots.

b. Points (0,18) and (4,16) are to be used
to calculate the rate of change.

$$m = \frac{y_2 - y_1}{x_2 - x_1}$$

$$m = \frac{16 - 18}{4 - 0}$$

$$m = \frac{-2}{4}$$

$$m = -\frac{1}{2}$$

The rate of change is $-\frac{1}{2}$ knot per foot.

c. $y = -\frac{1}{2}x + 18$

WRITING
103. Answers will vary.

REVIEW
105. CABLE TV

Analyze

• A 186-foot cable is cut into four pieces.

• Each successive piece is 3 feet longer than
the previous one.

• Find the length of each piece.

Assign

Let x = length of 1st piece in feet

$x + 3$ = length of 2nd piece in feet

$x + 6$ = length of 3rd piece in feet

$x + 9$ = length of 4th piece in feet

Form

The length of the 1st piece	plus	the length of the 2nd piece	plus	the length of the 3rd piece	plus	the length of the 4th piece	equals	the total length.
x	+	$x+3$	+	$x+6$	+	$x+9$	=	186

Solve

$$x + (x+3) + (x+6) + (x+9) = 186$$

$$4x + 18 = 186$$

$$4x + 18 - \mathbf{18} = 186 - \mathbf{18}$$

$$4x = 168$$

$$\frac{4x}{4} = \frac{168}{4}$$

$$x = 42$$

2nd piece	3rd piece	4th piece
$x+3 = \mathbf{42}+3$	$x+6 = \mathbf{42}+6$	$x+9 = \mathbf{42}+9$
$= 45$	$= 48$	$= 51$

State

Length of 1st piece is 42 ft.

Length of 2nd piece is 45 ft.

Length of 3rd piece is 48 ft.

Length of 4th piece is 51 ft.

Check

42 ft $+ 45$ ft $+ 48$ ft $+ 51$ ft $= 186$ ft

$$186 \text{ ft} = 186 \text{ ft}$$

The results check.

CHALLENGE PROBLEMS

107. If one would draw several lines such that
they went through quadrants I, II, and IV,
one would see that the **slopes** of all the
lines would be **negative** and all the
y-intercepts would be **positive**.

$$m < 0; b > 0$$

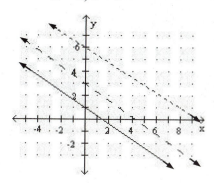

Section 3.5

SECTION 3.6
VOCABULARY

Fill in the blanks.

1. $y - y_1 = m(x - x_1)$ is called the **point–slope** form of the equation of a line. In words, we read this as y minus y **sub** one equals m **times** the quantity of x **minus** x sub **one**.

CONCEPTS

3. Determine in what form each equation is written.

   a. $y - 4 = 2(x - 5)$ **Point–slope**

   b. $y = 2x + 15$ **Slope–intercept**

5. Refer to the following graph of a line.

   a. What highlighted point does the line pass through?   **(– 2, – 3)**

   b. What is the slope of the line? $\dfrac{5}{6}$

   c. Write an equation of the line in point–slope form. $y + 3 = \dfrac{5}{6}(x + 2)$

7. Suppose you are asked to write an equation of the line in the scatter diagram below. What two points would you use to write the point–slope equation?

   (67, 170), (79, 220)

NOTATION

Complete the solution.

9. Find an equation of the line with slope –2 that passes through the point (–1, 5). Write the answer in slope–intercept form.

$$y - y_1 = m(x - x_1)$$
$$y - \mathbf{5} = -2[x - (\mathbf{-1})]$$
$$y - 5 = -2[x + 1]$$
$$y - 5 = -2x - \mathbf{2}$$
$$y = -2x + \mathbf{3}$$

11. Consider the steps below and then fill in the blanks:

$$y - 3 = 2(x + 1)$$
$$y - 3 = 2x + 2$$
$$y = 2x + 5$$

   The original equation was in **point–slope** form. After solving for y, we obtain an equation in **slope–intercept** form.

GUIDED PRACTICE

Use the point - slope form to write an equation of the line with the given slope and point.

13. Passes through $(2,\ 1)$, $m = 3$

$$x_1 = 2 \text{ and } y_1 = 1$$
$$y - y_1 = m(x - x_1)$$
$$y - 1 = 3(x - 2)$$

15. Passes through $(-5, -1)$, $m = \dfrac{4}{5}$

$$x_1 = -5 \text{ and } y_1 = -1$$
$$y - y_1 = m(x - x_1)$$
$$y - (-1) = \frac{4}{5}(x - (-5))$$
$$y + 1 = \frac{4}{5}(x + 5)$$

Use the point - slope form to find an equation of the line with the given slope and point. Then write the equation in slope - intercept form. See Example 1.

17. Passes through $(3,\ 5)$, $m = 2$

$$x_1 = 3 \text{ and } y_1 = 5$$
$$y - y_1 = m(x - x_1)$$
$$y - 5 = 2(x - 3) \text{ point-slope form}$$

$$y - 5 = 2(x - 3)$$
$$y - 5 = 2x - 6$$
$$y - 5 + \mathbf{5} = 2x - 6 + \mathbf{5}$$
$$y = 2x - 1 \text{ slope-intercept form}$$

19. Passes through $(-9,\ 8)$, $m = -5$

$$x_1 = -9 \text{ and } y_1 = 8$$
$$y - y_1 = m(x - x_1)$$
$$y - 8 = -5(x - (-9))$$
$$y - 8 = -5(x + 9) \text{ point-slope form}$$

$$y - 8 = -5(x + 9)$$
$$y - 8 = -5x - 45$$
$$y - 8 + \mathbf{8} = -5x - 45 + \mathbf{8}$$
$$y = -5x - 37 \text{ slope-intercept form}$$

21. Passes through $(0,\ 0)$, $m = -3$

$$x_1 = 0 \text{ and } y_1 = 0$$
$$y - y_1 = m(x - x_1)$$
$$y - 0 = -3(x - 0)$$
$$y = -3x \text{ slope-intercept form}$$

23. Passes through $(10, 1)$, $m = \dfrac{1}{5}$

$$x_1 = 10 \text{ and } y_1 = 1$$
$$y - y_1 = m(x - x_1)$$
$$y - 1 = \frac{1}{5}(x - 10) \text{ point-slope form}$$

$$y - 1 = \frac{1}{5}(x - 10)$$
$$y - 1 = \frac{1}{5}x - 2$$
$$y - 1 + \mathbf{1} = \frac{1}{5}x - 2 + \mathbf{1}$$
$$y = \frac{1}{5}x - 1 \text{ slope-intercept form}$$

25. Passes through $(6, -4)$, $m = -\dfrac{4}{3}$

$$x_1 = 6 \text{ and } y_1 = -4$$
$$y - y_1 = m(x - x_1)$$
$$y - (-4) = -\frac{4}{3}(x - 6)$$
$$y + 4 = -\frac{4}{3}(x - 6) \text{ point-slope form}$$

$$y + 4 = -\frac{4}{3}(x - 6)$$
$$y + 4 = -\frac{\mathbf{4}}{\mathbf{3}}x - \left(-\frac{\mathbf{4}}{\mathbf{3}}\right)6$$
$$y + 4 = -\frac{4}{3}x + 8$$
$$y + 4 - \mathbf{4} = -\frac{4}{3}x + 8 - \mathbf{4}$$
$$y = -\frac{4}{3}x + 4 \text{ slope-intercept form}$$

27. Passes through $(2, -6)$, $m = -\dfrac{11}{6}$

$$x_1 = 2 \text{ and } y_1 = -6$$
$$y - y_1 = m(x - x_1)$$
$$y - (-6) = -\frac{11}{6}(x - 2)$$
$$y + 6 = -\frac{11}{6}(x - 2) \text{ point-slope form}$$

$$y + 6 = -\frac{11}{6}(x - 2)$$
$$y + 6 = -\frac{\mathbf{11}}{\mathbf{6}}x - \left(-\frac{\mathbf{11}}{\mathbf{6}}\right)2$$
$$y + 6 = -\frac{11}{6}x + \frac{11}{3}$$
$$y + 6 - \mathbf{6} = -\frac{11}{6}x + \frac{11}{3} - \frac{\mathbf{18}}{\mathbf{3}}$$
$$y = -\frac{11}{6}x - \frac{7}{3} \text{ slope-intercept form}$$

Find an equation of the line that passes through the two given points. Write the equation in slope - intercept form. See Example 2.

29. $(1, 7)$ and $(-2, 1)$

(x_1, y_1) and (x_2, y_2)

$$m = \frac{y_2 - y_1}{x_2 - x_1}$$
$$= \frac{1 - 7}{-2 - 1}$$
$$= \frac{-6}{-3}$$
$$= 2$$

Passes through $(1, \ 7)$, $m = 2$

$$x_1 = 1 \text{ and } y_1 = 7$$
$$y - y_1 = m(x - x_1)$$
$$y - 7 = 2(x - 1)$$
$$y - 7 = 2x - 2$$
$$y - 7 + \mathbf{7} = 2x - 2 + \mathbf{7}$$
$$y = 2x + 5 \text{ slope-intercept form}$$

31. $(-4, 3)$ and $(2, 0)$
(x_1, y_1) and (x_2, y_2)

$$m = \frac{y_2 - y_1}{x_2 - x_1}$$

$$= \frac{0 - 3}{2 - (-4)}$$

$$= \frac{-3}{6}$$

$$= -\frac{1}{2}$$

Passes through $(2, 0)$, $m = -\frac{1}{2}$

$x_1 = 2$ and $y_1 = 0$

$$y - y_1 = m(x - x_1)$$

$$y - 0 = -\frac{1}{2}(x - 2)$$

$$y = -\frac{1}{2}x - \frac{1}{2}(-2)$$

$$y = -\frac{1}{2}x + 1 \text{ slope-intercept form}$$

33. $(5, 5)$ and $(7, 5)$
(x_1, y_1) and (x_2, y_2)

$$m = \frac{y_2 - y_1}{x_2 - x_1}$$

$$= \frac{5 - 5}{7 - 5}$$

$$= \frac{0}{2}$$

$$= 0$$

Passes through $(5, 5)$, $m = 0$

$x_1 = 5$ and $y_1 = 5$

$$y - y_1 = m(x - x_1)$$

$$y - 5 = 0(x - 5)$$

$$y - 5 = 0$$

$$y - 5 + 5 = 0 + 5$$

$$y = 5$$

35. $(5, 1)$ and $(-5, 0)$
(x_1, y_1) and (x_2, y_2)

$$m = \frac{y_2 - y_1}{x_2 - x_1}$$

$$= \frac{0 - 1}{-5 - 5}$$

$$= \frac{-1}{-10}$$

$$= \frac{1}{10}$$

Passes through $(5, 1)$, $m = \frac{1}{10}$

$x_1 = 5$ and $y_1 = 1$

$$y - y_1 = m(x - x_1)$$

$$y - 1 = \frac{1}{10}(x - 5)$$

$$y - 1 = \frac{1}{10}x - \frac{1}{10}(5)$$

$$y - 1 = \frac{1}{10}x - \frac{1}{2}$$

$$y - 1 + 1 = \frac{1}{10}x - \frac{1}{2} + \frac{2}{2}$$

$$y = \frac{1}{10}x + \frac{1}{2} \text{ slope-intercept form}$$

37. $(-8, 2)$ and $(-8, 17)$
(x_1, y_1) and (x_2, y_2)

$$m = \frac{y_2 - y_1}{x_2 - x_1}$$

$$= \frac{17 - 2}{-8 - (-8)}$$

$$= \frac{15}{0}$$

Undefined

Passes through $(-8, 2)$, $m = $ undefined

$x_1 = -8$ and $y_1 = 2$

$$x = -8$$

39. $\left(\dfrac{2}{3}, \dfrac{1}{3}\right)$ and $(0,0)$

(x_1, y_1) and (x_2, y_2)

$$m = \dfrac{y_2 - y_1}{x_2 - x_1}$$

$$= \dfrac{0 - \dfrac{1}{3}}{0 - \dfrac{2}{3}}$$

$$= \dfrac{-\dfrac{1}{3}}{-\dfrac{2}{3}}$$

$$= -\dfrac{1}{\cancel{3}}\left(-\dfrac{\overset{1}{\cancel{3}}}{2}\right)$$

$$= \dfrac{1}{2}$$

Passes through $(0,0)$, $m = \dfrac{1}{2}$

$x_1 = 0$ and $y_1 = 0$

$$y - y_1 = m(x - x_1)$$

$$y - 0 = \dfrac{1}{2}(x - 0)$$

$$y = \dfrac{1}{2}x \text{ \small slope-intercept form}$$

Write an equation of each line. See Example 3.

41. Vertical, passes through $(4,5)$
    The equation of a vertical line can be
    written in the form $x = a$.
    $$x = 4$$

43. Horizontal, passes through $(4,5)$
    The equation of a horizontal line can be
    written in the form $y = b$.
    $$y = 5$$

Graph the line that passes through the given point and has the given slope. See Example 4.

45. $(1, -2)$, slope $-1 = \dfrac{-1}{1}$

    Start at $(1, -2)$, go down 1, right 1

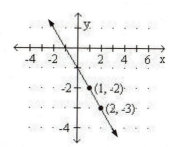

47. $(5, -3)$, $m = \dfrac{3}{4}$

    Start at $(5, -3)$, go up 3, right 4

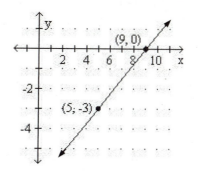

49. $(-2, -3)$, $m = 2 = \dfrac{2}{1}$

    Start at $(-2, -3)$, go up 2, right 1

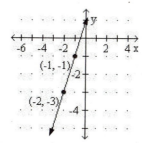

Section 3.6

51. $(4, -3)$, $m = -\dfrac{7}{8} = \dfrac{-7}{8}$

Start at $(4, -3)$, go down 7, right 8

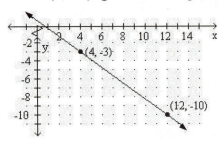

TRY IT YOURSELF
Use either the slope - intercept form (from Section 3.5) or the point - slope form (from Section 3.6) to find an equation of each line. Write each result in slope - intercept form, if possible.

53. $(5, 0)$ and $(-11, -4)$
$\quad (x_1, y_1)$ and $\quad (x_2, y_2)$

$$m = \dfrac{y_2 - y_1}{x_2 - x_1}$$

$$= \dfrac{-4 - 0}{-11 - 5}$$

$$= \dfrac{-4}{-16}$$

$$= \dfrac{1}{4}$$

Passes through $(5, 0)$, $m = \dfrac{1}{4}$

$x_1 = 5$ and $y_1 = 0$

$$y - y_1 = m(x - x_1)$$

$$y - 0 = \dfrac{1}{4}(x - 5)$$

$$y = \dfrac{\mathbf{1}}{\mathbf{4}}x - \dfrac{\mathbf{1}}{\mathbf{4}}(5)$$

$$y = \dfrac{1}{4}x - \dfrac{5}{4} \text{ slope-intercept form}$$

55. Horizontal, passes through $(-8, 12)$
The equation of a horizontal line can be written in the form $y = b$.
$$y = 12$$

57. Passes through y-intercept $\left(0, \dfrac{7}{8}\right)$, $m = -\dfrac{1}{4}$

$x_1 = 0$ and $y_1 = \dfrac{7}{8}$

$$y - y_1 = m(x - x_1)$$

$$y - \dfrac{7}{8} = -\dfrac{1}{4}(x - 0)$$

$$y - \dfrac{7}{8} = -\dfrac{1}{4}x$$

$$y - \dfrac{7}{8} + \dfrac{\mathbf{7}}{\mathbf{8}} = -\dfrac{1}{4}x + \dfrac{7}{\mathbf{8}}$$

$$y = -\dfrac{1}{4}x + \dfrac{7}{8} \text{ slope-intercept form}$$

59. Passes through $(3, 0)$, $m = -\dfrac{2}{3}$

$x_1 = 3$ and $y_1 = 0$

$$y - y_1 = m(x - x_1)$$

$$y - 0 = -\dfrac{2}{3}(x - 3)$$

$$y = -\dfrac{2}{3}(x - 3)$$

$$y = -\dfrac{\mathbf{2}}{\mathbf{3}}x - \left(-\dfrac{\mathbf{2}}{\mathbf{3}}\right)3$$

$$y = -\dfrac{2}{3}x + 2 \text{ slope-intercept form}$$

61. Passes through $(2, 20)$, $m = 8$
$x_1 = 2$ and $y_1 = 20$

$$y - y_1 = m(x - x_1)$$
$$y - 20 = 8(x - 2)$$
$$y - 20 = \mathbf{8}x - \mathbf{8}(2)$$
$$y - 20 = 8x - 16$$
$$y - 20 + \mathbf{20} = 8x - 16 + \mathbf{20}$$
$$y = 8x + 4 \text{ slope-intercept form}$$

63. Vertical, passes through $(-3, 7)$
The equation of a vertical line can be written in the form $x = a$.
$$x = -3$$

65. Passes through y-intercept $(0, -11)$, $m = 7$
$b = -11$
$$y = mx + b$$
$$y = 7x + (-11)$$
$$y = 7x - 11$$

67. $(-2,-1)$ and $(-1,-5)$

(x_1, y_1) and (x_2, y_2)

$$m = \frac{y_2 - y_1}{x_2 - x_1}$$

$$= \frac{-5 - (-1)}{-1 - (-2)}$$

$$= \frac{-4}{1}$$

$$= -4$$

Passes through $(-2, -1)$, $m = -4$

$x_1 = -2$ and $y_1 = -1$

$$y - y_1 = m(x - x_1)$$

$$y - (-1) = -4(x - (-2))$$

$$y + 1 = -4(x + 2)$$

$$y + 1 = -4x - 8$$

$$y + 1 - \mathbf{1} = -4x - 8 - \mathbf{1}$$

$$y = -4x - 9 \text{ slope-intercept form}$$

69. x-intercept $(7,0)$ and y-intercept $(0, -2)$

$(7,0)$ and $(0,-2)$

(x_1, y_1) and (x_2, y_2)

$$m = \frac{y_2 - y_1}{x_2 - x_1}$$

$$= \frac{-2 - 0}{0 - 7}$$

$$= \frac{-2}{-7}$$

$$= \frac{2}{7}$$

Passes through $(7,0)$, $m = \frac{2}{7}$

$x_1 = 7$ and $y_1 = 0$

$$y - y_1 = m(x - x_1)$$

$$y - 0 = \frac{2}{7}(x - 7)$$

$$y = \frac{2}{7}(x) - \frac{2}{7}(7)$$

$$y = \frac{2}{7}x - 2 \text{ slope-intercept form}$$

71. Passes through origin $(0,0)$, $m = \frac{1}{10}$

$x_1 = 0$ and $y_1 = 0$

$$y - y_1 = m(x - x_1)$$

$$y - 0 = \frac{1}{10}(x - 0)$$

$$y = \frac{1}{10}x \text{ slope-intercept form}$$

73. Passes through $\left(-\frac{1}{8}, 12\right)$, m is undefined

$x_1 = -\frac{1}{8}$ and $y_1 = 12$

$$x = -\frac{1}{8}$$

75. Passes through y-intercept $(0, -2.8)$, $m = 1.7$

$b = -2.8$

$$y = mx + b$$

$$y = 1.7x + (-2.8)$$

$$y = 1.7x - 2.8$$

APPLICATIONS

77. ANATOMY

In this problem, we are given two different letters r and h to use other than x and y. One has to determine which letter correlates with which letter. $x = r$ and $y = h$.

Find the slope which is given as 3.9 inches for each 1-inch for the radius.

$$m = \frac{\text{rise}}{\text{run}}$$

$$= \frac{3.9}{1}$$

$$= 3.9$$

Given: 64-inch tall woman is h_1.

Given: 9-inch-long radius bone is r_1.

Now use the point-slope formula.

$$h - h_1 = m(r - r_1)$$

$$h - 64 = 3.9(r - 9)$$

$$h - 64 = \mathbf{3.9}(r) - \mathbf{3.9}(9)$$

$$h - 64 = 3.9r - 35.1$$

$$h - 64 + \mathbf{64} = 3.9r - 35.1 + \mathbf{64}$$

$$h = 3.9r + 28.9 \text{ slope-intercept form}$$

Section 3.6

79. POLE VAULTING

Part 1: Given $(5, 2)$ and $(10, 0)$

Find the slope:

$$m = \frac{y_2 - y_1}{x_2 - x_1}$$

$$= \frac{2 - 0}{5 - 10}$$

$$= \frac{2}{-5}$$

$$= -\frac{2}{5}$$

Part 2: Now pick a point $(10, 0)$ and use

$$y - y_1 = m(x - x_1)$$

$$y - 0 = -\frac{2}{5}(x - 10)$$

$$y = -\frac{2}{5}x + 4 \text{ slope-intercept form}$$

Part 3: Given $(9, 7)$ and $(10, 0)$

Find the slope:

$$m = \frac{y_2 - y_1}{x_2 - x_1}$$

$$= \frac{7 - 0}{9 - 10}$$

$$= \frac{7}{-1}$$

$$= -7$$

Now pick a point $(10, 0)$ and use

$$y - y_1 = m(x - x_1)$$

$$y - 0 = -7(x - 10)$$

$$y = -7x + 70 \text{ slope-intercept form}$$

Part 4: Given $(10, 7)$ and $(10, 0)$

Find the slope:

$$m = \frac{y_2 - y_1}{x_2 - x_1}$$

$$= \frac{7 - 0}{10 - 10}$$

$$= \frac{7}{0}$$

Slope is undefined.

Undefined slope means the line is vertical and $x = a$ is the equation. $x = 10$

81. TOXIC CLEANUP

a. In this problem, we are given two different letters m (month) and y (yard) to use for x and y. One has to determine which letter correlates with which letter, $x = m$ and $y = y$.

Find the slope which is given as two ordered pairs $(3, 800)$ and $(5, 720)$. The reason for the numbers 3 and 5 is because after "3 months" and then "2 months later".

$$m = \frac{y_2 - y_1}{m_2 - m_1.}$$

$$= \frac{800 - 720}{3 - 5}$$

$$= \frac{80}{-2}$$

$$= -40$$

Now pick one point $(3, 800)$ and use

$$y - y_1 = m(m - m_1)$$
$$y - 800 = -40(m - 3)$$
$$y - 800 = -40m + 120$$
$$y - 800 + \mathbf{800} = -40m + 120 + \mathbf{800}$$
$$y = -40m + 920 \text{ slope-intercept form}$$

The remainder of the solution is on the next page.

b. To predict the number of cubic yards of waste on the site one year after the cleanup project began,
let $m = 1$ yr ,   1 year $= 12$ months.

$$y = -40(\mathbf{12}) + 920$$

$$= -480 + 920$$

$$= 440$$

440 cubic yards of waste that will still be on the site after 1 year.

83. TRAMPOLINES

In this problem, we are given two different letters r and l to use for x and y. One has to determine which letter correlates with which letter, $x = r$ and $y = l$.

Find the slope which is given as two ordered pairs (3, 19) and (7, 44).

$$m = \frac{l_2 - l_1}{r_2 - r_1}$$
$$= \frac{44 - 19}{7 - 3}$$
$$= \frac{25}{4}$$

Now pick one point (3, 19) and use

$$l - l_1 = m(r - r_1)$$
$$l - 19 = \frac{25}{4}(r - 3)$$
$$l - 19 = \frac{25}{4}r - \frac{75}{4}$$
$$l - 19 + \mathbf{19} = \frac{25}{4}r - \frac{75}{4} + \mathbf{19}$$
$$l = \frac{25}{4}r - \frac{75}{4} + \frac{\mathbf{76}}{\mathbf{4}}$$
$$l = \frac{25}{4}r + \frac{1}{4} \text{ slope-intercept form}$$

85. GOT MILK

a. Two points must be selected (yr, gal). (6, 26.5) and (21, 22.0) were selected because they both lie on the line. Now you must find the slope of the two selected ordered pairs.

$$m = \frac{y_2 - y_1}{x_2 - x_1}$$
$$= \frac{26.5 - 22.0}{6 - 21}$$
$$= \frac{4.5}{-15}$$
$$= -\frac{4.5 \cdot 10}{15 \cdot 10}$$
$$= -\frac{45}{150}$$
$$= -\frac{3}{10}$$

Now pick one point (21, 22.0) and use
$$y - y_1 = m(x - x_1)$$
$$y - 22.0 = -\frac{3}{10}(x - 21)$$
$$y - 22.0 = -\frac{3}{10}x + \frac{63}{10}$$
$$y - 22.0 + 22.0 = -\frac{3}{10}x + 6.3 + 22.0$$
$$y = -\frac{3}{10}x + 28.3 \text{ slope-intercept form}$$

or

$$y = -0.3x + 28.3 \text{ slope-intercept form}$$

or

$$y = -\frac{3}{10}x + \frac{283}{10} \text{ slope-intercept form}$$

b. Use $x = 40$ (the year 2020 correlates with 40) to calculate the number of gallons of milk to be consumed.
$$y = -0.3x + 28.3$$
$$y = -0.3(\mathbf{40}) + 28.3$$
$$y = -12 + 28.3$$
$$y = 16.3$$

The amount of milk that an average American will drink in 2020 is 16.3 gal.

WRITING

87-89. Answers will vary.

REVIEW

91. FRAMES

Analyze

- The length of a rectangular picture is 5 inches greater than twice the width.
- The perimeter is 112 inches.
- Find the dimensions of the frame.

Assign

Let x = width of frame in inches

$2x + 5$ = length of frame in inches

Form

$2 \cdot \boxed{\text{the width of the frame}}$ plus $2 \cdot \boxed{\text{the length of the frame}}$ equals $\boxed{\text{the perimeter.}}$

$$2x + 2(2x+5) = 112$$

Solve

$$2x + 2(2x+5) = 112$$
$$2x + 4x + 10 = 112$$
$$6x + 10 = 112$$
$$6x + 10 - \mathbf{10} = 112 - \mathbf{10}$$
$$6x = 102$$
$$\frac{6x}{\mathbf{6}} = \frac{102}{\mathbf{6}}$$
$$x = 17$$

$$\text{length}$$
$$2x + 5 = 2(\mathbf{17}) + 5$$
$$= 34 + 5$$
$$= 39$$

State

The dimensions of the frame are 17 inches by 39 inches.

Check

$$2x + 2(2x+5) = 112$$
$$2(\mathbf{17}) + 2(\mathbf{39}) = 112$$
$$34 + 78 = 112$$
$$112 = 112$$

The results check.

CHALLENGE PROBLEMS

93. Given (2, 5) Find the equation of a line parallel to $y = 4x - 7$.

First, find the slope from $y = 4x - 7$, $m = 4$.

The two lines are parallel and have the same slope, so use $m = 4$ and (2, 5) to write the equation of the line.

$$y - y_1 = m(x - x_1)$$
$$y - 5 = 4(x - 2)$$
$$y - 5 = 4x - 8$$
$$y - 5 + \mathbf{5} = 4x - 8 + \mathbf{5}$$
$$y = 4x - 3 \text{ slope-intercept form}$$

SECTION 3.7

VOCABULARY

Fill in the blanks.

1. $2x - y \leq 4$ is a linear **inequality** in two variables.

3. $(7, 2)$ is a solution of $x - y > 1$. We say that $(7, 2)$ **satisfies** the inequality.

5. In the graph, the line $2x - y = 4$ divides the coordinate plane into two **half-planes**.

CONCEPTS

7. Determine whether $(-3, -5)$ is a solution of $5x - 3y \geq 0$. **Yes**

$$5x - 3y \overset{?}{\geq} 0$$
$$5(-3) - 3(-5) \overset{?}{\geq} 0$$
$$-15 + 15 \overset{?}{\geq} 0$$
$$0 \geq 0 \quad \text{True}$$

9. Fill in the blanks: A **dashed** line indicates that points on the boundary are not solutions and a **solid** line indicates that points on the boundary are solutions.

11. If a false statement results when the coordinates of a test point are substituted into a linear inequality, which half-plane should be shaded to represent the solution of the inequality?

    The half-plane opposite that in which the test point lies

13. A linear inequality has been graphed. Determine whether each point satisfies the inequality.

    a. $(2, 1)$ **Yes**
    b. $(-2, -4)$ **No**
    c. $(4, -2)$ **No**
    d. $(-3, 4)$ **Yes**

NOTATION

15. Write the meaning of each symbol in words.

    a. $<$ **Is less than**
    b. $\geq$ **Is greater than or equal to**
    c. $\leq$ **Is less than or equal to**
    d. $\overset{?}{>}$ **Is possibly greater than**

17. Fill in the blanks: The inequality $4x + 2y \leq 9$ means $4x + 2y = 9$ or $4x + 2y < 9$.

GUIDED PRACTICE

Determine whether each ordered pair is a solution of the given inequality.

19. $2x + y > 6;\ (3,2)$

$$2(3) + 2 \overset{?}{>} 6$$
$$6 + 2 \overset{?}{>} 6$$
$$8 > 6 \quad \text{True}$$
$$(3, 2) \text{ is a solution.}$$

20. $4x - 2y \geq -6;\ (-2,1)$

$$4(-2) - 2(1) \overset{?}{\geq} -6$$
$$-8 - 2 \overset{?}{\geq} -6$$
$$-10 \geq -6 \quad \text{False}$$
$$(-2, 1) \text{ is not a solution.}$$

21. $-5x - 8y < 8;\ (-8,4)$

$$-5(-8) - 8(4) \overset{?}{<} 8$$
$$40 - 32 \overset{?}{<} 8$$
$$8 < 8 \quad \text{False}$$
$$(-8, 4) \text{ is not a solution.}$$

22. $x + 3y > 14;\ (-3,8)$

$$-3 + 3(8) \overset{?}{>} 14$$
$$-3 + 24 \overset{?}{>} 14$$
$$21 > 14 \quad \text{True}$$
$$(-3, 8) \text{ is a solution.}$$

23. $4x - y \leq 0;\ \left(\dfrac{1}{2}, 1\right)$

$$4\left(\dfrac{1}{2}\right) - 1 \overset{?}{\leq} 0$$
$$2 - 1 \overset{?}{\leq} 0$$
$$1 \leq 0 \quad \text{False}$$
$$\left(\dfrac{1}{2}, 1\right) \text{ is not a solution.}$$

25. $-5x + 2y > -4;\ (0.8, 0.6)$

$$-5(0.8) + 2(0.6) \overset{?}{>} -4$$
$$-4 + 1.2 \overset{?}{>} -4$$
$$-2.8 > -4 \quad \text{True}$$
$$(0.8, 0.6) \text{ is a solution.}$$

Section 3.7

Complete the graph by shading the correct side of the boundary.

27. $x - y \geq -2$, Boundary line is <u>solid</u> because of the $\geq$ symbol.

Select test point $(0,0)$ and substitute into

$$x - y \geq -2$$
$$\overset{?}{0 - 0 \geq -2}$$
$$0 \geq -2$$
True

The coordinates of every point on the same side of the line as the origin satisfy the inequality. To indicate this, we shade the half-plane that contains the test point $(0,0)$.

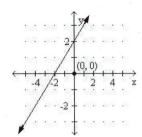 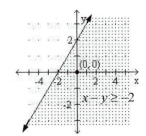

29. $y > 2x - 4$, Boundary line is <u>dashed</u> because of the $>$ symbol.

Select test point $(0,0)$ and substitute into

$$y > 2x - 4$$
$$?$$
$$0 > 2(0) - 4$$
$$0 > -4$$
True

The coordinates of every point on the same side of the line as the origin satisfy the inequality. To indicate this, we shade the half-plane that contains the test point $(0,0)$.

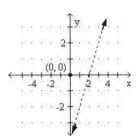 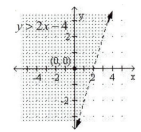

31. $x - 2y \geq 4$, Boundary line is <u>solid</u> because of the $\geq$ symbol.

Select test point $(0,0)$ and substitute into

$$x - 2y \geq 4$$
$$?$$
$$0 - 2(0) \geq 4$$
$$0 \geq 4$$
False

The coordinates of every point on the same side of the line as the origin *do not* satisfy the inequality. To indicate this, we shade the half-plane that *does not* contain the test point $(0, 0)$.

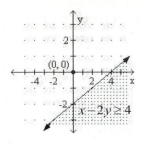 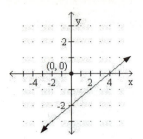

33. $y \leq 4x$, Boundary line is <u>solid</u> because of the $\leq$ symbol.

The line passes through the origin so select test point $(1,1)$ and substitute into

$$y \leq 4x$$
$$?$$
$$1 \leq 4(1)$$
$$1 \leq 4$$
True

The coordinates of every point on the same side of the line as $(1,1)$ satisfy the inequality. To indicate this, we shade the half-plane that contains the test point $(1,1)$.

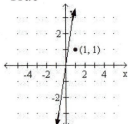 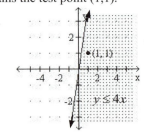

Graph each inequality.

35. $x + y \geq 3$, Boundary line is <u>solid</u> because of the $\geq$ symbol.

Graph as $x + y = 3$.

y-intercept:
If $x = 0$,
$$0 + y = 3$$
$$y = 3$$

x-intercept:
If $y = 0$,
$$x + 0 = 3$$
$$x = 3$$

The y-int is $(0, 3)$, and the x-int is $(3, 0)$.

Select test point $(0,0)$ and substitute into

$$x + y \geq 3$$
$$?$$
$$0 - 2(0) \geq 3$$
$$0 \geq 3$$
False

The coordinates of every point on the same side of the line as the origin *do not* satisfy the inequality. To indicate this, we shade the half-plane that *does not* contain the test point $(0, 0)$.

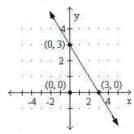 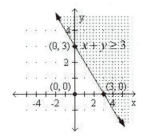

37. $3x - 4y > 12$, Boundary line is <u>dashed</u>

because of the $>$ symbol.
Graph as $3x - 4y = 12$.

y-intercept:

If $x = 0$,

$3(0) - 4y = 12$

$-4y = 12$

$\dfrac{-4y}{-4} = \dfrac{12}{-4}$

$y = -3$

x-intercept:

If $y = 0$,

$3x + 4(0) = 12$

$3x = 12$

$\dfrac{3x}{3} = \dfrac{12}{3}$

$x = 4$

The y-int is $(0, -3)$, and the x-int is $(4, 0)$.

Select test point $(0,0)$ and substitute into

$3x - 4y > 12$

$\overset{?}{3(0) - 4(0)} > 12$

$0 > 12$

False

The coordinates of every point on the same side of the line as the origin *do not* satisfy the inequality. To indicate this, we shade the half-plane that *does not* contain the test point $(0, 0)$.

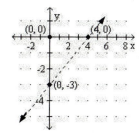 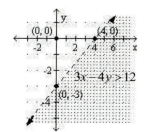

39. $2x + 3y \le -12$, Boundary line is <u>solid</u>
because of the $\le$ symbol.
Graph as $2x + 3y = -12$.

y-intercept:

If $x = 0$,

$2(0) + 3y = -12$

$3y = -12$

$\dfrac{3y}{3} = \dfrac{-12}{3}$

$y = -4$

x-intercept:

If $y = 0$,

$2x + 3(0) = -12$

$2x = -12$

$\dfrac{2x}{2} = \dfrac{-12}{2}$

$x = -6$

The y-int is $(0, -4)$, and the x-int is $(-6, 0)$.

Select test point $(0,0)$ and substitute into

$2x + 3y \le -12$

$\overset{?}{2(0) + 3(0)} \le -12$

$0 \le -12$

False

The coordinates of every point on the same side of the line as the origin *do not* satisfy the inequality. To indicate this, we shade the half-plane that *does not* contain the test point $(0, 0)$.

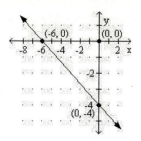 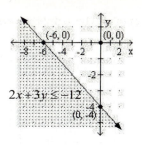

41. $y < 2x - 1$, Boundary line is <u>dashed</u>

because of the $<$ symbol.
Graph as $y = 2x - 1$.

y-intercept:

If $x = 0$,

$y = 2(0) - 1$

$y = -1$

x-intercept:

If $y = 0$,

$0 = 2x - 1$

$1 = 2x$

$\dfrac{1}{2} = \dfrac{2x}{2}$

$\dfrac{1}{2} = x$

The y-int is $(0, -1)$, and the x-int is $\left(\dfrac{1}{2}, 0\right)$.

Select test point $(0,0)$ and substitute into

$y < 2x - 1$

$\overset{?}{0} < 2(0) - 1$

$0 < -1$

False

The coordinates of every point on the same side of the line as the origin *do not* satisfy the inequality. To indicate this, we shade the half-plane that *does not* contain the test point $(0, 0)$.

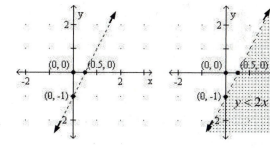

Section 3.7

43. $y < -3x + 2$, Boundary line is <u>dashed</u> --------- because of the $<$ symbol.

Graph as $y = -3x + 2$.

y-intercept:	x-intercept:
If $x = 0$,	If $y = 0$,
$y = -3(0) + 2$	$0 = -3x + 2$
$y = 2$	$0 - 2 = -3x + 2 - 2$
	$-2 = -3x$
	$\dfrac{-2}{-3} = \dfrac{-3x}{-3}$
	$\dfrac{2}{3} = x$

The y-int is (0, 2), and the x-int is $\left(\dfrac{2}{3}, 0\right)$.

Select test point (0,0) and substitute into

$y < -3x + 2$

$\overset{?}{0 < -3(0) + 2}$

$0 < 2$

True

The coordinates of every point on the same side of the line as the origin satisfy the inequality. To indicate this, we shade the half-plane that contains the test point (0,0).

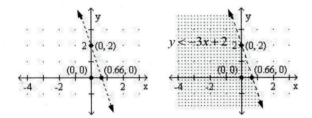

45. $y \geq -\dfrac{3}{2}x + 1$, Boundary line is <u>solid</u> because of the $\geq$ symbol.

Graph as $y = -\dfrac{3}{2}x + 1$.

y-intercept:	x-intercept:
If $x = 0$,	If $y = 0$,
$y = -\dfrac{3}{2}(0) + 1$	$0 = -\dfrac{3}{2}x + 1$
$y = 1$	$0 - 1 = -\dfrac{3}{2}x + 1 - 1$
	$-1 = -\dfrac{3}{2}x$
	$\left(-\dfrac{2}{3}\right) \cdot (-1) = \left(-\dfrac{2}{3}\right) - \dfrac{3}{2}x$
	$\dfrac{2}{3} = x$

The y-int is (0, 1), and the x-int is $\left(\dfrac{2}{3}, 0\right)$.

Select test point (0,0) and substitute into

$y \geq -\dfrac{3}{2}x + 1$

$\overset{?}{0 \geq -\dfrac{3}{2}(0) + 1}$

$0 \geq 1$

False

The coordinates of every point on the same side of the line as the origin **do not** satisfy the inequality. To indicate this, we shade the half-plane that **does not** contain the test point (0, 0).

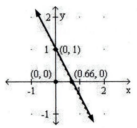 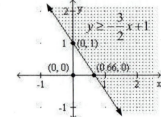

47. $x - 2y \geq 4$, Boundary line is <u>solid</u> because of the $\geq$ symbol.

Graph as $x - 2y = 4$.

y-intercept:	x-intercept:
If $x = 0$,	If $y = 0$,
$x - 2y = 4$	$x - 2y = 4$
$0 - 2y = 4$	$x - 2(0) = 4$
$-2y = 4$	$x = 4$
$\dfrac{-2y}{-2} = \dfrac{4}{-2}$	
$y = -2$	

The y-int is $(0, -2)$, and the x-int is $(4, 0)$.

Select test point (0,0) and substitute into

$x - 2y \geq 4$

$\overset{?}{0 - 2(0) \geq 4}$

$0 \geq 4$

False

The coordinates of every point on the same side of the line as the origin **do not** satisfy the inequality. To indicate this, we shade the half-plane that **does not** contain the test point (0, 0).

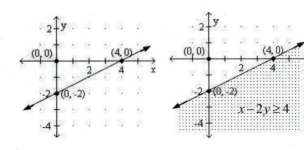

49. $2y - x < 8$, Boundary line is <u>dashed</u>
because of the $<$ symbol.
Graph as $2y - x = 8$.

y-intercept: $\qquad$ x-intercept:

If $x = 0$, $\qquad$ If $y = 0$,

$2y - \mathbf{0} = 8$ $\qquad$ $2(\mathbf{0}) - x = 8$

$\quad 2y = 8$ $\qquad$ $\quad -x = 8$

$\quad\; y = 4$ $\qquad$ $\qquad x = -8$

The y-int is $(0,4)$, and the x-int is $(-8,0)$.

Select test point $(0,0)$ and substitute into

$2y - x < 8$ $\qquad$ The coordinates of every
$\overset{?}{}$ $\qquad\qquad$ point on the same side of
$2(\mathbf{0}) - \mathbf{0} < 8$ $\qquad$ the line as the origin satisfy
$\qquad 0 < 8$ $\qquad$ the inequality. To indicate this,
$\qquad$ **True** $\qquad$ we shade the half-plane that
$\qquad\qquad\qquad$ contains the test point (0,0).

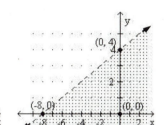

51. $7x - 2y < 21$, Boundary line is <u>dashed</u>
because of the $<$ symbol.
Graph as $7x - 2y = 21$.

y-intercept: $\qquad$ x-intercept:

If $x = 0$, $\qquad$ If $y = 0$,

$7(\mathbf{0}) - 2y = 21$ $\qquad$ $7x - 2(\mathbf{0}) = 21$

$\quad\;\; -2y = 21$ $\qquad$ $\qquad 7x = 21$

$\dfrac{-2y}{-\mathbf{2}} = \dfrac{21}{-\mathbf{2}}$ $\qquad$ $\dfrac{7x}{7} = \dfrac{21}{7}$

$\qquad y = -\dfrac{21}{2}$ $\qquad$ $\qquad x = 3$

The y-int is $\left(0, -\dfrac{21}{2}\right)$, and the x-int is $(3,0)$.

Select test point $(0,0)$ and substitute into

$7(\mathbf{0}) - 2(\mathbf{0}) < 21$ $\qquad$ The coordinates of every
$\qquad\qquad 0 < 21$ $\qquad$ point on the same side of
$\qquad\qquad$ **True** $\qquad$ the line as (0,0) satisfy the
$\qquad\qquad\qquad$ inequality. To indicate this,
$\qquad\qquad\qquad$ we shade the half-plane that
$\qquad\qquad\qquad$ contains the test point (0,0).

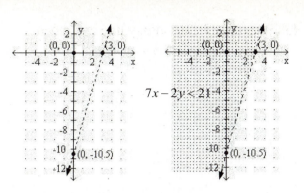

53. $2x - 3y \geq 4$, Boundary line is <u>solid</u>
because of the $\geq$ symbol.
Graph as $2x - 3y = 4$.

y-intercept: $\qquad$ x-intercept:

If $x = 0$, $\qquad$ If $y = 0$,

$2(\mathbf{0}) - 3y = 4$ $\qquad$ $2x - 3(\mathbf{0}) = 4$

$\quad\; -3y = 4$ $\qquad$ $\qquad 2x = 4$

$\dfrac{-3y}{-\mathbf{3}} = \dfrac{4}{-\mathbf{3}}$ $\qquad$ $\dfrac{2x}{2} = \dfrac{4}{2}$

$\qquad y = -\dfrac{4}{3}$ $\qquad$ $\qquad x = 2$

The y-int is $\left(0, -\dfrac{4}{3}\right)$, and the x-int is $(2,0)$.

Select test point $(0,0)$ and substitute into

$\qquad\qquad\qquad\qquad$ The coordinates of every
$2x - 3y \geq 4$ $\qquad\qquad$ point on the same side of
$\overset{?}{}$ $\qquad\qquad$ the line as the origin ***do not***
$2(\mathbf{0}) - 3(\mathbf{0}) \geq 4$ $\qquad$ satisfy the inequality. To
$\qquad\quad 0 \geq 4$ $\qquad\qquad$ indicate this, we shade the
$\qquad\quad$ **False** $\qquad\qquad$ half-plane that ***does not***
$\qquad\qquad\qquad\qquad$ contain the test point (0, 0).

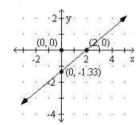 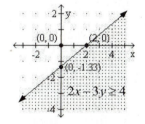

Graph each inequality.

55. $y \geq 2x$, Boundary line is <u>solid</u> because of the $\geq$ symbol.

Graph as $y = 2x$.

y-intercept:	x-intercept:
If $x = 0$,	If $y = 0$,
$y = 2(\mathbf{0})$	$\mathbf{0} = 2x$
$y = 0$	$0 = x$

The y-int is $(0,0)$, and the x-int is $(0,0)$.

The line passes through the origin so select test point $(0,2)$ and substitute into

$y > 2x$
$\overset{?}{}$
$\mathbf{2} > 2(\mathbf{0})$
$2 > 0$
True

The coordinates of every point on the same side of the line as $(0,2)$ satisfy the inequality. To indicate this, we shade the half-plane that contains the test point $(0,2)$.

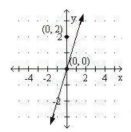

57. $y < -\dfrac{x}{2}$, Boundary line is <u>dashed</u> because of the $<$ symbol.

Graph as $y = -\dfrac{x}{2}$.

y-intercept:	x-intercept:
If $x = 0$,	If $y = 0$,
$y = -\dfrac{\mathbf{0}}{2}$	$0 = -\dfrac{x}{2}$
$y = 0$	$0 = x$

The y-int is $(0,0)$, and the x-int is $(0,0)$.

The line passes through the origin so select test point $(0,2)$ and substitute into

$y < -\dfrac{x}{2}$
$\mathbf{2} \overset{?}{<} -\dfrac{\mathbf{0}}{2}$
$2 < 0$
False

The coordinates of every point on the same side of the line as $(0, 2)$ ***do not*** satisfy the inequality. To indicate this, we shade the half-plane that ***does not*** contain the test point $(0, 2)$.

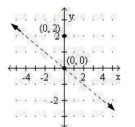

59. $y + x < 0$, Boundary line is <u>dashed</u> because of the $<$ symbol.

Graph as $y = -x$.

y-intercept:	x-intercept:
If $x = 0$,	If $y = 0$,
$y = -\mathbf{0}$	$\mathbf{0} = -x$
$y = 0$	$0 = x$

The y-int is $(0,0)$, and the x-int is $(0,0)$.

The line passes through the origin so select test point $(0,2)$ and substitute into

$y + x < 0$
$\overset{?}{}$
$\mathbf{2} + \mathbf{0} < 0$
$2 < 0$
False

The coordinates of every point on the same side of the line as $(0, 2)$ ***do not*** satisfy the inequality. To indicate this, we shade the half-plane that ***does not*** contain the test point $(0, 2)$.

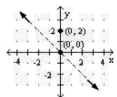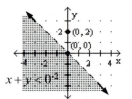

61. $5x + 3y < 0$, Boundary line is <u>dashed</u> because of the $<$ symbol.

Graph as $5x + 3y = 0$.

y-intercept:	x-intercept:
If $x = 0$,	If $y = 0$,
$5(\mathbf{0}) + 3y = 0$	$5x + 3(\mathbf{0}) = 0$
$3y = 0$	$5x = 0$
$\dfrac{3y}{3} = \dfrac{0}{3}$	$\dfrac{5x}{5} = \dfrac{0}{5}$
$y = 0$	$x = 0$

The y-int is $(0,0)$, and the x-int is $(0,0)$.

The line goes through the origin so select test point $(0,2)$ and substitute into

$$5x + 3y < 0$$
$$\overset{?}{5(0) + 3(2) < 0}$$
$$6 < 0$$
False

The coordinates of every point on the same side of the line as (0, 2) **do not** satisfy the inequality. To indicate this, we shade the half-plane that **does not** contain the test point (0, 2).

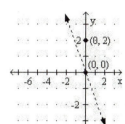 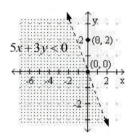

Graph each inequality.

63. $x < 2$, Boundary line is <u>dashed</u> because of the $<$ symbol.
Graph as $x = 2$.

There is no y-int, and the x-int is $(2, 0)$.
$x = 2$ is a line parallel to the y-axis.
Select test point $(0,0)$ and substitute into

$x < 2$
$\mathbf{0} < 2$
True

The coordinates of every point on the same side of the line as (0,0) satisfy the inequality. To indicate this, we shade the half-plane that contains the test point (0,0).

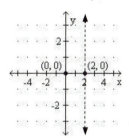 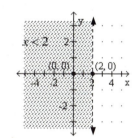

65. $y \le 1$, Boundary line is <u>solid</u> because of the $\le$ symbol.
Graph as $y = 1$.
The y-int is $(0,1)$, and there is no x-int.
$y = 1$ is a line parallel to the x-axis.
Select test point $(0,0)$ and substitute into

$y \le 1$
$\mathbf{0} \le 1$
True

The coordinates of every point on the same side of the line as (0,0) satisfy the inequality. To indicate this, we shade the half-plane that contains the test point (0,0).

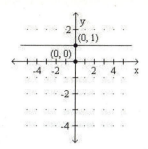 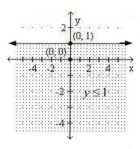

67. $y + 2.5 > 0$, Boundary line is <u>dashed</u> because of the $>$ symbol.
Graph as $y + 2.5 = 0$.
y-intercept:
If $x = 0$,
$y + 2.5 = 0$
$y = -2.5$

The y-int is $(0, -2.5)$, and no x-int.
$y = -2.5$ is parallel to the x-axis.
Select test point $(1,1)$ and substitute into

$y + 2.5 > 0$
$\overset{?}{1 + 2.5 > 0}$
$3.5 > 0$
True

The coordinates of every point on the same side of the line as (1, 1) satisfy the inequality. To indicate this, we shade the half-plane that contains the test point (1, 1).

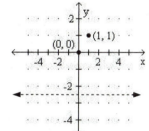 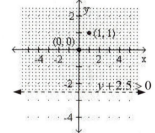

Section 3.7

69. $x \le 0$, Boundary line is <u>solid</u> because of the $\le$ symbol.
Graph as $x = 0$.

There are many y-int, and the x-int is $(0,0)$.
$x = 0$ is the same line as the y-axis.
Select test point $(1,1)$ and substitute into

$x \le 0$

$1 \le 0$

False

The coordinates of every point on the same side of the line as (1, 1) ***do not*** satisfy the inequality. To indicate this, we shade the half-plane that ***does not*** contain the test point (1, 1).

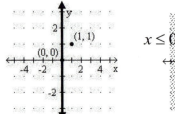
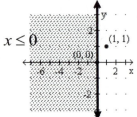

$x \le 0$

TRY IT YOURSELF

71. **a.** $5x - 3y \ge -15$, Boundary line is <u>solid</u> because of the $\ge$ symbol.
Graph as $5x - 3y = -15$.

y-intercept:

If $x = 0$,

$5(0) - 3y = -15$

$-3y = -15$

$\dfrac{-3y}{-3} = \dfrac{-15}{-3}$

$y = 5$

x-intercept:

If $y = 0$,

$5x - 3(0) = -15$

$5x = -15$

$\dfrac{5x}{5} = \dfrac{-15}{5}$

$x = -3$

The y-int is $(0,5)$, and the x-int is $(-3,0)$.

Select test point $(0,0)$ and substitute into

$5x - 3y \ge -15$

$5(0) - 3(0) \overset{?}{\ge} -15$

$0 \ge -15$

True

The coordinates of every point on the same side of the line as (0,0) satisfy the inequality. To indicate this, we shade the half-plane that contains the test point (0,0).

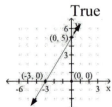
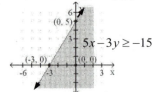

$5x - 3y \ge -15$

b. $5x - 3y < -15$, Boundary line is <u>dashed</u> because of the $<$ symbol.
Graph as $5x - 3y = -15$.

If one would notice the inequality symbol is the opposite of the original, then logically one would shade the opposite side of the boundary line and make it dashed instead of solid.

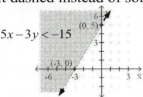

$5x - 3y < -15$

73. **a.** $y + 2x < 0$, Boundary line is <u>dashed</u> because of the $<$ symbol.
Graph as $y = -2x$.

y-intercept:

If $x = 0$,

$y = -2(0)$

$y = 0$

x-intercept:

If $y = 0$,

$0 = -2x$

$0 = x$

The y-int is $(0,0)$, and the x-int is $(0,0)$.

The line passes through the origin so select test point $(0,2)$ and substitute into

$y \overset{?}{<} -2x$

$2 \overset{?}{<} -2(0)$

$2 < 0$

False

The coordinates of every point on the same side of the line as (0, 2) ***do not*** satisfy the inequality. To indicate this, we shade the half-plane that ***does not*** contain the test point (0, 2).

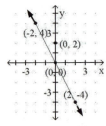
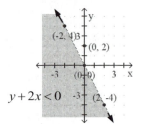

$y + 2x < 0$

b. $y + 2x \ge 0$, Boundary line is <u>solid</u> because of the $\ge$ symbol.
Graph as $y = -2x$.

If one would notice the inequality symbol is the opposite of the original, then logically one would shade the opposite side of the boundary line and make it solid instead of dashed.

$y + 2x \ge 0$

APPLICATIONS
75. DELIVERIES

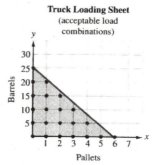

Truck Loading Sheet
(acceptable load combinations)

No, the truck can not deliver 4 pallets and 10 barrels in one trip.

FROM CAMPUS TO CAREERS
77. DENTAL ASSISTANT

Let x represents the number children and y represents the number of adults so let $c = x$ and $a = y$

$$\frac{3}{4}c + a \le 9$$

$$a \le -\frac{3}{4}c + 9$$

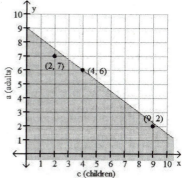

$(2,7),\ (4,6),\ (9,2)$; answers may vary.

79. PRODUCTION PLANNING

$3x + 4y \le 120$, Graph as $3x + 4y = 120$.
Boundary line is <u>solid</u>.

y-intercept:	x-intercept:
If $x = 0$,	If $y = 0$,
$3(\mathbf{0}) + 4y = 120$	$3x + 4(\mathbf{0}) = 120$
$4y = 120$	$3x = 120$
$\dfrac{4y}{4} = \dfrac{120}{4}$	$\dfrac{3x}{3} = \dfrac{120}{3}$
$y = 30$	$x = 40$

The y-int is $(0,30)$, and the x-int is $(40,0)$.
Select test point $(0,0)$ and substitute into

$$3(\mathbf{0}) + 4(\mathbf{0}) \le 120$$
$$0 \le 120$$
$$\text{True}$$

The coordinates of every point on the same side of the line as $(0,0)$ satisfy the inequality. To indicate this, we shade the half-plane that contains the test point $(0,0)$.

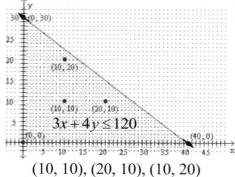

$3x + 4y \le 120$

$(10, 10), (20, 10), (10, 20)$
answers may vary

81. INVENTORIES

$100x + 88y \ge 4,400$,

Graph as $100x + 88y = 4,400$.
Boundary line is <u>solid</u>.

y-intercept:	x-intercept:
If $x = 0$,	If $y = 0$,
$100(\mathbf{0}) + 88y = 4,400$	$100x + 88(\mathbf{0}) = 4,400$
$88y = 4,400$	$100x = 4,400$
$\dfrac{88y}{88} = \dfrac{4,400}{88}$	$\dfrac{100x}{100} = \dfrac{4,400}{100}$
$y = 50$	$x = 44$

The y-int is $(0,50)$, and the x-int is $(44,0)$.

Select test point $(0,0)$ and substitute into

$$100(\mathbf{0}) + 88(\mathbf{0}) \ge 4,400$$
$$0 \ge 4,400$$
$$\text{False}$$

The coordinates of every point on the same side of the line as $(0,0)$ ***does not*** satisfy the inequality. To indicate this, we shade the half-plane that ***does not*** contain the test point $(0,0)$.

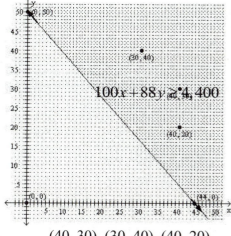

$100x + 88y \ge 4,400$

$(40, 30), (30, 40), (40, 20)$
answers may vary

Section 3.7

WRITING
83-85. Answers will vary.

REVIEW
87. Solve $A = P + Prt$ for t.

$$A = P + Prt$$

$$A - \mathbf{P} = P + Prt - \mathbf{P}$$

$$A - P = Prt$$

$$\frac{A - P}{\mathbf{Pr}} = \frac{Prt}{\mathbf{Pr}}$$

$$\frac{A - P}{Pr} = t$$

$$t = \frac{A - P}{Pr}$$

89. Simplify:

$$40\left(\frac{3}{8}x - \frac{1}{4}\right) + 40\left(\frac{4}{5}\right) = \mathbf{40}\left(\frac{3}{8}x\right) - \mathbf{40}\left(\frac{1}{4}\right) + \mathbf{40}\left(\frac{4}{5}\right)$$

$$= 15x - 10 + 32$$

$$= 15x + 22$$

CHALLENGE PROBLEMS
91. Find a linear inequality that has the graph shown.

slope $= \dfrac{3}{2}$

y-intercept $= (0, -3)$

Line is solid indicating $\leq$ or $\geq$.

Start with a line equation using the found information.

$$y = mx + b$$

$$y = \frac{3}{2}x - 3$$

$$\mathbf{2}y = \mathbf{2}\left(\frac{3}{2}x\right) - \mathbf{2}(3)$$

$$2y = 3x - 6$$

$$2y - \mathbf{3x} = 3x - 6 - \mathbf{3x}$$

$$-3x + 2y = -6$$

Now test the inequality: $-3x + 2y \leq -6$

Substitute test point $(0,0)$ in it.

$$-3x + 2y \leq -6$$

$$-3(\mathbf{0}) + 2(\mathbf{0}) \overset{?}{\leq} -6$$

$$0 \leq -6$$

$$\text{False}$$

The above false statement indicates this is the inequality with the given shaded half:

$$-3x + 2y \leq -6$$

or

$$3x - 2y \geq 6$$

SECTION 3.8

VOCABULARY

Fill in the blanks.

1. A set of ordered pairs is called a **relation** .

3. The set of all input values for a function is called the **domain**, and the set of all output values is called the **range**.

5. If $f(2) = -3$, we call -3 a function **value**.

CONCEPTS

7. **N** FEDERAL MINIMUM HOURLY

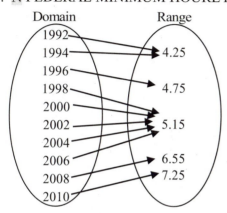

9. For the given input, what value will the function machine output?

$$f(x) = x^2 + 8, \quad x = -5$$
$$f(-5) = (-5)^2 + 8$$
$$f(-5) = 25 + 8$$
$$f(-5) = \mathbf{33}$$

NOTATION

Fill in the blanks.

11. We read $f(x) = 5x - 6$ as "f **of** x is $5x$ minus 6."

13. The notation $f(4) = 5$ indicates that when the x-value **4** is input into a function rule, the output is **5**. This fact can be shown graphically by plotting the ordered pair (**4** , **5**).

GUIDED PRACTICE

Find the domain and range of each relation.

15. **Domain:{–6, –1, 6, 8}; Range:{–10, –5, –1, 2}**

17. **Domain:{–8, 0, 6}; Range: {9, 50}**

Determine whether each arrow diagram or table defines y as a function of x. If a function is defined, give its domain and range. If it does not define a function, find two ordered pairs that show a value of x that is assigned more than one value of y.

19. **Yes; domain: {10,20,30}; range: {20,40,60}**

21. **No; (4, 2), (4, 4), (4, 6)** (Answers may vary)

23. **Yes; domain: {1, 2, 3, 4, 5};**
    range: {7, 8, 15, 16, 23}

25. **No; (–1, 0), (–1, 2)**

27. **No; (3, 4), (3, –4) or (4, 3), (4, –3)**

29. **Yes; domain: {–3, 1, 5, 6}; range: {–8, 0, 4, 9}**

31. **No; (3, 4), (3, –4) or (4, 3), (4, –3)**

33. **Yes; domain: {–2, –1, 0, 1};**
    range: {7, 10, 13, 16}

Find each function value.

35. $f(x) = 4x - 1$

   a. $f(x) = 4x - 1$
$$f(1) = 4(1) - 1$$
$$= 4 - 1$$
$$= 3$$
Thus, $f(1) = 3$

   b. $f(x) = 4x - 1$
$$f(-2) = 4(-2) - 1$$
$$= -8 - 1$$
$$= -9$$
Thus, $f(-2) = -9$

   c. $f(x) = 4x - 1$
$$f\left(\frac{1}{4}\right) = 4\left(\frac{1}{4}\right) - 1$$
$$= 1 - 1$$
$$= 0$$
Thus, $f\left(\frac{1}{4}\right) = 0$

   d. $f(x) = 4x - 1$
$$f(50) = 4(50) - 1$$
$$= 200 - 1$$
$$= 199$$
Thus, $f(50) = 199$

37. $f(x) = 2x^2$

   a. $f(x) = 2x^2$
$$f(0.4) = 2(0.4)^2$$
$$= 2(0.16)$$
$$= 0.32$$
Thus, $f(0.4) = 0.32$

b. $f(x) = 2x^2$

$f(-3) = 2(-3)^2$

$= 2(9)$

$= 18$

Thus, $f(-3) = 18$

c. $f(x) = 2x^2$

$f(1,000) = 2(1,000)^2$

$= 2(1,000,000)$

$= 2,000,000$

Thus, $f(1,000) = 2,000,000$

d. $f(x) = 2x^2$

$f\left(\dfrac{1}{8}\right) = 2\left(\dfrac{1}{8}\right)^2$

$= 2\left(\dfrac{1}{64}\right)$

$= \dfrac{1}{32}$

Thus, $f\left(\dfrac{1}{8}\right) = \left(\dfrac{1}{32}\right)$

39. $h(x) = |x - 7|$

a. $h(x) = |x - 7|$

$h(0) = |0 - 7|$

$= |-7|$

$= 7$

Thus, $h(0) = 7$

b. $h(x) = |x - 7|$

$h(-7) = |-7 - 7|$

$= |-14|$

$= 14$

Thus, $h(-7) = 14$

c. $h(x) = |x - 7|$

$h(7) = |7 - 7|$

$= |0|$

$= 0$

Thus, $h(7) = 0$

d. $h(x) = |x - 7|$

$h(8) = |8 - 7|$

$= |1|$

$= 1$

Thus, $h(8) = 1$

41. $g(x) = x^3 - x$

a. $g(x) = x^3 - x$

$g(1) = 1^3 - 1$

$= 1 - 1$

$= 0$

Thus, $g(1) = 0$

b. $g(x) = x^3 - x$

$g(10) = 10^3 - 10$

$= 1,000 - 10$

$= 990$

Thus, $g(10) = 990$

c. $g(x) = x^3 - x$

$g(-3) = (-3)^3 - (-3)$

$= -27 + 3$

$= -24$

Thus, $g(-3) = -24$

d. $g(x) = x^3 - x$

$g(6) = (6)^3 - 6$

$= 216 - 6$

$= 210$

Thus, $g(6) = 210$

43. $s(x) = (x + 3)^2$

a. $s(x) = (x + 3)^2$

$s(3) = (3 + 3)^2$

$= (6)^2$

$= 36$

Thus, $s(3) = 36$

b. $s(x) = (x + 3)^2$

$s(-3) = (-3 + 3)^2$

$= (0)^2$

$= 0$

Thus, $s(-3) = 0$

c. $s(x) = (x + 3)^2$

$s(0) = (0 + 3)^2$

$= (3)^2$

$= 9$

Thus, $s(0) = 9$

d. $s(x) = (x + 3)^2$

$s(-5) = (-5 + 3)^2$

$= (-2)^2$

$= 4$

Thus, $s(-5) = 4$

45. $f(x) = 3.4x^2 - 1.2x + 0.5$, find $f(-0.3)$

$f(-0.3) = 3.4(-0.3)^2 - 1.2(-0.3) + 0.5$

$= 3.4(0.09) - 1.2(-0.3) + 0.5$

$= 0.306 + 0.36 + 0.5$

$= 0.666 + 0.5$

$= 1.166$

Thus, $f(-0.3) = 1.166$

Complete each table of function values and then graph each function.

47. $f(x) = -3x - 2$

x	$f(x)$
-2	$f(-2) = -3(-2) - 2$ $= 6 - 2$ $= 4$
-1	$f(-1) = -3(-1) - 2$ $= 3 - 2$ $= 1$
0	$f(0) = -3(0) - 2$ $= 0 - 2$ $= -2$
1	$f(1) = -3(1) - 2$ $= -3 - 2$ $= -5$

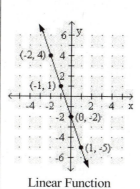

Linear Function

49. $h(x) = |1 - x|$

x	$h(x)$
-2	$h(-2) = |1 - (-2)|$ $= |1 + 2|$ $= |3|$ $= 3$
-1	$h(-1) = |1 - (-1)|$ $= |1 + 1|$ $= |2|$ $= 2$
0	$h(0) = |1 - 0|$ $= |1|$ $= 1$
1	$h(1) = |1 - 1|$ $= |0|$ $= 0$
2	$h(2) = |1 - 2|$ $= |-1|$ $= 1$
3	$h(3) = |1 - 3|$ $= |-2|$ $= 2$
4	$h(4) = |1 - 4|$ $= |-3|$ $= 3$

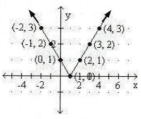

Absolute Value Function

Graph each function.

51. $f(x) = \dfrac{1}{2}x - 2$

x	$f(x)$
-2	$f(-2) = \dfrac{1}{2}(-2) - 2$ $= -1 - 2$ $= -3$
0	$f(0) = \dfrac{1}{2}(0) - 2$ $= 0 - 2$ $= -2$
2	$f(2) = \dfrac{1}{2}(2) - 2$ $= 1 - 2$ $= -1$

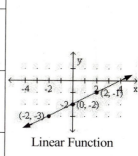

Linear Function

53. $h(x) = -|x|$

x	$h(x)$
-3	$h(-3) = -|-3|$ $= -(3)$ $= -3$
-2	$h(-2) = -|-2|$ $= -(2)$ $= -2$
-1	$h(-1) = -|-1|$ $= -(1)$ $= -1$
0	$h(0) = -|0|$ $= 0$
1	$h(1) = -|1|$ $= -(1)$ $= -1$
2	$h(2) = -|2|$ $= -(2)$ $= -2$
3	$h(3) = -|3|$ $= -(3)$ $= -3$

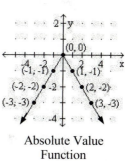

Absolute Value Function

Determine whether each graph is the graph of a function. If it is not, find ordered pairs that show a value of x that is assigned more than one value of y.

55. Yes

57. No; $(3, 4), (3, -1)$; Answers will vary.

59. No; $(0, 2), (0, -4)$; Answers will vary.

61. No; $(3, 0), (3, 1)$; Answers will vary.

APPLICATIONS

63. REFLECTIONS

$$f(x) = |x|$$

Section 3.8

65. VACATIONING

$$C(d) = 500 + 100(d - 3)$$
$$C(7) = 500 + 100(7 - 3)$$
$$= 500 + 100(4)$$
$$= 500 + 400$$
$$= 900$$

The cost will be $900.

67. LAWN SPRINKLERS

$$A(r) = \pi r^2, \quad \pi = 3.141592654$$
$$A(5) \approx (3.141592654)(5)^2$$
$$\approx (3.141592654)(25)$$
$$\approx 78.53$$
$$\approx 78.5$$

The area covered is about 78.5 ft^2.

$$A(r) = \pi r^2, \quad \pi = 3.141592654$$
$$A(20) \approx (3.141592654)(20)^2$$
$$\approx (3.141592654)(400)$$
$$\approx 1,256.63$$
$$\approx 1,256.6$$

The area covered is about $1,256.6 \text{ ft}^2$.

69. POSTAGE

Yes

WRITING

71-75. Answers will vary.

REVIEW PROBLEMS

77. COFFEE BLENDS

Analyze

• Regular coffee sells for $4/lb.
• Gourmet coffee sells for $7/lb.
• 40 lbs of gourmet coffee on hand.
• Blend to sell for $5/lb.
• How much regular coffee is needed?

Assign

Let x = the amount of regular coffee in lb.

Form

	Amount ·	Value	= Total value
Regular	x	4	$4x$
Gourmet	40	7	$7(40)$
Blend	$x + 40$	5	$5(x + 40)$

The value of the regular coffee	plus	the value of the gourmet coffee	equals	the value of the blend.
$4x$	+	280	=	$5(x+40)$

Solve

$$4x + 280 = 5(x + 40)$$
$$4x + 280 = 5x + 200$$
$$4x + 280 - \mathbf{4x} = 5x + 200 - \mathbf{4x}$$
$$280 = x + 200$$
$$280 - \mathbf{200} = x + 200 - \mathbf{200}$$
$$80 = x$$

State

80 pounds of regular coffee will be needed.

Check

The value of regular is 80($4), or $320.
The value of gourmet is 40($7), or $280.
The value of blend is 120($5), or $600.
Since the total was $320 + $280 = $600,
the result checks.

CHALLENGE PROBLEMS

79. No. The graph contains many values of x that are assigned more than one value of y, such as (2, 1) and (2, 2).

81. Let $f(x) = -2x + 5$. For what value of x is $f(x) = -7$?

If $f(x) = -2x + 5$ and $f(x) = -7$, then $-2x + 5 = -7$.

$$-2x + 5 = -7$$
$$-2x + 5 - \mathbf{5} = -7 - \mathbf{5}$$
$$-2x = -12$$
$$\frac{-2x}{-\mathbf{2}} = \frac{-12}{-\mathbf{2}}$$
$$x = 6$$

The value of x is 6.

CHAPTER 3 REVIEW

SECTION 3.1

Graphing Using the Rectangular Coordinate System

1. Graph the points with coordinates $(-1, 3)$, $(0, 1.5)$, $(-4, -4)$, $\left(2, \frac{7}{2}\right)$, and $(4, 0)$.

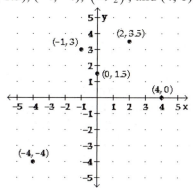

2. **HAWAII**
   Estimate the coordinates of Oahu using an ordered pair of the form (longitude, latitude). **(158, 21.5)**

3. In what quadrant does the point $(-3, -4)$ lie? **Quadrant III**

4. What are the coordinates of the origin? **(0, 0)**

5. **GEOMETRY**
   Three vertices (corners) of a square are $(-5, 4)$, $(-5, -2)$, and $(1, -2)$. Find the coordinates of the fourth vertex and then find the area of the square.

   The coordinates of the fourth vertex would be **(1, 4)**. Each side of the square is 6 units long.

   $$A = lw$$
   $$= (6)(6)$$
   $$= 36$$

   The area of the square is 36 unit^2.

6. **COLLEGE ENROLLMENTS**
   The graph gives the number of students enrolled at a college for the period from 4 weeks before to 5 weeks after the semester began.
   a. What was the maximum enrollment and when did it occur? **2,500; week 2**
   b. How many students had enrolled 2 weeks before the semester began? **1,000**
   c. When was the enrollment 2,250?
   1st week and 5th week

SECTION 3.2

Graphing Linear Equations

7. Is $(-3, -2)$ a solution of $y = 2x + 4$?
   $$y = 2x + 4$$
   $$(-2) \overset{?}{=} 2(-3) + 4$$
   $$-2 \overset{?}{=} -6 + 4$$
   $$-2 = -2$$
   True
   $(-3, -2)$ is a solution of $y = 2x + 4$.

8. Complete the table of solutions.
   $3x + 2y = -18$

x	y	(x, y)
-2	**-6**	**$(-2, -6)$**
-8	3	**$(-8, 3)$**

   $$3x + 2y = -18 \qquad\qquad 3x + 2y = -18$$
   $$3(-2) + 2y = -18 \qquad 3x + 2(3) = -18$$
   $$-6 + 2y = -18 \qquad\quad 3x + 6 = -18$$
   $$-6 + 2y + 6 = -18 + 6 \quad 3x + 6 - 6 = -18 - 6$$
   $$2y = -12 \qquad\qquad 3x = -24$$
   $$\frac{2y}{2} = \frac{-12}{2} \qquad\qquad \frac{3x}{3} = \frac{-24}{3}$$
   $$y = -6 \qquad\qquad x = -8$$

9. Which of the following equations are not linear equations?

   $$8x - 2y = 6 \qquad y = x^2 + 1 \qquad y = x$$

   $$3y = -x + 4 \qquad y - x^3 = 0$$

   $$y = x^2 + 1 \text{ and } y - x^3 = 0$$

10. The graph of a linear equation is shown.

    a. When the coordinates of point A are substituted into the equation, will a true or false statement result?
    True, because point A lies on the line and therefore is a solution of the equation.

    b. When the coordinates of point B are substituted into the equation, will a true or false statement result?

    False, because point B does not lie on the line and therefore is not a solution of the equation.

Graph each equation by constructing a table of solutions.

11. $y = 4x - 2$

x	y	(x, y)
-1	$y = 4x - 2$ $y = 4(\mathbf{-1}) - 2$ $y = -4 - 2$ $y = -6$	$(-1, -6)$
0	$y = 4x - 2$ $y = 4(\mathbf{0}) - 2$ $y = 0 - 2$ $y = -2$	$(0, -2)$
1	$y = 4x - 2$ $y = 4(\mathbf{1}) - 2$ $y = 4 - 2$ $y = 2$	$(1, 2)$

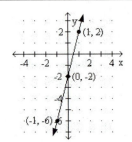

12. $y = \dfrac{3}{4}x$

x	y	(x, y)
-4	$y = \dfrac{3}{4}x$ $y = \dfrac{3}{4}(\mathbf{-4})$ $y = -3$	$(-4, -3)$
0	$y = \dfrac{3}{4}x$ $y = \dfrac{3}{4}(\mathbf{0})$ $y = 0$	$(0, 0)$
4	$y = \dfrac{3}{4}x$ $y = \dfrac{3}{4}(\mathbf{4})$ $y = 3$	$(4, 3)$

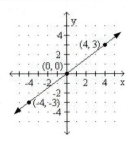

13. $5y = -5x + 15$

$$\dfrac{5y}{5} = \dfrac{-5x}{5} + \dfrac{15}{5}$$

$$y = -x + 3$$

x	y	(x, y)
-1	$y = -x + 3$ $y = -(\mathbf{-1}) + 3$ $y = 1 + 3$ $y = 4$	$(-1, 4)$
0	$y = -x + 3$ $y = -(\mathbf{0}) + 3$ $y = 0 + 3$ $y = 3$	$(0, 3)$
1	$y = -x + 3$ $y = -(\mathbf{1}) + 3$ $y = -1 + 3$ $y = 2$	$(1, 2)$

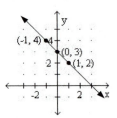

14. $6y = -4x$

$$\dfrac{6y}{6} = \dfrac{-4x}{6}$$

$$y = -\dfrac{2}{3}x$$

x	y	(x, y)
-3	$y = -\dfrac{2}{3}x$ $y = -\dfrac{2}{3}(\mathbf{-3})$ $y = 2$	$(-3, 2)$
0	$y = -\dfrac{2}{3}x$ $y = -\dfrac{2}{3}(\mathbf{0})$ $y = 0$	$(0, 0)$
3	$y = -\dfrac{2}{3}x$ $y = -\dfrac{2}{3}(\mathbf{3})$ $y = -2$	$(3, -2)$

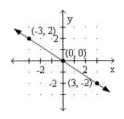

Chapter 3 Review and Chapter 3 Test   - 180 -

15. BIRTHDAY PARTIES

$$c = 8n + 50$$

Do not calculate the answer.

n	c	(n, c)
1	$c = 8n + 50$ $c = 8(1) + 50$ $c = 8 + 50$ $c = 58$	$(1, 58)$
5	$c = 8n + 50$ $c = 8(5) + 50$ $c = 40 + 50$ $c = 90$	$(5, 90)$
10	$c = 8n + 50$ $c = 8(10) + 50$ $c = 80 + 50$ $c = 130$	$(10, 130)$

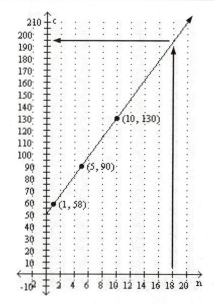

According to the graph, the cost of the party for 18 children is about $195.

16. Determine whether each statement is true or false.
   a. It takes three or more points to determine a line. **False**
   b. A linear equation in two variables has infinitely many solutions. **True**

SECTION 3.3 REVIEW
Intercepts

17. Identify the x- and y-intercepts of the graph.
   The x-intercept is $(-3, 0)$.

   The y-intercept is $(0, 2.5)$.

18. **DEPRECIATION**

   The y-intercept $(0, 25{,}000)$ indicates the equipment was originally valued at $25,000.

   The x-intercept $(10, 0)$ indicates in 10 years the sound equipment has no value.

Use the intercept method to graph each equation.

19. $-4x + 2y = 8$

   y-intercept:
   If $x = 0$,
   $-4(0) + 2y = 8$
   $2y = 8$
   $y = 4$

   x-intercept:
   If $y = 0$
   $-4x + 2(0) = 8$
   $-4x = 8$
   $x = -2$

   The y-intercept is $(0, 4)$, and the x-intercept is $(-2, 0)$.

   Check Point
   $-4x + 2y = 8$
   $-4(-1) + 2y = 8$
   $4 + 2y - 4 = 8 - 4$
   $2y = 4$
   $y = 2$
   $(-1, 2)$

20. $5x - 4y = 13$

   y-intercept:
   If $x = 0$,
   $5(0) - 4y = 13$
   $-4x = 13$
   $y = -13/4$

   x-intercept:
   If $y = 0$
   $5x - 4(0) = 13$
   $5x = 13$
   $x = 13/5$

   The y-intercept is $\left(0, -\dfrac{13}{4}\right)$, and the x-intercept is $\left(\dfrac{13}{5}, 0\right)$.

   Check Point
   $5x - 4y = 13$
   $5(1) - 4y = 13$
   $5 - 4y - 5 = 13 - 5$
   $-4y = 8$
   $y = -2$
   $(1, -2)$

Chapter 3 Review and Chapter 3 Test

21. $y = 4$

Let $x = -2$ | Let $x = 0$ | Let $x = 3$
$0x + y = 4$ | $0x + y = 4$ | $0x + y = 4$
$0(-2) + y = 4$ | $0(0) + y = 4$ | $0(3) + y = 4$
$y = 4$ | $y = 4$ | $y = 4$
$(-2, 4)$ | $(0, 4)$ | $(3, 4)$

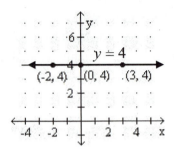

22. $x = -1$

Let $y = -3$ | Let $y = 0$ | Let $y = 2$
$x + 0y = -1$ | $x + 0y = -1$ | $x + 0y = -1$
$x + 0(-3) = -1$ | $x + 0(0) = -1$ | $x + 0(2) = -1$
$x = -1$ | $x = -1$ | $x = -1$
$(-1, -3)$ | $(-1, 0)$ | $(-1, 2)$

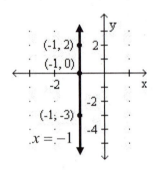

SECTION 3.4 REVIEW
Slope and Rate of Change
In each case, find the slope of the line.

23. $m = \dfrac{1}{4}$

24. $m = -\dfrac{7}{8}$

25. The line with this table of solutions

x	y	(x, y)
2	-3	$(2, -3)$
4	-17	$(4, -17)$

Let $(x_1, y_1) = (2, -3)$
Let $(x_2, y_2) = (4, -17)$

$$m = \frac{y_2 - y_1}{x_2 - x_1}$$

$$= \frac{-17 - (-3)}{4 - 2}$$

$$= \frac{-17 + 3}{4 - 2}$$

$$= \frac{-14}{2}$$

$$= -7$$

26. The line passing through the points $(1, -4)$ and $(3, -7)$

Let $(x_1, y_1) = (1, -4)$
Let $(x_2, y_2) = (3, -7)$

$$m = \frac{y_2 - y_1}{x_2 - x_1}$$

$$= \frac{-7 - (-4)}{3 - 1}$$

$$= \frac{-7 + 4}{3 - 1}$$

$$= -\frac{3}{2}$$

27. Draw a line having a slope that is

a. Positive

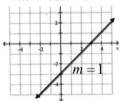

b. Negative

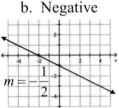

c. 0

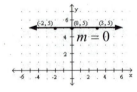

d. Undefined

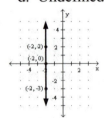

28. CARPENTRY

$$m = \frac{\text{rise}}{\text{run}}$$

$$= \frac{6}{8}$$

$$= \frac{\overset{1}{\cancel{2}} \cdot 3}{\underset{1}{\cancel{2}} \cdot 4}$$

$$= \frac{3}{4}$$

29. RAMPS

$$m = \frac{\text{rise}}{\text{run}}$$

$$= \frac{2}{24}$$

$$\approx 0.083$$

$$\approx 8.3\%$$

The grade is about 8.3%.

30. BOTTLED WATER

a. The y values would be the number of gallons and the x values would be the years.

$$m = \frac{\text{change in } y}{\text{change in } x}$$

$$= \frac{29 - 17}{2008 - 2000}$$

$$= \frac{12}{8}$$

$$= 1.5$$

The rate of change was 1.5 gallons per year.

31. Step 1: Find each slope.

Let $(x_1, y_1) = (6,6)$

Let $(x_2, y_2) = (4,2)$

Let $(x_1, y_1) = (2,-10)$

Let $(x_2, y_2) = (-2,-2)$

$$m_1 = \frac{y_2 - y_1}{x_2 - x_1} \qquad m_2 = \frac{y_2 - y_1}{x_2 - x_1}$$

$$= \frac{2-6}{4-6} \qquad\qquad = \frac{-2-(-10)}{-2-2}$$

$$= \frac{-4}{-2} \qquad\qquad = \frac{-2+10}{-2-2}$$

$$= 2 \qquad\qquad\qquad = \frac{8}{-4}$$

$$\qquad\qquad\qquad\qquad = -2$$

Step 2: Compare the 2 slopes.

The slopes are not equal.
The slopes are not negative reciprocals.
The lines are neither parallel nor perpendicular.

32. Step 1: Determine the slope.

Let $(x_1, y_1) = (-1, 9)$

Let $(x_2, y_2) = (-8, 4)$

$$m = \frac{y_2 - y_1}{x_2 - x_1}$$

$$= \frac{4-9}{-8-(-1)}$$

$$= \frac{4-9}{-8+1}$$

$$= \frac{-5}{-7}$$

$$= \frac{5}{7}$$

Step 2: Determine the negative reciprocal.

The negative reciprocal is $-\dfrac{7}{5}$.

This is the slope of the line perpendicular and passing through the given points.

SECTION 3.5 REVIEW
Slope - Intercept Form
Find the slope and the y - intercept of each line.

33. $y = \dfrac{3}{4}x - 2$

Since $m = \dfrac{3}{4}$ and $b = -2$, the slope is $\dfrac{3}{4}$

and the y-intercept is $(0, -2)$.

Chapter 3 Review and Chapter 3 Test

34. $y = -4x$

    $y = -4x + 0$

    Since $m = -4$ and $b = 0$, the slope is -4

    and the y-intercept is $(0,0)$.

35. $y = \dfrac{x}{8} + 10$

    $y = \dfrac{1}{8}x + 10$

    Since $m = \dfrac{1}{8}$ and $b = 10$, the slope is $\dfrac{1}{8}$

    and the y-intercept is $(0,10)$.

36. $\quad 7x + 5y = -21$

    $7x + 5y - 7x = -21 - 7x$

    $\quad 5y = -7x - 21$

    $\quad \dfrac{5y}{5} = \dfrac{-7x}{5} - \dfrac{21}{5}$

    $\quad y = -\dfrac{7}{5}x - \dfrac{21}{5}$

    Since $m = -\dfrac{7}{5}$ and $b = -\dfrac{21}{5}$, the slope is $-\dfrac{7}{5}$

    and the y-intercept is $\left(0, -\dfrac{21}{5}\right)$.

37. Slope 4, y-intercept $(0, -1)$

    The equation of the line is $y = 4x - 1$.

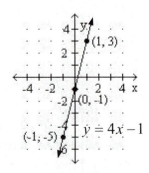

38. $m = \dfrac{3}{2}$, y-intercept is $(0, -3)$

    Equation of the line is $y = \dfrac{3}{2}x - 3$.

39. $\quad 9x - 3y = 15$

    $9x - 3y - 9x = 15 - 9x$

    $\quad -3y = -9x + 15$

    $\quad \dfrac{-3y}{-3} = \dfrac{-9x}{-3} + \dfrac{15}{-3}$

    $\quad y = 3x - 5$

    $m = 3$; y-intercept is $(0, -5)$.

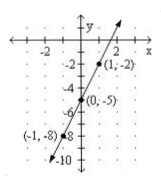

40. COPIERS

a. • Copies already made, 75,000

    • New copies, 300 per wk

    • Total copies is not known (c)

Total copies	equals	new copies of 300 per wk	plus	75,000 copies already made.
c	$=$	$300w$	$+$	$75,000$

b. Find the predicted copies after 52 weeks, replace w with 52 and then calculate the result.

    $c = 300w + 75,000$

    $c = 300(52) + 75,000$

    $c = 15,600 + 75,000$

    $c = 90,600$

    The total copies for 52 weeks is 90,600.

Without graphing, determine whether graphs of the given pairs of lines are be parallel, perpendicular, or neither.

41. $y = -\dfrac{2}{3}x + 6$ $\quad\bigg|\quad$ $y = -\dfrac{2}{3}x - 6$

    $m = -\dfrac{2}{3}$ $\quad\bigg|\quad$ $m = -\dfrac{2}{3}$

    The slopes are equal.

    The lines are parallel.

42.

$$x + 5y = -10$$
$$x + 5y - x = -10 - x$$
$$5y = -x - 10$$
$$\frac{5y}{5} = \frac{-x}{5} - \frac{10}{5}$$
$$y = -\frac{1}{5}x - 2$$
$$m = -\frac{1}{5}$$

$$y - 5x = 0$$
$$y - 5x + 5x = 0 + 5x$$
$$y = 5x$$
$$m = 5$$

The slopes are negative reciprocals.
The lines are perpendicular.

SECTION 3.6 REVIEW
Point - Slope Form

Find an equation of a line with the given slope that passes through the given point. Write the equation in slope - intercept form and graph it.

43. Passes through $(1, 5)$, $m = 3$

$$x_1 = 1 \text{ and } y_1 = 5$$
$$y - y_1 = m(x - x_1)$$
$$y - 5 = 3(x - 1) \text{ point-slope form}$$

$$y - 5 = 3(x - 1)$$
$$y - 5 = 3x - 3$$
$$y - 5 + 5 = 3x - 3 + 5$$
$$y = 3x + 2 \text{ slope-intercept form}$$

y-int $(0, 2)$, $m = 3 = \dfrac{3}{1}$

Start at $(0, 2)$, go up 3, right 1

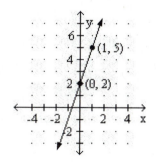

44. Passes through $(-4, -1)$, $m = -\dfrac{1}{2}$

$$x_1 = -4 \text{ and } y_1 = -1$$
$$y - y_1 = m(x - x_1)$$
$$y - (-1) = -\frac{1}{2}(x - (-4))$$
$$y + 1 = -\frac{1}{2}(x + 4) \text{ point-slope form}$$

$$y + 1 = -\frac{1}{2}(x + 4)$$
$$y + 1 = -\frac{1}{2}(x) - \frac{1}{2}(4)$$
$$y + 1 = -\frac{1}{2}x - 2$$
$$y + 1 - 1 = -\frac{1}{2}x - 2 - 1$$
$$y = -\frac{1}{2}x - 3 \text{ slope-intercept form}$$

y-int $(0, -3)$, $m = -\dfrac{1}{2} = \dfrac{-1}{2}$

Start at $(0, -3)$, go down 1, right 2

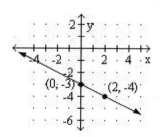

Write an equation of the line with the following characteristics. Write the equation in slope - intercept form.

45. Let $(x_1, y_1) = (3, 7)$

Let $(x_2, y_2) = (-6, 1)$

Step 1: Find the slope.

$$m = \frac{y_2 - y_1}{x_2 - x_1}$$
$$= \frac{1 - 7}{-6 - 3}$$
$$= \frac{-6}{-9}$$
$$= \frac{2}{3}$$

Chapter 3 Review and Chapter 3 Test

Passes through $(3,7)$, $m = \dfrac{2}{3}$

Step 2: $x_1 = 3$ and $y_1 = 7$

$$y - y_1 = m(x - x_1)$$

$$y - 7 = \frac{2}{3}(x - 3)$$

$$y - 7 = \frac{2}{3}x - \frac{2}{3}(3)$$

$$y - 7 = \frac{2}{3}x - 2$$

$$y - 7 + 7 = \frac{2}{3}x - 2 + 7$$

$$y = \frac{2}{3}x + 5 \text{ \scriptsize slope-intercept form}$$

46. Horizontal, passes through $(6, -8)$

The equation of a horizontal line can be written in the form $y = b$.

$$y = -8$$

47. CAR REGISTRATION

a. In this problem, we are told to let $x = $ # of years and $y = $ value. Now you must find the slope from the two ordered pairs:
$(2, 380)$ and $(4, 310)$

Let $(x_1, y_1) = (2, 380)$

Let $(x_2, y_2) = (4, 310)$

$$m = \frac{y_2 - y_1}{x_2 - x_1}$$

$$= \frac{310 - 380}{4 - 2}$$

$$= \frac{-70}{2}$$

$$= -35$$

Now pick one point $(2, 380)$ and use

$$y - y_1 = m(x - x_1)$$

$$y - 380 = -35(x - 2)$$

$$y - 380 = -35x + 70$$

$$y - 380 + 380 = -35x + 70 + 380$$

$$y = -35x + 450 \text{ \scriptsize slope-intercept form}$$

48. THE ATMOSPHERE

a. Let $(x_1, y_1) = (20, 340)$

Let $(x_2, y_2) = (40, 370)$

Step 1: Find the slope.

$$m = \frac{y_2 - y_1}{x_2 - x_1}$$

$$= \frac{370 - 340}{40 - 20}$$

$$= \frac{30}{20}$$

$$= \frac{3}{2}$$

Passes through $(20, 340)$, $m = \dfrac{3}{2}$

Step 2: $x_1 = 20$ and $y_1 = 340$

$$y - y_1 = m(x - x_1)$$

$$y - 340 = \frac{3}{2}(x - 20)$$

$$y - 340 = \frac{3}{2}x - 30$$

$$y - 340 + 340 = \frac{3}{2}x - 30 + 340$$

$$y = \frac{3}{2}x + 310 \text{ \scriptsize slope-intercept form}$$

b. Use 60 for x (the year $2020 - 1960 = 60$) to calculate the amount of carbon dioxide.

$$y = \frac{3}{2}x + 310$$

$$y = \frac{3}{2}(60) + 310$$

$$y = 90 + 310$$

$$y = 400$$

The amount of carbon dioxide in 2020 will be 400 parts per million.

SECTION 3.7 REVIEW
Graphing Linear Inequalities

49. Determine whether each ordered pair is a solution of $2x - y \leq -4$.

   a. $2x - y \leq -4;\ (0, 5)$

   $$2(\mathbf{0}) - (\mathbf{5}) \overset{?}{\leq} -4$$

   $$0 - 5 \overset{?}{\leq} -4$$

   $$-5 \leq -4 \quad \text{True}$$

   $(0, 5)$ is a solution.

   b. $2x - y \leq -4;\ (2, 8)$

   $$2(\mathbf{2}) - (\mathbf{8}) \overset{?}{\leq} -4$$

   $$4 - 8 \overset{?}{\leq} -4$$

   $$-4 \leq -4 \quad \text{True}$$

   $(2, 8)$ is a solution.

   c. $2x - y \leq -4;\ (-3, -2)$

   $$2(\mathbf{-3}) - (\mathbf{-2}) \overset{?}{\leq} -4$$

   $$-6 + 2 \overset{?}{\leq} -4$$

   $$-4 \leq -4 \quad \text{True}$$

   $(-3, -2)$ is a solution.

   d. $2x - y \leq -4;\ \left(\dfrac{1}{2}, -5\right)$

   $$2\left(\dfrac{\mathbf{1}}{\mathbf{2}}\right) - (\mathbf{-5}) \overset{?}{\leq} -4$$

   $$1 + 5 \overset{?}{\leq} -4$$

   $$6 \leq -4 \quad \text{False}$$

   $\left(\dfrac{1}{2}, -5\right)$ is not a solution.

50. Fill in the blanks: $2x - 3y \geq 6$

   means $2x - 3y \boxed{=} 6$ or $2x - 3y \boxed{>} 6$.

Graph each inequality.

51. $x - y < 5$, Boundary line is dashed because of the $<$ symbol.

   Graph as $x - y = 5$.

y-intercept:	x-intercept:
If $x = 0$,	If $y = 0$,
$\mathbf{0} - y = 5$	$x - \mathbf{0} = 5$
$-y = 5$	$x = 5$
$\dfrac{-y}{-1} = \dfrac{5}{-1}$	
$y = -5$	

   The y-int is $(0, -5)$, and the x-int is $(5, 0)$.

   Select test point $(0,0)$ and substitute into

   $x - y < 5$

   $$\mathbf{0} - \mathbf{0} \overset{?}{<} 5$$

   $$0 < 5$$

   True

   The coordinates of every point on the same side of the line as the origin satisfy the inequality. To indicate this, we shade the half-plane that contains the test point $(0,0)$.

   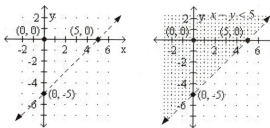

52. $2x - 3y \geq 6$, Boundary line is solid because of the $\geq$ symbol.
   Graph as $2x - 3y = 6$.

   The y-int is $(0, -2)$, and the x-int is $(3, 0)$.

   Select test point $(0,0)$ and substitute into

   $2x - 3y \geq 6$

   $$2(\mathbf{0}) - 3(\mathbf{0}) \overset{?}{\geq} 6$$

   $$0 \geq 6$$

   False

   The coordinates of every point on the same side of the line as the origin **does not** satisfy the inequality. To indicate this, we shade the half-plane that **does not** contains the test point $(0,0)$.

   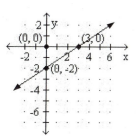
   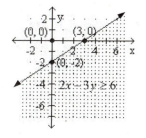

Chapter 3 Review and Chapter 3 Test

53. $y \leq -2x$, Boundary line is <u>solid</u>
because of the $\leq$ symbol.
Graph as $y = -2x$.

y-intercept:	x-intercept:
If $x = 0$,	If $y = 0$,
$y = -2(\mathbf{0})$	$\mathbf{0} = -2x$
$y = 0$	$0 = x$

The y-int is $(0, 0)$, and the x-int is $(0, 0)$.

The line goes through the origin so select test point $(0,2)$ and substitute into

$y \leq -2x$

$2 \overset{?}{\leq} -2(\mathbf{0})$

$2 \leq 0$

False

The coordinates of every point on the same side of the line as $(0,2)$ *does not* satisfy the inequality. To indicate this, we shade the half-plane that *does not* contain the test point $(0,2)$.

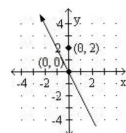 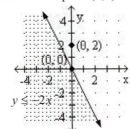

54. $y < -4$, Boundary line is <u>dashed</u>
because of the $<$ symbol.

Graph as $y = -4$.
The y-int is $(0, -4)$, and there is no x-int.
$y = -4$ is a line parallel to the x-axis.
Select test point $(0,0)$ and substitute into

$y < -4$

$0 < -4$

False

The coordinates of every point on the same side of the line as the origin *do not* satisfy the inequality. To indicate this, we shade the half-plane that *does not* contain the test point $(0, 0)$.

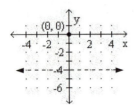 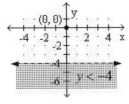

55. a. True   b. False   c. False

56. WORK SCHEDULES
$3x + 5y \leq 30$

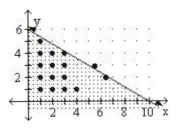

$(2,4)$, $(5,3)$, $(6,2)$
Answers will vary.

SECTION 3.8 REVIEW
An Introduction to Functions
Find the domain and range of each relation.

57. $\{(7, -3), (-5, 9), (4, 4), (0, -11)\}$

Domain: $\{-5, 0, 4, 7\}$

Range: $\{-11, -3, 4, 9\}$

58. $\{(2, -2), (15, -8), (-6, 9), (1, -8)\}$

Domain: $\{-6, 1, 2, 15\}$

Range: $\{-8, -2, 9\}$

Determine whether each arrow diagram or table defines y as a function of x. If a function is defined, give its domain and range. If it does not define a function, find ordered pairs that show a value of x that is assigned more than one value of y.

59. Yes; domain:$\{1, 4, 8\}$; range:$\{0, 6, 9\}$

60. Yes; domain:$\{2,3,5,6\}$; range:$\{1,4\}$

61. Yes; domain:$\{3,5,7,9\}$; range:$\{9,25,49,81\}$

62. No; $(-1, 2)$, $(-1, 4)$

63. Yes; domain:$\{-1, 0, 1, 2\}$; range:$\{5, 6\}$

64. No; $(4, 4)$, $(4, 6)$

Fill in the blanks.

65. The set of all input values for a function is called the **domain**, and the set of all output values is called the **range**.

66. Since $y = $ **$f(x)$**, the equations $y = 2x - 8$ and $f(x) = 2x - 8$ are equivalent.

For $f(x) = x^2 - 4x$, find each of the following function values.

67. $f(x) = x^2 - 4x$

$f(\mathbf{1}) = \mathbf{1}^2 - 4(\mathbf{1})$

$= 1 - 4$

$= -3$

Thus, $f(1) = -3$

68. $f(x) = x^2 - 4x$
$f(0) = 0^2 - 4(0)$
$= 0 - 0$
$= 0$
Thus, $f(0) = 0$

69. $f(x) = x^2 - 4x$
$f(-3) = (-3)^2 - 4(-3)$
$= 9 + 12$
$= 21$
Thus, $f(-3) = 21$

70. $f(x) = x^2 - 4x$
$f\left(\dfrac{1}{2}\right) = \left(\dfrac{1}{2}\right)^2 - 4\left(\dfrac{1}{2}\right)$
$= \dfrac{1}{4} - \dfrac{4}{2}$
$= \dfrac{1}{4} - \dfrac{8}{4}$
$= -\dfrac{7}{4}$
Thus, $f\left(\dfrac{1}{2}\right) = -\dfrac{7}{4}$

For $g(x) = 1 - 6x$, find each of the following function values.

71. $g(x) = 1 - 6x$
$g(1) = 1 - 6(1)$
$= 1 - 6$
$= -5$
Thus, $g(1) = -5$

72. $g(x) = 1 - 6x$
$g(-6) = 1 - 6(-6)$
$= 1 + 36$
$= 37$
Thus, $g(-6) = 37$

73. $g(x) = 1 - 6x$
$g(0.5) = 1 - 6(0.5)$
$= 1 - 3$
$= -2$
Thus, $g(0.5) = -2$

74. $g(x) = 1 - 6x$
$g\left(\dfrac{3}{2}\right) = 1 - 6\left(\dfrac{3}{2}\right)$
$= 1 - 9$
$= -8$

Thus, $g\left(\dfrac{3}{2}\right) = -8$

Determine whether each graph is the graph of a function. If it is not, find two ordered pairs that show a value of x that is assigned more than one value of y.

75. No: $(1, 0.5), (1, 4)$ (answers may vary)

76. Yes.

Complete the table of function values. Then graph the function.

77. $f(x) = 1 - |x|$

x	$f(x)$
0	$f(x) = 1 - |x|$ $f(0) = 1 - |0|$ $= 1 - 0$ $= 1$
1	$f(x) = 1 - |x|$ $f(1) = 1 - |1|$ $= 1 - 1$ $= 0$
3	$f(x) = 1 - |x|$ $f(3) = 1 - |3|$ $= 1 - 3$ $= -2$
−1	$f(x) = 1 - |x|$ $f(-1) = 1 - |-1|$ $= 1 - 1$ $= 0$
−3	$f(x) = 1 - |x|$ $f(-3) = 1 - |-3|$ $= 1 - 3$ $= -2$

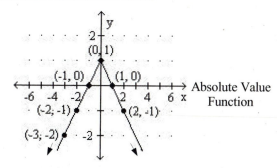

Absolute Value Function

78. ALUMINUM CANS
$V(r) = 15.7r^2$
$V(8) = 15.7(8)^2$
$= 15.7(64)$
$= 1{,}004.8$
The volume of the can is $1{,}004.8$ in^3.

Chapter 3 Review and Chapter 3 Test

CHAPTER 3 TEST

1. Fill in the blanks.

a. A rectangular coordinate system is formed by two perpendicular number lines called the x- **axis** and the y- **axis** .

b. A **solution** of an equation in two variables is an ordered pair of numbers that makes the equation a true statement.

c. $3x + y = 10$ is called a **linear** equation in two variables because its graph is a line.

d. The **slope** of a line is a measure of steepness.

e. A **function** is a set of ordered pairs in which to each first component there corresponds exactly one second component.

The graph shows the number of dogs being boarded in a kennel over a 3-day holiday weekend.

2. How many dogs were in the kennel 2 days before the holiday? **10**

3. What is the maximum number of dogs that were boarded on the holiday weekend? **60**

4. When were there 30 dogs in the kennel? **1 day before and the 3rd day of the holiday**

5. What information does the -intercept of the graph give? **50 dogs were in the kennel when the holiday began.**

6. Plot each point on a rectangular coordinate system:

$(1,3), (-2,4), (-3,-2), (3,-2), (-1,0),$

$(0,-1),$ and $\left(-\dfrac{1}{2}, \dfrac{7}{2}\right)$

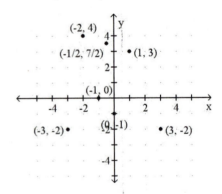

7. Find the coordinates of each point shown in the graph.
$A(2, 4), B(-3, 3), C(-2, -3), D(4, -3),$
$E(-4, 0), F(3.5, 1.5)$

8. In which quadrant is each point located?

a. $(-1,-5)$ **III**   b. $\left(6, -2\dfrac{3}{4}\right)$ **IV**

9. Is $(-3, -4)$ a solution of $3x - 4y = 7$?

$$3x - 4y = 7$$
$$3(-3) - 4(-4) \overset{?}{=} 7$$
$$-9 + 16 \overset{?}{=} 7$$
$$7 = 7$$
$$\text{True}$$

$(-3, -4)$ is a solution of $3x - 4y = 7$.

10. Complete the table of solutions for the linear equation.

x	y	(x, y)
2	$x + 4y = 6$ $2 + 4y = 6$ $2 + 4y - 2 = 6 - 2$ $4y = 4$ $\dfrac{4y}{4} = \dfrac{4}{4}$ $y = 1$	$(2,1)$
$x + 4y = 6$ $x + 4(3) = 6$ $x + 12 = 6$ $x + 12 - 12 = 6 - 12$ $x = -6$	3	$(-6,3)$

11. a. False, because point C does not lie on the line and therefore is not a solution of the equation.

   b. True, because point D lies on the line and therefore is a solution of the equation.

12. Graph: $y = \dfrac{x}{3}$

x	y	(x, y)
-3	$y = \dfrac{x}{3}$ $y = \dfrac{-3}{3}$ $y = -1$	$(-3, -1)$
0	$y = \dfrac{x}{3}$ $y = \dfrac{0}{3}$ $y = 0$	$(0,0)$
3	$y = \dfrac{x}{3}$ $y = \dfrac{3}{3}$ $y = 1$	$(3, 1)$

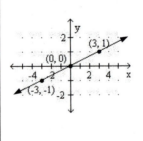

13. What are the *x*- and *y*-intercepts of the graph of $2x - 3y = 6$?

let $x = 0$	let $y = 0$
$2x - 3y = 6$	$2x - 3y = 6$
$2(0) - 3y = 6$	$2x - 3(0) = 6$
$-3y = 6$	$2x = 6$
$\dfrac{-3y}{-3} = \dfrac{6}{-3}$	$\dfrac{2x}{2} = \dfrac{6}{2}$
$y = -2$	$x = 3$
y-int: $(0, -2)$	*x*-int: $(3, 0)$

14. Graph: $8x + 4y = -24$.

let $x = 0$	let $y = 0$
$8x + 4y = -24$	$8x + 4y = -24$
$8(0) + 4y = -24$	$8x + 4(0) = -24$
$4y = -24$	$8x = -24$
$\dfrac{4y}{4} = \dfrac{-24}{4}$	$\dfrac{8x}{8} = \dfrac{-24}{8}$
$y = -6$	$x = -3$
y-int: $(0, -6)$	*x*-int: $(-3, 0)$

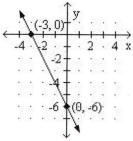

15. Find the slope of the line.

$$m = \frac{8}{7}$$

16. Find the slope of the line passing through

Let $(x_1, y_1) = (-1, 3)$

Let $(x_2, y_2) = (3, -1)$

$$m_1 = \frac{y_2 - y_1}{x_2 - x_1}$$

$$= \frac{-1 - 3}{3 - (-1)}$$

$$= \frac{-4}{3 + 1}$$

$$= \frac{-4}{4}$$

$$= -1$$

The slope of the line is -1.

17. What is the slope of a horizontal line?

The slope is zero or $m = 0$.

18. RAMPS

$$m = \frac{\text{rise}}{\text{run}}$$

$$= \frac{2}{20}$$

$$= 0.1$$

$$= 10\%$$

The grade is 10%.

19. One line passes through (9, 2) and (6, 4). Another line passes through (0, 7) and (2, 10). Without graphing, determine whether the lines are parallel, perpendicular, or neither.

Step 1: Find each slope.

Let $(x_1, y_1) = (9, 2)$	Let $(x_1, y_1) = (0, 7)$
Let $(x_2, y_2) = (6, 4)$	Let $(x_2, y_2) = (2, 10)$
$m_1 = \dfrac{y_2 - y_1}{x_2 - x_1}$	$m_2 = \dfrac{y_2 - y_1}{x_2 - x_1}$
$= \dfrac{4 - 2}{6 - 9}$	$= \dfrac{10 - 7}{2 - 0}$
$= \dfrac{2}{-3}$	$= \dfrac{3}{2}$
$= -\dfrac{2}{3}$	

Step 2: Compare the slopes.

The slopes are negative reciprocals.
The lines are perpendicular.

Chapter 3 Review and Chapter 3 Test

When graphed, are the lines $y = 2x + 6$ and $2x - y = 0$ parallel, perpendicular, or neither?

20.

$$y = 2x + 6$$

$$2x - y = 0$$
$$2x = y$$

x	y
-3	$y = 2x + 6$ $y = 2(-3) + 6$ $y = -6 + 6$ $y = 0$
-2	$y = 2x + 6$ $y = 2(-2) + 6$ $y = -4 + 6$ $y = 2$
-1	$y = 2x + 6$ $y = 2(-1) + 6$ $y = -2 + 6$ $y = 4$

x	y
-1	$y = 2x$ $y = 2(-1)$ $y = -2$
0	$y = 2x$ $y = 2(0)$ $y = 0$
1	$y = 2x$ $y = 2(1)$ $y = 2$

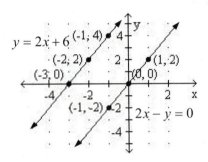

The lines are parallel.

21. Find the rate of change of the decline on which the woman is running.

(2, 250) and (12, 100)
(x_1, y_1) and (x_2, y_2)

$$m_{woman} = \frac{y_2 - y_1}{x_2 - x_1}$$

$$= \frac{100 - 250}{12 - 2}$$

$$= \frac{-150}{10}$$

$$= -15$$

Her rate of change is -15 feet per mi.

22. Find the rate of change of the incline on which the man is running.

(12, 100) and (20, 300)
(x_1, y_1) and (x_2, y_2)

$$m_{man} = \frac{y_2 - y_1}{x_2 - x_1}$$

$$= \frac{300 - 100}{20 - 12}$$

$$= \frac{200}{8}$$

$$= 25$$

His rate of change is 25 feet per mi.

23. Graph: $x = -4$.

x	y	(x, y)
-4	-3	$(-4, -3)$
-4	0	$(-4, 0)$
-4	2	$(-4, 2)$

24. Graph the line passing through $(-2, -4)$ having slope $\frac{2}{3}$.

Start at the given point $(-2, -4)$ and $m = \frac{2}{3}$.
Move up 2 spaces, then move right 3 spaces.

Start at the given point $(-2, -4)$ and $m = \frac{-2}{-3}$.
Move down 2 spaces, then move left 3 spaces.

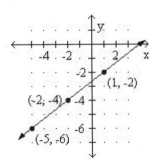

25. Find the slope and the y-intercept of the graph of $x + 2y = 8$.

$$x + 2y = 8$$

$$x + 2y - x = 8 - x$$

$$2y = -x + 8$$

$$\frac{2y}{2} = \frac{-x}{2} + \frac{8}{2}$$

$$y = -\frac{x}{2} + 4$$

The slope is $-\frac{1}{2}$. The y-intercept is $(0,4)$.

Find an equation of the line passing through $(-2, 5)$ with slope 7. Write the equation in slope - intercept form.

26. Passes through $(-2, 5)$, $m = 7$

$$x_1 = -2 \text{ and } y_1 = 5$$

$$y - y_1 = m(x - x_1)$$

$$y - 5 = 7(x - (-2))$$

$$y - 5 = 7(x + 2) \quad \text{point-slope form}$$

$$y - 5 = 7(x + 2)$$

$$y - 5 = 7x + 7(2)$$

$$y - 5 = 7x + 14$$

$$y - 5 + 5 = 7x + 14 + 5$$

$$y = 7x + 19 \quad \text{slope-intercept form}$$

27. Find an equation for the line shown. Write the equation in slope-intercept form.

Slope -2, y-intercept $(0, -5)$

The equation of the line is $y = -2x - 5$.

28. **DEPRECIATION**

a. computer cost new $15,000
loses value of $1,500 per yr
Write a linear equation that gives the resale value v of the computer x years after being purchased.

Resale value	equals	depreciation of $1,500 per yr	plus	the cost of $15,000.
v	$=$	$-1,500x$	$+$	$15,000$

b. Use your equation from part 'a' to predict the value of the computer 8 years after it is purchased.

$$v = -1,500x + 15,000$$

$$v = -1,500(8) + 15,000$$

$$v = -12,000 + 15,000$$

$$v = 3,000$$

The value of the computer after 8 years is $3,000.

29. **Determine whether (6, 1) is a solution of $2x - 4y \geq 8$.**

$$2x - 4y \geq 8$$

$$2(6) - 4(1) \overset{?}{\geq} 8$$

$$12 - 4 \overset{?}{\geq} 8$$

$$8 \geq 8$$

True

$(6,1)$ is a solution.

30. **WATER HEATERS**

Choose two ordered pairs.

$(140, 13)$ and $(180, 5)$

(T_1, y_1) and (T_2, y_2)

$$m = \frac{y_2 - y_1}{T_2 - T_1}$$

$$= \frac{5 - 13}{180 - 140}$$

$$= \frac{-8}{40}$$

$$= -\frac{1}{5}$$

Passes through $(180, 5)$, $m = -\frac{1}{5}$

$$T_1 = 180 \text{ and } y_1 = 5$$

$$y - y_1 = m(T - T_1)$$

$$y - 5 = -\frac{1}{5}(T - 180)$$

$$y - 5 = -\frac{1}{5}(T) - \frac{1}{5}(-180)$$

$$y - 5 = -\frac{1}{5}T + 36 \quad \text{point-slope form}$$

$$y - 5 + 5 = -\frac{1}{5}T + 36 + 5$$

$$y = -\frac{1}{5}T + 41 \quad \text{slope-intercept form}$$

Chapter 3 Review and Chapter 3 Test

31. Determine whether each point satisfies the inequality.

a. $(-2, 3)$, Yes, because the point lies within the shaded region of the graph and therefore is a solution of the inequality.

b. $(3, -4)$, No, because point does not lie within the shaded region of the graph and therefore is not a solution of the inequality.

c. $(0, 0)$, No, because the point lies on the the dashed boundary line of the graph and therefore is not a solution of the inequality.

32. Is $(-20, -2)$ a solution of

$$\frac{1}{2}x - 3y \geq -4?$$

$$\frac{1}{2}(-20) - 3(-2) \overset{?}{\geq} -4$$

$$-10 + 6 \overset{?}{\geq} -4$$

$$-4 \geq -4 \text{ Ture}$$

$(-20, -2)$ is a solution.

Graph the linear inequality $2x - 5y \leq -10$.

33. $2x - 5y \leq -10$, Boundary line is <u>solid</u> because of the $\leq$ symbol.

Graph as $2x - 5y = -10$.

y-intercept:	x-intercept:
If $x = 0$,	If $y = 0$,
$2(0) - 5y = -10$	$2x - 5(0) = -10$
$-5y = -10$	$2x = -10$
$\dfrac{-5y}{-5} = \dfrac{-10}{-5}$	$\dfrac{2x}{2} = \dfrac{-10}{2}$
$y = 2$	$x = -5$

The y-int is $(0, 2)$, and the x-int is $(-5, 0)$.

Select test point $(0, 0)$ and substitute into

$2x - 5y \leq -10$

$5(0) - 3(0) \overset{?}{\leq} -10$

$0 \leq -10$

False

The coordinates of every point on the same side of the line as $(0, 0)$ ***do not*** satisfy the inequality. To indicate this, we shade the half-plane that ***does not*** contain the test point $(0, 0)$.

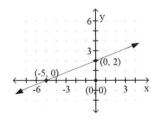
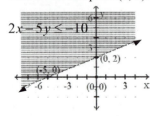

34. Find the domain and range of the relation: $\{(5, 3), (1, 12), (-4, 3), (0, -8)\}$.

Domain: $\{-4, 0, 1, 5\}$; Range: $\{-8, 3, 12\}$

Determine whether the relation defines y as a function of x. If a function is defined, give its domain and range. If it does not define a function, find ordered pairs that show a value of x that is assigned more than one value of y.

35. Yes; domain: $\{1, 2, 3, 4\}$; range: $\{1, 2, 3, 4\}$

36. No; domain: $(-3, 9), (-3, -7)$

37. Yes; domain: $\{6, 7, 8, 9, 10\}$; range: $\{5\}$

38. No; domain: $(2, 6), (2, 2)$

39. No; $(2, 3.5), (2, -3.5)$, (answers may vary)

40. No; $(-2, 2), (-2, -1)$, (answers may vary)

41. If $f(x) = 2x - 7$, find: $f(-3)$

$$f(-3) = 2(-3) - 7$$
$$= -6 - 7$$
$$= -13$$

Thus, $f(-3) = -13$

42. If $g(x) = 3.5x^3$, find $g(6)$.

$$g(x) = 3.5x^3$$
$$g(6) = 3.5(6)^3$$
$$= 3.5(216)$$
$$= 756$$

Thus, $g(6) = 756$

43. **TELEPHONE CALLS**

$C(n) = 0.30n + 15$, find $C(45)$.

$$C(n) = 0.30n + 15$$
$$C(45) = 0.30(45) + 15$$
$$= 13.5 + 15$$
$$= 28.5$$

It costs $28.50 to make 45 calls.

44. Graph: $f(x) = |x| - 1$.

x	$f(x)$
-3	$f(x) = |x| - 1$ $f(-3) = |-3| - 1$ $= 3 - 1$ $= 2$
-2	$f(x) = |x| - 1$ $f(-2) = |-2| - 1$ $= 2 - 1$ $= 1$
1	$f(x) = |x| - 1$ $f(-1) = |-1| - 1$ $= 1 - 1$ $= 0$
0	$f(x) = |x| - 1$ $f(0) = |0| - 1$ $= 0 - 1$ $= -1$
2	$f(x) = |x| - 1$ $f(2) = |2| - 1$ $= 2 - 1$ $= 1$
3	$f(x) = |x| - 1$ $f(3) = |3| - 1$ $= 3 - 1$ $= 2$

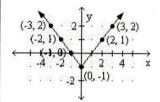

Absolute Value
Function

Chapter 3 Review and Chapter 3 Test

CHAPTER 3 CUMULATIVE REVIEW

1. Prime Factorization: [Section 1.2]
$$108 = 2 \cdot 54$$
$$= 2 \cdot 2 \cdot 27$$
$$= 2 \cdot 2 \cdot 3 \cdot 9$$
$$= 2 \cdot 2 \cdot 3 \cdot 3 \cdot 3$$
$$= 2^2 \cdot 3^3$$

2. Write $\dfrac{1}{250}$ as a decimal. [Section 1.3]

$$\frac{1}{250} = 250\overline{)1.000}^{\;0.004} = 0.004$$

3. Determine whether each statement is true or false. [Section 1.3]

   a. Every whole number is an integer. **True**

   b. Every integer is a real number. **True**

   c. 0 is a whole number, an integer, and a rational number. **True**

Perform the operations.

4. $-27 + 21 + (-9)$ [Section 1.4]
$$-27 + 21 + (-9) = -6 + (-9)$$
$$= -15$$

5. $-1.57 - (-0.8)$ [Section 1.5]
$$-1.57 - (-0.8) = -1.57 + 0.8$$
$$= -0.77$$

6. $-9(-7)(5)(-3)$ [Section 1.6]
$$-9(-7)(5)(-3) = 63(5)(-3)$$
$$= 63(5)(-3)$$
$$= 315(-3)$$
$$= -945$$

7. $\dfrac{-180}{-6}$ [Section 1.6]

$$\frac{-180}{-6} = \frac{\overset{1}{\cancel{-6}} \cdot 30}{\underset{1}{\cancel{-6}}}$$
$$= 30$$

8. Evaluate: $\left| \dfrac{(6-5)^4 - (-21)}{-27 + 4^2} \right|$. [Section 1.7]

$$\left| \frac{(6-5)^4 - (-21)}{-27 + 4^2} \right| = \left| \frac{(1)^4 + 21}{-27 + 16} \right|$$
$$= \left| \frac{1 + 21}{-11} \right|$$
$$= \left| \frac{22}{-11} \right|$$
$$= \left| -2 \right|$$
$$= 2$$

9. Evaluate: $b^2 - 4ac$ for $a = 2$, $b = -8$, and $c = 4$. [Section 1.8]
$$b^2 - 4ac = (-8)^2 - 4(2)(4)$$
$$= 64 - 32$$
$$= 32$$

10. Suppose x sheets from a 500-sheet ream of paper have been used. How many sheets are left? [Section 1.8]
$$500 - x$$

11. How many terms does the algebraic expression $3x^2 - 2x + 1$ have? What is the coefficient of the second term? [Section 1.8]

    3 terms

    -2 is the coefficient of the second term

12. Use the distributive property to remove parentheses. [Section 1.9]

    a. $2(x + 4) = \mathbf{2}x + \mathbf{2}(4)$
    $$= 2x + 8$$

    b. $-2(x - 4) = -\mathbf{2}x - \mathbf{2}(-4)$
    $$= -2x + 8$$

Simplify each expression. [Section 1.9]

13. $5a + 10 - a = (5 - 1)a + 10$
$$= 4a + 10$$

14. $-7(9t) = -63t$

15. $-2b^2 + 6b^2 = (-2 + 6)b^2$
$$= 4b^2$$

16. $5(-17)(0)(2) = 0$

17. $(a+2)-(a-2) = a+2-\mathbf{1}(a)-\mathbf{1}(-2)$
$\qquad = a+2-a+2$
$\qquad = (1-1)a+(2+2)$
$\qquad = 4$

18. $-4(-5)(-8a) = 20(-8a)$
$\qquad = 20(-8)a$
$\qquad = -160a$

19. $-y-y-y = [-1+(-1)+(-1)]y$
$\qquad = -3y$

20. $\dfrac{3}{2}(4x-8)+x = \dfrac{3}{2}(4x)-\dfrac{3}{2}(8)+x$
$\qquad = 6x-12+x$
$\qquad = (6+1)x-12$
$\qquad = 7x-12$

Solve each equation. [Sections 2.1 and 2.2]

21. $\qquad 3x-5 = 13$
$\qquad 3x-5+\mathbf{5} = 13+\mathbf{5}$
$\qquad\quad 3x = 18$
$\qquad\quad \dfrac{3x}{\mathbf{3}} = \dfrac{18}{\mathbf{3}}$
$\qquad\quad x = 6$
The solution is 6.

22. $\qquad 1.2-x = -1.7$
$\qquad 1.2-x-\mathbf{1.2} = -1.7-\mathbf{1.2}$
$\qquad\quad -x = -2.9$
$\qquad\quad \dfrac{-x}{\mathbf{-1}} = \dfrac{-2.9}{\mathbf{-1}}$
$\qquad\quad x = 2.9$
The solution is 2.9.

23. $\qquad \dfrac{2}{3}x-2 = 4$
$\qquad \dfrac{2}{3}x-2+\mathbf{2} = 4+\mathbf{2}$
$\qquad\quad \dfrac{2}{3}x = 6$
$\qquad \left(\dfrac{\mathbf{3}}{\mathbf{2}}\right)\dfrac{2}{3}x = \left(\dfrac{\mathbf{3}}{\mathbf{2}}\right)6$
$\qquad\quad x = 9$
The solution is 9.

24. $\qquad \dfrac{y-2}{7} = -3$
$\qquad \mathbf{7}\cdot\left(\dfrac{y-2}{7}\right) = \mathbf{7}(-3)$
$\qquad\quad y-2 = -21$
$\qquad y-2+\mathbf{2} = -21+\mathbf{2}$
$\qquad\quad y = -19$
The solution is -19.

25. $\qquad -3(2y-2)-y = 5$
$\qquad \mathbf{-3}(2y)-\mathbf{3}(-2)-y = 5$
$\qquad -6y+6-y = 5$
$\qquad -7y+6 = 5$
$\qquad -7y+6-\mathbf{6} = 5-\mathbf{6}$
$\qquad\quad -7y = -1$
$\qquad\quad \dfrac{-7y}{-7} = \dfrac{-1}{-7}$
$\qquad\quad y = \dfrac{1}{7}$
The solution is $\dfrac{1}{7}$.

26. $\qquad 9y-3 = 6y$
$\qquad 9y-3-\mathbf{9y} = 6y-\mathbf{9y}$
$\qquad\quad -3 = -3y$
$\qquad\quad \dfrac{-3}{-3} = \dfrac{-3y}{-3}$
$\qquad\quad 1 = y$
The solution is 1.

Chapter 3 Cumulative Review

27.

$$\frac{1}{3}+\frac{c}{5}=-\frac{3}{2}$$

$$\mathbf{30}\left(\frac{1}{3}\right)+\mathbf{30}\left(\frac{c}{5}\right)=\mathbf{30}\left(-\frac{3}{2}\right)$$

$$10+6c=-45$$

$$10+6c-\mathbf{10}=-45-\mathbf{10}$$

$$6c=-55$$

$$\frac{6c}{\mathbf{6}}=\frac{-55}{\mathbf{6}}$$

$$c=-\frac{55}{6}$$

The solution is $-\dfrac{55}{6}$.

28.

$$5(x+2)=5x-2$$

$$\mathbf{5}x+\mathbf{5}(2)=5x-2$$

$$5x+10=5x-2$$

$$5x+10-\mathbf{5x}=5x-2-\mathbf{5x}$$

$$10=-2$$

No solution, contradiction

29. $-x=99$

$$\frac{-x}{\mathbf{-1}}=\frac{99}{\mathbf{-1}}$$

$$x=-99$$

The solution is -99.

30.

$$3c-2=\frac{11(c-1)}{5}$$

$$\mathbf{5}(3c)-\mathbf{5}(2)=\mathbf{5}\left(\frac{11(c-1)}{5}\right)$$

$$15c-10=11(c-1)$$

$$15c-10=\mathbf{11}c-\mathbf{11}(1)$$

$$15c-10=11c-11$$

$$15c-10+\mathbf{10}=11c-11+\mathbf{10}$$

$$15c=11c-1$$

$$15c-\mathbf{11c}=11c-1-\mathbf{11c}$$

$$4c=-1$$

$$\frac{4c}{\mathbf{4}}=\frac{-1}{\mathbf{4}}$$

$$c=-\frac{1}{4}$$

The solution is $-\dfrac{1}{4}$.

31. PENNIES [Section 2.3]

79% of what number is 869?

$$0.79 \cdot \quad x \quad = 869$$

$$0.79x=869$$

$$\frac{0.79x}{\mathbf{0.79}}=\frac{869}{\mathbf{0.79}}$$

$$x=1{,}100$$

1,100 people were surveyed.

32. Solve for h: $S=2\pi rh+2\pi r^2$.
[Section 2.4]

$$S=2\pi rh+2\pi r^2$$

$$S-\mathbf{2\pi r^2}=2\pi rh+2\pi r^2-\mathbf{2\pi r^2}$$

$$S-2\pi r^2=2\pi rh$$

$$\frac{S-2\pi r^2}{\mathbf{2\pi r}}=\frac{2\pi rh}{\mathbf{2\pi r}}$$

$$\frac{S-2\pi r^2}{2\pi r}=h$$

$$h=\frac{S-2\pi r^2}{2\pi r}$$

33. BAND AIDS [Section 2.4]

$$P = 2w + 2l$$

$$= 2\left(\frac{13}{16}\right) + 2\left(\frac{3}{4}\right)$$

$$= \frac{13}{8} + \frac{3}{2}$$

$$= \frac{13}{8} + \frac{3}{2} \cdot \frac{4}{4}$$

$$= \frac{13 + 12}{8}$$

$$= \frac{25}{8}$$

$$= 3\frac{1}{8}$$

The perimeter is $3\frac{1}{8}$ inches.

$$A = lw$$

$$= \left(\frac{13}{16}\right)\left(\frac{3}{4}\right)$$

$$= \frac{39}{64}$$

The area is $\frac{39}{64}$ in^2.

34. HIGH HEELS [Section 2.5]

Analyze

• 3 angles of a triangle measure $180°$.
• Right angle is $90°$.
• Find the value of x.

Assign

Let x = the measure of one angle in degrees

Form

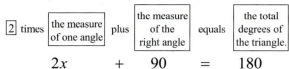

$$2x \quad + \quad 90 \quad = \quad 180$$

Solve

$$2x + 90 = 180$$

$$2x + 90 - \mathbf{90} = 180 - \mathbf{90}$$

$$2x = 90$$

$$\frac{2x}{\mathbf{2}} = \frac{90}{\mathbf{2}}$$

$$x = 45$$

State

$45°$ is the measure of each angle.

Check

The measure of each angle is $45°$.
The measure of the right angle is $90°$.
Since the total was $45° + 45° + 90° = 180°$,
the solution checks.

35. Complete the table. [Section 2.6]

	% acid •	Liters =	Amt of acid
50% solution	0.50	x	**0.50x**
25% solution	0.25	$13 - x$	**0.25(13–x)**
30% solution	0.30	13	**0.30(13)**

36. ROAD TRIPS [Section 2.6]

Analyze

• Bus and truck leave at the same time and from the same place.
• Bus travels at 60 mph.
• Truck travels at 50 mph.
• How long will it take for them to be 75 miles apart?

Assign

Let t = the number of hours traveled by each

	rate •	time =	distance
Bus	60	t	**60t**
Truck	50	t	**50t**
		Total:	75

Form

The distance traveled by the bus	minus	the distance traveled by the truck	equals	the total of 75 miles.
60t	–	50t	=	75

Solve

$$60t - 50t = 75$$

$$10t = 75$$

$$\frac{10t}{\mathbf{10}} = \frac{75}{\mathbf{10}}$$

$$t = 7.5$$

State

It will take 7.5 hours before the two are 75 miles apart.

Check

The bus's distance is $60(7.5) = 450$.
The truck's distance is $50(7.5) = 375$.
Since the difference was $450 - 375 = 75$,
the solution checks.

Chapter 3 Cumulative Review

37. MIXING CANDY [Section 2.6]

Analyze
- Candy corn sells for $2.85 per lb.
- Black gumdrop sells for $1.80 per lb.
- A blend of the two is to sell for $2.22 per lb.
- Mixture of 200 lb is needed.
- How many pounds of each is needed?

Assign
Let x = the pounds candy corn
$200 - x$ = the pounds black gumdrops

Form

	Number	•	Value	=	Total value
Candy corn	x		2.85		$2.85x$
Black gumdrops	$200 - x$		1.80		$1.80(200 - x)$
Blend	200		2.22		$200(2.22)$

The value of the candy corn	plus	the value of the gumdrops	equals	the total value of the blend.
$2.85x$	$+$	$1.80(200 - x)$	$=$	$200(2.22)$

Solve

$$2.85x + 1.80(200 - x) = 200(2.22)$$
$$2.85x + 360 - 1.80x = 444$$
$$1.05x + 360 = 444$$
$$1.05x + 360 - \mathbf{360} = 444 - \mathbf{360}$$
$$1.05x = 84$$
$$\frac{1.05x}{\mathbf{1.05}} = \frac{84}{\mathbf{1.05}}$$
$$x = 80$$

gumdrops
$$200 - x = 200 - \mathbf{80}$$
$$= 120$$

State
80 lb of candy corn will be needed.
120 lb of black gumdrops will be needed.

Check
The value of corn is 80($2.85), or $228.
The value of dumdrops is 120($1.80), or $216.
The value of the blend is 200($2.22), or $444.
Since the total was $228 + $216 = $444,
the results check.

Solve each inequality. Write the solution set in interval notation and graph it. [Section 2.7]

38.
$$-\frac{3}{16}x \geq -9$$
$$\left(-\frac{16}{3}\right)\left(-\frac{3}{16}x\right) \leq \left(-\frac{16}{3}\right)(-9)$$
$$x \leq 48$$
$$(-\infty, 48]$$

(number line graphed with endpoint at 48)

39.
$$8x + 4 > 3x + 4$$
$$8x + 4 - \mathbf{4} > 3x + 4 - \mathbf{4}$$
$$8x > 3x$$
$$8x - \mathbf{3x} > 3x - \mathbf{3x}$$
$$5x > 0$$
$$\frac{5x}{\mathbf{5}} > \frac{0}{\mathbf{5}}$$
$$x > 0$$
$$(0, \infty)$$

(number line graphed with open endpoint at 0)

40. In which quadrants are the second coordinates of points positive? [Section 3.1] **Quadrants I and II**

41. Is $(-2, 4)$ a solution of $y = 2x - 8$? [Section 3.2]
$$y = 2x - 8$$
$$\mathbf{4} \overset{?}{=} -2(\mathbf{-2}) - 8$$
$$4 \overset{?}{=} 4 - 8$$
$$4 = -4 \text{ False}$$
$(-2, 4)$ is not a solution.

Graph each equation.

42. $y = x$ [Section 3.2]

x	y	(x, y)
-2	-2	$(-2, -2)$
0	0	$(0, 0)$
2	2	$(2, 2)$

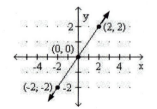

43. $4y + 2x = -8$ [Section 3.3]

y-intercept:	x-intercept:
If $x = 0$,	If $y = 0$
$4y + 2(0) = -8$	$4(0) + 2x = -8$
$4y = -8$	$2x = -8$
$y = -2$	$x = -4$

The y-intercept is $(0, -2)$, and the x-intercept is $(-4, 0)$.

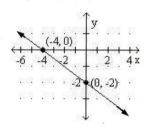

44. What is the slope of the graph of the line $y = 5$? [Section 3.4]

$y = 5$ is the graph of a horizontial line.

The slope of the line is zero.

45. What is the slope of the line passing through $(-2, 4)$ and $(5, -6)$? [Section 3.4]

Let $(x_1, y_1) = (-2, 4)$. Let $(x_2, y_2) = (5, -6)$.

$$m = \frac{y_2 - y_1}{x_2 - x_1}$$
$$= \frac{-6 - 4}{5 - (-2)}$$
$$= \frac{-6 - 4}{5 + 2}$$
$$= \frac{-10}{7}$$
$$= -\frac{10}{7}$$

46. ROOFING
Find the pitch of the roof. [Section 3.4]

$$m = \frac{\text{rise}}{\text{run}}$$
$$= \frac{7}{12}$$

47. Find the slope and the y-intercept of the graph of the line described by $4x - 6y = -12$. [Section 3.5]
Write the equation in $y = mx + b$ form.

$$4x - 6y = -12$$
$$4x - 6y - \mathbf{4x} = -12 - \mathbf{4x}$$
$$-6y = -4x - 12$$
$$\frac{-6y}{-6} = \frac{-4x}{-6} - \frac{12}{-6}$$
$$y = \frac{2}{3}x + 2$$

The slope is $\frac{2}{3}$. The y-intercept is $(0, 2)$.

48. Write an equation of the line that has slope -2 and y-int $(0, 1)$. [Section 3.5]
$$y = mx + b$$
$$y = -2x + 1$$

49. Write an equation of the line that has slope $-\frac{7}{8}$ and passes through $(2, -9)$.
Express the answer in point-slope form and slope-intercept form. [Section 3.6]

Passes through $(2, -9)$, $m = -\frac{7}{8}$

$x_1 = 2$ and $y_1 = -9$
$$y - y_1 = m(x - x_1)$$
$$y - (-9) = -\frac{7}{8}(x - 2)$$
$$y + 9 = -\frac{7}{8}(x - 2) \text{ point-slope form}$$
$$y + 9 = -\frac{7}{8}(x - 2)$$
$$y + 9 = -\frac{7}{8}x - \frac{7}{8}(-2)$$
$$y + 9 - \mathbf{9} = -\frac{7}{8}x + \frac{7}{4} - \frac{36}{4}$$
$$y = -\frac{7}{8}x - \frac{29}{4} \text{ slope-intercept form}$$

50. Is $(-2, -4)$ a solution of $x + y \le -6$? [Section 3.7]

$$x + y \le -6$$
$$\overset{?}{-2 + (-4) \le -6}$$
$$-6 \le -6$$

True

$(-2, -4)$ is a solution.

Chapter 3 Cumulative Review

51. $y \geq x+1$, Boundary line is <u>solid</u>
because of the $\geq$ symbol.

y-intercept:	x-intercept:
If $x = 0$,	If $y = 0$,
	$0 = x+1$
$y = 0+1$	$0-1 = x+1-1$
$y = 1$	$-1 = x$

The y-int is $(0, 1)$, and the x-int is $(-1, 0)$.

Select test point $(0,0)$ and substitute into

$y \geq x+1$

$\overset{?}{0 \geq 0+1}$

$0 \geq 1$

False

The coordinates of every point on the same side of the line as the origin *do not* satisfy the inequality. To indicate this, we shade the half-plane that *does not* contain the test point $(0, 0)$.

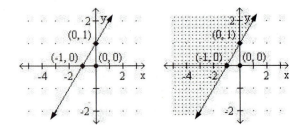

52. Graph $x < 4$ on a rectangular coordinate system. [Section 3.7]
Boundary line is <u>dashed</u> because of the $<$ symbol. Graph as $x = 4$.

There is no y-int, and the x-int is $(4,0)$.
$x = 4$ is a line parallel to the y-axis.
Select test point $(0,0)$ and substitute into

$x < 4$

$0 < 4$

True

The coordinates of every point on the same side of the line as $(0,0)$ satisfy the inequality. To indicate this, we shade the half-plane that contains the test point $(0,0)$.

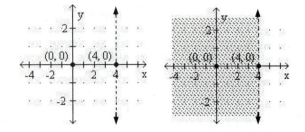

53. If $f(x) = x^4 + x$, find $f(-3)$.
[Section 3.8]

$f(x) = x^4 + x$

$f(-3) = (-3)^4 + (-3)$

$= 81 - 3$

$= 78$

Thus, $f(-3) = 78$

54. Is this the graph of a function?
[Section 3.8] **No, because it will not pass the vertical line test since the following points** $(1, 2)$ **and** $(1, -2)$ **have the same** x **- coordinate and different** y **- coordinates.**

SECTION 4.1 STUDY SET
VOCABULARY

Fill in the blanks.

1. The pair of equations $\begin{cases} x - y = -1 \\ 2x - y = 1 \end{cases}$ is called a **system** of equations.

3. The point of **intersection** of the lines graphed in part (a) of the following figure is (1, 2).

5. A system of equations that has at least one solution is called a **consistent** system. A system with no solution is called an **inconsistent** system.

CONCEPTS

7. Refer to the illustration.
   a. **True, because the point lies on the line.**
   b. **True, because the point lies on the line.**

9. a. To graph $5x - 2y = 10$, we can use the intercept method.

x	y	(x, y)
0	$5x - 2y = 10$ $5(0) - 2y = 10$ $-2y = 10$ $y = -5$	**(0, –5)**
$5x - 2y = 10$ $5x - 2(0) = 10$ $5x = 10$ $x = 2$	0	**(2, 0)**

   b. To graph $y = 3x - 2$, we can use the slope and y-intercept.

   slope: $\mathbf{3 = \dfrac{3}{1}}$

   y-intercept: **(0, –2)**

11. **No solution; independent, there are different graphs**

GUIDED PRACTICE

Determine whether the ordered pair is a solution of the given system.

13. (1,1), $\begin{cases} x + y = 2 \\ 2x - y = 1 \end{cases}$

$$x + y = 2 \qquad\qquad 2x - y = 1$$
$$1 + 1 \overset{?}{=} 2 \qquad\qquad 2(1) - 1 \overset{?}{=} 1$$
$$2 = 2 \ \text{True} \qquad\qquad 2 - 1 \overset{?}{=} 1$$
$$\qquad\qquad\qquad\qquad 1 = 1 \ \text{True}$$

Since (1, 1) satisfies both equations, it is a solution of the system.

15. (3,–2), $\begin{cases} 2x + y = 4 \\ y = 1 - x \end{cases}$

$$2x + y = 4 \qquad\qquad y = 1 - x$$
$$2(3) + (-2) \overset{?}{=} 4 \qquad\qquad -2 \overset{?}{=} 1 - 3$$
$$6 - 2 \overset{?}{=} 4 \qquad\qquad -2 = -2 \ \text{True}$$
$$4 = 4 \ \text{True}$$

Since $(3, -2)$ satisfies both equations, it is a solution of the system.

17. (12,0), $\begin{cases} x - 9y = 12 \\ y = 10 - x \end{cases}$

$$x - 9y = 12 \qquad\qquad y = 10 - x$$
$$12 - 9(0) \overset{?}{=} 12 \qquad\qquad 0 \overset{?}{=} 10 - (12)$$
$$12 - 0 \overset{?}{=} 12 \qquad\qquad 0 = -2 \ \text{False}$$
$$12 = 12 \ \text{True}$$

Since $(12, 0)$ does not satisfy both equations, it is not a solution of the system.

19. (–2,–4), $\begin{cases} 4x + 5y = -23 \\ -3x + 2y = 0 \end{cases}$

$$4x + 5y = -23 \qquad\qquad -3x + 2y = 0$$
$$4(-2) + 5(-4) \overset{?}{=} -23 \qquad -3(-2) + 2(-4) \overset{?}{=} 0$$
$$-8 - 20 \overset{?}{=} -23 \qquad\qquad 6 - 8 \overset{?}{=} 0$$
$$-28 = -23 \qquad\qquad\qquad -2 = 0$$
$$\text{False} \qquad\qquad\qquad\qquad \text{False}$$

Since $(-2, -4)$ does not satisfy both equations, it is not a solution of the system.

Section 4.1

21. $\left(\dfrac{1}{2}, 3\right)$, $\begin{cases} 2x+y=4 \\ 4x-11=3y \end{cases}$

$$\begin{array}{c|c} 2x+y=4 & 4x-11=3y \\ 2\left(\dfrac{1}{2}\right)+3\overset{?}{=}4 & 4\left(\dfrac{1}{2}\right)-11\overset{?}{=}3(3) \\ 1+3\overset{?}{=}4 & 2-11\overset{?}{=}9 \\ 4=4 & -9=9 \\ \text{True} & \text{False} \end{array}$$

Since $\left(\dfrac{1}{2},3\right)$ does not satisfy both equations, it is not a solution of the system.

23. $(2.5, 3.5)$, $\begin{cases} 4x-3=2y \\ 4y+1=6x \end{cases}$

$$\begin{array}{c|c} 4x-3=2y & 4y+1=6x \\ 4(2.5)-3\overset{?}{=}2(3.5) & 4(3.5)+1\overset{?}{=}6(2.5) \\ 10-3\overset{?}{=}7 & 14+1\overset{?}{=}15 \\ 7=7 & 15=15 \\ \text{True} & \text{True} \end{array}$$

Since $(2.5,\ 3.5)$ satisfies both equations, it is a solution of the system.

Solve each system by graphing. If a system has no solution or infinitely many, so state.

25. $\begin{cases} 2x+3y=12; & x\text{-int }(6,0);\ y\text{-int }(0,4) \\ 2x-y=4; & x\text{-int }(2,0);\ y\text{-int }(0,-4) \end{cases}$

The solution is $(3,2)$.

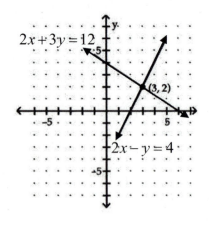

27. $\begin{cases} x+y=4; & x\text{-int }(4,0);\ y\text{-int }(0,4) \\ x-y=-6; & x\text{-int }(-6,0);\ y\text{-int }(0,6) \end{cases}$

The solution is $(-1,5)$.

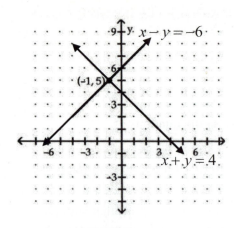

29. $\begin{cases} y=-\dfrac{1}{3}x-4; & m=-\dfrac{1}{3};\ y\text{-int }(0,-4) \\ x+3y=6; & x\text{-int }(6,0);\ y\text{-int }(0,2) \end{cases}$

The lines are parallel.
The system has no solution.

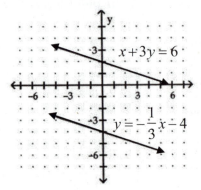

N31. $\begin{cases} y=3x; & m=3;\ y\text{-int }(0,0) \\ y-3x=-3; & x\text{-int }(1,0);\ y\text{-int }(0,-3) \end{cases}$

The lines are parallel.
The system has no solution.

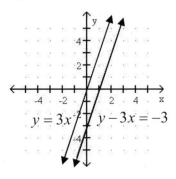

33. $\begin{cases} y = x - 1; \quad m = 1; \ y\text{-int } (0, -1) \\ 3x - 3y = 3; \quad x\text{-int } (1, 0); \ y\text{-int } (0, -1) \end{cases}$

The graphs are the same line.
The system has infinitely many solutions.

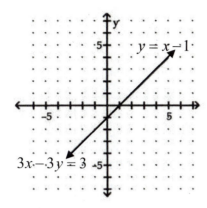

$N35.$ $\begin{cases} 4x + 6y = 12; \ x\text{-int } (3, 0); \ y\text{-int } (0, 2) \\ 2x + 3y = 6; \ x\text{-int } (3, 0); \ y\text{-int } (0, 2) \end{cases}$

The graphs are the same line.
The system has infinitely many solutions.

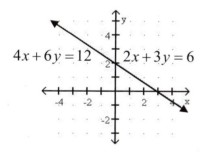

Find the slope and the y-intercept of the graph of each line in the system of equations. Then, use that information to determine the number of solutions of the system.

37. $\begin{cases} y = 6x - 7 \\ y = -2x + 1 \end{cases}$

$\begin{array}{l|l} y = 6x - 7 & y = -2x + 1 \\ m = 6 & m = -2 \\ y\text{-int } = (0, -7) & y\text{-int } = (0, 1) \end{array}$

Since the line graphs have different slopes, the lines are neither parallel nor identical. Therefore, they will intersect at one point and the system will have 1 solution.

39. $\begin{cases} 3x - y = -3 \\ y - 3x = 3 \end{cases}$

$\begin{array}{l|l} 3x - y = -3 & y - 3x = 3 \\ 3x - y + y = -3 + y & y - 3x + 3x = 3 + 3x \\ 3x = -3 + y & y = 3x + 3 \\ 3x + 3 = -3 + y + 3 & m = 3 \\ y = 3x + 3 & y\text{-int } = (0, 3) \\ m = 3 & \\ y\text{-int } = (0, 3) & \end{array}$

Since the line graphs are exactly the same line, the system has infinitely many solutions.

41. $\begin{cases} x + y = 6 \\ x + y = 8 \end{cases}$

$\begin{array}{l|l} x + y = 6 & x + y = 8 \\ x + y - x = 6 - x & x + y - x = 8 - x \\ y = -x + 6 & y = -x + 8 \\ m = -1 & m = -1 \\ y\text{-int } = (0, 6) & y\text{-int } = (0, 8) \end{array}$

Since the line graphs have the same slope, they are parallel and there is no solution.

43. $\begin{cases} 6x + y = 0 \\ 2x + 2y = 0 \end{cases}$

$\begin{array}{l|l} 6x + y = 0 & 2x + 2y = 0 \\ 6x + y - 6x = 0 - 6x & 2x + 2y - 2x = 0 - 2x \\ y = -6x + 0 & 2y = -2x + 0 \\ m = -6 & \dfrac{2y}{2} = \dfrac{-2x}{2} + \dfrac{0}{2} \\ y\text{-int } = (0, 0) & y = -x + 0 \\ & m = -1 \\ & y\text{-int } = (0, 0) \end{array}$

Since the line graphs have different slopes, the lines are neither parallel nor identical. Therefore, they will intersect at one point and the system will have 1 solution.

Use a graphing calculator to solve each system, if possible.

45. $\begin{cases} y = 4 - x \\ y = 2 + x \end{cases}$

The solution is $(1, 3)$.

47. $\begin{cases} 6x - 2y = 5 \\ 3x = y + 10 \end{cases}$

The lines are parallel. There is no solution.

Section 4.1

49. $\begin{cases} y = 3x + 6; & m = 3; \ y\text{-int } (0,6) \\ y = -2x - 4; & m = -2; \ y\text{-int } (0,-4) \end{cases}$

The solution is $(-2, 0)$.

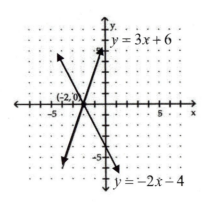

51. $\begin{cases} 2y = 3x + 2; & m = \dfrac{3}{2}; \ y\text{-int } (0,1) \\ 3x - 2y = 6; & m = \dfrac{3}{2}; \ y\text{-int } (0,-3) \end{cases}$

The lines are parallel.
The system has no solution.

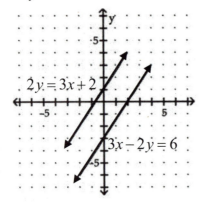

53. $\begin{cases} x + y = 2; & x\text{-int } (2,0); \ y\text{-int } (0,2) \\ y = x - 4; & m = 1; \ y\text{-int } (0,-4) \end{cases}$

The solution is $(3, -1)$.

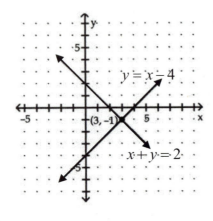

55. $\begin{cases} x = 3; & x\text{-int } (3,0); \ \text{no } y\text{-int} \\ 3y = 6 - 2x; & m = -\dfrac{2}{3}; \ y\text{-int } (0,2) \end{cases}$

The solution is $(3, 0)$.

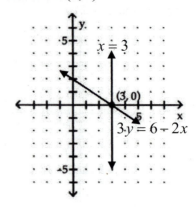

57. $\begin{cases} y = \dfrac{3}{4}x + 3; & m = \dfrac{3}{4}; \ y\text{-int } (0,3) \\ y = -\dfrac{x}{4} - 1; & m = -\dfrac{1}{4}; \ y\text{-int } (0,-1) \end{cases}$

The solution is $(-4, 0)$.

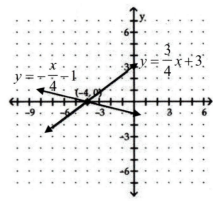

59. $\begin{cases} y = -x - 2; & m = -1; \ y\text{-int } (0,-2) \\ y = -3x + 6; & m = -3; \ y\text{-int } (0,6) \end{cases}$

The solution is $(4, -6)$.

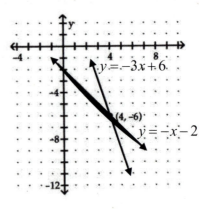

61. $\begin{cases} -x+3y=-11; \ x\text{-int } (11,0); \ y\text{-int } (0,-\frac{11}{3}) \\ 3x-y=17; \ x\text{-int } (\frac{17}{3},0); \ y\text{-int } (0,-17) \end{cases}$

The solution is $(5,-2)$.

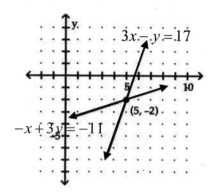

63. $\begin{cases} x+y=2; \ x\text{-int } (2,0); \ y\text{-int } (0,2) \\ y=x; \quad m=1; \ y\text{-int } (0,0) \end{cases}$

The solution is $(1,1)$.

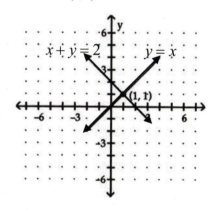

65. $\begin{cases} 4x-2y=8; \ x\text{-int } (2,0); \ y\text{-int } (0,-4) \\ y=2x-4; \ m=2; \ y\text{-int } (0,-4) \end{cases}$

The graphs are the same line.
The system has infinitely many solutions.

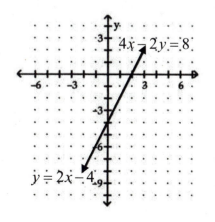

67. $\begin{cases} x+4y=-2; \ x\text{-int } (-2,0); \ y\text{-int } (0,-\frac{1}{2}) \\ y=-x-5; \quad m=-1; \ y\text{-int } (0,-5) \end{cases}$

The solution is $(-6,1)$.

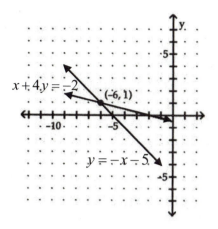

69. $\begin{cases} y=-3; \qquad \text{no } x\text{-int }; \ y\text{-int } (0,-3) \\ -x+2y=-4; \ x\text{-int } (4,0); \ y\text{-int } (0,-2) \end{cases}$

The solution is $(-2,-3)$.

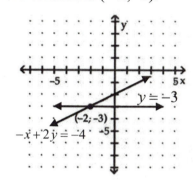

71. $\begin{cases} x+2y=-4; \ x\text{-int } (-4,0); \ y\text{-int } (0,-2) \\ x-\frac{1}{2}y=6; \ x\text{-int } (6,0); \ y\text{-int } (0,-12) \end{cases}$

The solution is $(4,-4)$.

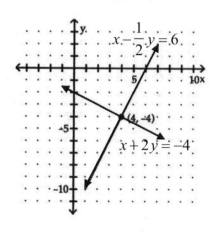

Section 4.1

APPLICATIONS

73. SOCIAL NETWORKS
The same number of visitors was in May '09 and it was about 70 million.

75. LATITUDE AND LONGITUDE
a. Houston, New Orleans, St. Augustine
b. St. Louis, Memphis, New Orleans
c. New Orleans

77. DAILY TRACKING POLL
a. The incumbent; 7%
b. November 2
c. The challenger; 3

FROM CAMPUS TO CAREERS
79. PHOTOGRAPHERS
$(6,4),(6,8),(12,4),(12,8);$ yes

WRITING
81-86. Answers will vary.

REVIEW
Solve each inequality. Write the solution set in interval notation and graph it.

87. $\quad -4(2y+2) \le 4y+28$
$$-8y-8-\mathbf{4y} \le 4y+28-\mathbf{4y}$$
$$-12y-8 \le 28$$
$$-12y-8+\mathbf{8} \le 28+\mathbf{8}$$
$$-12y \le 36$$
$$\frac{-12y}{-12} \ge \frac{36}{-12}$$
$$y \ge -3$$
$$[-3,\infty)$$

89. $\quad -1 \le -\dfrac{1}{2}n$
$$-\frac{1}{2}n \ge -1$$
$$-\frac{2}{1}\cdot\left(-\frac{1}{2}n\right) \le -\frac{2}{1}\cdot(-1)$$
$$n \le 2$$
$$(-\infty, 2]$$

CHALLENGE PROBLEMS

91. No. Suppose the equations are dependent, then they have the same graph, and would intersect at an infinite number of points. Therefore, the system would have at least one solution, and be consistent.

93. Answers will vary.
Given (2, 3) is a solution.
$$\begin{cases} x+y = 5 \\ x-y = 10 \end{cases}$$
There are many more systems than these two.

SECTION 4.2

VOCABULARY

Fill in the blanks.

1. To solve the system $\begin{cases} x = y+1 \\ 3x+2y=8 \end{cases}$ using the method discussed in this section, we begin by **substituting** $y+1$ for x in the second equation.

CONCEPTS

3. If the substitution method is used to solve $\begin{cases} 5x+y=2 \\ y=-3x \end{cases}$ which equation should be used as the substitution equation? $y = -3x$

5. Suppose $x-4$ is substituted for y in the equation $x + 3y = 8$. Insert parentheses in $x + 3x - 4 = 8$ to show the substitution.
$$x + 3(x-4) = 8$$

7. A student uses the substitution method to solve the system $\begin{cases} 4a+5b=2 \\ b=3a-11 \end{cases}$ and finds that $a = 3$.

What is the easiest way for her to determine the value of b?
Substitute 3 for a in the second equation.

9. Suppose $-2 = 1$ is obtained when a system is solved by the substitution method.
   a. Does the system have a solution? **No**
   b. Which of the following is a possible graph of the system? **ii, parallel lines have no solution point**

NOTATION

Complete the solution of the system.

11. Solve: $\begin{cases} y=3x \\ x-y=4 \end{cases}$

$x-y = 4$ This is the second equation. Substitute for y.

$x - (\mathbf{3x}) = 4$

$-2x = \mathbf{4}$

$x = \mathbf{-2}$

$y = 3x$ This is the first equation.

$y = 3(\mathbf{-2})$

$y = \mathbf{-6}$

The solution is $(\mathbf{-2}, \mathbf{-6})$.

GUIDED PRACTICE

Solve each system by substituion. See Example 1.

13. $\begin{cases} y=2x \\ x+y=6 \end{cases}$

$x + y = 6$ The 2nd equation Substitute for y.

$x + (\mathbf{2x}) = 6$

$3x = 6$

$\dfrac{3x}{3} = \dfrac{6}{3}$

$x = 2$

$y = 2x$ The 1st equation

$y = 2(\mathbf{2})$

$y = 4$

The solution is $(2,4)$.

Check:	Check:
$y = 2x$	$x + y = 6$
$\overset{?}{4 = 2(2)}$	$\overset{?}{2+4=6}$
$4 = 4$	$6 = 6$
True	True

15. $\begin{cases} y=2x-6 \\ 2x+y=6 \end{cases}$

$2x + y = 6$ The 2nd equation Substitute for y.

$2x + (\mathbf{2x-6}) = 6$

$4x - 6 = 6$

$4x - 6 + \mathbf{6} = 6 + \mathbf{6}$

$4x = 12$

$\dfrac{4x}{4} = \dfrac{12}{4}$

$x = 3$

$y = 2x - 6$ The 1st equation

$y = 2(\mathbf{3}) - 6$

$y = 6 - 6$

$y = 0$

The solution is $(3, 0)$.

Check:	Check:
$y = 2x - 6$	$2x + y = 6$
$\overset{?}{0 = 2(3) - 6}$	$\overset{?}{2(3) + 0 = 6}$
$0 = 0$	$6 = 6$
True	True

Section 4.2

Solve each system by substituion.
See Example 2.

17. $\begin{cases} x+3y=-4 \\ x=-5y \end{cases}$

$x+3y=-4$ The 1st equation
         Substitute for x.

$-5y+3y=-4$

$-2y=-4$

$\dfrac{-2y}{-2}=\dfrac{-4}{-2}$

$y=2$

$x=-5y$ The 2nd equation

$x=-5(2)$

$x=-10$

The solution is $(-10,2)$.

Check: | Check:

$x+3y=-4$ $x=-5y$

$\overset{?}{-10+3(2)=-4}$ $\overset{?}{-10=-5(2)}$

$-4=-4$ $-10=-10$

True   True

19. $\begin{cases} y=-5x \\ x+3y=-28 \end{cases}$

$x+3y=-28$ The 2nd equation
          Substitute for y.

$x+3(-5x)=-28$

$x-15x=-28$

$-14x=-28$

$\dfrac{-14x}{-14}=\dfrac{-28}{-14}$

$x=2$

$y=-5x$ The 1st equation

$y=-5(2)$

$y=-10$

The solution is $(2,-10)$.

Check: | Check:

$y=-5x$ $x+3y=-28$

$\overset{?}{-10=-5(2)}$ $\overset{?}{2+3(-10)=-28}$

$-10=-10$ $\overset{?}{2-30=-28}$

True $-28=-28$

                True

Solve each system by substituion.
See Example 3.

21. $\begin{cases} r+3s=9 \\ 3r+2s=13 \end{cases}$

Solve for r in the 1st equation.

$\begin{cases} r=9-3s \\ 3r+2s=13 \end{cases}$

$3r+2s=13$ The 2nd equation
          Substitute for x.

$3(9-3s)+2s=13$

$27-9s+2s=13$

$27-7s=13$

$27-7s-27=13-27$

$-7s=-14$

$\dfrac{-7s}{-7}=\dfrac{-14}{-7}$

$s=2$

$r=9-3s$ The 1st equation

$r=9-3(2)$

$r=9-6$

$r=3$

The solution is $(3,2)$.

Check: | Check:

$r+3s=9$ $3r+2s=13$

$\overset{?}{3+3(2)=9}$ $\overset{?}{3(3)+2(2)=13}$

$\overset{?}{3+6=9}$ $\overset{?}{9+4=13}$

$9=9$ $13=13$

True   True

23. $\begin{cases} 4x+y=-15 \\ 2x+3y=5 \end{cases}$

Solve for y in the 1st equation.

$\begin{cases} y=-4x-15 \\ 2x+3y=5 \end{cases}$

$2x+3y=5$ The 2nd equation
         Substitute for y.

$2x+3(-4x-15)=5$

$2x-12x-45=5$

$-10x-45=5$

$-10x-45+45=5+45$

$-10x=50$

$\dfrac{-10x}{-10}=\dfrac{50}{-10}$

$x=-5$

$y=-4x-15$ The 1st equation

$y=-4(-5)-15$

$y=20-15$

$y=5$

The solution is $(-5,5)$.

Check: | Check:

$4x+y=-15$ $2x+3y=5$

$\overset{?}{4(-5)+5=-15}$ $\overset{?}{2(-5)+3(5)=5}$

$\overset{?}{-20+5=-15}$ $\overset{?}{-10+15=5}$

$-15=-15$ $5=5$

True   True

Solve each system by substituion.
See Example 4.

25. $\begin{cases} 8x - 6y = 4 \\ 2x - y = -2 \end{cases}$

Solve for y in the 2nd equation.

$\begin{cases} 8x - 6y = 4 \\ y = 2x + 2 \end{cases}$

$8x - 6y = 4$ The 1st equation
Substitute for y.
$8x - 6(\mathbf{2x + 2}) = 4$
$8x - 12x - 12 = 4$
$-4x - 12 = 4$
$-4x - 12 + \mathbf{12} = 4 + \mathbf{12}$
$-4x = 16$
$\dfrac{-4x}{\mathbf{-4}} = \dfrac{16}{\mathbf{-4}}$
$x = -4$

$y = 2x + 2$ The 2nd equation
$y = 2(\mathbf{-4}) + 2$
$y = -8 + 2$
$y = -6$

The solution is $(-4, -6)$.

Check:
$8x - 6y = 4$
$8(\mathbf{-4}) - 6(\mathbf{-6}) \overset{?}{=} 4$
$-32 + 36 \overset{?}{=} 4$
$4 = 4$
True

Check:
$2x - y = -2$
$2(\mathbf{-4}) - (\mathbf{-6}) \overset{?}{=} -2$
$-8 + 6 \overset{?}{=} -2$
$-2 = -2$
True

27. $\begin{cases} 4x + 5y = 2 \\ 3x - y = 11 \end{cases}$

Solve for y in the 2nd equation.

$\begin{cases} 4x + 5y = 2 \\ y = 3x - 11 \end{cases}$

$4x + 5y = 2$ The 1st equation
Substitute for y.
$4x + 5(\mathbf{3x - 11}) = 2$
$4x + 15x - 55 = 2$
$19x - 55 = 2$
$19x - 55 + \mathbf{55} = 2 + \mathbf{55}$
$19x = 57$
$\dfrac{19x}{\mathbf{19}} = \dfrac{57}{\mathbf{19}}$
$x = 3$

$y = 3x - 11$ The 2nd equation
$y = 3(\mathbf{3}) - 11$
$y = 9 - 11$
$y = -2$

The solution is $(3, -2)$.

Check:
$4x + 5y = 2$
$4(3) + 5(-2) \overset{?}{=} 2$
$12 - 10 \overset{?}{=} 2$
$2 = 2$
True

Check:
$3x - y = 11$
$3(3) - (-2) \overset{?}{=} 11$
$9 + 2 \overset{?}{=} 11$
$11 = 11$
True

Solve each system by substituion.
See Example 5.

29. $\begin{cases} \dfrac{x}{4} + \dfrac{y}{4} = -\dfrac{1}{2} \\ x - 2y = y - 24 - x \end{cases}$

Clear 1st equation of fractions.
Write 2nd equation in standard form.

$\begin{cases} 4\left(\dfrac{x}{4}\right) + 4\left(\dfrac{y}{4}\right) = 4\left(-\dfrac{1}{2}\right) \\ x - 2y + x - y = y - 24 - x + x - y \end{cases}$

$\begin{cases} x + y = -2 \\ 2x - 3y = -24 \end{cases}$

Solve for x in the 1st equation.

$\begin{cases} x = -y - 2 \\ 2x - 3y = -24 \end{cases}$

$2x - 3y = -24$ The 2nd equation
Substitute for x.
$2(\mathbf{-y - 2}) - 3y = -24$
$-2y - 4 - 3y = -24$
$-5y - 4 = -24$
$-5y - 4 + \mathbf{4} = -24 + \mathbf{4}$
$-5y = -20$
$\dfrac{-5y}{\mathbf{-5}} = \dfrac{-20}{\mathbf{-5}}$
$y = 4$

$x = -y - 2$ The 1st equation
$x = -(\mathbf{4}) - 2$
$x = -6$

The solution is $(-6, 4)$.

Check:
$\dfrac{x}{4} + \dfrac{y}{4} = -\dfrac{1}{2}$
$\dfrac{-6}{4} + \dfrac{4}{4} \overset{?}{=} -\dfrac{1}{2}$
$\dfrac{-2}{4} \overset{?}{=} -\dfrac{1}{2}$
$-\dfrac{1}{2} = -\dfrac{1}{2}$
True

Check:
$x - 2y = y - 24 - x$
$-6 - 2(4) \overset{?}{=} 4 - 24 - (-6)$
$-6 - 8 \overset{?}{=} 4 - 24 + 6$
$-14 = -14$
True

Section 4.2

31. $\begin{cases} x - 5y = 20 - 4x - y \\ \dfrac{y}{3} = \dfrac{x}{2} - \dfrac{5}{2} \end{cases}$

Write 1st equation in standard form.
Clear 2nd equation of fractions.

$\begin{cases} x - 5y + \mathbf{4x} + \mathbf{y} = 20 - 4x - y + \mathbf{4x} + \mathbf{y} \\ \mathbf{6}\left(\dfrac{y}{3}\right) = \mathbf{6}\left(\dfrac{x}{2}\right) - \mathbf{6}\left(\dfrac{5}{2}\right) \end{cases}$

$\begin{cases} 5x - 4y = 20 \\ 2y = 3x - 15 \end{cases}$

Solve for y in the 2nd equation.

$\begin{cases} 5x - 4y = 20 \\ y = \dfrac{3x}{2} - \dfrac{15}{2} \end{cases}$

$5x - 4y = 20$ The 1st equation Substitute for y.

$5x - 4\left(\dfrac{3x}{2} - \dfrac{15}{2}\right) = 20$

$5x - 6x + 30 = 20$

$-x + 30 = 20$

$-x + 30 - \mathbf{30} = 20 - \mathbf{30}$

$-x = -10$

$\dfrac{-x}{-1} = \dfrac{-10}{-1}$

$x = 10$

$2y = 3x - 15$ The 2nd equation

$2y = 3(\mathbf{10}) - 15$

$2y = 30 - 15$

$2y = 15$

$\dfrac{2y}{\mathbf{2}} = \dfrac{15}{\mathbf{2}}$

$y = \dfrac{15}{2}$

The solution is $\left(10, \dfrac{15}{2}\right)$.

Solve each system by substituion.
See Example 6.

33. $\begin{cases} 2a + 4b = -24 \\ a = 20 - 2b \end{cases}$

$2a + 4b = -24$ The 1st equation Substitute for a.

$2(\mathbf{20} - \mathbf{2b}) + 4b = -24$

$40 - 4b + 4b = -24$ The variables drop out.

$40 = -24$

False

No Solution.

35. $\begin{cases} 6 - y = 4x \\ 2y = -8x - 20 \end{cases}$

Solve for y in the 1st equation.

$\begin{cases} 6 - 4x = y \\ 2y = -8x - 20 \end{cases}$

$2y = -8x - 20$ The 2nd equation Substitute for y.

$2(6 - 4x) = -8x - 20$

$12 - 8x = -8x - 20$

$12 - 8x + \mathbf{8x} = -8x - 20 + \mathbf{8x}$ The variables drop out.

$12 = -20$

False

No Solution.

Solve each system by substituion.
See Example 7.

37. $\begin{cases} y - 3x = -5 \\ 21x = 7y + 35 \end{cases}$

Solve for y in the 1st equation.

$\begin{cases} y = 3x - 5 \\ 21x = 7y + 35 \end{cases}$

$21x = 7y + 35$ The 2nd equation Substitute for y.

$21x = 7(\mathbf{3x} - \mathbf{5}) + 35$

$21x = 21x - 35 + 35$

$21x = 21x$

$21x - \mathbf{21x} = 21x - \mathbf{21x}$ The variables drop out.

$0 = 0$

True

The system has infinitely many solutions.

39. $\begin{cases} x = -3y + 6 \\ 2x + 4y = 6 + x + y \end{cases}$

Write the 2nd equation in standard form.

$\begin{cases} x = -3y + 6 \\ 2x + 4y - \mathbf{x} - \mathbf{y} = 6 + x + y - \mathbf{x} - \mathbf{y} \end{cases}$

$\begin{cases} x = -3y + 6 \\ x + 3y = 6 \end{cases}$

$x + 3y = 6$ The 2nd equation Substitute for x.

$(-3\mathbf{y} + \mathbf{6}) + 3y = 6$ The variables drop out.

$6 = 6$

True

The system has infinitely many solutions.

Solve each system by substitution. If a system has no solution or infinitely many solutions, so state.

41. $\begin{cases} -y = 11 - 3x \\ 2x + 5y = -4 \end{cases}$

Solve for y in the 1st equation.

$\begin{cases} y = 3x - 11 \\ 2x + 5y = -4 \end{cases}$

$2x + 5y = -4$ The 2nd equation
Substitute for y.

$2x + 5(3x - 11) = -4$
$2x + 15x - 55 = -4$
$17x - 55 = -4$
$17x - 55 + 55 = -4 + 55$
$17x = 51$
$\dfrac{17x}{17} = \dfrac{51}{17}$
$x = 3$

$y = 3x - 11$ The 1st equation
$y = 3(3) - 11$
$y = 9 - 11$
$y = -2$

The solution is $(3, -2)$.

Check:
$-y = 11 - 3x$
$-(-2) \overset{?}{=} 11 - 3(3)$
$2 \overset{?}{=} 11 - 9$
$2 = 2$
True

Check:
$2x + 5y = -4$
$2(3) + 5(-2) \overset{?}{=} -4$
$6 - 10 \overset{?}{=} -4$
$-4 = -4$
True

43. $\begin{cases} \dfrac{x}{2} + \dfrac{y}{6} = \dfrac{2}{3} \\ \dfrac{x}{3} - \dfrac{y}{4} = \dfrac{1}{12} \end{cases}$

Clear both equations of fractions.

$\begin{cases} 6\left(\dfrac{x}{2}\right) + 6\left(\dfrac{y}{6}\right) = 6\left(\dfrac{2}{3}\right) \\ 12\left(\dfrac{x}{3}\right) - 12\left(\dfrac{y}{4}\right) = 12\left(\dfrac{1}{12}\right) \end{cases}$

$\begin{cases} 3x + y = 4 \\ 4x - 3y = 1 \end{cases}$

Solve for y in the 1st equation.

$\begin{cases} y = 4 - 3x \\ 4x - 3y = 1 \end{cases}$

$4x - 3y = 1$ The 2nd equation
Substitute for y.

$4x - 3(4 - 3x) = 1$
$4x - 12 + 9x = 1$
$13x - 12 = 1$
$13x - 12 + 12 = 1 + 12$
$13x = 13$
$\dfrac{13x}{13} = \dfrac{13}{13}$
$x = 1$

$y = 4 - 3x$ The 1st equation
$y = 4 - 3(1)$
$y = 4 - 3$
$y = 1$

The solution is $(1, 1)$.

Check:
$\dfrac{x}{2} + \dfrac{y}{6} = \dfrac{2}{3}$
$\dfrac{1}{2} + \dfrac{1}{6} \overset{?}{=} \dfrac{2}{3}$
$\dfrac{3}{6} + \dfrac{1}{6} \overset{?}{=} \dfrac{2}{3}$
$\dfrac{4}{6} \overset{?}{=} \dfrac{2}{3}$
$\dfrac{2}{3} = \dfrac{2}{3}$
True

Check:
$\dfrac{x}{3} - \dfrac{y}{4} = \dfrac{1}{12}$
$\dfrac{1}{3} - \dfrac{1}{4} \overset{?}{=} \dfrac{1}{12}$
$\dfrac{4}{12} - \dfrac{3}{12} \overset{?}{=} \dfrac{1}{12}$
$\dfrac{1}{12} = \dfrac{1}{12}$
True

45. $\begin{cases} y + x = 2x + 2 \\ 6x - 4y = 21 - y \end{cases}$

Write both equations in standard form.

$\begin{cases} y + x - 2x = 2x + 2 - 2x \\ 6x - 4y + y = 21 - y + y \end{cases}$

$\begin{cases} -x + y = 2 \\ 6x - 3y = 21 \end{cases}$

Solve for y in the 1st equation.

$\begin{cases} y = 2 + x \\ 6x - 3y = 21 \end{cases}$

$6x - 3y = 21$ The 2nd equation
Substitute for y.

$6x - 3(2 + x) = 21$
$6x - 6 - 3x = 21$
$3x - 6 = 21$
$3x - 6 + 6 = 21 + 6$
$3x = 27$
$\dfrac{3x}{3} = \dfrac{27}{3}$
$x = 9$

Section 4.2

$y = 2 + x$ The 1st equation

$y = 2 + \mathbf{9}$

$y = 11$

The solution is $(9, 11)$.

Check:

$y + x = 2x + 2$

$11 + 9 \overset{?}{=} 2(9) + 2$

$20 \overset{?}{=} 18 + 2$

$20 = 20$

True

Check:

$6x - 4y = 21 - y$

$6(9) - 4(11) \overset{?}{=} 21 - 11$

$54 - 44 \overset{?}{=} 10$

$10 = 10$

True

47. $\begin{cases} y - 4 = 2x \\ y = 2x + 2 \end{cases}$

$y - 4 = 2x$ The 1st equation — Substitute for y.

$(\mathbf{2x + 2}) - 4 = 2x$

$2x - 2 = 2x$

$2x - 2 - \mathbf{2x} = 2x - \mathbf{2x}$ The variables drop out.

$-2 = 0$

False

No Solution.

49. $\begin{cases} a + b = 1 \\ a - 2b = -1 \end{cases}$

Solve for a in the 1st equation.

$\begin{cases} a = 1 - b \\ a - 2b = -1 \end{cases}$

$a - 2b = -1$ The 2nd equation — Substitute for a.

$(\mathbf{1 - b}) - 2b = -1$

$1 - 3b = -1$

$1 - 3b - \mathbf{1} = -1 - \mathbf{1}$

$-3b = -2$

$\dfrac{-3b}{-\mathbf{3}} = \dfrac{-2}{-\mathbf{3}}$

$b = \dfrac{2}{3}$

$a = 1 - b$ The 1st equation

$a = 1 - \dfrac{\mathbf{2}}{\mathbf{3}}$

$a = \dfrac{3}{3} - \dfrac{2}{3}$

$a = \dfrac{1}{3}$

The solution is $\left(\dfrac{1}{3}, \dfrac{2}{3} \right)$.

Check:

$a + b = 1$

$\dfrac{1}{3} + \dfrac{2}{3} \overset{?}{=} 1$

$\dfrac{3}{3} \overset{?}{=} 1$

$1 = 1$

True

Check:

$a - 2b = -1$

$\dfrac{1}{3} - 2\left(\dfrac{2}{3}\right) \overset{?}{=} -1$

$\dfrac{1}{3} - \dfrac{4}{3} \overset{?}{=} -1$

$-\dfrac{3}{3} \overset{?}{=} -1$

$-1 = -1$

True

51. $\begin{cases} 5x = \dfrac{1}{2}y - 1 \\ \dfrac{1}{4}y = 10x - 1 \end{cases}$

Clear both equations of fractions.

$\begin{cases} \mathbf{2}(5x) = \mathbf{2}\left(\dfrac{1}{2}y\right) - \mathbf{2}(1) \\ \mathbf{4}\left(\dfrac{1}{4}y\right) = \mathbf{4}(10x) - \mathbf{4}(1) \end{cases}$

$\begin{cases} 10x = y - 2 \\ y = 40x - 4 \end{cases}$

$10x = y - 2$ The 1st equation — Substitute for y.

$10x = (\mathbf{40x - 4}) - 2$

$10x = 40x - 6$

$10x - \mathbf{40x} = 40x - 6 - \mathbf{40x}$

$-30x = -6$

$\dfrac{-30x}{-30} = \dfrac{-6}{-30}$

$x = \dfrac{1}{5}$

$y = 40x - 4$ The 2st equation

$y = 40\left(\dfrac{\mathbf{1}}{\mathbf{5}}\right) - 4$

$y = 8 - 4$

$y = 4$

The solution is $\left(\dfrac{1}{5}, 4 \right)$.

Check:

$5x = \dfrac{1}{2}y - 1$

$5\left(\dfrac{1}{5}\right) \overset{?}{=} \dfrac{1}{2}(4) - 1$

$1 \overset{?}{=} 2 - 1$

$1 = 1$

True

Check:

$\dfrac{1}{4}y = 10x - 1$

$\dfrac{1}{4}(4) \overset{?}{=} 10\left(\dfrac{1}{5}\right) - 1$

$1 \overset{?}{=} 2 - 1$

$1 = 1$

True

53. $\begin{cases} x+2y=-6 \\ x=y \end{cases}$

$x+2y=-6$ The 1st equation
         Substitute for x.

$y+2y=-6$

$3y=-6$

$\dfrac{3y}{3}=\dfrac{-6}{3}$

$y=-2$

$x=y$ The 2nd equation

$x=-2$

The solution is $(-2,-2)$.

Check:
$x+2y=-6$
$-2+2(-2)\overset{?}{=}-6$
$-2+(-4)\overset{?}{=}-6$
$-6=-6$
True

Check:
$x=y$
$-2=-2$
True

55. $\begin{cases} b=\dfrac{2}{3}a \\ 8a-3b=3 \end{cases}$

$8a-3b=3$ The 2nd equation
         Substitute for b.

$8a-3\left(\dfrac{2}{3}a\right)=3$

$8a-2a=3$

$6a=3$

$\dfrac{6a}{6}=\dfrac{3}{6}$

$a=\dfrac{1}{2}$

$b=\dfrac{2}{3}a$ The 1st equation

$b=\dfrac{2}{3}\left(\dfrac{1}{2}\right)$

$b=\dfrac{1}{3}$

The solution is $\left(\dfrac{1}{2},\dfrac{1}{3}\right)$.

Check:
$b=\dfrac{2}{3}a$
$\dfrac{1}{3}\overset{?}{=}\dfrac{2}{3}\left(\dfrac{1}{2}\right)$
$\dfrac{1}{3}=\dfrac{1}{3}$
True

Check:
$8a-3b=3$
$8\left(\dfrac{1}{2}\right)-3\left(\dfrac{1}{3}\right)\overset{?}{=}3$
$4-1\overset{?}{=}3$
$3=3$
True

57. $\begin{cases} 6x-3y=5 \\ x+2y=0 \end{cases}$

Solve for x in the 2nd equation.

$\begin{cases} 6x-3y=5 \\ x=-2y \end{cases}$

$6x-3y=5$ The 1st equation
         Substitute for x.

$6(-2y)-3y=5$

$-12y-3y=5$

$-15y=5$

$\dfrac{-15y}{-15}=\dfrac{5}{-15}$

$y=-\dfrac{1}{3}$

$x=-2y$ The 2nd equation

$x=-2\left(-\dfrac{1}{3}\right)$

$x=\dfrac{2}{3}$

The solution is $\left(\dfrac{2}{3},-\dfrac{1}{3}\right)$.

Check:
$6x-3y=5$
$6\left(\dfrac{2}{3}\right)-3\left(-\dfrac{1}{3}\right)\overset{?}{=}5$
$4+1\overset{?}{=}5$
$5=5$
True

Check:
$x+2y=0$
$\dfrac{2}{3}+2\left(-\dfrac{1}{3}\right)\overset{?}{=}0$
$\dfrac{2}{3}-\dfrac{2}{3}\overset{?}{=}0$
$0=0$
True

59. $\begin{cases} 2x+3=-4y \\ x-6=-8y \end{cases}$

Solve for x in 2nd equation.

$\begin{cases} 2x+3=-4y \\ x=6-8y \end{cases}$

$2x+3=-4y$ The 1st equation
         Substitute for x.

$2(6-8y)+3=-4y$

$12-16y+3=-4y$

$-16y+15=-4y$

$-16y+15+16y=-4y+16y$

$15=12y$

$\dfrac{15}{12}=\dfrac{12y}{12}$

$\dfrac{5}{4}=y$

$x=6-8y$ The 2nd equation

$x=6-8\left(\dfrac{5}{4}\right)$

$x=6-10$

$x=-4$

The solution is $\left(-4,\dfrac{5}{4}\right)$.

Section 4.2

Check:
$$2x+3=-4y$$
$$2(-4)+3\overset{?}{=}-4\left(\frac{5}{4}\right)$$
$$-8+3\overset{?}{=}-5$$
$$-5=-5$$
True

Check:
$$x-6=-8y$$
$$-4-6\overset{?}{=}-8\left(\frac{5}{4}\right)$$
$$-10=-10$$
True

61. $$\begin{cases} 2x+5y=-2 \\ y=-\dfrac{x}{2} \end{cases}$$

Clear the fraction from the 2nd equation.

$$\begin{cases} 2x+5y=-2 \\ -2(y)=-2\left(-\dfrac{x}{2}\right) \end{cases}$$

$$\begin{cases} 2x+5y=-2 \\ -2y=x \end{cases}$$

$$2x+5y=-2 \quad \text{The } 1^{st} \text{ equation} \\ \text{Substitute for } x.$$

$$2(-2y)+5y=-2$$
$$-4y+5y=-2$$
$$y=-2$$

$$y=-\frac{x}{2} \quad \text{The } 2^{nd} \text{ equation}$$

$$-2=-\frac{x}{2}$$

$$-2(-2)=-2\left(-\frac{x}{2}\right)$$

$$4=x$$

The solution is $(4,-2)$.

Check:
$$2x+5y=-2$$
$$2(4)+5(-2)\overset{?}{=}-2$$
$$8-10\overset{?}{=}-2$$
$$-2=-2$$
True

Check:
$$y=-\frac{x}{2}$$
$$-2\overset{?}{=}-\left(\frac{4}{2}\right)$$
$$-2=-2$$
True

63. $$\begin{cases} 3x+4y=-19 \\ 2y-x=3 \end{cases}$$

Solve for x in the 2nd equation.

$$\begin{cases} 3x+4y=-19 \\ x=2y-3 \end{cases}$$

$$3x+4y=-19 \quad \text{The } 1^{st} \text{ equation} \\ \text{Substitute for } x.$$

$$3(2y-3)+4y=-19$$
$$6y-9+4y=-19$$
$$10y-9=-19$$
$$10y-9+9=-19+9$$
$$10y=-10$$
$$\frac{10y}{10}=\frac{-10}{10}$$
$$y=-1$$

$$x=2y-3 \quad \text{The } 2^{nd} \text{ equation}$$
$$x=2(-1)-3$$
$$x=-2-3$$
$$x=-5$$

The solution is $(-5,-1)$.

Check:
$$3x+4y=-19$$
$$3(-5)+4(-1)\overset{?}{=}-19$$
$$-15-4\overset{?}{=}-19$$
$$-19=-19$$
True

Check:
$$2y-x=3$$
$$2(-1)-(-5)\overset{?}{=}3$$
$$-2+5\overset{?}{=}3$$
$$3=3$$
True

65. $$\begin{cases} 3(x-1)+3=8+2y \\ 2(x+1)=8+y \end{cases}$$

Distribute in both equations.

$$\begin{cases} 3x-3+3=8+2y \\ 2x+2=8+y \end{cases}$$

Write the 1st equation in standard form.
Solve for y in the 2nd equation.

$$\begin{cases} 3x-2y=8+2y-2y \\ 2x+2-8=8+y-8 \end{cases}$$

$$\begin{cases} 3x-2y=8 \\ 2x-6=y \end{cases}$$

$$3x-2y=8 \quad \text{The } 1^{st} \text{ equation} \\ \text{Substitute for } y.$$

$$3x-2(2x-6)=8$$
$$3x-4x+12=8$$
$$-x+12=8$$
$$-x+12-12=8-12$$
$$-x=-4$$
$$\frac{-x}{-1}=\frac{-4}{-1}$$
$$x=4$$

$$2x-6=y \quad \text{The } 2^{nd} \text{ equation}$$
$$2(4)-6=y$$
$$8-6=y$$
$$2=y$$

The solution is $(4,2)$.

Check:
$$3(x-1)+3=8+2y$$
$$3(4-1)+3\overset{?}{=}8+2(2)$$
$$3(3)+3\overset{?}{=}8+4$$
$$9+3\overset{?}{=}12$$
$$12=12$$
True

Check:
$$2(x+1)=8+y$$
$$2(4+1)\overset{?}{=}8+2$$
$$2(5)\overset{?}{=}10$$
$$10=10$$
True

67. $\begin{cases} x = \dfrac{1}{3}y - 1 \\ x = y + 5 \end{cases}$

Clear the fraction from the 1st equation.

$\begin{cases} 3(x) = 3\left(\dfrac{1}{3}y\right) - 3(1) \\ x = y + 5 \end{cases}$

$\begin{cases} 3x = y - 3 \\ x = y + 5 \end{cases}$

$3x = y - 3$ The 1st equation

Substitute for x.

$3(y + 5) = y - 3$

$3y + 15 = y - 3$

$3y + 15 - 15 = y - 3 - 15$

$3y = y - 18$

$3y - y = y - 18 - y$

$2y = -18$

$\dfrac{2y}{2} = \dfrac{-18}{2}$

$y = -9$

$x = y + 5$ The 2nd equation

$x = -9 + 5$

$x = -4$

The solution is $(-4, -9)$.

Check:
$x = \dfrac{1}{3}y - 1$

$-4 \overset{?}{=} \dfrac{1}{3}(-9) - 1$

$-4 \overset{?}{=} -3 - 1$

$-4 = -4$

True

Check:
$x = y + 5$

$-4 \overset{?}{=} -9 + 5$

$-4 = -4$

True

69. $\begin{cases} 2a - 3b = -13 \\ -b = -2a - 7 \end{cases}$

Solve for b in 2nd equation.

$\begin{cases} 2a - 3b = -13 \\ b = 2a + 7 \end{cases}$

$2a - 3b = -13$ The 1st equation

Substitute for b.

$2a - 3(2a + 7) = -13$

$2a - 6a - 21 = -13$

$-4a - 21 = -13$

$-4a - 21 + 21 = -13 + 21$

$-4a = 8$

$\dfrac{-4a}{-4} = \dfrac{8}{-4}$

$a = -2$

$b = 2a + 7$ The 2nd equation

$b = 2(-2) + 7$

$b = -4 + 7$

$b = 3$

The solution is $(-2, 3)$.

Check:
$2a - 3b = -13$

$2(-2) - 3(3) \overset{?}{=} -13$

$-4 - 9 \overset{?}{=} -13$

$-13 = -13$

True

Check:
$-b = -2a - 7$

$-(3) \overset{?}{=} -2(-2) - 7$

$-3 \overset{?}{=} 4 - 7$

$-3 = -3$

True

71. $\begin{cases} x = 7y - 10 \\ 2x - 14y + 20 = 0 \end{cases}$

Put the 2nd equation in standard form.

$\begin{cases} x = 7y - 10 \\ 2x - 14y + 20 - 20 = 0 - 20 \end{cases}$

$\begin{cases} x = 7y - 10 \\ 2x - 14y = -20 \end{cases}$

$2x - 14y = -20$ The 2nd equation

Substitute for x.

$2(7y - 10) - 14y = -20$

$14y - 20 - 14y = -20$ The variables drop out.

$-20 = -20$

True

The system has infinitely many solutions.

73. $\begin{cases} 4x + 1 = 2x + 5 + y \\ 2x + 2y = 5x + y + 6 \end{cases}$

Write both equation in standard form.

$\begin{cases} 4x + 1 - 2x - y - 1 = 2x + 5 + y - 2x - y - 1 \\ 2x + 2y - 5x - y = 5x + y + 6 - 5x - y \end{cases}$

$\begin{cases} 2x - y = 4 \\ -3x + y = 6 \end{cases}$

Solve for y in the 1st equation.

$\begin{cases} 2x - y + y - 4 = 4 + y - 4 \\ -3x + y = 6 \end{cases}$

$\begin{cases} 2x - 4 = y \\ -3x + y = 6 \end{cases}$

$-3x + y = 6$ The 2nd equation

Substitute for y.

$-3x + (2x - 4) = 6$

$-x - 4 = 6$

$-x - 4 + 4 = 6 + 4$

$-x = 10$

$\dfrac{-x}{-1} = \dfrac{10}{-1}$

$x = -10$

$2x - 4 = y$ The 1st equation

$2(-10) - 4 = y$

$-20 - 4 = y$

$-24 = y$

The solution is $(-10, -24)$.

Check:
$4x + 1 = 2x + 5 + y$

$4(-10) + 1 \overset{?}{=} 2(-10) + 5 + (-24)$

$-40 + 1 \overset{?}{=} -20 + 5 - 24$

$-39 = -39$

True

Check:
$2x + 2y = 5x + y + 6$

$2(-10) + 2(-24) \overset{?}{=} 5(-10) + (-24) + 6$

$-20 - 48 \overset{?}{=} -50 - 24 + 6$

$-68 = -68$

True

Section 4.2

75. $\begin{cases} 2a+3b=7 \\ 6a-b=1 \end{cases}$

Solve for b in the 2nd equation.

$\begin{cases} 2a+3b=7 \\ 6a-b+\boldsymbol{b}-\boldsymbol{1}=1+\boldsymbol{b}-\boldsymbol{1} \end{cases}$

$\begin{cases} 2a+3b=7 \\ 6a-1=b \end{cases}$

$2a+3b=7$ The 1st equation
Substitute for b.

$2a+3(\boldsymbol{6a-1})=7$

$2a+18a-3=7$

$20a-3=7$

$20a-3+\boldsymbol{3}=7+\boldsymbol{3}$

$20a=10$

$\dfrac{20a}{\boldsymbol{20}}=\dfrac{10}{\boldsymbol{20}}$

$a=\dfrac{1}{2}$

$6a-1=b$ The 2nd equation

$6\left(\dfrac{1}{\boldsymbol{2}}\right)-1=b$

$3-1=b$

$2=b$

The solution is $\left(\dfrac{1}{2},2\right)$.

Check:
$2a+3b=7$
$2\left(\dfrac{1}{2}\right)+3(2)\overset{?}{=}7$

$1+6\overset{?}{=}7$
$7=7$
True

Check:
$6a-b=1$
$6\left(\dfrac{1}{2}\right)-2\overset{?}{=}1$

$3-2\overset{?}{=}1$
$1=1$
True

77. $\begin{cases} 2x-3y=-4 \\ x=-\dfrac{3}{2}y \end{cases}$

$2x-3y=-4$ The 1st equation
Substitute for x.

$2\left(-\dfrac{\boldsymbol{3}}{\boldsymbol{2}}y\right)-3y=-4$

$-3y-3y=-4$

$-6y=-4$

$\dfrac{-6y}{-6}=\dfrac{-4}{-6}$

$y=\dfrac{2}{3}$

$x=-\dfrac{3}{2}y$ The 2nd equation

$x=-\dfrac{3}{2}\left(\dfrac{\boldsymbol{2}}{\boldsymbol{3}}\right)$

$x=-1$

The solution is $\left(-1,\dfrac{2}{3}\right)$.

Check:
$2x-3y=-4$
$2(-1)-3\left(\dfrac{2}{3}\right)\overset{?}{=}-4$

$-2-2\overset{?}{=}-4$
$-4=-4$
True

Check:
$x=-\dfrac{3}{2}y$
$-1\overset{?}{=}-\dfrac{3}{2}\left(\dfrac{2}{3}\right)$

$-1=-1$
True

79. $\begin{cases} \dfrac{9x}{7}-\dfrac{3y}{7}=\dfrac{12}{7} \\ y-3x=-4 \end{cases}$

Clear 1st equations of fractions.
Solve for y in the 2nd equation.

$\begin{cases} \boldsymbol{7}\left(\dfrac{9x}{7}\right)-\boldsymbol{7}\left(\dfrac{3y}{7}\right)=\boldsymbol{7}\left(\dfrac{12}{7}\right) \\ y-3x+\boldsymbol{3x}=-4+\boldsymbol{3x} \end{cases}$

$\begin{cases} 9x-3y=12 \\ y=3x-4 \end{cases}$

$9x-3y=12$ The 1st equation
Substitute for y.

$9x-3(\boldsymbol{3x-4})=12$

$9x-9x+12=12$ The variables
drop out.

$12=12$
True

The system has infinitely many solutions.

81. $\begin{cases} 4x+5y+1=-16+x \\ x-3y+2=-3-x \end{cases}$

Write both equations in standard form.

$\begin{cases} 4x+5y+1-\boldsymbol{x}-\boldsymbol{1}=-16+x-\boldsymbol{x}-\boldsymbol{1} \\ x-3y+2+\boldsymbol{x}-\boldsymbol{2}=-3-x+\boldsymbol{x}-\boldsymbol{2} \end{cases}$

$\begin{cases} 3x+5y=-17 \\ 2x-3y=-5 \end{cases}$

Solve for x in the 1st equation.

$\begin{cases} 3x+5y=-17 \\ 2x-3y=-5 \end{cases}$

$$3x + 5y = -17$$
$$3x + 5y - \mathbf{5y} = -17 - \mathbf{5y}$$
$$3x = -17 - 5y$$
$$\frac{3x}{\mathbf{3}} = \frac{-17}{\mathbf{3}} - \frac{5y}{\mathbf{3}}$$
$$x = \frac{-17}{3} - \frac{5y}{3}$$

$$\begin{cases} x = \dfrac{-17}{3} - \dfrac{5y}{3} \\ 2x - 3y = -5 \end{cases}$$

$$2x - 3y = -5 \quad \text{The } 2^{\text{nd}} \text{ equation}$$
$$\text{Substitute for } x.$$

$$2\left(\frac{-17}{\mathbf{3}} - \frac{\mathbf{5y}}{\mathbf{3}}\right) - 3y = -5$$

$$\frac{-34}{3} - \frac{10y}{3} - 3y = -5$$

$$\mathbf{3}\left(\frac{-34}{3}\right) - \mathbf{3}\left(\frac{10y}{3}\right) - \mathbf{3}(3y) = \mathbf{3}(-5)$$

$$-34 - 10y - 9y = -15$$
$$-34 - 19y = -15$$
$$-34 - 19y + \mathbf{34} = -15 + \mathbf{34}$$
$$-19y = 19$$
$$\frac{-19y}{\mathbf{-19}} = \frac{19}{\mathbf{-19}}$$
$$y = -1$$

$$2x = -13 - 5y \quad \text{The } 1^{\text{st}} \text{ equation}$$
$$2x = -13 - 5(\mathbf{-1})$$
$$2x = -13 + 5$$
$$2x = -8$$
$$\frac{2x}{2} = \frac{-8}{2}$$
$$x = -4$$

The solution is $(-4, -1)$.

Check:
$$4x + 5y + 1 = -12 + 2x$$
$$4(-4) + 5(-1) + 1 \overset{?}{=} -12 + 2(-4)$$
$$-16 - 5 + 1 \overset{?}{=} -12 - 8$$
$$-20 = -20$$
$$\text{True}$$

Check:
$$x - 3y + 2 = -3 - x$$
$$-4 - 3(-1) + 2 \overset{?}{=} -3 - (-4)$$
$$-4 + 3 + 2 \overset{?}{=} -3 + 4$$
$$1 = 1$$
$$\text{True}$$

APPLICATIONS

83. OFFROADING

Let a = angle of approach in degrees
d = angle of departure in degrees
$$\begin{cases} a + d = 77 \\ a = d + 3 \end{cases}$$

$$a + d = 77 \quad \text{The } 1^{\text{st}} \text{ equation}$$
$$\text{Substitute for } a.$$
$$\mathbf{d} + \mathbf{3} + d = 77$$
$$2d + 3 = 77$$
$$2d + 3 - \mathbf{3} = 77 - \mathbf{3}$$
$$2d = 74$$
$$\frac{2d}{\mathbf{2}} = \frac{74}{\mathbf{2}}$$
$$d = 37$$

$$a = d + 3 \quad \text{The } 2^{\text{nd}} \text{ equation}$$
$$a = \mathbf{37} + 3$$
$$a = 40$$

The angle of approach is $40°$.
The angle of departure is $37°$.

Check: Check:
$$a + d = 77 \qquad a = d + 3$$
$$\overset{?}{} \qquad\qquad \overset{?}{}$$
$$40 + 37 = 77 \qquad 40 = 37 + 3$$
$$77 = 77 \qquad\quad 40 = 40$$
$$\text{True} \qquad\qquad \text{True}$$

85. GEOMETRY

Let x = measure of 1st angle in degrees
y = measure of 2nd angle in degrees
$$\begin{cases} x + y = 90 \\ y = 3x \end{cases}$$

$$x + y = 90 \quad \text{The } 1^{\text{st}} \text{ equation}$$
$$\text{Substitute for } y.$$
$$x + \mathbf{3x} = 90$$
$$4x = 90$$
$$\frac{4x}{\mathbf{4}} = \frac{90}{\mathbf{4}}$$
$$x = 22.5$$

$$y = 3x \quad \text{The } 2^{\text{nd}} \text{ equation}$$
$$y = 3(\mathbf{22.5})$$
$$y = 67.5$$

The measure of angle x is $22.5°$.
The measure of angle y is $67.5°$.

Check: Check:
$$x + y = 90 \qquad y = 3x$$
$$\overset{?}{} \qquad\qquad \overset{?}{}$$
$$22.5 + 67.5 = 90 \qquad 67.5 = 3(22.5)$$
$$90 = 90 \qquad\quad 67.5 = 67.5$$
$$\text{True} \qquad\qquad \text{True}$$

Section 4.2

WRITING

87-91. Answers will vary.

REVIEW

93. Find the prime factors of 189.

$$189 = 3 \cdot 63$$
$$= 3 \cdot 7 \cdot 9$$
$$= 3 \cdot 7 \cdot 3 \cdot 3$$
$$= 3^3 \cdot 7$$

95. Add:

$$\frac{5}{12} + \frac{1}{4} = \frac{5}{12} + \frac{1}{4}\left(\frac{3}{3}\right)$$
$$= \frac{5+3}{12}$$
$$= \frac{8}{12}$$
$$= \frac{\cancel{2} \cdot \cancel{2} \cdot 2}{\cancel{2} \cdot \cancel{2} \cdot 3}$$
$$= \frac{2}{3}$$

CHALLENGE PROBLEMS

97.
$$\begin{cases} \dfrac{6x-1}{3} - \dfrac{5}{3} = \dfrac{3y+1}{2} \\ \dfrac{1+5y}{4} + \dfrac{x+3}{4} = \dfrac{17}{2} \end{cases}$$

Clear both equations of fractions.

$$\begin{cases} \mathbf{6}\left(\dfrac{6x-1}{3}\right) - \mathbf{6}\left(\dfrac{5}{3}\right) = \mathbf{6}\left(\dfrac{3y+1}{2}\right) \\ \mathbf{4}\left(\dfrac{1+5y}{4}\right) + \mathbf{4}\left(\dfrac{x+3}{4}\right) = \mathbf{4}\left(\dfrac{17}{2}\right) \end{cases}$$

$$\begin{cases} 12x - 2 - 10 = 9y + 3 \\ 1 + 5y + x + 3 = 34 \end{cases}$$

Write the 1st equation in standard form.
Solve for x in the 2nd equation.

$$\begin{cases} 12x - 9y = 15 \\ x = 30 - 5y \end{cases}$$

$12x - 9y = 15$ The 1st equation
    Substitute for x.

$$12(\mathbf{30 - 5y}) - 9y = 15$$
$$360 - 60y - 9y = 15$$
$$360 - 69y = 15$$
$$360 - 69y - \mathbf{360} = 15 - \mathbf{360}$$
$$-69y = -345$$
$$\frac{-69y}{-69} = \frac{-345}{-69}$$
$$y = 5$$

$x = 30 - 5y$ The 2nd equation
$$x = 30 - 5(\mathbf{5})$$
$$x = 30 - 25$$
$$x = 5$$

The solution is (5, 5).

Check:

$$\frac{6x-1}{3} - \frac{5}{3} = \frac{3y+1}{2}$$
$$\frac{6(5)-1}{3} - \frac{5}{3} \overset{?}{=} \frac{3(5)+1}{2}$$
$$\frac{30-1}{3} - \frac{5}{3} \overset{?}{=} \frac{15+1}{2}$$
$$\frac{29}{3} - \frac{5}{3} \overset{?}{=} \frac{16}{2}$$
$$\frac{24}{3} \overset{?}{=} \frac{16}{2}$$
$$8 = 8$$
True

Check:

$$\frac{1+5y}{4} + \frac{x+3}{4} = \frac{17}{2}$$
$$\frac{1+5(5)}{4} + \frac{5+3}{4} \overset{?}{=} \frac{17}{2}$$
$$\frac{1+25}{4} + \frac{8}{4} \overset{?}{=} \frac{17}{2}$$
$$\frac{26}{4} + \frac{8}{4} \overset{?}{=} \frac{17}{2}$$
$$\frac{34}{4} \overset{?}{=} \frac{17}{2}$$
$$\frac{17}{2} = \frac{17}{2}$$
True

99. $\begin{cases} 2(2x+3y)=5 \\ 8x=3(1+3y) \end{cases}$

Distribute in both equations.

$\begin{cases} 4x+6y=5 \\ 8x=3+9y \end{cases}$

Solve 2nd for x.

$\begin{cases} 4x+6y=5 \\ x=\dfrac{9}{8}y+\dfrac{3}{8} \end{cases}$

$4x+6y=5$ The 1st equation
Substitute for x.

$4\left(\dfrac{9}{8}y+\dfrac{3}{8}\right)+6y=5$

$\dfrac{9}{2}y+\dfrac{3}{2}+6y=5$

$2\left(\dfrac{9}{2}y\right)+2\left(\dfrac{3}{2}\right)+2(6y)=2(5)$

$9y+3+12y=10$

$21y+3=10$

$21y+3-3=10-3$

$21y=7$

$\dfrac{21y}{21}=\dfrac{7}{21}$

$y=\dfrac{\cancel{7}^{1}}{3\cdot\cancel{7}_{1}}$

$y=\dfrac{1}{3}$

$x=\dfrac{9}{8}y+\dfrac{3}{8}$ The 2nd equation

$x=\dfrac{\cancel{9}^{3}}{8}\left(\dfrac{1}{\cancel{3}_{1}}\right)+\dfrac{3}{8}$

$x=\dfrac{3}{8}+\dfrac{3}{8}$

$x=\dfrac{6}{8}$

$x=\dfrac{\cancel{2}^{1}\cdot 3}{\cancel{2}_{1}\cdot 4}$

$x=\dfrac{3}{4}$

The solution is $\left(\dfrac{3}{4},\dfrac{1}{3}\right)$.

Check:

$2(2x+3y)=5$

$2\left(2\cdot\dfrac{3}{4}+3\cdot\dfrac{1}{3}\right)\overset{?}{=}5$

$2\left(\dfrac{3}{2}+1\right)\overset{?}{=}5$

$2\left(\dfrac{5}{2}\right)\overset{?}{=}5$

$5=5$

True

Check:

$8x=3(1+3y)$

$8\left(\dfrac{3}{4}\right)\overset{?}{=}3\left(1+3\cdot\dfrac{1}{3}\right)$

$2(3)\overset{?}{=}3(1+1)$

$6=6$

True

SECTION 4.3
VOCABULARY

Fill in the blanks.
1. The coefficients of $3x$ and $-3x$ are **opposites**.

CONCEPTS

3. In the following system, which terms have coefficients that are opposites? **$7y$ and $-7y$**

$$\begin{cases} 3x + 7y = -25 \\ 4x - 7y = 12 \end{cases}$$

5. Add each pair of equations.

a.
$$\begin{array}{l} 2a + 2b = -6 \\ \underline{3a - 2b = \ \ 2} \\ 5a \quad\quad = -4 \end{array}$$

b.
$$\begin{array}{l} x - 3y = 15 \\ \underline{-x - y = -14} \\ -4y = 1 \end{array}$$

7. If the elimination method is used to solve

$$\begin{cases} 3x + 12y = 4 \\ 6x - 4y = 7 \end{cases}$$

   a. By what would we multiply the first equation to eliminate x? **-2**

   b. By what would we multiply the second equation to eliminate y? **3**

9. What algebraic step should be performed to

   a. Clear $\dfrac{2}{3}x + 4y = -\dfrac{4}{5}$ of fractions?

   Multiply both sides by 15.

   b. Clear $0.2x - 0.9y = 6.4$ of decimals?

   Multiply both sides by 10.

NOTATION

Complete the solution to solve the system.

11. Solve: $\begin{cases} x + y = 5 \\ x - y = -3 \end{cases}$.

$$\begin{array}{l} x + y = 5 \\ \underline{x - y = -3} \quad \text{Add the equations.} \\ \boxed{2x} = 2 \\ \quad x = \boxed{1} \end{array}$$

$x + y = 5$ This is the first equation.

$\boxed{1} + y = 5$

$\quad y = \boxed{4}$

The solution is $(\boxed{1}, \boxed{4})$.

GUIDED PRACTICE

Use the elimination method to solve each system. See Example 1.

13. $\begin{cases} x + y = 5 \\ x - y = 1 \end{cases}$

Eliminate y.

$$\begin{array}{l} x + y = 5 \\ \underline{x - y = 1} \\ 2x \quad = 6 \\ \dfrac{2x}{2} = \dfrac{6}{2} \\ x = 3 \end{array}$$

$x + y = 5$ The 1st equation

$3 + y = 5$

$3 + y - 3 = 5 - 3$

$y = 2$

The solution is $(3, 2)$.

15. $\begin{cases} x + y = -5 \\ -x + y = -1 \end{cases}$

Eliminate x.

$$\begin{array}{l} x + y = -5 \\ \underline{-x + y = -1} \\ 2y = -6 \\ \dfrac{2y}{2} = \dfrac{-6}{2} \\ y = -3 \end{array}$$

$x + y = -5$ The 1st equation

$x + (-3) = -5$

$x - 3 + 3 = -5 + 3$

$x = -2$

The solution is $(-2, -3)$.

17. $\begin{cases} 4x + 3y = 24 \\ 4x - 3y = -24 \end{cases}$

Eliminate y.

$$\begin{array}{l} 4x + 3y = \ \ 24 \\ \underline{4x - 3y = -24} \\ 8x \quad\quad = 0 \\ \dfrac{8x}{8} = \dfrac{0}{8} \\ x = 0 \end{array}$$

$4x + 3y = 24$ The 1st equation

$4(0) + 3y = 24$

$3y = 24$

$\dfrac{3y}{3} = \dfrac{24}{3}$

$y = 8$

The solution is $(0, 8)$.

19. $\begin{cases} 2s + t = -2 \\ -2s - 3t = -6 \end{cases}$

Eliminate s.

$$2s + t = -2$$
$$\underline{-2s - 3t = -6}$$
$$-2t = -8$$
$$\frac{-2t}{-2} = \frac{-8}{-2}$$
$$t = 4$$

$2s + t = -2$ The 1st equation
$$2s + \mathbf{4} = -2$$
$$2s + 4 - \mathbf{4} = -2 - \mathbf{4}$$
$$2s = -6$$
$$\frac{2s}{\mathbf{2}} = \frac{-6}{\mathbf{2}}$$
$$s = -3$$

The solution is $(-3, 4)$.

Use the elimination method to solve each system. See Example 2.

21. $\begin{cases} x + 3y = -9 \\ x + 8y = -4 \end{cases}$

Eliminate x.

Multiply both sides of the 1st equation by -1.

$\begin{cases} \mathbf{-1}(x) - \mathbf{1}(3y) = \mathbf{-1}(-9) \\ x + 8y = -4 \end{cases}$

$$-x - 3y = 9$$
$$\underline{x + 8y = -4}$$
$$5y = 5$$
$$\frac{5y}{\mathbf{5}} = \frac{5}{\mathbf{5}}$$
$$y = 1$$

$x + 3y = -9$ The 1st equation
$$x + 3(\mathbf{1}) = -9$$
$$x + 3 - \mathbf{3} = -9 - \mathbf{3}$$
$$x = -12$$

The solution is $(-12, 1)$.

23. $\begin{cases} 7x - y = 10 \\ 8x - y = 13 \end{cases}$

Eliminate y.

Multiply both sides of the 1st equation by -1.

$\begin{cases} \mathbf{-1}(7x) - \mathbf{1}(-y) = \mathbf{-1}(10) \\ 8x - y = 13 \end{cases}$

$$-7x + y = -10$$
$$\underline{8x - y = 13}$$
$$x = 3$$

$7x - y = 10$ The 1st equation
$$7(\mathbf{3}) - y = 10$$
$$21 - y = 10$$
$$21 - y + \mathbf{y} = 10 + \mathbf{y}$$
$$21 - \mathbf{10} = 10 + y - \mathbf{10}$$
$$11 = y$$

The solution is $(3, 11)$

Use the elimination method to solve each system. See Example 3.

25. $\begin{cases} 7x + 4y - 14 = 0 \\ 3x = 2y - 20 \end{cases}$

Write both equations in standard form.

$\begin{cases} 7x + 4y - 14 + \mathbf{14} = 0 + \mathbf{14} \\ 3x - \mathbf{2y} = 2y - 20 - \mathbf{2y} \end{cases}$

$\begin{cases} 7x + 4y = 14 \\ 3x - 2y = -20 \end{cases}$

Eliminate y.

Multiply both sides of the 2nd equation by 2.

$\begin{cases} 7x + 4y = 14 \\ \mathbf{2}(3x) - \mathbf{2}(2y) = \mathbf{2}(-20) \end{cases}$

$$7x + 4y = 14$$
$$\underline{6x - 4y = -40}$$
$$13x = -26$$
$$\frac{13x}{\mathbf{13}} = \frac{-26}{\mathbf{13}}$$
$$x = -2$$

$7x + 4y = 14$ The 1st equation
$$7(\mathbf{-2}) + 4y = 14$$
$$-14 + 4y = 14$$
$$-14 + 4y + \mathbf{14} = 14 + \mathbf{14}$$
$$4y = 28$$
$$\frac{4y}{\mathbf{4}} = \frac{28}{\mathbf{4}}$$
$$y = 7$$

The solution is $(-2, 7)$.

Section 4.3

27.
$$\begin{cases} 7x - 50y + 43 = 0 \\ x = 4 - 3y \end{cases}$$
Write both equations in standard form.
$$\begin{cases} 7x - 50y + 43 - \mathbf{43} = 0 - \mathbf{43} \\ x + \mathbf{3}\mathbf{y} = 4 - 3y + \mathbf{3}\mathbf{y} \end{cases}$$
$$\begin{cases} 7x - 50y = -43 \\ x + 3y = 4 \end{cases}$$
Eliminate x.
Multiply both sides of the 2nd equation by -7.
$$\begin{cases} 7x - 50y = -43 \\ -\mathbf{7}(x) - \mathbf{7}(3y) = -\mathbf{7}(4) \end{cases}$$

$$\begin{array}{r} 7x - 50y = -43 \\ -7x - 21y = -28 \\ \hline -71y = -71 \end{array}$$
$$\frac{-71y}{-71} = \frac{-71}{-71}$$
$$y = 1$$
$$x + 3y = 4 \quad \text{The } 2^{nd} \text{ equation}$$
$$x + 3(\mathbf{1}) = 4$$
$$x + 3 - \mathbf{3} = 4 - \mathbf{3}$$
$$x = 1$$
The solution is $(1, 1)$.

Use the elimination method to solve each system. See Example 4.

29.
$$\begin{cases} 4x + 3y = 7 \\ 3x - 2y = -16 \end{cases}$$
Eliminate y.
Multiply both sides of the 1st equation by 2.
Multiply both sides of the 2nd equation by 3.
$$\begin{cases} \mathbf{2}(4x) + \mathbf{2}(3y) = \mathbf{2}(7) \\ \mathbf{3}(3x) - \mathbf{3}(2y) = \mathbf{3}(-16) \end{cases}$$

$$\begin{array}{r} 8x + 6y = 14 \\ 9x - 6y = -48 \\ \hline 17x = -34 \end{array}$$
$$\frac{17x}{17} = \frac{-34}{17}$$
$$x = -2$$

$$4x + 3y = 7 \quad \text{The } 1^{st} \text{ equation}$$
$$4(-\mathbf{2}) + 3y = 7$$
$$-8 + 3y + \mathbf{8} = 7 + \mathbf{8}$$
$$3y = 15$$
$$\frac{3y}{3} = \frac{15}{3}$$
$$y = 5$$
The solution is $(-2, 5)$

31.
$$\begin{cases} 5a + 8b = 2 \\ 11a - 3b = 25 \end{cases}$$
Eliminate b.
Multiply both sides of the 1st equation by 3.
Multiply both sides of the 2nd equation by 8.
$$\begin{cases} \mathbf{3}(5a + \mathbf{3}(8b) = \mathbf{3}(2) \\ \mathbf{8}(11a) - \mathbf{8}(3b) = \mathbf{8}(25) \end{cases}$$

$$\begin{array}{r} 15a + 24b = 6 \\ 88a - 24b = 200 \\ \hline 103a = 206 \end{array}$$
$$\frac{103a}{\mathbf{103}} = \frac{206}{\mathbf{103}}$$
$$a = 2$$

$$5a + 8b = 2 \quad \text{The } 1^{st} \text{ equation}$$
$$5(\mathbf{2}) + 8b = 2$$
$$10 + 8b - \mathbf{10} = 2 - \mathbf{10}$$
$$8b = -8$$
$$\frac{8b}{\mathbf{8}} = \frac{-8}{\mathbf{8}}$$
$$b = -1$$
The solution is $(2, -1)$

Use the elimination method to solve each system. See Example 5.

33.
$$\begin{cases} \dfrac{3}{5}s + \dfrac{4}{5}t = 1 \\ -\dfrac{1}{4}s + \dfrac{3}{8}t = 1 \end{cases}$$
Clear both equations of fractions.
$$\begin{cases} \mathbf{5}\left(\dfrac{3}{5}s\right) + \mathbf{5}\left(\dfrac{4}{5}t\right) = \mathbf{5}(1) \\ \mathbf{8}\left(-\dfrac{1}{4}s\right) + \mathbf{8}\left(\dfrac{3}{8}t\right) = \mathbf{8}(1) \end{cases}$$

$$\begin{cases} 3s + 4t = 5 \\ -2s + 3t = 8 \end{cases}$$
Eliminate s.
Multiply both sides of the 1st equation by 2.
Multiply both sides of the 2nd equation by 3..
$$\begin{cases} \mathbf{2}(3s) + \mathbf{2}(4t) = \mathbf{2}(5) \\ \mathbf{3}(-2s) + \mathbf{3}(3t) = \mathbf{3}(8) \end{cases}$$

$$\begin{cases} 6s + 8t = 10 \\ -6s + 9t = 24 \end{cases}$$

$$\begin{array}{r} 6s + 8t = 10 \\ -6s + 9t = 24 \\ \hline 17t = 34 \end{array}$$
$$\frac{17t}{17} = \frac{34}{17}$$
$$t = 2$$

$$3s + 4t = 5 \quad \text{The } 1^{\text{st}} \text{ equation}$$
$$3s + 4(\mathbf{2}) = 5$$
$$3s + 8 - \mathbf{8} = 5 - \mathbf{8}$$
$$3s = -3$$
$$\frac{3s}{\mathbf{3}} = \frac{-3}{\mathbf{3}}$$
$$s = -1$$

The solution is $(-1, 2)$.

35. $\begin{cases} \dfrac{1}{2}s - \dfrac{1}{4}t = 1 \\ \dfrac{1}{3}s + t = 3 \end{cases}$

Clear both equations of fractions.
Multiply both sides of the 1st equation by 4.
Multiply both sides of the 2nd equation by 3.

$$\begin{cases} \mathbf{4}\left(\dfrac{1}{2}s\right) - \mathbf{4}\left(\dfrac{1}{4}t\right) = \mathbf{4}(1) \\ \mathbf{3}\left(\dfrac{1}{3}s\right) + \mathbf{3}(t) = \mathbf{3}(3) \end{cases}$$

$$\begin{cases} 2s - t = 4 \\ s + 3t = 9 \end{cases}$$

Eliminate s.
Multiply both sides of the 2nd equation by -2.
$$2s - t = 4$$
$$\underline{-2s - 6t = -18}$$
$$-7t = -14$$
$$\frac{-7t}{-7} = \frac{-14}{-7}$$
$$t = 2$$

$$\frac{1}{3}s + t = 3 \quad \text{The } 2^{\text{nd}} \text{ equation}$$
$$\frac{1}{3}s + (\mathbf{2}) = 3$$
$$\frac{1}{3}s + 2 - \mathbf{2} = 3 - \mathbf{2}$$
$$\frac{1}{3}s = 1$$
$$\frac{\mathbf{3}}{\mathbf{1}}\left(\frac{1}{3}s\right) = \mathbf{3}(1)$$
$$s = 3$$

The solution is $(3, 2)$.

Use the elimination method to solve each system. If there is no solution, or infinitely many solutions, so indicate. See Example 6.

37. $\begin{cases} 3x - 5y = -29 \\ 3x - 5y = 15 \end{cases}$

Eliminate x.
Multiply both sides of the 1st equation by -1.
$$\begin{cases} \mathbf{-1}(3x) - \mathbf{1}(-5y) = \mathbf{-1}(-29) \\ 3x - 5y = 15 \end{cases}$$
$$-3x + 5y = 29$$
$$\underline{3x - 5y = 15}$$
$$0 = 44 \quad \text{Both variables are eliminated.}$$
$$\text{False}$$

No solution.

39. $\begin{cases} 3x - 16 = 5y \\ -3x + 5y - 33 = 0 \end{cases}$

Write both equations in standard form.
$$\begin{cases} 3x - 16 - \mathbf{5y} + \mathbf{16} = 5y - \mathbf{5y} + \mathbf{16} \\ -3x + 5y - 33 + \mathbf{33} = 0 + \mathbf{33} \end{cases}$$
$$\begin{cases} 3x - 5y = 16 \\ -3x + 5y = 33 \end{cases}$$

Eliminate x.
$$3x - 5y = 16$$
$$\underline{-3x + 5y = 33}$$
$$0 = 49 \quad \text{Both variables are eliminated.}$$
$$\text{False}$$

No solution.

Use the elimination method to solve each system. If there is no solution, or infinitely many solutions, so indicate. See Example 7.

41. $\begin{cases} 0.4x - 0.7y = -1.9 \\ -x + \dfrac{7}{4}y = \dfrac{19}{4} \end{cases}$

Multiply the 1st equation by 10.

Multiply the 2nd equation by 4.

$$\begin{cases} \mathbf{10}(0.4x) - \mathbf{10}(0.7y) = \mathbf{10}(-1.9) \\ \mathbf{4}(-x) + \mathbf{4}\left(\dfrac{7}{4}y\right) = \mathbf{4}\left(\dfrac{19}{4}\right) \end{cases}$$

$$\begin{cases} 4x - 7y = -19 \\ -4x + 7y = 19 \end{cases}$$
$$4x - 7y = -19$$
$$\underline{-4x + 7y = 19}$$
$$0 = 0 \quad \text{Both variables are eliminated.}$$
$$\text{True}$$

The system has infinitely
many solutions.

Section 4.3

43.
$$\begin{cases} \dfrac{x-6y}{2}=7 \\ -x+6y+14=0 \end{cases}$$

Write the 2nd equations in standard form.

$$\begin{cases} \dfrac{x-6y}{2}=7 \\ -x+6y+14-\mathbf{14}=0-\mathbf{14} \end{cases}$$

$$\begin{cases} \dfrac{x-6y}{2}=7 \\ -x+6y=-14 \end{cases}$$

Multiply both sides of the 1st equation by 2.

$$\begin{cases} \mathbf{2}\left(\dfrac{x-6y}{2}\right)=\mathbf{2}(7) \\ -x+6y=-14 \end{cases}$$

$$\begin{cases} x-6y=14 \\ -x+6y=-14 \end{cases}$$

$$\begin{aligned} x-6y&=14 \\ \underline{-x+6y}&=\underline{-14} \\ 0&=0 \quad \text{Both variables are eliminated.} \\ &\quad\text{True} \end{aligned}$$

The system has infinitely
   many solutions.

TRY IT YOURSELF

**Solve the system by either the substitution or the
elimination method if possible.**

45.
$$\begin{cases} y=-3x+9 \\ y=x+1 \end{cases}$$

$$y=-3x+9 \quad \text{The 1}^{\text{st}}\text{ equation}$$
$$\text{Substitute for } y.$$
$$(\mathbf{x+1})=-3x+9$$
$$x+1+\mathbf{3x}=-3x+9+\mathbf{3x}$$
$$4x+1=9$$
$$4x+1-\mathbf{1}=9-\mathbf{1}$$
$$4x=8$$
$$\dfrac{4x}{\mathbf{4}}=\dfrac{8}{\mathbf{4}}$$
$$x=2$$

$$y=x+1 \quad \text{The 2}^{\text{nd}}\text{ equation}$$
$$y=\mathbf{2}+1$$
$$y=3$$

The solution is (2, 3).

47.
$$\begin{cases} 4x+6y=5 \\ 8x-9y=3 \end{cases}$$

Eliminate y.

Multiply both sides of the 1st equation by 3.
Multiply both sides of the 2nd equation by 2.

$$\begin{aligned} 12x+18y&=15 \\ \underline{16x-18y}&=\underline{6} \\ 28x&=21 \end{aligned}$$

$$\dfrac{28x}{\mathbf{28}}=\dfrac{21}{\mathbf{28}}$$
$$x=\dfrac{3}{4}$$

$$4x+6y=5 \quad \text{The 1}^{\text{st}}\text{ equation}$$
$$4\left(\dfrac{\mathbf{3}}{\mathbf{4}}\right)+6y=5$$
$$3+6y=5$$
$$3+6y-\mathbf{3}=5-\mathbf{3}$$
$$6y=2$$
$$\dfrac{6y}{\mathbf{6}}=\dfrac{2}{\mathbf{6}}$$
$$y=\dfrac{1}{3}$$

The solution is $\left(\dfrac{3}{4},\ \dfrac{1}{3}\right)$.

49.
$$\begin{cases} 6x-3y=-7 \\ y+9x=6 \end{cases}$$

Write the 2nd equation in standard form.

$$\begin{cases} 6x-3y=-7 \\ 9x+y=6 \end{cases}$$

Eliminate y.

Multiply both sides of the 2nd equation by 3.

$$\begin{aligned} 6x-3y&=-7 \\ \underline{27x+3y}&=\underline{18} \\ 33x&=11 \end{aligned}$$

$$\dfrac{33x}{\mathbf{33}}=\dfrac{11}{\mathbf{33}}$$
$$x=\dfrac{1}{3}$$

$$6x - 3y = -7 \quad \text{1}^{\text{st}} \text{ equation}$$

$$6\left(\frac{1}{3}\right) - 3y = -7$$

$$2 - 3y = -7$$

$$2 - 3y - \mathbf{2} = -7 - \mathbf{2}$$

$$-3y = -9$$

$$\frac{-3y}{-3} = \frac{-9}{-3}$$

$$y = 3$$

The solution is $\left(\dfrac{1}{3}, 3\right)$.

51. $\begin{cases} x + y = 1 \\ x - y = 5 \end{cases}$

Eliminate y.

$$x + y = 1$$
$$\underline{x - y = 5}$$
$$2x \quad\;\; = 6$$

$$\frac{2x}{\mathbf{2}} = \frac{6}{\mathbf{2}}$$

$$x = 3$$

$$x + y = 1 \quad \text{The 1}^{\text{st}} \text{ equation}$$
$$\mathbf{3} + y = 1$$
$$3 + y - \mathbf{3} = 1 - \mathbf{3}$$
$$y = -2$$

The solution is $(3, -2)$.

53. $\begin{cases} 4(x - 2y) = 36 \\ 3x - 6y = 27 \end{cases}$

Distribute in 1st equation.

$\begin{cases} 4x - 8y = 36 \\ 3x - 6y = 27 \end{cases}$

Eliminate x.

Multiply both sides of the 1st equation by 3.

Multiply both sides of the 2nd equation by -4.

$\begin{cases} \mathbf{3}(4x) - \mathbf{3}(8y) = \mathbf{3}(36) \\ -\mathbf{4}(x) - \mathbf{4}(-6y) = -\mathbf{4}(27) \end{cases}$

$$12x - 24y = \;\;\;108$$
$$\underline{-12x + 24y = -108}$$
$$0 = 0 \;\; \text{Both variables are eliminated.}$$
$$\text{True}$$

The system has infinitely many solutions.

55. $\begin{cases} x = y \\ 0.1x + 0.2y = 1.0 \end{cases}$

Multiply both sides of the 2nd equation by 10.

$\begin{cases} x = y \\ x + 2y = 10 \end{cases}$

$$x + 2y = 10 \quad \begin{array}{l}\text{The 2}^{\text{nd}} \text{ equation} \\ \text{Substitute for } y.\end{array}$$

$$x + 2(\mathbf{x}) = 10$$
$$3x = 10$$
$$\frac{3x}{\mathbf{3}} = \frac{10}{\mathbf{3}}$$
$$x = \frac{10}{3}$$

$$x = y \quad \text{The 1}^{\text{st}} \text{ equation}$$
$$\frac{\mathbf{10}}{\mathbf{3}} = y$$

The solution is $\left(\dfrac{10}{3}, \dfrac{10}{3}\right)$.

57. $\begin{cases} 2x + 11y = -10 \\ 5x + 4y = 22 \end{cases}$

Eliminate x.

Multiply both sides of the 1st equation by 5.

Multiply both sides of the 2nd equation by -2.

$\begin{cases} \mathbf{5}(2x) + \mathbf{5}(11y) = \mathbf{5}(-10) \\ -\mathbf{2}(5x) + -\mathbf{2}(4y) = -\mathbf{2}(22) \end{cases}$

$$10x + 55y = -50$$
$$\underline{-10x - 8y \;\; = -44}$$
$$47y = -94$$
$$\frac{47y}{\mathbf{47}} = \frac{-94}{\mathbf{47}}$$
$$y = -2$$

$$5x + 4y = 22 \quad \text{The 2}^{\text{nd}} \text{ equation}$$
$$5x + 4(-\mathbf{2}) = 22$$
$$5x - 8 + \mathbf{8} = 22 + \mathbf{8}$$
$$5x = 30$$
$$\frac{5x}{\mathbf{5}} = \frac{30}{\mathbf{5}}$$
$$x = 6$$

The solution is $(6, -2)$.

Section 4.3

59.
$$\begin{cases} 7x = 21 - 6y \\ 4x + 5y = 12 \end{cases}$$

Write the 1st equations in standard form.

$$\begin{cases} 7x + 6y = 21 - 6y + 6y \\ 4x + 5y = 12 \end{cases}$$

$$\begin{cases} 7x + 6y = 21 \\ 4x + 5y = 12 \end{cases}$$

Eliminate x.

Multiply both sides of the 1st equation by 4.

Multiply both sides of the 2nd equation by -7.

$$\begin{cases} 4(7x) + 4(6y) = 4(21) \\ -7(4x) - 7(5y) = -7(12) \end{cases}$$

$$\begin{aligned} 28x + 24y &= 84 \\ -28x - 35y &= -84 \\ \hline -11y &= 0 \end{aligned}$$

$$\frac{-11y}{-11} = \frac{0}{-11}$$
$$y = 0$$

$7x = 21 - 6y$ The 1st equation
$7x = 21 - 6(0)$
$7x = 21$
$$\frac{7x}{7} = \frac{21}{7}$$
$$x = 3$$

The solution is (3, 0).

61.
$$\begin{cases} 9x - 10y = 0 \\ \dfrac{9x - 3y}{63} = 1 \end{cases}$$

Multiply both sides of the 2nd equation by 63.

$$\begin{cases} 9x - 10y = 0 \\ 63\left(\dfrac{9x - 3y}{63}\right) = 63(1) \end{cases}$$

$$\begin{cases} 9x - 10y = 0 \\ 9x - 3y = 63 \end{cases}$$

Eliminate x.

Multiply both sides of the 1st equation by -1.

$$\begin{cases} (-1)9x - (-1)(10y) = (-1)0 \\ 9x - 3y = 63 \end{cases}$$

$$\begin{aligned} -9x + 10y &= 0 \\ 9x - 3y &= 63 \\ \hline 7y &= 63 \end{aligned}$$

$$\frac{7y}{7} = \frac{63}{7}$$
$$y = 9$$

$9x - 10y = 0$ The 1st equation
$9x - 10(9) = 0$
$9x - 90 + 90 = 0 + 90$
$9x = 90$
$$\frac{9x}{9} = \frac{90}{9}$$
$$x = 10$$

The solution is (10, 9).

63.
$$\begin{cases} \dfrac{m}{4} + \dfrac{n}{3} = -\dfrac{1}{12} \\ \dfrac{m}{2} - \dfrac{5}{4}n = \dfrac{7}{4} \end{cases}$$

Clear both equations of fractions.

Multiply both sides of the 1st equation by 12.

Multiply both sides of the 2nd equation by 4.

$$\begin{cases} 12\left(\dfrac{m}{4}\right) + 12\left(\dfrac{n}{3}\right) = 12\left(-\dfrac{1}{12}\right) \\ 4\left(\dfrac{m}{2}\right) - 4\left(\dfrac{5}{4}n\right) = 4\left(\dfrac{7}{4}\right) \end{cases}$$

$$\begin{cases} 3m + 4n = -1 \\ 2m - 5n = 7 \end{cases}$$

Eliminate n.

Multiply both sides of the 1st equation by 5.

Multiply both sides of the 2nd equation by 4.

$$\begin{cases} (5)3m + (5)4n = (5)(-1) \\ (4)2m - (4)5n = (4)7 \end{cases}$$

$$\begin{aligned} 15m + 20n &= -5 \\ 8m - 20n &= 28 \\ \hline 23m \quad\quad &= 23 \end{aligned}$$

$$\frac{23m}{23} = \frac{23}{23}$$
$$m = 1$$

$2m - 5n = 7$ The 2nd equation
$2(1) - 5n = 7$
$2 - 5n - 2 = 7 - 2$
$-5n = 5$
$$\frac{-5n}{-5} = \frac{5}{-5}$$
$$n = -1$$

The solution is (1, −1).

65.
$$\begin{cases} x - \dfrac{4}{3}y = \dfrac{1}{3} \\ 2x + \dfrac{3}{2}y = \dfrac{1}{2} \end{cases}$$

Clear both equations of fractions.

Multiply both sides of the 1st equation by 3.

Multiply both sides of the 2nd equation by 2.

$$\begin{cases} 3(x) - 3\left(\dfrac{4}{3}y\right) = 3\left(\dfrac{1}{3}\right) \\ 2(2x) + 2\left(\dfrac{3}{2}y\right) = 2\left(\dfrac{1}{2}\right) \end{cases}$$

$$\begin{cases} 3x - 4y = 1 \\ 4x + 3y = 1 \end{cases}$$

Eliminate y.

Multiply both sides of the 1st equation by 3.

Multiply both sides of the 2nd equation by 4.

$$\begin{cases} (3)3x - (3)4y = (3)1 \\ (4)4x + (4)3y = (4)1 \end{cases}$$

$$\begin{aligned} 9x - 12y &= 3 \\ 16x + 12y &= 4 \\ \hline 25x \qquad &= 7 \end{aligned}$$

$$\dfrac{25x}{25} = \dfrac{7}{25}$$

$$x = \dfrac{7}{25}$$

$4x + 3y = 1$ The 2nd equation

$$4\left(\dfrac{7}{25}\right) + 3y = 1$$

$$\dfrac{28}{25} + 3y = 1$$

$$\dfrac{28}{25} + 3y - \dfrac{28}{25} = \dfrac{25}{25} - \dfrac{28}{25}$$

$$3y = -\dfrac{3}{25}$$

$$\dfrac{1}{3}(3y) = -\dfrac{3}{25}\left(\dfrac{1}{3}\right)$$

$$y = -\dfrac{1}{25}$$

The solution is $\left(\dfrac{7}{25}, -\dfrac{1}{25}\right)$.

67.
$$\begin{cases} 4x - 7y + 32 = 0 \\ 5x = 4y - 2 \end{cases}$$

Write both equations in standard form.

$$\begin{cases} 4x - 7y + 32 - 32 = 0 - 32 \\ 5x - 4y = 4y - 2 - 4y \end{cases}$$

$$\begin{cases} 4x - 7y = -32 \\ 5x - 4y = -2 \end{cases}$$

Eliminate x.

Multiply both sides of the 1st equation by 5.

Multiply both sides of the 2nd equation by -4.

$$\begin{cases} 5(4x) - 5(7y) = 5(-32) \\ (-4)5x(-4)(-4y) = (-4)(-2) \end{cases}$$

$$\begin{aligned} 20x - 35y &= -160 \\ -20x + 16y &= 8 \\ \hline -19y &= -152 \end{aligned}$$

$$\dfrac{-19y}{-19} = \dfrac{-152}{-19}$$

$$y = 8$$

$5x = 4y - 2$ The 2nd equation

$5x = 4(8) - 2$

$5x = 32 - 2$

$5x = 30$

$$\dfrac{5x}{5} = \dfrac{30}{5}$$

$$x = 6$$

The solution is $(6, 8)$.

69.
$$\begin{cases} 3(x + 4y) = -12 \\ x = 3y + 10 \end{cases}$$

Distribute in 1st equation.

Write the 2nd equation in standard form.

$$\begin{cases} 3(x) + 3(12y) = -12 \\ x - 3y = 3y + 10 - 3y \end{cases}$$

$$\begin{cases} 3x + 12y = -12 \\ x - 3y = 10 \end{cases}$$

Eliminate y.

Multiply both sides of the 2nd equation by 4.

$$\begin{cases} 3x + 12y = -12 \\ 4(x) - 4(3y) = 4(10) \end{cases}$$

$$\begin{aligned} 3x + 12y &= -12 \\ 4x - 12y &= 40 \\ \hline 7x \qquad &= 28 \end{aligned}$$

$$\dfrac{7x}{7} = \dfrac{28}{7}$$

$$x = 4$$

Section 4.3

$$x - 3y = 10 \quad \text{The 2}^{\text{nd}} \text{ equation.}$$
$$4 - 3y = 10$$
$$4 - 3y - \mathbf{4} = 10 - \mathbf{4}$$
$$-3y = 6$$
$$\frac{-3y}{-\mathbf{3}} = \frac{6}{-\mathbf{3}}$$
$$y = -2$$

The solution is $(4, -2)$.

71. $\begin{cases} 4a + 7b = 2 \\ 9a - 3b = 1 \end{cases}$

Eliminate b.

Multiply both sides of the 1st equation by 3.
Multiply both sides of the 2nd equation by 7.

$\begin{cases} \mathbf{3}(4a) + \mathbf{3}(7b) = \mathbf{3}(2) \\ \mathbf{7}(9a) - \mathbf{7}(3b) = \mathbf{7}(1) \end{cases}$

$$12a + 21b = 6$$
$$\underline{63a - 21b = 7}$$
$$75a \qquad = 13$$
$$\frac{75a}{\mathbf{75}} = \frac{13}{\mathbf{75}}$$
$$a = \frac{13}{75}$$

$$9a - 3b = 1 \quad \text{The 2}^{\text{nd}} \text{ equation}$$
$$9\left(\frac{\mathbf{13}}{\mathbf{75}}\right) - 3b = 1$$
$$\frac{39}{25} - 3b = 1$$
$$\frac{39}{25} - 3b - \frac{\mathbf{39}}{\mathbf{25}} = \frac{25}{25} - \frac{\mathbf{39}}{\mathbf{25}}$$
$$-3b = -\frac{14}{25}$$
$$-\frac{\mathbf{1}}{\mathbf{3}}(-3b) = -\frac{\mathbf{1}}{\mathbf{3}}\left(-\frac{14}{25}\right)$$
$$b = \frac{14}{75}$$

The solution is $\left(\frac{13}{75}, \frac{14}{75}\right)$.

73. $\begin{cases} 3a - b = 12.3 \\ 4a - b = 14.9 \end{cases}$

Eliminate b.

Multiply both sides of the 1st equation by -1.

$\begin{cases} -\mathbf{1}(3a) - \mathbf{1}(-b) = -\mathbf{1}(12.3) \\ 4a - b = 14.9 \end{cases}$

$$-3a + b = -12.3$$
$$\underline{4a - b = 14.9}$$
$$a \qquad = 2.6$$

$$3a - b = 12.3 \quad \text{The 1}^{\text{st}} \text{ equation}$$
$$3(\mathbf{2.6}) - b = 12.3$$
$$7.8 - b = 12.3$$
$$7.8 - b - \mathbf{7.8} = 12.3 - \mathbf{7.8}$$
$$-b = 4.5$$
$$\frac{-b}{-\mathbf{1}} = \frac{4.5}{-\mathbf{1}}$$
$$b = -4.5$$

The solution is $(2.6, -4.5)$.

75. $\begin{cases} 5x - 4y = 8 \\ -5x - 4y = 8 \end{cases}$

Eliminate x.

$$5x - 4y = 8$$
$$\underline{-5x - 4y = 8}$$
$$-8y = 16$$
$$\frac{-8y}{-\mathbf{8}} = \frac{16}{-\mathbf{8}}$$
$$y = -2$$

$$5x - 4y = 8 \quad \text{1}^{\text{st}} \text{ equation}$$
$$5x - 4(-\mathbf{2}) = 8$$
$$5x + 8 = 8$$
$$5x + 8 - \mathbf{8} = 8 - \mathbf{8}$$
$$5x = 0$$
$$\frac{5x}{\mathbf{5}} = \frac{0}{\mathbf{5}}$$
$$x = 0$$

The solution is $(0, -2)$.

77.
$$\begin{cases} 9a + 16b = -36 \\ 7a + 4b = 48 \end{cases}$$

Eliminate b.

Multiply both sides of the 2nd equation by -4.

$$\begin{cases} 9a + 16b = -36 \\ -4(7a) - 4(4b) = -4(48) \end{cases}$$

$$\begin{array}{r} 9a + 16b = -36 \\ -28a - 16b = -192 \\ \hline -19a \qquad = -228 \end{array}$$

$$\frac{-19a}{-19} = \frac{-228}{-19}$$

$$a = 12$$

$$7a + 4b = 48 \quad \text{The 2}^{nd} \text{ equation}$$
$$7(12) + 4b = 48$$
$$84 + 4b = 48$$
$$84 + 4b - 84 = 48 - 84$$
$$4b = -36$$
$$\frac{4b}{4} = \frac{-36}{4}$$
$$b = -9$$

The solution is $(12, -9)$.

79.
$$\begin{cases} 8x + 12y = -22 \\ 3x - 2y = 8 \end{cases}$$

Eliminate y.

Multiply both sides of the 2nd equation by 6.

$$\begin{cases} 8x + 12y = -22 \\ (6)3x - (6)2y = (6)8 \end{cases}$$

$$\begin{array}{r} 8x + 12y = -22 \\ 18x - 12y = 48 \\ \hline 26x \qquad = 26 \end{array}$$

$$\frac{26x}{26} = \frac{26}{26}$$

$$x = 1$$

$$3x - 2y = 8 \quad \text{The 2}^{nd} \text{ equation}$$
$$3(1) - 2y = 8$$
$$3 - 2y = 8$$
$$3 - 2y - 3 = 8 - 3$$
$$-2y = 5$$
$$\frac{-2y}{-2} = \frac{5}{-2}$$
$$y = -\frac{5}{2}$$

The solution is $\left(1, -\frac{5}{2}\right)$.

81.
$$\begin{cases} 6x + 5y + 29 = 0 \\ 0.02x = 0.03y - 0.05 \end{cases}$$

Write both equations in standard form.

$$\begin{cases} 6x + 5y = -29 \\ 0.02x - 0.03y = -0.05 \end{cases}$$

Multiply both sides of the 2nd equation by 100.

$$\begin{cases} 6x + 5y = -29 \\ 100(0.02x) - 100(0.03y) = 100(-0.05) \end{cases}$$

$$\begin{cases} 6x + 5y = -29 \\ 2x - 3y = -5 \end{cases}$$

$$\begin{cases} 6x + 5y = -29 \\ (-3)2x - (-3)3y = (-3)(-5) \end{cases}$$

$$\begin{array}{r} 6x + 5y = -29 \\ -6x + 9y = 15 \\ \hline 14y = -14 \end{array}$$

$$\frac{14y}{14} = \frac{-14}{14}$$

$$y = -1$$

$$2x - 3y = -5 \quad \text{The 2}^{nd} \text{ equation}$$
$$2x - 3(-1) = -5$$
$$2x + 3 = -5$$
$$2x + 3 - 3 = -5 - 3$$
$$2x = -8$$
$$\frac{2x}{2} = \frac{-8}{2}$$
$$x = -4$$

The solution is $(-4, -1)$.

83.
$$\begin{cases} c = d - 9 \\ 5c = 3d - 35 \end{cases}$$

Write both equations in standard form.

$$\begin{cases} c - d = -9 \\ 5c - 3d = -35 \end{cases}$$

Eliminate c.

Multiply both sides of the 1st equation by -5.

$$\begin{cases} (-5)c - (-5)d = (-5)(-9) \\ 5c - 3d = -35 \end{cases}$$

$$\begin{array}{r} -5c + 5d = 45 \\ 5c - 3d = -35 \\ \hline 2d = 10 \end{array}$$

$$\frac{2d}{2} = \frac{10}{2}$$

$$d = 5$$

$$c = d - 9 \quad \text{The 1}^{st} \text{ equation}$$
$$c = 5 - 9$$
$$c = -4$$

The solution is $(-4, 5)$.

Section 4.3

85. $\begin{cases} 0.9x + 2.1 = 0.3y \\ 0.4x = 0.7y + 1.9 \end{cases}$

Write both equations in standard form.

$\begin{cases} 0.9x - 0.3y = -2.1 \\ 0.4x - 0.7y = 1.9 \end{cases}$

Multiply both sides of both equations by 10.

$\begin{cases} 10(0.9x) - 10(0.3y) = 10(-2.1) \\ 10(0.4x) - 10(0.7y) = 10(1.9) \end{cases}$

$\begin{cases} 9x - 3y = -21 \\ 4x - 7y = 19 \end{cases}$

Eliminate y.

Multiply both sides of the 1st equation by 7.
Multiply both sides of the 2nd equation by -3.

$\begin{cases} 7(9x) - 7(3y) = 7(-21) \\ -3(4x) - 3(-7y) = -3(19) \end{cases}$

$\begin{array}{r} 63x - 21y = -147 \\ -12x + 21y = -57 \\ \hline 51x = -204 \end{array}$

$$\frac{51x}{51} = \frac{-204}{51}$$

$$x = -4$$

$9x + 21 = 3y$ The 1st equation

$9(-4) + 21 = 3y$

$-36 + 21 = 3y$

$-15 = 3y$

$$\frac{-15}{3} = \frac{3y}{3}$$

$-5 = y$

The solution is $(-4, -5)$.

87. $\begin{cases} 5c + 2d = -5 \\ 6c + 2d = -10 \end{cases}$

Eliminate d.

Multiply both sides of the 1st equation by -1.

$\begin{cases} -1(5c) - 1(2d) = -1(-5) \\ 6c + 2d = -10 \end{cases}$

$\begin{array}{r} -5c - 2d = 5 \\ 6c + 2d = -10 \\ \hline c = -5 \end{array}$

$5c + 2d = -5$ The 1st equation

$5(-5) + 2d = -5$

$-25 + 2d = -5$

$-25 + 2d + 25 = -5 + 25$

$2d = 20$

$$\frac{2d}{2} = \frac{20}{2}$$

$d = 10$

The solution is $(-5, 10)$.

89. $\begin{cases} \dfrac{2}{15}x - \dfrac{1}{5}y = \dfrac{1}{3} \\ \dfrac{2}{15}x - \dfrac{1}{5}y = \dfrac{1}{10} \end{cases}$

Clear both equations of fractions.

Multiply both sides of the 1st equation by 30.

Multiply both sides of the 2nd equation by -30.

$\begin{cases} 30\left(\dfrac{2}{15}x\right) - 30\left(\dfrac{1}{5}y\right) = 30\left(\dfrac{1}{3}\right) \\ -30\left(\dfrac{2}{15}x\right) - 30\left(-\dfrac{1}{5}y\right) = -30\left(\dfrac{1}{10}\right) \end{cases}$

$\begin{array}{r} 4x - 6y = 10 \\ -4x + 6y = -3 \\ \hline 0 = 7 \end{array}$ Both variables are eliminated.

False

No solution.

APPLICATIONS

91. EDUCATION

Let x = the number of years since 1980

y = the % less than high school

$\begin{cases} 9x + 11y = 352 \\ 5x - 11y = -198 \end{cases}$

Eliminate y.

$\begin{array}{r} 9x + 11y = 352 \\ 5x - 11y = -198 \\ \hline 14x = 154 \end{array}$

$$\frac{14x}{14} = \frac{154}{14}$$

$x = 11$

$x = 0$ implies 1980

$x = 11$ implies $1980 + 11$

The year when the percents are equal is 1991.

93. CFL BULBS

Let c = the operating cost of a CFL bulb

d = the number of days before savings start for the CFL bulb

$$\begin{cases} 60c - d = 96 \\ 15c - d = 6 \end{cases}$$

Eliminate d.

Multiply both sides of the 2nd equation by -1.

$$\begin{cases} 60c - d = 96 \\ \mathbf{-1}(15c) - \mathbf{1}(-d) = \mathbf{-1}(6) \end{cases}$$

$$\begin{array}{r} 60c - d = 96 \\ -15c + d = -6 \\ \hline 45c \quad\quad = 90 \end{array}$$

$$\frac{45c}{\mathbf{45}} = \frac{90}{\mathbf{45}}$$

$$c = 2$$

$60c - d = 96$ The 1st equation

$60(\mathbf{2}) - d = 96$

$120 - d - \mathbf{120} = 96 - \mathbf{120}$

$-d = -24$

$-1(-d) = -1(-24)$

$d = 24$

After 24 days the CFL bulb will begin to save money.

WRITING

95 - 97 Answers will vary.

REVIEW

99. Find an equation of the line with slope $-\dfrac{11}{6}$ that passes through $(2, -6)$. Write the equation in slope-intercept form.

$$y - y_1 = m(x - x_1)$$

$$y - (-6) = -\frac{11}{6}(x - 2)$$

$$y + 6 = -\frac{11}{6}x + \frac{22}{6}$$

$$y + 6 - \mathbf{6} = -\frac{11}{6}x + \frac{22}{6} - \frac{\mathbf{36}}{\mathbf{6}}$$

$$y = -\frac{11}{6}x - \frac{14}{6}$$

$$y = -\frac{11}{6}x - \frac{7}{3} \quad \text{slope-intercept form}$$

101. Evaluate: $-10(18 - 4^2)^3$

$$-10(18 - 4^2)^3 = -10(18 - 16)^3$$

$$= -10(2)^3$$

$$= -10(8)$$

$$= -80$$

CHALLENGE PROBLEMS

Use the elimination method to solve each system.

103. $\begin{cases} \dfrac{x-3}{2} = \dfrac{11}{6} - \dfrac{y+5}{3} \\ \dfrac{x+3}{3} - \dfrac{y+3}{4} = \dfrac{5}{12} \end{cases}$

Clear both equations of fractions

$$\begin{cases} \mathbf{6}\left(\dfrac{x-3}{2}\right) = \mathbf{6}\left(\dfrac{11}{6}\right) - \mathbf{6}\left(\dfrac{y+5}{3}\right) \\ \mathbf{12}\left(\dfrac{x+3}{3}\right) - \mathbf{12}\left(\dfrac{y+3}{4}\right) = \mathbf{12}\left(\dfrac{5}{12}\right) \end{cases}$$

$$\begin{cases} 3(x-3) = 11 - 2(y+5) \\ 4(x+3) - 3(y+3) = 5 \end{cases}$$

Distribute twice in both equations.
Collect like terms

$$\begin{cases} 3x - 9 = 1 - 2y \\ 4x - 3y + 3 = 5 \end{cases}$$

Write both equations in standard form.

$$\begin{cases} 3x - 9 + \mathbf{2y} + \mathbf{9} = 1 - 2y + \mathbf{2y} + \mathbf{9} \\ 4x - 3y + 3 - \mathbf{3} = 5 - \mathbf{3} \end{cases}$$

$$\begin{cases} 3x + 2y = 10 \\ 4x - 3y = 2 \end{cases}$$

Eliminate y.

Multiply both sides of the 1st equation by 3.

Multiply both sides of the 2nd equation by 2.

$$\begin{cases} \mathbf{3}(3x) + \mathbf{3}(2y) = \mathbf{3}(10) \\ \mathbf{2}(4x) - \mathbf{2}(3y) = \mathbf{2}(2) \end{cases}$$

$$\begin{array}{r} 9x + 6y = 30 \\ 8x - 6y = 4 \\ \hline 17x \quad\quad = 34 \end{array}$$

$$\frac{17x}{\mathbf{17}} = \frac{34}{\mathbf{17}}$$

$$x = 2$$

$3x + 2y = 10$ The 1st equation

$3(\mathbf{2}) + 2y = 10$

$6 + 2y - \mathbf{6} = 10 - \mathbf{6}$

$2y = 4$

$$\frac{2y}{\mathbf{2}} = \frac{4}{\mathbf{2}}$$

$$y = 2$$

The solution is $(2, 2)$.

Section 4.3

SECTION 4.4

VOCABULARY
Fill in the blanks.

1. Two angles are said to be **complementary** if the sum of their measures is 90°. Two angles are said to be **supplementary** if the sum of their measures is 180°.

CONCEPTS

3. A length of pipe is to be cut into two pieces. The longer piece is to be 1 foot less than twice the shorter piece. Write two equations that model the situation. $x + y = 20$, $y = 2x - 1$

5. Two angles are supplementary. The measure of the smaller angle is 25° less than the measure of the larger angle. Write two equations that model the situation. $x + y = 180$, $y = x - 25$

7. Let x = the cost of a chicken taco and y = the cost of a beef taco. Write an equation that models the offer shown in the advertisement.

$$5x + 2y = 15$$

9. For each case below, write an algebraic expression that represents the speed of the canoe in miles per hour if its speed in still water is x mph. $x + c$, $x - c$

11. a. If the contents of the two test tubes are poured into a third tube, how much solution will the third tube contain? (mL stands for milliliter. A milliliter is about two drops from an eyedropper.) $(x + y)$ mL

    b. Which of the following strengths could the mixture possibly be: 27%, 33%, or 44% acid solution? **33%**

GUIDED PRACTICE
13. COMPLEMENTARY ANGLES
Analyze
- Two angles are complementary (90°).
- One angle is 10° more than three times the measure of the other.
- Find the measure of each angle.

Assign
Let x = the measure of one $\angle$
    y = the measure of the other $\angle$

Form

The measure of one angle	plus	the measure of the other angle	is	complementary.
x	$+$	y	$=$	90

One angle	is	10°	more than	3 times the other angle.
x	$=$	10	$+$	$3y$

Solve
$$\begin{cases} x + y = 90 \\ x = 10 + 3y \end{cases}$$

Write the 2nd equation in standard form.

$$\begin{cases} x + y = 90 \\ x - 3y = 10 \end{cases}$$

Eliminate x.

Multply both sides of the 2nd equation by -1.

$$\begin{array}{r} x + y = 90 \\ -x + 3y = -10 \\ \hline 4y = 80 \end{array}$$

$$\frac{4y}{4} = \frac{80}{4}$$

$$y = 20$$

$x = 10 + 3y$ The 2nd equation
$x = 10 + 3(20)$
$x = 10 + 60$
$x = 70$

State
The measure of one angle is 70°.
The measure of the other angle is 20°.

Check

$70 + 20 = 90$	$70 = 10 + 3(20)$
	$70 = 70$

The results check.

15. SUPPLEMENTARY ANGLES

Analyze
- Two angles are supplementary (180º).
- The difference of the measures of two supplementary angles is 80º.
- Find the measure of each angle.

Assign

Let x = the measure of one $\angle$ (larger$\angle$)

y = the measure of the other $\angle$ (smaller$\angle$)

Form

| The measure of one angle | plus | the measure of the other angle | is | supplementary. |

$$x \ + \ y \ = \ 180$$

| The difference of the measures of the 2 angles | is | 80º. |

$$x - y \ = \ 80$$

Solve

$$\begin{cases} x + y = 180 \\ x - y = 80 \end{cases}$$

Eliminate y.

$$x + y = 180$$
$$\underline{x - y = 80}$$
$$2x \quad = 260$$
$$\frac{2x}{2} = \frac{260}{2}$$
$$x = 130$$

$$x + y = 180 \quad \text{The 1}^{\text{st}} \text{ equation}$$
$$\mathbf{130} + y = 180$$
$$130 + y - \mathbf{130} = 180 - \mathbf{130}$$
$$y = 50$$

State

The measure of one angle is 130º.
The measure of the other angle is 50º.

Check

$$130 + 50 = 180 \quad | \quad 130 - 50 = 80$$
$$180 = 180 \quad | \quad 80 = 80$$

The results check.

APPLICATION

Write a system of two equations in two variables to solve each problem.

17. TREE TRIMMING

Analyze
- The total length of both arms is 51 ft.
- The upper arm is 7 ft. shorter than the lower arm.
- Find the length of each arm.

Assign

Let x = length of upper arm in feet
   (shorter length)

y = length of lower arm in feet
   (longer length)

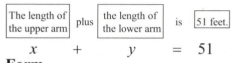

| The length of the upper arm | plus | the length of the lower arm | is | 51 feet. |

$$x \ + \ y \ = \ 51$$

Form

| The upper arm | is | 7 feet | shorter than | the lower arm. |

$$x \ = \ y \ - \ 7$$

Solve

$$\begin{cases} x + y = 51 \\ x = y - 7 \end{cases}$$

Write the 2nd equation in standard form.

$$\begin{cases} x + y = 51 \\ x - y = -7 \end{cases}$$

Eliminate y.

$$x + y = 51$$
$$\underline{x - y = -7}$$
$$2x \quad = 44$$
$$\frac{2x}{2} = \frac{44}{2}$$
$$x = 22$$

$$x + y = 51 \quad \text{The 1}^{\text{st}} \text{ equation}$$
$$\mathbf{22} + y = 51$$
$$22 + y - \mathbf{22} = 51 - \mathbf{22}$$
$$y = 29$$

State

The measure of the upper arm is 22 ft.
The measure of the lower arm is 29 ft.

Check

$$22 + 29 = 51 \quad | \quad 22 = 29 - 7$$
$$51 = 51 \quad | \quad 22 = 22$$

The results check.

Section 4.4

19. GOVERNMENT
Analyze
- Both salaries total $627,300.
- The president's makes $172,700 more than VP.
- Find the salary of each.

Assign
  Let x = president's salary (larger salary)
      y = vice-president salary (smaller salary)

Form

The salary of the President	plus	the salary of the VP	is	$627,300.
x	$+$	y	$=$	$627,300$

The President	makes	$172,700	more than	the VP.
x	$=$	$172,700$	$+$	y

Solve
$$\begin{cases} x + y = 627,300 \\ x = 172,700 + y \end{cases}$$

Write the 2nd equation in standard form

$$\begin{cases} x + y = 627,300 \\ x - y = 172,700 \end{cases}$$

Eliminate y.

$$\begin{aligned} x + y &= 627,300 \\ \underline{x - y} &= \underline{172,700} \\ 2x &= 800,000 \\ \frac{2x}{2} &= \frac{800,000}{2} \\ x &= 400,000 \end{aligned}$$

$$x + y = 627,300$$
$$\mathbf{400,000} + y = 627,300$$
$$400,000 + y - \mathbf{400,000} = 627,300 - \mathbf{400,000}$$
$$y = 227,300$$

State
The President's salary is $400,000.
The Vice-President's salary is $227,300.

Check
$$400,000 + 227,300 = 627,300 \mid 400,000 = 172,700 + 227,300$$
$$627,300 = 627,300 \mid 400,000 = 400,000$$
The results check.

GEOMETRY PROBLEMS
21. MONUMENTS
Analyze
- Two angles are supplementary (180°).
- The measure of $\angle 2$ is 15° less than twice the measure of $\angle 1$.
- Find the measure of each angle.

Assign
  Let x = the measure of $\angle 1$ (larger$\angle$)
      y = the measure of $\angle 2$ (smaller$\angle$)

Form

The measure of $\angle 1$	plus	the measure of the $\angle 2$	is	supplementary.
x	$+$	y	$=$	180

The measure of $\angle 2$	is	15°	less than	twice the measure of $\angle 1$.
x	$=$	$2y$	$-$	15

Solve
$$\begin{cases} x + y = 180 \\ y = 2x - 15 \end{cases}$$

$$x + y = 180 \quad \text{The 1}^{st}\text{ equation}$$
$$\text{Substitute for } x.$$
$$x + (\mathbf{2x - 15}) = 180$$
$$3x - 15 = 180$$
$$3x - 15 + \mathbf{15} = 180 + \mathbf{15}$$
$$\frac{3x}{\mathbf{3}} = \frac{195}{\mathbf{3}}$$
$$x = 65$$

$$y = 2x - 15 \quad \text{The 2}^{nd}\text{ equation}$$
$$y = 2(\mathbf{65}) - 15$$
$$y = 130 - 15$$
$$y = 115$$

State
The measure of $\angle 1$ is 65°.
The measure of $\angle 2$ is 115°.

Check
$$65 + 115 = 180 \qquad 115 = 2(65) - 15$$
$$180 = 180 \qquad 115 = 130 - 15$$
$$\qquad\qquad\qquad 115 = 115$$

The results check.

23. THEATER SCREENS

Analyze
- A giant rectangular movie screen has a width 26 feet less than its length.
- Its perimeter is 332 feet.
- Find the measures of the width and length.

Assign

Let w = width of screen in ft (shorter length)

l = length of screen in ft (longer length)

Form

The measure of two widths	plus	the measure of two lengths	is	the perimerter of 332 feet.
$2w$	$+$	$2l$	$=$	332

The width	is	26 ft	less than	the length.
w	$=$	l	$-$	26

Solve

$$\begin{cases} 2l + 2w = 332 \\ w = l - 26 \end{cases}$$

$2l + 2w = 332$ The 1ˢᵗ equation

Substitute for w.

$2l + 2(l - 26) = 332$

$2l + 2l - 52 = 332$

$4l - 52 = 332$

$4l - 52 + 52 = 332 + 52$

$4l = 384$

$\dfrac{4l}{4} = \dfrac{384}{4}$

$l = 96$

$w = l - 26$ The 2ⁿᵈ equation

$w = 96 - 26$

$w = 70$

State

The width of the screen is 70 feet.
The length of the screen is 96 feet.

Check

$2(70) + 2(96) = 332$	$70 = 96 - 26$
$140 + 192 = 332$	$70 = 70$
$332 = 332$	

The results check.

25. GEOMETRY

Analyze
- A 50-meter path surrounds a rectangular garden.
- The width of the garden is two-thirds its length.
- Find the measures of the width and length.

Assign

Let w = width of garden in m (shorter length)

l = length of garden in m (longer length)

Form

The measure of two widths	plus	the measure of two lengths	is	the perimeter of 50 meters.
$2w$	$+$	$2l$	$=$	50

The width	is	two-thirds	of	the length.
w	$=$	$\dfrac{2}{3}$	$\cdot$	l

Solve

$$\begin{cases} 2l + 2w = 50 \\ w = \dfrac{2}{3}l \end{cases}$$

Clear the fraction from the 2ⁿᵈ equation by multiplying both sides by 3.

$$\begin{cases} 2l + 2w = 50 \\ 2l = 3w \end{cases}$$

$2l + 2w = 50$ The 1ˢᵗ equation

Substitute for $2l$.

$3w + 2w = 50$

$5w = 50$

$\dfrac{5w}{5} = \dfrac{50}{5}$

$w = 10$

$2l + 2w = 50$ The 1ˢᵗ equation

$2l + 2(10) = 50$

$2l + 20 = 50$

$2l + 20 - 20 = 50 - 20$

$2l = 30$

$\dfrac{2l}{2} = \dfrac{30}{2}$

$l = 15$

State

The width of the garden is 10 m.
The length of the garden is 15 m.

Check

$2(15) + 2(10) = 50$	$3(10) = 2(15)$
$30 + 20 = 50$	$30 = 30$
$50 = 50$	

The results check.

NUMBER/VALUE PROBLEMS
27. EMPTY CARTRIDGES
Analyze
- $40 for 5 printer and 2 copier cartridges.
- $57 for 6 printer and 3 copier cartridges.
- How much is each type of cartridge?

Assign

Let x = value for a printer cartridge

y = value for a copier cartridge

	Number •	Value	= Total Value
Printer	5	x	$5x$
Copier	2	y	$2y$
		Total	40

Form

| The value of 5 printer cartridges | plus | the value of 2 copier cartridges | is | $40. |

$$5x \quad + \quad 2y \quad = \quad 40$$

	Number •	Value	= Total Value
Printer	6	x	$6x$
Copier	3	y	$3y$
		Total	57

| The value of 6 printer cartridges | plus | the value of 3 copier cartridges | is | $57. |

$$6x \quad + \quad 3y \quad = \quad 57$$

Solve

$$\begin{cases} 5x + 2y = 40 \\ 6x + 3y = 57 \end{cases}$$

Eliminate y.

Multiply both sides of the 1st equation by 3.

Multiply both sides of the 2nd equation by -2.

$$15x + 6y = 120$$
$$-12x - 6y = -114$$
$$\overline{3x = 6}$$

$$\frac{3x}{3} = \frac{6}{3}$$

$$x = 2$$

$$5x + 2y = 40 \quad \text{The } 1^{st} \text{ equation}$$
$$5(\mathbf{2}) + 2y = 40$$
$$10 + 2y - \mathbf{10} = 40 - \mathbf{10}$$
$$2y = 30$$
$$\frac{2y}{2} = \frac{30}{2}$$
$$y = 15$$

State

Bank was paid $2 for each printer cartridge.

Bank was paid $15 for each copier cartridge.

Check

$5(\$2) + 2(\$15) = \$40$	$6(\$2) + 3(\$15) = \$57$
$\$10 + \$30 = \$40$	$\$12 + \$45 = \$57$
$\$40 = \$40$	$\$57 = \$57$

The results check.

from CAMPUS TO CAREERS
29. PORTRAIT PHOTOGRAPHER
Analyze
- Pk 1: 1 10×14 and 10 8×10 photos cost $239.50.
- Pk 2: 1 10×14 and 5 8×10 photos cost $134.50.
- Find the cost of each size photo.

Assign

Let x = the cost of one 10x14 photo

y = the cost of one 8x10 photo

	Number •	Value	= Total Value
10x14	1	x	x
8x10	10	y	$10y$
		Total	239.50

Form

| The cost of one 10x14 photo | plus | the cost of ten 8x10 photos | is | $239.50. |

$$x \quad + \quad 10y \quad = \quad 239.50$$

	Number •	Value	= Total Value
10x14	1	x	x
8x10	5	y	$5y$
		Total	134.50

| The cost of one 10x14 photo | plus | the cost of five 8x10 photos | is | $134.50. |

$$x \quad + \quad 5y \quad = \quad 134.50$$

Solve

$$\begin{cases} x + 10y = 239.50 \\ x + 5y = 134.50 \end{cases}$$

Eliminate y.

Multiply both sides of the 2nd equation by -1.

$$x + 10y = 239.50$$
$$-x - 5y = -134.50$$
$$\overline{5y = 105}$$

$$\frac{5y}{5} = \frac{105}{5}$$

$$y = 21$$

$$x + 5y = 134.50 \quad \text{The } 2^{nd} \text{ equation}$$
$$x + 5(\mathbf{21}) = 134.50$$
$$x + 105 - \mathbf{105} = 134.50 - \mathbf{105}$$
$$x = 29.50$$

State

The cost of one 10×14 photo is $29.50.

The cost of one 8×10 photo is $21.

Check

$\$29.50 + 10(\$21) = \$239.50$	$\$29.50 + 5(\$21) = \$134.50$
$\$29.50 + \$210 = \$239.50$	$\$29.50 + \$105 = \$134.50$
$\$239.50 = \$239.50$	$\$134.50 = \$134.50$

The results check.

31. COLLECTING STAMPS

Analyze
- One Elvis stamp and one Liberty stamp cost a total of $0.63.
- A sheet of 40 Elvis stamps and a sheet of 20 Liberty stamps cost a total of $18.40.
- How much is each stamp?

Assign

Let x = the cost of one Elvis stamp
y = the cost of one Liberty stamp

	Number •	Value	= Total Value
Elvis	1	x	x
Liberty	1	y	y
		Total	0.63

Form

The value of one Elvis stamp	plus	the value of one Liberty stamp	is	$0.63.

$$x \quad + \quad y \quad = \quad 0.63$$

	Number •	Value	= Total Value
Elvis	40	x	$40x$
Liberty	20	y	$20y$
		Total	18.40

The value of 40 Elvis stamps	plus	the value of 20 Liberty stamps	is	$18.40.

$$40x \quad + \quad 20y \quad = \quad 18.40$$

Solve

$$\begin{cases} x + y = 0.63 \\ 40x + 20y = 18.40 \end{cases}$$

Eliminate y.
Multiply both sides of the 1st equation by -20.

$$-20x - 20y = -12.60$$
$$\underline{40x + 20y = 18.40}$$
$$20x = 5.80$$

$$\frac{20x}{20} = \frac{5.80}{20}$$

$$x = 0.29$$

$$x + y = 0.63 \quad \text{The 1}^{st}\text{ equation}$$
$$(0.29) + y = 0.63$$
$$0.29 + y - 0.29 = 0.63 - 0.29$$
$$y = 0.34$$

State
The cost of one Elvis stamp is $0.29.
The cost of one Liberty stamp is $0.34.

Check

$\$0.29 + \$0.34 = \$.63$	$40(\$0.29) + 20(\$0.34) = \$18.40$
$\$.63 = \$.63$	$\$11.60 + \$6.80 = \$18.40$
	$\$18.40 = \$18.40$

The results check.

33. SELLING ICE CREAM

Analyze
- Ice cream cones cost $1.80 and sundaes cost $3.30.
- The receipts for a total of 148 cones and sundaes were $360.90.
- How many of each was sold?

Assign

Let x = number of cones sold
y = number of sundaes sold

Form

The number of cones	plus	the number of sundaes	is	148.

$$x \quad + \quad y \quad = \quad 148$$

	Number •	Value	= Total Value
Cones	x	1.80	$1.80x$
Sundaes	y	3.30	$3.30y$
		Total	360.90

The value of x cones	plus	the value of y sundaes	is	$360.90.

$$1.80x \quad + \quad 3.30y \quad = \quad 360.90$$

Solve

$$\begin{cases} x + y = 148 \\ 1.80x + 3.30y = 360.90 \end{cases}$$

Solve the 1st equation for x.

$$\begin{cases} x = 148 - y \\ 1.80x + 3.30y = 360.90 \end{cases}$$

$$1.80x + 3.30y = 360.90 \quad \text{The 2}^{nd}\text{ equation}$$
$$1.80(148 - y) + 3.30y = 360.90 \quad \text{Substitute for } x.$$
$$266.40 - 1.80y + 3.30y = 360.90$$
$$266.40 + 1.50y = 360.90$$
$$266.40 + 1.50y - 266.40 = 360.90 - 266.40$$
$$1.50y = 94.50$$
$$\frac{1.50y}{1.50} = \frac{94.50}{1.50}$$
$$y = 63$$

$$x + y = 148 \quad \text{The 1}^{st}\text{ equation}$$
$$x + 63 = 148$$
$$x + 63 - 63 = 148 - 63$$
$$x = 85$$

State
The number of cones sold is 85.
The number of sunades sold is 63.

Check

$85 + 63 = 148$	$\$1.80(85) + \$3.30(63) = \$360.90$
$148 = 148$	$\$153 + \$207.90 = \$360.90$
	$\$360.90 = \$360.90$

The results check.

Section 4.4

INTEREST PROBLEMS
35. STUDENT LOANS
Analyze
- A college used $5,000 to make two student loans.
- The first at 5% annual interest to a nursing student.
- The second at 7% to a business major.
- College collected $310 in interest the first year.
- How much was loaned to each student?

Assign

Let x = amount loaned to nursing student

y = amount loaned to business student

Form

The amount loaned to nursing student	plus	the amount loaned to business student	is	$5,000.
x	$+$	y	$=$	5,000

	Principal	• Rate	• Time	= Interest
Nurse	x	5%	1 yr	$0.05x$
Business	y	7%	1 yr	$0.07y$

Total Interest = **$310**

The amount of interest of 5% loan	plus	the amount of interest of 7% loan	is	$310.
$0.05x$	$+$	$0.07y$	$=$	310

Solve

$$\begin{cases} x + y = 5,000 \\ 0.05x + 0.07y = 310 \end{cases}$$

Eliminate x.

Multiply both sides of the 1st equation by -5.

Multiply both sides of the 2nd equation by 100.

$$-5x - 5y = -25,000$$
$$\underline{5x + 7y = 31,000}$$
$$2y = 6,000$$
$$\frac{2y}{2} = \frac{6,000}{2}$$
$$y = 3,000$$

$$x + y = 5,000 \quad \text{The 1}^{st} \text{ equation}$$
$$x + 3,000 = 5,000$$
$$x + 3,000 - 3,000 = 5,000 - 3,000$$
$$x = 2,000$$

State

The nursing student loan is $2,000.

The business student loan is $3,000

Check

$$\begin{array}{c|c} \$2,000 + \$3,000 = \$5,000 & 0.05(\$2,000) + 0.07(\$3,000) = \$310 \\ \$5,000 = \$5,000 & \$100 + \$210 = \$310 \\ & \$310 = \$310 \end{array}$$

The results check.

37. INVESTING A BONUS
Analyze
- Bonus amount is $40,000.
- Invested part in a fund at 8% annual interest.
- The rest in an offshore bank at 9% annual interest.
- $3,415 in interest was earned the first year.
- How much was invested at each rate?

Assign

Let x = amount invested int fund at 8%

y = amount invested offshore at 9%

Form

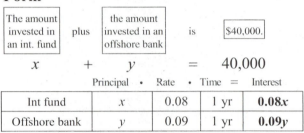

The amount invested in an int. fund	plus	the amount invested in an offshore bank	is	$40,000.
x	$+$	y	$=$	40,000

	Principal	• Rate	• Time	= Interest
Int fund	x	0.08	1 yr	$0.08x$
Offshore bank	y	0.09	1 yr	$0.09y$

Total Interest = **$3,415**

The amount of interest of 8% investment	plus	the amount of interest of 9% investment	is	$3,415.
$0.08x$	$+$	$0.09y$	$=$	3,415

Solve

$$\begin{cases} x + y = 40,000 \\ 0.08x + 0.09y = 3,415 \end{cases}$$

Eliminate x.

Multiply both sides of the 1st equation by -8.

Multiply both sides of the 2nd equation by 100.

$$-8x - 8y = -320,000$$
$$\underline{8x + 9y = 341,500}$$
$$y = 21,500$$

$$x + y = 40,000 \quad \text{The 1}^{st} \text{ equation}$$
$$x + 21,500 = 40,000$$
$$x + 21,500 - 21,500 = 40,000 - 21,500$$
$$x = 18,500$$

State

$18,500 was invested at 8% with Int fund.

$21,500 was invested at 9% with offshore.

Check

$$\begin{array}{c|c} \$18,500 + \$21,500 = \$40,000 & 0.08(\$18,500) + 0.09(\$21,500) = \$3,415 \\ \$40,000 = \$40,000 & \$1,480 + \$1,935 = \$3,415 \\ & \$3,415 = \$3,415 \end{array}$$

The results check.

39. LOSSES

Analyze the Problem
- CEO has $22,000 to invest.
- Invested part at 4% annual interest.
- Invested in biotech at 3% annual interest.
- Biotech showed a loss after first year.
- $110 in interest was earned the first year.
- Find the amount of each investment.

Assign

Let x = amount invested at 4%
$\quad\quad y$ = amount invested at 3%

Form

The amount invested in the 4% account	plus	the amount invested in the 3% biotech	is	$22,000.
x	$+$	y	$=$	$22,000$

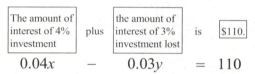

	Principal ·	Rate ·	Time =	Interest
account	x	0.04	1 yr	$0.04x$
biotech	y	0.03	1 yr	$-0.03y$

Total Interest = $110

The amount of interest of 4% investment	plus	the amount of interest of 3% investment lost	is	$110.
$0.04x$	$-$	$0.03y$	$=$	110

Solve

$$\begin{cases} x + y = 22{,}000 \\ 0.04x - 0.03y = 110 \end{cases}$$

Eliminate y.

Multiply both sides of the 1st equation by 3.
Multiply both sides of the 2nd equation by 100.

$$\begin{aligned} 3x + 3y &= 66{,}000 \\ 4x - 3y &= 11{,}000 \\ \hline 7x &= 77{,}000 \end{aligned}$$

$$\frac{7x}{7} = \frac{77{,}000}{7}$$

$$x = 11{,}000$$

$\quad\quad x + y = 22{,}000$ The 1st equation

$\quad\quad x + \mathbf{11{,}000} = 22{,}000$

$x + 11{,}000 - \mathbf{11{,}000} = 22{,}000 - \mathbf{11{,}000}$

$\quad\quad\quad x = 11{,}000$

State

$11,000 was invested at 4%.
$11,000 was invested at 3%.

Check

$\$11{,}000 + \$11{,}000 = \$22{,}000$ | $0.04(\$11{,}000) - 0.03(\$11{,}000) = \$110$
$\quad\quad\quad \$22{,}000 = \$22{,}000$ | $\quad\quad\quad \$440 - \$330 = \$110$
$\quad$ | $\quad\quad\quad\quad\quad \$110 = \$110$

The results check.

UNIFORM MOTION PROBLEMS
41. THE GULF STREAM

Analyze
- With the current, a ship traveled 300 mi in 10 hrs.
- Against the current, it took 15 hours for return trip.
- Find the speed of the ship in still water and the speed of the current.

Assign

Let x = speed of ship in still water in mph
$\quad\quad y$ = speed of current in mph

	Rate ·	Time =	Distance
With current	$x + y$	10	$10(x + y)$

Total Distance = 300

Form

The rate of the ship with the current	times	10 hours traveled	is	300 miles.
$(x + y)$	·	10	$=$	300

	Rate ·	Time =	Distance
Against current	$x - y$	15	$15(x - y)$

Total Distance = 300

The rate of the ship against the current	times	15 hours traveled	is	300 miles.
$(x - y)$	·	15	$=$	300

Solve

$$\begin{cases} 10(x + y) = 300 \\ 15(x - y) = 300 \end{cases}$$

Distribute in both equations.

$$\begin{cases} 10x + 10y = 300 \\ 15x - 15y = 300 \end{cases}$$

Eliminate y.

Multiply both sides of the 1st equation by 3.
Multiply both sides of the 2nd equation by 2.

$$\begin{aligned} 30x + 30y &= 900 \\ 30x - 30y &= 600 \\ \hline 60x &= 1{,}500 \end{aligned}$$

$$\frac{60x}{60} = \frac{1{,}500}{60}$$

$$x = 25$$

$\quad\quad 10(x + y) = 300$ The 1st equation

$\quad\quad 10(\mathbf{25} + y) = 300$

$\quad\quad 250 + 10y = 300$

$250 + 10y - \mathbf{250} = 300 - \mathbf{250}$

$\quad\quad\quad 10y = 50$

$$\frac{10y}{10} = \frac{50}{10}$$

$$y = 5$$

State

The speed of the ship in still water is 25 mph.
The speed of the current is 5 mph.

Check

$10(25 + 5) = 300$ | $15(25 - 5) = 300$
$\quad 10(30) = 300$ | $\quad 15(20) = 300$
$\quad\quad 300 = 300$ | $\quad\quad 300 = 300$

The results check.

Section 4.4

43. AVIATION

Analyze
- Airplane flys with the wind 800 mi in 4 hrs.
- The return trip against the wind takes 5 hours.
- Find the speed of the plane in still air and the speed of the wind current.

Assign

Let x = speed of plane in still air in mph
y = speed of wind in mph

	Rate •	Time =	Distance
With wind	$x + y$	4	$4(x + y)$

Total Distance = 800

Form

The rate of the plane with the wind	times	4 hours traveled	is	800 miles.

$$(x + y) \qquad \bullet \qquad 4 \qquad = 800$$

	Rate •	Time =	Distance
Against wind	$x - y$	5	$5(x - y)$

Total Distance = 800

The rate of the plane against the wind	times	5 hours traveled	is	800 miles.

$$(x - y) \qquad \bullet \qquad 5 \qquad = 800$$

Solve

$$\begin{cases} 4(x + y) = 800 \\ 5(x - y) = 800 \end{cases}$$

Distribute in both equations.

$$\begin{cases} 4x + 4y = 800 \\ 5x - 5y = 800 \end{cases}$$

Eliminate y.

Multiply both sides of the 1st equation by 5.

Multiply both sides of the 2nd equation by 4.

$$20x + 20y = 4,000$$
$$20x - 20y = 3,200$$

$$40x \qquad = 7,200$$

$$\frac{40x}{40} = \frac{7,200}{40}$$

$$x = 180$$

$$4(x + y) = 800 \quad \text{The } 1^{st} \text{ equation}$$
$$4(\mathbf{180} + y) = 800$$
$$720 + 4y = 800$$
$$720 + 4y - \mathbf{720} = 800 - \mathbf{720}$$
$$4y = 80$$
$$\frac{4y}{4} = \frac{80}{4}$$
$$y = 20$$

State

The speed of the plane in still air is 180 mph.
The speed of the wind is 20 mph.

Check

$$4(180 + 20) = 800 \quad \Big| \quad 5(180 - 20) = 800$$
$$4(200) = 800 \quad \Big| \quad 5(160) = 800$$
$$800 = 800 \quad \Big| \quad 800 = 800$$

The results check.

45. MARINE BIOLOGY

Analyze
- Set up an aquarium containing 3% salt water.
- 2 tanks contain 6% and 2% salt water.
- How much from each tank must he use to fill a 32-gallon aquarium with a 3% mixture?

Assign

Let x = amount of 2% salt water in gallons
y = amount of 6% salt water in gallons

Form

The # of gallons of 2% salt water	plus	the # of gallons of 6% salt water	is	32 gallons.

$$x \qquad + \qquad y \qquad = \qquad 32$$

	Amt •	Strength =	Amt of salt
Weak	x	0.02	$\mathbf{0.02x}$
Strong	y	0.06	$\mathbf{0.06y}$
Mix	32	0.03	$\mathbf{32(0.03)}$

The amount of salt in the 2% salt water	plus	the amount of salt in the 6% salt water	is equal to	the amount of salt in the 3% salt water.

$$0.02x \qquad + \qquad 0.06y \qquad = \qquad 32(0.03)$$

Solve

$$\begin{cases} x + y = 32 \\ 0.02x + 0.06y = 0.96 \end{cases}$$

Eliminate x.

Multiply both sides of the 1st equation by -2.

Multiply both sides of the 2nd equation by 100.

$$-2x - 2y = -64$$
$$2x + 6y = 96$$

$$4y = 32$$

$$\frac{4y}{4} = \frac{32}{4}$$

$$y = 8$$

$$x + y = 32 \quad \text{The } 1^{st} \text{ equation}$$
$$x + 8 = 32$$
$$x + 8 - \mathbf{8} = 32 - \mathbf{8}$$
$$x = 24$$

State

24 gallons of 2% salt water will be needed.
8 gallons of 6% salt water will be needed.

Check

$$24 + 8 = 32 \quad \Big| \quad 0.02(24) + 0.06(8) = 0.96$$
$$32 = 32 \quad \Big| \quad 0.48 + 0.48 = 0.96$$
$$\Big| \quad 0.96 = 0.96$$

The results check.

47. CLEANING FLOORS

Analyze
- Custodian mixes a 4% ammonia solution and a 12% ammonia solution to get 1 gallon (128 fluid ounces) of a 9% ammonia solution.
- How many fluid ounces of the 4% solution and the 12% solution should be used?

Assign

Let x = amount of 4% ammonia solution in oz

y = amount of 12% ammonia solution in oz

Form

The # of ounces of 4% ammonia	plus	the # of ounces of 12% ammonia	is	128 ounces.
x	$+$	y	$=$	128

	Amt $\cdot$	Strength =	Amt of salt
Weak	x	0.04	**0.04x**
Strong	y	0.12	**0.12y**
Mix	128	0.09	**128(0.09)**

The amount of ammonia in the 4% solution	plus	the amount of ammonia in the 12% solution	is equal to	the amount of ammonia in the 9% solution.
$0.04x$	$+$	$0.12y$	$=$	$128(0.09)$

Solve

$$\begin{cases} x + y = 128 \\ 0.04x + 0.12y = 11.52 \end{cases}$$

Eliminate x.
Multiply both sides of the 1st equation by -4.
Multiply both sides of the 2nd equation by 100.

$$\begin{array}{r} -4x - 4y = -512 \\ 4x + 12y = 1,152 \\ \hline 8y = 640 \end{array}$$

$$\frac{8y}{8} = \frac{640}{8}$$

$$y = 80$$

$$x + y = 128 \quad \text{The 1st equation}$$
$$x + 80 = 128$$
$$x + 80 - 80 = 128 - 80$$
$$x = 48$$

State
48 ounces of 4% solution will be needed.
80 ounces of 12% solution will be needed.

Check

$48 + 80 = 128$	$0.04(48) + 0.12(80) = 11.52$
$128 = 128$	$1.92 + 9.6 = 11.52$
	$11.52 = 11.52$

The results check.

49. COFFEE SALES

Analyze
- A 100 pound blend sells for $6.35 a pound.
- Brazilian coffee selling for $3.75 a pound.
- Columbian coffee selling for $8.75 a pound.
- How much of each type should be used?

Assign

Let x = amount of $8.75/lb coffee in lbs

y = amount of $3.75/lb coffee in lbs

Form

The # of lbs of $8.75/lb coffee	plus	the # of lbs of $3.75/lb coffee	is	100 pounds.
x	$+$	y	$=$	100

	Amt $\cdot$	Cost/lb =	Total Value
Columbian	x	8.75	**8.75x**
Brazilian	y	3.75	**3.75y**
Mixture	100	6.35	**100(6.35)**

The value of $8.75/lb coffee	plus	the value of $3.75/lb coffee	is equal to	the value of $6.35/lb coffee.
$8.75x$	$+$	$3.75y$	$=$	$6.35(100)$

Solve

$$\begin{cases} x + y = 100 \\ 8.75x + 3.75y = 635 \end{cases}$$

Eliminate y.
Multiply both sides of the 1st equation by -375.
Multiply both sides of the 2nd equation by 100.

$$\begin{array}{r} -375x - 375y = -37,500 \\ 875x + 375y = 63,500 \\ \hline 500x = 26,000 \end{array}$$

$$\frac{500x}{500} = \frac{26,000}{500}$$

$$x = 52$$

$$x + y = 100 \quad \text{The 1st equation}$$
$$52 + y = 100$$
$$52 + y - 52 = 100 - 52$$
$$y = 48$$

State
52 lb of $8.75/lb coffee will be needed.
48 lb of $3.75/lb coffee will be needed.

Check

$52 + 48 = 100$	$\$8.75(52) + \$3.75(48) = \$635$
$100 = 100$	$\$455 + \$180 = \$635$
	$\$635 = \$635$

The results check.

51. GOURMET FOODS

Analyze
- Stuffed Kalamata olives sell for $9 a pint.
- Marinated mushrooms sell for $12 a pint.
- How many pints of each are needed to get 20 pints of a mixture that will sell for $10 a pint?

Assign

Let x = pints of $9.00/pt olives
  y = pints of $12.00/pt mushrooms

Form

The # of pints of $9.00/pt olives	plus	the # of pints of $12.00/pt mushrooms	is	a total of 20 pints.
x	$+$	y	$=$	20

	Amt $\cdot$	Cost/pt $=$	Total Value
Olives	x	9.00	**9.00x**
Mushrooms	y	12.00	**12.00y**
Mixture	20	10.00	**10.00(20)**

The value of $9/pt olives	plus	the value of $12/pt mushrooms	is equal to	the value of $10/pt blend.
$9x$	$+$	$12y$	$=$	10(20)

Solve

$$\begin{cases} x + y = 20 \\ 9x + 12y = 200 \end{cases}$$

Eliminate x.

Multiply both sides of the 1st equation by -9.

$$\begin{array}{r} -9x - 9y = -180 \\ 9x + 12y = 200 \\ \hline 3y = 20 \end{array}$$

$$\frac{3y}{3} = \frac{20}{3}$$

$$y = 6\frac{2}{3}$$

$x + y = 20$ The 1st equation

$$x + 6\frac{2}{3} = 20$$

$$x + 6\frac{2}{3} - 6\frac{2}{3} = 19\frac{3}{3} - 6\frac{2}{3}$$

$$x = 13\frac{1}{3}$$

State

$13\frac{1}{3}$ pt of $9.00/pt olives will be needed.

$6\frac{2}{3}$ lb of $12.0/pt mushrooms will be needed.

Check

$$13\frac{1}{3} + 6\frac{2}{3} = 20 \qquad \$9\left(\frac{40}{3}\right) + \$12\left(\frac{20}{3}\right) = \$200$$

$$20 = 20 \qquad\qquad \$120 + \$80 = \$200$$

$$\$200 = \$200$$

The results check.

WRITING

53. Answers will vary.

REVIEW

Graph each inequality. Then describe the graph using interval notation.

55. $\qquad x < 4$

$(-\infty, 4)$

57. $\qquad -1 < x \le 2$

$(-1, 2]$

CHALLENGE PROBLEMS

59. SCALE

Analyze
- 3 nails and 1 bolt is the same as 3 nuts.
- 1 bolt and 1 nut is the same as 5 nails.
- How many nails will it take to balance 1 nut?

Assign

Let bolt = number of bolts
  nut = number of nuts
  nail = number of nails

Form

Three nails	plus	one bolt	is	3 nuts.
3 nails	$+$	1 bolt	$=$	3 nuts

One bolt	plus	one nut	is	5 nails.
1 bolt	$+$	1 nut	$=$	5 nails

Solve

$$\begin{cases} 3 \text{ nails} + 1 \text{ bolt} = 3 \text{ nuts} \\ 1 \text{ bolt} + 1 \text{ nut} = 5 \text{ nails} \end{cases}$$

Solve the 2nd equation for 1 bolt.

$$\begin{cases} 3 \text{ nails} + 1 \text{ bolt} = 3 \text{ nuts} \\ 1 \text{ bolt} = 5 \text{ nails} - 1 \text{ nut} \end{cases}$$

$3 \text{ nails} + 1 \text{ bolt} = 3 \text{ nuts}$ The 1st equation
  Substitute for 1 bolt.

$3 \text{ nails} + (\mathbf{5 \text{ nails}} - \mathbf{1 \text{ nut}}) = 3 \text{ nuts}$

$8 \text{ nails} - 1 \text{ nut} = 3 \text{ nuts}$

$8 \text{ nails} - 1 \text{ nut} + \mathbf{1 \text{ nut}} = 3 \text{ nut} + \mathbf{1 \text{ nut}}$

$8 \text{ nails} = 4 \text{ nuts}$

$$\frac{8 \text{ nails}}{4} = \frac{4 \text{ nuts}}{4}$$

$2 \text{ nails} = 1 \text{ nut}$

State

It will take 2 nails to balance 1 nut.

Check

The results check.

VOCABULARY
Fill in the blanks.

1. $\begin{cases} x+y>2 \\ x+y<4 \end{cases}$ is a system of linear **inequalities**.

3. To find the solutions of a system of two linear inequalities graphically, look for the **intersection**, or overlap, of the two shaded regions.

CONCEPTS

5. a. What is the equation of the boundary line of the graph of $3x-y<5$? **$3x-y=5$**

   b. Is the boundary a solid or dashed line?
      Dashed

7. Find the slope and the y-intercept of the line whose equation is $y=4x-3$. **Slope: $4=\frac{4}{1}$,**

   y-intercept: $(0,-3)$

9. The boundary of the graph of $2x+y>4$ is shown.
   a. Does the point $(0, 0)$ make the inequality true? **No**

   b. Should the region above or below the boundary be shaded? **Above**

11. The graph of a system of two linear inequalities is shown. Determine whether each point is a solution of the system.
    a. $(4,-2)$ **Yes**

    b. $(1, 3)$ **No**

    c. the origin **No**

13. Match each equation, inequality, or system with the graph of its solution.
    a. $x+y=2$ **ii**   b. $x+y\geq 2$ **iii**

    c. $\begin{cases} x+y=2 \\ x-y=2 \end{cases}$ **iv**   d. $\begin{cases} x+y\geq 2 \\ x-y\leq 2 \end{cases}$ **i**

GUIDED PRACTICE
Graph the solutions of each system.
See Example 1.

15. $\begin{cases} x+2y\leq 3 \\ 2x-y\geq 1 \end{cases}$

Step 1: Graph $x+2y\leq 3$. We begin by graphing the boundary line $x+2y=3$. Since the inequality contains a $\leq$ symbol, the boundary is a solid line. We check test point $(0,0)$ to determine which side of the line to shade.

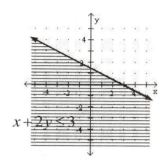

Step 2: Graph $2x-y\geq 1$. We begin by graphing the boundary line $2x-y=1$. Since the inequality contains a $\geq$ symbol, the boundary is a solid line. We check test point $(0,0)$ to determine which side of the line to shade.

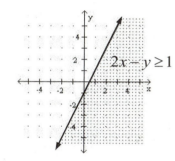

Step 3: We now superimpose the two graphs so that we can determine the region that the graphs have in common.

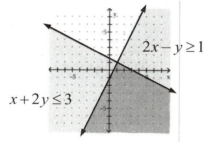

Step 4: Check one point in the double shaded region.

Check: $(2,0)$ Check: $(2,0)$

$x+2y\leq 3$ $2x-y\geq 1$

$2+2(0)\leq 3$ $2(2)-0\geq 1$

$2\leq 3$ $4\geq 1$

True   True

17. $\begin{cases} x+y < -1 \\ x-y > -1 \end{cases}$

Step 1: Graph $x+y < -1$. We begin by graphing the boundary line $x+y = -1$. Since the inequality contains a $<$ symbol, the boundary is a dashed line. We check test point $(0,0)$ to determine which side of the line to shade.

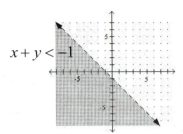

$x+y < -1$

Step 2: Graph $x-y > -1$. We begin by graphing the boundary line $x-y = -1$. Since the inequality contains a $>$ symbol, the boundary is a dashed line. We check test point $(0,0)$ to determine which side of the line to shade.

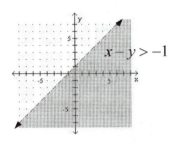

$x-y > -1$

Step 3: We now superimpose the two graphs so that we can determine the region that the graphs have in common.

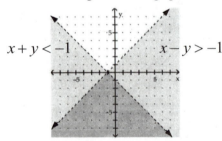

$x+y < -1$ $x-y > -1$

Step 4: Check one point in the double shaded region.

Check: $(0,-2)$ Check: $(0,-2)$

$x+y < -1$ $x-y > -1$

$0+(-2) < -1$ $0-(-2) > -1$

$-2 < -1$ $2 > -1$

True   True

Graph the solutions of each system. See Example 2.

19. $\begin{cases} y > 2x \\ x+2y < 6 \end{cases}$

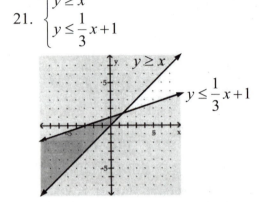

$y > 2x$

$x+2y < 6$

21. $\begin{cases} y \geq x \\ y \leq \dfrac{1}{3}x+1 \end{cases}$

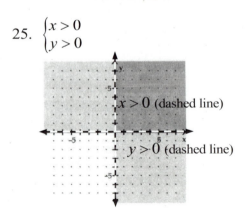

$y \geq x$

$y \leq \dfrac{1}{3}x+1$

Graph the solutions of each system. See Example 3.

23. $\begin{cases} x \geq 2 \\ y \leq 3 \end{cases}$

$x \geq 2$

$y \leq 3$

25. $\begin{cases} x > 0 \\ y > 0 \end{cases}$

$x > 0$ (dashed line)

$y > 0$ (dashed line)

Graph the solutions of each system. See Example 4.
With these systems, one extra boundary is drawn
before identifying the intersection of the system.

27. $\begin{cases} x \geq 0 \\ y \geq 0 \\ x + y \leq 3 \end{cases}$

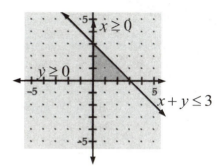

29. $\begin{cases} x - y < 4 \\ y \leq 0 \\ x \geq 0 \end{cases}$

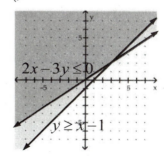

TRY IT YOURSELF

Graph the solutions of each system.

31. $\begin{cases} 2x - 3y \leq 0 \\ y \geq x - 1 \end{cases}$

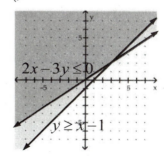

33. $\begin{cases} x + y < 2 \\ x + y \leq 1 \end{cases}$

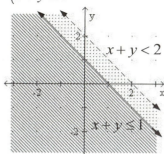

35. $\begin{cases} 3x + 4y \geq -7 \\ 2x - 3y \geq 1 \end{cases}$

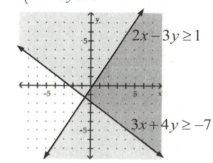

37. $\begin{cases} 2x + y < 7 \\ y > 2 - 2x \end{cases}$

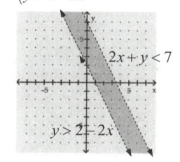

39. $\begin{cases} 2(x - 2y) > -6 \\ 3x + y \geq 5 \end{cases}$

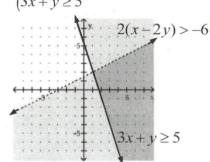

- 247 -

Section 4.5

41. $\begin{cases} 3x - y + 4 \le 0 \\ 3y > -2x - 10 \end{cases}$

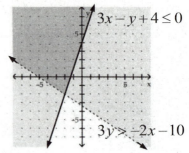

43. $\begin{cases} x \ge -1 \\ y \le -x \\ x - y \le 3 \end{cases}$

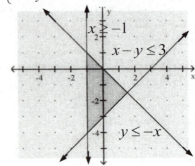

45. $\begin{cases} x + y > 0 \\ y - x < -2 \end{cases}$

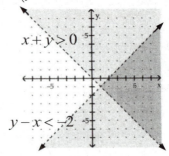

LOOK ALIKES

In part a, graph the solution of each system. Use your answer to part a to determine the solution of the system of equations in part b.

47. a. $\begin{cases} x + y > -1 \\ y \ge x - 3 \end{cases}$ b. $\begin{cases} x + y = -1 \\ y = x - 3 \end{cases}$

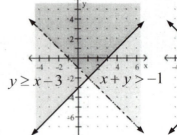
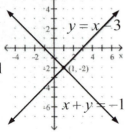

APPLICATIONS

49. BIRDS OF PREY

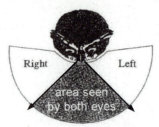

Graph each system of inequalities and give two possible solutions. See Example 5.

51. BUYING COMPACT DISCS

$\begin{cases} 10x + 15y \ge 30 \\ 10x + 15y \le 60 \end{cases}$

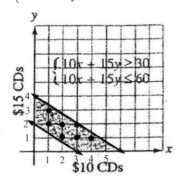

1 $10 CD and 2 $15 CD's
1 $10 CD and 3 $15 CD's
2 $10 CD's and 1 $15 CD
2 $10 CD's and 2 $15 CD's
3 $10 CD's and 1 $15 CD
3 $10 CD's and 2 $15 CD's
4 $10 CD's and 1 $15 CD

53. FURNITURE

$$\begin{cases} 150x + 100y \le 90 \\ y > x \end{cases}$$

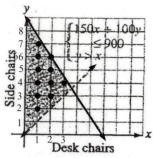

1 desk chair and 2 side chairs
1 desk chair and 3 side chairs
1 desk chair and 4 side chairs
1 desk chair and 5 side chairs
1 desk chair and 6 side chairs
1 desk chair and 7 side chairs
2 desk chair and 3 side chairs
2 desk chair and 4 side chairs
2 desk chair and 5 side chairs
2 desk chair and 6 side chairs
3 desk chair and 4 side chairs

from CAMPUS TO CAREERS
55. PHOTOGRAPHERS

$$\begin{cases} y \le \dfrac{1}{4}x + 2 \\ y \ge -\dfrac{1}{4}x + 2 \end{cases}$$

$y \le \dfrac{1}{4}x + 2$

$y \ge -\dfrac{1}{4}x + 2$

WRITING
57-59. Answers will vary.

REVIEW
Simplify each expression.

61. $8\left(\dfrac{3}{4}t\right) = \dfrac{8 \cdot 3t}{4}$

$$= \dfrac{2 \cdot \cancel{4} \cdot 3t}{\cancel{4}}$$

$$= 6t$$

---

63. $-\dfrac{7}{16}x - \dfrac{3}{16}x = \dfrac{-7x - 3x}{16}$

$$= \dfrac{-10x}{16}$$

$$= -\dfrac{\cancel{2} \cdot 5x}{\cancel{2} \cdot 8}$$

$$= -\dfrac{5}{8}x$$

CHALLENGE PROBLEMS
Graph the solutions of each system.

65. $\begin{cases} \dfrac{x}{3} - \dfrac{y}{2} < -3 \\ \dfrac{x}{3} + \dfrac{y}{2} > -1 \end{cases}$

Clear both inequalities of fractions.

$$\begin{cases} 2x - 3y < -18 \\ 2x + 3y > -6 \end{cases}$$

Graph the boundaries of both.

Identify the intersection.

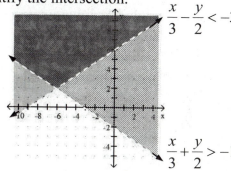

$\dfrac{x}{3} - \dfrac{y}{2} < -3$

$\dfrac{x}{3} + \dfrac{y}{2} > -1$

67. $\begin{cases} 2x + 3y \le 6 \\ 3x + y \le 1 \\ x \le 0 \end{cases}$

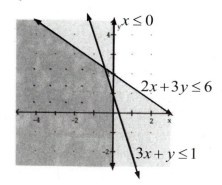

$x \le 0$

$2x + 3y \le 6$

$3x + y \le 1$

Section 4.5

Solving System Equations by Graphing
Determine whether the ordered pair is a solution
of the system.

1. $(2, -3)$ $\begin{cases} 3x - 2y = 12 \\ 2x + 3y = -5 \end{cases}$

$$3x - 2y = 12 \qquad\qquad 2x + 3y = -5$$
$$3(2) - 2(-3) \overset{?}{=} 12 \qquad 2(2) + 3(-3) \overset{?}{=} -5$$
$$6 + 6 \overset{?}{=} 12 \qquad\qquad 4 - 9 \overset{?}{=} -5$$
$$12 = 12 \qquad\qquad -5 = -5$$
$$\text{True} \qquad\qquad\qquad \text{True}$$

Since $(2, -3)$ satisfies both equations,
it is a solution of the system.

2. $\left(\dfrac{7}{2}, -\dfrac{2}{3}\right)$ $\begin{cases} 3y = 2x - 9 \\ 2x + 3y = 6 \end{cases}$

$$3y = 2x - 9 \qquad\qquad 2x + 3y = 6$$
$$3\left(\dfrac{-2}{3}\right) \overset{?}{=} 2\left(\dfrac{7}{2}\right) - 9 \quad 2\left(\dfrac{7}{2}\right) + 3\left(\dfrac{-2}{3}\right) \overset{?}{=} 6$$
$$-2 \overset{?}{=} 7 - 9 \qquad\qquad 7 - 2 \overset{?}{=} 6$$
$$-2 = -2 \qquad\qquad 5 = 6$$
$$\text{True} \qquad\qquad\qquad \text{False}$$

Since $\left(\dfrac{7}{2}, -\dfrac{2}{3}\right)$ does not satisfy both

equations, it is not a solution of the system.

Use the graphing method to solve each system.

3. $\begin{cases} x + y = 7 \\ 2x - y = 5 \end{cases}$

The solution is $(4, 3)$.

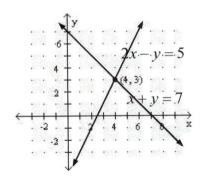

4. $\begin{cases} 2x + y = 5 \\ y = -\dfrac{x}{3} \end{cases}$

The solution is $(3, -1)$.

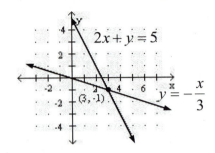

5. $\begin{cases} 3x + 6y = 6 \\ x + 2y - 2 = 0 \end{cases}$

The graphs are the same line.
The system has infinitely many solutions.

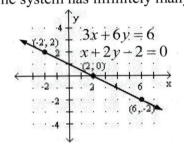

6. $\begin{cases} 6x + 3y = 12 \\ y = -2x + 2 \end{cases}$

The lines are parallel.
The system has no solution.

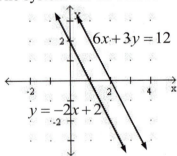

Find the slope and the y-intercept of the graph of each line in the system. Then, use that information to determine the number of solutions of the system.

7. $\begin{cases} y = -2x + 1 \\ 8x + 4y = 3 \end{cases}$

$y = -2x + 1$

$m = -2$

$y\text{-int} = (0, 1)$

$8x + 4y = 3$

$8x + 4y - 8x = 3 - 8x$

$4y = 3 - 8x$

$y = -2x + \dfrac{3}{4}$

$m = -2$

$y\text{-int} = \left(0, \dfrac{3}{4}\right)$

Since the line graphs have the same slope, they are parallel and there is no solution.

8. **BACHELOR'S DEGREE**
   Estimate the point of intersection of the graphs. Explain its significance.
   (1980, 480,000)
   In 1980, the same number of men as women awarded Bachelor' Degrees in the U.S. ws the same 480,000.

SECTION 4.2 REVIEW EXERCISES
SOLVING SYSTEMS OF EQUATIONS
BY SUBSTITUTION

Use the substitution method to solve each system.

9. $\begin{cases} y = 15 - 3x \\ 7y + 3x = 15 \end{cases}$

$7y + 3x = 15$ The 2nd equation. Substitute for y.

$7(15 - 3x) + 3x = 15$

$105 - 21x + 3x = 15$

$105 - 18x = 15$

$105 - 18x - 105 = 15 - 105$

$-18x = -90$

$\dfrac{-18x}{-18} = \dfrac{-90}{-18}$

$x = 5$

$y = 15 - 3x$ The 1st equation

$y = 15 - 3(5)$

$y = 15 - 15$

$y = 0$

The solution is (5,0).

10. $\begin{cases} x = y \\ 5x - 4y = 3 \end{cases}$

$5x - 4y = 3$ The 2nd equation. Substitute for x.

$5(y) - 4y = 3$

$y = 3$

$x = y$

$x = 3$

The solution is (3,3).

11. $\begin{cases} 6x + 2y = 8 - y + x \\ 3x = 2 - y \end{cases}$

Write the 1st equation in standard form. Solve the 2nd equation for y.

$\begin{cases} 5x + 3y = 8 \\ y = 2 - 3x \end{cases}$

$5x + 3y = 8$ The 1st equation. Substitute for y.

$5x + 3(2 - 3x) = 8$

$5x + 6 - 9x = 8$

$-4x + 6 = 8$

$-4x + 6 - 6 = 8 - 6$

$-4x = 2$

$\dfrac{-4x}{-4} = \dfrac{2}{-4}$

$x = -\dfrac{1}{2}$

$y = 2 - 3x$ The 2nd equation

$y = 2 - 3\left(-\dfrac{1}{2}\right)$

$y = 2 + \dfrac{3}{2}$

$y = \dfrac{4}{2} + \dfrac{3}{2}$

$y = \dfrac{7}{2}$

The solution is $\left(-\dfrac{1}{2}, \dfrac{7}{2}\right)$.

Chapter 4 Review and Test

12. $\begin{cases} r = 3s + 7 \\ r = 2s + 5 \end{cases}$

$r = 2s + 5$ The 2nd equation. Substitute for r.

$(3s + 7) = 2s + 5$

$3s + 7 - \mathbf{2s} = 2s + 5 - \mathbf{2s}$

$s + 7 = 5$

$s + 7 - \mathbf{7} = 5 - \mathbf{7}$

$s = -2$

$r = 3s + 7$ The 1st equation

$r = 3(\mathbf{-2}) + 7$

$r = -6 + 7$

$r = 1$

The solution is $(1, -2)$.

13. $\begin{cases} 9x + 3y - 5 = 0 \\ 3x + y = \dfrac{5}{3} \end{cases}$

Solve the 2nd equation for y.

$\begin{cases} 9x + 3y - 5 = 0 \\ y = \dfrac{5}{3} - 3x \end{cases}$

$9x + 3y - 5 = 0$ The 1st equation. Substitute for y.

$9x + 3\left(\dfrac{\mathbf{5}}{\mathbf{3}} - \mathbf{3x}\right) - 5 = 0$

$9x + 5 - 9x - 5 = 0$ The variables drop out.

$0 = 0$

True

The system has infinitely many solutions.

14. $\begin{cases} \dfrac{x}{2} + \dfrac{y}{2} = 11 \\ \dfrac{5x}{16} - \dfrac{3y}{16} = \dfrac{15}{8} \end{cases}$

Clear both equations of fractions.

$\begin{cases} \mathbf{2}\left(\dfrac{x}{2}\right) + \mathbf{2}\left(\dfrac{y}{2}\right) = \mathbf{2}(11) \\ \mathbf{16}\left(\dfrac{5x}{16}\right) - \mathbf{16}\left(\dfrac{3y}{16}\right) = \mathbf{16}\left(\dfrac{15}{8}\right) \end{cases}$

$\begin{cases} x + y = 22 \\ 5x - 3y = 30 \end{cases}$

Solve the 1st equation for x.

$\begin{cases} x = 22 - y \\ 5x - 3y = 30 \end{cases}$

$5x - 3y = 30$ The 2nd equation. Substitute for x.

$5(\mathbf{22 - y}) - 3y = 30$

$110 - 5y - 3y = 30$

$-8y + 110 = 30$

$-8y + 110 - \mathbf{110} = 30 - \mathbf{110}$

$-8y = -80$

$\dfrac{-8y}{-\mathbf{8}} = \dfrac{-80}{-\mathbf{8}}$

$y = 10$

$x = 22 - y$ The 1st equation

$x = 22 - (\mathbf{10})$

$x = 12$

The solution is $(12, 10)$.

15. When solving a system using the substitution method, suppose you obtain the result $8 = 9$.

a. How many solutions does the system have?
   No solution

b. Describe the graph of the system.
   Two parallel lines

c. What term is used to describe the system?
   Inconsistent system

16. Fill in the blank. With the substitution method, the objective is to use an appropriate substitution to obtain one equation in **one** variable.

SECTION 4.3 REVIEW EXERCISES
SOLVING SYSTEMS OF EQUATIONS
BY ELIMINATION (ADDITION)

17. Write each equation of the system in general $Ax + By = C$ form.

$\begin{cases} 4x + 2y - 7 = 0 \rightarrow \\ 3y = 5x + 6 \quad\rightarrow \end{cases}$ $\begin{cases} \mathbf{4x + 2y = 7} \\ \mathbf{5x - 3y = -6} \end{cases}$

18. Fill in the blank. With the elimination method, the basic objective is to obtain two equations whose sum will be one equation in **one** variable.

Solve each system using the elimination (addition) method.

19. $\begin{cases} 2x+y=1 \\ 5x-y=20 \end{cases}$

Eliminate y.

$$\begin{array}{r} 2x+y= 1 \\ 5x-y=20 \\ \hline 7x\quad\ =21 \end{array}$$

$$\frac{7x}{7}=\frac{21}{7}$$
$$x=3$$

$2x+y=1$ The 1st equation
$2(3)+y=1$
$6+y-6=1-6$
$\quad y=-5$

The solution is $(3,\ -5)$.

20. $\begin{cases} x+8y=7 \\ x-4y=1 \end{cases}$

Eliminate x.
Multiply both sides of the 2nd equation by -1.

$$\begin{array}{r} x+8y= 7 \\ -x+4y=-1 \\ \hline 12y= 6 \end{array}$$

$$\frac{12y}{12}=\frac{6}{12}$$
$$y=\frac{1}{2}$$

$x+8y=7$ The 1st equation

$x+8\left(\dfrac{1}{2}\right)=7$

$x+4-4=7-4$
$\quad x=3$

The solution is $\left(3,\ \dfrac{1}{2}\right)$.

21. $\begin{cases} 5a+b=2 \\ 3a+2b=11 \end{cases}$

Eliminate b.
Multiply both sides of the 1st equation by -2.

$$\begin{array}{r} -10a-2b=-4 \\ 3a+2b=11 \\ \hline -7a\quad\ =7 \end{array}$$

$$\frac{-7a}{-7}=\frac{7}{-7}$$
$$a=-1$$

$5a+b=2$ The 1st equation
$5(-1)+b=2$
$-5+b+5=2+5$
$\quad b=7$

The solution is $(-1,\ 7)$.

22. $\begin{cases} 11x+3y=27 \\ 8x+4y=36 \end{cases}$

Eliminate y.
Multiply both sides of the 1st equation by 4.
Multiply both sides of the 2nd equation by -3.

$$\begin{array}{r} 44x+12y= 108 \\ -24x-12y=-108 \\ \hline 20x\quad\quad =0 \end{array}$$

$$\frac{20x}{20}=\frac{0}{20}$$
$$x=0$$

$8x+4y=36$ The 2nd equation
$8(0)+4y=36$
$\quad 4y=36$

$$\frac{4y}{4}=\frac{36}{4}$$
$$y=9$$

The solution is $(0,\ 9)$.

23. $\begin{cases} 9x+3y=15 \\ 3x=5-y \end{cases}$

Write the 2nd equation in standard form.
$\begin{cases} 9x+3y=15 \\ 3x+y=5 \end{cases}$

Eliminate y.
Multiply both sides of the 2nd equation by -3.

$$\begin{array}{r} 9x+3y= 15 \\ -9x-3y=-15 \\ \hline 0=0 \end{array}$$ Both variables are eliminated.
$\qquad\qquad$ True

The system has infinitely
$\qquad$ many solutions.

Chapter 4 Review and Test

24. $\begin{cases} 0.02x + 0.05y = 0 \\ 0.3x - 0.2y = -1.9 \end{cases}$

Clear both equations of decimals.
Multiply both sides of the 1st equation by 100.
Multiply both sides of the 2nd equation by 10.

$\begin{cases} 100(0.02x) + 100(0.05y) = 100(0) \\ 10(0.3x) - 10(0.2y) = 10(-1.9) \end{cases}$

$\begin{cases} 2x + 5y = 0 \\ 3x - 2y = -19 \end{cases}$

Eliminate y.
Multiply both sides of the 1st equation by 2.
Multiply both sides of the 2nd equation by 5.

$\begin{array}{r} 4x + 10y = 0 \\ 15x - 10y = -95 \\ \hline 19x = -95 \end{array}$

$\dfrac{19x}{\mathbf{19}} = \dfrac{-95}{\mathbf{19}}$

$x = -5$

$4x + 10y = 0$ The 1st equation
$4(\mathbf{-5}) + 10y = 0$
$-20 + 10y + \mathbf{20} = 0 + \mathbf{20}$
$10y = 20$
$\dfrac{10y}{\mathbf{10}} = \dfrac{20}{\mathbf{10}}$
$y = 2$

The solution is $(-5, 2)$.

25. $\begin{cases} -\dfrac{a}{4} - \dfrac{b}{3} = \dfrac{1}{12} \\ \dfrac{a}{2} - \dfrac{5b}{4} = \dfrac{7}{4} \end{cases}$

Clear both equations of fractions.

$\begin{cases} \mathbf{12}\left(-\dfrac{a}{4}\right) - \mathbf{12}\left(\dfrac{b}{3}\right) = \mathbf{12}\left(\dfrac{1}{12}\right) \\ \mathbf{4}\left(\dfrac{a}{2}\right) - \mathbf{4}\left(\dfrac{5b}{4}\right) = \mathbf{4}\left(\dfrac{7}{4}\right) \end{cases}$

$\begin{cases} -3a - 4b = 1 \\ 2a - 5b = 7 \end{cases}$

Eliminate a.
Multiply both sides of the 1st equation by 2.
Multiply both sides of the 2nd equation by 3.

$\begin{array}{r} -6a - 8b = 2 \\ 6a - 15b = 21 \\ \hline -23b = 23 \end{array}$

$\dfrac{-23b}{\mathbf{-23}} = \dfrac{23}{\mathbf{-23}}$

$b = -1$

$2a - 5b = 7$ The 2nd equation
$2a - 5(\mathbf{-1}) = 7$
$2a + 5 = 7$
$2a + 5 - \mathbf{5} = 7 - \mathbf{5}$
$2a = 2$
$\dfrac{2a}{\mathbf{2}} = \dfrac{2}{\mathbf{2}}$
$a = 1$

The solution is $(1, -1)$.

26. $\begin{cases} -\dfrac{1}{4}x = 1 - \dfrac{2}{3}y \\ 6x - 18y = 5 - 2y \end{cases}$

Clear the 1st equation of fractions.

$\begin{cases} \mathbf{12}\left(-\dfrac{1}{4}x\right) = \mathbf{12}(1) - \mathbf{12}\left(\dfrac{2}{3}y\right) \\ 6x - 18y = 5 - 2y \end{cases}$

$\begin{cases} -3x = 12 - 8y \\ 6x - 18y = 5 - 2y \end{cases}$

Write both equations in standard form.

$\begin{cases} -3x + 8y = 12 \\ 6x - 16y = 5 \end{cases}$

Eliminate x.
Multiply both sides of the 1st equation by 2.

$\begin{array}{r} -6x + 16y = 24 \\ 6x - 16y = 5 \\ \hline 0 = 29 \end{array}$ Both variables are eliminated.

False
No solution.

For each system, determine which method, substitution or elimination (addition), would be easier to use to solve the system and explain why.

27. $\begin{cases} 6x + 2y = 5 \\ 3x - 3y = -4 \end{cases}$ **Elimination; no variables have a coefficient of 1 or −1.**

28. $\begin{cases} x = 5 - 7y \\ 3x - 3y = -4 \end{cases}$ **Substitution; equation 1 is solved for x.**

SECTION 4.4 REVIEW EXERCISES
PROBLEM SOLVING USING SYSTEMS
OF EQUATIONS

29. ELEVATIONS
Analyze
- Elevation of Las Vegas is 20 times greater than that of Baltimore.
- The sum of their elevations is 2,100 feet.
- Find the elevation of each city.

Assign
Let x = the elevation of Las Vegas in feet
    y = the elevation of Baltimore in feet

Form

The elevation of Las Vegas	is	twenty	times greater than	the elevation of Baltimore.
x	=	20	$\bullet$	y

The elevation of Las Vegas	plus	the elevation of Baltimore	is	2,100 feet.
x	+	y	=	2,100

Solve
$$\begin{cases} x = 20y \\ x + y = 2{,}100 \end{cases}$$

$x + y = 2{,}100$ The 2nd equation. Substitute for x.

$(20y) + y = 2{,}100$

$21y = 2{,}100$

$\dfrac{21y}{21} = \dfrac{2{,}100}{21}$

$y = 100$

$x = 20y$ The 1st equation

$x = 20(100)$

$x = 2{,}000$

State
The elevation of Las Vegas is 2,000 feet.
The elevation of Baltimore is 100 feet.

Check

$2{,}000 = 20(100)$	$2{,}000 + 100 = 2{,}100$
$2{,}000 = 2{,}000$	$2{,}100 = 2{,}100$

The results check.

30. PAINTING EQUIPMENT
Analyze
- Fully extended, a ladder is 35 feet in length.
- The extension is 7 feet shorter than the base.
- How long is each part of the ladder?

Assign
Let x = the base in feet (longer section)
    y = the extension in feet (shorter section)

Form

The length of the bases	plus	the length of the bases extension	is	35 feet.
x	+	y	=	35

The extension	is	7 feet	shorter than	the base.
y	=	x	−	7

Solve
$$\begin{cases} x + y = 35 \\ y = x - 7 \end{cases}$$

$x + y = 35$ The 1st equation. Substitute for y.

$x + (x - 7) = 35$

$2x - 7 = 35$

$2x - 7 + 7 = 35 + 7$

$2x = 42$

$\dfrac{2x}{2} = \dfrac{42}{2}$

$x = 21$

$y = x - 7$ The 2nd equation

$y = 21 - 7$

$y = 14$

State
The length of the base is 21 feet.
The length of the extension is 14 feet.

Check
The results check.

31. GEOMETRY

Analyze
- Two angles are complementary (90°).
- One is 15° more than twice the measure of the other.
- Find the measure of each angle.

Assign
Let x = the measure of one $\angle$
y = the measure of the other $\angle$

Form

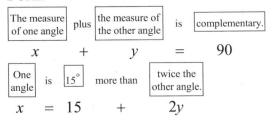

The measure of one angle	plus	the measure of the other angle	is	complementary.
x	$+$	y	$=$	90

One angle	is	$15°$	more than	twice the other angle.
x	$=$	15	$+$	$2y$

Solve

$$\begin{cases} x + y = 90 \\ x = 15 + 2y \end{cases}$$

$x + y = 90$ The 1ˢᵗ equation. Substitute for x.

$(15 + 2y) + y = 90$

$3y + 15 - 15 = 90 - 15$

$3y = 75$

$\dfrac{3y}{3} = \dfrac{75}{3}$

$y = 25$

$x + y = 90$ The 1ˢᵗ equation

$x + 25 = 90$

$x + 25 - 25 = 90 - 25$

$x = 65$

State
The measure of one angle is 65°.
The measure of the other angle is 25°.

Check

$65 + 25 = 90$	$65 = 15 + 2(25)$
	$65 = 15 + 50$
	$65 = 65$

The results check.

32. CRASH INVESTIGATION

Analyze
- 420 yards of tape is used to seal off a rectangular-shaped area.
- Width is three-fourths of the length.
- What is the area of the rectangle?

Assign
Let w = width of rectangle in yds (shorter length)
l = length of rectangle in yds (longer length)

Form

The measure of two widths	plus	the measure of two lengths	is	the perimerter of 420 feet.
$2w$	$+$	$2l$	$=$	420

The width	is	three-fourth	of	the length.
w	$=$	$\dfrac{3}{4}$	$\cdot$	l

Solve

$$\begin{cases} 2l + 2w = 420 \\ w = \dfrac{3}{4}l \end{cases}$$

$2w + 2l = 420$ The 1ˢᵗ equation. Substitute for w.

$2\left(\dfrac{3}{4}l\right) + 2l = 420$

$\dfrac{3}{2}l + 2l = 420$

$\dfrac{3}{2}l + \dfrac{4}{2}l = 420$

$\dfrac{7}{2}l = 420$

$\dfrac{2}{7}\left(\dfrac{7}{2}l\right) = \dfrac{2}{7}(420)$

$l = 120$

$w = \dfrac{3}{4}l$ The 2ⁿᵈ equation

$w = \dfrac{3}{4}(120)$

$w = 3(30)$

$w = 90$

State
The area of the rectangle is
120 yd $\cdot$ 90 yd $= 10,800$ yd^2.

Check
The results check.

33. Complete each table.

a.

	Amt •	Strength =	Amt of antifreeze
Weak	x	0.02	**0.02x**
Strong	y	0.09	**0.09y**
Mixture	100	0.08	**0.08(100)**

b.

	Rate •	Time =	Distance
With the wind	$s + w$	5	**5(s + w)**
Against the wind	$s - w$	7	**7(s − w)**

c.

	Principal •	Rate •	Time =	Interest
Mack Financial	x	0.11	1 yr	**0.11x**
Union Savings	y	0.06	1 yr	**0.06y**

d.

	Amount •	Price =	Total Value
Carmel corn	x	4	**4x**
Peanuts	y	8	**8y**
Mixture	10	5	**5(10)**

34. CANDY STORE

Analyze
- Gummy bears worth \$3 per pound.
- Gummy worms worth \$6 per pound.
- 30 pound mixture sells for \$4.20 per pound.
- How much of each type should be used?

Assign

Let x = amount of \$3/lb bears in lbs

y = amount of \$6/lb worms in lbs

Form

The # of lbs of \$3/lb bears	plus	the # of lbs of \$6/lb worms	is	30 pounds.
x	$+$	y	$=$	30

	Amt •	Cost/lb =	Total Value
Bears	x	3	**3x**
Worms	y	6	**6y**
Mixture	30	4.20	**4.20(30)**

The value of \$3/lb bears	plus	the value of \$6/lb worms	is equal to	the value of \$4.20/lb mixture.
$3x$	$+$	$6y$	$=$	4.20(30)

Solve

$$\begin{cases} x + y = 30 \\ 3x + 6y = 126 \end{cases}$$

Eliminate x.

Multiply both sides of the 1st equation by -3.

$$-3x - 3y = -90$$
$$\underline{3x + 6y = 126}$$
$$3y = 36$$
$$\frac{3y}{\mathbf{3}} = \frac{36}{\mathbf{3}}$$
$$y = 12$$

$$x + y = 30 \quad \text{The 1st equation}$$
$$x + \mathbf{12} = 30$$
$$x + 12 - \mathbf{12} = 30 - \mathbf{12}$$
$$x = 18$$

State

18 lb of \$3/lb gummy bears will be needed.
12 lb of \$6/lb gummy worms will be needed.

Check

The results check.

Chapter 4 Review and Test

35. BOATING

Analyze
- A boat can travel 56 miles downriver in 4 hours.
- The return trip takes 3 hours longer (7 hours).
- Find the speed of the current.

Assign

Let x = speed of boat in still water in mph

y = speed of river in mph

Form

	Rate •	Time =	Distance
With current	$x + y$	4	$4(x + y)$

Total Distance = 56

The rate of the boat with the current	times	4 hours traveled	is	56 miles.
$(x + y)$	•	4	=	56

	Rate •	Time =	Distance
Against current	$x - y$	7	$7(x - y)$

Total Distance = 56

The rate of the boat against the current	times	7 hours traveled	is	56 miles.
$(x - y)$	•	7	=	56

Solve

$$\begin{cases} 4(x + y) = 56 \\ 7(x - y) = 56 \end{cases}$$

Distribute in both equations.

$$\begin{cases} 4x + 4y = 56 \\ 7x - 7y = 56 \end{cases}$$

Eliminate x.

Multiply both sides of the 1st equation by 7.
Multiply both sides of the 2nd equation by -4.

$$28x + 28y = 392$$
$$\underline{-28x + 28y = -224}$$
$$56y = 168$$
$$\frac{56y}{56} = \frac{168}{56}$$
$$y = 3$$

State

The speed of the river is 3 mph.

Check

The results check.

36. SHOPPING

Analyze
- 2 bottles of cleaner and 3 bottles of soaking solution cost $63.40
- 3 bottles of cleaner and 3 bottles of soaking solution cost $69.60.
- How much is each bottle of solution?

Assign

Let x = cost of a bottle of cleaner

y = cost of a bottle of soaking solution

Form

	Amount •	Price	= Total Value
Cleaner	2	x	$2x$
Soaking	3	y	$3y$

Total 63.40

The value of 2 bottles of cleaner	plus	the value of 3 bottles of soaking	is	$63.40.
$2x$	+	$3y$	=	63.40

	Amount •	Price	= Total Value
Cleaner	3	x	$3x$
Soaking	2	y	$2y$

Total 69.60

The value of 3 bottles of cleaner	plus	the value of 2 bottles of soaking	is	$69.60.
$3x$	+	$2y$	=	69.60

Solve

$$\begin{cases} 2x + 3y = 63.40 \\ 3x + 2y = 69.60 \end{cases}$$

Eliminate x.

Multiply both sides of the 1st equation by 3.
Multiply both sides of the 2nd equation by -2.

$$6x + 9y = 190.20$$
$$\underline{-6x - 4y = -139.20}$$
$$5y = 51$$
$$\frac{5y}{5} = \frac{51}{5}$$
$$y = 10.20$$

$$2x + 3y = 63.40 \quad \text{The 1st equation}$$
$$2x + 3(\mathbf{10.20}) = 63.40$$
$$2x + 30.60 - \mathbf{30.60} = 63.40 - \mathbf{30.60}$$
$$2x = 32.80$$
$$\frac{2x}{2} = \frac{32.80}{2}$$
$$x = 16.40$$

State

A bottle of cleaning solution cost $16.40.
A bottle of soaking solution cost $10.20.

Check

The results check.

37. INVESTING

Analyze
- Investing $3,000.
- Passbook account paying 6% annual interest.
- Certificate account paying 10% annually.
- Total interest for first year was $270.
- How much was invested at 6%?

Assign

Let x = amt invested in passbook account

$\quad\ y$ = amt invested in certificate account

Form

The amount invested in the passbook	plus	the amount invested in the certificate	is	$3,000.
x	$+$	y	$=$	$3,000$

	Principal	$\cdot$ Rate	$\cdot$ Time	$=$ Interest
Passbook	x	0.06	1 yr	**0.06x**
Certificate	y	0.10	1 yr	**0.10y**

Total Interest = $270

The interest earned by the passbook	plus	the interest earned by the certificate	is	$270.
$0.06x$	$+$	$0.10\,y$	$=$	270

Solve

$$\begin{cases} x + y = 3,000 \\ 0.06x + 0.10\,y = 270 \end{cases}$$

Eliminate y.

Multiply both sides of the 1st equation by -10.

Multiply both sides of the 2nd equation by 100.

$$\begin{aligned} -10x - 10y &= -30,000 \\ 6x + 10y &= 27,000 \\ \hline -4x \qquad\ &= -3,000 \end{aligned}$$

$$\frac{-4x}{-4} = \frac{-3,000}{-4}$$

$$x = 750$$

State

$750 was invested at 6%.

Check

The results check.

38. ANTIFREEZE

Analyze
- A 40% antifreeze solution.
- A 70% antifreeze solution.
- How much of each is needed to make a 20 gal mixture that is 50% antifreeze solution?

Assign

Let x = amount of 40% antifreeze in gallons

$\quad\ y$ = amount of 70% antifreeze in gallons

Form

The # of gallons of 40% antifreeze	plus	the # of gallons of 70% antifreezes	is	20 gallons.
x	$+$	y	$=$	20

	Amt $\cdot$	Strength $=$	Amt of antifreeze
Weak	x	0.40	**0.4x**
Strong	y	0.70	**0.7y**
Mix	20	0.50	**20(0.5)**

The amount of antifreeze in the 40% solution	plus	the amount of antifreeze in the 70% solution	is equal to	the amount of antifreeze 50% solution.
$0.4x$	$+$	$0.7y$	$=$	$20(0.5)$

Solve

$$\begin{cases} x + y = 20 \\ 0.4x + 0.7\,y = 10 \end{cases}$$

Eliminate x.

Multiply both sides of the 1st equation by -4.

Multiply both sides of the 2nd equation by 10.

$$\begin{aligned} -4x - 4y &= -80 \\ 4x + 7y &= 100 \\ \hline 3y &= 20 \end{aligned}$$

$$\frac{3y}{3} = \frac{20}{3}$$

$$y = 6\frac{2}{3}$$

$$x + y = 20 \quad \text{The 1st equation}$$

$$x + 6\frac{2}{3} = 20$$

$$x + \frac{20}{3} - \frac{20}{3} = \frac{60}{3} - \frac{20}{3}$$

$$x = \frac{40}{3}$$

$$x = 13\frac{1}{3}$$

State

$13\frac{1}{3}$ gallons of 40% antifreeze will be needed.

$6\frac{2}{3}$ gallons of 70% antifreeze will be needed.

Check

The results check.

Chapter 4 Review and Test

39. $\begin{cases} 5x+3y<15 \\ 3x-y>3 \end{cases}$

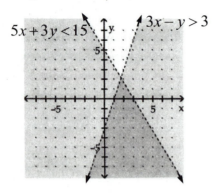

40. $\begin{cases} 3y \le x \\ y > 3x \end{cases}$

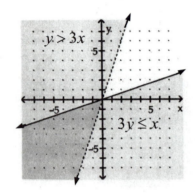

41. $\begin{cases} x \le 0 \\ y < 0 \end{cases}$

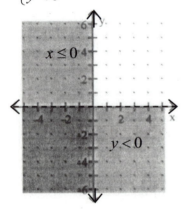

42. $\begin{cases} y \ge x \\ y \le \dfrac{1}{3}x+1 \\ x > -3 \end{cases}$

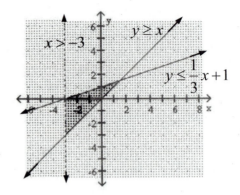

Use a check to determine whether each ordered pair is a solution of the system.

43. a. $(5, -4)$ $\begin{cases} x+2y \le 3 \\ 2x-y>1 \end{cases}$

$$x+2y \le 3 \qquad\qquad 2x-y>1$$
$$\overset{?}{5+2(-4) \le 3} \qquad \overset{?}{2(5)-(-4)>1}$$
$$\overset{?}{5-8 \le 3} \qquad\qquad \overset{?}{10+4>1}$$
$$-3 \le 3 \qquad\qquad 14>3$$
$$\text{True} \qquad\qquad\qquad \text{True}$$

$(5,-4)$ is a solution of the system.

b. $(-1, -3)$ $\begin{cases} x+2y \le 3 \\ 2x-y>1 \end{cases}$

$$x+2y \le 3 \qquad\qquad 2x-y>1$$
$$\overset{?}{-1+2(-3) \le 3} \qquad \overset{?}{2(-1)-(-3)>1}$$
$$\overset{?}{-1-6 \le 3} \qquad\qquad \overset{?}{-2+3>1}$$
$$-7 \le 3 \qquad\qquad 1>3$$
$$\text{True} \qquad\qquad\qquad \text{False}$$

$(-1,-3)$ is not a solution of the system.

44. GIFT SHOPPING

$$\begin{cases} 10x + 20y \geq 40 \\ 10x + 20y \leq 60 \end{cases}$$

Let x = # of T-shirts

y = # of pants

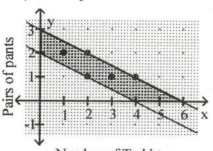

Number of T-shirts

1 T-shirt and 2 pairs of pants

3 T-shirts and 1 pair of pants

There are other answers.

CHAPTER 4 TEST

Determine whether the ordered pair is a solution of the system.

1. $(5,3)$ $\begin{cases} 3x + 2y = 21 \\ x + y = 8 \end{cases}$

$3x + 2y = 21$	$x + y = 8$
$\overset{?}{}$	$\overset{?}{}$
$3(5) + 2(3) \overset{?}{=} 21$	$5 + 3 \overset{?}{=} 8$
$15 + 6 \overset{?}{=} 21$	$8 = 8$
$21 = 21$	True
True	

Since $(5,3)$ satisfies both equations, it is a solution of the system.

2. $(-2,-1)$ $\begin{cases} 4x + y = -9 \\ 2x - 3y = -7 \end{cases}$

$4x + y = -9$	$2x - 3y = -7$
$\overset{?}{}$	$\overset{?}{}$
$4(-2) + (-1) \overset{?}{=} -9$	$2(-2) - 3(-1) \overset{?}{=} -7$
$-8 - 1 \overset{?}{=} -9$	$-4 + 3 \overset{?}{=} -7$
$-9 = -9$	$-1 = -7$
True	False

Since $(-2,-1)$ does not satisfies both equations, it is not a solution of the system.

3. **Fill in the blanks.**

a. A **solution** of a system of linear equations is an ordered pair that satisfies each equation.

b. A system of equations that has at least one solution is called a **consistent** system.

c. A system of equations that has no solution is called an **inconsistent** system.

d. Equations with different graphs are called **independent** equations.

e. A system of **dependent** equations has an infinite number of solutions.

f. Two angles are said to be **supplementary** if the sum of their measures is $180°$.

4. **ENERGY**

(2005, 770); In 2005, the amount of electricity generated by natural gas and nuclear sources was the same, about 770 billion kilowatts hours each.

Solve each system by graphing.

5. $\begin{cases} y = 2x - 1 \\ x - 2y = -4 \end{cases}$

The solution is $(2,3)$.

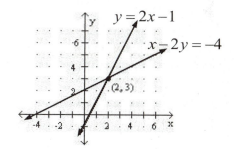

6. $\begin{cases} x + y = 5 \\ y = -x \end{cases}$

The lines are parallel.
The system has no solution.

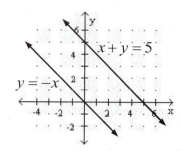

7. Find the slope and the y-intercept of the graph of each line in the system $\begin{cases} y = 4x - 10 \\ x - 2y = -16 \end{cases}$.

Then, use that information to determine the *number of solutions* of the system. **Do not solve the system.**

$$\begin{cases} y = 4x - 10 \\ x - 2y = -16 \end{cases}$$

$y = 4x - 10$ $x - 2y = -16$

$m = 4$ $x - 2y - x = -16 - x$

y-int $= (0, -10)$ $\dfrac{-2y}{-2} = \dfrac{-x}{-2} - \dfrac{16}{-2}$

$$y = \frac{1}{2}x + 8$$

$$m = \frac{1}{2}$$

y-int $= (0, 8)$

Since the lines have different slopes, they will intersect at one point. The system has 1 solution.

8. Infinitely many solutions. **(0, 2), (3, 1), (6, 0).**

Solve each system by substitution.

9. $\begin{cases} y = x - 1 \\ 2x + y = -7 \end{cases}$

$2x + y = -7$ The 2nd equation.

$2x + (\boldsymbol{x - 1}) = -7$ Substitute for y.

$3x - 1 = -7$

$3x - 1 + \boldsymbol{1} = -7 + \boldsymbol{1}$

$3x = -6$

$\dfrac{3x}{\boldsymbol{3}} = \dfrac{-6}{\boldsymbol{3}}$

$x = -2$

$y = x - 1$ The 1st equation

$y = -\boldsymbol{2} - 1$

$y = -3$

The solution is $(-2, -3)$.

---

10. $\begin{cases} 3x + 6y = -15 \\ x + 2y = -5 \end{cases}$

Solve the 2nd equation for x.

$\begin{cases} 3x + 6y = -15 \\ x = -5 - 2y \end{cases}$

$3x + 6y = -15$ The 1st equation. Substitute for x.

$3(\boldsymbol{-5 - 2y}) + 6y = -15$

$-15 - 6y + 6y = -15$

$-15 = -15$ The variables drop out.

True

The system has infinitely many solutions.

Solve each system using elimination (addition).

11. $\begin{cases} 3x - y = 2 \\ 2x + y = 8 \end{cases}$

Eliminate y.

$\begin{array}{r} 3x - y = 2 \\ 2x + y = 8 \\ \hline 5x \quad\;\; = 10 \end{array}$

$\dfrac{5x}{\boldsymbol{5}} = \dfrac{10}{\boldsymbol{5}}$

$x = 2$

$2x + y = 8$ The 2nd equation

$2(\boldsymbol{2}) + y = 8$

$4 + y - \boldsymbol{4} = 8 - \boldsymbol{4}$

$y = 4$

The solution is $(2, 4)$.

12. $\begin{cases} 4x + 3y = -3 \\ -3x = -4y + 21 \end{cases}$

Write the 2nd equation in standard form.

$\begin{cases} 4x + 3y = -3 \\ -3x + 4y = 21 \end{cases}$

Eliminate x.

Multiply both sides of the 1st equation by 3.
Multiply both sides of the 2nd equation by 4.

$\begin{array}{r} 12x + 9y \;\; = -9 \\ -12x + 16y = 84 \\ \hline 25y = 75 \end{array}$

$\dfrac{25y}{\boldsymbol{25}} = \dfrac{75}{\boldsymbol{25}}$

$y = 3$

$4x + 3y = -3$ The 1st equation
$4x + 3(3) = -3$
$4x + 9 - 9 = -3 - 9$
$4x = -12$
$$\frac{4x}{4} = \frac{-12}{4}$$
$x = -3$
The solution is $(-3, 3)$.

Solve each system using substitution or elimination (addition).

13. $\begin{cases} 3x - 5y - 16 = 0 \\ \dfrac{x}{2} - \dfrac{5}{6}y = \dfrac{1}{3} \end{cases}$

Write the 1st equation in standard form.
Clear the 2nd equation of fractions.

$\begin{cases} 3x - 5y - 16 + 16 = 0 + 16 \\ 6\left(\dfrac{x}{2}\right) - 6\left(\dfrac{5}{6}y\right) = 6\left(\dfrac{1}{3}\right) \end{cases}$

$\begin{cases} 3x - 5y = 16 \\ 3x - 5y = 2 \end{cases}$

Eliminate x.
Multiply both sides of the 1st equation by -1.

$-3x + 5y = -16$
$\underline{3x - 5y = 2}$
$0 = -14$ Both variables are eliminated.

False
No solution.

14. $\begin{cases} 3a + 4b = -7 \\ 2b - a = -1 \end{cases}$

Rewrite the 2nd equation in standard form.

$\begin{cases} 3a + 4b = -7 \\ -a + 2b = -1 \end{cases}$

Eliminate a.
Multiply both sides of the 2nd equation by 3.

$3a + 4b = -7$
$\underline{-3a + 6b = -3}$
$10b = -10$
$$\frac{10b}{10} = \frac{-10}{10}$$
$b = -1$

$-a + 2b = -1$ The 2nd equation
$-a + 2(-1) = -1$
$-a - 2 + 2 = -1 + 2$
$-a = 1$
$$\frac{-a}{-1} = \frac{1}{-1}$$
$x = -1$
The solution is $(-1, -1)$.

15. $\begin{cases} y = 3x - 1 \\ y = 2x + 4 \end{cases}$

$y = 2x + 4$ The 2nd equation. Substitute for y.
$(3x - 1) = 2x + 4$
$3x - 1 + 1 = 2x + 4 + 1$
$3x = 2x + 5$
$3x - 2x = 2x + 5 - 2x$
$x = 5$

$y = 3x - 1$ The 1st equation
$y = 3(5) - 1$
$y = 15 - 1$
$y = 14$
The solution is $(5, 14)$.

16. $\begin{cases} 0.6c + 0.5d = 0 \\ 0.02c + 0.09d = 0 \end{cases}$

Eliminate c.
Multiply both sides of the 1st equation by -10.
Multiply both sides of the 2nd equation by 300.

$-6c - 5d = 0$
$\underline{6c + 27d = 0}$
$22d = 0$
$$\frac{22d}{22} = \frac{0}{22}$$
$d = 0$

$0.6c + 0.5d = 0$ The 1st equation
$0.6c + 0.5(0) = 0$
$0.6c = 0$
$$\frac{0.6c}{0.6} = \frac{0}{0.6}$$
$c = 0$
The solution is $(0, 0)$.

Chapter 4 Review and Test

17. $\begin{cases} a-1=2b \\ 3a+1=-10b \end{cases}$

Rewrite both equations in standard form.

$\begin{cases} a-2b=1 \\ 3a+10b=-1 \end{cases}$

Eliminate b.

Multiply both sides of the 1st equation by 5.

$\begin{cases} 5a-10b=5 \\ 3a+10b=-1 \end{cases}$

$\begin{array}{r} 5a-10b=5 \\ \underline{3a+10b=-1} \\ 8a=4 \end{array}$

$\dfrac{8a}{8} = \dfrac{4}{8}$

$a = \dfrac{1}{2}$

$a-2b=1$ The 1st equation

$\dfrac{1}{2}-2b=1$

$\dfrac{1}{2}-2b-\dfrac{1}{2}=1-\dfrac{1}{2}$

$-2b=\dfrac{2}{2}-\dfrac{1}{2}$

$-2b=\dfrac{1}{2}$

$-\dfrac{1}{2}(-2b)=-\dfrac{1}{2}\left(\dfrac{1}{2}\right)$

$b=-\dfrac{1}{4}$

The solution is $\left(\dfrac{1}{2},-\dfrac{1}{4}\right)$.

18. $\begin{cases} \dfrac{x+2}{6}=\dfrac{y+3}{4} \\ \dfrac{x}{5}=\dfrac{3y-3}{6} \end{cases}$

Clear both equations of fractions.

$\begin{cases} 12\left(\dfrac{x+2}{6}\right)=12\left(\dfrac{y+3}{4}\right) \\ 30\left(\dfrac{x}{5}\right)=30\left(\dfrac{3y-3}{6}\right) \end{cases}$

$\begin{cases} 2(x+2)=3(y+3) \\ 6x=5(3y-3) \end{cases}$

Distribute both equations.

$\begin{cases} 2x+4=3y+9 \\ 6x=15y-15 \end{cases}$

Rewrite both equations in standard form.

$\begin{cases} 2x-3y=5 \\ 6x-15y=-15 \end{cases}$

Eliminate x.

Multiply both sides of the 1st equation by -3.

$\begin{cases} -6x+9y=-15 \\ 6x-15y=-15 \end{cases}$

$\begin{array}{r} -6x+9y=-15 \\ \underline{6x-15y=-15} \\ -6y=-30 \end{array}$

$\dfrac{-6y}{-6}=\dfrac{-30}{-6}$

$y=5$

$2x-3y=5$ The 1st equation

$2x-3(5)=5$

$2x-15=5$

$2x-15+15=5+15$

$2x=20$

$\dfrac{2x}{2}=\dfrac{20}{2}$

$x=10$

The solution is $(10, 5)$.

19. $\begin{cases} 4(a-2)+5y=19 \\ 3(a-2)-y=0 \end{cases}$

Distribute both equations.

$\begin{cases} 4a-8+5y=19 \\ 3a-6-y=0 \end{cases}$

Rewrite both equations in standard form.

$\begin{cases} 4a-8+5y+\mathbf{8}=19+\mathbf{8} \\ 3a-6-y+\mathbf{6}=0+\mathbf{6} \end{cases}$

$\begin{cases} 4a+5y=27 \\ 3a-y=6 \end{cases}$

Eliminate y.

Multiply both sides of the 2nd equation by 5.

$\begin{cases} 4a+5y=27 \\ 15a-5y=30 \end{cases}$

$\begin{array}{r} 4a+5y=27 \\ 15a-5y=30 \\ \hline 19a=57 \end{array}$

$\dfrac{19a}{\mathbf{19}}=\dfrac{57}{\mathbf{19}}$

$a=3$

$3a-y=6$ The 2nd equation.
Substitute for x.

$3(\mathbf{3})-y=6$

$9-y=6$

$9-y-\mathbf{9}=6-\mathbf{9}$

$-y=-3$

$\dfrac{-y}{\mathbf{-1}}=\dfrac{-3}{\mathbf{-1}}$

$y=3$

The solution is (3, 3).

20. $\begin{cases} 3x+1=2x-4y+2 \\ 7x-y-1=9y+5x+10 \end{cases}$

Rewrite both equations in standard form.

$\begin{cases} 3x+1-\mathbf{1}-\mathbf{2x}+\mathbf{4y}=2x-4y+2-\mathbf{1}-\mathbf{2x}+\mathbf{4y} \\ 7x-y-1+\mathbf{1}-\mathbf{5x}-\mathbf{9y}=9y+5x+10+\mathbf{1}-\mathbf{5x}-\mathbf{9y} \end{cases}$

$\begin{cases} x+4y=1 \\ 2x-10y=11 \end{cases}$

Solve the 1st equation for x.

Substitute for x in the 2nd equation.

$\begin{cases} x=1-4y \\ 2x-10y=11 \end{cases}$

$2(\mathbf{1}-\mathbf{4y})-10y=11$ The 2nd equation.
Substitute for x.

$2-8y-10y=11$

$2-18y=11$

$2-18y-\mathbf{2}=11-\mathbf{2}$

$-18y=9$

$\dfrac{-18y}{\mathbf{-18}}=\dfrac{9}{\mathbf{-18}}$

$y=-\dfrac{1}{2}$

$x=1-4y$

$x=1-4\left(-\dfrac{\mathbf{1}}{\mathbf{2}}\right)$

$x=1+2$

$x=3$

The solution is $\left(3,-\dfrac{1}{2}\right)$.

Write a system of two equations in two variables to solve each problem.

21. CHILD CARE

Analyze

- 22-mile commute, one way.
- First part of the trip is 6 miles less than the second part.
- How long is each part of her commute?

Assign

Let x = the 1st part in miles

$\quad\;\; y$ = the 2nd part in miles

Form

| The first part | plus | the second part | is | 22 miles. |

$$x \quad + \quad y \quad = \quad 22$$

| The first part | is | 6 miles | less than | the second part. |

$$x \quad = \quad y \quad - \quad 6$$

Solve

$$\begin{cases} x + y = 22 \\ x = y - 6 \end{cases}$$

$$x + y = 22 \quad \text{The } 1^{st} \text{ equation.}$$
$$\text{Substitute for } x.$$

$$(y - 6) + y = 22$$

$$2y - 6 = 22$$

$$2y - 6 + 6 = 22 + 6$$

$$2y = 28$$

$$\frac{2y}{2} = \frac{28}{2}$$

$$y = 14$$

$$x = y - 6 \quad \text{The } 2^{nd} \text{ equation}$$

$$x = (14) - 6$$

$$x = 8$$

State

The first part is 8 miles.

The second part is 14 miles.

Check

$$8 + 14 = 22 \quad | \quad 8 = 14 - 6$$
$$22 = 22 \quad | \quad 8 = 8$$

The results check.

22. VACATIONING

Analyze

- Cost $219 for a family of 7.
- Adult tickets cost $37.00.
- Child tickets cost $27.00.
- How many of each type were bought?

Assign

Let x = number of child tickets

$\quad\;\; y$ = number of adult tickets

Form

| The number of children tickets | plus | the number of adult tickets | is | 7. |

$$x \quad + \quad y \quad = \quad 7$$

	Number	•	Value	=	Total Value
Child	x		$27		$27x$
Adult	y		$37		$37y$
			Total		219

| The value of children tickets | plus | the value of adult tickets | is | $219. |

$$27x \quad + \quad 37y \quad = \quad 219$$

Solve

$$\begin{cases} x + y = 7 \\ 27x + 33y = 219 \end{cases}$$

Eliminate x.

Multiply both sides of the 1st equation by -27.

$$-27x - 27y = -189$$
$$\underline{27x + 37y = \;\; 219}$$
$$10y = 30$$

$$\frac{10y}{10} = \frac{30}{10}$$

$$y = 3$$

$$x + y = 7 \quad \text{The } 1^{st} \text{ equation}$$

$$x + 3 = 7$$

$$x + 3 - 3 = 7 - 3$$

$$x = 4$$

State

4 child tickets were bought.

3 adult tickets were bought.

Check

The results check.

23. FINANCIAL PLANNING

Analyze

- Investing $10,000.
- Invest part at 8% annual interest.
- Invest the rest at 9% annual interest.
- Combined interest for first year was $840.
- How much was invested at each rate?

Assign

Let x = amt invested at 8%

y = amt invested at 9%

Form

The amount invested at 8%	plus	the amount invested at 9%	is	$10,000.
x	$+$	y	$=$	$10,000$

	Principal	• Rate	• Time	= Interest
8%	x	8%	1 yr	**0.08x**
9%	y	9%	1 yr	**0.09y**

Total Interest = $840

The interest earned by 8% investment	plus	the interest earned by 9% investment	is	$840.
$0.08x$	$+$	$0.09y$	$=$	840

Solve

$$\begin{cases} x + y = 10,000 \\ 0.08x + 0.09y = 840 \end{cases}$$

Eliminate x.

Multiply both sides of the 1st equation by -8.

Multiply both sides of the 2nd equation by 100.

$$-8x - 8y = -80,000$$
$$\underline{8x + 9y = \ \ 84,000}$$
$$y = 4,000$$

$$x + y = 10,000 \quad \text{The 1}^{st}\text{ equation}$$
$$x + \textbf{4,000} = 10,000$$
$$x + 4,000 - \textbf{4,000} = 10,000 - \textbf{4,000}$$
$$x = 6,000$$

State

$6,000 was invested at 8%.

$4,000 was invested at 9%.

Check

$6,000 + \$4,000 = \$10,000$	$0.08(\$6,000) + 0.09(\$4,000) = \$840$
$10,000 = \$10,000$	$480 + \$360 = \$840$
	$840 = \$840$

The results check.

24. TAILWINDS/HEADWINDS

Analyze

- Plane flies 450 miles in 2.5 hours with a tailwind.
- Plane flies 450 miles in 3 hours into a headwind.
- Find the speed of the plane in calm air and the speed of the wind.

Assign

Let x = speed of plane in calm air in mph

y = speed of wind in mph

Form

	Rate •	Time =	Distance
With tailwind	$x + y$	2.5	**2.5(x + y)**

Total Distance = 450

The rate of the plane with tailwind	times	2.5 hours traveled	is	450 miles.
$(x + y)$	•	2.5	=	450

	Rate •	Time =	Distance
Into headwind	$x - y$	3	**3(x − y)**

Total Distance = 450

The rate of the plnae into headwind	times	3 hours traveled	is	450 miles.
$(x - y)$	•	3	=	450

Solve

$$\begin{cases} 2.5(x + y) = 450 \\ 3(x - y) = 450 \end{cases}$$

Distribute in both equations.

$$\begin{cases} 2.5x + 2.5y = 450 \\ 3x - 3y = 450 \end{cases}$$

Eliminate y.

Multiply both sides of the 1st equation by 30.

Multiply both sides of the 2nd equation by 25.

$$75x + 75y = 13,500$$
$$\underline{75x - 75y = 11,250}$$
$$150x \qquad = 24,750$$
$$\frac{150x}{\textbf{150}} = \frac{24,750}{\textbf{150}}$$
$$x = 165$$

$$3x - 3y = 450 \quad \text{The 2}^{nd}\text{ equation}$$
$$3(\textbf{165}) - 3y = 450$$
$$495 - 3y - \textbf{495} = 450 + \textbf{495}$$
$$-3y = -45$$
$$\frac{-3y}{\textbf{-3}} = \frac{-45}{\textbf{-3}}$$
$$y = 15$$

State

The speed of the plane is 165 mph.

The speed of the wind is 15 mph.

Check

The results check.

Chapter 4 Review and Test

25. TETHER BALL

Analyze
- Two angles are complementary (90°).
- Larger angle is 10° more than 3 times the measure of the smaller angle.
- Find the measure of each angle.

Assign
Let x = the measure of smaller $\angle$
y = the measure of larger $\angle$

Form

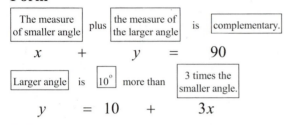

The measure of smaller angle	plus	the measure of the larger angle	is	complementary.
x	$+$	y	$=$	90

Larger angle	is	$10°$	more than	3 times the smaller angle.
y	$=$	10	$+$	$3x$

Solve

$$\begin{cases} x + y = 90 \\ y = 10 + 3x \end{cases}$$

$x + y = 90$ The 1st equation.
Substitute for y.

$x + (\mathbf{10 + 3x}) = 90$

$4x + 10 = 90$

$4x + 10 - \mathbf{10} = 90 - \mathbf{10}$

$4x = 80$

$\dfrac{4x}{\mathbf{4}} = \dfrac{80}{\mathbf{4}}$

$x = 20$

$y = 10 + 3x$ The 2nd equation

$y = 10 + 3(\mathbf{20})$

$y = 10 + 60$

$y = 70$

State
The measure of the smaller angle is 20°.
The measure of the larger angle is 70°.

Check

$20 + 70 = 90$	$70 = 10 + 3(20)$
$90 = 90$	$70 = 70$

The results check.

26. ANTIFREEZE

Analyze
- A 5% antifreeze solution.
- A 20% antifreeze solution.
- How much of each is needed to make a 12 pint solution that is 15% antifreeze?

Assign
Let x = amount of 5% antifreeze in pints
y = amount of 20% antifreeze in pints

Form

The # of pints of 5% antifreeze	plus	the # of pints of 20% antifreezes	is	12 pints.
x	$+$	y	$=$	12

	Amt $\bullet$	Strength $=$	Amt of pure antifreeze
Weak	x	0.05	**0.05x**
Strong	y	0.20	**0.20y**
Mixture	12	0.15	**12(0.15)**

The amount of pure antifreeze in the 5% solution	plus	the amount of pure antifreeze in the 20% solution	is equal to	the amount of pure antifreeze in the 15% solution.
$0.05x$	$+$	$0.20y$	$=$	$12(0.15)$

Solve

$$\begin{cases} x + y = 12 \\ 0.05x + 0.20y = 1.8 \end{cases}$$

Eliminate x.
Multiply both sides of the 1st equation by -5.
Multiply both sides of the 2nd equation by 100.

$$\begin{array}{r} -5x - 5y = -60 \\ 5x + 20y = 180 \\ \hline 15y = 120 \end{array}$$

$\dfrac{15y}{\mathbf{15}} = \dfrac{120}{\mathbf{15}}$

$y = 8$

$x + y = 12$ The 1st equation

$x + \mathbf{8} = 12$

$x + \mathbf{8} - \mathbf{8} = 12 - \mathbf{8}$

$x = 4$

State
4 pints of 5% antifreeze will be needed.
8 pints of 20% antifreeze will be needed.

Check
The results check.

27. SUNSCREEN

Analyze

- $1.50/ounce sunscreen.
- $0.80/ounce sunscreen.
- How much of each is needed to make 10 ounces that sells for $1.01?

Assign

Let x = amount of $1.50/oz sunscreen in oz

y = amount of $0.80/oz sunscreen in oz

Form

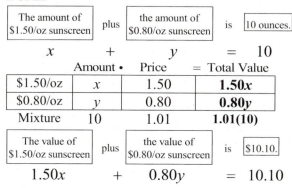

The amount of $1.50/oz sunscreen	plus	the amount of $0.80/oz sunscreen	is	10 ounces.
x	$+$	y	$=$	10

	Amount •	Price	= Total Value
$1.50/oz	x	1.50	**1.50x**
$0.80/oz	y	0.80	**0.80y**
Mixture	10	1.01	**1.01(10)**

The value of $1.50/oz sunscreen	plus	the value of $0.80/oz sunscreen	is	$10.10.
1.50x	$+$	0.80y	$=$	10.10

Solve

$$\begin{cases} x+y=10 \\ 1.50x+0.80y=10.10 \end{cases}$$

Eliminate y.

Multiply both sides of the 1st equation by -80.

Multiply both sides of the 2nd equation by 100.

$$-80x-80y=-800$$
$$150x+80y=1010$$
$$\overline{70x=210}$$

$$\frac{70x}{70}=\frac{210}{70}$$

$$x=3$$

$x+y=10$ The 1st equation

$3+y=10$

$3+y-3=10-3$

$y=7$

State

3 ounces of $1.50/oz sunscreen is needed.

7 ounces of $0.80/oz sunscreen is needed.

Check

$3+7=10$	$1.50(3)+\$0.80(7)=\$10.10$
$10=10$	$4.50+\$5.60=\$10.10$
	$10.10=\$10.10$

The results check.

Use a check to determine whether (3, 1) is a solution of the system :

28. $\begin{cases} y \le 2x-1 \\ x+3y > 6 \end{cases}$

$y \le 2x-1$	$x+3y > 6$
$\overset{?}{\le}$	$\overset{?}{>}$
$1 \le 2(3)-1$	$(3)+3(1)>6$
$\overset{?}{\le}$	$\overset{?}{>}$
$1 \le 6-1$	$3+3>6$
$1 \le 5$	$6 > 6$
True	False

(3,1) is not a solution of the system.

Solve the system by graphing.

29. $\begin{cases} 3x+2y \le 6 \\ y \ge x+1 \end{cases}$

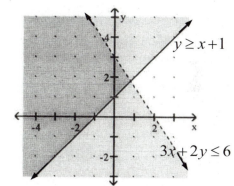

30. $\begin{cases} x-y < 3 \\ y \le 0 \\ x \ge 0 \end{cases}$

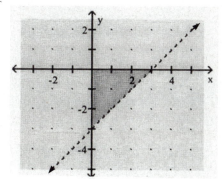

Chapter 4 Review and Test

31. CLOTHES SHOPPING

$$\begin{cases} 20x + 40y \geq 80 \\ 20x + 40y \leq 120 \end{cases}$$

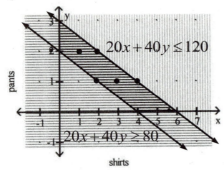

1 shirt, 2 pants

2 shirts, 1 pants

2 shirts, 2 pants

32. Match each equation, inequality, or system with the graph of its solution.

a. $2x + y = 2$ **iii**

b. $2x + y \geq 2$ **ii**

c. $\begin{cases} 2x + y = 2 \\ 2x - y = 2 \end{cases}$ **i**

d. $\begin{cases} 2x + y \geq 2 \\ 2x - y \leq 2 \end{cases}$ **iv**

CUMULATIVE REVIEW
CHAPTERS 1 - 4
Fill in the blanks.

1. The answer to an addition problem is called a **sum**. The answer to a subtraction problem is called a **difference**. The answer to a multiplication problem is called a **product**. The answer to a division problem is called a **quotient**. [Section 1.1]

2. Prime Factorization: [Section 1.2]
$$100 = 2 \cdot 50$$
$$= 2 \cdot 2 \cdot 25$$
$$= 2 \cdot 2 \cdot 5 \cdot 5$$
$$= 2^2 \cdot 5^2$$

3. Divide: [Section 1.2]
$$\frac{3}{4} \div \frac{6}{5} = \frac{3}{4} \cdot \frac{5}{6}$$
$$= \frac{\overset{1}{\cancel{3}} \cdot 5}{4 \cdot 2 \cdot \underset{1}{\cancel{3}}}$$
$$= \frac{5}{8}$$

4. Subtract: [Section 1.2]
$$LCD = 2 \cdot 5 \cdot 7 = 70$$
$$\frac{7}{10} - \frac{1}{14} = \frac{7}{10} \cdot \left(\frac{7}{7}\right) - \frac{1}{14} \cdot \left(\frac{5}{5}\right)$$
$$= \frac{49 - 5}{70}$$
$$= \frac{44}{70}$$
$$= \frac{\overset{1}{\cancel{2}} \cdot 2 \cdot 11}{\underset{1}{\cancel{2}} \cdot 5 \cdot 7}$$
$$= \frac{22}{35}$$

5. Rational or Irrational: [Section 1.3]
π is irrational.

6. Graph: [Section 1.3]
$$\left\{-2\frac{1}{4}, \sqrt{2}, -1.75, \frac{7}{2}, 0.5\right\}$$

7. Write as a decimal: [Section 1.3]
$$\frac{2}{3} = 3\overline{)2.000} \xrightarrow{\;0.666\;} = 0.6\bar{6}$$

8. What property of real numbers is illustrated? [Section 1.6]
$$3(2x) = (3 \cdot 2)x$$
Associative property of multiplication

Evaluate each expression.

9. $-3^2 + \left|4^2 - 5^2\right|$ [Section 1.7]
$$-3^2 + \left|4^2 - 5^2\right| = -(3)(3) + |16 - 25|$$
$$= -9 + |-9|$$
$$= -9 + 9$$
$$= 0$$

10. $(4-5)^{20}$ [Section 1.7]
$$(4-5)^{20} = (-1)^{20}$$
$$= 1$$

11. $\dfrac{-3-(-7)}{2^2 - 3}$ [Section 1.7]
$$\frac{-3-(-7)}{2^2 - 3} = \frac{-3+7}{4-3}$$
$$= \frac{4}{1}$$
$$= 4$$

12. $12 - 2[1 - (-8 + 2)]$ [Section 1.7]
$$12 - 2[1 - (-8 + 2)] = 12 - 2[1 - (-6)]$$
$$= 12 - 2[1 + 6]$$
$$= 12 - 2(7)$$
$$= 12 - 14$$
$$= -2$$

13. RACING [Section 1.8]
$$250 - x$$

14. Value of d dimes? [Section 1.8]
$10d$ cents or $0.10d

Simplify each expression.

15. $13r - 12r$ [Section 1.9]

$$13r - 12r = (13 - 12)r$$
$$= r$$

16. $27\left(\dfrac{2}{3}x\right)$ [Section 1.9]

$$27\left(\dfrac{2}{3}x\right) = \dfrac{27 \cdot 2}{3}x$$

$$= \dfrac{\overset{1}{\cancel{3}} \cdot 9 \cdot 2}{\underset{1}{\cancel{3}}}x$$

$$= 18x$$

17. $4(d-3) - (d-1)$ [Section 1.9]

$$4(d-3) - (d-1) = \mathbf{4}d - \mathbf{4}(3) - d + 1$$
$$= 4d - 12 - d + 1$$
$$= (4-1)d + (-12+1)$$
$$= 3d - 11$$

18. $(13c - 3)(-6)$ [Section 1.9]

$$(13c - 3)(-6) = -\mathbf{6}(13c) - \mathbf{6}(-3)$$
$$= -78c + 18$$

Solve each equation. Check each result.

19. $3(x-5) + 2 = 2x$ [Section 2.2]

$$3(x-5) + 2 = 2x$$
$$\mathbf{3}(x) - \mathbf{3}(5) + 2 = 2x$$
$$3x - 15 + 2 = 2x$$
$$3x - 13 = 2x$$
$$3x - 13 + \mathbf{13} = 2x + \mathbf{13}$$
$$3x = 2x + 13$$
$$3x - \mathbf{2x} = 2x + 13 - \mathbf{2x}$$
$$x = 13$$

Check: $x = 13$

$$3(x-5) + 2 = 2x$$
$$3(\mathbf{13} - 5) + 2 \overset{?}{=} 2(\mathbf{13})$$
$$3(8) + 2 \overset{?}{=} 26$$
$$24 + 2 \overset{?}{=} 26$$
$$26 = 26 \quad \text{True}$$

The solution is 13.

20. $\dfrac{x-5}{3} - 5 = 7$ [Section 2.2]

$$\dfrac{x-5}{3} - 5 = 7$$

$$3\left(\dfrac{x-5}{3} - 5\right) = \mathbf{3}(7)$$

$$\mathbf{3}\left(\dfrac{x-5}{3}\right) - \mathbf{3}(5) = \mathbf{3}(7)$$

$$x - 5 - 15 = 21$$
$$x - 20 = 21$$
$$x - 20 + \mathbf{20} = 21 + \mathbf{20}$$
$$x = 41$$

Check: $x = 41$

$$\dfrac{x-5}{3} - 5 = 7$$

$$\dfrac{\mathbf{41} - 5}{3} - 5 \overset{?}{=} 7$$

$$\dfrac{36}{3} - 5 \overset{?}{=} 7$$

$$12 - 5 \overset{?}{=} 7$$

$$7 = 7 \quad \text{True}$$

The solution is 41.

21. $\dfrac{2}{5}x + 1 = \dfrac{1}{3} + x$ [Section 2.2]

$$\dfrac{2}{5}x + 1 = \dfrac{1}{3} + x$$

$$\mathbf{15}\left(\dfrac{2}{5}x + 1\right) = \mathbf{15}\left(\dfrac{1}{3} + x\right)$$

$$\mathbf{15}\left(\dfrac{2}{5}x\right) + \mathbf{15}(1) = \mathbf{15}\left(\dfrac{1}{3}\right) + \mathbf{15}(x)$$

$$6x + 15 = 5 + 15x$$
$$6x + 15 - \mathbf{5} = 5 + 15x - \mathbf{5}$$
$$6x + 10 = 15x$$
$$6x + 10 - \mathbf{6x} = 15x - \mathbf{6x}$$
$$10 = 9x$$
$$\dfrac{10}{\mathbf{9}} = \dfrac{9x}{\mathbf{9}}$$
$$\dfrac{10}{9} = x$$

Check: $x = \dfrac{10}{9}$

$$\frac{2}{5}\left(\frac{\mathbf{10}}{\mathbf{9}}\right)+1\overset{?}{=}\frac{1}{3}+\frac{\mathbf{10}}{\mathbf{9}}$$

$$\frac{4}{9}+\frac{9}{9}\overset{?}{=}\frac{3}{9}+\frac{10}{9}$$

$$\frac{13}{9}=\frac{13}{9} \quad \text{True}$$

The solution is $\dfrac{10}{9}$.

22. $-\dfrac{5}{8}h=15$ [Section 2.2]

$$-\frac{5}{8}h=15$$

$$-\frac{\mathbf{8}}{\mathbf{5}}\left(-\frac{5}{8}h\right)=-\frac{\mathbf{8}}{\mathbf{5}}(15)$$

$$h=-24$$

Check: $h=-24$

$$-\frac{5}{8}h=15$$

$$-\frac{5}{8}(-\mathbf{24})\overset{?}{=}15$$

$$15=15 \quad \text{True}$$

The solution is -24.

23. GYMNASTICS [Section 2.3]
85% of what number is 119.

$$0.85 \;\cdot\; x \;=119$$
$$0.85x=119$$

$$\frac{0.85x}{\mathbf{0.85}}=\frac{119}{\mathbf{0.85}}$$

$$x=140$$

Check: $x=140$
$$0.85x=119$$
$$0.85(\mathbf{140})\overset{?}{=}119$$
$$119=119 \quad \text{True}$$

The maximum number of children is 140.

24. GREEN HOUSE [Section 2.3]

Let $x=$ amt of emissions from transportation
What number is 27.9% of 6,957 teragrams?

$$x \;=0.279 \cdot 6,957 \text{ teragrams}$$
$$x=1,941.003 \text{ teragrams}$$

1,941 teragrams comes from transportation.

25. Solve: $A=\dfrac{1}{2}h(b+B)$ for h. [Section 2.4]

$$A=\frac{1}{2}h(b+B)$$

$$\mathbf{2}(A)=\mathbf{2}\left[\frac{1}{2}h(b+B)\right]$$

$$2A=h(b+B)$$

$$\frac{2A}{\mathbf{b+B}}=\frac{h(b+B)}{\mathbf{b+B}}$$

$$\frac{2A}{b+B}=h$$

$$h=\frac{2A}{b+B}$$

26. CANCER [Section 2.5]
Analyze
• Total number of people with cancer is 219,000.
• 13,000 more cases dealing with men than women.
• How many cases for each gender are there?
Assign

Let $x=$ number of cases for women
$x+13,000=$ number of cases for men

Form

The number of cancer cases for women	plus	the number of cancer cases for men	equals	the total number of cases.
x	$+$	$x+13,000$	$=$	219,000

Solve

$$x+x+13,000=219,000$$
$$2x+13,000=219,000$$
$$2x+13,000-\mathbf{13,000}=219,000-\mathbf{13,000}$$
$$2x=206,000$$
$$\frac{2x}{\mathbf{2}}=\frac{206,000}{\mathbf{2}}$$
$$x=103,000$$

men
$$x+13,000=103,000+13,000$$
$$=116,000$$

State
Number of women with cancer is 103,000.
Number of men with cancer is 116,000.
Check
$$103,000+113,000=219,000$$

27. MIXING CANDY [Section 2.6]

Analyze
- Lemon gumdrops sell for $4.40 per lb.
- Red licorice bits sell for $3.80 per lb.
- The blend is to sell for $4 per lb.
- A blend of 30 lb is needed.
- How many pounds of each is needed?

Assign

Let x = the number of pounds lemon gumdrops
$30 - x$ = the number of pounds red licorice

	Number	• Value =	Total value
Lemon	x	4.40	**4.40x**
Licorice	**30 − x**	3.80	**3.80(30 − x)**
Blend	30	4.00	**120**

Form

The value of the lemon gumdrops	plus	the value of the licorice	equals	the total value of the blend.

$$4.40x \;+\; 3.80(30-x) \;=\; 120$$

Solve

$$4.40x + 3.80(30-x) = 120$$
$$4.40x + \mathbf{3.80}(30) - \mathbf{3.80}(x) = 120$$
$$4.40x + 114 - 3.80x = 120$$
$$0.60x + 114 = 120$$
$$0.60x + 114 - \mathbf{114} = 120 - \mathbf{114}$$
$$0.60x = 6$$
$$\frac{0.60x}{\mathbf{0.60}} = \frac{6}{\mathbf{0.60}}$$
$$x = 10$$

licorice
$$30 - x = 30 - \mathbf{10}$$
$$= 20$$

State

10 lb of the $4.40/lb lemon drops will be needed.
20 lb of the $3.80/lb licorice will be needed.

Check

The value of the lemon drops is 10($4.40), or $44.
The value of the licorice is 20($3.80), or $76.
The value of the blend is 30($4), or $120.
Since the total was $44 + $76 = $120, the answers check.

28. Solve: $8(4+x) > 10(6+x)$ [Section 2.7]

$$8(4+x) > 10(6+x)$$
$$\mathbf{8}(4) + \mathbf{8}(x) > \mathbf{10}(6) + \mathbf{10}(x)$$
$$32 + 8x > 60 + 10x$$
$$32 + 8x - \mathbf{32} > 60 + 10x - \mathbf{32}$$
$$8x > 10x + 28$$
$$8x - \mathbf{10x} > 10x + 28 - \mathbf{10x}$$
$$-2x > 28$$
$$\frac{-2x}{\mathbf{-2}} < \frac{28}{\mathbf{-2}}$$
$$x < -14$$
$$(-\infty, -14)$$

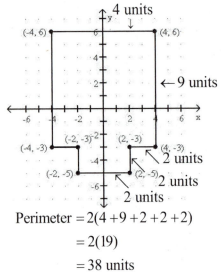

29. NEW YORK [Section 3.1]
(Ninth Avenue, 44th Street)

30. PERIMETER [Section 3.1]

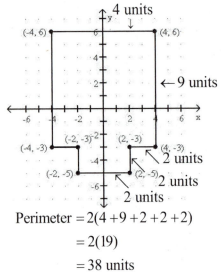

$$\text{Perimeter} = 2(4+9+2+2+2)$$
$$= 2(19)$$
$$= 38 \text{ units}$$

The perimeter is 38 units.

31. In what quadrant does $(-3.5, 6)$ lie?
[Section 3.1] **Quadrant II**

32. Is $(-2, 8)$ a solution of $y = -2x + 3$?
[Section 3.2]

$$y = -2x + 3$$
$$\overset{?}{8} = -2(-2) + 3$$
$$\overset{?}{8} = 4 + 3$$
$$8 = 7 \text{ False}$$
$$(-2, 8) \text{ is not a solution.}$$

Graph each equation.

33. $x = 4$ [Section 3.3]

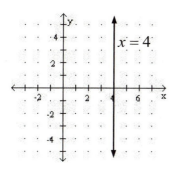

34. $4x - 3y = 12$ [Section 3.3]

y-intercept: | x-intercept:
If $x = 0$, | If $y = 0$
$4(\mathbf{0}) - 3y = 12$ | $4x - 3(\mathbf{0}) = 12$
$-3y = 12$ | $4x = 12$
$y = -4$ | $x = 3$

The y-intercept is $(0, -4)$, and the x-intercept is $(3, 0)$.

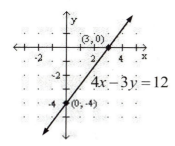

Find the slope of the line with the given properties.

35. Passing through $(-2, 4)$ and $(6, 8)$ [Section 3.4]

$(-2, 4)$ and $(6, 8)$
(x_1, y_1) and (x_2, y_2)

$m = \dfrac{y_2 - y_1}{x_2 - x_1}$

$\quad = \dfrac{8 - 4}{6 - (-2)}$

$\quad = \dfrac{4}{6 + 2}$

$\quad = \dfrac{4}{8}$

$\quad = \dfrac{1}{2}$

36. A line that is horizontial [Section 3.4]

$m = 0$

37. An equation of $2x - 3y = 12$ [Section 3.5]

$2x - 3y = 12$
$2x - 3y - \mathbf{2x} = 12 - \mathbf{2x}$
$-3y = -2x + 12$
$\dfrac{-3y}{-3} = \dfrac{-2x}{-3} + \dfrac{12}{-3}$
$y = \dfrac{2}{3}x - 4$

The slope is $\dfrac{2}{3}$.

38. Are the graphs of the lines parallel or perpendicular? [Section 3.5]

$y = -\dfrac{3}{4}x + \dfrac{15}{4}$ | $4x - 3y = 25$
 | $4x - 3y - \mathbf{4x} = 25 - \mathbf{4x}$
$m = -\dfrac{3}{4}$ | $-3y = -4x + 25$
 | $\dfrac{-3y}{-3} = \dfrac{-4x}{-3} + \dfrac{25}{-3}$
 | $y = \dfrac{4}{3}x - \dfrac{25}{3}$
 | $m = \dfrac{4}{3}$

The slopes are negative reciprocals. The lines are perpendicular.

39. Find the slope and the x- and y-intercepts of the line.

The slope is 2.
The x-intercept is $(-1, 0)$.
The y-intercept is $(0, 2)$.

40. NEWSPAPERS [Section 3.5]

$A(1999, \ 44\%)$ and $B(2009, \ 28\%)$
$\quad (x_1, y_1) \qquad\qquad (x_2, y_2)$

$m = \dfrac{y_2 - y_1}{x_2 - x_1}$

$m = \dfrac{28\% - 44\%}{2009 - 1999}$

$m = \dfrac{-16\%}{10}$

$m = -1.6\%$

There is a decrease of 1.6% per year.

Cumulative Review, Chapters 1 - 4

Find an equation of the line with the following properties. Write the equation in slope - intercept form.

41. Slope $= \dfrac{2}{3}$, y-intercept $= (0,5)$
    [Section 3.5]

    $y = mx + b$

    $y = \dfrac{2}{3}x + 5$

42. Passing through $(-2,4)$ and $(6,10)$
    [Section 3.6]

    Step 1: $(-2,4)$ and $(6,10)$
    $\qquad (x_1, y_1)$ and (x_2, y_2)

    $m = \dfrac{y_2 - y_1}{x_2 - x_1}$

    $\quad = \dfrac{10 - 4}{6 - (-2)}$

    $\quad = \dfrac{6}{6 + 2}$

    $\quad = \dfrac{6}{8}$

    $\quad = \dfrac{3}{4}$

    Step 2: $y - y_1 = m(x - x_1)$

    $y - 10 = \dfrac{3}{4}(x - 6)$

    $y - 10 = \dfrac{3}{4}x - \dfrac{3}{4}(6)$

    $y - 10 = \dfrac{3}{4}x - \dfrac{9}{2}$

    $y - 10 + \mathbf{10} = \dfrac{3}{4}x - \dfrac{9}{2} + \dfrac{\mathbf{20}}{\mathbf{2}}$

    $y = \dfrac{3}{4}x + \dfrac{11}{2}$

43. A horizontial line passing through $(2,4)$
    [Section 3.6]

    $y = 4$

44. Graph: $y < \dfrac{x}{3} - 1$ [Section 3.7]

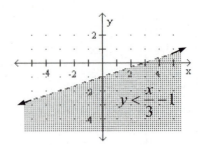

45. If $f(x) = -2x^2 - 3x^3$, find $f(-1)$.
    [Section 3.8]

    $f(x) = -2x^2 - 3x^3$

    $f(-1) = -2(-1)^2 - 3(-1)^3$

    $f(-1) = -2(1) - 3(-1)$

    $f(-1) = -2 + 3$

    $f(-1) = 1$

46. Is this a graph of a function?
    [Section 3.8]

    No ; $(1, \; 2)$ and $(1, \; -2)$
    There are other points.

Solve each system by graphing.

47. $\begin{cases} x + 4y = -2 \\ y = -x - 5 \end{cases}$ [Section 4.1]

    The solution is $(-6, \; 1)$

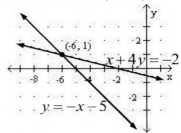

48. $\begin{cases} 2x - 3y < 0 \\ y > x - 1 \end{cases}$ [Section 4.5]

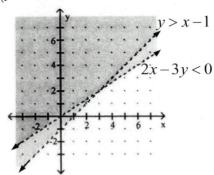

49. Solve $\begin{cases} x - 2y = 2 \\ 2x + 3y = 11 \end{cases}$ by substitution

[Section 4.2]

$\begin{cases} x - 2y = 2 \\ 2x + 3y = 11 \end{cases}$

Solve the 1st equation for x.

$\begin{cases} x = 2y + 2 \\ 2x + 3y = 11 \end{cases}$

$2x + 3y = 11$ The 2nd equation.
                  Substitute for x.

$2(2y + 2) + 3y = 11$

$4y + 4 + 3y = 11$

$7y + 4 = 11$

$7y + 4 - 4 = 11 - 4$

$7y = 7$

$\dfrac{7y}{7} = \dfrac{7}{7}$

$y = 1$

$x = 2y + 2$ The 1st equation

$x = 2(1) + 2$

$x = 4$

The solution is $(4, 1)$.

50. $\begin{cases} \dfrac{3}{2}x - \dfrac{2}{3}y = 0 \\ \dfrac{3}{4}x + \dfrac{4}{3}y = \dfrac{5}{2} \end{cases}$

Clear both equations of fractions.

$\begin{cases} 6\left(\dfrac{3}{2}x\right) - 6\left(\dfrac{2}{3}y\right) = 6(0) \\ 12\left(\dfrac{3}{4}x\right) + 12\left(\dfrac{4}{3}y\right) = 12\left(\dfrac{5}{2}\right) \end{cases}$

$\begin{cases} 9x - 4y = 0 \\ 9x + 16y = 30 \end{cases}$

Eliminate x.

Multiply both sides of the 1st equation by -1.

$\begin{aligned} -9x + 4y &= 0 \\ 9x + 16y &= 30 \\ \hline 20y &= 30 \end{aligned}$

$\dfrac{20y}{20} = \dfrac{30}{20}$

$y = \dfrac{3}{2}$

$9x - 4y = 0$ The 1st equation

$9x - 4\left(\dfrac{3}{2}\right) = 0$

$9x - 6 = 0$

$9x - 6 + 6 = 0 + 6$

$9x = 6$

$\dfrac{9x}{9} = \dfrac{6}{9}$

$x = \dfrac{2}{3}$

The solution is $\left(\dfrac{2}{3}, \dfrac{3}{2}\right)$.

51. NUTRITION [Section 4.4]

Analyze
- Protein in 1 serving of egg noodles: 5 g
- Protein in 1 serving of rice pilaf: 4 g
- Wants to consume only 22 g of protein.
- Fat in 1 serving of egg noodles: 3 g
- Fat in 1 serving of rice pilaf: 5 g
- Wants to consume only 21 g of fat.

Assign

Let x = # of servings of egg noodles
y = # of servings of rice pilaf

Form

	Grams •	Servings =	Total Protein
Noodles	5	x	$5x$
Rice	4	y	$4y$
		Total	22

The grams of protein in egg noodles	plus	the grams of protein in rice pilaf	is	22 grams.
$5x$	$+$	$4y$	$=$	22

	Grams •	Servings =	Total Fat
Noodles	3	x	$3x$
Rice	5	y	$5y$
		Total	21

The grams of fat in egg noodles	plus	the grams of fat in rice pilaf	is	21 grams.
$3x$	$+$	$5y$	$=$	21

Solve

$$\begin{cases} 5x + 4y = 22 \\ 3x + 5y = 21 \end{cases}$$

Eliminate x.

Multiply both sides of the 1st equation by -3.
Multiply both sides of the 2nd equation by 5.

$$-15x - 12y = -66$$
$$\underline{15x + 25y = \ 105}$$
$$13y = 39$$
$$\frac{13y}{13} = \frac{39}{13}$$
$$y = 3$$

$$5x + 4y = 22 \quad \text{The } 1^{st} \text{ equation}$$
$$5x + 4(\mathbf{3}) = 22$$
$$5x + 12 - \mathbf{12} = 22 - \mathbf{12}$$
$$\frac{5x}{5} = \frac{10}{5}$$
$$x = 2$$

State
2 servings of egg noodles.
3 servings of rice pilaf.

Check
The results check.

52. INVESTMENT CLUBS

Analyze
- $8,000 to invest.
- Invested part at 10% annual interest.
- The rest at 12% annual interest.
- $900 in interest was earned the first year.
- How much was invested at each rate?

Assign

Let x = amount invested at 10%
y = amount invested at 12%

Form

The amount invested at 10%	plus	the amount invested at 12%	is	$8,000.
x	$+$	y	$=$	8,000

	Principal •	Rate •	Time =	Interest
10% part	x	0.10	1 yr	$0.10x$
12% part	y	0.12	1 yr	$0.12y$
			Total Interest =	$900

The amount of interest of 10% investment	plus	the amount of interest of 12% investment	is	$900.
$0.10x$	$+$	$0.12y$	$=$	900

Solve

$$\begin{cases} x + y = 8,000 \\ 0.10x + 0.12y = 900 \end{cases}$$

Eliminate x.
Multiply both sides of the 1st equation by -10.
Multiply both sides of the 2nd equation by 100.

$$-10x - 10y = -80,000$$
$$\underline{10x + 12y = \ \ 90,000}$$
$$2y = \ \ 10,000$$
$$\frac{2y}{2} = \frac{10,000}{2}$$
$$y = 5,000$$

$$x + y = 8,000 \quad \text{The } 1^{st} \text{ equation}$$
$$x + \mathbf{5,000} = 8,000$$
$$x + 5,000 - \mathbf{5,000} = 8,000 - \mathbf{5,000}$$
$$x = 3,000$$

State
$3,000 was invested at 10%.
$5,000 was invested at 12%.

Check
The results check.

SECTION 5.1

VOCABULARY

Fill in the blanks.

1. Expressions such as x^4, 10^3, and $(5t)^2$ are called **exponential** expressions.

CONCEPTS

Fill in the blanks.

3. a. $(3x)^4 = \mathbf{3x \cdot 3x \cdot 3x \cdot 3x}$

   b. $(-5y)(-5y)(-5y) = \mathbf{(-5y)^3}$

5. To simplify each expression, determine whether you add, subtract, multiply, or divide the exponents.

   a. $\dfrac{x^8}{x^2}$ **Subtract**   b. $b^6 \cdot b^9$ **Add**

   c. $(n^8)^4$ **Multiply**   d. $(a^4 b^2)^5$ **Multiply**

Simplify each expression, if possible.

7. a. $x^2 + x^2 = (1+1)x^2$
   $= \mathbf{2x^2}$

   b. $x^2 \cdot x^2 = x^{2+2}$
   $= \mathbf{x^4}$

   c. $x^2 + x$, Doesn't simplify.

   d. $x^2 \cdot x = x^{2+1}$
   $= \mathbf{x^3}$

NOTATION

Complete each solution to simplify each expression.

9. $\left(x^4 x^2\right)^3 = \left(x^6\right)^3 = x^{\mathbf{18}}$

11. a. We read 9^4 as "nine to the fourth **power**."

   b. We read $(a^2 b^6)(a^4 b^5)$ as "the **quantity** of of $a^2 b^6$ times the **quantity** of $a^4 b^5$."

   c. We read $(3^6)^9$ as "3 to the **sixth** power, raised to the **ninth** power."

GUIDED PRACTICE

Identify the base and the exponent in each expression. See Example 1.

Look Alikes . . .

13. a. 4^3; 4, 3   b. -4^3; 4, 3   c. $(-4)^3$; -4, 3

15. a. $(-3x)^2$; $-3x$, 2   b. -3^2; x, 2

   c. $-(-3x)^2$; $-3x$, 2

17. a. $9m^{12}$; m, 12   b. $(9m)^{12}$; $9m$, 12

   c. $-9m^{12}$; , 12

Write each expression in an equivalent form using an exponent. See Example 2.

19. $4t \cdot 4t \cdot 4t \cdot 4t = (4t)^4$

21. $-4 \cdot t \cdot t \cdot t \cdot t \cdot t = -4t^5$

23. $\dfrac{t}{2} \cdot \dfrac{t}{2} \cdot \dfrac{t}{2} = \left(\dfrac{t}{2}\right)^3$

25. $(x-y)(x-y) = (x-y)^2$

Use the product rule for exponents to simplify each expression. Write the results using exponents. See Example 3.

27. $5^3 \cdot 5^4 = 5^{3+4}$
   $= 5^7$

29. $bb^2 b^3 = b^{1+2+3}$
   $= b^6$

31. $(y-2)^5 (y-2)^2 = (y-2)^{5+2}$
   $= (y-2)^7$

33. $(a^2 b^3)(a^3 b^3) = (a^2 a^3)(b^3 b^3)$
   $= a^{2+3} b^{3+3}$
   $= a^5 b^6$

Find an expression that represents the area or volume of each figure. Recall that the formula for the volume of a rectangular solid is $V = $ length $\bullet$ width $\bullet$ height. See Example 4.

35. $A = lw$
   $= a^5 \cdot a^5$
   $= a^{5+5}$
   $= a^{10}$

   The area is a^{10} mi^2.

37. $V = lwh$
   $= x^4 \cdot x^3 \cdot x^2$
   $= x^{4+3+2}$
   $= x^9$

   The volume is x^9 ft^3.

Use the quotient rule for exponents to simplify each expression. Write the results using exponents. See Example 5.

39. $\dfrac{8^{12}}{8^4} = 8^{12-4}$
   $= 8^8$

41. $\dfrac{x^{15}}{x^3} = x^{15-3}$
   $= x^{12}$

43. $\dfrac{(3.7p)^7}{(3.7p)^2} = (3.7p)^{7-2}$
   $= (3.7p)^5$

45. $\dfrac{c^3 d^7}{cd} = \dfrac{c^3}{c} \cdot \dfrac{d^7}{d}$
   $= c^{3-1} d^{7-1}$
   $= c^2 d^6$

Use the product and quotient rules for exponents to simplify each expression. See Example 6.

47. $\dfrac{y^3 y^4}{y y^2} = \dfrac{y^{3+4}}{y^{1+2}}$

$= \dfrac{y^7}{y^3}$

$= y^{7-3}$

$= y^4$

49. $\dfrac{a^2 a^3 a^4}{a^8} = \dfrac{a^{2+3+4}}{a^8}$

$= \dfrac{a^9}{a^8}$

$= a^{9-8}$

$= a^1$

$= a$

Use the power rule for exponents to simplify each expression. Write the results using exponents. See Example 7.

51. $\left(3^2\right)^4 = 3^{2 \cdot 4}$

$= 3^8$

53. $\left[(-4.3)^3\right]^8 = (-4.3)^{3 \cdot 8}$

$= (-4.3)^{24}$

55. $\left(m^{50}\right)^{10} = m^{50 \cdot 10}$

$= m^{500}$

57. $\left(y^5\right)^3 = y^{5 \cdot 3}$

$= y^{15}$

Use the product and power rules for exponents to simplify each expression. See Example 8.

59. $\left(x^2 x^3\right)^5 = (x^{2+3})^5$

$= (x^5)^5$

$= x^{5 \cdot 5}$

$= x^{25}$

61. $\left(p^2 p^3\right)^5 = \left(p^{2+3}\right)^5$

$= \left(p^5\right)^5$

$= p^{5 \cdot 5}$

$= p^{25}$

63. $\left(t^3\right)^4 \left(t^2\right)^3 = \left(t^{3 \cdot 4}\right)\left(t^{2 \cdot 3}\right)$

$= \left(t^{12}\right)\left(t^6\right)$

$= t^{12+6}$

$= t^{18}$

65. $\left(u^4\right)^2 \left(u^3\right)^2 = \left(u^{4 \cdot 2}\right)\left(u^{3 \cdot 2}\right)$

$= \left(u^8\right)\left(u^6\right)$

$= u^{8+6}$

$= u^{14}$

Use the power of a product rule for exponents to simplify each expression. See Example 8.

67. $(6a)^2 = 6^2 a^2$

$= 36a^2$

69. $(5y)^4 = 5^4 y^4$

$= 625y^4$

71. $\left(-2r^2 s^3\right)^3 = (-2)^3 \left(r^2\right)^3 \left(s^3\right)^3$

$= -8 r^{2 \cdot 3} s^{3 \cdot 3}$

$= -8 r^6 s^9$

73. $\left(-\dfrac{1}{3} y^2 z^4\right)^5 = \left(-\dfrac{1}{3}\right)^5 \left(y^2\right)^5 \left(z^4\right)^5$

$= -\dfrac{1}{243} y^{2 \cdot 5} z^{4 \cdot 5}$

$= -\dfrac{1}{243} y^{10} z^{20}$

Use rules for exponents to simplify each expression. See Example 10.

75. $\dfrac{\left(ab^2\right)^3}{a^2 b^2} = \dfrac{\left(a^1\right)^3 \left(b^2\right)^3}{a^2 b^2}$

$= \dfrac{\left(a^{1 \cdot 3}\right)\left(b^{2 \cdot 3}\right)}{a^2 b^2}$

$= \dfrac{a^3 b^6}{a^2 b^2}$

$= a^{3-2} b^{6-2}$

$= a^1 b^4$

$= ab^4$

77. $\dfrac{\left(r^4 s^3\right)^4}{r^3 s^9} = \dfrac{\left(r^4\right)^4 \left(s^3\right)^4}{r^3 s^9}$

$= \dfrac{\left(r^{4 \cdot 4}\right)\left(s^{3 \cdot 4}\right)}{r^3 s^9}$

$= \dfrac{r^{16} s^{12}}{r^3 s^9}$

$= r^{16-3} s^{12-9}$

$= r^{13} s^3$

Use rules for exponents to simplify each expression. See Example 11.

79. $\dfrac{(6k)^7}{(6k)^4} = (6k)^{7-4}$

$= (6k)^3$

$= 6^3 k^3$

$= 216k^3$

81. $\dfrac{(3q)^5}{(3q)^3} = (3q)^{5-3}$

$= (3q)^2$

$= 3^2 q^2$

$= 9q^2$

Use the power of a quotient rule for exponents to simplify each expression. See Example 12.

83. $\left(\dfrac{a}{b}\right)^3 = \dfrac{a^3}{b^3}$

85. $\left(\dfrac{8a^2}{11b^5}\right)^2 = \dfrac{\left(8a^2\right)^2}{\left(11b^5\right)^2}$

$= \dfrac{(8)^2\left(a^2\right)^2}{(11)^2\left(b^5\right)^2}$

$= \dfrac{8^2 a^{2\cdot2}}{11^2 b^{5\cdot2}}$

$= \dfrac{64a^4}{121b^{10}}$

TRY IT YOURSELF
Simplify each expression, if possible

87. $\left(\dfrac{x^2}{y^3}\right)^5 = \dfrac{\left(x^2\right)^5}{\left(y^3\right)^5}$

$= \dfrac{x^{2\cdot5}}{y^{3\cdot5}}$

$= \dfrac{x^{10}}{y^{15}}$

89. $y^3 y^2 y^4 = y^{3+2+4}$

$= y^9$

91. $\dfrac{15^9}{15^6} = 15^{9-6}$

$= 15^3$

93. $\dfrac{t^5 t^6 t}{t^2 t^3} = \dfrac{t^5 t^6 t^1}{t^2 t^3}$

$= \dfrac{t^{5+6+1}}{t^{2+3}}$

$= \dfrac{t^{12}}{t^5}$

$= t^{12-5}$

$= t^7$

95. $\dfrac{(k-2)^{15}}{(k-2)} = \dfrac{(k-2)^{15}}{(k-2)^1}$

$= (k-2)^{15-1}$

$= (k-2)^{14}$

97. $cd^4 \cdot cd = cc \cdot dd^4$

$= c^{1+1} \cdot d^{1+4}$

$= c^2 d^5$

99. $\left(\dfrac{y^3 y^5}{yy^2}\right)^3 = \left(\dfrac{y^{3+5}}{y^{1+2}}\right)^3$

$= \left(\dfrac{y^8}{y^3}\right)^3$

$= \left(y^{8-3}\right)^3$

$= \left(y^5\right)^3$

$= y^{5\cdot3}$

$= y^{15}$

101. $\dfrac{s^2 s^2 s^2}{s^3 s} = \dfrac{s^2 s^2 s^2}{s^3 s^1}$

$= \dfrac{s^{2+2+2}}{s^{3+1}}$

$= \dfrac{s^6}{s^4}$

$= s^{6-4}$

$= s^2$

103. $\left(-6a^3 b^2\right)^3 = (-6)^3\left(a^3\right)^3\left(b^2\right)^3$

$= -216a^{3\cdot3}b^{2\cdot3}$

$= -216a^9 b^6$

105. $\left(\dfrac{3m^4}{2n^5}\right)^5 = \dfrac{\left(3m^4\right)^5}{\left(2n^5\right)^5}$

$= \dfrac{(3)^5\left(m^4\right)^5}{(2)^5\left(n^5\right)^5}$

$= \dfrac{3^5 m^{4\cdot5}}{2^5 n^{5\cdot5}}$

$= \dfrac{243m^{20}}{32n^{25}}$

107. $\dfrac{(a^2 b^2)^{15}}{(ab)^9} = \dfrac{\left(a^2\right)^{15}\left(b^2\right)^{15}}{\left(a^1\right)^9\left(b^1\right)^9}$

$= \dfrac{\left(a^{2\cdot15}\right)\left(b^{2\cdot15}\right)}{\left(a^{1\cdot9}\right)\left(b^{1\cdot9}\right)}$

$= \dfrac{a^{30}b^{30}}{a^9 b^9}$

$= a^{30-9}b^{30-9}$

$= a^{21}b^{21}$

Section 5.1

109.
$$\left(n^4 n\right)^3 \left(n^3\right)^6 = \left(n^{4+1}\right)^3 \left(n^3\right)^6$$
$$= \left(n^5\right)^3 \left(n^3\right)^6$$
$$= \left(n^{5\cdot3}\right)\left(n^{3\cdot6}\right)$$
$$= \left(n^{15}\right)\left(n^{18}\right)$$
$$= n^{15+18}$$
$$= n^{33}$$

111.
$$\frac{(6h)^8}{(6h)^6} = (6h)^{8-6}$$
$$= (6h)^2$$
$$= 6^2 h^2$$
$$= 36h^2$$

113.
$$\frac{x^4 y^7}{xy^3} = \frac{x^4}{x^1} \cdot \frac{y^7}{y^3}$$
$$= x^{4-1} y^{7-3}$$
$$= x^3 y^4$$

115.
$$\left(\frac{m}{3}\right)^4 = \frac{m^4}{3^4}$$
$$= \frac{m^4}{81}$$

(All new) LOOK ALIKES ...

117. a. $a^3 \cdot a^3 = a^{3+3}$
$$= a^6$$
b. $\left(a^3\right)^3 = a^{3\cdot3}$
$$= a^9$$
c. $a^3 + a^3 = 1a^3 + 1a^3$
$$= 2a^3$$

119. a. $b^3 b^2 b^4 = b^{3+2+4}$
$$= b^9$$
b. $\left(b^3 b^2\right)^4 = \left(b^{3+2}\right)^4$
$$= \left(b^5\right)^4$$
$$= b^{5\cdot4}$$
$$= b^{20}$$
c. $\dfrac{b^3 b^2}{b^4} = \dfrac{b^{3+2}}{b^4}$
$$= \frac{b^5}{b^4}$$
$$= b^{5-4}$$
$$= b$$

APPLICATIONS

121. ART HISTORY

a. $A = lw$
$$= (5x)(5x)$$
$$= 25x^{1+1}$$
$$= 25x^2$$
The area of the square is $25x^2$ ft^2.

b. $A = \pi r^2$
$$= (3a)(3a)\pi$$
$$= 9a^{1+1}\pi$$
$$= 9a^2 \pi$$
The area of the circle is $9a^2 \pi$ ft^2.

123. CHILDBIRTH
$$\left(\frac{1}{2}\right)^{13} = \frac{1^{13}}{2^{13}}$$
$$= \frac{1}{8,192}$$
The probability is $\dfrac{1}{8,192}$.

WRITING
125. Answers will vary.

REVIEW
Match each equation with its graph below.
127. $y = 2x - 1$ **c**   **129.** $y = 3$ **d**

CHALLENGE PROBLEMS
131. Simplify each expression. The variables represent natural numbers.

a. $x^{2m} x^{3m} = x^{2m+3m}$
$$= x^{5m}$$
b. $\left(y^{5c}\right)^4 = y^{5c\cdot4}$
$$= y^{20c}$$
c. $\dfrac{m^{8x}}{m^{4x}} = m^{8x-4x}$
$$= m^{4x}$$
d. $\left(2a^{6y}\right)^4 = \left(2^1\right)^4 \left(a^{6y}\right)^4$
$$= \left(2^{1\cdot4}\right)\left(a^{6y\cdot4}\right)$$
$$= 2^4 a^{24y}$$
$$= 16a^{24y}$$

SECTION 5.2

VOCABULARY

Fill in the blanks.

1. In the expression 5^{-1}, the exponent is a **negative** integer.

3. We read a^0 as "a to the **zero** power".

CONCEPTS

5. Complete the table.

Expression	Base	Exponent
4^{-2}	**4**	**−2**
$6x^{-5}$	**x**	**−5**
$\left(\frac{3}{y}\right)^{-8}$	$\frac{3}{y}$	**−8**
-7^{-1}	**7**	**−1**
$(-2)^{-3}$	**−2**	**−3**
$10a^0$	**a**	**0**

7. Complete the table.

x	3^x
2	$3^2 = 3\cdot 3$ $=\mathbf{9}$
1	$3^1 = \mathbf{3}$
0	$3^0 = \mathbf{1}$
−1	$3^{-1} = \dfrac{\mathbf{1}}{\mathbf{3}}$
−2	$3^{-2} = \dfrac{1}{3^2}$ $=\dfrac{1}{3\cdot 3}$ $=\dfrac{1}{9}$

9. Fill in the blanks.

a. $2^{-3} = \dfrac{1}{2^{\mathbf{3}}}$

b. $\dfrac{1}{t^{-6}} = t^{\mathbf{6}}$

c. A factor can be moved from the denominator to the numerator or from the numerator to the denominator of a fraction if the **sign** of its exponent is changed.

$$\frac{5^{-2}}{6^{-3}} = \frac{6^{\mathbf{3}}}{5^{\mathbf{2}}}$$

d. A fraction raised to a power is equal to the **reciprocal** of the fraction raise to the opposite power.

$$\left(\frac{3}{d}\right)^{-2} = \left(\frac{d}{3}\right)^{\mathbf{2}}$$

NOTATION

Complete each solution to simplify each expression.

11. $(y^5 y^3)^{-5} = (y^{\mathbf{8}})^{-5}$

$= y^{-40}$

$= \dfrac{1}{y^{40}}$

GUIDED PRACTICE

Simplify each expression. See Example 1.

13. $7^0 = 1$

15. $\left(\dfrac{1}{4}\right)^0 = 1$

17. $2x^0 = 2\cdot 1$ $= 2$

19. $\dfrac{5}{2x^0} = \dfrac{5}{2\cdot 1}$ $= \dfrac{5}{2}$

Express using positive exponents and simplify, if possible. See Example 2.

21. $2^{-2} = \dfrac{1}{2^2}$ $= \dfrac{1}{4}$

23. $6^{-1} = \dfrac{1}{6^1}$ $= \dfrac{1}{6}$

25. $b^{-5} = \dfrac{1}{b^5}$

27. $(-5)^{-1} = \dfrac{1}{(-5)^1}$ $= -\dfrac{1}{5}$

29. $2^{-2} + 4^{-1} = \dfrac{1}{2^2} + \dfrac{1}{4^1}$ $= \dfrac{1}{4} + \dfrac{1}{4}$ $= \dfrac{2}{4}$ $= \dfrac{1}{2}$

31. $9^0 - 9^{-1} = 1 - \dfrac{1}{9^1}$ $= \dfrac{9}{9} - \dfrac{1}{9}$ $= \dfrac{8}{9}$

Simplify. Do not use negative exponents in the answer. See Example 3.

33. $15g^{-6} = 15\cdot g^{-6}$ $= 15\cdot \dfrac{1}{g^6}$ $= \dfrac{15}{g^6}$

35. $5x^{-3} = 5\cdot x^{-3}$ $= 5\cdot \dfrac{1}{x^3}$ $= \dfrac{5}{x^3}$

Section 5.2

37. $-3^{-3} = -1 \cdot 3^{-3}$

$\quad\quad = -1 \cdot \dfrac{1}{3^3}$

$\quad\quad = -\dfrac{1}{27}$

39. $-8^{-2} = -1 \cdot 8^{-2}$

$\quad\quad = -1 \cdot \dfrac{1}{8^2}$

$\quad\quad = -\dfrac{1}{64}$

Simplify. Do not use negative exponents in the answer. See Example 4.

41. $\dfrac{1}{5^{-3}} = 5^3$

$\quad\quad = 125$

43. $\dfrac{8}{s^{-1}} = 8s$

45. $\dfrac{2^{-4}}{3^{-1}} = \dfrac{3^1}{2^4}$

$\quad\quad = \dfrac{3}{16}$

47. $\dfrac{-4d^{-1}}{p^{-10}} = \dfrac{-4p^{10}}{d^1}$

$\quad\quad = -\dfrac{4p^{10}}{d}$

Simplify. See Example 5.

49. $\left(\dfrac{1}{6}\right)^{-2} = \left(\dfrac{6^1}{1^1}\right)^2$

$\quad\quad = \dfrac{6^{1 \cdot 2}}{1^{1 \cdot 2}}$

$\quad\quad = \dfrac{6^2}{1}$

$\quad\quad = 36$

51. $\left(\dfrac{1}{2}\right)^{-3} = \left(\dfrac{2^1}{1^1}\right)^3$

$\quad\quad = \dfrac{2^{1 \cdot 3}}{1^{1 \cdot 3}}$

$\quad\quad = \dfrac{2^3}{1}$

$\quad\quad = 8$

53. $\left(\dfrac{c}{d}\right)^{-8} = \left(\dfrac{d^1}{c^1}\right)^8$

$\quad\quad = \dfrac{d^{1 \cdot 8}}{c^{1 \cdot 8}}$

$\quad\quad = \dfrac{d^8}{c^8}$

55. $\left(\dfrac{3}{m}\right)^{-4} = \left(\dfrac{m^1}{3^1}\right)^4$

$\quad\quad = \dfrac{m^{1 \cdot 4}}{3^{1 \cdot 4}}$

$\quad\quad = \dfrac{m^4}{3^4}$

$\quad\quad = \dfrac{m^4}{81}$

Simplify. Do not use negative exponents in the answer. See Example 6.

57. $y^8 \cdot y^{-2} = y^{8+(-2)}$

$\quad\quad = y^6$

59. $b^{-7} \cdot b^{14} = b^{-7+14}$

$\quad\quad = b^7$

61. $\dfrac{y^4}{y^5} = y^{4-5}$

$\quad\quad = y^{-1}$

$\quad\quad = \dfrac{1}{y}$

63. $\dfrac{h^{-5}}{h^2} = h^{-5-2}$

$\quad\quad = h^{-7}$

$\quad\quad = \dfrac{1}{h^7}$

65. $(x^4)^{-3} = x^{-12}$

$\quad\quad = \dfrac{1}{x^{12}}$

67. $(b^2)^{-4} = b^{-8}$

$\quad\quad = \dfrac{1}{b^8}$

69. $\left(6s^4t^{-7}\right)^2 = \left(6^1 s^4 t^{-7}\right)^2$

$\quad\quad = 6^2 \left(s^4\right)^2 \left(t^{-7}\right)^2$

$\quad\quad = 36 s^{4 \cdot 2} t^{-7 \cdot 2}$

$\quad\quad = 36 s^8 t^{-14}$

$\quad\quad = \dfrac{36 s^8}{t^{14}}$

71. $\left(\dfrac{4}{x^3}\right)^{-3} = \left(\dfrac{x^3}{4^1}\right)^3$

$\quad\quad = \dfrac{\left(x^3\right)^3}{4^3}$

$\quad\quad = \dfrac{x^{3 \cdot 3}}{64}$

$\quad\quad = \dfrac{x^9}{64}$

Simplify. Do not use negative exponents in the answer. See Example 7.

73. $\dfrac{y^{-3}}{y^{-4}y^{-2}} = \dfrac{y^{-3}}{y^{-4+(-2)}}$

$\quad\quad = \dfrac{y^{-3}}{y^{-6}}$

$\quad\quad = y^{-3-(-6)}$

$\quad\quad = y^{-3+6}$

$\quad\quad = y^3$

75. $\dfrac{a^{-5}a^{-9}}{a^{-8}} = \dfrac{a^{-5+(-9)}}{a^{-8}}$

$\quad\quad = \dfrac{a^{-14}}{a^{-8}}$

$\quad\quad = a^{-14-(-8)}$

$\quad\quad = a^{-14+8}$

$\quad\quad = a^{-6}$

$\quad\quad = \dfrac{1}{a^6}$

77. $\dfrac{2^{-1}a^4b^2}{3^{-2}a^2b^4} = \dfrac{3^2 a^4 b^2}{2a^2 b^4}$

$= \dfrac{9a^{4-2}b^{2-4}}{2}$

$= \dfrac{9a^2 b^{-2}}{2}$

$= \dfrac{9a^2}{2b^2}$

79. $\left(\dfrac{y^3 z^{-2}}{y^{-4} z^3}\right)^2 = \left(y^{3-(-4)} z^{-2-3}\right)^2$

$= \left(y^{3+4} z^{-5}\right)^2$

$= \left(y^7 z^{-5}\right)^2$

$= y^{7\cdot 2} z^{-5\cdot 2}$

$= y^{14} z^{-10}$

$= \dfrac{y^{14}}{z^{10}}$

TRY IT YOURSELF

Simplify. Do not use negative exponents in the answer. Assume that no variables are 0.

81. $\left(\dfrac{a^4}{2b}\right)^{-3} = \left(\dfrac{2^1 b^1}{a^4}\right)^3$

$= \dfrac{\left(2^1 b^1\right)^3}{\left(a^4\right)^3}$

$= \dfrac{2^3 b^3}{a^{4\cdot 3}}$

$= \dfrac{8b^3}{a^{12}}$

83. $\dfrac{r^{-50}}{r^{-70}} = \dfrac{r^{70}}{r^{50}}$

$= r^{70-50}$

$= r^{20}$

85. $\dfrac{a^{-5}}{b^{-2}} = \dfrac{b^2}{a^5}$

87. $(-10)^{-3} = \dfrac{1}{(-10)^3}$

$= \dfrac{1}{(-10)(-10)(-10)}$

$= -\dfrac{1}{1,000}$

89. $\left(\dfrac{a^2 b^3}{ab^4}\right)^0 = 1$

91. $\dfrac{9^{-2} s^6 t}{4^{-3} s^4 t^5} = \dfrac{4^3 s^6 t^1}{9^2 s^4 t^5}$

$= \dfrac{64 s^{6-4} t^{1-5}}{81}$

$= \dfrac{64 s^2 t^{-4}}{81}$

$= \dfrac{64 s^2}{81 t^4}$

93. $\left(2u^{-2} v^5\right)^5 = \left(2^1 u^{-2} v^5\right)^5$

$= 2^5 \left(u^{-2}\right)^5 \left(v^5\right)^5$

$= 32 u^{-2\cdot 5} v^{5\cdot 5}$

$= 32 u^{-10} v^{25}$

$= \dfrac{32 v^{25}}{u^{10}}$

95. $\left(\dfrac{y^4}{3}\right)^{-2} = \left(\dfrac{3^1}{y^4}\right)^2$

$= \dfrac{3^2}{\left(y^4\right)^2}$

$= \dfrac{9}{y^{4\cdot 2}}$

$= \dfrac{9}{y^8}$

97. $-15x^0 y = -15 \cdot 1y$

$= -15y$

99. $\left(\dfrac{4}{h^{10}}\right)^{-2} = \left(\dfrac{h^{10}}{4^1}\right)^2$

$= \dfrac{h^{10\cdot 2}}{4^{1\cdot 2}}$

$= \dfrac{h^{10\cdot 2}}{4^2}$

$= \dfrac{h^{20}}{16}$

Section 5.2

101. $x^{-3} \cdot x^{-3} = x^{-3+(-3)}$

$\qquad = x^{-6}$

$\qquad = \dfrac{1}{x^6}$

103. $\left(\dfrac{c^3 d^{-4}}{c^{-1} d^5}\right)^3 = \left(c^{3-(-1)} d^{-4-5}\right)^3$

$\qquad = \left(c^{3+1} d^{-9}\right)^3$

$\qquad = \left(c^4 d^{-9}\right)^3$

$\qquad = c^{4 \cdot 3} d^{-9 \cdot 3}$

$\qquad = c^{12} d^{-27}$

$\qquad = \dfrac{c^{12}}{d^{27}}$

105. $15(-6x)^0 = 15 \cdot 1$

$\qquad = 15$

107. $\dfrac{2^{-2} g^{-2} h^{-3}}{9^{-1} h^{-3}} = \dfrac{9^1 h^{-3-(-3)}}{2^2 g^2}$

$\qquad = \dfrac{9 h^{-3+3}}{4 g^2}$

$\qquad = \dfrac{9 h^0}{4 g^2}$

$\qquad = \dfrac{9 \cdot 1}{4 g^2}$

$\qquad = \dfrac{9}{4 g^2}$

109. $\left(5d^{-2}\right)^3 = \left(5^1 d^{-2}\right)^3$

$\qquad = 5^3 \left(d^{-2}\right)^3$

$\qquad = 125 d^{-2 \cdot 3}$

$\qquad = 125 d^{-6}$

$\qquad = \dfrac{125}{d^6}$

111. $\left(\dfrac{x^2 y^{-2}}{x^{-5} y^3}\right)^4 = \left(x^{2-(-5)} y^{-2-3}\right)^4$

$\qquad = \left(x^{2+5} y^{-5}\right)^4$

$\qquad = \left(x^7 y^{-5}\right)^4$

$\qquad = x^{7 \cdot 4} y^{-5 \cdot 4}$

$\qquad = x^{28} y^{-20}$

$\qquad = \dfrac{x^{28}}{y^{20}}$

113. $\left(2x^3 y^{-2}\right)^5 = \left(2^1 x^3 y^{-2}\right)^5$

$\qquad = 2^5 \left(x^3\right)^5 \left(y^{-2}\right)^5$

$\qquad = 32 x^{3 \cdot 5} y^{-2 \cdot 5}$

$\qquad = 32 x^{15} y^{-10}$

$\qquad = \dfrac{32 x^{15}}{y^{10}}$

115. $\dfrac{t(t^{-2})^{-2}}{t^{-5}} = \dfrac{t^1 (t^{-2 \cdot -2})}{t^{-5}}$

$\qquad = \dfrac{t^1 (t^4)}{t^{-5}}$

$\qquad = \dfrac{t^{1+4}}{t^{-5}}$

$\qquad = \dfrac{t^5}{t^{-5}}$

$\qquad = t^{5-(-5)}$

$\qquad = t^{5+5}$

$\qquad = t^{10}$

117. $\dfrac{-4 s^{-5}}{t^{-2}} = \dfrac{-4 t^2}{s^5}$

$\qquad = -\dfrac{4 t^2}{s^5}$

119. $\left(x^{-4} x^3\right)^3 = \left(x^{-4+3}\right)^3$

$\qquad = \left(x^{-1}\right)^3$

$\qquad = x^{-1 \cdot 3}$

$\qquad = x^{-3}$

$\qquad = \dfrac{1}{x^3}$

121. a. $8^{-1} = \dfrac{1}{8}$ b. $-8^{-1} = -\dfrac{1}{8}$

c. $(-8)^{-1} = \dfrac{1}{(-8)^1}$ d. $-(-8)^{-1} = -\dfrac{1}{(-8)^1}$

$= -\dfrac{1}{8}$ $= \dfrac{1}{8}$

123. a. $4xy^{-2} = \dfrac{4x}{y^2}$ b. $(4xy)^{-2} = \dfrac{1}{(4xy)^2}$

$= \dfrac{1}{16x^2 y^2}$

c. $4x^{-2}y = \dfrac{4y}{x^2}$ d. $4^{-2}xy = \dfrac{xy}{16}$

APPLICATIONS
from Campus to Careers

125. SOUND ENGINEERING TECHNICIAN

Type of sound	Intensity
Front row rock concert	10^{-1}
Normal conversation	10^{-6}
Vacuum cleaner	10^{-4}
Military jet takeoff	10^{2}
Whisper	10^{-10}

WRITING
127. Answers will vary.

REVIEW
Find the slope of the line that passes through the given points.

129. Let $(x_1, y_1) = (1, -4)$
Let $(x_2, y_2) = (3, -7)$

$m = \dfrac{y_2 - y_1}{x_2 - x_1}$

$= \dfrac{-7 - (-4)}{3 - 1}$

$= \dfrac{-7 + 4}{2}$

$= \dfrac{-3}{2}$

$= -\dfrac{3}{2}$

131. Write an equation of the line having slope $\dfrac{3}{4}$ and y-intercept -5.

$y = mx + b$

$y = \dfrac{3}{4}x - 5$

CHALLENGE PROBLEMS
133. **Simplify each expression. Do not use negative exponents in the answer. The variable m represents a positive integer.**

a. $r^{5m} r^{-6m} = r^{5m + (-6m)}$

$= r^{-m}$

$= \dfrac{1}{r^m}$

b. $\dfrac{x^{3m}}{x^{6m}} = x^{3m - (6m)}$

$= x^{3m + (-6m)}$

$= x^{-3m}$

$= \dfrac{1}{x^{3m}}$

SECTION 5.3
VOCABULARY
Fill in the blanks.

1. 4.84×10^5 is written in **scientific** notation.
484,000 is written in **standard** notation.

CONCEPTS
Fill in the blanks.

3. When we multiply a decimal by 10^5, the decimal point moves 5 places to the **right**. When we multiply a decimal by 10^{-7}, the decimal point moves 7 places to the **left**.

5. a. When a real number greater than 10 is written in scientific notation, the exponent on 10 is a **positive** integer.

   b. When a real number between 0 and 1 is written in scientific notation, the exponent on 10 is a **negative** integer.

Fill in the blanks to write each number in scientific notation.

7. a. $7,700 = \mathbf{7.7} \times 10^3$

   b. $500,000 = \mathbf{5.0} \times 10^5$

   c. $114,000,000 = 1.44 \times 10^{\mathbf{8}}$

9. Write each expression so that the decimal numbers are grouped together and the powers of ten are grouped together.

   a. $\left(5.1 \times 10^9\right)\left(1.5 \times 10^{22}\right) = \left(\mathbf{5.1 \times 1.5}\right)\left(\mathbf{10^9 \times 10^{22}}\right)$

   b. $\dfrac{8.8 \times 10^{30}}{2.2 \times 10^{19}} = \dfrac{\mathbf{8.8}}{\mathbf{2.2}} \times \dfrac{\mathbf{10^{30}}}{\mathbf{10^{19}}}$

NOTATION

11. Fill in the blanks. A positive number is written in scientific notation when it is written in the form $N \times 10^n$, where $\mathbf{1 \le N < 10}$ and n is an **integer**.

GUIDED PRACTICE
Convert each number to standard notation. See Example 1.

13. The exponent is 2.
Move the decimal 2 places to the right.
$$2.3 \times 10^2 = 230$$

15. The exponent is 5.
Move the decimal 5 places to the right.
$$8.12 \times 10^5 = 812,000$$

17. The exponent is -3.
Move the decimal 3 places to the left.
$$1.15 \times 10^{-3} = 0.00115$$

19. The exponent is -4.
Move the decimal 4 places to the left.
$$9.76 \times 10^{-4} = 0.000976$$

21. The exponent is 6.
Move the decimal 6 places to the right.
$$6.001 \times 10^6 = 6,001,000$$

23. The exponent is 0.
Do not move the decimal.
$$2.718 \times 10^0 = 2.718$$

25. The exponent is -2.
Move the decimal 2 places to the left.
$$6.798 \times 10^{-2} = 0.06798$$

27. The exponent is -5.
Move the decimal 5 places to the left.
$$2.0 \times 10^{-5} = 0.00002$$

Write each number in scientific notation. See Example 2.

29. Move the decimal 4 places to the left.
$$23,000 = 2.3 \times 10^4$$

31. Move the decimal 6 places to the left.
$$1,700,000 = 1.7 \times 10^6$$

33. Move the decimal 2 places to the right.
$$0.062 = 6.2 \times 10^{-2}$$

35. Move the decimal 6 places to the right.
$$0.0000051 = 5.1 \times 10^{-6}$$

37. Move the decimal 9 places to the left.
$$5,000,000,000 = 5.0 \times 10^9$$

39. Move the decimal 7 places to the right.
$$0.0000003 = 3.0 \times 10^{-7}$$

41. Move the decimal 8 places to the left.
$$909,000,000 = 9.09 \times 10^8$$

43. Move the decimal 2 places to the right.
$$0.0345 = 3.45 \times 10^{-2}$$

45. Do not move the decimal.
$$9 = 9.0 \times 10^0$$

47. Move the decimal 1 place to the left.
$$11 = 1.1 \times 10^1$$

49. Move the decimal 18 places to the left.
$$1,718,000,000,000,000,000 = 1.718 \times 10^{18}$$

51. Move the decimal 14 places to the right.
$$0.0000000000000123 = 1.23 \times 10^{-14}$$

53. Move the decimal 1 place to the left.
$$73 \times 10^4 = 7.3 \times 10^{4+1}$$
$$= 7.3 \times 10^5$$

55. Move the decimal 2 places to the left.
$$201.8 \times 10^{15} = 2.018 \times 10^{15+2}$$
$$= 2.018 \times 10^{17}$$

57. Move the decimal 2 places to the right.
$$0.073 \times 10^{-3} = 7.3 \times 10^{-3+(-2)}$$
$$= 7.3 \times 10^{-5}$$

59. Move the decimal 1 place to the left.
$$36.02 \times 10^{-20} = 3.602 \times 10^{-20+1}$$
$$= 3.602 \times 10^{-19}$$

Use scientific notation to perform the calculations. Give all answers in scientific notation and standard notation. See Examples 3 and 4.

61. $(3.4 \times 10^2)(2.1 \times 10^3)$
$$= (3.4 \cdot 2.1) \times (10^2 \times 10^3)$$
$$= (3.4 \cdot 2.1) \times (10^{2+3})$$
$$= 7.14 \times 10^5$$
$$= 714,000$$

63. $(8.4 \times 10^{-13})(4.8 \times 10^9)$
$$= (8.4 \cdot 4.8) \times (10^{-13} \times 10^9)$$
$$= (8.4 \cdot 4.8) \times (10^{-13+9})$$
$$= 40.32 \times 10^{-4}$$
$$= 4.032 \times 10^{-4+1}$$
$$= 4.032 \times 10^{-3}$$
$$= 0.004032$$

65. $\dfrac{2.24 \times 10^4}{5.6 \times 10^7} = \dfrac{2.24}{5.6} \times \dfrac{10^4}{10^7}$
$$= \dfrac{2.24}{5.6} \times 10^{4-7}$$
$$= 0.4 \times 10^{-3}$$
$$= 4.0 \times 10^{-3+(-1)}$$
$$= 4.0 \times 10^{-4}$$
$$= 0.0004$$

67. $\dfrac{9.3 \times 10^2}{3.1 \times 10^{-2}} = \dfrac{9.3}{3.1} \times \dfrac{10^2}{10^{-2}}$
$$= \dfrac{9.3}{3.1} \times 10^{2-(-2)}$$
$$= 3 \times 10^{2+2}$$
$$= 3.0 \times 10^4$$
$$= 30,000$$

69. $\dfrac{0.00000129}{0.0003} = \dfrac{1.29 \times 10^{-6}}{3.0 \times 10^{-4}}$
$$= \dfrac{1.29}{3.0} \times \dfrac{10^{-6}}{10^{-4}}$$
$$= \dfrac{1.29}{3.0} \times 10^{-6-(-4)}$$
$$= 0.43 \times 10^{-6+4}$$
$$= 0.43 \times 10^{-2}$$
$$= 4.3 \times 10^{-2+(-1)}$$
$$= 4.3 \times 10^{-3}$$
$$= 0.0043$$

71. $(0.0000000056)(5,500,000)$
$$= (5.6 \times 10^{-9})(5.5 \times 10^6)$$
$$= (5.6 \cdot 5.5) \times (10^{-9} \times 10^6)$$
$$= (5.6 \cdot 5.5) \times (10^{-9+6})$$
$$= 30.8 \times 10^{-3}$$
$$= 3.08 \times 10^{-3+1}$$
$$= 3.08 \times 10^{-2}$$
$$= 0.0308$$

73. $\dfrac{96,000}{(12,000)(0.00004)} = \dfrac{9.6 \times 10^4}{(1.2 \times 10^4)(4.0 \times 10^{-5})}$
$$= \dfrac{9.6 \times 10^4}{(1.2 \cdot 4.0) \times (10^4 \times 10^{-5})}$$
$$= \dfrac{9.6 \times 10^4}{(1.2 \cdot 4.0) \times (10^{4+(-5)})}$$
$$= \dfrac{9.6 \times 10^4}{4.8 \times 10^{-1}}$$
$$= \dfrac{9.6}{4.8} \times \dfrac{10^4}{10^{-1}}$$
$$= \dfrac{9.6}{4.8} \times 10^{4-(-1)}$$
$$= \dfrac{9.6}{4.8} \times 10^{4+1}$$
$$= 2.0 \times 10^5$$
$$= 200,000$$

Section 5.3

75.

$$\frac{2{,}475}{(132{,}000{,}000{,}000{,}000)(0.25)}$$

$$=\frac{2.475\times10^3}{(1.32\times10^{14})(2.5\times10^{-1})}$$

$$=\frac{2.475\times10^3}{(1.32\cdot2.5)\times(10^{14}\times10^{-1})}$$

$$=\frac{2.475\times10^3}{(1.32\cdot2.5)\times(10^{14+(-1)})}$$

$$=\frac{2.475\times10^3}{3.3\times10^{13}}$$

$$=\frac{2.475}{3.3}\times\frac{10^3}{10^{13}}$$

$$=\frac{2.475}{3.3}\times10^{3-13}$$

$$=\frac{2.475}{3.3}\times10^{3+(-13)}$$

$$=0.75\times10^{-10}$$

$$=7.5\times10^{-10+(-1)}$$

$$=7.5\times10^{-11}$$

$$=0.000000000075$$

Find each power. These exact answers were computed using MS Excel.

77. $(456.4)^6 = 9{,}038{,}030{,}747{,}579{,}110$
$$= 9.03803074757911\times10^{15}$$

79. $225^{-5} = 0.00000000000173415299158326$
$$= 1.73415299158326\times10^{-12}$$

APPLICATIONS

81. ASTRONOMY
Move the decimal 13 places to the left.

$$25{,}700{,}000{,}000{,}000 = 2.57\times10^{13}$$

The distance is about 2.57×10^{13} miles.

83. EARTH, SUN, MOON
Surface area of the Earth

The exponent is 8.
Move the decimal 8 places to the right.

$$1.97\times10^8 = 197{,}000{,}000$$

The surface area is $197{,}000{,}000$ mi^2.

Surface area of the sun

The exponent is 17.
Move the decimal 17 places to the right.

$$1.09\times10^{17} = 109{,}000{,}000{,}000{,}000{,}000$$

The surface area is
$109{,}000{,}000{,}000{,}000{,}000$ mi^2.

Surface area of the moon

The exponent is 7.
Move the decimal 7 places to the right.

$$1.46\times10^7 = 14{,}600{,}000$$

The surface area is $14{,}600{,}000$ mi^2.

85. SAND
Move the decimal 10 places to the right.

$$0.00000000045 = 4.5\times10^{-10}$$

The mass of one grain of beach sand is approximately 4.5×10^{-10} oz.

87. WAVELENGTHS

gamma ray	treating cancer	8.9×10^{-14}
x-ray	medical	2.3×10^{-11}
ultraviolet	sun lamp	6.1×10^{-8}
visible light	lighting	9.3×10^{-6}
infrared	photography	3.7×10^{-5}
microwave	cooking	1.1×10^{-2}
radio wave	communication	3.0×10^{2}

89. from CAMPUS TO CAREERS
SOUND ENGINEERING TECHNICIAN

$$\frac{3.3\times10^4}{(100)(1000)} = \frac{3.3\times10^4}{(1.0\times10^2)(1.0\times10^3)}$$

$$=\frac{3.3\times10^4}{(1.0\times1.0)(10^2\times10^3)}$$

$$=\frac{3.3}{1.0}\cdot\frac{10^4}{10^{2+3}}$$

$$=3.3\times10^{4-5}$$

$$=3.3\times10^{-1}$$

The speed of sound in air is approximately 3.3×10^{-1} km/sec.

91. LIGHT YEARS

$$(5.87 \times 10^{12})(5.28 \times 10^{3}) = (5.87 \times 5.28)(10^{12} \times 10^{3})$$
$$= 30.9936 \times 10^{12+3}$$
$$= 30.9936 \times 10^{15}$$
$$= 3.09936 \times 10^{15+1}$$
$$= 3.09936 \times 10^{16}$$

One light year is about 3.09936×10^{16} feet.

93. INSURED DEPOSITS

$$I = Prt$$
$$= (7.6 \times 10^{12})(0.04)(1)$$
$$= (7.6 \times 10^{12})(4.0 \times 10^{-2})(1)$$
$$= (7.6 \times 4.0 \times 1)(10^{12} \times 10^{-2})$$
$$= 30.4 \times 10^{12+(-2)}$$
$$= 30.4 \times 10^{10}$$
$$= 3.04 \times 10^{10+1}$$
$$= 3.04 \times 10^{11}$$

The amount of interest would be about 3.04×10^{11} dollars.

95. POWERS OF 10

One million:
$$1,000,000 = 1.0 \times 10^{6}$$
One billion:
$$1,000,000,000 = 1.0 \times 10^{9}$$
One trillion:
$$1,000,000,000,000 = 1.0 \times 10^{12}$$
One quadrillion:
$$1,000,000,000,000,000 = 1.0 \times 10^{15}$$
One quintillion:
$$1,000,000,000,000,000,000 = 1.0 \times 10^{18}$$

WRITING
97-99. Answers will vary.
REVIEW
101. If $y = -1$, find the value of $-5y^{55}$.
$$-5y^{55} = -5(-1)^{55}$$
$$= -5(-1)$$
$$= 5$$

103. COUNSELING

1st year, counselor had 75 clients.

2nd year, she had 105 clients.

t years is the x-axis.

c clints is the y-axis.

Write a linear equation.

In this problem, we are given two different letters t(years) and c(clients) to use for x and y. One has to determine which letter correlates with which letter, $x = t$ and $y = c$.

Find the slope which is given as two ordered pairs (1, 75) and (2, 105). The reason for the numbers 1 and 2 is because after "1 year" and then after "2 years".

$$\text{Let } (t_1, c_1) = (1, 75)$$
$$\text{Let } (t_2, c_2) = (2, 105)$$
$$m = \frac{c_2 - c_1}{t_2 - t_1}$$
$$= \frac{105 - 75}{2 - 1}$$
$$= \frac{30}{1}$$
$$= 30$$

Now pick one point (1, 75) and use
$$c - c_1 = m(t - t_1)$$
$$c - 75 = 30(t - 1)$$
$$c - 75 = 30t - 30$$
$$c - 75 - \mathbf{75} = 30t - 30 + \mathbf{75}$$
$$c = 30t + 45 \quad \text{slope-intercept form}$$

CHALLENGE PROBLEMS

105. Consider 2.5×10^{-4}. Answer the following questions in scientific notation form.

a. What is its opposite?

Its opposite is -2.5×10^{-4}.

b. What is its reciprocal?

$$\frac{1}{2.5 \times 10^{-4}} = \frac{1}{2.5} \times \frac{1}{10^{-4}}$$
$$= 0.4 \times 10^{4}$$
$$= 4.0 \times 10^{4+(-1)}$$
$$= 4.0 \times 10^{3}$$

Its reciprocal is 4.0×10^{3}.

Section 5.3

SECTION 5.4
VOCABULARY

Fill in the blanks.

1. A **polynomial** is a term or a sum of terms in which all variables have whole-number exponents and no variable appears in a denominator.

3. $x^3 - 6x^2 + 9x - 2$ is a polynomial in **one** variable, and is written in **descending** powers of x and $c^3 + 2c^2d - d^2$ is a polynomial in **two** variables and is written in **ascending** powers of d.

5. A **monomial** is a polynomial with exactly one term. A **binomial** is a polynomial with exactly two terms. A **trinomial** is a polynomial with exactly three terms.

7. To **evaluate** the polynomial $x^2 - 2x + 1$ for $x = 6$, we substitute 6 for x and follow the rules for the order of operations.

CONCEPTS

Determine whether each expression is a polynomial.

9. a. $x^3 - 5x^2 - 2$ **Yes**   b. $x^{-4} - 5x$ **No**

   c. $x^2 - \dfrac{1}{2x} + 3$ **No**   d. $x^3 - 1$ **Yes**

   e. $x^2 - y^2$ **Yes**   f. $a^4 + a^3 + a^2 + a$ **Yes**

Make a term-coefficient-degree table like that shown in Example 1 for each polynomial.

11. $8x^2 + x - 7$

Term	Coefficient	Degree
$8x^2$	8	2
x	1	1
-7	-7	0

Degree of the polynomial: **2**

13. $8a^6b^3 - 27ab$

Term	Coefficient	Degree
$8a^6b^3$	8	9
$-27ab$	-27	2

Degree of the polynomial: **9**

NOTATION

15. a. Write $x - 9 + 3x^2 + 5x^3$ in descending powers of x.
$$5x^3 + 3x^2 + x - 9$$
   b. Write $-2xy + y^2 + x^2$ in ascending powers of y.
$$x^2 - 2xy + y^2$$

GUIDED PRACTICE

Classify each polynomial as a monomial, a binomial, a trinomial, or none of these. See Example 1.

	Monomial one term	Binomial two terms	Trinomial three terms	none of these
17. $3x + 7$		X		
19. $y^2 + 4y + 3$			X	
21. $\dfrac{3}{2}z^2$	X			
23. $t - 32$		X		
25. $s^2 - 23s + 31$			X	
27. $6x^5 - x^4 - 3x^3 + 7$				X
29. $3m^3n - 4m^2n^2 + mn - 1$				X
31. $2a^2 - 3ab + b^2$			X	

Find the degree of each polynomial. See Example 1.

33. $3x^4$ **4th**

35. $-2x^2 + 3x + 1$ **2nd**

37. $\dfrac{1}{3}x - 5$ **1st**

39. $-5r^2s^2 - r^3s + 3$ **4th**

41. $x^{12} + 3x^2y^3$ **12th**

43. 38   **0th**

45. $\dfrac{3}{2}m^7 - \dfrac{3}{4}m^{18}$ **18th**

47. $5.5tw - 6.5t^2w - 7.5t^3$ **3rd**

Evaluate each expression. See Examples 2 and 3.

49. $x^2 - x + 1$
   a. For $x = 2$
$$x^2 - x + 1 = (2)^2 - (2) + 1$$
$$= 4 - 2 + 1$$
$$= 2 + 1$$
$$= 3$$

   b. For $x = -3$
$$x^2 - x + 1 = (-3)^2 - (-3) + 1$$
$$= 9 + 3 + 1$$
$$= 12 + 1$$
$$= 13$$

51. $4t^2 + 2t - 8$
   a. For $t = -1$
$$4t^2 + 2t - 8 = 4(-1)^2 + 2(-1) - 8$$
$$= 4(1) + 2(-1) - 8$$
$$= 4 - 2 - 8$$
$$= 2 - 8$$
$$= 2 + (-8)$$
$$= -6$$

b. For $t = 0$

$$4t^2 + 2t - 8 = 4(\mathbf{0})^2 + 2(\mathbf{0}) - 8$$
$$= 4(0) + 2(0) - 8$$
$$= 0 + 0 - 8$$
$$= -8$$

53. $\dfrac{1}{2}a^2 - \dfrac{1}{4}a$

a. For $a = 4$

$$\frac{1}{2}a^2 - \frac{1}{4}a = \frac{1}{2}(\mathbf{4})^2 - \frac{1}{4}(\mathbf{4})$$
$$= \frac{1}{2}(16) - \frac{1}{4}(4)$$
$$= 8 - 1$$
$$= 7$$

b. For $a = -8$

$$\frac{1}{2}a^2 - \frac{1}{4}a = \frac{1}{2}(\mathbf{-8})^2 - \frac{1}{4}(\mathbf{-8})$$
$$= \frac{1}{2}(64) - \frac{1}{4}(-8)$$
$$= 32 + 2$$
$$= 34$$

55. $-9.2x^2 + x - 1.4$

a. For $x = -1$

$$-9.2x^2 + x - 1.4 = -9.2(\mathbf{-1})^2 + (\mathbf{-1}) - 1.4$$
$$= -9.2(1) - 1 - 1.4$$
$$= -9.2 - 1 - 1.4$$
$$= -10.2 - 1.4$$
$$= -11.6$$

b. For $x = -2$

$$-9.2x^2 + x - 1.4 = -9.2(\mathbf{-2})^2 + (\mathbf{-2}) - 1.4$$
$$= -9.2(4) - 2 - 1.4$$
$$= -36.8 - 2 - 1.4$$
$$= -38.8 - 1.4$$
$$= -40.2$$

57. $x^3 + 3x^2 + 2x + 4$

a. For $x = 2$

$$x^3 + 3x^2 + 2x + 4 = (\mathbf{2})^3 + 3(\mathbf{2})^2 + 2(\mathbf{2}) + 4$$
$$= 8 + 3(4) + 4 + 4$$
$$= 8 + 12 + 4 + 4$$
$$= 20 + 4 + 4$$
$$= 24 + 4$$
$$= 28$$

b. For $x = -2$

$$x^3 + 3x^2 + 2x + 4$$
$$= (\mathbf{-2})^3 + 3(\mathbf{-2})^2 + 2(\mathbf{-2}) + 4$$
$$= -8 + 3(4) - 4 + 4$$
$$= -8 + 12 - 4 + 4$$
$$= 4 - 4 + 4$$
$$= 0 + 4$$
$$= 4$$

59. $y^4 - y^3 + y^2 + 2y - 1$

a. For $y = 1$

$$y^4 - y^3 + y^2 + 2y - 1$$
$$= (\mathbf{1})^4 - (\mathbf{1})^3 + (\mathbf{1})^2 + 2(\mathbf{1}) - 1$$
$$= 1 - 1 + 1 + 2 - 1$$
$$= 0 + 1 + 2 - 1$$
$$= 1 + 2 - 1$$
$$= 3 - 1$$
$$= 2$$

b. For $y = -1$

$$y^4 - y^3 + y^2 + 2y - 1$$
$$= (\mathbf{-1})^4 - (\mathbf{-1})^3 + (\mathbf{-1})^2 + 2(\mathbf{-1}) - 1$$
$$= 1 - (-1) + 1 + 2(-1) - 1$$
$$= 1 + 1 + 1 - 2 - 1$$
$$= 2 + 1 - 2 - 1$$
$$= 3 - 2 - 1$$
$$= 1 - 1$$
$$= 0$$

Evaluate each polynomial for $a = -2$ and $b = 3$. See Example 4.

61. $6a^2b$, For $a = -2$, $b = 3$.

$$6a^2b = 6(\mathbf{-2})^2(\mathbf{3})$$
$$= 6(4)(3)$$
$$= 72$$

63. $a^3 + b^3$, For $a = -2$, $b = 3$.

$$a^3 + b^3 = (\mathbf{-2})^3 + (\mathbf{3})^3$$
$$= -8 + 27$$
$$= 19$$

65. $a^2 + 5ab - b^2$, For $a = -2$, $b = 3$.

$$a^2 + 5ab - b^2 = (\mathbf{-2})^2 + 5(\mathbf{-2})(\mathbf{3}) - (\mathbf{3})^2$$
$$= 4 - 30 - 9$$
$$= -26 - 9$$
$$= -35$$

Section 5.4

67. $5ab^3 - ab - b + 10$, For $a = -2$, $b = 3$.

$5ab^3 - ab - b + 10$

$$= 5(-2)(3)^3 - (-2)(3) - (3) + 10$$
$$= 5(-2)27 + 6 - 3 + 10$$
$$= -270 + 6 - 3 + 10$$
$$= -270 + 6 - 3 + 10$$
$$= -264 - 3 + 10$$
$$= -267 + 10$$
$$= -257$$

Construct a table of solutions and then graph the equation. See Examples 5-7.

69. $y = x^2 + 1$

x	y	(x, y)
-3	$y = (-3)^2 + 1$ $= 9 + 1$ $= 10$	$(-3, 10)$
-2	$y = (-2)^2 + 1$ $= 4 + 1$ $= 5$	$(-2, 5)$
-1	$y = (-1)^2 + 1$ $= 1 + 1$ $= 2$	$(-1, 2)$
0	$y = (0)^2 + 1$ $= 0 + 1$ $= 1$	$(0, 1)$
1	$y = (1)^2 + 1$ $= 1 + 1$ $= 2$	$(1, 2)$
2	$y = (2)^2 + 1$ $= 4 + 1$ $= 5$	$(2, 5)$
3	$y = (3)^2 + 1$ $= 9 + 1$ $= 10$	$(3, 10)$

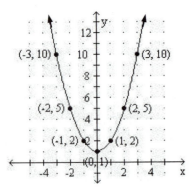

71. $y = -x^2 - 2$

x	y	(x, y)
-3	$y = -(-3)^2 - 2$ $= -(9) - 2$ $= -9 + (-2)$ $= -11$	$(-3, -11)$
-2	$y = -(-2)^2 - 2$ $= -(4) - 2$ $= -4 + (-2)$ $= -6$	$(-2, -6)$
-1	$y = -(-1)^2 - 2$ $= -(1) - 2$ $= -1 + (-2)$ $= -3$	$(-1, -3)$
0	$y = -(0)^2 - 2$ $= -(0) - 2$ $= 0 + (-2)$ $= -2$	$(0, -2)$
1	$y = -(1)^2 - 2$ $= -(1) - 2$ $= -1 + (-2)$ $= -3$	$(1, -3)$
2	$y = -(2)^2 - 2$ $= -(4) - 2$ $= -4 + (-2)$ $= -6$	$(2, -6)$
3	$y = -(3)^2 - 2$ $= -(9) - 2$ $= -9 + (-2)$ $= -11$	$(3, -11)$

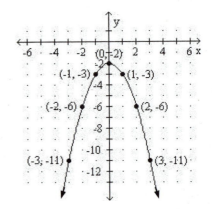

73. $y = 2x^2 - 3$

x	y	(x, y)
-3	$y = 2(-3)^2 - 3$ $= 2(9) - 3$ $= 18 + (-3)$ $= 15$	$(-3, 15)$
-2	$y = 2(-2)^2 - 3$ $= 2(4) - 3$ $= 8 + (-3)$ $= 5$	$(-2, 5)$
-1	$y = 2(-1)^2 - 3$ $= 2(1) - 3$ $= 2 + (-3)$ $= -1$	$(-1, -1)$
0	$y = 2(0)^2 - 3$ $= 2(0) - 3$ $= 0 + (-3)$ $= -3$	$(0, -3)$
1	$y = 2(1)^2 - 3$ $= 2(1) - 3$ $= 2 + (-3)$ $= -1$	$(1, -1)$
2	$y = 2(2)^2 - 3$ $= 2(4) - 3$ $= 8 + (-3)$ $= 5$	$(2, 5)$
3	$y = 2(3)^2 - 3$ $= 2(9) - 3$ $= 18 + (-3)$ $= 15$	$(3, 15)$

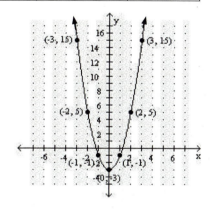

75. $y = x^3 + 2$

x	y	(x, y)
-3	$y = (-3)^3 + 2$ $= -27 + 2$ $= -25$	$(-3, -25)$
-2	$y = (-2)^3 + 2$ $= -8 + 2$ $= -6$	$(-2, -6)$
-1	$y = (-1)^3 + 2$ $= -1 + 2$ $= 1$	$(-1, 1)$
0	$y = (0)^3 + 2$ $= 0 + 2$ $= 2$	$(0, 2)$
1	$y = (1)^3 + 2$ $= 1 + 2$ $= 3$	$(1, 3)$
2	$y = (2)^3 + 2$ $= 8 + 2$ $= 10$	$(2, 10)$
3	$y = (3)^3 + 2$ $= 27 + 2$ $= 29$	$(3, 29)$

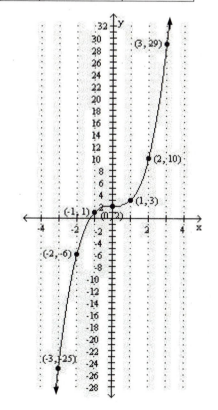

Section 5.4

77. $y = x^3 - 3$

x	y	(x, y)
-3	$y = (-3)^3 - 3$ $= -27 + (-3)$ $= -30$	$(-3, -30)$
-2	$y = (-2)^3 - 3$ $= -8 + (-3)$ $= -11$	$(-2, -11)$
-1	$y = (-1)^3 - 3$ $= -1 + (-3)$ $= -4$	$(-1, -4)$
0	$y = (0)^3 - 3$ $= 0 + (-3)$ $= -3$	$(0, -3)$
1	$y = (1)^3 - 3$ $= 1 + (-3)$ $= -2$	$(1, -2)$
2	$y = (2)^3 - 3$ $= 8 + (-3)$ $= 5$	$(2, 5)$
3	$y = (3)^3 - 3$ $= 27 + (-3)$ $= 24$	$(3, 24)$

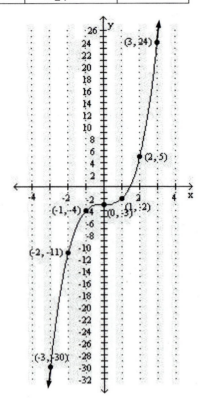

79. $y = -x^3 - 1$

x	y	(x, y)
-3	$y = -(-3)^3 - 1$ $= -(-27) - 1$ $= 27 + (-1)$ $= 26$	$(-3, 26)$
-2	$y = -(-2)^3 - 1$ $= -(-8) - 1$ $= 8 + (-1)$ $= 7$	$(-2, 7)$
-1	$y = -(-1)^3 - 1$ $= -(-1) - 1$ $= 1 + (-1)$ $= 0$	$(-1, 0)$
0	$y = -(0)^3 - 1$ $= -(0) - 1$ $= 0 + (-1)$ $= -1$	$(0, -1)$
1	$y = -(1)^3 - 1$ $= -(1) - 1$ $= -1 + (-1)$ $= -2$	$(1, -2)$
2	$y = -(2)^3 - 1$ $= -(8) - 1$ $= -8 + (-1)$ $= -9$	$(2, -9)$
3	$y = -(3)^3 - 1$ $= -(27) - 1$ $= -27 + (-1)$ $= -28$	$(3, -28)$

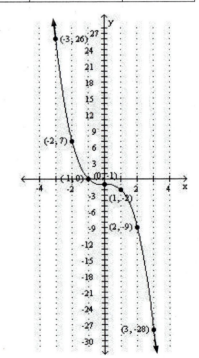

APPLICATION

81. SUPERMARKETS

$$\frac{1}{3}c^3 + \frac{1}{2}c^2 + \frac{1}{6}c = \frac{1}{3}(6)^3 + \frac{1}{2}(6)^2 + \frac{1}{6}(6)$$
$$= \frac{1}{3}(216) + \frac{1}{2}(36) + \frac{1}{6}(6)$$
$$= 72 + 18 + 1$$
$$= 91$$

91 cantaloupes will be used.

83. STOPPING DISTANCE

$$0.04v^2 + 0.9v = 0.04(30)^2 + 0.9(30)$$
$$= 0.04(900) + 0.9(30)$$
$$= 36 + 27$$
$$= 63$$

The stopping distance is 63 feet.

from CAMPUS TO CAREERS

85. SOUND ENGINEERING TECHNICIAN

Let $x = 0$, when the year is 2004.

Let $x = 1$, when the year is 2005.

Let $x = 10$, when the year is 2014.

$$0.32x^2 - 0.36x + 0.21$$
$$= 0.32(10)^2 - 0.36(10) + 0.21$$
$$= 0.32(100) - 0.36(10) + 0.21$$
$$= 32 - 3.6 + 0.21$$
$$= 28.61$$

The approximate number of iTunes downloads as of January 2014 will be about 28.6 billion.

87. SCIENCE HISTORY

Time (seconds)	Distance (ft)	(s, f)
0	0	$(0, 0)$
$\frac{1}{4}$	$\frac{1}{2}$	$\left(\frac{1}{4}, \frac{1}{2}\right)$
$\frac{1}{2}$	2	$\left(\frac{1}{2}, 2\right)$
$\frac{3}{4}$	$4\frac{1}{2}$	$\left(\frac{3}{4}, 4\frac{1}{2}\right)$
1	8	$(1, 8)$
$1\frac{1}{4}$	$12\frac{1}{2}$	$\left(1\frac{1}{4}, 12\frac{1}{2}\right)$
$1\frac{1}{2}$	18	$\left(1\frac{1}{2}, 18\right)$

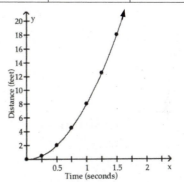

WRITING

89-91. Answers will vary.

REVIEW

Solve each inequality. Write the solution set in interval notation and graph it.

93. $\quad -4(3y + 2) \le 28$

$$-4(3y) - 4(2) \le 28$$
$$-12y - 8 \le 28$$
$$-12y - 8 + 8 \le 28 + 8$$
$$-12y \le 36$$
$$\frac{-12y}{-12} \ge \frac{36}{-12}$$
$$y \ge -3$$
$$[-3, \infty)$$

Section 5.4

Simplify each expression. Do not use negative exponents in the answer.

95. $\left(x^2 x^4\right)^3 = \left(x^{2+4}\right)^3$

$$= \left(x^6\right)^3$$

$$= x^{6 \cdot 3}$$

$$= x^{18}$$

97. $\left(\dfrac{y^2 y^5}{y^4}\right)^3 = \left(\dfrac{y^{2+5}}{y^4}\right)^3$

$$= \left(\dfrac{y^7}{y^4}\right)^3$$

$$= \left(y^{7-4}\right)^3$$

$$= \left(y^3\right)^3$$

$$= y^{3 \cdot 3}$$

$$= y^9$$

CHALLENGE PROBLEMS

99. Find a three-term polynomial of degree 2 whose value will be 1 when it is evaluated for $x = 2$. Answers will vary.

Example: $x^2 - x - 1 = (2)^2 - (2) - 1$

$$= 4 - 2 - 1$$

$$= 2 - 1$$

$$= 1$$

SECTION 5.5
VOCABULARY

Fill in the blanks.

1. $(b^3 - b^2 - 9b + 1) + (b^3 - b^2 - 9b + 1)$ is the sum of two **polynomials**.

3. **Like** terms have the same variables with the same exponents.

CONCEPTS

Fill in the blanks.

5. To add polynomials, **combine** their like terms.

7. Simplify each polynomial, if possible.

a. $2x^2 + 3x^2 = (2+3)x^2$
$= 5x^2$

b. $15m^3 - m^3 = (15-1)m^3$
$= 14m^3$

c. $8a^3b - a^3b = (8-1)a^3b$
$= 7a^3b$

d. $6cd + 4c^2d$; Will not simplify.

9. Write without parentheses.

a. $-(5x^2 - 8x + 23)$
$\mathbf{-5x^2 + 8x - 23}$

b. $-(-5y^4 + 3y^2 - 7)$
$\mathbf{5y^4 - 3y^2 + 7}$

NOTATION

Fill in the blanks to add (subtract) the polynomials.

11. $(6x^2 + 2x + 3) + (4x^2 - 7x + 1)$
$= (6x^2 + \mathbf{4x^2}) + (\mathbf{2x} - 7x) + (3 + \mathbf{1})$
$= \mathbf{10x^2} - 5x + \mathbf{4}$

GUIDED PRACTICE

Simplify each polynomial and write it in descending powers of one variable. See Example 1.

13. $8t^2 + 4t^2 = (8+4)t^2$
$= 12t^2$

15. $18x^2 - 19x + 2x^2 = (18+2)x^2 - 19x$
$= 20x^2 - 19x$

17. $10x^2 - 8x + 9x - 9x^2 = (10-9)x^2 + (-8+9)x$
$= x^2 + x$

19. $\frac{1}{5}x^2 - \frac{3}{8}x + \frac{2}{3}x^2 + \frac{1}{4}x$

$= \left(\frac{1}{5} + \frac{2}{3}\right)x^2 + \left(-\frac{3}{8} + \frac{1}{4}\right)x$

$= \left(\frac{1}{5} \cdot \frac{3}{3} + \frac{2}{3} \cdot \frac{5}{5}\right)x^2 + \left(-\frac{3}{8} + \frac{1}{4} \cdot \frac{2}{2}\right)x$

$= \left(\frac{3+10}{15}\right)x^2 + \left(\frac{-3+2}{8}\right)x$

$= \left(\frac{13}{15}\right)x^2 + \left(\frac{-1}{8}\right)x$

$= \frac{13}{15}x^2 - \frac{1}{8}x$

21. $0.6x^3 + 0.8x^4 + 0.7x^3 + (-0.8x^4)$
$= [0.8 + (-0.8)]x^4 + (0.6 + 0.7)x^3$
$= 0x^4 + (1.3)x^3$
$= 1.3x^3$

23. $\frac{1}{2}st + \frac{3}{2}st = \left(\frac{1}{2} + \frac{3}{2}\right)st$

$= \frac{4}{2}st$

$= 2st$

25. $-4ab + 4ab - ab = (-4 + 4 - 1)ab$
$= -ab$

27. $4x^2y + 5 - 6x^3y - 3x^2y + 2x^3y$
$= (-6+2)x^3y + (4-3)x^2y + 5$
$= -4x^3y + x^2y + 5$

Add the polynomials. See Example 2.

29. $(3q^2 - 5q + 7) + (2q^2 + q - 12)$
$= (3q^2 + 2q^2) + (-5q + q) + (7 - 12)$
$= 5q^2 - 4q - 5$

31. $\left(\dfrac{2}{3}y^3 + \dfrac{3}{4}y^2 + \dfrac{1}{2}\right) + \left(\dfrac{1}{3}y^3 + \dfrac{1}{5}y^2 - \dfrac{1}{6}\right)$

$= \left(\dfrac{2}{3} + \dfrac{1}{3}\right)y^3 + \left(\dfrac{3}{4} + \dfrac{1}{5}\right)y^2 + \left(\dfrac{1}{2} - \dfrac{1}{6}\right)$

$= \left(\dfrac{2+1}{3}\right)y^3 + \left(\dfrac{3}{4}\cdot\dfrac{5}{5} + \dfrac{1}{5}\cdot\dfrac{4}{4}\right)y^2 + \left(\dfrac{1}{2}\cdot\dfrac{3}{3} - \dfrac{1}{6}\right)$

$= \left(\dfrac{3}{3}\right)y^3 + \left(\dfrac{15+4}{20}\right)y^2 + \left(\dfrac{3-1}{6}\right)$

$= y^3 + \dfrac{19}{20}y^2 + \dfrac{\overset{1}{\cancel{2}}}{\underset{1}{\cancel{2}}\cdot 3}$

$= y^3 + \dfrac{19}{20}y^2 + \dfrac{1}{3}$

33. $(0.3p + 2.1q) + (0.4p - 3q)$

$= (0.3 + 0.4)p + (2.1 - 3)q$

$= 0.7p - 0.9q$

35. $(2x^2 + xy + 3y^2) + (5x^2 - y^2)$

$= (2+5)x^2 + xy + (3-1)y^2$

$= 7x^2 + xy + 2y^2$

Find a polynomial that represents the perimeter of the figure. See Example 3.

37. $(x^2 + 3x + 1) + (x^2 + 3x + 1) + (x^2 - 4)$

$= (1+1+1)x^2 + (3+3)x + (1+1-4)$

$= 3x^2 + 6x - 2$

The perimeter is $(3x^2 + 6x - 2)$ yd.

39. $(2x^2 - 7) + (x + 6) + (5x^2 + 3x + 1) + (x + 6)$

$= (2+5)x^2 + (1+3+1)x + (-7+6+1+6)$

$= 7x^2 + 5x + 6$

The perimeter is $(7x^2 + 5x + 6)$ mi.

Use vertical form to add the polynomials. See Example 4.

41. $\begin{array}{r} 3x^2 + 4x + 5 \\ \underline{2x^2 - 3x + 6} \\ 5x^2 + x + 11 \end{array}$

43. $\begin{array}{r} 6a^2 + 7a + 9 \\ \underline{-\ 9a^2\qquad -2} \\ -3a^2 + 7a + 7 \end{array}$

45. $\begin{array}{r} z^3 + 6z^2 - 7z + 16 \\ \underline{9z^3 - 6z^2 + 8z - 18} \\ 10z^3 \qquad + z\ -\ 2 \end{array}$

47. $\begin{array}{r} -3x^3y^2 + 4x^2y - 4x + 9 \\ \underline{2x^3y^2 \qquad\quad +9x - 3} \\ -x^3y^2 + 4x^2y + 5x + 6 \end{array}$

Subtract the polynomials. See Example 5.

49. $(3a^2 - 2a + 4) - (a^2 - 3a + 7)$

$= 3a^2 - 2a + 4 - a^2 + 3a - 7$

$= (3-1)a^2 + (-2+3)a + (4-7)$

$= 2a^2 + a - 3$

51. $(-4h^3 + 5h^2 + 15) - (h^3 - 15)$

$= -4h^3 + 5h^2 + 15 - h^3 + 15$

$= (-4-1)h^3 + 5h^2 + (15+15)$

$= -5h^3 + 5h^2 + 30$

53. $\left(\dfrac{3}{8}s^8 - \dfrac{3}{4}s^7\right) - \left(\dfrac{1}{3}s^8 + \dfrac{1}{5}s^7\right)$

$= \dfrac{3}{8}s^8 - \dfrac{3}{4}s^7 - \dfrac{1}{3}s^8 - \dfrac{1}{5}s^7$

$= \left(\dfrac{3}{8} - \dfrac{1}{3}\right)s^8 + \left(-\dfrac{3}{4} - \dfrac{1}{5}\right)s^7$

$= \left(\dfrac{3}{8}\cdot\dfrac{3}{3} - \dfrac{1}{3}\cdot\dfrac{8}{8}\right)s^8 + \left(-\dfrac{3}{4}\cdot\dfrac{5}{5} - \dfrac{1}{5}\cdot\dfrac{4}{4}\right)s^7$

$= \dfrac{9-8}{24}s^8 + \dfrac{-15-4}{20}s^7$

$= \dfrac{1}{24}s^8 - \dfrac{19}{20}s^7$

55. $(5ab + 2b^2) - (2 + ab + b^2)$

$= 5ab + 2b^2 - 2 - ab - b^2$

$= (2-1)b^2 + (5-1)ab - 2$

$= b^2 + 4ab - 2$

Use vertical form to subtract the polynomials. See Example 6.

57. $\begin{array}{r} 3x^2 + 4x + 5 \\ \underline{-(2x^2 - 2x + 3)} \end{array}$ $\xrightarrow[\text{and add}]{\text{Change signs}}$ $\begin{array}{r} 3x^2 + 4x + 5 \\ \underline{-2x^2 + 2x - 3} \\ x^2 + 6x + 2 \end{array}$

59. $\begin{array}{r} 5s^2 \quad + 9 \\ \underline{-(s^2 + 4s + 2)} \end{array}$ $\xrightarrow[\text{and add}]{\text{Change signs}}$ $\begin{array}{r} 5s^2 \quad + 9 \\ \underline{-s^2 - 4s - 2} \\ 4s^2 - 4s + 7 \end{array}$

61. $\begin{array}{r} 17a^3 \quad + 25a - 10 \\ \underline{-(8a^3 + 8a^2 - 3a + 1)} \end{array}$ $\xrightarrow[\text{and add}]{\text{Change signs}}$ $\begin{array}{r} 17a^3 \qquad\quad + 25a - 10 \\ \underline{-8a^3 - 8a^2 + 3a - \ 1} \\ 9a^3 - 8a^2 + 28a - 11 \end{array}$

63.

$$0.8x^3 \qquad\quad -2.3x+0.6$$
$$-(0.2x^3-1.2x^2-3.6x+0.9)$$

Change the signs of the 2nd line and add

$$0.8x^3 \qquad\quad -2.3x+0.6$$
$$\underline{-0.2x^3+1.2x^2+3.6x-0.9}$$
$$0.6x^3+1.2x^2+1.3x-0.3$$

Perform the operations. See Example 7.

65. $(-2x^2-7x+1)+(-4x^2+8x-1)$

$$=(-2-4)x^2+(-7+8)x+(1-1)$$
$$=-6x^2+x$$
$$(-6x^2+x)-(3x^2+4x-7)$$
$$=-6x^2+x-3x^2-4x+7$$
$$=(-6-3)x^2+(1-4)x+7$$
$$=-9x^2-3x+7$$

67. $(3t^3+t^2)+(-t^3+6t-3)$

$$=(3-1)t^3+t^2+6t-3$$
$$=2t^3+t^2+6t-3$$
$$(2t^3+t^2+6t-3)-(t^3-2t^2+2)$$
$$=2t^3+t^2+6t-3-t^3+2t^2-2$$
$$=(2-1)t^3+(1+2)t^2+6t+(-3-2)$$
$$=t^3+3t^2+6t-5$$

TRY IT YOURSELF

Perform the operations.

69. $(9a^2+3a)-(2a-4a^2)=9a^2+3a-2a+4a^2$

$$=(9+4)a^2+(3-2)a$$
$$=13a^2+a$$

71. $(2y^5-y^4)-(-y^5+5y^4-1.2)$

$$=2y^5-y^4+y^5-5y^4+1.2$$
$$=(2+1)y^5+(-1-5)y^4+1.2$$
$$=3y^5-6y^4+1.2$$

73. $3r^4-4r+7r^4=(3+7)r^4-4r$

$$=10r^4-4r$$

75. $(0.03f^2+0.25f+0.91)-(0.17f^2-1.18)$

$$=0.03f^2+0.25f+0.91-0.17f^2+1.18$$
$$=(0.03-0.17)f^2+0.25f+(0.91+1.18)$$
$$=-0.14f^2+0.25f+2.09$$

77. $\left(\dfrac{7}{8}r^4+\dfrac{5}{9}r^2-\dfrac{9}{4}\right)-\left(-\dfrac{3}{8}r^4-\dfrac{2}{3}r^2-\dfrac{1}{4}\right)$

$$=\dfrac{7}{8}r^4+\dfrac{5}{9}r^2-\dfrac{9}{4}+\dfrac{3}{8}r^4+\dfrac{2}{3}r^2+\dfrac{1}{4}$$
$$=\left(\dfrac{7}{8}+\dfrac{3}{8}\right)r^4+\left(\dfrac{5}{9}+\dfrac{2}{3}\right)r^2+\left(-\dfrac{9}{4}+\dfrac{1}{4}\right)$$
$$=\left(\dfrac{7}{8}+\dfrac{3}{8}\right)r^4+\left(\dfrac{5}{9}+\dfrac{2}{3}\cdot\dfrac{3}{3}\right)r^2+\left(-\dfrac{9}{4}+\dfrac{1}{4}\right)$$
$$=\dfrac{7+3}{8}r^4+\dfrac{5+6}{9}r^2+\dfrac{-9+1}{4}$$
$$=\dfrac{10}{8}r^4+\dfrac{11}{9}r^2+\dfrac{-8}{4}$$
$$=\dfrac{\overset{1}{\cancel{2}}\cdot5}{\underset{1}{\cancel{2}}\cdot4}r^4+\dfrac{11}{9}r^2-2$$
$$=\dfrac{5}{4}r^4+\dfrac{11}{9}r^2-2$$

79.

$$8c^2-4c-5 \qquad\qquad 8c^2-4c-5$$
$$-(-c^2+2c+9) \;\xrightarrow[\text{and add}]{\text{Change signs}}\; \underline{c^2-2c-9}$$
$$9c^2-6c-14$$

81. $(12.1h^3+9.9h^2)+(7.3h^3+1.1h^2)$

$$=(12.1+7.3)h^3+(9.9+1.1)h^2$$
$$=19.4h^3+11h^2$$

83. $(20-4rt-5r^2t)+(10-5rt)$

$$=-5r^2t+(-4rt-5rt)+(20+10)$$
$$=-5r^2t+(-4-5)rt+(30)$$
$$=-5r^2t-9rt+30$$

85. $(3x^2-3x-2)+(3x^2+4x-3)$

$$=3x^2-3x-2+3x^2+4x-3$$
$$=(3+3)x^2+(-3+4)x+(-2-3)$$
$$=6x^2+x-5$$

Section 5.5

87. $\dfrac{2}{3}d^2 - \dfrac{1}{4}c^2 + \dfrac{5}{6}c^2 - \dfrac{1}{2}cd + \dfrac{1}{3}d^2$

$= \left(-\dfrac{1}{4} + \dfrac{5}{6}\right)c^2 - \dfrac{1}{2}cd + \left(\dfrac{2}{3} + \dfrac{1}{3}\right)d^2$

$= \left(-\dfrac{1}{4} \cdot \dfrac{\mathbf{3}}{\mathbf{3}} + \dfrac{5}{6} \cdot \dfrac{\mathbf{2}}{\mathbf{2}}\right)c^2 - \dfrac{1}{2}cd + \left(\dfrac{2}{3} + \dfrac{1}{3}\right)d^2$

$= \left(\dfrac{-3+10}{12}\right)c^2 - \dfrac{1}{2}cd + \left(\dfrac{2+1}{3}\right)d^2$

$= \dfrac{7}{12}c^2 - \dfrac{1}{2}cd + \dfrac{3}{3}d^2$

$= \dfrac{7}{12}c^2 - \dfrac{1}{2}cd + d^2$

89. $(3x+7) + (4x-3) = (3+4)x + (7-3)$

$= 7x + 4$

91. $(-2.7t^2 + 2.1t - 1.7) + (3.1t^2 - 2.5t + 2.3)$

$= (-2.7 + 3.1)t^2 + (2.1 - 2.5)t + (-1.7 + 2.3)$

$= 0.4t^2 - 0.4t + 0.6$

$(0.4t^2 - 0.4t + 0.6) - (1.7t^2 - 1.1t)$

$= 0.4t^2 - 0.4t + 0.6 - 1.7t^2 + 1.1t$

$= (0.4 - 1.7)t^2 + (-0.4 + 1.1)t + 0.6$

$= -1.3t^2 + 0.7t + 0.6$

93. $-32u^3 - 16u^3 = (-32 - 16)u^3$

$= -48u^3$

95. $(9d^2 + 6d) + (8d - 4d^2)$

$= (9 - 4)d^2 + (6 + 8)d$

$= 5d^2 + 14d$

97. $\quad 3x^3y^2 + 4x^2y + 7x + 12$

$\underline{-(-4x^3y^2 + 6x^2y + 9x - 3)}$

Change the signs of the 2nd line and add

$3x^3y^2 + 4x^2y + 7x + 12$

$\underline{4x^3y^2 - 6x^2y - 9x \quad +3}$

$7x^3y^2 - 2x^2y - 2x + 15$

99. $(2x^2 - 3x + 1) - (4x^2 - 3x + 2)$

$\qquad\qquad\qquad\qquad + (2x^2 + 3x + 2)$

$= 2x^2 - 3x + 1 - 4x^2 + 3x - 2 + 2x^2 + 3x + 2$

$= (2 - 4 + 2)x^2 + (-3 + 3 + 3)x + (1 - 2 + 2)$

$= 0x^2 + 3x + 1$

$= 3x + 1$

101. $\quad 4x^3 + 4x^2 - 3x + 10$

$\underline{+(5x^3 - 2x^2 - 4x - 4)}$

$\quad 9x^3 + 2x^2 - 7x + \ 6$

LOOK ALIKES...

103. **a.** $(-8x^2 - 3x) + (-11x^2 + 6x + 10)$

$= (-8 - 11)x^2 + (-3 + 6)x + 10$

$= -19x^2 + 3x + 10$

b. $(-8x^2 - 3x) - (-11x^2 + 6x + 10)$

$= -8x^2 - 3x + 11x^2 - 6x - 10$

$= (-8 + 11)x^2 + (-3 - 6)x - 10$

$= 3x^2 - 9x - 10$

APPLICATIONS

105. GREEK ARCHITECTURE

a. $(x^2 - 3x + 2) - (5x - 10)$

$= x^2 - 3x + 2 - 5x + 10$

$= x^2 + (-3 - 5)x + (2 + 10)$

$= x^2 - 8x + 12$

The difference in the heights is

$(x^2 - 8x + 12)$ ft.

b. $(x^2 - 3x + 2) + (5x - 10)$

$= x^2 - 3x + 2 + 5x - 10$

$= x^2 + (-3 + 5)x + (2 - 10)$

$= x^2 + 2x - 8$

The sum of the heights is

$(x^2 + 2x - 8)$ ft.

107. PIÑATAS

$(4a^2 + 6a - 1) - (2a^2 - 6)$

$= 4a^2 + 6a - 1 - 2a^2 + 6$

$= (4 - 2)a^2 + 6a + (-1 + 6)$

$= 2a^2 + 6a + 5$

The length of the rope is $(2a^2 + 6a + 5)$ in.

109. NAVAL OPERATIONS

a. $(-16t^2 + 150t + 40) - (-16t^2 + 128t + 20)$

$= -16t^2 + 150t + 40 + 16t^2 - 128t - 20$

$= (-16 + 16)t^2 + (150 - 128)t + (40 - 20)$

$= 0t^2 + 22t + 20$

$= 22t + 20$

The difference in the heights is
$(22t + 20)$ ft.

b. $22t + 20 = 22(\mathbf{4}) + 20$

$= 88 + 20$

$= 108$

The second flare is 108 ft higher.

WRITING
111-115. Answers will vary.

REVIEW
117. What is the sum of the measures of
the angles of a triangle? **180°**

119. Graph: $y = -\dfrac{1}{2}x + 2$

y-intercept is $(0,2)$, $m = \dfrac{-1}{2}$

Start at $(0,2)$, go down 1, right 2

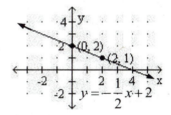

CHALLENGE PROBLEMS
121. $(6x^2 - 7x - 8) - (2x^2 - x + 3)$

$= 6x^2 - 7x - 8 - 2x^2 + x - 3$

$= (6 - 2)x^2 + (-7 + 1)x + (-8 - 3)$

$= 4x^2 - 6x - 11$

The needed polynomial is $(4x^2 - 6x - 11)$.

SECTION 5.6

VOCABULARY

Fill in the blanks.

1. $(2x^3)(3x^4)$ is the product of two **monomials** and $(2a - 4)(3a + 5)$ is the product of two **binomials**.

3. In the acronym FOIL, F stands for **first** terms, O for **outer** terms, I for **inner** terms, and L for **last** terms.

CONCEPTS

Fill in the blanks.

5. a. To multiply two polynomials, multiply **each** term of one polynomial by **each** term of the other polynomial, and then combine like terms.

   b. When multiplying three polynomials, we begin by multiplying **any** two of them, and then we multiply that result by the **third** polynomial.

7. Simplify each polynomial by combining like terms.

   a. $6x^2 - 8x + 9x - 12$ $\mathbf{6x^2 + x - 12}$

   b. $5x^4 + 3ax^2 + 5ax^2 + 3a^2$ $\mathbf{5x^4 + 8ax^2 + 3a^2}$

NOTATION

Complete each solution.

9. $(9n^3)(8n^2) = (9 \cdot \mathbf{8})(n^3 \cdot n^2) = \mathbf{72}n^\mathbf{5}$

11. $(2x + 5)(3x - 2) = 2x(3x) - \mathbf{2x}(2) + \mathbf{5}(3x) - \mathbf{5}(2)$
$$= 6x^2 - \mathbf{4x} + \mathbf{15x} - 10$$
$$= 6x^2 + \mathbf{11x} - 10$$

GUIDED PRACTICE

Multiply. See Example 1.

13. $5m(m) = 5m^{1+1}$
$$= 5m^2$$

15. $(3x^2)(4x^3) = (3 \cdot 4)(x^2 x^3)$
$$= 12(x^{2+3})$$
$$= 12x^5$$

17. $(1.2c^3)(5c^3) = (1.2 \cdot 5)(c^3 c^3)$
$$= 6(c^{3+3})$$
$$= 6c^6$$

19. $(3b^2)(-2b)(4b^3) = (3 \cdot -2 \cdot 4)(b^2 b b^3)$
$$= -24(b^{2+1+3})$$
$$= -24b^6$$

21. $(2x^2 y^3)(4x^3 y^2) = (2 \cdot 4)(x^2 x^3)(y^3 y^2)$
$$= 8(x^{2+3})(y^{3+2})$$
$$= 8x^5 y^5$$

23. $(8a^5)\left(-\dfrac{1}{4}a^6\right) = \left(8 \cdot -\dfrac{1}{4}\right)(a^5 a^6)$
$$= -2(a^{5+6})$$
$$= -2a^{11}$$

Multiply. See Example 2.

25. $3x(x + 4) = \mathbf{3x}(x) + \mathbf{3x}(4)$
$$= 3(x^{1+1}) + 12x$$
$$= 3x^2 + 12x$$

27. $-4t(t^2 - 7) = -\mathbf{4t}(t^2) - \mathbf{4t}(-7)$
$$= -4(t^{1+2}) + 28t$$
$$= -4t^3 + 28t$$

29. $-2x^3(3x^2 - x + 1)$
$$= -\mathbf{2x^3}(3x^2) - \mathbf{2x^3}(-x) - \mathbf{2x^3}(1)$$
$$= -6(x^{3+2}) + 2(x^{3+1}) - 2x^3$$
$$= -6x^5 + 2x^4 - 2x^3$$

31. $\dfrac{5}{8}t^2(t^6 + 8t^2) = \dfrac{\mathbf{5}}{\mathbf{8}}t^2(t^6) + \dfrac{\mathbf{5}}{\mathbf{8}}t^2(8t^2)$
$$= \dfrac{5}{8}(t^{2+6}) + 5(t^{2+2})$$
$$= \dfrac{5}{8}t^8 + 5t^4$$

33. $-4x^2 z(3x^2 + z^2 + xz - 1)$
$$= -\mathbf{4x^2 z}(3x^2) - \mathbf{4x^2 z}(z^2) - \mathbf{4x^2 z}(xz) - \mathbf{4x^2 z}(-1)$$
$$= -12(x^{2+2}z) - 4(x^2 z^{1+2}) - 4(x^{2+1}z^{1+1}) + 4x^2 z$$
$$= -12x^4 z - 4x^2 z^3 - 4x^3 z^2 + 4x^2 z$$

35. $(x^2 - 12x)(6x^{12}) = x^2(\mathbf{6x^{12}}) - 12x(\mathbf{6x^{12}})$
$$= 6(x^{2+12}) - 72(x^{1+12})$$
$$= 6x^{14} - 72x^{13}$$

Find a polynomial that represents the area of the parallelogram or rectangle. See Example 3.

37. $A = bh$
$$= (7h + 3)h$$
$$= (7h)\mathbf{h} + (3)\mathbf{h}$$
$$= 7h^{1+1} + 3h$$
$$= 7h^2 + 3h$$

The area is $(7h^2 + 3h)$ in^2.

39. $A = bh$
$= (4w - 2)w$
$= (4w)w - (2)w$
$= 4w^{1+1} - 2w$
$= 4w^2 - 2w$

The area is $(4w^2 - 2w)$ ft^2.

Multiply. See Examples 4 and 5.

41. $(y + 3)(y + 5) = \textbf{y}(y) + \textbf{y}(5) + \textbf{3}(y) + \textbf{3}(5)$
$= y^2 + 5y + 3y + 15$
$= y^2 + 8y + 15$

43. $(m + 6)(m - 9) = \textbf{m}(m) + \textbf{m}(-9) + \textbf{6}(m) + \textbf{6}(-9)$
$= m^2 - 9m + 6m - 54$
$= m^2 - 3m - 54$

45. $(4y - 5)(y + 7) = \textbf{4y}(y) + \textbf{4y}(7) - \textbf{5}(y) - \textbf{5}(7)$
$= 4y^2 + 28y - 5y - 35$
$= 4y^2 + 23y - 35$

47. $(2x - 3)(6x - 5)$
$= \textbf{2x}(6x) + \textbf{2x}(-5) - \textbf{3}(6x) - \textbf{3}(-5)$
$= 12x^2 - 10x - 18x + 15$
$= 12x^2 - 28x + 15$

49. $(3.8y - 1)(2y - 1)$
$= \textbf{3.8y}(2y) + \textbf{3.8y}(-1) - \textbf{1}(2y) - \textbf{1}(-1)$
$= 7.6y^2 - 3.8y - 2y + 1$
$= 7.6y^2 - 5.8y + 1$

51. $\left(6m - \dfrac{2}{3}\right)\left(3m - \dfrac{4}{3}\right)$

$= \textbf{6m}(3m) + \textbf{6m}\left(-\dfrac{4}{3}\right) + \left(-\dfrac{2}{3}\right)3m + \left(-\dfrac{2}{3}\right)\left(-\dfrac{4}{3}\right)$

$= 18m^2 - 8m - 2m + \dfrac{8}{9}$

$= 18m^2 - 10m + \dfrac{8}{9}$

53. $(t^2 - 3)(t^2 - 4) = \textbf{t}^2(t^2) + \textbf{t}^2(-4) - \textbf{3}(t^2) - \textbf{3}(-4)$
$= t^4 - 4t^2 - 3t^2 + 12$
$= t^4 - 7t^2 + 12$

55. $(3a - 2b)(4a + b)$
$= \textbf{3a}(4a) + \textbf{3a}(b) - \textbf{2b}(4a) - \textbf{2b}(b)$
$= 12a^2 + 3ab - 8ab - 2b^2$
$= 12a^2 - 5ab - 2b^2$

Multiply. See Example 6.

57. $(x + 2)(x^2 - 2x + 3)$
$= \textbf{x}(x^2 - 2x + 3) + \textbf{2}(x^2 - 2x + 3)$
$= \textbf{x}(x^2) + \textbf{x}(-2x) + \textbf{x}(3) + \textbf{2}(x^2) + \textbf{2}(-2x) + \textbf{2}(3)$
$= x^3 - 2x^2 + 3x + 2x^2 - 4x + 6$
$= x^3 + (-2 + 2)x^2 + (3 - 4)x + 6$
$= x^3 + 0x^2 - x + 6$
$= x^3 - x + 6$

59. $(4t + 3)(t^2 + 2t + 3)$
$= \textbf{4t}(t^2 + 2t + 3) + \textbf{3}(t^2 + 2t + 3)$
$= \textbf{4t}(t^2) + \textbf{4t}(2t) + \textbf{4t}(3) + \textbf{3}(t^2) + \textbf{3}(2t) + \textbf{3}(3)$
$= 4t^3 + 8t^2 + 12t + 3t^2 + 6t + 9$
$= 4t^3 + (8 + 3)t^2 + (12 + 6)t + 9$
$= 4t^3 + 11t^2 + 18t + 9$

61. $(x^2 + 6x + 7)(2x - 5)$
$= (\textbf{x}^2 + \textbf{6x} + \textbf{7})2x - (\textbf{x}^2 + \textbf{6x} + \textbf{7})5$
$= \textbf{x}^2(2x) + \textbf{6x}(2x) + \textbf{7}(2x) - [\textbf{x}^2(5) + \textbf{6x}(5) + \textbf{7}(5)]$
$= 2x^3 + 12x^2 + 14x - [5x^2 + 30x + 35]$
$= 2x^3 + 12x^2 + 14x - 5x^2 - 30x - 35$
$= 2x^3 + (12 - 5)x^2 + (14 - 30)x - 35$
$= 2x^3 + 7x^2 - 16x - 35$

63. $(r^2 - r + 3)(r^2 - 4r - 5)$
$= \textbf{r}^2(r^2 - 4r - 5) - \textbf{r}(r^2 - 4r - 5) + \textbf{3}(r^2 - 4r - 5)$
$= \textbf{r}^2(r^2) + \textbf{r}^2(-4r) + \textbf{r}^2(-5) - \textbf{r}(r^2) - \textbf{r}(-4r) - \textbf{r}(-5)$
$\quad + \textbf{3}(r^2) + \textbf{3}(-4r) + \textbf{3}(-5)$
$= r^4 - 4r^3 - 5r^2 - r^3 + 4r^2 + 5r + 3r^2 - 12r - 15$
$= r^4 + (-4 - 1)r^3 + (-5 + 4 + 3)r^2 + (5 - 12)r - 15$
$= r^4 - 5r^3 + 2r^2 - 7r - 15$

Multiply using vertical form. See Example 7.

65.
$$
\begin{array}{r}
x^2 - 2x + 1 \\
x + 2 \\
\hline
2x^2 - 4x + 2 \\
x^3 - 2x^2 + x \\
\hline
x^3 - 3x + 2
\end{array}
$$

67.
$$
\begin{array}{r}
4x^2 + 3x - 4 \\
3x + 2 \\
\hline
8x^2 + 6x - 8 \\
12x^3 + 9x^2 - 12x \\
\hline
12x^3 + 17x^2 - 6x - 8
\end{array}
$$

Multiply. See Example 8.

69. $4x(2x + 1)(x - 2)$
$= 4x[\textbf{2x}(x) + \textbf{2x}(-2) + \textbf{1}(x) + \textbf{1}(-2)]$
$= 4x[2x^2 - 4x + x - 2]$
$= 4x[2x^2 - 3x - 2]$
$= \textbf{4x}(2x^2) + \textbf{4x}(-3x) + \textbf{4x}(-2)$
$= 8x^3 - 12x^2 - 8x$

71. $-3a(a + b)(a - b)$
$= -3a[\textbf{a}(a) + \textbf{a}(-b) + \textbf{b}(a) + \textbf{b}(-b)]$
$= -3a[a^2 - ab + ab - b^2]$
$= -3a[a^2 + 0ab - b^2]$
$= \textbf{-3a}(a^2) - \textbf{3a}(-b^2)$
$= -3a^3 + 3ab^2$

Section 5.6

73. $(-2a^2)(-3a^3)(3a-2)$
$= (-2)(-3)(a^{2+3})(3a-2)$
$= 6a^5(3a-2)$
$= 6a^5(3a) + 6a^5(-2)$
$= (6)(3)a^{5+1} + (6)(-2)a^5$
$= 18a^6 - 12a^5$

75. $(x-4)(x+1)(x-3)$
$= (x-4)[x(x) + x(-3) + 1(x) + 1(-3)]$
$= (x-4)[x^2 - 3x + x - 3]$
$= (x-4)[x^2 - 2x - 3]$
$= x(x^2) - x(2x) - x(3) - 4(x^2) - 4(-2x) - 4(-3)$
$= x^{2+1} - 2x^{1+1} - 3x - 4x^2 + 8x + 12$
$= x^3 - 2x^2 - 3x - 4x^2 + 8x + 12$
$= x^3 + (-2-4)x^2 + (-3+8)x + 12$
$= x^3 - 6x^2 + 5x + 12$

Try It Yourself

Multiply.

77. $(5x-2)(6x-1)$
$= 5x(6x) + 5x(-1) - 2(6x) - 2(-1)$
$= 30x^2 - 5x - 12x + 2$
$= 30x^2 + (-5-12)x + 2$
$= 30x^2 - 17x + 2$

79. $(3x^2 + 4x - 7)(2x^2)$
$= 3x^2(2x^2) + 4x(2x^2) - 7(2x^2)$
$= 6x^{2+2} + 8x^{1+2} - 14x^2$
$= 6x^4 + 8x^3 - 14x^2$

81. $2(t+4)(t-3) = 2[t(t) + t(-3) + 4(t) + 4(-3)]$
$= 2(t^2 - 3t + 4t - 12)$
$= 2(t^2 + t - 12)$
$= 2t^2 + 2t - 24$

83.
$$
\begin{array}{r}
2a^2 + 3a + 1 \\
3a^2 - 2a + 4 \\
\hline
8a^2 + 12a + 4 \\
-4a^3 - 6a^2 - 2a \\
6a^4 + 9a^3 + 3a^2 \\
\hline
6a^4 + 5a^3 + 5a^2 + 10a + 4
\end{array}
$$

85. $(t+2s)(9t-3s)$
$= t(9t) + t(-3s) + 2s(9t) + 2s(-3s)$
$= 9t^2 - 3st + 18st - 6s^2$
$= 9t^2 + 15st - 6s^2$

87. $\left(\dfrac{1}{2}a\right)(4a^4)(a^5) = \left(\dfrac{1}{2} \cdot 4\right)(a \cdot a^4 \cdot a^5)$
$= 2a^{1+4+5}$
$= 2a^{10}$

89. $\left(4a - \dfrac{5}{4}r\right)\left(4a + \dfrac{3}{4}r\right)$
$= 4a(4a) + 4a\left(\dfrac{3}{4}r\right) + \left(-\dfrac{5}{4}r\right)4a + \left(-\dfrac{5}{4}r\right)\left(\dfrac{3}{4}r\right)$
$= 16a^2 + 3ar - 5ar - \dfrac{15}{16}r^2$
$= 16a^2 + (3-5)ar - \dfrac{15}{16}r^2$
$= 16a^2 - 2ar - \dfrac{15}{16}r^2$

91. $(a+b)(a+b) = a(a) + a(b) + b(a) + b(b)$
$= a^2 + ab + ab + b^2$
$= a^2 + 2ab + b^2$

93. $(x+6)(x^3 + 5x^2 - 4x - 4)$
$= x(x^3 + 5x^2 - 4x - 4) + 6(x^3 + 5x^2 - 4x - 4)$
$= x(x^3) + x(5x^2) + x(-4x) + x(-4)$
$\qquad + 6(x^3) + 6(5x^2) + 6(-4x) + 6(-4)$
$= x^4 + 5x^3 - 4x^2 - 4x + 6x^3 + 30x^2 - 24x - 24$
$= x^4 + (5+6)x^3 + (-4+30)x^2 + (-4-24)x - 24$
$= x^4 + 11x^3 + 26x^2 - 28x - 24$

95. $9x^2(x^2 - 2x + 6)$
$= 9x^2(x^2) + 9x^2(-2x) + 9x^2(6)$
$= 9(x^{2+2}) - 18(x^{2+1}) + 54x^2$
$= 9x^4 - 18x^3 + 54x^2$

97. $4y(y+3)(y+7)$
$= 4y[y(y) + y(7) + 3(y) + 3(7)]$
$= 4y[y^2 + 7y + 3y + 21]$
$= 4y[y^2 + 10y + 21]$
$= 4y(y^2) + 4y(10y) + 4y(21)$
$= 4y^3 + 40y^2 + 84y$

99. $0.3p^5(0.4p^4 - 6p^2)$
$= 0.3p^5(0.4p^4) + 0.3p^5(-6p^2)$
$= 0.12(p^{5+4}) - 1.8(p^{5+2})$
$= 0.12p^9 - 1.8p^7$

101. $8.2pq(2pq - 3p + 5q)$
$= \mathbf{8.2pq}(2pq) - \mathbf{8.2pq}(3p) + \mathbf{8.2pq}(5q)$
$= 16.4p^2q^2 - 24.6p^2q + 41pq^2$

103. $(-3x + y)(x^2 - 8xy + 16y^2)$
$= \mathbf{-3x}(x^2 - 8xy + 16y^2) + \mathbf{y}(x^2 - 8xy + 16y^2)$
$= \mathbf{-3x}(x^2) - \mathbf{3x}(-8xy) - \mathbf{3x}(16y^2)$
$\qquad + \mathbf{y}(x^2) + \mathbf{y}(-8xy) + \mathbf{y}(16y^2)$
$= -3x^3 + 24x^2y - 48xy^2 + x^2y - 8xy^2 + 16y^3$
$= -3x^3 + (24 + 1)x^2y + (-48 - 8)xy^2 + 16y^3$
$= -3x^3 + 25x^2y - 56xy^2 + 16y^3$

LOOK ALIKES . . .
Perform the indicated operations to simplify each expression, if possible.

105. a. $(x - 2) + (x^2 + 2x + 4)$
$\qquad = x^2 + (1 + 2)x + (-2 + 4)$
$\qquad = x^2 + 3x + 2$

105. b.
$$\begin{array}{r} x^2 + 2x + 4 \\ x - 2 \\ \hline -2x^2 - 4x - 8 \\ x^3 + 2x^2 + 4x \\ \hline x^3 - 8 \end{array}$$

107. a. $(6x^2z^5) - (-3xz^3)$; Does not simplify.
b. $(6x^2z^5)(-3xz^3) = (6 \cdot -3)(x^2x)(z^5z^3)$
$\qquad = -18x^{2+1}z^{5+3}$
$\qquad = -18x^3z^8$

109. a. $(2x^2 - x) - (3x^2 - 3x)$
$\qquad = 2x^2 - x - 3x^2 + 3x$
$\qquad = (2 - 3)x^2 + (-1 + 3)x$
$\qquad = -x^2 + 2x$

b. $(2x^2 - x)(3x^2 - 3x)$
$= \mathbf{2x^2}(3x^2 - 3x) - \mathbf{x}(3x^2 - 3x)$
$= \mathbf{2x^2}(3x^2) + \mathbf{2x^2}(-3x) - \mathbf{x}(3x^2) - \mathbf{x}(-3x)$
$= 6x^{2+2} - 6x^{2+1} - 3x^{2+1} + 3x^{1+1}$
$= 6x^4 - 6x^3 - 3x^3 + 3x^2$
$= 6x^4 - 9x^3 + 3x^2$

111. a. $3a + (4a - 1) + (6a + 2)$
$\qquad = (3 + 4 + 6)a + (-1 + 2)$
$\qquad = 13a + 1$

b. $3a(4a - 1)(6a + 2)$
$= 3a\left[\mathbf{4a}(6a + 2) - \mathbf{1}(6a + 2)\right]$
$= 3a\left[\mathbf{4a}(6a) + \mathbf{4a}(2) - \mathbf{1}(6a) - \mathbf{1}(2)\right]$
$= 3a\left[24a^2 + 8a - 6a - 2\right]$
$= 3a\left[24a^2 + 2a - 2\right]$
$= \mathbf{3a}(24a^2) + \mathbf{3a}(2a) - \mathbf{3a}(2)$
$= 72a^3 + 6a^2 - 6a$

APPLICATIONS
113. STAMPS
$A = lw$
$\quad = (3x - 1)(2x + 1)$
$\quad = \mathbf{3x}(2x) + \mathbf{3x}(1) - \mathbf{1}(2x) - \mathbf{1}(1)$
$\quad = 6x^2 + 3x - 2x - 1$
$\quad = 6x^2 + (3 - 2)x - 1$
$\quad = 6x^2 + x - 1$
The area is $(6x^2 + x - 1)$ cm^2.

115. SUNGLASSES
$A = 0.785ab$
$\quad = 0.785(x + 1)(x - 1)$
$\quad = 0.785[\mathbf{x}(x) - \mathbf{x}(1) + \mathbf{1}(x) - \mathbf{1}(1)]$
$\quad = 0.785[x^2 + x - x - 1]$
$\quad = 0.785[x^2 + (1 - 1)x - 1]$
$\quad = 0.785(x^2 - 1)$
$\quad = 0.785x^2 - 0.785$
The area is $(0.785x^2 - 0.785)$ in^2.

117. LUGGAGE
$V = lwh$
$\quad = (x - 3)(x)(2x + 2)$
$\quad = x(x - 3)(2x + 2)$
$\quad = x[\mathbf{x}(2x) + \mathbf{x}(2) - \mathbf{3}(2x) - \mathbf{3}(2)]$
$\quad = x[2x^2 + 2x - 6x - 6]$
$\quad = x[2x^2 + (2 - 6)x - 6]$
$\quad = x(2x^2 - 4x - 6)$
$\quad = \mathbf{x}(2x^2) - \mathbf{x}(4x) - \mathbf{x}(6)$
$\quad = 2x^{1+2} - 4x^{1+1} - 6x$
$\quad = 2x^3 - 4x^2 - 6x$
The volume is $(2x^3 - 4x^2 - 6x)$ in^3.

WRITING

REVIEW

125. a. $m = \dfrac{\text{rise}}{\text{run}}$

Start at point $(-2, 0)$. Move up the axis one space (rise), then move to the right one space (run) to arrive at the blue line again.

$$m = \frac{1}{1}$$
$$= 1$$

The slope of the blue line is 1.

b. Line 2 is a vertical line. Its slope is always undefined.

c. $m = \dfrac{\text{rise}}{\text{run}}$

Start at point $(-1, 0)$. Move down the axis two spaces (rise), then move to the right three spaces (run) to arrive at the red line again.

$$m = \frac{-2}{3}$$
$$= -\frac{2}{3}$$

The slope of the red line is $-\dfrac{2}{3}$.

d. The x-axis is a horizontal line. Its slope is always zero.

CHALLENGE PROBLEMS

127. a. Find each of the following products.

   i. $(x - 1)(x + 1) = x^2 - 1$

   ii. $(x - 1)(x^2 + x + 1)$
   $= x(x^2 + x + 1) - 1(x^2 + x + 1)$
   $= x^3 + x^2 + x - x^2 - x - 1$
   $= x^3 + (1-1)x^2 + (1-1)x - 1$
   $= x^3 - 1$

   iii. $(x - 1)(x^3 + x^2 + x + 1)$
   $= x(x^3 + x^2 + x + 1) - 1(x^3 + x^2 + x + 1)$
   $= x^4 + x^3 + x^2 + x - x^3 - x^2 - x - 1$
   $= x^4 + (1-1)x^3 + (1-1)x^2 + (1-1)x - 1$
   $= x^4 - 1$

   b. Write a product of two polynomials such that the result is $x^5 - 1$.

   $(x - 1)(x^4 + x^3 + x^2 + x + 1)$

SECTION 5.7

VOCABULARY

Fill in the blanks.

1. Expressions of the form $(x + y)^2$, $(x - y)^2$, and $(x + y)(x - y)$ occur so frequently in algebra that they are called special **products**.

CONCEPTS

3. Complete each special product.

a. $(x + y)^2 = x + 2xy + y^2$

The **square** of the second term

Twice the product of the first and second terms

The square of the **first** term

b. $(x + y)(x - y) = x^2 - y^2$.

The square of the **second** term

The **square** of the first term

NOTATION

Complete each solution to find the product.

5. $(x + 4)^2 = \mathbf{x^2} + 2(x)(4) + \mathbf{4}^2$
$ = x^2 + \mathbf{8x} + 16$

7. $(s + 5)(s - 5) = s^2 - \mathbf{5}^2$
$ = s^2 - \mathbf{25}$

GUIDED PRACTICE

Find each product. See Example 1.

9. $(x + 1)^2 = x^2 + 2(x)(1) + (1)^2$
$ = x^2 + 2x + 1$

11. $(m - 6)^2 = m^2 + 2(m)(-6) + (-6)^2$
$ = m^2 - 12m + 36$

13. $(4x + 5)^2 = (4x)^2 + 2(4x)(5) + (5)^2$
$ = 16x^2 + 40x + 25$

15. $(7m - 2)^2 = (7m)^2 + 2(7m)(-2) + (-2)^2$
$ = 49m^2 - 28m + 4$

17. $(1 - 3y)^2 = (1)^2 + 2(1)(-3y) + (-3y)^2$
$ = 1 - 6y + 9y^2$

19. $(y + 0.9)^2 = y^2 + 2(y)(0.9) + (0.9)^2$
$ = y^2 + 1.8y + 0.81$

21. $(a^2 + b^2)^2 = (a^2)^2 + 2(a^2)(b^2) + (b^2)^2$
$ = a^{2\cdot2} + 2a^2b^2 + b^{2\cdot2}$
$ = a^4 + 2a^2b^2 + b^4$

23. $\left(s + \dfrac{3}{4}\right)^2 = s^2 + 2(s)\left(\dfrac{3}{4}\right) + \left(\dfrac{3}{4}\right)^2$
$\phantom{\left(s + \dfrac{3}{4}\right)^2} = s^2 + \dfrac{3}{2}s + \dfrac{9}{16}$

Find each product. See Example 2.

25. $(x + 3)(x - 3) = x^2 - (3)^2$
$ = x^2 - 9$

27. $(2p + 7)(2p - 7) = (2p)^2 - (7)^2$
$ = 4p^2 - 49$

29. $(3n + 1)(3n - 1) = (3n)^2 - (1)^2$
$ = 9n^2 - 1$

31. $\left(c + \dfrac{3}{4}\right)\left(c - \dfrac{3}{4}\right) = c^2 - \left(\dfrac{3}{4}\right)^2$
$\phantom{\left(c + \dfrac{3}{4}\right)\left(c - \dfrac{3}{4}\right)} = c^2 - \dfrac{9}{16}$

33. $(0.4 - 9m^2)(0.4 + 9m^2) = (0.4)^2 - (9m^2)^2$
$ = 0.16 - 81m^{2\cdot2}$
$ = 0.16 - 81m^4$

35. $(5 - 6g)(5 + 6g) = 5^2 - (6g)^2$
$ = 25 - 36g^2$

Expand each binomial. See Example 3.

37. $(x + 4)^3$
$ = (x + 4)^2(x + 4)$
$ = [x^2 + 2(x)(4) + (4)^2](x + 4)$
$ = (x^2 + 8x + 16)(x + 4)$
$ = \mathbf{x^2}(x) + \mathbf{x^2}(4) + \mathbf{8x}(x) + \mathbf{8x}(4)$
$ + \mathbf{16}(x) + \mathbf{16}(4)$
$ = x^3 + 4x^2 + 8x^2 + 32x + 16x + 64$
$ = x^3 + (\mathbf{4 + 8})x^2 + (\mathbf{32 + 16})x + 64$
$ = x^3 + 12x^2 + 48x + 64$

Section 5.7

39. $(n-6)^3$

$= (n-6)^2(n-6)$

$= [n^2 + 2(n)(-6) + (-6)^2](n-6)$

$= (n^2 - 12n + 36)(n-6)$

$= \boldsymbol{n^2}(n) + \boldsymbol{n^2}(-6) - \boldsymbol{12n}(n) - \boldsymbol{12n}(-6)$
$\qquad\qquad + \boldsymbol{36}(n) + \boldsymbol{36}(-6)$

$= n^3 - 6n^2 - 12n^2 + 72n + 36n - 216$

$= n^3 + (\boldsymbol{-6-12})n^2 + (\boldsymbol{72+36})n - 216$

$= n^3 - 18n^2 + 108n - 216$

41. $(2g-3)^3$

$= (2g-3)^2(2g-3)$

$= [(2g)^2 + 2(2g)(-3) + (-3)^2](2g-3)$

$= (4g^2 - 12g + 9)(2g-3)$

$= \boldsymbol{4g^2}(2g) + \boldsymbol{4g^2}(-3) - \boldsymbol{12g}(2g) - \boldsymbol{12g}(-3)$
$\qquad\qquad + \boldsymbol{9}(2g) + \boldsymbol{9}(-3)$

$= 8g^3 - 12g^2 - 24g^2 + 36g + 18g - 27$

$= 8g^3 + (\boldsymbol{-12-24})g^2 + (\boldsymbol{36+18})g - 27$

$= 8g^3 - 36g^2 + 54g - 27$

43. $(a+b)^3$

$= (a+b)^2(a+b)$

$= [a^2 + 2(a)(b) + (b)^2](a+b)$

$= (a^2 + 2ab + b^2)(a+b)$

$= \boldsymbol{a^2}(a) + \boldsymbol{a^2}(b) + \boldsymbol{2ab}(a) + \boldsymbol{2ab}(b)$
$\qquad\qquad + \boldsymbol{b^2}(a) + \boldsymbol{b^2}(b)$

$= a^3 + a^2b + 2a^2b + 2ab^2 + ab^2 + b^3$

$= a^3 + (\boldsymbol{1+2})a^2b + (\boldsymbol{2+1})ab^2 + b^3$

$= a^3 + 3a^2b + 3ab^2 + b^3$

Perform the operations.

45. $2(x^2 + 7x - 1) - 3(x^2 - 2x + 2)$

$= \boldsymbol{2}(x^2) + \boldsymbol{2}(7x) + \boldsymbol{2}(-1)$
$\qquad\qquad - \boldsymbol{3}(x^2) - \boldsymbol{3}(-2x) - \boldsymbol{3}(2)$

$= 2x^2 + 14x - 2 - 3x^2 + 6x - 6$

$= (\boldsymbol{2-3})x^2 + (\boldsymbol{14+6})x + (\boldsymbol{-2-6})$

$= -x^2 + 20x - 8$

47. $(3x+4)(2x-2) - (2x+1)(x+3)$

$= [\boldsymbol{3x}(2x) - \boldsymbol{3x}(2) + \boldsymbol{4}(2x) - \boldsymbol{4}(2)] -$
$\qquad\qquad [\boldsymbol{2x}(x) + \boldsymbol{2x}(3) + \boldsymbol{1}(x) + \boldsymbol{1}(3)]$

$= [6x^2 - 6x + 8x - 8] - [2x^2 + 6x + x + 3]$

$= [6x^2 + (\boldsymbol{-6+8})x - 8] - [2x^2 + (\boldsymbol{6+1})x + 3]$

$= [6x^2 + 2x - 8] - [2x^2 + 7x + 3]$

$= 6x^2 + 2x - 8 - 2x^2 - 7x - 3$

$= (\boldsymbol{6-2})x^2 + (\boldsymbol{2-7})x + (\boldsymbol{-8-3})$

$= 4x^2 - 5x - 11$

49. $-5d(4d-1)^2$

$= -5d[(4d)^2 + 2(4d)(-1) + (-1)^2]$

$= -5d[16d^2 - 8d + 1]$

$= \boldsymbol{-5d}(16d^2) - \boldsymbol{5d}(-8d) - \boldsymbol{5d}(1)$

$= -80d^3 + 40d^2 - 5d$

51. $4d(d^2 + g^3)(d^2 - g^3) = 4d\left[(d^2)^2 - (g^3)^2\right]$

$\qquad\qquad = 4d\left[d^{2\cdot2} - g^{3\cdot2}\right]$

$\qquad\qquad = 4d\left[d^4 - g^6\right]$

$\qquad\qquad = \boldsymbol{4d}(d^4) + \boldsymbol{4d}(-g^6)$

$\qquad\qquad = 4(d^{1+4}) - 4(dg^6)$

$\qquad\qquad = 4d^5 - 4dg^6$

Find a polynomial that represents the area of the figure. Leave π in your answer.

53. $A = \dfrac{1}{2}bh$

$= \dfrac{1}{2}(2x+2)(2x-2)$

$= \dfrac{1}{2}[(2x)^2 - 2^2]$

$= \dfrac{1}{2}(4x^2 - 4)$

$= \dfrac{\boldsymbol{1}}{\boldsymbol{2}}(4x^2) + \dfrac{\boldsymbol{1}}{\boldsymbol{2}}(-4)$

$= 2x^2 - 2$

The area is $(2x^2 - 2)$ yd^2.

55. $A = lw$

$= (3x+1)(3x+1)$

$= (3x)^2 + 2(3x)(1) + 1^2$

$= 9x^2 + 6x + 1$

The area is $(9x^2 + 6x + 1)$ ft^2.

TRY IT YOURSELF
Perform the operations.

57. $(2v^3 - 8)^2 = (2v^3)^2 + 2(2v^3)(-8) + (-8)^2$
$= 4v^{3\cdot2} - 32v^3 + 64$
$= 4v^6 - 32v^3 + 64$

59. $3x(2x+3)(2x+3)$
$= 3x[(2x)^2 + 2(2x)(3) + (3)^2]$
$= 3x[4x^2 + 12x + 9]$
$= 3x(4x^2) + 3x(12x) + 3x(9)$
$= 12x^3 + 36x^2 + 27x$

61. $(4f + 0.4)(4f - 0.4) = (4f)^2 - (0.4)^2$
$= 16f^2 - 0.16$

63. $(r^2 + 10s)^2 = (r^2)^2 + 2(r^2)(10s) + (10s)^2$
$= r^{2\cdot2} + 20r^2s + 100s^2$
$= r^4 + 20r^2s + 100s^2$

65. $2(x+3) + 4(x-2)$
$= 2(x) + 2(3) + 4(x) + 4(-2)$
$= 2x + 6 + 4x - 8$
$= (2+4)x + (6-8)$
$= 6x - 2$

67. $\left(d^4 + \dfrac{1}{4}\right)^2 = (d^4)^2 + 2(d^4)\left(\dfrac{1}{4}\right) + \left(\dfrac{1}{4}\right)^2$
$= d^{4\cdot2} + \dfrac{1}{2}d^4 + \dfrac{1}{16}$
$= d^8 + \dfrac{1}{2}d^4 + \dfrac{1}{16}$

69. $(d+7)(d-7) = d^2 - (7)^2$
$= d^2 - 49$

71. $(2a - 3b)^2 = (2a)^2 + 2(2a)(-3b) + (-3b)^2$
$= 4a^2 - 12ab + 9b^2$

73. $(n+6)(n-6) = n^2 - 6^2$
$= n^2 - 36$

75. $(m+10)^2 - (m-8)^2$
$= (m)^2 + 2(m)(10) + (10)^2$
$\qquad -[(m)^2 + 2(m)(-8) + (-8)^2]$
$= m^2 + 20m + 100 - [m^2 - 16m + 64]$
$= m^2 + 20m + 100 - 1(m^2) - 1(-16m) - 1(64)$
$= m^2 + 20m + 100 - m^2 + 16m - 64$
$= (1-1)m^2 + (20+16)m + (100-64)$
$= 36m + 36$

77. $(2m+n)^3$
$= (2m+n)^2(2m+n)$
$= [(2m)^2 + 2(2m)(n) + (n)^2](2m+n)$
$= (4m^2 + 4mn + n^2)(2m+n)$
$= 4m^2(2m) + 4m^2(n) + 4mn(2m) + 4mn(n)$
$\qquad\qquad + n^2(2m) + n^2(n)$
$= 8m^3 + 4m^2n + 8m^2n + 4mn^2 + 2mn^2 + n^3$
$= 8m^3 + (4+8)m^2n + (4+2)mn^2 + n^3$
$= 8m^3 + 12m^2n + 6mn^2 + n^3$

79. $\left(5m - \dfrac{6}{5}\right)^2 = (5m)^2 + 2(5m)\left(-\dfrac{6}{5}\right) + \left(-\dfrac{6}{5}\right)^2$
$= 25m^2 - 12m + \dfrac{36}{25}$

81. $(r^2 - s^2)^2 = (r^2)^2 + 2(r^2)(-s^2) + (-s^2)^2$
$= r^{2\cdot2} - 2r^2s^2 + s^{2\cdot2}$
$= r^4 - 2r^2s^2 + s^4$

83. $(x-2)^2 = (x)^2 + 2(x)(-2) + (-2)^2$
$= x^2 - 4x + 4$

85. $(r+2)^2 = r^2 + 2(r)(2) + (2)^2$
$= r^2 + 4r + 4$

Section 5.7

87. $(n-2)^4$

$$= (n-2)^2(n-2)^2$$
$$= [(n)^2 + 2(n)(-2) + (-2)^2]$$
$$[(n)^2 + 2(n)(-2) + (-2)^2]$$
$$= (n^2 - 4n + 4)(n^2 - 4n + 4)$$
$$= \mathbf{n^2}(n^2) + \mathbf{n^2}(-4n) + \mathbf{n^2}(4)$$
$$\qquad - \mathbf{4n}(n^2) - \mathbf{4n}(-4n) - \mathbf{4n}(4)$$
$$\qquad\qquad + \mathbf{4}(n^2) + \mathbf{4}(-4n) + \mathbf{4}(4)$$
$$= n^4 - 4n^3 + 4n^2 - 4n^3 + 16n^2 - 16n$$
$$\qquad\qquad + 4n^2 - 16n + 16$$
$$= n^4 + (\mathbf{-4-4})n^3 + (\mathbf{4+16+4})n^2$$
$$\qquad\qquad + (\mathbf{-16-16})n + 16$$
$$= n^4 - 8n^3 + 24n^2 - 32n + 16$$

89. $5(y^2 - 2y - 6) + 6(2y^2 + 2y - 5)$

$$= 5(y^2) - 5(2y) - 5(6) + 6(2y^2) + 6(2y - 6(5)$$
$$= 5y^2 - 10y - 30 + 12y^2 + 12y - 30$$
$$= (5+12)y^2 + (-10+12)y + (-30-30)$$
$$= 17y^2 + 2y - 60$$

91. $(3x-2)^2 + (2x+1)^2$

$$= (3x)^2 + 2(3x)(-2) + (-2)^2$$
$$\qquad\qquad + (2x)^2 + 2(2x)(1) + (1)^2$$
$$= 9x^2 - 12x + 4 + [4x^2 + 4x + 1]$$
$$= 9x^2 - 12x + 4 + 4x^2 + 4x + 1$$
$$= (\mathbf{9+4})x^2 + (\mathbf{-12+4})x + (\mathbf{4+1})$$
$$= 13x^2 - 8x + 5$$

93. $(f-8)^2 = f^2 + 2(f)(-8) + (-8)^2$
$$= f^2 - 16f + 64$$

95. $\left(6b + \dfrac{1}{2}\right)\left(6b - \dfrac{1}{2}\right) = (6b)^2 - \left(\dfrac{1}{2}\right)^2$
$$= 36b^2 - \dfrac{1}{4}$$

97. $3y(y+2) + (y+1)(y-1)$

$$= \left[\mathbf{3y}(y+2)\right] + \left[(y+1)(y-1)\right]$$
$$= \left[\mathbf{3y}(y) + \mathbf{3y}(2)\right] + \left[(y^2 - 1^2)\right]$$
$$= 3y^2 + 6y + y^2 - 1$$
$$= (\mathbf{3+1})y^2 + 6y - 1$$
$$= 4y^2 + 6y - 1$$

99. $(6 - 2d^3)^2 = 6^2 + 2(6)(-2d^3) + (-2d^3)^2$
$$= 36 - 24d^3 + 4d^{3 \cdot 2}$$
$$= 36 - 24d^3 + 4d^6$$

101. $(2e+1)^3$

$$= (2e+1)^2(2e+1)$$
$$= [(2e)^2 + 2(2e)(1) + (1)^2](2e+1)$$
$$= (4e^2 + 4e + 1)(2e+1)$$
$$= \mathbf{4e^2}(2e) + \mathbf{4e^2}(1) + \mathbf{4e}(2e) + \mathbf{4e}(1)$$
$$\qquad\qquad + \mathbf{1}(2e) + \mathbf{1}(1)$$
$$= 8e^3 + 4e^2 + 8e^2 + 4e + 2e + 1$$
$$= 8e^3 + (\mathbf{4+8})e^2 + (\mathbf{4+2})e + 1$$
$$= 8e^3 + 12e^2 + 6e + 1$$

103. $(8x+3)^2 = (8x)^2 + 2(8x)(3) + (3)^2$
$$= 64x^2 + 48x + 9$$

LOOK ALIKES...
Perform the indicated operations.

105. a. $(xy)^2 = x^{1 \cdot 2} y^{1 \cdot 2}$
$$= x^2 y^2$$
  b. $(x+y)^2 = x^2 + 2(x)(y) + y^2$
$$= x^2 + 2xy + y^2$$

107. a. $(2b^2 d)^2 = 2^2 b^{2 \cdot 2} d^{1 \cdot 2}$
$$= 4b^4 d^2$$
  b. $(2b^2 + d)^2 = (2b^2)^2 + 2(2b^2)(d) + d^2$
$$= 2^2 b^{2 \cdot 2} + 4b^2 d + d^2$$
$$= 4b^4 + 4bd + d^2$$

APPLICATIONS

109. PLAYPENS

$$(x+6)^2 = (x)^2 + 2(x)(6) + (6)^2$$
$$= x^2 + 12x + 36$$

The area is $(x^2 + 12x + 36)$ in.^{2}

111. PAPER TOWELS

$$\pi h(R+r)(R-r) = \pi h(R^2 - r^2)$$
$$= \pi h R^2 - \pi h r^2$$

WRITING

113-115. Answers will vary.

REVIEW

117. Simplify:

$$\frac{30}{36} = \frac{5 \cdot 6}{6 \cdot 6}$$

$$= \frac{5 \cdot \cancel{6}^{\,1}}{6 \cdot \cancel{6}_{\,1}}$$

$$= \frac{5}{6}$$

119. Multiply:

$$\frac{7}{8} \cdot \frac{3}{5} = \frac{7 \cdot 3}{8 \cdot 5}$$
$$= \frac{21}{40}$$

120. Divide:

$$\frac{1}{3} \div \frac{4}{5} = \frac{1}{3} \cdot \frac{5}{4}$$
$$= \frac{1 \cdot 5}{3 \cdot 4}$$
$$= \frac{5}{12}$$

CHALLENGE PROBLEMS

121. a. Find two binomials whose product is a binomial.

$$(x + 1)(x - 1) = x^2 - 1$$

b. Find two binomials whose product is a trinomial.

$$(x + 1)(x + 1) = x^2 + 2x + 1$$

c. Find two binomials whose product is a four-term polynomial.

$$(x + a)(x + b) = x^2 + bx + ax + ab$$

Section 5.7

VOCABULARY

Fill in the blanks.

1. The expression $\frac{18x^7}{9x^4}$ is a monomial divided by a

   monomial.

3. The expression $\frac{x^2-8x+12}{x-6}$ is a trinomial divided

   by a **binomial**.

CONCEPTS

Fill in the blanks.

5. The long division method is a series of four steps
   that are repeated. Put them in the correct order:
   subtract   multiply   bring down   divide

   Divide, multiply, subtract, bring down

7. **Fill in the blanks:**   To check an answer of a long
   division, we use the fact that
   divisor • **quotient** + remainder = **dividend**

NOTATION

Complete each solution.

9. $\dfrac{28x^5 - x^3 + 5x^2}{7x^2} = \dfrac{28x^5}{\boxed{7x^2}} - \dfrac{\boxed{x^3}}{7x^2} + \dfrac{5x^2}{\boxed{7x^2}}$

   $\qquad = 4x^{\boxed{5}-\boxed{2}} - \dfrac{x^{3-2}}{\boxed{7}} + 5x^{\boxed{2}-\boxed{2}}$

   $\qquad = \boxed{4x^3} - \dfrac{x}{7} + \dfrac{\boxed{5}}{\boxed{7}}$

11. Write the polynomial $2x^2 - 1 + 5x^4$ in
    descending powers of x and insert
    placeholders for each missing term.

    $5x^4 + 0x^3 + 2x^2 + 0x - 1$

GUIDED PRACTICE

Divide the monomials. See Example 1.

13. $\dfrac{x^5}{x^2} = x^{5-2}$

    $\qquad = x^3$

15. $\dfrac{12h^8}{9h^6} = \dfrac{4h^{8-6}}{3}$

    $\qquad = \dfrac{4h^2}{3}$

17. $\dfrac{-3d^4}{15d^8} = -\dfrac{1}{5}d^{4-8}$

    $\qquad = -\dfrac{1}{5}d^{-4}$

    $\qquad = -\dfrac{1}{5d^4}$

19. $\dfrac{10s^2}{s^3} = 10s^{2-3}$

    $\qquad = 10s^{-1}$

    $\qquad = \dfrac{10}{s}$

21. $\dfrac{8x^3y^2}{40xy^6} = \dfrac{1x^{3-1}y^{2-6}}{5}$

    $\qquad = \dfrac{1x^2y^{-4}}{5}$

    $\qquad = \dfrac{x^2}{5y^4}$

23. $\dfrac{-16r^3y^2}{-4r^2y^7} = 4r^{3-2}y^{2-7}$

    $\qquad = 4r^1y^{-5}$

    $\qquad = \dfrac{4r}{y^5}$

Divide the polynomial by the monomial. See Example 2.

25. $\dfrac{6x+3}{3} = \dfrac{6x}{3} + \dfrac{3}{3}$

    $\qquad = 2x+1$

27. $\dfrac{a-a^3+a^4}{a^4} = \dfrac{a}{a^4} - \dfrac{a^3}{a^4} + \dfrac{a^4}{a^4}$

    $\qquad = a^{1-4} - a^{3-4} + a^{4-4}$

    $\qquad = a^{-3} - a^{-1} + a^0$

    $\qquad = \dfrac{1}{a^3} - \dfrac{1}{a} + 1$

29. $\dfrac{6h^{12}+48h^9}{24h^{10}} = \dfrac{6h^{12}}{24h^{10}} + \dfrac{48h^9}{24h^{10}}$

    $\qquad = \dfrac{h^{12-10}}{4} + 2h^{9-10}$

    $\qquad = \dfrac{h^2}{4} + 2h^{-1}$

    $\qquad = \dfrac{h^2}{4} + \dfrac{2}{h}$

31. $\dfrac{9s^8-18s^5+12s^4}{3s^3} = \dfrac{9s^8}{3s^3} - \dfrac{18s^5}{3s^3} + \dfrac{12s^4}{3s^3}$

    $\qquad = 3s^{8-3} - 6s^{5-3} + 4s^{4-3}$

    $\qquad = 3s^5 - 6s^2 + 4s^1$

    $\qquad = 3s^5 - 6s^2 + 4s$

33. $\dfrac{7c^5 + 21c^4 - 14c^3 - 35c}{7c^2}$

$$= \frac{7c^5}{7c^2} + \frac{21c^4}{7c^2} - \frac{14c^3}{7c^2} - \frac{35c}{7c^2}$$

$$= c^{5-2} + 3c^{4-2} - 2c^{3-2} - 5c^{1-2}$$

$$= c^3 + 3c^2 - 2c^1 - 5c^{-1}$$

$$= c^3 + 3c^2 - 2c - \frac{5}{c^1}$$

$$= c^3 + 3c^2 - 2c - \frac{5}{c}$$

35. $-25x^2y^3 + 30xy^2 - 5xy$

$$= \frac{-25x^2y^3}{-5x^2y^2} + \frac{30xy^2}{-5x^2y^2} - \frac{5xy}{-5x^2y^2}$$

$$= 5x^{2-2}y^{3-2} - 6x^{1-2}y^{2-2} + x^{1-2}y^{1-2}$$

$$= 5x^0y^1 - 6x^{-1}y^0 + x^{-1}y^{-1}$$

$$= 5y - \frac{6}{x} + \frac{1}{xy}$$

Perform each division. See Examples 3 and 4.

37.
$$\begin{array}{r} x + 6 \\ x+2\overline{)x^2 + 8x + 12} \\ \underline{-(x^2 + 2x)} \\ 6x + 12 \\ \underline{-(6x + 12)} \\ 0 \end{array}$$

39.
$$\begin{array}{r} x - 2 \\ x-3\overline{)x^2 - 5x + 6} \\ \underline{-(x^2 - 3x)} \\ -2x + 6 \\ \underline{-(-2x + 6)} \\ 0 \end{array}$$

41.
$$\begin{array}{r} x + 1 + \frac{-1}{2x+3} \\ 2x+3\overline{)2x^2 + 5x + 2} \\ \underline{-(2x^2 + 3x)} \\ 2x + 2 \\ \underline{-(2x + 3)} \\ -1 \end{array}$$

43.
$$\begin{array}{r} 2x - 3 + \frac{-1}{3x-1} \\ 3x-1\overline{)6x^2 - 11x + 2} \\ \underline{-(6x^2 - 2x)} \\ -9x + 2 \\ \underline{-(-9x + 3)} \\ -1 \end{array}$$

Perform each division. See Example 5.

45.
$$\begin{array}{r} 2x - 1 \\ x+2\overline{)2x^2 + 3x - 2} \\ \underline{-(2x^2 + 4x)} \\ -x - 2 \\ \underline{-(-x - 2)} \\ 0 \end{array}$$

47.
$$\begin{array}{r} 2x + 1 \\ 5x+3\overline{)10x^2 + 11x + 3} \\ \underline{-(10x^2 + 6x)} \\ 5x + 3 \\ \underline{-(5x + 3)} \\ 0 \end{array}$$

Perform each division. See Example 6.

49.
$$\begin{array}{r} a - 5 \\ a+5\overline{)a^2 + \mathbf{0}a - 25} \\ \underline{-(a^2 + 5a)} \\ -5a - 25 \\ \underline{-(-5a - 25)} \\ 0 \end{array}$$

51.
$$\begin{array}{r} x + 1 \\ x-1\overline{)x^2 + \mathbf{0}x - 1} \\ \underline{-(x^2 - x)} \\ x - 1 \\ \underline{-(x - 1)} \\ 0 \end{array}$$

53.
$$\begin{array}{r} 2x - 3 \\ 2x+3\overline{)4x^2 + \mathbf{0}x - 9} \\ \underline{-(4x^2 + 6x)} \\ -6x - 9 \\ \underline{-(-6x - 9)} \\ 0 \end{array}$$

55.
$$9b - 7 \overline{\smash{\big)}\, 81b^2 + \mathbf{0b} - 49}$$
$$\underline{-(81b^2 - 63b)}$$
$$63b - 49$$
$$\underline{-(63b - 49)}$$
$$0$$

quotient: $9b + 7$

65.
$$2a + 3 \overline{\smash{\big)}\, 6a^2 + 5a - 6}$$
$$\underline{-(6a^2 + 9a)}$$
$$-4a - 6$$
$$\underline{-(-4a - 6)}$$
$$0$$

quotient: $3a - 2$

TRY IT YOURSELF
Perform each division.

57.
$$y + 1 \overline{\smash{\big)}\, y^2 + 13y + 13}$$
$$\underline{-(y^2 + y)}$$
$$12y + 13$$
$$\underline{-(12y + 12)}$$
$$1$$

quotient: $y + 12 + \dfrac{1}{y+1}$

59.
$$\frac{15a^8 b^2 - 10a^2 b^5}{5a^3 b^2} = \frac{15a^8 b^2}{5a^3 b^2} - \frac{10a^2 b^5}{5a^3 b^2}$$
$$= \frac{3a^{8-3} b^{2-2}}{1} - \frac{2a^{2-3} b^{5-2}}{1}$$
$$= \frac{3a^5 b^0}{1} - \frac{2a^{-1} b^3}{1}$$
$$= 3a^5 - \frac{2b^3}{a}$$

61.
$$3x + 2 \overline{\smash{\big)}\, 6x^3 + 10x^2 + 7x + 2}$$
$$\underline{-(6x^3 + 4x^2)}$$
$$6x^2 + 7x$$
$$\underline{-(6x^2 + 4x)}$$
$$3x + 2$$
$$\underline{-(3x + 2)}$$
$$0$$

quotient: $2x^2 + 2x + 1$

63.
$$\frac{8x^9 - 32x^6}{4x^4} = \frac{8x^9}{4x^4} - \frac{32x^6}{4x^4}$$
$$= 2x^{9-4} - 8x^{6-4}$$
$$= 2x^5 - 8x^2$$

67.
$$\frac{45m^{10}}{9m^5} = 5 \cdot m^{10-5}$$
$$= 5m^5$$

69.
$$3b + 2 \overline{\smash{\big)}\, 3b^2 + 11b + 6}$$
$$\underline{-(3b^2 + 2b)}$$
$$9b + 6$$
$$\underline{-(9b + 6)}$$
$$0$$

quotient: $b + 3$

71.
$$2x - 7 \overline{\smash{\big)}\, 2x^2 - x - 21}$$
$$\underline{-(2x^2 - 7x)}$$
$$6x - 21$$
$$\underline{-(6x - 21)}$$
$$0$$

quotient: $x + 3$

73.
$$x + 1 \overline{\smash{\big)}\, x^3 + \mathbf{0x^2} + \mathbf{0x} + 1}$$
$$\underline{-(x^3 + x^2)}$$
$$-x^2 + 0x$$
$$\underline{-(-x^2 - x)}$$
$$x + 1$$
$$\underline{-(x + 1)}$$
$$0$$

quotient: $x^2 - x + 1$

75.
$$\frac{-65rs^2}{15r^2 s^5} = \frac{-13r^{1-2} s^{2-5}}{3}$$
$$= \frac{-13r^{-1} s^{-3}}{3}$$
$$= -\frac{13}{3rs^3}$$

77. $\dfrac{-18w^6-9}{9w^4}=\dfrac{-18w^6}{9w^4}-\dfrac{-9}{9w^4}$

$\qquad\qquad =-2w^{6-4}-\dfrac{1}{w^4}$

$\qquad\qquad =-2w^2-\dfrac{1}{w^4}$

79. $\dfrac{9m-6}{m}=\dfrac{9m}{m}-\dfrac{6}{m}$

$\qquad\qquad =9m^{1-1}-\dfrac{6}{m}$

$\qquad\qquad =9m^0-\dfrac{6}{m}$

$\qquad\qquad =9-\dfrac{6}{m}$

81.
$$
\begin{array}{r}
y^2+2y+5+\dfrac{10}{y-2}\\[2pt]
y-2\overline{)\,y^3+\mathbf{0}\mathbf{y^2}+y+\mathbf{0}}\\
\underline{-(y^3-2y^2)}\\
2y^2+y\\
\underline{-(2y^2-4y)}\\
5y+0\\
\underline{-(5y-10)}\\
10
\end{array}
$$

83. $\dfrac{5x^4-10x}{25x^3}=\dfrac{5x^4}{25x^3}-\dfrac{10x}{25x^3}$

$\qquad\qquad =\dfrac{x^{4-3}}{5}-\dfrac{2x^{1-3}}{5}$

$\qquad\qquad =\dfrac{x}{5}-\dfrac{2x^{-2}}{5}$

$\qquad\qquad =\dfrac{x}{5}-\dfrac{2}{5x^2}$

85.
$$
\begin{array}{r}
x^2-2x+1\\
4x+3\overline{)\,4x^3-5x^2-2x+3}\\
\underline{-(4x^3+3x^2)}\\
-8x^2-2x\\
\underline{-(-8x-6x)}\\
4x+3\\
\underline{-(4x+3)}\\
0
\end{array}
$$

87.
$$
\begin{array}{r}
x+1+\dfrac{10}{x+5}\\
x+5\overline{)\,x^2+6x+15}\\
\underline{-(x^2+5x)}\\
x+15\\
\underline{-(x+5)}\\
10
\end{array}
$$

89. $\dfrac{12x^3y^2-8x^2y-4x}{4xy}$

$\qquad =\dfrac{12x^3y^2}{4xy}-\dfrac{8x^2y}{4xy}-\dfrac{4x}{4xy}$

$\qquad =3x^{3-1}y^{2-1}-2x^{2-1}y^{1-1}-\dfrac{x^{1-1}}{y}$

$\qquad =3x^2y^1-2x^1y^0-\dfrac{x^0}{y}$

$\qquad =3x^2y-2x-\dfrac{1}{y}$

91.
$$
\begin{array}{r}
a-12+\dfrac{4}{a-5}\\
a-5\overline{)\,a^2-17a+64}\\
\underline{-(a^2-5a)}\\
-12a+64\\
\underline{-(-12a+60)}\\
4
\end{array}
$$

93.
$$
\begin{array}{r}
a^2+a+1\\
a-1\overline{)\,a^3+\mathbf{0}\mathbf{a^2}+\mathbf{0}\mathbf{a}-1}\\
\underline{-(a^3-a^2)}\\
a^2+0a\\
\underline{-(a^2-a)}\\
a-1\\
\underline{-(a-1)}\\
0
\end{array}
$$

95.
$$
\begin{array}{r}
2x^2+x+1+\dfrac{2}{3x-1}\\
3x-1\overline{)\,6x^3+x^2+2x+1}\\
\underline{-(6x^3-2x^2)}\\
3x^2+2x\\
\underline{-(3x^2-x)}\\
3x+1\\
\underline{-(3x-1)}\\
2
\end{array}
$$

Section 5.8

97.
$$\frac{8x^{17}y^{20}}{16x^{15}y^{30}} = \frac{x^{17-15}y^{20-30}}{2}$$
$$= \frac{x^2 y^{-10}}{2}$$
$$= \frac{x^2}{2y^{10}}$$

99.

$$\begin{array}{r} 3m-8 \\ 2m+5{\overline{\smash{\big)}\,6m^2 - m - 40}} \\ \underline{-(6m^2 +15m)} \\ -16m-40 \\ \underline{-(-16m-40)} \\ 0 \end{array}$$

LOOK ALIKES…
Perform the indicated operations.

101. a.
$$\frac{16x^2 -16x -5}{4x} = \frac{16x^2}{4x} - \frac{16x}{4x} - \frac{5}{4x}$$
$$= 4x^{2-1} - 4x^{1-1} - \frac{5}{4x}$$
$$= 4x - 4 - \frac{5}{4x}$$

b.
$$\begin{array}{r} 4x-5 \\ 4x+1{\overline{\smash{\big)}\,16x^2 -16x -5}} \\ \underline{-(16x^2 +4x)} \\ -20x-5 \\ \underline{-(-20x-5)} \\ 0 \end{array}$$

APPLICATIONS
103. FURNACE FILTERS

$$\begin{array}{r} x-6 \\ x+4{\overline{\smash{\big)}\,x^2 -2x -24}} \\ \underline{-(x^2 +4x)} \\ -6x-24 \\ \underline{-(-6x-24)} \\ 0 \end{array}$$

The length is $(x-6)$ in.

105. POOL
$$\frac{6x^2 -3x +9}{3} = \frac{6x^2}{3} - \frac{3x}{3} + \frac{+9}{3}$$
$$= \frac{2 \cdot \overset{1}{\cancel{3}}\, x^2}{\underset{1}{\cancel{3}}} - \frac{\overset{1}{\cancel{3}}\, x}{\underset{1}{\cancel{3}}} + \frac{\overset{1}{\cancel{3}} \cdot 3}{\underset{1}{\cancel{3}}}$$
$$= 2x^2 - x + 3$$

The length of one side is $(2x^2 - x +3)$ in.

WRITING
107-109.   Answers will vary.

REVIEW
111. Write an equation of the line with slope $-\frac{11}{6}$ that passes through (2, –6). Write the answer in slope–intercept form.

$$y - y_1 = m(x - x_1)$$
$$y - (-6) = -\frac{11}{6}(x - 2)$$
$$y + 6 = -\frac{11}{6}(x) - \frac{11}{6}(-2)$$
$$y + 6 = -\frac{11}{6}x + \frac{11}{3}$$
$$y + 6 - \mathbf{6} = -\frac{11}{6}x + \frac{11}{3} - \mathbf{6}$$
$$y = -\frac{11}{6}x + \frac{11}{3} - \frac{\mathbf{18}}{\mathbf{3}}$$
$$y = -\frac{11}{6}x - \frac{7}{3}$$

CHALLENGE PROBLEMS
Perform each division.

113. $\dfrac{6a^3 -17a^2b +14ab^2 -3b^3}{2a -3b}$

$$\begin{array}{r} 3a^2 -4ab + b^2 \\ 2a-3b{\overline{\smash{\big)}\,6a^3 -17a^2b +14ab^2 -3b^3}} \\ \underline{-(6a^3 -9a^2b)} \\ -8a^2b +14ab^2 \\ \underline{-(-8a^2b +12ab^2)} \\ 2ab^2 -3b^3 \\ \underline{-(2ab^2 -3b^3)} \\ 0 \end{array}$$

115. $(x^6 + 2x^4 - 6x^2 - 9) \div (x^2 + 3)$

$$
\begin{array}{r}
x^4 \qquad\qquad -x^2 \;-\; 3 \\[2pt]
x^2 + 0x + 3 \overline{)\; x^6 + 0x^5 + 2x^4 + 0x^3 - 6x^2 + 0x - 9} \\[2pt]
\underline{-(x^6 - 0x^5 + 3x^4)\qquad\qquad\qquad\qquad} \\[2pt]
-x^4 + 0x^3 - 6x^2 \qquad\qquad \\[2pt]
\underline{-(-x^4 + 0x^3 - 3x^2)\qquad\qquad} \\[2pt]
-3x^2 + 0x - 9 \\[2pt]
\underline{-(-3x^2 - 0x - 9)} \\[2pt]
0
\end{array}
$$

117. $\dfrac{a^8 + a^6 - 4a^4 + 5a^2 - 3}{a^4 + 2a^2 - 3}$

$$
\begin{array}{r}
a^4 \;-\; a^2 \;+\; 1 \\[2pt]
a^4 + 2a^2 - 3 \overline{)\; a^8 + \; a^6 - 4a^4 + 5a^2 - 3} \\[2pt]
\underline{-(a^8 + 2a^6 - 3a^4)\qquad\qquad\qquad} \\[2pt]
-a^6 - \; a^4 + 5a^2 \qquad \\[2pt]
\underline{-(-a^6 - 2a^4 + 3a^2)\qquad} \\[2pt]
a^4 + 2a^2 - 3 \\[2pt]
\underline{-(a^4 + 2a^2 - 3)} \\[2pt]
0
\end{array}
$$

Section 5.8

CHAPTER 5 REVIEW
SECTION 5.1
Rules for Exponents

1. Identify the base and the exponent in each expression.
   a. n^{12} base n, exponent 12
   b. $(2x)^6$ base $2x$, exponent 6
   c. $3r^4$ base r, exponent 4
   d. $(y-7)^3$ base $y-7$, exponent 3

2. Write each expression in an equivalent form using an exponent.
   a. $m \cdot m \cdot m \cdot m \cdot m = \boldsymbol{m^5}$

   b. $-3 \cdot x \cdot x \cdot x \cdot x = \boldsymbol{-3x^4}$

   c. $(x+8)(x+8) = \boldsymbol{(x+8)^2}$

   d. $\left(\dfrac{1}{2}pq\right)\left(\dfrac{1}{2}pq\right)\left(\dfrac{1}{2}pq\right) = \boldsymbol{\left(\dfrac{1}{2}pq\right)^3}$

Simplify each expression. Assume there are no divisions by 0.

3. $7^4 \cdot 7^8 = 7^{4+8}$
   $\qquad = 7^{12}$

4. $mmnn = m^2n^2$

5. $(y^7)^3 = y^{7 \cdot 3}$
   $\qquad = y^{21}$

6. $(3x)^4 = 81x^4$

7. $\dfrac{b^{12}}{b^3} = b^{12-3}$
   $\qquad = b^9$

8. $-b^3b^4b^5 = -b^{12}$

9. $(-16s^3)^2 s^4 = (-16)^2 s^{3 \cdot 2} s^4$
   $\qquad\qquad = 256s^6 s^4$
   $\qquad\qquad = 256s^{6+4}$
   $\qquad\qquad = 256s^{10}$

10. $(2.1x^2y)^2 = 4.41x^4y^2$

11. $[(-9)^3]^5 = (-9)^{3 \cdot 5}$
    $\qquad\quad = (-9)^{15}$

12. $(a^5)^3(a^2)^4 = (a^{15})(a^8)$
    $\qquad\qquad = a^{23}$

13. $\left(\dfrac{1}{2}x^2x^3\right)^3 = \left(\dfrac{1}{2}\right)^3 (x^{2+3})^3$
    $\qquad\quad = \dfrac{1}{8}(x^5)^3$
    $\qquad\quad = \dfrac{1}{8}x^{5 \cdot 3}$
    $\qquad\quad = \dfrac{1}{8}x^{15}$

14. $\left(\dfrac{x^7}{3xy}\right)^2 = \left(\dfrac{x^{7-1}}{3y}\right)^2$
    $\qquad\quad = \dfrac{x^{6 \cdot 2}}{3^2 y^2}$
    $\qquad\quad = \dfrac{x^{12}}{9y^2}$

15. $\dfrac{(m-25)^{16}}{(m-25)^4} = (m-25)^{16-4}$
    $\qquad\qquad = (m-25)^{12}$

16. $\dfrac{(5y^2z^3)^3}{(yz)^5} = \dfrac{5^3 y^{2 \cdot 3} z^{3 \cdot 3}}{y^5 z^5}$
    $\qquad\quad = \dfrac{125y^6 z^9}{y^5 z^5}$
    $\qquad\quad = 125y^{6-5} z^{9-5}$
    $\qquad\quad = 125yz^4$

17. $\dfrac{a^5 a^4 a^5}{a^2 a} = \dfrac{a^{5+4+5}}{a^{2+1}}$
    $\qquad\quad = \dfrac{a^{14}}{a^3}$
    $\qquad\quad = a^{14-3}$
    $\qquad\quad = a^{11}$

18. $\dfrac{(cd)^9}{(cd)^4} = (cd)^{9-4}$
    $\qquad\quad = (cd)^5$
    $\qquad\quad = c^5d^5$

Find an expression that represents the area or the volume of each figure, whichever is appropriate.

19. $V = s^3$

$= (4x^4)^3$

$= 4^3 x^{4 \cdot 3}$

$= 64x^{12}$

The volume is $64x^{12}$ in.^{3}

20. $A = s^2$

$= (y^2)^2$

$= y^{2 \cdot 2}$

$= y^4$

The area is y^4 ft^2.

SECTION 5.2 REVIEW EXERCISES
Zero and Negative Exponents

Simplify each expression. Do not use negative exponents in the answer.

21. $x^0 = 1$

22. $(3x^2 y^2)^0 = 1$

23. $3x^0 = 3(1)$

$= 3$

24. $10^{-3} = \dfrac{1}{10^3}$

$= \dfrac{1}{1,000}$

25. $-5^{-2} = \dfrac{-1}{5^2}$

$= -\dfrac{1}{25}$

26. $\dfrac{t^4}{t^{10}} = t^{4-10}$

$= t^{-6}$

$= \dfrac{1}{t^6}$

27. $\dfrac{8}{x^{-5}} = 8x^5$

28. $-6y^4 y^{-5} = -6y^{4-5}$

$= -6y^{-1}$

$= -\dfrac{6}{y}$

29. $\dfrac{7^{-2}}{2^{-3}} = \dfrac{2^3}{7^2}$

$= \dfrac{8}{49}$

30. $(x^{-3} x^{-4})^{-2} = (x^{-7})^{-2}$

$= x^{-7 \cdot -2}$

$= x^{14}$

31. $\left(\dfrac{-3r^4 r^{-3}}{r^{-3} r^7}\right)^3 = \left(\dfrac{-3r^{4-3}}{r^{-3+7}}\right)^3$

$= \left(\dfrac{-3r^1}{r^4}\right)^3$

$= \left(\dfrac{-3r^{1-4}}{1}\right)^3$

$= \left(\dfrac{-3r^{-3}}{1}\right)^3$

$= \left(\dfrac{-3}{r^3}\right)^3$

$= \dfrac{(-3)^3}{r^{3 \cdot 3}}$

$= -\dfrac{27}{r^9}$

32. $\left(\dfrac{4z^4}{z^3}\right)^{-2} = \left(\dfrac{z^3}{4z^4}\right)^2$

$= \left(\dfrac{1}{4z^{4-3}}\right)^2$

$= \left(\dfrac{1}{4z}\right)^2$

$= \dfrac{1^2}{4^2 z^2}$

$= \dfrac{1}{16z^2}$

Chapter 5 Review and Test

33.
$$\frac{3^{-2}c^3 d^3}{2^{-3}c^2 d^8} = \frac{2^3 c^{3-2} d^{3-8}}{3^2}$$
$$= \frac{8c^1 d^{-5}}{9}$$
$$= \frac{8c}{9d^5}$$

34.
$$\frac{t^{-30}}{t^{-60}} = t^{-30-(-60)}$$
$$= t^{-30+60}$$
$$= t^{30}$$

35.
$$\frac{w(w^{-3})^{-4}}{w^{-9}} = \frac{w(w^{-3\cdot-4})}{w^{-9}}$$
$$= \frac{w(w^{12})}{w^{-9}}$$
$$= \frac{w^{1+12+9}}{1}$$
$$= w^{22}$$

36.
$$\left(\frac{4}{f^4}\right)^{-10} = \left(\frac{f^4}{4}\right)^{10}$$
$$= \frac{f^{4\cdot10}}{4^{10}}$$
$$= \frac{f^{40}}{4^{10}}$$

SECTION 5.3 REVIEW EXERCISES
Scientific Notation
Write each number in scientific notation.

37. Move the decimal 8 places to the left.
$$720,000,000 = 7.2 \times 10^8$$

38. Move the decimal 15 places to the left.
$$9,370,000,000,000,000 = 9.37 \times 10^{15}$$

39. Move the decimal 9 places to the right.
$$0.00000000942 = 9.42 \times 10^{-9}$$

40. Move the decimal 4 places to the right.
$$0.00013 = 1.3 \times 10^{-4}$$

41. Move the decimal 2 places to the right.
$$0.018 \times 10^{-2} = 1.8 \times 10^{-2+(-2)}$$
$$= 1.8 \times 10^{-4}$$

42. Move the decimal 2 places to the left.
$$853 \times 10^3 = 8.53 \times 10^{3+2}$$
$$= 8.53 \times 10^5$$

Write each number in standard notation.

43. The exponent is 5.
Move the decimal 5 places to the right.
$$1.26 \times 105 = 126,000$$

44. The exponent is -8.
Move the decimal 8 places to the left.
$$3.919 \times 10^{-8} = 0.00000003919$$

45. The exponent is 0.
Do not move the decimal.
$$2.68 \times 10^0 = 2.68$$

46. The exponent is 1.
Move the decimal 1 place to the right.
$$5.76 \times 10^1 = 57.6$$

Evaluate each expression by first writing each number in scientific notation. Express the result in scientific notation and standard notation.

47.
$$\frac{(0.000012)(0.000004)}{0.00000016} = \frac{(1.2 \times 10^{-5})(4.0 \times 10^{-6})}{1.6 \times 10^{-7}}$$
$$= \frac{(1.2 \cdot 4.0)}{1.6} \times \frac{10^{-5+(-6)}}{10^{-7}}$$
$$= \frac{(1.2 \cdot 4.0)}{1.6} \times \frac{10^{-11}}{10^{-7}}$$
$$= \frac{(1.2 \cdot 4.0)}{1.6} \times 10^{-11-(-7)}$$
$$= 3.0 \times 10^{-11+7}$$
$$= 3.0 \times 10^{-4}$$
$$= 0.0003$$

48.
$$\frac{(4,800,000)(20,000,000)}{600,000}$$
$$= \frac{(4.8 \times 10^6)(2.0 \times 10^7)}{6.0 \times 10^5}$$
$$= \frac{(4.8 \cdot 2.0)}{6} \times \frac{10^{13}}{10^5}$$
$$= \frac{(4.8 \cdot 2.0)}{6} \times 10^{13-5}$$
$$= 1.6 \times 10^8$$
$$= 160,000,000$$

49. WORLD POPULATION
6.57 billion $= 6,570,000,000$
Move the decimal 9 places to the left.
$$6,570,000,000 = 6.57 \times 10^9$$

50. ATOMS
Move the decimal 5 places to the right.
$$1.0 \times 10^5 = 100,000$$

SECTION 5.4 REVIEW EXERCISES
Polynomials

51. Consider the polynomial $3x^3 - x^2 + x + 10$.

  a. How many terms does the polynomial have?
  4

  b. What is the lead term? **$3x^3$**

  c. What is the coefficient of each term?

  3, –1, 1, 10

  d. What is the constant term? **10**

52. Find the degree of each polynomial and classify it as a monomial, binomial, trinomial, or none of these.

  a. $13x^7$ **7th, monomial**
  b. $-16a^2b$ **3rd, monomial**
  c. $5^3x + x^2$ **2nd, binomial**
  d. $-3x^5 + x - 1$ **5th, trinomial**
  e. $9xy^2 + 21x^3y^3$ **6th, binomial**
  f. $4s^4 - 3s^2 + 5s + 4$ **4th, none of these**

53. Evaluate $-x^5 - 3x^4 + 3$
  for $x = 0$ and $x = -2$.
  $-x^5 - 3x^4 + 3 = -(0)^5 - 3(0)^4 + 3$
  $$= 3$$

  $-x^5 - 3x^4 + 3 = -(-2)^5 - 3(-2)^4 + 3$
  $$= -(-32) - 3(16) + 3$$
  $$= 32 - 48 + 3$$
  $$= -16 + 3$$
  $$= -13$$

54. DIVING
  $0.1875x^2 - 0.0078125x^3$
  $$= 0.1875(8)^2 - 0.0078125(8)^3$$
  $$= 0.1875(64) - 0.0078125(512)$$
  $$= 12 - 4$$
  $$= 8$$
  The amount of deflection is 8 in.

55. $y = x^2$

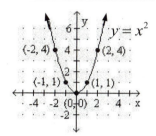

x	y	(x, y)
-2	$y = (-2)^2$ $= 4$	$(-2, 4)$
-1	$y = (-1)^2$ $= 1$	$(-1, 1)$
0	$y = (0)^2$ $= 0$	$(0, 0)$
1	$y = (1)^2$ $= 1$	$(1, 1)$
2	$y = (2)^2$ $= 4$	$(2, 4)$

56. $y = x^3 + 1$

x	y	(x, y)
-2	$y = (-2)^3 + 1$ $= -8 + 1$ $= -7$	$(-2, -7)$
-1	$y = (-1)^3 + 1$ $= -1 + 1$ $= 0$	$(-1, 0)$
0	$y = 0^3 + 1$ $= 1$	$(0, 1)$
1	$y = 1^3 + 1$ $= 1 + 1$ $= 2$	$(1, 2)$
2	$y = 2^3 + 1$ $= 8 + 1$ $= 9$	$(2, 9)$

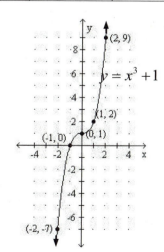

Chapter 5 Review and Test

Adding and Subtracting Polynomials

Simplify each polynomial and write the result in descending powers of one variable.

57. $6y^3 + 8y^4 + 7y^3 + (-8y^4)$

$$= (8-8)y^4 + (6+7)y^3$$
$$= 0y^4 + 13y^3$$
$$= 13y^3$$

58. $4a^2b + 5 - 6a^3b - 3a^2b + 2a^3b + 1$

$$= (-6+2)a^3b + (4-3)a^2b + (5+1)$$
$$= -4a^3b + a^2b + 6$$

59. $\dfrac{5}{6}x^2 + \dfrac{1}{3}y^2 - \dfrac{1}{4}x^2 - \dfrac{3}{4}xy + \dfrac{2}{3}y^2$

$$= \left(\dfrac{5}{6} - \dfrac{1}{4}\right)x^2 - \dfrac{3}{4}xy + \left(\dfrac{1}{3} + \dfrac{2}{3}\right)y^2$$
$$= \left(\dfrac{5}{6}\cdot\dfrac{2}{2} - \dfrac{1}{4}\cdot\dfrac{3}{3}\right)x^2 - \dfrac{3}{4}xy + \left(\dfrac{1}{3} + \dfrac{2}{3}\right)y^2$$
$$= \left(\dfrac{10-3}{12}\right)x^2 - \dfrac{3}{4}xy + \left(\dfrac{1+2}{3}\right)y^2$$
$$= \left(\dfrac{7}{12}\right)x^2 - \dfrac{3}{4}xy + \left(\dfrac{3}{3}\right)y^2$$
$$= \dfrac{7}{12}x^2 - \dfrac{3}{4}xy + y^2$$

60. $7.6c^5 - 2.1c^3 - 0.9c^5 + 8.1c^4$

$$= (7.6 - 0.9)c^5 + 8.1c^4 - 2.1c^3$$
$$= 6.7c^5 + 8.1c^4 - 2.1c^3$$

Perform the operations.

61. $(2r^6 + 14r^3) + (23r^6 - 5r^3 + 5r)$

$$= (2+23)r^6 + (14-5)r^3 + 5r$$
$$= 25r^6 + 9r^3 + 5r$$

62. $(7.1a^2 + 2.2a - 5.8) - (3.4a^2 - 3.9a + 11.8)$

$$= 7.1a^2 + 2.2a - 5.8 - 3.4a^2 + 3.9a - 11.8$$
$$= (7.1 - 3.4)a^2 + (2.2 + 3.9)a + (-5.8 - 11.8)$$
$$= 3.7a^2 + 6.1a - 17.6$$

63. $(3r^3s + r^2s^2 - 3rs^3 - 3s^4)$
$$+ (r^3s - 8r^2s^2 - 4rs^3 + s^4)$$
$$= (3+1)r^3s + (1-8)r^2s^2 + (-3-4)rs^3$$
$$+ (-3+1)s^4$$
$$= 4r^3s - 7r^2s^2 - 7rs^3 - 2s^4$$

64. $\left(\dfrac{7}{8}m^4 - \dfrac{1}{5}m^3\right) - \left(\dfrac{1}{4}m^4 + \dfrac{1}{5}m^3\right) - \dfrac{3}{5}m^3$

$$= \dfrac{7}{8}m^4 - \dfrac{1}{5}m^3 - \dfrac{1}{4}m^4 - \dfrac{1}{5}m^3 - \dfrac{3}{5}m^3$$
$$= \left(\dfrac{7}{8} - \dfrac{1}{4}\right)m^4 + \left(-\dfrac{1}{5} - \dfrac{1}{5} - \dfrac{3}{5}\right)m^3$$
$$= \left(\dfrac{7}{8} - \dfrac{1}{4}\cdot\dfrac{2}{2}\right)m^4 + \left(-\dfrac{5}{5}\right)m^3$$
$$= \left(\dfrac{7}{8} - \dfrac{2}{8}\right)m^4 + (-1)m^3$$
$$= \dfrac{5}{8}m^4 - m^3$$

65. Step 1:

$$(2z^2 + 3z - 7) + (-4z^3 - 2z - 3)$$
$$= -4z^3 + 2z^2 + (3-2)z + (-7-3)$$
$$= -4z^3 + 2z^2 + z - 10$$

Step 2:

$$(-4z^3 + 2z^2 + z - 10) - (-3z^3 - 4z + 7)$$
$$= -4z^3 + 2z^2 + z - 10 + 3z^3 + 4z - 7$$
$$= (-4+3)z^3 + 2z^2 + (1+4)z + (-10-7)$$
$$= -z^3 + 2z^2 + 5z - 17$$

66. **GARDENING**

$$(2x^2 + x + 1) - (x^2 - 2) = 2x^2 + x + 1 - x^2 + 2$$
$$= (2-1)x^2 + x + (1+2)$$
$$= x^2 + x + 3$$

The length of the handle is $(x^2 + x + 3)$ in.

67.
$$\begin{array}{r} 3x^2 + 5x + 2 \\ + \ \underline{x^2 - 3x + 6} \\ 4x^2 + 2x + 8 \end{array}$$

68.
$$\begin{array}{r} 20x^3 \qquad + 12x \\ - \underline{(12x^3 + 7x^2 - 7x)} \end{array} \xrightarrow[\text{and add}]{\text{Change signs}} \begin{array}{r} 20x^3 \qquad + 12x \\ \underline{-12x^3 - 7x^2 + \ 7x} \\ 8x^3 - 7x^2 + 19x \end{array}$$

Multiply.

69. $(2x^2)(5x) = (2 \cdot 5)(x^{2+1})$
$$= 10x^3$$

70. $(-6x^4 z^3)(x^6 z^2) = -6x^{4+6} z^{3+2}$
$$= -6x^{10} z^5$$

71. $5b^3 \cdot 6b^2 \cdot 4b^6 = (5 \cdot 6 \cdot 4)(b^{3+2+6})$
$$= 120 b^{11}$$

72. $\dfrac{2}{3} h^5 (3h^9 + 12h^6) = \dfrac{2}{3} h^5 (3h^9) + \dfrac{2}{3} h^5 (12 h^6)$
$$= \left(\dfrac{2}{3} \cdot 3 \right) h^{5+9} + \left(\dfrac{2}{3} \cdot 12 \right) h^{5+6}$$
$$= 2h^{14} + 8h^{11}$$

73. $3n^2 (3n^2 - 5n + 2)$
$$= 3n^2 (3n^2) + 3n^2 (-5n) + 3n^2 (2)$$
$$= (3 \cdot 3) n^{2+2} + (3 \cdot -5) n^{2+1} + (3 \cdot 2) n^2$$
$$= 9n^4 - 15n^3 + 6n^2$$

74. $x^2 y(y^2 - xy) = x^2 y(y^2) + x^2 y(-xy)$
$$= x^2 y^{1+2} - x^{2+1} y^{1+1}$$
$$= x^2 y^3 - x^3 y^2$$

75. $2x(3x^4)(x+2) = (2 \cdot 3)(x^{1+4})(x+2)$
$$= 6x^5 (x+2)$$
$$= 6x^5 (x) + 6x^5 (2)$$
$$= 6x^{5+1} + (6 \cdot 2) x^5$$
$$= 6x^6 + 12x^5$$

76. $-a^2 b^2 (-a^4 b^2 + a^3 b^3 - ab^4 + 7a)$
$$= -a^2 b^2 (-a^4 b^2) - a^2 b^2 (a^3 b^3)$$
$$\qquad - a^2 b^2 (-ab^4) - a^2 b^2 (7a)$$
$$= a^{2+4} b^{2+2} - a^{2+3} b^{2+3} + a^{2+1} b^{2+4} - 7a^{2+1} b^2$$
$$= a^6 b^4 - a^5 b^5 + a^3 b^6 - 7a^3 b^2$$

77. $(x+3)(x+2) = x(x) + x(2) + 3(x) + 3(2)$
$$= x^2 + 2x + 3x + 6$$
$$= x^2 + 5x + 6$$

78. $(2x+1)(x-1)$
$$= 2x(x) + 2x(-1) + 1(x) + 1(-1)$$
$$= 2x^2 - 2x + x - 1$$
$$= 2x^2 + (-2+1)x - 1$$
$$= 2x^2 - x - 1$$

79. $(3t-3)(2t+2)$
$$= 3t(2t) + 3t(2) - 3(2t) - 3(2)$$
$$= 6t^2 + 6t - 6t - 6$$
$$= 6t^2 - 6$$

80. $(3n^4 - 5n^2)(2n^4 - n^2)$
$$= 3n^4 (2n^4) + 3n^4 (-n^2) - 5n^2 (2n^4) - 5n^2 (-n^2)$$
$$= 6n^{4+4} - 3n^{4+2} - 10n^{2+4} + 5n^{2+2}$$
$$= 6n^8 - 3n^6 - 10n^6 + 5n^4$$
$$= 6n^8 + (-3 - 10)n^6 + 5n^4$$
$$= 6n^8 - 13n^6 + 5n^4$$

81. $-a^5 (a^2 - b)(5a^2 + b)$
$$= -a^5 [a^2 (5a^2) + a^2 (b) - b(5a^2) - b(b)]$$
$$= -a^5 [5a^4 + a^2 b - 5a^2 b - b^2]$$
$$= -a^5 [5a^4 - 4a^2 b - b^2]$$
$$= -a^5 (5a^4) - a^5 (-4a^2 b) - a^5 (-b^2)$$
$$= -5a^9 + 4a^7 b + a^5 b^2$$

82. $6.6(a-1)(a+1) = 6.6(a^2 - 1)$
$$= 6.6a^2 - 6.6$$

83. $\left(3t - \dfrac{1}{3} \right) \left(6t + \dfrac{5}{3} \right)$
$$= 3t(6t) + 3t\left(\dfrac{5}{3} \right) + \left(-\dfrac{1}{3} \right) 6t + \left(-\dfrac{1}{3} \right) \left(\dfrac{5}{3} \right)$$
$$= 3t(6t) + \dfrac{\overset{1}{\cancel{3}} \cdot 5t}{\underset{1}{\cancel{3}}} - \dfrac{\overset{1}{\cancel{3}} \cdot 2t}{\underset{1}{\cancel{3}}} - \dfrac{1 \cdot 5}{3 \cdot 3}$$
$$= 18t^2 + 5t - 2t - \dfrac{5}{9}$$
$$= 18t^2 + (5-2)t - \dfrac{5}{9}$$
$$= 18t^2 + 3t - \dfrac{5}{9}$$

84. $(5.5 - 6b)(2 - 4b)$
$$= 5.5(2) + 5.5(-4b) - 6b(2) - 6b(-4b)$$
$$= 11 - 22b - 12b + 24b^2$$
$$= 11 + (-22 - 12)b + 24b^2$$
$$= 11 - 34b + 24b^2$$

85. $(2a-3)(4a^2+6a+9)$
$= 2a(4a^2+6a+9) - 3(4a^2+6a+9)$
$= 2a(4a^2) + 2a(6a) + 2a(9) - 3(4a^2) - 3(6a) - 3(9)$
$= 8a^3 + 12a^2 + 18a - 12a^2 - 18a - 27$
$= 8a^3 + (12-12)a^2 + (18-18)a - 27$
$= 8a^3 + 0a^2 + 0a - 27$
$= 8a^3 - 27$

86.
$$\begin{array}{r} 8x^2 + x - 2 \\ 7x^2 + x - 1 \\ \hline -8x^2 - x + 2 \\ 8x^3 + x^2 - 2x \\ 56x^4 + 7x^3 - 14x^2 \\ \hline 56x^4 + 15x^3 - 21x^2 - 3x + 2 \end{array}$$

87.
$$\begin{array}{r} 4x^2 - 2x + 1 \\ 2x + 1 \\ \hline 4x^2 - 2x + 1 \\ 8x^3 - 4x^2 + 2x \\ \hline 8x^3 + 1 \end{array}$$

88. APPLIANCES
   a. Find the perimeter of the base of the dishwasher.
   $P = 2w + 2l$
   $= 2(x+6) + 2(2x-1)$
   $= 2x + 2(6) + 2(2x) - 2(1)$
   $= 2x + 12 + 4x - 2$
   $= (2+4)x + (12-2)$
   $= 6x + 10$
   The perimeter is **(6x + 10) in.**
   b. Find the area of the base of the dishwasher.
   $A = wl$
   $= (x+6)(2x-1)$
   $= x(2x) + x(-1) + 6(2x) + 6(-1)$
   $= 2x^2 - x + 12x - 6$
   $= 2x^2 + (-1+12)x - 6$
   $= 2x^2 + 11x - 6$
   The area is **(2x² + 11x − 6) in.²**
   c. Find the volume occupied by the dishwasher.
   $V = lwh$
   $= (x+6)(2x-1)(3x)$
   $= [x(2x) + x(-1) + 6(2x) + 6(-1)](3x)$
   $= [2x^2 - x + 12x - 6](3x)$
   $= [2x^2 + (-1+12)x - 6](3x)$
   $= [2x^2 + 11x - 6](3x)$
   $= 3x(2x^2) + 3x(11x) - 3x(6)$
   $= (3 \cdot 2)x^{1+2} + (3 \cdot 11)x^{1+1} - (3 \cdot 6)x$
   $= 6x^3 + 33x^2 - 18x$
   The volume is **(6x³ + 33x² − 18x) in.³**

SECTION 5.7 REVIEW EXERCISES
Special Products
Find each product.

89. $(a-3)^2 = a^2 + 2(a)(-3) + (-3)^2$
$= a^2 - 6a + 9$

90. $(m+2)^3 = m^3 + 6m^2 + 12m + 8$

91. $(x+7)(x-7) = (x)^2 - (7)^2$
$= x^2 - 49$

92. $(2x-0.9)(2x+0.9) = 4x^2 - 0.81$

93. $(2y+1)^2 = (2y)^2 + 2(2y)(1) + (1)^2$
$= 4y^2 + 4y + 1$

94. $(y^2+1)(y^2-1) = y^4 - 1$

95. $(6r^2+10s)^2 = (6r^2)^2 + 2(6r^2)(10s) + (10s)^2$
$= 36r^4 + 120r^2s + 100s^2$

96. $-(8a-3c)^2 = -(64a^2 - 48ac + 9c^2)$
$= -64a^2 + 48ac - 9c^2$

97. $80s(r^2+s^2)(r^2-s^2) = 80s[(r^2)^2 - (s^2)^2]$
$= 80s[r^4 - s^4]$
$= 80s(r^4) + 80s(-s^4)$
$= 80r^4s - 80s^5$

98. $4b(3b-4)^2 = 4b[(3b)^2 + 2(3b)(-4) + (-4)^2]$
$= 4b(9b^2 - 24b + 16)$
$= 36b^3 - 96b^2 + 64b$

99. $\left(t - \dfrac{3}{4}\right)^2 = t^2 + 2(t)\left(-\dfrac{3}{4}\right) + \left(-\dfrac{3}{4}\right)^2$
$= t^2 - \dfrac{3}{2}t + \dfrac{9}{16}$

100. $\left(x + \dfrac{4}{3}\right)^2 = x^2 + 2(x)\left(\dfrac{4}{3}\right) + \left(\dfrac{4}{3}\right)^2$
$= x^2 + \dfrac{8}{3}x + \dfrac{16}{9}$

Perform the operations.

101. $3(9x^2+3x+7)-2(11x^2-5x+9)$

$=3(9x^2)+3(3x)+3(7)-2(11x^2)-2(-5x)-2(9)$

$=27x^2+9x+21-22x^2+10x-18$

$=(27-22)x^2+(9+10)x+(21-18)$

$=5x^2+19x+3$

102. $(5c-1)^2-(c+6)(c-6)$

$=(5c)^2+2(5c)(-1)+(-1)^2-(c^2-36)$

$=25c^2-10c+1-c^2+36$

$=(25-1)c^2-10c+(1+36)$

$=24c^2-10c+37$

103. GRAPHIC ARTS

$[(x+3)-1][(x-1)-1]$

$=[x+(3-1)][x+(-1-1)]$

$=(x+2)(x-2)$

$=x^2-(2)^2$

$=x^2-4$

The area of the picture is (x^2-4) in.^{2}

104. Find a polynomial that represents the area of the triangle.

$A=\dfrac{1}{2}bh$

$=\dfrac{1}{2}(10x+4)(10x-4)$

$=\dfrac{1}{2}(100x^2-16)$

$=\dfrac{1}{2}(100x^2)+\dfrac{1}{2}(-16)$

$=50x^2-8$

The area of the triangle is $(50x^2-8)$ in.^{2}

SECTION 5.8 REVIEW EXERCISES
Dividing Polynomials

Divide. Do not use negative exponents in the answer.

105. $\dfrac{16n^8}{8n^5}=2n^{8-5}$

$=2n^3$

106. $\dfrac{-14x^2y}{21xy^3}=-\dfrac{14x^{2-1}y^{1-3}}{21}$

$=-\dfrac{2\cdot\cancel{7}\,x^1y^{-2}}{3\cdot\cancel{7}}$

$=-\dfrac{2x}{3y^2}$

107. $\dfrac{a^{15}-24a^8}{6a^{12}}=\dfrac{a^{15}}{6a^{12}}-\dfrac{24a^8}{6a^{12}}$

$=\dfrac{a^{15-12}}{6}-4a^{8-12}$

$=\dfrac{a^3}{6}-4a^{-4}$

$=\dfrac{a^3}{6}-\dfrac{4}{a^4}$

108. $\dfrac{15a^5b+ab^2-25b}{5a^2b}$

$=\dfrac{15a^5b}{5a^2b}+\dfrac{ab^2}{5a^2b}-\dfrac{25b}{5a^2b}$

$=\dfrac{3\cdot\cancel{5}\,a^{5-2}b^{1-1}}{\cancel{5}}+\dfrac{a^{1-2}b^{2-1}}{5}-\dfrac{5\cdot\cancel{5}\,b^{1-1}}{\cancel{5}\,a^2}$

$=3a^3b^0+\dfrac{a^{-1}b^1}{5}-\dfrac{5b^0}{a^2}$

$=3a^3+\dfrac{b}{5a}-\dfrac{5}{a^2}$

109.
$$\begin{array}{r}
x-5 \\
x-1\overline{\smash{)}x^2-6x+5} \\
\underline{-(x^2-x)} \\
-5x+5 \\
\underline{-(-5x+5)} \\
0
\end{array}$$

110.
$$\begin{array}{r}
2x+1 \\
x+3\overline{\smash{)}2x^2+7x+3} \\
\underline{-(2x^2+6x)} \\
x+3 \\
\underline{-(x+3)} \\
0
\end{array}$$

$$111. \quad 3x+2 \overline{\smash{\big)}\, 15x^2 - 8x - 8} \qquad 5x-6+\frac{4}{3x+2}$$
$$\underline{-(15x^2+10x)}$$
$$-18x-8$$
$$\underline{-(-18x-12)}$$
$$4$$

$$112. \quad 5y+3 \overline{\smash{\big)}\, 25y^2 + \mathbf{0}y - 9} \qquad 5y-3$$
$$\underline{-(25y^2+15y)}$$
$$-15y-9$$
$$\underline{-(-15y-9)}$$
$$0$$

$$113. \quad 3x+1 \overline{\smash{\big)}\, 9x^3 + \mathbf{0}x^2 - 13x - 4} \qquad 3x^2 \quad -x-4$$
$$\underline{-(9x^3+3x^2)}$$
$$-3x^2-13x$$
$$\underline{-(-3x^2-\quad x)}$$
$$-12x-4$$
$$\underline{-(-12x-4)}$$
$$0$$

$$114. \quad 2x-1 \overline{\smash{\big)}\, 6x^3 + \quad x^2 + \mathbf{0}x + 1} \qquad 3x^2+2x+1+\frac{2}{2x-1}$$
$$\underline{-(6x^3-3x^2)}$$
$$4x^2+0x$$
$$\underline{-(4x^2-2x)}$$
$$2x+1$$
$$\underline{-(2x-1)}$$
$$2$$

$$115. \quad (y+3)(3y+2) = \mathbf{y}(3y)+\mathbf{y}(2)+\mathbf{3}(3y)+\mathbf{3}(2)$$
$$= 3y^2+2y+9y+6$$
$$= 3y^2+(2+9)y+6$$
$$= 3y^2+11y+6$$

116. BEDDING
$$2x+3 \overline{\smash{\big)}\, 4x^3 + 12x^2 + x - 12} \qquad 2x^2+3x-4$$
$$\underline{-(4x^3+6x^2)}$$
$$6x^2+\quad x$$
$$\underline{-(6x^2+9x)}$$
$$-8x-12$$
$$\underline{-(-8x-12)}$$
$$0$$

The length of the bed is $(2x^2+3x-4)$ in.

CHAPTER 5 TEST

1. **Fill in the blanks.**
   a. In the expression y^{10}, the **base** is y and the **exponent** is 10.

   b. We call a polynomial with exactly one term a **monomia** , with exactly two terms a **binomial**, and with exactly three terms a **trinomial**.

   c. The **degree** of a term of a polynomial in one variable is the value of the exponent on the variable.

   d. $(x+y)^2$, $(x-y)^2$, and $(x+y)(x-y)$ are called **special** products .

2. Use exponents to rewrite $2xxxyyyy$. $\mathbf{2x^3y^4}$

Simplify each expression. Do not use negative exponent in the answer.

3. $y^2(yy^3) = y^{2+1+3}$
$$= y^6$$

4. $\left(\dfrac{1}{2}x^3\right)^5 (x^2)^3 = \left(\dfrac{1^5}{2^5}x^{3\cdot5}\right)(x^{2\cdot3})$
$$= \left(\dfrac{1}{32}x^{15}\right)(x^6)$$
$$= \dfrac{1}{32}x^{15+6}$$
$$= \dfrac{1}{32}x^{21}$$

5. $3.5x^0 = 3.5(1)$
$$= 3.5$$

6. $2y^{-5}y^2 = 2y^{-5+2}$
$= 2y^{-3}$
$= \dfrac{2}{y^3}$

7. $5^{-3} = \dfrac{1}{5^3}$
$= \dfrac{1}{125}$

8. $\dfrac{(x+1)^{15}}{(x+1)^6} = (x+1)^{15-6}$
$= (x+1)^9$

9. $\dfrac{(y^{-5})^{-4}}{yy^{-2}} = \dfrac{y^{-5\cdot-4}}{y^{1-2}}$
$= \dfrac{y^{20}}{y^{-1}}$
$= y^{20-(-1)}$
$= y^{20+1}$
$= y^{21}$

10. $\left(\dfrac{a^2 b^{-1}}{4a^3 b^{-2}}\right)^3 = \left(\dfrac{a^{2-3} b^{-1-(-2)}}{4}\right)^3$
$= \left(\dfrac{a^{-1} b^{-1+2}}{4}\right)^3$
$= \left(\dfrac{b}{4a}\right)^3$
$= \dfrac{b^3}{4^3 a^3}$
$= \dfrac{b^3}{64a^3}$

11. $\left(\dfrac{8}{m^6}\right)^{-2} = \left(\dfrac{m^6}{8}\right)^2$
$= \dfrac{m^{6\cdot2}}{8^2}$
$= \dfrac{m^{12}}{64}$

12. $\dfrac{-6a}{b^{-9}} = -6ab^9$

13. $V = (10y^4)^3$
$= 10^3 y^{4\cdot3}$
$= 1,000y^{12}$

The volume of the cube is $1,000y^{12}$ in.^{3}

14. ELECTRICITY
Move the decimal 18 places to the left.
$6,250,000,000,000,000,000 = 6.25 \times 10^{18}$

15. The exponent is -5.
Move the decimal 5 places to the left.
$9.3 \times 10^{-5} = 0.000093$

16. $(2.3 \times 10^{18})(4.0 \times 10^{-15})$
$= (2.3 \cdot 4.0) \times (10^{18} \times 10^{-15})$
$= (2.3 \cdot 4.0) \times (10^{18+(-15)})$
$= 9.2 \times 10^3$
$= 9,200$

17. $x^4 + 8x^2 - 12$ **Trinomial**

Term	Coefficient	Degree
x^4	1	4
$8x^2$	8	2
-12	-12	0

Degree of the polynomial: **4**

18. **Find the degree of the polynomial**
$3x^3 y + 2x^2 y^3 - 5xy^2 - 6y$ **5th degree**

Complete the table of solutions and then graph the equation.

19. $y = x^2 + 2$

x	y	(x, y)
-2	$y = (-2)^2 + 2$ $= 4 + 2$ $= 6$	$(-2, 6)$
-1	$y = (-1)^2 + 2$ $= 1 + 2$ $= 3$	$(-1, 3)$
0	$y = (0)^2 + 2$ $= 0 + 2$ $= 2$	$(0, 2)$
1	$y = (1)^2 + 2$ $= 1 + 2$ $= 3$	$(1, 3)$
2	$y = (2)^2 + 2$ $= 4 + 2$ $= 6$	$(2, 6)$

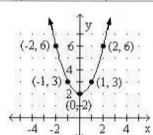

Chapter 5 Review and Test

20. FREE FALL

Evaluate $-16t^2 + 5{,}184$ when $t = 18$.

$$-16t^2 + 5{,}184 = -16(\mathbf{18})^2 + 5{,}184$$
$$= -16(324) + 5{,}184$$
$$= -5{,}184 + 5{,}184$$
$$= 0$$

0 ft; The rock hits the canyon floor 18 seconds after being dropped.

Simplify each polynomial.

21. $\dfrac{3}{5}x^2 + 6 + \dfrac{1}{4}x - 8 - \dfrac{1}{2}x^2 + \dfrac{1}{3}x$

$$= \left(\dfrac{3}{5} - \dfrac{1}{2}\right)x^2 + \left(\dfrac{1}{4} + \dfrac{1}{3}\right)x + (6 - 8)$$

$$= \left(\dfrac{3}{5}\cdot\dfrac{2}{2} - \dfrac{1}{2}\cdot\dfrac{5}{5}\right)x^2 + \left(\dfrac{1}{4}\cdot\dfrac{3}{3} + \dfrac{1}{3}\cdot\dfrac{4}{4}\right)x + (6 - 8)$$

$$= \left(\dfrac{6 - 5}{10}\right)x^2 + \left(\dfrac{3 + 4}{12}\right)x - 2$$

$$= \dfrac{1}{10}x^2 + \dfrac{7}{12}x - 2$$

22. $4a^2b + 5 - 6a^3b - 3a^2b + 2a^3b$

$$= (-6 + 2)a^3b + (4 - 3)a^2b + 5$$
$$= -4a^3b + a^2b + 5$$

Perform the operations.

23. $(12.1h^3 - 9.9h^2 + 9.5)$
$$+ (7.3h^3 - 1.2h^2 - 10.1)$$
$$= 12.1h^3 - 9.9h^2 + 9.5 + 7.3h^3 - 1.2h^2 - 10.1$$
$$= (12.1 + 7.3)h^3 + (-9.9 - 1.2)h^2 + (9.5 - 10.1)$$
$$= 19.4h^3 - 11.1h^2 - 0.6$$

24. $(6b^3c - 3bc) + (b^3c - 2bc) - (b^3c - 3bc + 12)$
$$= 6b^3c - 3bc + b^3c - 2bc - b^3c + 3bc - 12$$
$$= (6 + 1 - 1)b^3c + (-3 - 2 + 3)bc - 12$$
$$= 6b^3c - 2bc - 12$$

25. $-5y^3 + 4y^2 + 3$
$-(-2y^3 - 14y^2 + 17y - 32)$

Change the signs of the 2nd line and add

$$\begin{aligned}-5y^3 + 4y^2 + 3 \\ \underline{2y^3 + 14y^2 - 17y + 32} \\ -3y^3 + 18y^2 - 17y + 35\end{aligned}$$

26. $A = 2l + 2w$
$$= 2(5a^2 + 3a - 1) + 2(a - 9)$$
$$= \mathbf{2}(5a^2) + \mathbf{2}(3a) - \mathbf{2}(1) + \mathbf{2}(a) - \mathbf{2}(9)$$
$$= 10a^2 + 6a - 2 + 2a - 18$$
$$= 10a^2 + (6 + 2)a + (-2 - 18)$$
$$= 10a^2 + 8a - 20$$

The perimeter of the rectangle is $(10a^2 + 8a - 20)$ in.

Multiply.

27. $(2x^3y^3)(5x^2y^8) = (2 \cdot 5)(x^{3+2})(y^{3+8})$
$$= 10x^5y^{11}$$

28. $9b^3(8b^4)(-b) = (9 \cdot 8 \cdot -1)b^{3+4+1}$
$$= -72b^8$$

29. $3y^2(y^2 - 2y + 3)$
$$= \mathbf{3y^2}(y^2) + \mathbf{3y^2}(-2y) + \mathbf{3y^2}(3)$$
$$= 3y^{2+2} - 6y^{2+1} + 9y^2$$
$$= 3y^4 - 6y^3 + 9y^2$$

30. $0.6p^5(0.4p^6 - 0.9p^3)$
$$= \mathbf{0.6p^5}(0.4p^6) - \mathbf{0.6p^5}(0.9p^3)$$
$$= (0.6 \cdot 0.4)(p^{5+6}) - (0.6 \cdot 0.9)(p^{5+3})$$
$$= 0.24p^{11} - 0.54p^8$$

31. $\dfrac{3}{4}s^3t^9(s^4t^8 + 16st)$

$$= \dfrac{3}{4}\mathbf{s^3t^9}(s^4t^8) + \dfrac{3}{4}\mathbf{s^3t^9}(16st)$$

$$= \dfrac{3}{4}(s^{3+4}t^{9+8}) + \left(\dfrac{3}{4}\cdot 16\right)(s^{3+1}t^{9+1})$$

$$= \dfrac{3}{4}s^7t^{17} + 12s^4t^{10}$$

32. $(x - 5)(3x + 4) = 3x^2 + 4x - 15x - 20$
$$= 3x^2 + (4 - 15)x - 20$$
$$= 3x^2 - 11x - 20$$

33. $\left(6t + \dfrac{1}{2}\right)\left(2t - \dfrac{3}{2}\right)$

$$= 6t(2t) + 6t\left(-\dfrac{3}{2}\right) + \left(\dfrac{1}{2}\right)2t + \left(\dfrac{1}{2}\right)\left(-\dfrac{3}{2}\right)$$

$$= 6t(2t) - \dfrac{\overset{1}{\cancel{2}} \cdot 3t \cdot 3}{\underset{1}{\cancel{2}}} + \dfrac{\overset{1}{\cancel{2}} \cdot t}{\underset{1}{\cancel{2}}} - \dfrac{1 \cdot 3}{2 \cdot 2}$$

$$= 12t^2 - 9t + t - \dfrac{3}{4}$$

$$= 12t^2 - 8t - \dfrac{3}{4}$$

34. $(3.8m - 1)(2m - 1)$

$$= 3.8m(2m) - 3.8m(1) - 1(2m) - 1(-1)$$
$$= 7.6m^2 - 3.8m - 2m + 1$$
$$= 7.6m^2 - 5.8m + 1$$

35. $(a^3 - 6)(a^3 + 7)$

$$= a^3(a^3) + a^3(7) - 6(a^3) - 6(7)$$
$$= a^6 + 7a^3 - 6a^3 - 42$$
$$= a^6 + (7 - 6)a^3 - 42$$
$$= a^6 + a^3 - 42$$

36.
$$
\begin{array}{r}
x^2 - 2x + 4 \\
2x - 3 \\
\hline
-3x^2 + 6x - 12 \\
2x^3 - 4x^2 + 8x \\
\hline
2x^3 - 7x^2 + 14x - 12
\end{array}
$$

37. $(1 + 10c)(1 - 10c) = (1)^2 - (10c)^2$

$$= 1 - 100c^2$$

38. $(7b^3 - 3t)^2 = (7b^3)^2 + 2(7b^3)(-3t) + (-3t)^2$

$$= 49b^6 - 42b^3t + 9t^2$$

39. $(2.2a)(a + 5)(a - 3)$

$$= 2.2a[a(a) + a(-3) + 5(a) + 5(-3)]$$
$$= 2.2a[a^2 - 3a + 5a - 15]$$
$$= 2.2a[a^2 + 2a - 15]$$
$$= 2.2a(a^2) + 2.2a(2a) + 2.2a(-15)$$
$$= 2.2a^3 + 4.4a^2 - 33a$$

40. $(x + y)(x - y) + (x + y)^2$

$$= x^2 - y^2 + x^2 + 2xy + y^2$$
$$= (1 + 1)x^2 + (-1 + 1)y^2 + 2xy$$
$$= 2x^2 + 2xy$$

Divide.

41. $\dfrac{6a^2 - 12b^2}{24ab} = \dfrac{6a^2}{24ab} - \dfrac{12b^2}{24ab}$

$$= \dfrac{a^{2-1}}{4b} - \dfrac{b^{2-1}}{2a}$$

$$= \dfrac{a}{4b} - \dfrac{b}{2a}$$

42.
$$
\begin{array}{r}
x - 2 \\
x + 3 \overline{) x^2 + \ x - 6} \\
\underline{-(x^2 + 3x)} \\
-2x - 6 \\
\underline{-(-2x - 6)} \\
0
\end{array}
$$

43.
$$
\begin{array}{r}
3x^2 + 2x + 1 + \dfrac{2}{2x - 1} \\
2x - 1 \overline{) 6x^3 + \ x^2 + \ 0x + 1} \\
\underline{-(6x^2 - 3x^2)} \\
4x^2 + \ 0x \\
\underline{-(4x^2 - \ 2x)} \\
2x + 1 \\
\underline{-(2x - 1)} \\
2
\end{array}
$$

44.
$$
\begin{array}{r}
x - 5 \\
x - 1 \overline{) x^2 - 6x + 5} \\
\underline{-(x^2 - x)} \\
-5x + 5 \\
\underline{-(-5x + 5)} \\
0
\end{array}
$$

The width of the rectangle is $(x - 5)$ ft.

45. $(5m + 1)(m - 6)$

$$= 5m(m) + 5m(-6) + 1(m) + 1(-6)$$
$$= 5m^2 - 30m + m - 6$$
$$= 5m^2 + (-30 + 1)m - 6$$
$$= 5m^2 - 29m - 6$$

Yes, it checks.

46. Is $(a + b)^2 = a^2 + b^2$?

No; $(a + b)^2 = a^2 + 2ab + b^2$

Chapter 5 Review and Test

CHAPTER 5 CUMULATIVE REVIEW

1. Prime Factorization: [Section 1.2]

$270 = 2 \cdot 135$
$= 2 \cdot 5 \cdot 27$
$= 2 \cdot 5 \cdot 3 \cdot 9$
$= 2 \cdot 5 \cdot 3 \cdot 3 \cdot 3$
$= 2 \cdot 3^3 \cdot 5$

2. a. Commutative Property of Addition: [Section 1.4]

$a + b = \boldsymbol{b} + \boldsymbol{a}$

b. Associative Property of Multiplication: [Section 1.6]

$(xy)z = \boldsymbol{x}(\boldsymbol{yz})$

Evaluate each expression. [Section 1.7]

3. $3 - 4[-10 - 4(-5)] = 3 - 4(-10 + 20)$
$= 3 - 4(10)$
$= 3 - 40$
$= 3 + (-40)$
$= -37$

4. $\dfrac{|-45| - 2(-5) + 1^5}{2 \cdot 9 - 2^4} = \dfrac{45 - 2(-5) + 1}{2 \cdot 9 - 16}$
$= \dfrac{45 + 10 + 1}{18 - 16}$
$= \dfrac{56}{2}$
$= 28$

Simplify each expression. [Section 1.9]

5. $27\left(\dfrac{2}{3}x\right) = \dfrac{27 \cdot 2}{3}x$
$= \dfrac{3 \cdot 9 \cdot 2}{3}x$
$= \dfrac{\overset{1}{\cancel{3}} \cdot 9 \cdot 2}{\underset{1}{\cancel{3}}}x$
$= 18x$

6. $3x^2 + 2x^2 - 5x^2 = (3 + 2 - 5)x^2$
$= 0$

Solve each equation. [Section 2.2]

7. $2 - (4x + 7) = 3 + 2(x + 2)$
$2 - 4x - 7 = 3 + 2x + 4$
$-4x - 5 = 2x + 7$
$-4x - 5 + \boldsymbol{4x} = 2x + \boldsymbol{4x} + 7$
$-5 = 6x + 7$
$-5 - \boldsymbol{7} = 6x + 7 - \boldsymbol{7}$
$-12 = 6x$
$\dfrac{-12}{6} = \dfrac{6x}{6}$
$-2 = x$

The solution is -2.

8. $\dfrac{2}{5}y + 3 = 9$

$\boldsymbol{5}\left(\dfrac{2}{5}y + 3\right) = \boldsymbol{5}(9)$

$\boldsymbol{5}\left(\dfrac{2}{5}y\right) + \boldsymbol{5}(3) = \boldsymbol{5}(9)$

$2y + 15 = 45$
$2y + 15 - \boldsymbol{15} = 45 - \boldsymbol{15}$
$2y = 30$
$\dfrac{2y}{\boldsymbol{2}} = \dfrac{30}{\boldsymbol{2}}$
$y = 15$

The solution is 15.

9. CANDY SALES [Section 2.3]

34% of $6.5 billions is what number?

$0.34 \quad \cdot \quad 6.5 \qquad = x$
$0.34(6.5) = x$
$2.21 = x$

The candy sales is $2.21 billion.

10. AIR CONDITIONING [Section 2.4]

One will need to change the radius of 3 inches into part of a foot before using the data in the formula. 12 inches in a foot.

$V = \pi r^2 h$

$= 3.14\left(\dfrac{3}{12}\right)^2 (6)$

$= 3.14(0.25)^2(6)$

$= 3.14(0.0625)(6)$

$= 3.14(0.375)$

$= 1.1775$

The volume of air is about 1.2 ft^3.

11. ANGLE OF ELEVATION [Section 2.5]

Analyze

- $x°$ is the measure of one acute angle.
- $2x°$ is the measure of other acute angle.
- $90°$ is the measure of the right angle.
- What is the value of x?

Assign

Let x = the measure of one acute angle
$2x$ = the measure of other acute angle

Form

Angle 1	plus	angle 2	plus	angle 3	equals	the total of 180.
x	+	$2x$	+	90	=	180

Solve

$$x + 2x + 90 = 180$$
$$3x + 90 = 180$$
$$3x + 90 - 90 = 180 - 90$$
$$3x = 90$$
$$\frac{3x}{3} = \frac{90}{3}$$
$$x = 30$$

State

The value of x is 30.

Check

$$30 + 2(30) + 90 = 180$$
$$30 + 60 + 90 = 180$$
$$180 = 180$$

The result checks.

12. LIVESTOCK AUCTION [Section 2.5]

Analyze

- Wants to make $6,000.
- 4% commission.
- What is the selling price?

Assign

Let x = selling price of the hog

Form

The selling price of the hog	minus	the amount of the commission	equals	the amount made on the auction.
x	−	$0.04x$	=	6,000

Solve

$$x - 0.04x = 6,000$$
$$0.96x = 6,000$$
$$\frac{0.96x}{0.96} = \frac{6,000}{0.96}$$
$$x = 6,250$$

State

$6,250 is the selling price.

Check

$$x - 0.04x = 6,000$$
$$6,250 - 0.04(6,250) = 6,000$$
$$6,250 - 250 = 6,000$$
$$6,000 = 6,000$$

The result checks.

13. STOCK MARKET [Section 2.6]

Analyze

- $45,000 is the total investment.
- Mutual fund at 12% annual interest.
- Treasury bond at 6.5% annual interest.
- $4,300 is the first-year combined income.
- How much is invested in each account?

Assign

Let x = amount invested at 12%
$45,000 - x$ = amount invested at 6.5%

Form

	Principal	• rate	• time =	interest
Mutual	x	0.12	1	**0.12x**
Treasury	$45,000 - x$	0.065	1	**0.065(45,000 − x)**
			Total:	**4,300**

The interest earned at 12%	plus	the interest earned at 6.5%	equals	the total interest.
$0.12x$	+	$0.065(45,000 - x)$	=	4,300

Solve

$$0.12x + 0.065(45,000 - x) = 4,300$$
$$\mathbf{1,000}[0.12x + 0.065(45,000 - x)] = \mathbf{1,000}(4,300)$$
$$1,000(0.12x) + 1,000[0.065(45,000 - x)] = 1,000(4,300)$$
$$120x + 65(45,000 - x) = 4,300,000$$
$$120x + 2,925,000 - 65x = 4,300,000$$
$$55x + 2,925,000 = 4,300,000$$
$$55x + 2,925,000 - \mathbf{2,925,000} = 4,300,000 - \mathbf{2,925,000}$$
$$55x = 1,375,000$$
$$\frac{55x}{55} = \frac{1,375,000}{55}$$
$$x = 25,000$$

6.5% investment
$$45,000 - x = 45,000 - 25,000$$
$$= 20,000$$

State

$25,000 was invested in the mutual fund.
$20,000 was invested in treasury bonds.

Check

The results check.

Chapter 5 Cumulative Review

14. Solve: [Section 2.7]

$$-4x + 6 > 17$$

$$-4x + 6 - \mathbf{6} > 17 - \mathbf{6}$$

$$-4x > 11$$

$$\frac{-4x}{-4} < \frac{11}{-4}$$

$$x < -\frac{11}{4}$$

$$\left(-\infty, -\frac{11}{4}\right)$$

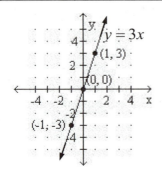

Graph each equation.

15. $y = 3x$ [Section 3.2]

x	y	(x,y)
-1	$y = 3x$ $= 3(-1)$ $= -3$	$(-1, -3)$
0	$y = 3x$ $= 3(0)$ $= 0$	$(0, 0)$
1	$y = 3x$ $= 3(1)$ $= 3$	$(1, 3)$

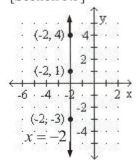

16. $x = -2$ [Section 3.3]

17. Find the slope of the line passing through $(6, -2)$ and $(-3, 2)$. [Section 3.4]

Let $(x_1, y_1) = (6, -2)$

Let $(x_2, y_2) = (-3, 2)$

$$m = \frac{y_2 - y_1}{x_2 - x_1}$$

$$= \frac{2 - (-2)}{-3 - 6}$$

$$= \frac{2 + 2}{-3 + (-6)}$$

$$= -\frac{4}{9}$$

The slope of the line is $-\dfrac{4}{9}$.

18. DIABETES [Section 3.4]

Select any two points that lie on the line.

Let $(x_1, y_1) = (1996,\ 8.8)$

Let $(x_2, y_2) = (2008,\ 17.8)$

$$m = \frac{y_2 - y_1}{x_2 - x_1}$$

$$= \frac{17.8 - 8.8}{2008 - 1996}$$

$$= \frac{9}{12}$$

$$= \frac{3}{4}$$

0.75 million people per year or

$\dfrac{3}{4}$ million people per year.

19. Find the slope and y-intercept of the line. Then write the equation of the line. [Section 3.5]

$$m = \frac{3}{1}$$

$$= 3$$

$$y\text{-intercept} = (0, -2)$$

$$y = mx + b$$

$$= 3x + (-2)$$

$$= 3x - 2$$

The equation of the line is $y = 3x - 2$.

20. Determine if the graphs of $y = \dfrac{3}{2}x - 1$ and $2x + 3y = 10$ are parallel, perpendicular, or neither. [Section 3.5]

$$y = \dfrac{3}{2}x - 1 \qquad \bigg| \qquad 2x + 3y = 10$$

$$m = \dfrac{3}{2} \qquad \bigg| \qquad 3y = -2x + 10$$

$$\bigg| \qquad y = -\dfrac{2}{3}x + \dfrac{10}{3}$$

$$\bigg| \qquad m = -\dfrac{2}{3}$$

The slopes are negative reciprocals.
The lines are perpendicular.

Find an equation of the line with the following properties.
Write the equation in slope - intercept form.

21. Passing through $(-2, 10)$ with slope -4.
    [Section 3.6]

$$(-2, 10), \quad m = -4$$
$$y - y_1 = m(x - x_1)$$
$$y - 10 = -4[x - (-2)]$$
$$y - 10 = -4(x + 2)$$
$$y - 10 = -4x - 8$$
$$y - 10 + \mathbf{10} = -4x - 8 + \mathbf{10}$$
$$y = -4x + 2$$

22. Is $(-2, 1)$ a solution of $2x - 3y \geq -6$?
    [Section 3.7]

$$2x - 3y \geq -6$$
$$2(\mathbf{-2}) - 3(\mathbf{1}) \overset{?}{\geq} -6$$
$$-4 - 3 \overset{?}{\geq} -6$$
$$-7 \geq -6$$
$$\text{False}$$
$$(-2, 1) \text{ is not a solution.}$$

23. If $f(x) = 2x^2 + 3x - 9$, find $f(-5)$.
    [Section 3.8]

$$f(x) = 2x^2 + 3x - 9$$
$$f(\mathbf{-5}) = 2(\mathbf{-5})^2 + 3(\mathbf{-5}) - 9$$
$$f(-5) = 2(25) + (-15) - 9$$
$$f(-5) = 50 + (-15) + (-9)$$
$$f(-5) = 35 + (-9)$$
$$f(-5) = 26$$

24. Is the graph of function? [Section 3.8]

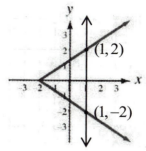

It is not a graph of a function, because the vertical line intersects the graph at points $(1,2)$ and $(1,-2)$ which have the same x-coordinate. There are many more points that demonstrate this same concept.

25. $\left(\dfrac{2}{3}, -1\right)$, $\begin{cases} y = -3x + 1 \\ 3x + 3y = -2 \end{cases}$ [Section 4.1]

$$y = -3x + 1 \qquad \bigg| \qquad 3x + 3y = -2$$
$$-1 \overset{?}{=} -3\left(\dfrac{2}{3}\right) + 1 \quad \bigg| \quad 3\left(\dfrac{2}{3}\right) + 3(-1) \overset{?}{=} -2$$
$$-1 \overset{?}{=} -2 + 1 \qquad \bigg| \qquad 2 + (-3) \overset{?}{=} -2$$
$$-1 = -1 \qquad \bigg| \qquad -1 = -2$$
$$\text{True} \qquad \bigg| \qquad \text{False}$$

Since $\left(\dfrac{2}{3}, -1\right)$ does not satisfies both equations, it is not a solution of the system.

26. Solve the system $\begin{cases} 3x + 2y = 14 \\ y = \dfrac{1}{4}x \end{cases}$ by graphing. [Section 4.1]

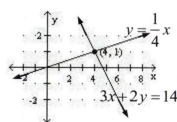

The solution is $(4,1)$.

27. Solve the system $\begin{cases} 2b - 3a = 18 \\ a + 3b = 5 \end{cases}$

by substitution. [Section 4.2]

$\begin{cases} 2b - 3a = 18 \\ a + 3b = 5 \end{cases}$

Solve for a in 2nd equation.

$\begin{cases} 2b - 3a = 18 \\ a = 5 - 3b \end{cases}$

$2b - 3a = 18$ The 1st equation

        Substitute for a.

$2b - 3(\mathbf{5 - 3b}) = 18$

$2b - 15 + 9b = 18$

$11b - 15 = 18$

$11b - 15 + \mathbf{15} = 18 + \mathbf{15}$

$11b = 33$

$\dfrac{11b}{\mathbf{11}} = \dfrac{33}{\mathbf{11}}$

$b = 3$

$a = 5 - 3b$ The 2nd equation

$a = 5 - 3(3)$

$a = 5 + (-9)$

$a = -4$

The solution is $(-4, 3)$.

28. Solve the system $\begin{cases} 8s + 10t = 24 \\ 11s - 3t = -34 \end{cases}$

by elimination (addition). [Section 4.3]

$\begin{cases} 8s + 10t = 24 \\ 11s - 3t = -34 \end{cases}$

Eliminate t.

Multiply both sides of the 1st equation by 3.

Multiply both sides of the 2nd equation by 10.

$\begin{array}{r} 24s + 30t = 72 \\ 110s - 30t = -340 \\ \hline 134s \quad\quad = -268 \end{array}$

$\dfrac{134s}{\mathbf{134}} = \dfrac{-268}{\mathbf{134}}$

$s = -2$

$8s + 10t = 24$ The 1st equation

$8(\mathbf{-2}) + 10t = 24$

$-16 + 10t = 24$

$-16 + 10t + \mathbf{16} = 24 + \mathbf{16}$

$10t = 40$

$\dfrac{10t}{\mathbf{10}} = \dfrac{40}{\mathbf{10}}$

$t = 4$

The solution is $(-2, 4)$.

29. VACATION [Section 4.4]

Analyze
- Tickets for 2 adults, 3 children cost $275.
- Tickets for 3 adults, 3 children cost $336.
- What does an adult and child ticket cost?

Assign

Let x = cost of adult ticket

     y = cost of child ticket

Form

	Number •	Value =	Total Value
Adult	2	$x	2x
Child	3	$y	3y
		Total	275

The cost of 2 adults tickets	plus	the cost of 3 children tickets	is	275.
$2x$	$+$	$3y$	$=$	275

	Number •	Value =	Total Value
Adult	3	$x	3x
Child	3	$y	3y
		Total	336

The cost of 3 adults tickets	plus	the cost of 3 children tickets	is	336.
$3x$	$+$	$3y$	$=$	336

Solve

$\begin{cases} 2x + 3y = 275 \\ 3x + 3y = 336 \end{cases}$

Eliminate y.

Multiply both sides of the 1st equation by -1.

$\begin{array}{r} -2x - 3y = -275 \\ 3x + 3y = 336 \\ \hline x = 61 \end{array}$

$2x + 3y = 275$ The 1st equation

$2(\mathbf{61}) + 3y = 275$

$122 + 3y - \mathbf{122} = 275 - \mathbf{122}$

$3y = 153$

$\dfrac{3y}{\mathbf{3}} = \dfrac{153}{\mathbf{3}}$

$y = 51$

State

$61 is the cost of an adult ticket.

$51 is the cost of a child ticket.

Check

The results check.

30. Graph: $\begin{cases} y \le 2x - 1 \\ x + 3y > 6 \end{cases}$ [Section 4.5]

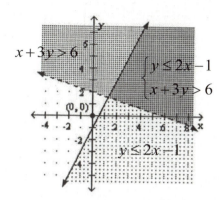

Simplify: Do not use negative exponents in the answer. [31-34 Section 5.1]

31. $(-3x^2y^4)^2 = (-3)^2(x^2)^2(y^4)^2$
$$= 9x^{2\cdot2}y^{4\cdot2}$$
$$= 9x^4y^8$$

32. $(v^5)^2(v^3)^4 = (v^{5\cdot2})(v^{3\cdot4})$
$$= v^{10}v^{12}$$
$$= v^{10+12}$$
$$= v^{22}$$

33. $ab^3c^4 \cdot ab^4c^2 = aa \cdot b^3b^4 \cdot c^4c^2$
$$= a^{1+1} \cdot b^{3+4} \cdot c^{4+2}$$
$$= a^2b^7c^6$$

34. $\left(\dfrac{4t^3t^4t^5}{3t^2t^6}\right)^3 = \left(\dfrac{4t^{3+4+5}}{3t^{2+6}}\right)^3$
$$= \left(\dfrac{4t^{12}}{3t^8}\right)^3$$
$$= \left(\dfrac{4t^{12-8}}{3}\right)^3$$
$$= \left(\dfrac{4t^4}{3}\right)^3$$
$$= \dfrac{4^3t^{4\cdot3}}{3^3}$$
$$= \dfrac{64t^{12}}{27}$$

[35-38 Section 5.2]

35. $(2y)^{-4} = \dfrac{1}{(2y)^4}$
$$= \dfrac{1}{2^4y^4}$$
$$= \dfrac{1}{16y^4}$$

36. $\dfrac{a^4b^0}{a^{-3}} = a^{4-(-3)} \cdot 1$
$$= a^{4+3}$$
$$= a^7$$

37. $-5^{-2} = \dfrac{-1}{5^2}$
$$= -\dfrac{1}{25}$$

38. $\left(\dfrac{a}{x}\right)^{-10} = \left(\dfrac{x}{a}\right)^{10}$
$$= \dfrac{x^{10}}{a^{10}}$$

Write each number in scientific notation.

39. 615,000 [Section 5.3]
Move the decimal 5 places to the left.
$615,000 = 6.15 \times 10^5$

40. 0.0000013 [Section 5.3]
Move the decimal 6 places to the right.
$0.0000013 = 1.3 \times 10^{-6}$

41. Graph: $y = x^2$. [Section 5.4]

x	y	(x, y)
-3	$y = (-3)^2$ $= 9$	$(-3, 9)$
-2	$y = (-2)^2$ $= 4$	$(-2, 4)$
-1	$y = (-1)^2$ $= 1$	$(-1, 1)$
0	$y = (0)^2$ $= 0$	$(0, 0)$
1	$y = (1)^2$ $= 1$	$(1, 1)$
2	$y = (2)^2$ $= 4$	$(2, 4)$
3	$y = (3)^2$ $= 9$	$(3, 9)$

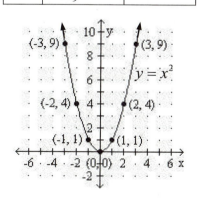

Chapter 5 Cumulative Review

42. MUSICAL INSTRUMENT [Section 5.4]

$0.01875x^4 - 0.15x^3 + 1.2x$
$$= 0.01875(2)^4 - 0.15(2)^3 + 1.2(2)$$
$$= 0.01875(16) - 0.15(8) + 1.2(2)$$
$$= 0.3 - 1.2 + 2.4$$
$$= 1.5$$

The amount of deflection is 1.5 in.

Perform the operations.

43. $(4c^2 + 3c - 2) + (3c^2 + 4c + 2)$ [Section 5.5]
$$= (4c^2 + 3c^2) + (3c + 4c) + (-2 + 2)$$
$$= (4 + 3)c^2 + (3 + 4)c + (-2 + 2)$$
$$= 7c^2 + 7c$$

44. Subtract: [Section 5.5]

$$\begin{array}{r} 17x^4 - 3x^2 - 65x - 12 \\ -(23x^4 + 14x^2 + 3x - 23) \end{array}$$

Change the signs of the 2nd line and then add.

$$\begin{array}{r} 17x^4 - 3x^2 - 65x - 12 \\ -23x^4 - 14x^2 - 3x + 23 \\ \hline -6x^4 - 17x^2 - 68x + 11 \end{array}$$

45. $(2t + 3s)(3t - s)$ [Section 5.6]
$$= 2t(3t) + 2t(-s) + 3s(3t) + 3s(-s)$$
$$= 6t^2 - 2st + 9st - 3s^2$$
$$= 6t^2 + (-2 + 9)st - 3s^2$$
$$= 6t^2 + 7st - 3s^2$$

46. $3x(2x + 3)^2$ [Section 5.7]

$$= 3x[(2x)^2 + 2(2x)(3) + (3)^2]$$
$$= 3x(4x^2 + 12x + 9)$$
$$= 3x(4x^2) + 3x(12x) + 3x(9)$$
$$= (3 \cdot 4)x^{1+2} + (3 \cdot 12)x^{1+1} + (3 \cdot 9)x$$
$$= 12x^3 + 36x^2 + 27x$$

47. [Section 5.8]

$$\begin{array}{r} 2x + 1 \\ 5x+3{\overline{\smash{\big)}\,10x^2 + 11x + 3}} \\ \underline{-(10x^2 + 6x)} \\ 5x + 3 \\ \underline{-(5x + 3)} \\ 0 \end{array}$$

48. [Section 5.8]

$$\frac{2x - 32}{16x} = \frac{2x}{16x} - \frac{32}{16x}$$

$$= \frac{\cancel{2}\,\cancel{x}}{\cancel{2} \cdot 8\,\cancel{x}} - \frac{2 \cdot \cancel{16}}{\cancel{16}\,x}$$

$$= \frac{1}{8} - \frac{2}{x}$$

SECTION 6.1
VOCABULARY
Fill in the blanks.

1. To **factor** a polynomial means to express it as a product of two (or more) polynomials.

3. To factor $m^3 + 3m^2 + 4m + 12$ by **grouping**, we begin by writing $m^2(m + 3) + 4(m + 3)$.

CONCEPTS

5. Complete each factorization.
   a. $6x = 2 \cdot \mathbf{3} \cdot x$

   b. $35h^2 = 5 \cdot \mathbf{7} \cdot h \cdot \mathbf{h}$

   c. $18y^3z = 2 \cdot \mathbf{3} \cdot 3 \cdot \mathbf{y} \cdot y \cdot \mathbf{y} \cdot z$

7.a. Write a binomial such that the GCF of its terms is 2.   **$2x + 4$ (answers may vary)**

   b. Write a trinomial such that the GCF of its terms is x.   **$x^3 + x^2 + x$ (answers may vary)**

Fill in the blanks to complete each factorization.

9. $2x^2 + 6x = \mathbf{2x} \cdot x + \mathbf{2x} \cdot 3$
   $ = 2x(\,\mathbf{x + 3}\,)$

11. Consider the polynomial $2k - 8 + hk - 4h$.
   a. How many terms does the polynomial have?
      4
   b. Is there a common factor of all the terms, other than 1? **No**
   c. What is the GCF of the first two terms and what is the GCF of the last two terms? **2; h**

NOTATION

Complete each factorization.

13. $8m^2 - 32m + 16 = \mathbf{8}(m^2 - 4m + 2)$

15. $b^3 - 6b^2 + 2b - 12 = \mathbf{b^2}(b - 6) + \mathbf{2}(b - 6)$
   $ = (\mathbf{b - 6})(b^2 + 2)$

GUIDED PRACTICE

Find the GCF of each list of numbers. See Example 1.

17. $6 = \mathbf{2} \cdot \mathbf{3}$
    $10 = \mathbf{2} \cdot \mathbf{5}$
    GCF is 2.

19. $18 = \mathbf{2} \cdot \mathbf{3} \cdot \mathbf{3}$
    $24 = \mathbf{2} \cdot 2 \cdot 2 \cdot \mathbf{3}$
    GCF is $2 \cdot 3 = 6$.

21. $14 = 2 \cdot \mathbf{7}$
    $21 = 3 \cdot \mathbf{7}$
    $42 = 2 \cdot 3 \cdot \mathbf{7}$
    GCF is 7.

23. $40 = \mathbf{2} \cdot \mathbf{2} \cdot \mathbf{2} \cdot 5$
    $32 = \mathbf{2} \cdot \mathbf{2} \cdot \mathbf{2} \cdot 2 \cdot 2$
    $24 = \mathbf{2} \cdot \mathbf{2} \cdot \mathbf{2} \cdot 3$
    GCF is $2 \cdot 2 \cdot 2 = 8$.

Find the GCF of each list of terms. See Example 2.

25. $m^4 = \mathbf{m} \cdot \mathbf{m} \cdot \mathbf{m} \cdot m$
    $m^3 = \mathbf{m} \cdot \mathbf{m} \cdot \mathbf{m}$
    GCF is $m \cdot m \cdot m = m^3$.

27. $15x = 3 \cdot \mathbf{5} \cdot x$
    $25 = \mathbf{5} \cdot 5$
    GCF is 5.

29. $20c^2 = \mathbf{2} \cdot \mathbf{2} \cdot 5 \cdot c \cdot c$
    $12c = \mathbf{2} \cdot \mathbf{2} \cdot 3 \cdot \mathbf{c}$
    GCF is $2 \cdot 2 \cdot c = 4c$.

31. $18a^4 = 2 \cdot \mathbf{3} \cdot \mathbf{3} \cdot \mathbf{a} \cdot \mathbf{a} \cdot \mathbf{a} \cdot a$
    $9a^3 = \mathbf{3} \cdot \mathbf{3} \cdot \mathbf{a} \cdot \mathbf{a} \cdot \mathbf{a}$
    $27a^3 = \mathbf{3} \cdot \mathbf{3} \cdot 3 \cdot \mathbf{a} \cdot \mathbf{a} \cdot \mathbf{a}$
    GCF is $3 \cdot 3 \cdot a \cdot a \cdot a = 9a^3$.

33. $24a^2 = \mathbf{2} \cdot \mathbf{2} \cdot 2 \cdot 3 \cdot \mathbf{a} \cdot a$
    $16a^3b = \mathbf{2} \cdot \mathbf{2} \cdot 2 \cdot 2 \cdot \mathbf{a} \cdot a \cdot a \cdot b$
    $40ab = \mathbf{2} \cdot \mathbf{2} \cdot 2 \cdot 5 \cdot \mathbf{a} \cdot b$
    GCF is $2 \cdot 2 \cdot 2 \cdot a = 8a$.

35. $6m^4n = 2 \cdot \mathbf{3} \cdot \mathbf{m} \cdot \mathbf{m} \cdot \mathbf{m} \cdot m \cdot \mathbf{n}$
    $12m^3n^2 = 2 \cdot 2 \cdot \mathbf{3} \cdot \mathbf{m} \cdot \mathbf{m} \cdot \mathbf{m} \cdot \mathbf{n} \cdot n$
    $9m^3n^3 = \mathbf{3} \cdot 3 \cdot \mathbf{m} \cdot \mathbf{m} \cdot \mathbf{m} \cdot \mathbf{n} \cdot n \cdot n$
    GCF is $3 \cdot m \cdot m \cdot m \cdot n = 3m^3n$.

37. $4(x + 7) = 2 \cdot 2 \cdot \mathbf{(x + 7)}$
    $9(x + 7) = 3 \cdot 3 \cdot \mathbf{(x + 7)}$
    GCF is $x + 7$.

39. $4(p - t) = 2 \cdot 2 \cdot \mathbf{(p - t)}$
    $p(p - t) = p \cdot \mathbf{(p - t)}$
    GCF is $p - t$.

Factor out the GCF. See Example 3.

41. $3x + 6 = \mathbf{3} \cdot x + \mathbf{3} \cdot 2$
    $ = \mathbf{3}(x + 2)$

43. $18m - 9 = \mathbf{9} \cdot 2m - \mathbf{9} \cdot 1$
    $ = \mathbf{9}(2m - 1)$

45. $d^2 - 7d = \mathbf{d} \cdot d - \mathbf{d} \cdot 7$
    $ = \mathbf{d}(d - 7)$

47. $15c^3 + 25 = \mathbf{5} \cdot 3c^3 + \mathbf{5} \cdot 5$
    $ = \mathbf{5}(3c^3 + 5)$

49. $24a - 16a^2 = \mathbf{8a} \cdot 3 - \mathbf{8a} \cdot 2a$
    $ = \mathbf{8a}(3 - 2a)$

51. $14x^2 - 7x - 7 = \mathbf{7} \cdot 2x^2 - \mathbf{7} \cdot x - \mathbf{7} \cdot 1$
    $ = \mathbf{7}(2x^2 - x - 1)$

53. $t^4 + t^3 + 2t^2 = \mathbf{t^2} \cdot t^2 + \mathbf{t^2} \cdot t + \mathbf{t^2} \cdot 2$
    $ = \mathbf{t^2}(t^2 + t + 2)$

55. $21x^2y^3 + 3xy^2 = \mathbf{3xy^2} \cdot 7xy + \mathbf{3xy^2} \cdot 1$
    $ = \mathbf{3xy^2}(7xy + 1)$

Section 6.1

Factor out –1 from each polynomial. See Example 3.

57. $-a-b = (-1)a + (-1)b$
$= (-1)(a+b)$
$= -(a+b)$

59. $-x^2 - x + 16 = (-1)x^2 + (-1)(x) + (-1)(-16)$
$= (-1)(x^2 + x - 16)$
$= -(x^2 + x - 16)$

61. $5 - x = (-1)(-5) + (-1)(x)$
$= (-1)(-5 + x)$
$= -(-5 + x) \text{ or } -(x - 5)$

63. $9 - 4a = (-1)(-9) + (-1)(4a)$
$= (-1)(-9 + 4a)$
$= -(-9 + 4a) \text{ or } -(4a - 9)$

Factor each polynomial by factoring out the opposite of the GCF. See Example 5.

65. $-3x^2 - 6x = (-3x)(x) + (-3x)(2)$
$= -3x(x + 2)$

67. $-4a^2b + 12a^3 = (-4a^2)(b) + (-4a^2)(-3a)$
$= -4a^2(b - 3a)$

69. $-24x^4 - 48x^3 + 36x^2$
$= (-12x^2)2x^2 + (-12x^2)(4x) + (-12x^2)(-3)$
$= -12x^2(2x^2 + 4x - 3)$

71. $-4a^3b^2 + 14a^2b^2 - 10ab^2$
$= (-2ab^2)2a^2 + (-2ab^2)(-7a) + (-2ab^2)(5)$
$= -2ab^2(2a^2 - 7a + 5)$

Factor each expression. See Example 6.

73. $y(x+2) + 3(x+2) = y(x+2) + 3(x+2)$
$= (x+2)(y+3)$

75. $m(p-q) - 5(p-q) = m(p-q) - 5(p-q)$
$= (p-q)(m-5)$

Factor by grouping. See Example 7.

77. $2x + 2y + ax + ay = (2x + 2y) + (ax + ay)$
$= 2(x+y) + a(x+y)$
$= (x+y)(2+a)$

79. $rs - ru + 8sw - 8uw = (rs - ru) + (8sw - 8uw)$
$= r(s-u) + 8w(s-u)$
$= (s-u)(r+8w)$

81. $7m^3 - 2m^2 + 14m - 4$
$= (7m^2m - 2m^2) + (2 \cdot 7m - 2 \cdot 2)$
$= m^2(7m - 2) + 2(7m - 2)$
$= (7m - 2)(m^2 + 2)$

83. $5x^3 - x^2 + 10x - 2$
$= (5x^2x - x^2 \cdot 1) + (2 \cdot 5x - 2 \cdot 1)$
$= x^2(5x - 1) + 2(5x - 1)$
$= (5x - 1)(x^2 + 2)$

Factor by grouping. See Example 8.

85. $ab + ac + b + c = (ab + ac) + (1b + 1c)$
$= a(b+c) + 1(b+c)$
$= (b+c)(a+1)$

87. $rs + 4s^2 - r - 4s = (rs + 4ss) + (-r - 4s)$
$= s(r + 4s) - 1(r + 4s)$
$= (r + 4s)(s - 1)$

89. $2ax + 2bx - 3a - 3b$
$= (2ax + 2bx) + (-3a - 3b)$
$= 2x(a+b) - 3(a+b)$
$= (a+b)(2x - 3)$
or
$= (2ax - 3a) + (2bx - 3b)$
$= a(2x - 3) + b(2x - 3)$
$= (2x - 3)(a + b)$

91. $mp - np - mq + nq$
$= (mp - np) + (-mq + nq)$
$= (mp - np) - 1(mq - nq)$
$= p(m-n) - q(m-n)$
$= (m-n)(p-q)$

Factor by grouping. See Example 9.

93. $5m^3 + 6 + 5m^2 + 6m = 5m^3 + 5m^2 + 6m + 6$
$= (5m^3 + 5m^2) + (6m + 6)$
$= (5m^2m^1 + 5m^2) + (6m + 6)$
$= 5m^2(m + 1) + 6(m + 1)$
$= (m + 1)(5m^2 + 6)$

95. $y^3 - 12 + 3y - 4y^2 = y^3 - 4y^2 + 3y - 12$
$= (y^3 - 4y^2) + (3y - 12)$
$= (y^2y^1 - 4y^2) + (3y - 3 \cdot 4)$
$= y^2(y - 4) + 3(y - 4)$
$= (y - 4)(y^2 + 3)$

Factor by grouping. Remember to factor out the GCF first. See Example 10.

97. $ax^3 - 2ax^2 + 5ax - 10a$
$= ax^3 - 2ax^2 + 5ax - 10a$
$= a(x^3 - 2x^2 + 5x - 10)$
$= a[(x^2x - 2x^2) + (5x - 5 \cdot 2)]$
$= a[x^2(x - 2) + 5(x - 2)]$
$= a(x - 2)(x^2 + 5)$

99. $6x^3 - 6x^2 + 12x - 12$

$$= \mathbf{6}x^3 - \mathbf{6}x^2 + \mathbf{6} \cdot 2x - \mathbf{6} \cdot 2$$
$$= \mathbf{6}(x^3 - x^2 + 2x - 2)$$
$$= \mathbf{6}[(x^2x - x^2 \cdot 1) + (2x - 2 \cdot 1)]$$
$$= \mathbf{6}[x^2(x-1) + 2(x-1)]$$
$$= \mathbf{6}(x-1)(x^2 + 2)$$

TRY IT YOURSELF
Factor.

101. $h^2(14+r) + 5(14+r)$

$$= h^2(\mathbf{14+r}) + 5(\mathbf{14+r})$$
$$= (\mathbf{14+r})(h^2 + 5)$$

103. $22a^3 - 33a^2 = \mathbf{11} \cdot 2a^2a - \mathbf{11} \cdot 3a^2$

$$= \mathbf{11}a^2(2a - 3)$$

105. $ax + bx - a - b = (ax + b\mathbf{x}) + (-\mathbf{1}a - \mathbf{1}b)$

$$= \mathbf{x}(a+b) - \mathbf{1}(a+b)$$
$$= (a+b)(\mathbf{x}-\mathbf{1})$$

107. $15r^8 - 18r^6 - 30r^5$

$$= \mathbf{3} \cdot 5r^5r^3 - \mathbf{3} \cdot 6r^5r^1 - \mathbf{3} \cdot 10r^5$$
$$= \mathbf{3}r^5(5r^3 - 6r - 10)$$

109. $27mp + 9mq - 9np - 3nq$

$$= \mathbf{3} \cdot 9mp + \mathbf{3} \cdot 3mq - \mathbf{3} \cdot 3np - \mathbf{3} \cdot nq$$
$$= \mathbf{3}(9mp + 3mq - 3np - nq)$$
$$= \mathbf{3}[(\mathbf{3} \cdot 3mp + \mathbf{3} \cdot mq) + (-3np - nq)]$$
$$= \mathbf{3}[\mathbf{3}m(3p+q) - n(3p+q)]$$
$$= \mathbf{3}(3p+q)(\mathbf{3}m-n)$$

111. $-60p^2t^2 - 80pt^3 = -\mathbf{20} \cdot 3ppt^2 - \mathbf{20} \cdot 4pt^2t$

$$= -\mathbf{20}pt^2(3p + 4t)$$

113. $-2x + 5 = (-\mathbf{1})2x + (-\mathbf{1})(-5)$

$$= (-\mathbf{1})(2x - 5)$$
$$= -(2x - 5)$$

115. $6x^2 - 2xy - 15x + 5y$

$$= (\mathbf{2} \cdot 3x\mathbf{x} - \mathbf{2} \cdot xy) + (-\mathbf{5} \cdot 3x + \mathbf{5}y)$$
$$= \mathbf{2}x(3x - y) - \mathbf{5}(3x - y)$$
$$= (3x - y)(\mathbf{2}x - \mathbf{5})$$

117. $2x^3z - 4x^2z + 32xz - 64z$

$$= \mathbf{2}x^3z - \mathbf{2} \cdot 2x^2z + \mathbf{2} \cdot 16xz - \mathbf{2} \cdot 32z$$
$$= \mathbf{2}z(x^3 - 2x^2 + 16x - 32)$$
$$= \mathbf{2}z[(x^2x - 2x^2) + (16x - 16 \cdot 2)]$$
$$= \mathbf{2}z[x^2(x - 2) + 16(x - 2)]$$
$$= \mathbf{2}z(x - 2)(x^2 + 16)$$

119. $12uvw^3 - 54uv^2w^2 = \mathbf{6} \cdot 2uvw^2w - \mathbf{6} \cdot 9uvvw^2$

$$= \mathbf{6}uvw^2(2w - 9v)$$

121. $x^3 + x^2 + x + 1 = (x^2 \cdot x + x^2) + (x + 1)$

$$= x^2(x+1) + \mathbf{1}(x+1)$$
$$= (x+1)(x^2 + \mathbf{1})$$

123. $-3r + 2s - 3 = (-\mathbf{1})3r + (-\mathbf{1})(-2s) + (-\mathbf{1})3$

$$= (-\mathbf{1})(3r - 2s + 3)$$
$$= -(3r - 2s + 3)$$

LOOK ALIKES ...

125. a. $5t^3 + 6t^2 + 15t + 18$

$$= (5t^3 + 6t^2) + (15t + 18)$$
$$= (5t^2t^1 + 6t^2) + (\mathbf{3} \cdot 5t + \mathbf{3} \cdot 6)$$
$$= t^2(5t + 6) + \mathbf{3}(5t + 6)$$
$$= (5t + 6)(t^2 + \mathbf{3})$$

b. $3t^3 + 6t^2 + 15t + 18$

$$= 3t^3 + (\mathbf{3} \cdot 2)t^2 + (\mathbf{3} \cdot 5)t + (\mathbf{3} \cdot 6)$$
$$= 3(t^3 + 2t^2 + 5t + 6)$$

APPLICATIONS

127. GEOMETRY

$$x^3 + 4x^2 + 5x + 20$$

$$= (x^2 \cdot x + 4 \cdot x^2) + (\mathbf{5} \cdot x + \mathbf{5} \cdot 4)$$
$$= x^2(x + 4) + \mathbf{5}(x + 4)$$
$$= (x + 4)(x^2 + 5)$$

The length is $(x^2 + 5)$ ft.

The width is $(x + 4)$ ft.

from Campus to Careers

129. **ELEMENTARY SCHOOL TEACHER**

a. Rewrite the formula in factored form.

$$V = \pi r^2 h_1 + \frac{1}{3}\pi r^2 h_2$$

$$= \pi r^2\left(h_1 + \frac{1}{3}h_2\right)$$

The formula in factored form is

$$V = \pi r^2\left(h_1 + \frac{1}{3}h_2\right).$$

b. Evaluate the factored formula if $h_1 = 1\frac{5}{6}$ in.

$h_2 = \frac{1}{2}$ in. and $r = \frac{1}{4}$ in.

Section 6.1

$$V = \pi r^2 \left(h_1 + \frac{1}{3} h_2 \right)$$

$$= 3.14 \left(\frac{1}{4} \right)^2 \left(1\frac{5}{6} + \frac{1}{3} \cdot \frac{1}{2} \right)$$

$$= 3.14 \left(\frac{1}{16} \right) \left(\frac{11}{6} + \frac{1}{6} \right)$$

$$= 3.14 \left(\frac{1}{16} \right) \left(\frac{12}{6} \right)$$

$$= 3.14 \left(\frac{1}{16} \right) (2)$$

$$= 0.3925$$

20 students will each make 5 crayons which is 100 crayons. Mutiply 100 by 0.3925 and obtain the total volume of wax that will be needed. 39.25 or about 40 in.³ of wax.

Depending on the value of π that was used the answers will vary.

WRITING
131-133. Answers will vary.

REVIEW
135. INSURANCE COSTS

Step 1: $\$1,050 - \$925 = \$125$

This is the amount of decrease.

Step 2:   125 is what percent of 1,050?

$$125 = \quad x \quad \cdot 1,050$$

$$\frac{125}{1,050} = \frac{1,050x}{1,050}$$

$$0.119 \approx x$$

$$0.12 \approx x$$

The percent of decrease is about 12% .

CHALLENGE PROBLEMS
137. $6x^{4m}y^n + 21x^{3m}y^{2n} - 15x^{2m}y^{3n}$
$= 3x^{2m}y^n \cdot 2x^{2m} + 3x^{2m}y^n \cdot 7x^m y^n - 3x^{2m}y^n \cdot 5y^{2n}$
$= 3x^{2m}y^n (2x^{2m} + 7x^m y^n - 5y^{2n})$

SECTION 6.2
VOCABULARY

Fill in the blanks.

1. The trinomial $x^2 - x - 12$ **factors** as the product of two binomials: $(x - 4)(x + 3)$.

3. The **leading** coefficient of $x^2 - 3x + 2$ is 1.

CONCEPTS

Fill in the blanks.

5. a. Before attempting to factor a trinomial, be sure that it is written in **descending** powers of a variable.

   b. Before attempting to factor a trinomial into two binomials, always factor out any **common** factors first.

7. $x^2 + 5x + 3$ cannot be factored because we cannot find two integers whose product is **3** and whose sum is **5**.

9. Check to determine whether each factorization is correct.
   a. $x^2 - x - 20 = (x + 5)(x - 4)$ **No**
   b. $4a^2 + 12a - 16 = 4(a - 1)(a + 4)$ **Yes**

11. Consider a trinomial of the form $x^2 + bx + c$.
   a. If c is positive, what can be said about the two integers that should be chosen for the factorization? **They are both positive or both negative.**

   b. If c is negative, what can be said about the two integers that should be chosen for the factorization? **One will be positive, the other negative.**

NOTATION

13. $(x + \textbf{3})(x - \textbf{2})$

GUIDED PRACTICE

Factor each trinomial. See Example 1.

15. The positive factors of 2 whose sum is 3 are 1 and 2.
$$x^2 + 3x + 2 = (x + 2)(x + 1)$$

17. The positive factors of 12 whose sum is 7 are 3 and 4.
$$z^2 + 7z + 12 = (z + 4)(z + 3)$$

Factor each trinomial. See Example 2.

19. The negative factors of 6 whose sum is -5 are -2 and -3.
$$m^2 - 5m + 6 = (m - 3)(m - 2)$$

21. The negative factors of 28 whose sum is -11 are -4 and -7.
$$t^2 - 11t + 28 = (t - 7)(t - 4)$$

Factor each trinomial. See Example 3.

23. Two different sign factors of -24 whose sum is $+5$ are $+8$ and -3.
$$x^2 + 5x - 24 = (x + 8)(x - 3)$$

25. Two different sign factors of -48 whose sum is $+13$ are $+16$ and -3.
$$t^2 + 13t - 48 = (t - 3)(t + 16)$$

Factor each trinomial. See Example 4.

27. Two different sign factors of -16 whose sum is -6 are -8 and $+2$.
$$a^2 - 6a - 16 = (a - 8)(a + 2)$$

29. Two different sign factors of -36 whose sum is -9 are -12 and $+3$.
$$b^2 - 9b - 36 = (b - 12)(b + 3)$$

Factor each trinomial. See Example 5.

31. Factor out a -1.
$$-x^2 - 7x - 10 = -1(x^2 + 7x + 10)$$
   The positive factors of 10 whose sum is 7 are 2 and 5.
$$-x^2 - 7x - 10 = -1(x^2 + 7x + 10)$$
$$= -(x + 5)(x + 2)$$

33. Factor out a -1.
$$-t^2 - t + 30 = -1(t^2 + t - 30)$$
   Two different sign factors of -30 whose sum is $+1$ are $+6$ and -5.
$$-t^2 - t + 30 = -1(t^2 + t - 30)$$
$$= -(t + 6)(t - 5)$$

35. Factor out a -1.
$$-r^2 - 3r + 54 = -1(r^2 + 3r - 54)$$
   Two different sign factors of -54 whose sum is $+3$ are $+9$ and -6.
$$-r^2 - 3r + 54 = -1(r^2 + 3r - 54)$$
$$= -(r + 9)(r - 6)$$

37. Factor out a -1.
$$-m^2 + 18m - 77 = -1(m^2 - 18m + 77)$$
   The negative factors of 77 whose sum is -18 are -7 and -11.
$$-m^2 + 18m - 77 = -1(m^2 - 18m + 77)$$
$$= -(m - 7)(m - 11)$$

Factor each trinomial. See Example 6.

39. The positive factors of 3 whose sum is 4 are 1 and 3.
$$a^2 + 4ab + 3b^2 = (a + 3b)(a + b)$$

41. Two different sign factors of -7 whose sum is -6 are -7 and $+1$.
$$x^2 - 6xy - 7y^2 = (x - 7y)(x + y)$$

- 343 -

43. Two different sign factors of -2 whose sum is $+1$ are $+2$ and -1.
$$r^2 + rs - 2s^2 = (r + 2s)(r - s)$$

45. Two negative factors of 6 whose sum is -5 are -3 and -2.
$$a^2 - 5ab + 6b^2 = (a - 3b)(a - 2b)$$

Factor completely. See Example 7.

47. Factor out a 2.
$$2x^2 + 10x + 12 = 2(x^2 + 5x + 6)$$
The positive factors of 6 whose sum is 5 are 2 and 3.
$$2x^2 + 10x + 12 = 2(x^2 + 5x + 6)$$
$$= 2(x + 2)(x + 3)$$

49. Factor out a 6.
$$6a^2 - 30a + 24 = 6(a^2 - 5a + 4)$$
The negative factors of 4 whose sum is -5 are -1 and -4.
$$6a^2 - 30a + 24 = 6(a^2 - 5a + 4)$$
$$= 6(a - 1)(a - 4)$$

51. Factor out a 5.
$$5a^2 - 25a + 30 = 5(a^2 - 5a + 6)$$
The negative factors of 6 whose sum is -5 are -2 and -3.
$$5a^2 - 25a + 30 = 5(a^2 - 5a + 6)$$
$$= 5(a - 2)(a - 3)$$

53. Factor out a $-z$.
$$-z^3 + 29z^2 - 100z = -z(z^2 - 29z + 100)$$
The negative factors of 100 whose sum is -29 are -4 and -25.
$$-z^3 + 29z^2 - 100z = -z(z^2 - 29z + 100)$$
$$= -z(z - 4)(z - 25)$$

Write each trinomial in descending powers of one variable and factor. See Example 8.

55. $80 - 24x + x^2 = x^2 - 24x + 80$
The negative factors of 80 whose sum is -24 are -20 and -4.
$$80 - 24x + x^2 = x^2 - 24x + 80$$
$$= (x - 20)(x - 4)$$

57. $10y + 9 + y^2 = y^2 + 10y + 9$
The positive factors of 9 whose sum is 10 are 1 and 9.
$$10y + 9 + y^2 = y^2 + 10y + 9$$
$$= (y + 1)(y + 9)$$

59. $r^3 - 16r + 6r^2 = r^3 + 6r^2 - 16r$
Factor out a r.
Two different sign factors of -16 whose sum is $+6$ are $+8$ and -2.
$$r^3 - 16r + 6r^2 = r^3 + 6r^2 - 16r$$
$$= r(r^2 + 6r - 16)$$
$$= r(r + 8)(r - 2)$$

61. $4r^2x + r^3 + 3rx^2 = r^3 + 4r^2x + 3rx^2$
Factor out a r.
The positive factors of 3 whose sum is 4 are 1 and 3.
$$4r^2x + r^3 + 3rx^2 = r^3 + 4r^2x + 3rx^2$$
$$= r(r^2 + 4rx + 3x^2)$$
$$= r(r + x)(r + 3x)$$

Factor each trinomial. See Example 9.

63. There are no two integers whose product is 15 and whose sum is 10, the trinomial $u^2 + 10u + 15$ cannot be factored and is a *prime trinomial*.

65. There are no two diffferent sign integers whose product is -4 and whose sum is 2, the trinomial $r^2 + 2r - 4$ cannot be factored and is a *prime trinomial*.

TRY IT YOURSELF

Choose the correct method from Section 6.1 or Section 6.2 to factor each of the following.

67. $5x + 15 + xy + 3y = (5x + 5 \cdot 3) + (xy + 3y)$
$$= 5(x + 3) + y(x + 3)$$
$$= (x + 3)(5 + y)$$

69. $26n^2 - 8n = 2 \cdot 13nn - 2 \cdot 4n$
$$= 2n(13n - 4)$$

71. Two different sign factors of -5 whose sum is -4 are -5 and $+1$.
$$a^2 - 4a - 5 = (a - 5)(a + 1)$$

73. Factor out a -1.
$$-x^2 + 21x + 22 = -1(x^2 - 21x - 22)$$
Two different sign factors of -22 whose sum is -21 are -22 and $+1$.
$$-x^2 + 21x + 22 = -1(x^2 - 21x - 22)$$
$$= -(x - 22)(x + 1)$$

75. $4xy - 4x + 28y - 28$
$$= 4xy - 4x + 4 \cdot 7y - 4 \cdot 7$$
$$= 4(xy - x + 7y - 7)$$
$$= 4[(xy - x \cdot 1) + (7y - 7 \cdot 1)]$$
$$= 4[x(y - 1) + 7(y - 1)]$$
$$= 4(y - 1)(x + 7)$$

77. Factor out a $12b^2$.

$24b^4 - 48b^3 - 36b^2$
$= \mathbf{12} \cdot 2 \cdot \mathbf{b^2} \cdot b^2 - \mathbf{12} \cdot 4 \cdot \mathbf{b^2} \cdot b - \mathbf{12} \cdot 3 \cdot \mathbf{b^2}$
$= 12b^2(2b^2 - 4b - 3)$

79. The negative factors of 18 whose sum is -9 are -3 and -6.

$r^2 - 9r + 18 = (r-3)(r-6)$

81. Factor out a $-n^2$.

$-n^4 + 28n^3 + 60n^2 = -n^2(n^2 - 28n - 60)$

Two different sign factors of -60 whose sum is -28 are -30 and $+2$.

$-n^4 + 28n^3 + 60n^2 = -n^2(n^2 - 28n - 60)$
$= -n^2(n-30)(n+2)$

83. The positive factors of 4 whose sum is 4 are 2 and 2.

$x^2 + 4xy + 4y^2 = (x+2y)(x+2y)$
$= (x + 2y)^2$

85. Two different sign factors of -12 whose sum is -4 are -6 and $+2$.

$a^2 - 4ab - 12b^2 = (a-6b)(a+2b)$

87. Factor out a $4x^2$.

$4x^4 + 16x^3 + 16x^2 = 4x^2(x^2 + 4x + 4)$

The positive factors of 4 whose sum is 4 are 2 and 2.

$4x^4 + 16x^3 + 16x^2 = 4x^2(x^2 + 4x + 4)$
$= 4x^2(x+2)(x+2)$
$= 4x^2(x+2)^2$

89. The negative factors of 45 whose sum is -46 are -1 and -45.

$a^2 - 46a + 45 = (a-45)(a-1)$

91. There are no two integers whose product is 4 and whose sum is -2, the trinomial $r^2 - 2r + 4$ cannot be factored and is a *prime trinomial*.

93. $t(x+2) + 7(x+2) = (x+2)(t+7)$

95. Factor out a s^2.

$s^4 + 11s^3 - 26s^2 = s^2(s^2 + 11s - 26)$

Two different sign factors of -26 whose sum is $+11$ are $+13$ and -2.

$s^4 + 11s^3 - 26s^2 = s^2(s^2 + 11s - 26)$
$= s^2(s+13)(s-2)$

97. $15s^3 + 75 = \mathbf{15} \cdot s^3 + \mathbf{15} \cdot 5$
$= 15(s^3 + 5)$

99. $-13y + y^2 - 14 = y^2 - 13y - 14$

Two different sign factors of -14 whose sum is -13 are -14 and $+1$.

$-13y + y^2 - 14 = y^2 - 13y - 14$
$= (y-14)(y+1)$

101. Factor out a 2.

$2x^2 - 12x + 16 = 2(x^2 - 6x + 8)$

The negative factors of 8 whose sum is -6 are -4 and -2.

$2x^2 - 12x + 16 = 2(x^2 - 6x + 8)$
$= 2(x-4)(x-2)$

LOOK ALIKES…

103. The negative factors of 24 whose sum is -10 are -4 and -6.

a. $x^2 - 10x + 24 = (x-4)(x-6)$

Two different sign factors of -24 whose sum is -10 are -12 and $+2$.

b. $x^2 - 10x - 24 = (x-12)(x+2)$

APPLICATIONS

105. PETS

Factor out a x.
$x^3 + 12x^2 + 27x = x(x^2 + 12x + 27)$

The positive factors of 27 whose sum is 12 are 9 and 3.
$x^3 + 12x^2 + 27x = x(x^2 + 12x + 27)$
$= x(x+9)(x+3)$

Its length is $(x+9)$ in.
Its width is x in.
Its height is $(x+3)$ in.

WRITING

107-111. Answers will vary.

REVIEW

Simplify each expression. Write each answer without negative exponents.

113. $\dfrac{x^{12}x^{-7}}{x^3 x^4} = \dfrac{x^{12-7}}{x^{3+4}}$

$= \dfrac{x^5}{x^7}$

$= \dfrac{x^{5-7}}{1}$

$= \dfrac{x^{-2}}{1}$

$= \dfrac{1}{x^2}$

Section 6.2

115.
$$(x^{-3}x^{-2})^2 = (x^{-3+(-2)})^2$$
$$= (x^{-5})^2$$
$$= \left(\frac{1}{x^5}\right)^2$$
$$= \frac{1}{x^{5\cdot2}}$$
$$= \frac{1}{x^{10}}$$

CHALLENGE PROBLEMS

Factor completely.

117. The negative factors of $\dfrac{9}{25}$

whose sum is $-\dfrac{6}{5}$ are $-\dfrac{3}{5}$ and $-\dfrac{3}{5}$.

$$x^2 - \frac{6}{5}x + \frac{9}{25} = \left(x - \frac{3}{5}\right)\left(x - \frac{3}{5}\right)$$
$$= \left(x - \frac{3}{5}\right)^2$$

119. Two different sign factors of -45
whose sum is -12 are -15 and $+3$.
$$x^{2m} - 12x^m - 45 = (x^m - 15)(x^m + 3)$$

121. Find all positive integer values of c that
make $n^2 + 6n + c$ factorable.   **5, 8, 9**

SECTION 6.3

VOCABULARY
Fill in the blanks.

1. The **leading** coefficient of $3x^2 - x - 12$ is 3.

3. The first terms of the binomial factors $(5y + 1)$ $(y + 3)$ are **5y** and **y**. The second terms of the binomial factors are **1** and **3**.

CONCEPTS

5. If $10x^2 - 27x + 5$ is to be factored as the product of two binomials, what are the possible first terms of the binomial factors?

   10x and x, 5x and 2x

7. a. Fill in the blanks. When factoring a trinomial, we write it in **descending** powers of the variable. Then we factor out any **GCF** (including -1 if that is necessary to make the lead **coefficient** positive).

   b. What is the GCF of the terms of $6s^4 + 33s^3 + 36s^2$? **$3s^2$**

   c. Factor out -1 from $-2d^2 + 19d - 8$.
   $-(2d^2 - 19d + 8)$

A trinomial has been partially factored. Complete each statement that describes the type of integers we should consider for the blanks.

9. $5y^2 - 13y + 6 = (5y \boxed{})(y \boxed{})$

   Since the last term of the trinomial is positive and the middle term is negative, the integers must be **negative** factors of 6.

11. $5y^2 - 7y - 6 = (5y \boxed{})(y \boxed{})$

   Since the last term of the trinomial is negative, the signs of the integers will be **different**.

13. Complete the key number table.

Negative factors of 12	Sum of the negative factors of 12
$-1(-12)$	**-13**
$-2(-6)$	**-8**
$-3(-4)$	-7

NOTATION

15. a. Suppose we wish to factor $12b^2 + 20b - 9$ by grouping. Identify a, b, and c. **12, 20, -9**

   b. What is the key number, ac? **-108**

Complete each step of the factorization of the trinomial by grouping.

17. $12t^2 + 17t + 6 = 12t^2 + 9t + 8t + 6$
$$= \mathbf{3t}(4t + 3) + \mathbf{2}(4t + 3)$$
$$= (\mathbf{4t + 3})(3t + 2)$$

GUIDED PRACTICE
Factor. See Example 1.

19. $2x^2 + 3x + 1 = (2x + 1)(x + 1)$

21. $3a^2 + 10a + 3 = (3a + 1)(a + 3)$

23. $5x^2 + 7x + 2 = (5x + 2)(x + 1)$

25. $7x^2 + 18x + 11 = (7x + 11)(x + 1)$

Factor. See Example 2.

27. $4x^2 - 8x + 3 = (2x - 3)(2x - 1)$

29. $8x^2 - 22x + 5 = (4x - 1)(2x - 5)$

31. $15t^2 - 26t + 7 = (5t - 7)(3t - 1)$

33. $6y^2 - 13y + 2 = (6y - 1)(y - 2)$

Factor. See Example 3.

35. $3\underset{a}{x^2} - 2\underset{b}{x} - \underset{c}{21}$

In $3x^2 - 2x - 21$, we have $a = 3$, $b = -2$, and $c = -21$. The key number is $ac = 3(-21) = -63$. We must find a factorization of -63 in which the sum of the factors is $b = -2$. Since the factors must have a negative product, their signs must be different. The pairs of factors are -9 and $+7$.

$$3x^2 - 2x - 21 = 3x^2 - 9x + 7x - 21$$
$$= (3x^2 - 9x) + (7x - 21)$$
$$= 3x(x - 3) + 7(x - 3)$$
$$= (x - 3)(3x + 7)$$

37. $5\underset{a}{m^2} - 7\underset{b}{m} - \underset{c}{6}$

In $5m^2 - 7m - 6$, we have $a = 5$, $b = -7$, and $c = -6$. The key number is $ac = 5(-6) = -30$. We must find a factorization of -30 in which the sum of the factors is $b = -7$. Since the factors must have a negative product, their signs must be different. The pairs of factors are -10 and $+3$.

$$5m^2 - 7m - 6 = 5m^2 - 10m + 3m - 6$$
$$= (5m^2 - 10m) + (3m - 6)$$
$$= 5m(m - 2) + 3(m - 2)$$
$$= (m - 2)(5m + 3)$$

39. $7 \underset{a}{y^2} + 55 \underset{b}{y} - \underset{c}{8}$

In $7y^2 + 55y - 8$, we have $a = 7$, $b = 55$ and $c = -8$. The key number is $ac = 7(-8) = -56$. We must find a factorization of -56 in which the sum of the factors is $b = 55$. Since the factors must have a negative product, their signs must be different. The pairs of factors are -1 and $+56$.

$$
\begin{aligned}
7y^2 + 55y - 8 &= 7y^2 - y + 56y - 8 \\
&= (7y^2 - y) + (56y - 8) \\
&= y(7y - 1) + 8(7y - 1) \\
&= (7y - 1)(y + 8)
\end{aligned}
$$

41. $11 \underset{a}{y^2} + \underset{b}{7} y - \underset{c}{4}$

In $11y^2 + 7y - 4$, we have $a = 11$, $b = 7$ and $c = -4$. The key number is $ac = 11(-4) = -44$. We must find a factorization of -44 in which the sum of the factors is $b = 7$. Since the factors must have a negative product, their signs must be different. The pairs of factors are -4 and $+11$.

$$
\begin{aligned}
11y^2 + 7y - 4 &= 11y^2 - 4y + 11y - 4 \\
&= (11y^2 - 4y) + (11y - 4) \\
&= y(11y - 4) + 1(11y - 4) \\
&= (11y - 4)(y + 1)
\end{aligned}
$$

Factor. See Example 4.

43. $6 \underset{a}{r^2} + \underset{b}{1} rs - 2 \underset{c}{s^2}$

In $6r^2 + 1rs - 2s^2$, we have $a = 6$, $b = 1$ and $c = -2$. The key number is $ac = 6(-2) = -12$. We must find a factorization of -12 in which the sum of the factors is $b = 1$. Since the factors must have a negative product, their signs must be different. The pairs of factors are -3 and $+4$.

$$
\begin{aligned}
6r^2 + rs - 2s^2 &= 6r^2 - 3rs + 4rs - 2s^2 \\
&= (6r^2 - 3rs) + (4rs - 2s^2) \\
&= 3r(2r - s) + 2s(2r - s) \\
&= (2r - s)(3r + 2s)
\end{aligned}
$$

45. $4 \underset{a}{x^2} + \underset{b}{8} xy + 3 \underset{c}{y^2}$

In $4x^2 + 8xy + 3y^2$, we have $a = 4$, $b = 8$ and $c = 3$. The key number is $ac = 4(3) = 12$. We must find a factorization of 12 in which the sum of the factors is $b = 8$. Since the factors must have a positive product, their signs must be the same. The pairs of factors are $+2$ and $+6$.

$$
\begin{aligned}
4x^2 + 8xy + 3y^2 &= 4x^2 + 2xy + 6xy + 3y^2 \\
&= (4x^2 + 2xy) + (6xy + 3y^2) \\
&= 2x(2x + y) + 3y(2x + y) \\
&= (2x + y)(2x + 3y)
\end{aligned}
$$

47. $8 \underset{a}{m^2} + 91 \underset{b}{mn} + 33 \underset{c}{n^2}$

In $8m^2 + 91mn + 33n^2$, we have $a = 8$, $b = 91$ and $c = 33$. The key number is $ac = 8(33) = 264$. We must find a factorization of 264 in which the sum of the factors is $b = 91$. Since the factors must have a positive product, their signs must be the same. The pairs of factors are $+3$ and $+88$.

$$
\begin{aligned}
8m^2 + 91mn + 33n^2 &= 8m^2 + 88mn + 3mn + 33n^2 \\
&= (8m^2 + 88mn) + (3mn + 33n^2) \\
&= 8m(m + 11n) + 3n(m + 11n) \\
&= (m + 11n)(8m + 3n)
\end{aligned}
$$

49. $15 \underset{a}{x^2} - \underset{b}{1} xy - 6 \underset{c}{y^2}$

In $15x^2 - 1xy - 6y^2$, we have $a = 15$, $b = -1$ and $c = -6$. The key number is $ac = 15(-6) = -90$. We must find a factorization of -90 in which the sum of the factors is $b = -1$. Since the factors must have a negative product, their signs must be different. The pairs of factors are -10 and $+9$.

$$
\begin{aligned}
15x^2 - xy - 6y^2 &= 15x^2 - 10xy + 9xy - 6y^2 \\
&= (15x^2 - 10xy) + (9xy - 6y^2) \\
&= 5x(3x - 2y) + 3y(3x - 2y) \\
&= (3x - 2y)(5x + 3y)
\end{aligned}
$$

Factor. See Example 7.

51.
$$
\begin{aligned}
-26x + 6x^2 - 20 &= 6x^2 - 26x - 20 \\
&= 2(3x^2 - 13x - 10)
\end{aligned}
$$

$$3 \underset{a}{x^2} - 13 \underset{b}{x} - \underset{c}{10}$$

In $3x^2 - 13x - 10$, we have $a = 3$, $b = -13$ and $c = -10$. The key number is $ac = 3(-10) = -30$. We must find a factorization of -30 in which the sum of the factors is $b = -13$. Since the factors must have a negative product, their signs must be different. The pairs of factors are -15 and $+2$.

$$
\begin{aligned}
2(3x^2 - 13x - 10) &= 2(3x^2 - 15x + 2x - 10) \\
&= 2[(3x^2 - 15x) + (2x - 10)] \\
&= 2[3x(x - 5) + 2(x - 5)] \\
&= 2(x - 5)(3x + 2)
\end{aligned}
$$

53.
$$15a + 8a^3 - 26a^2 = 8a^3 - 26a^2 + 15a$$
$$= a(8a^2 - 26a + 15)$$

$$\underset{a}{8a^2} - \underset{b}{26a} + \underset{c}{15}$$

In $8a^2 - 26a + 15$, we have $a = 8$, $b = -26$ and $c = 15$. The key number is $ac = 8(15) = 120$. We must find a factorization of 120 in which the sum of the factors is $b = -26$. Since the factors must have a positive product, their signs must be the same. The pairs of factors are -6 and -20.

$$a(8a^2 - 26a + 15) = a(8a^2 - 6a - 20a + 15)$$
$$= a[(8a^2 - 6a) + (-20a + 15)]$$
$$= a[2a(4a - 3) - 5(4a - 3)]$$
$$= a(4a - 3)(2a - 5)$$

55.
$$2u^2 - 6v^2 - uv = 2u^2 - uv - 6v^2$$

$$\underset{a}{2u^2} - \underset{b}{1uv} - \underset{c}{6v^2}$$

In $2u^2 - 1uv - 6v^2$, we have $a = 2$, $b = -1$ and $c = -6$. The key number is $ac = 2(-6) = -12$. We must find a factorization of -12 in which the sum of the factors is $b = -1$. Since the factors must have a negative product, their signs must be different. The pairs of factors are -4 and 3.

$$2u^2 - uv - 6v^2 = (2u^2 - 4uv + 3uv - 6v^2)$$
$$= (2u^2 - 4uv) + (3uv - 6v^2)$$
$$= 2u(u - 2v) + 3v(u - 2v)$$
$$= (u - 2v)(2u + 3v)$$

57.
$$36y^2 - 88y + 32 = 4(9y^2 - 22y + 8)$$

$$\underset{a}{9y^2} - \underset{b}{22y} + \underset{c}{8}$$

In $9y^2 - 22y + 8$, we have $a = 9$, $b = -22$ and $c = 8$. The key number is $ac = 9(8) = 72$. We must find a factorization of 72 in which the sum of the factors is $b = -22$. Since the factors must have a positive product, their signs must be the same. The pairs of factors are -4 and -18.

$$4(9y^2 - 22y + 8) = 4(9y^2 - 4y - 18y + 8)$$
$$= 4[(9y^2 - 4y) + (-18y + 8)]$$
$$= 4[y(9y - 4) - 2(9y - 4)]$$
$$= 4(9y - 4)(y - 2)$$

TRY IT YOURSELF

Factor. If an expression is *prime*, so indicate.

59. $\underset{a}{6t^2} - \underset{b}{7t} - \underset{c}{20}$

In $6t^2 - 7t - 20$, we have $a = 6$, $b = -7$ and $c = -20$. The key number is $ac = 6(-20) = -120$. We must find a factorization of -120 in which the sum of the factors is $b = -7$. Since the factors must have a negative product, their signs must be different. The pairs of factors are -15 and 8.

$$6t^2 - 7t - 20 = 6t^2 - 15t + 8t - 20$$
$$= (6t^2 - 15t) + (8t - 20)$$
$$= 3t(2t - 5) + 4(2t - 5)$$
$$= (2t - 5)(3t + 4)$$

61. $\underset{a}{15p^2} - \underset{b}{2}pq - \underset{c}{1}q^2$

In $15p^2 - 2pq - 1q^2$, we have $a = 15$, $b = -2$ and $c = -1$. The key number is $ac = 15(-1) = -15$. We must find a factorization of -15 in which the sum of the factors is $b = -2$. Since the factors must have a negative product, their signs must be different. The pairs of factors are -5 and 3.

$$15p^2 - 2pq - q^2 = 15p^2 - 5pq + 3pq - q^2$$
$$= (15p^2 - 5pq) + (3pq - q^2)$$
$$= 5p(3p - q) + q(3p - q)$$
$$= (3p - q)(5p + q)$$

63. $\underset{a}{4t^2} - \underset{b}{16t} + \underset{c}{7}$

In $4t^2 - 16t + 7$, we have $a = 4$, $b = -16$ and $c = 7$. The key number is $ac = 4(7) = 28$. We must find a factorization of 28 in which the sum of the factors is $b = -16$. Since the factors must have a positive product, their signs must be the same. The pairs of factors are -2 and -14.

$$4t^2 - 16t + 7 = 4t^2 - 2t - 14t + 7$$
$$= (4t^2 - 2t) + (-14t + 7)$$
$$= 2t(2t - 1) - 7(2t - 1)$$
$$= (2t - 1)(2t - 7)$$

65. $130r^2 + 20r - 110 = 10(13r^2 + 2r - 11)$

$$\underset{a}{13r^2} + \underset{b}{2}r - \underset{c}{11}$$

In $13r^2 + 2r - 11$, we have $a = 13$, $b = 2$ and $c = -11$. The key number is $ac = 13(-11) = -143$. We must find a factorization of -143 in which the sum of the factors is $b = 2$. Since the factors must have a negative product, their signs must be different. The pairs of factors are -11 and 13.

$$10(13r^2 + 2r - 11) = 10(13r^2 - 11r + 13r - 11)$$
$$= 10[(13r^2 - 11r) + (13r - 11)]$$
$$= 10[r(13r - 11) + 1(13r - 11)]$$
$$= 10(13r - 11)(r + 1)$$

Section 6.3

67. $8y^2 - \underset{b}{2}y - \underset{c}{1}$

In $8y^2 - 2y - 1$, we have $a = 8$, $b = -2$ and $c = -1$. The key number is $ac = 8(-1) = -8$. We must find a factorization of -8 in which the sum of the factors is $b = -2$. Since the factors must have a negative product, their signs must be different. The pairs of factors are -4 and 2.

$$8y^2 - 2y - 1 = 8y^2 - 4y + 2y - 1$$
$$= (8y^2 - 4y) + (2y - 1)$$
$$= 4y(2y - 1) + 1(2y - 1)$$
$$= (2y - 1)(4y + 1)$$

69. $18\underset{a}{x^2} + 31\underset{b}{x} - \underset{c}{10}$

In $18x^2 + 31x - 10$, we have $a = 18$, $b = 31$ and $c = -10$. The key number is $ac = 18(-10) = -180$. We must find a factorization of -180 in which the sum of the factors is $b = 31$. Since the factors must have a negative product, their signs must be different. The pairs of factors are -5 and 36.

$$18x^2 + 31x - 10 = 18x^2 - 5x + 36x - 10$$
$$= (18x^2 - 5x) + (36x - 10)$$
$$= x(18x - 5) + 2(18x - 5)$$
$$= (18x - 5)(x + 2)$$

71. $-y^3 - 13y^2 - 12y = -y(y^2 + 13y + 12)$
$$= -y(y + 12)(y + 1)$$

73. $10\underset{a}{u^2} - 13\underset{b}{u} - \underset{c}{6}$

In $10u^2 - 13u - 6$, we have $a = 10$, $b = -13$ and $c = -6$. The key number is $ac = 10(-6) = -60$. We must find a factorization of -60 in which the sum of the factors is $b = -13$. Since the factors must have a negative product, their signs must be different. The pairs of factors are none. This trinomial is prime.

$$10u^2 - 13u - 6 \text{ is prime.}$$

75. $-6x^4 + 15x^3 + 9x^2 = -3x^2(2x^2 - 5x - 3)$

In $2x^2 - 5x - 3$, we have $a = 2$, $b = -5$ and $c = -3$. The key number is $ac = 2(-3) = -6$. We must find a factorization of -6 in which the sum of the factors is $b = -5$. Since the factors must have a negative product, their signs must be different. The pairs of factors are -6 and 1.

$$-3x^2(2x^2 - 5x - 3) = -3x^2(2x^2 - 6x + 1x - 3)$$
$$= -3x^2[(2x^2 - 6x) + (1x - 3)]$$
$$= -3x^2[2x(x - 3) + 1(x - 3)]$$
$$= -3x^2(x - 3)(2x + 1)$$

77. $6\underset{a}{p^2} + \underset{b}{1}pq - \underset{c}{1}q^2$

In $6p^2 + 1pq - 1q^2$, we have $a = 6$, $b = 1$ and $c = -1$. The key number is $ac = 6(-1) = -6$. We must find a factorization of -6 in which the sum of the factors is $b = 1$. Since the factors must have a negative product, their signs must be different. The pairs of factors are -2 and 3.

$$6p^2 + 1pq - 1q^2 = 6p^2 - 2pq + 3pq - 1q^2$$
$$= (6p^2 - 2pq) + (3pq - 1q^2)$$
$$= 2p(3p - q) + q(3p - q)$$
$$= (3p - q)(2p + q)$$

79. $30r^5 + 63r^4 - 30r^3 = 3r^3(10r^2 + 21r - 10)$

In $10r^2 + 21r - 10$, we have $a = 10$, $b = 21$ and $c = -10$. The key number is $ac = 10(-10) = -100$. We must find a factorization of -100 in which the sum of the factors is $b = 21$. Since the factors must have a negative product, their signs must be different. The pairs of factors are -4 and 25.

$$3r^3(10r^2 + 21r - 10)$$
$$= 3r^3(10r^2 - 4r + 25r - 10)$$
$$= 3r^3[(10r^2 - 4r) + (25r - 10)]$$
$$= 3r^3[2r(5r - 2) + 5(5r - 2)]$$
$$= 3r^3(5r - 2)(2r + 5)$$

81. $16m^3n + 20m^2n^2 + 6mn^3$
$$= 2mn(8m^2 + 10mn + 3n^2)$$

$$8\underset{a}{m^2} + 10\underset{b}{mn} + 3\underset{c}{n^2}$$

In $8m^2 + 10mn + 3n^2$, we have $a = 8$, $b = 10$ and $c = 3$. The key number is $ac = 8(3) = 24$. We must find a factorization of 24 in which the sum of the factors is $b = 10$. Since the factors must have a positive product, their signs must be the same. The pairs of factors are 6 and 4.

$$2mn(8m^2 + 10mn + 3n^2)$$
$$= 2mn(8m^2 + 6mn + 4mn + 3n^2)$$
$$= 2mn[(8m^2 + 6mn) + (4mn + 3n^2)]$$
$$= 2mn[2m(4m + 3n) + n(4m + 3n)]$$
$$= 2mn(4m + 3n)(2m + n)$$

83. $3\underset{a}{x^2} + \underset{b}{1}x + \underset{c}{6}$

In $3x^2 + 1x + 6$, we have $a = 3$, $b = 1$ and $c = 6$. The key number is $ac = 3(6) = 18$. We must find a factorization of 18 in which the sum of the factors is $b = 1$. Since the factors must have a positive product, their signs must be the same. The pairs of factors are none. This trinomial is prime.

$$3x^2 + x + 6 \text{ is prime.}$$

85. $-12y^2 - 12 + 25y = -12y^2 + 25y - 12$
$$= -(12y^2 - 25y + 12)$$

$\underset{a}{12}y^2 \underset{b}{-25}y \underset{c}{+12}$

In $12y^2 - 25y + 12$, we have $a = 12$, $b = -25$ and $c = 12$. The key number is $ac = 12(12) = 144$. We must find a factorization of 144 in which the sum of the factors is $b = -25$. Since the factors must have a positive product, their signs must be the same. The pairs of factors are -9 and -16.

$$-(12y^2 - 25y + 12)$$
$$= -[12y^2 - 16y - 9y + 12]$$
$$= -[(12y^2 - 16y) + (-9y + 12)]$$
$$= -[4y(3y - 4) - 3(3y - 4)]$$
$$= -(3y - 4)(4y - 3)$$

Choose the correct method from Sections 6.1, 6.2, or 6.3 to factor each of the following.

87. $m^2 + 3m - 28 = (m + 7)(m - 4)$

89. $6a^3 + 15a^2 = 3a^2(2a + 5)$

91. $x^3 - 2x^2 + 5x - 10 = (x^3 - 2x^2) + (5x - 10)$
$$= x^2(x - 2) + 5(x - 2)$$
$$= (x - 2)(x^2 + 5)$$

93. $5y^2 + 3 - 8y = 5y^2 - 8y + 3$

$\underset{a}{5}y^2 \underset{b}{-8}y \underset{c}{+3}$

In $5y^2 - 8y + 3$, we have $a = 5$, $b = -8$ and $c = 3$. The key number is $ac = 5(3) = 15$. We must find a factorization of 15 in which the sum of the factors is $b = -8$. Since the factors must have a positive product, their signs must be the same. The pairs of factors are -3 and -5.

$$5y^2 - 8y + 3 = 5y^2 - 5y - 3y + 3$$
$$= (5y^2 - 5y) + (-3y + 3)$$
$$= 5y(y - 1) - 3(y - 1)$$
$$= (y - 1)(5y - 3)$$

95. $-2x^2 - 10x - 12 = -2(x^2 + 5x + 6)$
$$= -2(x + 2)(x + 3)$$

97. $12x^3y^3 - 18x^2y^3 + 15x^2y^2$
$$= 3x^2y^2(4xy - 6y + 5)$$

99. $a^2 - 7ab + 10b^2 = (a - 5b)(a - 2b)$

101. $9u^6 - 71u^5 - 8u^4 = u^4(9u^2 - 71u - 8)$

$\underset{a}{9}u^2 \underset{b}{-71}u \underset{c}{-8}$

In $9u^2 - 71u - 8$, we have $a = 9$, $b = -71$ and $c = -8$. The key number is $ac = 9(-8) = -72$. We must find a factorization of -72 in which the sum of the factors is $b = -71$. Since the factors must have a negative product, their signs must be different. The pairs of factors are 1 and -72.

$$u^4(9u^2 - 71u - 8) = u^4(9u^2 - 72u + 1u - 8)$$
$$= u^4[(9u^2 - 72u) + (1u - 8)]$$
$$= u^4[9u(u - 8) + 1(u - 8)]$$
$$= u^4(u - 8)(9u + 1)$$

APPLICATIONS
from CAMPUS to CAREERS
103. ELEMENTARY SCHOOL TEACHER

$\underset{a}{4}x^2 \underset{b}{+20}x \underset{c}{-11}$

In $4x^2 + 20x - 11$, we have $a = 4$, $b = 20$ and $c = -11$. The key number is $ac = 4(-11) = -44$. We must find a factorization of -44 in which the sum of the factors is $b = 20$. Since the factors must have a negative product, their signs must be different. The pairs of factors are -2 and 22.

$$4x^2 + 20x - 11 = 4x^2 + 22x - 2x - 11$$
$$= (4x^2 + 22x) + (-2x - 11)$$
$$= 2x(2x + 11) - 1(2x + 11)$$
$$= (2x + 11)(2x - 1)$$

The length is $(2x + 11)$ in.
The width is $(2x - 1)$ in.

WRITING
105-107. Answers will vary.

REVIEW
Evaluate each expression.
109. $-7^2 = -(7)(7)$
$$= -49$$

111. $7^0 = 1$

113. $\dfrac{1}{7^{-2}} = 7^2$
$$= 49$$

CHALLENGE PROBLEMS

Factor.

115. $6a^{10} + 5a^5 - 21$
$\quad\quad\underset{a}{} \quad \underset{b}{} \quad \underset{c}{}$

In $6a^{10} + 5a^5 - 21$, we have $a = 6$, $b = 5$ and $c = -21$.
The key number is $ac = 6(-21) = -126$. We must find a
factorization of -126 in which the sum of the factors is
$b = 5$. Since the factors must have a negative product,
their signs must be different. The pairs of factors are 14
and -9.

$$\begin{aligned}
6a^{10} + 5a^5 - 21 &= 6a^{10} - 9a^5 + 14a^5 - 21 \\
&= (6a^{10} - 9a^5) + (14a^5 - 21) \\
&= 3a^5(2a^5 - 3) + 7(2a^5 - 3) \\
&= (2a^5 - 3)(3a^5 + 7)
\end{aligned}$$

117. $8x^2(c^2 + c - 2) - 2x(c^2 + c - 2)$
$$\begin{aligned}
&\quad\quad\quad\quad\quad\quad\quad - 1(c^2 + c - 2) \\
&= (c^2 + c - 2)(8x^2 - 2x - 1) \\
&= (c + 2)(c - 1)(4x + 1)(2x - 1)
\end{aligned}$$

VOCABULARY

Fill in the blanks.

1. $x^2 + 6x + 9$ is a **perfect** -square trinomial because it is the square of the binomial $x + 3$.

CONCEPTS

3. Consider $25x^2 + 30x + 9$.

   a. The first term is the square of **5x**.

   b. The last term is the square of **3**.

   c. The middle term is twice the product of **5x** and **3**.

5. a. $x^2 + 2xy + y^2 = (\mathbf{x} + \mathbf{y})^2$

   b. $x^2 - 2xy + y^2 = (x - \mathbf{y})^2$

   c. $x^2 - y^2 = (x + y)(\mathbf{x} - \mathbf{y})$

7. List the squares of the integers from 1 through 20.

   1, 4, 9, 16, 25, 36, 49, 64, 81, 100, 121, 144, 169, 196, 225, 256, 289, 324, 361, 400

NOTATION

Complete each factorization.

9. $x^2 + 10x + 25 = (x + 5)^{\mathbf{2}}$

11. $x^2 - 64 = (x + 8)(x - 8)$

GUIDED PRACTICE

Determine whether each of the following is a perfect-square trinomial. See Example 1.

13. $x^2 + 18x + 81$

   $2 \cdot x \cdot 9 = 18x$ **Yes**

15. $y^2 + 2y + 4$

   $2 \cdot y \cdot 2 = 4y$ **No**

17. $9n^2 - 30n - 25$

   The last term is negative. **No**

19. $4y^2 - 12y + 9$

   $2 \cdot 2y \cdot 3 = 12y$ **Yes**

Factor. See Example 2.

21. $x^2 + 6x + 9$

   The first term x^2 is the square of **x**.
   The last term 9 is the square of **3**.
   The middle term is twice the product of x and 3: $2(x)(3) = \mathbf{6x}$.
   $$x^2 + 6x + 9 = (x + 3)(x + 3)$$
   $$= (x + 3)^2$$

23. $b^2 + 2b + 1$

   The first term b^2 is the square of **b**.
   The last term 1 is the square of **1**.
   The middle term is twice the product of b and 1: $2(b)(1) = \mathbf{2b}$.
   $$b^2 + 2b + 1 = (b + 1)(b + 1)$$
   $$= (b + 1)^2$$

25. $c^2 - 12c + 36$

   The first term c^2 is the square of **c**.
   The last term 36 is the square of **−6**.
   The middle term is twice the product of c and -6: $2(c)(-6) = \mathbf{-12c}$.
   $$c^2 - 12c + 36 = (c - 6)(c - 6)$$
   $$= (c - 6)^2$$

27. $9 + 4x^2 + 12x = 4x^2 + 12x + 9$

   The first term $4x^2$ is the square of **2x**.
   The last term 9 is the square of **3**.
   The middle term is twice the product of $2x$ and 3: $2(2x)(3) = \mathbf{12x}$.
   $$4x^2 + 12x + 9 = (2x + 3)(2x + 3)$$
   $$= (2x + 3)^2$$

29. $36m^2 + 60mn + 25n^2$

   The first term $36m^2$ is the square of **6m**.
   The last term $25n^2$ is the square of **5n**.
   The middle term is twice the product of $6m$ and $5n$: $2(6m)(5n) = \mathbf{60mn}$.
   $$36m^2 + 60mn + 25n^2 = (6m + 5n)(6m + 5n)$$
   $$= (6m + 5n)^2$$

31. $81x^2 - 72xy + 16y^2$

   The first term $81x^2$ is the square of **9x**.
   The last term $16y^2$ is the square of **−4y**.
   The middle term is twice the product of $9x$ and $-4y$: $2(9x)(-4y) = \mathbf{-72xy}$.
   $$81x^2 - 72xy + 16y^2 = (9x - 4y)(9x - 4y)$$
   $$= (9x - 4y)^2$$

Factor. See Example 3.

33. $3u^2 - 18u + 27 = 3(u^2 - 6u + 9)$

The first term u^2 is the square of **u**.
The last term 9 is the square of **−3**.
The middle term is twice the product of
u and -3: $2(u)(-3) = -6u$.

$$3(u^2 - 6u + 9) = 3(u - 3)(u - 3)$$
$$= 3(u - 3)^2$$

35. $36x^3 + 12x^2 + x = x(36x^2 + 12x + 1)$

The first term $36x^2$ is the square of **6x**.
The last term 1 is the square of **1**.
The middle term is twice the product of
$6x$ and 1: $2(6x)(1) = 12x$.

$$x(36x^2 + 12x + 1) = x(6x + 1)(6x + 1)$$
$$= x(6x + 1)^2$$

Factor. If a polynomial can't be factored, write "prime." See Example 4.

37. $x^2 - 4$

$$F^2 - L^2 = (F+L)(F-L)$$
$$\downarrow \quad \downarrow \qquad \downarrow \quad \downarrow \quad \downarrow \quad \downarrow$$
$$x^2 - 2^2 = (x + 2)\ (x - 2)$$

39. $x^2 - 16$

$$F^2 - L^2 = (F+L)(F-L)$$
$$\downarrow \quad \downarrow \qquad \downarrow \quad \downarrow \quad \downarrow \quad \downarrow$$
$$x^2 - 4^2 = (x + 4)\ (x - 4)$$

41. $36 - y^2$

$$F^2 - L^2 = (F+L)(F-L)$$
$$\downarrow \quad \downarrow \qquad \downarrow \quad \downarrow \quad \downarrow \quad \downarrow$$
$$6^2 - y^2 = (6 + y)\ (6 - y)$$

43. $t^2 - 25$

$$(F^2 - L^2) = (F+L)(F-L)$$
$$\downarrow \quad \downarrow \qquad \downarrow \quad \downarrow \quad \downarrow \quad \downarrow$$
$$(t^2 - 5^2) = (t + 5)(t - 5)$$

45. $a^2 + b^2$ is prime.

47. $y^2 - 63$ is prime.

Factor. See Example 5.

49. $25t^2 - 64$

$$F^2 - L^2 = (F+L)(F-L)$$
$$\downarrow \quad \downarrow \qquad \downarrow \quad \downarrow \quad \downarrow \quad \downarrow$$
$$(5t)^2 - 8^2 = (5t + 8)\ (5t - 8)$$

51. $81y^2 - 1$

$$F^2 - L^2 = (F+L)\ (F-L)$$
$$\downarrow \quad \downarrow \qquad \downarrow \quad \downarrow \quad \downarrow \quad \downarrow$$
$$(9y)^2 - 1^2 = (9y + 1)\ (9y - 1)$$

53. $9x^4 - y^2$

$$F^2 - L^2 = (F + L)\ (F - L)$$
$$\downarrow \quad \downarrow \qquad \downarrow \quad \downarrow \quad \downarrow \quad \downarrow$$
$$(3x^2)^2 - y^2 = (3x^2 + y)\ (3x^2 - y)$$

55. $-49d^4 + 16c^2 = 16c^2 - 49d^4$

$$F^2 - L^2 = (F + L)\ (F - L)$$
$$\downarrow \quad \downarrow \qquad \downarrow \quad \downarrow \quad \downarrow \quad \downarrow$$
$$(4c)^2 - (7d^2)^2 = (4c + 7d^2)\ (4c - 7d^2)$$

Factor. See Example 6.

57. $8x^2 - 32y^2 = 8(x^2 - 4y^2)$
$$= 8(x + 2y)(x - 2y)$$

59. $63a^2 - 7 = 7(9a^2 - 1)$
$$= 7(3a + 1)(3a - 1)$$

Factor. See Example 7.

61. $81 - s^4 = 9^2 - (s^2)^2$
$$= (9 + s^2)(9 - s^2)$$
$$= (9 + s^2)(3 + s)(3 - s)$$

63. $b^4 - 256 = (b^2)^2 - (16)^2$
$$= (b^2 + 16)(b^2 - 16)$$
$$= (b^2 + 16)(b + 4)(b - 4)$$

TRY IT YOURSELF

Factor.

65. $a^4 - 144b^2 = (a^2)^2 - (12b)^2$
$$= (a^2 + 12b)(a^2 - 12b)$$

67. $9x^2y^2 + 30xy + 25$

The first term $9x^2y^2$ is the square of **3xy**.
The last term 25 is the square of **5**.
The middle term is twice the product of
$3xy$ and 5: $2(3xy)(5) = 30xy$.

$$9x^2y^2 + 30xy + 25 = (3xy + 5)(3xy + 5)$$
$$= (3xy + 5)^2$$

69. $16t^4 - 16s^4 = 16(t^4 - s^4)$
$$= 16[(t^2)^2 - (s^2)^2]$$
$$= 16(t^2 + s^2)(t^2 - s^2)$$
$$= 16(t^2 + s^2)(t + s)(t - s)$$

71. $t^2 - 20t + 100$

The first term t^2 is the square of t.
The last term 100 is the square of -10.
The middle term is twice the product of
t and -10: $2(t)(-10) = -20t$.
$$t^2 - 20t + 100 = (t-10)(t-10)$$
$$= (t-10)^2$$

73. $9y^2 - 24y + 16$

The first term $9y^2$ is the square of $3y$.
The last term 16 is the square of -4.
The middle term is twice the product of
$3y$ and -4: $2(3y)(-4) = -24y$.
$$9y^2 - 24y + 16 = (3y-4)(3y-4)$$
$$= (3y-4)^2$$

75. $z^2 - 64 = (z+8)(z-8)$

77. $25m^4 - 25 = 25(m^4 - 1)$
$$= 25[(m^2)^2 - 1]$$
$$= 25(m^2 + 1)(m^2 - 1)$$
$$= 25(m^2 + 1)(m+1)(m-1)$$

79. $18a^5 + 84a^4b + 98a^3b^2$
$$= 2a^3(9a^2 + 42ab + 49b^2)$$

The first term $9a^2$ is the square of $3a$.
The last term $49b^2$ is the square of $7b$.
The middle term is twice the product of
$3a$ and $7b$: $2(3a)(7b) = 42ab$.
$$2a^3(9a^2 + 42ab + 49b^2)$$
$$= 2a^3(3a + 7b)(3a + 7b)$$
$$= 2a^3(3a + 7b)^2$$

81. $x^3 - 144x = x(x^2 - 144)$
$$= x(x+12)(x-12)$$

83. $49t^2 - 28ts + 4s^2$

The first term $49t^2$ is the square of $7t$.
The last term $4s^2$ is the square of $-2s$.
The middle term is twice the product of
$7t$ and $-2s$: $2(7t)(-2s) = -28ts$.
$$49t^2 - 28ts + 4s^2 = (7t - 2s)(7t - 2s)$$
$$= (7t - 2s)^2$$

85. $3m^4 - 3n^4 = 3(m^4 - n^4)$
$$= 3[(m^2)^2 - (n^2)^2]$$
$$= 3(m^2 + n^2)(m^2 - n^2)$$
$$= 3(m^2 + n^2)(m + n)(m - n)$$

87. $25m^2 + 70m + 49$

The first term $25m^2$ is the square of $5m$.
The last term 49 is the square of 7.
The middle term is twice the product of
$5m$ and 7: $2(5m)(7) = 70m$.
$$25m^2 + 70m + 49 = (5m + 7)(5m + 7)$$
$$= (5m + 7)^2$$

89. $-100t^2 + 20t - 1 = -(100t^2 - 20t + 1)$

The first term $100t^2$ is the square of $10t$.
The last term 1 is the square of -1.
The middle term is twice the product of
$10t$ and -1: $2(10t)(-1) = -20t$.
$$-(100t^2 - 20t + 1) = -(10t - 1)(10t - 1)$$
$$= -(10t - 1)^2$$

91. $6x^4 - 6x^2y^2 = 6x^2(x^2 - y^2)$
$$= 6x^2(x + y)(x - y)$$

93. $100a^2 + 81$ is prime.

95. $-169 + 25x^2 = -(169 - 25x^2)$
$$= -(13 + 5x)(13 - 5x)$$

or $-169 + 25x^2 = 25x^2 - 169$
$$= (5x + 13)(5x - 13)$$

Choose the correct method from Section 6.1, Section 6.2, Section 6.3, or Section 6.4 to factor each of the following:

97. Two different sign factors of -42
whose sum is $+1$ are $+7$ and -6.
$$x^2 + x - 42 = (x + 7)(x - 6)$$

99. $x^2 - 9 = (x + 3)(x - 3)$

101. $24a^3b - 16a^2b = 8a^2b(3a - 2)$

103. $-2r^2 + 28r - 80 = -2(r^2 - 14r + 40)$
The negative factors of 40
whose sum is -14 are -10 and -4.
$$= -2(r - 10)(r - 4)$$

Section 6.4

105. $x^3 + 3x^2 + 4x + 12 = (x^3 + 3x^2) + (4x + 12)$
$$= (x^2x^1 + 3x^2) + (4x + 4 \cdot 3)$$
$$= x^2(x+3) + 4(x+3)$$
$$= (x+3)(x^2+4)$$

107. $4b^2 - 20b + 25$

The first term $4b^2$ is the square of $2b$.
The last term 25 is the square of -5.
The middle term is twice the product of
$2b$ and -5: $2(2b)(-5) = -20b$.
$$4b^2 - 20b + 25 = (2b-5)(2b-5)$$
$$= (2b-5)^2$$

APPLICATIONS

109. GENETICS
$$p^2 + 2pq + q^2 = (p+q)(p+q)$$
$$= (p+q)^2$$

111. PHYSICS
$$0.5gt_1^2 - 0.5gt_2^2 = 0.5g\left(t_1^2 - t_2^2\right)$$
$$= 0.5g\left(t_1 + t_2\right)\left(t_1 - t_2\right)$$
The distance is $0.5g\left(t_1 + t_2\right)\left(t_1 - t_2\right)$.

WRITING
113-115. Answers will vary.

REVIEW
Perform each division.

117. $\dfrac{-30c^2d^2 - 15c^2d - 10cd^2}{-10cd}$

$$= \frac{-30c^2d^2}{-10cd} - \frac{15c^2d}{-10cd} - \frac{10cd^2}{-10cd}$$

$$= \frac{3c^{2-1}d^{2-1}}{1} + \frac{3c^{2-1}d^{1-1}}{2} + \frac{1c^{1-1}d^{2-1}}{1}$$

$$\frac{3c^1d^1}{1} + \frac{3c^1d^0}{2} + \frac{1c^0d^1}{1}$$

$$= 3cd + \frac{3c}{2} + d$$

CHALLENGE PROBLEMS
119. For what value of c does $80x^2 - c$ factor
as $5(4x + 3)(4x - 3)$? **5·3·3 = 45**

Factor completely.

121. $81x^6 + 36x^3y^2 + 4y^4$

The first term $81x^6$ is the square of $9x^3$.
The last term $4y^4$ is the square of $2y^2$.
The middle term is twice the product of
$9x^3$ and $2y^2$: $2(9x^3)(2y^2) = 36x^3y^2$.
$$81x^6 + 36x^3y^2 + 4y^4 = (9x^3 + 2y^2)(9x^3 + 2y^2)$$
$$= (9x^3 + 2y^2)^2$$

123. $c^2 + 1.6c + 0.64$

The first term c^2 is the square of c.
The last term 0.64 is the square of 0.8.
The middle term is twice the product of
c and 0.8: $2(c)(0.8) = 1.6c$.

$$c^2 + 1.6c + 0.64 = (c + 0.8)(c + 0.8)$$
$$= (c + 0.8)^2$$

125. $(x+5)^2 - y^2 = (x+5+y)(x+5-y)$

127. $c^2 - \dfrac{1}{16} = c^2 - \left(\dfrac{1}{4}\right)^2$

$$= \left(c + \frac{1}{4}\right)\left(c - \frac{1}{4}\right)$$

VOCABULARY
Fill in the blanks.

1. $x^3 + 27$ is the **sum** of two cubes and $a^3 - 125$ is the difference of two **cubes**.

CONCEPTS
Fill in the blanks.

3. a. $F^3 + L^3 = (\mathbf{F} + \mathbf{L})(F^2 - FL + L^2)$

   b. $F^3 - L^3 = (F - L)(\mathbf{F^2} + FL + \mathbf{L^2})$

5. $216n^3 - 125$

   ↑   ↑

   This is   This is
   6n cubed.   **5** cubed.

7. List the first ten positive integer cubes.

   1, 8, 27, 64, 125, 216, 343, 512, 729, 1,000

9. Use multiplication to determine if the factorization is correct.

   $b^3 + 27 = (b + 3)(b^2 + 3b + 9)$

$$b^2 + 3b + 9$$
$$\underline{\qquad\qquad b + 3}$$
$$3b^2 + 9b + 27$$
$$\underline{b^3 + 3b^2 + 9b \qquad\quad}$$
$$b^3 + 6b^2 + 18b + 27$$

No

NOTATION
Complete each factorization.

11. $a^3 + 8 = (a + 2)(a^2 - \mathbf{2a} + 4)$

13. $b^3 + 27 = (\mathbf{b + 3})(b^2 - 3b + 9)$

Give an example of each type of expression.

15. a. the sum of two cubes $\mathbf{x^3 + 8}$

   b. the cube of a sum $\mathbf{(x + 8)^3}$

GUIDED PRACTICE
Factor. See Example 1.

17. $y^3 + 125 = y^3 + 5^3$
$$= (y + 5)(y^2 - y \cdot 5 + 5^2)$$
$$= (y + 5)(y^2 - 5y + 25)$$

19. $a^3 + 64 = a^3 + 4^3$
$$= (a + 4)(a^2 - a \cdot 4 + 4^2)$$
$$= (a + 4)(a^2 - 4a + 16)$$

21. $n^3 + 512 = n^3 + 8^3$
$$= (n + 8)(n^2 - n \cdot 8 + 8^2)$$
$$= (n + 8)(n^2 - 8n + 64)$$

23. $8 + t^3 = 2^3 + t^3$
$$= (2 + t)(2^2 - 2 \cdot t + t^2)$$
$$= (2 + t)(4 - 2t + t^2)$$

25. $a^3 + 1,000b^3$
$$= a^3 + (10b)^3$$
$$= (a + 10b)[a^2 - a \cdot 10b + (10b)^2]$$
$$= (a + 10b)(a^2 - 10ab + 100b^2)$$

27. $125c^3 + 27d^3$
$$= (5c)^3 + (3d)^3$$
$$= (5c + 3d)[(5c)^2 - 5c \cdot 3d + (3d)^2]$$
$$= (5c + 3d)(25c^2 - 15cd + 9d^2)$$

Factor. See Example 2.

29. $a^3 - 27 = a^3 - 3^3$
$$= (a - 3)(a^2 + a \cdot 3 + 3^2)$$
$$= (a - 3)(a^2 + 3a + 9)$$

31. $m^3 - 343 = m^3 - 7^3$
$$= (m - 7)(m^2 + m \cdot 7 + 7^2)$$
$$= (m - 7)(m^2 + 7m + 49)$$

33. $216 - v^3 = 6^3 - v^3$
$$= (6 - v)(6^2 + 6 \cdot v + v^2)$$
$$= (6 - v)(36 + 6v + v^2)$$

35. $8s^3 - t^3 = (2s)^3 - t^3$
$$= (2s - t)[(2s)^2 + 2s \cdot t + t^2)]$$
$$= (2s - t)(4s^2 + 2st + t^2)$$

37. $1,000a^3 - w^3$
$$= (10a)^3 - w^3$$
$$= (10a - w)[(10a)^2 + 10a \cdot w + (w)^2]$$
$$= (10a - w)(100a^2 + 10aw + w^2)$$

39. $64x^3 - 27y^3$
$$= (4x)^3 - (3y)^3$$
$$= (4x - 3y)[(4x)^2 + 4x \cdot 3y + (3y)^2]$$
$$= (4x - 3y)(16x^2 + 12xy + 9y^2)$$

Factor. See Example 3.

41. $2x^3 + 2 = 2(x^3 + 1)$
$$= 2(x^3 + 1^3)$$
$$= 2(x + 1)(x^2 - x \cdot 1 + 1^2)$$
$$= 2(x + 1)(x^2 - x + 1)$$

43. $3d^3 + 81 = 3(d^3 + 27)$
$$= 3(d^3 + 3^3)$$
$$= 3(d+3)(d^2 - d \cdot 3 + 3^2)$$
$$= 3(d+3)(d^2 - 3d + 9)$$

45. $x^4 - 216x = x(x^3 - 216)$
$$= x(x^3 - 6^3)$$
$$= x(x-6)(x^2 + x \cdot 6 + 6^2)$$
$$= x(x-6)(x^2 + 6x + 36)$$

47. $64m^3 x - 8n^3 x$
$$= 8x(8m^3 - n^3)$$
$$= 8x((2m)^3 - n^3)$$
$$= 8x(2m - n)[(2m)^2 + 2m \cdot n + n^2]$$
$$= 8x(2m - n)(4m^2 + 2mn + n^2)$$

TRY IT YOURSELF

Choose the correct method from Section 6.1 through Section 6.5 and factor completely.

49. $x^2 + 8x + 16$

The first term x^2 is the square of x.
The last term 16 is the square of **4**.
The middle term is twice the product of
x and 4: $2(x)(4) = \mathbf{8x}$.
$$x^2 + 8x + 16 = (x+4)(x+4)$$
$$= (x+4)^2$$

51. $9r^2 - 16s^2 = (3r + 4s)(3r - 4s)$

53. $xy - ty + sx - st = (xy - ty) + (sx - st)$
$$= y(x-t) + s(x-t)$$
$$= (x-t)(y+s)$$

55. $4p^3 + 32q^3 = 4(p^3 + 8q^3)$
$$= 4[(p^3 + (2q)^3]$$
$$= 4[(p+2q)(p^2 - p \cdot 2q + (2q)^2]$$
$$= 4(p+2q)(p^2 - 2pq + 4q^2)$$

57. $16c^3 t^2 + 20c^2 t^3 + 6ct^4$
$$= \mathbf{2} \cdot 8\mathbf{c^1} c^2 \mathbf{t^2} + \mathbf{2} \cdot 10 \mathbf{c^1} c^1 t^1 \mathbf{t^2} + \mathbf{2} \cdot 3 \mathbf{c^1 t^2 t^2}$$
$$= 2ct^2(8c^2 + 10ct + 3t^2)$$

$$\underset{a}{8c^2} + \underset{b}{10ct} + \underset{c}{3t^2}$$

In $8c^2 + 10ct + 3t^2$, we have $a = 8$, $b = 10$ and $c = 3$.
The key number is $ac = 8(3) = 24$. We must find a
factorization of 24 in which the sum of the factors is

$b = 10$. Since the factors must have a positive product,
their signs must be the same. The pairs of factors are +4
and +6.
$$2ct^2(8c^2 + 10ct + 3t^2)$$
$$= 2ct^2[8c^2 + 4ct + 6ct + 3t^2]$$
$$= 2ct^2[(8c^2 + 4ct) + (6ct + 3t^2)]$$
$$= 2ct^2[4c(2c+t) + 3t(2c+t)]$$
$$= 2ct^2(2c+t)(4c+3t)$$

59. $36e^4 - 36 = 36(e^4 - 1)$
$$= 36[(e^2)^2 - 1)]$$
$$= 36(e^2 + 1)(e^2 - 1)$$
$$= 36(e^2 + 1)(e+1)(e-1)$$

61. $35a^3 b^2 - 14a^2 b^3 + 14a^3 b^3$
$$= \mathbf{7} \cdot 5\mathbf{a^2} a^1 \mathbf{b^2} - \mathbf{7} \cdot 2a^2 b^1 \mathbf{b^2} + \mathbf{7} \cdot 2a^1 \mathbf{a^2} b^1 \mathbf{b^2}$$
$$= 7a^2 b^2 (5a - 2b + 2ab)$$

63. $36r^2 + 60rs + 25s^2$

The first term $36r^2$ is the square of $\mathbf{6r}$.
The last term $25s^2$ is the square of $\mathbf{5s}$.
The middle term is twice the product of
$6r$ and $5s$: $2(6r)(5s) = \mathbf{60rs}$.
$$36r^2 + 60rs + 25s^2 = (6r + 5s)(6r + 5s)$$
$$= (6r + 5s)^2$$

LOOK ALIKES . . .

65. a. $x^2 - 1 = (x+1)(x-1)$
   b. $x^3 - 1 = (x-1)(x^2 + x + 1)$

67. a. $x^2 + 2x = x(x+2)$
   b. $x^2 + 2x + 1 = (x+1)(x+1)$
$$= (x+1)^2$$

APPLICATIONS

69. MAILING BREAKABLES

Find volume of the larger box.
$$V = lwh$$
$$= 10 \cdot 10 \cdot 10$$
$$= 1,000$$

The volume is 1,000 in^3.

Find volume of the smaller box.
$$V = lwh$$
$$= x \cdot x \cdot x$$
$$= x^3$$

The volume is x^3 in^3.
Subtract the samller volume from the larger volume.

$$V_L - V_S = 1,000 - x^3$$

The volume of packing is $(1,000 - x^3)$ in^3.

$$1,000 - x^3 = 10^3 - x^3$$
$$= (10 - x)(10^2 + 10 \cdot x + x^2)$$
$$= (10 - x)(100 + 10x + x^2)$$

WRITING

71. Answers will vary.

REVIEW

73. $\dfrac{7}{9} = 0.\overline{7}$, repeating decimal

75. Solve.

$$2x + 2 = \frac{2}{3}x - 2$$

$$3 \cdot (2x + 2) = 3 \cdot \left(\frac{2}{3}x - 2\right)$$

$$6x + 6 = 2x - 6$$

$$6x + 6 - \mathbf{6} = 2x - 6 - \mathbf{6}$$

$$6x = 2x - 12$$

$$6x - \mathbf{2x} = 2x - 12 - \mathbf{2x}$$

$$4x = -12$$

$$\frac{4x}{\mathbf{4}} = \frac{-12}{\mathbf{4}}$$

$$x = -3$$

CHALLENGE PROBLEMS

77. a) $(x^3)^2 - 1^2$
$$= (x^3 + 1)(x^3 - 1)$$
$$= (x + 1)(x^2 - x + 1)$$
$$(x - 1)(x^2 + x + 1)$$

Factor.

79. $x^6 - y^9 = (x^2)^3 - (y^3)^3$
$$= (x^2 - y^3)[(x^2)^2 + x^2 \cdot y^3 + (y^3)^2]$$
$$= (x^2 - y^3)(x^4 + x^2 y^3 + y^6)$$

81. $64x^{12} + y^{15}z^{18}$
$$= (4x^4)^3 + (y^5 z^6)^3$$
$$= (4x^4 + y^5 z^6)[(4x^4)^2 - 4x^4 \cdot y^5 z^6 + (y^5 z^6)^2]$$
$$= (4x^4 + y^5 z^6)(16x^8 - 4x^4 y^5 z^6 + y^{10} z^{12})$$

Section 6.5

SECTION 6.6
VOCABULARY

Fill in the blanks.

1. To factor a polynomial means to express it as a **product** of two (or more) polynomials.

CONCEPTS

For each of the following polynomials, which factoring method would you use first?

3. $2x^5y - 4x^3y$

Factor out the GCF

5. $x^2 + 18x + 81$

Perfect square trinomial

7. $x^3 + 27$

Sum of two cubes

9. $m^2 + 3mn + 2n^2$

Trinomial factoring

11. What is the first question that should be asked when using the strategy of this section to factor a polynomial? **Is there a common factor?**

NOTATION

Complete each factorization.

13. $6m^3 - 28m^2 + 16m = 2m(3m^2 - \mathbf{14m} + 8)$
$$= 2m(3m - 2)(\mathbf{m} - 4)$$

TRY IT YOURSELF

The following is a list of random factoring problems. Factor each expression. If an expression is not factorable, write "prime." See Examples 1–5.

15. $2b^2 + 8b - 24 = 2(b^2 + 4b - 12)$

Two different sign factors of -12 whose sum is $+4$ are $+6$ and -2.
$$2(b^2 + 4b - 12) = 2(b + 6)(b - 2)$$

17. $8p^3q^7 + 4p^2q^3 = \mathbf{4 \cdot 2\, p^2}\, pq^3q^4 + \mathbf{4p^2q^3}$
$$= 4p^2q^3(2pq^4 + 1)$$

19. $2 + 24y + 40y^2 = 2(20y^2 + 12y + 1)$

The positive factors of 20 whose sum is 12 are 10 and 2.
$$2(20y^2 + 12y + 1) = 2(2y + 1)(10y + 1)$$

21. $8x^4 - 8 = 8(x^4 - 1)$
$$= 8(x^2 + 1)(x^2 - 1)$$
$$= 8(x^2 + 1)(x + 1)(x - 1)$$

23. $14c - 147 + c^2 = c^2 + 14c - 147$

Two different sign factors of -147 whose sum is $+14$ are $+21$ and -7.
$$c^2 + 14c - 147 = (c + 21)(c - 7)$$

25. $x^2 + 7x + 1$ is prime.

27. $-2x^5 + 128x^2 = -2x^2(x^3 - 64)$
$$= -2x^2(x^3 - 4^3)$$
$$= -2x^2(x - 4)(x^2 + x \cdot 4 + 4^2)$$
$$= -2x^2(x - 4)(x^2 + 4x + 16)$$

29. $a^2c + a^2d^2 + bc + bd^2$
$$= (\mathbf{a^2c + a^2d^2}) + (\mathbf{bc + bd^2})$$
$$= a^2(c + d^2) + b(c + d^2)$$
$$= (c + d^2)(a^2 + b)$$

31. $-9x^2 + 6x - 1 = -(9x^2 - 6x + 1)$

$-(9x^2 - 6x + 1)$

The first term $9x^2$ is the square of $\mathbf{3x}$.
The last term 1 is the square of $\mathbf{-1}$.
The middle term is twice the product of $3x$ and -1: $2(3x)(-1) = \mathbf{-6x}$.
$$-(9x^2 - 6x + 1) = -(3x - 1)(3x - 1)$$
$$= -(3x - 1)^2$$

33. $-20m^3 - 100m^2 - 125m$
$$= -5m(4m^2 + 20m + 25)$$

$4m^2 + 20m + 25$

The first term $4m^2$ is the square of $\mathbf{2m}$.
The last term 25 is the square of $\mathbf{5}$.
The middle term is twice the product of $2m$ and 5: $2(2m)(5) = \mathbf{20m}$.

$-5m(4m^2 + 20m + 25)$
$$= -5m(2m + 5)(2m + 5)$$
$$= -5m(2m + 5)^2$$

35. $2c^2 - 5cd - 3d^2$
$\quad\; a \quad\;\; b \quad\;\; c$

In $2c^2 - 5cd - 3c^2$, we have $a = 2$, $b = -5$, and $c = -3$. The key number is $ac = 2(-3) = -6$. We must find a factorization of -6 in which the sum of the factors is $b = -5$. Since the factors must have a negative product, their signs must be different. The pairs of factors are -6 and 1.

$2c^2 - 5cd - 3d^2 = 2c^2 - 6cd + cd - 3d^2$
$$= (2c^2 - 6cd) + (cd - 3d^2)$$
$$= 2c(c - 3d) + d(c - 3d)$$
$$= (c - 3d)(2c + d)$$

37. $p^4 - 2p^3 - 8p + 16$
$$= (p^4 - 2p^3) + (-8p + 16)$$
$$= p^3(p - 2) - 8(p - 2)$$
$$= (p - 2)(p^3 - 8)$$
$$= (p - 2)(p^3 - 2^3)$$
$$= (p - 2)(p - 2)(p^2 + p \cdot 2 + 2^2)$$
$$= (p - 2)(p - 2)(p^2 + 2p + 4)$$
$$= (p - 2)^2(p^2 + 2p + 4)$$

39. $a^2(x - a) - b^2(x - a) = (x - a)(a^2 - b^2)$
$$= (x - a)(a + b)(a - b)$$

41. $a^2b^2 - 144 = (ab + 12)(ab - 12)$

43. $2x^3 + 10x^2 + x + 5 = (2x^3 + 10x^2) + (x + 5)$
$$= 2x^2(x + 5) + 1(x + 5)$$
$$= (x + 5)(2x^2 + 1)$$

45. $8v^2 - 14v^3 + v^4 = v^4 - 14v^3 + 8v^2$
$$= v^2(v^2 - 14v + 8)$$

47. $18a^2 - 6ab + 42ac - 14bc$
$$= 2 \cdot 9a^2 - 2 \cdot 3ab + 2 \cdot 21ac - 2 \cdot 7bc$$
$$= 2(9a^2 - 3ab + 21ac - 7bc)$$
$$= 2\left[(9a^2 - 3ab) + (21ac - 7bc)\right]$$
$$= 2\left[(3 \cdot 3a^1a^1 - 3a^1b) + (7 \cdot 3ac - 7bc)\right]$$
$$= 2\left[3a(3a - b) + 7c(3a - b)\right]$$
$$= 2(3a - b)(3a + 7c)$$

49. $8a^2x^3 - 2b^2x = 2x(4a^2x^2 - b^2)$
$$= 2x(2ax + b)(2ax - b)$$

51. $6x^2 - 14x + 8 = 2(3x^2 - 7x + 4)$

$\underset{a}{3}x^2 - \underset{b}{7}x + \underset{c}{4}$

In $3x^2 - 7x + 4$, we have $a = 3$, $b = -7$, and $c = 4$. The key number is $ac = 3(4) = 12$. We must find a factorization of 12 in which the sum of the factors is $b = -7$. Since the factors must have a positive product, their signs must be the same. The pairs of factors are -3 and -4.

$2(3x^2 - 7x + 4) = 2(3x^2 - 4x - 3x + 4)$
$$= 2[(3x^2 - 4x) + (-3x + 4)]$$
$$= 2[x(3x - 4) - 1(3x - 4)]$$
$$= 2(3x - 4)(x - 1)$$

53. $4x^2y^2 + 4xy^2 + y^2 = y^2(4x^2 + 4x + 1)$

$4x^2 + 4x + 1$

The first term $4x^2$ is the square of **2x**.
The last term 1 is the square of **1**.
The middle term is twice the product of $2x$ and 1: $2(2x)(1) = \mathbf{4x}$.

$y^2(4x^2 + 4x + 1) = y^2(2x + 1)(2x + 1)$
$$= y^2(2x + 1)^2$$

55. $4m^5 + 500m^2 = 4m^2(m^3 + 125)$
$$= 4m^2(m^3 + 5^3)$$
$$= 4m^2(m + 5)(m^2 - m \cdot 5 + 5^2)$$
$$= 4m^2(m + 5)(m^2 - 5m + 25)$$

57. $a^3 - 24 - 4a + 6a^2$
$$= a^3 + 6a^2 - 4a - 24$$
$$= (a^2a^1 + 6a^2) + (-4a - 4 \cdot 6)$$
$$= a^2(a + 6) - 4(a + 6)$$
$$= (a + 6)(a^2 - 4)$$
$$= (a + 6)(a + 2(a - 2))$$

59. $4x^2 + 9y^2$ is prime.

61. $16a^5 - 54a^2 = 2a^2(8a^3 - 27)$
$$= 2a^2[(2a)^3 - 3^3]$$
$$= 2a^2[2a - 3][(2a)^2 + 2a \cdot 3 + 3^2]$$
$$= 2a^2(2a - 3)(4a^2 + 6a + 9)$$

63. $27x - 27y - 27z = 27(x - y - z)$

65. $xy - ty + xs - ts = (xy - ty) + (xs - ts)$
$$= y(x - t) + s(x - t)$$
$$= (x - t)(y + s)$$

67. $35x^8 - 2x^7 - x^6 = x^6(35x^2 - 2x - 1)$
$$= x^6(7x + 1)(5x - 1)$$

69. $5(x - 2) + 10y(x - 2) = (x - 2)(5 + 10y)$
$$= (x - 2)5(1 + 2y)$$
$$= 5(x - 2)(1 + 2y)$$

Section 6.6

71. $49p^2 + 28pq + 4q^2$

The first term $49p^2$ is the square of **7p**.
The last term $4q^2$ is the square of **2q**.
The middle term is twice the product of
7p and 2q: $2(7p)(2q) = \mathbf{28pq}$.
$$49p^2 + 28pq + 4q^2 = (7p + 2q)(7p + 2q)$$
$$= (7p + 2q)^2$$

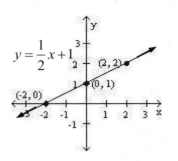

73. $4t^2 + 36 = 4(t^2 + 9)$

75. $m^2n^2 - 9m^2 + 3n^2 - 27$
$$= (m^2n^2 - 9m^2) + (3n^2 - 27)$$
$$= m^2(n^2 - 9) + 3(n^2 - 9)$$
$$= (n^2 - 9)(m^2 + 3)$$
$$= (n + 3)(n - 3)(m^2 + 3)$$

WRITING
77-79. Answers will vary.

REVIEW
81. Graph the real numbers $-3, 0, 2,$
and $-\dfrac{3}{2}$ on a number line.

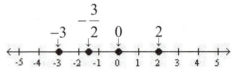

83. Graph: $y = \dfrac{1}{2}x + 1$.

x	y	(x, y)
-2	$y = \dfrac{1}{2}x + 1$ $= \dfrac{-2}{2} + 1$ $= -1 + 1$ $= 0$	$(-2, 0)$
0	$y = \dfrac{1}{2}x + 1$ $= \dfrac{0}{2} + 1$ $= 1$	$(0, 1)$
2	$y = \dfrac{1}{2}x + 1$ $= \dfrac{2}{2} + 1$ $= 1 + 1$ $= 2$	$(2, 2)$

CHALLENGE PROBLEMS
Factor using rational numbers.

85. $x^6 - 4x^3 - 12 = (x^3 + 2)(x^3 - 6)$

87. $24 - x^3 + 8x^2 - 3x = 24 - 3x + 8x^2 - x^3$
$$= (24 - 3x) + (8x^2 - x^3)$$
$$= 3(8 - x) + x^2(8 - x)$$
$$= (8 - x)(3 + x^2)$$

89. $x^9 + y^6 = (x^3)^3 + (y^2)^3$
$$= (x^3 + y^2)[(x^3)^2 - x^3 \cdot y^2 + (y^2)^2]$$
$$= (x^3 + y^2)(x^6 - x^3y^2 + y^4)$$

91. $x^4 - 13x^2 + 36 = (x^2 - 9)(x^2 - 4)$
$$= (x + 3)(x - 3)(x + 2)(x - 2)$$

93. $x^2y^2 - 6xy - 16 = (xy + 2)(xy - 8)$

SECTION 6.7
VOCABULARY

Fill in the blanks.

1. $2x^2 + 3x - 1 = 0$ and $x^2 - 36 = 0$ are examples of **quadratic** equations.

3. The **zero-factor** property states that if the product of two numbers is 0, at least one of them is 0: If $ab = 0$, then $a = \underline{\mathbf{0}}$ or $b = \underline{\mathbf{0}}$.

CONCEPTS

5. Which of the following are quadratic equations?
   a. $x^2 + 2x - 10 = 0$ **Yes** b. $2x - 10 = 0$ **No**
   c. $x^2 = 15x$ **Yes**   d. $x^3 + x^2 + 2x = 0$ **No**

7. Set $5x + 4$ equal to 0 and solve for x. $-\dfrac{4}{5}$

9. What step (or steps) should be performed first before factoring is used to solve each equation?
   a. $x^2 + 7x = -6$ **Add 6 to both sides.**
   b. $x(x + 7) = 3$ **Distribute the multiplication by x and subtract 3 from both sides.**

NOTATION

Complete each solution to solve the equation.

11. $\quad (x - 1)(x + 7) = 0$
$$x - 1 = \boxed{\mathbf{0}} \quad \text{or} \quad \boxed{\mathbf{x + 7}} = 0$$
$$x = 1 \quad | \quad x = \boxed{\mathbf{-7}}$$

13. $\quad\quad p^2 - p - 6 = 0$
$$(\boxed{\mathbf{p}} - 3)(p + 2) = 0$$
$$\boxed{\mathbf{p - 3}} = 0 \quad \text{or} \quad p + 2 = \boxed{\mathbf{0}}$$
$$p = \boxed{\mathbf{3}} \quad | \quad p = \boxed{\mathbf{-2}}$$

GUIDED PRACTICE

Solve each equation. See Example 1.

15. $(x - 3)(x - 2) = 0$
$$x - 3 = 0 \quad \text{or} \quad x - 2 = 0$$
$$x = 3 \quad | \quad x = 2$$
The solutions are 3 and 2.

17. $(x + 7)(x - 7) = 0$
$$x + 7 = 0 \quad \text{or} \quad x - 7 = 0$$
$$x = -7 \quad | \quad x = 7$$
The solutions are -7 and 7.

19. $6x(2x - 5) = 0$
$$6x = 0 \quad \text{or} \quad 2x - 5 = 0$$
$$\frac{6x}{6} = \frac{0}{6} \quad\quad\quad 2x = 5$$
$$x = 0 \quad\quad\quad\quad x = \frac{5}{2}$$
The solutions are 0 and $\dfrac{5}{2}$.

21. $-7a(3a + 10) = 0$
$$-7a = 0 \quad \text{or} \quad 3a + 10 = 0$$
$$a = 0 \quad\quad\quad\quad 3a = -10$$
$$a = -\frac{10}{3}$$
The solutions are 0 and $-\dfrac{10}{3}$.

23. $t(t - 6)(t + 8) = 0$
$$t = 0 \quad \text{or} \quad t - 6 = 0 \quad \text{or} \quad t + 8 = 0$$
$$\quad\quad\quad\quad t = 6 \quad\quad\quad\quad t = -8$$
The solutions are 0, 6 and -8.

25. $(x - 1)(x + 2)(x - 3) = 0$
$$x - 1 = 0 \quad \text{or} \quad x + 2 = 0 \quad \text{or} \quad x - 3 = 0$$
$$x = 1 \quad | \quad x = -2 \quad | \quad x = 3$$
The solutions are 1, -2 and 3.

Solve each equation. See Example 2.

27. $x^2 - 13x + 12 = 0$
$$(x - 12)(x - 1) = 0$$
$$x - 12 = 0 \quad \text{or} \quad x - 1 = 0$$
$$x = 12 \quad | \quad x = 1$$
The solutions are 12 and 1.

29. $x^2 - 4x - 21 = 0$
$$(x + 3)(x - 7) = 0$$
$$x + 3 = 0 \quad \text{or} \quad x - 7 = 0$$
$$x = -3 \quad | \quad x = 7$$
The solutions are -3 and 7.

31. $x^2 - 9x + 8 = 0$
$$(x - 8)(x - 1) = 0$$
$$x - 8 = 0 \quad \text{or} \quad x - 1 = 0$$
$$x = 8 \quad | \quad x = 1$$
The solutions are 8 and 1.

33. $a^2 + 8a + 15 = 0$
$$(a + 3)(a + 5) = 0$$
$$a + 3 = 0 \quad \text{or} \quad a + 5 = 0$$
$$a = -3 \quad | \quad a = -5$$
The solutions are -3 and -5.

Solve each equation. See Example 3.

35. $\quad\quad x^2 - 81 = 0$
$$(x + 9)(x - 9) = 0$$
$$x + 9 = 0 \quad \text{or} \quad x - 9 = 0$$
$$x = -9 \quad | \quad x = 9$$

Section 6.7

The solutions are -9 and 9.

37. $$t^2 - 25 = 0$$
$$(t+5)(t-5) = 0$$

$t + 5 = 0$ or $t - 5 = 0$
$t = -5$ $t = 5$

The solutions are -5 and 5.

39. $$4x^2 - 1 = 0$$
$$(2x+1)(2x-1) = 0$$

$2x + 1 = 0$ or $2x - 1 = 0$
$2x = -1$ $2x = 1$
$x = -\dfrac{1}{2}$ $x = \dfrac{1}{2}$

The solutions are $-\dfrac{1}{2}$ and $\dfrac{1}{2}$.

41. $$9y^2 - 49 = 0$$
$$(3y+7)(3y-7) = 0$$

$3y + 7 = 0$ or $3y - 7 = 0$
$3y = -7$ $3y = 7$
$y = -\dfrac{7}{3}$ $y = \dfrac{7}{3}$

The solutions are $-\dfrac{7}{3}$ and $\dfrac{7}{3}$.

Solve each equation. See Example 4.

43. $$w^2 = 7w$$
$$w^2 - 7w = 7w - 7w$$
$$w^2 - 7w = 0$$
$$w(w-7) = 0$$

$w = 0$ or $w - 7 = 0$
                    $w = 7$

The solutions are 0 and 7.

45. $$s^2 = 16s$$
$$s^2 - 16s = 16s - 16s$$
$$s^2 - 16s = 0$$
$$s(s-16) = 0$$

$s = 0$ or $s - 16 = 0$
                    $s = 16$

The solutions are 0 and 16.

47. $$4y^2 = 12y$$
$$4y^2 - 12y = 12y - 12y$$
$$4y^2 - 12y = 0$$
$$4y(y-3) = 0$$

$4y = 0$ or $y - 3 = 0$
$y = 0$ $y = 3$

The solutions are 0 and 3.

49. $$3x^2 = -8x$$
$$3x^2 + 8x = -8x + 8x$$
$$3x^2 + 8x = 0$$
$$x(3x+8) = 0$$

$x = 0$ or $3x + 8 = 0$
                    $3x = -8$
                    $x = -\dfrac{8}{3}$

The solutions are 0 and $-\dfrac{8}{3}$.

Solve each equation. See Example 5.

51. $$3x^2 + 5x = 2$$
$$3x^2 + 5x - 2 = 2 - 2$$
$$3x^2 + 5x - 2 = 0$$
$$(3x-1)(x+2) = 0$$

$3x - 1 = 0$ or $x + 2 = 0$
$3x = 1$ $x = -2$
$x = \dfrac{1}{3}$

The solutions are $\dfrac{1}{3}$ and -2.

53. $$2x^2 + x = 3$$
$$2x^2 + x - 3 = 3 - 3$$
$$2x^2 + x - 3 = 0$$
$$(2x+3)(x-1) = 0$$

$2x + 3 = 0$ or $x - 1 = 0$
$2x = -3$ $x = 1$
$x = -\dfrac{3}{2}$

The solutions are $-\dfrac{3}{2}$ and 1.

55. $$5x^2 + 1 = 6x$$
$$5x^2 + 1 - 6x = 6x - 6x$$
$$5x^2 - 6x + 1 = 0$$
$$(5x-1)(x-1) = 0$$

$5x - 1 = 0$ or $x - 1 = 0$
$5x = 1$ $x = 1$
$x = \dfrac{1}{5}$

The solutions are $\dfrac{1}{5}$ and 1.

57.
$$2x^2 - 3x = 20$$
$$2x^2 - 3x - \mathbf{20} = 20 - \mathbf{20}$$
$$2x^2 - 3x - 20 = 0$$
$$(2x + 5)(x - 4) = 0$$

$2x + 5 = 0$ or $x - 4 = 0$
$$2x = -5 \qquad\qquad x = 4$$
$$x = -\frac{5}{2}$$

The solutions are $-\dfrac{5}{2}$ and 4.

Solve each equation. See Example 6.

59.
$$4r(r + 7) = -49$$
$$4r^2 + 28r = -49$$
$$4r^2 + 28r + \mathbf{49} = -49 + \mathbf{49}$$
$$4r^2 + 28r + 49 = 0$$
$$(2r + 7)(2r + 7) = 0$$

$2r + 7 = 0$ or $2r + 7 = 0$
$$2r = -7 \qquad\qquad 2r = -7$$
$$r = -\frac{7}{2} \qquad\qquad r = -\frac{7}{2}$$

There is a repeated solution of $-\dfrac{7}{2}$.

61.
$$9a(a - 3) = 3a - 25$$
$$9a^2 - 27a = 3a - 25$$
$$9a^2 - 27a - \mathbf{3a} = 3a - 25 - \mathbf{3a}$$
$$9a^2 - 30a = -25$$
$$9a^2 - 30a + \mathbf{25} = -25 + \mathbf{25}$$
$$9a^2 - 30a + 25 = 0$$
$$(3a - 5)(3a - 5) = 0$$

$3a - 5 = 0$ or $3a - 5 = 0$
$$3a = 5 \qquad\qquad 3a = 5$$
$$a = \frac{5}{3} \qquad\qquad a = \frac{5}{3}$$

There is a repeated solution of $\dfrac{5}{3}$.

Solve each equation. See Example 7.

63.
$$x^3 + 3x^2 + 2x = 0$$
$$x(x^2 + 3x + 2) = 0$$
$$x(x + 1)(x + 2) = 0$$

$x = 0$ or $x + 1 = 0$ or $x + 2 = 0$
$$x = -1 \qquad\qquad x = -2$$

The solutions are 0, -1 and -2.

65.
$$k^3 - 27k - 6k^2 = 0$$
$$k^3 - 6k^2 - 27k = 0$$
$$k(k^2 - 6k - 27) = 0$$
$$k(k - 9)(k + 3) = 0$$

$k = 0$ or $k - 9 = 0$ or $k + 3 = 0$
$$k = 9 \qquad\qquad k = -3$$

The solutions are 0, 9, and -3.

TRY IT YOURSELF.
Solve each equation.

67.
$$4x^2 = 81$$
$$4x^2 - \mathbf{81} = 81 - \mathbf{81}$$
$$4x^2 - 81 = 0$$
$$(2x + 9)(2x - 9) = 0$$

$2x + 9 = 0$ or $2x - 9 = 0$
$$2x = -9 \qquad\qquad 2x = 9$$
$$x = -\frac{9}{2} \qquad\qquad x = \frac{9}{2}$$

The solutions are $-\dfrac{9}{2}$ and $\dfrac{9}{2}$.

69. $x^2 - 16x + 64 = 0$
$$(x - 8)(x - 8) = 0$$

$x - 8 = 0$ or $x - 8 = 0$
$$x = 8 \qquad\qquad x = 8$$

There is a repeated solution of 8.

71. $(2s - 5)(s + 6) = 0$

$2s - 5 = 0$ or $s + 6 = 0$
$$2s = 5 \qquad\qquad s = -6$$
$$s = \frac{5}{2}$$

The solutions are $\dfrac{5}{2}$ and -6.

73.
$$3b^2 - 30b = 6b - 60$$
$$3b^2 - 30b - \mathbf{6b} = 6b - 60 - \mathbf{6b}$$
$$3b^2 - 30b - 6b = -60$$
$$3b^2 - 30b - 6b + \mathbf{60} = -60 + \mathbf{60}$$
$$3b^2 - 30b - 6b + 60 = 0$$
$$3(b^2 - 10b - 2b + 20) = 0$$
$$3[(b^2 - 10b) + (-2b + 20)] = 0$$
$$3[b(b - 10) - 2(b - 10)] = 0$$
$$3(b - 10)(b - 2) = 0$$

$b - 10 = 0$ or $b - 2 = 0$
$$b = 10 \qquad\qquad b = 2$$

- 365 -

Section 6.7

The solutions are 10 and 2.

75. $k^3 + k^2 - 20k = 0$

$k(k^2 + k - 20) = 0$

$k(x+5)(x-4) = 0$

$k = 0$ or $k+5 = 0$ or $k-4 = 0$

$k = -5$ | $k = 4$

The solutions are 0, -5 and 4.

77. $x^2 - 100 = 0$

$(x+10)(x-10) = 0$

$x+10 = 0$ or $x-10 = 0$

$x = -10$ | $x = 10$

The solutions are -10 and 10.

79. $z(z-7) = -12$

$z^2 - 7z = -12$

$z^2 - 7z + 12 = -12 + 12$

$z^2 - 7z + 12 = 0$

$(z-3)(z-4) = 0$

$z-3 = 0$ or $z-4 = 0$

$z = 3$ | $z = 4$

The solutions are 3 and 4.

81. $3y^2 - 14y - 5 = 0$

$(3y+1)(y-5) = 0$

$3y+1 = 0$ or $y-5 = 0$

$3y = -1$ | $y = 5$

$y = -\dfrac{1}{3}$

The solutions are $-\dfrac{1}{3}$ and 5.

83. $(x-2)(x^2 - 8x + 7) = 0$

$(x-2)(x-7)(x-1) = 0$

$x-2 = 0$ or $x-7 = 0$ or $x-1 = 0$

$x = 2$ | $x = 7$ | $x = 1$

The solutions are 2, 7 and 1.

85. $(n+8)(n-3) = -30$

$n^2 + 5n - 24 = -30$

$n^2 + 5n - 24 + 30 = -30 + 30$

$n^2 + 5n + 6 = 0$

$(n+3)(n+2) = 0$

$n+3 = 0$ or $n+2 = 0$

$n = -3$ | $n = -2$

The solutions are -3 and -2.

87. $x^3 - 6x^2 = -9x$

$x^3 - 6x^2 + 9x = -9x + 9x$

$x^3 - 6x^2 + 9x = 0$

$x(x^2 - 6x + 9) = 0$

$x(x-3)(x-3) = 0$

$x = 0$ or $x-3 = 0$

$x = 3$

The solutions are 0 and a repeated answer of 3.

89. $4a^2 + 1 = 8a + 1$

$4a^2 + 1 - 8a = 8a + 1 - 8a$

$4a^2 - 8a + 1 = 1$

$4a^2 - 8a + 1 - 1 = 1 - 1$

$4a^2 - 8a = 0$

$4a(a-2) = 0$

$4a = 0$ or $a-2 = 0$

$a = 0$ | $a = 2$

The solutions are 0 and 2.

91. $2b(6b + 13) = -12$

$12b^2 + 26b = -12$

$12b^2 + 26b + 12 = -12 + 12$

$12b^2 + 26b + 12 = 0$

$2(6b^2 + 13b + 6) = 0$

$2(2b+3)(3b+2) = 0$

$2b+3 = 0$ or $3b+2 = 0$

$2b = -3$ | $3b = -2$

$b = -\dfrac{3}{2}$ | $b = -\dfrac{2}{3}$

The solutions are $-\dfrac{3}{2}$ and $-\dfrac{2}{3}$.

93. $3a^3 + 4a^2 + a = 0$

$a(3a^2 + 4a + 1) = 0$

$a(3a+1)(a+1) = 0$

$a = 0$ or $3a+1 = 0$ or $a+1 = 0$

$3a = -1$ | $a = -1$

$a = -\dfrac{1}{3}$

The solutions are 0, $-\dfrac{1}{3}$ and -1.

95.
$$2x^3 = 2x(x+2)$$
$$2x^3 = 2x^2 + 4x$$
$$2x^3 - \mathbf{2x^2} = 2x^2 + 4x - \mathbf{2x^2}$$
$$2x^3 - 2x^2 = 4x$$
$$2x^3 - 2x^2 - \mathbf{4x} = 4x - \mathbf{4x}$$
$$2x^3 - 2x^2 - 4x = 0$$
$$2x(x^2 - x - 2) = 0$$
$$x(x+1)(x-2) = 0$$

$$x = 0 \quad \text{or} \quad x+1 = 0 \quad \text{or} \quad x-2 = 0$$
$$\qquad\qquad x = -1 \qquad\qquad x = 2$$

The solutions are 0, -1 and 2.

97.
$$-15x^2 + 2 + 7x = 0$$
$$-15x^2 + 7x + 2 = 0$$
$$-(15x^2 - 7x - 2) = 0$$
$$-(5x + 1)(3x - 2) = 0$$

$$5x + 1 = 0 \quad \text{or} \quad 3x - 2 = 0$$
$$5x = -1 \qquad\qquad 3x = 2$$
$$x = -\frac{1}{5} \qquad\qquad x = \frac{2}{3}$$

The solutions are $-\dfrac{1}{5}$ and $\dfrac{2}{3}$.

99.
$$4p^2 - 121 = 0$$
$$(2p + 11)(2p - 11) = 0$$

$$2p + 11 = 0 \quad \text{or} \quad 2p - 11 = 0$$
$$2p = -11 \qquad\qquad 2p = 11$$
$$p = -\frac{11}{2} \qquad\qquad p = \frac{11}{2}$$

The solutions are $-\dfrac{11}{2}$ and $\dfrac{11}{2}$.

101.
$$d(8d - 9) = -1$$
$$8d^2 - 9d = -1$$
$$8d^2 - 9d + \mathbf{1} = -1 + \mathbf{1}$$
$$8d^2 - 9d + 1 = 0$$
$$(8d - 1)(d - 1) = 0$$

$$8d - 1 = 0 \quad \text{or} \quad d - 1 = 0$$
$$8d = 1 \qquad\qquad d = 1$$
$$d = \frac{1}{8}$$

The solutions are $\dfrac{1}{8}$ and 1.

LOOK ALIKES...

Factor the expression in part *a* and solve the equation in part *b*.

103. a. $x^2 + 4x - 21 = (x + 7)(x - 3)$

b. $x^2 + 4x - 21 = 0$
$$(x + 7)(x - 3) = 0$$
$$x + 7 = 0 \quad \text{or} \quad x - 3 = 0$$
$$x = -7 \qquad\qquad x = 3$$
The solutions are -7 and 3.

105. a. $12n^2 - 5n - 2 = (4n + 1)(3n - 2)$

b. $12n^2 - 5n - 2 = 0$
$$(4n + 1)(3n - 2) = 0$$
$$4n + 1 = 0 \quad \text{or} \quad 3n - 2 = 0$$
$$4n = -1 \qquad\qquad 3n = 2$$
$$n = -\frac{1}{4} \qquad\qquad n = \frac{2}{3}$$

The solutions are $-\dfrac{1}{4}$ and $\dfrac{2}{3}$.

WRITING
107-111. Answers will vary.

REVIEW
113. EXERCISE
$$15 \text{ min} \le t < 30 \text{ min}$$

CHALLENGE PROBLEMS
Solve each equation.

115.
$$x^4 - 625 = 0$$
$$(x^2)^2 - 25^2 = 0$$
$$(x^2 + 25)(x^2 - 25) = 0$$
$$(x^2 + 25)(x + 5)(x - 5) = 0$$
$$x^2 + 25 = 0 \quad \text{or} \quad x + 5 = 0 \quad \text{or} \quad x - 5 = 0$$
$$x^2 = -25 \qquad\qquad x = -5 \qquad\qquad x = 5$$
$$\text{not real}$$

The solutions are -5 and 5.

117.
$$(x - 3)^2 = 2x + 9$$
$$x^2 - 6x + 9 = 2x + 9$$
$$x^2 - 6x + 9 - \mathbf{2x} = 2x + 9 - \mathbf{2x}$$
$$x^2 - 8x + 9 = 9$$
$$x^2 - 8x + 9 - \mathbf{9} = 9 - \mathbf{9}$$
$$x^2 - 8x = 0$$
$$x(x - 8) = 0$$
$$x = 0 \quad \text{or} \quad x - 8 = 0$$
$$\qquad\qquad x = 8$$
The solutions are 0 and 8.

Section 6.7

SECTION 6.8
VOCABULARY

Fill in the blanks.

1. Integers that follow one another, such as 6 and 7, are called **consecutive** integers.

3. The longest side of a right triangle is the **hypotenuse**. The remaining two sides are the **legs** of the triangle.

CONCEPTS

5. **ii**

7. $20 = b(b + 5)$

9. a. What kind of triangle is shown below?

   A right triangle

   b. What are the lengths of the legs of the triangle? **x ft; $(x + 1)$ ft**

   c. How much longer is the hypotenuse than the shorter leg? **9 ft**

NOTATION

Complete the solution to solve the equation.

11.
$$0 = -16t^2 + 32t + 48$$
$$0 = \boxed{-16}(t^2 - 2t - 3)$$
$$0 = -16(t - 3)(t + \boxed{1})$$
$$t - 3 = \boxed{0} \quad \text{or} \quad t + 1 = \boxed{0}$$
$$t = \boxed{3} \quad | \quad t = \boxed{-1}$$

APPLICATIONS
GEOMETRY

13. FLAGS

Analyze

· Length is twice as long as it's width.
· Area is 18 sq. ft.
· Find the dimensions.

Assign

Let w = width of flag in ft
   $2w$ = length of flag in ft

Form

The area of the rectangle	equals	the length	times	the width.
18	=	$2w$	·	w

Solve

$$2w(w) = 18$$
$$2w^2 = 18$$
$$2w^2 - \mathbf{18} = 18 - \mathbf{18}$$
$$2w^2 - 18 = 0$$
$$2(w^2 - 9) = 0$$
$$2(w + 3)(w - 3) = 0$$

$$w + 3 = 0 \quad \text{or} \quad w - 3 = 0$$
$$\cancel{w = -3} \quad | \quad w = 3$$

State

The width of the flag is 3 feet.
The length of the flag is $2(3) = 6$ feet.

Check

A rectangle with dimensions 3 feet by 6 feet has an area of 18 ft^2, and the length is twice the width. The results check.

14. BILLIARDS
    Width: 5 ft; length: 10 ft

15. X-RAYS

Analyze

· Length is 2 inches more than the width.
· Area is 80 sq. in.
· Find the dimensions.

Assign

Let w = width of x-ray in inches
   $w + 2$ = length of x-ray in inches

Form

The area of the rectangle	equals	the length	times	the width.
80	=	$w + 2$	·	w

Solve

$$w(w + 2) = 80$$
$$w^2 + 2w = 80$$
$$w^2 + 2w - \mathbf{80} = 80 - \mathbf{80}$$
$$w^2 + 2w - 80 = 0$$
$$(w + 10)(w - 8) = 0$$

$$w + 10 = 0 \quad \text{or} \quad w - 8 = 0$$
$$\cancel{w = -10} \quad | \quad w = 8$$

State

The width of the film is 8 inches.
The length of the film is $8 + 2 = 10$ inches.

Check

A rectangle with dimensions 8 in by 10 in has an area of 80 in^2, and the length is two inches more the width. The results check.

from CAMPUS TO CAREERS

17. ELEMENTARY TEACHERS

Analyze
· Length is 3 times the width.
· Area of wall is 90 sq. ft.
· Fire code only allows for 30% coverage.
· Find the dimensions.

Assign
Let w = width of board in feet
   $3w$ = length of board in feet

   First calculate 30% of the wall area.
$$90(30\%) = 90(0.30)$$
$$= 27$$

Form

The area of the rectangle	equals	the length	times	the width.
27	=	$3w$	·	w

Solve
$$3w(w) = 27$$
$$3w^2 = 27$$
$$3w^2 - 27 = 27 - 27$$
$$3w^2 - 27 = 0$$
$$3(w^2 - 9) = 0$$
$$3(w + 3)(w - 3) = 0$$

$w + 3 = 0$ or $w - 3 = 0$
~~$w = -3$~~ $w = 3$

State
The width of the board is 3 feet.
The length of the board is $3(3) = 9$ feet.

Check
A rectangle with dimensions 3 ft by 9 ft
has an area of 27 ft^2, and the length is three
times the width. The results check.

19. JEANS

Analyze
· Height of the triangle is 1 cm less than the base.
· Area is 15 sq. cm.
· Find the lengths of the base and height.

Assign
Let b = base of triangle in cm
   $b - 1$ = height of triangle in cm

Form

The area of the triangle	equals	$\frac{1}{2}$	times	the base	times	the height.
15	=	$\frac{1}{2}$	·	b	·	$b - 1$

Solve
$$\frac{1}{2}b(b-1) = 15$$
$$(2)\frac{1}{2}b(b-1) = (2)15$$
$$b(b-1) = 30$$
$$b^2 - b = 30$$
$$b^2 - b - 30 = 30 - 30$$
$$b^2 - b - 30 = 0$$
$$(b + 5)(b - 6) = 0$$

$b + 5 = 0$ or $b - 6 = 0$
~~$b = -5$~~ $b = 6$

State
The base is 6 cm.
The height is $6 - 1 = 5$ cm.

Check
A triangle with dimensions 5 cm by 6 cm
has an area of 15 cm^2, and the height is one
cm less than the base. The results check.

Section 6.8

21. SAILBOATS

Analyze

· Length of the luff is 3 times longer than the length of the foot.
· Area is 24 sq. ft.
· Find the lengths of the luff and foot.

Assign

Let f = length of foot (base) in feet
$3f$ = length of luff (height) in feet

Form

The area of the triangle	equals one half of	the base	times	the height.

$$24 \quad = \quad \frac{1}{2} \cdot f \cdot 3f$$

Solve

$$\frac{1}{2}f(3f) = 24$$
$$\frac{3}{2}f^2 = 24$$
$$2 \cdot \frac{3}{2}f^2 = 2 \cdot 24$$
$$3f^2 = 48$$
$$3f^2 - 48 = 48 - 48$$
$$3f^2 - 48 = 0$$
$$3(f^2 - 16) = 0$$
$$3(f+4)(f-4) = 0$$

$$f + 4 = 0 \qquad \text{or} \qquad f - 4 = 0$$
$$\bcancel{f = -4} \qquad\qquad f = 4$$

State

The foot is 4 ft.
The luff is $3(4) = 12$ ft.

Check

A triangle with dimensions 4 ft by 12 ft has an area of 24 ft^2, and the luff is three times the foot. The results check.

CONSECUTIVE INTEGERS

23. NASCAR

Analyze

· Product of two consecutive positive integers is 90.
· Kahne's car number is the smaller.
· Riggs' car number is the larger.
· What is the number of each car?

Assign

Let x = Kahne's car number
$x + 1$ = Riggs' car number

Form

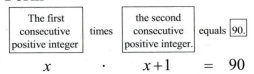

The first consecutive positive integer	times	the second consecutive positive integer.	equals	90.

$$x \quad \cdot \quad x+1 \quad = \quad 90$$

Solve

$$x(x+1) = 90$$
$$x^2 + x = 90$$
$$x^2 + x - \mathbf{90} = 90 - \mathbf{90}$$
$$x^2 + x - 90 = 0$$
$$(x+10)(x-9) = 0$$

$$x + 10 = 0 \qquad \text{or} \qquad x - 9 = 0$$
$$\bcancel{x = -10} \qquad\qquad\qquad x = 9$$

State

Kahne's car number is 9.
Riggs' car number is $9 + 1 = 10$.

Check

Since 9 and 10 are consecutive positive integers, and since $9 \cdot 10 = 90$. The results check.

25. CUSTOMER SERVICE

Analyze

· Product of two consecutive positive integers is 156.
· Ticket being served now is the smaller.
· Next ticket is the larger.
· What is the ticket number being served now?

Assign

Let x = Ticket being served now
$x+1$ = Next ticket being served

Form

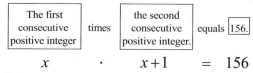

| The first consecutive positive integer | times | the second consecutive positive integer. | equals | 156. |

$$x \quad \cdot \quad x+1 \quad = \quad 156$$

Solve

$$x(x+1) = 156$$
$$x^2 + x = 156$$
$$x^2 + x - \mathbf{156} = 156 - \mathbf{156}$$
$$x^2 + x - 156 = 0$$
$$(x+13)(x-12) = 0$$

$$x + 13 = 0 \qquad \text{or} \qquad x - 12 = 0$$
$$\cancel{x = -13} \qquad\qquad\qquad x = 12$$

State

Ticket being served now is 12.

Check

Since 12 and 13 are consecutive positive integers, and since $12 \cdot 13 = 156$. The results check.

27. PLOTTING POINTS

Analyze

· Product of two consecutive odd positive integers is 143.
· The x-coordinate is the smaller.
· The y-coordinate is the larger.
· What is the coordinate of the point?

Assign

Let x = x-coordinate
$x+2$ = y-coordinate

Form

| The first consecutive odd integer | times | the second consecutive odd integer. | equals | 143. |

$$x \quad \cdot \quad x+2 \quad = \quad 143$$

Solve

$$x(x+2) = 143$$
$$x^2 + 2x = 143$$

$$x^2 + 2x - \mathbf{143} = 143 - \mathbf{143}$$
$$x^2 + 2x - 143 = 0$$
$$(x+13)(x-11) = 0$$

$$x + 13 = 0 \qquad \text{or} \qquad x - 11 = 0$$
$$\cancel{x = -13} \qquad\qquad\qquad x = 11$$

State

The x-coordinate is 11.
The y-coordinate is 13.

Check

Since 11 and 13 are consecutive odd positive integers, and since $11 \cdot 13 = 143$. The results check.

PYTHAGOREAN THEOREM PROBLEMS
29. HIGH-ROPES ADVENTURES COURSES

Analyze

· Anchor is 8 yds from the base of the pole. (longer leg)
· Tie off point is 6 yds up the pole. (shorter leg)
· How long is the cable? (hypotenuse)

Assign

Let a = length of shorter leg in yd
b = length of longer leg in yd

Form

$$\left(\begin{array}{c}\text{The length of}\\\text{the shorter leg}\end{array}\right)^2 \text{ plus } \left(\begin{array}{c}\text{the length of}\\\text{the longer leg}\end{array}\right)^2 \text{ equals } \left(\begin{array}{c}\text{the length of}\\\text{the hypotenuse.}\end{array}\right)^2$$

$$a^2 \quad + \quad b^2 \quad = \quad c^2$$

Solve

$$a^2 + b^2 = c^2$$
$$6^2 + 8^2 = c^2$$
$$36 + 64 = c^2$$
$$100 = c^2$$
$$c^2 = 100$$
$$c^2 - \mathbf{100} = 100 - \mathbf{100}$$
$$c^2 - 100 = 0$$
$$(c+10)(c-10) = 0$$

$$c + 10 = 0 \qquad \text{or} \qquad c - 10 = 0$$
$$\cancel{c = -10} \qquad\qquad\qquad c = 10$$

State

The length of cable is 10 yards long.

Check

Since $6^2 + 8^2 = 10^2$. The results check.

Section 6.8

31. MOTO X

Analyze
- 15 ft is the longer leg.
- x ft is the shorter leg.
- 17 ft is the hypotenuse.
- What is the height of the ramp (x)?

Assign

Let x = height of the ramp in ft

Form

$$\left(\begin{array}{c}\text{The length of}\\\text{the shorter leg}\end{array}\right)^2 \quad \text{plus} \quad \left(\begin{array}{c}\text{the length of}\\\text{the longer leg}\end{array}\right)^2 \quad \text{equals} \quad \left(\begin{array}{c}\text{the length of}\\\text{the hypotenuse.}\end{array}\right)^2$$

$$x^2 \quad + \quad 15^2 \quad = \quad 17^2$$

Solve

$$x^2 + 15^2 = 17^2$$
$$x^2 + 225 = 289$$
$$x^2 + 225 - \mathbf{289} = 289 - \mathbf{289}$$
$$x^2 - 64 = 0$$
$$(x+8)(x-8) = 0$$

$$x + 8 = 0 \qquad \text{or} \qquad x - 8 = 0$$
$$\xcancel{x = -8} \qquad \qquad \qquad x = 8$$

State
The height of the ramp is 8 ft.

Check
Since $8^2 + 15^2 = 17^2$. The results check.

33. BOATING

Analyze
- x m is the shorter leg (rise).
- Run is 7m longer than the rise.
- Ramp is 8 m longer than the rise.
- How long is each side?

Assign

Let x = length of shorter leg in meters

$x + 7$ = length of longer leg in meters

$x + 8$ = length of hypotenuse in meters

Form

$$\left(\begin{array}{c}\text{The length of}\\\text{the shorter leg}\end{array}\right)^2 \quad \text{plus} \quad \left(\begin{array}{c}\text{the length of}\\\text{the longer leg}\end{array}\right)^2 \quad \text{equals} \quad \left(\begin{array}{c}\text{the length of}\\\text{the hypotenuse.}\end{array}\right)^2$$

$$x^2 \quad + \quad (x+7)^2 \quad = \quad (x+8)^2$$

Solve

$$x^2 + (x+7)^2 = (x+8)^2$$
$$x^2 + x^2 + 14x + 49 = x^2 + 16x + 64$$
$$2x^2 + 14x + 49 = x^2 + 16x + 64$$
$$2x^2 + 14x + 49 - \mathbf{x^2} = x^2 + 16x + 64 - \mathbf{x^2}$$
$$x^2 + 14x + 49 = 16x + 64$$
$$x^2 + 14x + 49 - \mathbf{16x} = 16x + 64 - \mathbf{16x}$$
$$x^2 - 2x + 49 = 64$$
$$x^2 - 2x + 49 - \mathbf{64} = 64 - \mathbf{64}$$
$$x^2 - 2x - 15 = 0$$
$$(x+3)(x-5) = 0$$

$$x + 3 = 0 \qquad \text{or} \qquad x - 5 = 0$$
$$\xcancel{x = -3} \qquad \qquad \qquad x = 5$$

State
The rise is 5 m.
The run is $5 + 7 = 12$ m.
The ramp is $5 + 8 = 13$ m.

Check
Since $5^2 + 12^2 = 13^2$. The results check.

QUADRATIC EQUATION MODEL PROBLEMS

35. THRILL RIDES

Analyze
- When the sunglasses hit the ground, its height will be 0 feet. To find the time that it takes for the glasses to hit the ground, we set h equal to 0, and solve the quadratic equation for t.

Assign

Let t = time in seconds

Form

$$h = -16t^2 + 64t + 80$$
$$0 = -16t^2 + 64t + 80$$

Solve

$$-16t^2 + 64t + 80 = 0$$
$$-16(t^2 - 4t - 5) = 0$$
$$-16(t+1)(t-5) = 0$$

$$\xcancel{-16 = 0} \quad t + 1 = 0 \qquad \text{or} \qquad t - 5 = 0$$
$$\xcancel{t = -1} \qquad \qquad t = 5$$

State
The equation has two solutions, -1 and 5. Since t represents time, and, in this case, time cannot be negative, we discard -1. The second solution, 5, indicates that the glasses hits the ground 5 seconds after falling off.

Check
Check this result by substituting 5 for t in $h = -16t^2 + 64t + 80$. You should get $h = 0$.

37. SOFTBALL

Analyze

· When the softball hits the ground, its height will be 0 feet. To find the time that it takes for the ball to hit the ground, we set h equal to 0, and solve the quadratic equation for t.

Assign

Let t = time in seconds

Form

$$h = -16t^2 + 63t + 4$$
$$0 = -16t^2 + 63t + 4$$

Solve

$$-16t^2 + 63t + 4 = 0$$
$$-1(16t^2 - 63t - 4) = 0$$
$$-(16t + 1)(t - 4) = 0$$

$\cancel{-1 = 0}$ $16t + 1 = 0$ or $t - 4 = 0$

$t = \cancel{-\dfrac{1}{16}}$ | $t = 4$

State

The equation has two solutions, $-\dfrac{1}{16}$ and 4. Since t represents time, and, in this case, time cannot be negative, we discard $-\dfrac{1}{16}$. The second solution, 4, indicates that the ball hits the ground 4 seconds after being thrown.

Check

Check this result by substituting 4 for t in $h = -16t^2 + 63t + 4$. You should get $h = 0$.

39. DOLPHINS

Analyze

· The height h in feet reached by a dolphin t seconds after breaking the surface of the water is given by $h = -16t^2 + 32t$.
· Height is 16 ft.

Assign

Let t = time in seconds

Form

$$h = -16t^2 + 32t$$
$$16 = -16t^2 + 32t$$

Solve

$$-16t^2 + 32t = 16$$
$$-16t^2 + 32t - \mathbf{16} = 16 - \mathbf{16}$$
$$-16t^2 + 32t - 16 = 0$$
$$-16(t^2 - 2t + 1) = 0$$
$$-16(t - 1)(t - 1) = 0$$

$\cancel{-16 = 0}$ or $t - 1 = 0$

| $t = 1$

State

The equation has two repeating solutions, 1. The solution, 1, indicates that the dolphin will touch the trainer's hand 1 second after breaking the surface of the water.

Check

Check this result by substituting 1 for t in $-16t^2 + 32t = 16$. You should get $16 = 16$.

41. CHOREOGRAPHY

Analyze

· 36 dancers are to assemble in a triangular-shaped series of rows, where each row has one more dancer than the previous row. The relationship between the number of rows r and the number of dancers d is given by

$$d = \frac{1}{2}r(r + 1).$$

· What is the number of rows in the formation?

Assign

Let r = the number of rows

Form

$$d = \frac{1}{2}r(r + 1)$$
$$36 = \frac{1}{2}r(r + 1)$$

Solve

$$\frac{1}{2}r(r + 1) = 36$$
$$(\mathbf{2})\frac{1}{2}r(r + 1) = (\mathbf{2})(36)$$
$$r(r + 1) = 72$$
$$r^2 + r = 72$$
$$r^2 + r - \mathbf{72} = 72 - \mathbf{72}$$
$$r^2 + r - 72 = 0$$
$$(r + 9)(r - 8) = 0$$

$r + 9 = 0$ or $r - 8 = 0$

$\cancel{r = -9}$ | $r = 8$

State

The equation has two solutions, -9 and 8. Since r represents number of rows, and, in this case, number of rows cannot be negative, we discard -9. The second solution, 8, indicates that the number of rows is 8.

Check

Check this result by substituting 8 for r in $\frac{1}{2}r(r + 1) = 36$. You should get $36 = 36$.

Section 6.8

WRITING

43-45. Answers will vary.

REVIEW

Find each special product.

47. $(5b - 2)^2 = (5b)^2 + 2 \cdot 5b \cdot (-2) + (-2)^2$
$= 25b^2 - 20b + 4$

49. $(s^2 + 4)^2 = (s^2)^2 + 2 \cdot s^2 \cdot (4) + 4^2$
$= s^4 + 8s^2 + 16$

51. $(9x + 6)(9x - 6) = (9x)^2 - (6)^2$
$= 81x^2 - 36$

CHALLENGE PROBLEMS

53. POOL BORDERS

Analyze

· $25 + 2w$ is the new length in meters.

· $10 + 2w$ is the new width in meters.

· w is the border in meters.

· Pool is 10 m by 25 m.

· Area of border pool minus area of pool is 74

Assign

Let $w =$ width of border in m

Form

The area of the outside of the pool	minus	the area of the pool	equals	the area of border.
$(2w + 25)(2w + 10)$	$-$	$(10)(25)$	$=$	74

Solve

$$(2w + 25)(2w + 10) - (10)(25) = 74$$
$$4w^2 + 70w + 250 - 250 = 74$$
$$4w^2 + 70w = 74$$
$$4w^2 + 70w - \mathbf{74} = 74 - \mathbf{74}$$
$$4w^2 + 70w - 74 = 0$$
$$2(2w^2 + 35w - 37) = 0$$
$$2(2w + 37)(w - 1) = 0$$

$2w + 37 = 0$ or $w - 1 = 0$

$w = -\dfrac{37}{2}$ $\quad\quad$ $w = 1$

State

The equation has two solutions, $-\dfrac{37}{2}$ and 1.

Since w represents the width of the border, and, in this case, the border cannot be negative, we discard $-\dfrac{37}{2}$. The second solution, 1, indicates that the border is 1 m.

Check

Check this result by substituting 1 for w in
$(2w + 25)(2w + 10) - (10)(25) = 324 - 250 = 74$.
You should get $74 = 74$.

CHAPTER 6 REVIEW

SECTION 6.1
The Greatest Common Factor;
Factoring by Grouping

Find the prime-factorization of each number.

1. $35 = 5 \cdot 7$

2. $96 = 2 \cdot 48$
   $= 2 \cdot 2 \cdot 24$
   $= 2 \cdot 2 \cdot 2 \cdot 12$
   $= 2 \cdot 2 \cdot 2 \cdot 2 \cdot 6$
   $= 2 \cdot 2 \cdot 2 \cdot 2 \cdot 2 \cdot 3$
   $= 2^5 \cdot 3$

Find the GCF of each list.

3. 28 and 35
$$\left. \begin{array}{l} 28 = 2 \cdot 2 \cdot \mathbf{7} \\ 35 = 5 \cdot \mathbf{7} \end{array} \right\} \text{GCF} = 7$$

4. $36a^4$, $54a^3$, and $126a^6$
$$\left. \begin{array}{l} 36a^4 = \mathbf{2} \cdot 2 \cdot \mathbf{3 \cdot 3} \quad \cdot \mathbf{a \cdot a} \cdot a \cdot a \\ 54a^3 = \mathbf{2} \quad \cdot \mathbf{3 \cdot 3} \cdot 3 \cdot \mathbf{a \cdot a} \cdot a \\ 126a^6 = \mathbf{2} \cdot \quad \mathbf{3 \cdot 3} \cdot 7 \cdot \mathbf{a \cdot a} \cdot a \cdot a \cdot a \cdot a \end{array} \right\}$$

$$\text{GCF} = \mathbf{2 \cdot 3 \cdot 3 \cdot a \cdot a \cdot a} = 18a^3$$

Factor.

5. $3x + 9y = \mathbf{3} \cdot x + \mathbf{3} \cdot 3y$
   $= \mathbf{3}(x + 3y)$

6. $5ax^2 + 15a = \mathbf{5 \cdot a} \cdot x \cdot x + 3 \cdot \mathbf{5 \cdot a}$
   $= 5a(x^2 + 3)$

7. $7s^5 + 14s^3 = \mathbf{7 \cdot s^2 \cdot s^3} + \mathbf{7} \cdot 2s^3$
   $= \mathbf{7s^3}(s^2 + 2)$

8. $\pi ab - \pi ac = \mathbf{\pi \cdot a} \cdot b - \mathbf{\pi \cdot a} \cdot c$
   $= \pi a(b - c)$

9. $24x^3 + 60x^2 - 48x$
   $= \mathbf{12} \cdot 2x^2 \cdot \mathbf{x} + \mathbf{12} \cdot 5x \cdot \mathbf{x} - \mathbf{12} \cdot 4\mathbf{x}$
   $= \mathbf{12x}(2x^2 + 5x - 4)$

10. $x^5y^3z^2 + xy^5z^3 - xy^3z^2$
    $= \mathbf{x} \cdot x^4 \cdot \mathbf{y^3} \cdot \mathbf{z^2} + \mathbf{x} \cdot \mathbf{y^3} \cdot y^2 \cdot \mathbf{z^2} \cdot z - \mathbf{x} \cdot \mathbf{y^3} \cdot \mathbf{z^2}$
    $= xy^3z^2(x^4 + y^2z - 1)$

11. $-5ab^2 + 10a^2b - 15ab$
    $= (-\mathbf{5} \cdot \mathbf{ab} \cdot \mathbf{b}) + (-\mathbf{5} \cdot -\mathbf{2a} \cdot \mathbf{ab}) + (-\mathbf{5} \cdot 3\mathbf{ab})$
    $= -\mathbf{5ab}(b - 2a + 3)$

12. $4(x-2) - x(x-2) = 4(\mathbf{x-2}) - x(\mathbf{x-2})$
    $= (x-2)(4-x)$

Factor out -1.

13. $-a - 7 = (\mathbf{-1})a + (\mathbf{-1})(7)$
    $= (\mathbf{-1})(a + 7)$
    $= -(a + 7)$

14. $-4t^2 + 3t - 1 = (\mathbf{-1})4t^2 - (\mathbf{-1})3t + (\mathbf{-1})1$
    $= -(4t^2 - 3t + 1)$

Factor.

15. $2c + 2d + ac + ad = (\mathbf{2c + 2d}) + (\mathbf{ac + ad})$
    $= \mathbf{2}(c+d) + \mathbf{a}(c+d)$
    $= (c+d)(\mathbf{2} + \mathbf{a})$

16. $3xy + 18x - 5y - 30$
    $= (3xy + 18x) + (-5y - 30)$
    $= (\mathbf{3xy + 3x} \cdot 6) + [\mathbf{-5}y + (\mathbf{-5} \cdot 6)]$
    $= 3x(y + 6) - 5(y + 6)$
    $= (y + 6)(3x - 5)$

17. $2a^3 + 2a^2 - a - 1$
    $= (\mathbf{2a^2} \cdot a + \mathbf{2a^2}) + [\mathbf{-1} \cdot a + (\mathbf{-1} \cdot 1)]$
    $= 2a^2(a + 1) - 1(a + 1)$
    $= (a + 1)(2a^2 - 1)$

18. $4m^2n + 12m^2 - 8mn - 24m$
    $= 4m(mn + 3m - 2n - 6)$
    $= 4m[(mn + 3m) + (-2n - 6)]$
    $= 4m[m(n + 3) - 2(n + 3)]$
    $= 4m(n + 3)(m - 2)$

SECTION 6.2 REVIEW EXERCISES
Factoring Trinomials of the Form $x^2 + bx + c$

19. What is the lead coefficient of
    $x^2 + 8x - 9$?   **1**

20. Complete the table.

Factors of 6	Sum of the factors of 6
1(6)	**7**
2(3)	**5**
−1(−6)	**−7**
−2(−3)	**−5**

Chapter 6 Review and Test

21. Two different sign factors of -24
    whose sum is $+2$ are $+6$ and -4.
    $$x^2 + 2x - 24 = (x+6)(x-4)$$

22. Two different sign factors of -40
    whose sum is -18 are -20 and $+2$.
    $$x^2 - 18x - 40 = (x-20)(x+2)$$

23. The negative factors of 45
    whose sum is -14 are -5 and -9.
    $$x^2 - 14x + 45 = (x-5)(x-9)$$

24. $t^2 + 10t + 15$ is prime.

25. Factor out a -1.
    $$-y^2 + 15y - 56 = -(y^2 - 15y + 56)$$
    The negative factors of 56
    whose sum is -15 are -8 and -7.
    $$-(y^2 - 15y + 56) = -(y-8)(y-7)$$

26. $10y + 9 + y^2 = y^2 + 10y + 9$
    The positive factors of 9
    whose sum is $+10$ are $+9$ and $+1$.
    $$y^2 + 10y + 9 = (y+9)(y+1)$$

27. Two different sign factors of -10
    whose sum is $+3$ are $+5$ and -2.
    $$c^2 + 3cd - 10d^2 = (c+5d)(c-2d)$$

28. $-3mn + m^2 + 2n^2 = m^2 - 3mn + 2n^2$
    The negative factors of 2
    whose sum is -3 are -2 and -1.
    $$m^2 - 3mn + 2n^2 = (m-2n)(m-n)$$

29. Explain how we can check to
    determine whether $(x-4)(x+5)$
    is the factorization of $x^2 + x - 20$.
    Multiply

30. Explain why $x^2 + 7x + 11$ is prime.
    **There are no two integers whose
    product is 11 and whose sum is 7.**

Completely factor each trinomial.

31. Factor out GCF $5a^3$.
    $$5a^5 + 45a^4 - 50a^3 = 5a^3(a^2 + 9a - 10)$$
    Two different sign factors of -10
    whose sum is $+9$ are $+10$ and -1.
    $$5a^5 + 45a^4 - 50a^3 = 5a^3(a^2 + 9a - 10)$$
    $$= 5a^3(a+10)(a-1)$$

32. Rearrange and factor out GCF $-4x$.
    $$-4x^2y - 4x^3 + 24xy^2 = -4x^3 - 4x^2y + 24xy^2$$
    $$= -4x(x^2 + xy - 6y^2)$$
    Two different sign factors of -6
    whose sum is $+1$ are $+3$ and -2.
    $$-4x(x^2 + xy - 6y^2) = -4x(x+3y)(x-2y)$$

SECTION 6.3 REVIEW EXERCISES
Factoring Trinomials of the Form $ax^2 + bx + c$
Factor each trinomial completely, if possible.

33. $\underset{a}{2x^2} - \underset{b}{5x} - \underset{c}{3}$
    In $2x^2 - 5x - 3$, we have $a = 2$, $b = -5$, and $c = -3$.
    The key number is $ac = 2(-3) = -6$. We must find a
    factorization of -6 in which the sum of the factors is
    $b = -5$. Since the factors must have a negative product,
    their signs must be different. The pairs of factors are -6
    and $+1$.
    $$2x^2 - 5x - 3 = 2x^2 - 6x + x - 3$$
    $$= (2x^2 - 6x) + (x - 3)$$
    $$= 2x(x-3) + 1(x-3)$$
    $$= (x-3)(2x+1)$$

34. $\underset{a}{35y^2} + \underset{b}{11y} - \underset{c}{10}$
    In $35y^2 + 11y - 10$, we have $a = 35$, $b = 11$, and $c = -10$.
    The key number is $ac = 35(-10) = -350$. We must find a
    factorization of -350 in which the sum of the factors is
    $b = +11$. Since the factors must have a negative product,
    their signs must be different. The pairs of factors are -14
    and $+25$.
    $$35y^2 + 11y - 10 = 35y^2 - 14y + 25y - 10$$
    $$= (35y^2 - 14y) + (25y - 10)$$
    $$= 7y(5y - 2) + 5(5y - 2)$$
    $$= (5y - 2)(7y + 5)$$

35. $-3x^2 + 13x + 30 = -(3x^2 - 13x - 30)$
    $$\underset{a}{3x^2} - \underset{b}{13x} - \underset{c}{30}$$
    In $3x^2 - 13x - 30$, we have $a = 3$, $b = -13$ and $c = -30$.
    The key number is $ac = 3(-30) = -90$. We must find a
    factorization of -90 in which the sum of the factors is
    $b = -13$. Since the factors must have a negative product,
    their signs must be different. The pairs of factors are -18
    and $+5$.
    $$-(3x^2 - 13x - 30) = -(3x^2 - 18x + 5x - 30)$$
    $$= -[(3x^2 - 18x) + (5x - 30)]$$
    $$= -[3x(x-6) + 5(x-6)]$$
    $$= -(x-6)(3x+5)$$

36. $-33p^2 - 6p + 18p^3 = 18p^3 - 33p^2 - 6p$
$$= 3p(6p^2 - 11p - 2)$$

$$3p(\underset{a}{6p^2} \underset{b}{-11p} \underset{c}{-2})$$

In $6p^2 - 11p - 2$, we have $a = 6$, $b = -11$ and $c = -2$. The key number is $ac = 6(-2) = -12$. We must find a factorization of -12 in which the sum of the factors is $b = -11$. Since the factors must have a negative product, their signs must be different. The pairs of factors are -12 and $+1$.

$$3p(6p^2 - 11p - 2) = 3p(6p^2 - 12p + 1p - 2)$$
$$= 3p[(6p^2 - 12p) + (1p - 2)]$$
$$= 3p[6p(p - 2) + 1(p - 2)]$$
$$= 3p(p - 2)(6p + 1)$$

37. $\underset{a}{4b^2} \underset{b}{-17bc} + \underset{c}{4c^2}$

In $4b^2 - 17bc + 4c^2$, we have $a = 4$, $b = -17$ and $c = 4$. The key number is $ac = 4(4) = 16$. We must find a factorization of 16 in which the sum of the factors is $b = -17$. Since the factors must have a positive product, their signs must be the same. The pairs of factors are -1 and -16.

$$4b^2 - 17bc + 4c^2 = 4b^2 - 16bc - bc + 4c^2$$
$$= (4b^2 - 16bc) + (-bc + 4c^2)$$
$$= 4b(b - 4c) - c(b - 4c)$$
$$= (b - 4c)(4b - c)$$

38. $7y^2 + 7y - 18$ is prime.

39. ENTERTAINING

$$\underset{a}{12x^2} \underset{b}{-1x} \underset{c}{-1}$$

In $12x^2 - 1x - 1$, we have $a = 12$, $b = -1$ and $c = -1$. The key number is $ac = 12(-1) = -12$. We must find a factorization of -12 in which the sum of the factors is $b = -1$. Since the factors must have a negative product, their signs must be different. The pairs of factors are -4 and $+3$.

$$12x^2 - x - 1 = 12x^2 - 4x + 3x - 1$$
$$= (12x^2 - 4x) + (3x - 1)$$
$$= 4x(3x - 1) + 1(3x - 1)$$
$$= (3x - 1)(4x + 1)$$

The length is $(4x + 1)$ in.
The width is $(3x - 1)$ in.

40. In the following work, a student began to factor $5x^2 - 8x + 3$. Explain his mistake.

$$(5x - \quad)(x + \quad)$$

The signs of the second terms must be negative.

Factoring Perfect-Square Trinomials and the Difference of Two Squares
Factor completely, if possible.

41. $x^2 + 10x + 25$

The first term x^2 is the square of **x**.
The last term 25 is the square of **5**.
The middle term is twice the product of x and 5: $2(x)(5) = \mathbf{10x}$.
$$x^2 + 10x + 25 = (x + 5)(x + 5)$$
$$= (x + 5)^2$$

42. $9y^2 + 16 - 24y = 9y^2 - 24y + 16$

The first term $9y^2$ is the square of **$3y$**.
The last term 16 is the square of **-4**.
The middle term is twice the product of $3y$ and -4: $2(3y)(-4) = \mathbf{-24y}$.
$$9y^2 - 24y + 16 = (3y - 4)(3y - 4)$$
$$= (3y - 4)^2$$

43. $-z^2 + 2z - 1 = -(z^2 - 2z + 1)$

The first term z^2 is the square of **z**.
The last term 1 is the square of **-1**.
The middle term is twice the product of z and -1: $2(z)(-1) = \mathbf{-2z}$.
$$-(z^2 - 2z + 1) = -(z - 1)(z - 1)$$
$$= -(z - 1)^2$$

44. $25a^2 + 20ab + 4b^2$

The first term $25a^2$ is the square of **$5a$**.
The last term $4b^2$ is the square of **$2b$**.
The middle term is twice the product of $5a$ and $2b$: $2(5a)(2b) = \mathbf{20ab}$.
$$25a^2 + 20ab + 4b^2 = (5a + 2b)(5a + 2b)$$
$$= (5a + 2b)^2$$

45. $x^2 - 9$

$$F^2 - L^2 = (F + L)(F - L)$$
$$\downarrow \quad \downarrow \qquad \downarrow \quad \downarrow \quad \downarrow \quad \downarrow$$
$$x^2 - 3^2 = (x + 3)(x - 3)$$

46. $49t^2 - 121y^2$

$$F^2 - L^2 = (F + L)(F - L)$$
$$\downarrow \qquad \downarrow \qquad \downarrow \quad \downarrow \quad \downarrow \quad \downarrow$$
$$(7t)^2 - (11y)^2 = (7t + 11y)(7t - 11y)$$

47. $x^2y^2 - 400$

$$
\begin{array}{ccccc}
F^2 & - & L^2 & = (F & + & L)\ (F & - & L)\\
\downarrow & & \downarrow & & \downarrow & & \downarrow & & \downarrow
\end{array}
$$

$$(xy)^2 - 20^2 = (xy + 20)(xy - 20)$$

48. $8at^2 - 32a = 8a(t^2 - 4)$

$$
\begin{array}{ccccc}
F^2 & - & L^2 & = & (F + L)(F - L)\\
\downarrow & & \downarrow & & \downarrow\ \ \ \downarrow\ \ \ \downarrow\ \ \ \downarrow
\end{array}
$$

$$8a(t^2 - 2^2) = 8a(t + 2)\ (t - 2)$$

49. $c^4 - 256$

$$
\begin{array}{ccccc}
F^2 & - & L^2 & = (F & + & L)\ (F & - & L)\\
\downarrow & & \downarrow & & \downarrow & & \downarrow & & \downarrow
\end{array}
$$

$$(c^2)^2 - 16^2 = (c^2 + 16)(c^2 - 16)$$

$c^2 - 16$

$$
\begin{array}{ccccc}
F^2 & - & L^2 & = (F + L)(F - L)\\
\downarrow & & \downarrow & & \downarrow\ \ \ \downarrow\ \ \ \downarrow\ \ \ \downarrow
\end{array}
$$

$$c^2 - 4^2 = (c + 4)\ (c - 4)$$

$$c^4 - 256 = (c^2 + 16)(c + 4)(c - 4)$$

50. $h^2 + 36$ is prime.

SECTION 6.5 REVIEW EXERCISES
Factoring the Sum and Difference of Two Cubes
Factor each polynomial completely.

51. $b^3 + 1 = b^3 + 1^3$
$$= (b + 1)(b^2 - b \cdot 1 + 1^2)$$
$$= (b + 1)(b^2 - b + 1)$$

52. $x^3 - 216 = x^3 - 6^3$
$$= (x - 6)(x^2 + x \cdot 6 + 6^2)$$
$$= (x - 6)(x^2 + 6x + 36)$$

53. $p^3 + 125q^3 = p^3 + (5q)^3$
$$= (p + 5q)[p^2 - p \cdot 5q + (5q)^2]$$
$$= (p + 5q)(p^2 - 5pq + 25q^2)$$

54. $16x^5 - 54x^2y^3$
$$= 2x^2(8x^3 - 27y^3)$$
$$= 2x^2[(2x)^3 - (3y)^3]$$
$$= 2x^2(2x - 3y)[(2x)^2 + (2x)(3y) + (3y)^2]$$
$$= 2x^2(2x - 3y)(4x^2 + 6xy + 9y^2)$$

SECTION 6.6 REVIEW EXERCISES
A Factoring Strategy
Factor each polynomial completely, if possible.

55. $14y^3 + 6y^4 - 40y^2 = 6y^4 + 14y^3 - 40y^2$
$$= 2y^2(3y^2 + 7y - 20)$$

$$3\underset{a}{y^2} + 7\underset{b}{y} - \underset{c}{20}$$

In $3y^2 + 7y - 20$, we have $a = 3$, $b = 7$, and $c = -20$. The key number is $ac = 3(-20) = -60$. We must find a factorization of -60 in which the sum of the factors is $b = 7$. Since the factors must have a negative product, their signs must be different. The pairs of factors are -5 and $+12$.

$$2y^2(3y^2 + 7y - 20) = 2y^2(3y^2 - 5y + 12y - 20)$$
$$= 2y^2[y(3y - 5) + 4(3y - 5)]$$
$$= 2y^2(3y - 5)(y + 4)$$

56. $5s^2t + 5s^2u^2 + 5tv + 5u^2v$
$$= 5(s^2t + s^2u^2 + tv + u^2v)$$
$$= 5[(s^2t + s^2u^2) + (tv + u^2v)]$$
$$= 5[s^2(t + u^2) + v(t + u^2)]$$
$$= 5(t + u^2)(s^2 + v)$$

57. $j^4 - 16$

$$
\begin{array}{ccccc}
F^2 & - & L^2 & = (F + L)(F - L)\\
\downarrow & & \downarrow & & \downarrow\ \ \ \downarrow\ \ \ \downarrow\ \ \ \downarrow
\end{array}
$$

$$(j^2)^2 - 4^2 = (j^2 + 4)(j^2 - 4)$$

$$
\begin{array}{ccccc}
F^2 & - & L^2 & = (F + L)(F - L)\\
\downarrow & & \downarrow & & \downarrow\ \ \ \downarrow\ \ \ \downarrow\ \ \ \downarrow
\end{array}
$$

$$j^2 - 2^2 = (j + 2)\ (j - 2)$$

$$j^4 - 16 = (j^2 + 4)(j + 2)(j - 2)$$

58. $-3j^3 - 24 = -3(j^3 + 8)$
$$= -3[(j^3 + 2^3]$$
$$= -3[(j + 2)(j^2 - j \cdot 2 + 2^2)]$$
$$= -3(j + 2)(j^2 - 2j + 4)$$

59. $400x + 400 - m^2x - m^2$
$$= (400x + 400) + (-m^2x - m^2)$$
$$= 400(x + 1) - m^2(x + 1)$$
$$= (x + 1)(400 - m^2)$$
$$= (x + 1)(20 + m)(20 - m)$$

60. $12w^4 - 36w^3 + 27w^2 = 3w^2(4w^2 - 12w + 9)$
$$= 3w^2(2w - 3)(2w - 3)$$
$$= 3w^2(2w - 3)^2$$

61. $2t^3 + 10 = 2t^3 + 2 \cdot 5$
$$= 2(t^3 + 5)$$

62. $121p^2 + 36q^2$ is prime.

63. $x^2z + 64y^2z + 16xyz = x^2z + 16xyz + 64y^2z$
$\qquad = z(x^2 + 16xy + 64y^2)$

$x^2 + 16xy + 64y^2$
The first term x^2 is the square of x.
The last term $64y^2$ is the square of $8y$.
The middle term is twice the product of
x and $8y$: $2(x)(8y) = 16xy$.
$z(x^2 + 16xy + 64y^2) = z(x + 8y)(x + 8y)$
$\qquad = z(x + 8y)^2$

64. $18c^3d^2 - 12c^3d - 24c^2d$
$\qquad = 6 \cdot 3c^2cdd - 6 \cdot 2c^2cd - 6 \cdot 4c^2d$
$\qquad = 6c^2d(3cd - 2c - 4)$

SECTION 6.7 REVIEW EXERCISES
Solving Quadratic Equations by Factoring
Solve each equation by factoring.

65. $8x(x - 6) = 0$

$\quad 8x = 0 \qquad$ or $\qquad x - 6 = 0$
$\quad x = 0 \qquad\qquad\qquad x = 6$

The solutions are 0 and 6.

66. $(4x - 7)(x + 1) = 0$

$\quad 4x - 7 = 0 \qquad$ or $\qquad x + 1 = 0$
$\quad 4x = 7 \qquad\qquad\qquad x = -1$
$\quad x = \dfrac{7}{4}$

The solutions are $\dfrac{7}{4}$ and -1.

67. $x^2 + 2x = 0$

$x(x + 2) = 0$
$\quad x = 0 \qquad$ or $\qquad x + 2 = 0$
$\qquad\qquad\qquad\qquad x + 2 - 2 = 0 - 2$
$\qquad\qquad\qquad\qquad\qquad x = -2$

The solutions are 0 and -2.

68. $\qquad x^2 - 9 = 0$

$(x + 3)(x - 3) = 0$
$\quad x + 3 = 0 \qquad$ or $\qquad x - 3 = 0$
$\quad x = -3 \qquad\qquad\qquad x = 3$

The solutions are -3 and 3.

69. $\qquad 144x^2 - 25 = 0$

$(12x + 5)(12x - 5) = 0$
$\quad 12x + 5 = 0 \qquad$ or $\qquad 12x - 5 = 0$
$\quad 12x + 5 - 5 = 0 - 5 \qquad\; 12x - 5 + 5 = 0 + 5$
$\quad\quad 12x = -5 \qquad\qquad\qquad 12x = 5$
$\quad\quad \dfrac{12x}{12} = \dfrac{-5}{12} \qquad\qquad\quad \dfrac{12x}{12} = \dfrac{5}{12}$
$\quad\quad x = -\dfrac{5}{12} \qquad\qquad\qquad x = \dfrac{5}{12}$

The solutions are $-\dfrac{5}{12}$ and $\dfrac{5}{12}$.

70. $a^2 - 7a + 12 = 0$

$(a - 3)(a - 4) = 0$
$\quad a - 3 = 0 \qquad$ or $\qquad a - 4 = 0$
$\quad a = 3 \qquad\qquad\qquad a = 4$

The solutions are 3 and 4.

71. $2t^2 + 28t + 98 = 0$

$2(t^2 + 14t + 49) = 0$
$2(t + 7)(t + 7) = 0$

$\quad t + 7 = 0$
$\quad\quad t = -7$

The solution is a repeated answer of -7.

72. $\qquad 2x - x^2 = -24$

$2x - x^2 + x^2 = -24 + x^2$
$\quad\quad 2x = x^2 - 24$
$\quad 2x - 2x = x^2 - 24 - 2x$
$\quad\quad 0 = x^2 - 2x - 24$
$\quad\quad 0 = (x + 4)(x - 6)$
$\quad x + 4 = 0 \qquad$ or $\qquad x - 6 = 0$
$\quad x = -4 \qquad\qquad\qquad x = 6$

The solutions are -4 and 6.

73. $5a^2 - 6a + 1 = 0$

$(5a - 1)(a - 1) = 0$
$\quad 5a - 1 = 0 \qquad$ or $\qquad a - 1 = 0$
$\quad 5a - 1 + 1 = 0 + 1 \qquad\; a - 1 + 1 = 0 + 1$
$\quad\quad 5a = 1 \qquad\qquad\qquad\; a = 1$
$\quad\quad \dfrac{5a}{5} = \dfrac{1}{5}$
$\quad\quad a = \dfrac{1}{5}$

The solutions are $\dfrac{1}{5}$ and 1.

74.

$$2p^3 = 2p(p+2)$$
$$2p^3 = 2p^2 + 4p$$
$$2p^3 - \mathbf{2p^2} = 2p^2 + 4p - \mathbf{2p^2}$$
$$2p^3 - 2p^2 = 4p$$
$$2p^3 - 2p^2 - \mathbf{4p} = 4p - \mathbf{4p}$$
$$2p^3 - 2p^2 - 4p = 0$$
$$2p(p^2 - p - 2) = 0$$
$$2p(p+1)(p-2) = 0$$

$$p = 0 \quad \text{or} \quad p+1 = 0 \quad \text{or} \quad p-2 = 0$$
$$\qquad\qquad p = -1 \qquad\qquad p = 2$$

The solutions are 0, -1 and 2.

SECTION 6.8 REVIEW EXERCISES
Applications of Quadratic Equations
75. SANDPAPER
Analyze
· Length is 2 inches longer than it is wide.
· Area is 99 sq. in.
· Find the width.
Assign
Let w = width of sandpaper in inches
$\quad w+2$ = length of sandpaper in inches
Form

The area of the rectangle	equals	the length	times	the width.
99	=	$w+2$	·	w

Solve
$$w(w+2) = 99$$
$$w^2 + 2w = 99$$
$$w^2 + 2w - \mathbf{99} = 99 - \mathbf{99}$$
$$w^2 + 2w - 99 = 0$$
$$(w+11)(h-9) = 0$$
$$w+11 = 0 \qquad \text{or} \quad w-9 = 0$$
$$w+11 - \mathbf{11} = 0 - \mathbf{11} \qquad\qquad w = 9$$
$$\cancel{w = -11}$$

State
The width is 9 inches.
The length is $(9+2) = 11$ inches.
Check
A recangle with dimensions 11 in by 9 in
has an area of 99 in^2. The results check.

76. CONSTRUCTION
Analyze
· Base is 3 m longer than twice its height.
· Area is 45 sq. m.
· Find the length of the base.
Assign
Let h = height of triangle in m
$\quad 2h+3$ = base of triangle in m
Form

The area of the triangle	equals	$\frac{1}{2}$	times	the base	times	the height.
45	=	$\frac{1}{2}$	·	$2h+3$	·	h

Solve
$$\frac{1}{2}h(2h+3) = 45$$
$$(\mathbf{2})\frac{1}{2}h(2h+3) = (\mathbf{2})45$$
$$h(2h+3) = 90$$
$$2h^2 + 3h = 90$$
$$2h^2 + 3h - \mathbf{90} = 90 - \mathbf{90}$$
$$2h^2 + 3h - 90 = 0$$
$$(2h+15)(h-6) = 0$$
$$2h+15 = 0 \qquad\qquad \text{or} \qquad h-6 = 0$$
$$2h+15 - \mathbf{15} = 0 - \mathbf{15} \qquad\qquad\qquad h = 6$$
$$2h = -15$$
$$\frac{2h}{2} = \frac{-15}{2}$$
$$\cancel{h = -\frac{15}{2}}$$

State
The height is 6 m.
The base is $2(6)+3 = 15$ m.

Check the Results
A triangle with dimensions 15 m by 6 m
has an area of 45 cm^2, and the base is twice
the height plus 3 m. The results check.

77. Fill in the blanks.

If we let $x =$ the first integer, then:

☐ two consecutive integers are x and $\boxed{x+1}$

☐ two consecutive even integers are x and $\boxed{x+2}$

☐ two consecutive odd integers are x and $\boxed{x+2}$

78. MUSIC AWARDS

Analyze

· Product of two consecutive even integers is 120. West has been nominated less.

· The first even integer represents number of times West was nominated.

· The second even integer represents number of times Jackson was nominated.

Assign

Let $x =$ West's number of nominations

$x + 2 =$ Jackson's number of nominations

Form

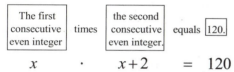

The first consecutive even integer	times	the second consecutive even integer.	equals	$\boxed{120.}$
x	·	$x+2$	=	120

Solve

$$x(x+2) = 120$$
$$x^2 + 2x = 120$$
$$x^2 + 2x - \mathbf{120} = 120 - \mathbf{120}$$
$$x^2 + 2x - 120 = 0$$
$$(x+12)(x-10) = 0$$

$x + 12 = 0$ or $x - 10 = 0$

$\cancel{x = -12}$ | $x = 10$

State

West has been nominated 10 times.

Jackson has been nominated $10 + 2 = 12$ times.

Check

Since 10 and 12 are consecutive even integers, and since $10 \cdot 12 = 120$, the results check.

79. TIGHTROPE WALKERS

Analyze

· The longer leg is $(x+7)$ meters long.

· The shorter leg is x meters long.

· The hypotenuse is $(x+8)$ meters long.

· How high above the ground is the platform?

Assign

Let $x =$ length of shorter leg in m

$x + 7 =$ length of longer leg in m

$x + 8 =$ length of hypotenuse in m

Form

The length of the shorter leg^2	plus	the length of the longer leg^2	equals	the length of the hypotenuse.^{2}
x^2	+	$(x+7)^2$	=	$(x+8)^2$

Solve

$$x^2 + (x+7)^2 = (x+8)^2$$
$$x^2 + x^2 + 14x + 49 = x^2 + 16x + 64$$
$$2x^2 + 14x + 49 = x^2 + 16x + 64$$
$$2x^2 + 14x + 49 - \mathbf{x^2} = x^2 + 16x + 64 - \mathbf{x^2}$$
$$x^2 + 14x + 49 = 16x + 64$$
$$x^2 + 14x + 49 - \mathbf{16x} = 16x + 64 - \mathbf{16x}$$
$$x^2 - 2x + 49 = 64$$
$$x^2 - 2x + 49 - \mathbf{64} = 64 - \mathbf{64}$$
$$x^2 - 2x - 15 = 0$$
$$(x+3)(x-5) = 0$$

$x + 3 = 0$ or $x - 5 = 0$

$\cancel{x = -3}$ | $x = 5$

State

The platform is 5 meters above the ground.

Check

Since $5^2 + 12^2 = 13^2$, the results check.

80. BALLOONING

$h = -16t^2 + 1,600,$ the camera hitting the ground means h is zero.

$$h = -16t^2 + 1,600$$
$$0 = -16t^2 + 1,600$$
$$0 = -16(t^2 - 100)$$
$$0 = -16(t+10)(t-10)$$

$t + 10 = 0$ or $t - 10 = 0$

$\cancel{t = -10}$ | $t = 10$

The camera will hit the ground after 10 sec.

CHAPTER 6 TEST

1. Fill in the blanks.
   a. The letters GCF stand for **greatest common factor**.
   b. To factor a polynomial means to express it as a **product** of two (or more) polynomials.
   c. The **Pythagorean** theorem provides a formula relating the lengths of the three sides of a right triangle.
   d. $y^2 - 25$ is a **difference** of two squares.
   e. The trinomial $x^2 + x - 6$ factors as the product of two **binomials**: $(x + 3)(x - 2)$.
2. a. Find the prime factorizations of 45 and 30.
   $$45 = 3^2 \cdot 5; \quad 30 = 2 \cdot 3 \cdot 5$$
   b. Find the greatest common factor of $45x^4$ and $30x^3$. $15x^3$

Factor. If an expression cannot be factored, write "prime."

3. $4x + 16 = 4(x + 4)$

4. $q^2 - 81 = (q + 9)(q - 9)$

5. $30a^2b^3 - 20a^3b^2 + 5ab = 5ab(6ab^2 - 4a^2b + 1)$

6. $x^2 + 9$ Prime

7. $2x(x + 1) + 3(x + 1) = (x + 1)(2x + 3)$

8. $x^2 + 4x + 3 = (x + 3)(x + 1)$

9. $-x^2 + 9x + 22 = -(x - 11)(x + 2)$

10. $60x^2 - 32x^3 + x^4 = x^2(x - 30)(x - 2)$

11. $9a - 9b + ax - bx = (a - b)(9 + x)$

12. $2a^2 + 5a - 12 = (2a - 3)(a + 4)$

13. $18x^2 + 60xy + 50y^2 = 2(3x + 5y)^2$

14. $x^3 + 8 = (x + 2)(x^2 - 2x + 4)$

15. $60m^8 - 45m^6 = 15m^6(4m^2 - 3)$

16. $3a^3 - 81 = 3(a - 3)(a^2 + 3a + 9)$

17. $16x^4 - 81 = (4x^2 + 9)(2x + 3)(2x - 3)$

18. $a^3 + 5a^2 + a + 5 = (a + 5)(a^2 + 1)$

19. $a^2 - 24 - 4a + 6a^3 = (a + 6)(a^3 - 4)$

20. $3d - 4 + 10d^2 = (5d + 4)(2d - 1)$

21. $8m^2 - 800 = 8(m + 10)(m - 10)$

22. $36n^2 - 84n + 49 = (6n - 7)^2$

23. $8r^2 - 14r + 3 = (4r - 1)(2r - 3)$

24. $t^2 - 6t + 10$ Prime

25. CHECKERS
$(25x^2 - 40x + 16) = (5x - 4)(5x - 4)$
The length of a side of the checkerboard is $(5x - 4)$ in.

26. Factor $x^2 - 3x - 54$. Show a check of your answer.
$$(x - 9)(x + 6) = x^2 + 6x - 9x - 54$$
$$= x^2 - 3x - 54$$

Solve each equation.

27. $(x + 3)(x - 2) = 0$

$\quad x + 3 = 0 \qquad$ or $\qquad x - 2 = 0$

$\qquad x = -3 \qquad\qquad\qquad x = 2$

The solutions are –3 and 2.

28. $\qquad\quad x^2 - 25 = 0$

$\quad (x + 5)(x - 5) = 0$

$\quad x + 5 = 0 \qquad$ or $\qquad x + 5 = 0$

$\qquad x = -5 \qquad\qquad\qquad x = 5$

The solutions are –5 and 5.

29. $\quad 36x^2 - 6x = 0$

$\quad 6x(6x - 1) = 0$

$\quad 6x = 0 \qquad$ or $\qquad 6x - 1 = 0$

$\quad x = 0 \qquad\qquad\qquad 6x - 1 + 1 = 0 + 1$

$\qquad\qquad\qquad\qquad\qquad 6x = 1$

$\qquad\qquad\qquad\qquad\qquad \dfrac{6x}{6} = \dfrac{1}{6}$

$\qquad\qquad\qquad\qquad\qquad x = \dfrac{1}{6}$

The solutions are 0 and $\dfrac{1}{6}$.

30. $\qquad\quad x^2 + 6x = -9$

$\quad x^2 + 6x + \mathbf{9} = -9 + \mathbf{9}$

$\quad x^2 + 6x + 9 = 0$

$\quad (x + 3)(x + 3) = 0$

$\qquad\quad x + 3 = 0$

$\qquad\qquad x = -3$

The solution repeats and is –3.

31.
$$6x^2 + x - 1 = 0$$
$$(3x-1)(2x+1) = 0$$

$$3x - 1 = 0 \quad \text{or} \quad 2x + 1 = 0$$
$$3x - 1 + 1 = 0 + 1 \qquad 2x + 1 - 1 = 0 - 1$$
$$3x = 1 \qquad\qquad 2x = -1$$
$$\frac{3x}{3} = \frac{1}{3} \qquad\qquad \frac{2x}{2} = -\frac{1}{2}$$
$$x = \frac{1}{3} \qquad\qquad x = -\frac{1}{2}$$

The solutions are $\dfrac{1}{3}$ and $-\dfrac{1}{2}$.

32.
$$a(a-7) = 18$$
$$a^2 - 7a = 18$$
$$a^2 - 7a - \mathbf{18} = 18 - \mathbf{18}$$
$$a^2 - 7a - 18 = 0$$
$$(a-9)(a+2) = 0$$

$$a - 9 = 0 \quad \text{or} \quad a + 2 = 0$$
$$a = 9 \qquad\qquad a = -2$$

The solutions are 9, –2.

33.
$$x^3 + 7x^2 = -6x$$
$$x^3 + 7x^2 + 6x = -6x + 6x$$
$$x^3 + 7x^2 + 6x = 0$$
$$x(x^2 + 7x + 6) = 0$$
$$x(x+1)(x+6) = 0$$

$$x = 0 \;\text{or} \quad x + 1 = 0 \quad \text{or} \quad x + 6 = 0$$
$$\qquad\qquad x + 1 - 1 = 0 - 1 \qquad x + 6 - 6 = 0 - 6$$
$$\qquad\qquad x = -1 \qquad\qquad x = -6$$

The solutions are 0, –1, –6.

34. DRIVING SAFETY
Analyze
· Length is 3 feet longer than its width.
· Area is 54 sq. ft.
· Find the dimensions.

Assign
Let w = width of blind spot in ft
$w + 3$ = length of blind spot in ft

Form

The area of the rectangle	equals	the length	times	the width.
54	=	$w + 3$	·	w

Solve
$$w(w+3) = 54$$
$$w^2 + 3w = 54$$
$$w^2 + 3w - \mathbf{54} = 54 - \mathbf{54}$$
$$w^2 + 3w - 54 = 0$$
$$(w+9)(w-6) = 0$$

$$w + 9 = 0 \quad \text{or} \quad w - 6 = 0$$
$$\cancel{w = -9} \qquad\qquad w = 6$$

State
The width of the blind spot is 6 feet.
The length of the blind spot is $6 + 3 = 9$ feet.

Check
A rectangle with dimensions 6 feet by 9 feet has an area of 54 ft^2, and the length is 3 feet longer than the width. The results check.

35. ROCKETRY
Analyze
· When the rocket hits the ground, its height will be 0 feet. To find the time that it takes for the ball to hit the ground, we set h equal to 0, and solve the quadratic equation for t.

Assign
Let t = time in seconds

Form
$$h = -16t^2 + 80t$$
$$0 = -16t^2 + 80t$$

Solve
$$-16t^2 + 80t = 0$$
$$-16t(t-5) = 0$$

$$-16t = 0 \quad \text{or} \quad t - 5 = 0$$
$$t = 0 \qquad\qquad t = 5$$

State
The equation has two solutions, 0 and 5. Since t represents time, and, in this case, time cannot be zero, we discard 0. The second solution, 5, indicates that the rockets hits the ground 5 seconds after being launched.

Check
Check this result by substituting 5 for t in $h = -16t^2 + 80t$. You should get $h = 0$.

36. ATV'S

Analyze
· Area is 33 sq. in.
· Height is 1 inch less than twice the base
· Find the lengths of the base and height.

Assign
Let x = length of base in inches
    $2x - 1$ = length of height in inches

Form

The area of the triangle	equals one half of	the base	times	the height.

$$33 \quad = \quad \frac{1}{2} \cdot x \quad \cdot \quad 2x - 1$$

Solve

$$\frac{1}{2}x(2x-1) = 33$$
$$2 \cdot \frac{1}{2}x(2x-1) = 2 \cdot 33$$
$$x(2x-1) = 66$$
$$2x^2 - x = 66$$
$$2x^2 - x - 66 = 66 - 66$$
$$2x^2 - x - 66 = 0$$
$$(2x+11)(x-6) = 0$$

$$2x+11 = 0 \qquad \text{or} \qquad x - 6 = 0$$
$$2x = -11 \qquad\qquad\qquad x = 6$$
$$x = \cancel{\frac{-11}{2}}$$

State
The base is 6 inches.
The height is $2(6) - 1 = 11$ inches.

Check
A triangle with dimensions 6 in by 11 in has an area of 33 in^2, and the height is 1 less than twice base. The results check.

37. CONSECUTIVE NUMBERS

Analyze
· Product of two consecutive positive integers is 156.
· What are the two integers?

Assign

Let $x = 1^{st}$ consecutive positive even integer
    $x + 1 = 2^{nd}$ consecutive positive even integer

Form

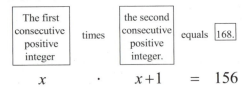

The first consecutive positive integer	times	the second consecutive positive integer.	equals	168.

$$x \qquad \cdot \qquad x+1 \qquad = \qquad 156$$

Solve

$$x(x+1) = 156$$
$$x^2 + x = 156$$
$$x^2 + x - 156 = 156 - 156$$
$$x^2 + x - 156 = 0$$
$$(x+13)(x-12) = 0$$

$$x + 13 = 0 \qquad \text{or} \qquad x - 12 = 0$$
$$\cancel{x = -13} \qquad\qquad x = 12$$

State
1st consecutive positive integer is 12.
2nd consecutive positive integer is 13.

Check
Since 12 and 13 are consecutive positive integers, and since $12 \cdot 13 = 156$, the results check.

38. RIGHT TRIANGLE

Analyze

· The longer leg is $(x-2)$ units long.
· The shorter leg is $(x-4)$ units long.
· The hypotenuse is x units long.
· Find the length of the hypotenuse.

Assign

Let $x =$ length of hypotenuse
$x-2 =$ length of longer leg
$x-4 =$ length of in shorter leg

Form

$$\boxed{\begin{array}{c}\text{The length of}\\\text{the shorter leg}\end{array}}^2 \text{ plus } \boxed{\begin{array}{c}\text{the length of}\\\text{the longer leg}\end{array}}^2 \text{ equals } \boxed{\begin{array}{c}\text{the length of}\\\text{the hypotenuse.}\end{array}}^2$$

$$(x-4)^2 \quad + \quad (x-2)^2 \quad = \quad x^2$$

Solve

$$(x-4)^2 + (x-2)^2 = x^2$$
$$x^2 - 8x + 16 + x^2 - 4x + 4 = x^2$$
$$2x^2 - 12x + 20 = x^2$$
$$2x^2 - 12x + 20 - x^2 = x^2 - x^2$$
$$x^2 - 12x + 20 = 0$$
$$(x-2)(x-10) = 0$$

$$x - 2 = 0 \qquad \text{or} \qquad x - 10 = 0$$
$$x = 2 \qquad\qquad\qquad x = 10$$

State

The solution of 2 will not work for the longer leg would be 0 and the shorter leg would be -2.

The length of the hypotenuse is 10 units long.

Check

Since $4^2 + 8^2 = 10^2$, the results check.

39. What is a quadratic equation? Give an example.

   A quadratic equation is an equation that can be written in the form $ax^2 + bx + c = 0$; $x^2 - 2x + 1 = 0$. (Answers may vary.)

40. If the product of two numbers is 0, what conclusion can be drawn about the numbers? **At least one of them is 0.**

CHAPTER 6 CUMULATIVE REVIEW

1. HEART RATES [Section 1.1]
   185 − 150 = 35. The difference is which is 35 beats/min.

2. Find the prime factorization of 250. [Section 1.2]
$$250 = 2 \cdot 125$$
$$= 2 \cdot 5 \cdot 25$$
$$= 2 \cdot 5 \cdot 5 \cdot 5$$
$$= 2 \cdot 5^3$$

3. Find the quotient: [Section 1.2]
$$\frac{16}{5} \div \frac{10}{3} = \frac{16}{5} \cdot \frac{3}{10}$$
$$= \frac{16 \cdot 3}{5 \cdot 10}$$
$$= \frac{\overset{1}{\cancel{2}} \cdot 8 \cdot 3}{5 \cdot \underset{1}{\cancel{2}} \cdot 5}$$
$$= \frac{24}{25}$$

4. Write as a decimal. [Section 1.3]
$$\frac{124}{125} = 0.992$$

5. Determine whether each statement is true or false. [Section 1.2]

   a. Every integer is a whole number. **False**
   b. Every integer is a rational number. **True**
   c. π is a real number. **True**

6. Which division is undefined, $\dfrac{0}{5}$ or $\dfrac{5}{0}$?

   [Section 1.6] $\quad \dfrac{5}{0}$

Evaluate each expression.

7. $3 + 2[-1 - 4(5)]$ [Section 1.7]
$$3 + 2[-1 - 4(5)] = 3 + 2[-1 - 20]$$
$$= 3 + 2[-1 + (-20)]$$
$$= 3 + 2(-21)$$
$$= 3 + (-42)$$
$$= -39$$

8. $\dfrac{|-25| - 2(-5)}{9 - 2^4}$ [Section 1.7]
$$\frac{|-25| - 2(-5)}{9 - 2^4} = \frac{25 + 10}{9 - 16}$$
$$= \frac{35}{9 + (-16)}$$
$$= \frac{35}{-7}$$
$$= -5$$

9. What is -3 cubed? [Section 1.7]
$$-3 \text{ cubed} = (-3)(-3)(-3)$$
$$= -27$$

10. What is the value of x twenty-dollar bills? [Section 1.8] $\quad \$20x$

11. Evaluate $\dfrac{-x - a}{y - b}$ for $x = -2$, $y = 1$, $a = 5$, and $b = 2$. [Section 1.8]
$$\frac{-x - a}{y - b} = \frac{-(-2) - 5}{1 - 2}$$
$$= \frac{2 + (-5)}{1 + (-2)}$$
$$= \frac{-3}{-1}$$
$$= 3$$

12. Identify the coefficient of each term in expression $8x^2 - x + 9$. [Section 1.8]
    The coefficients are 8, −1, 9.

Simplify each expression. [Section 1.9]

13. $-8y^2 - 5y^2 + 6 = (-8 - 5)y^2 + 6$
$$= -13y^2 + 6$$

14. $3z + 2(y - z) + y = 3z + 2y - 2z + y$
$$= (2 + 1)y + (3 - 2)z$$
$$= 3y + z$$

Solve each equation. [15-18 Section 2.2]

15. $-(3a+1)+a = 2$

$-3a-1+a = 2$

$(-3+1)a-1 = 2$

$-2a-1 = 2$

$-2a-1+\mathbf{1} = 2+\mathbf{1}$

$-2a = 3$

$\dfrac{-2a}{-2} = \dfrac{3}{-2}$

$a = -\dfrac{3}{2}$

16. $2-(4x+7) = 3+2(x+2)$

$2-4x-7 = 3+2x+4$

$-4x+(2-7) = 2x+(3+4)$

$-4x-5 = 2x+7$

$-4x-5-\mathbf{2x} = 2x+7-\mathbf{2x}$

$-6x-5 = 7$

$-6x-5+\mathbf{5} = 7+\mathbf{5}$

$-6x = 12$

$\dfrac{-6x}{-6} = \dfrac{12}{-6}$

$x = -2$

17. $\dfrac{3t-21}{2} = t-6$

$\mathbf{2}\cdot\dfrac{3t-21}{2} = \mathbf{2}(t-6)$

$3t-21 = 2t-12$

$3t-21-\mathbf{2t} = 2t-12-\mathbf{2t}$

$t-21 = -12$

$t-21+\mathbf{21} = -12+\mathbf{21}$

$t = 9$

18. $-\dfrac{1}{3}-\dfrac{x}{5} = \dfrac{3}{2}$

$\mathbf{30}\cdot\left(-\dfrac{1}{3}\right)+\mathbf{30}\cdot\left(-\dfrac{x}{5}\right) = \mathbf{30}\left(\dfrac{3}{2}\right)$

$-10-6x = 45$

$-6x-10+\mathbf{10} = 45+\mathbf{10}$

$-6x = 55$

$\dfrac{-6x}{-6} = \dfrac{55}{-6}$

$x = -\dfrac{55}{6}$

19. WATERMELON [Section 2.3]

92% of 270 is what number?

$0.92 \bullet 270 = \quad x$

$248.4 = x$

$248 \approx x$

The approximate weight is 248 lbs.

20. Find the distance traveled by a truck traveling for 5½ hours at a rate of 60 miles per hour. [Section 2.4]

$d = rt$

$= 60\cdot 5\dfrac{1}{2}$

$= 60\cdot 5.5$

$= 330$

The distance traveled is 330 miles.

21. What is the formula for simple interest? [Section 2.4] $I = Prt$

22. GEOMETRY TOOLS [Section 2.4]

$A = \pi r^2,\ \ \pi = 3.141592654$

$\approx 3.141592654\cdot 2^2$

$\approx 3.141592654\cdot 4$

≈ 12.56637061

≈ 12.6

The area is about 12.6 in^2.

23. Solve $A = P+Prt$ for t. [Section 2.4]

$A = P+Prt$

$A-\mathbf{P} = P+Prt-\mathbf{P}$

$A-P = Prt$

$\dfrac{A-P}{\mathbf{Pr}} = \dfrac{Prt}{\mathbf{Pr}}$

$\dfrac{A-P}{Pr} = t$

$t = \dfrac{A-P}{Pr}$

Chapter 6 Cumulative Review

24. HISTORY [Section 2.5]

Analyze

· Consecutive integers represent the order of all the presidents.

· Cleveland's two terms add up to be 46.

· What are the two numbers of his 2 terms?

Assign

Let x = number of his 1st term

$x+2$ = number of his 2nd term

Form

The 1st term	plus	the 2nd term	equals	the total of 46.
x	$+$	$x+2$	$=$	46

Solve

$$x+x+2=46$$
$$2x+2=46$$
$$2x+2-\mathbf{2}=46-\mathbf{2}$$
$$2x=44$$
$$\frac{2x}{\mathbf{2}}=\frac{44}{\mathbf{2}}$$
$$x=22$$

$$2^{nd}\text{ term}$$
$$x+2=22+2$$
$$=24$$

State

He was the 22th and 24th president.

Check

$22+24=46$

The solutions check.

25. ANTIQUE SHOWS [Section 2.5]

Analyze

· 17 week tour to 3 cities.

· Will stay in LA 2 weeks longer than in LV.

· Will stay in Dallas 1 week less than twice in LV.

· How long did they spend in each city?

Assign

Let x = number of weeks in Las Vegas

$x+2$ = number of weeks in Los Angles

$2x-1$ = number of weeks in Dallas

Form

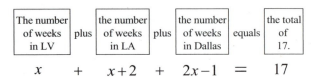

The number of weeks in LV	plus	the number of weeks in LA	plus	the number of weeks in Dallas	equals	the total of 17.
x	$+$	$x+2$	$+$	$2x-1$	$=$	17

Solve

$$x+(x+2)+(2x-1)=17$$
$$4x+1=17$$
$$4x+1-\mathbf{1}=17-\mathbf{1}$$
$$4x=16$$
$$\frac{4x}{\mathbf{4}}=\frac{16}{\mathbf{4}}$$
$$x=4$$

LA $\qquad$ Dallas

$x+2=\mathbf{4}+2 \qquad 2x-1=2(\mathbf{4})-1$

$=6 \qquad\qquad =7$

State

The show stays 4 weeks in Las Vegas, 6 weeks in Los Angles, and 7 weeks in Dallas.

Check

$4+6+7=17$

The solutions check.

26. PHOTOGRAPHIC CHEMICALS
[Section 2.6]

Analyze
· Weak 6 liter acetic acid solution is 5%.
· Strong acetic solution is 10%.
· A mixture of the two is to be 7%.
· How many liters of 10% solution are needed?

Assign
Let x = the amount of 10% solution in liters

Amount • Strength = Amount of pure acid

Form

	Amount	Strength	Amount of pure acid
Weak	6	0.05	**6(0.05)**
Strong	x	0.10	**0.10 x**
Mixture	$x+6$	0.07	**0.07(x + 6)**

The acetic acid in the 5% solution	plus	the acetic acid in the 10% solution	equals	the acetic acid in the 7% solution.
0.30	+	0.10x	=	0.07(x+6)

Solve

$$0.30 + 0.10x = 0.07(x + 6)$$
$$\mathbf{100}[0.30 + 0.10x] = \mathbf{100}[0.07(x + 6)]$$
$$100(0.30) + 100(0.10x) = 100[0.07(x + 6)]$$
$$30 + 10x = 7(x + 6)$$
$$30 + 10x = 7x + 42$$
$$30 + 10x - \mathbf{7x} = 7x + 42 - \mathbf{7x}$$
$$30 + 3x = 42$$
$$30 + 3x - \mathbf{30} = 42 - \mathbf{30}$$
$$3x = 12$$
$$\frac{3x}{3} = \frac{12}{3}$$
$$x = 4$$

State
4 liters of the 10% acetic acid solution will be needed.

Check
The acid in the 5% solution is 6(0.05), or 0.3 L.
The acid in the 10% solution is 4(0.10), or 0.4 L.
The acid in the 7% solution is 10(0.07), or 0.7 L.
Since the total was 0.3 L + 0.4 L = 0.7 L,
the solutions check.

27. DRIED FRUIT [Section 2.6]

Analyze
· Dried apple slices cost $4.60 per lb.
· Dried banana chips cost $3.40 per lb.
· 10 lb mixture sells for $4.00 per lb.
· How many pounds of each are needed?

Assign
Let x = the pounds of apples slices
$10 - x$ = the pounds of banana chips

Form

	Amount	Price	Value
Apple	x	4.60	**4.60x**
Banana	**10 − x**	3.40	**3.40(10−x)**
Mixture	10	4.00	4.00(10)

The value of the apple slices	plus	the value of the banana chips	equals	the value of the mixture.
4.60x	+	3.40(10−x)	=	4.00(10)

Solve

$$4.60x + 3.40(10 - x) = 4.00(10)$$
$$\mathbf{100}[4.60x + 3.40(10 - x)] = \mathbf{100}[4.00(10)]$$
$$100(4.60x) + 100[3.40(10 - x)] = 100[4.00(10)]$$
$$460x + 340(10 - x) = 400(10)$$
$$460x + 3,400 - 340x = 4,000$$
$$120x + 3,400 = 4,000$$
$$120x + 3,400 - \mathbf{3,400} = 4,000 - \mathbf{3,400}$$
$$120x = 600$$
$$\frac{120x}{120} = \frac{600}{120}$$
$$x = 5$$

State
5 lbs of apple slices will be needed.
5 lbs of banana chips will be needed.

Check
The value of the apple is 5(4.60), or $23.
The value of the banana is 5(3.40), or $17.
The value of the mixture is 10(4.00), or $40.
Since the total was $23 + $17 = $40,
the solutions check.

Chapter 6 Cumulative Review

28. Solve the inequality. Write the solution set in interval notation and graph it.
[Section 2.8]

$$-\frac{x}{2}+4>5$$

$$-\frac{x}{2}+4-4>5-4$$

$$-\frac{x}{2}>1$$

$$-2\cdot\left(-\frac{x}{2}\right)<-2\cdot1$$

$$x<-2$$

$$(-\infty,-2)$$

29. Is $(-2,5)$ a solution of $3x+2y=4$?
[Section 3.2]

$$3x+2y=4$$

$$3(-2)+2(5)\overset{?}{=}4$$

$$-6+10\overset{?}{=}4$$

$$4=4 \text{ True}$$

$(-2, 5)$ is a solution.

30. Graph: $y=2x-3$ [Section 3.2]

x	y	(x, y)
-2	$y=2x-3$ $=2(-2)-3$ $=-4-3$ $=-7$	$(-2,-7)$
0	$y=2x-3$ $=2(0)-3$ $=0-3$ $=-3$	$(0,-3)$
2	$y=2x-3$ $=2(2)-3$ $=4-3$ $=1$	$(2,1)$

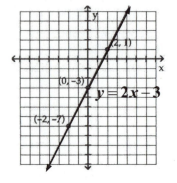

31. Is the graph of $x=3$ a vertical or horizontal line? [Section 3.3]
It is a vertical line.

32. If two lines are parallel, what can be said about their slopes? [Section 3.4]
They are the same.

33. ENCYCLOPEDIAS [Section 3.4]
Select any two points that lie on the line.
Let $(x_1, y_1) = (2005, 470{,}000)$
Let $(x_2, y_2) = (2010, 3{,}150{,}000)$

$$m=\frac{y_2-y_1}{x_2-x_1}$$

$$=\frac{3{,}150{,}000-470{,}000}{2010-2005}$$

$$=\frac{2{,}680{,}000}{5}$$

$$=536{,}000$$

The rate of change is an increase of 536,000 articles per year.

34. Find the slope and the y-intercept of the graph of $3x-3y=6$. [Section 3.5]

$$3x-3y=6$$

$$3x-3y-3x=6-3x$$

$$-3y=-3x+6$$

$$\frac{-3y}{-3}=\frac{-3x}{-3}+\frac{6}{-3}$$

$$y=x-2$$

The slope is 1. The y-intercept is $(0,-2)$.

35. Find an equation of the line passing through $(-2, 5)$ and $(-3,-2)$. Write the equation in slope-intercept form. [Section 3.5]

$$(-2, 5) \text{ and } (-3, -2)$$

$$(x_1,y_1) \text{ and } (x_2, y_2)$$

$$m=\frac{y_2-y_1}{x_2-x_1}$$

$$=\frac{-2-5}{-3-(-2)}$$

$$=\frac{-7}{-1}$$

$$=7$$

Passes through $(-2, 5)$, $m = 7$

$x_1 = -2$ and $y_1 = 5$

$$y - y_1 = m(x - x_1)$$
$$y - 5 = 7[x - (-2)]$$
$$y - 5 = 7(x + 2)$$
$$y - 5 = 7x + 14$$
$$y - 5 + \mathbf{5} = 7x + 14 + \mathbf{5}$$
$$y = 7x + 19 \quad \text{slope-intercept form}$$

36. Graph the line passing through $(-4, 1)$ that has slope -3 [Section 3.6]

$(-4, 1)$, $m = -3$

Start at $(-4, 1)$, go down 3, right 1

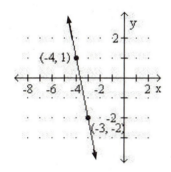

37. Graph: $8x + 4y \geq -24$ [Section 3.7]

Graph as $8x + 4y = -24$.

Boundary line is <u>solid</u>.

y-intercept:	x-intercept:
If $x = 0$,	If $y = 0$,
$8x + 4y = -24$	$8x + 4y = -24$
$\mathbf{0} + 4y = -24$	$8x + 2(\mathbf{0}) = -24$
$4y = -24$	$8x = -24$
$\dfrac{4y}{\mathbf{4}} = \dfrac{-24}{\mathbf{4}}$	$\dfrac{8x}{\mathbf{8}} = \dfrac{-24}{\mathbf{8}}$
$y = -6$	$x = -3$

The y-int is $(0, -6)$, and the x-int is $(-3, 0)$.

Select test point $(0, 0)$ and substitute into

$8x + 4y \geq -24$

$8(\mathbf{0}) + 4(\mathbf{0}) \overset{?}{\geq} -24$

$0 \geq -24$

True

The coordinates of every point on the same side of the line as $(0, 0)$ satisfy the inequality. To indicate this, we shade the half-plane that contains the test point $(0, 0)$.

$8x + 4y = -24$

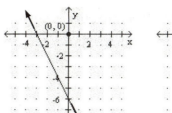 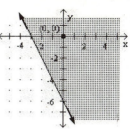

$8x + 4y \geq -24$

38. If $f(x) = 3x^2 - 2x + 1$, find $f(-2)$. [Section 3.8]

$$f(x) = 3x^2 - 2x + 1$$
$$f(-\mathbf{2}) = 3(-\mathbf{2})^2 - 2(-\mathbf{2}) + 1$$
$$= 3(4) + 4 + 1$$
$$= 12 + 5$$
$$= 17$$

Thus, $f(-2) = 17$

39. Is $\left(\dfrac{1}{2}, 1\right)$ a solution of the system

$\begin{cases} 4x - y = 1 \\ 2x + y = 2 \end{cases}$? [Section 4.1]

$4x - y = 1$	$2x + y = 2$
$4\left(\dfrac{1}{2}\right) - 1 \overset{?}{=} 1$	$2\left(\dfrac{1}{2}\right) + 1 \overset{?}{=} 2$
$2 - 1 \overset{?}{=} 1$	$1 + 1 \overset{?}{=} 2$
$1 = 1$	$2 = 2$
True	True

Since $(\frac{1}{2}, 3)$ does satisfy both equations, it is a solution of the system.

40. Solve the system $\begin{cases} 3x - 2y = 6 \\ x - y = 1 \end{cases}$ by graphing. [Section 4.1]

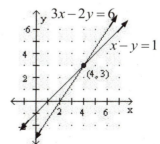

The solution is $(4, 3)$.

41. Solve the system $\begin{cases} y = -4x + 1 \\ 4x - y = 5 \end{cases}$ by substitution. [Section 4.2]

$4x - y = 5$ The 2nd equation
Substitute for y.

$4x - (-4x + 1) = 5$

$4x + 4x - 1 = 5$

$8x - 1 + 1 = 5 + 1$

$8x = 6$

$\dfrac{8x}{8} = \dfrac{6}{8}$

$x = \dfrac{3}{4}$

$y = -4x + 1$ The 1st equation

$y = -4\left(\dfrac{3}{4}\right) + 1$

$y = -3 + 1$

$y = -2$

The solution is $\left(\dfrac{3}{4}, -2\right)$.

Check: | Check:
$y = -4x + 1$ | $4x - y = 5$
$-2 \overset{?}{=} -4\left(\dfrac{3}{4}\right) + 1$ | $4\left(\dfrac{3}{4}\right) - (-2) \overset{?}{=} 5$
$-2 \overset{?}{=} -3 + 1$ | $3 + 2 \overset{?}{=} 5$
$-2 = -2$ | $5 = 5$
True | True

40. Solve the system $\begin{cases} 5a + 3b = -8 \\ 2a + 9b = 2 \end{cases}$ by elimination (addition) [Section 4.3]

$\begin{cases} 5a + 3b = -8 \\ 2a + 9b = 2 \end{cases}$

Eliminate b.

Multiply both sides of the 1st equation by -3.

$-15a - 9b = 24$
$\underline{2a + 9b = 2}$
$-13a = 26$

$\dfrac{-13a}{-13} = \dfrac{26}{-13}$

$a = -2$

$2a + 9b = 2$ The 2nd equation

$2(-2) + 9b = 2$

$-4 + 9b = 2$

$-4 + 9b + 4 = 2 + 4$

$9b = 6$

$\dfrac{9b}{9} = \dfrac{6}{9}$

$b = \dfrac{2}{3}$

The solution is $\left(-2, \dfrac{2}{3}\right)$.

43. FUNDRAISING [Section 4.4]

Analyze

- Rotary Club earned $356 by recycling a total of 14 tons of newspaper and cardboard.
- They were paid $31 per ton for newspaper.
- They were paid $18 per ton for cardboard.
- How many ton of each were collected?

Assign

Let x = the tons of newspaper
y = the tons of cardboard

Form

The number of tons of newsaper	plus	the number of tons of cardboard	equals	the total of 14 tons.
x	$+$	y	$=$	14

	Number $\cdot$	Value	= Total Value
Newspaper	x	31	**$31x$**
Cardboard	y	18	**$18y$**
		Total	**356**

The value of the newspaper	plus	the value of the cardboard	equals	the total value of $356.00.
$31x$	$+$	$18y$	$=$	356

Solve

$\begin{cases} x + y = 14 \\ 31x + 18y = 356 \end{cases}$

Eliminate y.

Multiply both sides of the 1st equation by -18.

$-18x - 18y = -252$
$\underline{31x + 18y = 356}$
$13x = 104$

$\dfrac{13x}{13} = \dfrac{104}{13}$

$x = 8$

$x + y = 14$ The 1st equation
$\mathbf{8} + y = 14$
$8 + y - \mathbf{8} = 14 - \mathbf{8}$
$y = 6$

State
They collected 8 tons of newspaper.
They collected 6 tons of cardboard.

Check
The solutions check.

44. Graph $\begin{cases} 4x + 3y \geq 12 \\ y < 4 \end{cases}$ [Section 4.5]

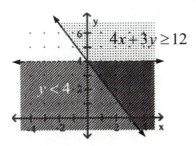

Simplify each expression. Write each answer without negative exponents.

45. [Section 5.1]

$-y^2(4y^3) = -4y^{2+3}$
$\qquad\qquad = -4y^5$

46. [Section 5.1]

$\dfrac{(x^2 y^5)^5}{(x^3 y)^2} = \dfrac{x^{2 \cdot 5} y^{5 \cdot 5}}{x^{3 \cdot 2} y^2}$

$\qquad = \dfrac{x^{10} y^{25}}{x^6 y^2}$

$\qquad = x^{10-6} y^{25-2}$

$\qquad = x^4 y^{23}$

47. [Section 5.2]

$\dfrac{b^5}{b^{-2}} = b^{5-(-2)}$

$\qquad = b^{5+2}$

$\qquad = b^7$

48. [Section 5.2]

$2x^0 = 2 \cdot 1$
$\qquad = 2$

49. Write in scientific notation. [Section 5.3]
Move the decimal 5 places to the right.

$0.00009011 = 9.011 \times 10^{-5}$

50. Write in scientific notation. [Section 5.3]
Move the decimal 6 places to the left.

$1,700,000 = 1.7 \times 10^6$

51. Find the degree of $7y^3 + 4y^2 + y + 3$. [Section 5.4]

The degree is 3.

52. Graph: $y = x^3 + 2$ [Section 5.4]

x	y	(x, y)
-3	$y = (-3)^3 + 2$ $= -27 + 2$ $= -25$	$(-3, -25)$
-2	$y = (-2)^3 + 2$ $= -8 + 2$ $= -6$	$(-2, -6)$
-1	$y = (-1)^3 + 2$ $= -1 + 2$ $= 1$	$(-1, 1)$
0	$y = (0)^3 + 2$ $= 0 + 2$ $= 2$	$(0, 2)$
1	$y = (1)^3 + 2$ $= 1 + 2$ $= 3$	$(1, 3)$
2	$y = (2)^3 + 2$ $= 8 + 2$ $= 10$	$(2, 10)$
3	$y = (3)^3 + 2$ $= 27 + 2$ $= 29$	$(3, 29)$

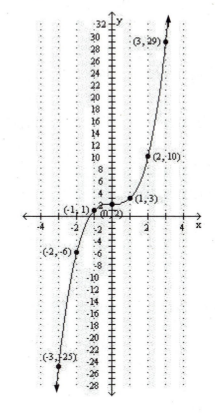

Chapter 6 Cumulative Review

Perform the operations.

53. $(x^2 - 3x + 8) - (3x^2 + x + 3)$ [Section 5.5]

$$= x^2 - 3x + 8 - 3x^2 - x - 3$$
$$= (1-3)x^2 + (-3-1)x + (8-3)$$
$$= -2x^2 - 4x + 5$$

54. $4b^3(2b^2 - 2b)$ [Section 5.6]

$$= \mathbf{4b^3}(2b^2) + \mathbf{4b^3}(-2b)$$
$$= 4(2)b^{3+2} + 4(-2)b^{3+1}$$
$$= 8b^5 - 8b^4$$

55. $(3x - 2)(x + 4)$ [Section 5.6]

$$= \mathbf{3x}(x) + \mathbf{3x}(4) - \mathbf{2}(x) - \mathbf{2}(4)$$
$$= 3x^2 + 12x - 2x - 8$$
$$= 3x^2 + 10x - 8$$

56. $(y - 6)^2$ [Section 5.7]

$$= y^2 + 2(y)(-6) + (-6)^2$$
$$= y^2 - 12y + 36$$

57. $\dfrac{12a^2b^2 - 8a^2b - 4ab}{4ab}$ [Section 5.8]

$$= \frac{12a^2b^2}{4ab} - \frac{8a^2b}{4ab} - \frac{4ab}{4ab}$$
$$= 3a^{2-1}b^{2-1} - 2a^{2-1}b^{1-1} - a^{1-1}b^{1-1}$$
$$= 3a^1b^1 - 2a^1b^0 - a^0b^0$$
$$= 3ab - 2a - 1$$

58. [Section 5.8]

$$\begin{array}{r} 2x+1 \\ x-3{\overline{\smash{\big)}\,2x^2-5x-3}} \\ \underline{-(2x^2-6x)} \\ x-3 \\ \underline{-(x-3)} \\ 0 \end{array}$$

59. PLAYPENS

a. perimeter of the playpen [Section 5.5]

$$P = 2l + 2w$$
$$= 2(x+3) + 2(x+1)$$
$$= 2x + 6 + 2x + 2$$
$$= (2+2)x + (6+2)$$
$$= 4x + 8$$

The perimeter of the playpen is $(4x + 8)$ in.

b. area of the floor of the playpen [Section 5.6]

$$A = lw$$
$$= (x+3)(x+1)$$
$$= x^2 + x + 3x + 3$$
$$= x^2 + (1+3)x + 3$$
$$= x^2 + 4x + 3$$

The area of the floor of the playpen is $(x^2 + 4x + 3)$ in^2.

c. volume of the playpen [Section 5.6]

$$V = lwh$$
$$= (x+3)(x+1)x$$
$$= (x^2 + x + 3x + 3)x$$
$$= (x^2 + (1+3)x + 3)x$$
$$= (x^2 + 4x + 3)x$$
$$= x^3 + 4x^2 + 3x$$

The volume of the playpen is $(x^3 + 4x^2 + 3x)$ in^3.

60. Find the GCF. [Section 6.1]

$$24x^5y^8 = 2 \cdot 2 \cdot 2 \cdot 3 \cdot x \cdot x \cdot x \cdot x \cdot x \quad \mathbf{y} \cdot y^7$$
$$54x^6y = 2 \cdot 3 \cdot 3 \cdot 3 \cdot x \cdot x \cdot x \cdot x \cdot x \cdot x \cdot \mathbf{y}$$

GCF is $\mathbf{2 \cdot 3 \cdot x^5 \cdot y} = 6x^5y$.

Factor.

61. [Section 6.1]

$$9b^3 - 27b^2 = \mathbf{9} \cdot \mathbf{b^2}b - \mathbf{9} \cdot 3\mathbf{b^2}$$
$$= 9b^2(b-3)$$

62. [Section 6.1]

$$ax + bx + ay + by = (ax + bx) + (ay + by)$$
$$= \mathbf{x}(a+b) + \mathbf{y}(a+b)$$
$$= (a+b)(x+y)$$

63. $u^2 - 3 + 2u$ [Section 6.2]

$$u^2 + 2u - 3$$

Two different sign factors of -3 whose sum is $+2$ are $+3$ and -1.
$$u^2 + 2u - 3 = (u+3)(u-1)$$

64. $10x^2 + x - 2$ [Section 6.3]

$$10x^2 + \underset{b}{1}x \underset{c}{-2}$$
$$ a$$

In $10x^2 + x - 2$, we have $a = 10$, $b = 1$ and $c = -2$. The key number is $ac = 10(-2) = -20$. We must find a factorization of -20 in which the sum of the factors is $b = 1$. Since the factors must have a negative product, their signs must be different. The pairs of factors are

−4 and 5.

$$10x^2 + x - 2 = 10x^2 - 4x + 5x - 2$$
$$= (10x^2 - 4x) + (5x - 2)$$
$$= 2x(5x - 2) + 1(5x - 2)$$
$$= (5x - 2)(2x + 1)$$

65. $4a^2 - 12a + 9$ [Section 6.4]

· The first term $4a^2$ is the square of **2a**.
· The last term 9 is the square of **−3**.
· The middle term is twice the product of
$2a$ and -3: $2(2a)(-3) = -12a$.

$$4a^2 - 12a + 9 = (2a - 3)(2a - 3)$$
$$= (2a - 3)^2$$

66. $9z^2 - 1$ [Section 6.4]

$$\begin{array}{cccccc} F^2 & - & L^2 & = (F+L) & (F-L) \\ \downarrow & & \downarrow & \downarrow \ \downarrow & \downarrow \ \downarrow \end{array}$$
$$(3z)^2 - 1^2 \ = (3z + 1) \ (3z - 1)$$

67. $t^3 - 8$ [Section 6.5]

$$= t^3 - 2^3$$
$$= (t - 2)(t^2 + t \cdot 2 + 2^2)$$
$$= (t - 2)(t^2 + 2t + 4)$$

68. $3a^2b^2 - 6a^2 - 3b^2 + 6$ [Section 6.6]

$$= 3(a^2b^2 - 2a^2 - b^2 + 2)$$
$$= 3[(a^2b^2 - 2a^2) + (-b^2 + 2)]$$
$$= 3[a^2(b^2 - 2) - 1(b^2 - 2)]$$
$$= 3(b^2 - 2)(a^2 - 1)$$
$$= 3(b^2 - 2)(a + 1)(a - 1)$$

Solve each equation.

69. $15s^2 - 20s = 0$ [Section 6.7]

$$15s^2 - 20s = 0$$
$$5s(3s - 4) = 0$$

$$5s = 0 \quad \text{or} \quad 3s - 4 = 0$$
$$s = 0 \quad\quad\quad\quad 3s = 4$$
$$s = \frac{4}{3}$$

The solutions are 0 and $\dfrac{4}{3}$.

70. $2x^2 - 5x = -2$ [Section 6.7]

$$2x^2 - 5x = -2$$
$$2x^2 - 5x + \mathbf{2} = -2 + \mathbf{2}$$
$$2x^2 - 5x + 2 = 0$$
$$(2x - 1)(x - 2) = 0$$

$$2x - 1 = 0 \quad \text{or} \quad x - 2 = 0$$
$$2x = 1 \quad\quad\quad\quad x = 2$$
$$x = \frac{1}{2}$$

The solutions are $\dfrac{1}{2}$ and 2.

71. $x^3 + 3x^2 + 2x = 0$ [Section 6.7]

$$x^3 + 3x^2 + 2x = 0$$
$$x(x^2 + 3x + 2) = 0$$
$$x(x + 1)(x + 2) = 0$$

$$x = 0 \quad \text{or} \quad x + 1 = 0 \quad \text{or} \quad x + 2 = 0$$
$$x = -1 \quad\quad\quad x = -2$$

The solutions are $0, -1,$ and -2.

72. **CAMPING**

Analyze
· Length is 3 inches longer than it is wide.
· Area is 108 sq. in.
· Find the dimensions.

Assign
Let w = width of surface in inches
$w + 3$ = length of surface in inches

Form

The area of the rectangle	equals	the length	times	the width.
108	=	$w + 3$	·	w

Solve

$$w(w + 3) = 108$$
$$w^2 + 3w = 108$$
$$w^2 + 3w - \mathbf{108} = 108 - \mathbf{108}$$
$$w^2 + 3w - 108 = 0$$
$$(w + 12)(h - 9) = 0$$

$$w + 12 = 0 \quad \text{or} \quad w - 9 = 0$$
$$\cancel{w = -12} \quad\quad\quad w = 9$$

State
The width is 9 inches.
The length is $(9 + 3) = 12$ inches.

Check
A recangle with dimensions 9 in by 12 in
has an area of 108 in^2. The solutions check.

Chapter 6 Cumulative Review

SECTION 7.1 STUDY SET
VOCABULARY

Fill in the blanks.

1. A quotient of two polynomials, such as $\frac{x^2+x}{x^2-3x}$, is called a **rational** expression.

3. Because of the division by 0, the expression $\frac{8}{0}$ is **undefined**.

CONCEPTS

5. When we simplify $\frac{x^2+5x}{4x+20}$, the result is $\frac{x}{4}$. These equivalent expressions have the same value for all real numbers, except $x=-5$. Show that they have the same value for $x=1$.

$$\frac{x^2+5x}{4x+20} \overset{?}{=} \frac{x}{4}$$

$$\frac{1^2+5(1)}{4(1)+20} \overset{?}{=} \frac{1}{4}$$

$$\frac{1+5}{4+20} \overset{?}{=} \frac{1}{4}$$

$$\frac{6}{24} \overset{?}{=} \frac{1}{4}$$

$$\frac{\overset{1}{\cancel{6}}}{\underset{1}{\cancel{6}}\cdot 4} \overset{?}{=} \frac{1}{4}$$

$$\frac{1}{4} = \frac{1}{4}$$

7. Simplify each expression, if possible.

a. $\frac{x-8}{x-8}$ **1**   b. $\frac{x-8}{8-x}$ **−1**

c. $\frac{x+8}{8+x}$ **1**   d. $\frac{x+8}{x}$ **Does not simplify**

NOTATION
Complete the solution to simplify the rational expression.

9. $\frac{x^2+2x+1}{x^2+4x+3} = \frac{(x+1)(x+1)}{(x+3)(x+1)}$

$$= \frac{(x+1)(\overset{1}{\cancel{x+1}})}{(x+3)(\underset{1}{\cancel{x+1}})}$$

$$= \frac{x+1}{x+3}$$

GUIDED PRACTICE
Evaluate each expression for $x=6$. See Example 1.

11. $\frac{x-2}{x-5} = \frac{6-2}{6-5}$

$$= \frac{4}{1}$$

$$= 4$$

13. $\frac{x^2-4x-12}{x^2+x-2} = \frac{6^2-4(6)-12}{6^2+6-2}$

$$= \frac{36-24-12}{36+6-2}$$

$$= \frac{0}{40}$$

$$= 0$$

15. $\frac{-x+1}{x^2-5x-6} = \frac{-(6)+1}{6^2-5(6)-6}$

$$= \frac{-6+1}{36-30-6}$$

$$= \frac{-5}{0}$$

Undefined

Evaluate each expression for $y=-3$. See Example 1.

17. $\frac{y+5}{3y-2} = \frac{(-3)+5}{3(-3)-2}$

$$= \frac{2}{-9-2}$$

$$= -\frac{2}{11}$$

19. $\frac{-y}{y^2-y+6} = \frac{-(-3)}{(-3)^2-(-3)+6}$

$$= \frac{3}{9+3+6}$$

$$= \frac{3}{18}$$

$$= \frac{1}{6}$$

21. $\dfrac{y^2+9}{9-y^2}=\dfrac{(-3)^2+9}{9-(-3)^2}$

$\qquad =\dfrac{9+9}{9-9}$

$\qquad =\dfrac{18}{0}$

Undefined

Find all real numbers for which the rational expression is undefined. See Example 2.

23. $\dfrac{15}{x-2}$ $\qquad\qquad x-2=0$

$\qquad\qquad\qquad\quad x-2+\mathbf{2}=0+\mathbf{2}$

$\qquad\qquad\qquad\qquad\quad x=2$

The rational expression is undefined for $x=2$.

25. $\dfrac{x+5}{8x}$ $\qquad\quad 8x=0$

$\qquad\qquad\qquad\quad x=0$

The rational expression is undefined for $x=0$.

27. $\dfrac{15x+2}{x^2+6}$

No matter what real number is substituted for x, the denominator, x^2+6, will not be equal to 0. Thus, no real numbers make the denominator undefined.

29. $\dfrac{x+1}{2x-1}$ $\qquad\qquad 2x-1=0$

$\qquad\qquad\qquad 2x-1+\mathbf{1}=0+\mathbf{1}$

$\qquad\qquad\qquad\qquad\quad 2x=1$

$\qquad\qquad\qquad\qquad\quad \dfrac{2x}{\mathbf{2}}=\dfrac{1}{\mathbf{2}}$

$\qquad\qquad\qquad\qquad\quad x=\dfrac{1}{2}$

The rational expression is undefined for $x=\dfrac{1}{2}$.

31. $\dfrac{x^2-6x}{9}$

Since the denominator does not contain a variable. the denominator can never be equal to 0. Thus, no real numbers make the expression undefined.

33. $\dfrac{30x}{x^2-36}$ $\qquad\qquad x^2-36=0$

$\qquad\qquad\qquad (x+6)(x-6)=0$

$\quad x+6=0 \qquad\qquad\quad x-6=0$

$x+6-\mathbf{6}=0-\mathbf{6} \quad\bigg|\quad x-6+\mathbf{6}=0+\mathbf{6}$

$\qquad x=-6 \qquad\qquad\qquad x=6$

The rational expression is undefined for $x=-6$ and $x=6$.

35. $\dfrac{15}{x^2+x-2}$ $\qquad\qquad x^2+x-2=0$

$\qquad\qquad\qquad (x+2)(x-1)=0$

$\quad x+2=0 \qquad\qquad\quad x-1=0$

$x+2-\mathbf{2}=0-\mathbf{2} \quad\bigg|\quad x-1+\mathbf{1}=0+\mathbf{1}$

$\qquad x=-2 \qquad\qquad\qquad x=1$

The rational expression is undefined for $x=-2$ and $x=1$.

37. $\dfrac{16}{20-x}$ $\qquad\qquad 20-x=0$

$\qquad\qquad\qquad 20-x+\mathbf{x}=0+\mathbf{x}$

$\qquad\qquad\qquad\qquad 20=x$

The rational expression is undefined for $x=20$.

Simplify each expression. See Example 3.

39. $\dfrac{45}{9a}=\dfrac{5\cdot\overset{1}{\cancel{9}}}{\cancel{9}\,a}$

$\qquad =\dfrac{5}{a}$

41. $\dfrac{6x^4}{4x^2}=\dfrac{\overset{1}{\cancel{2}}\cdot 3x^{4-2}}{\underset{1}{\cancel{2}}\cdot 2}$

$\qquad =\dfrac{3x^2}{2}$

Simplify. See Example 4.

43. $\dfrac{6x+3}{9}=\dfrac{3(2x+1)}{9}$

$\qquad =\dfrac{\overset{1}{\cancel{3}}(2x+1)}{3\cdot\underset{1}{\cancel{3}}}$

$\qquad =\dfrac{2x+1}{3}$

45. $\dfrac{x+3}{3x+9}=\dfrac{x+3}{3(x+3)}$

$\qquad =\dfrac{\overset{1}{\cancel{x+3}}}{3\,\underset{1}{\cancel{(x+3)}}}$

$\qquad =\dfrac{1}{3}$

47. $\dfrac{x^2-4}{x^2-6x+8}=\dfrac{(x+2)(x-2)}{(x-2)(x-4)}$

$\qquad =\dfrac{(x+2)\,\overset{1}{\cancel{(x-2)}}}{\underset{1}{\cancel{(x-2)}}(x-4)}$

$\qquad =\dfrac{x+2}{x-4}$

49. $\dfrac{x^2 + 5x + 4}{x^2 + 4x} = \dfrac{(x+1)(x+4)}{x(x+4)}$

$$= \dfrac{(x+1)\overset{1}{\cancel{(x+4)}}}{x\underset{1}{\cancel{(x+4)}}}$$

$$= \dfrac{x+1}{x}$$

Simplify. See Example 5.

51. $\dfrac{m^2 - 2mn + n^2}{7m^2 - 7n^2} = \dfrac{(m-n)(m-n)}{7(m+n)(m-n)}$

$$= \dfrac{\overset{1}{\cancel{(m-n)}}(m-n)}{7(m+n)\underset{1}{\cancel{(m-n)}}}$$

$$= \dfrac{m-n}{7(m+n)} \text{ or } \dfrac{m-n}{7m+7n}$$

53. $\dfrac{4b^2 + 4b + 1}{(2b+1)^3} = \dfrac{(2b+1)(2b+1)}{(2b+1)(2b+1)(2b+1)}$

$$= \dfrac{\overset{1}{\cancel{(2b+1)}}\,\overset{1}{\cancel{(2b+1)}}}{\underset{1}{\cancel{(2b+1)}}\,\underset{1}{\cancel{(2b+1)}}(2b+1)}$$

$$= \dfrac{1}{2b+1}$$

Simplify. See Example 6.

55. $\dfrac{10(c-3)+10}{3(c-3)+3} = \dfrac{10c - 30 + 10}{3c - 9 + 3}$

$$= \dfrac{10c - 20}{3c - 6}$$

$$= \dfrac{10(c-2)}{3(c-2)}$$

$$= \dfrac{10\,\overset{1}{\cancel{(c-2)}}}{3\,\underset{1}{\cancel{(c-2)}}}$$

$$= \dfrac{10}{3}$$

57. $\dfrac{6(x+3)-18}{3x-18} = \dfrac{6x + 18 - 18}{3x - 18}$

$$= \dfrac{6x}{3(x-6)}$$

$$= \dfrac{2 \cdot \overset{1}{\cancel{3}} \cdot x}{\underset{1}{\cancel{3}}(x-6)}$$

$$= \dfrac{2x}{x-6}$$

Simplify. See Example 7.

59. $\dfrac{2x-7}{7-2x} = \dfrac{2x-7}{-2x+7}$

$$= \dfrac{2x-7}{-1(2x-7)}$$

$$= \dfrac{\overset{1}{\cancel{2x-7}}}{-1\underset{1}{\cancel{(2x-7)}}}$$

$$= -1$$

61. $\dfrac{3-4t}{8t-6} = \dfrac{-4t+3}{8t-6}$

$$= \dfrac{-1(4t-3)}{2(4t-3)}$$

$$= \dfrac{-1\,\overset{1}{\cancel{(4t-3)}}}{2\,\underset{1}{\cancel{(4t-3)}}}$$

$$= -\dfrac{1}{2}$$

63. $\dfrac{2-a}{a^2 - a - 2} = \dfrac{-a+2}{(a-2)(a+1)}$

$$= \dfrac{-1(a-2)}{(a-2)(a+1)}$$

$$= \dfrac{-1\,\overset{1}{\cancel{(a-2)}}}{\underset{1}{\cancel{(a-2)}}(a+1)}$$

$$= -\dfrac{1}{a+1}$$

65. $\dfrac{25-5m}{m^2-25}=\dfrac{-5m+25}{(m+5)(m-5)}$

$\qquad =\dfrac{-5(m-5)}{(m+5)(m-5)}$

$\qquad =\dfrac{-5\;\overset{1}{\cancel{(m-5)}}}{(m+5)\;\underset{1}{\cancel{(m-5)}}}$

$\qquad =-\dfrac{5}{m+5}$

TRY IT YOURSELF

Simplify, if an expression cannot be simplified, write "Does not simplify."

67. $\dfrac{a^3-a^2}{a^4-a^3}=\dfrac{a^2(a-1)}{a^3(a-1)}$

$\qquad =\dfrac{a^{2-3}\;\overset{1}{\cancel{(a-1)}}}{\underset{1}{\cancel{(a-1)}}}$

$\qquad =\dfrac{a^{-1}}{1}$

$\qquad =\dfrac{1}{a}$

69. $\dfrac{4-x^2}{x^2-x-2}=\dfrac{-x^2+4}{x^2-x-2}$

$\qquad =\dfrac{-1(x^2-4)}{x^2-x-2}$

$\qquad =\dfrac{-1(x+2)(x-2)}{(x-2)(x+1)}$

$\qquad =\dfrac{-1(x+2)\;\overset{1}{\cancel{(x-2)}}}{\underset{1}{\cancel{(x-2)}}(x+1)}$

$\qquad =-\dfrac{x+2}{x+1}$

71. $\dfrac{6x-30}{5-x}=\dfrac{6x-30}{-x+5}$

$\qquad =\dfrac{6(x-5)}{-1(x-5)}$

$\qquad =\dfrac{6\;\overset{1}{\cancel{(x-5)}}}{-1\;\underset{1}{\cancel{(x-5)}}}$

$\qquad =-6$

73. $\dfrac{x^2+3x+2}{x^2+x-2}=\dfrac{(x+2)(x+1)}{(x+2)(x-1)}$

$\qquad =\dfrac{\overset{1}{\cancel{(x+2)}}(x+1)}{\underset{1}{\cancel{(x+2)}}(x-1)}$

$\qquad =\dfrac{x+1}{x-1}$

75. $\dfrac{15x^2y}{5xy^2}=\dfrac{3\cdot5\cdot x^{2-1}\cdot y^{1-2}}{5}$

$\qquad =\dfrac{3\cdot\overset{1}{\cancel{5}}\cdot x^1\cdot y^{-1}}{\underset{1}{\cancel{5}}}$

$\qquad =\dfrac{3x}{y}$

77. $\dfrac{x^8+9x^7}{9+x}=\dfrac{x^7(x+9)}{x+9}$

$\qquad =\dfrac{x^7\;\overset{1}{\cancel{(x+9)}}}{\underset{1}{\cancel{x+9}}}$

$\qquad =x^7$

79. $\dfrac{x(x-8)+16}{16-x^2}=\dfrac{x^2-8x+16}{-x^2+16}$

$\qquad =\dfrac{(x-4)(x-4)}{-1(x^2-16)}$

$\qquad =\dfrac{(x-4)(x-4)}{-1(x+4)(x-4)}$

$\qquad =\dfrac{(x-4)\;\overset{1}{\cancel{(x-4)}}}{-1(x+4)\;\underset{1}{\cancel{(x-4)}}}$

$\qquad =-\dfrac{x-4}{x+4}$ or $\dfrac{4-x}{4+x}$

81. $\dfrac{4c+4d}{d+c}=\dfrac{4(c+d)}{c+d}$

$\qquad =\dfrac{4\;\overset{1}{\cancel{(c+d)}}}{\underset{1}{\cancel{c+d}}}$

$\qquad =4$

Section 7.1

83. $\dfrac{3x^2-27}{2x^2-5x-3} = \dfrac{3(x^2-9)}{(2x+1)(x-3)}$

$\quad = \dfrac{3(x+3)(x-3)}{(2x+1)(x-3)}$

$\quad = \dfrac{3(x+3)\ \cancel{(x-3)}^{1}}{(2x+1)\ \cancel{(x-3)}_{1}}$

$\quad = \dfrac{3(x+3)}{2x+1}$ or $\dfrac{3x+9}{2x+1}$

85. $\dfrac{-3x^2+10x+77}{x^2-4x-21} = \dfrac{-(3x^2-10x-77)}{x^2-4x-21}$

$\quad = \dfrac{-(3x+11)(x-7)}{(x+3)(x-7)}$

$\quad = \dfrac{-(3x+11)\ \cancel{(x-7)}^{1}}{(x+3)\ \cancel{(x-7)}_{1}}$

$\quad = -\dfrac{3x+11}{x+3}$

87. $\dfrac{42c^3d}{18cd^3} = \dfrac{6\cdot 7\cdot c^{3-1}\cdot d^{1-3}}{3\cdot 6}$

$\quad = \dfrac{\cancel{6}^{1}\cdot 7\cdot c^2\cdot d^{-2}}{3\cdot \cancel{6}_{1}}$

$\quad = \dfrac{7c^2}{3d^2}$

89. $\dfrac{16a^2-1}{4a+4} = \dfrac{(4a+1)(4a-1)}{4(a+1)}$

$\quad$ Does not simplify.

91. $\dfrac{8u^2-2u-15}{4u^4+5u^3} = \dfrac{(2u-3)(4u+5)}{u^3(4u+5)}$

$\quad = \dfrac{(2u-3)\ \cancel{(4u+5)}^{1}}{u^3\ \cancel{(4u+5)}}$

$\quad = \dfrac{2u-3}{u^3}$

93. $\dfrac{(2x+3)^4}{4x^2+12x+9}$

$\quad = \dfrac{(2x+3)^4}{(2x+3)(2x+3)}$

$\quad = \dfrac{\cancel{(2x+3)}^{1}\ \cancel{(2x+3)}^{1}(2x+3)(2x+3)}{\cancel{(2x+3)}_{1}\ \cancel{(2x+3)}_{1}}$

$\quad = (2x+3)^2$

95. $\dfrac{6a+3(a+2)+12}{a+2} = \dfrac{6a+3a+6+12}{a+2}$

$\quad = \dfrac{9a+18}{a+2}$

$\quad = \dfrac{9(a+2)}{a+2}$

$\quad = \dfrac{9\ \cancel{(a+2)}^{1}}{\cancel{a+2}_{1}}$

$\quad = 9$

97. $\dfrac{15x-3x^2}{25y-5xy} = \dfrac{3x(5-x)}{5y(5-x)}$

$\quad = \dfrac{3x\ \cancel{(5-x)}^{1}}{5y\ \cancel{(5-x)}_{1}}$

$\quad = \dfrac{3x}{5y}$

99. $\dfrac{2x^2}{x+2}$ Does not simplify.

101. $\dfrac{18+2x}{x^2-81} = \dfrac{2(9+x)}{(x+9)(x-9)}$

$\quad = \dfrac{2\ \cancel{(x+9)}^{1}}{\cancel{(x+9)}_{1}(x-9)}$

$\quad = \dfrac{2}{x-9}$

103. ORGAN PIPES

$$n = \frac{512}{L}$$

$$= \frac{512}{6}$$

$$= \frac{\overset{1}{\cancel{2}} \cdot 256}{\cancel{2} \cdot 3}$$

$$= \frac{256}{3}$$

$$= 85\frac{1}{3}$$

The pipe will vibrate $85\frac{1}{3}$ times per sec.

105. MEDICAL DOSAGES

$$c = \frac{4t}{t^2 + 1} \text{ , at } 1:00$$

$$= \frac{4(1)}{1^2 + 1}$$

$$= \frac{4}{2}$$

$$= 2$$

At 1:00 there is 2 mg per liter.

$$c = \frac{4t}{t^2 + 1} \text{ , at } 2:00$$

$$= \frac{4(2)}{2^2 + 1}$$

$$= \frac{8}{5}$$

$$= 1.6$$

At 2:00 there is 1.6 mg per liter.

$$c = \frac{4t}{t^2 + 1} \text{ , at } 3:00$$

$$= \frac{4(3)}{3^2 + 1}$$

$$= \frac{12}{10}$$

$$= 1.2$$

At 3:00 there is 1.2 mg per liter.

WRITING

107-111. Answers will vary.

REVIEW

State each property using the variables *a*, *b*, and when necessary, *c*.

113. a. The associative property of addition

$$(a + b) + c = a + (b + c)$$

  b. The commutative property of multiplication

$$ab = ba$$

CHALLENGE PROBLEMS

Simplify each expression.

115. $\dfrac{(x^2 + 2x + 1)(x^2 - 2x + 1)}{(x^2 - 1)^2}$

$$= \frac{(x+1)(x+1)(x-1)(x-1)}{(x^2 - 1)(x^2 - 1)}$$

$$= \frac{(x+1)(x+1)(x-1)(x-1)}{(x+1)(x-1)(x+1)(x-1)}$$

$$= \frac{\overset{1}{\cancel{(x+1)}}\ \overset{1}{\cancel{(x+1)}}\ \overset{1}{\cancel{(x-1)}}\ \overset{1}{\cancel{(x-1)}}}{\underset{1}{\cancel{(x+1)}}\ \underset{1}{\cancel{(x-1)}}\ \underset{1}{\cancel{(x+1)}}\ \underset{1}{\cancel{(x-1)}}}$$

$$= 1$$

117. $\dfrac{x^3 - 27}{x^3 - 9x} = \dfrac{(x-3)(x^2 + 3x + 9)}{x(x^2 - 9)}$

$$= \frac{(x-3)(x^2 + 3x + 9)}{x(x+3)(x-3)}$$

$$= \frac{\overset{1}{\cancel{(x-3)}}(x^2 + 3x + 9)}{x(x+3)\underset{1}{\cancel{(x-3)}}}$$

$$= \frac{x^2 + 3x + 9}{x(x+3)}$$

119. $\dfrac{m^3 + 64}{m^3 + 4m^2 + 3m + 12}$

$$= \frac{(m+4)(m^2 - 4m + 16)}{(m^3 + 4m^2) + (3m + 12)}$$

$$= \frac{(m+4)(m^2 - 4m + 16)}{m^2(m+4) + 3(m+4)}$$

$$= \frac{(m+4)(m^2 - 4m + 16)}{(m+4)(m^2 + 3)}$$

$$= \frac{\overset{1}{\cancel{(m+4)}}(m^2 - 4m + 16)}{\underset{1}{\cancel{(m+4)}}(m^2 + 3)}$$

$$= \frac{m^2 - 4m + 16}{m^2 + 3}$$

SECTION 7.2 STUDY SET
VOCABULARY
Fill in the blanks.

1. The **reciprocal** of $\frac{x^2+6x+1}{10x}$ is $\frac{10x}{x^2+6x+1}$.

CONCEPTS
Fill in the blanks.

3. a. To multiply rational expressions, multiply their **numerators** and multiply their **denominators.** To divide two rational expressions, multiply the first by the **reciprocal** of the second. In symbols,

   b. $\frac{A}{B} \cdot \frac{C}{D} = \frac{AC}{BD}$ and $\frac{A}{B} \div \frac{C}{D} = \frac{A}{B} \cdot \frac{D}{C}$

Simplify each expression.

5. $\dfrac{y \cdot y \cdot y(15-y)}{y(y-15)(y+1)}$

$$\dfrac{\cancel{y} \cdot y \cdot y(15-y)}{\cancel{y} \cdot -1(15-y)(y+1)}$$

$$-\dfrac{y^2}{y+1}$$

7. Find the product of the rational expression and its reciprocal.

$$\frac{3}{x+2} \cdot \frac{x+2}{3} = \frac{\cancel{3}}{\cancel{x+2}} \cdot \frac{\cancel{x+2}}{\cancel{3}}$$

$$= 1$$

NOTATION

9. What units are common to the numerator and denominator?   **ft**

$$\frac{45 \text{ ft}}{1} \cdot \frac{1 \text{ yd}}{3 \text{ ft}}$$

GUIDED PRACTICE
Multiply, and then simplify, if possible. See Example 1.

11. $\dfrac{3}{7} \cdot \dfrac{y}{2} = \dfrac{3y}{14}$

13. $\dfrac{y+2}{y} \cdot \dfrac{3}{y^2} = \dfrac{(y+2)3}{y^{1+2}}$

$$= \dfrac{3(y+2)}{y^3} \text{ or } \dfrac{3y+6}{y^3}$$

15. $\dfrac{35n}{12} \cdot \dfrac{16}{7n^2} = \dfrac{35n \cdot 16}{12 \cdot 7n^2}$

$$= \dfrac{5 \cdot \cancel{7} \cdot \cancel{n} \cdot \cancel{4} \cdot 4}{3 \cdot \cancel{4} \cdot \cancel{7} \cdot \cancel{n} \cdot n}$$

$$= \dfrac{20}{3n}$$

17. $\dfrac{2x^2y}{3xy} \cdot \dfrac{3xy^2}{2} = \dfrac{2x^2y \cdot 3xy^2}{3xy \cdot 2}$

$$= \dfrac{\cancel{2} \cdot x^2 \cdot \cancel{3} \cdot \cancel{3} \cdot \cancel{x} \cdot y^2}{\cancel{3} \cdot \cancel{x} \cdot \cancel{3} \cdot \cancel{2}}$$

$$= x^2y^2$$

Multiply, and then simplify, if possible. See Example 2.

19. $\dfrac{x+5}{5} \cdot \dfrac{x}{x+5} = \dfrac{(x+5)x}{5(x+5)}$

$$= \dfrac{\cancel{(x+5)} \cdot x}{5 \cdot \cancel{(x+5)}}$$

$$= \dfrac{x}{5}$$

21. $\dfrac{2x+6}{x+3} \cdot \dfrac{3}{4x} = \dfrac{(2x+6)3}{(x+3)4x}$

$$= \dfrac{2 \cdot (x+3) \cdot 3}{(x+3) \cdot 2 \cdot 2 \cdot x}$$

$$= \dfrac{\cancel{2} \cdot \cancel{(x+3)} \cdot 3}{\cancel{(x+3)} \cdot \cancel{2} \cdot 2 \cdot x}$$

$$= \dfrac{3}{2x}$$

23. $\dfrac{(x+1)^2}{x+2} \cdot \dfrac{x+2}{x+1} = \dfrac{(x+1)^2(x+2)}{(x+2)(x+1)}$

$$= \dfrac{(x+1)\cancel{(x+1)}\cancel{(x+2)}}{\cancel{(x+2)}\cancel{(x+1)}}$$

$$= x+1$$

25. $\dfrac{x^2-x}{x} \cdot \dfrac{3x-6}{3-3x} = \dfrac{(x^2-x)(3x-6)}{x(3-3x)}$

$= \dfrac{x(x-1)\cdot 3(x-2)}{x\cdot 3(1-x)}$

$= \dfrac{x(x-1)\cdot 3(x-2)}{x\cdot 3(-x+1)}$

$= \dfrac{\cancel{x}\cdot(\cancel{x-1})\cdot\cancel{3}\cdot(x-2)}{\cancel{x}\cdot\cancel{3}\cdot -1\cdot(\cancel{x-1})}$

$= -(x-2) \text{ or } -x+2$

27. $\dfrac{x^2+x-6}{5x} \cdot \dfrac{5x-10}{x+3} = \dfrac{(x^2+x-6)(5x-10)}{5x(x+3)}$

$= \dfrac{(x+3)(x-2)\cdot 5(x-2)}{5x(x+3)}$

$= \dfrac{(\cancel{x+3})(x-2)\cdot\cancel{5}\cdot(x-2)}{\cancel{5}\cdot x\cdot(\cancel{x+3})}$

$= \dfrac{(x-2)^2}{x}$

29. $\dfrac{m^2-2m-3}{2m+4} \cdot \dfrac{m^2-4}{m^2+3m+2}$

$= \dfrac{(m^2-2m-3)(m^2-4)}{(2m+4)(m^2+3m+2)}$

$= \dfrac{(m-3)(m+1)(m+2)(m-2)}{2(m+2)(m+2)(m+1)}$

$= \dfrac{(m-3)\,(\cancel{m+1})\,(\cancel{m+2})\,(m-2)}{2(m+2)\,(\cancel{m+2})\,(\cancel{m+1})}$

$= \dfrac{(m-2)(m-3)}{2(m+2)}$

Multiply, and then simplify, if possible. See Example 3.

31. $7m\left(\dfrac{5}{m}\right) = \dfrac{7m}{1}\left(\dfrac{5}{m}\right)$

$= \dfrac{7m\cdot 5}{1\cdot m}$

$= \dfrac{7\cdot\cancel{m}\cdot 5}{1\cdot\cancel{m}}$

$= 35$

33. $15x\left(\dfrac{x+1}{5x}\right) = \dfrac{15x}{1}\left(\dfrac{x+1}{5x}\right)$

$= \dfrac{15x(x+1)}{1\cdot 5x}$

$= \dfrac{3\cdot\cancel{5}\cdot\cancel{x}\cdot(x+1)}{1\cdot\cancel{5}\cdot\cancel{x}}$

$= 3(x+1) \text{ or } 3x+3$

35. $12y\left(\dfrac{5y-8}{6y}\right) = \dfrac{12y}{1}\left(\dfrac{5y-8}{6y}\right)$

$= \dfrac{12y(5y-8)}{1\cdot 6y}$

$= \dfrac{2\cdot\cancel{6}\cdot\cancel{y}\cdot(5y-8)}{1\cdot\cancel{6}\cdot\cancel{y}}$

$= 2(5y-8) \text{ or } 10y-16$

37. $24\left(\dfrac{3a-5}{2a}\right) = \dfrac{24}{1}\left(\dfrac{3a-5}{2a}\right)$

$= \dfrac{24(3a-5)}{1\cdot 2a}$

$= \dfrac{\cancel{2}\cdot 12\cdot(3a-5)}{1\cdot\cancel{2}\cdot a}$

$= \dfrac{12(3a-5)}{a} \text{ or } \dfrac{36a-60}{a}$

Divide, and then simplify, if possible. See Example 4.

39. $\dfrac{2}{y} \div \dfrac{4}{3} = \dfrac{2}{y}\cdot\dfrac{3}{4}$

$= \dfrac{2\cdot 3}{y\cdot 2\cdot 2}$

$= \dfrac{\cancel{2}\cdot 3}{y\cdot 2\cdot\cancel{2}}$

$= \dfrac{3}{2y}$

41. $\dfrac{3a}{25} \div \dfrac{1}{5} = \dfrac{3a}{25}\cdot\dfrac{5}{1}$

$= \dfrac{3\cdot a\cdot 5}{5\cdot 5\cdot 1}$

$= \dfrac{3\cdot a\cdot\cancel{5}}{5\cdot\cancel{5}\cdot 1}$

$= \dfrac{3a}{5}$

42. $\dfrac{3y}{8} \div \dfrac{3}{2} = \dfrac{y}{4}$

- 403 -

43.
$$\frac{x^3}{18y} \div \frac{x}{6y} = \frac{x^3}{18y} \cdot \frac{6y}{x}$$
$$= \frac{x \cdot x^2 \cdot 6 \cdot y}{3 \cdot 6 \cdot y \cdot x}$$
$$= \frac{\overset{1}{\cancel{x}} \cdot x^2 \cdot \overset{1}{\cancel{6}} \cdot \overset{1}{\cancel{y}}}{3 \cdot \underset{1}{\cancel{6}} \cdot \underset{1}{\cancel{y}} \cdot \underset{1}{\cancel{x}}}$$
$$= \frac{x^2}{3}$$

45.
$$\frac{27p^4}{35q} \div \frac{9p}{21q} = \frac{27p^4}{35q} \cdot \frac{21q}{9p}$$
$$= \frac{3 \cdot 9 \cdot p \cdot p^3 \cdot 3 \cdot 7 \cdot q}{5 \cdot 7 \cdot q \cdot 9 \cdot p}$$
$$= \frac{3 \cdot \overset{1}{\cancel{9}} \cdot \overset{1}{\cancel{p}} \cdot p^3 \cdot 3 \cdot \overset{1}{\cancel{7}} \cdot \overset{1}{\cancel{q}}}{5 \cdot \underset{1}{\cancel{7}} \cdot \underset{1}{\cancel{q}} \cdot \underset{1}{\cancel{9}} \cdot \underset{1}{\cancel{p}}}$$
$$= \frac{9p^3}{5}$$

Divide, and then simplify, if possible. See Example 5.

47.
$$\frac{9a-18}{28} \div \frac{9a^3}{35} = \frac{(9a-18) \cdot 35}{28 \cdot 9a^3}$$
$$= \frac{9 \cdot (a-2) \cdot 5 \cdot 7}{4 \cdot 7 \cdot 9 \cdot a^3}$$
$$= \frac{\overset{1}{\cancel{9}} \cdot (a-2) \cdot 5 \cdot \overset{1}{\cancel{7}}}{4 \cdot \underset{1}{\cancel{7}} \cdot \underset{1}{\cancel{9}} \cdot a^3}$$
$$= \frac{5(a-2)}{4a^3} \text{ or } \frac{5a-10}{4a^3}$$

49.
$$\frac{x^2-4}{3x+6} \div \frac{2-x}{x+2} = \frac{x^2-4}{3x+6} \cdot \frac{x+2}{2-x}$$
$$= \frac{(x+2)(x-2)(x+2)}{3(x+2)(-x+2)}$$
$$= \frac{(x+2) \cdot \overset{1}{\cancel{(x-2)}} \cdot (x+2)}{3 \cdot \underset{1}{\cancel{(x+2)}} \cdot -1 \cdot \underset{1}{\cancel{(x-2)}}}$$
$$= -\frac{x+2}{3}$$

51.
$$\frac{x^2+7x}{5x-10} \div \frac{(x+7)^2}{15x-30}$$
$$= \frac{x(x+7)}{5(x-2)} \cdot \frac{15(x-2)}{(x+7)(x+7)}$$
$$= \frac{x\overset{1}{\cancel{(x+7)}} \cdot \overset{1}{\cancel{5}} \cdot 3 \overset{1}{\cancel{(x-2)}}}{\underset{1}{\cancel{5}}\underset{1}{\cancel{(x-2)}}\underset{1}{\cancel{(x+7)}}(x+7)}$$
$$= \frac{3x}{x+7}$$

53.
$$\frac{m^2+m-20}{m} \div \frac{4-m}{m} = \frac{m^2+m-20}{m} \cdot \frac{m}{4-m}$$
$$= \frac{(m+5) \cdot (m-4) \cdot m}{m \cdot (-m+4)}$$
$$= \frac{(m+5) \cdot \overset{1}{\cancel{(m-4)}} \cdot \overset{1}{\cancel{m}}}{\underset{1}{\cancel{m}} \cdot -1 \cdot \underset{1}{\cancel{(m-4)}}}$$
$$= -(m+5) \text{ or } -m-5$$

55.
$$\frac{t^2+5t-14}{t} \div \frac{t-2}{t} = \frac{t^2+5t-14}{t} \cdot \frac{t}{t-2}$$
$$= \frac{(t+7)(t-2)t}{t(t-2)}$$
$$= \frac{(t+7) \cdot \overset{1}{\cancel{(t-2)}} \cdot \overset{1}{\cancel{t}}}{\underset{1}{\cancel{t}} \cdot \underset{1}{\cancel{(t-2)}}}$$
$$= t+7$$

57.
$$\frac{x^2-2x-35}{3x^2+27x} \div \frac{3x^2+17x+10}{18x^2+12x}$$
$$= \frac{x^2-2x-35}{3x^2+27x} \cdot \frac{18x^2+12x}{3x^2+17x+10}$$
$$= \frac{(x^2-2x-35)(18x^2+12x)}{(3x^2+27x)(3x^2+17x+10)}$$
$$= \frac{(x-7)(x+5) \cdot 6x(3x+2)}{3x(x+9)(3x+2)(x+5)}$$
$$= \frac{(x-7)(x+5) \cdot 2 \cdot \overset{1}{\cancel{3}} \cdot \overset{1}{\cancel{x}} \overset{1}{\cancel{(3x+2)}}}{\underset{1}{\cancel{3}} \cdot \underset{1}{\cancel{x}}(x+9)\underset{1}{\cancel{(3x+2)}}\underset{1}{\cancel{(x+5)}}}$$
$$= \frac{2(x-7)}{x+9}$$

Divide, and then simplify, if possible. See Example 6.

59. $\dfrac{x^2-1}{3x-3} \div (x+1) = \dfrac{x^2-1}{3x-3} \div \dfrac{(x+1)}{1}$

$= \dfrac{x^2-1}{3x-3} \cdot \dfrac{1}{(x+1)}$

$= \dfrac{(x+1)(x-1)\cdot 1}{3(x-1)(x+1)}$

$= \dfrac{\cancel{(x+1)}\ \cancel{(x-1)}\cdot 1}{3\ \cancel{(x-1)}\ \cancel{(x+1)}}$

$= \dfrac{1}{3}$

61. $\dfrac{n^2-10n+9}{n-9} \div (n-1) = \dfrac{n^2-10n+9}{n-9} \div \dfrac{(n-1)}{1}$

$= \dfrac{n^2-10n+9}{n-9} \cdot \dfrac{1}{(n-1)}$

$= \dfrac{\cancel{(n-9)}\cdot \cancel{(n-1)}\cdot 1}{\cancel{(n-9)}\cdot \cancel{(n-1)}}$

$= 1$

63. $\dfrac{2r-3s}{12} \div (4r^2-12rs+9s^2)$

$= \dfrac{2r-3s}{12} \div \dfrac{(4r^2-12rs+9s^2)}{1}$

$= \dfrac{2r-3s}{12} \cdot \dfrac{1}{(4r^2-12rs+9s^2)}$

$= \dfrac{(2r-3s)\cdot 1}{12(4r^2-12rs+9s^2)}$

$= \dfrac{\cancel{(2r-3s)}\cdot 1}{12\ \cancel{(2r-3s)}\ (2r-3s)}$

$= \dfrac{1}{12(2r-3s)}$

65. $24n^2 \div \dfrac{18n^3}{n-1} = \dfrac{24n^2}{1} \div \dfrac{18n^3}{n-1}$

$= \dfrac{24n^2}{1} \cdot \dfrac{n-1}{18n^3}$

$= \dfrac{4\cdot\cancel{6}\cdot\cancel{n^2}\,(n-1)}{1\cdot 3\cdot\cancel{6}\cdot n\cdot\cancel{n}}$

$= \dfrac{4(n-1)}{3n}$

Complete each unit conversion. See Examples 7 and 8.

67. $\dfrac{150\text{ yd}}{1} \cdot \dfrac{3\text{ ft}}{1\text{ yd}} = \dfrac{150\ \cancel{\text{yd}}}{1} \cdot \dfrac{3\text{ ft}}{1\ \cancel{\text{yd}}}$

$= 450\text{ ft}$

69. $\dfrac{6\text{ pints}}{1} \cdot \dfrac{1\text{ gallon}}{8\text{ pints}} = \dfrac{\cancel{2}\cdot 3\ \cancel{\text{pints}}}{1} \cdot \dfrac{1\text{ gallon}}{\cancel{2}\cdot 4\ \cancel{\text{pints}}}$

$= \dfrac{3}{4}\text{ gallon}$

71. $\dfrac{30\text{ miles}}{1\text{ hr}} \cdot \dfrac{1\text{ hr}}{60\text{ min}} = \dfrac{\cancel{30}\text{ miles}}{1\ \cancel{\text{hr}}} \cdot \dfrac{1\ \cancel{\text{hr}}}{2\cdot \cancel{30}\text{ min}}$

$= \dfrac{1}{2}\text{ mile per min}$

73. $\dfrac{30\text{ meters}}{1\text{ sec}} \cdot \dfrac{60\text{ sec}}{1\text{ min}} = \dfrac{30\text{ meters}}{1\ \cancel{\text{sec}}} \cdot \dfrac{60\ \cancel{\text{sec}}}{1\text{ min}}$

$= 1{,}800\text{ meters per min}$

TRY IT YOURSELF
Perform the operations and simplify, if possible.

75. $\dfrac{b^2-5b+6}{b^2-10b+16} \div \dfrac{b^2+2b}{b^2-6b-16}$

$= \dfrac{b^2-5b+6}{b^2-10b+16} \cdot \dfrac{b^2-6b-16}{b^2+2b}$

$= \dfrac{(b^2-5b+6)(b^2-6b-16)}{(b^2-10b+16)(b^2+2b)}$

$= \dfrac{\cancel{(b-2)}(b-3)\ \cancel{(b-8)}\ \cancel{(b+2)}}{\cancel{(b-8)}\ \cancel{(b-2)}\ b\ \cancel{(b+2)}}$

$= \dfrac{b-3}{b}$

Section 7.2

77. $\dfrac{5x-5}{25} \cdot \dfrac{5}{(x-1)^3} = \dfrac{(5x-5)5}{25(x-1)^3}$

$$= \dfrac{\cancel{5} \cdot \cancel{5} \, \cancel{(x-1)}}{\cancel{5} \cdot \cancel{5} \, \cancel{(x-1)}(x-1)^2}$$

$$= \dfrac{1}{(x-1)^2}$$

79. $\dfrac{6a^2}{a^2+6a+9} \cdot \dfrac{(a+3)^4}{4a^5} = \dfrac{6a^2(a+3)^4}{(a^2+6a+9)4a^5}$

$$= \dfrac{6a^2(a+3)^4}{4a^5(a+3)^2}$$

$$= \dfrac{\cancel{2} \cdot 3 \cdot a^{2-5}(a+3)^{4-2}}{\cancel{2} \cdot 2}$$

$$= \dfrac{3 \cdot a^{-3}(a+3)^2}{2}$$

$$= \dfrac{3(a+3)^2}{2a^3}$$

81. $\dfrac{36c^2-49d^2}{3d^3} \div \dfrac{12c+14d}{d^4}$

$$= \dfrac{36c^2-49d^2}{3d^3} \cdot \dfrac{d^4}{12c+14d}$$

$$= \dfrac{(6c+7d)(6c-7d)d \cdot d^3}{3 \cdot d^3 \cdot 2(6c+7d)}$$

$$= \dfrac{\cancel{(6c+7d)}\,(6c-7d)d \cdot \cancel{d^4}}{3 \cdot \cancel{d^3} \cdot 2 \, \cancel{(6c+7d)}}$$

$$= \dfrac{d(6c-7d)}{6}$$

83. $10h\left(\dfrac{5h-3}{2h}\right) = \dfrac{10h}{1}\left(\dfrac{5h-3}{2h}\right)$

$$= \dfrac{\cancel{2} \cdot 5 \, \cancel{h}(5h-3)}{1 \cdot \cancel{2} \, \cancel{h}}$$

$$= 5(5h-3) \text{ or } 25h-15$$

85. $\dfrac{n^2-9}{n^2-3n} \div \dfrac{n+3}{n^2-n} = \dfrac{n^2-9}{n^2-3n} \cdot \dfrac{n^2-n}{n+3}$

$$= \dfrac{(n^2-9)(n^2-n)}{(n^2-3n)(n+3)}$$

$$= \dfrac{\cancel{(n+3)}\,\cancel{(n-3)}\,\cancel{n}(n-1)}{\cancel{n}\,\cancel{(n-3)}\,\cancel{(n+3)}}$$

$$= n-1$$

87. $\dfrac{10r^2s}{6rs^2} \cdot \dfrac{3r^3}{2rs} = \dfrac{10r^2s \cdot 3r^3}{6rs^2 \cdot 2rs}$

$$= \dfrac{\cancel{2} \cdot 5 \, \cancel{r} \cdot \cancel{r} \cdot \cancel{s} \cdot \cancel{3}\, r^3}{\cancel{2} \cdot \cancel{3} \, \cancel{r} \cdot s^2 \cdot 2 \, \cancel{r} \, \cancel{s}}$$

$$= \dfrac{5r^3}{2s^2}$$

89. $\dfrac{7}{3p^3} \cdot \dfrac{p+2}{p} = \dfrac{7(p+2)}{3p^{3+1}}$

$$= \dfrac{7(p+2)}{3p^4} \text{ or } \dfrac{7p+14}{3p^4}$$

91. $\dfrac{5x^2+13x-6}{x+3} \div \dfrac{5x^2-17x+6}{x-2}$

$$= \dfrac{5x^2+13x-6}{x+3} \cdot \dfrac{x-2}{5x^2-17x+6}$$

$$= \dfrac{(5x^2+13x-6)(x-2)}{(x+3)(5x^2-17x+6)}$$

$$= \dfrac{\cancel{(5x-2)}\,\cancel{(x+3)}(x-2)}{\cancel{(x+3)}\,\cancel{(5x-2)}(x-3)}$$

$$= \dfrac{x-2}{x-3}$$

93. $\dfrac{4x^2-12xy+9y^2}{x^3y^2} \cdot \dfrac{x^3y}{4x^2-9y^2}$

$$= \dfrac{(4x^2-12xy+9y^2)x^3y}{x^3y^2(4x^2-9y^2)}$$

$$= \dfrac{\cancel{x^3}\,\cancel{y}(2x-3y)\,\cancel{(2x-3y)}}{\cancel{x^3}\,\cancel{y}\,y(2x+3y)\,\cancel{(2x-3y)}}$$

$$= \dfrac{2x-3y}{y(2x+3y)}$$

95.
$$\frac{x-2}{x} \cdot \frac{2x}{2-x} = \frac{(x-2)2x}{x(2-x)}$$

$$= \frac{(x-2)2x}{x(-x+2)}$$

$$= \frac{\overset{1}{\cancel{(x-2)}} \cdot 2 \cdot \overset{1}{\cancel{x}}}{\underset{1}{\cancel{x}} \cdot -1 \underset{1}{\cancel{(x-2)}}}$$

$$= -2$$

LOOK ALIKES ...

97. a.
$$\frac{3x+6}{4} \cdot \frac{4x+8}{3} = \frac{3(x+2)}{4} \cdot \frac{4(x+2)}{3}$$

$$= \frac{\overset{1}{\cancel{3}} \cdot \overset{1}{\cancel{4}}(x+2)(x+2)}{\underset{1}{\cancel{3}} \cdot \underset{1}{\cancel{4}}}$$

$$= (x+2)^2$$

b.
$$\frac{3x+6}{4} \div \frac{4x+8}{3} = \frac{3(x+2)}{4} \div \frac{4(x+2)}{3}$$

$$= \frac{3(x+2)}{4} \cdot \frac{3}{4(x+2)}$$

$$= \frac{3 \cdot 3 \overset{1}{\cancel{(x+2)}}}{4 \cdot 4 \underset{1}{\cancel{(x+2)}}}$$

$$= \frac{9}{16}$$

99. a.
$$\frac{x^2-5x+6}{2x-4} \cdot \frac{2x-6}{x-2}$$

$$= \frac{(x-2)(x-3)}{2(x-2)} \cdot \frac{2(x-3)}{(x-2)}$$

$$= \frac{\overset{1}{\cancel{2}} \overset{1}{\cancel{(x-2)}}(x-3)(x-3)}{\underset{1}{\cancel{2}} \underset{1}{\cancel{(x-2)}}(x-2)}$$

$$= \frac{(x-3)^2}{x-2}$$

b.
$$\frac{x^2-5x+6}{2x-4} \div \frac{2x-6}{x-2}$$

$$= \frac{(x-2)(x-3)}{2(x-2)} \div \frac{2(x-3)}{(x-2)}$$

$$= \frac{(x-2)(x-3)}{2(x-2)} \cdot \frac{(x-2)}{2(x-3)}$$

$$= \frac{\overset{1}{\cancel{(x-2)}}(x-2)\overset{1}{\cancel{(x-3)}}}{2 \cdot 2 \underset{1}{\cancel{(x-2)}}\underset{1}{\cancel{(x-3)}}}$$

$$= \frac{x-2}{4}$$

APPLICATIONS
101. GEOMETRY
$$A = lw$$

$$= \left(\frac{x^2-7x}{5}\right)\left(\frac{x}{2x-14}\right)$$

$$= \frac{(x^2-7x)x}{5(2x-14)}$$

$$= \frac{x\,\overset{1}{\cancel{(x-7)}}\,x}{5 \cdot 2\,\underset{1}{\cancel{(x-7)}}}$$

$$= \frac{x^2}{10}$$

The area of the rectangle is $\dfrac{x^2}{10}$ ft^2.

103. TALKING
$$\frac{12{,}000 \text{ words}}{1 \text{ day}} \cdot \frac{365 \text{ days}}{1 \text{ year}}$$

$$= \frac{12{,}000 \text{ words}}{1 \overset{}{\cancel{\text{day}}}} \cdot \frac{365 \overset{1}{\cancel{\text{days}}}}{1 \text{ year}}$$

$$= 4{,}380{,}000 \text{ words per year}$$

The average speaks $4{,}380{,}000$ words per year.

105. NATURAL LIGHT
$$1 \text{ yd}^2 = 3 \text{ ft} \cdot 3 \text{ ft} = 9 \text{ ft}^2$$

$$\frac{72 \text{ ft}^2}{1} \cdot \frac{1 \text{ yd}^2}{9 \text{ ft}^2} = \frac{\overset{1}{\cancel{9}} \cdot 8 \; \overset{1}{\cancel{\text{ft}^2}}}{1} \cdot \frac{1 \text{ yd}^2}{\underset{1}{\cancel{9}} \; \underset{1}{\cancel{\text{ft}^2}}}$$

$$= 8 \text{ yd}^2$$

The average classroom has 8 yd^2 of windows.

107. BEARS

$$\frac{30 \text{ miles}}{1 \text{ hr}} \cdot \frac{1 \text{ hr}}{60 \text{ min}} = \frac{\overset{1}{\cancel{30}} \text{ miles}}{1 \cancel{\text{hr}}} \cdot \frac{1 \cancel{\text{hr}}}{2 \cdot \underset{1}{\cancel{30}} \text{ min}}$$

$$= \frac{1}{2} \text{ mile per min}$$

A bear can run $\frac{1}{2}$ mile per minute.

109. TV TRIVIA

$$\frac{160 \text{ acres}}{1} \cdot \frac{1 \text{ mi}^2}{640 \text{ acres}}$$

$$= \frac{\overset{1}{\cancel{160}} \text{ acres}}{1} \cdot \frac{1 \text{ mi}^2}{4 \cdot \underset{1}{\cancel{160}} \text{ acres}}$$

$$= \frac{1}{4} \text{ mi}^2$$

The farm was $\frac{1}{4}$ mi^2.

WRITING
111- 113. Answers will vary.

REVIEW
115. HARDWARE
Analyze
- A leg of the triangle is 8 inches long. (longer leg)
- The width of the shelf is the other leg. (shorter leg)
- The brace is 2 inches less than twice the width of the shelf. (hypotenuse)
- Find the width of the shelf and the length of the brace.

Assign
Let x = width of the shelf in inches
$2x - 2$ = length of the brace in inches

Form

$$\left(\begin{array}{c}\text{The length of}\\\text{the shorter leg}\end{array}\right)^2 \text{ plus } \left(\begin{array}{c}\text{the length of}\\\text{the longer leg}\end{array}\right)^2 \text{ equals } \left(\begin{array}{c}\text{the length of}\\\text{the hypotenuse.}\end{array}\right)^2$$

$$x^2 \quad + \quad 8^2 \quad = \quad (2x-2)^2$$

Solve

$$x^2 + 8^2 = (2x-2)^2$$
$$x^2 + 64 = 4x^2 - 8x + 4$$
$$x^2 + 64 - \boldsymbol{x^2} = 4x^2 - 8x + 4 - \boldsymbol{x^2}$$
$$64 = 3x^2 - 8x + 4$$
$$64 - \boldsymbol{64} = 3x^2 - 8x + 4 - \boldsymbol{64}$$
$$0 = 3x^2 - 8x - 60$$
$$0 = (3x + 10)(x - 6)$$

$$3x + 10 = 0 \qquad \text{or} \qquad x - 6 = 0$$
$$ \qquad\qquad\qquad x = 6$$
$$x = \cancel{-\frac{10}{3}}$$

State
The width of the shelf is 6 inches.
The length of the brace is $2(6) - 2 = 10$ inches.

Check
Since $6^2 + 8^2 = 10^2$, the answers check.

CHALLENGE PROBLEMS
Perform the operations and simplify, if possible.

117. $\dfrac{c^3 - 2c^2 + 5c - 10}{c^2 - c - 2} \cdot \dfrac{c^3 + c^2 - 5c - 5}{c^4 - 25}$

$$= \frac{c^2(c-2) + 5(c-2)}{(c-2)(c+1)} \cdot \frac{c^2(c+1) - 5(c+1)}{(c^2-5)(c^2+5)}$$

$$= \frac{\overset{1}{\cancel{(c-2)}}\,\overset{1}{\cancel{(c^2+5)}}\,\overset{1}{\cancel{(c+1)}}\,\overset{1}{\cancel{(c^2-5)}}}{\underset{1}{\cancel{(c-2)}}\,\underset{1}{\cancel{(c+1)}}\,\underset{1}{\cancel{(c^2-5)}}\,\underset{1}{\cancel{(c^2+5)}}}$$

$$= 1$$

119. $\dfrac{-x^3 + x^2 + 6x}{3x^3 + 21x^2} \div \left(\dfrac{2x+4}{3x^2} \div \dfrac{2x+14}{x^2 - 3x}\right)$

$$= \frac{-x^3 + x^2 + 6x}{3x^3 + 21x^2} \div \left(\frac{2x+4}{3x^2} \cdot \frac{x^2 - 3x}{2x+14}\right)$$

$$= \frac{-x(x^2 - x - 6)}{3x^2(x+7)} \div \left(\frac{2(x+2)x(x-3)}{3x^2 \cdot 2(x+7)}\right)$$

$$= \frac{-x(x-3)(x+2)}{3x^2(x+7)} \cdot \left(\frac{2 \cdot 3x^2(x+7)}{2x(x+2)(x-3)}\right)$$

$$= -\frac{\overset{1}{\cancel{2}} \cdot \overset{1}{\cancel{3}} \cdot \overset{1}{\cancel{x}} \cdot \overset{1}{\cancel{x^2}}\,\cancel{(x-3)}\,\cancel{(x+2)}\,\cancel{(x+7)}}{\underset{1}{\cancel{2}} \cdot \underset{1}{\cancel{3}}\,\underset{1}{\cancel{x}} \cdot \underset{1}{\cancel{x^2}}\,\cancel{(x+7)}\,\cancel{(x+2)}\,\cancel{(x-3)}}$$

$$= -1$$

SECTION 7.3 STUDY SET
VOCABULARY
Fill in the blanks.

1. The rational expressions $\frac{7}{6n}$ and $\frac{n+1}{6n}$ have the common **denominator** $6n$.

3. To **build** a rational expression, we multiply it by a form of 1. For example, $\frac{2}{n^2} \cdot \frac{8n}{8n} = \frac{16n}{8n^3}$.

CONCEPTS
Fill in the blanks.

5. To add or subtract rational expressions that have the same denominator, add or subtract the **numerators**, and write the sum or difference over the common **denominator**. In symbols,

$$\frac{A}{D} + \frac{B}{D} = \frac{A+B}{D} \quad \text{and} \quad \frac{A}{D} - \frac{B}{D} = \frac{A-B}{D}$$

7. The sum of two rational expressions is $\frac{4x+4}{5(x+1)}$. Factor the numerator and then simplify the result. $\quad \dfrac{4}{5}$

9. Consider the following factorizations.
$$18x - 36 = 2 \cdot 3 \cdot 3 \cdot (x - 2)$$
$$3x - 6 = 3(x - 2)$$

   a. What is the greatest number of times the factor 3 appears in any one factorization? **Twice**

   b. What is the greatest number of times the factor $x - 2$ appears in any one factorization? **Once**

NOTATION
Complete the solution.

11. $\dfrac{6a-1}{4a+1} + \dfrac{2a+3}{4a+1} = \dfrac{6a-1+(2a+\boxed{3})}{\boxed{4a+1}}$

$$= \dfrac{8a+\boxed{2}}{4a+1}$$

$$= \dfrac{2\boxed{4a+1}}{4a+1}$$

$$= \boxed{2}$$

GUIDED PRACTICE
Add and simplify the result, if possible. See Example 1.

13. $\dfrac{9}{x} + \dfrac{2}{x} = \dfrac{9+2}{x}$

$$= \dfrac{11}{x}$$

15. $\dfrac{x}{18} + \dfrac{5}{18} = \dfrac{x+5}{18}$

17. $\dfrac{a-5}{3a^3} + \dfrac{5}{3a^3} = \dfrac{a-5+5}{3a^3}$

$$= \dfrac{a}{3a^3}$$

$$= \dfrac{\cancel{a}^{1}}{3\cancel{a} \cdot a^2}_{1}$$

$$= \dfrac{1}{3a^2}$$

19. $\dfrac{x+3}{2y} + \dfrac{x+5}{2y} = \dfrac{x+3+x+5}{2y}$

$$= \dfrac{2x+8}{2y}$$

$$= \dfrac{\cancel{2}^{1}(x+4)}{\cancel{2} \cdot y}_{1}$$

$$= \dfrac{x+4}{y}$$

Add and simplify the result, if possible. See Example 2.

21. $\dfrac{2}{r^2-3r-10} + \dfrac{r}{r^2-3r-10} = \dfrac{r+2}{r^2-3r-10}$

$$= \dfrac{\cancel{r+2}^{1}}{\cancel{(r+2)}(r-5)}_{1}$$

$$= \dfrac{1}{r-5}$$

23. $\dfrac{3x-5}{x-2} + \dfrac{6x-13}{x-2} = \dfrac{3x-5+6x-13}{x-2}$

$$= \dfrac{9x-18}{x-2}$$

$$= \dfrac{9\cancel{(x-2)}^{1}}{\cancel{(x-2)}}_{1}$$

$$= 9$$

Subtract and simplify the result, if possible. See Example 3.

Section 7.3

25. $\dfrac{2x}{25} - \dfrac{x}{25} = \dfrac{2x-x}{25}$

   $\qquad\qquad = \dfrac{x}{25}$

27. $\dfrac{m-1}{6m^2} - \dfrac{5}{6m^2} = \dfrac{m-1-5}{6m^2}$

   $\qquad\qquad\quad = \dfrac{m-6}{6m^2}$

29. $\dfrac{t}{t^2+t-2} - \dfrac{1}{t^2+t-2} = \dfrac{t-1}{t^2+t-2}$

   $\qquad\qquad\qquad\quad = \dfrac{\cancel{t-1}^{\,1}}{\cancel{(t-1)}\,(t+2)}$

   $\qquad\qquad\qquad\quad = \dfrac{1}{t+2}$

31. $\dfrac{11w+6}{3w(w-9)} - \dfrac{11w}{3w(w-9)} = \dfrac{11w+6-11w}{3w(w-9)}$

   $\qquad\qquad\qquad\qquad = \dfrac{2 \cdot \cancel{3}^{\,1}}{\cancel{3} \cdot w(w-9)}$

   $\qquad\qquad\qquad\qquad = \dfrac{2}{w(w-9)}$

Subtract and simplify the result, if possible. See Example 4.

33. $\dfrac{3y-2}{2y+6} - \dfrac{2y-5}{2y+6} = \dfrac{3y-2-(2y-5)}{2y+6}$

   $\qquad\qquad\qquad\quad = \dfrac{3y-2-2y+5}{2y+6}$

   $\qquad\qquad\qquad\quad = \dfrac{y+3}{2y+6}$

   $\qquad\qquad\qquad\quad = \dfrac{\cancel{y+3}^{\,1}}{2\,\cancel{(y+3)}}$

   $\qquad\qquad\qquad\quad = \dfrac{1}{2}$

35. $\dfrac{6x^2}{3x+2} - \dfrac{11x+10}{3x+2} = \dfrac{6x^2-(11x+10)}{3x+2}$

   $\qquad\qquad\qquad\qquad = \dfrac{6x^2-11x-10}{3x+2}$

   $\qquad\qquad\qquad\qquad = \dfrac{(2x-5)\,\cancel{(3x+2)}^{\,1}}{\cancel{3x+2}_{\,1}}$

   $\qquad\qquad\qquad\qquad = 2x-5$

37. $\dfrac{6x-5}{3xy} - \dfrac{3x-5}{3xy} = \dfrac{6x-5-(3x-5)}{3xy}$

   $\qquad\qquad\qquad\quad = \dfrac{6x-5-3x+5}{3xy}$

   $\qquad\qquad\qquad\quad = \dfrac{3x}{3xy}$

   $\qquad\qquad\qquad\quad = \dfrac{\cancel{3}^{\,1}\,\cancel{x}^{\,1}}{\cancel{3}\cdot\cancel{x}\cdot y}$

   $\qquad\qquad\qquad\quad = \dfrac{1}{y}$

39. $\dfrac{2-p}{p^2-p} - \dfrac{-p+2}{p^2-p} = \dfrac{2-p-(-p+2)}{p^2-p}$

   $\qquad\qquad\qquad\qquad = \dfrac{2-p+p-2}{p^2-p}$

   $\qquad\qquad\qquad\qquad = \dfrac{0}{p^2-p}$

   $\qquad\qquad\qquad\qquad = 0$

Find the LCD of each pair of rational expressions. See Example 5.

41. $\dfrac{1}{2x}, \dfrac{9}{6x}$

   $2x = 2 \cdot x$

   $6x = 2 \cdot 3 \cdot x$

   $\text{LCD} = 2 \cdot 3 \cdot x$

   $\qquad\quad = 6x$

43. $\dfrac{33}{15a^3}, \dfrac{9}{10a}$

   $15a^3 = 3 \cdot 5 \cdot a \cdot a \cdot a$

   $10a = 2 \cdot 5 \cdot a$

   $\text{LCD} = 2 \cdot 3 \cdot 5 \cdot a \cdot a \cdot a$

   $\qquad\quad = 30a^3$

45. $\dfrac{35}{3a^2b}, \dfrac{23}{a^2b^3}$

   $3a^2b = 3 \cdot a \cdot a \cdot b$

   $a^2b^3 = a \cdot a \cdot b \cdot b \cdot b$

   $\text{LCD} = 3 \cdot a \cdot a \cdot b \cdot b \cdot b$

   $\qquad\quad = 3a^2b^3$

47. $\dfrac{8}{c}, \dfrac{8-c}{c+2}$

$c = c$

$c + 2 = c + 2$

$LCD = c(c+2)$

Find the LCD of each pair of rational expressions. See Example 6.

49. $\dfrac{3x+1}{3x-3}, \dfrac{3x}{4x-4}$

$3x - 3 = 3(x-1)$

$4x - 4 = 4(x-1)$

$LCD = 3 \cdot 4 \cdot (x-1)$

$\qquad = 12(x-1)$

51. $\dfrac{b-9}{4b+8}, \dfrac{b}{6}$

$4b + 8 = 2 \cdot 2 \cdot (b+2)$

$6 = 2 \cdot 3$

$LCD = 2 \cdot 2 \cdot 3 \cdot (b+2)$

$\qquad = 12(b+2)$

53. $\dfrac{6-k}{2k+4}, \dfrac{11}{8k}$

$2k + 4 = 2 \cdot (k+2)$

$8k = 2 \cdot 2 \cdot 2 \cdot k$

$LCD = 2 \cdot 2 \cdot 2 \cdot k(k+2)$

$\qquad = 8k(k+2)$

55. $\dfrac{-2x}{x^2-1}, \dfrac{5x}{x+1}$

$x^2 - 1 = (x+1)(x-1)$

$x + 1 = x + 1$

$LCD = (x+1)(x-1)$

57. $\dfrac{4x-5}{x^2-4x-5}, \dfrac{3x+1}{x^2-25}$

$x^2 - 4x - 5 = (x+1)(x-5)$

$x^2 - 25 = (x+5)(x-5)$

$LCD = (x+1)(x+5)(x-5)$

58.

59. $\dfrac{5n^2-16}{2n^2+13n+20}, \dfrac{3n^2}{n^2+8n+16}$

$2n^2 + 13n + 20 = (2n+5)(n+4)$

$n^2 + 8n + 16 = (n+4)(n+4)$

$LCD = (2n+5)(n+4)(n+4)$

$\qquad = (2n+5)(n+4)^2$

Build each rational expression into an equivalent expression with the given denominator. See Example 7.

61. $\dfrac{5}{r} = \dfrac{5}{r} \cdot \dfrac{\mathbf{10}}{\mathbf{10}}$

$\qquad = \dfrac{50}{10r}$

63. $\dfrac{8}{x} = \dfrac{8}{x} \cdot \dfrac{\boldsymbol{xy}}{\boldsymbol{xy}}$

$\qquad = \dfrac{8xy}{x^2 y}$

65. $\dfrac{9}{4b} = \dfrac{9}{4b} \cdot \dfrac{\mathbf{3b}}{\mathbf{3b}}$

$\qquad = \dfrac{27b}{12b^2}$

67. $\dfrac{3x}{x+1} = \dfrac{3x}{(x+1)} \cdot \dfrac{\boldsymbol{(x+1)}}{\boldsymbol{(x+1)}}$

$\qquad = \dfrac{3x^2+3x}{(x+1)^2}$

Build each rational expression into an equivalent expression with the given denominator. See Example 8.

69. $\dfrac{x+9}{x^2+5x} = \dfrac{x+9}{x(x+5)}$

$\qquad = \dfrac{x+9}{x(x+5)} \cdot \dfrac{\boldsymbol{x}}{\boldsymbol{x}}$

$\qquad = \dfrac{x^2+9x}{x^2(x+5)}$

71. $\dfrac{t+5}{4t+8} = \dfrac{t+5}{4(t+2)}$

$\qquad = \dfrac{t+5}{4(t+2)} \cdot \dfrac{\boldsymbol{(t+9)}}{\boldsymbol{(t+9)}}$

$\qquad = \dfrac{t^2+14t+45}{4(t+2)(t+9)}$

73. $\dfrac{y+3}{y^2-5y+6} = \dfrac{y+3}{(y-2)(y-3)}$

$\qquad = \dfrac{y+3}{(y-2)(y-3)} \cdot \dfrac{\boldsymbol{4y}}{\boldsymbol{4y}}$

$\qquad = \dfrac{4y^2+12y}{4y(y-2)(y-3)}$

75.
$$\frac{12-h}{h^2-81} = \frac{12-h}{(h+9)(h-9)}$$
$$= \frac{12-h}{(h+9)(h-9)} \cdot \frac{3}{3}$$
$$= \frac{36-3h}{3(h+9)(h-9)}$$

83.
$$\frac{17a}{2a+4} - \frac{7a}{2a+4} = \frac{17a-7a}{2a+4}$$
$$= \frac{10a}{2a+4}$$
$$= \frac{\overset{1}{\cancel{2}}\cdot 5a}{\underset{1}{\cancel{2}}(a+2)}$$
$$= \frac{5a}{a+2}$$

TRY IT YOURSELF
Perform the operations. Then simplify, if possible.

77.
$$\frac{3t}{t^2-8t+7} - \frac{3}{t^2-8t+7} = \frac{3t-3}{t^2-8t+7}$$
$$= \frac{3\,\cancel{(t-1)}}{(t-7)\,\cancel{(t-1)}}$$
$$= \frac{3}{t-7}$$

79.
$$\frac{c}{c^2-d^2} - \frac{d}{c^2-d^2} = \frac{c-d}{c^2-d^2}$$
$$= \frac{\cancel{c-d}}{(c+d)\,\cancel{(c-d)}}$$
$$= \frac{1}{c+d}$$

81.
$$\frac{a^2+a}{4a^2-8a} + \frac{2a^2-7a}{4a^2-8a} = \frac{a^2+a+2a^2-7a}{4a^2-8a}$$
$$= \frac{3a^2-6a}{4a^2-8a}$$
$$= \frac{3\,\cancel{a}\,\cancel{(a-2)}}{4\,\cancel{a}\,\cancel{(a-2)}}$$
$$= \frac{3}{4}$$

85.
$$\frac{8}{9-3x^2} - \frac{-6x+8}{9-3x^2} = \frac{8-(-6x+8)}{9-3x^2}$$
$$= \frac{8+6x-8}{9-3x^2}$$
$$= \frac{6x}{9-3x^2}$$
$$= \frac{2\cdot\overset{1}{\cancel{3}}\cdot x}{\underset{1}{\cancel{3}}(3-x^2)}$$
$$= \frac{2x}{3-x^2}$$

87.
$$\frac{11n}{(n+4)(n-2)} - \frac{4n-1}{(n-2)(n+4)}$$
$$= \frac{11n-(4n-1)}{(n+4)(n-2)}$$
$$= \frac{11n-4n+1}{(n+4)(n-2)}$$
$$= \frac{7n+1}{(n+4)(n-2)}$$

89.
$$\frac{5r-27}{3r^2-9r} + \frac{4r}{3r^2-9r} = \frac{5r-27+4r}{3r^2-9r}$$
$$= \frac{9r-27}{3r^2-9r}$$
$$= \frac{9(r-3)}{3r(r-3)}$$
$$= \frac{3\cdot\overset{1}{\cancel{3}}\,\overset{1}{\cancel{(r-3)}}}{\underset{1}{\cancel{3}}\cdot r\,\underset{1}{\cancel{(r-3)}}}$$
$$= \frac{3}{r}$$

91. $\dfrac{11}{36y}+\dfrac{9}{36y}=\dfrac{11+9}{36y}$

$=\dfrac{20}{36y}$

$=\dfrac{5\cdot\cancel{4}}{9\cdot\cancel{4}\,y}$

$=\dfrac{5}{9y}$

93. $\dfrac{-4x}{3x^2-7x+2}-\dfrac{-3x-2}{3x^2-7x+2}=\dfrac{-4x-(-3x-2)}{3x^2-7x+2}$

$=\dfrac{-4x+3x+2}{3x^2-7x+2}$

$=\dfrac{-x+2}{3x^2-7x+2}$

$=\dfrac{-\cancel{(x-2)}}{(3x-1)\,\cancel{(x-2)}}$

$=-\dfrac{1}{3x-1}$

95. $\dfrac{3x^2}{x+1}-\dfrac{-x+2}{x+1}=\dfrac{3x^2-(-x+2)}{x+1}$

$=\dfrac{3x^2+x-2}{x+1}$

$=\dfrac{(3x-2)(x+1)}{x+1}$

$=\dfrac{(3x-2)\,\cancel{(x+1)}}{\cancel{x+1}}$

$=3x-2$

LOOK ALIKES...

97. a. $\dfrac{t}{12}+\dfrac{5t}{12}=\dfrac{t+5t}{12}$

$=\dfrac{\cancel{6}\,t}{\cancel{6}\cdot 2}$

$=\dfrac{t}{2}$

b. $\dfrac{t}{12}\cdot\dfrac{5t}{12}=\dfrac{5t^{1+1}}{12\cdot 12}$

$=\dfrac{5t^2}{144}$

c. $\dfrac{t}{12}\div\dfrac{5t}{12}=\dfrac{\cancel{t}}{\cancel{12}}\cdot\dfrac{\cancel{12}}{5\cancel{t}}$

$=\dfrac{1}{5}$

99. a. $\dfrac{m+6}{5}-\dfrac{m+2}{5}=\dfrac{m+6-(m+2)}{5}$

$=\dfrac{m+6-m-2}{5}$

$=\dfrac{4}{5}$

b. $\dfrac{m+6}{5}\cdot\dfrac{m+2}{5}=\dfrac{(m+6)(m+2)}{5\cdot 5}$

$=\dfrac{m^2+8m+12}{25}$

c. $\dfrac{m+6}{5}\div\dfrac{m+2}{5}=\dfrac{m+6}{\cancel{5}}\cdot\dfrac{\cancel{5}}{m+2}$

$=\dfrac{m+6}{m+2}$

APPLICATIONS
101. GEOMETRY

$\dfrac{5x+11}{x+2}-\dfrac{3x+5}{x+2}=\dfrac{5x+11-(3x+5)}{x+2}$

$=\dfrac{5x+11-3x-5}{x+2}$

$=\dfrac{2x+6}{x+2}$

The difference between the length and the width is $\dfrac{2x+6}{x+2}$ feet.

WRITING
103-107. Answers will vary.
REVIEW
Give the formula for . . .
109. a. simple interest

$I=Prt$

b. the area of a triangle

$A=\dfrac{1}{2}bh$

c. the perimeter of a rectangle

$P=2l+2w$

Section 7.3

CHALLENGE PROBLEMS
Perform the operations. Simplify the results, if possible.

111. $\dfrac{3xy}{x-y} - \dfrac{x(3y-x)}{x-y} - \dfrac{x(x-y)}{x-y}$

$$= \frac{3xy - x(3y-x) - x(x-y)}{x-y}$$

$$= \frac{3xy - 3xy + x^2 - x^2 + xy}{x-y}$$

$$= \frac{xy}{x-y}$$

113. $\dfrac{2a^2+2}{a^3+8} + \dfrac{a^3+a}{a^3+8} = \dfrac{2a^2+2+a^3+a}{a^3+8}$

$$= \frac{a^3+2a^2+a+2}{a^3+8}$$

$$= \frac{(a^3+2a^2)+(a+2)}{a^3+8}$$

$$= \frac{a^2(a+2)+1(a+2)}{(a+2)(a^2-2a+4)}$$

$$= \frac{\overset{1}{\cancel{(a+2)}}(a^2+1)}{\underset{1}{\cancel{(a+2)}}(a^2-2a+4)}$$

$$= \frac{a^2+1}{a^2-2a+4}$$

SECTION 7.4 STUDY SET
VOCABULARY

Fill in the blanks.

1. $\frac{x}{x-7}$ and $\frac{1}{x-7}$ have like denominators. $\frac{x+5}{x-7}$ and $\frac{4x}{x+7}$ have **unlike** denominators.

CONCEPTS

3. a. $\frac{x+1}{20x^2}$, $\quad 20x^2 = \mathbf{2 \cdot 2 \cdot 5 \cdot x \cdot x}$

 b. $\frac{3x^2-4}{x^2+4x-12}$, $\quad x^2+4x-12 = \mathbf{(x+6)(x-2)}$

5. $\mathbf{(x+6)(x+3)}$

Fill in the blanks.

7. To build $\dfrac{x}{x+2}$ so that it has a denominator of $5(x+2)$, we multiply it by 1 in the form of $\dfrac{\mathbf{5}}{\mathbf{5}}$.

NOTATION
Complete the solution.

9. $\dfrac{2}{5} + \dfrac{7}{3x} = \dfrac{2}{5} \cdot \dfrac{\mathbf{3x}}{\mathbf{3x}} + \dfrac{7}{3x} \cdot \dfrac{\mathbf{5}}{\mathbf{5}}$

$= \dfrac{\mathbf{6x}}{\mathbf{15x}} + \dfrac{35}{\mathbf{15x}}$

$= \dfrac{\mathbf{6x+35}}{15x}$

GUIDED PRACTICE
Perform the operations. Simplify, if possible. See Example 1.

11. $\left.\begin{array}{l} 3=3 \\ 7=7 \end{array}\right\}$ LCD $= 3 \cdot 7 = 21$

$\dfrac{x}{3} + \dfrac{2x}{7} = \dfrac{x}{3} \cdot \dfrac{\mathbf{7}}{\mathbf{7}} + \dfrac{2x}{7} \cdot \dfrac{\mathbf{3}}{\mathbf{3}}$

$= \dfrac{7x}{21} + \dfrac{6x}{21}$

$= \dfrac{7x+6x}{21}$

$= \dfrac{13x}{21}$

13. $\left.\begin{array}{l} 8=2 \cdot 2 \cdot 2 \\ 5=5 \end{array}\right\}$ LCD $= 2 \cdot 2 \cdot 2 \cdot 5 = 40$

$\dfrac{7a}{8} + \dfrac{4a}{5} = \dfrac{7a}{8} \cdot \dfrac{\mathbf{5}}{\mathbf{5}} + \dfrac{4a}{5} \cdot \dfrac{\mathbf{8}}{\mathbf{8}}$

$= \dfrac{35a}{40} + \dfrac{32a}{40}$

$= \dfrac{35a+32a}{40}$

$= \dfrac{67a}{40}$

Perform the operations. Simplify, if possible. See Example 2.

15. $\left.\begin{array}{l} m^2 = m \cdot m \\ m = m \end{array}\right\}$ LCD $= m \cdot m = m^2$

$\dfrac{7}{m^2} - \dfrac{2}{m} = \dfrac{7}{m^2} - \dfrac{2}{m} \cdot \dfrac{\mathbf{m}}{\mathbf{m}}$

$= \dfrac{7}{m^2} - \dfrac{2m}{m^2}$

$= \dfrac{7-2m}{m^2}$

17. $\left.\begin{array}{l} 5p^2 = 5 \cdot p \cdot p \\ 10p = 2 \cdot 5 \cdot p \end{array}\right\}$ LCD $= 2 \cdot 5 \cdot p \cdot p = 10p^2$

$\dfrac{3}{5p^2} - \dfrac{5}{10p} = \dfrac{3}{5p^2} \cdot \dfrac{\mathbf{2}}{\mathbf{2}} - \dfrac{5}{10p} \cdot \dfrac{\mathbf{p}}{\mathbf{p}}$

$= \dfrac{6}{10p^2} - \dfrac{5p}{10p^2}$

$= \dfrac{6-5p}{10p^2}$

19. $\left.\begin{array}{l} 6t = 2 \cdot 3 \cdot t \\ 8t^3 = 2 \cdot 2 \cdot 2 \cdot t \cdot t \cdot t \end{array}\right\}$ LCD $= 2 \cdot 2 \cdot 2 \cdot 3 \cdot t \cdot t \cdot t = 24t^3$

$\dfrac{1}{6t} - \dfrac{11}{8t^3} = \dfrac{1}{6t} \cdot \dfrac{\mathbf{4t^2}}{\mathbf{4t^2}} - \dfrac{11}{8t^3} \cdot \dfrac{\mathbf{3}}{\mathbf{3}}$

$= \dfrac{4t^2}{24t^3} - \dfrac{33}{24t^3}$

$= \dfrac{4t^2-33}{24t^3}$

- 415 -

21.
$$\left.\begin{array}{l} 6c^4 = 2 \cdot 3 \cdot c \cdot c \cdot c \cdot c \\ 9c^2 = 3 \cdot 3 \cdot c \cdot c \end{array}\right\}$$

$$LCD = 2 \cdot 3 \cdot 3 \cdot c \cdot c \cdot c \cdot c$$
$$= 18c^4$$

$$\frac{1}{6c^4} - \frac{8}{9c^2} = \frac{1}{6c^4} \cdot \mathbf{\frac{3}{3}} - \frac{8}{9c^2} \cdot \mathbf{\frac{2c^2}{2c^2}}$$
$$= \frac{3}{18a^4} - \frac{16c^2}{18a^4}$$
$$= \frac{3 - 16c^2}{18a^4}$$

Perform the operations. Simplify, if possible. See Example 3.

23.
$$\left.\begin{array}{l} 2a + 4 = 2(a+2) \\ a^2 - 4 = (a+2)(a-2) \end{array}\right\}$$

$$LCD = 2(a+2)(a-2)$$

$$\frac{1}{2a+4} + \frac{5}{a^2-4}$$
$$= \frac{1}{2(a+2)} + \frac{5}{(a+2)(a-2)}$$
$$= \frac{1}{2(a+2)} \cdot \mathbf{\frac{(a-2)}{(a-2)}} + \frac{5}{(a+2)(a-2)} \cdot \mathbf{\frac{2}{2}}$$
$$= \frac{a-2}{2(a+2)(a-2)} + \frac{10}{2(a+2)(a-2)}$$
$$= \frac{a-2+10}{2(a+2)(a-2)}$$
$$= \frac{a+8}{2(a+2)(a-2)}$$

25.
$$\left.\begin{array}{l} 3a - 2 = 3a - 2 \\ 9a^2 - 4 = (3a+2)(3a-2) \end{array}\right\}$$

$$LCD = (3a+2)(3a-2)$$

$$\frac{2}{3a-2} + \frac{5}{9a^2-4}$$
$$= \frac{2}{3a-2} + \frac{5}{(3a+2)(3a-2)}$$
$$= \frac{2}{(3a-2)} \cdot \mathbf{\frac{(3a+2)}{(3a+2)}} + \frac{5}{(3a+2)(3a-2)}$$
$$= \frac{2(3a+2)+5}{(3a+2)(3a-2)}$$
$$= \frac{6a+4+5}{(3a+2)(3a-2)}$$
$$= \frac{6a+9}{(3a+2)(3a-2)}$$

27.
$$\left.\begin{array}{l} a + 2 = a + 2 \\ a^2 + 4a + 4 = (a+2)(a+2) \end{array}\right\}$$

$$LCD = (a+2)(a+2)$$

$$\frac{4}{a+2} - \frac{7}{a^2+4a+4}$$
$$= \frac{4}{a+2} - \frac{7}{(a+2)(a+2)}$$
$$= \frac{4}{(a+2)} \cdot \mathbf{\frac{(a+2)}{(a+2)}} - \frac{7}{(a+2)(a+2)}$$
$$= \frac{4(a+2)-7}{(a+2)(a+2)}$$
$$= \frac{4a+8-7}{(a+2)(a+2)}$$
$$= \frac{4a+1}{(a+2)^2}$$

29.
$$\left.\begin{array}{l} 5m - 5 = 5(m-1) \\ 5m^2 - 5m = 5m(m-1) \end{array}\right\} LCD = 5m(m-1)$$

$$\frac{6}{5m^2-5m} - \frac{3}{5m-5}$$
$$= \frac{6}{5m(m-1)} - \frac{3}{5(m-1)}$$
$$= \frac{6}{5m(m-1)} - \frac{3}{5(m-1)} \cdot \mathbf{\frac{m}{m}}$$
$$= \frac{6-3m}{5m(m-1)}$$

Perform the operations. Simplify, if possible. See Example 4.

31.
$$\left.\begin{array}{l} t + 3 = t + 3 \\ t + 2 = t + 2 \end{array}\right\} LCD = (t+3)(t+2)$$

$$\frac{9}{t+3} + \frac{8}{t+2}$$
$$= \frac{9}{(t+3)} \cdot \mathbf{\frac{(t+2)}{(t+2)}} + \frac{8}{(t+2)} \cdot \mathbf{\frac{(t+3)}{(t+3)}}$$
$$= \frac{9t+18+8t+24}{(t+3)(t+2)}$$
$$= \frac{17t+42}{(t+3)(t+2)}$$

33.
$$\left.\begin{array}{l}2x-1=2x-1\\2x+3=2x+3\end{array}\right\}\text{LCD}=(2x-1)(2x+3)$$

$$\frac{3x}{2x-1}-\frac{2x}{2x+3}$$

$$=\frac{3x}{(2x-1)}\cdot\frac{(2x+3)}{(2x+3)}-\frac{2x}{(2x+3)}\cdot\frac{(2x-1)}{(2x-1)}$$

$$=\frac{6x^2+9x-(4x^2-2x)}{(2x-1)(2x+3)}$$

$$=\frac{6x^2+9x-4x^2+2x}{(2x-1)(2x+3)}$$

$$=\frac{2x^2+11x}{(2x-1)(2x+3)}$$

35.
$$\left.\begin{array}{l}s+3=s+3\\s+7=s+7\end{array}\right\}\text{LCD}=(s+3)(s+7)$$

$$\frac{s+7}{s+3}-\frac{s-3}{s+7}$$

$$=\frac{(s+7)}{(s+3)}\cdot\frac{(s+7)}{(s+7)}-\frac{(s-3)}{(s+7)}\cdot\frac{(s+3)}{(s+3)}$$

$$=\frac{(s+7)(s+7)-(s-3)(s+3)}{(s+3)(s+7)}$$

$$=\frac{s^2+14s+49-(s^2-9)}{(s+3)(s+7)}$$

$$=\frac{s^2+14s+49-s^2+9}{(s+3)(s+7)}$$

$$=\frac{14s+58}{(s+3)(s+7)}$$

37.
$$\left.\begin{array}{l}m-2=m-2\\m+5=m+5\end{array}\right\}\text{LCD}=(m-2)(m+5)$$

$$\frac{3m}{m-2}-\frac{m-3}{m+5}$$

$$=\frac{3m}{(m-2)}\cdot\frac{(m+5)}{(m+5)}-\frac{(m-3)}{(m+5)}\cdot\frac{(m-2)}{(m-2)}$$

$$=\frac{3m(m+5)-(m-3)(m-2)}{(m-2)(m+5)}$$

$$=\frac{3m^2+15m-(m^2-5m+6)}{(m-2)(m+5)}$$

$$=\frac{3m^2+15m-m^2+5m-6}{(m-2)(m+5)}$$

$$=\frac{2m^2+20m-6}{(m-2)(m+5)}$$

Perform the operations. Simplify, if possible. See Example 5.

39.
$$\left.\begin{array}{l}s^2+5s+4=(s+1)(s+4)\\s^2+2s+1=(s+1)(s+1)\end{array}\right\}$$

$$\text{LCD}=(s+4)(s+1)(s+1)$$

$$\frac{4}{s^2+5s+4}+\frac{s}{s^2+2s+1}$$

$$=\frac{4}{(s+1)(s+4)}\cdot\frac{(s+1)}{(s+1)}+\frac{s}{(s+1)(s+1)}\cdot\frac{(s+4)}{(s+4)}$$

$$=\frac{4s+4+s^2+4s}{(s+4)(s+1)(s+1)}$$

$$=\frac{s^2+8s+4}{(s+4)(s+1)(s+1)}$$

41.
$$\left.\begin{array}{l}x^2-9x+8=(x-8)(x-1)\\x^2-6x-16=(x-8)(x+2)\end{array}\right\}$$

$$\text{LCD}=(x-1)(x-8)(x+2)$$

$$\frac{5}{x^2-9x+8}-\frac{3}{x^2-6x-16}$$

$$=\frac{5}{(x-1)(x-8)}\cdot\frac{(x+2)}{(x+2)}-\frac{3}{(x-8)(x+2)}\cdot\frac{(x-1)}{(x-1)}$$

$$=\frac{5(x+2)-3(x-1)}{(x-1)(x-8)(x+2)}$$

$$=\frac{5x+10-3x+3}{(x-1)(x-8)(x+2)}$$

$$=\frac{2x+13}{(x-1)(x-8)(x+2)}$$

43.
$$\left.\begin{array}{l}a^2+4a+3=(a+3)(a+1)\\a+3=a+3\end{array}\right\}$$

$$\text{LCD}=(a+1)(a+3)$$

$$\frac{2}{a^2+4a+3}+\frac{1}{a+3}$$

$$=\frac{2}{(a+1)(a+3)}+\frac{1}{(a+3)}\cdot\frac{(a+1)}{(a+1)}$$

$$=\frac{2+a+1}{(a+1)(a+3)}$$

$$=\frac{\overset{1}{\cancel{a+3}}}{(a+1)\cancel{(a+3)}}_{\,1}$$

$$=\frac{1}{a+1}$$

Section 7.4

45.

$$\left.\begin{array}{l} y^2 - 16 = (y+4)(y-4) \\ y^2 - y - 12 = (y-4)(y+3) \end{array}\right\}$$

$$\text{LCD} = (y+4)(y-4)(y+3)$$

$$\frac{8}{y^2-16} - \frac{7}{y^2-y-12}$$

$$= \frac{8}{(y+4)(y-4)} \cdot \frac{(y+3)}{(y+3)} - \frac{7}{(y-4)(y+3)} \cdot \frac{(y+4)}{(y+4)}$$

$$= \frac{8y+24-7y-28}{(y+4)(y-4)(y+3)}$$

$$= \frac{\overset{1}{\cancel{y-4}}}{(y+4)\,\cancel{(y-4)}\,(y+3)}$$

$$= \frac{1}{(y+4)(y+3)}$$

Perform the operations. Simplify, if possible. See Example 6.

47.

$$\frac{9y}{x-4} + y = \frac{9y}{x-4} + \frac{y}{1} \cdot \frac{x-4}{x-4}$$

$$= \frac{9y + y(x-4)}{x-4}$$

$$= \frac{9y + xy - 4y}{x-4}$$

$$= \frac{(9-4)y + xy}{x-4}$$

$$= \frac{5y + xy}{x-4}$$

49.

$$\frac{8}{x} + z = \frac{8}{x} + \frac{z}{1} \cdot \frac{x}{x}$$

$$= \frac{8 + xz}{x}$$

Perform the operations. Simplify, if possible. See Example 7.

51.

$$\frac{7}{a-4} + \frac{5}{4-a} = \frac{7}{a-4} + \frac{5}{4-a} \cdot \frac{-1}{-1}$$

$$= \frac{7}{a-4} + \frac{-5}{-4+a}$$

$$= \frac{7}{a-4} + \frac{-5}{a-4}$$

$$= \frac{7+(-5)}{a-4}$$

$$= \frac{2}{a-4}$$

53.

$$\frac{c}{7c-d} - \frac{d}{d-7c} = \frac{c}{7c-d} - \frac{d}{d-7c} \cdot \frac{-1}{-1}$$

$$= \frac{c}{7c-d} - \frac{-d}{-d+7c}$$

$$= \frac{c}{7c-d} - \frac{-d}{7c-d}$$

$$= \frac{c-(-d)}{7c-d}$$

$$= \frac{c+d}{7c-d}$$

TRY IT YOURSELF

Perform the operations. Then simplify, if possible.

55.

$$\frac{x-7}{x^2+4x-5} - \frac{x-9}{x^2+3x-10}$$

$$= \frac{x-7}{(x+5)(x-1)} - \frac{x-9}{(x+5)(x-2)}$$

$$= \frac{(x-7)}{(x+5)(x-1)} \cdot \frac{(x-2)}{(x-2)} - \frac{(x-9)}{(x+5)(x-2)} \cdot \frac{(x-1)}{(x-1)}$$

$$= \frac{(x-7)(x-2) - (x-9)(x-1)}{(x+5)(x-1)(x-2)}$$

$$= \frac{x^2 - 9x + 14 - (x^2 - 10x + 9)}{(x+5)(x-1)(x-2)}$$

$$= \frac{x^2 - 9x + 14 - x^2 + 10x - 9}{(x+5)(x-1)(x-2)}$$

$$= \frac{(1-1)x^2 + (-9+10)x + (14-9)}{(x+5)(x-1)(x-2)}$$

$$= \frac{x+5}{(x+5)(x-1)(x-2)}$$

$$= \frac{\overset{1}{\cancel{x+5}}}{\cancel{(x+5)}(x-1)(x-2)}$$

$$= \frac{1}{(x-1)(x-2)}$$

57. $\dfrac{3d-3}{d-9}-\dfrac{3d}{9-d}=\dfrac{3d-3}{d-9}-\dfrac{3d}{9-d}\cdot\dfrac{-1}{-1}$

$=\dfrac{3d-3}{d-9}-\dfrac{-3d}{-9+d}$

$=\dfrac{3d-3}{d-9}-\dfrac{-3d}{d-9}$

$=\dfrac{3d-3-(-3d)}{d-9}$

$=\dfrac{3d-3+3d}{d-9}$

$=\dfrac{6d-3}{d-9}$

59. $\dfrac{10}{x-1}+y=\dfrac{10}{x-1}+\dfrac{y}{1}\cdot\dfrac{(x-1)}{(x-1)}$

$=\dfrac{10+y(x-1)}{x-1}$

$=\dfrac{10+xy-y}{x-1}$

$=\dfrac{xy-y+10}{x-1}$

61. $\dfrac{b}{b+1}-\dfrac{b-1}{b+2}=\dfrac{b}{(b+1)}\cdot\dfrac{(b+2)}{(b+2)}-\dfrac{(b-1)}{(b+2)}\cdot\dfrac{(b+1)}{(b+1)}$

$=\dfrac{b(b+2)-(b-1)(b+1)}{(b+1)(b+2)}$

$=\dfrac{b^2+2b-(b^2-1)}{(b+1)(b+2)}$

$=\dfrac{b^2+2b-b^2+1}{(b+1)(b+2)}$

$=\dfrac{2b+1}{(b+1)(b+2)}$

63. $\dfrac{g}{g^2-4}+\dfrac{2}{4-g^2}=\dfrac{g}{g^2-4}+\dfrac{2}{4-g^2}\cdot\dfrac{-1}{-1}$

$=\dfrac{g}{g^2-4}+\dfrac{-2}{-4+g^2}$

$=\dfrac{g}{g^2-4}+\dfrac{-2}{g^2-4}$

$=\dfrac{g+(-2)}{g^2-4}$

$=\dfrac{\cancel{g-2}^{1}}{(g+2)\,\cancel{(g-2)}_{1}}$

$=\dfrac{1}{g+2}$

65. $\dfrac{5y}{6}+\dfrac{5y}{3}=\dfrac{5y}{6}+\dfrac{5y}{3}\cdot\dfrac{2}{2}$

$=\dfrac{5y}{6}+\dfrac{10y}{6}$

$=\dfrac{5y+10y}{6}$

$=\dfrac{15y}{6}$

$=\dfrac{\cancel{3}^{1}\cdot 5y}{\cancel{3}_{1}\cdot 2}$

$=\dfrac{5y}{2}$

67. $\dfrac{1}{5x}+\dfrac{7x}{x+5}=\dfrac{1}{5x}\cdot\dfrac{(x+5)}{(x+5)}+\dfrac{7x}{(x+5)}\cdot\dfrac{(5x)}{(5x)}$

$=\dfrac{x+5+35x^2}{5x(x+5)}$

$=\dfrac{35x^2+x+5}{5x(x+5)}$

69. $\dfrac{11}{5x}-\dfrac{5}{6x}=\dfrac{11}{5x}\cdot\dfrac{6}{6}-\dfrac{5}{6x}\cdot\dfrac{5}{5}$

$=\dfrac{66-25}{30x}$

$=\dfrac{41}{30x}$

71. $\dfrac{x}{x+1}+\dfrac{x-1}{x}=\dfrac{x}{(x+1)}\cdot\dfrac{(x)}{(x)}+\dfrac{(x-1)}{x}\cdot\dfrac{(x+1)}{(x+1)}$

$\qquad\qquad =\dfrac{x^2+x^2-1}{x(x+1)}$

$\qquad\qquad =\dfrac{2x^2-1}{x(x+1)}$

73. $\dfrac{y}{y-1}-\dfrac{4}{1-y}=\dfrac{y}{y-1}-\dfrac{4}{1-y}\cdot\dfrac{-1}{-1}$

$\qquad\qquad =\dfrac{y}{y-1}-\dfrac{-4}{-1+y}$

$\qquad\qquad =\dfrac{y}{y-1}-\dfrac{-4}{y-1}$

$\qquad\qquad =\dfrac{y-(-4)}{y-1}$

$\qquad\qquad =\dfrac{y+4}{y-1}$

75. $\dfrac{n}{5}-\dfrac{n-2}{15}=\dfrac{n}{5}\cdot\dfrac{3}{3}-\dfrac{n-2}{15}$

$\qquad\qquad =\dfrac{3n}{15}-\dfrac{n-2}{15}$

$\qquad\qquad =\dfrac{3n-(n-2)}{15}$

$\qquad\qquad =\dfrac{3n-n+2}{15}$

$\qquad\qquad =\dfrac{2n+2}{15}$

77. $\dfrac{y+2}{5y^2}+\dfrac{y+4}{15y}=\dfrac{(y+2)}{5y^2}\cdot\dfrac{3}{3}+\dfrac{(y+4)}{15y}\cdot\dfrac{y}{y}$

$\qquad\qquad =\dfrac{3(y+2)+y(y+4)}{15y^2}$

$\qquad\qquad =\dfrac{3y+6+y^2+4y}{15y^2}$

$\qquad\qquad =\dfrac{y^2+7y+6}{15y^2}$

79. $\dfrac{x}{x-2}+\dfrac{4+2x}{x^2-4}=\dfrac{x}{(x-2)}\cdot\dfrac{(x+2)}{(x+2)}+\dfrac{4+2x}{(x+2)(x-2)}$

$\qquad\qquad =\dfrac{x(x+2)+4+2x}{(x+2)(x-2)}$

$\qquad\qquad =\dfrac{x^2+2x+4+2x}{(x+2)(x-2)}$

$\qquad\qquad =\dfrac{x^2+4x+4}{(x+2)(x-2)}$

$\qquad\qquad =\dfrac{\overset{1}{\cancel{(x+2)}}(x+2)}{\underset{1}{\cancel{(x+2)}}(x-2)}$

$\qquad\qquad =\dfrac{x+2}{x-2}$

81. $b-\dfrac{3}{a^2}=\dfrac{b}{1}\cdot\dfrac{a^2}{a^2}-\dfrac{3}{a^2}$

$\qquad\qquad =\dfrac{a^2b-3}{a^2}$

83. $\dfrac{7}{3a}+\dfrac{1}{a-2}=\dfrac{7}{3a}\cdot\dfrac{(a-2)}{(a-2)}+\dfrac{1}{(a-2)}\cdot\dfrac{(3a)}{(3a)}$

$\qquad\qquad =\dfrac{7(a-2)+3a}{3a(a-2)}$

$\qquad\qquad =\dfrac{7a-14+3a}{3a(a-2)}$

$\qquad\qquad =\dfrac{10a-14}{3a(a-2)}$

85. $\dfrac{3}{x^2}+\dfrac{17}{x}=\dfrac{3}{x^2}+\dfrac{17}{x}\cdot\dfrac{x}{x}$

$\qquad\qquad =\dfrac{3}{x^2}+\dfrac{17x}{x^2}$

$\qquad\qquad =\dfrac{17x+3}{x^2}$

87.
$$\frac{x+2}{x+1} - 5 = \frac{x+2}{x+1} - \frac{5}{1} \cdot \frac{(x+1)}{(x+1)}$$

$$= \frac{x+2-5(x+1)}{x+1}$$

$$= \frac{x+2-5x-5}{x+1}$$

$$= \frac{-4x-3}{x+1}$$

$$= \frac{-(4x+3)}{x+1}$$

$$= -\frac{4x+3}{x+1} \text{ or } \frac{-4x-3}{x+1}$$

89.
$$\frac{4b}{3} - \frac{5b}{12} = \frac{4b}{3} \cdot \frac{4}{4} - \frac{5b}{12}$$

$$= \frac{16b}{12} - \frac{5b}{12}$$

$$= \frac{16b-5b}{12}$$

$$= \frac{11b}{12}$$

LOOK ALIKES ...
Perform the operations and simplify, if possible.

91. a.
$$\frac{5}{2x} + \frac{4x}{15} = \frac{5}{2x} \cdot \frac{15}{15} + \frac{4x}{15} \cdot \frac{2x}{2x}$$

$$= \frac{75+8x^2}{30x}$$

b.
$$\frac{5}{2x} \cdot \frac{4x}{15} = \frac{5 \cdot 4x}{2x \cdot 15}$$

$$= \frac{\cancel{5} \cdot \cancel{2} \cdot 2 \cancel{x}}{\cancel{2} \cancel{x} \cdot 3 \cdot \cancel{5}}$$

$$= \frac{2}{3}$$

93. a.
$$\frac{t}{t-5} - \frac{t}{t^2-25} = \frac{t}{(t-5)} - \frac{t}{(t+5)(t-5)}$$

$$= \frac{t}{(t-5)} \cdot \frac{(t+5)}{(t+5)} - \frac{t}{(t+5)(t-5)}$$

$$= \frac{t(t+5)-t}{(t-5)(t+5)}$$

$$= \frac{t^2+5t-t}{(t-5)(t+5)}$$

$$= \frac{t^2+4t}{(t-5)(t+5)}$$

b.
$$\frac{t}{t-5} \div \frac{t}{t^2-25} = \frac{t}{(t-5)} \div \frac{t}{(t+5)(t-5)}$$

$$= \frac{t}{(t-5)} \cdot \frac{(t+5)(t-5)}{t}$$

$$= \frac{\cancel{t}}{\cancel{(t-5)}} \cdot \frac{(t+5)(\cancel{t-5})}{\cancel{t}}$$

$$= t+5$$

APPLICATIONS

95.
$$\frac{3}{2x^2} + \frac{10}{3x} = \frac{3}{2x^2} \cdot \frac{3}{3} + \frac{10}{3x} \cdot \frac{2x}{2x}$$

$$= \frac{9}{6x^2} + \frac{20x}{6x^2}$$

$$= \frac{20x+9}{6x^2}$$

The total height is $\dfrac{20x+9}{6x^2}$ cm.

WRITING
97-99. Answers will vary.

REVIEW
101. Find the slope and y-intercept of the graph $y = 8x+2$.

$$y = mx+b$$
$$m = \text{slope of line}$$
$$b = y\text{-intercept, } (0,y)$$
$$m = 8$$
$$y\text{-int} = (0,2)$$

103. What is the slope of teh graph of $y = 2$?

$y = mx + b$

$m = $ slope of line

$b = y$-intercept, $(0, y)$

$y = 2$

This is the equation of a horizontal line.

$m = 0$

CHALLENGE PROBLEMS
Perform the operations and simplify the results, if possible.

105.
$$\left.\begin{array}{l} a - 1 = a - 1 \\ a + 2 = a + 2 \\ a^2 + a - 2 = (a-1)(a+2) \end{array}\right\}$$

$$\text{LCD} = (a-1)(a+2)$$

$$\frac{a}{a-1} - \frac{2}{a+2} + \frac{3(a-2)}{a^2 + a - 2}$$

$$= \frac{a}{(a-1)} \cdot \frac{(a+2)}{(a+2)} - \frac{2}{(a+2)} \cdot \frac{(a-1)}{(a-1)}$$

$$+ \frac{3(a-2)}{a^2 + a - 2}$$

$$= \frac{a(a+2) - 2(a-1) + 3(a-2)}{(a-1)(a+2)}$$

$$= \frac{a^2 + 2a - 2a + 2 + 3a - 6}{(a-1)(a+2)}$$

$$= \frac{a^2 + 3a - 4}{(a-1)(a+2)}$$

$$= \frac{(a-1)(a+4)}{(a-1)(a+2)}$$

$$= \frac{\overset{1}{\cancel{(a-1)}}(a+4)}{\underset{1}{\cancel{(a-1)}}(a+2)}$$

$$= \frac{a+4}{a+2}$$

107. $\dfrac{1}{a+1} + \dfrac{a^2 - 7a + 10}{2a^2 - 2a - 4} \cdot \dfrac{2a^2 - 50}{a^2 + 10a + 25}$

$$= \frac{1}{a+1} + \frac{a^2 - 7a + 10}{2(a^2 - a - 2)} \cdot \frac{2(a^2 - 25)}{a^2 + 10a + 25}$$

$$= \frac{1}{a+1} + \frac{(a-2)(a-5)}{2(a-2)(a+1)} \cdot \frac{2(a+5)(a-5)}{(a+5)(a+5)}$$

$$= \frac{1}{a+1} + \frac{\overset{1}{\cancel{2}}\overset{1}{\cancel{(a-2)}}(a-5)\overset{1}{\cancel{(a+5)}}(a-5)}{\underset{1}{\cancel{2}}\underset{1}{\cancel{(a-2)}}(a+1)\underset{1}{\cancel{(a+5)}}(a+5)}$$

$$= \frac{1}{a+1} + \frac{(a-5)(a-5)}{(a+1)(a+5)}$$

$$\left.\begin{array}{l} a + 1 = a + 1 \\ (a+1)(a+5) = (a+1)(a+5) \end{array}\right\}$$

$$\text{LCD} = (a+1)(a+5)$$

$$= \frac{1}{a+1} + \frac{(a-5)(a-5)}{(a+1)(a+5)}$$

$$= \frac{1}{(a+1)} \cdot \frac{(a+5)}{(a+5)} + \frac{(a-5)(a-5)}{(a+1)(a+5)}$$

$$= \frac{1(a+5) + (a-5)(a-5)}{(a+1)(a+5)}$$

$$= \frac{a + 5 + a^2 - 10a + 25}{(a+1)(a+5)}$$

$$= \frac{a^2 - 9a + 30}{(a+1)(a+5)}$$

SECTION 7.5 STUDY SET

VOCABULARY

Fill in the blanks.

1. The expression $\dfrac{\frac{2}{3} - \frac{1}{x}}{\frac{x-3}{4}}$ is called a **complex** rational expression or a **complex** fraction.

CONCEPTS

Fill in the blanks.

3. Method 1: To simplify a complex fraction, write its numerator and denominator as **single** rational expression. Then perform the indicated **division** by multiplying the numerator of the complex fraction by the **reciprocal** of the denominator.

5. a. $\dfrac{x-3}{4}$, **Yes**

   b. $\dfrac{1}{12} - \dfrac{x}{6}$, **No**

NOTATION

Fill in the blanks to simplify each complex fraction.

7. $\dfrac{\frac{12}{y^2}}{\frac{4}{y^3}} = \dfrac{12}{y^2} \boxed{\div} \dfrac{4}{y^3}$

GUIDED PRACTICE

Simplify each complex fraction. See Example 1.

9. $\dfrac{\frac{2}{3}}{\frac{3}{4}} = \dfrac{2}{3} \div \dfrac{3}{4}$

$= \dfrac{2}{3} \cdot \dfrac{4}{3}$

$= \dfrac{2 \cdot 4}{3 \cdot 3}$

$= \dfrac{8}{9}$

11. $\dfrac{\frac{x}{2}}{\frac{6}{5}} = \dfrac{x}{2} \div \dfrac{6}{5}$

$= \dfrac{x}{2} \cdot \dfrac{5}{6}$

$= \dfrac{x \cdot 5}{2 \cdot 6}$

$= \dfrac{5x}{12}$

13. $\dfrac{\frac{x}{y}}{\frac{1}{x}} = \dfrac{x}{y} \div \dfrac{1}{x}$

$= \dfrac{x}{y} \cdot \dfrac{x}{1}$

$= \dfrac{x \cdot x}{y \cdot 1}$

$= \dfrac{x^2}{y}$

15. $\dfrac{\frac{n}{8}}{\frac{1}{n^2}} = \dfrac{n}{8} \div \dfrac{1}{n^2}$

$= \dfrac{n}{8} \cdot \dfrac{n^2}{1}$

$= \dfrac{n \cdot n^2}{8}$

$= \dfrac{n^3}{8}$

17. $\dfrac{\frac{4a}{11}}{\frac{6a^4}{55}} = \dfrac{4a}{11} \div \dfrac{6a^4}{55}$

$= \dfrac{4a}{11} \cdot \dfrac{55}{6a^4}$

$= \dfrac{4a \cdot 55}{11 \cdot 6a^4}$

$= \dfrac{\cancel{2} \cdot 2 \cdot 5 \cdot \cancel{11} \cdot \cancel{a}}{\cancel{11} \cdot \cancel{2} \cdot 3 \cdot \cancel{a} \cdot a^3}$

$= \dfrac{10}{3a^3}$

19. $\dfrac{-\frac{x^4}{30}}{\frac{7x^2}{15}} = -\dfrac{x^4}{30} \div \dfrac{7x^2}{15}$

$= -\dfrac{x^4}{30} \cdot \dfrac{15}{7x^2}$

$= -\dfrac{x^4 \cdot 15}{30 \cdot 7x^2}$

$= -\dfrac{\cancel{3} \cdot \cancel{5} \cdot \cancel{x^2} \cdot x^2}{2 \cdot \cancel{3} \cdot \cancel{5} \cdot 7 \cdot \cancel{x^2}}$

$= -\dfrac{x^2}{14}$

Simplify each complex fraction. See Examples 2 or 4.

21. $\left.\begin{array}{l} 2=2 \\ 4=2\cdot 2 \end{array}\right\}$ LCD $= 2\cdot 2 = 4$

$$\dfrac{\dfrac{1}{2}+\dfrac{3}{4}}{\dfrac{3}{2}+\dfrac{1}{4}} = \dfrac{\dfrac{1}{2}+\dfrac{3}{4}}{\dfrac{3}{2}+\dfrac{1}{4}}\cdot\dfrac{\mathbf{4}}{\mathbf{4}}$$

$$= \dfrac{\left(\dfrac{1}{2}+\dfrac{3}{4}\right)4}{\left(\dfrac{3}{2}+\dfrac{1}{4}\right)4}$$

$$= \dfrac{\dfrac{1}{2}(4)+\dfrac{3}{4}(4)}{\dfrac{3}{2}(4)+\dfrac{1}{4}(4)}$$

$$= \dfrac{2+3}{6+1}$$

$$= \dfrac{5}{7}$$

23. $\left.\begin{array}{l} 4=2\cdot 2 \\ y=y \\ 3=3 \\ 2=2 \end{array}\right\}$ LCD $= 2\cdot 2\cdot 3\cdot y = 12y$

$$\dfrac{\dfrac{1}{4}+\dfrac{1}{y}}{\dfrac{y}{3}-\dfrac{1}{2}} = \dfrac{\dfrac{1}{4}+\dfrac{1}{y}}{\dfrac{y}{3}-\dfrac{1}{2}}\cdot\dfrac{\mathbf{12y}}{\mathbf{12y}}$$

$$= \dfrac{\left(\dfrac{1}{4}+\dfrac{1}{y}\right)12y}{\left(\dfrac{y}{3}-\dfrac{1}{2}\right)12y}$$

$$= \dfrac{\dfrac{1}{4}(12y)+\dfrac{1}{y}(12y)}{\dfrac{y}{3}(12y)-\dfrac{1}{2}(12y)}$$

$$= \dfrac{3y+12}{4y^2-6y}\ \text{ or }\ \dfrac{3(y+4)}{2y(2y-3)}$$

25. $\left.\begin{array}{l} y=y \\ 2=2 \end{array}\right\}$ LCD $= 2\cdot y = 2y$

$$\dfrac{\dfrac{1}{y}-\dfrac{5}{2}}{\dfrac{3}{y}} = \dfrac{\dfrac{1}{y}-\dfrac{5}{2}}{\dfrac{3}{y}}\cdot\dfrac{\mathbf{2y}}{\mathbf{2y}}$$

$$= \dfrac{\left(\dfrac{1}{y}-\dfrac{5}{2}\right)2y}{\left(\dfrac{3}{y}\right)2y}$$

$$= \dfrac{\dfrac{1}{y}(2y)-\dfrac{5}{2}(2y)}{\dfrac{3}{y}(2y)}$$

$$= \dfrac{2-5y}{6}$$

27. $\left.\begin{array}{l} c=c \\ 6=2\cdot 3 \end{array}\right\}$ LCD $= 2\cdot 3\cdot c = 6c$

$$\dfrac{\dfrac{4}{c}-\dfrac{c}{6}}{\dfrac{2}{c}} = \dfrac{\dfrac{4}{c}-\dfrac{c}{6}}{\dfrac{2}{c}}\cdot\dfrac{\mathbf{6c}}{\mathbf{6c}}$$

$$= \dfrac{\left(\dfrac{4}{c}-\dfrac{c}{6}\right)6c}{\left(\dfrac{2}{c}\right)6c}$$

$$= \dfrac{\dfrac{4}{c}(6c)-\dfrac{c}{6}(6c)}{\dfrac{2}{c}(6c)}$$

$$= \dfrac{24-c^2}{12}$$

29. LCD $=3$

$$\dfrac{\dfrac{2}{3}+1}{\dfrac{1}{3}+1}=\dfrac{\dfrac{2}{3}+1}{\dfrac{1}{3}+1}\cdot\dfrac{3}{3}$$

$$=\dfrac{\left(\dfrac{2}{3}+1\right)3}{\left(\dfrac{1}{3}+1\right)3}$$

$$=\dfrac{\dfrac{2}{3}(3)+1(3)}{\dfrac{1}{3}(3)+1(3)}$$

$$=\dfrac{2+3}{1+3}$$

$$=\dfrac{5}{4}$$

31. LCD $=x$

$$\dfrac{\dfrac{1}{x}-3}{\dfrac{5}{x}+2}=\dfrac{\dfrac{1}{x}-3}{\dfrac{5}{x}+2}\cdot\dfrac{x}{x}$$

$$=\dfrac{\left(\dfrac{1}{x}-3\right)x}{\left(\dfrac{5}{x}+2\right)x}$$

$$=\dfrac{\dfrac{1}{x}(x)-3(x)}{\dfrac{5}{x}(x)+2(x)}$$

$$=\dfrac{1-3x}{5+2x}$$

33. LCD $=x$

$$\dfrac{\dfrac{2}{x}+2}{\dfrac{4}{x}+2}=\dfrac{\dfrac{2}{x}+2}{\dfrac{4}{x}+2}\cdot\dfrac{x}{x}$$

$$=\dfrac{\left(\dfrac{2}{x}+2\right)x}{\left(\dfrac{4}{x}+2\right)x}$$

$$=\dfrac{\dfrac{2}{x}(x)+2(x)}{\dfrac{4}{x}(x)+2(x)}$$

$$=\dfrac{2+2x}{4+2x}$$

$$=\dfrac{\overset{1}{\cancel{2}}(1+x)}{\underset{1}{\cancel{2}}(2+x)}$$

$$=\dfrac{x+1}{x+2}$$

35. LCD $=x$

$$\dfrac{\dfrac{3y}{x}-y}{y-\dfrac{y}{x}}=\dfrac{\dfrac{3y}{x}-y}{y-\dfrac{y}{x}}\cdot\dfrac{x}{x}$$

$$=\dfrac{\left(\dfrac{3y}{x}-y\right)x}{\left(y-\dfrac{y}{x}\right)x}$$

$$=\dfrac{\dfrac{3y}{x}(x)-y(x)}{y(x)-\dfrac{y}{x}(x)}$$

$$=\dfrac{3y-xy}{xy-y}$$

$$=\dfrac{\overset{1}{\cancel{y}}(3-x)}{\underset{1}{\cancel{y}}(x-1)}$$

$$=\dfrac{3-x}{x-1}$$

37. $\left.\begin{array}{l}6=2\cdot3\\x=x\end{array}\right\}$ LCD $=2\cdot3\cdot x=6x$

$$\dfrac{\dfrac{1}{6}-\dfrac{2}{x}}{\dfrac{1}{6}+\dfrac{1}{x}}=\dfrac{\dfrac{1}{6}-\dfrac{2}{x}}{\dfrac{1}{6}+\dfrac{1}{x}}\cdot\dfrac{6x}{6x}$$

$$=\dfrac{\left(\dfrac{1}{6}-\dfrac{2}{x}\right)6x}{\left(\dfrac{1}{6}+\dfrac{1}{x}\right)6x}$$

$$=\dfrac{\dfrac{1}{6}(6x)-\dfrac{2}{x}(6x)}{\dfrac{1}{6}(6x)+\dfrac{1}{x}(6x)}$$

$$=\dfrac{x-12}{x+6}$$

39. $\left.\begin{array}{l}7=7\\a=a\end{array}\right\}$ LCD $=7\cdot a=7a$

$$\dfrac{\dfrac{a}{7}-\dfrac{7}{a}}{\dfrac{1}{a}+\dfrac{1}{7}}=\dfrac{\dfrac{a}{7}-\dfrac{7}{a}}{\dfrac{1}{a}+\dfrac{1}{7}}\cdot\dfrac{7a}{7a}$$

$$=\dfrac{\left(\dfrac{a}{7}-\dfrac{7}{a}\right)7a}{\left(\dfrac{1}{a}+\dfrac{1}{7}\right)7a}$$

$$=\dfrac{\dfrac{a}{7}(7a)-\dfrac{7}{a}(7a)}{\dfrac{1}{a}(7a)+\dfrac{1}{7}(7a)}$$

$$=\dfrac{a^2-49}{7+a}$$

$$=\dfrac{(a-7)\overset{1}{\cancel{(a+7)}}}{\underset{1}{\cancel{7+a}}}$$

$$=a-7$$

Section 7.5

Simplify each complex fraction. See Example 5.

41.
$$\left.\begin{array}{l} 2 = 2 \\ 4 = 2\cdot 2 \\ 5 = 5 \end{array}\right\} \text{LCD} = 2\cdot 2\cdot 5 = 20$$

$$\dfrac{\dfrac{d^2}{4} + \dfrac{4d}{5}}{\dfrac{d+1}{2}} = \dfrac{\dfrac{d^2}{4} + \dfrac{4d}{5}}{\dfrac{d+1}{2}} \cdot \dfrac{\mathbf{20}}{\mathbf{20}}$$

$$= \dfrac{\left(\dfrac{d^2}{4} + \dfrac{4d}{5}\right)20}{\left(\dfrac{d+1}{2}\right)20}$$

$$= \dfrac{\dfrac{d^2}{4}(20) + \dfrac{4d}{5}(20)}{\left(\dfrac{d+1}{2}\right)(20)}$$

$$= \dfrac{5d^2 + 16d}{10(d+1)}$$

$$= \dfrac{5d^2 + 16d}{10d + 10}$$

43.
$$\left.\begin{array}{l} x = x \\ y = y \end{array}\right\} \text{LCD} = x\cdot y = xy$$

$$\dfrac{\dfrac{2}{x}}{\dfrac{2}{y} - \dfrac{4}{x}} = \dfrac{\dfrac{2}{x}}{\dfrac{2}{y} - \dfrac{4}{x}} \cdot \dfrac{xy}{xy}$$

$$= \dfrac{\left(\dfrac{2}{x}\right)xy}{\left(\dfrac{2}{y} - \dfrac{4}{x}\right)xy}$$

$$= \dfrac{\left(\dfrac{2}{x}\right)(xy)}{\dfrac{2}{y}(xy) - \dfrac{4}{x}(xy)}$$

$$= \dfrac{2y}{2x - 4y}$$

$$= \dfrac{\overset{1}{\cancel{2}}\, y}{\underset{1}{\cancel{2}}(x - 2y)}$$

$$= \dfrac{y}{x - 2y}$$

Simplify each complex fraction. See Example 6.

45. $\text{LCD} = x+1$

$$\dfrac{\dfrac{1}{x+1}}{1 + \dfrac{1}{x+1}} = \dfrac{\dfrac{1}{x+1}}{1 + \dfrac{1}{x+1}} \cdot \dfrac{\mathbf{(x+1)}}{\mathbf{(x+1)}}$$

$$= \dfrac{\left(\dfrac{1}{x+1}\right)(x+1)}{\left(1 + \dfrac{1}{x+1}\right)(x+1)}$$

$$= \dfrac{\left(\dfrac{1}{x+1}\right)(x+1)}{1(x+1) + \left(\dfrac{1}{x+1}\right)(x+1)}$$

$$= \dfrac{1}{x+1+1}$$

$$= \dfrac{1}{x+2}$$

47. $\text{LCD} = x+2$

$$\dfrac{\dfrac{x}{x+2}}{\dfrac{x}{x+2} + x} = \dfrac{\dfrac{x}{x+2}}{\dfrac{x}{x+2} + x} \cdot \dfrac{\mathbf{(x+2)}}{\mathbf{(x+2)}}$$

$$= \dfrac{\left(\dfrac{x}{x+2}\right)(x+2)}{\left(\dfrac{x}{x+2} + x\right)(x+2)}$$

$$= \dfrac{\left(\dfrac{x}{x+2}\right)(x+2)}{\left(\dfrac{x}{x+2}\right)(x+2) + x(x+2)}$$

$$= \dfrac{x}{x + x^2 + 2x}$$

$$= \dfrac{x}{x^2 + 3x}$$

$$= \dfrac{\overset{1}{\cancel{x}}}{\underset{1}{\cancel{x}}(x+3)}$$

$$= \dfrac{1}{x+3}$$

TRY IT YOURSELF
Simplify each complex fraction.

49. $\text{LCD} = pq$

$$\dfrac{\dfrac{1}{p}+\dfrac{1}{q}}{\dfrac{1}{p}} = \dfrac{\dfrac{1}{p}+\dfrac{1}{q}}{\dfrac{1}{p}}\cdot\dfrac{\boldsymbol{pq}}{\boldsymbol{pq}}$$

$$= \dfrac{\left(\dfrac{1}{p}+\dfrac{1}{q}\right)pq}{\left(\dfrac{1}{p}\right)pq}$$

$$= \dfrac{\dfrac{1}{p}(pq)+\dfrac{1}{q}(pq)}{\dfrac{1}{p}(pq)}$$

$$= \dfrac{q+p}{q}$$

51. $$\dfrac{\dfrac{40x^2}{20x}}{9} = \dfrac{40x^2}{1}\div\dfrac{20x}{9}$$

$$= \dfrac{40x^2}{1}\cdot\dfrac{9}{20x}$$

$$= \dfrac{40\cdot 9\cdot x^2}{20x}$$

$$= \dfrac{2\cdot \overset{1}{\cancel{20}}\cdot 9\cdot \overset{1}{\cancel{x}}\cdot x}{\underset{1}{\cancel{20}}\,\underset{1}{\cancel{x}}}$$

$$= 18x$$

53. $\left.\begin{array}{l} 2 = 2 \\ 4 = 2\cdot 2 \\ c = c \\ c^2 = c\cdot c \end{array}\right\} \text{LCD} = 2\cdot 2\cdot c\cdot c = 4c^2$

$$\dfrac{\dfrac{1}{c}+\dfrac{1}{2}}{\dfrac{1}{c^2}-\dfrac{1}{4}} = \dfrac{\dfrac{1}{c}+\dfrac{1}{2}}{\dfrac{1}{c^2}-\dfrac{1}{4}}\cdot\dfrac{\boldsymbol{4c^2}}{\boldsymbol{4c^2}}$$

$$= \dfrac{\left(\dfrac{1}{c}+\dfrac{1}{2}\right)4c^2}{\left(\dfrac{1}{c^2}-\dfrac{1}{4}\right)4c^2}$$

$$= \dfrac{\dfrac{1}{c}(4c^2)+\dfrac{1}{2}(4c^2)}{\dfrac{1}{c^2}(4c^2)-\dfrac{1}{4}(4c^2)}$$

$$= \dfrac{4c+2c^2}{4-c^2}$$

$$= \dfrac{2c\,\overset{1}{\cancel{(2+c)}}}{\cancel{(2+c)}\,(2-c)}$$

$$= \dfrac{2c}{2-c}$$

55. $\text{LCD} = (r+1)(r-1)$

$$\dfrac{\dfrac{1}{r+1}+1}{\dfrac{3}{r-1}+1} = \dfrac{\dfrac{1}{r+1}+1}{\dfrac{3}{r-1}+1}\cdot\dfrac{\boldsymbol{(r+1)(r-1)}}{\boldsymbol{(r+1)(r-1)}}$$

$$= \dfrac{\left(\dfrac{1}{r+1}+1\right)(r+1)(r-1)}{\left(\dfrac{3}{r-1}+1\right)(r+1)(r-1)}$$

$$= \dfrac{\left(\dfrac{1}{r+1}\right)(r+1)(r-1)+1(r+1)(r-1)}{\left(\dfrac{3}{r-1}\right)(r+1)(r-1)+1(r+1)(r-1)}$$

$$= \dfrac{(r-1)+(r+1)(r-1)}{3(r+1)+(r+1)(r-1)}$$

$$= \dfrac{r-1+r^2-1}{3r+3+r^2-1}$$

$$= \dfrac{r^2+r-2}{r^2+3r+2}$$

$$= \dfrac{\overset{1}{\cancel{(r+2)}}\,(r-1)}{\underset{1}{\cancel{(r+2)}}\,(r+1)}$$

$$= \dfrac{r-1}{r+1}$$

57. $\dfrac{\dfrac{b^2-81}{18a^2}}{\dfrac{4b-36}{9a}} = \dfrac{b^2-81}{18a^2} \div \dfrac{4b-36}{9a}$

$\quad = \dfrac{b^2-81}{18a^2} \cdot \dfrac{9a}{4b-36}$

$\quad = \dfrac{(b^2-81)\cdot 9a}{18a^2(4b-36)}$

$\quad = \dfrac{\overset{1}{\cancel{9}}\cdot \overset{1}{\cancel{a}}\,(b\cancel{-9})(b+9)}{2\cdot \cancel{9}\,\cancel{a}\cdot a\cdot 4\,(b\cancel{-9})}$

$\quad = \dfrac{b+9}{8a}$

59. $\dfrac{\dfrac{10x}{x-3}}{\dfrac{6}{x-3}} = \dfrac{10x}{x-3} \div \dfrac{6}{x-3}$

$\quad = \dfrac{10x}{x-3} \cdot \dfrac{x-3}{6}$

$\quad = \dfrac{10x\cdot(x-3)}{(x-3)\cdot 6}$

$\quad = \dfrac{\overset{1}{\cancel{2}}\cdot 5\cdot x\,\overset{1}{\cancel{(x-3)}}}{\cancel{2}\cdot 3\,\cancel{(x-3)}}$

$\quad = \dfrac{5x}{3}$

61. $\left.\begin{array}{l} 8h = 2\cdot 2\cdot 2\cdot h \\ 4h = 2\cdot 2\cdot h \end{array}\right\} \text{LCD} = 2\cdot 2\cdot 2\cdot h = 8h$

$\dfrac{4-\dfrac{1}{8h}}{12+\dfrac{3}{4h}} = \dfrac{4-\dfrac{1}{8h}}{12+\dfrac{3}{4h}} \cdot \dfrac{\boldsymbol{8h}}{\boldsymbol{8h}}$

$\quad = \dfrac{\left(4-\dfrac{1}{8h}\right)8h}{\left(12+\dfrac{3}{4h}\right)8h}$

$\quad = \dfrac{4(8h)-\dfrac{1}{8h}(8h)}{12(8h)+\dfrac{3}{4h}(8h)}$

$\quad = \dfrac{32h-1}{96h+6}$

63. $\left.\begin{array}{l} n = n \\ m = m \end{array}\right\} \text{LCD} = n\cdot m = mn$

$\dfrac{\dfrac{m}{n}+\dfrac{n}{m}}{\dfrac{m}{n}-\dfrac{n}{m}} = \dfrac{\dfrac{m}{n}+\dfrac{n}{m}}{\dfrac{m}{n}-\dfrac{n}{m}} \cdot \dfrac{\boldsymbol{mn}}{\boldsymbol{mn}}$

$\quad = \dfrac{\left(\dfrac{m}{n}+\dfrac{n}{m}\right)mn}{\left(\dfrac{m}{n}-\dfrac{n}{m}\right)mn}$

$\quad = \dfrac{\dfrac{m}{n}(mn)+\dfrac{n}{m}(mn)}{\dfrac{m}{n}(mn)-\dfrac{n}{m}(mn)}$

$\quad = \dfrac{m^2+n^2}{m^2-n^2}$

65. $\left.\begin{array}{l} 4 = 2\cdot 2 \\ c = c \\ c^2 = c\cdot c \end{array}\right\} \text{LCD} = 2\cdot 2\cdot c\cdot c = 4c^2$

$\dfrac{\dfrac{2}{c^2}}{\dfrac{1}{c}+\dfrac{5}{4}} = \dfrac{\dfrac{2}{c^2}}{\dfrac{1}{c}+\dfrac{5}{4}} \cdot \dfrac{\boldsymbol{4c^2}}{\boldsymbol{4c^2}}$

$\quad = \dfrac{\left(\dfrac{2}{c^2}\right)4c^2}{\left(\dfrac{1}{c}+\dfrac{5}{4}\right)4c^2}$

$\quad = \dfrac{\left(\dfrac{2}{c^2}\right)(4c^2)}{\dfrac{1}{c}(4c^2)+\dfrac{5}{4}(4c^2)}$

$\quad = \dfrac{8}{4c+5c^2}$

67. $\dfrac{\dfrac{4t-8}{t^2}}{\dfrac{8t-16}{t^5}} = \dfrac{4t-8}{t^2} \div \dfrac{8t-16}{t^5}$

$\quad = \dfrac{4t-8}{t^2} \cdot \dfrac{t^5}{8t-16}$

$\quad = \dfrac{(4t-8)\cdot t^5}{t^2(8t-16)}$

$\quad = \dfrac{4t^5(t-2)}{8t^2(t-2)}$

$\quad = \dfrac{\overset{1}{\cancel{4}}\cdot \overset{1}{\cancel{t^2}}\cdot t^3\,\overset{1}{\cancel{(t-2)}}}{2\cdot \underset{1}{\cancel{4}}\cdot \underset{1}{\cancel{t^2}}\,\underset{1}{\cancel{(t-2)}}}$

$\quad = \dfrac{t^3}{2}$

69. $\left.\begin{array}{l} s=s \\ s^2 = s\cdot s \\ s^3 = s\cdot s\cdot s \end{array}\right\}$ LCD $= s\cdot s\cdot s = s^3$

$\dfrac{\dfrac{2}{s}-\dfrac{2}{s^2}}{\dfrac{4}{s^3}+\dfrac{4}{s^2}} = \dfrac{\dfrac{2}{s}-\dfrac{2}{s^2}}{\dfrac{4}{s^3}+\dfrac{4}{s^2}}\cdot \dfrac{s^3}{s^3}$

$\quad = \dfrac{\left(\dfrac{2}{s}-\dfrac{2}{s^2}\right)s^3}{\left(\dfrac{4}{s^3}+\dfrac{4}{s^2}\right)s^3}$

$\quad = \dfrac{\dfrac{2}{s}(s^3)-\dfrac{2}{s^2}(s^3)}{\dfrac{4}{s^3}(s^3)+\dfrac{4}{s^2}(s^3)}$

$\quad = \dfrac{2s^2-2s}{4+4s}$

$\quad = \dfrac{\overset{1}{\cancel{2}}\,s(s-1)}{2\cdot \underset{1}{\cancel{2}}(1+s)}$

$\quad = \dfrac{s(s-1)}{2(s+1)}$ or $\dfrac{s^2-s}{2+2s}$

71. $\left.\begin{array}{l} t=t \\ t^2 = t\cdot t \end{array}\right\}$ LCD $= t\cdot t = t^2$

$\dfrac{1+\dfrac{6}{t}+\dfrac{8}{t^2}}{1+\dfrac{1}{t}-\dfrac{12}{t^2}} = \dfrac{1+\dfrac{6}{t}+\dfrac{8}{t^2}}{1+\dfrac{1}{t}-\dfrac{12}{t^2}}\cdot \dfrac{t^2}{t^2}$

$\quad = \dfrac{\left(1+\dfrac{6}{t}+\dfrac{8}{t^2}\right)t^2}{\left(1+\dfrac{1}{t}-\dfrac{12}{t^2}\right)t^2}$

$\quad = \dfrac{1(t^2)+\dfrac{6}{t}(t^2)+\dfrac{8}{t^2}(t^2)}{1(t^2)+\dfrac{1}{t}(t^2)-\dfrac{12}{t^2}(t^2)}$

$\quad = \dfrac{t^2+6t+8}{t^2+t-12}$

$\quad = \dfrac{(t+2)\,\overset{1}{\cancel{(t+4)}}}{(t-3)\,\underset{1}{\cancel{(t+4)}}}$

$\quad = \dfrac{t+2}{t-3}$

73. LCD $= xy$

$\dfrac{\dfrac{1}{1}}{\dfrac{1}{x}+\dfrac{1}{y}} = \dfrac{1}{\dfrac{1}{x}+\dfrac{1}{y}}\cdot \dfrac{xy}{xy}$

$\quad = \dfrac{(1)xy}{\left(\dfrac{1}{x}+\dfrac{1}{y}\right)xy}$

$\quad = \dfrac{xy}{\dfrac{1}{x}(xy)+\dfrac{1}{y}(xy)}$

$\quad = \dfrac{xy}{y+x}$

Section 7.5

75. $\dfrac{-\dfrac{25}{16x^2}}{\dfrac{15}{32x^5}} = -\dfrac{25}{16x^2} \div \dfrac{15}{32x^5}$

$\qquad = -\dfrac{25}{16x^2} \cdot \dfrac{32x^5}{15}$

$\qquad = -\dfrac{25 \cdot 32x^5}{16x^2 \cdot 15}$

$\qquad = -\dfrac{5 \cdot \overset{1}{\cancel{5}} \cdot 2 \cdot \overset{1}{\cancel{16}} \cdot \overset{1}{\cancel{x^2}} \cdot x^3}{3 \cdot \underset{1}{\cancel{5}} \cdot \underset{1}{\cancel{16}} \cdot \underset{1}{\cancel{x^2}}}$

$\qquad = -\dfrac{10x^3}{3}$

77. LCD $= x - 1$

$\dfrac{3 + \dfrac{3}{x-1}}{3 - \dfrac{3}{x-1}} = \dfrac{3 + \dfrac{3}{x-1}}{3 - \dfrac{3}{x-1}} \cdot \dfrac{(x-1)}{(x-1)}$

$\qquad = \dfrac{\left(3 + \dfrac{3}{x-1}\right)(x-1)}{\left(3 - \dfrac{3}{x-1}\right)(x-1)}$

$\qquad = \dfrac{3(x-1) + \left(\dfrac{3}{x-1}\right)(x-1)}{3(x-1) - \left(\dfrac{3}{x-1}\right)(x-1)}$

$\qquad = \dfrac{3x - 3 + 3}{3x - 3 - 3}$

$\qquad = \dfrac{3x}{3x - 6}$

$\qquad = \dfrac{\overset{1}{\cancel{3}} x}{\underset{1}{\cancel{3}}(x-2)}$

$\qquad = \dfrac{x}{x-2}$

79. $\left. \begin{array}{l} d = d \\ d^2 = d \cdot d \end{array} \right\}$ LCD $= d \cdot d = d^2$

$\dfrac{1 - \dfrac{9}{d^2}}{2 + \dfrac{6}{d}} = \dfrac{1 - \dfrac{9}{d^2}}{2 + \dfrac{6}{d}} \cdot \dfrac{\boldsymbol{d^2}}{\boldsymbol{d^2}}$

$\qquad = \dfrac{\left(1 - \dfrac{9}{d^2}\right)d^2}{\left(2 + \dfrac{6}{d}\right)d^2}$

$\qquad = \dfrac{1(d^2) - \dfrac{9}{d^2}(d^2)}{2(d^2) + \dfrac{6}{d}(d^2)}$

$\qquad = \dfrac{d^2 - 9}{2d^2 + 6d}$

$\qquad = \dfrac{(d-3)\,\overset{1}{\cancel{(d+3)}}}{2d\,\underset{1}{\cancel{(d+3)}}}$

$\qquad = \dfrac{d-3}{2d}$

81. $\left. \begin{array}{l} a^2b = a \cdot a \cdot b \\ ab = a \cdot b \\ ab^2 = a \cdot b \cdot b \end{array} \right\}$ LCD $= a^2 \cdot b^2 = a^2b^2$

$\dfrac{\dfrac{1}{a^2b} - \dfrac{5}{ab}}{\dfrac{3}{ab} - \dfrac{7}{ab^2}} = \dfrac{\dfrac{1}{a^2b} - \dfrac{5}{ab}}{\dfrac{3}{ab} - \dfrac{7}{ab^2}} \cdot \dfrac{\boldsymbol{a^2b^2}}{\boldsymbol{a^2b^2}}$

$\qquad = \dfrac{\left(\dfrac{1}{a^2b} - \dfrac{5}{ab}\right)a^2b^2}{\left(\dfrac{3}{ab} - \dfrac{7}{ab^2}\right)a^2b^2}$

$\qquad = \dfrac{\dfrac{1}{a^2b}(a^2b^2) - \dfrac{5}{ab}(a^2b^2)}{\dfrac{3}{ab}(a^2b^2) - \dfrac{7}{ab^2}(a^2b^2)}$

$\qquad = \dfrac{b - 5ab}{3ab - 7a}$ or $\dfrac{b(1 - 5a)}{a(3b - 7)}$

83. $\text{LCD} = 2m+1$

$$\dfrac{m - \dfrac{1}{2m+1}}{1 - \dfrac{m}{2m+1}} = \dfrac{m - \dfrac{1}{2m+1}}{1 - \dfrac{m}{2m+1}} \cdot \dfrac{(2m+1)}{(2m+1)}$$

$$= \dfrac{\left(m - \dfrac{1}{2m+1}\right)(2m+1)}{\left(1 - \dfrac{m}{2m+1}\right)(2m+1)}$$

$$= \dfrac{m(2m+1) - \left(\dfrac{1}{2m+1}\right)(2m+1)}{1(2m+1) - \left(\dfrac{m}{2m+1}\right)(2m+1)}$$

$$= \dfrac{2m^2 + m - 1}{2m+1-m}$$

$$= \dfrac{2m^2 + m - 1}{m+1}$$

$$= \dfrac{\overset{1}{\cancel{(m+1)}}(2m-1)}{\underset{1}{\cancel{m+1}}}$$

$$= 2m - 1$$

APPLICATIONS

85. SLOPE

$$\left.\begin{array}{l} 2 = 2 \\ 3 = 3 \\ 4 = 2 \cdot 2 \\ 8 = 2 \cdot 2 \cdot 2 \end{array}\right\} \text{LCD} = 2 \cdot 2 \cdot 2 \cdot 3 = 24$$

$$m = \dfrac{\dfrac{5}{8} - \dfrac{1}{3}}{\dfrac{3}{4} - \dfrac{1}{2}}$$

$$= \dfrac{\left(\dfrac{5}{8} - \dfrac{1}{3}\right)\mathbf{24}}{\left(\dfrac{3}{4} - \dfrac{1}{2}\right)\mathbf{24}}$$

$$= \dfrac{\left(\dfrac{5}{8}\right)24 - \left(\dfrac{1}{3}\right)24}{\left(\dfrac{3}{4}\right)24 - \left(\dfrac{1}{2}\right)24}$$

$$= \dfrac{15 - 8}{18 - 12}$$

$$= \dfrac{7}{6}$$

87. ELECTRONICS

$\text{LCD} = R_1 R_2$

$$\text{Total resistance} = \dfrac{1}{\dfrac{1}{R_1} + \dfrac{1}{R_2}}$$

$$= \dfrac{1(R_1 R_2)}{\left(\dfrac{1}{R_1} + \dfrac{1}{R_2}\right)R_1 R_2}$$

$$= \dfrac{R_1 R_2}{\left(\dfrac{1}{R_1}\right)R_1 R_2 + \left(\dfrac{1}{R_2}\right)R_1 R_2}$$

$$= \dfrac{R_1 R_2}{R_2 + R_1}$$

WRITING

89-92. Answers will vary.

REVIEW

Simplify each expression. Write each answer without negative exponents.

93. $(8x)^0 = 1$

95.
$$\left(\dfrac{4x^3}{5x^{-3}}\right)^{-2} = \left(\dfrac{4x^{3-(-3)}}{5}\right)^{-2}$$

$$= \left(\dfrac{4x^{3+3}}{5}\right)^{-2}$$

$$= \left(\dfrac{4x^6}{5}\right)^{-2}$$

$$= \left(\dfrac{5}{4x^6}\right)^{2}$$

$$= \dfrac{5^2}{4^2 x^{6(2)}}$$

$$= \dfrac{25}{16x^{12}}$$

CHALLENGE PROBLEMS

Simplify.

97.
$$\left.\begin{array}{l} h+1 = h+1 \\ h+2 = h+2 \\ h^2 + 3h + 2 = (h+1)(h+2) \end{array}\right\}$$

$$\text{LCD} = (h+1)(h+2)$$

$$\frac{\dfrac{h}{h^2+3h+2}}{\dfrac{4}{h+2}-\dfrac{4}{h+1}} = \frac{\left(\dfrac{h}{(h+1)(h+2)}\right)}{\left(\dfrac{4}{h+2}-\dfrac{4}{h+1}\right)} \cdot \frac{(h+1)(h+2)}{(h+1)(h+2)}$$

$$= \frac{\left(\dfrac{h}{(h+1)(h+2)}\right)(h+1)(h+2)}{\left(\dfrac{4}{h+2}\right)(h+1)(h+2)-\left(\dfrac{4}{h+1}\right)(h+1)(h+2)}$$

$$= \frac{h}{4(h+1)-4(h+2)}$$

$$= \frac{h}{4h+4-4h-8}$$

$$= \frac{h}{-4}$$

$$= -\frac{h}{4}$$

99. $\text{LCD} = a+1$

$$a + \frac{a}{1+\dfrac{a}{a+1}} = a + \frac{a}{\left(1+\dfrac{a}{a+1}\right)} \cdot \frac{(a+1)}{(a+1)}$$

$$= a + \frac{a(a+1)}{1(a+1)+\left(\dfrac{a}{a+1}\right)(a+1)}$$

$$= a + \frac{a(a+1)}{a+1+a}$$

$$= a + \frac{a^2+a}{2a+1}$$

$\text{LCD} = 2a+1$

$$= a \cdot \frac{(2a+1)}{(2a+1)} + \frac{a^2+a}{2a+1}$$

$$= \frac{a(2a+1)+a^2+a}{2a+1}$$

$$= \frac{2a^2+a+a^2+a}{2a+1}$$

$$= \frac{3a^2+2a}{2a+1}$$

SECTION 7.6 STUDY SET

VOCABULARY

Fill in the blanks.

1. Equations that contain one or more rational expressions, such as $\frac{x}{x+2} = 4 + \frac{10}{x+2}$, are called **rational** equations.

3. To **clear** a rational equation of fractions, multiply both sides by the LCD of all rational expressions in the equation.

CONCEPTS

5. a.
$$\frac{1}{x-1} = 1 - \frac{3}{x-1}$$
$$\frac{1}{5-1} \overset{?}{=} 1 - \frac{3}{5-1}$$
$$\frac{1}{4} \overset{?}{=} 1 - \frac{3}{4}$$
$$\frac{1}{4} \overset{?}{=} \frac{4}{4} - \frac{3}{4}$$
$$\frac{1}{4} = \frac{1}{4}$$
Yes

b.
$$\frac{x}{x-5} = 3 + \frac{5}{x-5}$$
$$\frac{5}{5-5} \overset{?}{=} 3 + \frac{5}{5-5}$$
$$\frac{5}{0} \overset{?}{=} 3 + \frac{5}{0}$$
Undefined
No

7. $\dfrac{x}{x-3} = \dfrac{1}{x} + \dfrac{2}{x-3}$

  a. $x = 0$ and $x - 3 = 0$
$$x = 3$$

  b. **0, 3**

  c. **0, 3**

By what should both sides of the equation be multiplied to clear it of fractions?

9. a. y

  b. $(x+2)(x-2)$

11. a. $4x\left(\dfrac{3}{4x}\right) = \dfrac{\overset{1}{\cancel{4}} \cdot \overset{1}{\cancel{x}} \cdot 3}{\underset{1}{\cancel{4}} \cdot \underset{1}{\cancel{x}}}$
$$= 3$$

  b. $(x+6)(x-2)\left(\dfrac{3}{x-2}\right)$
$$= \dfrac{3(x+6)\,\overset{1}{\cancel{(x-2)}}}{\underset{1}{\cancel{x-2}}}$$
$$= 3(x+6) \text{ or } 3x+18$$

NOTATION

13.
$$\frac{2}{a} + \frac{1}{2} = \frac{7}{2a}$$
$$\boxed{2a}\left(\frac{2}{a} + \frac{1}{2}\right) = \boxed{2a}\left(\frac{7}{2a}\right)$$
$$\boxed{2a}\left(\frac{2}{a}\right) + \boxed{2a}\left(\frac{1}{2}\right) = \boxed{2a}\left(\frac{7}{2a}\right)$$
$$\boxed{4} + a = \boxed{7}$$
$$4 + a - 4 = 7 - \boxed{4}$$
$$a = \boxed{3}$$

GUIDED PRACTICE

Solve each equation and check the result. If an equation has no solution, so indicate.
Only a selected few of the rational equations are checked due to the complex nature of checking them. See Example 1.

15. $\left.\begin{array}{l} 2 = 2 \\ 3 = 3 \\ 6 = 2 \cdot 3 \end{array}\right\}$ LCD $= 2 \cdot 3 = 6$

$$\frac{2}{3} = \frac{1}{2} + \frac{x}{6}$$
$$6\left(\frac{2}{3}\right) = 6\left(\frac{1}{2} + \frac{x}{6}\right)$$
$$6\left(\frac{2}{3}\right) = 6\left(\frac{1}{2}\right) + 6\left(\frac{x}{6}\right)$$
$$2 \cdot \overset{1}{\cancel{3}}\left(\frac{2}{\underset{1}{\cancel{3}}}\right) = 3 \cdot \overset{1}{\cancel{2}}\left(\frac{1}{\underset{1}{\cancel{2}}}\right) + \overset{1}{\cancel{6}}\left(\frac{x}{\underset{1}{\cancel{6}}}\right)$$
$$4 = 3 + x$$
$$4 - 3 = 3 + x - 3$$
$$1 = x$$

check: $\dfrac{2}{3} = \dfrac{1}{2} + \dfrac{x}{6}$ $\dfrac{2}{3} \overset{?}{=} \dfrac{4}{6}$

$\dfrac{2}{3} \overset{?}{=} \dfrac{1}{2} + \dfrac{1}{6}$ $\dfrac{2}{3} = \dfrac{2}{3}$

$\dfrac{2}{3} \overset{?}{=} \dfrac{3}{6} + \dfrac{1}{6}$

This checks, so the solution is 1.

17. $\left.\begin{array}{l} 2 = 2 \\ 4 = 2 \cdot 2 \\ 12 = 2 \cdot 2 \cdot 3 \end{array}\right\}$ LCD $= 2 \cdot 2 \cdot 3 = 12$

$$\frac{s}{12} - \frac{s}{2} = \frac{5s}{4}$$

$$12\left(\frac{s}{12} - \frac{s}{2}\right) = 12\left(\frac{5s}{4}\right)$$

$$12\left(\frac{s}{12}\right) - 12\left(\frac{s}{2}\right) = 12\left(\frac{5s}{4}\right)$$

$$\overset{1}{\cancel{12}}\left(\frac{s}{\cancel{12}}\right) - 6\cdot\cancel{2}\left(\frac{s}{\cancel{2}}\right) = 3\cdot\cancel{4}\left(\frac{5s}{\cancel{4}}\right)$$

$$s - 6s = 15s$$
$$-5s = 15s$$
$$-5s + 5s = 15s + 5s$$
$$0 = 20s$$
$$\frac{0}{20} = \frac{20s}{20}$$
$$0 = s$$

This checks, so the solution is 0.

19. $\left.\begin{array}{r} 2 = 2 \\ 3 = 3 \\ 18 = 2\cdot3\cdot3 \end{array}\right\} \mathrm{LCD} = 2\cdot3\cdot3 = 18$

$$\frac{x}{18} = \frac{1}{3} - \frac{x}{2}$$

$$18\left(\frac{x}{18}\right) = 18\left(\frac{1}{3} - \frac{x}{2}\right)$$

$$18\left(\frac{x}{18}\right) = 18\left(\frac{1}{3}\right) - 18\left(\frac{x}{2}\right)$$

$$\overset{1}{\cancel{18}}\left(\frac{x}{\cancel{18}}\right) = 6\cdot\cancel{3}\left(\frac{1}{\cancel{3}}\right) - 9\cdot\cancel{2}\left(\frac{x}{\cancel{2}}\right)$$

$$x = 6 - 9x$$
$$x + 9x = 6 - 9x + 9x$$
$$10x = 6$$
$$\frac{10x}{10} = \frac{6}{10}$$
$$x = \frac{\cancel{2}\cdot3}{\cancel{2}\cdot5}$$
$$x = \frac{3}{5}$$

check: $\dfrac{x}{18} = \dfrac{1}{3} - \dfrac{x}{2}$

$x = \dfrac{3}{5}$. Convert to a decimal to make the

checking easier. $\dfrac{3}{5} = 0.6$

$$\frac{x}{18} = \frac{1}{3} - \frac{x}{2}$$

$$\frac{0.6}{18} \overset{?}{=} \frac{1}{3} - \frac{0.6}{2}$$

$$0.03\overline{3} \overset{?}{=} 0.33\overline{3} - 0.3$$

$$0.03\overline{3} \approx 0.03\overline{3}$$

This checks, so the solution is $\dfrac{3}{5}$.

21. $\left.\begin{array}{r} 4 = 2\cdot2 \\ 2 = 2 \\ 3 = 3 \end{array}\right\} \mathrm{LCD} = 2\cdot2\cdot3 = 12$

$$\frac{b}{4} + \frac{1}{2} = \frac{b}{3} - \frac{1}{4}$$

$$12\left(\frac{b}{4} + \frac{1}{2}\right) = 12\left(\frac{b}{3} - \frac{1}{4}\right)$$

$$12\left(\frac{b}{4}\right) + 12\left(\frac{1}{2}\right) = 12\left(\frac{b}{3}\right) - 12\left(\frac{1}{4}\right)$$

$$3\cdot\cancel{4}\left(\frac{b}{\cancel{4}}\right) + 6\cdot\cancel{2}\left(\frac{1}{\cancel{2}}\right) = 4\cdot\cancel{3}\left(\frac{b}{\cancel{3}}\right) - 3\cdot\cancel{4}\left(\frac{1}{\cancel{4}}\right)$$

$$3b + 6 = 4b - 3$$
$$3b + 6 - 3b = 4b - 3 - 3b$$
$$6 = b - 3$$
$$6 + 3 = b - 3 + 3$$
$$9 = b$$

Check: $\dfrac{b}{4} + \dfrac{1}{2} = \dfrac{b}{3} - \dfrac{1}{4}$

$$\frac{9}{4} + \frac{1}{2} \overset{?}{=} \frac{9}{3} - \frac{1}{4}$$

$$\frac{9}{4} + \frac{2}{4} \overset{?}{=} \frac{36}{12} - \frac{3}{12}$$

$$\frac{11}{4} \overset{?}{=} \frac{33}{12}$$

$$\frac{11}{4} = \frac{11}{4}$$

This checks, so the solution is 9.

Solve each equation and check the result. If an equation has no solution, so indicate. See Example 2.

23.
$$\left.\begin{array}{l} k = k \\ 3k = 3 \cdot k \end{array}\right\} \text{LCD} = 3 \cdot k = 3k$$

$$\frac{5}{3k} + \frac{1}{k} = -2$$

$$3k\left(\frac{5}{3k} + \frac{1}{k}\right) = 3k(-2)$$

$$3k\left(\frac{5}{3k}\right) + 3k\left(\frac{1}{k}\right) = 3k(-2)$$

$$\overset{1}{\cancel{3k}}\left(\frac{5}{\cancel{3k}}\right) + 3 \cdot \overset{1}{\cancel{k}}\left(\frac{1}{\cancel{k}}\right) = -6k$$

$$5 + 3 = -6k$$

$$8 = -6k$$

$$\frac{8}{-6} = \frac{-6k}{-6}$$

$$-\frac{\overset{}{\cancel{2}} \cdot 4}{\underset{1}{\cancel{2}} \cdot 3} = k$$

$$k = -\frac{4}{3}$$

This checks, so the solution is $-\dfrac{4}{3}$.

25.
$$\left.\begin{array}{l} 4 = 2 \cdot 2 \\ 6 = 2 \cdot 3 \\ a = a \end{array}\right\} \text{LCD} = 2 \cdot 2 \cdot 3 \cdot a = 12a$$

$$\frac{1}{4} - \frac{5}{6} = \frac{1}{a}$$

$$12a\left(\frac{1}{4} - \frac{5}{6}\right) = 12a\left(\frac{1}{a}\right)$$

$$12a\left(\frac{1}{4}\right) - 12a\left(\frac{5}{6}\right) = 12a\left(\frac{1}{a}\right)$$

$$3a \cdot \overset{1}{\cancel{4}}\left(\frac{1}{\cancel{4}}\right) - 2a \cdot \overset{1}{\cancel{6}}\left(\frac{5}{\cancel{6}}\right) = 12 \cdot \overset{1}{\cancel{a}}\left(\frac{1}{\cancel{a}}\right)$$

$$3a - 10a = 12$$

$$-7a = 12$$

$$\frac{-7a}{-7} = \frac{12}{-7}$$

$$a = -\frac{12}{7}$$

This checks, so the solution is $-\dfrac{12}{7}$.

27.
$$\left.\begin{array}{l} 8 = 2 \cdot 2 \cdot 2 \\ 12 = 2 \cdot 2 \cdot 3 \\ b = b \end{array}\right\} \text{LCD} = 2 \cdot 2 \cdot 2 \cdot 3 \cdot b = 24b$$

$$\frac{1}{8} + \frac{2}{b} - \frac{1}{12} = 0$$

$$24b\left(\frac{1}{8} + \frac{2}{b} - \frac{1}{12}\right) = 24b(0)$$

$$24b\left(\frac{1}{8}\right) + 24b\left(\frac{2}{b}\right) - 24b\left(\frac{1}{12}\right) = 24b(0)$$

$$3b \cdot \overset{}{\cancel{8}}\left(\frac{1}{\underset{1}{\cancel{8}}}\right) + 24\,\overset{}{\cancel{b}}\left(\frac{2}{\underset{1}{\cancel{b}}}\right) - 2b \cdot \overset{}{\cancel{12}}\left(\frac{1}{\underset{1}{\cancel{12}}}\right) = 0$$

$$3b + 48 - 2b = 0$$

$$b + 48 = 0$$

$$b + 48 - 48 = 0 - 48$$

$$b = -48$$

This checks, so the solution is -48.

29.
$$\left.\begin{array}{l} 5 = 5 \\ 10x = 2 \cdot 5 \cdot x \\ 15 = 3 \cdot 5 \end{array}\right\} \text{LCD} = 2 \cdot 3 \cdot 5 \cdot x = 30x$$

$$\frac{4}{5} - \frac{1}{10x} = \frac{7}{15}$$

$$30x\left(\frac{4}{5} - \frac{1}{10x}\right) = 30x\left(\frac{7}{15}\right)$$

$$30x\left(\frac{4}{5}\right) - 30x\left(\frac{1}{10x}\right) = 30x\left(\frac{7}{15}\right)$$

$$6x \cdot \overset{}{\cancel{5}}\left(\frac{4}{\underset{1}{\cancel{5}}}\right) - 3 \cdot \overset{}{\cancel{10x}}\left(\frac{1}{\underset{1}{\cancel{10x}}}\right) = 2x \cdot \overset{}{\cancel{15}}\left(\frac{7}{\underset{1}{\cancel{15}}}\right)$$

$$24x - 3 = 14x$$

$$24x - 3 - 24x = 14x - 24x$$

$$-3 = -10x$$

$$\frac{-3}{-10} = \frac{-10x}{-10}$$

$$x = \frac{3}{10}$$

This checks, so the solution is $\dfrac{3}{10}$.

Solve each equation and check the result. If an equation has no solution, so indicate. See Example 3.

31. $LCD = x$

$$x + \frac{8}{x} = 6$$

$$x\left(x + \frac{8}{x}\right) = x(6)$$

$$x(x) + x\left(\frac{8}{x}\right) = x(6)$$

$$x^2 + \overset{1}{\cancel{x}}\left(\frac{8}{\cancel{x}}\right) = 6x$$

$$x^2 + 8 = 6x$$

$$x^2 + 8 - 6x = 6x - 6x$$

$$x^2 - 6x + 8 = 0$$

$$(x-2)(x-4) = 0$$

$$x - 2 = 0 \qquad \text{or} \qquad x - 4 = 0$$
$$x - 2 + 2 = 0 + 2 \qquad\qquad x - 4 + 4 = 0 + 4$$
$$x = 2 \qquad\qquad\qquad x = 4$$

check: $x + \dfrac{8}{x} = 6$ \qquad check: $x + \dfrac{8}{x} = 6$

$$2 + \frac{8}{2} \overset{?}{=} 6 \qquad\qquad 4 + \frac{8}{4} \overset{?}{=} 6$$

$$2 + 4 \overset{?}{=} 6 \qquad\qquad 4 + 2 \overset{?}{=} 6$$

$$6 = 6 \qquad\qquad\qquad 6 = 6$$

Both check, so the solutions are 2 and 4.

33. $LCD = t$

$$\frac{10}{t} - t = 3$$

$$t\left(\frac{10}{t} - t\right) = t(3)$$

$$t\left(\frac{10}{t}\right) - t(t) = t(3)$$

$$\overset{1}{\cancel{t}}\left(\frac{10}{\cancel{t}}\right) - t^2 = 3t$$

$$10 - t^2 = 3t$$

$$10 - t^2 - 3t = 3t - 3t$$

$$-t^2 - 3t + 10 = 0$$

$$-(t^2 + 3t - 10) = 0$$

$$-(t-2)(t+5) = 0$$

$$t - 2 = 0 \qquad \text{or} \qquad t + 5 = 0$$
$$t - 2 + 2 = 0 + 2 \qquad\qquad t + 5 - 5 = 0 - 5$$
$$t = 2 \qquad\qquad\qquad t = -5$$

check: $\dfrac{10}{t} - t = 3$ \qquad check: $\dfrac{10}{t} - t = 3$

$$\frac{10}{2} - 2 \overset{?}{=} 3 \qquad\qquad \frac{10}{-5} - (-5) \overset{?}{=} 3$$

$$5 - 2 \overset{?}{=} 3 \qquad\qquad -2 + 5 \overset{?}{=} 3$$

$$3 = 3 \qquad\qquad\qquad 3 = 3$$

Both check, so the solutions are 2 and -5.

35. $LCD = c$

$$\frac{20}{c} + c = -9$$

$$c\left(\frac{20}{c} + c\right) = c(-9)$$

$$c\left(\frac{20}{c}\right) + c(c) = c(-9)$$

$$\overset{1}{\cancel{c}}\left(\frac{20}{\cancel{c}}\right) + c^2 = -9c$$

$$20 + c^2 = -9c$$

$$20 + c^2 + 9c = -9c + 9c$$

$$c^2 + 9c + 20 = 0$$

$$(c+4)(c+5) = 0$$

$$c + 4 = 0 \qquad \text{or} \qquad c + 5 = 0$$
$$c + 4 - 4 = 0 - 4 \qquad\qquad c + 5 - 5 = 0 - 5$$
$$c = -4 \qquad\qquad\qquad c = -5$$

Both check, so the solutions are -4 and -5.

37. $\text{LCD} = p$

$$4 + \frac{15}{p} = 3p$$

$$p\left(4 + \frac{15}{p}\right) = p(3p)$$

$$p(4) + p\left(\frac{15}{p}\right) = p(3p)$$

$$4p + \overset{1}{\cancel{p}}\left(\frac{15}{\cancel{p}}\right) = 3p^2$$

$$4p + 15 = 3p^2$$
$$4p + 15 - \mathbf{4p} = 3p^2 - \mathbf{4p}$$
$$15 = 3p^2 - 4p$$
$$15 - \mathbf{15} = 3p^2 - 4p - \mathbf{15}$$
$$0 = 3p^2 - 4p - 15$$
$$0 = (3p + 5)(p - 3)$$

$$3p + 5 = 0 \qquad \text{or} \qquad p - 3 = 0$$
$$3p + 5 - 5 = 0 - 5 \qquad\quad p - 3 + 3 = 0 + 3$$
$$\frac{3p}{3} = \frac{-5}{3} \qquad\qquad\qquad p = 3$$
$$p = -\frac{5}{3}$$

Both check, so the solutions are $-\dfrac{5}{3}$ and 3.

Solve each equation and check the result. If an equation has no solution, so indicate. See Example 4.

39. $\text{LCD} = x - 5$

$$\frac{x}{x-5} = 3 + \frac{5}{x-5}$$

$$(x-5)\left(\frac{x}{x-5}\right) = (x-5)\left(3 + \frac{5}{x-5}\right)$$

$$(x-5)\left(\frac{x}{x-5}\right) = (x-5)(3) + (x-5)\left(\frac{5}{x-5}\right)$$

$$\overset{1}{\cancel{(x-5)}}\left(\frac{x}{\cancel{x-5}}\right) = 3x - 15 + \overset{1}{\cancel{(x-5)}}\left(\frac{5}{\cancel{x-5}}\right)$$

$$x = 3x - 15 + 5$$
$$x - \mathbf{x} = 3x - 10 - \mathbf{x}$$
$$0 = 2x - 10$$
$$0 + \mathbf{10} = 2x - 10 + \mathbf{10}$$
$$10 = 2x$$
$$\frac{10}{2} = \frac{2x}{2}$$
$$5 = x$$

check: $\dfrac{x}{x-5} = 3 + \dfrac{5}{x-5}$

$$\frac{x}{5-5} = 3 + \frac{5}{5-5}$$

$$\frac{x}{\mathbf{0}} = 3 + \frac{5}{\mathbf{0}}$$

5 makes the denominators of the original equation 0. No solution. 5 is extraneous.

41. $\text{LCD} = a + 2$

$$\frac{a^2}{a+2} - a = \frac{4}{a+2}$$

$$(a+2)\left(\frac{a^2}{a+2} - a\right) = (a+2)\left(\frac{4}{a+2}\right)$$

$$(a+2)\left(\frac{a^2}{a+2}\right) - (a+2)(a) = (a+2)\left(\frac{4}{a+2}\right)$$

$$\overset{1}{\cancel{(a+2)}}\left(\frac{a^2}{\cancel{a+2}}\right) - (a+2)(a) = \overset{1}{\cancel{(a+2)}}\left(\frac{4}{\cancel{a+2}}\right)$$

$$a^2 - a(a+2) = 4$$
$$a^2 - a^2 - 2a = 4$$
$$-2a = 4$$
$$\frac{-2a}{-2} = \frac{4}{-2}$$
$$a = -2$$

check: $\dfrac{a^2}{a+2} - a = \dfrac{4}{a+2}$

$$\frac{a^2}{-2+2} - a = \frac{4}{-2+2}$$

$$\frac{a^2}{\mathbf{0}+2} - a = \frac{4}{\mathbf{0}+2}$$

-2 makes the denominators of the original equation 0. No solution. -2 is extraneous.

Section 7.6

Solve each equation and check the result. If an equation has no solution, so indicate. See Example 5.

43. $LCD = (x+4)(x-3)$

$$\frac{x+6}{x+4} + \frac{1}{x^2+x-12} = 1$$

$$\frac{x+6}{x+4} + \frac{1}{(x+4)(x-3)} = 1$$

$$(x+4)(x-3)\left(\frac{x+6}{x+4} + \frac{1}{(x+4)(x-3)}\right)$$
$$= (x+4)(x-3)(1)$$

$$(x+4)(x-3)\left(\frac{x+6}{x+4}\right) + (x+4)(x-3)\left(\frac{1}{(x+4)(x-3)}\right)$$
$$= (x+4)(x-3)$$

$$(x-3)\overset{1}{(\cancel{x+4})}\left(\frac{x+6}{\cancel{x+4}}\right) + \overset{1}{(\cancel{x+4})}\overset{1}{(\cancel{x-3})}\left(\frac{1}{\underset{1}{\cancel{(x+4)}}\underset{1}{\cancel{(x-3)}}}\right)$$
$$= (x+4)(x-3)$$

$$(x-3)(x+6)+1=(x+4)(x-3)$$
$$x^2+3x-18+1=x^2+x-12$$
$$x^2+3x-17=x^2+x-12$$
$$x^2+3x-17-x^2=x^2+x-12-x^2$$
$$3x-17=x-12$$
$$3x-17-x=x-12-x$$
$$2x-17=-12$$
$$2x-17+17=-12+17$$
$$2x=5$$
$$\frac{2x}{2}=\frac{5}{2}$$
$$x=\frac{5}{2}$$

This checks, so the solution is $\frac{5}{2}$.

45. $\left.\begin{array}{l}x+2=x+2\\x^2+x-2=(x+2)(x-1)\end{array}\right\}LCD=(x+2)(x-1)$

$$\frac{2x}{x^2+x-2} + \frac{2}{x+2} = 1$$

$$(x+2)(x-1)\cdot\left(\frac{2x}{(x+2)(x-1)}\right) + (x+2)(x-1)\cdot\left(\frac{2}{x+2}\right)$$
$$= (x+2)(x-1)\cdot 1$$

$$\frac{2x\,\overset{1}{\cancel{(x+2)}}\,\overset{1}{\cancel{(x-1)}}}{\underset{1}{\cancel{(x+2)}}\,\underset{1}{\cancel{(x-1)}}} + \frac{2\,\overset{1}{\cancel{(x+2)}}(x-1)}{\underset{1}{\cancel{(x+2)}}} = (x+2)(x-1)$$

$$2x+2(x-1)=(x+2)(x-1)$$
$$2x+2x-2=x^2+x-2$$
$$4x-2=x^2+x-2$$
$$4x-2-4x=x^2+x-2-4x$$
$$-2=x^2-3x-2$$
$$-2+2=x^2-3x-2+2$$
$$0=x^2-3x$$
$$0=x(x-3)$$

$$x=0 \quad \text{or} \quad \begin{array}{l}x-3=0\\x-3+3=0+3\\x=3\end{array}$$

Both check, so the solutions are 0 and 3.

Solve each formula for the indicated variable. See Example 6.

47. $LCD = b+d$

$$h = \frac{2A}{b+d}$$

$$(b+d)(h) = (b+d)\left(\frac{2A}{b+d}\right)$$

$$(b+d)(h) = \overset{1}{\cancel{(b+d)}}\left(\frac{2A}{\underset{1}{\cancel{b+d}}}\right)$$

$$h(b+d) = 2A$$

$$\frac{h(b+d)}{2} = \frac{\overset{1}{\cancel{2}}A}{\underset{1}{\cancel{2}}}$$

$$A = \frac{h(b+d)}{2}$$

49. $LCD = R+r$

$$I = \frac{E}{R+r}$$

$$(R+r)(I) = (R+r)\left(\frac{E}{R+r}\right)$$

$$(R+r)(I) = \overset{1}{\cancel{(R+r)}}\left(\frac{E}{\underset{1}{\cancel{R+r}}}\right)$$

$$I(R+r) = E$$
$$IR + Ir = E$$
$$IR + Ir - IR = E - IR$$
$$Ir = E - IR$$

$$\frac{\overset{1}{\cancel{I}}r}{\underset{1}{\cancel{I}}} = \frac{E-IR}{I}$$

$$r = \frac{E-IR}{I}$$

51. LCD $= xyz$

$$\frac{5}{x} - \frac{4}{y} = \frac{5}{z}$$

$$(xyz)\left(\frac{5}{x} - \frac{4}{y}\right) = (xyz)\left(\frac{5}{z}\right)$$

$$\overset{1}{\cancel{x}}\, yz\left(\frac{5}{\cancel{x}_1}\right) - x\,\overset{1}{\cancel{y}}\, z\left(\frac{4}{\cancel{y}_1}\right) = xy\,\overset{1}{\cancel{z}}\left(\frac{5}{\cancel{z}_1}\right)$$

$$5yz - 4xz = 5xy$$

$$5yz - 4xz + 4xz = 5xy + 4xz$$

$$5yz = x(5y + 4z)$$

$$\frac{5yz}{5y + 4z} = \frac{x\,\overset{1}{\cancel{(5y+4z)}}}{\cancel{5y+4z}_1}$$

$$x = \frac{5yz}{5y + 4z}$$

53. LCD $= rst$

$$\frac{1}{r} + \frac{1}{s} = \frac{1}{t}$$

$$(rst)\left(\frac{1}{r} + \frac{1}{s}\right) = (rst)\left(\frac{1}{t}\right)$$

$$\overset{1}{\cancel{r}}\, st\left(\frac{1}{\cancel{r}_1}\right) + r\,\overset{1}{\cancel{s}}\, t\left(\frac{1}{\cancel{s}_1}\right) = rs\,\overset{1}{\cancel{t}}\left(\frac{1}{\cancel{t}_1}\right)$$

$$st + rt = rs$$

$$st + rt - rt = rs - rt$$

$$st = r(s - t)$$

$$\frac{st}{s - t} = \frac{r\,\overset{1}{\cancel{(s-t)}}}{\cancel{s-t}_1}$$

$$r = \frac{st}{s - t}$$

Solve each formula for the indicated variable. See Example 7.

55. LCD $= n$

$$\frac{P}{n} = rt$$

$$n\left(\frac{P}{n}\right) = n(rt)$$

$$\overset{1}{\cancel{n}}\left(\frac{P}{\cancel{n}_1}\right) = nrt$$

$$P = nrt$$

57. LCD $= bd$

$$\frac{a}{b} = \frac{c}{d}$$

$$bd\left(\frac{a}{b}\right) = bd\left(\frac{c}{d}\right)$$

$$d\,\overset{1}{\cancel{b}}\left(\frac{a}{\cancel{b}_1}\right) = b\,\overset{1}{\cancel{d}}\left(\frac{c}{\cancel{d}_1}\right)$$

$$ad = bc$$

$$\frac{\overset{1}{\cancel{a}}\, d}{\cancel{a}_1} = \frac{bc}{a}$$

$$d = \frac{bc}{a}$$

59. LCD $= ab$

$$\frac{1}{a} + \frac{1}{b} = 1$$

$$(ab)\left(\frac{1}{a} + \frac{1}{b}\right) = (ab)(1)$$

$$\overset{1}{\cancel{(a)}}\, b\left(\frac{1}{\cancel{a}_1}\right) + a\,\overset{1}{\cancel{(b)}}\left(\frac{1}{\cancel{b}_1}\right) = (ab)(1)$$

$$b + a = ab$$

$$b + a - a = ab - a$$

$$b = ab - a$$

$$b = a(b - 1)$$

$$\frac{b}{b - 1} = \frac{a\,\overset{1}{\cancel{(b-1)}}}{\cancel{b-1}_1}$$

$$a = \frac{b}{b - 1}$$

61.

$$\left.\begin{array}{l} 2 = 2 \\ 6d = 2 \cdot 3 \cdot d \end{array}\right\} \text{LCD} = 2 \cdot 3 \cdot d = 6d$$

$$F = \frac{L^2}{6d} + \frac{d}{2}$$

$$(6d)(F) = (6d)\left(\frac{L^2}{6d} + \frac{d}{2}\right)$$

$$6d(F) = \overset{1}{\cancel{6d}}\left(\frac{L^2}{\cancel{6d}_1}\right) + 3 \cdot \overset{1}{\cancel{2}}d\left(\frac{d}{\cancel{2}_1}\right)$$

$$6dF = L^2 + 3d^2$$

$$6dF - 3d^2 = L^2 + 3d^2 - 3d^2$$

$$6dF - 3d^2 = L^2$$

$$L^2 = 6dF - 3d^2$$

TRY IT YOURSELF

Solve each equation and check the result. If an equation has no solution, so indicate. Only a selected few of the rational equations are checked due to the complex nature of checking them.

63. $\text{LCD} = 3(x-3)$

$$\frac{1}{3} + \frac{2}{x-3} = 1$$

$$3(x-3)\left(\frac{1}{3} + \frac{2}{x-3}\right) = 3(x-3)(1)$$

$$3(x-3)\left(\frac{1}{3}\right) + 3(x-3)\left(\frac{2}{x-3}\right) = 3(x-3)(1)$$

$$\overset{1}{\cancel{3}}(x-3)\left(\frac{1}{\cancel{3}_1}\right) + 3\overset{1}{(\cancel{x-3})}\left(\frac{2}{\cancel{x-3}_1}\right) = 3x-9$$

$$x-3+6 = 3x-9$$

$$x+3-x = 3x-9-x$$

$$3+9 = 2x-9+9$$

$$\frac{\overset{1}{\cancel{2}} \cdot 6}{\cancel{2}_1} = \frac{\overset{1}{\cancel{2}}x}{\cancel{2}_1}$$

$$6 = x$$

This checks, so the solution is 6.

65.

$$\left.\begin{array}{l} q-2 = q-2 \\ q+1 = q+1 \\ q^2 - q - 2 = (q-2)(q+1) \end{array}\right\}$$

$$\text{LCD} = (q-2)(q+1)$$

$$\frac{7}{q^2 - q - 2} + \frac{1}{q+1} = \frac{3}{q-2}$$

$$\frac{7}{(q-2)(q+1)} \cdot (q-2)(q+1) + \frac{1}{q+1} \cdot (q-2)(q+1) = \frac{3}{q-2} \cdot (q-2)(q+1)$$

$$\frac{7\overset{1}{\cancel{(q-2)}}\overset{1}{\cancel{(q+1)}}}{\cancel{(q-2)}\cancel{(q+1)}_1} + \frac{1(q-2)\overset{1}{\cancel{(q+1)}}}{\cancel{q+1}_1} = \frac{3\overset{1}{\cancel{(q-2)}}(q+1)}{\cancel{q-2}_1}$$

$$7 + q - 2 = 3q + 3$$

$$q + 5 - 3 = 3q + 3 - 3$$

$$q + 2 - q = 3q - q$$

$$\frac{\overset{1}{\cancel{2}}}{\cancel{2}_1} = \frac{\overset{1}{\cancel{2}}q}{\cancel{2}_1}$$

$$1 = q$$

This checks, so the solution is 1.

67. $\text{LCD} = (3-t)(t+3)$

$$\frac{2}{3-t} = \frac{-t}{t+3}$$

$$(3-t)(t+3)\left(\frac{2}{3-t}\right) = (3-t)(t+3)\left(\frac{-t}{t+3}\right)$$

$$\overset{1}{\cancel{(3-t)}}(t+3)\left(\frac{2}{\cancel{3-t}_1}\right) = (3-t)\overset{1}{\cancel{(t+3)}}\left(\frac{-t}{\cancel{t+3}_1}\right)$$

$$2(t+3) = -t(3-t)$$

$$2t + 6 = -3t + t^2$$

$$2t + 6 - 2t = -3t + t^2 - 2t$$

$$6 = t^2 - 5t$$

$$6 - 6 = t^2 - 5t - 6$$

$$0 = t^2 - 5t - 6$$

$$0 = (t-6)(t+1)$$

$$
\begin{array}{ccc}
t - 6 = 0 & \text{or} & t + 1 = 0 \\
t - 6 + 6 = 0 + 6 & & t + 1 - 1 = 0 - 1 \\
t = 6 & & t = -1
\end{array}
$$

check: $\dfrac{2}{3-t} = \dfrac{-t}{t+3}$

$$\frac{2}{3-6} \overset{?}{=} \frac{-6}{6+3}$$

$$-\frac{2}{3} \overset{?}{=} \frac{2 \cdot \overset{1}{\cancel{3}}}{3 \cdot \cancel{3}_1}$$

$$-\frac{2}{3} = -\frac{2}{3}$$

check: $\dfrac{2}{3-t} = \dfrac{-t}{t+3}$

$$\frac{2}{3-(-1)} \overset{?}{=} \frac{-(-1)}{-1+3}$$

$$\frac{\overset{1}{\cancel{2}}}{2 \cdot \cancel{2}_1} \overset{?}{=} \frac{1}{2}$$

$$\frac{1}{2} = \frac{1}{2}$$

These check, so the solutions are -1 and 6.

$$\left.\begin{array}{l} 8 = 2 \cdot 2 \cdot 2 \\ 69. \ 10 = 2 \cdot 5 \\ y = y \end{array}\right\} \text{LCD} = 2 \cdot 2 \cdot 2 \cdot 5 \cdot y = 40y$$

$$\frac{1}{8} + \frac{2}{y} = \frac{1}{y} + \frac{1}{10}$$

$$40y\left(\frac{1}{8} + \frac{2}{y}\right) = 40y\left(\frac{1}{y} + \frac{1}{10}\right)$$

$$40y\left(\frac{1}{8}\right) + 40y\left(\frac{2}{y}\right) = 40y\left(\frac{1}{y}\right) + 40y\left(\frac{1}{10}\right)$$

$$5 \cdot \overset{1}{\cancel{8}} \, y\left(\frac{1}{\cancel{8}}\right) + 40 \cdot \overset{1}{\cancel{y}}\left(\frac{2}{\cancel{y}}\right) = 40 \cdot \overset{1}{\cancel{y}}\left(\frac{1}{\cancel{y}}\right) + 4 \cdot \overset{1}{\cancel{10}} \, y\left(\frac{1}{\cancel{10}}\right)$$

$$5y + 80 = 40 + 4y$$
$$5y + 80 - 4y = 40 + 4y - 4y$$
$$y + 80 - 80 = 40 - 80$$
$$y = -40$$

This checks, so the solution is -40.

71. $\text{LCD} = x + 1$

$$4 - \frac{8}{x+1} = \frac{8x}{x+1}$$

$$(x+1)\left(4 - \frac{8}{x+1}\right) = (x+1)\left(\frac{8x}{x+1}\right)$$

$$(x+1)(4) - (x+1)\left(\frac{8}{x+1}\right) = (x+1)\left(\frac{8x}{x+1}\right)$$

$$4x + 4 - \overset{1}{\cancel{(x+1)}}\left(\frac{8}{\cancel{x+1}}\right) = \overset{1}{\cancel{(x+1)}}\left(\frac{8x}{\cancel{x+1}}\right)$$

$$4x + 4 - 8 = 8x$$
$$4x - 4 - 4x = 8x - 4x$$
$$-4 = 4x$$
$$\frac{-\cancel{4}}{\cancel{4}} = \frac{\cancel{4}x}{\cancel{4}}$$
$$-1 = x$$

check: $4 - \dfrac{8}{x+1} = \dfrac{8x}{x+1}$

-1 makes the denominators of the original equation 0. No solution. -1 is extraneous.

73. $\text{LCD} = a + 1$

$$\frac{5a}{a+1} - 4 = \frac{3}{a+1}$$

$$(a+1)\left(\frac{5a}{a+1} - 4\right) = (a+1)\left(\frac{3}{a+1}\right)$$

$$(a+1)\left(\frac{5a}{a+1}\right) - (a+1)(4) = (a+1)\left(\frac{3}{a+1}\right)$$

$$\overset{1}{\cancel{(a+1)}}\left(\frac{5a}{\cancel{a+1}}\right) - (a+1)(4) = \overset{1}{\cancel{(a+1)}}\left(\frac{3}{\cancel{a+1}}\right)$$

$$5a - 4(a+1) = 3$$
$$5a - 4a - 4 = 3$$
$$a - 4 = 3$$
$$a - 4 + 4 = 3 + 4$$
$$a = 7$$

check: $\dfrac{5a}{a+1} - 4 = \dfrac{3}{a+1}$

$$\frac{5(7)}{7+1} - 4 \overset{?}{=} \frac{3}{7+1}$$

$$\frac{35}{8} - 4 \cdot \frac{8}{8} \overset{?}{=} \frac{3}{8}$$

$$\frac{35 - 32}{8} \overset{?}{=} \frac{3}{8}$$

$$\frac{3}{8} = \frac{3}{8}$$

This checks, so the solution is 7.

75. $\text{LCD} = y + 1$

$$\frac{2}{y+1} + 5 = \frac{12}{y+1}$$

$$(y+1)\left(\frac{2}{y+1} + 5\right) = (y+1)\left(\frac{12}{y+1}\right)$$

$$(y+1)\left(\frac{2}{y+1}\right) + (y+1)(5) = (y+1)\left(\frac{12}{y+1}\right)$$

$$\overset{1}{\cancel{(y+1)}}\left(\frac{2}{\cancel{y+1}}\right) + (y+1)(5) = \overset{1}{\cancel{(y+1)}}\left(\frac{12}{\cancel{y+1}}\right)$$

$$2 + 5(y+1) = 12$$
$$2 + 5y + 5 = 12$$
$$5y + 7 - 7 = 12 - 7$$
$$5y = 5$$
$$\frac{5y}{5} = \frac{5}{5}$$
$$y = 1$$

This checks, so the solution is 1.

Section 7.6

77. $\text{LCD} = 2(x+1)$

$$\frac{3}{x+1} = \frac{x-2}{x+1} + \frac{x-2}{2}$$

$$2(x+1)\left(\frac{3}{x+1}\right) = 2(x+1)\left(\frac{x-2}{x+1} + \frac{x-2}{2}\right)$$

$$2\,\overset{1}{\cancel{(x+1)}}\left(\frac{3}{\underset{1}{\cancel{x+1}}}\right) = 2\,\overset{1}{\cancel{(x+1)}}\left(\frac{x-2}{\underset{1}{\cancel{x+1}}}\right) + \overset{1}{\cancel{2}}(x+1)\left(\frac{x-2}{\underset{1}{\cancel{2}}}\right)$$

$$6 = 2(x-2) + (x+1)(x-2)$$

$$6 = 2x - 4 + x^2 - x - 2$$

$$6 - \mathbf{6} = x^2 + x - 6 - \mathbf{6}$$

$$0 = x^2 + x - 12$$

$$0 = (x-3)(x+4)$$

$$\begin{array}{c|c} x - 3 = 0 & x + 4 = 0 \\ x - 3 + 3 = 0 + 3 & x + 4 - 4 = 0 - 4 \\ x = 3 & x = -4 \end{array}$$

This checks, so the solutions are 3 and -4.

79. $\text{LCD} = (z-3)(z+1)$

$$\frac{z-4}{z-3} = \frac{z+2}{z+1}$$

$$(z-3)(z+1)\left(\frac{z-4}{z-3}\right) = (z-3)(z+1)\left(\frac{z+2}{z+1}\right)$$

$$\overset{1}{\cancel{(z-3)}}(z+1)\left(\frac{z-4}{\underset{1}{\cancel{z-3}}}\right) = (z-3)\,\overset{1}{\cancel{(z+1)}}\left(\frac{z+2}{\underset{1}{\cancel{z+1}}}\right)$$

$$(z+1)(z-4) = (z-3)(z+2)$$

$$z^2 - 3z - 4 = (z-3)(z+2)$$

$$z^2 - 3z - 4 = z^2 - z - 6$$

$$z^2 - 3z - 4 - \mathbf{z^2} = z^2 - z - 6 - \mathbf{z^2}$$

$$-3z - 4 = -z - 6$$

$$-3z - 4 + \mathbf{3z} = -z - 6 + \mathbf{3z}$$

$$-4 = 2z - 6$$

$$-4 + \mathbf{6} = 2z - 6 + \mathbf{6}$$

$$2 = 2z$$

$$\frac{2}{\mathbf{2}} = \frac{2z}{\mathbf{2}}$$

$$1 = z$$

This checks, so the solution is 1.

81. $\text{LCD} = x$

$$\frac{3}{x} + 2 = 3$$

$$x\left(\frac{3}{x} + 2\right) = x(3)$$

$$x\left(\frac{3}{x}\right) + x(2) = x(3)$$

$$\overset{1}{\cancel{x}}\left(\frac{3}{\underset{1}{\cancel{x}}}\right) + 2x = 3x$$

$$3 + 2x - \mathbf{2x} = 3x - \mathbf{2x}$$

$$3 = x$$

check: $\dfrac{3}{x} + 2 = 3$

$$\frac{3}{3} + 2 \overset{?}{=} 3$$

$$1 + 2 \overset{?}{=} 3$$

$$3 = 3$$

This checks, so the solution is 3.

83. $\left.\begin{array}{l} y + 2 = y + 2 \\ y - 2 = y - 2 \\ y^2 - 4 = (y+2)(y-2) \end{array}\right\} \text{LCD} = (y+2)(y-2)$

$$\frac{4}{y^2 - 4} = \frac{1}{y-2} + \frac{1}{y+2}$$

$$(y+2)(y-2) \cdot \frac{4}{(y+2)(y-2)}$$

$$= (y+2)(y-2) \cdot \frac{1}{y-2} + (y+2)(y-2) \cdot \frac{1}{y+2}$$

$$\frac{4\,\overset{1}{\cancel{(y+2)}}\,\overset{1}{\cancel{(y-2)}}}{\underset{1}{\cancel{(y+2)}}\,\underset{1}{\cancel{(y-2)}}} = \frac{\overset{1}{\cancel{(y-2)}}(y+2)}{\underset{1}{\cancel{(y-2)}}} + \frac{\overset{1}{\cancel{(y+2)}}(y-2)}{\underset{1}{\cancel{(y+2)}}}$$

$$4 = y + 2 + y - 2$$

$$4 = 2y$$

$$\frac{4}{\mathbf{2}} = \frac{2y}{\mathbf{2}}$$

$$2 = y$$

check: $\dfrac{4}{y^2 - 4} = \dfrac{1}{y-2} + \dfrac{1}{y+2}$

$$\frac{4}{(2)^2 - 4} = \frac{1}{2 - 2} + \frac{1}{2 + 2}$$

$$\frac{4}{\mathbf{0}} = \frac{1}{\mathbf{0}} + \frac{1}{4}$$

2 makes two denominators of the original equation 0. No solution. 2 is extraneous.

85. $\left.\begin{array}{l} 3=3 \\ 5d=5\cdot d \\ 10d=2\cdot 5\cdot d \end{array}\right\}$ LCD $=2\cdot 3\cdot 5\cdot d = 30d$

$$\frac{3}{5d}+\frac{4}{3}=\frac{9}{10d}$$

$$30d\left(\frac{3}{5d}+\frac{4}{3}\right)=30d\left(\frac{9}{10d}\right)$$

$$30d\left(\frac{3}{5d}\right)+30d\left(\frac{4}{3}\right)=30d\left(\frac{9}{10d}\right)$$

$$6\cdot \overset{1}{\cancel{5d}}\left(\frac{3}{\cancel{5d}_{1}}\right)+10\cdot \overset{1}{\cancel{3}}d\left(\frac{4}{\cancel{3}_{1}}\right)=3\cdot \overset{1}{\cancel{10d}}\left(\frac{9}{\cancel{10d}_{1}}\right)$$

$$18+40d=27$$

$$18+40d-\mathbf{18}=27-\mathbf{18}$$

$$\frac{\overset{1}{\cancel{40}}d}{\cancel{40}_{1}}=\frac{9}{40}$$

$$d=\frac{9}{40}$$

This checks, so the solution is $\dfrac{9}{40}$.

87. $\left.\begin{array}{l} n+3=n+3 \\ n-3=n-3 \\ n^2-9=(n+3)(n-3) \end{array}\right\}$ LCD $=(n+3)(n-3)$

$$\frac{n}{n^2-9}+\frac{n+8}{n+3}=\frac{n-8}{n-3}$$

$$(n+3)(n-3)\cdot \frac{n}{(n+3)(n-3)}+(n+3)(n-3)\cdot \frac{n+8}{n+3}=(n+3)(n-3)\cdot \frac{n-8}{n-3}$$

$$\frac{n\,\overset{1}{\cancel{(n+3)}}\,\overset{1}{\cancel{(n-3)}}}{\cancel{(n+3)}_1\,\cancel{(n-3)}_1}+\frac{(n+8)\,\overset{1}{\cancel{(n+3)}}(n-3)}{\cancel{(n+3)}_1}=\frac{(n-8)(n+3)\,\overset{1}{\cancel{(n-3)}}}{\cancel{(n-3)}_1}$$

$$n+n^2+5n-24=n^2-5n-24$$

$$n^2+6n-24-\mathbf{n^2}=n^2-5n-24-\mathbf{n^2}$$

$$6n-24+\mathbf{24}=-5n-24+\mathbf{24}$$

$$6n+\mathbf{5n}=-5n+\mathbf{5n}$$

$$11n=0$$

$$\frac{11n}{\mathbf{11}}=\frac{0}{\mathbf{11}}$$

$$n=0$$

check: $\dfrac{n}{n^2-9}+\dfrac{n+8}{n+3}=\dfrac{n-8}{n-3}$

$$\frac{0}{0^2-9}+\frac{0+8}{0+3}\overset{?}{=}\frac{0-8}{0-3}$$

$$\frac{8}{3}\overset{?}{=}\frac{-8}{-3}$$

$$\frac{8}{3}=\frac{8}{3}$$

This checks, so the solution is 0.

89. $\left.\begin{array}{l} x=x \\ x-2=x-2 \\ x^2-2x=x(x-2) \end{array}\right\}$ LCD $=x(x-2)$

$$\frac{3}{x-2}+\frac{1}{x}=\frac{6x+4}{x^2-2x}$$

$$x(x-2)\cdot \frac{3}{x-2}+x(x-2)\cdot \frac{1}{x}=x(x-2)\cdot \frac{6x+4}{x(x-2)}$$

$$\frac{3x\,\overset{1}{\cancel{(x-2)}}}{\cancel{(x-2)}_1}+\frac{1\,\overset{1}{\cancel{x}}(x-2)}{\cancel{x}_1}=\frac{(6x+4)\,\overset{1}{\cancel{x}}\,\overset{1}{\cancel{(x-2)}}}{\cancel{x}_1\,\cancel{(x-2)}_1}$$

$$3x+x-2=6x+4$$

$$4x-2=6x+4$$

$$4x-2-\mathbf{4x}=6x+4-\mathbf{4x}$$

$$-2=2x+4$$

$$-2-4=2x+4-\mathbf{4}$$

$$-6=2x$$

$$\frac{-6}{\mathbf{2}}=\frac{2x}{\mathbf{2}}$$

$$-3=x$$

check:

$$\frac{3}{x-2}+\frac{1}{x}=\frac{6x+4}{x^2-2x}$$

$$\frac{3}{-3-2}+\frac{1}{-3}\overset{?}{=}\frac{6(-3)+4}{(-3)^2-2(-3)}$$

$$\frac{-3}{5}+\frac{-1}{3}\overset{?}{=}-\frac{14}{15}$$

$$\frac{-3}{5}\cdot\frac{3}{3}+\frac{-1}{3}\cdot\frac{5}{5}\overset{?}{=}-\frac{14}{15}$$

$$\frac{-9-5}{15}\overset{?}{=}-\frac{14}{15}$$

$$-\frac{14}{15}=-\frac{14}{15}$$

This checks, so the solution is -3.

Section 7.6

91.

$$\left.\begin{array}{l}3=3\\3y-9=3(y-3)\end{array}\right\}LCD=3(y-3)$$

$$y+\frac{2}{3}=\frac{2y-12}{3y-9}$$

$$\mathbf{3(y-3)}\cdot\left(y+\frac{2}{3}\right)=\mathbf{3(y-3)}\cdot\left(\frac{2y-12}{3y-9}\right)$$

$$y\cdot3(y-3)+\left(\frac{2}{3}\right)\cdot3(y-3)=\left(\frac{2y-12}{3y-9}\right)\cdot3(y-3)$$

$$3y^2-9y+\frac{2\cdot\overset{1}{\cancel{3}}(y-3)}{\cancel{3}_1}=\frac{(2y-12)\overset{1}{\cancel{3(y-3)}}}{\cancel{3(y-3)}_1}$$

$$3y^2-9y+2y-6=2y-12$$

$$3y^2-7y-6-\mathbf{2y}+\mathbf{12}=2y-12-\mathbf{2y}+\mathbf{12}$$

$$3y^2-9y+6=0$$

$$3(y^2-3y+2)=0$$

$$3(y-2)(y-1)=0$$

$$\begin{array}{lcl}y-2=0 & \text{or} & y-1=0\\y-2+2=0+2 & & y-1+1=0+1\\y=2 & & y=1\end{array}$$

Both check, so the solutions are 2 and 1.

93.

$$\left.\begin{array}{l}2=2\\7=7\\14=2\cdot7\end{array}\right\}LCD=2\cdot7=14$$

$$\frac{a-1}{7}-\frac{a-2}{14}=\frac{1}{2}$$

$$\mathbf{14}\left(\frac{a-1}{7}-\frac{a-2}{14}\right)=\mathbf{14}\left(\frac{1}{2}\right)$$

$$14\left(\frac{a-1}{7}\right)-14\left(\frac{a-2}{14}\right)=14\left(\frac{1}{2}\right)$$

$$2\cdot\overset{1}{\cancel{7}}\left(\frac{a-1}{\cancel{7}_1}\right)-\cancel{14}\left(\frac{a-2}{\cancel{14}_1}\right)=7\cdot\overset{1}{\cancel{2}}\left(\frac{1}{\cancel{2}_1}\right)$$

$$2(a-1)-(a-2)=7$$

$$2a-2-a+2=7$$

$$a=7$$

This checks, so the solution is 7.

LOOK ALIKES ...

For each expression, perform the indicated operations and then simplify, if possible. Solve each equation and check the results.

95. a.
$$\frac{a}{3}+\frac{3}{5}+\frac{a}{15}=\frac{\mathbf{5}}{\mathbf{5}}\cdot\frac{a}{3}+\frac{\mathbf{3}}{\mathbf{3}}\cdot\frac{3}{5}+\frac{a}{15}$$

$$=\frac{5a+9+a}{15}$$

$$=\frac{6a+9}{15}$$

$$=\frac{\overset{1}{\cancel{3}}(2a+3)}{\cancel{3}_1\cdot5}$$

$$=\frac{2a+3}{5}$$

b.
$$\frac{a}{3}+\frac{3}{5}=\frac{a}{15}$$

$$\mathbf{15}\left(\frac{a}{3}+\frac{3}{5}\right)=\mathbf{15}\cdot\frac{a}{15}$$

$$5\cdot\overset{1}{\cancel{3}}\left(\frac{a}{\cancel{3}_1}\right)+3\cdot\overset{1}{\cancel{5}}\left(\frac{3}{\cancel{5}_1}\right)=\overset{1}{\cancel{15}}\cdot\frac{a}{\cancel{15}_1}$$

$$5a+9=a$$

$$5a+9-\mathbf{5a}=a-\mathbf{5a}$$

$$9=-4a$$

$$\frac{9}{-4}=\frac{-4a}{-4}$$

$$a=-\frac{9}{4}$$

This checks, so the solution is $-\dfrac{9}{4}$.

97. a.
$$\frac{x}{x-2}-\frac{1}{x-3}$$

$$=\frac{\mathbf{(x-3)}}{\mathbf{(x-3)}}\cdot\frac{x}{(x-2)}-\frac{\mathbf{(x-2)}}{\mathbf{(x-2)}}\cdot\frac{1}{(x-3)}$$

$$=\frac{x^2-3x-(x-2)}{(x-2)(x-3)}$$

$$=\frac{x^2-3x-(x-2)}{(x-2)(x-3)}$$

$$=\frac{x^2-3x-x+2}{(x-2)(x-3)}$$

$$=\frac{x^2-4x+2}{(x-2)(x-3)}$$

b.

$$\frac{x}{x-2} - \frac{1}{x-3} = 1$$

$$(x-2)(x-3)\left(\frac{x}{x-2} - \frac{1}{x-3}\right)$$
$$= (x-2)(x-3)1$$

$$(x-3)(\overset{1}{\cancel{x-2}})\left(\frac{x}{\cancel{x-2}}\right) - (x-2)(\overset{1}{\cancel{x-3}})\left(\frac{1}{\cancel{x-3}}\right)$$
$$= (x-2)(x-3)$$

$$x(x-3) - (x-2) = (x-2)(x-3)$$
$$x^2 - 3x - x + 2 = x^2 - 5x + 6$$
$$x^2 - 4x + 2 - x^2 = x^2 - 5x + 6 - x^2$$
$$-4x + 2 + 5x = -5x + 6 + 5x$$
$$x + 2 - 2 = 6 - 2$$
$$x = 4$$

This checks, so the solution is 4.

APPLICATION
99. MEDICINE

$$LCD = R + B$$

$$H = \frac{RB}{R+B}$$

$$(R+B) \cdot H = (R+B) \cdot \frac{RB}{R+B}$$

$$(R+B) \cdot H = (\overset{1}{\cancel{R+B}}) \cdot \frac{RB}{\cancel{R+B}}$$

$$H(R+B) = RB$$

$$HR + HB = RB$$

$$HR + HB - HR = RB - HR$$

$$HB = R(B-H)$$

$$\frac{HB}{B-H} = \frac{R(\overset{1}{\cancel{B-H}})}{\cancel{B-H}}$$

$$\frac{HB}{B-H} = R$$

$$R = \frac{HB}{B-H}$$

101. ELECTRONICS

$$LCD = rr_1r_2$$

$$\frac{1}{r} = \frac{1}{r_1} + \frac{1}{r_2}$$

$$(rr_1r_2) \cdot \frac{1}{r} = (rr_1r_2) \cdot \left(\frac{1}{r_1} + \frac{1}{r_2}\right)$$

$$\overset{1}{\cancel{r}}r_1r_2 \cdot \frac{1}{\cancel{r}} = r\,\overset{1}{\cancel{r_1}}\,r_2\frac{1}{\cancel{r_1}} \cdot + rr_1\,\overset{1}{\cancel{r_2}} \cdot \frac{1}{\cancel{r_2}}$$

$$r_1r_2 = rr_2 + rr_1$$

$$r_1r_2 = r(r_2 + r_1)$$

$$\frac{r_1r_2}{r_2 + r_1} = \frac{r\,(\overset{1}{\cancel{r_2+r_1}})}{\cancel{(r_2+r_1)}}$$

$$\frac{r_1r_2}{r_2 + r_1} = r$$

$$r = \frac{r_1r_2}{r_2 + r_1}$$

WRITING
103-105. Answers will vary.
REVIEW
107. UNIFORMS
Analyze
- Cost per uniform is $18.50
- One time set up fee is $75.00
- Total cost is $445.00
- Find the total number of uniforms

Assign
Let x = the total number of uniforms
Form

	Number	• Value	+ Fee =	Total value
Uniforms	x	18.50	75	445

The number of uniforms	times	the value of one unifirm	plus	the fee	equals	the total value.
x	·	18.50	+	75	=	445

Solve

$$18.50x + 75 = 445$$
$$18.50x + 75 - 75 = 445 - 75$$
$$\frac{18.50x}{18.50} = \frac{370}{18.50}$$
$$x = 20$$

State

The number of uniforms widgets is 20.

Section 7.6

Check

The results check.

CHALLENGE PROBLEMS

Solve each equation and check the results. If an equation has not solution, so indicate.

109. $LCD = (x-3)$

$$\frac{x-4}{x-3} + \frac{x-2}{x-3} = x-3$$

$$(\boldsymbol{x-3})\cdot\left(\frac{x-4}{x-3}\right) + (\boldsymbol{x-3})\cdot\left(\frac{x-2}{x-3}\right) = (\boldsymbol{x-3})(x-3)$$

$$\frac{\overset{1}{(\cancel{x-3})}(x-4)}{\underset{1}{(\cancel{x-3})}} + \frac{\overset{1}{(\cancel{x-3})}(x-2)}{\underset{1}{(\cancel{x-3})}} = x^2 - 6x + 9$$

$$(x-4)+(x-2) = x^2 - 6x + 9$$

$$2x - 6 = x^2 - 6x + 9$$

$$2x - 6 - \boldsymbol{2x} = x^2 - 6x + 9 - \boldsymbol{2x}$$

$$-6 = x^2 - 8x + 9$$

$$-6 + \boldsymbol{6} = x^2 - 8x + 9 + \boldsymbol{6}$$

$$0 = x^2 - 8x + 15$$

$$0 = (x-5)(x-3)$$

$x - 5 = 0$	or	$x - 3 = 0$
$x-5+\boldsymbol{5}=0+\boldsymbol{5}$		$x-3+\boldsymbol{3}=0+\boldsymbol{3}$
$x = 5$		$x = 3$

check: $\dfrac{x-4}{x-3}+\dfrac{x-2}{x-3}=x-3$

$$\frac{\boldsymbol{5}-4}{\boldsymbol{5}-3} + \frac{\boldsymbol{5}-2}{\boldsymbol{5}-3} \overset{?}{=} \boldsymbol{5}-3$$

$$\frac{1}{2} + \frac{3}{2} \overset{?}{=} 2$$

$$\frac{4}{2} \overset{?}{=} 2$$

$$2 = 2$$

This checks, so the solution is 5.

check: $\dfrac{x-4}{x-3}+\dfrac{x-2}{x-3}=x-3$

$$\frac{\boldsymbol{3}-4}{\boldsymbol{3}-3} - \frac{\boldsymbol{3}-2}{\boldsymbol{3}-3} \overset{?}{=} \boldsymbol{3}-3$$

$$\frac{-1}{\boldsymbol{0}} - \frac{1}{\boldsymbol{0}} \overset{?}{=} 0$$

3 makes the denominators of the original equation 0. Not a solution. 3 is extraneous.

The only solution is 5.

111. $\left.\begin{array}{l} x = x \\ x^2 = x\cdot x \end{array}\right\} LCD = x\cdot x = x^2$

Solve

$$x^{-2} + 2x^{-1} + 1 = 0$$

$$\frac{1}{x^2} + \frac{2}{x} + 1 = 0$$

$$\left(\frac{1}{x^2} + \frac{2}{x} + 1\right)\cdot \boldsymbol{x^2} = 0\cdot \boldsymbol{x^2}$$

$$\frac{1}{x^2}\left(x^2\right) + \frac{2}{x}\left(x^2\right) + 1\left(x^2\right) = 0$$

$$1 + 2x + x^2 = 0$$

$$x^2 + 2x + 1 = 0$$

$$(x+1)(x+1) = 0$$

$$x + 1 = 0$$

$$x+1-1 = 0-1$$

$$x = -1$$

This checks, so the solution is -1.

SECTION 7.7 STUDY SET
VOCABULARY

Fill in the blanks.

1. In this section, problems that involve:
   - moving vehicles are called uniform **motion** problems.
   - depositing money are called **investment** problems.
   - people completing jobs are called shared-**work** problems.

CONCEPTS

3. $\dfrac{5+x}{8+x} = \dfrac{2}{3}$ **(iii)**

5. a. $\dfrac{1}{45}$ of the job per minute   b. $\dfrac{x}{4}$

7. *a.* $d = rt$
$$\dfrac{d}{r} = \dfrac{rt}{r}$$
$$\dfrac{d}{r} = t$$
$$t = \dfrac{d}{r}$$

b. $I = Prt$
$$\dfrac{I}{rt} = \dfrac{Prt}{rt}$$
$$\dfrac{I}{rt} = P$$
$$P = \dfrac{I}{rt}$$

9.

	Rate	· Time	= Work completed
1st printer	$\dfrac{1}{15}$	x	$\dfrac{x}{15}$
2nd printer	$\dfrac{1}{8}$	x	$\dfrac{x}{8}$

NOTATION

11. $\dfrac{55}{9} = 9\overline{)55} = 6\dfrac{1}{9}$ days
$$\dfrac{54}{1}$$

GUIDED PRACTICE

Solve each of these number problems. See Example 1.

13.

Analyze

- Begin with the fraction $\dfrac{2}{5}$.

- Add the same number to the numerator and denominator.

- The results is $\dfrac{2}{3}$.

- Find the number.

Assign

Let n = the unknown number.

Form
$$\dfrac{2+n}{5+n} = \dfrac{2}{3}$$

Solve
$$\dfrac{2+n}{5+n} = \dfrac{2}{3}$$
$$LCD = 3(n+5)$$
$$3(n+5)\left(\dfrac{2+n}{5+n}\right) = 3(n+5)\left(\dfrac{2}{3}\right)$$
$$3(n+5)\left(\dfrac{2+n}{5+n}\right) = 3(n+5)\left(\dfrac{2}{3}\right)$$
$$3(n+2) = 2(n+5)$$
$$3n+6 = 2n+10$$
$$3n+6-2n = 2n+10-2n$$
$$n+6 = 10$$
$$n+6-6 = 10-6$$
$$n = 4$$

State

The number is 4.

Check
$$\dfrac{2+n}{5+n} = \dfrac{2+4}{5+4} = \dfrac{6}{9} = \dfrac{2\cdot 3}{3\cdot 3} = \dfrac{2}{3}$$

The result checks.

Section 7.7

15.

Analyze
- Begin with the fraction $\frac{3}{4}$.
- Double the numerator.
- Add a number to the denominator.
- The results is 1.
- Find the number.

Assign
Let n = the unknown number.

Form
$$\frac{3(2)}{4+n} = 1$$

Solve
$$\frac{6}{4+n} = 1$$
$$\text{LCD} = n+4$$
$$(n+4)\left(\frac{6}{n+4}\right) = (n+4)(1)$$
$$(\overset{1}{\cancel{n+4}})\left(\frac{6}{\cancel{n+4}}\right) = n+4$$
$$6 = n+4$$
$$6-\mathbf{4} = n+4-\mathbf{4}$$
$$2 = n$$

State
The number is 2.

Check
$$\frac{6}{4+n} = \frac{6}{4+\mathbf{2}} = \frac{6}{6} = 1$$
The result checks.

17.

Analyze
- Begin with the fraction $\frac{3}{4}$.
- Add a number to the numerator.
- Twice as much is added to the denominator.
- The results is $\frac{4}{7}$.
- Find the number.

Assign
Let n = the unknown number.

Form
$$\frac{3+n}{4+2n} = \frac{4}{7}$$

Solve
$$\frac{3+n}{4+2n} = \frac{4}{7}$$
$$\text{LCD} = 7(4+2n)$$
$$7(4+2n)\left(\frac{3+n}{4+2n}\right) = 7(4+2n)\left(\frac{4}{7}\right)$$
$$7(\cancel{4+2n})\left(\frac{3+n}{\cancel{4+2n}}\right) = \cancel{7}(4+2n)\left(\frac{4}{\cancel{7}}\right)$$
$$7(3+n) = (4+2n)4$$
$$21+7n = 16+8n$$
$$21+7n-\mathbf{7n} = 16+8n-\mathbf{7n}$$
$$21 = 16+n$$
$$21-\mathbf{16} = 16+n-\mathbf{16}$$
$$5 = n$$

State
The number is 5.

Check
$$\frac{3+n}{4+2n} = \frac{3+\mathbf{5}}{4+2\cdot\mathbf{5}} = \frac{3+5}{4+10} = \frac{8}{14} = \frac{4\cdot\overset{1}{\cancel{2}}}{7\cdot\underset{1}{\cancel{2}}} = \frac{4}{7}$$

The result checks.

19.

Analyze

- The sum of a number and its reciprocal is $\frac{13}{6}$.

- Find the numbers.

Assign

Let n = the unknown number.

$\frac{1}{n}$ = the reciprocal

Form

$$n + \frac{1}{n} = \frac{13}{6}$$

Solve

$$n + \frac{1}{n} = \frac{13}{6}$$
$$\text{LCD} = 6n$$
$$6n\left(n + \frac{1}{n}\right) = 6n\left(\frac{13}{6}\right)$$
$$6n(n) + 6\,\overset{1}{\cancel{n}}\left(\frac{1}{\underset{1}{\cancel{n}}}\right) = \overset{1}{\cancel{6}}\,n\left(\frac{13}{\underset{1}{\cancel{6}}}\right)$$
$$6n^2 + 6 = 13n$$
$$6n^2 + 6 - \mathbf{13n} = 13n - \mathbf{13n}$$
$$6n^2 - 13n + 6 = 0$$
$$(3n - 2)(2n - 3) = 0$$

$3n - 2 = 0$	$2n - 3 = 0$
$3n - 2 + 2 = 0 + 2$	$2n - 3 + 3 = 0 + 3$
$3n = 2$	$2n = 3$
$\frac{3n}{3} = \frac{2}{3}$	$\frac{2n}{2} = \frac{3}{2}$
$n = \frac{2}{3}$	$n = \frac{3}{2}$

or

State

The numbers are $\frac{2}{3}$ and $\frac{3}{2}$.

Check

$$x + \frac{1}{x} = \frac{13}{6}$$
$$\frac{2}{3} + \frac{1}{\frac{2}{3}} \overset{?}{=} \frac{13}{6}$$
$$\frac{2}{3} + \frac{3}{2} \overset{?}{=} \frac{13}{6}$$
$$\frac{2}{3} \cdot \frac{\mathbf{2}}{\mathbf{2}} + \frac{3}{2} \cdot \frac{\mathbf{3}}{\mathbf{3}} \overset{?}{=} \frac{13}{6}$$
$$\frac{4 + 9}{6} \overset{?}{=} \frac{13}{6}$$
$$\frac{13}{6} = \frac{13}{6}$$

$$x + \frac{1}{x} = \frac{13}{6}$$
$$\frac{3}{2} + \frac{1}{\frac{3}{2}} \overset{?}{=} \frac{13}{6}$$
$$\frac{3}{2} + \frac{2}{3} \overset{?}{=} \frac{13}{6}$$
$$\frac{3}{2} \cdot \frac{\mathbf{3}}{\mathbf{3}} + \frac{2}{3} \cdot \frac{\mathbf{2}}{\mathbf{2}} \overset{?}{=} \frac{13}{6}$$
$$\frac{9 + 4}{6} \overset{?}{=} \frac{13}{6}$$
$$\frac{13}{6} = \frac{13}{6}$$

Both results check.

APPLICATIONS

21. COOKING

Analyze

- Begin with the fraction $\frac{1}{4}$.

- Add the same number to the numerator and denominator.

- The results is $\frac{3}{4}$.

- Find the number.

Assign

Let n = the unknown number.

Form

$$\frac{1 + n}{4 + n} = \frac{3}{4}$$

Solve

$$\frac{1 + n}{4 + n} = \frac{3}{4}$$
$$\text{LCD} = 4(4 + n)$$
$$4(4 + n)\left(\frac{1 + n}{4 + n}\right) = 4(4 + n)\left(\frac{3}{4}\right)$$
$$4\,\cancel{(4 + n)}\left(\frac{1 + n}{\cancel{4 + n}}\right) = \cancel{4}(4 + n)\left(\frac{3}{\cancel{4}}\right)$$
$$4(1 + n) = 3(4 + n)$$
$$4 + 4n = 12 + 3n$$
$$4 + 4n - \mathbf{3n} = 12 + 3n - \mathbf{3n}$$
$$4 + n = 12$$
$$4 + n - \mathbf{4} = 12 - \mathbf{4}$$
$$n = 8$$

State

The number is 8.

Check

$$\frac{1 + n}{4 + n} = \frac{1 + \mathbf{8}}{4 + \mathbf{8}} = \frac{9}{12} = \frac{3 \cdot \overset{1}{\cancel{3}}}{4 \cdot \underset{1}{\cancel{3}}} = \frac{3}{4}$$

The result checks.

- 449 -

Section 7.7

23. TOUR de FRANCE

Analyze

• Garin rode 80 miles.

• Armstrong rode 130 miles.

• Times are the same.

• Armstrong was 10 mph faster than Garin.

• What was each cyclist average speed?

Assign

Let r = Garin's speed in mph

$r + 10$ = Armstrong's speed in mph

Form

	Rate	•	Time	=	Distance
Garin	r		$\dfrac{80}{r}$		80
Armstrong	$r + 10$		$\dfrac{130}{r + 10}$		130

Time took Garin to cycle 80 miles	equals	Time took Armstrong to cycle 130 miles
$\dfrac{80}{r}$	$=$	$\dfrac{130}{r + 10}$

Solve

$$\frac{80}{r} = \frac{130}{r + 10}$$

$$\text{LCD} = r(r + 10)$$

$$r(r + 10)\left(\frac{80}{r}\right) = r(r + 10)\left(\frac{130}{r + 10}\right)$$

$$\overset{1}{\cancel{r}}(r + 10)\left(\frac{80}{\cancel{r}}\right) = r\,\overset{1}{\cancel{(r + 10)}}\left(\frac{130}{\cancel{r + 10}}\right)$$

$$80(r + 10) = 130r$$

$$80r + 800 = 130r$$

$$80r + 800 - \mathbf{80r} = 130r - \mathbf{80r}$$

$$800 = 50r$$

$$\frac{800}{\mathbf{50}} = \frac{50r}{\mathbf{50}}$$

$$16 = r$$

State

Garin's speed was 16 mph.

Armstrong's speed was 26 mph.

Check

The results check.

25. PACKAGING FRUIT

Analyze

• Shorter belt is 100 feet.

• Longer belt is 300 feet.

• Times are the same.

• Longer belt 1 foot per second faster.

• What is each belt's speed?

Assign

Let r = Shorter belt's speed in fps

$r + 1$ = Longer belt's speed in fps

Form

	Rate	•	Time	=	Distance
Shorter	r		$\dfrac{100}{r}$		100
Longer	$r + 1$		$\dfrac{300}{r + 1}$		300

Time for shorter to travel 100 ft	equals	Time for longer to travel 300 ft
$\dfrac{100}{r}$	$=$	$\dfrac{300}{r + 1}$

Solve

$$\frac{100}{r} = \frac{300}{r + 1}$$

$$\text{LCD} = r(r + 1)$$

$$r(r + 1)\left(\frac{100}{r}\right) = r(r + 1)\left(\frac{300}{r + 1}\right)$$

$$\overset{1}{\cancel{r}}(r + 1)\left(\frac{100}{\cancel{r}}\right) = r \cdot \overset{1}{\cancel{(r + 1)}}\left(\frac{300}{\cancel{r + 1}}\right)$$

$$100(r + 1) = 300r$$

$$100r + 100 = 300r$$

$$100r + 100 - \mathbf{100r} = 300r - \mathbf{100r}$$

$$100 = 200r$$

$$\frac{100}{\mathbf{200}} = \frac{200r}{\mathbf{200}}$$

$$\frac{1}{2} = r$$

State

Longer belt's speed is $1\dfrac{1}{2}$ feet per second.

Shorter belt's speed is $\dfrac{1}{2}$ foot per second.

Check

The results check.

27. BIRDS IN FLIGHT

Analyze
- Canada goose can fly 120 miles.
- Great Blue heron can fly 80 miles.
- Times are the same.
- Goose flies 10 mph faster.
- What are the speeds of both?

Assign

Let r = Heron's speed in mph

$r + 10$ = Goose's speed in mph

Form

	Rate	Time	Distance
Heron	r	$\dfrac{80}{r}$	80
Goose	$r + 10$	$\dfrac{120}{r+10}$	120

Time took heron to fly 80 miles	equals	Time took goose to fly 120 miles

$$\frac{80}{r} = \frac{120}{r+10}$$

Solve

$$\frac{80}{r} = \frac{120}{r+10}$$

$$\text{LCD} = r(r+10)$$

$$r(r+10)\left(\frac{80}{r}\right) = r(r+10)\left(\frac{120}{r+10}\right)$$

$$\overset{1}{\cancel{r}}(r+10)\left(\frac{80}{\cancel{r}_{1}}\right) = r \cdot \overset{1}{\cancel{(r+10)}}\left(\frac{120}{\cancel{r+10}_{1}}\right)$$

$$80(r+10) = 120r$$

$$80r + 800 = 120r$$

$$80r + 800 - \mathbf{80r} = 120r - \mathbf{80r}$$

$$800 = 40r$$

$$\frac{800}{40} = \frac{40r}{40}$$

$$20 = r$$

State

The Great Blue heron's speed is 20 mph.
The Canada goose's speed is 30 mph.

Check

The results check.

29. WIND SPEED

Analyze
- Rate of plane in still air is 255 mph.
- Flies 300 miles downwind.
- Flies 210 miles upwind.
- Times are the same for both directions.
- What is the speed of the wind?

Assign

Let r = Wind's speed in mph

Form

	Rate	Time	Distance
Downwind	$255 + r$	$\dfrac{300}{255+r}$	300
Upwind	$255 - r$	$\dfrac{210}{255-r}$	210

Time took plane to fly 300 miles downwind.	equals	time took plane to fly 210 miles upwind.

$$\frac{300}{255+r} = \frac{210}{255-r}$$

Solve

$$\frac{300}{255+r} = \frac{210}{255-r}$$

$$\text{LCD} = (255+r)(255-r)$$

$$(255+r)(255-r)\left(\frac{300}{255+r}\right) = (255+r)(255-r)\left(\frac{210}{255-r}\right)$$

$$\overset{1}{\cancel{(255+r)}}(255-r)\left(\frac{300}{\cancel{255+r}_{1}}\right) = (255+r)\overset{1}{\cancel{(255-r)}}\left(\frac{210}{\cancel{255-r}_{1}}\right)$$

$$300(255-r) = 210(255+r)$$

$$76,500 - 300r = 53,550 + 210r$$

$$76,500 - 300r + \mathbf{300r} = 53,550 + 210r + \mathbf{300r}$$

$$76,500 = 53,550 + 510r$$

$$76,500 - \mathbf{53,550} = 53,550 + 510r - \mathbf{53,550}$$

$$22,950 = 510r$$

$$\frac{22,950}{510} = \frac{510r}{510}$$

$$45 = r$$

State

The wind's speed is 45 mph.

Check

The result checks.

Section 7.7

31. ROOFING HOUSES

Analyze
- Homeowner takes 7 days to roof house.
- Professional takes 4 days to roof house.
- How long will it take both of them, working together, to roof the house?

Assign

Let x = the number of days it will take both working together to roof the house.

Form

	Rate	· Time	= Work Completed
Homeowner	$\dfrac{1}{7}$	x	$\dfrac{x}{7}$
Professional	$\dfrac{1}{4}$	x	$\dfrac{x}{4}$

The part of job done by homeowner.	plus	The part of job done by professional.	equals	1 job completed
$\dfrac{x}{7}$	$+$	$\dfrac{x}{4}$	$=$	1

Solve

$$\frac{x}{7} + \frac{x}{4} = 1$$
$$LCD = 28$$
$$28\left(\frac{x}{7} + \frac{x}{4}\right) = 28(1)$$
$$\overset{1}{\cancel{7}} \cdot 4\left(\frac{x}{\underset{1}{\cancel{7}}}\right) + \overset{1}{\cancel{4}} \cdot 7\left(\frac{x}{\underset{1}{\cancel{4}}}\right) = 28$$
$$4x + 7x = 28$$
$$11x = 28$$
$$\frac{11x}{11} = \frac{28}{11}$$
$$x = \frac{28}{11}$$
$$x = 2\frac{6}{11}$$

State

It takes both working together $2\dfrac{6}{11}$ days to roof the house.

Check
The result checks.

---

33. RECREATION DIRECTOR

Analyze
- 1st pipe can fill pool in 12 hours.
- 2nd pipe can fill pool in 18 hours.
- Will the pool be filled for a 2:00 p.m. opening if both are turned on a 8:00 a.m. that day?

Assign

Let x = the number of hours it will take both pipes working together to fill the pool.

Form

	Rate	· Time	= Work Completed
1st pipe	$\dfrac{1}{12}$	x	$\dfrac{x}{12}$
2nd pipe	$\dfrac{1}{18}$	x	$\dfrac{x}{18}$

The part of job done by 1st pipe.	plus	The part of job done by 2nd pipe.	equals	1 job completed
$\dfrac{x}{12}$	$+$	$\dfrac{x}{18}$	$=$	1

Solve

$$\frac{x}{12} + \frac{x}{18} = 1$$
$$LCD = 36$$
$$36\left(\frac{x}{12} + \frac{x}{18}\right) = 36(1)$$
$$\overset{}{\cancel{12}} \cdot 3\left(\frac{x}{\underset{1}{\cancel{12}}}\right) + \overset{}{\cancel{18}} \cdot 2\left(\frac{x}{\underset{1}{\cancel{18}}}\right) = 36$$
$$3x + 2x = 36$$
$$5x = 36$$
$$\frac{5x}{5} = \frac{36}{5}$$
$$x = \frac{36}{5}$$
$$x = 7\frac{1}{5}$$

State

It takes both working together $7\dfrac{1}{5}$ hours to fill the pool. The pool opens in 6 hours. No, the pool will not be filled in time.

Check
The result checks.

---

35. FILLING A POOL
Analyze
- Inlet pipe can fill pool in 4 hours.
- Outlet pipe can drain pool in 8 hours.
- How long will it take the inlet pipe to fill the pool, if the outlet pipe is left open?

Assign

Let x = the number of hours it will take the inlet pipe to fill the pool if the outlet drain is left open.

Form

Rate $\cdot$ Time $=$ Work Completed

Inlet pipe	$\dfrac{1}{4}$	x	$\dfrac{x}{4}$
Outlet pipe	$\dfrac{1}{8}$	x	$\dfrac{x}{8}$

The part of job done by inlet pipe.	minus	The part of job done by outlet pipe.	equals	1 job completed
$\dfrac{x}{4}$	$-$	$\dfrac{x}{8}$	$=$	1

Solve

$$\frac{x}{4} - \frac{x}{8} = 1$$

$$LCD = 8$$

$$8\left(\frac{x}{4} - \frac{x}{8}\right) = 8(1)$$

$$2 \cdot \overset{1}{\cancel{4}}\left(\frac{x}{\cancel{4}_1}\right) - \overset{1}{\cancel{8}}\left(\frac{x}{\cancel{8}_1}\right) = 8$$

$$2x - x = 8$$

$$x = 8$$

State

It will take the inlet pipe 8 hours to fill the pool with the outlet drain left oepn.

Check

The result checks.

37. GRADING PAPERS
Analyze
- Teacher takes 30 minutes.
- Aide takes 60 minutes.
- How long will it take both of them, working together, to grade a set of papers?

Assign

Let x = the number of minutes it will take both working together to grade tests.

Form

Rate $\cdot$ Time $=$ Work Completed

Teacher	$\dfrac{1}{30}$	x	$\dfrac{x}{30}$
Aide	$\dfrac{1}{60}$	x	$\dfrac{x}{60}$

The part of job done by teacher.	plus	The part of job done by aide.	equals	1 job completed
$\dfrac{x}{30}$	$+$	$\dfrac{x}{60}$	$=$	1

Solve

$$\frac{x}{30} + \frac{x}{60} = 1$$

$$LCD = 60$$

$$60\left(\frac{x}{30} + \frac{x}{60}\right) = 60(1)$$

$$\overset{1}{\cancel{30}} \cdot 2\left(\frac{x}{\cancel{30}_1}\right) + \overset{1}{\cancel{60}}\left(\frac{x}{\cancel{60}_1}\right) = 60$$

$$2x + x = 60$$

$$3x = 60$$

$$\frac{3x}{3} = \frac{60}{3}$$

$$x = 20$$

State

It takes both working together 20 minutes to grade one set of tests.

Check

The result checks.

39. PRINTERS

Analyze
- One printer takes 4 hours.
- Other printer takes 6 hours.
- How long will it take both of them, working together, to print $\frac{3}{4}$ of the schedules?

Assign

Let x = the number of hours it will take both working together to print schedules.

Form

	Rate	· Time	= Work Completed
Fast printer	$\frac{1}{4}$	x	$\frac{x}{4}$
Slow Printer	$\frac{1}{6}$	x	$\frac{x}{6}$

The part of job done by fast printer.	plus	The part of job done by slow printer.	equals	$\frac{3}{4}$ job completed
$\frac{x}{4}$	$+$	$\frac{x}{6}$	$=$	$\frac{3}{4}$

Solve

$$\frac{x}{4}+\frac{x}{6}=\frac{3}{4}$$
$$\text{LCD}=12$$
$$12\left(\frac{x}{4}+\frac{x}{6}\right)=12\left(\frac{3}{4}\right)$$
$$\overset{1}{\cancel{4}}\cdot3\left(\frac{x}{\cancel{4}}\right)+\overset{1}{\cancel{6}}\cdot2\left(\frac{x}{\cancel{6}}\right)=\overset{1}{\cancel{4}}\cdot3\left(\frac{3}{\cancel{4}}\right)$$
$${\scriptstyle 1}{\scriptstyle 1}{\scriptstyle 1}$$
$$3x+2x=9$$
$$5x=9$$
$$\frac{5x}{5}=\frac{9}{5}$$
$$x=1\frac{4}{5}$$

State

It takes both working together $1\frac{4}{5}$ or 1.8 hours to print $\frac{3}{4}$ of the schedules.

Check

The result checks.

41. COMPARING INVESTMENTS

Analyze
- Tax-free bonds earns $300 interest.
- Credit union earns $200 interest.
- Credit union rate is 2% less.
- Time for both is 1 year.
- Principals are the same amount for both.
- What are the two rates?

Assign

Let r = Bond's interest rate as a percent

$r-2$ = C Union's interest rate as a percent

Form

	Principal	· Rate	· Time	= Interest
Bonds	$\frac{300}{r}$	r	1	300
Credit	$\frac{200}{r-2}$	$r-2$	1	200

Principal invested in bonds.	equals	Principal invested with credit union.
$\frac{300}{r}$	$=$	$\frac{200}{r-2}$

Solve

$$\frac{300}{r}=\frac{200}{r-2}$$
$$\text{LCD}=r(r-2)$$
$$r(r-2)\left(\frac{300}{r}\right)=r(r-2)\left(\frac{200}{r-2}\right)$$
$$\overset{1}{\cancel{r}}\cdot(r-2)\left(\frac{300}{\cancel{r}}\right)=r\cdot\overset{1}{(\cancel{r-2})}\left(\frac{200}{\cancel{r-2}}\right)$$
$${\scriptstyle 1}{\scriptstyle 1}$$
$$300(r-2)=r(200)$$
$$300r-600=200r$$
$$300r-600+\mathbf{600}=200r+\mathbf{600}$$
$$300r=200r+600$$
$$300r-\mathbf{200r}=200r+600-\mathbf{200r}$$
$$\frac{100r}{\mathbf{100}}=\frac{600}{\mathbf{100}}$$
$$r=6$$

State

Bond's rate = 6%.
Credit Union's rate = $r-2=6-2=4\%$

Check

The result checks.

43. COMPARING INVESTMENTS

Analyze

- 1st CD earns $175 interest.
- 2nd CD earns $200 interest.
- 2nd CD investment rate is 1% more.
- Time for both is 1 year.
- Principals are the same amount for both.
- What are the two rates?

Assign

Let $r = 1^{st}$ CD interest rate as a percent

$r + 1 = 2^{nd}$ CD interest rate as a percent

Form

	Principal	• Rate	• Time	= Interest
1st CD	$\dfrac{175}{r}$	r	1	175
2nd CD	$\dfrac{200}{r+1}$	$r+1$	1	200

Principal invested in 1st CD.	equals	Principal invested in 2nd CD.
$\dfrac{175}{r}$	$=$	$\dfrac{200}{r+1}$

Solve

$$\frac{175}{r} = \frac{200}{r+1}$$

$$\text{LCD} = r(r+1)$$

$$r(r+1)\left(\frac{175}{r}\right) = r(r+1)\left(\frac{200}{r+1}\right)$$

$$\overset{1}{\cancel{r}} \cdot (r+1)\left(\frac{175}{\cancel{r}}\right) = r \cdot \overset{1}{\cancel{(r+1)}}\left(\frac{200}{\cancel{r+1}}\right)$$

$$175(r+1) = r(200)$$

$$175r + 175 = 200r$$

$$175r + 175 - \mathbf{175r} = 200r - \mathbf{175r}$$

$$175 = 25r$$

$$\frac{175}{\mathbf{25}} = \frac{25r}{\mathbf{25}}$$

$$7 = r$$

State

1st CD's rate = 7%.

2nd CD's rate = $r + 1 = 7 + 1 = 8\%$

Check

The results check.

WRITING

45. Answers will vary.

REVIEW

47. Solve using substitution:

$$\begin{cases} x + y = 4 \\ y = 3x \end{cases}$$

$x + y = 4$ The 1st equation
           Substitute for y.

$$x + 3x = 4$$

$$4x = 4$$

$$\frac{4x}{4} = \frac{4}{4}$$

$$x = 1$$

$y = 3x$ The 2nd equation

$$y = 3(1)$$

$$y = 3$$

The solution is $(1, 3)$.

49. $x + 20 = 4x - 1 + 2x$

$$\frac{21}{5} + 20 \overset{?}{=} 4\left(\frac{21}{5}\right) - 1 + 2\left(\frac{21}{5}\right)$$

$$\frac{21}{5} + \frac{100}{5} \overset{?}{=} \frac{84}{5} - \frac{5}{5} + \frac{42}{5}$$

$$\frac{121}{5} = \frac{121}{5}$$

          True

CHALLENGE PROBLEMS

51. RIVER TOUR

Analyze

- Tour's distance is 60 miles one way.
- Rate of current is 5 mph.
- Total time is 5 hours.
- What is the boat's still water speed?

Assign

Let $r =$ boat's still water speed in mph

Form

	Rate	$\cdot$	Time	=	Distance
Upstream	$r-5$		$\dfrac{60}{r-5}$		60
Downstream	$r+5$		$\dfrac{60}{r+5}$		60

Time took boat to go 60 miles upstream.	plus	Time took boat to go 60 miles downstream.	equals	Total time of 5 hours.

$$\frac{60}{r-5} \;+\; \frac{60}{r+5} \;=\; 5$$

Solve

$$\frac{60}{r-5}+\frac{60}{r+5}=5$$
$$LCD=(r-5)(r+5)$$

$$(r-5)(r+5)\left(\frac{60}{r-5}\right)+(r-5)(r+5)\left(\frac{60}{r+5}\right)=(r-5)(r+5)(5)$$

$$(r-5)(r+5)\left(\frac{60}{r-5}\right)+(r-5)(r+5)\left(\frac{60}{r+5}\right)=5(r^2-25)$$

$$60(r+5)+60(r-5)=5(r^2-25)$$
$$60r+300+60r-300=5r^2-125$$
$$120r=5r^2-125$$
$$5r^2-125=120r$$
$$5r^2-125-120r=120r-120r$$
$$5r^2-120r-125=0$$
$$5(r^2-24r-25)=0$$
$$5(r-25)(r+1)=0$$

$$\text{or}$$

$r-25=0$	$r+1=0$
$r-25+25=0+25$	$r+1-1=0-1$
$r=25$	$r=-1$

State

Boat's rate should be 25 mph.

Check

The result checks.

---

53. SALES

Analyze

- Spent $1,200 on some radios.
- Gave away 6 radios.
- Sold each for $10 more than paid.
- How many radios did she buy?
- Broke even.

Assign

Let $x =$ number of radios bought

Form

	Quanity	$\cdot$	Value of 1	=	Total Value
Bought	x		$\dfrac{1,200}{x}$		1,200
Selling	$x-6$		$\dfrac{1,200}{x-6}$		1,200

Buyer's cost per radio.	plus	Profit per radio.	equals	Selling price per radio.

$$\frac{1,200}{x} \;+\; 10 \;=\; \frac{1,200}{x-6}$$

Solve

$$\frac{1,200}{x}+10=\frac{1,200}{x-6}$$
$$LCD=x(x-6)$$

$$x(x-6)\left(\frac{1,200}{x}+10\right)=x(x-6)\left(\frac{1,200}{x-6}\right)$$

$$x(x-6)\left(\frac{1,200}{x}\right)+x(x-6)(10)=x(x-6)\left(\frac{1,200}{x-6}\right)$$

$$1,200(x-6)+10x(x-6)=1,200x$$
$$1,200x-7,200+10x^2-60x=1,200x$$
$$10x^2+1,140x-7,200=1,200x$$
$$10x^2+1,140x-7,200-\mathbf{1,200x}=1,200x-\mathbf{1,200x}$$
$$10x^2-60x-7,200=0$$
$$10(x^2-6x-720)=0$$
$$10(x-30)(x+24)=0$$

$$\text{or}$$

$x-30=0$	$x+24=0$
$x-30+30=0+30$	$x+24-24=0-24$
$x=30$	$x=-24$

State

The number of radios is 30.

Check

The result checks.

SECTION 7.8, STUDY SET

VOCABULARY

Fill in the blanks.

1. A **ratio** is the quotient of two numbers or the quotient of two quantities with the same units. A **rate** is a quotient of two quantities that have different units.

3. In $\frac{50}{3} = \frac{x}{9}$, the terms 50 and 9 are called the **extremes** and the terms 3 and x are called the **means** of the proportion.

5. Examples of **unit** prices are $1.65 per gallon, 17¢ per day, and $50 per foot.

CONCEPTS

7. Fill in the blanks: In a proportion, the product of the extremes is **equal** to the product of the means. In symbols, If $\frac{a}{b} = \frac{c}{d}$, then $ad = bc$.

9. SNACKS

$$\frac{\text{Number of bags} \rightarrow \mathbf{25}}{\text{Number underweight} \rightarrow \mathbf{2}} = \frac{\mathbf{1{,}000} \leftarrow \text{Number of bags}}{x \leftarrow \text{Number underweight}}$$

11. KLEENEX

$$\frac{\text{Price}}{\text{Number of sheets}} = \frac{\text{Price}}{\text{Number of sheets}}$$

$$\frac{\mathbf{2.19}}{85} = \frac{x}{\mathbf{1}}$$

NOTATION

13. Solve for x.

$$\frac{12}{18} = \frac{x}{24}$$

$$12 \cdot 24 = 18 \cdot x$$

$$\mathbf{288} = 18x$$

$$\frac{288}{\mathbf{18}} = \frac{18x}{\mathbf{18}}$$

$$16 = x$$

15. Fill in the blanks: The proportion $\frac{20}{1.6} = \frac{100}{8}$ can be read: 20 is to 1.6 **as** 100 is **to** 8.

GUIDED PRACTICE

Translate each ratio into a fraction in simplest form. See Example 1.

17. $\dfrac{4 \text{ boxes}}{15 \text{ boxes}} = \dfrac{4 \cancel{\text{ boxes}}}{15 \cancel{\text{ boxes}}} = \dfrac{4}{15}$

19. $\dfrac{18 \text{ watts}}{24 \text{ watts}} = \dfrac{3 \cdot \overset{1}{\cancel{6}} \cancel{\text{ watts}}}{4 \cdot \cancel{6} \cancel{\text{ watts}}} = \dfrac{3}{4}$

21. $\dfrac{30 \text{ days}}{24 \text{ days}} = \dfrac{5 \cdot \overset{1}{\cancel{6}} \cancel{\text{ days}}}{4 \cdot \cancel{6} \cancel{\text{ days}}} = \dfrac{5}{4}$

23. If 1 hour = 60 minutes, then 3 hours = 180 minutes.

$$\frac{90 \text{ minutes}}{3 \text{ hours}} = \frac{90 \text{ minutes}}{180 \text{ minutes}}$$

$$\frac{90 \text{ minutes}}{180 \text{ minutes}} = \frac{1 \cdot \overset{1}{\cancel{90}} \cancel{\text{ minutes}}}{2 \cdot \cancel{90} \cancel{\text{ minutes}}} = \frac{1}{2}$$

25. If 1 gallon = 4 quarts, then 4 gallons = 16 quarts.

$$\frac{8 \text{ quarts}}{4 \text{ gallons}} = \frac{8 \text{ quarts}}{16 \text{ quarts}}$$

$$\frac{8 \text{ quarts}}{16 \text{ quarts}} = \frac{1 \cdot \overset{1}{\cancel{8}} \cancel{\text{ quarts}}}{2 \cdot \cancel{8} \cancel{\text{ quarts}}} = \frac{1}{2}$$

27. 1 mile = 5,280 feet

$$\frac{6{,}000 \text{ feet}}{1 \text{ mile}} = \frac{6{,}000 \text{ feet}}{5{,}280 \text{ feet}}$$

$$\frac{6{,}000 \text{ feet}}{5{,}280 \text{ feet}} = \frac{25 \cdot \overset{1}{\cancel{240}} \cancel{\text{ feet}}}{22 \cdot \cancel{240} \cancel{\text{ feet}}} = \frac{25}{22}$$

Determine whether each equation is a true proportion. See Example 2.

29. $\dfrac{7}{3} = \dfrac{14}{6}$

$7 \cdot 6 = 14 \cdot 3$

$42 = 42$

Since the cross products are equal, it is a proportion.

31. $\dfrac{5}{8} = \dfrac{12}{19.4}$

$5 \cdot 19.4 = 12 \cdot 8$

$97 = 96$

Since the cross products are not equal, it is not a proportion.

Solve each proportion. See Example 3.

Section 7.8

33. $\dfrac{2}{3} = \dfrac{x}{6}$

check:

$2 \cdot 6 = 3x$

$\dfrac{2}{3} = \dfrac{x}{6}$

$12 = 3x$

$\dfrac{2}{3} \overset{?}{=} \dfrac{4}{6}$

$\dfrac{12}{3} = \dfrac{3x}{3}$

$2 \cdot 6 \overset{?}{=} 3 \cdot 4$

$4 = x$

$12 = 12$

---

35. $\dfrac{63}{g} = \dfrac{9}{2}$

check:

$63 \cdot 2 = 9g$

$\dfrac{63}{g} = \dfrac{9}{2}$

$126 = 9g$

$\dfrac{63}{14} \overset{?}{=} \dfrac{9}{2}$

$\dfrac{126}{9} = \dfrac{9g}{9}$

$63 \cdot 2 \overset{?}{=} 9 \cdot 14$

$14 = g$

$126 = 126$

---

37. $\dfrac{x+1}{5} = \dfrac{3}{15}$

check:

$15(x+1) = 3 \cdot 5$

$\dfrac{x+1}{5} = \dfrac{3}{15}$

$15x + 15 = 15$

$\dfrac{0+1}{5} \overset{?}{=} \dfrac{3}{15}$

$15x + 15 - \mathbf{15} = 15 - \mathbf{15}$

$15x = 0$

$\dfrac{1}{5} \overset{?}{=} \dfrac{3}{15}$

$\dfrac{15x}{\mathbf{15}} = \dfrac{0}{\mathbf{15}}$

$1 \cdot 15 = 5 \cdot 3$

$x = 0$

$15 = 15$

---

39. $\dfrac{5-x}{17} = \dfrac{13}{34}$

check:

$34(5-x) = 13 \cdot 17$

$\dfrac{5-x}{17} = \dfrac{13}{34}$

$170 - 34x = 221$

let $-\dfrac{3}{2} = -1.5$

$170 - 34x - \mathbf{170} = 221 - \mathbf{170}$

$\dfrac{5-(-1.5)}{17} \overset{?}{=} \dfrac{13}{34}$

$-34x = 51$

$\dfrac{6.5}{17} \overset{?}{=} \dfrac{13}{34}$

$\dfrac{-34x}{\mathbf{-34}} = \dfrac{51}{\mathbf{-34}}$

$6.5 \cdot 34 \overset{?}{=} 17 \cdot 13$

$x = -\dfrac{3 \cdot \cancel{17}^{1}}{2 \cdot \cancel{17}_{1}}$

$221 = 221$

$x = -\dfrac{3}{2}$

---

41. $\dfrac{15}{7b+5} = \dfrac{5}{2b+1}$

check:

$15(2b+1) = 5(7b+5)$

$\dfrac{15}{7b+5} = \dfrac{5}{2b+1}$

$30b + 15 = 35b + 25$

$\dfrac{15}{7(-2)+5} \overset{?}{=} \dfrac{5}{2(-2)+1}$

$30b + 15 - \mathbf{25} = 35b + 25 - \mathbf{25}$

$30b - 10 = 35b$

$\dfrac{15}{-14+5} \overset{?}{=} \dfrac{5}{-4+1}$

$30b - 10 - \mathbf{30b} = 35b - \mathbf{30b}$

$\dfrac{15}{-9} \overset{?}{=} \dfrac{5}{-3}$

$-10 = 5b$

$\dfrac{-10}{\mathbf{5}} = \dfrac{5b}{\mathbf{5}}$

$15 \cdot -3 \overset{?}{=} -9 \cdot 5$

$-2 = b$

$-45 = -45$

---

43. $\dfrac{8x}{3} = \dfrac{11x+9}{4}$

check:

$8x(4) = 3(11x+9)$

$\dfrac{8x}{3} = \dfrac{11x+9}{4}$

$32x = 33x + 27$

$\dfrac{8(-27)}{3} \overset{?}{=} \dfrac{11(-27)+9}{4}$

$32x - \mathbf{33x} = 33x + 27 - \mathbf{33x}$

$\dfrac{-216}{3} \overset{?}{=} \dfrac{-297+9}{4}$

$-x = 27$

$\dfrac{-216}{3} \overset{?}{=} \dfrac{-288}{4}$

$(\mathbf{-1})(-x) = (\mathbf{-1})(27)$

$-216 \cdot 4 \overset{?}{=} -288 \cdot 3$

$x = -27$

$-864 = -864$

Solve each proportion. See Example 4.

45. $\dfrac{2}{3x} = \dfrac{x}{6}$

$2(6) = 3x(x)$

$12 = 3x^2$

$12 - \mathbf{12} = 3x^2 - \mathbf{12}$

$0 = 3x^2 - 12$

$0 = 3(x^2 - 4)$

$0 = 3(x+2)(x-2)$

$\begin{array}{c|c} & \text{or} \\ x+2 = 0 & x-2 = 0 \\ x+2-\mathbf{2} = 0-\mathbf{2} & x-2+\mathbf{2} = 0+\mathbf{2} \\ x = -2 & x = 2 \end{array}$

The solutions are -2 and 2.

check:

$\dfrac{2}{3x} = \dfrac{x}{6}$

$\dfrac{2}{3(-2)} \overset{?}{=} \dfrac{-2}{6}$

$\dfrac{2}{-6} \overset{?}{=} \dfrac{-2}{6}$

$-2 \cdot 6 \overset{?}{=} 6 \cdot -2$

$-12 = -12$

check:

$\dfrac{2}{3x} = \dfrac{x}{6}$

$\dfrac{2}{3(2)} \overset{?}{=} \dfrac{2}{6}$

$\dfrac{2}{6} \overset{?}{=} \dfrac{2}{6}$

$2 \cdot 6 \overset{?}{=} 2 \cdot 6$

$12 = 12$

Both solutions check.

47.
$$\frac{b-5}{3}=\frac{2}{b}$$
$$b(b-5)=2\cdot 3$$
$$b^2-5b=6$$
$$b^2-5b-\mathbf{6}=6-\mathbf{6}$$
$$b^2-5b-6=0$$
$$(b-6)(b+1)=0$$

$$b-6=0 \qquad \text{or} \qquad b+1=0$$
$$b-6+\mathbf{6}=0+\mathbf{6} \qquad \quad b+1-\mathbf{1}=0-\mathbf{1}$$
$$b=6 \qquad\qquad\qquad b=-1$$

The solutions are 6 and -1.

check:
$$\frac{b-5}{3}=\frac{2}{b}$$
$$\frac{6-5}{3}\overset{?}{=}\frac{2}{6}$$
$$\frac{1}{3}\overset{?}{=}\frac{2}{6}$$
$$1\cdot 6\overset{?}{=}3\cdot 2$$
$$6=6$$

check:
$$\frac{b-5}{3}=\frac{2}{b}$$
$$\frac{-1-5}{3}\overset{?}{=}\frac{2}{-1}$$
$$\frac{-6}{3}\overset{?}{=}\frac{2}{-1}$$
$$-6\cdot -1\overset{?}{=}3\cdot 2$$
$$6=6$$

Both solutions check.

49.
$$\frac{a-4}{a}=\frac{15}{a+4}$$
$$(a-4)(a+4)=15a$$
$$a^2-16=15a$$
$$a^2-16-\mathbf{15a}=15a-\mathbf{15a}$$
$$a^2-15a-16=0$$
$$(a-16)(a+1)=0$$

$$a-16=0 \qquad \text{or} \qquad a+1=0$$
$$a-16+\mathbf{16}=0+\mathbf{16} \qquad a+1-\mathbf{1}=0-\mathbf{1}$$
$$a=16 \qquad\qquad\qquad a=-1$$

The solutions are 16 and -1.

check:
$$\frac{a-4}{a}=\frac{15}{a+4}$$
$$\frac{16-4}{16}\overset{?}{=}\frac{15}{16+4}$$
$$\frac{12}{16}\overset{?}{=}\frac{15}{20}$$
$$12\cdot 20\overset{?}{=}16\cdot 15$$
$$240=240$$

check:
$$\frac{a-4}{a}=\frac{15}{a+4}$$
$$\frac{-1-4}{-1}\overset{?}{=}\frac{15}{-1+4}$$
$$\frac{-5}{-1}\overset{?}{=}\frac{15}{3}$$
$$-5\cdot 3\overset{?}{=}-1\cdot 15$$
$$-15=-15$$

Both solutions check.

51.
$$\frac{t+3}{t+5}=\frac{-1}{2t}$$
$$2t(t+3)=-1(t+5)$$
$$2t^2+6t=-t-5$$
$$2t^2+6t+\mathbf{t}=-t-5+\mathbf{t}$$
$$2t^2+7t=-5$$
$$2t^2+7t+\mathbf{5}=-5+\mathbf{5}$$
$$2t^2+7t+5=0$$
$$(2t+5)(t+1)=0$$

$$2t+5=0 \qquad \text{or} \qquad t+1=0$$
$$2t+5-\mathbf{5}=0-\mathbf{5} \qquad t+1-\mathbf{1}=0-\mathbf{1}$$
$$2t=-5 \qquad\qquad\qquad t=-1$$
$$\frac{2t}{\mathbf{2}}=\frac{-5}{\mathbf{2}}$$
$$t=-\frac{5}{2}$$

The solutions are $-\dfrac{5}{2}$ and -1.

check:
$$\frac{t+3}{t+5}=\frac{-1}{2t}$$
$$\text{let } -\frac{5}{2}=-2.5$$
$$\frac{-2.5+3}{-2.5+5}\overset{?}{=}\frac{-1}{2(-2.5)}$$
$$\frac{0.5}{2.5}\overset{?}{=}\frac{-1}{-5}$$
$$0.5\cdot -5\overset{?}{=}2.5\cdot -1$$
$$-2.5=-2.5$$

check:
$$\frac{t+3}{t+5}=\frac{-1}{2t}$$
$$\frac{-1+3}{-1+5}\overset{?}{=}\frac{-1}{2(-1)}$$
$$\frac{2}{4}\overset{?}{=}\frac{-1}{-2}$$
$$2\cdot -2\overset{?}{=}4\cdot -1$$
$$-4=-4$$

Both solutions check.

Section 7.8

Each pair of triangles is similar. Find the missing side length. See Example 8.

53. $\dfrac{4}{12} = \dfrac{5}{x}$

$4x = 5 \cdot 12$

$4x = 60$

$\dfrac{4x}{4} = \dfrac{60}{4}$

$x = 15$

The solution is 15.

check:

$\dfrac{4}{12} = \dfrac{5}{x}$

$\dfrac{4}{12} = \dfrac{5}{\mathbf{15}}$

$4 \cdot 15 = 12 \cdot 5$

$60 = 60$

55. $\dfrac{9}{12} = \dfrac{6}{x}$

$9x = 6 \cdot 12$

$9x = 72$

$\dfrac{9x}{9} = \dfrac{72}{9}$

$x = 8$

The solution is 8.

check:

$\dfrac{9}{12} = \dfrac{6}{x}$

$\dfrac{9}{12} \overset{?}{=} \dfrac{6}{\mathbf{8}}$

$72 = 72$

TRY IT YOURSELF

Solve each proportion.

57. $\dfrac{x-1}{x+1} = \dfrac{2}{3x}$

$3x(x-1) = 2(x+1)$

$3x^2 - 3x = 2x + 2$

$3x^2 - 3x - \mathbf{2x} = 2x + 2 - \mathbf{2x}$

$3x^2 - 5x = 2$

$3x^2 - 5x - \mathbf{2} = 2 - \mathbf{2}$

$3x^2 - 5x - 2 = 0$

$(3x+1)(x-2) = 0$

$3x + 1 = 0 \qquad \text{or} \qquad x - 2 = 0$

$3x + 1 - \mathbf{1} = 0 - \mathbf{1} \qquad x - 2 + \mathbf{2} = 0 + \mathbf{2}$

$3x = -1 \qquad\qquad x = 2$

$\dfrac{3x}{3} = \dfrac{-1}{3}$

$x = -\dfrac{1}{3}$

The solutions are $-\dfrac{1}{3}$ and 2.

check:

$\dfrac{x-1}{x+1} = \dfrac{2}{3x}$

$\dfrac{-\dfrac{1}{3}-1}{-\dfrac{1}{3}+1} \overset{?}{=} \dfrac{2}{3 \cdot -\dfrac{1}{3}}$

$\dfrac{-\dfrac{4}{3}}{\dfrac{2}{3}} \overset{?}{=} \dfrac{2}{-1}$

$-\dfrac{4}{3} \cdot -1 \overset{?}{=} \dfrac{2}{3} \cdot 2$

$\dfrac{4}{3} = \dfrac{4}{3}$

check:

$\dfrac{x-1}{x+1} = \dfrac{2}{3x}$

$\dfrac{\mathbf{2}-1}{\mathbf{2}+1} \overset{?}{=} \dfrac{2}{3(\mathbf{2})}$

$\dfrac{1}{3} \overset{?}{=} \dfrac{2}{6}$

$1 \cdot 6 \overset{?}{=} 3 \cdot 2$

$6 = 6$

Both solutions check.

59. $\dfrac{x+1}{4} = \dfrac{3x}{8}$

$8(x+1) = 3x(4)$

$8x + 8 = 12x$

$8x + 8 - \mathbf{8x} = 12x - \mathbf{8x}$

$8 = 4x$

$\dfrac{8}{4} = \dfrac{4x}{4}$

$2 = x$

The solution is 2.

check:

$\dfrac{x+1}{4} = \dfrac{3x}{8}$

$\dfrac{\mathbf{2}+1}{4} \overset{?}{=} \dfrac{3(\mathbf{2})}{8}$

$\dfrac{3}{4} \overset{?}{=} \dfrac{6}{8}$

$3 \cdot 8 \overset{?}{=} 4 \cdot 6$

$24 = 24$

The solution checks.

61. $\dfrac{y-4}{y+1} = \dfrac{y+3}{y+6}$

$(y-4)(y+6) = (y+1)(y+3)$

$y^2 + 2y - 24 = y^2 + 4y + 3$

$y^2 + 2y - 24 - \mathbf{y^2} = y^2 + 4y + 3 - \mathbf{y^2}$

$2y - 24 = 4y + 3$

$2y - 24 - \mathbf{2y} = 4y + 3 - \mathbf{2y}$

$-24 = 2y + 3$

$-24 - \mathbf{3} = 2y + 3 - \mathbf{3}$

$-27 = 2y$

$\dfrac{-27}{2} = \dfrac{2y}{2}$

$-\dfrac{27}{2} = y$

The solution is $-\dfrac{27}{2}$.

check:

$$\dfrac{y-4}{y+1} = \dfrac{y+3}{y+6}$$

let $-\dfrac{27}{2} = -13.5$

$$\dfrac{-13.5-4}{-13.5+1} \overset{?}{=} \dfrac{-13.5+3}{-13.5+6}$$

$$\dfrac{-17.5}{-12.5} \overset{?}{=} \dfrac{-10.5}{-7.5}$$

$$-17.5 \cdot -7.5 \overset{?}{=} -12.5 \cdot -10.5$$

$$131.25 = 131.25$$

The solution checks.

63.

$$\dfrac{c}{10} = \dfrac{10}{c}$$

$$c \cdot c = 10 \cdot 10$$

$$c^2 = 100$$

$$c^2 - 100 = 100 - 100$$

$$c^2 - 100 = 0$$

$$(c+10)(c-10) = 0$$

$c+10 = 0$ or $c-10 = 0$

$c+10-10 = 0-10$ $c-10+10 = 0+10$

$\qquad c = -10$ $c = 10$

The solutions are -10 and 10.

check: check:

$$\dfrac{c}{10} = \dfrac{10}{c} \qquad\qquad \dfrac{c}{10} = \dfrac{10}{c}$$

$$\dfrac{-10}{10} \overset{?}{=} \dfrac{10}{-10} \qquad\quad \dfrac{10}{10} \overset{?}{=} \dfrac{10}{10}$$

$$-10 \cdot -10 \overset{?}{=} 10 \cdot 10 \qquad 10 \cdot 10 \overset{?}{=} 10 \cdot 10$$

$$100 = 100 \qquad\qquad\quad 100 = 100$$

Both solutions check.

65.

$$\dfrac{m}{3} = \dfrac{4}{m+1}$$

$$m(m+1) = 3 \cdot 4$$

$$m^2 + m = 12$$

$$m^2 + m - 12 = 12 - 12$$

$$m^2 + m - 12 = 0$$

$$(m+4)(m-3) = 0$$

$m+4 = 0$ or $m-3 = 0$

$m+4-4 = 0-4$ $m-3+3 = 0+3$

$\qquad m = -4$ $m = 3$

The solutions are -4 and 3.

check: check:

$$\dfrac{m}{3} = \dfrac{4}{m+1} \qquad\qquad \dfrac{m}{3} = \dfrac{4}{m+1}$$

$$\dfrac{-4}{3} \overset{?}{=} \dfrac{4}{-4+1} \qquad\quad \dfrac{3}{3} \overset{?}{=} \dfrac{4}{3+1}$$

$$\dfrac{-4}{3} \overset{?}{=} \dfrac{4}{-3} \qquad\qquad \dfrac{3}{3} \overset{?}{=} \dfrac{4}{4}$$

$$-4 \cdot -3 \overset{?}{=} 3 \cdot 4 \qquad\qquad 3 \cdot 4 \overset{?}{=} 3 \cdot 4$$

$$12 = 12 \qquad\qquad\qquad 12 = 12$$

Both solutions check.

67.

$$\dfrac{3}{3b+4} = \dfrac{2}{5b-6}$$

$$3(5b-6) = 2(3b+4)$$

$$15b - 18 = 6b + 8$$

$$15b - 18 - 6b = 6b + 8 - 6b$$

$$9b - 18 = 8$$

$$9b - 18 + 18 = 8 + 18$$

$$9b = 26$$

$$\dfrac{9b}{9} = \dfrac{26}{9}$$

$$b = \dfrac{26}{9}$$

The solution is $\dfrac{26}{9}$.

Section 7.8

check:

$$\frac{3}{3b+4}=\frac{2}{5b-6}$$

$$\frac{3}{3\left(\dfrac{26}{9}\right)+4}\overset{?}{=}\frac{2}{5\left(\dfrac{26}{9}\right)-6}$$

$$\frac{3}{\dfrac{26}{3}+4}\overset{?}{=}\frac{2}{\dfrac{130}{9}-6}$$

$$\frac{3}{\dfrac{38}{3}}\overset{?}{=}\frac{2}{\dfrac{76}{9}}$$

$$3\cdot\frac{76}{9}\overset{?}{=}\frac{38}{3}\cdot 2$$

$$\frac{76}{3}=\frac{76}{3}$$

The solution checks.

LOOK ALIKES...
Solve each equation.

69. a.
$$\frac{-2}{5}=\frac{3}{4x}$$
$$-2(4x)=3(5)$$
$$-8x=15$$
$$\frac{-8x}{-8}=\frac{15}{-8}$$
$$x=-\frac{15}{8}$$

b.
$$\frac{4}{x}-\frac{2}{5}=\frac{3}{4x}$$
$$20x\left(\frac{4}{x}-\frac{2}{5}\right)=20x\left(\frac{3}{4x}\right)$$
$$20\,\overset{1}{\cancel{x}}\left(\frac{4}{\cancel{x}}\right)-4x\cdot\overset{1}{\cancel{5}}\left(\frac{2}{\cancel{5}}\right)=5\cdot\overset{1}{\cancel{4x}}\left(\frac{3}{\cancel{4x}}\right)$$
$$80-8x=15$$
$$80-8x-\mathbf{80}=15-\mathbf{80}$$
$$-8x=-65$$
$$\frac{-8x}{-8}=\frac{-65}{-8}$$
$$x=\frac{65}{8}$$

71. a.
$$\frac{3}{a-1}=\frac{8}{a}$$
$$a(3)=8(a-1)$$
$$3a=8a-8$$
$$3a-\mathbf{8a}=8a-8-\mathbf{8a}$$
$$-5a=8$$
$$\frac{-5a}{-5}=\frac{8}{-5}$$
$$a=-\frac{8}{5}$$

b.
$$\frac{3}{a-1}+\frac{8}{a}=3$$
$$a(a-1)\left(\frac{3}{a-1}+\frac{8}{a}\right)=a(a-1)(3)$$
$$a\cdot\overset{1}{\cancel{(a-1)}}\left(\frac{3}{\cancel{a-1}}\right)+(a-1)\cdot\overset{1}{\cancel{a}}\left(\frac{8}{\cancel{a}}\right)=(a^2-a)(3)$$
$$3a+8a-8=3a^2-3a$$
$$11a-8=3a^2-3a$$
$$11a-8-\mathbf{11a}=3a^2-3a-\mathbf{11a}$$
$$-8=3a^2-14a$$
$$-8+\mathbf{8}=3a^2-14a+\mathbf{8}$$
$$0=3a^2-14a+8$$
$$0=(3a-2)(a-4)$$

$$
\begin{array}{ll}
3a-2=0 \quad \text{or} & a-4=0\\
3a=2 & a-4+\mathbf{4}=0+\mathbf{4}\\
a=\dfrac{2}{3} & a=4
\end{array}
$$

The solutions are $\dfrac{2}{3}$ and 4.

APPLICATIONS

73. SHOPPING FOR CLOTHES

Analyze
- We know the cost of two shirts is $25.
- How much do five shirts cost?

Assign

Let c = cost of 5 shirts

Form

2 shirts is to $25 as 5 shirts is to $$c$.

$$\text{2 shirts} \rightarrow \frac{2}{25} = \frac{5}{c} \leftarrow \text{5 shirts}$$

Cost of 2 shirts $\rightarrow$ $\leftarrow$ Cost of 5 shirts

Solve

$$\frac{2}{25} = \frac{5}{c}$$
$$2c = 125$$
$$\frac{2c}{2} = \frac{125}{2}$$
$$c = 62.5$$

State

Five shirts cost $62.50.

Check

$$\frac{2}{25} = \frac{5}{c}$$
$$\frac{2}{25} \overset{?}{=} \frac{5}{\mathbf{62.50}}$$
$$125 = 125$$

The result checks.

75. CPR

Analyze
- We know the ratio of chest compressions to breaths should be 5:2.
- How many breaths are needed for 210 compressions?

Assign

Let x = # of breaths needed

Form

5 compressions is to 2 breaths
as 210 compressions is to x breaths.

5 compressions $\rightarrow \dfrac{5}{2} = \dfrac{210}{x} \leftarrow$ 210 compressions

2 breaths $\rightarrow$ $\leftarrow$ number of breaths

Solve

$$\frac{5}{2} = \frac{210}{x}$$
$$5x = 420$$
$$\frac{5x}{5} = \frac{420}{5}$$
$$x = 84$$

State

84 breaths are needed with 210 compressions.

Check

$$\frac{5}{2} = \frac{210}{x}$$
$$\frac{5}{2} \overset{?}{=} \frac{210}{\mathbf{84}}$$
$$420 = 420$$

The result checks.

from CAMPUS TO CAREERS

77. RECREATION DIRECTOR

a.

Analyze
- We know there is total of 966 boys and girls.
- If 504 are boys how many are girls?

Assign

Let x = number of girls

Form

The # of boys plus the # of girls is 966.

Solve

$$504 + x = 966$$
$$504 + x - \mathbf{504} = 966 - \mathbf{504}$$
$$x = 462$$

State

The are 462 girls.

Check

$$504 + x = 966$$
$$504 + 462 \overset{?}{=} 966$$
$$966 = 966$$

The result checks.

b.
- Find the ratio of girls to boys.

$$\frac{\text{\# of girls}}{\text{\# of boys}} = \frac{462}{504}$$

$$= \frac{11 \cdot \cancel{42}^{1}}{12 \cdot \cancel{42}_{1}}$$

$$= \frac{11}{12} \quad \text{or} \quad 11:12$$

The ratio of girls to boys is 11:12.

79. COMPUTING A PAYCHECK

Analyze
- We know Billie earns $412 for 40 hours.
- She missed 10 hours of work.
- How much did she earn?

Section 7.8

Assign

Let x = amount earned for 30 hours

Form

40 hours is to $412 as 30 hours is to $$x$.

40 hrs worked → $\dfrac{40}{412} = \dfrac{30}{x}$ ← 30 hrs worked

Earned 40 hrs. →   ← Earned 30 hrs

Solve

$$\frac{40}{412} = \frac{30}{x}$$

$$40x = 12,360$$

$$\frac{40x}{40} = \frac{12,360}{40}$$

$$x = 309$$

State

Billie earns $309 for 30 hours of work.

Check

$$\frac{40}{412} = \frac{30}{x}$$

$$\frac{40}{412} \overset{?}{=} \frac{30}{309}$$

$$12,360 = 12,360$$

The result checks.

81. TWITTER

Analyze

• 7,500 tweets are sent every 10 seconds.

• How many tweets are sent in one minute?

Assign

Let x = number of tweets sent in one minute

Form

7,500 tweets is to 10 seconds as
x tweets is to 60 seconds.

tweets → $\dfrac{7,500}{10} = \dfrac{x}{60}$ ← tweets

seconds →   ← seconds

Solve

$$\frac{7,500}{10} = \frac{x}{60}$$

$$450,000 = 10x$$

$$\frac{450,000}{10} = \frac{10x}{10}$$

$$x = 45,000$$

State

45,000 tweets are sent in one minute.

Check

$$\frac{7,500}{10} = \frac{x}{60}$$

$$\frac{7,500}{10} \overset{?}{=} \frac{\mathbf{45,000}}{60}$$

$$450,000 = 450,000$$

The result checks.

83. NUTRITION

Analyze

• We know a 10-oz milkshake contains 355 calories, 8 gm of fat and 9 gm of protein.

• What are the amounts in a 16-oz milkshake?

Assign

Let c = # of calories in a 16-oz shake

Form

335 cal is to a 10-oz shake as c cal is to a 16-oz shake.

335 calories → $\dfrac{335}{10} = \dfrac{c}{16}$ ← number of calories

10-oz shake →   ← 16-oz shake

Assign

Let f = grams of fat in a 16-oz shake

Form

8 gm is to a 10-oz shake as f gm is to a 16-oz shake.

8 grams of fat → $\dfrac{8}{10} = \dfrac{f}{16}$ ← grams of fat

10-oz shake →   ← 16-oz shake

Assign

Let p = grams of protein in a 16-oz shake

Form

9 gm is to a 10-oz shake as p gm is to a 16-oz shake.

9 gm of protein → $\dfrac{9}{10} = \dfrac{p}{16}$ ← grams of protein

10-oz shake →   ← 16-oz shake

Solve

calories	fat	protein
$\dfrac{355}{10} = \dfrac{c}{16}$	$\dfrac{8}{10} = \dfrac{f}{16}$	$\dfrac{9}{10} = \dfrac{p}{16}$
$5680 = 10c$	$128 = 10f$	$144 = 10p$
$\dfrac{5680}{10} = \dfrac{10c}{10}$	$\dfrac{128}{10} = \dfrac{10f}{10}$	$\dfrac{144}{10} = \dfrac{10p}{10}$
$568 = c$	$12.8 = f$	$14.4 = p$
	$13 = f$	$14 = p$

State

In a 16-oz mikeshake there are 568 calories, 13 grams of fat and 14 grams of protein.

Check

$$\frac{355}{10} = \frac{c}{16} \qquad \frac{8}{10} = \frac{f}{16} \qquad \frac{9}{10} = \frac{p}{16}$$

$$\frac{355}{10} \overset{?}{=} \frac{\mathbf{568}}{16} \qquad \frac{8}{10} \overset{?}{=} \frac{\mathbf{12.8}}{16} \qquad \frac{9}{10} \overset{?}{=} \frac{\mathbf{14.4}}{16}$$

$$5680 = 5680 \qquad 128 = 128 \qquad 144 = 144$$

The results check.

85. MIXING FUEL

Analyze

- We know the ratio is 50 to 1.
- We know the amount of gasoline is 6 gallons and the amount of oil is 6 ounces.
- The number of ounces of gasoline is $6 \cdot 128 = 768$.
- Are the amounts correct?

Assign

Let x = the number of ounces of oil

Form

50 parts gasoline is to 1 part oil as 768 parts of gasoline is to x parts of oil.

$$50 \text{ parts gas} \rightarrow \frac{50}{1} = \frac{768}{x} \leftarrow 768 \text{ parts gas}$$
$$1 \text{ part oil} \rightarrow \qquad\qquad \leftarrow x \text{ parts oil}$$

Solve

$$\frac{50}{1} = \frac{768}{x}$$
$$50x = 768$$
$$\frac{50x}{50} = \frac{768}{50}$$
$$x = 15.36$$

State

The calculated amount of 15.36 ounces is close to the stated amount of 16 ounces.

Check

$$\frac{50}{1} = \frac{768}{x}$$
$$\frac{50}{1} \overset{?}{=} \frac{768}{\mathbf{15.36}}$$
$$768 = 768$$

The result checks.

87. CAPTURE-RELEASE METHOD

Analyze

- We know 12 tagged squirrels were released.
- 35 were captured with 3 being tagged.
- How many squirrels are on the acreage?

Assign

Let x = the number of squirrels

Form

35 squirrels is to 3 tagged squirrels as x squirrels is to 12 tagged squirrels.

$$35 \text{ squirrels} \rightarrow \frac{35}{3} = \frac{x}{12} \leftarrow x \text{ squirrels}$$
$$3 \text{ tagged} \rightarrow \qquad\qquad \leftarrow 12 \text{ tagged}$$

Solve

$$\frac{35}{3} = \frac{x}{12}$$
$$420 = 3x$$
$$\frac{420}{3} = \frac{3x}{3}$$
$$140 = x$$

State

There are 140 squirrels on the acreage.

Check

$$\frac{35}{3} = \frac{x}{12}$$
$$\frac{35}{3} \overset{?}{=} \frac{\mathbf{140}}{12}$$
$$420 = 420$$

The result checks.

89. MODEL RAILROADS
Analyze
- We know the HO scale is 1:87.
- How long is a real engine in inches and feet if a model one measures 6 inches?

Assign
Let x = the length of a real engine in inches

Form
1 is to 87 as 6 inches is to x inches.

$$\begin{array}{l} 1 \text{ part} \rightarrow \\ 87 \text{ parts} \rightarrow \end{array} \frac{1}{87} = \frac{6}{x} \begin{array}{l} \leftarrow 6 \text{ inch model} \\ \leftarrow x \text{ inches real engine} \end{array}$$

Solve
$$\frac{1}{87} = \frac{6}{x}$$
$$x = 87(6)$$
$$x = 522$$

State
A real engine would be 522 inches long

or $\dfrac{522 \text{ in}}{1} \cdot \dfrac{1 \text{ ft}}{12 \text{ in}} = 43.5$ feet.

Check
$$\frac{1}{87} = \frac{6}{x}$$
$$\frac{1}{87} \overset{?}{=} \frac{6}{\mathbf{522}}$$
$$522 = 522$$
The result checks.

91. BLUEPRINTS
Analyze
- We know that $\frac{1}{4}$ inch is equal to 1 foot.
- How long is the real kitchen if the length is $2\frac{1}{2}$ inches on the drawing?

Assign
Let x = the length of the real kitchen

Form
$\frac{1}{4}$ inch $= 0.25$ inches

$2\frac{1}{2}$ inches $= 2.5$ inches

0.25 inch is to 1 foot as 2.5 inches is to x feet.

$$\begin{array}{l} 0.25 \text{ inch} \rightarrow \\ 1 \text{ foot} \rightarrow \end{array} \frac{0.25}{1} = \frac{2.5}{x} \begin{array}{l} \leftarrow 2.5 \text{ inches} \\ \leftarrow x \text{ feet} \end{array}$$

Solve
$$\frac{0.25}{1} = \frac{2.5}{x}$$
$$0.25x = 2.5$$
$$\frac{0.25x}{0.25} = \frac{2.5}{0.25}$$
$$x = 10$$

State
The kitchen's real length is 10 feet.

Check
$$\frac{0.25}{1} = \frac{2.5}{x}$$
$$\frac{0.25}{1} \overset{?}{=} \frac{2.5}{\mathbf{10}}$$
$$2.5 = 2.5$$
The result checks.

For each of the following purchases, determine the better buy. See Example 7.

93. TRUMPET LESSONS
$25 for 45 minutes
45 minutes = 0.75 hour

$$\text{cost} \rightarrow \frac{25}{0.75} = \frac{x}{1} \leftarrow \text{cost}$$
$$0.75 \text{ hour} \rightarrow \quad\quad \leftarrow 1 \text{ hour}$$
$$\frac{25}{0.75} = x$$
$$33.\overline{3} = x$$

The cost per hour is $33.33.

$35 for 60 minutes,
The cost per hour is $35.00.

The lesson of $25 for 45 minutes is the better rate.

95. BUSINESS CARDS
100 cards for $9.99.

$$\text{cost} \rightarrow \frac{9.99}{100} = \frac{x}{1} \leftarrow \text{cost}$$
$$100 \text{ cards} \rightarrow \quad\quad \leftarrow 1 \text{ card}$$
$$\frac{9.99}{100} = x$$
$$0.0999 = x$$

The cost per card is $0.0999.

150 cards for $12.99.

$$\text{cost} \rightarrow \frac{12.99}{150} = \frac{x}{1} \leftarrow \text{cost}$$
$$150 \text{ cards} \rightarrow \quad\quad \leftarrow 1 \text{ card}$$
$$\frac{12.99}{150} = x$$
$$0.0866 = x$$

The cost per card is $0.0866.

The 150 cards for $12.99 is the better buy.

97. SOFT DRINKS
6 cans for $1.50.

$$\text{cost} \rightarrow \frac{1.50}{6} = \frac{x}{1} \leftarrow \text{cost}$$
$$6 \text{ cans} \rightarrow \quad\quad \leftarrow 1 \text{ can}$$
$$\frac{1.50}{6} = x$$
$$0.25 = x$$

The cost per can is $0.25.

24 cans for $6.25.

$$\text{cost} \rightarrow \frac{6.25}{24} = \frac{x}{1} \leftarrow \text{cost}$$
$$24 \text{ cans} \rightarrow \quad\quad \leftarrow 1 \text{ can}$$
$$\frac{6.25}{24} = x$$
$$0.2604 \approx x$$

The cost per can is $0.2604.

The 6 - pack for $1.50 is the better buy.

99. AQUACLEAR WATER

12 8-oz bottles for $1.79.

$$\text{cost} \rightarrow \frac{1.79}{96} = \frac{x}{1} \leftarrow \text{cost}$$
$$96 \text{ ounces} \rightarrow \quad\quad \leftarrow 1 \text{ ounce}$$
$$\frac{1.79}{96} = x$$
$$0.0186 \approx x$$

The cost per ounce is about $0.0186.

24 12-oz bottles for $4.49.

$$\text{cost} \rightarrow \frac{4.49}{288} = \frac{x}{1} \leftarrow \text{cost}$$
$$288 \text{ ounces} \rightarrow \quad\quad \leftarrow 1 \text{ ounce}$$
$$\frac{4.49}{288} = x$$
$$0.0156 \approx x$$

The cost per ounce is about $0.0156.

The 24-12 oz bottles at $4.49 is the better buy.

101. HEIGHT OF A TREE

Analyze
- Similar triangles determined by the tree and its shadow and the man and his shadow.
- What is the height of the tree?

Assign

Let h = the height of the tree

Form

6 feet is to 4 feet as h feet is to 26 feet.

$$\text{man's height} \rightarrow \frac{6}{4} = \frac{h}{26} \leftarrow \text{tree's height}$$
$$\text{man's shadow} \rightarrow \quad\quad\quad\quad \leftarrow \text{tree's shadow}$$

Solve

$$\frac{6}{4} = \frac{h}{26}$$
$$156 = 4h$$
$$\frac{156}{4} = \frac{4h}{4}$$
$$39 = h$$

State

A height of the tree is 39 feet.

Check

$$\frac{6}{4} = \frac{h}{26}$$
$$\frac{6}{4} \overset{?}{=} \frac{39}{26}$$
$$156 = 156$$

The result checks.

103. SURVEYING

Analyze
- Similar triangles are determined by the layout.
- What is the width of the river?

Assign

Let w = the width of the river

Form a Proportion

20 feet is to 32 feet as w feet is to 75 feet.

$$\text{dis to bank} \rightarrow \frac{20}{32} = \frac{w}{75} \leftarrow \text{width of river}$$
$$\text{dis along river} \rightarrow \quad\quad\quad\quad \leftarrow \text{dis along river}$$

Solve

$$\frac{20}{32} = \frac{w}{75}$$
$$1,500 = 32w$$
$$\frac{1,500}{32} = \frac{32w}{32}$$
$$46.875 = w$$
$$46\frac{7}{8} = x$$

State

A width of the river is $46\frac{7}{8}$ feet.

Check

$$\frac{20}{32} = \frac{x}{75}$$
$$\frac{20}{32} \overset{?}{=} \frac{46.875}{75}$$
$$1,500 = 1,500$$

The result checks.

105. SLOPE

Analyze

- The rise to run is 14 to 21 of the given large triangle.
- What is the rise of the smaller triangle with a run of 12?

Assign

Let x = the rise of the smaller triangle

Form

14 is to 21 as x is to 12.

$$\text{rise large} \rightarrow \frac{14}{21} = \frac{x}{12} \leftarrow \text{rise small}$$
$$\text{run large} \rightarrow \qquad \qquad \leftarrow \text{run small}$$

Solve

$$\frac{14}{21} = \frac{x}{12}$$
$$168 = 21x$$
$$\frac{168}{21} = \frac{21x}{21}$$
$$8 = x$$

State

The rise of the smaller triangle is 8.

Check

$$\frac{14}{21} = \frac{x}{12}$$
$$\frac{14}{21} \overset{?}{=} \frac{8}{12}$$
$$168 = 168$$

The result checks.

WRITING

107-109. Answers will vary.

REVIEW

111. $\dfrac{9}{10} = 10\overline{)9.0}^{\,0.9} \Rightarrow 0.9 \cdot 100 = 90\%$

113. 30% of $1{,}600 = 0.30 \cdot 1{,}600$
$$= 480$$

CHALLENGE PROBLEMS

115. $\dfrac{a}{c} = \dfrac{b}{d}, \qquad \dfrac{b}{a} = \dfrac{d}{c}, \qquad \dfrac{c}{a} = \dfrac{d}{b}$

1. Find undefined values.

$$\frac{x-1}{x^2-16}=\frac{x-1}{(x+4)(x-4)}$$

$$x+4=0 \quad \text{or} \quad x-4=0$$

$$x+4-4=0-4 \quad | \quad x-4+4=0+4$$

$$x=-4 \quad | \quad x=4$$

The rational expression is undefined for $x=-4$ and $x=4$.

2. Evaluate: $\dfrac{x^2-1}{x-5}$ for $x=-2$.

$$\frac{x^2-1}{x-5}=\frac{(-2)^2-1}{-2-5}$$

$$=\frac{4-1}{-7}$$

$$=-\frac{3}{7}$$

Simplify each rational expression, if possible.
Assume that no denominators are zero.

3. $\dfrac{3x^2}{6x^3}=\dfrac{3\cdot x^2}{2\cdot 3\cdot x\cdot x^2}=\dfrac{1}{2x}$

4. $\dfrac{5xy^2}{2x^2y^2}=\dfrac{5\cdot x\cdot y^2}{2x\cdot x\cdot y^2}=\dfrac{5}{2x}$

5. $\dfrac{x^2}{x^2+x}=\dfrac{x^2}{x(x+1)}=\dfrac{x\cdot x}{x(x+1)}=\dfrac{x}{x+1}$

6. $\dfrac{a^2-4}{a+2}=\dfrac{(a+2)(a-2)}{a+2}=\dfrac{(a+2)(a-2)}{a+2}=a-2$

7. $\dfrac{3p-2}{2-3p}=\dfrac{3p-2}{-1(-2+3p)}=\dfrac{3p-2}{-1(3p-2)}=-1$

8. $\dfrac{8-x}{x^2-5x-24}=\dfrac{8-x}{(x-8)(x+3)}=\dfrac{-1(x-8)}{(x-8)(x+3)}=-\dfrac{1}{x+3}$

9. $\dfrac{2x^2-16x}{2x^2-18x+16}=\dfrac{2x(x-8)}{2(x^2-9x+8)}=\dfrac{2x(x-8)}{2(x-1)(x-8)}=\dfrac{2\cdot x(x-8)}{2(x-1)(x-8)}=\dfrac{x}{x-1}$

10. $\dfrac{x^2+x-2}{x^2-x-2}=\dfrac{(x+2)(x-1)}{(x-2)(x+1)}$

This will not simplify.

11. $\dfrac{x^2-2xy+y^2}{(x-y)^3}=\dfrac{(x-y)(x-y)}{(x-y)(x-y)(x-y)}=\dfrac{(x-y)(x-y)}{(x-y)(x-y)(x-y)}=\dfrac{1}{x-y}$

12. $\dfrac{4(t+3)+8}{3(t+3)+6}=\dfrac{4t+12+8}{3t+9+6}=\dfrac{4t+20}{3t+15}=\dfrac{4(t+5)}{3(t+5)}=\dfrac{4(t+5)}{3(t+5)}=\dfrac{4}{3}$

13. $\dfrac{x+1}{x} = \dfrac{\overset{1}{\cancel{x}}+1}{\cancel{x}} = \dfrac{2}{1} = 2$

x is not a common factor of the numerator and the denominator; x is a term of the numerator.

14. DOSAGES

$C = \dfrac{D(A+1)}{24}$

$C = \dfrac{300(11+1)}{24}$

$= \dfrac{300(12)}{24}$

$= \dfrac{3,600}{24}$

$= 150$

The 11 year old's dosage is 150 milligrams.

SECTION 7.2
Multiplying and Dividing Rational Expressions

Multiply and simplify, if possible.

15. $\dfrac{3xy}{2x} \cdot \dfrac{4x}{2y^2} = \dfrac{3xy \cdot 4x}{2x \cdot 2y^2}$

$= \dfrac{3 \cdot \overset{1}{\cancel{4}} \cdot x \cdot \overset{1}{\cancel{x}} \cdot \overset{1}{\cancel{y}}}{\cancel{4} \cdot \cancel{x} \cdot \cancel{y} \cdot y}$

$= \dfrac{3x}{y}$

16. $56x\left(\dfrac{12}{7x}\right) = \dfrac{56x \cdot 12}{7x}$

$= \dfrac{\overset{1}{\cancel{7}} \cdot 8 \cdot 12 \cdot \overset{1}{\cancel{x}}}{\cancel{7} \cdot \cancel{x}}$

$= 96$

17. $\dfrac{x^2-1}{x^2+2x} \cdot \dfrac{x}{x+1} = \dfrac{(x+1)(x-1)x}{x(x+2)(x+1)}$

$= \dfrac{\overset{1}{\cancel{x}}\,\overset{1}{\cancel{(x+1)}}(x-1)}{\cancel{x}(x+2)\,\cancel{(x+1)}}$

$= \dfrac{x-1}{x+2}$

18. $\dfrac{x^2+x}{3x-15} \cdot \dfrac{6x-30}{x^2+2x+1} = \dfrac{x(x+1)6(x-5)}{3(x-5)(x+1)(x+1)}$

$= \dfrac{2 \cdot \overset{1}{\cancel{3}} \cdot x \,\overset{1}{\cancel{(x+1)}}\,\overset{1}{\cancel{(x-5)}}}{\cancel{3}\,\cancel{(x-5)}\,\cancel{(x+1)}(x+1)}$

$= \dfrac{2x}{x+1}$

Divide and simplify, if possible.

19. $\dfrac{3x^2}{5x^2y} \div \dfrac{6x}{15xy^2} = \dfrac{3x^2}{5x^2y} \cdot \dfrac{15xy^2}{6x}$

$= \dfrac{3x^2 \cdot 15xy^2}{5x^2y \cdot 6x}$

$= \dfrac{3 \cdot \overset{1}{\cancel{3}} \cdot \overset{1}{\cancel{5}} \cdot \overset{1}{\cancel{x^2}}\, y\, \overset{1}{\cancel{y}}}{2 \cdot \cancel{3} \cdot \cancel{5} \cdot \cancel{x^2}\, \cancel{y}}$

$= \dfrac{3y}{2}$

20. $\dfrac{x^2-x-6}{1-2x} \div \dfrac{x^2-2x-3}{2x^2+x-1}$

$= \dfrac{x^2-x-6}{1-2x} \cdot \dfrac{2x^2+x-1}{x^2-2x-3}$

$= \dfrac{(x-3)(x+2)(2x-1)(x+1)}{-1(-1+2x)(x-3)(x+1)}$

$= -\dfrac{\overset{1}{\cancel{(x-3)}}(x+2)\,\overset{1}{\cancel{(2x-1)}}\,\overset{1}{\cancel{(x+1)}}}{\cancel{(2x-1)}\,\cancel{(x-3)}\,\cancel{(x+1)}}$

$= -(x+2)$ or $-x-2$

21. a. Yes b. No c. Yes d. Yes

22. TRAFFIC SIGN

$\dfrac{20\text{ miles}}{1\text{ hour}} \cdot \dfrac{1\text{ hour}}{60\text{ minutes}} = \dfrac{20\text{ miles}}{1\text{ \cancel{hour}}} \cdot \dfrac{1\text{ \cancel{hour}}}{60\text{ minutes}}$

$= \dfrac{\overset{1}{\cancel{2}} \cdot \overset{1}{\cancel{2}} \cdot \overset{1}{\cancel{5}}\text{ miles}}{3 \cdot \cancel{2} \cdot \cancel{2} \cdot \cancel{5}\text{ minutes}}$

$= \dfrac{1}{3}\text{ mile per minute}$

The conversion is $\dfrac{1}{3}$ mile per minute.

SECTION 7.3

Adding and Subtracting with Like Denominators; Least Common Denominators

Add or subtract and simplify, if possible.

23. $\dfrac{13}{15d} - \dfrac{8}{15d} = \dfrac{13-8}{15d}$

$$= \dfrac{\overset{1}{\cancel{5}}}{3 \cdot \cancel{5} \, d}$$

$$= \dfrac{1}{3d}$$

24. $\dfrac{x}{x+y} + \dfrac{y}{x+y} = \dfrac{x+y}{x+y}$

$$= \dfrac{\overset{1}{\cancel{x+y}}}{\underset{1}{\cancel{x+y}}}$$

$$= 1$$

25. $\dfrac{3x}{x-7} - \dfrac{x-2}{x-7} = \dfrac{3x-(x-2)}{x-7}$

$$= \dfrac{3x-x+2}{x-7}$$

$$= \dfrac{2x+2}{x-7}$$

$$= \dfrac{2(x+1)}{x-7} \text{ or } \dfrac{2x+2}{x-7}$$

26. $\dfrac{a}{a^2-2a-8} + \dfrac{2}{a^2-2a-8} = \dfrac{a+2}{a^2-2a-8}$

$$= \dfrac{a+2}{(a-4)(a+2)}$$

$$= \dfrac{\overset{1}{\cancel{a+2}}}{(a-4)\,\underset{1}{\cancel{(a+2)}}}$$

$$= \dfrac{1}{a-4}$$

Find the LCD of each pair of rational expressions.

27. $\dfrac{12}{x}, \dfrac{1}{9}$

$\text{LCD} = 9x$

28. $\dfrac{1}{2x^3}, \dfrac{5}{8x}$

$2x^3 = 2 \cdot x \cdot x \cdot x$

$8x = 2 \cdot 2 \cdot 2 \cdot x$

$\text{LCD} = 2 \cdot 2 \cdot 2 \cdot x \cdot x \cdot x = 8x^3$

29. $\dfrac{7}{m}, \dfrac{m+2}{m-8}$

$\text{LCD} = m(m-8)$

30. $\dfrac{x}{5x+1}, \dfrac{5x}{5x-1}$

$\text{LCD} = (5x+1)(5x-1)$

31. $\dfrac{6-a}{a^2-25}, \dfrac{a^2}{a-5}$

$a^2-25 = (a+5)(a-5)$

$a-5 = a-5$

$\text{LCD} = (a+5)(a-5)$

32. $\dfrac{4t+25}{t^2+10t+25}, \dfrac{t^2-7}{2t^2+17t+35}$

$t^2+10t+25 = (t+5)(t+5)$

$2t^2+17t+35 = (2t+7)(t+5)$

$\text{LCD} = (2t+7)(t+5)^2$

Build each rational expression into an equivalent fraction having the given denominator.

33. $\dfrac{9}{a} \cdot \dfrac{7}{7} = \dfrac{63}{7a}$

34. $\dfrac{2y+1}{x-9} \cdot \dfrac{x}{x} = \dfrac{x(2y+1)}{x(x-9)}$

$$= \dfrac{2xy+x}{x(x-9)}$$

35. $\dfrac{b+7}{3b-15} = \dfrac{b+7}{3(b-5)} \cdot \dfrac{2}{2}$

$$= \dfrac{2(b+7)}{6(b-5)}$$

$$= \dfrac{2b+14}{6(b-5)}$$

36. $\dfrac{9r}{r^2+6r+5} = \dfrac{9r}{(r+1)(r+5)} \cdot \dfrac{r-4}{r-4}$

$\qquad = \dfrac{9r(r-4)}{(r+1)(r+5)(r-4)}$

$\qquad = \dfrac{9r^2-36r}{(r+1)(r+5)(r-4)}$

SECTION 7.4
Adding and Subtracting with Unlike Denominators
Add or subtract and simplify, if possible.

37. $\text{LCD} = 7a$

$\dfrac{1}{7} - \dfrac{1}{a} = \dfrac{1}{7} \cdot \dfrac{a}{a} - \dfrac{1}{a} \cdot \dfrac{7}{7}$

$\qquad = \dfrac{a-7}{7a}$

38. $\text{LCD} = x(x-1)$

$\dfrac{x}{x-1} + \dfrac{1}{x} = \dfrac{x}{x-1} \cdot \dfrac{x}{x} + \dfrac{1}{x} \cdot \dfrac{(x-1)}{(x-1)}$

$\qquad = \dfrac{x^2+x-1}{x(x-1)}$

39. $\left.\begin{array}{l} t^2+2t+1=(t+1)(t+1) \\ (t+1)=t+1 \end{array}\right\}$

$\text{LCD} = (t+1)(t+1)$

$\dfrac{2t+2}{t^2+2t+1} - \dfrac{1}{t+1} = \dfrac{2t+2}{(t+1)(t+1)} - \dfrac{1}{(t+1)} \cdot \dfrac{(t+1)}{(t+1)}$

$\qquad = \dfrac{2t+2-(t+1)}{(t+1)(t+1)}$

$\qquad = \dfrac{2t+2-t-1}{(t+1)(t+1)}$

$\qquad = \dfrac{\overset{1}{\cancel{t+1}}}{(t+1)\,\underset{1}{\cancel{(t+1)}}}$

$\qquad = \dfrac{1}{t+1}$

40. $\left.\begin{array}{l} 2x=2x \\ x^2=x\cdot x \end{array}\right\}\text{LCD}=2x^2$

$\dfrac{x+2}{2x} - \dfrac{2-x}{x^2} = \dfrac{x+2}{2x} \cdot \dfrac{x}{x} - \dfrac{2-x}{x^2} \cdot \dfrac{2}{2}$

$\qquad = \dfrac{x(x+2)-2(2-x)}{2x^2}$

$\qquad = \dfrac{x^2+2x-4+2x}{2x^2}$

$\qquad = \dfrac{x^2+4x-4}{2x^2}$

41. $\text{LCD}=b-1$

$\dfrac{6}{b-1} - \dfrac{b}{1-b} = \dfrac{6}{b-1} - \dfrac{b}{-1(-1+b)}$

$\qquad = \dfrac{6}{b-1} - \dfrac{-b}{b-1}$

$\qquad = \dfrac{6}{b-1} + \dfrac{b}{b-1}$

$\qquad = \dfrac{b+6}{b-1}$

42. $\text{LCD}=c$

$\dfrac{8}{c} + 6 = \dfrac{8}{c} + \dfrac{6}{1} \cdot \dfrac{c}{c}$

$\qquad = \dfrac{8}{c} + \dfrac{6c}{c}$

$\qquad = \dfrac{6c+8}{c}$

43. $\left.\begin{array}{l} n+3=n+3 \\ n+7=n+7 \end{array}\right\}\text{LCD}=(n+3)(n+7)$

$\dfrac{n+7}{n+3} - \dfrac{n-3}{n+7} = \dfrac{(n+7)}{(n+3)} \cdot \dfrac{(n+7)}{(n+7)} - \dfrac{(n-3)}{(n+7)} \cdot \dfrac{(n+3)}{(n+3)}$

$\qquad = \dfrac{(n+7)(n+7)-(n-3)(n+3)}{(n+3)(n+7)}$

$\qquad = \dfrac{n^2+14n+49-(n^2-9)}{(n+3)(n+7)}$

$\qquad = \dfrac{n^2+14n+49-n^2+9}{(n+3)(n+7)}$

$\qquad = \dfrac{14n+58}{(n+3)(n+7)}$

Chapter 7 Review and Chapter 7 Test

44.

$$\left.\begin{array}{l} t+2=t+2 \\ (t+2)^2=(t+2)^2 \end{array}\right\} \text{LCD} = (t+2)^2$$

$$\frac{4}{t+2} - \frac{7}{(t+2)^2} = \frac{4}{(t+2)} \cdot \frac{(t+2)}{(t+2)} - \frac{7}{(t+2)^2}$$

$$= \frac{4(t+2)-7}{(t+2)^2}$$

$$= \frac{4t+8-7}{(t+2)^2}$$

$$= \frac{4t+1}{(t+2)^2}$$

45.

$$\left.\begin{array}{l} a^2-9=(a+3)(a-3) \\ a^2-a-6=(a-3)(a+2) \end{array}\right\}$$

$$\text{LCD} = (a+3)(a-3)(a+2)$$

$$\frac{6}{a^2-9} - \frac{5}{a^2-a-6}$$

$$= \frac{6}{(a+3)(a-3)} - \frac{5}{(a-3)(a+2)}$$

$$= \frac{6}{(a+3)(a-3)} \cdot \frac{(a+2)}{(a+2)} - \frac{5}{(a-3)(a+2)} \cdot \frac{(a+3)}{(a+3)}$$

$$= \frac{6(a+2)-5(a+3)}{(a+3)(a-3)(a+2)}$$

$$= \frac{6a+12-5a-15}{(a+3)(a-3)(a+2)}$$

$$= \frac{a-3}{(a+3)(a-3)(a+2)}$$

$$= \frac{\overset{1}{\cancel{a-3}}}{(a+3)\,\underset{1}{\cancel{(a-3)}}\,(a+2)}$$

$$= \frac{1}{(a+3)(a+2)}$$

46.

$$\left.\begin{array}{l} 3y-6=3(y-2) \\ 4y+8=4(y+2) \end{array}\right\} \text{LCD} = 12(y-2)(y+2)$$

$$\frac{2}{3y-6} + \frac{3}{4y+8}$$

$$= \frac{2}{3(y-2)} + \frac{3}{4(y+2)}$$

$$= \frac{2}{3(y-2)} \cdot \frac{4(y+2)}{4(y+2)} + \frac{3}{4(y+2)} \cdot \frac{3(y-2)}{3(y-2)}$$

$$= \frac{8(y+2)+9(y-2)}{12(y-2)(y+2)}$$

$$= \frac{8y+16+9y-18}{12(y-2)(y+2)}$$

$$= \frac{17y-2}{12(y-2)(y+2)}$$

47. $\dfrac{-5n^3-7}{3n(n+6)} = \dfrac{-(5n^3+7)}{3n(n+6)} = -\dfrac{5n^3+7}{3n(n+6)}$

Yes, they are equivalent.

48. DIGITAL VIDEO CAMERAS

$$P = 2L + 2W \qquad \text{LCD} = (x+6)(x-1)$$

$$P = 2\left(\frac{4}{x+6}\right) + 2\left(\frac{3}{x-1}\right)$$

$$= \frac{8}{x+6} + \frac{6}{x-1}$$

$$= \frac{8}{(x+6)} \cdot \frac{(x-1)}{(x-1)} + \frac{6}{x-1} \cdot \frac{(x+6)}{(x+6)}$$

$$= \frac{8(x-1)+6(x+6)}{(x+6)(x-1)}$$

$$= \frac{8x-8+6x+36}{(x+6)(x-1)}$$

$$= \frac{14x+28}{(x+6)(x-1)}$$

The perimeter is $\dfrac{14x+28}{(x+6)(x-1)}$ units.

AREA $\quad A = LW$

$$= \left(\frac{4}{x+6}\right)\left(\frac{3}{x-1}\right)$$

$$= \frac{12}{(x+6)(x-1)}$$

The area is $\dfrac{12}{(x+6)(x-1)}$ square units.

SECTION 7.5
Simplifying Complex Fractions
Simplify each complex fraction.

49. $\dfrac{\dfrac{n^4}{30}}{\dfrac{7n}{15}} = \dfrac{n^4}{30} \div \dfrac{7n}{15}$

$= \dfrac{n^4}{30} \cdot \dfrac{15}{7n}$

$= \dfrac{15n^4}{30 \cdot 7n}$

$= \dfrac{\overset{1}{\cancel{5}} \cdot \overset{1}{\cancel{3}} \cdot \overset{1}{\cancel{n}} \cdot n^3}{2 \cdot \underset{1}{\cancel{3}} \cdot \underset{1}{\cancel{5}} \cdot 7 \cdot \underset{1}{\cancel{n}}}$

$= \dfrac{n^3}{14}$

50. $\dfrac{\dfrac{r^2-81}{18s^2}}{\dfrac{4r-36}{9s}} = \dfrac{r^2-81}{18s^2} \div \dfrac{4r-36}{9s}$

$= \dfrac{r^2-81}{18s^2} \cdot \dfrac{9s}{4r-36}$

$= \dfrac{9s(r+9)(r-9)}{18s^2 \cdot 4(r-9)}$

$= \dfrac{\overset{1}{\cancel{3}} \cdot \overset{1}{\cancel{3}} \cdot \overset{1}{\cancel{s}}(r+9)\,\overset{1}{\cancel{(r-9)}}}{2 \cdot \underset{1}{\cancel{3}} \cdot \underset{1}{\cancel{3}} \cdot 4 \cdot \underset{1}{\cancel{s}} \cdot s \,\underset{1}{\cancel{(r-9)}}}$

$= \dfrac{r+9}{8s}$

51. LCD $= y$

$\dfrac{\dfrac{1}{y}+1}{\dfrac{1}{y}-1} = \dfrac{\left(\dfrac{1}{y}+1\right)y}{\left(\dfrac{1}{y}-1\right)y}$

$= \dfrac{\dfrac{1}{y}(y)+1(y)}{\dfrac{1}{y}(y)-1(y)}$

$= \dfrac{1+y}{1-y}$

52. LCD $= 3a^2$

$\dfrac{\dfrac{7}{a^2}}{\dfrac{1}{a}+\dfrac{10}{3}} = \dfrac{\left(\dfrac{7}{a^2}\right)3a^2}{\left(\dfrac{1}{a}+\dfrac{10}{3}\right)3a^2}$

$= \dfrac{\dfrac{7}{a^2}(3a^2)}{\dfrac{1}{a}(3a^2)+\dfrac{10}{3}(3a^2)}$

$= \dfrac{21}{3a+10a^2}$

53. LCD $= (x+1)(x-1)$

$\dfrac{\dfrac{2}{x-1}+\dfrac{x-1}{x+1}}{\dfrac{1}{x^2-1}} = \dfrac{\dfrac{2}{x-1}+\dfrac{x-1}{x+1}}{\dfrac{1}{(x+1)(x-1)}} \cdot \dfrac{(x+1)(x-1)}{(x+1)(x-1)}$

$= \dfrac{\left(\dfrac{2}{x-1}\right)\cdot(x+1)(x-1)+\left(\dfrac{x-1}{x+1}\right)\cdot(x+1)(x-1)}{\left(\dfrac{1}{(x+1)(x-1)}\right)\cdot(x+1)(x-1)}$

$= \dfrac{2(x+1)+(x-1)(x-1)}{1}$

$= 2x+2+x^2-2x+1$

$= x^2+3$

54. $\left.\begin{array}{l} x^2y = x \cdot x \cdot y \\ xy = x \cdot y \\ xy^2 = x \cdot y \cdot y \end{array}\right\}$ LCD $= x^2y^2$

$\dfrac{\dfrac{1}{x^2y}-\dfrac{5}{xy}}{\dfrac{3}{xy}-\dfrac{7}{xy^2}} = \dfrac{\dfrac{1}{x^2y}-\dfrac{5}{xy}}{\dfrac{3}{xy}-\dfrac{7}{xy^2}} \cdot \dfrac{x^2y^2}{x^2y^2}$

$= \dfrac{\left(\dfrac{1}{x^2y}-\dfrac{5}{xy}\right)(x^2y^2)}{\left(\dfrac{3}{xy}-\dfrac{7}{xy^2}\right)(x^2y^2)}$

$= \dfrac{\left(\dfrac{1}{x^2y}\right)(x^2y^2)-\left(\dfrac{5}{xy}\right)(x^2y^2)}{\left(\dfrac{3}{xy}\right)(x^2y^2)-\left(\dfrac{7}{xy^2}\right)(x^2y^2)}$

$= \dfrac{y-5xy}{3xy-7x}$

Chapter 7 Review and Chapter 7 Test

SECTION 7.6

Solving Rational Equations

Solve each equation and check the results. If an equation has no solution, so indicate.

55. $\text{LCD} = x(x-1) \quad x \neq 0, 1$

$$\frac{3}{x} = \frac{2}{x-1}$$

$$x(x-1)\left(\frac{3}{x}\right) = x(x-1)\left(\frac{2}{x-1}\right)$$

$$\overset{1}{\cancel{x}}(x-1)\left(\frac{3}{\cancel{x}_1}\right) = x\,\overset{1}{\cancel{(x-1)}}\left(\frac{2}{\cancel{x-1}_1}\right)$$

$$3(x-1) = x(2)$$
$$3x - 3 = 2x$$
$$3x - 3 - 2x = 2x - 2x$$
$$x - 3 = 0$$
$$x - 3 + 3 = 0 + 3$$
$$x = 3$$

Check: $\dfrac{3}{x} = \dfrac{2}{x-1}$

$$\frac{3}{3} \overset{?}{=} \frac{2}{3-1}$$

$$\frac{3}{3} \overset{?}{=} \frac{2}{2}$$

$$1 = 1$$

The solution is 3.
The result checks.

56. $\text{LCD} = a - 5 \; ; \; a \neq 5$

$$\frac{a}{a-5} = 3 + \frac{5}{a-5}$$

$$(a-5)\left(\frac{a}{a-5}\right) = (a-5)\left(3 + \frac{5}{a-5}\right)$$

$$(a-5)\left(\frac{a}{a-5}\right) = (a-5)(3) + (a-5)\left(\frac{5}{a-5}\right)$$

$$\overset{1}{\cancel{(a-5)}}\left(\frac{a}{\cancel{a-5}_1}\right) = (a-5)(3) + \overset{1}{\cancel{(a-5)}}\left(\frac{5}{\cancel{a-5}_1}\right)$$

$$a = 3(a-5) + 5$$
$$a = 3a - 15 + 5$$
$$a = 3a - 10$$
$$a - a = 3a - 10 - a$$
$$0 = 2a - 10$$
$$0 + 10 = 2a - 10 + 10$$
$$10 = 2a$$
$$\frac{10}{2} = \frac{2a}{2}$$
$$5 = a$$

The denominator becomes zero. No solution. 5 is extraneous.

57. $\left.\begin{array}{l} 3t = 3t \\ t = t \\ 9 = 3 \cdot 3 \end{array}\right\} \text{LCD} = 9t \; ; \; t \neq 0$

$$\frac{2}{3t} + \frac{1}{t} = \frac{5}{9}$$

$$9t\left(\frac{2}{3t}\right) + 9t\left(\frac{1}{t}\right) = 9t\left(\frac{5}{9}\right)$$

$$3 \cdot \overset{1}{\cancel{3t}}\left(\frac{2}{\cancel{3t}_1}\right) + 9\,\overset{1}{\cancel{t}}\left(\frac{1}{\cancel{t}_1}\right) = \overset{1}{\cancel{9}}\,t\left(\frac{5}{\cancel{9}_1}\right)$$

$$3(2) + 9 = t(5)$$
$$6 + 9 = 5t$$
$$15 = 5t$$
$$\frac{15}{5} = \frac{5t}{5}$$
$$3 = t$$

The result checks.

58.

$$\left.\begin{array}{c} 4a-24=4(a-6) \\ 4=4 \end{array}\right\}$$

$$\text{LCD}=4(a-6)\ ;\ \ a\neq 6$$

$$a=\frac{3a-50}{4a-24}-\frac{3}{4}$$

$$\mathbf{4(a-6)}(a)=\mathbf{4(a-6)}\left(\frac{3a-50}{4(a-6)}-\frac{3}{4}\right)$$

$$4a(a-6)=4(a-6)\left(\frac{3a-50}{4(a-6)}\right)-4(a-6)\left(\frac{3}{4}\right)$$

$$4a(a-6)=\cancel{4}\ \cancel{(a-6)}\left(\frac{3a-50}{\cancel{4}\ \cancel{(a-6)}}\right)-\cancel{4}(a-6)\left(\frac{3}{\cancel{4}}\right)$$

$$4a(a-6)=3a-50-3(a-6)$$

$$4a^2-24a=3a-50-3a+18$$

$$4a^2-24a=-32$$

$$4a^2-24a+\mathbf{32}=-32+\mathbf{32}$$

$$4a^2-24a+32=0$$

$$4(a^2-6a+8)=0$$

$$4(a-2)(a-4)=0$$

$$\begin{array}{c|c} a-2=0 \quad \overset{\text{or}}{} & a-4=0 \\ a-2+\mathbf{2}=0+\mathbf{2} & a-4+\mathbf{4}=0+\mathbf{4} \\ a=2 & a=4 \end{array}$$

The solutions are 2 and 4.
The results check.

59.

$$\left.\begin{array}{c} x+2=x+2 \\ x+3=x+3 \\ (x^2+5x+6)=(x+2)(x+3) \end{array}\right\}$$

$$\text{LCD}=(x+2)(x+3)\ ;\ \ x\neq -2,-3$$

$$\frac{4}{x+2}-\frac{3}{x+3}=\frac{6}{(x^2+5x+6)}$$

$$(x+2)(x+3)\left(\frac{4}{x+2}-\frac{3}{x+3}\right)=(x+2)(x+3)\left(\frac{6}{(x+2)(x+3)}\right)$$

$$(x+2)(x+3)\left(\frac{4}{x+2}\right)-(x+2)(x+3)\left(\frac{3}{x+3}\right)=(x+2)(x+3)\left(\frac{6}{(x+2)(x+3)}\right)$$

$$\cancel{(x+2)}(x+3)\left(\frac{4}{\cancel{x+2}}\right)-(x+2)\cancel{(x+3)}\left(\frac{3}{\cancel{x+3}}\right)=\cancel{(x+2)}\ \cancel{(x+3)}\left(\frac{6}{\cancel{(x+2)}\ \cancel{(x+3)}}\right)$$

$$4(x+3)-3(x+2)=6$$

$$4x+12-3x-6=6$$

$$x+6=6$$

$$x+6-\mathbf{6}=6-\mathbf{6}$$

$$x=0$$

The solution is 0.
The result checks.

60. $\text{LCD}=2(x+1)\ ;\ \ x\neq -1$

$$\frac{3}{x+1}-\frac{x-2}{2}=\frac{x-2}{x+1}$$

$$\mathbf{2(x+1)}\left(\frac{3}{x+1}-\frac{x-2}{2}\right)=\mathbf{2(x+1)}\left(\frac{x-2}{x+1}\right)$$

$$2(x+1)\left(\frac{3}{x+1}\right)-2(x+1)\left(\frac{x-2}{2}\right)=2(x+1)\left(\frac{x-2}{x+1}\right)$$

$$2\ \cancel{(x+1)}\left(\frac{3}{\cancel{x+1}}\right)-\cancel{2}(x+1)\left(\frac{x-2}{\cancel{2}}\right)=2\ \cancel{(x+1)}\left(\frac{x-2}{\cancel{x+1}}\right)$$

$$2(3)-(x+1)(x-2)=2(x-2)$$

$$6-(x^2-x-2)=2x-4$$

$$6-x^2+x+2=2x-4$$

$$-x^2+x+8=2x-4$$

$$-x^2+x+8-\mathbf{2x}=2x-4-\mathbf{2x}$$

$$-x^2+x+8-2x=-4$$

$$-x^2-x+8+\mathbf{4}=-4+\mathbf{4}$$

$$-x^2-x+12=0$$

$$-(x^2+x-12)=0$$

$$-(x+4)(x-3)=0$$

$$\begin{array}{c|c} x+4=0 \quad \overset{\text{or}}{} & x-3=0 \\ x+4-\mathbf{4}=0-\mathbf{4} & x-3+\mathbf{3}=0+\mathbf{3} \\ x=-4 & x=3 \end{array}$$

The solutions are -4 and 3.
The results check.

61. ENGINEERING

$$E=1-\frac{T_2}{T_1}\ ;\ \ \text{LCD}=T_1$$

$$E=1-\frac{T_2}{T_1}$$

$$\mathbf{T_1}E=\mathbf{T_1}(1)-\mathbf{T_1}\left(\frac{T_2}{T_1}\right)$$

$$T_1E=T_1-\cancel{T_1}\left(\frac{T_2}{\cancel{T_1}}\right)$$

$$T_1E=T_1-T_2$$

$$T_1E-\mathbf{T_1}=T_1-T_2-\mathbf{T_1}$$

$$T_1E-T_1=-T_2$$

$$T_1(E-1)=-T_2$$

$$\frac{T_1(E-1)}{\mathbf{E-1}}=\frac{-T_2}{\mathbf{E-1}}$$

$$T_1=-\frac{T_2}{E-1}\ \text{ or }\ \frac{T_2}{1-E}$$

62. LCD $= xyz$; Solve for y.

$$\frac{1}{x} = \frac{1}{y} + \frac{1}{z}$$

$$xyz\left(\frac{1}{x}\right) = xyz\left(\frac{1}{y} + \frac{1}{z}\right)$$

$$xyz\left(\frac{1}{x}\right) = xyz\left(\frac{1}{y}\right) + xyz\left(\frac{1}{z}\right)$$

$$\overset{1}{\cancel{x}}\,yz\left(\frac{1}{\underset{1}{\cancel{x}}}\right) = x\,\overset{1}{\cancel{y}}\,z\left(\frac{1}{\underset{1}{\cancel{y}}}\right) + xy\,\overset{1}{\cancel{z}}\left(\frac{1}{\underset{1}{\cancel{z}}}\right)$$

$$yz = xz + xy$$

$$yz - xy = xz + xy - xy$$

$$yz - xy = xz$$

$$y(z - x) = xz$$

$$\frac{y(z-x)}{z-x} = \frac{xz}{z-x}$$

$$y = \frac{xz}{z-x}$$

SECTION 7.7
Problem Solving Using Rational Equations

63. NUMBER PROBLEM

Analyze

• Begin with the fraction $\dfrac{4}{5}$.

• Subtract a number from the denominator.

• Twice as much is added to the numerator.

• The results is 5.

• Find the number.

Assign

Let $n =$ the unknown number.

Form

$$\frac{4+2n}{5-n} = 5$$

Solve

$$\frac{4+2n}{5-n} = 5$$

$$\text{LCD} = 5 - n$$

$$(5-n)\left(\frac{4+2n}{5-n}\right) = (5-n)5$$

$$\overset{1}{\cancel{(5-n)}}\left(\frac{4+2n}{\underset{1}{\cancel{5-n}}}\right) = 5(5-n)$$

$$4 + 2n = 25 - 5n$$

$$4 + 2n + 5n = 25 - 5n + 5n$$

$$4 + 7n = 25$$

$$4 + 7n - 4 = 25 - 4$$

$$7n = 21$$

$$\frac{7n}{7} = \frac{21}{7}$$

$$n = 3$$

State

The number is 3.

Check

$$\frac{4+2n}{5-n} = \frac{4+2(3)}{5-3} = \frac{4+6}{2} = \frac{10}{2} = 5$$

The result checks.

64. EXERCISE

Analyze
- Jogger bikes 30 miles.
- She jogs 10 miles.
- Times are the same.
- Biking was 10 mph faster than jogging.
- How fast can she jog?

Assign

Let r = jogging speed in mph

$r + 10$ = biking speed in mph

Form

	Rate	$\cdot$ Time	= Distance
Jogging	r	$\dfrac{10}{r}$	10
Biking	$r + 10$	$\dfrac{30}{r+10}$	30

Time it took to jog 10 miles	equals	time it took to bike 30 miles.

$$\frac{10}{r} = \frac{30}{r+10}$$

Solve

$$\frac{10}{r} = \frac{30}{r+10}$$

$$LCD = r(r+10)$$

$$r(r+10)\left(\frac{10}{r}\right) = r(r+10)\left(\frac{30}{r+10}\right)$$

$$\cancel{r}(r+10)\left(\frac{10}{\cancel{r}}\right) = r\,\cancel{(r+10)}\left(\frac{30}{\cancel{r+10}}\right)$$

$$10(r+10) = 30r$$

$$10r + 100 = 30r$$

$$10r + 100 - 10r = 30r - 10r$$

$$100 = 20r$$

$$\frac{100}{20} = \frac{20r}{20}$$

$$5 = r$$

State

Her jogging speed is 5 mph.

Check

The result checks.

65. HOUSE CLEANING

The rate of work is $\dfrac{1}{4}$ of the job per hour.

66. HOUSE PAINTING

Analyze
- Homeowner takes 14 days.
- Professional takes 10 days.
- How long will it take both of them, working together, to paint the house?

Assign

Let x = the number of days it will take both working together to paint the house.

Form

	Rate	$\cdot$ Time	= Work Completed
Homeowner	$\dfrac{1}{14}$	x	$\dfrac{x}{14}$
Professional	$\dfrac{1}{10}$	x	$\dfrac{x}{10}$

The part of job done by homeowner	plus	the part of job done by professional	equals	1 job completed.

$$\frac{x}{14} \quad + \quad \frac{x}{10} \quad = \quad 1$$

Solve

$$\frac{x}{14} + \frac{x}{10} = 1 \ , \ LCD = 70$$

$$70\left(\frac{x}{14} + \frac{x}{10}\right) = 70(1)$$

$$\cancel{14} \cdot 5\left(\frac{x}{\cancel{14}}\right) + \cancel{10} \cdot 7\left(\frac{x}{\cancel{10}}\right) = 70$$

$$5x + 7x = 70$$

$$12x = 70$$

$$\frac{12x}{12} = \frac{70}{12}$$

$$x = 5\frac{5}{6}$$

State

It takes both working together $5\dfrac{5}{6}$ days to paint the house.

Check

The result checks.

Chapter 7 Review and Chapter 7 Test

67. INVESTMENTS
Analyze
- S and L earns $100 interest.
- Credit Union earns $120 interest.
- Credit Union rate is 1% more.
- Time for both is 1 year.
- Principals are the same amount for both.
- What are the two rates?

Assign
Let $r = $ S and L interest rate as a percent
$r + 1 = $ Credit Union interest rate as a percent

Form

	Principal	• Rate	• Time	= Interest
S and L	$\dfrac{100}{r}$	r	1	100
Credit Union	$\dfrac{120}{r+1}$	$r+1$	1	120

Principal invested in S and L	equals	principal invested in Credit Union.

$$\frac{100}{r} = \frac{120}{r+1}$$

Solve

$$\frac{100}{r} = \frac{120}{r+1}$$

$$\text{LCD} = r(r+1)$$

$$r(r+1)\left(\frac{100}{r}\right) = r(r+1)\left(\frac{120}{r+1}\right)$$

$$\overset{1}{\cancel{r}}\cdot(r+1)\left(\frac{100}{\underset{1}{\cancel{r}}}\right) = r\cdot\overset{1}{\cancel{(r+1)}}\left(\frac{120}{\underset{1}{\cancel{r+1}}}\right)$$

$$100(r+1) = r(120)$$
$$100r + 100 = 120r$$
$$100r + 100 - \mathbf{100r} = 120r - \mathbf{100r}$$
$$100 = 20r$$
$$\frac{100}{\mathbf{20}} = \frac{20r}{\mathbf{20}}$$
$$5 = r$$

State
Savings and Loan's rate is 5%.

Check
The result checks.

68. WIND SPEED
Analyze
- Rate of plane in still air is 360 mph.
- Flies 400 miles with the wind.
- Flies 320 miles against the wind.
- Times are the same for both directions.
- What is the speed of the wind?

Assign
Let $r = $ Wind's speed in mph

Form

	Rate	• Time	= Distance
With wind	$360 + r$	$\dfrac{400}{360+r}$	400
Against wind	$360 - r$	$\dfrac{320}{360-r}$	320

The time it took the plane to fly 400 miles with the wind	equals	the time it took the plane to fly 320 miles against the wind.

$$\frac{400}{360+r} = \frac{320}{360-r}$$

Solve

$$\frac{400}{360+r} = \frac{320}{360-r}$$

$$\text{LCD} = (360+r)(360-r)$$

$$(360+r)(360-r)\left(\frac{400}{360+r}\right) = (360+r)(360-r)\left(\frac{320}{360-r}\right)$$

$$\overset{1}{\cancel{(360+r)}}(360-r)\left(\frac{400}{\underset{1}{\cancel{360+r}}}\right) = (360+r)\overset{1}{\cancel{(360-r)}}\left(\frac{320}{\underset{1}{\cancel{360-r}}}\right)$$

$$400(360-r) = 320(360+r)$$
$$144,000 - 400r = 115,200 + 320r$$
$$144,000 - 400r + \mathbf{400r} = 115,200 + 320r + \mathbf{400r}$$
$$144,000 = 115,200 + 720r$$
$$144,000 - \mathbf{115,200} = 115,200 + 720r - \mathbf{115,20}$$
$$28,800 = 720r$$
$$\frac{28,800}{\mathbf{720}} = \frac{720r}{\mathbf{720}}$$
$$40 = r$$

State
The wind's speed is 40 mph.

Check
The result checks.

Determine whether each equation is a true proportion.

69. $\dfrac{4}{7} = \dfrac{20}{34}$ 70. $\dfrac{5}{7} = \dfrac{30}{42}$

$\overset{?}{4 \cdot 34 = 7 \cdot 20}$ $\overset{?}{5 \cdot 42 = 7 \cdot 30}$

$136 \neq 140$ $210 = 210$

    No   Yes

Solve each proportion.

71. $\dfrac{3}{x} = \dfrac{6}{9}$ Check: $\dfrac{3}{x} = \dfrac{6}{9}$

$3 \cdot 9 = 6x$ $\dfrac{3}{\frac{9}{2}} \overset{?}{=} \dfrac{6}{9}$

$27 = 6x$

$\dfrac{27}{\mathbf{6}} = \dfrac{6x}{\mathbf{6}}$ $3 \cdot \dfrac{2}{9} \overset{?}{=} \dfrac{6}{9}$

$\dfrac{9}{2} = x$ $\dfrac{6}{9} = \dfrac{6}{9}$ True

$\dfrac{9}{2}$ is the solution.

72. $\dfrac{x}{3} = \dfrac{x}{5}$ Check: $\dfrac{x}{3} = \dfrac{x}{5}$

$5x = 3x$ $\dfrac{0}{3} \overset{?}{=} \dfrac{0}{5}$

$5x - \mathbf{3x} = 3x - \mathbf{3x}$ $0 = 0$ True

$2x = 0$ 0 is the solution.

$\dfrac{2x}{\mathbf{2}} = \dfrac{0}{\mathbf{2}}$

$x = 0$

73. $\dfrac{x-2}{5} = \dfrac{x}{7}$ Check: $\dfrac{x-2}{5} = \dfrac{x}{7}$

$7(x-2) = 5x$ $\dfrac{7-2}{5} \overset{?}{=} \dfrac{7}{7}$

$7x - 14 = 5x$ $\dfrac{5}{5} \overset{?}{=} \dfrac{7}{7}$

$7x - 14 - \mathbf{7x} = 5x - \mathbf{7x}$

$-14 = -2x$ $1 = 1$ True

$\dfrac{-14}{\mathbf{-2}} = \dfrac{-2x}{\mathbf{-2}}$ 7 is the solution.

$7 = x$

74. $\dfrac{2x}{x+4} = \dfrac{3}{x-1}$

$2x(x-1) = 3(x+4)$

$2x^2 - 2x = 3x + 12$

$2x^2 - 2x - 3x = 3x + 12 - 3x$

$2x^2 - 5x = 12$

$2x^2 - 5x - \mathbf{12} = 12 - \mathbf{12}$

$2x^2 - 5x - 12 = 0$

$(2x+3)(x-4) = 0$

$2x + 3 = 0$ or $x - 4 = 0$

$2x + 3 - \mathbf{3} = 0 - \mathbf{3}$ $x - 4 + \mathbf{4} = 0 + \mathbf{4}$

$2x = -3$ $x = 4$

$\dfrac{2x}{\mathbf{2}} = \dfrac{-3}{\mathbf{2}}$

$x = -\dfrac{3}{2}$

The solutions are $-\dfrac{3}{2}$ and 4.

The results check.

75. DENTISTRY
Analyze
• 3 out of 4 will develop gum disease.
• How many of his 340 adult patients will develop gum disease?

Assign
Let g = number of gum disease patients

Form
3 adults is to 4 as g adults is to 340.

3 gum disease → $\dfrac{3}{4} = \dfrac{g}{340}$ ← # of gum disease

4 patients →   ← Total # of patients

Solve

$\dfrac{3}{4} = \dfrac{g}{340}$

$1{,}020 = 4g$

$\dfrac{1{,}020}{\mathbf{4}} = \dfrac{4g}{\mathbf{4}}$

$255 = g$

State
255 patients will develop gum disease.

Check
The result checks.

76. UTILITY POLES

Analyze

• Similar triangles determined by the pole and its shadow and the man and his shadow.

• What is the height of the pole?

Assign

Let h = the height of the pole

Form

6 feet is to 3.6 feet as h feet is to 12 feet.

$$\text{man's height} \rightarrow \frac{6}{3.6} = \frac{h}{12} \leftarrow \text{pole's height}$$
$$\text{man's shadow} \rightarrow \qquad\qquad \leftarrow \text{pole's shadow}$$

Solve

$$\frac{6}{3.6} = \frac{h}{12}$$
$$72 = 3.6h$$
$$\frac{72}{3.6} = \frac{3.6h}{3.6}$$
$$20 = h$$

State

A height of the pole is 20 feet.

Check

The result checks.

77. PORCELAIN FIGURINES

Analyze the Problem

• We know the scale is 1:12.

• How tall is a real flutist in feet if a model one measures 5.5 inches?

Assign

Let x = the height of a real flutist

Form

1 is to 12 as 5.5 inches is to x inches.

$$\text{1 part} \rightarrow \frac{1}{12} = \frac{5.5}{x} \leftarrow \text{5.5 inch model}$$
$$\text{12 parts} \rightarrow \qquad\qquad \leftarrow x \text{ inches real flutist}$$

Solve

$$\frac{1}{12} = \frac{5.5}{x}$$
$$x = 66$$

State

A real flutist would be 66 inches tall or 5 feet 6 inches tall.

Check

The result checks.

78. COMPARISON SHOPPING

150 compact discs for $60.

$$\text{cost} \rightarrow \frac{60}{150} = \frac{x}{1} \leftarrow \text{cost}$$
$$\text{150 discs} \rightarrow \qquad\qquad \leftarrow \text{1 disc}$$
$$\frac{60}{150} = x$$
$$0.4 = x$$

The cost per disc is $0.40.

250 compact discs for $98.

$$\text{cost} \rightarrow \frac{98}{250} = \frac{x}{1} \leftarrow \text{cost}$$
$$\text{250 discs} \rightarrow \qquad\qquad \leftarrow \text{1 disc}$$
$$\frac{98}{250} = x$$
$$0.392 = x$$

The cost per disc is about $0.392.

The 250 discs for $98 is the better buy.

CHAPTER 7 TEST

1. Fill in the blanks.

a. A quotient of two polynomials, such as $\dfrac{x+7}{x^2+2x}$, is called a **rational** expression.

b. Two triangles with the same shape, but not necessarily the same size, are called **similar** triangles.

c. A **proportion** is a mathematical statement that two ratios or two rates are equal.

d. To **build** a rational expression, we multiply it by a form of 1. For example, $\frac{2}{5x} \cdot \frac{8}{8} = \frac{16}{40x}$.

e. To simplify $\dfrac{x-3}{(x+3)(x-3)}$, we remove common **factors** of the numerator and denominator.

2. MEMORY

$$n = \frac{35 + 5d}{d}$$

$$= \frac{35 + 5(7)}{7}$$

$$= \frac{35 + 35}{7}$$

$$= \frac{70}{7}$$

$$= 10$$

After one week, 10 words will be remembered.

For what real numbers are each rational expression undefined?

3. $\dfrac{6x - 9}{5x}$

$$5x = 0$$

$$\frac{5x}{5} = \frac{0}{5}$$

$$x = 0$$

The rational expression is undefined for $x = 0$.

4. $\dfrac{x}{x^2 + x - 6} = \dfrac{x}{(x+3)(x-2)}$

$x + 3 = 0$ or $x - 2 = 0$

$x + 3 - 3 = 0 - 3$ $x - 2 + 2 = 0 + 2$

     $x = -3$ $x = 2$

The rational expression is undefined for $x = -3$ and $x = 2$.

5. THE INTERNET

$$\frac{56K \text{ bits}}{1 \text{ second}} \cdot \frac{60 \text{ seconds}}{1 \text{ minute}} = \frac{56K \text{ bits}}{1 \text{ second}} \cdot \frac{60 \text{ seconds}}{1 \text{ minute}}$$

$$= \frac{3,360K \text{ bits}}{1 \text{ minute}}$$

The modem can transmit 3,360K bits per minute or 3,360,000 bits per minute.

6. $\dfrac{x+5}{5} = \dfrac{\cancel{x+5}}{\cancel{5}}$

$$= x + 1$$

5 is not a comman factor of the numerator, and therefore can not be <u>removed</u>. 5 is a term of the numerator.

Simplify each rational expression.

7. $\dfrac{48x^2 y}{54xy^2} = \dfrac{8 \cdot \cancel{6}\,\cancel{x}\, x\, \cancel{y}}{9 \cdot \cancel{6}\,\cancel{x}\,\cancel{y}\, y}$

$$= \frac{8x}{9y}$$

8. $\dfrac{7m - 49}{7 - m} = \dfrac{7(m - 7)}{7 - m}$

$$= \frac{7(m - 7)}{-1(-7 + m)}$$

$$= -\frac{7\,\cancel{(m-7)}}{\cancel{(m-7)}}$$

$$= -7$$

9. $\dfrac{2x^2 - x - 3}{4x^2 - 9} = \dfrac{(2x - 3)(x + 1)}{(2x + 3)(2x - 3)}$

$$= \frac{\cancel{(2x-3)}(x + 1)}{(2x + 3)\,\cancel{(2x-3)}}$$

$$= \frac{x + 1}{2x + 3}$$

10. $\dfrac{3(x + 2) - 3}{6x + 5 - (3x + 2)} = \dfrac{3x + 6 - 3}{6x + 5 - 3x - 2}$

$$= \frac{3x + 3}{3x + 3}$$

$$= \frac{\cancel{3x + 3}}{\cancel{3x + 3}}$$

$$= 1$$

Find the LCD of each pair of rational expressions.

11. $\dfrac{19}{3c^2 d}, \dfrac{6}{c^2 d^3};$ LCD $= 3c^2 d^3$

12. $\dfrac{4n + 25}{n^2 - 4n - 5}, \dfrac{6n}{n^2 - 25}$

$$\frac{4n + 25}{(n - 5)(n + 1)}, \frac{6n}{(n + 5)(n - 5)}$$

LCD $= (n + 5)(n - 5)(n + 1)$

Perform the operations. Simplify, if possible.

Chapter 7 Review and Chapter 7 Test

13. $\dfrac{12x^2y}{15xy} \cdot \dfrac{25y^2}{16x} = \dfrac{12 \cdot 25 x^2 y y^2}{15 \cdot 16 x^2 y}$

$= \dfrac{\overset{1}{\cancel{3}} \cdot \overset{1}{\cancel{4}} \cdot \overset{1}{\cancel{5}} \cdot 5\, \overset{1}{\cancel{x^2}}\, \overset{1}{\cancel{y}}\, y^2}{\cancel{3} \cdot \cancel{5} \cdot \cancel{4} \cdot 4\, \cancel{x^2}\, \cancel{y}}$
$\qquad\quad {}_{1}\; {}_{1}\; {}_{1}\qquad {}_{1}\; {}_{1}$

$= \dfrac{5y^2}{4}$

14. $\dfrac{x^2 + 3x + 2}{3x + 9} \cdot \dfrac{x+3}{x^2 - 4} = \dfrac{(x^2 + 3x + 2)(x+3)}{(3x+9)(x^2-4)}$

$= \dfrac{(x+2)(x+1)(x+3)}{3(x+3)(x+2)(x-2)}$

$= \dfrac{\overset{1}{\cancel{(x+2)}}(x+1)\overset{1}{\cancel{(x+3)}}}{3\,\underset{1}{\cancel{(x+3)}}\,\underset{1}{\cancel{(x+2)}}(x-2)}$

$= \dfrac{x+1}{3(x-2)}$

15. $\dfrac{x - x^2}{3x^2 + 6x} \div \dfrac{3x-3}{3x^3 + 6x^2} = \dfrac{x-x^2}{3x^2+6x} \cdot \dfrac{3x^3 + 6x^2}{3x-3}$

$= \dfrac{(x-x^2)(3x^3 + 6x^2)}{(3x^2+6x)(3x-3)}$

$= \dfrac{x(1-x)3x^2(x+2)}{3x(x+2)3(x-1)}$

$= \dfrac{3x^2 x \cdot -1(-1+x)(x+2)}{3 \cdot 3x(x+2)(x-1)}$

$= -\dfrac{\overset{1}{\cancel{3}}\, x^2\, \overset{1}{\cancel{x}}\, \overset{1}{\cancel{(x-1)}}\, \overset{1}{\cancel{(x+2)}}}{3 \cdot \underset{1}{\cancel{3}}\, \underset{1}{\cancel{x}}\, \underset{1}{\cancel{(x+2)}}\, \underset{1}{\cancel{(x-1)}}}$

$= -\dfrac{x^2}{3}$

16. $\dfrac{a^2 - 16}{a - 4} \div (6a + 24) = \dfrac{a^2 - 16}{a-4} \cdot \dfrac{1}{(6a+24)}$

$= \dfrac{(a+4)(a-4)}{(a-4)6(a+4)}$

$= \dfrac{\overset{1}{\cancel{(a+4)}}\, \overset{1}{\cancel{(a-4)}}}{6\, \underset{1}{\cancel{(a-4)}}\, \underset{1}{\cancel{(a+4)}}}$

$= \dfrac{1}{6}$

17. $\dfrac{3y+7}{2y+3} - \dfrac{-3y-2}{2y+3} = \dfrac{3y+7-(-3y-2)}{2y+3}$

$= \dfrac{3y+7+3y+2}{2y+3}$

$= \dfrac{6y+9}{2y+3}$

$= \dfrac{3\,\overset{1}{\cancel{(2y+3)}}}{\underset{1}{\cancel{2y+3}}}$

$= 3$

18. LCD $= 10m$

$\dfrac{2n}{5m} - \dfrac{n}{2} = \dfrac{2n}{5m} \cdot \dfrac{\mathbf{2}}{\mathbf{2}} - \dfrac{n}{2} \cdot \dfrac{\mathbf{5m}}{\mathbf{5m}}$

$= \dfrac{4n - 5mn}{10m}$

19. LCD $= x(x+1)$

$\dfrac{x+1}{x} + \dfrac{x-1}{x+1} = \dfrac{x+1}{x} \cdot \dfrac{\mathbf{x+1}}{\mathbf{x+1}} + \dfrac{x-1}{x+1} \cdot \dfrac{\mathbf{x}}{\mathbf{x}}$

$= \dfrac{(x+1)(x+1) + (x-1)x}{x(x+1)}$

$= \dfrac{x^2 + 2x + 1 + x^2 - x}{x(x+1)}$

$= \dfrac{2x^2 + x + 1}{x(x+1)}$

20. LCD $= (a-1)$

$\dfrac{a+3}{a-1} - \dfrac{a+4}{1-a} = \dfrac{a+3}{a-1} - \dfrac{a+4}{-1(-1+a)}$

$= \dfrac{a+3}{a-1} + \dfrac{a+4}{a-1}$

$= \dfrac{a+3+a+4}{a-1}$

$= \dfrac{(1+1)a + (3+4)}{a-1}$

$= \dfrac{2a+7}{a-1}$

21. $\text{LCD} = c - 4$

$$\frac{9}{c-4} + c = \frac{9}{c-4} + \frac{c}{1}$$

$$= \frac{9}{c-4} + \frac{c}{1} \cdot \frac{c-4}{c-4}$$

$$= \frac{9 + c(c-4)}{c-4}$$

$$= \frac{9 + c^2 - 4c}{c-4}$$

$$= \frac{c^2 - 4c + 9}{c-4}$$

22. $\text{LCD} = (t+3)(t-3)(t+2)$

$$\frac{6}{t^2-9} - \frac{5}{t^2-t-6}$$

$$= \frac{6}{(t+3)(t-3)} - \frac{5}{(t-3)(t+2)}$$

$$= \frac{6}{(t+3)(t-3)} \cdot \frac{t+2}{t+2} - \frac{5}{(t-3)(t+2)} \cdot \frac{t+3}{t+3}$$

$$= \frac{6(t+2) - 5(t+3)}{(t+3)(t-3)(t+2)}$$

$$= \frac{6t + 12 - 5t - 15}{(t+3)(t-3)(t+2)}$$

$$= \frac{t-3}{(t+3)(t-3)(t+2)}$$

$$= \frac{\overset{1}{\cancel{t-3}}}{(t+3)\,\underset{1}{\cancel{(t-3)}}\,(t+2)}$$

$$= \frac{1}{(t+3)(t+2)}$$

Simplify each complex fraction.

23. $$\frac{\dfrac{3m-9}{8m}}{\dfrac{5m-15}{32}} = \frac{3m-9}{8m} \div \frac{5m-15}{32}$$

$$= \frac{3m-9}{8m} \cdot \frac{32}{5m-15}$$

$$= \frac{3(m-3) \cdot 32}{8m \cdot 5(m-3)}$$

$$= \frac{3 \cdot \overset{1}{\cancel{8}} \cdot 4 \,\overset{1}{\cancel{(m-3)}}}{\underset{1}{\cancel{8}} \cdot 5m \,\underset{1}{\cancel{(m-3)}}}$$

$$= \frac{12}{5m}$$

24. $\text{LCD} = a^2 s^2$

$$\frac{\dfrac{3}{as^2} + \dfrac{6}{a^2 s}}{\dfrac{6}{a} - \dfrac{9}{s^2}} = \frac{\dfrac{3}{as^2} + \dfrac{6}{a^2 s}}{\dfrac{6}{a} - \dfrac{9}{s^2}} \cdot \frac{a^2 s^2}{a^2 s^2}$$

$$= \frac{\left(\dfrac{3}{as^2} + \dfrac{6}{a^2 s}\right) a^2 s^2}{\left(\dfrac{6}{a} - \dfrac{9}{s^2}\right) a^2 s^2}$$

$$= \frac{\dfrac{3}{as^2}(a^2 s^2) + \dfrac{6}{a^2 s}(a^2 s^2)}{\dfrac{6}{a}(a^2 s^2) - \dfrac{9}{s^2}(a^2 s^2)}$$

$$= \frac{3a + 6s}{6as^2 - 9a^2}$$

$$= \frac{3(a + 2s)}{3a(2s^2 - 3a)}$$

$$= \frac{\overset{1}{\cancel{3}}(a+2s)}{\underset{1}{\cancel{3}}\,a(2s^2 - 3a)}$$

$$= \frac{a+2s}{a(2s^2 - 3a)} \quad \text{or} \quad \frac{a+2s}{2as^2 - 3a^2}$$

Solve each equation. If an equation has no solution, so indicate.

25. $\text{LCD} = 3y$; $y \neq 0$

$$\frac{1}{3} + \frac{4}{3y} = \frac{5}{y}$$

$$3y\left(\frac{1}{3} + \frac{4}{3y}\right) = 3y\left(\frac{5}{y}\right)$$

$$\overset{1}{\cancel{3}}y\left(\frac{1}{\underset{1}{\cancel{3}}}\right) + \overset{1}{\cancel{3}}\,\overset{1}{\cancel{y}}\left(\frac{4}{\underset{1}{\cancel{3}}\,\underset{1}{\cancel{y}}}\right) = 3\,\overset{1}{\cancel{y}}\left(\frac{5}{\underset{1}{\cancel{y}}}\right)$$

$$y + 4 = 15$$

$$y + 4 - 4 = 15 - 4$$

$$y = 11$$

The solution is 11.

26. $LCD = n - 6; \quad n \neq 6$

$$\frac{9n}{n-6} = 3 + \frac{54}{n-6}$$

$$(n-6)\left(\frac{9n}{n-6}\right) = (n-6)\left(3 + \frac{54}{n-6}\right)$$

$$(n-6)\left(\frac{9n}{n-6}\right) = (n-6)(3) + (n-6)\left(\frac{54}{n-6}\right)$$

$$9n = 3(n-6) + 54$$

$$9n = 3n - 18 + 54$$

$$9n = 3n + 36$$

$$9n - 3n = 3n + 36 - 3n$$

$$6n = 36$$

$$\frac{6n}{6} = \frac{36}{6}$$

$$n = 6$$

The denominator becomes zero.
No solution. 6 is extraneous.

27. $LCD = (q+1)(q-2); \quad q \neq -1, 2$

$$\frac{7}{q^2 - q - 2} + \frac{1}{q+1} = \frac{3}{q-2}$$

$$\frac{7}{(q-2)(q+1)} + \frac{1}{q+1} = \frac{3}{q-2}$$

$$(q-2)(q+1)\left(\frac{7}{(q-2)(q+1)} + \frac{1}{q+1}\right) = (q-2)(q+1)\left(\frac{3}{q-2}\right)$$

$$(q-2)(q+1)\left(\frac{7}{(q-2)(q+1)}\right) + (q-2)(q+1)\left(\frac{1}{q+1}\right) = (q-2)(q+1)\left(\frac{3}{q-2}\right)$$

$$7 + q - 2 = 3(q+1)$$

$$q + 5 = 3q + 3$$

$$q + 5 - q = 3q + 3 - q$$

$$5 = 2q + 3$$

$$5 - 3 = 2q + 3 - 3$$

$$2 = 2q$$

$$\frac{2}{2} = \frac{2q}{2}$$

$$1 = q$$

The solution is 1.

28. $LCD = 3(c-3); \quad c \neq 3$

$$\frac{2}{3} = \frac{2c-12}{3c-9} - c$$

$$3(c-3)\left(\frac{2}{3}\right) = 3(c-3)\left(\frac{2c-12}{3(c-3)} - c\right)$$

$$3(c-3)\left(\frac{2}{3}\right) = 3(c-3)\left(\frac{2c-12}{3(c-3)}\right) - 3(c-3)(c)$$

$$2(c-3) = 2c - 12 - 3c(c-3)$$

$$2c - 6 = 2c - 12 - 3c^2 + 9c$$

$$2c - 6 = -3c^2 + 11c - 12$$

$$2c - 6 - 2c = -3c^2 + 11c - 12 - 2c$$

$$-6 = -3c^2 + 9c - 12$$

$$-6 + 6 = -3c^2 + 9c - 12 + 6$$

$$0 = -3c^2 + 9c - 6$$

$$0 = -3(c^2 - 3c + 2)$$

$$0 = -3(c-2)(c-1)$$

$$\begin{array}{c|c} c - 2 = 0 & \text{or} \quad c - 1 = 0 \\ c - 2 + 2 = 0 + 2 & c - 1 + 1 = 0 + 1 \\ c = 2 & c = 1 \end{array}$$

The solutions are 1 and 2.

29. $LCD = y(y-1); \quad y \neq 0, 1$

$$\frac{y}{y-1} = \frac{y-2}{y}$$

$$y(y-1)\left(\frac{y}{y-1}\right) = y(y-1)\left(\frac{y-2}{y}\right)$$

$$y(y-1)\left(\frac{y}{y-1}\right) = y(y-1)\left(\frac{y-2}{y}\right)$$

$$y^2 = (y-1)(y-2)$$

$$y^2 = y^2 - 3y + 2$$

$$y^2 - y^2 = y^2 - 3y + 2 - y^2$$

$$0 = -3y + 2$$

$$0 + 3y = -3y + 2 + 3y$$

$$3y = 2$$

$$\frac{3y}{3} = \frac{2}{3}$$

$$y = \frac{2}{3}$$

The solution is $\frac{2}{3}$.

30. $\text{LCD} = (a+3)(a-3); \quad a \neq -3, 3$

$$\frac{a}{a-3} + \frac{4}{a+3} = \frac{18}{a^2-9}$$

$$(a+3)(a-3)\left(\frac{a}{a-3} + \frac{4}{a+3}\right) = (a+3)(a-3)\left(\frac{18}{(a+3)(a-3)}\right)$$

$$(a+3)(a-3)\left(\frac{a}{a-3}\right) + (a+3)(a-3)\left(\frac{4}{a+3}\right)$$

$$= (a+3)(a-3)\left(\frac{18}{(a+3)(a-3)}\right)$$

$$a(a+3) + 4(a-3) = 18$$
$$a^2 + 3a + 4a - 12 = 18$$
$$a^2 + 7a - 12 = 18$$
$$a^2 + 7a - 12 - \mathbf{18} = 18 - \mathbf{18}$$
$$a^2 + 7a - 30 = 0$$
$$(a+10)(a-3) = 0$$

$$a + 10 = 0 \quad \text{or} \quad a - 3 = 0$$
$$a = -10 \qquad \quad \cancel{a=3}$$

The solution is -10.
3 is an extraneous solution because it makes one denominator zero.

31. $H = \dfrac{RB}{R+B}$; Solve for B.

$\text{LCD} = R + B$

$$H = \frac{RB}{R+B}$$
$$(R+B)H = (R+B)\frac{RB}{R+B}$$
$$H(R+B) = (R+B)\frac{RB}{R+B}$$
$$HR + HB = RB$$
$$HR + HB - HB = RB - HB$$
$$HR = RB - HB$$
$$HR = B(R-H)$$
$$\frac{HR}{R-H} = \frac{B(R-H)}{R-H}$$
$$\frac{HR}{R-H} = B$$
$$B = \frac{HR}{R-H}$$

32. $\text{LCD} = rst;$ Solve for s.

$$\frac{1}{r} + \frac{1}{s} = \frac{1}{t}$$

$$(rst)\left(\frac{1}{r} + \frac{1}{s}\right) = (rst)\left(\frac{1}{t}\right)$$

$$r\,st\left(\frac{1}{r}\right) + r\,s\,t\left(\frac{1}{s}\right) = rs\,t\left(\frac{1}{t}\right)$$

$$st + rt = rs$$
$$st + rt - st = rs - st$$
$$rt = s(r-t)$$
$$\frac{rt}{r-t} = \frac{s(r-t)}{(r-t)}$$
$$s = \frac{rt}{r-t}$$

33. HEALTH RISKS

$$\frac{114}{120} = \frac{6 \cdot 19}{6 \cdot 20}$$

$$= \frac{19}{20}$$

Yes, the patient falls within the range.

34. CURRENCY EXCHANGE RATES

Let $x =$ number of pounds

$$\frac{51}{100} = \frac{x}{3,500}$$
$$51 \cdot 3,500 = 100x$$
$$\frac{51 \cdot 3,500}{100} = \frac{100x}{100}$$
$$1,785 = x$$

The traveler receives 1,785 pounds.

35. TV TOWERS

Let $h =$ the height of the tower

man's height $\rightarrow$ $\dfrac{6}{4} = \dfrac{h}{114}$ $\leftarrow$ tower's height
man's shadow $\rightarrow$ $\qquad\qquad\quad$ $\leftarrow$ tower's shadow

$$6 \cdot 114 = 4h$$
$$\frac{6 \cdot 114}{4} = \frac{4h}{4}$$
$$171 = h$$

The height of the tower is 171 feet.

Chapter 7 Review and Chapter 7 Test

36. COMPARISON SHOPPING

80 sheets for $3.89.

$$\underset{80 \text{ sheets} \rightarrow}{\overset{\text{cost} \rightarrow}{}} \frac{3.89}{80} = \frac{x}{1} \begin{array}{l} \leftarrow \text{cost} \\ \leftarrow 1 \text{ sheet} \end{array}$$

$$\frac{3.89}{80} = x$$

$$0.0486 \simeq x$$

The cost per sheet is about $0.0486.

120 sheets for $6.19.

$$\underset{120 \text{ sheets} \rightarrow}{\overset{\text{cost} \rightarrow}{}} \frac{6.19}{120} = \frac{x}{1} \begin{array}{l} \leftarrow \text{cost} \\ \leftarrow 1 \text{ card} \end{array}$$

$$\frac{6.19}{120} = x$$

$$0.0516 \simeq x$$

The cost per sheet is about $0.0516.

The 80 sheets for $3.89 is the better buy.

37. CLEANING HIGHWAYS

Analyze
- Highway worker takes 7 hours.
- Helper takes 9 hours.
- How long will it take both of them, working together, to pick up trash?

Assign

Let x = the number of hours it will take both working together to pick up the trash.

Form

	Rate	· Time	= Work Completed
Worker	$\frac{1}{7}$	x	$\frac{x}{7}$
Helper	$\frac{1}{9}$	x	$\frac{x}{9}$

The part of job done by worker	plus	the part of job done by helper	equals	1 job completed.
$\frac{x}{7}$	$+$	$\frac{x}{9}$	$=$	1

Solve

$$\frac{x}{7} + \frac{x}{9} = 1$$

$$LCD = 63$$

$$63\left(\frac{x}{7} + \frac{x}{9}\right) = 63(1)$$

$$\overset{1}{\cancel{7}} \cdot 9 \left(\frac{x}{\cancel{7}}\right) + \overset{1}{\cancel{9}} \cdot 7 \left(\frac{x}{\cancel{9}}\right) = 63$$

$$9x + 7x = 63$$

$$16x = 63$$

$$\frac{16x}{16} = \frac{63}{16}$$

$$x = 3\frac{15}{16}$$

State

It takes both working together $3\frac{15}{16}$ hours to pick up the trash.

Check

The result checks.

38. PHYSICAL FITNESS

Analyze
- He roller blades 5 miles.
- He jogs 2 miles.
- Times are the same.
- Roller is 6 mph faster than jogging.
- How fast can he jog?

Assign

Let r = jogging speed in mph

$r + 6$ = roller speed in mph

Form

	Rate	Time	= Distance
Jogging	r	$\dfrac{2}{r}$	2
Roller Blade	$r + 6$	$\dfrac{5}{r+6}$	5

Time took to jog 2 miles	equals	time took to blade 5 miles.
$\dfrac{2}{r}$	$=$	$\dfrac{5}{r+6}$

Solve

$$\frac{2}{r} = \frac{5}{r+6}$$

$$\text{LCD} = r(r+6)$$

$$r(r+6)\left(\frac{2}{r}\right) = r(r+6)\left(\frac{5}{r+6}\right)$$

$$\overset{1}{\cancel{r}}(r+6)\left(\frac{2}{\cancel{r}}\right) = r \cdot \overset{1}{\cancel{(r+6)}}\left(\frac{5}{\cancel{r+6}}\right)$$

$$2(r+6) = 5r$$

$$2r + 12 = 5r$$

$$2r + 12 - 2r = 5r - 2r$$

$$12 = 3r$$

$$\frac{12}{3} = \frac{3r}{3}$$

$$4 = r$$

State

His jogging speed is 4 mph.

Check

The result checks.

39. NUMBER PROBLEM

Analyze

- Begin with the fraction $\dfrac{5}{8}$.
- Subtract a number from the numerator.
- Twice as much is added to the denominator.
- The results is $\dfrac{1}{4}$.
- Find the number.

Assign

Let n = the unknown number.

Form

$$\frac{5-n}{8+2n} = \frac{1}{4}$$

Solve

$$\frac{5-n}{8+2n} = \frac{1}{4}$$

$$\text{LCD} = 4(4+n)$$

$$4(4+n)\left(\frac{5-n}{8+2n}\right) = 4(4+n)\frac{1}{4}$$

$$2 \cdot \overset{1}{\cancel{2}} \, \overset{1}{\cancel{(4+n)}}\left(\frac{5-n}{\underset{1}{\cancel{2}} \, \underset{1}{\cancel{(4+n)}}}\right) = \overset{1}{\cancel{4}}(4+n)\frac{1}{\underset{1}{\cancel{4}}}$$

$$2(5-n) = 4+n$$

$$10 - 2n = 4 + n$$

$$10 - 2n + 2n = 4 + n + 2n$$

$$10 = 4 + 3n$$

$$10 - 4 = 4 + 3n - 4$$

$$6 = 3n$$

$$\frac{6}{3} = \frac{3n}{3}$$

$$2 = n$$

State

The number is 2.

Check

$$\frac{5-n}{8+2n} = \frac{5-2}{8+2(2)} = \frac{3}{12} = \frac{1}{4}$$

The result checks.

40. **Explain the diffence between the procedure used to simplify $\dfrac{1}{x} + \dfrac{1}{4}$ and the procedure used to solve $\dfrac{1}{x} + \dfrac{1}{4} = \dfrac{1}{2}$.**

To simplify $\dfrac{1}{x} + \dfrac{1}{4}$, we build each fraction to have the LCD of $4x$. To solve $\dfrac{1}{x} + \dfrac{1}{4} = \dfrac{1}{2}$, we multiply both sides by the LCD $4x$ to eliminate the denominators.

Chapter 7 Review and Chapter 7 Test

CHAPTERS 1-7
CUMULATIVE REVIEW

1. Determine whether each statement is true or false. [Section 1.3]

   a. Every integer is a whole number. **False**

   b. 0 is not a rational number. **False**

   c. π is an irrational number. **True**

   d. The set of integers is the set of whole numbers and their opposites. **True**

2. Insert the proper symbol, $<$ or $>$, in the blank to make a true statement. [Section 1.5]

$$|2-4|\boxed{?}-(-6)$$
$$|2+(-4)|\boxed{?}-(-6)$$
$$|-2|\boxed{?}\ 6$$
$$2 < 6$$
$$|2-4|\boxed{<}-(-6)$$

3. Evaluate: $9^2 - 3[45 - 3(6+4)]$ [Section 1.7]

$$9^2 - 3[45 - 3(6+4)] = 81 - 3[45 - 3(10)]$$
$$= 81 - 3[45 - 30]$$
$$= 81 - 3[15]$$
$$= 81 - 45$$
$$= 36$$

4. Find the average (mean) test score of a student in a history class with scores of 80, 73, 61, 73, and 98. [Section 1.7]

$$Mean\ test\ score = \frac{80 + 73 + 61 + 73 + 98}{5}$$
$$= \frac{385}{5}$$
$$= 77$$

The mean test score is 77.

5. Simplify: $8(c+7) - 2(c-3)$. [Section 1.9]

$$8(c+7) - 2(c-3) = \mathbf{8}(c) + \mathbf{8}(7) - \mathbf{2}(c) - \mathbf{2}(-3)$$
$$= 8c + 56 - 2c + 6$$
$$= (8-2)c + (56+6)$$
$$= 6c + 62$$

6. Solve: $\frac{4}{5}d = -4$ [Section 2.1]

$$\left(\frac{5}{4}\right) \cdot \frac{4}{5}d = \left(\frac{5}{4}\right) \cdot (-4)$$
$$d = -5$$

The solution is -5.

7. Solve: $2 - 3(x-5) = 4(x-1)$. [Section 2.2]

$$2 - \mathbf{3}(x) - \mathbf{3}(-5) = \mathbf{4}(x) - \mathbf{4}(1)$$
$$2 - 3x + 15 = 4x - 4$$
$$-3x + 17 = 4x - 4$$
$$-3x + 17 + \mathbf{3x} = 4x - 4 + \mathbf{3x}$$
$$17 = 7x - 4$$
$$17 + \mathbf{4} = 7x - 4 + \mathbf{4}$$
$$21 = 7x$$
$$\frac{21}{7} = \frac{7x}{7}$$
$$3 = x$$

The solution is 3.

8. GRAND KING SIZE BEDS [Section 2.3]

1,600 is what % of 6,240?

$$1,600 = 6,240x$$
$$\frac{1,600}{\mathbf{6,240}} = \frac{6,240x}{\mathbf{6,240}}$$
$$0.256 \approx x$$
$$0.26 \approx x$$
$$26\% \approx x$$

The increase in area is about 26%.

9. Solve: $A - c = 2B + r$ for B. [Section 2.4]

$$A - c - \mathbf{r} = 2B + r - \mathbf{r}$$
$$A - c - r = 2B$$
$$\frac{A-c-r}{\mathbf{2}} = \frac{2B}{\mathbf{2}}$$
$$\frac{A-c-r}{2} = B$$
$$B = \frac{A-c-r}{2}$$

10. Change $40°$ C to degrees Fahrenheit. [Section 2.4]

$$F = \frac{9}{5}C + 32$$

$$F = \frac{9}{5}(40) + 32$$

$$F = \frac{9}{\cancel{5}}(\cancel{40}^{\,8}) + 32$$

$$F = 9(8) + 32$$

$$F = 72 + 32$$

$$F = 104$$

The Fahrenheit temperature is $104°$.

11. Find the volume of a pyramid that has a square base, measuring 6 feet on a side, and whose height is 20 feet. [Section 2.4]

$$V = \frac{1}{3}Bh$$

$$V = \frac{1}{3}(6^2)(20)$$

$$V = \frac{1}{3}(36)(20)$$

$$V = \frac{1}{\cancel{3}}(\cancel{36}^{\,12})(20)$$

$$V = 12(20)$$

$$V = 240$$

The volume is 240 ft^3.

12. BLENDING TEA [Section 2.6]

Analyze

- 1st tea cost \$6.40 per pound.
- 2nd tea cost \$4.00 per pound.
- 20 pound mixture worth \$5.44 per pound.
- How much of each grade of tea will be needed?

Assign

Let $x =$ the amount of 1st grade tea

$20 - x =$ the amount of 2nd grade tea

Form

	Amount	$\cdot$ Price	$=$ Total Value
1st Tea	x	6.40	**6.40x**
2nd Tea	$20 - x$	4.00	**4.00(20 − x)**
Mixture	20	5.44	**5.44(20)**

Value of 1st tea.	plus	Value of 2nd tea.	equals	Value of the mixture.
$6.40x$	$+$	$4.00(20 - x)$	$=$	$5.44(20)$

Solve

$$6.40x + 4.00(20 - x) = 5.44(20)$$

$$6.40x + 80 - 4.00x = 108.80$$

$$2.40x + 80 = 108.80$$

$$2.40x + 80 - \mathbf{80} = 108.80 - \mathbf{80}$$

$$2.40x = 28.80$$

$$\frac{2.40x}{\mathbf{2.40}} = \frac{28.80}{\mathbf{2.40}}$$

$$x = 12$$

State

12 lb. of the \$6.40 tea. 8 lb of the \$4 tea.

Check

$$6.40x + 4.00(20 - x) = 5.44(20)$$

$$6.40(12) + 4.00(20 - 12) \overset{?}{=} 5.44(20)$$

$$76.8 + 32 \overset{?}{=} 108.8$$

$$108.8 = 108.8 \ \text{True}$$

12 pounds of \$6.40 tea and 8 pounds of \$4.00 tea are the correct solutions.

13. SPEED OF A PLANE [Section 2.6]

Analyze
- Two planes are 6,000 miles a part.
- Faster plane is 200 mph faster.
- They travel towards each other.
- They meet in 5 hours.
- What is the speed of the slower plane?

Assign

Let x = the speed of slower plane in mph
$x + 200$ = the speed of faster plane in mph

Form

Rate $\cdot$ Time = Distance

Slower	x	5	**5x**
Faster	$x + 200$	5	**5(x + 200)**

Distance of slower.	plus	Distance of faster.	equals	Total distance.
$5x$	+	$5(x + 200)$	=	6,000

Solve

$$5x + 5(x + 200) = 6,000$$
$$5x + 5x + 1,000 = 6,000$$
$$10x + 1,000 = 6,000$$
$$10x + 1,000 - \mathbf{1,000} = 6,000 - \mathbf{1,000}$$
$$10x = 5,000$$
$$\frac{10x}{\mathbf{10}} = \frac{5,000}{\mathbf{10}}$$
$$x = 500$$

State

The speed of the slower plane is 500 mph.

Check

The result checks.

14. Solve: $7x + 2 \geq 4x - 1$ [Section 2.7]

$$7x + 2 - \mathbf{4x} \geq 4x - 1 - \mathbf{4x}$$
$$3x + 2 \geq -1$$
$$3x + 2 - \mathbf{2} \geq -1 - \mathbf{2}$$
$$3x \geq -3$$
$$\frac{3x}{\mathbf{3}} \geq \frac{-3}{\mathbf{3}}$$
$$x \geq -1$$

Interval Notation: $[-1, \infty)$

Graph:

-1

15. Graph: $y = 2x - 3$ [Section 3.2]

x	y	(x, y)
-2	$\begin{aligned} 2x - 3 \\ = 2(-2) - 3 \\ = -4 - 3 \\ = -7 \end{aligned}$	$(-2, -7)$
0	$\begin{aligned} 2x - 3 \\ = 2(0) - 3 \\ = 0 - 3 \\ = -3 \end{aligned}$	$(0, -3)$
3	$\begin{aligned} 2x - 3 \\ = 2(3) - 3 \\ = 6 - 3 \\ = 3 \end{aligned}$	$(3, 3)$

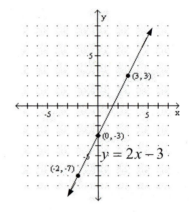

16. Find the slope of the line passing through $(-1, 3)$ and $(3, -1)$. [Section 3.4]

$\qquad (x_1, y_1) \qquad (x_2, y_2)$

$$m = \frac{y_2 - y_1}{x_2 - x_1}$$
$$m = \frac{-1 - 3}{3 - (-1)}$$
$$m = \frac{-4}{4}$$
$$m = -1$$

17. CUTTING STEEL. [Section 3.4]

$(0, 0)$ and $(50, 0.4)$
$(x_1, y_1) \qquad (x_2, y_2)$

$$m = \frac{y_2 - y_1}{x_2 - x_1}$$
$$m = \frac{0.4 - 0}{50 - 0}$$
$$m = \frac{0.4}{50}$$
$$m = 0.008$$

The rate of change is 0.008 mm/m.

18. What is the slope of a line perpendicular to the line $y = -\dfrac{7}{8}x - 6$. [Section 3.5]

$y = mx + b$, where m is the slope of the line.

$-\dfrac{7}{8}$ is the slope of the given line.

Perpendicular slope is the negative reciprocal of the given slope.

$\dfrac{8}{7}$ is the slope of the perpendicular line.

$-\dfrac{7}{8} \cdot \left(\dfrac{8}{7}\right) = -1$

19. Write the equation of a line in slope - intercept form that has slope 3 and passes through the point (1, 5). [Section 3.6]

$$y - y_1 = m(x - x_1)$$
$$y - 5 = 3(x - 1)$$
$$y - 5 = 3x - 3$$
$$y - 5 + 5 = 3x - 3 + 5$$
$$y = 3x + 2$$

20. Graph: $3x - 2y \le 6$ [Section 3.7]

• Graph the line $3x - 2y = 6$.

$$y = \dfrac{3}{2}x - 3$$

x	y	(x, y)
-2	$\dfrac{3}{2}x - 3$ $= \dfrac{3}{2}(-2) - 3$ $= -3 - 3$ $= -6$	$(-2, -6)$
0	$\dfrac{3}{2}x - 3$ $= \dfrac{3}{2}(0) - 3$ $= 0 - 3$ $= -3$	$(0, -3)$
	$\dfrac{3}{2}x - 3$ $= \dfrac{3}{2}(2) - 3$ $= 3 - 3$ $= 0$	

• Test (0,0) in the inequality $3x - 2y \le 6$.
$$3(0) - 2(0) \le 6$$
$$0 \le 6$$
True

• Shade in the side that is true.

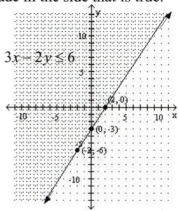

21. If $f(x) = -3x^2 - 6x$, find $f(-2)$. [Section 3.8]

$$f(x) = -3x^2 - 6x$$
$$f(-2) = -3(-2)^2 - 6(-2)$$
$$f(-2) = -3(4) + 12$$
$$f(-2) = -12 + 12$$
$$f(-2) = 0$$

The output is 0.

22. Fill in the blanks. [Section 3.8]

23. Solve the system by graphing. [Section 4.1]

$$\begin{cases} x + y = 1 \\ y = x + 5 \end{cases}$$

The method of choice to graph each line is left up to the student.

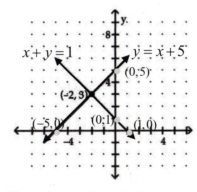

The solution is $(-2, 3)$.

Chapters 1- 7 Cumulative Review

24. Solve the system by substitution.
[Section 4.2]

$$\begin{cases} x = 3y - 1 \\ 2x - 3y = 4 \end{cases}$$

$2x - 3y = 4$ The 2nd equation
               Substitute for x.

$2(\mathbf{3y - 1}) - 3y = 4$

$6y - 2 - 3y = 4$

$3y - 2 = 4$

$3y - 2 + \mathbf{2} = 4 + \mathbf{2}$

$3y = 6$

$\dfrac{3y}{\mathbf{3}} = \dfrac{6}{\mathbf{3}}$

$y = 2$

$x = 3y - 1$ The 1st equation

$x = 3(\mathbf{2}) - 1$

$x = 5$

The solution is (5, 2).

Check:
$x = 3y - 1$
$5 \overset{?}{=} 3(2) - 1$
$5 = 5$
True

Check:
$2x - 3y = 4$
$2(5) - 3(2) \overset{?}{=} 4$
$10 - 6 \overset{?}{=} 4$
$4 = 4$
True

25. Solve the system by elimination (addition).
[Section 4.3]

$$\begin{cases} 2x + 3y = -1 \\ 3x + 5y = -2 \end{cases}$$

Eliminate x.

Multiply both sides of the 1st equation by 3.

Multiply both sides of the 2nd equation by -2.

$6x + 9y = -3$
$\underline{-6x - 10y = 4}$
$-y = 1$

$\dfrac{-y}{\mathbf{-1}} = \dfrac{1}{\mathbf{-1}}$

$y = -1$

$2x + 3y = -1$ The 1st equation

$2x + 3(\mathbf{-1}) = -1$

$2x - 3 + \mathbf{3} = -1 + \mathbf{3}$

$2x = 2$

$\dfrac{2x}{\mathbf{2}} = \dfrac{2}{\mathbf{2}}$

$x = 1$

The solution is (1, −1)

26. POKER [Section 4.4]

Analyze

- Red chips worth $5 each.
- Blue chips worth $10 each.
- Ended with $190 and 23 chips.
- How many of each chips did he have?

Assign

Let x = number of red chips
      y = number of blue chips

Form

The number of red chips	plus	the number of blue chips	is	23.
x	$+$	y	$=$	23

	Number •	Value	= Total Value
Red chips	x	5	**5x**
Blue chips	y	10	**10y**
		Total	190

The value of red chips	plus	the value of blue chips	is	$190.
$5x$	$+$	$10y$	$=$	190

Solve

$$\begin{cases} x + y = 23 \\ 5x + 10y = 190 \end{cases}$$

Eliminate x.

Multiply both sides of the 1st equation by -5.

$-5x - 5y = -115$
$\underline{5x + 10y = 190}$
$5y = 75$

$\dfrac{5y}{\mathbf{5}} = \dfrac{75}{\mathbf{5}}$

$y = 15$

$x + y = 23$ The 1st equation

$x + 15 = 23$

$x + 15 - \mathbf{15} = 23 - \mathbf{15}$

$x = 8$

State

He had 8 red chips.
He had 15 blue chips.

Check

$x + y = 23$
$8 + 15 \overset{?}{=} 23$
$23 = 23$
True

$5x + 10y = 190$
$5(8) + 10(15) \overset{?}{=} 190$
$40 + 150 \overset{?}{=} 190$
$190 = 190$
True

The results check.

Simplify each expression. Write each answer without using negative exponents. [Section 5.1]

27. $x^4 x^3 = x^{4+3}$
$$= x^7$$

28. $(x^2 x^3)^5 = (x^{2+3})^5$
$$= (x^5)^5$$
$$= x^{5\cdot5}$$
$$= x^{25}$$

29. $\left(\dfrac{y^3 y}{2yy^2}\right)^3 = \left(\dfrac{y^{3+1}}{2y^{1+2}}\right)^3$
$$= \left(\dfrac{y^4}{2y^3}\right)^3$$
$$= \left(\dfrac{y^{4-3}}{2}\right)^3$$
$$= \left(\dfrac{y^1}{2^1}\right)^3$$
$$= \dfrac{y^{1\cdot3}}{2^{1\cdot3}}$$
$$= \dfrac{y^3}{8}$$

30. $\left(\dfrac{-2a}{b}\right)^5 = \dfrac{(-2)^5 a^5}{b^5}$
$$= -\dfrac{32a^5}{b^5}$$

[Section 5.2]

31. $(a^{-2}b^3)^{-4} = a^{-2\cdot-4}b^{3\cdot-4}$
$$= a^8 b^{-12}$$
$$= \dfrac{a^8}{b^{12}}$$

32. $\dfrac{9b^0 b^3}{3b^{-3}b^4} = \dfrac{9b^{0+3}}{3b^{-3+4}}$
$$= \dfrac{9b^3}{3b^1}$$
$$= \dfrac{\overset{1}{\cancel{3}}\cdot3b^{3-1}}{\underset{1}{\cancel{3}}}$$
$$= 3b^2$$

33. Write $290,000$ in scientific notation. [Section 5.3]
Move the decimal 5 places to the left.
$$290,000 = 2.9 \times 10^5$$

34. What is the degree of the polynomial $5x^3 - 4x + 16$? [Section 5.4]

The degree is 3, because the degree of a polynomial is the same as the highest degree of any term of the polynomial.

35. Graph: $y = -x^3$ [Section 5.4]

x	y	(x, y)
-2	$-x^3 = -(-2)^3$ $= -(-8)$ $= 8$	$(-2, 8)$
-1	$-x^3 = -(-1)^3$ $= -(-1)$ $= 1$	$(-1, 1)$
0	$-x^3 = -(0)^3$ $= 0$	$(0, 0)$
1	$-x^3 = -(1)^3$ $= -1$	$(1, -1)$
	$-x^3 = -(2)^3$ $= -8$	$(2, -8)$

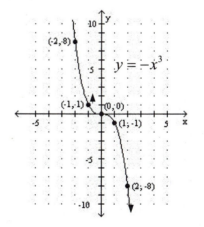

36. CONCENTRIC CIRCLES [Section 5.7]
The area of the ring between the two concentric circles of radius r and R is given by the formula $A = \pi(R+r)(R-r)$. Do the multiplication on the right-hand side of the equation.

$$A = \pi(R+r)(R-r)$$
$$A = \pi(R^2 - r^2)$$
$$A = \pi R^2 - \pi r^2$$

Perform the operations. [Section 5.5]

37. $(3x^2 - 3x - 2) + (3x^2 + 4x - 3)$
$$= 3x^2 - 3x - 2 + 3x^2 + 4x - 3$$
$$= (3+3)x^2 + (-3+4)x + (-2-3)$$
$$= 6x^2 + x - 5$$

38. $\left(\dfrac{1}{16}t^3 + \dfrac{1}{2}t^2 - \dfrac{1}{6}t\right) - \left(\dfrac{9}{16}t^3 + \dfrac{9}{4}t^2 - \dfrac{1}{12}t\right)$

$$= \dfrac{1}{16}t^3 + \dfrac{1}{2}t^2 - \dfrac{1}{6}t - \dfrac{9}{16}t^3 - \dfrac{9}{4}t^2 + \dfrac{1}{12}t$$

$$= \left(\dfrac{1}{16} - \dfrac{9}{16}\right)t^3 + \left(\dfrac{1}{2} - \dfrac{9}{4}\right)t^2 + \left(-\dfrac{1}{6} + \dfrac{1}{12}\right)t$$

$$= \left(-\dfrac{8}{16}\right)t^3 + \left(\dfrac{1}{2} - \dfrac{9}{4}\right)t^2 + \left(-\dfrac{1}{6} + \dfrac{1}{12}\right)t$$

$$= \left(-\dfrac{1 \cdot \overset{1}{\cancel{8}}}{2 \cdot \underset{1}{\cancel{8}}}\right)t^3 + \left(\dfrac{1}{2} \cdot \dfrac{2}{2} - \dfrac{9}{4}\right)t^2 + \left(-\dfrac{1}{6} \cdot \dfrac{2}{2} + \dfrac{1}{12}\right)t$$

$$= \left(-\dfrac{1}{2}\right)t^3 + \left(\dfrac{2-9}{4}\right)t^2 + \left(\dfrac{-2+1}{12}\right)t$$

$$= -\dfrac{1}{2}t^3 - \dfrac{7}{4}t^2 - \dfrac{1}{12}t$$

[Section 5.6]

39. $(2x^2 y^3)(3x^2 y^2) = (2 \cdot 3)(x^2 x^2)(y^3 y^2)$
$$= 6x^{2+2} y^{3+2}$$
$$= 6x^4 y^5$$

40. $(2y-5)(3y+7) = 6y^2 + 14y - 15y - 35$
$$= 6y^2 - y - 35$$

41. $-4x^2 z(3x^2 - z) = -4x^2 z(3x^2) - 4x^2 z(-z)$
$$= -4 \cdot 3 x^{2+2} z - 4 \cdot -1 x^2 z^{1+1}$$
$$= -12x^4 z + 4x^2 z^2$$

[Section 5.7]

42. $(3a-4)^2 = (3a-4)(3a-4)$
$$= 9a^2 - 12a - 12a + 16$$
$$= 9a^2 - 24a + 16$$

[Section 5.8]

43. $\dfrac{6x+9}{3} = \dfrac{6x}{3} + \dfrac{9}{3}$

$$= \dfrac{2 \cdot \overset{1}{\cancel{3}} x}{\underset{1}{\cancel{3}}} + \dfrac{3 \cdot \overset{1}{\cancel{3}}}{\underset{1}{\cancel{3}}}$$

$$= 2x + 3$$

44.
$$\require{enclose}
\begin{array}{r}
x^2 + 2x - 1 \\
2x+3 \enclose{longdiv}{2x^3 + 7x^2 + 4x - 3} \\
\underline{-(2x^3 + 3x^2)} \\
4x^2 + 4x \\
\underline{-(4x^2 + 6x)} \\
-2x - 3 \\
\underline{-(-2x - 3)} \\
0
\end{array}$$

Factor each polynomial completely, if possible. [Section 6.1]

45. $k^3 t - 3k^2 t$, GCF $= k^2 t$
$$k^3 t - 3k^2 t = k^2 t \cdot k - k^2 t \cdot 3$$
$$= k^2 t(k-3)$$

46. $2ab + 2ac + 3b + 3c$
$$= (2ab + 2ac) + (3b + 3c)$$
$$= 2a(b+c) + 3(b+c)$$
$$= (b+c)(2a+3)$$

[Section 6.2]

47. $u^2 - 18u + 81 = (u-9)(u-9)$
$$= (u-9)^2$$

48. $-r^2 + 2 + r = -r^2 + r + 2$
$$= -(r^2 - r - 2)$$
$$= -(r-2)(r+1)$$

49. $u^2 + 10u + 15$
    prime

[Section 6.3]

50. $6x^2 - 63 - 13x = 6x^2 - 13x - 63$
$= (2x - 9)(3x + 7)$

[Section 6.4]

51. $2a^2 - 200b^2 = 2(a^2 - 100b^2)$
$= 2(a + 10b)(a - 10b)$

[Section 6.5]

52. $b^3 + 125 = b^3 + 5^3$
$= (b + 5)(b^2 - 5b + 25)$

Solve each equation by factoring.
[Section 6.7]

53. $5x^2 + x = 0$

$x(5x + 1) = 0$

$x = 0$ or $5x + 1 = 0$

$5x + 1 - \mathbf{1} = 0 - \mathbf{1}$

$5x = -1$

$\dfrac{5x}{\mathbf{5}} = \dfrac{-1}{\mathbf{5}}$

$x = -\dfrac{1}{5}$

The solutions are 0 and $-\dfrac{1}{5}$.

54. $6x^2 - 5x = -1$

$6x^2 - 5x + \mathbf{1} = -1 + \mathbf{1}$

$6x^2 - 5x + 1 = 0$

$(2x - 1)(3x - 1) = 0$

$2x - 1 = 0$ or $3x - 1 = 0$

$2x - 1 + \mathbf{1} = 0 + \mathbf{1}$ | $3x - 1 + \mathbf{1} = 0 + \mathbf{1}$

$2x = 1$ | $3x = 1$

$\dfrac{2x}{\mathbf{2}} = \dfrac{1}{\mathbf{2}}$ | $\dfrac{3x}{\mathbf{3}} = \dfrac{1}{\mathbf{3}}$

$x = \dfrac{1}{2}$ | $x = \dfrac{1}{3}$

The solutions are $\dfrac{1}{2}$ and $\dfrac{1}{3}$.

55. COOKING [Section 6.7]

Analyze
- Cooking surface is 160 sq. in.
- Length is $w + 6$.
- Width is w.
- Griddle is a rectangle.
- Find the length and width.

Assign
Let w = Width of griddle in inches
$w + 6$ = Length of griddle in inches

Form

Width of griddle	times	length of griddle	equals	area of griddle.
w	$\bullet$	$w + 6$	$=$	160

Solve

$$w(w + 6) = 160$$
$$w^2 + 6w = 160$$
$$w^2 + 6w - \mathbf{160} = 160 - \mathbf{160}$$
$$w^2 + 6w - 160 = 0$$
$$(w + 16)(w - 10) = 0$$

$w + 16 = 0$ or $w - 10 = 0$

$w + 16 - \mathbf{16} = 0 - \mathbf{16}$ | $w - 10 + \mathbf{10} = 0 + \mathbf{10}$

$w = \cancel{-16}$ | $w = 10$

State
The width is 10 inches.
The length is 16 inches.

Check
The result checks.

56. For what values of $\dfrac{3x^2}{x^2 - 25}$ is the rational expression undefined? [Section 7.1]

$$\dfrac{3x^2}{x^2 - 25} = \dfrac{3x^2}{(x + 5)(x - 5)}$$

$x + 5 = 0$ or $x - 5 = 0$

$x + 5 - \mathbf{5} = 0 - \mathbf{5}$ | $x - 5 + \mathbf{5} = 0 + \mathbf{5}$

$x = -5$ | $x = 5$

The rational expression is undefined for $x = 5$ and $x = -5$.

Chapters 1- 7 Cumulative Review

Perform the operations. Simplify, if possible.
[Section 7.1]

57. $\dfrac{2x^2 - 8x}{x^2 - 6x + 8} = \dfrac{2x(x-4)}{(x-2)(x-4)}$

$$= \dfrac{2x \, \cancel{(x-4)}^{\;1}}{(x-2) \, \cancel{(x-4)}_{\;1}}$$

$$= \dfrac{2x}{x-2}$$

[Section 7.2]

58. $\dfrac{x^2 - 16}{4 - x} \div \dfrac{3x + 12}{x^3} = \dfrac{x^2 - 16}{4 - x} \cdot \dfrac{x^3}{3x + 12}$

$$= \dfrac{(x^2 - 16)x^3}{(4 - x)(3x + 12)}$$

$$= \dfrac{x^3(x+4)(x-4)}{-1(-4+x)3(x+4)}$$

$$= -\dfrac{x^3 \, \cancel{(x+4)}^{\;1} \cdot \cancel{(x-4)}^{\;1}}{3 \, \cancel{(x-4)}_{\;1} \, \cancel{(x+4)}_{\;1}}$$

$$= -\dfrac{x^3}{3}$$

59. LCD $= 2m + 5$ [Section 7.3]

$\dfrac{8m^2}{2m+5} - \dfrac{4m^2 + 25}{2m+5} = \dfrac{8m^2 - (4m^2 + 25)}{2m+5}$

$$= \dfrac{8m^2 - 4m^2 - 25}{2m+5}$$

$$= \dfrac{4m^2 - 25}{2m+5}$$

$$= \dfrac{(2m+5)(2m-5)}{2m+5}$$

$$= \dfrac{\cancel{(2m+5)}^{\;1}(2m-5)}{\cancel{2m+5}_{\;1}}$$

$$= 2m - 5$$

60. LCD $= x - 3$ [Section 7.4]

$\dfrac{4}{x-3} + \dfrac{5}{3-x} = \dfrac{4}{x-3} + \dfrac{5}{-1(-3+x)}$

$$= \dfrac{4}{x-3} - \dfrac{5}{x-3}$$

$$= \dfrac{4-5}{x-3}$$

$$= \dfrac{-1}{x-3}$$

$$= -\dfrac{1}{x-3}$$

[Section 7.4]

61. $\left. \begin{array}{l} m^2 + 5m + 6 = (m+2)(m+3) \\ m^2 + 3m + 2 = (m+2)(m+1) \end{array} \right\}$

$$\text{LCD} = (m+2)(m+3)(m+1)$$

$\dfrac{m}{m^2 + 5m + 6} - \dfrac{2}{m^2 + 3m + 2}$

$$= \dfrac{m}{(m+2)(m+3)} - \dfrac{2}{(m+2)(m+1)}$$

$$= \dfrac{m}{(m+2)(m+3)} \cdot \dfrac{(m+1)}{(m+1)} - \dfrac{2}{(m+2)(m+1)} \cdot \dfrac{(m+3)}{(m+3)}$$

$$= \dfrac{m(m+1) - 2(m+3)}{(m+2)(m+3)(m+1)}$$

$$= \dfrac{m^2 + m - 2m - 6}{(m+2)(m+3)(m+1)}$$

$$= \dfrac{m^2 - m - 6}{(m+2)(m+3)(m+1)}$$

$$= \dfrac{(m+2)(m-3)}{(m+2)(m+3)(m+1)}$$

$$= \dfrac{\cancel{(m+2)}^{\;1}(m-3)}{\cancel{(m+2)}_{\;1}(m+3)(m+1)}$$

$$= \dfrac{(m-3)}{(m+3)(m+1)}$$

Simplify. [Section 7.5]

62. LCD $= x(x+1)$

$$\frac{2-\dfrac{2}{x+1}}{2+\dfrac{2}{x}} = \frac{2-\dfrac{2}{x+1}}{2+\dfrac{2}{x}} \cdot \frac{\boldsymbol{x(x+1)}}{\boldsymbol{x(x+1)}}$$

$$= \frac{\left(2-\dfrac{2}{x+1}\right)x(x+1)}{\left(2+\dfrac{2}{x}\right)x(x+1)}$$

$$= \frac{2x(x+1)-x(x+1)\left(\dfrac{2}{x+1}\right)}{2x(x+1)+x(x+1)\left(\dfrac{2}{x}\right)}$$

$$= \frac{2x^2+2x-x\,\overset{1}{(\cancel{x+1})}\left(\dfrac{2}{\cancel{x+1}}\right)}{2x^2+2x+\cancel{x}(x+1)\left(\dfrac{2}{\cancel{x}}\right)}$$

$$= \frac{2x^2+2x-x(2)}{2x^2+2x+(x+1)(2)}$$

$$= \frac{2x^2+2x-2x}{2x^2+2x+2x+2}$$

$$= \frac{2x^2}{2x^2+4x+2}$$

$$= \frac{2x^2}{2(x^2+2x+1)}$$

$$= \frac{\overset{1}{\cancel{2}}\,x^2}{\underset{1}{\cancel{2}}(x^2+2x+1)}$$

$$= \frac{x^2}{x^2+2x+1} \quad \text{or} \quad \frac{x^2}{(x+1)^2}$$

Solve each equation. [Section 7.6]

63. LCD $= 60x$; $x \neq 0$

$$\frac{7}{5x}-\frac{1}{2} = \frac{5}{6x}+\frac{1}{3}$$

$$\boldsymbol{60x}\left(\frac{7}{5x}-\frac{1}{2}\right) = \boldsymbol{60x}\left(\frac{5}{6x}+\frac{1}{3}\right)$$

$$60x\left(\frac{7}{5x}\right)-60x\left(\frac{1}{2}\right) = 60x\left(\frac{5}{6x}\right)+60x\left(\frac{1}{3}\right)$$

$$12\cdot\overset{1}{\cancel{5}}\,\overset{1}{\cancel{x}}\left(\frac{7}{\underset{1\,1}{\cancel{5}\,\cancel{x}}}\right)-30\cdot\overset{1}{\cancel{2}}\,x\left(\frac{1}{\underset{1}{\cancel{2}}}\right) = 10\cdot\overset{1}{\cancel{6}}\,\overset{1}{\cancel{x}}\left(\frac{5}{\underset{1\,1}{\cancel{6}\,\cancel{x}}}\right)+20\cdot\overset{1}{\cancel{3}}\,x\left(\frac{1}{\underset{1}{\cancel{3}}}\right)$$

$$12(7)-30x = 10(5)+20x$$

$$84-30x = 50+20x$$

$$84-30x+\boldsymbol{30x} = 50+20x+\boldsymbol{30x}$$

$$84 = 50+50x$$

$$84-\boldsymbol{50} = 50+50x-\boldsymbol{50}$$

$$34 = 50x$$

$$\frac{34}{\boldsymbol{50}} = \frac{50x}{\boldsymbol{50}}$$

$$\frac{\overset{1}{\cancel{2}}\cdot 17}{\underset{1}{\cancel{2}}\cdot 25} = x$$

$$\frac{17}{25} = x$$

The solution is $\dfrac{17}{25}$.

64. LCD $= u(u-1)$; $u \neq 0,1$

$$\frac{u}{u-1}+\frac{1}{u} = \frac{u^2+1}{u^2-u}$$

$$\boldsymbol{u(u-1)}\left(\frac{u}{u-1}+\frac{1}{u}\right) = \boldsymbol{u(u-1)}\left(\frac{u^2+1}{u^2-u}\right)$$

$$u(u-1)\left(\frac{u}{u-1}\right)+u(u-1)\left(\frac{1}{u}\right) = u(u-1)\left(\frac{u^2+1}{u(u-1)}\right)$$

$$u\,\overset{1}{(\cancel{u-1})}\left(\frac{u}{\cancel{u-1}}\right)+\overset{1}{\cancel{u}}(u-1)\left(\frac{1}{\cancel{u}}\right) = \overset{1}{\cancel{u}}\,\overset{1}{(\cancel{u-1})}\left(\frac{u^2+1}{\underset{1}{\cancel{u}}\underset{1}{(\cancel{u-1})}}\right)$$

$$u^2+u-1 = u^2+1$$

$$u^2+u-1-\boldsymbol{u^2} = u^2+1-\boldsymbol{u^2}$$

$$u-1 = 1$$

$$u-1+\boldsymbol{1} = 1+\boldsymbol{1}$$

$$u = 2$$

The solution is 2.

Chapters 1- 7 Cumulative Review

65. DRAINING A TANK [Section 7.7]

Analyze
- 1st Outlet pipe can drain pool in 24 hours.
- 2nd Outlet pipe can drain pool in 36 hours.
- How long will it take both pipes to drain the pool?

Assign

Let x = the number of hours it will take both outlet pipes to drain the pool.

Form

	Rate	· Time	= Work Completed
1st Outlet pipe	$\dfrac{1}{24}$	x	$\dfrac{x}{24}$
2nd Outlet pipe	$\dfrac{1}{36}$	x	$\dfrac{x}{36}$

The part of job done by 1st outlet pipe	plus	the part of job done by 2nd outlet pipe	equals	1 job completed.
$\dfrac{x}{24}$	$+$	$\dfrac{x}{36}$	$=$	1

Solve

$$\frac{x}{24} + \frac{x}{36} = 1$$

$$LCD = 72$$

$$72\left(\frac{x}{24} + \frac{x}{36}\right) = 72(1)$$

$$\overset{1}{\cancel{24}} \cdot 3\left(\frac{x}{\underset{1}{\cancel{24}}}\right) + \overset{1}{\cancel{36}} \cdot 2\left(\frac{x}{\underset{1}{\cancel{36}}}\right) = 72$$

$$3x + 2x = 72$$

$$5x = 72$$

$$\frac{5x}{5} = \frac{72}{5}$$

$$x = 14\frac{2}{5}$$

State

It will take both outlet pipes $14\frac{2}{5}$ hours to empty the pool.

Check

The result checks.

66. HEIGHT OF A TREE [Section 7.8]

Analyze
- Similar triangles determined by the tree and its shadow and the yardstick and its shadow.
- What is the height of the tree?

Assign

Let h = the height of the tree

Form

1 yard is to 2.5 feet as h feet is to 29 feet.
We will convert 1 yard to 3 feet so all units will be the same.
3 feet is to 2.5 feet as h feet is to 29 feet.

$$\text{yardstick} \rightarrow \frac{3}{2.5} = \frac{h}{29} \leftarrow \text{tree's height}$$
$$\text{yardstick shadow} \rightarrow \qquad\qquad \leftarrow \text{tree's shadow}$$

Solve

$$\frac{3}{2.5} = \frac{h}{29}$$

$$87 = 2.5h$$

$$\frac{87}{2.5} = \frac{2.5h}{2.5}$$

$$34.8 = h$$

State

A height of the tree is 34.8 feet.

Check

$$\frac{3}{2.5} = \frac{h}{29}$$

$$\frac{3}{2.5} \overset{?}{=} \frac{34.8}{29}$$

$$87 = 87$$

The result checks.

VOCABULARY

1. An **equation** is a statement indicating that two expressions are equal.

3. A number that makes an equation true when substituted for the variable is called a **solution**.

5. An equation that is made true by any permissible replacement value for the variable is called an **identity**.

7. $<, >, \leq$, and $\geq$ are **inequality** symbols.

CONCEPTS

9. a. Adding the **same** number to, or subtracting the same number from, both sides of an equation does not change its solution.
   b. Multiplying or dividing both sides of an equation by the **same** nonzero number does not change its solution.

11. a. No.
$$5(2x+7)=2x-4$$
$$5(2\cdot-5+7)\overset{?}{=}2(-5)-4$$
$$5(-10+7)\overset{?}{=}-10-4$$
$$5(-3)\overset{?}{=}-14$$
$$-15 \neq -14$$

   b. Yes.
$$3x+6\leq-9$$
$$3(-5)+6\overset{?}{\leq}-9$$
$$-15+6\overset{?}{\leq}-9$$
$$-9\leq-9$$

NOTATION

13. a. iii
   b. i
   c. ii

15.
$$4x+1=13$$
$$4x+1-1=13-1$$
$$4x=12$$
$$x=3$$

17.
$$3(x+1)=15$$
$$3x+3=15$$
$$3x+3-3=15-3$$
$$3x=12$$
$$x=4$$

19.
$$2x+6(2x+3)=-10$$
$$2x+12x+18=-10$$
$$14x+18=-10$$
$$14x+18-18=-10-18$$
$$14x=-28$$
$$x=-2$$

21.
$$7(a+2)=11a+17-7a$$
$$7a+14=4a+17$$
$$7a+14-4a=4a+17-4a$$
$$3a+14=17$$
$$3a+14-14=17-14$$
$$3a=3$$
$$a=1$$

23.
$$\frac{1}{2}x-4=-1+2x$$
$$2\left(\frac{1}{2}x-4\right)=2(-1+2x)$$
$$x-8=-2+4x$$
$$x-8-4x=-2+4x-4x$$
$$-3x-8=-2$$
$$-3x-8+8=-2+8$$
$$-3x=6$$
$$x=-2$$

25.

$$\frac{x}{2} - \frac{x}{3} = 4$$

$$6\left(\frac{x}{2} - \frac{x}{3}\right) = 6(4)$$

$$3x - 2x = 24$$

$$x = 24$$

27.

$$\frac{1}{6}(x+12) + 1 = \frac{x}{3}$$

$$6\left[\frac{1}{6}(x+12) + 1\right] = 6\left[\frac{x}{3}\right]$$

$$1(x+12) + 6 = 2x$$

$$x + 12 + 6 = 2x$$

$$x + 18 = 2x$$

$$x + 18 - x = 2x - x$$

$$18 = x$$

29.

$$\frac{1}{5}(x+6) = \frac{5}{8}(2x-1) + 2$$

$$\frac{1}{5}x + \frac{6}{5} = \frac{10}{8}x - \frac{5}{8} + 2$$

$$40\left(\frac{1}{5}x + \frac{6}{5}\right) = 40\left(\frac{10}{8}x - \frac{5}{8} + 2\right)$$

$$8x + 48 = 50x - 25 + 80$$

$$8x + 48 = 50x + 55$$

$$8x + 48 - 50x = 50x + 55 - 50x$$

$$-42x + 48 = 55$$

$$-42x + 48 - 48 = 55 - 48$$

$$-42x = 7$$

$$x = -\frac{1}{6}$$

31.

$$2x - 6 = -2x + 4(x-2)$$

$$2x - 6 = -2x + 4x - 8$$

$$2x - 6 = 2x - 8$$

$$2x - 6 - 2x = 2x - 8 - 2x$$

$$-6 \neq -8$$

$$\varnothing; \text{ contradiction}$$

33.

$$2y + 1 = 5(0.2y + 1) - (4 - y)$$

$$2y + 1 = y + 5 - 4 + y$$

$$2y + 1 = 2y + 1$$

$$2y + 1 - 2y = 2y + 1 - 2y$$

$$1 = 1$$

$$\mathbb{R}; \text{ identity}$$

all real numbers

35.

$$\frac{7}{2}(y-1) + \frac{1}{2} = \frac{1}{2}(7y - 6)$$

$$2\left[\frac{7}{2}(y-1) + \frac{1}{2}\right] = 2\left[\frac{1}{2}(7y - 6)\right]$$

$$7(y-1) + 1 = 1(7y - 6)$$

$$7y - 7 + 1 = 7y - 6$$

$$7y - 6 = 7y - 6$$

$$7y - 6 - 7y = 7y - 6 - 7y$$

$$-6 = -6$$

$$\mathbb{R}; \text{ identity}$$

all real numbers

37.

$$0.3(x-4) + 0.6 = -0.2(x+4) + 0.5x$$

$$10\left[0.3(x-4) + 0.6\right] = 10\left[-0.2(x+4) + 0.5x\right]$$

$$3(x-4) + 6 = -2(x+4) + 5x$$

$$3x - 12 + 6 = -2x - 8 + 5x$$

$$3x - 6 = 3x - 8$$

$$3x - 6 - 3x = 3x - 8 - 3x$$

$$-6 \neq -8$$

$$\varnothing; \text{ no solution}$$

39.

$$p = 2\ell + 2w$$

$$p - 2\ell = 2\ell + 2w - 2\ell$$

$$p - 2\ell = 2w$$

$$\frac{p - 2\ell}{2} = \frac{2w}{2}$$

$$\frac{p - 2\ell}{2} = w$$

$$w = \frac{p - 2\ell}{2}$$

41.

$$V = \frac{1}{3}Bh$$

$$3(V) = 3\left(\frac{1}{3}Bh\right)$$

$$3V = Bh$$

$$\frac{3V}{h} = \frac{Bh}{h}$$

$$\frac{3V}{h} = B$$

$$B = \frac{3V}{h}$$

43.

$$T - W = ma$$

$$T - W - T = ma - T$$

$$-W = ma - T$$

$$\frac{-W}{-1} = \frac{ma - T}{-1}$$

$$W = -ma + T$$

$$W = T - ma$$

45.

$$z = \frac{x - \mu}{\sigma}$$

$$\sigma(z) = \sigma\left(\frac{x - \mu}{\sigma}\right)$$

$$z\sigma = x - \mu$$

$$z\sigma + \mu = x - \mu + \mu$$

$$z\sigma + \mu = x$$

47.

$$S = \frac{n(a + l)}{2}$$

$$2(S) = 2\left(\frac{n(a + l)}{2}\right)$$

$$2S = n(a + l)$$

$$2S = na + nl$$

$$2S - na = na + nl - na$$

$$2S - na = nl$$

$$\frac{2S - na}{n} = \frac{nl}{n}$$

$$\frac{2S - na}{n} = l$$

$$l = \frac{2S - na}{n} \quad \text{or} \quad l = \frac{2S}{n} - a$$

49.

$$l = a + (n - 1)d$$

$$l - a = a + (n - 1)d - a$$

$$l - a = (n - 1)d$$

$$\frac{l - a}{n - 1} = \frac{(n - 1)d}{n - 1}$$

$$\frac{l - a}{n - 1} = d$$

$$d = \frac{l - a}{n - 1}$$

51.

$$5x - 3 > 7$$

$$5x - 3 + 3 > 7 + 3$$

$$5x > 10$$

$$x > 2$$

$$(2, \infty)$$

53.

$$9a + 11 \le 29$$

$$9a + 11 - 11 \le 29 - 11$$

$$9a \le 18$$

$$a \le 2$$

$$(-\infty, 2]$$

55.

$$3(z - 2) \le 2(z + 7)$$

$$3z - 6 \le 2z + 14$$

$$3z - 6 - 2z \le 2z + 14 - 2z$$

$$z - 6 \le 14$$

$$z - 6 + 6 \le 14 + 6$$

$$z \le 20$$

$$(-\infty, 20]$$

57.
$$2x+4+6x>2-3x+2$$
$$8x+4>4-3x$$
$$8x+4+3x>4-3x+3x$$
$$11x+4>4$$
$$11x+4-4>4-4$$
$$11x>0$$
$$x>0$$
$$(0,\infty)$$

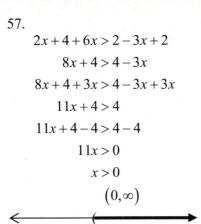

59.
$$-3x-1\le 5$$
$$-3x-1+1\le 5+1$$
$$-3x\le 6$$
$$x\ge -2$$
$$[-2,\infty)$$

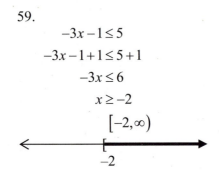

61.
$$-5t+3\le 5$$
$$-5t+3-3\le 5-3$$
$$-5t\le 2$$
$$t\ge -\frac{2}{5}$$
$$\left[-\frac{2}{5},\infty\right)$$

63.
$$-7y+5>-5y-1$$
$$-7y+5+5y>-5y-1+5y$$
$$-2y+5>-1$$
$$-2y+5-5>-1-5$$
$$-2y>-6$$
$$y<3$$
$$(-\infty,3)$$

65.
$$t+1-3t\ge t-20$$
$$-2t+1\ge t-20$$
$$-2t+1-t\ge t-20-t$$
$$-3t+1\ge -20$$
$$-3t+1-1\ge -20-1$$
$$-3t\ge -21$$
$$t\le 7$$
$$(-\infty,7]$$

67.
$$2(5x-6)>4x-15+6x$$
$$10x-12>10x-15$$
$$10x-12-10x>10x-15-10x$$
$$-12>-15$$
$$(-\infty,\infty);\ \mathbb{R}$$

69.
$$\frac{3b+7}{3}\le \frac{2b-9}{2}$$
$$6\left(\frac{3b+7}{3}\right)\le 6\left(\frac{2b-9}{2}\right)$$
$$2(3b+7)\le 3(2b-9)$$
$$6b+14-6b\le 6b-27-6b$$
$$14\nleq -27$$
No Solution; $\varnothing$

TRY IT YOURSELF

71.
$$2r-5=1-r$$
$$2r-5+r=1-r+r$$
$$3r-5=1$$
$$3r-5+5=1+5$$
$$3r=6$$
$$r=2$$

73.

$$0.2(a-5)-0.1(3a+1)=0$$
$$10\left[0.2(a-5)-0.1(3a+1)\right]=10\left[0\right]$$
$$2(a-5)-(3a+1)=0$$
$$2a-10-3a-1=0$$
$$-a-11=0$$
$$-a-11+11=0+11$$
$$-a=11$$
$$a=-11$$

75.

$$-\frac{4}{5}s=2$$
$$5\left(-\frac{4}{5}s\right)=5(2)$$
$$-4s=10$$
$$s=-\frac{10}{4}$$
$$s=-\frac{5}{2}$$

77.

$$\frac{1}{2}(3y+2)-\frac{5}{8}=\frac{3}{4}y$$
$$8\left[\frac{1}{2}(3y+2)-\frac{5}{8}\right]=8\left[\frac{3}{4}y\right]$$
$$4(3y+2)-5=6y$$
$$12y+8-5=6y$$
$$12y+3=6y$$
$$12y+3-12y=6y-12y$$
$$3=-6y$$
$$-\frac{1}{2}=y$$

79.

$$8x+3(2-x)=5x+6$$
$$8x+6-3x=5x+6$$
$$5x+6=5x+6$$
$$5x+6-5x=5x+6-5x$$
$$6=6$$

All real numbers; $\mathbb{R}$

identity

81.

$$12+3(x-4)-21=5\left[5-4(4-x)\right]$$
$$12+3x-12-21=5\left[5-16+4x\right]$$
$$3x-21=5\left[-11+4x\right]$$
$$3x-21=-55+20x$$
$$3x-21-20x=-55+20x-20x$$
$$-17x-21=-55$$
$$-17x-21+21=-55+21$$
$$-17x=-34$$
$$x=2$$

83.

$$\frac{3+p}{3}-4p=1-\frac{p+7}{2}$$
$$6\left(\frac{3+p}{3}-4p\right)=6\left(1-\frac{p+7}{2}\right)$$
$$2(3+p)-24p=6-3(p+7)$$
$$6+2p-24p=6-3p-21$$
$$6-22p=-15-3p$$
$$6-22p+3p=-15-3p+3p$$
$$6-19p=-15$$
$$6-19p-6=-15-6$$
$$-19p=-21$$
$$p=\frac{21}{19}$$

85.

$$5x+10=5x$$
$$5x+10-5x=5x-5x$$
$$10\neq0$$

No solution; $\varnothing$

contradiction

87.

$$0.06(a+200)+0.1a=172$$
$$0.06a+12+0.1a=172$$
$$12+0.16a=172$$
$$0.16a=160$$
$$a=1,000$$

Section 8.1

89.

$$-4\big[p-(3-p)\big]=3(6p-2)$$
$$-4\big[p-3+p\big]=18p-6$$
$$-4\big[2p-3\big]=18p-6$$
$$-8p+12=18p-6$$
$$-8p+12-18p=18p-6-18p$$
$$-26p+12=-6$$
$$-26p+12-12=-6-12$$
$$-26p=-18$$
$$p=\frac{18}{26}$$
$$p=\frac{9}{13}$$

91.

$$-3(a+2)>2(a+1)$$
$$-3a-6>2a+2$$
$$-3a-6-2a>2a+2-2a$$
$$-5a-6>2$$
$$-5a-6+6>2+6$$
$$-5a>8$$
$$a<-\frac{8}{5}$$
$$\left(-\infty,-\frac{8}{5}\right)$$

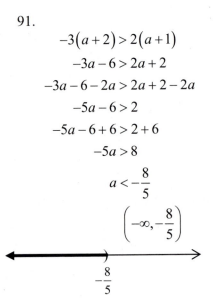

93.

$$\frac{x-7}{2}-\frac{x-1}{5}\le-\frac{x}{4}$$
$$20\left(\frac{x-7}{2}-\frac{x-1}{5}\right)\le20\left(-\frac{x}{4}\right)$$
$$10(x-7)-4(x-1)\le-5x$$
$$10x-70-4x+4\le-5x$$
$$6x-66\le-5x$$
$$6x-66-6x\le-5x-6x$$
$$-66\le-11x$$
$$6\ge x$$
$$x\le6$$
$$(-\infty,6]$$

95.

$$5(2n+2)-n>3n-3(1-2n)$$
$$10n+10-n>3n-3+6n$$
$$9n+10>9n-3$$
$$9n+10-9n>9n-3-9n$$
$$10>-3$$
$$(-\infty,\infty);\ \mathbb{R}$$

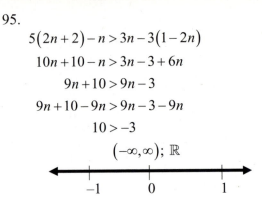

97.

$$0.4x+0.4\le0.1x+0.85$$
$$100(0.4x+0.4)\le100(0.1x+0.85)$$
$$40x+40\le10x+85$$
$$40x+40-10x\le10x+85-10x$$
$$30x+40\le85$$
$$30x+40-40\le85-40$$
$$30x\le45$$
$$x\le1.5$$
$$(-\infty,1.5]$$

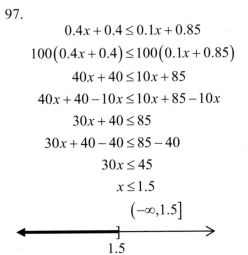

99.

$$\frac{1}{2}y+2\ge\frac{1}{3}y-4$$
$$6\left(\frac{1}{2}y+2\right)\ge6\left(\frac{1}{3}y-4\right)$$
$$3y+12\ge2y-24$$
$$3y+12-2y\ge2y-24-2y$$
$$y+12\ge-24$$
$$y+12-12\ge-24-12$$
$$y\ge-36$$
$$[-36,\infty)$$

101.

$$7 < \frac{5}{3}a - 3$$

$$3(7) < 3\left(\frac{5}{3}a - 3\right)$$

$$21 < 5a - 9$$

$$21 + 9 < 5a - 9 + 9$$

$$30 < 5a$$

$$6 < a$$

$$a > 6$$

$$(6, \infty)$$

103. a.

$$\frac{1}{2}(6x + 8) - 10 - \frac{2}{3}(6x - 9)$$

$$= 3x + 4 - 10 - 4x + 6$$

$$= -x$$

b.

$$\frac{1}{2}(6x + 8) - 10 = -\frac{2}{3}(6x - 9)$$

$$3x + 4 - 10 = -4x + 6$$

$$3x - 6 = -4x + 6$$

$$3x - 6 + 4x = -4x + 6 + 4x$$

$$7x - 6 = 6$$

$$7x - 6 + 6 = 6 + 6$$

$$7x = 12$$

$$x = \frac{12}{7}$$

105. a.

$$12x - 33.16 \leq 5.84$$

$$12x - 33.16 + 33.16 \leq 5.84 + 33.16$$

$$12x \leq 39$$

$$x \leq 3.25$$

$$(-\infty, 3.25]$$

b.

$$12x - 33.16 > 5.84$$

$$(3.25, \infty)$$

APPLICATIONS

107. SPRING TOURS
Let x = the # of students they supervise.

$$1,810 - 15.5x = 1,500$$

$$1,810 - 15.5x - 1,810 = 1,500 - 1,810$$

$$-15.5x = -310$$

$$x = 20$$

He or she must supervise 20 students to get the reduced cost of $1,500.

109. MOVING EXPENSES
Let x = the number of miles he drives the truck.

$$41.50 + 0.35x = 150$$

$$100(41.50 + 0.35x) = 100(150)$$

$$4,150 + 35x = 15,000$$

$$4,150 + 35x - 4,150 = 15,000 - 4,150$$

$$35x = 10,850$$

$$x = 310$$

He can drive 310 miles.

111. FENCING PENS
There are 5 sides of the pen that are x ft and 2 sides that are $(x + 5)$ ft. He has 150 feet of fencing.

$$5x + 2(x + 5) = 150$$

$$5x + 2x + 10 = 150$$

$$7x + 10 = 150$$

$$7x + 10 - 10 = 150 - 10$$

$$7x = 140$$

$$x = 20$$

The width of the pen is $x = 20$.
The length of the pen is
$x + (x + 5) = 2x + 5 = 2(20) + 5 = 45$.
The dimensions are 20 ft by 45 ft.

113. WEBMASTER
Let x = number of views on April 30.

$$\frac{650,568,999 + x}{30} = 22,000,000$$

$$30\left(\frac{650,568,999 + x}{30}\right) = 30(22,000,000)$$

$$650,568,999 + x = 660,000,000$$

$$x = 9,431,001$$

The site would need 9,431,001 views on April 30.

Section 8.1

115. FUND-RAISING

Let x = the number of hours the tank can be rented.

$$85 + 19.50x = 185$$

$$85 + 19.50x - 85 = 185 - 85$$

$$19.50x = 100$$

$$x \approx 5$$

He can rent the tank for a total of 5 hours + the 3 first hours = 8 hours.

117. SCHEDULING EQUIPMENT

Let x = the number of hours the operator can use the backhoe.

$$300x + 500(40 - x) \leq 18,500$$

$$300x + 20,000 - 500x \leq 18,500$$

$$-200x + 20,000 \leq 18,500$$

$$-200x \leq -1,500$$

$$x \leq 7.5$$

$$x = 7$$

They can use the backhoe 7 hours.

WRITING

119. Answers will vary.

REVIEW

121.

$$\left(\frac{t^3 t^5 t^{-6}}{t^2 t^{-4}}\right)^{-3} = \left(\frac{t^{3+5+(-6)}}{t^{2+(-4)}}\right)^{-3}$$

$$= \left(\frac{t^2}{t^{-2}}\right)^{-3}$$

$$= \left(t^{2-(-2)}\right)^{-3}$$

$$= \left(t^4\right)^{-3}$$

$$= \frac{1}{\left(t^4\right)^3}$$

$$= \frac{1}{t^{4\cdot 3}}$$

$$= \frac{1}{t^{12}}$$

CHALLENGE PROBLEMS

123. Let $x = 4$.

$$k + 3(4) - 6 = 3k(4) - k + 16$$

$$k + 12 - 6 = 12k - k + 16$$

$$k + 6 = 11k + 16$$

$$6 = 10k + 16$$

$$-10 = 10k$$

$$-1 = k$$

125. a.

$$\frac{1}{3} > \frac{1}{x}$$

$$\frac{1}{3} > \frac{1}{-1}$$

$$\frac{1}{3} > -1$$

b. Multiplying both sides of an inequality by the same positive number does not change the solutions. In this case, however, both sides were multiplied by $3x$, an expression which, depending on the value of x, could be negative or 0.

VOCABULARY

1. A set of ordered pairs is called a **relation**. The set of all first components of the ordered pairs is called the **domain** and the set of all second components is called the **range**.

3. Given a relation in x and y, if to each value of x in the domain there corresponds exactly one value of y in the range, y is said to be a **function** of x. We call x the independent **variable** and y the **dependent** variable.

5. A **linear** function is a function that can be written in the form $f(x) = mx + b$. A polynomial function is a function whose equation is defined by a polynomial in **one** variable.

CONCEPTS

7. a. {(2000, 63), (2001, 56), (2002, 54), (2003, 50), (2004, 52), (2005, 51), (2006, 51)}
    b. D: {2000, 2001, 2002, 2003, 2004, 2005, 2006}
        R: {50, 51, 52, 54, 56, 63}
    c.

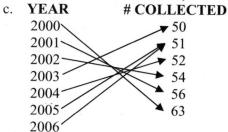

9. If $x = -4$, then the denominator of $\dfrac{1}{x+4}$ is 0 and undefined.

11. a. $(0, -4)$
    b. $\left(-\dfrac{2}{3}, 0\right)$

NOTATION

13. a. We read $f(x) = 5x - 6$ as "f [of] x is equal to $5x$ minus 6."
    b. We read $g(t) = t + 9$ as "g [of] t is equal to t plus 9."

15.
    *   If $y = 5x + 1$, find the value of y when $x = 8$.
    *   If $f(x) = 5x + 1$, find $\boxed{f(8)}$.

17. When graphing the function $f(x) = -x + 5$, the vertical axis of the rectangular coordinate system can be labeled $\boxed{f(x)}$ or $\boxed{y}$.

GUIDED PRACTICE

19. D:{−5, −1, 7, 8}; R:{−11, −6, −1, 3}

21. D:{−23, 0, 7}; R:{1, 35}

23. yes

25. no; (4, 2), (4, 4) and (4, 6)

27. no; (3, 4) and (3, −4) or (4, 3) and (4, −3)

29. yes

31. yes

33. no; (−1, 0) and (−1, 2)

35. yes

37. yes

39. no; (1, 1) and (1, −1)

41. yes

43. yes

45. no; (1, 1), (1, −1)

47. $f(x) = 3x$
    $\quad f(3) = 3(3) \qquad\qquad f(-1) = 3(-1)$
    $\qquad\quad = 9 \qquad\qquad\qquad\quad = -3$

Section 8.2

49. $f(x) = 2x - 3$

$$f(3) = 2(3) - 3 \qquad f(-1) = 2(-1) - 3$$
$$= 6 - 3 \qquad\qquad = -2 - 3$$
$$= 3 \qquad\qquad\quad = -5$$

51. $g(x) = x^2 - 10$

$$g(2) = \mathbf{2}^2 - 10 \qquad g(3) = \mathbf{3}^2 - 10$$
$$= 4 - 10 \qquad\qquad = 9 - 10$$
$$= -6 \qquad\qquad\quad = -1$$

53. $g(x) = -x^3 + x$

$$g(2) = -\mathbf{2}^3 + \mathbf{2} \qquad g(3) = -\mathbf{3}^3 + \mathbf{3}$$
$$= -8 + 2 \qquad\qquad = -27 + 3$$
$$= -6 \qquad\qquad\quad = -24$$

55. $g(x) = (x + 1)^2$

$$g(2) = (\mathbf{2} + 1)^2 \qquad g(3) = (\mathbf{3} + 1)^2$$
$$= 3^2 \qquad\qquad\quad = 4^2$$
$$= 9 \qquad\qquad\quad = 16$$

57. $g(x) = 2x^2 - x + 1$

$$g(2) = 2(\mathbf{2})^2 - \mathbf{2} + 1 \quad g(3) = 2(\mathbf{3})^2 - \mathbf{3} + 1$$
$$= 2(4) - 2 + 1 \qquad = 2(9) - 3 + 1$$
$$= 8 - 2 + 1 \qquad\quad = 18 - 3 + 1$$
$$= 7 \qquad\qquad\qquad = 16$$

59. $h(x) = |x| + 2$

$$h(5) = |\mathbf{5}| + 2 \qquad h(-2) = |\mathbf{-2}| + 2$$
$$= 5 + 2 \qquad\qquad = 2 + 2$$
$$= 7 \qquad\qquad\quad = 4$$

61. $h(x) = \dfrac{1}{x + 3}$

$$h(5) = \dfrac{1}{\mathbf{5} + 3} \qquad h(-2) = \dfrac{1}{\mathbf{-2} + 3}$$
$$= \dfrac{1}{8} \qquad\qquad\quad = \dfrac{1}{1}$$
$$\qquad\qquad\qquad\qquad = 1$$

63. $h(x) = \dfrac{x}{x - 3}$

$$h(5) = \dfrac{\mathbf{5}}{\mathbf{5} - 3} \qquad h(-2) = \dfrac{\mathbf{-2}}{\mathbf{-2} - 3}$$
$$= \dfrac{5}{2} \qquad\qquad\qquad = \dfrac{-2}{-5}$$
$$\qquad\qquad\qquad\qquad\quad = \dfrac{2}{5}$$

65. $h(x) = \dfrac{x^2 + 2x - 35}{x^2 + 5x + 6}$

$$h(5) = \dfrac{(\mathbf{5})^2 + 2(\mathbf{5}) - 35}{(\mathbf{5})^2 + 5(\mathbf{5}) + 6}$$
$$= \dfrac{25 + 10 - 35}{25 + 25 + 6}$$
$$= \dfrac{0}{56}$$
$$= 0$$

$$h(-2) = \dfrac{(\mathbf{-2})^2 + 2(\mathbf{-2}) - 35}{(\mathbf{-2})^2 + 5(\mathbf{-2}) + 6}$$
$$= \dfrac{4 - 4 - 35}{4 - 10 + 6}$$
$$= \dfrac{-35}{0}$$
$$= \text{undefined}$$

67. $f(t) = |t - 2|$

t	$f(x) = |t - 2|$
-1.7	$|\mathbf{-1.7} - 2| = |-3.7|$ $= 3.7$
0.9	$|\mathbf{0.9} - 2| = |-1.1|$ $= 1.1$
5.4	$|\mathbf{5.4} - 2| = |3.4|$ $= 3.4$

69. $g(a) = a^3$

Input	Output
$-\dfrac{3}{4}$	$\left(-\dfrac{3}{4}\right)^3 = -\dfrac{3}{4}\left(-\dfrac{3}{4}\right)\left(-\dfrac{3}{4}\right)$ $= -\dfrac{27}{64}$
$\dfrac{1}{6}$	$\left(\dfrac{1}{6}\right)^3 = \dfrac{1}{6}\left(\dfrac{1}{6}\right)\left(\dfrac{1}{6}\right)$ $= \dfrac{1}{216}$
$\dfrac{5}{2}$	$\left(\dfrac{5}{2}\right)^3 = \dfrac{5}{2}\left(\dfrac{5}{2}\right)\left(\dfrac{5}{2}\right)$ $= \dfrac{125}{8}$

71. $g(x) = 2x$

$g(w) = 2(w)$ $\qquad$ $g(w+1) = 2(w+1)$
$\quad = 2w$ $\qquad\qquad\quad = 2w + 2$

73. $g(x) = 3x - 5$

$g(w) = 3(w) - 5$ $\quad$ $g(w+1) = 3(w+1) - 5$
$\quad = 3w - 5$ $\qquad\qquad\quad = 3w + 3 - 5$
$\qquad\qquad\qquad\qquad\qquad\quad = 3x - 2$

75.

$$f(x) = -2x + 5$$
$$f(x) = 5$$
$$-2x + 5 = 5$$
$$-2x + 5 - 5 = 5 - 5$$
$$-2x = 0$$
$$x = 0$$

77.

$$f(x) = \frac{3}{2}x - 2$$
$$f(x) = -\frac{1}{2}$$
$$\frac{3}{2}x - 2 = -\frac{1}{2}$$
$$2\left(\frac{3}{2}x - 2\right) = 2\left(-\frac{1}{2}\right)$$
$$3x - 4 = -1$$
$$3x - 4 + 4 = -1 + 4$$
$$3x = 3$$
$$x = 1$$

79. a. The set of real numbers
b. The set of real numbers except 4

81. a. The set of real numbers
b. The set of real numbers except $-\dfrac{1}{2}$

83. $f(x) = 2x - 1$

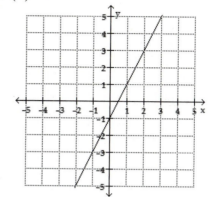

85. $f(x) = -\dfrac{3}{2}x - 3$

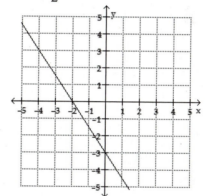

87. $f(x) = x$

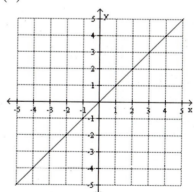

Section 8.2

89. $f(x) = -4$

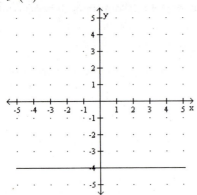

91. $g(x) = 0.75x$

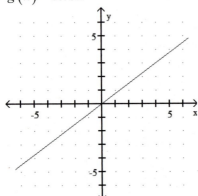

93. $s(x) = \dfrac{7}{8}x + 2$

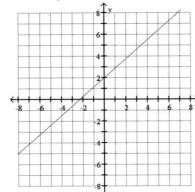

95. $m = 5, b = -3$

$$y = mx + b$$
$$y = 5x - 3$$
$$f(x) = 5x - 3$$

97. $m = \dfrac{1}{5}, x_1 = 10, y_1 = 1$

$$y - y_1 = m(x - x_1)$$
$$y - 1 = \frac{1}{5}(x - 10)$$
$$y - 1 = \frac{1}{5}x - 2$$
$$y - 1 + 1 = \frac{1}{5}x - 2 + 1$$
$$y = \frac{1}{5}x - 1$$
$$f(x) = \frac{1}{5}x - 1$$

99. First find the slope of the line:

$$m = \frac{y_2 - y_1}{x_2 - x_1}$$
$$= \frac{1 - 7}{-2 - 1}$$
$$= \frac{-6}{-3}$$
$$= 2$$

Use $m = 2, x_1 = 1, y_1 = 7$

$$y - y_1 = m(x - x_1)$$
$$y - 7 = 2(x - 1)$$
$$y - 7 = 2x - 2$$
$$y - 7 + 7 = 2x - 2 + 7$$
$$y = 2x + 5$$
$$f(x) = 2x + 5$$

101. The slope of the given line is $-\dfrac{2}{3}$. Since the lines are parallel, then use the same slope for the new line.

Use $m = -\dfrac{2}{3}, x_1 = 3 \; y_1 = 0$

$$y - y_1 = m(x - x_1)$$
$$y - 0 = -\frac{2}{3}(x - 3)$$
$$y = -\frac{2}{3}x + 2$$
$$f(x) = -\frac{2}{3}x + 2$$

103. The slope of the given line is $-\frac{1}{6}$. The slope of the line perpendicular is the reciprocal with the opposite sign so slope would be 6.
Use $m = 6$, $x_1 = 1$ $y_1 = 2$
$$y - y_1 = m(x - x_1)$$
$$y - 2 = 6(x - 1)$$
$$y - 2 = 6x - 6$$
$$y - 2 + 2 = 6x - 6 + 2$$
$$y = 6x - 4$$
$$f(x) = 6x - 4$$

105. A horizontal line through the point $(-8, 12)$ has the equation $f(x) = 12$.

APPLICATIONS

107. WEBMASTER
$$w(t) = 12.67t + 29.15$$
$$w(6) = 12.67(6) + 29.15$$
$$= 76.02 + 29.15$$
$$= 105.17 \text{ million websites}$$

109. CONCESSIONAIRES
a. $p(b) = 4.75b - 125$
b. $\quad p(b) = 4.75b - 125$
$$p(110) = 4.75(110) - 125$$
$$= 522.50 - 125$$
$$= \$397.50$$

111. NURSES
Write the two ordered pairs using the first coordinate as the years after 2000 and the second coordinate as the number of nurses: (5, 2,175,500) and (15, 2,586,500).
Find the slope of the line between those points:
$$m = \frac{y_2 - y_1}{x_2 - x_1}$$
$$= \frac{2,586,500 - 2,175,500}{15 - 5}$$
$$= \frac{411,000}{10}$$
$$= 41,100$$

a. Use $m = 41,100$, $x_1 = 5$, $y_1 = 2,175,500$
$$y - y_1 = m(x - x_1)$$
$$y - 2,175,500 = 41,100(x - 5)$$
$$y - 2,175,500 = 41,100x - 205,500$$
$$y = 41,100x + 1,970,000$$
$$N(t) = 41,100t + 1,970,000$$

b. Let $t = 25$ since 2025 is 25 years after 2000.
$$N(25) = 41,100(25) + 1,970,000$$
$$= 1,027,500 + 1,970,000$$
$$= 2,997,500$$

113. BREATHING CAPACITY
Write the information as ordered pairs: (35, 90) and (55, 66).
a. Find the slope of the line between those points:
$$m = \frac{y_2 - y_1}{x_2 - x_1}$$
$$= \frac{66 - 90}{55 - 35}$$
$$= \frac{-24}{20}$$
$$= -1.2$$
Use $m = -1.2$, $x_1 = 55$, $y_1 = 66$
$$y - y_1 = m(x - x_1)$$
$$y - 66 = -1.2(x - 55)$$
$$y - 66 = -1.2x + 66$$
$$y = -1.2x + 132$$
$$L(a) = -1.2a + 132$$

b. Let $a = 80$.
$$L(80) = -1.2(80) + 132$$
$$= -96 + 132$$
$$= 36\%$$

115. TAXES
a. $\quad T(a) = 837.50 + 0.15(a - 8,375)$
$$T(25,000) = 837.50 + 0.15(25,000 - 8,375)$$
$$= 837.50 + 0.15(16,625)$$
$$= 837.50 + 2,493.75$$
$$= \$3,331.25$$
The tax on an adjusted gross income of $25,000 is $3,331.25.
b.
$$T(a) = 4,681.25 + 0.25(a - 34,000)$$

- 513 -

117. ROLLER COASTERS

$$f(x) = 0.001x^3 - 0.12x^2 + 3.6x + 10$$

$$f(0) = 0.001(0)^3 - 0.12(0)^2 + 3.6(0) + 10$$
$$= 0 - 0 + 0 + 10$$
$$= 10 \text{ m}$$

$$f(20) = 0.001(20)^3 - 0.12(20)^2 + 3.6(20) + 10$$
$$= 0.001(8,000) - 0.12(400) + 3.6(20) + 10$$
$$= 8 - 48 + 72 + 10$$
$$= 42 \text{ m}$$

$$f(40) = 0.001(40)^3 - 0.12(40)^2 + 3.6(40) + 10$$
$$= 0.001(64,000) - 0.12(1,600) + 3.6(40) + 10$$
$$= 64 - 192 + 144 + 10$$
$$= 26 \text{ m}$$

$$f(60) = 0.001(60)^3 - 0.12(60)^2 + 3.6(60) + 10$$
$$= 0.001(216,000) - 0.12(3,600) + 3.6(60) + 10$$
$$= 216 - 432 + 216 + 10$$
$$= 10 \text{ m}$$

119. RAIN GUTTERS

$$f(x) = -240x^2 + 1,440x$$

$$f(3) = -240(3)^2 + 1,440(3)$$
$$= -240(9) + 1,440(3)$$
$$= -2,160 + 4,320$$
$$= 2,160 \text{ in.}^3$$

CHALLENGE PROBLEMS

127.

$$f(8) = 4(8) + 6$$
$$= 32 + 6$$
$$= 38$$
$$g(8) = 4$$
$$h(8) = 6$$

$$\frac{f(8) + g(8)}{h(8)} = \frac{38 + 4}{6}$$
$$= \frac{42}{6}$$
$$= 7$$

WRITING

121. Answers will vary.

123. Answers will vary.

REVIEW

125.

$$-2(t + 4) + 5t + 1 = 3(t - 4) + 7$$
$$-2t - 8 + 5t + 1 = 3t - 12 + 7$$
$$3t - 7 = 3t - 5$$
$$3t - 7 - 3t = 3t - 5 - 3t$$
$$-7 \neq -5$$
$$\varnothing$$

No Solution; contradiction

VOCABULARY

1. Functions whose graphs are not lines are called **nonlinear** functions.

3. The set of **nonnegative** real numbers is the set of real numbers greater than or equal to 0.

CONCEPTS

5. a. the squaring function

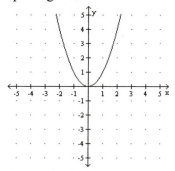

b. the cubing function

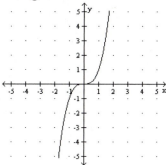

c. the absolute value function

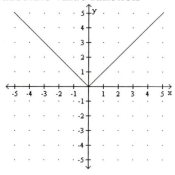

7. The graph of $g(x) = -x^2$ is the **reflection** of the graph of $f(x) = x^2$ about the x-axis.

9. a.

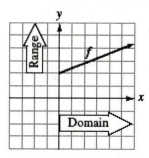

b. D: the set of nonnegative real numbers
   R: the set of all real numbers greater than or equal to 2

11. a. h is the x-value of the vertex: 4
    b. h is the x-value of the vertex: 0
    c. h is the x-value of the vertex: –2

13. a. (–2, 4) and (–2, –4)
    b. No, since the x-value of –2 corresponds to more than one y-value, 4 and –4.

NOTATION

15. a. The graph of $f(x) = (x + 4)^3$ is the same as the graph of $f(x) = x^3$ except that it is shifted **4** units to the **left**.
    b. The graph of $f(x) = x^3 + 4$ is the same as the graph of $f(x) = x^3$ except that it is shifted **4** units **up**.

GUIDED PRACTICE

17. a. –4
    b. 0
    c. 2
    d. –1

19. a. 2
    b. 2
    c. –1
    d. –3 and 1

21. D: the set of all real numbers
    R: the set of all real numbers

23. D: the set of all real numbers
    R: the set of all real numbers less than or
       equal to 5

25. D: the set of all real numbers
    R: the set of all real numbers greater than
       or equal to –4

27. D: the set of nonnegative real numbers
    R: the set of nonnegative real numbers

29. $f(x) = x^2 + 2$

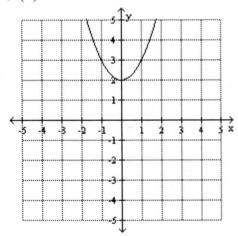

    D: the set of all real numbers
    R: the set of all real numbers greater than
       or equal to 2

31. $f(x) = x^3 - 3$

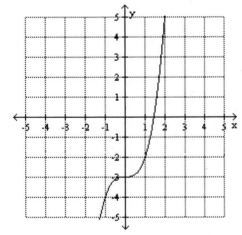

    D: the set of all real numbers
    R: the set of all real numbers

33. $f(x) = |x - 1|$

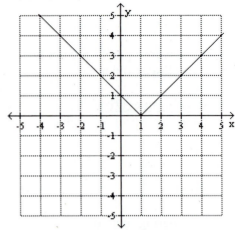

    D: the set of all real numbers
    R: the set of nonnegative real numbers

35. $f(x) = (x + 4)^2$

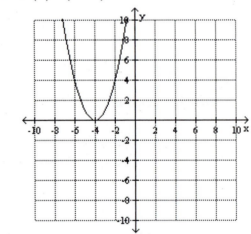

    D: the set of all real numbers
    R: the set of nonnegative real numbers

37. $g(x) = |x| - 2$

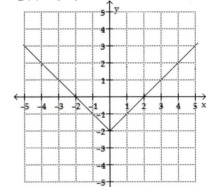

    D: the set of real numbers
    R: the set of real numbers greater than
       or equal to –2

39. $f(x) = (x + 1)^3$

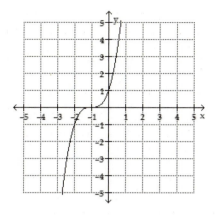

D: the set of real numbers
R: the set of real numbers

41. $g(x) = x^2 - 3$

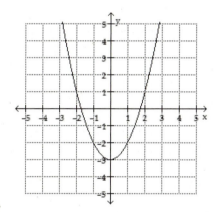

D: the set of real numbers
R: the set of real numbers greater than
    or equal to –3

43. $g(x) = (x - 4)^3$

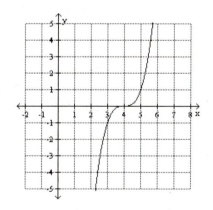

D: the set of real numbers
R: the set of real numbers

45. $g(x) = x^3 + 4$

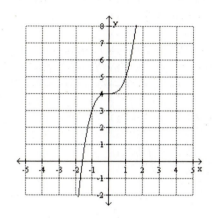

D: the set of real numbers
R: the set of real numbers

47. $g(x) = (x + 4)^2$

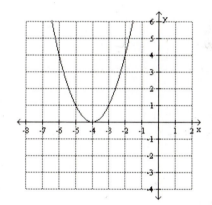

D: the set of real numbers
R: the set of nonnegative real numbers

49. $g(x) = |x - 2| - 1$

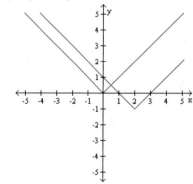

Section 8.3

51. $g(x) = (x+1)^3 - 2$

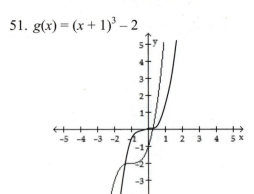

59. $g(x) = -x^2$

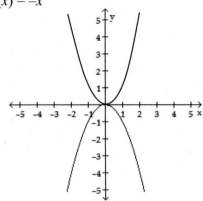

53. $g(x) = (x-2)^2 + 4$

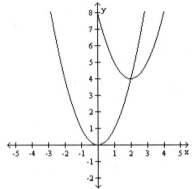

61. $g(x) = -|x+5|$

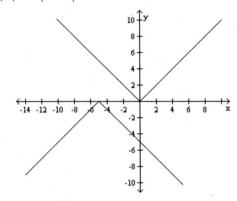

55. $g(x) = |x+3| + 5$

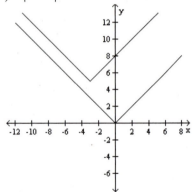

63. $g(x) = -x^2 + 3$

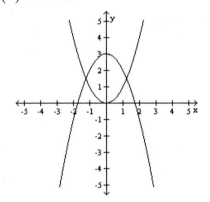

57. $g(x) = -x^3$

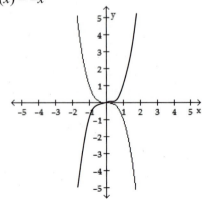

65. a. $f(1) = 15$
    b. $f(-3) = 5$
    c. $x = 0, -2, -4$
    d. $D = (-\infty, \infty); \ R = (-\infty, \infty)$

67. a. $g(1) = 3$
    b. $g(-4) = -5$
    c. $x = -2, 2$
    d. $D: = (-\infty, \infty); \ R = (-\infty, 4]$

69. no; $(0, 2), (0, -2)$

71. yes

73. yes

75. no; (3, 0), (3, 1)

77. $f(x) = x^2 + 8$

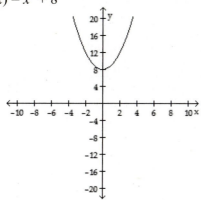

79. $f(x) = |x + 5|$

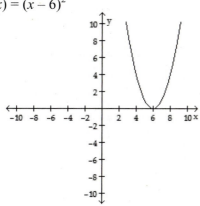

81. $f(x) = (x - 6)^2$

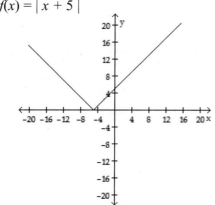

83. $f(x) = x^3 + 8$

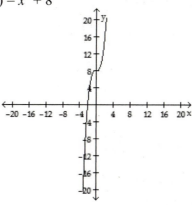

APPLICATIONS

85. OPTICS
$f(x) = |x|$

87. CENTER OF GRAVITY
a parabola

89. LABOR STATISTICS
   a. $J(9) \approx 12$; In 2009, there were about 12 million manufacturing jobs in the U.S.
   b. $x = 2$; In 2002, there were about 16.5 million manufacturing jobs in the U.S.

WRITING

91. Answers will vary.

93. Answers will vary.

95. Answers will very.

REVIEW

97.
$$T - W = ma$$
$$T - W - T = ma - T$$
$$-W = ma - T$$
$$\frac{-W}{-1} = \frac{ma - T}{-1}$$
$$W = -ma + T$$
$$W = T - ma$$

Section 8.3

99.

$$s = \frac{1}{2}gt^2 + vt$$

$$s - vt = \frac{1}{2}gt^2 + vt - vt$$

$$s - vt = \frac{1}{2}gt^2$$

$$2(s - vt) = 2\left(\frac{1}{2}gt^2\right)$$

$$2(s - vt) = gt^2$$

$$\frac{2(s - vt)}{t^2} = \frac{gt^2}{t^2}$$

$$\frac{2(s - vt)}{t^2} = g$$

CHALLENGE PROBLEMS

101. $f(x) = \begin{cases} |x| & \text{for } x \geq 0 \\ x^3 & \text{for } x < 0 \end{cases}$

x	y
–3	$(-3)^3 = -27$
–2	$(-2)^3 = -8$
–1	$(-1)^3 = -1$
0	$|0| = 0$
1	$|1| = 1$
2	$|2| = 2$
3	$|3| = 3$

105. $f(x) = 2x^3 - 3x^2 - 11x + 6$

x	$f(x)$
–3	$2(-3)^3 - 3(-3)^2 - 11(-3) + 6$ $= 2(-27) - 3(9) + 33 + 6$ $= -42$
–2	$2(-2)^3 - 3(-2)^2 - 11(-2) + 6$ $= 2(-8) - 3(4) + 22 + 6$ $= 0$
–1	$2(-1)^3 - 3(-1)^2 - 11(-1) + 6$ $= 2(-1) - 3(1) + 11 + 6$ $= 12$
0	$2(0)^3 - 3(0)^2 - 11(0) + 6$ $= 2(0) - 3(0) - 0 + 6$ $= 6$
1	$2(1)^3 - 3(1)^2 - 11(1) + 6$ $= 2(1) - 3(1) - 11 + 6$ $= -6$
2	$2(2)^3 - 3(2)^2 - 11(2) + 6$ $= 2(8) - 3(4) - 22 + 6$ $= -12$
3	$2(3)^3 - 3(3)^2 - 11(3) + 6$ $= 2(27) - 3(9) - 33 + 6$ $= 0$
4	$2(4)^3 - 3(4)^2 - 11(4) + 6$ $= 2(64) - 3(16) - 44 + 6$ $= 42$

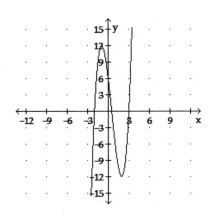

D: $(-\infty, \infty)$;   R: $(-\infty, \infty)$

103   a.   D: the set of all real numbers from –4 to 4; R: {–2, 1, 3}

b.   D: the set of all real numbers except –3 and 1; R: the set of all real numbers

SECTION 8.4

VOCABULARY

1. The **intersection** of two sets is the set of elements that are common to both sets and the **union** of two sets is the set of elements that are in one set, or the other, or both.

3. $-6 < x + 1 \le 1$ is a **double** linear inequality.

CONCEPTS

5. a. The solution set of a compound inequality containing the word *and* includes all numbers that make **both** inequalities true.
   b. The solution set of a compound inequality containing the word *or* includes all numbers that make **one**, or the other, or **both** inequalities true.

7. a. When solving a compound inequality containing the word *and*, the solution set is the **intersection** of the solution sets of the inequalities.
   b. When solving a compound inequality containing the word *or*, the solution set is the **union** of the solution sets of the inequalities.

9. a. Let $x = -3$.

$$\frac{-3}{3} + 1 \overset{?}{\ge} 0 \quad \text{and} \quad 2(-3) - 3 \overset{?}{<} -10$$

$$-1 + 1 \overset{?}{\ge} 0 \qquad\qquad -6 - 3 \overset{?}{<} -10$$

$$0 \overset{?}{\ge} 0 \qquad\qquad -9 < -10$$

$$\text{true} \qquad\qquad\qquad \text{false}$$

No. Since both inequalities are not true, -3 is **not a solution** of the compound inequality.

   b. Let $x = -3$.

$$2(-3) \overset{?}{\le} 0 \quad \text{or} \quad -3(-3) \overset{?}{<} -5$$

$$-6 \overset{?}{\le} 0 \qquad\qquad 9 \overset{?}{<} -5$$

$$\text{true} \qquad\qquad\qquad \text{false}$$

Yes. Since one inequality is true, -3 **is a solution** of the compound inequality.

11. a. $[-2, 1)$
    b. $[2, 2]$
    c. $\varnothing$

NOTATION

13. We read $\cup$ as **union** and $\cap$ as **intersection**.

15. all real numbers

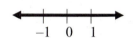

GUIDED PRACTICE

17. $\{4, 6\}$

19. $\{-3, 1, 2\}$

21. $\{-3, -1, 0, 1, 2, 4, 6, 8, 10\}$

23. $\{-3, 0, 1, 2, 3, 4, 5, 6, 8\}$

25. $x > -2$ and $x \le 5$
    $(-2, 5]$

27. $\quad 2x - 1 > 3 \quad$ and $\quad x + 8 \le 11$
    $\quad 2x - 1 + 1 > 3 + 1 \qquad x + 8 - 8 \le 11 - 8$
    $\qquad 2x > 4 \qquad\qquad\qquad x \le 3$
    $\qquad \dfrac{2x}{2} > \dfrac{4}{2}$
    $\qquad x > 2$

    $(2, 3]$

29. $\quad 6x + 1 < 5x - 3 \quad$ and $\quad \dfrac{x}{2} + 9 \le 6$

    $6x + 1 - 5x < 5x - 3 - 5x \qquad \dfrac{x}{2} + 9 - 9 \le 6 - 9$

    $\qquad x + 1 < -3 \qquad\qquad\qquad \dfrac{x}{2} \le -3$

    $\qquad x + 1 - 1 < -3 - 1$

    $\qquad x < -4 \qquad\qquad\qquad 2\left(\dfrac{x}{2}\right) \le 2(-3)$

    $\qquad\qquad\qquad\qquad\qquad x \le -6$

    $(-\infty, -6]$

31. $x + 2 < -\dfrac{1}{3}x$ and $-6x < 9x$

$3(x+2) < 3\left(-\dfrac{1}{3}x\right)$ $\quad$ $-6x - 9x < 9x - 9x$

$3x + 6 < -x$ $\quad\quad\quad$ $-15x < 0$

$3x + 6 - 3x < -x - 3x$ $\quad$ $\dfrac{-15x}{-15} > \dfrac{0}{-15}$

$6 < -4x$ $\quad\quad\quad\quad$ $x > 0$

$\dfrac{6}{-4} > \dfrac{-4x}{-4}$

$-\dfrac{3}{2} > x$

$x < -\dfrac{3}{2}$

No solution; $\varnothing$. There are no numbers that are less than $-\frac{3}{2}$ and greater than 0.

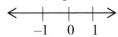

33. $\quad 4 \leq x + 3 \leq 7$

$4 - 3 \leq x + 3 - 3 \leq 7 - 3$

$1 \leq x \leq 4$

$[1, \ 4]$

(number line with solid bracket from 1 to 4)

35. $\quad 0.9 < 2x - 0.7 < 1.5$

$10(0.9) < 10(2x - 0.7) < 10(1.5)$

$9 < 20x - 7 < 15$

$9 + 7 < 20x - 7 + 7 < 15 + 7$

$16 < 20x < 22$

$\dfrac{16}{20} < \dfrac{20x}{20} < \dfrac{22}{20}$

$0.8 < x < 1.1$

$(0.8, \ 1.1)$

(number line with open interval from 0.8 to 1.1)

37. $x \leq -2$ or $x > 6$

$(-\infty, -2] \cup (6, \infty)$

(number line with bracket at −2 and open paren at 6)

39. $\quad x - 3 < -4 \quad$ or $\quad -x + 2 < 0$

$x - 3 + 3 < -4 + 3 \quad -x + 2 - 2 < 0 - 2$

$x < -1 \quad\quad\quad\quad -x < -2$

$\quad\quad\quad\quad\quad \dfrac{-x}{-1} > \dfrac{-2}{-1}$

$\quad\quad\quad\quad\quad\quad x > 2$

$(-\infty, -1) \cup (2, \infty)$

(number line with open parens at −1 and 2)

41. $\quad 3x + 2 < 8 \quad$ or $\quad 2x - 3 > 11$

$3x + 2 - 2 < 8 - 2 \quad 2x - 3 + 3 > 11 + 3$

$3x < 6 \quad\quad\quad\quad 2x > 14$

$x < 2 \quad\quad\quad\quad x > 7$

$(-\infty, 2) \cup (7, \infty)$

(number line with open parens at 2 and 7)

43. $\quad 2x > x + 3 \quad$ or $\quad \dfrac{x}{8} + 1 < \dfrac{13}{8}$

$2x - x > x + 3 - x \quad 8\left(\dfrac{x}{8} + 1\right) < 8\left(\dfrac{13}{8}\right)$

$x > 3 \quad\quad\quad\quad\quad x + 8 < 13$

$\quad\quad\quad\quad\quad\quad x + 8 - 8 < 13 - 8$

$\quad\quad\quad\quad\quad\quad\quad x < 5$

$(-\infty, \infty)$

All real numbers are either greater than 3 or less than 5 or both.

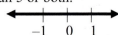

TRY IT YOURSELF

45.

$-4(x + 2) \geq 12 \quad$ or $\quad 3x + 8 < 11$

$-4x - 8 \geq 12 \quad\quad 3x + 8 - 8 < 11 - 8$

$-4x - 8 + 8 \geq 12 + 8 \quad\quad 3x < 3$

$-4x \geq 20 \quad\quad\quad\quad \dfrac{3x}{3} < \dfrac{3}{3}$

$\dfrac{-4x}{-4} \leq \dfrac{20}{-4} \quad\quad\quad\quad x < 1$

$x \leq -5$

$(-\infty, 1)$

(number line with open interval extending left of 1)

47.

$$2.2x < -19.8 \quad \text{and} \quad -4x < 40$$

$$\frac{2.2x}{2.2} < \frac{-19.8}{2.2} \qquad \frac{-4x}{-4} > \frac{40}{-4}$$

$$x < -9 \qquad\qquad x > -10$$

$$-10 < x < -9$$

$$(-10, -9)$$

49.

$$-2 < -b + 3 < 5$$

$$-2 - 3 < -b + 3 - 3 < 5 - 3$$

$$-5 < -b < 2$$

$$\frac{-5}{-1} > \frac{-b}{-1} > \frac{2}{-1}$$

$$5 > b > -2$$

$$-2 < b < 5$$

$$(-2, 5)$$

51.

$$4.5x - 2 > 2.5 \quad \text{or} \quad \frac{1}{2}x \le 1$$

$$4.5x - 2 + 2 > 2.5 + 2 \qquad 2\left(\frac{1}{2}x\right) \le 2(1)$$

$$4.5x > 4.5 \qquad\qquad x \le 2$$

$$\frac{4.5x}{4.5} > \frac{4.5}{4.5}$$

$$x > 1$$

$$(-\infty, \infty)$$

All real numbers are either greater than 1 or less than 2 or both.

53.

$$5(x - 2) \ge 0 \quad \text{and} \quad -3x < 9$$

$$5x - 10 \ge 0 \qquad \frac{-3x}{-3} > \frac{9}{-3}$$

$$5x - 10 + 10 \ge 0 + 10 \qquad x > -3$$

$$5x \ge 10$$

$$x \ge 2$$

$$[2, \infty)$$

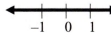

55.

$$-x < -2x \qquad \text{and} \qquad 3x > 2x$$

$$-x + 2x < -2x + 2x \qquad 3x - 2x > 2x - 2x$$

$$x < 0 \qquad\qquad x > 0$$

no solution; $\varnothing$

There are no numbers that are both less than and greater than 0.

57.

$$-6 < -3(x - 4) \le 24$$

$$-6 < -3x + 12 \le 24$$

$$-6 - 12 < -3x + 12 - 12 \le 24 - 12$$

$$-18 < -3x \le 12$$

$$\frac{-18}{-3} > \frac{-3x}{-3} \ge \frac{12}{-3}$$

$$6 > x \ge -4$$

$$-4 \le x < 6$$

$$[-4, 6)$$

59.

$$2x + 1 \ge 5 \quad \text{and} \quad -3(x + 1) \ge -9$$

$$2x + 1 - 1 \ge 5 - 1 \qquad -3x - 3 \ge -9$$

$$2x \ge 4 \qquad -3x - 3 + 3 \ge -9 + 3$$

$$x \ge 2 \qquad\qquad -3x \ge -6$$

$$x \le 2$$

$$[2, 2]$$

2 is the only number less than or equal to 2 AND greater than or equal to 2.

61.

$$\frac{4.5x-12}{2} < x \quad \text{or} \quad -15.3 > -3(x-1.4)$$

$$2\left(\frac{4.5x-12}{2}\right) < 2(x) \qquad -15.3 > -3x + 4.2$$

$$4.5x - 12 < 2x \qquad -15.3 - 4.2 > -3x + 4.2 - 4.2$$

$$4.5x - 12 + 12 < 2x + 12 \qquad -19.5 > -3x$$

$$4.5x < 2x + 12 \qquad \frac{-19.5}{-3} > \frac{-3x}{-3}$$

$$4.5x - 2x < 2x - 2x + 12 \qquad 6.5 < x$$

$$2.5x < 12 \qquad x > 6.5$$

$$\frac{2.5x}{2.5} < \frac{12}{2.5}$$

$$x < 4.8$$

$$(-\infty, 4.8) \cup (6.5, \infty)$$

4.8   6.5

63.

$$\frac{x}{0.7} + 5 > 4 \quad \text{and} \quad -4.8 \le \frac{3x}{-0.125}$$

$$\frac{x}{0.7} + 5 - 5 > 4 - 5 \quad -0.125(-4.8) \ge -0.125\left(\frac{3x}{-0.125}\right)$$

$$\frac{x}{0.7} > -1 \qquad 0.6 \ge 3x$$

$$0.7\left(\frac{x}{0.7}\right) > 0.7(-1) \qquad \frac{0.6}{3} \ge \frac{3x}{3}$$

$$x > -0.7 \qquad 0.2 \ge x$$

$$x \le 0.2$$

$$(-0.7,\ 0.2]$$

−0.7   0.2

65.

$$-24 < \frac{3}{2}x - 6 \le -15$$

$$-24 + 6 < \frac{3}{2}x - 6 + 6 \le -15 + 6$$

$$-18 < \frac{3}{2}x \le -9$$

$$2(-18) < 2\left(\frac{3}{2}x\right) \le 2(-9)$$

$$-36 < 3x \le -18$$

$$-12 < x \le -6$$

$$(-12, -6]$$

−12   −6

67.

$$\frac{x}{3} - \frac{x}{4} > \frac{1}{6} \quad \text{or} \quad \frac{x}{2} + \frac{2}{3} \le \frac{3}{4}$$

$$12\left(\frac{x}{3} - \frac{x}{4}\right) > 12\left(\frac{1}{6}\right) \qquad 12\left(\frac{x}{2} + \frac{2}{3}\right) \le 12\left(\frac{3}{4}\right)$$

$$4x - 3x > 2 \qquad 6x + 8 \le 9$$

$$x > 2 \qquad 6x + 8 - 8 \le 9 - 8$$

$$6x \le 1$$

$$\frac{6x}{6} \le \frac{1}{6}$$

$$x \le \frac{1}{6}$$

$$\left(-\infty, \frac{1}{6}\right] \cup (2, \infty)$$

$\frac{1}{6}$ 2

69.

$$0 \le \frac{4-x}{3} \le 2$$

$$3(0) \le 3\left(\frac{4-x}{3}\right) \le 3(2)$$

$$0 \le 4 - x \le 6$$

$$0 - 4 \le 4 - x - 4 \le 6 - 4$$

$$-4 \le -x \le 2$$

$$\frac{-4}{-1} \ge \frac{-x}{-1} \ge \frac{2}{-1}$$

$$4 \ge x \ge -2$$

$$-2 \le x \le 4$$

$$[-2, 4]$$

−2   4

71.

$$x \le 6 - \frac{1}{2}x \quad \text{and} \quad \frac{1}{2}x + 1 \ge 3$$

$$2(x) \le 2\left(6 - \frac{1}{2}x\right) \quad 2\left(\frac{1}{2}x + 1\right) \ge 2(3)$$

$$2x \le 12 - x \qquad x + 2 \ge 6$$

$$2x + x \le 12 - x + x \qquad x + 2 - 2 \ge 6 - 2$$

$$3x \le 12 \qquad x \ge 4$$

$$x \le 4$$

$$[4, 4]$$

4

4 is the only number less than or equal to 4 AND greater than or equal to 4.

73.

$$-6 < f(x) \le 0$$
$$-6 < 3x - 9 \le 0$$
$$-6 + 9 < 3x - 9 + 9 \le 0 + 9$$
$$3 < 3x \le 9$$
$$\frac{3}{3} < \frac{3x}{3} \le \frac{9}{3}$$
$$1 < x \le 3$$
$$(1, 3]$$

75.

$$f(x) > 29 \quad \text{and} \quad g(x) < 20$$
$$5x + 14 > 29 \qquad 2x + 8 < 20$$
$$5x + 14 - 14 > 29 - 14 \quad 2x + 8 - 8 < 20 - 8$$
$$5x > 15 \qquad 2x < 12$$
$$x > 3 \qquad x < 6$$
$$(3, 6)$$

77. a.

$$3x - 2 \ge 4 \quad \text{and} \quad x + 6 \ge 12$$
$$3x - 2 + 2 \ge 4 + 2 \quad x + 6 - 6 \ge 12 - 6$$
$$3x \ge 6 \qquad x \ge 6$$
$$x \ge 2$$
$$[6, \infty)$$

b.

$$3x - 2 \ge 4 \quad \text{or} \quad x + 6 \ge 12$$
$$3x - 2 + 2 \ge 4 + 2 \quad x + 6 - 6 \ge 12 - 6$$
$$3x \ge 6 \qquad x \ge 6$$
$$x \ge 2$$
$$[2, \infty)$$

79. a.

$$2x + 1 \le 7 \quad \text{and} \quad 3x + 5 \ge 23$$
$$2x + 1 - 1 \le 7 - 1 \qquad 3x + 5 - 5 \ge 23 - 5$$
$$2x \le 6 \qquad 3x \ge 18$$
$$x \le 3 \qquad x \ge 6$$

no solution; $\varnothing$

There are no numbers that are both less than 3 and greater than 6.

b.

$$2x + 1 \le 7 \quad \text{or} \quad 3x + 5 \ge 23$$
$$2x + 1 - 1 \le 7 - 1 \qquad 3x + 5 - 5 \ge 23 - 5$$
$$2x \le 6 \qquad 3x \ge 18$$
$$x \le 3 \qquad x \ge 6$$
$$(-\infty, 3] \cup [6, \infty)$$

APPLICATIONS

81. BABY FURNITURE
  a. $128 \le 4s \le 192$
  b.
$$128 \le 4s \le 192$$
$$\frac{128}{4} \le \frac{4s}{4} \le \frac{192}{4}$$
$$32 \le s \le 48$$

83. from Campus to Careers

  a. $(67, 77)$
   5 degrees below is $72 - 5 = 67$
   5 degrees above is $75 + 5 = 77$
  b. $(62, 82)$
   10 degrees below is $72 - 10 = 62$
   10 degrees above is $72 + 10 = 82$

85. U.S. HEALTH CARE
  a. 2004
  b. 2004, 2005, 2004, 2007
  c. 2006 and 2007
  d. 2007

87. STREET INTERSECTIONS

   a. intersection

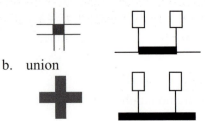

   b. union

99.

$$x - 12 < 4x < 2x + 16$$

$$x - 12 < 4x \quad \text{and} \quad 4x < 2x + 16$$

$$-12 < 3x \qquad\qquad 2x < 16$$

$$-4 < x \qquad\qquad\quad x < 8$$

$$x > -4$$

$$(-4, 8)$$

WRITING

89. Answers will vary.

91. Answers will vary.

93. Answers will vary.

REVIEW

95. Let x = number the B747 seats.
    Then $x - 125$ = number the B777 seats.

$$x + (x - 125) = 681$$

$$2x - 125 = 681$$

$$2x - 125 + 125 = 681 + 125$$

$$2x = 806$$

$$x = 403 \quad \text{and} \quad x - 125 = 278$$

    The B747 holds 403 passengers and the
    B777 holds 278 passengers.

CHALLENGE PROBLEMS

97.

$$-5 < \frac{x+2}{-2} < 0 \qquad \text{or} \qquad 2x + 10 \geq 30$$

$$-2(-5) > -2\left(\frac{x+2}{-2}\right) > -2(0) \quad 2x + 10 - 10 \geq 30 - 10$$

$$10 > x + 2 > 0 \qquad\qquad 2x \geq 20$$

$$10 - 2 > x + 2 - 2 > 0 - 2 \qquad \frac{2x}{2} \geq \frac{20}{2}$$

$$8 > x > -2 \qquad\qquad x \geq 10$$

$$(-2, 8) \cup [10, \infty)$$

SECTION 8.5

VOCABULARY

1. The **absolute value** of a number is its distance from 0 on a number line.

3. To **isolate** the absolute value in $|3-x|-4=5$, we add 4 to both sides.

5. When two equations are joined by the word *or*, such as $x+1=5$ *or* $x+1=-5$, we call the statement a **compound** equation.

CONCEPTS

7. To solve absolute value equations and inequalities, we write and solve equivalent **compound** equations and inequalities.

9. a. -2 and 2 are exactly 2 units from 0 on a number line.
   b. $-1.99, -1, 0, 1,$ and 1.99 are less than 2 units from 0 on the number line.
   c. $-3, -2.01, 2.01,$ and 3 are more than 2 units from 0 on the number line.

11. a.
$$|x-7|=8$$
is equivalent to
$$x-7=\boxed{8} \text{ or } x-7=\boxed{-8}$$
   b.
$$|x+10|=|x-3|$$
is equivalent to
$$x+10=\boxed{x-3} \text{ or } x+10=\boxed{-(x-3)}$$

13. a. $x=8$ or $x=-8$
   b. $x\le-8$ or $x\ge8$
   c. $-8\le x\le8$
   d. $5x-1=x+3$ or $5x-1=-(x+3)$

15. a. $|7x+6|=-8$ has no solution since the absolute value will never equal a negative number.
   b. $|7x+6|\le-8$ has no solution since the absolute value will never be less than or equal to a negative number.

c. $|7x+6|\ge-8$ has solutions that include all real numbers since every absolute value will be greater than a negative number.

NOTATION

17. a. ii
   b. iii
   c. i

GUIDED PRACTICE

19.
$$|x|=23$$
$$x=23 \text{ or } x=-23$$
$$\boxed{x=23,-23}$$

21.
$$|x-5|=8$$
$$x-5=8 \qquad \text{or} \qquad x-5=-8$$
$$x-5+5=8+5 \qquad x-5+5=-8+5$$
$$x=13 \qquad\qquad x=-3$$
$$\boxed{x=13,-3}$$

23.
$$|3x+2|=16$$
$$3x+2=16 \qquad \text{or} \qquad 3x+2=-16$$
$$3x+2-2=16-2 \quad 3x+2-2=-16-2$$
$$3x=14 \qquad\qquad 3x=-18$$
$$x=\frac{14}{3} \qquad\qquad x=-6$$
$$\boxed{x=\frac{14}{3},-6}$$

25.
$$\left|\frac{x}{5}\right|=10$$
$$\frac{x}{5}=10 \quad \text{or} \quad \frac{x}{5}=-10$$
$$5\left(\frac{x}{5}\right)=5(10) \quad 5\left(\frac{x}{5}\right)=5(-10)$$
$$x=50 \qquad\qquad x=-50$$
$$\boxed{x=50,-50}$$

27.

$$|2x+3.6|=9.8$$

$2x+3.6=9.8$ or $2x+3.6=-9.8$

$2x+3.6-3.6=9.8-3.6$ $2x+3.6-3.6=-9.8-3.6$

     $2x=6.2$ $2x=-13.4$

      $x=3.1$ $x=-6.7$

$$\boxed{x=3.1,-6.7}$$

29.

$$\left|\frac{7}{2}x+3\right|=-5$$

Since an absolute value can never be negative, there are no real numbers x that make $\left|\dfrac{7}{2}x+3\right|=-5$ true. The equation has no solution and the solution set is $\varnothing$.

31.

$$|x-3|-19=3$$

$|x-3|-19+19=3+19$

      $|x-3|=22$

      $x-3=22$ or $x-3=-22$

  $x-3+3=22+3$ $x-3+3=-22+3$

       $x=25$ $x=-19$

$$\boxed{x=25,-19}$$

33.

$$|3x-7|+8=22$$

$|3x-7|+8-8=22-8$

     $|3x-7|=14$

     $3x-7=14$ or $3x-7=-14$

  $3x-7+7=14+7$ $3x-7+7=-14+7$

      $3x=21$ $3x=-7$

       $x=7$ $x=-\dfrac{7}{3}$

$$\boxed{x=7,-\frac{7}{3}}$$

35.

$$|3-4x|+1=6$$

$|3-4x|+1-1=6-1$

     $|3-4x|=5$

     $3-4x=5$ or $3-4x=-5$

  $3-4x-3=5-3$ $3-4x-3=-5-3$

     $-4x=2$ $-4x=-8$

      $x=-\dfrac{1}{2}$ $x=2$

$$\boxed{x=-\frac{1}{2},2}$$

37.

$$\left|\frac{7}{8}x+5\right|-2=7$$

$\left|\dfrac{7}{8}x+5\right|-2+2=7+2$

     $\left|\dfrac{7}{8}x+5\right|=9$

     $\dfrac{7}{8}x+5=9$ or $\dfrac{7}{8}x+5=-9$

  $\dfrac{7}{8}x+5-5=9-5$ $\dfrac{7}{8}x+5-5=-9-5$

     $\dfrac{7}{8}x=4$ $\dfrac{7}{8}x=-14$

  $\dfrac{8}{7}\left(\dfrac{7}{8}x\right)=\dfrac{8}{7}(4)$ $\dfrac{8}{7}\left(\dfrac{7}{8}x\right)=\dfrac{8}{7}(-14)$

      $x=\dfrac{32}{7}$ $x=-16$

$$\boxed{x=\frac{32}{7},-16}$$

39.

$$\left|\frac{1}{5}x+2\right|-8=-8$$

$$\left|\frac{1}{5}x+2\right|-8+8=-8+8$$

$$\left|\frac{1}{5}x+2\right|=0$$

$$\frac{1}{5}x+2=0$$

$$\frac{1}{5}x+2-2=0-2$$

$$\frac{1}{5}x=-2$$

$$5\left(\frac{1}{5}x\right)=5(-2)$$

$$\boxed{x=-10}$$

41.

$$2\left|3x+24\right|=0$$

$$\frac{2\left|3x+24\right|}{2}=\frac{0}{2}$$

$$\left|3x+24\right|=0$$

$$3x+24=0$$

$$3x+24-24=0-24$$

$$3x=-24$$

$$\boxed{x=-8}$$

43.

$$-5\left|2x-9\right|+14=14$$

$$-5\left|2x-9\right|+14-14=14-14$$

$$-5\left|2x-9\right|=0$$

$$\frac{-5\left|2x-9\right|}{-5}=\frac{0}{-5}$$

$$\left|2x-9\right|=0$$

$$2x-9=0$$

$$2x-9+9=0+9$$

$$2x=9$$

$$\boxed{x=\frac{9}{2}}$$

45.

$$6-3\left|10x+5\right|=6$$

$$6-3\left|10x+5\right|-6=6-6$$

$$-3\left|10x+5\right|=0$$

$$\frac{-3\left|10x+5\right|}{-3}=\frac{0}{-3}$$

$$\left|10x+5\right|=0$$

$$10x+5=0$$

$$10x+5-5=0-5$$

$$10x=-5$$

$$x=-\frac{5}{10}$$

$$\boxed{x=-\frac{1}{2}}$$

47.

$$\left|5x-12\right|=\left|4x-16\right|$$

$5x-12=4x-16$ or $5x-12=-(4x-16)$

$\qquad 5x-12=4x-16 \qquad\qquad 5x-12=-4x+16$

$5x-12-4x=4x-16-4x \quad 5x-12+4x=-4x+16+4x$

$\qquad\qquad x-12=-16 \qquad\qquad\qquad 9x-12=16$

$\quad x-12+12=-16+12 \qquad 9x-12+12=16+12$

$\qquad\qquad x=-4 \qquad\qquad\qquad\qquad 9x=28$

$\qquad\qquad x=-4 \qquad\qquad\qquad\qquad x=\frac{28}{9}$

$$\boxed{x=-4,\ \frac{28}{9}}$$

49.

$$\left|10x\right|=\left|x-18\right|$$

$10x=x-18$ or $10x=-(x-18)$

$\qquad 10x=x-18 \qquad\qquad 10x=-x+18$

$10x-x=x-18-x \quad 10x+x=-x+18+x$

$\qquad\qquad 9x=-18 \qquad\qquad\qquad 11x=18$

$\qquad\qquad x=-2 \qquad\qquad\qquad x=\frac{18}{11}$

$$\boxed{x=-2,\ \frac{18}{11}}$$

51.

$$|2-x|=|3x+2|$$

$$2-x=3x+2 \quad \text{or} \quad 2-x=-(3x+2)$$

$$2-x=3x+2 \qquad\qquad 2-x=-3x-2$$

$$2-x-3x=3x+2-3x \quad 2-x+3x=-3x-2+3x$$

$$2-4x=2 \qquad\qquad 2+2x=-2$$

$$2-4x-2=2-2 \qquad 2+2x-2=-2-2$$

$$-4x=0 \qquad\qquad 2x=-4$$

$$x=0 \qquad\qquad x=-2$$

$$\boxed{x=0,-2}$$

53.

$$|5x-7|=|4(x+1)|$$

$$5x-7=4(x+1) \quad \text{or} \quad 5x-7=-4(x+1)$$

$$5x-7=4x+4 \qquad\qquad 5x-7=-4x-4$$

$$5x-7+7=4x+4+7 \quad 5x-7+7=-4x-4+7$$

$$5x=4x+11 \qquad\qquad 5x=-4x+3$$

$$5x-4x=4x+11-4x \quad 5x+4x=-4x+3+4x$$

$$x=11 \qquad\qquad 9x=3$$

$$x=11 \qquad\qquad x=\frac{1}{3}$$

$$\boxed{x=11,\frac{1}{3}}$$

55.

$$|x|<4$$

$$-4<x<4$$

$$(-4,4)$$

57.

$$|x+9|\le 12$$

$$-12\le x+9\le 12$$

$$-12-9\le x+9-9\le 12-9$$

$$-21\le x\le 3$$

$$[-21,3]$$

59.

$$|3x-2|<10$$

$$-10<3x-2<10$$

$$-10+2<3x-2+2<10+2$$

$$-8<3x<12$$

$$-\frac{8}{3}<x<4$$

$$\left(-\frac{8}{3},4\right)$$

61.

$$|5x-12|<-5$$

No solution; $\varnothing$. Since $|5x-12|$ can never be negative, there are no real numbers x that can make the equation true.

63.

$$|x|>3$$

$$x<-3 \quad \text{or} \quad x>3$$

$$(-\infty,-3)\cup(3,\infty)$$

65.

$$|x-12|>24$$

$$x-12<-24 \quad \text{or} \quad x-12>24$$

$$x-12+12<-24+12 \quad x-12+12>24+12$$

$$x<-12 \qquad\qquad x>36$$

$$(-\infty,-12)\cup(36,\infty)$$

67.

$$|5x - 1| - 2 \geq 0$$
$$|5x - 1| - 2 + 2 \geq 0 + 2$$
$$|5x - 1| \geq 2$$

$$5x - 1 \leq -2 \quad \text{or} \quad 5x - 1 \geq 2$$
$$5x - 1 + 1 \leq -2 + 1 \quad 5x - 1 + 1 \geq 2 + 1$$
$$5x \leq -1 \quad\quad\quad 5x \geq 3$$
$$x \leq -\frac{1}{5} \quad\quad\quad x \geq \frac{3}{5}$$

$$\left(-\infty, -\frac{1}{5}\right] \cup \left[\frac{3}{5}, \infty\right)$$

$$-\frac{1}{5} \quad \frac{3}{5}$$

69.

$$|4x + 3| > -5$$
$$(-\infty, \infty)$$

$$0$$

Since $|4x + 3|$ is always greater than or equal to 0 for any real number x, then this absolute value inequality is true for all real numbers, $\mathbb{R}$.

71.

$$f(x) = |x + 3|$$
$$3 = |x + 3|$$
$$3 = x + 3 \quad \text{or} \quad -3 = x + 3$$
$$3 - 3 = x + 3 - 3 \quad -3 - 3 = x + 3 - 3$$
$$0 = x \quad\quad\quad -6 = x$$

$$\boxed{x = 0, -6}$$

73.

$$f(x) = |2(x - 1) + 4|$$
$$|2(x - 1) + 4| < 4$$
$$|2x - 2 + 4| < 4$$
$$|2x + 2| < 4$$
$$-4 < 2x + 2 < 4$$
$$-4 - 2 < 2x + 2 - 2 < 4 - 2$$
$$-6 < 2x < 2$$
$$\frac{-6}{2} < \frac{2x}{2} < \frac{2}{2}$$
$$-3 < x < 1$$
$$(-3, 1)$$

75.

$$|3x + 2| + 1 > 15$$
$$|3x + 2| + 1 - 1 > 15 - 1$$
$$|3x + 2| > 14$$
$$3x + 2 < -14 \quad \text{or} \quad 3x + 2 > 14$$
$$3x + 2 - 2 < -14 - 2 \quad 3x + 2 - 2 > 14 - 2$$
$$3x < -16 \quad\quad\quad 3x > 12$$
$$x < -\frac{16}{3} \quad\quad\quad x > 4$$

$$\left(-\infty, -\frac{16}{3}\right) \cup (4, \infty)$$

$$-\frac{16}{3} \quad 4$$

77.

$$6\left|\frac{x - 2}{3}\right| \leq 24$$
$$\frac{6\left|\dfrac{x - 2}{3}\right|}{6} \leq \frac{24}{6}$$
$$\left|\frac{x - 2}{3}\right| \leq 4$$
$$-4 \leq \frac{x - 2}{3} \leq 4$$
$$3(-4) \leq 3\left(\frac{x - 2}{3}\right) \leq 3(4)$$
$$-12 \leq x - 2 \leq 12$$
$$-12 + 2 \leq x - 2 + 2 \leq 12 + 2$$
$$-10 \leq x \leq 14$$
$$[-10, 14]$$

$$-10 \quad 14$$

79.

$$-7 = 2 - |0.3x - 3|$$
$$-7 - 2 = 2 - |0.3x - 3| - 2$$
$$-9 = -|0.3x - 3|$$
$$\frac{-9}{-1} = \frac{-|0.3x - 3|}{-1}$$
$$9 = |0.3x - 3|$$
$$|0.3x - 3| = 9$$

$$0.3x - 3 = 9 \quad \text{or} \quad 0.3x - 3 = -9$$
$$0.3x - 3 + 3 = 9 + 3 \quad 0.3x - 3 + 3 = -9 + 3$$
$$0.3x = 12 \qquad\qquad 0.3x = -6$$
$$x = 40 \qquad\qquad\quad x = -20$$

$$\boxed{x = 40, -20}$$

81.

$$|2 - 3x| \geq -8$$
$$(-\infty, \infty)$$

Since $|2 - 3x|$ is always greater than -8 for any real number x, then this absolute value inequality is true for all real numbers, $\mathbb{R}$.

83.

$$|7x + 12| = |x - 6|$$

$$7x + 12 = x - 6 \quad \text{or} \quad 7x + 12 = -(x - 6)$$
$$7x + 12 = x - 6 \qquad\qquad 7x + 12 = -x + 6$$
$$7x + 12 - 12 = x - 6 - 12 \quad 7x + 12 - 12 = -x + 6 - 12$$
$$7x = x - 18 \qquad\qquad 7x = -x - 6$$
$$7x - x = x - 18 - x \qquad 7x + x = -x - 6 + x$$
$$6x = -18 \qquad\qquad 8x = -6$$
$$x = -3 \qquad\qquad\quad x = -\frac{3}{4}$$

$$\boxed{x = -3, -\frac{3}{4}}$$

85.

$$3|2 - 3x| + 2 \leq 2$$
$$3|2 - 3x| + 2 - 2 \leq 2 - 2$$
$$3|2 - 3x| \leq 0$$
$$\frac{3|2 - 3x|}{3} \leq \frac{0}{3}$$
$$|2 - 3x| \leq 0$$
$$2 - 3x \leq 0$$
$$2 - 3x - 2 \leq 0 - 2$$
$$-3x \leq -2$$
$$\frac{-3x}{-3} \geq \frac{-2}{-3}$$
$$x \geq \frac{2}{3}; \qquad \left[\frac{2}{3}, \infty\right)$$

87. $-14 = |x - 3|$

Since an absolute value can never be negative, there are no real numbers x that make $-14 = |x - 3|$ true. The equation has no solution and the solution set is $\varnothing$.

89.

$$\frac{6}{5} = \left|\frac{3x}{5} + \frac{x}{2}\right|$$

$$\frac{3x}{5} + \frac{x}{2} = \frac{6}{5} \quad \text{or} \quad \frac{3x}{5} + \frac{x}{2} = -\frac{6}{5}$$
$$10\left(\frac{3x}{5} + \frac{x}{2}\right) = 10\left(\frac{6}{5}\right) \quad 10\left(\frac{3x}{5} + \frac{x}{2}\right) = 10\left(-\frac{6}{5}\right)$$
$$6x + 5x = 12 \qquad\qquad 6x + 5x = -12$$
$$11x = 12 \qquad\qquad 11x = -12$$
$$x = \frac{12}{11} \qquad\qquad x = -\frac{12}{11}$$

$$\boxed{x = \frac{12}{11}, -\frac{12}{11}}$$

91.

$$-|2x-3| < -7$$

$$\frac{-|2x-3|}{-1} > \frac{-7}{-1}$$

$$|2x-3| > 7$$

$$2x-3 < -7 \quad \text{or} \quad 2x-3 > 7$$

$$2x-3+3 < -7+3 \quad 2x-3+3 > 7+3$$

$$2x < -4 \qquad\qquad 2x > 10$$

$$x < -2 \qquad\qquad x > 5$$

$$(-\infty,-2) \cup (5,\infty)$$

93.

$$|0.5x+1| < -23$$

No solution; $\varnothing$. Since $|0.5x+1|$ can never be negative, there are no real numbers x that can make the equation true.

LOOK ALIKES

95. a.

$$\frac{x}{10} - 1 = 1$$

$$\frac{x}{10} - 1 + 1 = 1 + 1$$

$$\frac{x}{10} = 2$$

$$10\left(\frac{x}{10}\right) = 10(2)$$

$$x = 20$$

b.

$$\left|\frac{x}{10} - 1\right| = 1$$

$$\frac{x}{10} - 1 = 1 \quad \text{and} \quad \frac{x}{10} - 1 = -1$$

$$\frac{x}{10} = 2 \qquad\qquad \frac{x}{10} = 0$$

$$x = 20 \qquad\qquad x = 0$$

$$x = 20, 0$$

c.

$$\frac{x}{10} - 1 > 1$$

$$\frac{x}{10} > 2$$

$$x > 20$$

$$(20,\infty)$$

d.

$$\left|\frac{x}{10} - 1\right| > 1$$

$$\frac{x}{10} - 1 > 1 \quad \text{and} \quad \frac{x}{10} - 1 < -1$$

$$\frac{x}{10} > 2 \qquad\qquad \frac{x}{10} < 0$$

$$x > 20 \qquad\qquad x < 0$$

$$(-\infty,0) \cup (20,\infty)$$

Section 8.5

97. a.
$$0.9 - 0.3x = 8.4$$
$$0.9 - 0.3x - 0.9 = 8.4 - 0.9$$
$$-0.3x = 7.5$$
$$x = -25$$

b.
$$|0.9 - 0.3x| = 8.4$$
$$0.9 - 0.3x = 8.4 \quad \text{and} \quad 0.9 - 0.3x = -8.4$$
$$0.9 - 0.3x - 0.9 = 8.4 - 0.9 \quad 0.9 - 0.3x - 0.9 = -8.4 - 0.9$$
$$-0.3x = 7.5 \qquad\qquad -0.3x = -9.3$$
$$x = -25 \qquad\qquad x = 31$$
$$x = -25, 31$$

c.
$$0.9 - 0.3x > 8.4$$
$$0.9 - 0.3x - 0.9 > 8.4 - 0.9$$
$$-0.3x > 7.5$$
$$x < -25$$
$$(-\infty, -25)$$

d.
$$|0.9 - 0.3x| > 8.4$$
$$0.9 - 0.3x > 8.4 \quad \text{and} \quad 0.9 - 0.3x < -8.4$$
$$0.9 - 0.3x - 0.9 > 8.4 - 0.9 \quad 0.9 - 0.3x - 0.9 < -8.4 - 0.9$$
$$-0.3x > 7.5 \qquad\qquad -0.3x < -9.3$$
$$x < -25 \qquad\qquad x > 31$$
$$(-\infty, -25) \cup (31, \infty)$$

99. a.
$$|8x - 40| \le 16$$
$$-16 \le 8x - 40 \le 16$$
$$-16 + 40 \le 8x - 40 + 40 \le 16 + 40$$
$$24 \le 8x \le 56$$
$$3 \le x \le 7$$
$$[3, 7]$$

b.
$$|8x - 40| \ge 16$$
$$8x - 40 \ge 16 \quad \text{or} \quad 8x - 40 \le -16$$
$$8x \ge 56 \qquad\qquad 8x \le 24$$
$$x \ge 7 \qquad\qquad x \le 3$$
$$(-\infty, 3] \cup [7, \infty)$$

101. a.
$$\left|\frac{4x - 4}{3}\right| - 1 > 11$$
$$\left|\frac{4x - 4}{3}\right| - 1 + 1 > 11 + 1$$
$$\left|\frac{4x - 4}{3}\right| > 12$$
$$\frac{4x - 4}{3} > 12 \quad \text{or} \quad \frac{4x - 4}{3} < -12$$
$$3\left(\frac{4x - 4}{3}\right) > 3(12) \quad 3\left(\frac{4x - 4}{3}\right) < 3(-12)$$
$$4x - 4 > 36 \qquad\qquad 4x - 4 < -36$$
$$4x > 40 \qquad\qquad 4x < -32$$
$$x > 10 \qquad\qquad x < -8$$
$$(-\infty, -8) \cup (10, \infty)$$

b.
$$\left|\frac{4x - 4}{3}\right| - 1 \le 11$$
$$\left|\frac{4x - 4}{3}\right| - 1 + 1 \le 11 + 1$$
$$\left|\frac{4x - 4}{3}\right| \le 12$$
$$-12 \le \frac{4x - 4}{3} \le 12$$
$$3(-12) \le 3\left(\frac{4x - 4}{3}\right) \le 3(12)$$
$$-36 \le 4x - 4 \le 36$$
$$-36 + 4 \le 4x - 4 + 4 \le 36 + 4$$
$$-32 \le 4x \le 40$$
$$-8 \le x \le 10$$
$$[-8, 10]$$

APPLICATIONS

103.   TEMPERATURE RANGES

$$|t - 78| \leq 8$$

$$-8 \leq t - 78 \leq 8$$

$$-8 + 78 \leq t - 78 + 78 \leq 8 + 78$$

$$70° \leq t \leq 86°$$

105.   AUTO MECHANICS

a. $|c - 0.6°| \leq 0.5°$

b. $|c - 0.6°| \leq 0.5°$

$$-0.5 \leq c - 0.6 \leq 0.5$$

$$-0.5 + 0.6 \leq c - 0.6 + 0.6 \leq 0.5 + 0.6$$

$$0.1 \leq c \leq 1.1$$

$$\left[0.1°, 1.1° \right]$$

107.   ERROR ANALYSIS

a.   Trial 1: $p = 22.91\%$

$$|22.91 - 25.46| \overset{?}{\leq} 1.00$$

$$|-2.55| \overset{?}{\leq} 1.00$$

$$2.55 \not\leq 1.00$$

no

Trial 2: $p = 26.45\%$

$$|26.45 - 25.46| \overset{?}{\leq} 1.00$$

$$|0.99| \overset{?}{\leq} 1.00$$

$$0.99 \leq 1.00$$

yes

Trial 3: $p = 26.49\%$

$$|26.49 - 25.46| \overset{?}{\leq} 1.00$$

$$|1.03| \overset{?}{\leq} 1.00$$

$$1.03 \not\leq 1.00$$

no

Trial 4: $p = 24.76\%$

$$|24.76 - 25.46| \overset{?}{\leq} 1.00$$

$$|-0.70| \overset{?}{\leq} 1.00$$

$$0.70 \leq 1.00$$

yes

b.   It is less than or equal to 1%.

WRITING

109.   Answers will vary.

111.   Answers will vary.

REVIEW

113.   The angles x and y are supplementary (their sum is 180°).

$$\begin{cases} x + y = 180 \\ y = 30 + 2x \end{cases}$$

$$x + y = 180$$

$$x + (30 + 2x) = 180$$

$$3x + 30 = 180$$

$$3x + 30 - 30 = 180 - 30$$

$$3x = 150$$

$$x = 50°$$

$$y = 30 + 2(50)$$

$$y = 130°$$

CHALLENGE PROBLEMS

115.   a. $k < 0$

b. $k = 0$

SECTION 8.6

VOCABULARY

1. When we write $2x + 4$ as $2(x + 2)$, we say that we have **factored** $2x + 4$.

3. The abbreviation GCF stands for **greatest common factor**.

5. To factor $ab + 6a + 2b + 12$ by **grouping**, we begin by factoring out a from the first two terms and 2 from the last two terms.

7. The **leading** coefficient of the trinomial $x^2 - 3x + 2$ is 1, the **coefficient** of the middle term is -3, and the last term is **2**.

CONCEPTS

9. $\text{GCF} = 2 \cdot 3 \cdot x \cdot y \cdot y = 6xy^2$

11.

Factors of 8	Sum of the factors of 8
$1(8) = 8$	$1 + 8 = 9$
$2(4) = 8$	$2 + 4 = 6$
$-1(-8) = 8$	$-1 + (-8) = -9$
$-2(-4) = 8$	$-2 + (-4) = -6$

13.

Negative factors of 12	Sum of factors of 12
$-1(-12) = 12$	$-1 + (-12) = -13$
$-2(-6) = 12$	$-2 + (-6) = -8$
$-3(-4) = 12$	$-3 + (-4) = -7$

NOTATION

15.
$$15c^3 d^4 - 25c^2 d^4 + 5c^3 d^6 = \boxed{5c^2 d^4}\left(3c - 5 + cd^2\right)$$

17.
$$6m^2 + 7m - 3 = \left(\boxed{3m} - 1\right)\left(2m + \boxed{3}\right)$$

GUIDED PRACTICE

19.
$$2x^2 - 6x = 2x(x - 3)$$

21.
$$15x^2 y - 10x^2 y^2 = 5x^2 y(3 - 2y)$$

23.
$$27z^3 + 12z^2 + 3z = 3z\left(9z^2 + 4z + 1\right)$$

25.
$$24s^3 - 12s^2 t + 6st^2 = 6s\left(4s^2 - 2st + t^2\right)$$

27.
$$11x^3 - 12y \quad \text{is prime}$$

29.
$$23a^2 b^3 + 4x^3 y^2 \quad \text{is prime}$$

31.
$$-8a - 16 = -8(a + 2)$$

33.
$$-6x^2 - 3xy = -3x(2x + y)$$

35.
$$-18a^2 b + 12ab^2 = -6ab(3a - 2b)$$

37.
$$-8a^4 c^8 + 28a^3 c^8 - 20a^2 c^9 = -4a^2 c^8\left(2a^2 - 7a + 5c\right)$$

39.
$$(x + y)u + (x + y)v = (x + y)(u + v)$$

41.
$$5(a - b + c) - t(a - b + c)$$
$$= (a - b + c)(5 - t)$$

43.
$$ax + bx + ay + by = x(a + b) + y(a + b)$$
$$= (a + b)(x + y)$$

45.
$$x^2 + yx - x - y = x(x + y) - 1(x + y)$$
$$= (x + y)(x - 1)$$

47.
$$t^3 - 3t^2 - 7t + 21 = t^2(t - 3) - 7(t - 3)$$
$$= (t - 3)(t^2 - 7)$$

49.

$$a^2 - 4b + ab - 4a = a^2 + ab - 4a - 4b$$
$$= a(a+b) - 4(a+b)$$
$$= (a+b)(a-4)$$

51.

$$6x^3 - 6x^2 + 12x - 12 = 6(x^3 - x^2 + 2x - 2)$$
$$= 6[x^2(x-1) + 2(x-1)]$$
$$= 6(x-1)(x^2+2)$$

53.

$$28a^3b^3c + 14a^3c - 4b^3c - 2c$$
$$= 2c(14a^3b^3 + 7a^3 - 2b^3 - 1)$$
$$= 2c[7a^3(2b^3+1) - (2b^3+1)]$$
$$= 2c(2b^3+1)(7a^3-1)$$

55.

$$2g = ch + dh$$
$$2g = h(c+d)$$
$$\frac{2g}{c+d} = \frac{h(c+d)}{c+d}$$
$$\frac{2g}{c+d} = h$$
$$h = \frac{2g}{c+d}$$

57.

$$r_1 r_2 = rr_2 + rr_1$$
$$r_1 r_2 - rr_1 = rr_2 + rr_1 - rr_1$$
$$r_1(r_2 - r) = rr_2$$
$$\frac{r_1(r_2 - r)}{r_2 - r} = \frac{rr_2}{r_2 - r}$$
$$r_1 = \frac{rr_2}{r_2 - r}$$

59.

$$b^2x^2 + a^2y^2 = a^2b^2$$
$$b^2x^2 + a^2y^2 - a^2y^2 = a^2b^2 - a^2y^2$$
$$b^2x^2 = a^2b^2 - a^2y^2$$
$$b^2x^2 = a^2(b^2 - y^2)$$
$$\frac{b^2x^2}{b^2 - y^2} = \frac{a^2(b^2 - y^2)}{b^2 - y^2}$$
$$\frac{b^2x^2}{b^2 - y^2} = a^2$$
$$a^2 = \frac{b^2x^2}{b^2 - y^2}$$

61.

$$Sn = (n-2)180$$
$$Sn = 180n - 360$$
$$Sn - 180n = 180n - 360 - 180n$$
$$n(S - 180) = -360$$
$$\frac{n(S-180)}{S-180} = \frac{-360}{S-180}$$
$$n = \frac{-360}{S-180} \cdot \frac{-1}{-1}$$
$$n = \frac{360°}{180° - S}$$

63. $x^2 - 5x + 6 = (x-2)(x-3)$

65. $x^2 + x - 30 = (x+6)(x-5)$

67. $3x^2 + 12xy - 63y^2 = 3(x^2 + 4xy - 21y^2)$
$$= 3(x+7y)(x-3y)$$

69. $6a^2 - 30ab + 24b^2 = 6(a^2 - 5ab + 4b^2)$
$$= 6(a-b)(a-4b)$$

71. $n^4 - 28n^3t - 60n^2t^2 = n^2(n^2 - 28nt - 60t^2)$
$$= n^2(n-30t)(n+2t)$$

73. $-3x^2 + 15xy - 18y^2 = -3(x^2 - 5xy + 6y^1)$
$$= -3(x-2y)(x-3y)$$

75. $5x^2 + 13x + 6 = (5x+3)(x+2)$

77. $7a^2 + 12a + 5 = (7a+5)(a+1)$

79. $11y^2 + 32y - 3 = (11y-1)(y+3)$

81. $8x^2 - 22x + 5 = (4x-1)(2x-5)$

83. $6y^2 - 13y + 6 = (3y - 2)(2y - 3)$

85. $15b^2 + 4b - 4 = (5b - 2)(3b + 2)$

87. $30x^4 - 25x^2 - 20 = 5(6x^4 - 5x^2 - 4)$
$= 5(3x^2 - 4)(2x^2 + 1)$

89. $32x^4 - 96x^2 + 72 = 8(4x^4 - 12x^2 + 9)$
$= 8(2x^2 - 3)^2$

91. $64h^5 - 4h + 24h^3 = 64h^5 + 24h^3 - 4h$
$= 4h(16h^4 + 6h^2 - 1)$
$= 4h(8h^2 - 1)(2h^2 + 1)$

93. $-3a^4 - 5a^2b^2 - 2b^4 = -(3a^4 + 5a^2b^2 + 2b^4)$
$= -(3a^2 + 2b^2)(a^2 + b^2)$

95. $(a + b)^2 - 2(a + b) - 24$
Let $x = (a + b)$
$x^2 - 2x - 24 = (x - 6)(x + 4)$
Substitute $(a + b)$ for x.
$= (a + b - 6)(a + b + 4)$

97. $(x + a)^2 + 2(x + a) + 1$
Let $y = (x + a)$
$y^2 + 2y + 1 = (y + 1)(y + 1)$
$= (y + 1)^2$
Substitute $(x + a)$ for y.
$= (x + a + 1)^2$

TRY IT YOURSELF

99.
$3(m + n + p) + x(m + n + p) = (m + n + p)(3 + x)$

101.
$-63u^3v^6 + 28u^2v^7 - 21u^3v^3$
$= -7u^2v^3(9uv^3 - 4v^4 + 3u)$

103.
$b^4x^2 - 12b^2x^2 + 35x^2 = x^2(b^4 - 12b^2 + 35)$
$= x^2(b^2 - 5)(b^2 - 7)$

105.
$1 - m + mn - n = 1(1 - m) + n(m - 1)$
$= 1(1 - m) - n(1 - m)$
$= (1 - m)(1 - n)$

107.
$-x^2 + 4xy + 21y^2 = -(x^2 - 4xy - 21y^2)$
$= -(x - 7y)(x + 3y)$

109.
$a^2x + bx - a^2 - b = x(a^2 + b) - 1(a^2 + b)$
$= (a^2 + b)(x - 1)$

111.
$4y^2 + 4y + 1 = (2y + 1)(2y + 1)$
$= (2y + 1)^2$

113.
$b^2 + 8b + 18$ is prime

115.
$13r + 3r^2 - 10 = 3r^2 + 13r - 10$
$= (3r - 2)(r + 5)$

117.
$y^3 - 12 + 3y - 4y^2 = y^3 - 4y^2 + 3y - 12$
$= y^2(y - 4) + 3(y - 4)$
$= (y - 4)(y^2 + 3)$

119.
$2y^5 - 26y^3 + 60y = 2y(y^4 - 13y^2 + 30)$
$= 2y(y^2 - 10)(y^2 - 3)$

121. $14(q - r)^2 - 17(q - r) - 6$
Let $x = (q - r)$.
$14x^2 - 17x - 6 = (2x - 3)(7x + 2)$
Substitute $(q - r)$ for x.
$= [2(q - r) - 3][7(q - r) + 2]$
$= (2q - 2r - 3)(7q - 7r + 2)$

APPLICATIONS

123. CRAYONS
$V = \pi r^2 h_1 + \frac{1}{3}\pi r^2 h_2$
$= \pi r^2 \left(h_1 + \frac{1}{3} h_2 \right)$

125.   ICE

Area of square $= s^2$

$$s^2 = 6x^2 + 36x + 54$$
$$= 6\left(x^2 + 6x + 9\right)$$
$$= 6\left(x + 3\right)\left(x + 3\right)$$
$$= 6\left(x + 3\right)^2$$

The length of an edge of the block is $x + 3$.

WRITING

127.   Answers will vary.

REVIEW

129.   a. $s(-3) = 2$
      b. $s(3) = 4$
      c. $x = -1$, 1, and 5
      d. $x = 2$ and 4

CHALLENGE PROBLEMS

131.

$$t^5 + 4t^{-6} = t^{-3}\left(t^8 + 4t^{-3}\right)$$

133.

$$x^{2n} + 2x^n + 1 = \left(x^n + 1\right)\left(x^n + 1\right)$$
$$= \left(x^n + 1\right)^2$$

135.

$$x^{4n} + 2x^{2n}y^{2n} + y^{4n} = \left(x^{2n} + y^{2n}\right)\left(x^{2n} + y^{2n}\right)$$
$$= \left(x^{2n} + y^{2n}\right)^2$$

SECTION 8.7

VOCABULARY

1. When the polynomial $4x^2 - 25$ is written as $(2x)^2 - (5)^2$, we see that it is the difference of two **squares**.

CONCEPTS

3. a. 1, 4, 9, 16, 25, 36, 49, 64, 81, 100
   b. 1, 8, 27, 64, 125, 216, 343, 512, 729, 1,000

5. a. $F^2 - L^2 = (F + L)(F - L)$
   b. $F^3 + L^3 = (F + L)(F^2 - FL + L^2)$
   c. $F^3 - L^3 = (F - L)(F^2 + FL + L^2)$

NOTATION

7. Answers will vary.
   a. $x^2 - 4$
   b. $(x - 4)^2$
   c. $x^2 + 4$
   d. $x^3 + 8$
   e. $(x + 8)^3$

GUIDED PRACTICE

9.
$$x^2 - 16 = (x)^2 - (4)^2$$
$$= (x + 4)(x - 4)$$

11.
$$9y^2 - 64 = (3y)^2 - (8)^2$$
$$= (3y + 8)(3y - 8)$$

13.
$$144 - c^2 = (12)^2 - (c)^2$$
$$= (12 + c)(12 - c)$$

15.
$$100m^2 - 1 = (10m)^2 - (1)^2$$
$$= (10m + 1)(10m - 1)$$

17.
$$81a^2 - 49b^2 = (9a)^2 - (7b)^2$$
$$= (9a - 7b)(9a + 7b)$$

19.
$$x^2 + 25 \text{ is prime}$$

21.
$$9r^4 - 121s^2 = (3r^2)^2 - (11s)^2$$
$$= (3r^2 - 11s)(3r^2 + 11s)$$

23.
$$16t^2 - 25w^4 = (4t)^2 - (5w^2)^2$$
$$= (4t + 5w^2)(4t - 5w^2)$$

25.
$$100r^2s^4 - t^4 = (10rs^2)^2 - (t^2)^2$$
$$= (10rs^2 + t^2)(10rs^2 - t^2)$$

27.
$$36x^4y^2 - 49z^4 = (6x^2y)^2 - (7z^2)^2$$
$$= (6x^2y - 7z^2)(6x^2y + 7z^2)$$

29.
$$x^4 - y^4 = (x^2)^2 - (y^2)^2$$
$$= (x^2 + y^2)(x^2 - y^2)$$
$$= (x^2 + y^2)(x + y)(x - y)$$

31.
$$16a^4 - 81b^4 = (4a^2)^2 - (9b^2)^2$$
$$= (4a^2 + 9b^2)(4a^2 - 9b^2)$$
$$= (4a^2 + 9b^2)(2a + 3b)(2a - 3b)$$

33.
$$(x + y)^2 - z^2 = (x + y + z)(x + y - z)$$

35.
$$(r - s)^2 - t^4 = (r - s + t^2)(r - s - t^2)$$

37.

$$2x^2 - 288 = 2(x^2 - 144)$$
$$= 2(x + 12)(x - 12)$$

39.

$$3x^3 - 243x = 3x(x^2 - 81)$$
$$= 3x(x + 9)(x - 9)$$

41.

$$5ab^4 - 5a = 5a(b^4 - 1)$$
$$= 5a(b^2 + 1)(b^2 - 1)$$
$$= 5a(b^2 + 1)(b + 1)(b - 1)$$

43.

$$64b - 4b^5 = 4b(16 - b^4)$$
$$= 4b(4 + b^2)(4 - b^2)$$
$$= 4b(4 + b^2)(2 + b)(2 - b)$$

45.

$$c^2 - d^2 + c + d = (c^2 - d^2) + (c + d)$$
$$= (c - d)(c + d) + 1(c + d)$$
$$= (c + d)(c - d + 1)$$

47.

$$a^2 - b^2 + 2a - 2b = (a^2 - b^2) + (2a - 2b)$$
$$= (a + b)(a - b) + 2(a - b)$$
$$= (a - b)(a + b + 2)$$

49.

$$x^2 + 12x + 36 - y^2 = (x + 6)^2 - y^2$$
$$= (x + 6 - y)(x + 6 + y)$$

51.

$$x^2 - 2x + 1 - 9z^2 = (x^2 - 2x + 1) - 9z^2$$
$$= (x - 1)^2 - 9z^2$$
$$= (x - 1 - 3z)(x - 1 + 3z)$$

53.

$$a^3 + 125 = (a + 5)(a^2 - 5a + 5^2)$$
$$= (a + 5)(a^2 - 5a + 25)$$

55.

$$8r^3 + s^3 = (2r + s)((2r)^2 - 2rs + s^2)$$
$$= (2r + s)(4r^2 - 2rs + s^2)$$

57.

$$64t^6 - 27v^3 = (4t^2 - 3v)((4t^2)^2 + (4t^2)(3v) + (3v)^2)$$
$$= (4t^2 - 3v)(16t^4 + 12t^2v + 9v^2)$$

59.

$$x^3 - 216y^6 = (x - 6y^2)(x^2 + 6xy^2 + (6y^2)^2)$$
$$= (x - 6y^2)(x^2 + 6xy^2 + 36y^4)$$

61.

$$(a - b)^3 + 27$$
$$= (a - b)^3 + (3)^3$$
$$= ((a - b) + 3)((a - b)^2 - 3(a - b) + 9)$$
$$= (a - b + 3)(a^2 - 2ab + b^2 - 3a + 3b + 9)$$

63.

$$64 - (a + b)^3$$
$$= (4)^3 - (a + b)^3$$
$$= (4 - (a + b))(4^2 + 4(a + b) + (a + b)^2)$$
$$= (4 - a - b)(16 + 4a + 4b + a^2 + 2ab + b^2)$$

65.

$$x^6 - 1 = (x^3 + 1)(x^3 - 1)$$
$$= (x + 1)(x^2 - x + 1)(x - 1)(x^2 + x + 1)$$

67.

$$x^{12} - y^6$$
$$= (x^6 - y^3)(x^6 + y^3)$$
$$= (x^2 - y)(x^4 + x^2y + y^2)(x^2 + y)(x^4 - x^2y + y^2)$$

69.

$$5x^3 + 625 = 5(x^3 + 125)$$
$$= 5(x + 5)(x^2 - 5x + 25)$$

Section 8.7

71.

$$4x^5 - 256x^2 = 4x^2\left(x^3 - 64\right)$$
$$= 4x^2\left(x - 4\right)\left(x^2 + 4x + 16\right)$$

TRY IT YOURSELF

73.

$$64a^3 - 125b^6$$
$$= \left(4a\right)^3 - \left(5b^2\right)^3$$
$$= \left(4a - 5b^2\right)\left(\left(4a\right)^2 + \left(4a\right)\left(5b^2\right) + \left(5b^2\right)^2\right)$$
$$= \left(4a - 5b^2\right)\left(16a^2 + 20ab^2 + 25b^4\right)$$

75.

$$288b^2 - 2b^6 = 2b^2\left(144 - b^4\right)$$
$$= 2b^2\left(12 + b^2\right)\left(12 - b^2\right)$$

77.

$$x^2 - y^2 + 8x + 8y = \left(x^2 - y^2\right) + \left(8x + 8y\right)$$
$$= \left(x - y\right)\left(x + y\right) + 8\left(x + y\right)$$
$$= \left(x + y\right)\left(x - y + 8\right)$$

79.

$$x^9 + y^9 = \left(x^3\right)^3 + \left(y^3\right)^3$$
$$= \left(x^3 + y^3\right)\left(x^6 - x^3 y^3 + y^6\right)$$
$$= \left(x + y\right)\left(x^2 - xy + y^2\right)\left(x^6 - x^3 y^3 + y^6\right)$$

81.

$$144a^2 t^2 - 169b^6 = \left(12at\right)^2 - \left(13b^3\right)^2$$
$$= \left(12at + 13b^3\right)\left(12at - 13b^3\right)$$

83.

$$100a^2 + 9b^2 \text{ is prime}$$

85.

$$81c^4 d^4 - 16t^4 = \left(9c^2 d^2\right)^2 - \left(4t^2\right)$$
$$= \left(9c^2 d^2 + 4t^2\right)\left(9c^2 d^2 - 4t^2\right)$$
$$= \left(9c^2 d^2 + 4t^2\right)\left(3cd + 2t\right)\left(3cd - 2t\right)$$

87.

$$128u^2 v^3 - 2t^3 u^2 = 2u^2\left(64v^3 - t^3\right)$$
$$= 2u^2\left(4v - t\right)\left(16v^2 + 4tv + t^2\right)$$

89.

$$y^2 - \left(2x - t\right)^2 = \left(y + \left(2x - t\right)\right)\left(y - \left(2x - t\right)\right)$$
$$= \left(y + 2x - t\right)\left(y - 2x + t\right)$$

91.

$$x^2 + 20x + 100 - 9z^2 = \left(x + 10\right)^2 - 9z^2$$
$$= \left(x + 10 - 3z\right)\left(x + 10 + 3z\right)$$

93.

$$\left(c - d\right)^3 + 216$$
$$= \left(c - d\right)^3 + 6^3$$
$$= \left(c - d + 6\right)\left(\left(c - d\right)^2 - 6\left(c - d\right) + 6^2\right)$$
$$= \left(c - d + 6\right)\left(c^2 - 2cd + d^2 - 6c + 6d + 36\right)$$

95.

$$\frac{1}{36} - y^4 = \left(\frac{1}{6}\right)^2 - \left(y^2\right)^2$$
$$= \left(\frac{1}{6} + y^2\right)\left(\frac{1}{6} - y^2\right)$$

97.

$$m^6 - 64$$
$$= \left(m^3\right)^2 - 8^2$$
$$= \left(m^3 + 8\right)\left(m^3 - 8\right)$$
$$= \left(m + 2\right)\left(m^2 - 2m + 4\right)\left(m - 2\right)\left(m^2 + 2m + 4\right)$$

99.

$$\left(a + b\right)x^3 + 27\left(a + b\right)$$
$$= \left(a + b\right)\left(x^3 + 27\right)$$
$$= \left(a + b\right)\left(x + 3\right)\left(x^2 - 3x + 9\right)$$

101.

$$x^9 - y^{12} z^{15} = \left(x^3\right)^3 - \left(y^4 z^5\right)^3$$
$$= \left(x^3 - y^4 z^5\right)\left(x^6 + x^3 y^4 z^5 + y^8 z^{10}\right)$$

LOOK ALIKES

103. a. $q^2 - 64 = (q+8)(q-8)$

    b. $q^3 - 64 = (q-4)(q^2+4q+16)$

105. a. $d^2 - 25 = (d-5)(d+5)$

    b. $d^3 - 125 = (d-5)(d^2+5d+25)$

107. a. $a^6 - b^3 = (a^2-b)(a^4+a^2b+b^2)$

    b. $a^6 + b^3 = (a^2+b)(a^4-a^2b+b^2)$

109.

    a. $125m^3 + 8n^3 = (5m+2n)(25m^2-10mn+4n^2)$

    b. $125m^3 - 8n^3 = (5m-2n)(25m^2+10mn+4n^2)$

121. $a^{3b} - c^{3b} = (a^b)^3 - (c^b)^3$
$$= (a^b - c^b)(a^{2b} + a^b c^b + c^{2b})$$

123.

$x^{32} - y^{32}$

$$= (x^{16} + y^{16})(x^{16} - y^{16})$$

$$= (x^{16} + y^{16})(x^8 + y^8)(x^8 - y^8)$$

$$= (x^{16} + y^{16})(x^8 + y^8)(x^4 + y^4)(x^4 - y^4)$$

$$= (x^{16} + y^{16})(x^8 + y^8)(x^4 + y^4)(x^2 + y^2)(x^2 - y^2)$$

$$= (x^{16} + y^{16})(x^8 + y^8)(x^4 + y^4)(x^2 + y^2)(x+y)(x-y)$$

APPLICATIONS

111. CANDY

$$V = \frac{4}{3}\pi r_1^3 - \frac{4}{3}\pi r_2^3$$

$$= \frac{4}{3}\pi(r_1^3 - r_2^3)$$

$$= \frac{4}{3}\pi(r_1 - r_2)(r_1^2 + r_1 r_2 + r_2^2)$$

WRITING

113. Answers will vary.

115. Answers will vary.

REVIEW

117. $\frac{25}{45} = 0.55$ \$/min ; $\frac{35}{60} = 0.58$ \$/min

    45 minutes for \$25 is the better buy.

CHALLENGE PROBLEMS

119. $4x^{2n} - 9y^{2n} = (2x^n)^2 - (3y^n)^2$
$$= (2x^n - 3y^n)(2x^n + 3y^n)$$

Section 8.7

SECTION 8.8

VOCABULARY

1. A quotient of two polynomials, such as $\frac{x^2+x}{x^2-3x}$, is called a **rational** expression.

3. The **domain** of a function is the set of all permissible input values for the variable.

5. To simplify a rational expression, we remove factors **common** to the numerator and denominator.

7. In the rational expression $\frac{(x+2)(3x-1)}{(x+2)(4x+2)}$, $x+2$ is a common **factor** of the numerator and the denominator.

9. To **build** a rational expression, we multiply it by a form of 1. For example, $\frac{2}{n^2} \cdot \frac{8}{8} = \frac{16}{8n^2}$.

CONCEPTS

11. a. $f(0) = \dfrac{2(0)+1}{(0)^2+3(0)-4}$

    $= \dfrac{0+1}{0+0-4}$

    $= -\dfrac{1}{4}$

    b. $f(2) = \dfrac{2(2)+1}{(2)^2+3(2)-4}$

    $= \dfrac{4+1}{4+6-4}$

    $= \dfrac{5}{6}$

    c. $f(1) = \dfrac{2(1)+1}{(1)^2+3(1)-4}$

    $= \dfrac{2+1}{1+3-4}$

    $= \dfrac{3}{0}$; undefined

13. a. $f(1) = -1$

    b. $f(4) = 2$

    c. $x = 2$

    d. $x = 5$

15. To multiply rational expressions, multiply their **numerators** and multiply their **denominators**. To divide two rational expressions, multiply the first by the **reciprocal** of the second.

17. To find the least common denominator of several rational expressions, **factor** each denominator completely. The LCD is a product that uses each different factor the **greatest** number of times it appears in any one factorization.

19. a. twice
    b. once

21. a. $3a$
    b. $(x-2)(x+5)$

NOTATION

23. $\dfrac{5x^2+35x}{1}$

GUIDED PRACTICE

25.
$$f(x) = \frac{2}{x}$$
$$x \neq 0$$
$$(-\infty, 0) \cup (0, \infty)$$

27.
$$f(x) = \frac{2x}{x+2}$$
$$x + 2 \neq 0$$
$$x \neq -2$$
$$(-\infty, -2) \cup (-2, \infty)$$

29.
$$f(x) = \frac{3x-1}{x-x^2}$$
$$x - x^2 \neq 0$$
$$x(1-x) \neq 0$$
$$x \neq 0 \quad \text{and} \quad 1-x \neq 0$$
$$x \neq 0 \qquad\qquad 1 \neq x$$
$$(-\infty, 0) \cup (0, 1) \cup (1, \infty)$$

31.
$$f(x) = \frac{x^2 + 3x + 2}{x^2 - x - 56}$$
$$x^2 - x - 56 \neq 0$$
$$(x-8)(x+7) \neq 0$$
$$x - 8 \neq 0 \text{ and } x + 7 \neq 0$$
$$x \neq 8 \qquad x \neq -7$$
$$(-\infty, -7) \cup (-7, 8) \cup (8, \infty)$$

33.
$$\frac{15a^2}{25a^8} = \frac{3 \cdot \cancel{5} \cdot \cancel{a} \cdot \cancel{a}}{5 \cdot \cancel{5} \cdot \cancel{a} \cdot \cancel{a} \cdot a \cdot a \cdot a \cdot a \cdot a \cdot a}$$
$$= \frac{3}{5a^6}$$

35.
$$\frac{24x^3 y^4}{54x^4 y^3} = \frac{\cancel{2} \cdot 2 \cdot 2 \cdot \cancel{3} \cdot \cancel{x} \cdot \cancel{x} \cdot \cancel{x} \cdot \cancel{y} \cdot \cancel{y} \cdot \cancel{y} \cdot y}{\cancel{2} \cdot \cancel{3} \cdot 3 \cdot 3 \cdot \cancel{x} \cdot \cancel{x} \cdot \cancel{x} \cdot x \cdot \cancel{y} \cdot \cancel{y} \cdot \cancel{y}}$$
$$= \frac{4y}{9x}$$

37.
$$\frac{5x^2 - 10x}{x^2 - 4x + 4} = \frac{5x \cancel{(x-2)}}{(x-2)\cancel{(x-2)}}$$
$$= \frac{5x}{x-2}$$

39.
$$\frac{6x^2 - 7x - 5}{2x^2 + 5x + 2} = \frac{(3x-5)\cancel{(2x+1)}}{\cancel{(2x+1)}(x+2)}$$
$$= \frac{3x-5}{x+2}$$

41.
$$\frac{4 - x^2}{x^2 - x - 2} = \frac{(2-x)(2+x)}{(x-2)(x+1)}$$
$$= \frac{-\cancel{(x-2)}(x+2)}{\cancel{(x-2)}(x+1)}$$
$$= \frac{-(x+2)}{(x+1)}$$
$$= -\frac{(x+2)}{(x+1)}$$

43.
$$\frac{p^3 + p^2 q - 2pq^2}{pq^2 + p^2 q - 2p^3} = \frac{p(p^2 + pq - 2q^2)}{p(q^2 + pq - 2p^2)}$$
$$= \frac{p(p+2q)(p-q)}{p(q-p)(q+2p)}$$
$$= \frac{-1\cancel{p}(p+2q)\cancel{(q-p)}}{\cancel{p}\cancel{(q-p)}(q+2p)}$$
$$= -\frac{p+2q}{q+2p} \quad \text{or} \quad \frac{-p-2q}{q+2p}$$

45
$$\frac{10a^2}{3b^4} \cdot \frac{12b^3}{5a^2} = \frac{2 \cdot \cancel{5} \cdot \cancel{a} \cdot \cancel{a} \cdot 2 \cdot 2 \cdot \cancel{3} \cdot \cancel{b} \cdot \cancel{b} \cdot \cancel{b}}{\cancel{3} \cdot \cancel{b} \cdot \cancel{b} \cdot \cancel{b} \cdot b \cdot \cancel{5} \cdot \cancel{a} \cdot \cancel{a}}$$
$$= \frac{8}{b}$$

47.
$$\frac{3p^2}{6p+24} \cdot \frac{p^2 - 16}{6p} = \frac{\cancel{3} \cdot \cancel{p} \cdot p}{6\cancel{(p+4)}} \cdot \frac{\cancel{(p+4)}(p-4)}{2 \cdot \cancel{3} \cdot \cancel{p}}$$
$$= \frac{p(p-4)}{12}$$

49.
$$\frac{x^2 + 2x + 1}{9x} \cdot \frac{2x^2 - 2x}{2x^2 - 2} = \frac{(x^2 + 2x + 1)(2x^2 - 2x)}{9x(2x^2 - 2)}$$
$$= \frac{(x+1)(x+1)2x(x-1)}{9x(2)(x^2 - 1)}$$
$$= \frac{\cancel{(x+1)}(x+1)\cancel{(2)}\cancel{(x)}\cancel{(x-1)}}{9\cancel{x}\cancel{(2)}\cancel{(x-1)}\cancel{(x+1)}}$$
$$= \frac{x+1}{9}$$

51.
$$\frac{2x^2 - x - 3}{x^2 - 1} \cdot \frac{x^2 + x - 2}{2x^2 + x - 6}$$
$$= \frac{(2x^2 - x - 3)(x^2 + x - 2)}{(x^2 - 1)(2x^2 + x - 6)}$$
$$= \frac{\cancel{(2x-3)}\cancel{(x+1)}\cancel{(x+2)}\cancel{(x-1)}}{\cancel{(x+1)}\cancel{(x-1)}\cancel{(2x-3)}\cancel{(x+2)}}$$
$$= 1$$

Section 8.8

53.

$$\left(6a - a^2\right) \cdot \frac{a^3}{a^3 - 6a^2 + 3a - 18}$$

$$= \frac{a(6-a)}{1} \cdot \frac{a^3}{(a^2+3)(a-6)}$$

$$= -\frac{a(a-6)}{1} \cdot \frac{a^3}{(a^2+3)(a-6)}$$

$$= -\frac{a^4}{a^2+3}$$

55.

$$\left(x^2 + x - 2cx - 2c\right) \cdot \frac{x^2 + 3x + 2}{4c^2 - x^2}$$

$$= \frac{\left(x^2 + x - 2cx - 2c\right)\left(x^2 + 3x + 2\right)}{4c^2 - x^2}$$

$$= \frac{(x(x+1) - 2c(x+1))(x+2)(x+1)}{(2c+x)(2c-x)}$$

$$= -\frac{(x+1)(x-2c)(x+2)(x+1)}{(2c+x)(x-2c)}$$

$$= -\frac{(x+1)^2(x+2)}{(2c+x)}$$

57.

$$\frac{m^2 n}{4} \div \frac{mn^3}{6} = \frac{m^2 n}{4} \cdot \frac{6}{mn^3}$$

$$= \frac{m \cdot m \cdot n \cdot 2 \cdot 3}{2 \cdot 2 \cdot m \cdot n \cdot n \cdot n}$$

$$= \frac{3m}{2n^2}$$

59.

$$\frac{x^2 - 16}{x^2 - 25} \div \frac{x+4}{x-5} = \frac{\left(x^2 - 16\right)(x-5)}{\left(x^2 - 25\right)(x+4)}$$

$$= \frac{(x+4)(x-4)(x-5)}{(x+5)(x-5)(x+4)}$$

$$= \frac{x-4}{x+5}$$

61.

$$\frac{5c+1}{6} \div \frac{125c^3 + 1}{6c+6}$$

$$= \frac{5c+1}{6} \cdot \frac{6c+6}{125c^3 + 1}$$

$$= \frac{5c+1}{6} \cdot \frac{6(c+1)}{(5c+1)(25c^2 - 5c + 1)}$$

$$= \frac{c+1}{25c^2 - 5c + 1}$$

63.

$$\frac{3n^2 + 5n - 2}{12n^2 - 13n + 3} \div \frac{n^2 + 3n + 2}{4n^2 + 5n - 6}$$

$$= \frac{3n^2 + 5n - 2}{12n^2 - 13n + 3} \cdot \frac{4n^2 + 5n - 6}{n^2 + 3n + 2}$$

$$= \frac{\left(3n^2 + 5n - 2\right)\left(4n^2 + 5n - 6\right)}{\left(12n^2 - 13n + 3\right)\left(n^2 + 3n + 2\right)}$$

$$= \frac{(3n-1)(n+2)(4n-3)(n+2)}{(4n-3)(3n-1)(n+2)(n+1)}$$

$$= \frac{n+2}{n+1}$$

65.

$$\frac{3x}{x^2 - 9} - \frac{9}{x^2 - 9} = \frac{3x - 9}{x^2 - 9}$$

$$= \frac{3(x-3)}{(x-3)(x+3)}$$

$$= \frac{3}{x+3}$$

67.

$$\frac{3y-2}{2y+6} - \frac{2y-5}{2y+6} = \frac{3y - 2 - (2y-5)}{2y+6}$$

$$= \frac{3y - 2 - 2y + 5}{2y+6}$$

$$= \frac{y+3}{2y+6}$$

$$= \frac{y+3}{2(y+3)}$$

$$= \frac{1}{2}$$

69.

$$\frac{3}{4x} + \frac{2}{3x} = \frac{3}{4x}\left(\frac{3}{3}\right) + \frac{2}{3x}\left(\frac{4}{4}\right)$$

$$= \frac{9}{12x} + \frac{8}{12x}$$

$$= \frac{9+8}{12x}$$

$$= \frac{17}{12x}$$

71.

$$\frac{8}{9y^2} + \frac{1}{6y^4} = \frac{8}{9y^2}\left(\frac{2y^2}{2y^2}\right) + \frac{1}{6y^4}\left(\frac{3}{3}\right)$$

$$= \frac{16y^2}{18y^4} + \frac{3}{18y^4}$$

$$= \frac{16y^2 + 3}{18y^4}$$

73.

$$\frac{3}{4ab^2} - \frac{5}{2a^2b} = \frac{3}{4ab^2}\left(\frac{a}{a}\right) - \frac{5}{2a^2b}\left(\frac{2b}{2b}\right)$$

$$= \frac{3a}{4a^2b^2} - \frac{10b}{4a^2b^2}$$

$$= \frac{3a - 10b}{4a^2b^2}$$

75.

$$\frac{y-7}{y^2} - \frac{y+7}{2y} = \frac{y-7}{y^2}\left(\frac{2}{2}\right) - \frac{y+7}{2y}\left(\frac{y}{y}\right)$$

$$= \frac{2y-14}{2y^2} - \frac{y^2+7y}{2y^2}$$

$$= \frac{2y-14-\left(y^2+7y\right)}{2y^2}$$

$$= \frac{2y-14-y^2-7y}{2y^2}$$

$$= \frac{-y^2-5y-14}{2y^2}$$

$$= -\frac{y^2+5y+14}{2y^2}$$

77.

$$\frac{3}{x+2} + \frac{5}{x-4} = \frac{3}{x+2}\left(\frac{x-4}{x-4}\right) + \frac{5}{x-4}\left(\frac{x+2}{x+2}\right)$$

$$= \frac{3(x-4)}{(x+2)(x-4)} + \frac{5(x+2)}{(x+2)(x-4)}$$

$$= \frac{3x-12}{(x+2)(x-4)} + \frac{5x+10}{(x+2)(x-4)}$$

$$= \frac{3x-12+5x+10}{(x+2)(x-4)}$$

$$= \frac{8x-2}{(x+2)(x-4)}$$

$$= \frac{2(4x-1)}{(x+2)(x-4)}$$

79.

$$\frac{x+2}{x+5} - \frac{x-3}{x+7}$$

$$= \frac{x+2}{x+5}\left(\frac{x+7}{x+7}\right) - \frac{x-3}{x+7}\left(\frac{x+5}{x+5}\right)$$

$$= \frac{(x+2)(x+7)}{(x+5)(x+7)} - \frac{(x-3)(x+5)}{(x+5)(x+7)}$$

$$= \frac{x^2+7x+2x+14}{(x+5)(x+7)} - \frac{x^2+5x-3x-15}{(x+5)(x+7)}$$

$$= \frac{x^2+9x+14}{(x+5)(x+7)} - \frac{x^2+2x-15}{(x+5)(x+7)}$$

$$= \frac{x^2+9x+14-\left(x^2+2x-15\right)}{(x+5)(x+7)}$$

$$= \frac{x^2+9x+14-x^2-2x+15}{(x+5)(x+7)}$$

$$= \frac{7x+29}{(x+5)(x+7)}$$

Section 8.8

81.

$$\frac{x}{x^2+5x+6}+\frac{x}{x^2-4}$$

$$=\frac{x}{(x+2)(x+3)}+\frac{x}{(x+2)(x-2)}$$

$$=\frac{x}{(x+2)(x+3)}\left(\frac{x-2}{x-2}\right)+\frac{x}{(x+2)(x-2)}\left(\frac{x+3}{x+3}\right)$$

$$=\frac{x^2-2x}{(x+2)(x-2)(x+3)}+\frac{x^2+3x}{(x+2)(x-2)(x+3)}$$

$$=\frac{x^2-2x+x^2+3x}{(x+2)(x-2)(x+3)}$$

$$=\frac{2x^2+x}{(x+2)(x-2)(x+3)}$$

$$=\frac{x(2x+1)}{(x+2)(x-2)(x+3)}$$

83.

$$\frac{4}{x^2-2x-3}-\frac{x}{3x^2-7x-6}$$

$$=\frac{4}{(x-3)(x+1)}-\frac{x}{(3x+2)(x-3)}$$

$$=\frac{4}{(x-3)(x+1)}\left(\frac{3x+2}{3x+2}\right)-\frac{x}{(3x+2)(x-3)}\left(\frac{x+1}{x+1}\right)$$

$$=\frac{12x+8}{(x-3)(x+1)(3x+2)}-\frac{x^2+x}{(3x+2)(x-3)(x+1)}$$

$$=\frac{12x+8-(x^2+x)}{(x-3)(x+1)(3x+2)}$$

$$=\frac{12x+8-x^2-x}{(x-3)(x+1)(3x+2)}$$

$$=\frac{-x^2+11x+8}{(x-3)(x+1)(3x+2)}$$

85.

$$\frac{6}{5d^2-5d}-\frac{3}{5d-5}=\frac{6}{5d(d-1)}-\frac{3}{5(d-1)}$$

$$=\frac{6}{5d(d-1)}-\frac{3}{5(d-1)}\left(\frac{d}{d}\right)$$

$$=\frac{6}{5d(d-1)}-\frac{3d}{5d(d-1)}$$

$$=\frac{6-3d}{5d(d-1)}$$

87.

$$\frac{m}{m^2+9m+20}-\frac{4}{m^2+7m+12}$$

$$=\frac{m}{(m+4)(m+5)}-\frac{4}{(m+3)(m+4)}$$

$$=\frac{m}{(m+4)(m+5)}\left(\frac{m+3}{m+3}\right)-\frac{4}{(m+3)(m+4)}\left(\frac{m+5}{m+5}\right)$$

$$=\frac{m^2+3m}{(m+3)(m+4)(m+5)}-\frac{4m+20}{(m+3)(m+4)(m+5)}$$

$$=\frac{m^2+3m-(4m+20)}{(m+3)(m+4)(m+5)}$$

$$=\frac{m^2+3m-4m-20}{(m+3)(m+4)(m+5)}$$

$$=\frac{m^2-m-20}{(m+3)(m+4)(m+5)}$$

$$=\frac{(m-5)\cancel{(m+4)}}{(m+3)\cancel{(m+4)}(m+5)}$$

$$=\frac{m-5}{(m+3)(m+5)}$$

89.

$$\frac{3}{y}+\frac{7}{2y}=13$$

$$2y\left(\frac{3}{y}+\frac{7}{2y}\right)=2y(13)$$

$$2(3)+7=26y$$

$$6+7=26y$$

$$13=26y$$

$$\frac{13}{26}=y$$

$$\frac{1}{2}=y$$

91.

$$\frac{2}{x} + \frac{1}{2} = \frac{9}{4x} - \frac{1}{2x}$$

$$4x\left(\frac{2}{x} + \frac{1}{2}\right) = 4x\left(\frac{9}{4x} - \frac{1}{2x}\right)$$

$$4(2) + 2x(1) = 9 - 2(1)$$

$$8 + 2x = 9 - 2$$

$$8 + 2x = 7$$

$$2x = -1$$

$$x = -\frac{1}{2}$$

93.

$$\frac{a}{2} = \frac{a-6}{3a-9} - \frac{1}{3}$$

$$\frac{a}{2} = \frac{a-6}{3(a-3)} - \frac{1}{3}$$

$$6(a-3)\left(\frac{a}{2}\right) = 6(a-3)\left(\frac{a-6}{3(a-3)} - \frac{1}{3}\right)$$

$$3a(a-3) = 2(a-6) - 2(a-3)$$

$$3a^2 - 9a = 2a - 12 - 2a + 6$$

$$3a^2 - 9a = -6$$

$$3a^2 - 9a + 6 = 0$$

$$3(a^2 - 3a + 2) = 0$$

$$3(a-1)(a-2) = 0$$

$$a - 1 = 0 \quad \text{or} \quad a - 2 = 0$$

$$a = 1 \qquad\qquad a = 2$$

95.

$$\frac{2}{5x-5} + \frac{x-2}{15} = \frac{4}{5x-5}$$

$$\frac{2}{5(x-1)} + \frac{x-2}{15} = \frac{4}{5(x-1)}$$

$$15(x-1)\left(\frac{2}{5(x-1)} + \frac{x-2}{15}\right) = 15(x-1)\left(\frac{4}{5(x-1)}\right)$$

$$3(2) + (x-1)(x-2) = 3(4)$$

$$6 + x^2 - 2x - x + 2 = 12$$

$$x^2 - 3x + 8 = 12$$

$$x^2 - 3x - 4 = 0$$

$$(x-4)(x+1) = 0$$

$$x - 4 = 0 \quad \text{or} \quad x + 1 = 0$$

$$x = 4 \qquad\qquad x = -1$$

97.

$$f(x) = \frac{x^2 + 6x - 16}{x^2 - 4}$$

$$= \frac{(x+8)(x-2)}{(x+2)(x-2)}$$

$$= \frac{x+8}{x+2}; x \neq -2 \text{ and } x \neq 2$$

99.

$$g(x) = \frac{x^3 + 64}{x^3 + 4x^2 + 3x + 12}$$

$$= \frac{(x)^3 + (4)^3}{x^2(x+4) + 3(x+4)}$$

$$= \frac{(x+4)(x^2 - 4x + 16)}{(x+4)(x^2 + 3)}$$

$$= \frac{x^2 - 4x + 16}{x^2 + 3}; \ x \neq -4$$

101.

$$\frac{p^3 - q^3}{q^2 - p^2} \cdot \frac{q^2 + pq}{p^3 + p^2q + pq^2}$$

$$= \frac{(p^3 - q^3)(q^2 + pq)}{(q^2 - p^2)(p^3 + p^2q + pq^2)}$$

$$= \frac{(p-q)(p^2 + pq + q^2)(q)(q+p)}{(q+p)(q-p)(p)(p^2 + pq + q^2)}$$

$$= -\frac{(q-p)(p^2 + pq + q^2)(q)(q+p)}{(q+p)(q-p)(p)(p^2 + pq + q^2)}$$

$$= -\frac{q}{p}$$

Section 8.8

103.

$$\frac{t}{t^2+5t+6}-\frac{2}{t^2+3t+2}$$

$$=\frac{t}{(t+2)(t+3)}-\frac{2}{(t+2)(t+1)}$$

$$=\frac{t}{(t+2)(t+3)}\left(\frac{t+1}{t+1}\right)-\frac{2}{(t+2)(t+1)}\left(\frac{t+3}{t+3}\right)$$

$$=\frac{t^2+t}{(t+1)(t+2)(t+3)}-\frac{2t+6}{(t+1)(t+2)(t+3)}$$

$$=\frac{t^2+t-(2t+6)}{(t+1)(t+2)(t+3)}$$

$$=\frac{t^2+t-2t-6}{(t+1)(t+2)(t+3)}$$

$$=\frac{t^2-t-6}{(t+1)(t+2)(t+3)}$$

$$=\frac{(t-3)\cancel{(t+2)}}{(t+1)\cancel{(t+2)}(t+3)}$$

$$=\frac{t-3}{(t+1)(t+3)}$$

105.

$$\frac{4}{m^2-9}+\frac{5}{m^2-m-12}=\frac{7}{m^2-7m+12}$$

$$\frac{4}{(m-3)(m+3)}+\frac{5}{(m-4)(m+3)}=\frac{7}{(m-3)(m-4)}$$

Multiply each term by the LCD of $(m+3)(m-3)(m-4)$.

$$4(m-4)+5(m-3)=7(m+3)$$
$$4m-16+5m-15=7m+21$$
$$9m-31=7m+21$$
$$9m-31-7m=7m+21-7m$$
$$2m-31=21$$
$$2m-31+31=21+31$$
$$2m=52$$
$$m=26$$

107.

$$\frac{2}{x-1}-\frac{2x}{x^2-1}-\frac{x}{x^2+2x+1}$$

$$=\frac{2}{x-1}-\frac{2x}{(x+1)(x-1)}-\frac{x}{(x+1)(x+1)}$$

$$=\frac{2}{x-1}\left(\frac{(x+1)(x+1)}{(x+1)(x+1)}\right)-\frac{2x}{(x+1)(x-1)}\left(\frac{x+1}{x+1}\right)-\frac{x}{(x+1)(x+1)}\left(\frac{x-1}{x-1}\right)$$

$$=\frac{2(x^2+2x+1)}{(x-1)(x+1)^2}-\frac{2x(x+1)}{(x-1)(x+1)^2}-\frac{x(x-1)}{(x-1)(x+1)^2}$$

$$=\frac{2x^2+4x+2}{(x-1)(x+1)^2}-\frac{2x^2+2x}{(x-1)(x+1)^2}-\frac{x^2-x}{(x-1)(x+1)^2}$$

$$=\frac{2x^2+4x+2-(2x^2+2x)-(x^2-x)}{(x-1)(x+1)^2}$$

$$=\frac{2x^2+4x+2-2x^2-2x-x^2+x}{(x-1)(x+1)^2}$$

$$=\frac{-x^2+3x+2}{(x-1)(x+1)^2}$$

109.

$$(2x^2-15x+25)\div\frac{2x^2-3x-5}{x+1}$$

$$=\frac{2x^2-15x+25}{1}\cdot\frac{x+1}{2x^2-3x-5}$$

$$=\frac{(2x-5)(x-5)}{1}\cdot\frac{(x+1)}{(2x-5)(x+1)}$$

$$=x-5$$

111.

$$\frac{y^3-x^3}{2x^2+2xy+x+y}\cdot\frac{2x^2-5x-3}{yx-3y-x^2+3x}$$

$$=\frac{(y-x)(y^2+xy+x^2)}{2x(x+y)+1(x+y)}\cdot\frac{(2x+1)(x-3)}{y(x-3)-x(x-3)}$$

$$=\frac{\cancel{(y-x)}(y^2+xy+x^2)}{(x+y)\cancel{(2x+1)}}\cdot\frac{\cancel{(2x+1)}\cancel{(x-3)}}{\cancel{(x-3)}\cancel{(y-x)}}$$

$$=\frac{y^2+xy+x^2}{x+y}$$

113.

$$\frac{24n^4}{16n^4+24n^3}=\frac{24n^4}{8n^3(2n+3)}$$

$$=\frac{\cancel{2}\cdot\cancel{2}\cdot\cancel{2}\cdot3\cdot\cancel{n}\cdot\cancel{n}\cdot\cancel{n}\cdot n}{\cancel{2}\cdot\cancel{2}\cdot\cancel{2}\cdot\cancel{n}\cdot\cancel{n}\cdot\cancel{n}(2n+3)}$$

$$=\frac{3n}{2n+3}$$

115.

$$1 + x - \frac{x}{x-5} = 1\left(\frac{x-5}{x-5}\right) + x\left(\frac{x-5}{x-5}\right) - \frac{x}{x-5}$$

$$= \frac{x-5}{x-5} + \frac{x^2-5x}{x-5} - \frac{x}{x-5}$$

$$= \frac{x-5+x^2-5x-x}{x-5}$$

$$= \frac{x^2-5x-5}{x-5}$$

117.

$$\frac{ax+by+ay+bx}{a^2-b^2} = \frac{(ax+bx)+(ay+by)}{(a+b)(a-b)}$$

$$= \frac{x(a+b)+y(a+b)}{(a+b)(a-b)}$$

$$= \frac{(x+y)\,\cancel{(a+b)}}{\cancel{(a+b)}(a-b)}$$

$$= \frac{x+y}{a-b}$$

119.

$$\frac{3}{s-2} + \frac{s-14}{2s^2-3s-2} - \frac{4}{2s+1} = 0$$

$$\frac{3}{s-2} + \frac{s-14}{(s-2)(2s+1)} - \frac{4}{2s+1} = 0$$

Multiply each term by the LCD of $(s-2)(2s+1)$.

$$\frac{3}{s-2} + \frac{s-14}{(s-2)(2s+1)} - \frac{4}{2s+1} = 0$$

$$3(2s+1)+(s-14)-4(s-2) = 0$$

$$6s+3+s-14-4s+8 = 0$$

$$3s-3 = 0$$

$$3s = 3$$

$$s = 1$$

121.

$$\frac{5}{x+4} + \frac{1}{x+4} = x-1$$

$$(x+4)\left(\frac{5}{x+4} + \frac{1}{x+4}\right) = (x+4)(x-1)$$

$$5+1 = x^2-x+4x-4$$

$$6 = x^2+3x-4$$

$$0 = x^2+3x-10$$

$$0 = (x+5)(x-2)$$

$$x+5=0 \quad \text{or} \quad x-2=0$$

$$x=-5 \qquad x=2$$

123.

$$\frac{5x}{x-3} + \frac{4x}{3-x} = \frac{5x}{x-3} - \frac{4x}{x-3}$$

$$= \frac{5x-4x}{x-3}$$

$$= \frac{x}{x-3}$$

APPLICATIONS

125. ENVIRONMENTAL CLEANUP
   a.
$$f(50) = \frac{50,000(50)}{100-50}$$

$$= \frac{2,500,000}{50}$$

$$= \$50,000$$

   b.
$$f(80) = \frac{50,000(80)}{100-80}$$

$$= \frac{4,000,000}{20}$$

$$= \$200,000$$

127. UTILITY COSTS
   a. $c(n) = 0.09n + 7.50$

   b. $c(n) = \dfrac{0.09n + 7.50}{n}$

   c.
$$c(775) = \frac{0.09(775)+7.50}{775}$$

$$= \frac{69.75+7.50}{775}$$

$$= \frac{77.25}{775}$$

$$\approx \$0.10$$

WRITING

129. Answers will vary.

131. Answers will vary.

Section 8.8

REVIEW

133.

$$-10|16x+4|-3=-3$$

$$-10|16x+4|-3+3=-3+3$$

$$-10|16x+4|=0$$

$$\frac{-10|16x+4|}{-10}=\frac{0}{-10}$$

$$|16x+4|=0$$

$$16x+4=0$$

$$16x=-4$$

$$x=-\frac{4}{16}$$

$$x=-\frac{1}{4}$$

CHALLENGE PROBLEMS

135.

$$\frac{a^6-64}{\left(a^2+2a+4\right)\left(a^2-2a+4\right)}$$

$$=\frac{\left(a^3+8\right)\left(a^3-8\right)}{\left(a^2+2a+4\right)\left(a^2-2a+4\right)}$$

$$=\frac{(a+2)\left(a^2-2a+4\right)(a-2)\left(a^2+2a+4\right)}{\left(a^2+2a+4\right)\left(a^2-2a+4\right)}$$

$$=(a+2)(a-2)$$

137.

$$\left[\left(x^{-1}+1\right)^{-1}+1\right]^{-1}=\frac{1}{\left[\left(x^{-1}+1\right)^{-1}+1\right]}$$

$$=\frac{1}{\dfrac{1}{\left(x^{-1}+1\right)}+1}$$

$$=\frac{1}{\dfrac{1}{\dfrac{1}{x}+1}+1}$$

$$=\frac{1}{\dfrac{1}{\dfrac{1}{x}+\dfrac{1}{1}\cdot\dfrac{x}{x}}+1}$$

$$=\frac{1}{\dfrac{1}{\dfrac{1}{x}+\dfrac{x}{x}}+1}$$

$$=\frac{1}{\dfrac{1}{\dfrac{x+1}{x}}+1}$$

$$=\frac{1}{\dfrac{x}{x+1}+1}$$

$$=\frac{1}{\dfrac{x}{x+1}+\dfrac{1}{1}\cdot\dfrac{x+1}{x+1}}$$

$$=\frac{1}{\dfrac{x}{x+1}+\dfrac{x+1}{x+1}}$$

$$=\frac{1}{\dfrac{2x+1}{x+1}}$$

$$=\frac{x+1}{2x+1}$$

139. $f(x) = \dfrac{1}{x-1}$

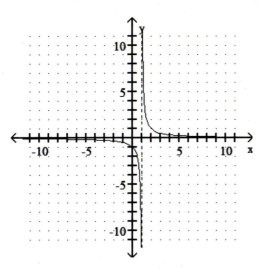

141.

$$\frac{6a^2-7a-3}{2a^2-2} \div \frac{4a^2-12a+9}{a^2-1} \cdot \frac{2a^2-a-3}{3a^2-2a-1}$$

$$= \frac{6a^2-7a-3}{2a^2-2} \cdot \frac{a^2-1}{4a^2-12a+9} \cdot \frac{2a^2-a-3}{3a^2-2a-1}$$

$$= \frac{(2a-3)(3a+1)}{2(a-1)(a+1)} \cdot \frac{(a-1)(a+1)}{(2a-3)(2a-3)} \cdot \frac{(2a-3)(a+1)}{(3a+1)(a-1)}$$

$$= \frac{a+1}{2(a-1)}$$

143.

$$\frac{h}{\dfrac{h^2+3h+2}{4}} = \frac{h}{\dfrac{(h+2)(h+1)}{4}}$$
$$\dfrac{4}{h+2} - \dfrac{4}{h+1} \qquad \dfrac{4}{h+2} - \dfrac{4}{h+1}$$

$$= \frac{\dfrac{h}{(h+2)(h+1)}}{\dfrac{4}{h+2} - \dfrac{4}{h+1}} \cdot \frac{(h+2)(h+1)}{(h+2)(h+1)}$$

$$= \frac{\dfrac{h(h+2)(h+1)}{(h+2)(h+1)}}{\dfrac{4(h+2)(h+1)}{h+2} - \dfrac{4(h+2)(h+1)}{h+1}}$$

$$= \frac{h}{4(h+1)-4(h+2)}$$

$$= \frac{h}{4h+4-4h-8}$$

$$= \frac{h}{-4}$$

$$= -\frac{h}{4}$$

Section 8.8

SECTION 8.9

VOCABULARY

1. The equation $y = kx$ defines **direct** variation: As x increased, y **increases**.

3. The equation $y = kxz$ defines **joint** variation, and $y = \dfrac{kx}{z}$ defines **combined** variation.

CONCEPTS

5. a.

b.

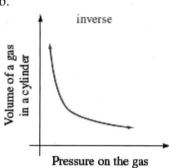

7. direct

9. inverse

11. direct

13. inverse

NOTATION

15. a. yes
  b. no
  c. no
  d. yes

GUIDED PRACTICE

17. $A = kp^2$

19. $z = \dfrac{k}{t^3}$

21. $C = kxyz$

23. $P = \dfrac{ka^2}{j^3}$

25. r varies directly as t

27. b varies inversely as h

29. U varies jointly with r, the square of s, and t.

31. P varies directly with m and inversely with n.

APPLICATIONS

33. WEBMASTER
  a. false
  b. false
  c. true

35. GRAVITY
  Let f = force and m = mass.
  Find k if $f = 49$ and $m = 5$.
  $$f = km$$
  $$49 = k(5)$$
  $$9.8 = k$$
  Find f if $m = 12$ and $k = 9.8$
  $$f = km$$
  $$f = 9.8(12)$$
  $$f = 117.6$$
  The force is 117.6 newtons.

37. FINDING DISTANCE

Let d = distance driven and g = gallons of gasoline.

Find k if $d = 288$ and $g = 12$.

$$d = kg$$

$$288 = k(12)$$

$$288 = 12k$$

$$24 = k$$

Find d if $k = 24$ and $g = 18$.

$$d = kg$$

$$d = 24(18)$$

$$d = 432$$

The car can go 432 miles.

39. FARMING

Let d = days feed will last and a = number of animals.

Find k if $d = 10$ and $a = 25$.

$$d = \frac{k}{a}$$

$$10 = \frac{k}{25}$$

$$25(10) = 25\left(\frac{k}{25}\right)$$

$$250 = k$$

Find d if $k = 250$ and $a = 10$.

$$d = \frac{k}{a}$$

$$d = \frac{250}{10}$$

$$d = 25$$

The feed will last 25 days.

41. GAS PRESSURE

Let v = volume of gas and p = pressure.

Find k when $v = 20$ and $p = 6$.

$$v = \frac{k}{p}$$

$$20 = \frac{k}{6}$$

$$6(20) = 6\left(\frac{k}{6}\right)$$

$$120 = k$$

Find v when $k = 120$ and $p = 10$.

$$v = \frac{k}{p}$$

$$v = \frac{120}{10}$$

$$v = 12 \text{ in.}^3$$

The pressure would be 12 in.^3

43. TRUCKING COSTS

Let c = costs, t = # of trucks, and h = # of hours.

Find k if $c = 1,800$, $t = 4$ and $h = 6$.

$$c = thk$$

$$1,800 = (4)(6)k$$

$$1,800 = 24k$$

$$75 = k$$

Find c if $k = 75$, $t = 10$, and $h = 12$.

$$c = thk$$

$$c = (10)(12)(75)$$

$$c = 120(75)$$

$$c = 9,000$$

The cost is $9,000.

45. ELECTRONICS

Let v = voltage, r = resistance, and c = current.

$$v = rc$$

$$6 = r \cdot 2$$

$$6 = 2r$$

$$3 = r$$

$$r = 3 \text{ ohms}$$

Section 8.9

47. STRUCTURAL ENGINEERING

Let F = force of deflection, w = width, d = depth, and k = constant. Find k using the given information that $F = 1.1$, $w = 4$, and $d = 4$.

$$F = \frac{k}{wd^3}$$

$$1.1 = \frac{k}{4(4)^3}$$

$$1.1 = \frac{k}{4(64)}$$

$$1.1 = \frac{k}{256}$$

$$256(1.1) = 256\left(\frac{k}{256}\right)$$

$$281.6 = k$$

Let $k = 281.6$, $w = 2$, and $d = 8$ to find the force of deflection.

$$F = \frac{k}{wd^3}$$

$$= \frac{281.6}{2(8)^3}$$

$$= \frac{281.6}{2(512)}$$

$$= \frac{281.6}{1,024}$$

$$= 0.275 \text{ in.}$$

The deflection is 0.275 in.

49. ELECTRONICS

Let R = resistance, l = length, and d = diameter of the wire, and k = constant. Find k using the given information that $R = 11.2$, $l = 80$, and $d = 0.01$.

$$R = \frac{kl}{d^2}$$

$$11.2 = \frac{80k}{(0.01)^2}$$

$$11.2 = \frac{80k}{0.0001}$$

$$0.00112 = 80k$$

$$0.000014 = k$$

Let $k = 0.000014$, $l = 160$, and $d = 0.04$ to find the resistance.

$$R = \frac{kl}{d^2}$$

$$R = \frac{160(0.000014)}{(0.04)^2}$$

$$R = \frac{0.00224}{0.0016}$$

$$R = 1.4$$

51. TENSION IN A STRING

Let T = tension of the string, s = speed of the ball, and r = radius of the circle, and k = constant.

Find k when $s = 6$, $T = 6$, and $r = 3$.

$$T = \frac{ks^2}{r}$$

$$6 = \frac{k \cdot (6)^2}{3}$$

$$6 = \frac{k \cdot 36}{3}$$

$$6 = 12k$$

$$0.5 = k$$

Find T given $k = 0.5$, $s = 8$ and $r = 2.5$.

$$T = \frac{ks^2}{r}$$

$$= \frac{0.5 \cdot (8)^2}{2.5}$$

$$= \frac{0.5(64)}{2.5}$$

$$= \frac{32}{2.5}$$

$$= 12.8 \text{ lb}$$

The tension is 12.8 lb.

WRITING

53. Answers will vary.

REVIEW

55.

$$2a + 10 < 7a \quad \text{and} \quad 5a - 15 < 2a$$
$$10 < 5a \qquad\qquad -15 < -3a$$
$$2 < a \qquad\qquad 5 > a$$
$$a > 2 \qquad\qquad a < 5$$

$$(2, 5)$$

57.

$$\frac{n}{2} < \frac{n}{6} + 2 \quad \text{or} \quad 0.1n + 0.1 > 1$$
$$6\left(\frac{n}{2}\right) < 6\left(\frac{n}{6} + 2\right) \quad 10(0.1n + 0.1) > 10(1)$$
$$3n < n + 12 \qquad\qquad n + 1 > 10$$
$$2n < 12 \qquad\qquad n > 9$$
$$n < 6$$

$$(-\infty, 6) \cup (9, \infty)$$

CHALLENGE PROBLEMS

59. No; change times cost is not a constant.

SECTION 8.1
Review of Solving Linear Equations,
Formulas, and Linear Inequalities

1.
$$5x + 12 = 0$$
$$5x + 12 - 12 = 0 - 12$$
$$5x = -12$$
$$\frac{5x}{5} = \frac{-12}{5}$$
$$x = -\frac{12}{5}$$

2.
$$-3x - 7 + x = 6x + 20 - 5x$$
$$-2x - 7 = x + 20$$
$$-2x - 7 - x = x + 20 - x$$
$$-3x - 7 = 20$$
$$-3x - 7 + 7 = 20 + 7$$
$$-3x = 27$$
$$\frac{-3x}{-3} = \frac{27}{-3}$$
$$x = -9$$

3.
$$4(y - 1) = 28$$
$$4y - 4 = 28$$
$$4y - 4 + 4 = 28 + 4$$
$$4y = 32$$
$$\frac{4y}{4} = \frac{32}{4}$$
$$y = 8$$

4.
$$2 - 13(x - 1) = 4 - 6x$$
$$2 - 13x + 13 = 4 - 6x$$
$$15 - 13x = 4 - 6x$$
$$15 - 13x + 6x = 4 - 6x + 6x$$
$$15 - 7x = 4$$
$$15 - 7x - 15 = 4 - 15$$
$$-7x = -11$$
$$\frac{-7x}{-7} = \frac{-11}{-7}$$
$$x = \frac{11}{7}$$

5.
$$\frac{8}{3}(x - 5) = \frac{2}{5}(x - 4)$$
$$15 \cdot \frac{8}{3}(x - 5) = 15 \cdot \frac{2}{5}(x - 4)$$
$$40(x - 5) = 6(x - 4)$$
$$40x - 200 = 6x - 24$$
$$40x - 200 - 6x = 6x - 24 - 6x$$
$$34x - 200 = -24$$
$$34x - 200 + 200 = -24 + 200$$
$$34x = 176$$
$$\frac{34x}{34} = \frac{176}{34}$$
$$x = \frac{88}{17}$$

6.
$$\frac{3y}{4} - 14 = -\frac{y}{3} - 1$$
$$12\left(\frac{3y}{4} - 14\right) = 12\left(-\frac{y}{3} - 1\right)$$
$$9y - 168 = -4y - 12$$
$$9y - 168 + 4y = -4y - 12 + 4y$$
$$13y - 168 = -12$$
$$13y - 168 + 168 = -12 + 168$$
$$13y = 156$$
$$\frac{13y}{13} = \frac{156}{13}$$
$$y = 12$$

7.
$$2x + 4 = 2(x + 3) - 2$$
$$2x + 4 = 2x + 6 - 2$$
$$2x + 4 = 2x + 4$$
$$4 = 4$$
$$\mathbb{R}; \text{ identity}$$

8.
$$3x - 2 - x = 2(x - 4)$$
$$2x - 2 = 2x - 8$$
$$-2 \neq -8$$
$$\varnothing; \text{ contradiction}$$

9.

$$-\frac{5}{4}p = 10$$

$$-\frac{4}{5}\left(-\frac{5}{4}p\right) = -\frac{4}{5}(10)$$

$$p = -8$$

10.

$$\frac{4t+1}{3} - \frac{t+5}{6} = \frac{t-3}{6}$$

$$6\left(\frac{4t+1}{3} - \frac{t+5}{6}\right) = 6\left(\frac{t-3}{6}\right)$$

$$2(4t+1) - (t+5) = t-3$$

$$8t + 2 - t - 5 = t - 3$$

$$7t - 3 = t - 3$$

$$6t - 3 = -3$$

$$6t = 0$$

$$t = 0$$

11.

$$V = \pi r^2 h$$

$$\frac{V}{\pi r^2} = \frac{\pi r^2 h}{\pi r^2}$$

$$\frac{V}{\pi r^2} = h$$

12.

$$v = \frac{1}{6}ab(x+y)$$

$$6 \cdot v = 6 \cdot \frac{1}{6}ab(x+y)$$

$$6v = ab(x+y)$$

$$\frac{6v}{ab} = \frac{ab(x+y)}{ab}$$

$$\frac{6v}{ab} = x + y$$

$$\frac{6v}{ab} - y = x + y - y$$

$$\frac{6v}{ab} - y = x$$

$$x = \frac{6v}{ab} - y \quad \text{or}$$

$$x = \frac{6v - aby}{ab}$$

13.

$$0.3x - 0.4 \geq 1.2 - 0.1x$$

$$10(0.3x - 0.4) \geq 10(1.2 - 0.1x)$$

$$3x - 4 \geq 12 - x$$

$$3x - 4 + x \geq 12 - x + x$$

$$4x - 4 \geq 12$$

$$4x - 4 + 4 \geq 12 + 4$$

$$4x \geq 16$$

$$x \geq 4; \quad [4, \infty)$$

14.

$$\frac{7}{4}(x+3) < \frac{3}{8}(x-3)$$

$$8 \cdot \frac{7}{4}(x+3) < 8 \cdot \frac{3}{8}(x-3)$$

$$14(x+3) < 3(x-3)$$

$$14x + 42 < 3x - 9$$

$$14x + 42 - 3x < 3x - 9 - 3x$$

$$11x + 42 < -9$$

$$11x + 42 - 42 < -9 - 42$$

$$11x < -51$$

$$x < -\frac{51}{11}; \quad \left(-\infty, -\frac{51}{11}\right)$$

15.

$$-16 < -\frac{4}{5}x$$

$$-\frac{5}{4}(-16) > -\frac{5}{4}\left(-\frac{4}{5}x\right)$$

$$20 > x$$

$$x < 20; \quad (-\infty, 20)$$

Chapter 8 Review

16.

$$5(2n+2)-n>3n-3(1-2n)$$
$$10n+10-n>3n-3+6n$$
$$9n+10>9n-3$$
$$10>-3$$
$$(-\infty,\infty); \ \mathbb{R}$$

17. Let x = length of one piece.
Then $3x$ = length of the other piece.
$$x+3x=20$$
$$4x=20$$
$$x=5$$
He should cut the board 5 ft from one end.

18. Let width = x. Then length = $x+4$.
$$P=2l+2w$$
$$28=2(x+4)+2(x)$$
$$28=2x+8+2x$$
$$28=4x+8$$
$$20=4x$$
$$5=x$$
The width = 5 meters and the
length = $5+4=9$ meters.

SECTION 8.2
Functions

19. D: $\{-4, -1, 2, 5\}$
R: $\{-2, 0, 16\}$

20. a. A **function** is a set of ordered pairs (a
relation) where to each first
component there corresponds exactly
one second component.
b. Given a relation in x and y, if to each
value of x in the domain there
corresponds exactly one value of y in
the range, y is said to be a **function** of
x. We call x the independent **variable**
and y the **dependent** variable.

21. a. yes
b. no; $(-1, 8), (-1, 9)$
c. yes

22. $f(x)=-3x^3-7x^2+4$
$$f(-2)=-3(\mathbf{-2})^3-7(\mathbf{-2})^2+4$$
$$=-3(-8)-7(4)+4$$
$$=24-28+4$$
$$=0$$

23. yes

24. yes

25. no; $(25, 5)$ and $(25, -5)$

26. no; $(3, 4)$ and $(3, -4)$

27. $f(x)=3x+2$
$$f(-3)=3(-3)+2$$
$$=-9+2$$
$$=-7$$

28. $g(a)=\dfrac{a^2-4a+4}{2}$
$$g(8)=\dfrac{(\mathbf{8})^2-4(\mathbf{8})+4}{2}$$
$$=\dfrac{64-32+4}{2}$$
$$=\dfrac{36}{2}$$
$$=18$$

29. $g(a)=\dfrac{a^2-4a+4}{2}$
$$g(-2)=\dfrac{(\mathbf{-2})^2-4(\mathbf{-2})+4}{2}$$
$$=\dfrac{4+8+4}{2}$$
$$=\dfrac{16}{2}$$
$$=8$$

30. $f(x)=3x+2$
$$f(t+2)=3(\mathbf{t}+2)+2$$
$$=3t+6+2$$
$$=3t+8$$

31. $f(x) = -8$

$-5x + 7 = -8$

$-5x + 7 - 7 = -8 - 7$

$-5x = -15$

$x = 3$

32. $g(t) = \dfrac{3}{4}t - 1$

$\dfrac{3}{4}x - 1 = 0$

$4\left(\dfrac{3}{4}x - 1\right) = 4(0)$

$3x - 4 = 0$

$3x - 4 + 4 = 0 + 4$

$3x = 4$

$x = \dfrac{4}{3}$

33. D: the set of real numbers

34. D: the set of real numbers

35. D: the set of real numbers except 2
(2 makes the denominator 0 and
undefined)

36. D: the set of real numbers except –5

37. When written in function form, the
coefficient of x is the slope and the
constant is the y-intercept. So, for
$g(x) = -2x - 16$, the slope is –2 and the y-
intercept is (0, –16).

38. $f(x) = \dfrac{2}{3}x - 2$

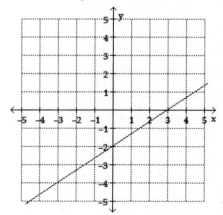

39. $m = \dfrac{9}{10}$ and $b = \dfrac{7}{8}$

$y = mx + b$

$y = \dfrac{9}{10}x + \dfrac{7}{8}$

40. $m = \dfrac{1}{5}$, $x_1 = 10, y_1 = 1$

$y - y_1 = m(x - x_1)$

$y - 1 = \dfrac{1}{5}(x - 10)$

$y - 1 = \dfrac{1}{5}x - 2$

$y - 1 + 1 = \dfrac{1}{5}x - 2 + 1$

$y = \dfrac{1}{5}x - 1$

$f(x) = \dfrac{1}{5}x - 1$

41. A horizontal line through (–8, 12) has an
equation of $y = 12$, so $f(x) = 12$.

42. The slope of $g(x) = 4x - 70$ is 4. Parallel
lines have identical slopes, so use $m = 4$,
$x_1 = 2$, and $y_1 = 5$.

$y - y_1 = m(x - x_1)$

$y - 5 = 4(x - 2)$

$y - 5 = 4x - 8$

$y - 5 + 5 = 4x - 8 + 5$

$y = 4x - 3$

$f(x) = 4x - 3$

Chapter 8 Review

43. The slope of $g(x) = -3x - 120$ is -3. The slope of a line perpendicular has an opposite sign and is the reciprocal, so use $m = \dfrac{1}{3}$, $x_1 = -6$, and $y_1 = 3$.

$$y - y_1 = m(x - x_1)$$

$$y - 3 = \frac{1}{3}(x + 6)$$

$$y - 3 = \frac{1}{3}x + 2$$

$$y - 3 + 3 = \frac{1}{3}x + 2 + 3$$

$$y = \frac{1}{3}x + 5$$

$$f(x) = \frac{1}{3}x + 5$$

44. a. Find the slope using the ordered pairs (10, 5.25) and (30, 5.65).

$$m = \frac{y_2 - y_1}{x_2 - x_1}$$

$$= \frac{5.65 - 5.25}{30 - 10}$$

$$= \frac{0.4}{20}$$

$$= 0.02$$

$$y - y_1 = m(x - x_1)$$

$$y - 5.25 = 0.02(x - 10)$$

$$y - 5.25 = 0.02x - 0.2$$

$$y = 0.02x + 5.05$$

$$R(t) = 0.02t + 5.05$$

b. $R(100) = 0.02(100) + 5.05$

$$= 2 + 5.05$$

$$= 7.05 \text{ milliohms}$$

45. $V(r) = 4.19r^3 + 25.13r^2$

$$V(2) = 4.19(2)^3 + 25.13(2)^2$$

$$= 4.19(8) + 25.13(4)$$

$$= 33.52 + 100.52$$

$$\approx 134 \text{ in.}^3$$

46. MISSING SOMETHING

$$f(x) = 1 + x + \frac{x^2}{2} + \frac{x^3}{6} + \frac{x^4}{24}$$

SECTION 8.3
Graphs of Functions

47. a. $f(-2) = -4$
    b. $f(3) = 3$
    c. $x = 1$

48. a. $g(0) = 4$
    b. $g(-3) = 1$
    c. $x = -4$ and $x = 2$

49. a. $f(0) = -1$
    b. $x = -2, -1$, and 1
    c. D: $(-\infty, \infty)$; R: $[-1, \infty)$

50. The real numbers greater than or equal to 0.

51. D: the set of real numbers
    R: the set of real numbers

52. D: the set of real numbers
    R: the set of real numbers greater than or equal to 1

53. D: the set of real numbers
    R: the set of nonnegative real numbers

54. a. The graph of $f(x) = x^2 + 6$ is the same as the graph of $f(x) = x^2$ except that it is shifted **6** units **up**.
    b. The graph of $f(x) = (x + 6)^2$ is the same as the graph of $f(x) = x^2$ except that it is shifted **6** units **left**.

55. $g(x) = x^2 - 3$

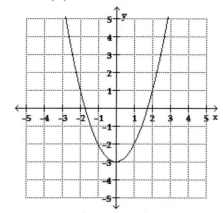

D: the set of real numbers
R: the sent of real numbers greater than or equal to –3

56. $g(x) = |x - 4|$

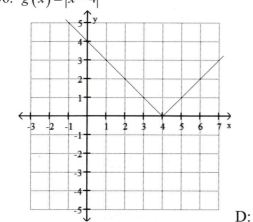

D: the set of real numbers
R: the sent of real numbers greater than or equal to –3

57. $g(x) = (x - 2)^3 + 1$

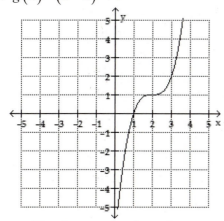

D: the set of real numbers
R: the set of real numbers

58. $g(x) = -x^3$

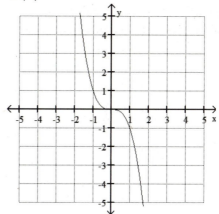

D: the set of real numbers
R: the set of real numbers

59. function

60. not a function; (0, 2) and (0, 3)

LESSON 8.4
Solving Compound Inequalities

61. {–3, 3}

62. {–6, –5, –3, 0, 3, 6, 8}

63. yes
$$x < 0 \qquad x > -5$$
$$-4 < 0 \qquad -4 > -5$$

64. no
$$x + 3 < -3x - 1 \qquad \text{and} \qquad 4x - 3 > 3x$$
$$-4 + 3 \overset{?}{<} -3(-4) - 1 \qquad 4(-4) - 3 \overset{?}{>} 3(-4)$$
$$-1 \overset{?}{<} 12 - 1 \qquad -16 - 3 \overset{?}{>} -12$$
$$-1 < 11 \qquad -19 \ngtr -12$$

65. $(-3, 3) \cup [1, 6]$

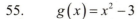

66. $(-\infty, 2] \cap [1, 4)$

$\begin{array}{c}\longleftarrow\overset{[\underline{\quad\quad}]}{\underset{1\qquad 2}{}}\longrightarrow\end{array}$

67.

$$-2x > 8 \quad \text{and} \quad x+4 \geq -6$$

$$\frac{-2x}{-2} < \frac{8}{-2} \qquad x+4-4 \geq -6-4$$

$$x < -4 \qquad\qquad x \geq -10$$

$$-10 \leq x < -4$$

$$[-10,-4)$$

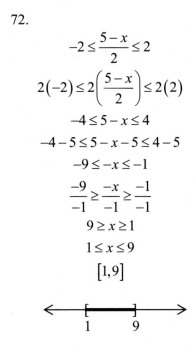

68.

$$5(x+2) \leq 4(x+1) \quad \text{and} \quad 11+x < 0$$

$$5x+10 \leq 4x+4 \qquad\qquad x < -11$$

$$x+10 \leq 4$$

$$x \leq -6$$

$$(-\infty,-11)$$

69.

$$\frac{2}{5}x - 2 < -\frac{4}{5} \quad \text{and} \quad \frac{x}{-3} < -1$$

$$5\left(\frac{2}{5}x-2\right) < 5\left(-\frac{4}{5}\right) \quad -3\left(\frac{x}{-3}\right) > -3(-1)$$

$$2x-10 < -4 \qquad\qquad x > 3$$

$$2x-10+10 < -4+10$$

$$2x < 6$$

$$x < 3$$

$$\text{no solution;} \varnothing$$

70.

$$4\left(x-\frac{1}{4}\right) \leq 3x-1 \quad \text{and} \quad x \geq 0$$

$$4x-1 \leq 3x-1$$

$$4x-1-3x \leq 3x-1-3x$$

$$x-1 \leq -1$$

$$x-1+1 \leq -1+1$$

$$x \leq 0$$

$$[0,0]$$

71.

$$3 < 3x+4 < 10$$

$$3-4 < 3x+4-4 < 10-4$$

$$-1 < 3x < 6$$

$$\frac{-1}{3} < \frac{3x}{3} < \frac{6}{3}$$

$$-\frac{1}{3} < x < 2$$

$$\left(-\frac{1}{3}, 2\right)$$

72.

$$-2 \leq \frac{5-x}{2} \leq 2$$

$$2(-2) \leq 2\left(\frac{5-x}{2}\right) \leq 2(2)$$

$$-4 \leq 5-x \leq 4$$

$$-4-5 \leq 5-x-5 \leq 4-5$$

$$-9 \leq -x \leq -1$$

$$\frac{-9}{-1} \geq \frac{-x}{-1} \geq \frac{-1}{-1}$$

$$9 \geq x \geq 1$$

$$1 \leq x \leq 9$$

$$[1,9]$$

73. Yes.

$$-4 < 1.6 \quad \text{or} \quad -4 \not> -3.9$$

$$\text{true} \qquad\qquad \text{false}$$

74. No.

$$-4+1 \overset{?}{<} 2(-4)-1 \quad \text{or} \quad 4(-4)-3 \overset{?}{>} 3(-4)$$

$$-3 \not< -9 \qquad\qquad -19 \not> -12$$

$$\text{false} \qquad\qquad\qquad \text{false}$$

75.

$$x+1<-4 \quad \text{or} \quad x-4>0$$
$$x+1-1<-4-1 \quad x-4+4>0+4$$
$$x<-5 \qquad\qquad x>4$$
$$(-\infty-5)\cup(4,\infty)$$

76.

$$\frac{x}{2}+3>-2 \quad \text{or} \quad 4-x>4$$
$$2\left(\frac{x}{2}+3\right)>2(-2) \quad 4-x-4>4-4$$
$$x+6>-4 \qquad\qquad -x>0$$
$$x+6-6>-4-6 \qquad \frac{-x}{-1}<\frac{0}{-1}$$
$$x>-10 \qquad\qquad x<0$$
$$(-\infty,\infty)$$

77. Area = length · width
$$= l \cdot 4$$
$$= 4l$$
$$17 \le 4l \le 25$$
$$\frac{17}{4} \le \frac{4l}{4} \le \frac{25}{4}$$
$$4.25 \le l \le 6.25$$

78. a.   ii and iv
b.   i and iii

79.

$$f(x)<-48 \qquad \text{or} \qquad f(x)\ge 32$$
$$\frac{5}{4}x-140<-48 \qquad\qquad \frac{5}{4}x-140\ge 32$$
$$4\left(\frac{5}{4}x-140\right)<4(-48) \quad 4\left(\frac{5}{4}x-140\right)\ge 4(32)$$
$$5x-560<-192 \qquad\qquad 5x-560\ge 128$$
$$5x<368 \qquad\qquad 5x\ge 688$$
$$x<73.6 \qquad\qquad x\ge 137.6$$
$$(-\infty,73.6)\cup[137.6,\infty)$$

80.

$$-4 \ge f(x) > -12$$
$$-4 \ge 3x-5 > -12$$
$$-4+5 \ge 3x-5+5 > -12+5$$
$$1 \ge 3x > -7$$
$$\frac{1}{3} \ge x > -\frac{7}{3}$$
$$-\frac{7}{3} < x \le \frac{1}{3}$$
$$\left(-\frac{7}{3},\frac{1}{3}\right]$$

SECTION 8.5
Solving Absolute Value Equations and Inequalities.

81.

$$|4x|=8$$
$$4x=8 \quad \text{or} \quad 4x=-8$$
$$x=2 \qquad\qquad x=-2$$

82.

$$2|3x+1|-1=19$$
$$2|3x+1|-1+1=19+1$$
$$2|3x+1|=20$$
$$\frac{2|3x+1|}{2}=\frac{20}{2}$$
$$|3x+1|=10$$
$$3x+1=10 \quad \text{or} \quad 3x+1=-10$$
$$3x+1-1=10-1 \quad 3x+1-1=-10-1$$
$$3x=9 \qquad\qquad 3x=-11$$
$$x=3 \qquad\qquad x=-\frac{11}{3}$$

Chapter 8 Review

83.

$$\left|\frac{3}{2}x - 4\right| - 10 = -1$$

$$\left|\frac{3}{2}x - 4\right| - 10 + 10 = -1 + 10$$

$$\left|\frac{3}{2}x - 4\right| = 9$$

$$\frac{3}{2}x - 4 = 9 \quad \text{or} \quad \frac{3}{2}x - 4 = -9$$

$$2\left(\frac{3}{2}x - 4\right) = 2(9) \quad 2\left(\frac{3}{2}x - 4\right) = 2(-9)$$

$$3x - 8 = 18 \qquad\qquad 3x - 8 = -18$$

$$3x - 8 + 8 = 18 + 8 \quad 3x - 8 + 8 = -18 + 8$$

$$3x = 26 \qquad\qquad 3x = -10$$

$$x = \frac{26}{3} \qquad\qquad x = -\frac{10}{3}$$

84.

$$\left|\frac{2-x}{3}\right| = -4$$

No solution. Since an absolute value can never be negative, there are no real numbers x that can make the equation true.

85.

$$\left|-4(2x - 6)\right| = 0$$

$$-4(2x - 6) = 0$$

$$-8x + 24 = 0$$

$$-8x + 24 - 24 = 0 - 24$$

$$-8x = -24$$

$$\frac{-8x}{-8} = \frac{-24}{-8}$$

$$x = 3$$

$$\boxed{x = 3}$$

86.

$$\left|\frac{3}{8} + \frac{x}{3}\right| = \frac{5}{12}$$

$$\frac{3}{8} + \frac{x}{3} = \frac{5}{12} \quad \text{or} \quad \frac{3}{8} + \frac{x}{3} = -\frac{5}{12}$$

$$24\left(\frac{3}{8} + \frac{x}{3}\right) = 24\left(\frac{5}{12}\right) \quad 24\left(\frac{3}{8} + \frac{x}{3}\right) = 24\left(-\frac{5}{12}\right)$$

$$9 + 8x = 10 \qquad\qquad 9 + 8x = -10$$

$$9 + 8x - 9 = 10 - 9 \qquad 9 + 8x - 9 = -10 - 9$$

$$8x = 1 \qquad\qquad 8x = -19$$

$$x = \frac{1}{8} \qquad\qquad x = -\frac{19}{8}$$

$$\boxed{x = \frac{1}{8}, -\frac{19}{8}}$$

87.

$$|3x + 2| = |2x - 3|$$

$$3x + 2 = 2x - 3 \quad \text{or} \quad 3x + 2 = -(2x - 3)$$

$$3x + 2 = 2x - 3 \qquad\qquad 3x + 2 = -2x + 3$$

$$3x + 2 - 2x = 2x - 3 - 2x \quad 3x + 2 + 2x = -2x + 3 + 2x$$

$$x + 2 = -3 \qquad\qquad 5x + 2 = 3$$

$$x + 2 - 2 = -3 - 2 \qquad 5x + 2 - 2 = 3 - 2$$

$$x = -5 \qquad\qquad 5x = 1$$

$$x = -5 \qquad\qquad x = \frac{1}{5}$$

88.

$$\left|\frac{2(1-x)+1}{2}\right| = \left|\frac{3x-2}{3}\right|$$

$$\left|\frac{2-2x+1}{2}\right| = \left|\frac{3x-2}{3}\right|$$

$$\left|\frac{-2x+3}{2}\right| = \left|\frac{3x-2}{3}\right|$$

$$\frac{-2x+3}{2} = \frac{3x-2}{3} \quad \text{or} \quad \frac{-2x+3}{2} = -\left(\frac{3x-2}{3}\right)$$

$$6\left(\frac{-2x+3}{2}\right) = 6\left(\frac{3x-2}{3}\right) \quad 6\left(\frac{-2x+3}{2}\right) = -6\left(\frac{3x-2}{3}\right)$$

$$3(-2x+3) = 2(3x-2) \qquad 3(-2x+3) = -2(3x-2)$$

$$-6x + 9 = 6x - 4 \qquad\qquad -6x + 9 = -6x + 4$$

$$-6x + 9 - 6x = 6x - 4 - 6x \quad -6x + 9 + 6x = -6x + 4 + 6x$$

$$-12x + 9 = -4 \qquad\qquad 9 \ne 4$$

$$-12x + 9 - 9 = -4 - 9$$

$$-12x = -13$$

$$x = \frac{13}{12}$$

$$\boxed{x = \frac{13}{12}}$$

89.

$$|x| \le 3$$
$$-3 \le x \le 3$$
$$[-3, 3]$$

90.

$$|2x + 7| < 3$$
$$-3 < 2x + 7 < 3$$
$$-3 - 7 < 2x + 7 - 7 < 3 - 7$$
$$-10 < 2x < -4$$
$$-5 < x < -2$$
$$(-5, -2)$$

91.

$$2|5 - 3x| \le 28$$
$$\frac{2|5 - 3x|}{2} \le \frac{28}{2}$$
$$|5 - 3x| \le 14$$
$$-14 \le 5 - 3x \le 14$$
$$-14 - 5 \le 5 - 3x - 5 \le 14 - 5$$
$$-19 \le -3x \le 9$$
$$\frac{-19}{-3} \ge \frac{-3x}{-3} \ge \frac{9}{-3}$$
$$\frac{19}{3} \ge x \ge -3$$
$$-3 \le x \le \frac{19}{3}$$
$$\left[-3, \frac{19}{3}\right]$$

92.

$$\left|\frac{2}{3}x + 14\right| + 6 < 6$$
$$\left|\frac{2}{3}x + 14\right| + 6 - 6 < 6 - 6$$
$$\left|\frac{2}{3}x + 14\right| < 0$$

No solution. Since an absolute value can never be negative (less than zero), there are no real numbers x that can make the equation true.

93.

$$|x| > 1$$
$$x > 1 \quad \text{or} \quad x < -1$$
$$(-\infty, -1) \cup (1, \infty)$$

94.

$$\left|\frac{1 - 5x}{3}\right| \ge 7$$

$$\frac{1 - 5x}{3} \le -7 \qquad \text{or} \qquad \frac{1 - 5x}{3} \ge 7$$

$$3\left(\frac{1 - 5x}{3}\right) \le 3(-7) \qquad 3\left(\frac{1 - 5x}{3}\right) \ge 3(7)$$

$$1 - 5x \le -21 \qquad\qquad 1 - 5x \ge 21$$

$$1 - 5x - 1 \le -21 - 1 \qquad 1 - 5x - 1 \ge 21 - 1$$

$$-5x \le -22 \qquad\qquad -5x \ge 20$$

$$x \ge \frac{22}{5} \qquad\qquad\qquad x \le -4$$

$$(-\infty, -4] \cup \left[\frac{22}{5}, \infty\right)$$

Chapter 8 Review

95.

$$|3x-8|-4 \geq 0$$
$$|3x-8|-4+4 \geq 0+4$$
$$|3x-8| \geq 4$$

$$3x-8 \leq -4 \quad \text{or} \quad 3x-8 \geq 4$$
$$3x-8+8 \leq -4+8 \quad 3x-8+8 \geq 4+8$$
$$3x \leq 4 \qquad\qquad 3x \geq 12$$
$$x \leq \frac{4}{3} \qquad\qquad x \geq 4$$

$$\left(-\infty, \frac{4}{3}\right] \cup [4, \infty)$$

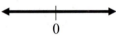

$$\frac{4}{3} \qquad 4$$

96.

$$\left|\frac{3}{2}x-14\right| \geq 0$$

$$(-\infty, \infty)$$

0

Since $\left|\frac{3}{2}x-14\right|$ is always greater than or equal to 0 for any real number x, then this absolute value inequality is true for all real numbers, $\mathbb{R}$.

97. a. $|w-8| \leq 2$

b.
$$|w-8| \leq 2$$
$$-2 \leq w-8 \leq 2$$
$$-2+8 \leq w-8+8 \leq 2+8$$
$$6 \leq w \leq 10$$
$$[6,10]$$

98.
$$f(x) = \frac{1}{3}|6x|-1$$
$$\frac{1}{3}|6x|-1 = 5$$
$$\frac{1}{3}|6x|-1+1 = 5+1$$
$$\frac{1}{3}|6x| = 6$$
$$3 \cdot \frac{1}{3}|6x| = 3 \cdot 6$$
$$|6x| = 18$$
$$6x = 18 \quad \text{or} \quad 6x = -18$$
$$x = 3 \qquad\qquad x = -3$$
$$\boxed{x = 3, -3}$$

99. $f(x) = 2|3(x+4)|+1.5$
$$2|3(x+4)|+1.5 = 25.5$$
$$2|3(x+4)| = 24$$
$$|3(x+4)| = 12$$
$$|3x+12| = 12$$
$$3x+12 = 12 \quad \text{and} \quad 3x+12 = -12$$
$$3x = 0 \qquad\qquad 3x = -24$$
$$x = 0 \qquad\qquad x = -8$$
$$\boxed{x = 0, -8}$$

100. $f(x) = |7-x|$
$$|7-x| < 5$$
$$-5 < 7-x < 5$$
$$-5-7 < 7-x-7 < 5-7$$
$$-12 < -x < -2$$
$$12 > x > 2$$
$$2 < x < 12$$
$$(2,12)$$

101. Since $|0.04x-8.8|$ is always greater than or equal to 0 for any real number x, this absolute value inequality has no solution.

102. Since $\left|\dfrac{3x}{50}+\dfrac{1}{45}\right|$ is always greater than

or equal to 0 for any real number x, this absolute value inequality is true

for all real numbers, $\mathbb{R}$.

SECTION 8.6
Review of Factoring Methods: GCF, Grouping, Trinomials

103.
$$z^2 - 11z + 30 = (z-5)(z-6)$$

104.
$$x^4 + 4x^2 + x^2 y + 4y = x^2\left(x^2+4\right) + y\left(x^2+4\right)$$
$$= \left(x^2 + y\right)\left(x^2 + 4\right)$$

105.
$$4a^2 - 5a + 1 = (4a-1)(a-1)$$

106.
$$27x^3 y^3 z^3 + 81x^4 y^5 z^2 - 90x^2 y^3 z^7$$
$$= 9x^2 y^3 z^2 \left(3xz + 9x^2 y^2 - 10z^5\right)$$

107.
$$15b^2 + 4b - 4 = (5b-2)(3b+2)$$

108.
$$-x^2 - 3x + 28 = -\left(x^2 + 3x - 28\right)$$
$$= -(x+7)(x-4)$$

109.
$$15x^2 - 57xy - 12y^2 = 3\left(5x^2 - 19xy - 4y^2\right)$$
$$= 3(5x+y)(x-4y)$$

110.
$$w^8 - w^4 - 90 = \left(w^4 - 10\right)\left(w^4 + 9\right)$$

111.
$$r^2 y - ar - ry + a + r - 1$$
$$= \left(r^2 y - ar + r\right) + (-ry + a - 1)$$
$$= r(ry - a + 1) - (ry - a + 1)$$
$$= (ry - a + 1)(r-1)$$

112.
$$49a^6 + 84a^3 b^2 + 36b^4 = \left(7a^3 + 6b^2\right)^2$$

113.
$$3b^2 + 2b + 1$$
prime

114.
$$2a^4 + 4a^3 - 6a^2 = 2a^2\left(a^2 + 2a - 3\right)$$
$$= 2a^2(a+3)(a-1)$$

115.
$$(s+t)^2 - 2(s+t) + 1$$
Let $x = (s+t)$
$$x^2 - 2x + 1 = (x-1)^2$$
Substitute $(s+t)$ for x in the answer.
$$(x-1)^2 = (s+t-1)^2$$

116.
$$m_1 m_2 = mm_2 + mm_1$$
$$m_1 m_2 - mm_1 = mm_2 + mm_1 - mm_1$$
$$m_1 m_2 - mm_1 = mm_2$$
$$m_1\left(m_2 - m\right) = mm_2$$
$$\frac{m_1\left(m_2 - m\right)}{m_2 - m} = \frac{mm_2}{m_2 - m}$$
$$m_1 = \frac{mm_2}{m_2 - m}$$

SECTION 8.7
The Difference of Two Squares; the Sum and Difference of Two Cubes

117. $\quad z^2 - 16 = (z+4)(z-4)$

118. $\quad x^2 y^4 - 64z^6 = (xy^2 + 8z^3)(xy^2 - 8z^3)$

119. $\quad a^2 b^2 + c^2$
prime

120. $\quad c^2 - (a+b)^2 = [c + (a+b)][c - (a+b)]$
$$= (c + a + b)(c - a - b)$$

121.

$$32a^4c - 162b^4c$$
$$= 2c\left(16a^4 - 81b^4\right)$$
$$= 2c\left(4a^2 + 9b^2\right)\left(4a^2 - 9b^2\right)$$
$$= 2c\left(4a^2 + 9b^2\right)\left(2a + 3b\right)\left(2a - 3b\right)$$

122.

$$k^2 + 2k + 1 - 9m^2 = \left(k^2 + 2k + 1\right) - 9m^2$$
$$= \left(k + 1\right)^2 - 9m^2$$
$$= \left(k + 1 + 3m\right)\left(k + 1 - 3m\right)$$

123.

$$m^2 - n^2 - m - n = \left(m^2 - n^2\right) - \left(m + n\right)$$
$$= \left(m + n\right)\left(m - n\right) - 1\left(m + n\right)$$
$$= \left(m + n\right)\left(m - n - 1\right)$$

124.

$$t^3 + 64 = \left(t + 4\right)\left(t^2 - 4t + 16\right)$$

125.

$$8a^3 - 125b^9$$
$$= \left(2a - 5b^3\right)\left(\left(2a\right)^2 + \left(2a\right)\left(5b^3\right) + \left(5b^3\right)^2\right)$$
$$= \left(2a - 5b^3\right)\left(4a^2 + 10ab^3 + 25b^6\right)$$

126.

$$V = \frac{\pi}{2}r_1^2 h - \frac{\pi}{2}r_2^2 h$$
$$= \frac{\pi}{2}h\left(r_1^2 - r_2^2\right)$$
$$= \frac{\pi}{2}h\left(r_1 + r_2\right)\left(r_1 - r_2\right)$$

SECTION 8.8
Review of Rational Expressions and Rational Equations

127. a. $f(2) = 1$

  b. $f(-1) = -\dfrac{1}{2}$

  c. $x = 0$

128. $f(x) = \dfrac{2x^2 + 8x}{x^2 + 2x - 24}$

$$x^2 + 2x - 24 = 0$$
$$(x + 6)(x - 4) = 0$$
$$x + 6 = 0 \quad \text{or} \quad x - 4 = 0$$
$$x = -6 \qquad\qquad x = 4$$
$$\left(-\infty, -6\right) \cup \left(-6, 4\right) \cup \left(4, \infty\right)$$

The domain is the set of all real numbers except –6 and 4.

129. $n(t) = \dfrac{28t}{t^2 + 1}$

$$n(3) = \frac{28(3)}{(3)^2 + 1}$$
$$= \frac{84}{9 + 1}$$
$$= \frac{84}{10}$$
$$= 8.4$$

Three hours after the injection, the concentration of pain medication in the patient's bloodstream was 8.4 milligrams per liter.

130. $f(x) = \dfrac{4x^2 - 6x + 7}{3x^2 - 27}$

$$f(3) = \frac{4(3)^2 - 6(3) + 7}{3(3)^2 - 27}$$
$$= \frac{4(9) - 6(3) + 7}{3(9) - 27}$$
$$= \frac{36 - 18 + 7}{27 - 27}$$
$$= \frac{25}{0}$$
$$= \text{undefined}$$

131.

$$\frac{62x^2 y}{144xy^2} = \frac{\cancel{2} \cdot 31 \cdot \cancel{x} \cdot x \cdot \cancel{y}}{\cancel{2} \cdot 2 \cdot 2 \cdot 2 \cdot 3 \cdot 3 \cdot \cancel{x} \cdot \cancel{y} \cdot y}$$
$$= \frac{31x}{72y}$$

132.

$$\frac{2m-2n}{n-m} = \frac{2(m-n)}{n-m}$$

$$= -\frac{2(m-n)}{m-n}$$

$$= -2$$

133.

$$\frac{3x^3y^4}{c^2d} \cdot \frac{c^3d^2}{21x^5y^4} = \frac{\cancel{3}\cancel{x}\cancel{x}\cancel{x}\cancel{y}\cancel{y}\cancel{y}\cancel{y}\cancel{c}\cancel{c}cdd}{\cancel{3}\cdot 7\cancel{c}\cancel{c}\cancel{d}\cancel{x}\cancel{x}\cancel{x}xx\cancel{y}\cancel{y}\cancel{y}\cancel{y}}$$

$$= \frac{cd}{7x^2}$$

134.

$$\frac{2a^2-5a-3}{a^2-9} \div \frac{2a^2+5a+2}{2a^2+5a-3}$$

$$= \frac{2a^2-5a-3}{a^2-9} \cdot \frac{2a^2+5a-3}{2a^2+5a+2}$$

$$= \frac{(2a+1)(a-3)}{(a-3)(a+3)} \cdot \frac{(2a-1)(a+3)}{(2a+1)(a+2)}$$

$$= \frac{2a-1}{a+2}$$

135.

$$\frac{m^2+3m+9}{m^2+mp+mr+pr} \div \frac{m^3-27}{am+ar+bm+br}$$

$$= \frac{m^2+3m+9}{m^2+mp+mr+pr} \cdot \frac{am+ar+bm+br}{m^3-27}$$

$$= \frac{m^2+3m+9}{m(m+p)+r(m+p)} \cdot \frac{a(m+r)+b(m+r)}{(m-3)(m^2+3m+9)}$$

$$= \frac{m^2+3m+9}{(m+r)(m+p)} \cdot \frac{(a+b)(m+r)}{(m-3)(m^2+3m+9)}$$

$$= \frac{a+b}{(m+p)(m-3)}$$

136.

$$\frac{x^3+3x^2+2x}{2x^2-2x-12} \cdot \frac{3x^2-3x}{x^3-3x^2-4x} \div \frac{x^2+3x+2}{2x^2-4x-16}$$

$$= \frac{x^3+3x^2+2x}{2x^2-2x-12} \cdot \frac{3x^2-3x}{x^3-3x^2-4x} \cdot \frac{2x^2-4x-16}{x^2+3x+2}$$

$$= \frac{x(x^2+3x+2)}{2(x^2-x-6)} \cdot \frac{3x(x-1)}{x(x^2-3x-4)} \cdot \frac{2(x^2-2x-8)}{x^2+3x+2}$$

$$= \frac{x(x+1)(x+2)}{2(x-3)(x+2)} \cdot \frac{3x(x-1)}{x(x-4)(x+1)} \cdot \frac{2(x-4)(x+2)}{(x+1)(x+2)}$$

$$= \frac{3x(x-1)}{(x-3)(x+1)}$$

137.

$$\frac{d^2}{c^3-d^3} + \frac{c^2+cd}{c^3-d^3} = \frac{d^2+c^2+cd}{c^3-d^3}$$

$$= \frac{c^2+cd+d^2}{(c-d)(c^2+cd+d^2)}$$

$$= \frac{1}{c-d}$$

138.

$$\frac{4}{t-3} + \frac{6}{3-t} = \frac{4}{t-3} - \frac{6}{t-3}$$

$$= \frac{4-6}{t-3}$$

$$= \frac{-2}{t-3}$$

$$= -\frac{2}{t-3}$$

139.

$$\frac{5x}{14z^2} + \frac{y^2}{16z} = \frac{5x}{14z^2}\left(\frac{8}{8}\right) + \frac{y^2}{16z}\left(\frac{7z}{7z}\right)$$

$$= \frac{40x}{112z^2} + \frac{7y^2z}{112z^2}$$

$$= \frac{40x+7y^2z}{112z^2}$$

Chapter 8 Review

140.

$$\frac{4}{3xy-6y}-\frac{4}{10-5x}=\frac{4}{3y(x-2)}-\frac{4}{5(2-x)}$$

$$=\frac{4}{3y(x-2)}+\frac{4}{5(x-2)}$$

$$=\frac{4}{3y(x-2)}\left(\frac{5}{5}\right)+\frac{4}{5(x-2)}\left(\frac{3y}{3y}\right)$$

$$=\frac{20}{15y(x-2)}+\frac{12y}{15y(x-2)}$$

$$=\frac{20+12y}{15y(x-2)}$$

141.

$$\frac{y+7}{y+3}-\frac{y-3}{y+7}=\frac{y+7}{y+3}\left(\frac{y+7}{y+7}\right)-\frac{y-3}{y+7}\left(\frac{y+3}{y+3}\right)$$

$$=\frac{y^2+14y+49}{(y+3)(y+7)}-\frac{y^2-9}{(y+3)(y+7)}$$

$$=\frac{y^2+14y+49-(y^2-9)}{(y+3)(y+7)}$$

$$=\frac{y^2+14y+49-y^2+9}{(y+3)(y+7)}$$

$$=\frac{14y+58}{(y+3)(y+7)}$$

142.

$$\frac{2x}{x+1}+\frac{3x}{x+2}+\frac{4x}{x^2+3x+2}$$

$$=\frac{2x}{x+1}+\frac{3x}{x+2}+\frac{4x}{(x+1)(x+2)}$$

$$=\frac{2x}{x+1}\left(\frac{x+2}{x+2}\right)+\frac{3x}{x+2}\left(\frac{x+1}{x+1}\right)+\frac{4x}{(x+1)(x+2)}$$

$$=\frac{2x^2+4x}{(x+1)(x+2)}+\frac{3x^2+3x}{(x+1)(x+2)}+\frac{4x}{(x+1)(x+2)}$$

$$=\frac{2x^2+4x+3x^2+3x+4x}{(x+1)(x+2)}$$

$$=\frac{5x^2+11x}{(x+1)(x+2)}$$

143.

$$\frac{4}{x}-\frac{1}{10}=\frac{7}{2x}$$

$$10x\left(\frac{4}{x}-\frac{1}{10}\right)=10x\left(\frac{7}{2x}\right)$$

$$40-x=35$$

$$40-x-40=35-40$$

$$-x=-5$$

$$x=5$$

144.

$$\frac{3}{y}-\frac{2}{y+1}=\frac{1}{2}$$

$$2y(y+1)\left(\frac{3}{y}-\frac{2}{y+1}\right)=2y(y+1)\left(\frac{1}{2}\right)$$

$$3(2)(y+1)-2(2y)=y(y+1)$$

$$6y+6-4y=y^2+y$$

$$2y+6=y^2+y$$

$$0=y^2-y-6$$

$$0=(y-3)(y+2)$$

$$y-3=0 \quad\text{or}\quad y+2=0$$

$$y=3 \qquad\qquad y=-2$$

145.

$$\frac{2}{3x+15}-\frac{1}{18}=\frac{1}{3x+12}$$

$$\frac{2}{3(x+5)}-\frac{1}{18}=\frac{1}{3(x+4)}$$

$$18(x+5)(x+4)\left(\frac{2}{3(x+5)}-\frac{1}{18}\right)=18(x+5)(x+4)\left(\frac{1}{3(x+4)}\right)$$

$$6(x+4)2-(x+5)(x+4)=6(x+5)$$

$$12(x+4)-(x^2+4x+5x+20)=6x+30$$

$$12x+48-x^2-4x-5x-20=6x+30$$

$$-x^2+3x+28=6x+30$$

$$-x^2+3x+28+x^2-3x-28=6x+30+x^2-3x-28$$

$$0=x^2+3x+2$$

$$0=(x+1)(x+2)$$

$$x+1=0 \quad\text{or}\quad x+2=0$$

$$x=-1 \qquad\qquad x=-2$$

146

$$\frac{3}{x+2}=\frac{1}{2-x}+\frac{2}{x^2-4}$$

$$\frac{3}{x+2}=\frac{-1}{x-2}+\frac{2}{(x-2)(x+2)}$$

$$(x-2)(x+2)\left(\frac{3}{x+2}\right)=(x-2)(x+2)\left(\frac{-1}{x-2}+\frac{2}{(x-2)(x+2)}\right)$$

$$3(x-2)=-(x+2)+2$$

$$3x-6=-x-2+2$$

$$3x-6=-x$$

$$-6=-4x$$

$$\frac{3}{2}=x$$

147.

$$\frac{x+3}{x-5}+\frac{2x^2+6}{x^2-7x+10}=\frac{3x}{x-2}$$

$$\frac{x+3}{x-5}+\frac{2x^2+6}{(x-2)(x-5)}=\frac{3x}{x-2}$$

$$(x-5)(x-2)\left(\frac{x+3}{x-5}+\frac{2x^2+6}{(x-2)(x-5)}\right)=(x-2)(x-5)\left(\frac{3x}{x-2}\right)$$

$$(x-2)(x+3)+2x^2+6=3x(x-5)$$

$$x^2+3x-2x-6+2x^2+6=3x^2-15x$$

$$3x^2+x=3x^2-15x$$

$$3x^2+x-3x^2=3x^2-15x-3x^2$$

$$x=-15x$$

$$x+15x=-15x+15x$$

$$16x=0$$

$$x=0$$

148.

$$\frac{5a}{a-3}-7=\frac{15}{a-3}$$

$$(a-3)\left(\frac{5a}{a-3}-7\right)=(a-3)\left(\frac{15}{a-3}\right)$$

$$5a-7(a-3)=15$$

$$5a-7a+21=15$$

$$-2a+21=15$$

$$-2a=-6$$

$$a=\cancel{3};\ \text{No solution}$$

3 is extraneous

SECTION 8.9
Variation

149. Let t = property tax, k = constant, and a = assessed valuation.

$$t=ka$$

$$1,575=k(90,000)$$

$$1,575=90,000k$$

$$0.0175=k$$

Find t if $a=312,000$ and $k=0.0175$.

$$t=0.0175(312,000)$$

$$t=\$5,460$$

150. Let c = current, k = constant, and r = resistance.

$$c=\frac{k}{r}$$

$$2.5=\frac{k}{150}$$

$$150(2.5)=150\left(\frac{k}{150}\right)$$

$$375=k$$

Find c if $k=375$ and $r=2(150)=300$.

$$c=\frac{k}{r}$$

$$=\frac{375}{300}$$

$$=1.25\text{ amps}$$

151.

$$y=kxz$$

$$3=k(24)(4)$$

$$3=96k$$

$$\frac{3}{96}=k$$

$$\frac{1}{32}=k$$

$$k=\frac{1}{32}$$

Chapter 8 Review

152. f = force of wind, k = constant, A = area of sign, and v = velocity of wind.

$$f = kAv^2$$

$$1.98 = k(3 \cdot 1.5)(10)^2$$

$$1.98 = k(4.5)(100)$$

$$1.98 = 450k$$

$$0.0044 = k$$

Find f if $k = 0.0044$ and $v = 80$.

$$f = kAv^2$$

$$f = 0.0044(1.5 \cdot 3)(80)^2$$

$$f = 0.0044(4.5)(6,400)$$

$$f = 126.72 \text{ lb}$$

153. inverse variation

154.

$$x_1 = \frac{kt^3}{x_2}$$

$$1.6 = \frac{k(8)^3}{64}$$

$$1.6 = \frac{512k}{64}$$

$$1.6 = 8k$$

$$0.2 = k$$

1. a. To **solve** an equation means to find all of the values of the variable that make the equation true.
    b. $<, >, \leq$, and $\geq$ are **inequality** symbols.
    c. The statement $x^2 - x - 12 = (x - 4)(x + 3)$ shows that the trinomial $x^2 - x - 12$ **factors** as the product of two binomials.
    d. The **reciprocal** of $\dfrac{x+1}{x-7}$ is $\dfrac{x-7}{x+1}$.
    e. Given a relation in x and y, if to each value of x in the domain there corresponds exactly one value of y in the range, y is said to be a **function** of x.
    f. For a function, the set of all possible values that can be used for the independent variable is called the **domain**. The set of all values of the dependent variable is called the **range**.

2. Yes, it is a solution.
$$1.6y + (-3) = y + 1.02$$
$$1.6(6.7) + (-3) \overset{?}{=} (6.7) + 1.02$$
$$10.72 + (-3) \overset{?}{=} 7.72$$
$$7.72 = 7.72$$

3.
$$t + 18 = 5t - 3 + t$$
$$t + 18 = 6t - 3$$
$$-5t + 18 = -3$$
$$-5t = -21$$
$$t = \frac{-21}{-5}$$
$$t = \frac{21}{5}$$

4.
$$\frac{2}{3}(2s + 2) = \frac{1}{6}(5s + 29) - 4$$
$$6\left[\frac{2}{3}(2s + 2)\right] = 6\left[\frac{1}{6}(5s + 29) - 4\right]$$
$$4(2s + 2) = 1(5s + 29) - 24$$
$$8s + 8 = 5s + 29 - 24$$
$$8s + 8 = 5s + 5$$
$$3s + 8 = 5$$
$$3s = -3$$
$$s = -1$$

5.
$$6 - (x - 3) - 5x = 3[1 - 2(x + 2)]$$
$$6 - x + 3 - 5x = 3[1 - 2x - 4]$$
$$-6x + 9 = 3[-2x - 3]$$
$$-6x + 9 = -6x - 9$$
$$-6x + 9 + 6x = -6x - 9 + 6x$$
$$9 \neq -9$$

$\varnothing$ (no solution); contradiction

6.
$$y - y_1 = m(x - x_1)$$
$$y - y_1 = mx - mx_1$$
$$y - y_1 - mx = mx - mx_1 - mx$$
$$y - y_1 - mx = -mx_1$$
$$\frac{y - y_1 - mx}{-m} = \frac{-mx_1}{-m}$$
$$\frac{-y + y_1 + mx}{m} = x_1$$
$$x_1 = \frac{y_1 + mx - y}{m}$$

7. Let x = the number of passes.
$$0.9375 - 0.03125x = 0.6875$$
$$0.9375 - 0.03125x - 0.9375 = 0.6875 - 0.9375$$
$$-0.03125x = -0.25$$
$$x = 8$$

8. Let x = grade on Exam 5.

$$\frac{70 + 79 + 85 + 88 + x}{5} > 80$$

$$\frac{322 + x}{5} > 80$$

$$5\left(\frac{322 + x}{5}\right) > 5(80)$$

$$322 + x > 400$$

$$322 + x - 322 > 400 - 322$$

$$x > 78$$

She must make higher than 78.

9. a.   No; $(1, -8)$ and $(1, 6)$
    b.   Yes
    c.   Yes
    d.   No; $(4, -4)$ and $(4, 4)$

10. $f(x) = \dfrac{15}{12 - 2x}$

$$12 - 2x = 0$$

$$-2x = -12$$

$$x = 6$$

The set of all real numbers except 6.

11. $$f(x) = -\frac{4}{5}x - 12$$

$$-\frac{4}{5}x - 12 = 4$$

$$5\left(-\frac{4}{5}x - 12\right) = 5(4)$$

$$-4x - 60 = 20$$

$$-4x = 80$$

$$x = -20$$

12.

$$f(x) = 8x - 9$$

$$f(x) = mx + b$$

The slope is 8 and the y-intercept is $(0, -9)$.

13. If the graph is perpendicular to

$g(x) = \dfrac{4}{5}x + \dfrac{1}{9}$, then its slope would be

$-\dfrac{5}{4}$ since perpendicular lines have slopes

that are negative reciprocals.

$$y - y_1 = m(x - x_1)$$

$$y - 0 = -\frac{5}{4}(x - 2)$$

$$y = -\frac{5}{4}x + \frac{5}{2}$$

$$f(x) = -\frac{5}{4}x + \frac{5}{2}$$

14. a.   Use the ordered pairs (1980, 14.5) and (2005, 10.5) to find the slope.

$$m = \frac{y_2 - y_1}{x_2 - x_1}$$

$$= \frac{10.5 - 14.5}{2005 - 1980}$$

$$= \frac{-4}{25}$$

$$= -0.16$$

$$y - y_1 = m(x - x_1)$$

$$y - 10.5 = -0.16(x - 2005)$$

$$y - 10.5 = -0.16x + 320.8$$

$$y = -0.16x + 331.3$$

$$T(m) = -0.16m + 331.3$$

   b.   Let $m = 2015$.

$$T(2015) = -0.16(2015) + 331.3$$

$$= 8.9 \text{ sec}$$

15. $f(x) = 3x + 1$

$$f(3) = 3(3) + 1$$

$$= 9 + 1$$

$$= 10$$

16. $g(t) = t^2 - 2t + 1$

$$g(-6) = (-6)^2 - 2(-6) + 1$$

$$= 36 + 12 + 1$$

$$= 49$$

17. $g(t) = t^2 - 2t + 1$

$$g\left(\frac{1}{4}\right) = \left(\frac{1}{4}\right)^2 - 2\left(\frac{1}{4}\right) + 1$$

$$= \frac{1}{16} - \frac{1}{2} + 1$$

$$= \frac{1}{16} - \frac{8}{16} + \frac{16}{16}$$

$$= \frac{9}{16}$$

18. $f(x) = 3x + 1$

$$f(r+8) = 3(r+8) + 1$$

$$= 3r + 24 + 1$$

$$= 3r + 25$$

19. $h(t) = -16t^2 + 80t + 10$

$$h(2.5) = -16(2.5)^2 + 80(2.5) + 10$$

$$= -16(6.25) + 80(2.5) + 10$$

$$= -100 + 200 + 10$$

$$= 110$$

20. a. $f(4) = 0$
    b. $x = 2$ and $x = 6$
    c. D: $(-\infty, \infty)$; R: $(-\infty, 2]$

21. Function

22. Not a function; (2,2) and (2, −2)

23. $f(x) = |x| + 3$

D: the set of real numbers
R: the set of real numbers greater than or
   equal to 3

24. $g(x) = (x-4)^3 + 1$

D: the set of real numbers
R: the set of real numbers

25.
$$-2(2x+3) \geq 14$$
$$-4x - 6 \geq 14$$
$$-4x - 6 + 6 \geq 14 + 6$$
$$-4x \geq 20$$
$$x \leq -5; \quad (-\infty, -5]$$

−5

26.
$$-2 < \frac{x-4}{3} < 4$$

$$3(-2) < 3\left(\frac{x-4}{3}\right) < 3(4)$$

$$-6 < x - 4 < 12$$

$$-6 + 4 < x - 4 + 4 < 12 + 4$$

$$-2 < x < 16$$

$$(-2, 16)$$

−2   16

27.
$$3x \geq -2x + 5 \quad \text{and} \quad 7 \geq 4x - 2$$
$$3x + 2x \geq -2x + 5 + 2x \quad 7 + 2 \geq 4x - 2 + 2$$
$$5x \geq 5 \qquad\qquad 9 \geq 4x$$
$$x \geq 1 \qquad\qquad \frac{9}{4} \geq x$$

$$\left[1, \frac{9}{4}\right]$$

1     $\frac{9}{4}$

Chapter 8 Test

28.

$$3x < -9 \quad \text{or} \quad -\frac{x}{4} < -2$$

$$\frac{3x}{3} < \frac{-9}{3} \quad -4\left(-\frac{x}{4}\right) > -4(-2)$$

$$x < -3 \qquad\qquad x > 8$$

$$(-\infty, -3) \cup (8, \infty)$$

29.

$$|2x - 4| > 22$$

$$2x - 4 < -22 \quad \text{or} \quad 2x - 4 > 22$$

$$2x - 4 + 4 < -22 + 4 \quad 2x - 4 + 4 > 22 + 4$$

$$2x < -18 \qquad\qquad 2x > 26$$

$$x < -9 \qquad\qquad x > 13$$

$$(-\infty, -9) \cup (13, \infty)$$

30.

$$2|3(x - 2)| \le 4$$

$$2|3x - 6| \le 4$$

$$\frac{2|3x - 6|}{2} \le \frac{4}{2}$$

$$|3x - 6| \le 2$$

$$-2 \le 3x - 6 \le 2$$

$$-2 + 6 \le 3x - 6 + 6 \le 2 + 6$$

$$4 \le 3x \le 8$$

$$\frac{4}{3} \le x \le \frac{8}{3}$$

$$\left[\frac{4}{3}, \frac{8}{3}\right]$$

31.

$$|2x + 3| - 19 = 0$$

$$|2x + 3| = 19$$

$$2x + 3 = 19 \quad \text{or} \quad 2x + 3 = -19$$

$$2x = 16 \qquad\qquad 2x = -22$$

$$x = 8 \qquad\qquad x = -11$$

32.

$$|3x + 4| = |x + 12|$$

$$3x + 4 = x + 12 \quad \text{or} \quad 3x + 4 = -(x + 12)$$

$$3x + 4 = x + 12 \qquad\qquad 3x + 4 = -x - 12$$

$$3x + 4 - 4 = x + 12 - 4 \quad 3x + 4 - 4 = -x - 12 - 4$$

$$3x = x + 8 \qquad\qquad 3x = -x - 16$$

$$3x - x = x + 8 - x \qquad 3x + x = -x - 16 + x$$

$$2x = 8 \qquad\qquad 4x = -16$$

$$x = 4 \qquad\qquad x = -4$$

33.

$$12a^3b^2c - 3a^2b^2c^2 + 6abc^3$$

$$= 3abc\left(4a^2b - abc + 2c^2\right)$$

34.

$$4y^4 - 64 = 4\left(y^4 - 16\right)$$

$$= 4\left(y^2 + 4\right)\left(y^2 - 4\right)$$

$$= 4\left(y^2 + 4\right)(y + 2)(y - 2)$$

35.

$$b^3 + 125 = (b)^3 + (5)^3$$

$$= (b + 5)\left(b^2 - 5b + 25\right)$$

36.

$$6u^2 + 9u - 6 = 3\left(2u^2 + 3u - 2\right)$$

$$= 3(2u - 1)(u + 2)$$

37.

$$ax - xy + ay - y^2 = x(a - y) + y(a - y)$$

$$= (a - y)(x + y)$$

38.

$$25m^8 - 60m^4n + 36n^2 = \left(5m^4 - 6n\right)^2$$

39.

$$144b^2 + 25$$

prime

40.

$$x^2 + 6x + 9 - y^2 = \left(x^2 + 6x + 9\right) - y^2$$

$$= (x + 3)^2 - y^2$$

$$= (x + 3 - y)(x + 3 + y)$$

41.

$$64a^3 - 125b^6$$

$$= \left(4a - 5b^2\right)\left[(4a)^2 + (4a)(5b^2) + (5b^2)^2\right]$$

$$= \left(4a - 5b^2\right)\left(16a^2 + 20ab^2 + 25b^4\right)$$

42.

$$(x-y)^2 + 3(x-y) - 10$$

$$= \left[(x-y)+5\right]\left[(x-y)-2\right]$$

$$= (x-y+5)(x-y-2)$$

43. a. $f(5) = 1$

   b. $f(-1) = -\dfrac{1}{2}$

   c. $f(1) = -1$

   d. $x = 2$

44. $p(t) = \dfrac{200t}{t+1}$

$$p(7) = \dfrac{200(7)}{(7)+1}$$

$$= \dfrac{1,400}{8}$$

$$= 175$$

45. No.

$$-\dfrac{x+4}{x^2+x-18} = \dfrac{-x-4}{x^2+x-18}$$

$$\neq \dfrac{x-4}{x^2+x-18}$$

46. $f(x) = \dfrac{x^2+6x+5}{x-x^2}$

$$x - x^2 = 0$$

$$x(1-x) = 0$$

$$x = 0 \quad \text{and} \quad 1-x = 0$$

$$x = 0 \qquad\qquad 1 = x$$

$$(-\infty, 0) \cup (0,1) \cup (1, \infty)$$

The set of all real numbers except 0 and 1.

47.

$$\dfrac{3y-6z}{2z-y} = \dfrac{3(y-2z)}{2z-y}$$

$$= \dfrac{3(y-2z)}{-1(y-2z)}$$

$$= -3$$

48.

$$\dfrac{2x^2+7xy+3y^2}{4xy+12y^2} = \dfrac{(2x+y)(x+3y)}{4y(x+3y)}$$

$$= \dfrac{2x+y}{4y}$$

49.

$$\dfrac{x^3+y^3}{4} \div \dfrac{x^2-xy+y^2}{2x+2y} = \dfrac{x^3+y^3}{4} \cdot \dfrac{2x+2y}{x^2-xy+y^2}$$

$$= \dfrac{(x+y)(x^2-xy+y^2)}{2 \cdot 2} \cdot \dfrac{2(x+y)}{x^2-xy+y^2}$$

$$= \dfrac{(x+y)^2}{2}$$

50.

$$\dfrac{xu+2u+3x+6}{u^2-9} \cdot \dfrac{13u-39}{x^2+3x+2}$$

$$= \dfrac{u(x+2)+3(x+2)}{(u+3)(u-3)} \cdot \dfrac{13(u-3)}{(x+1)(x+2)}$$

$$= \dfrac{(u+3)(x+2)}{(u+3)(u-3)} \cdot \dfrac{13(u-3)}{(x+1)(x+2)}$$

$$= \dfrac{13}{x+1}$$

51.

$$\dfrac{-3t+4}{t^2+t-20} + \dfrac{6+5t}{t^2+t-20} = \dfrac{-3t+4+6+5t}{t^2+t-20}$$

$$= \dfrac{2t+10}{t^2+t-20}$$

$$= \dfrac{2(t+5)}{(t+5)(t-4)}$$

$$= \dfrac{2}{t-4}$$

52.

$$\frac{a+3}{a^2-a-2}-\frac{a-4}{a^2-2a-3}$$

$$=\frac{a+3}{(a-2)(a+1)}-\frac{a-4}{(a-3)(a+1)}$$

$$=\frac{a+3}{(a-2)(a+1)}\left(\frac{a-3}{a-3}\right)-\frac{a-4}{(a-3)(a+1)}\left(\frac{a-2}{a-2}\right)$$

$$=\frac{a^2-9}{(a-2)(a+1)(a-3)}-\frac{a^2-2a-4a+8}{(a-2)(a+1)(a-3)}$$

$$=\frac{a^2-9}{(a-2)(a+1)(a-3)}-\frac{a^2-6a+8}{(a-2)(a+1)(a-3)}$$

$$=\frac{a^2-9-\left(a^2-6a+8\right)}{(a-2)(a+1)(a-3)}$$

$$=\frac{a^2-9-a^2+6a-8}{(a-2)(a+1)(a-3)}$$

$$=\frac{6a-17}{(a-2)(a+1)(a-3)}$$

53.

$$\frac{34}{x^2}+\frac{13}{20x}=\frac{3}{2x}$$

$$20x^2\left(\frac{34}{x^2}+\frac{13}{20x}\right)=20x^2\left(\frac{3}{2x}\right)$$

$$680+13x=30x$$

$$680=17x$$

$$40=x$$

54.

$$\frac{u-2}{u-3}+3=u+\frac{u-4}{3-u}$$

$$\frac{u-2}{u-3}+3=u-\frac{u-4}{u-3}$$

$$(u-3)\left(\frac{u-2}{u-3}+3\right)=(u-3)\left(u-\frac{u-4}{u-3}\right)$$

$$u-2+3(u-3)=u(u-3)-(u-4)$$

$$u-2+3u-9=u^2-3u-u+4$$

$$4u-11=u^2-4u+4$$

$$0=u^2-8u+15$$

$$0=(u-5)(u-3)$$

$$u-5=0 \quad\text{and}\quad u-3=0$$

$$u=5 \qquad\qquad u=\cancel{3}$$

$$u=5;\ 3\text{ is extraneous}$$

55.

$$\frac{3}{x-2}=\frac{x+3}{2x}$$

$$2x(x-2)\left(\frac{3}{x-2}\right)=2x(x-2)\left(\frac{x+3}{2x}\right)$$

$$2x(3)=(x-2)(x+3)$$

$$6x=x^2+3x-2x-6$$

$$0=x^2-5x-6$$

$$0=(x-6)(x+1)$$

$$x-6=0 \quad\text{and}\quad x+1=0$$

$$x=6 \qquad\qquad x=-1$$

56.

$$\frac{4}{m^2-9}+\frac{5}{m^2-m-12}=\frac{7}{m^2-7m+12}$$

$$\frac{4}{(m-3)(m+3)}+\frac{5}{(m-4)(m+3)}=\frac{7}{(m-4)(m-3)}$$

Multiply by LCD of $(m-3)(m+3)(m-4)$.

$$4(m-4)+5(m-3)=7(m+3)$$

$$4m-16+5m-15=7m+21$$

$$9m-31=7m+21$$

$$2m-31=21$$

$$2m=52$$

$$m=26$$

57. Find k when $x=30$ and $y=4$.

$$x=ky$$

$$30=4k$$

$$\frac{30}{4}=\frac{4k}{4}$$

$$\frac{15}{2}=k$$

Find y when $x=9$.

$$x=ky$$

$$9=\frac{15}{2}y$$

$$\frac{2}{15}(9)=\frac{2}{15}\left(\frac{15}{2}y\right)$$

$$\frac{6}{5}=y$$

58. Let L = loudness and d = distance.

$$L = \frac{k}{d^2}$$

$$100 = \frac{k}{30^2}$$

$$100 = \frac{k}{900}$$

$$90,000 = k$$

Find L if $k = 90,000$ and $d = 60$.

$$L = \frac{k}{d^2}$$

$$= \frac{90,000}{60^2}$$

$$= \frac{90,000}{3,600}$$

$$= 25 \text{ decibels}$$

VOCABULARY

1. $5x^2$ is the **square** root of $25x^4$ because $(5x^2)^2 = 25x^4$. The **cube** root of 216 is 6 because $6^3 = 216$.

3. The radical symbol $\sqrt{}$ represents the **positive** or principle square root of a number.

5. The number 100 has two square roots. The positive or **principal** square root of 100 is 10.

7. When we write $\sqrt{b^4} = b^2$, we say that we have **simplified** the radical expression.

9. $f(x) = \sqrt{x}$ and $g(x) = \sqrt[3]{x}$ are **radical** functions.

CONCEPTS

11 b is a square root of a if $b^2 = \boxed{a}$.

13. $\sqrt{-4}$ is not a real number, because no real number **squared** equals -4.

15. $\sqrt{x^2} = \boxed{|x|}$ and $\sqrt[3]{x^3} = \boxed{x}$.

17. a. $f(11) = 3$
    b. $f(2) = 0$
    c. $f(-1)$ is undefined
    d. $x = 6$
    e. none
    f. D: $[2, \infty)$ and R: $[0, \infty)$

19. a. $f(-8) = -5$
    b. $f(0) = -3$
    c. $x = 1$
    d. D: $(-\infty, \infty)$ and R: $(-\infty, \infty)$

NOTATION

21. a. $\sqrt{x^2} = |x|$
    b. $\sqrt[3]{x^3} = x$

23.
$$\sqrt{100} = 10$$

25.
$$-\sqrt{64} = -1(8)$$
$$= -8$$

27.
$$\sqrt{\frac{1}{9}} = \frac{\sqrt{1}}{\sqrt{9}}$$
$$= \frac{1}{3}$$

29.
$$\sqrt{0.25} = 0.5$$

31.
$$\sqrt{-81} \text{ is not real}$$

33.
$$\sqrt{121} = 11$$

35.
$$\sqrt{12} = 3.4641$$

37.
$$\sqrt{679.25} = 26.0624$$

39.
$$\sqrt{4x^2} = \sqrt{(2x)^2}$$
$$= |2x|$$
$$= 2|x|$$

41.
$$\sqrt{81h^4} = \sqrt{(9h^2)^2}$$
$$= |9h^2|$$
$$= 9h^2$$

43.
$$\sqrt{36s^6} = \sqrt{(6s^3)^2}$$
$$= |6s^3|$$
$$= 6|s^3|$$

45.
$$\sqrt{144m^8} = \sqrt{(12m^4)^2}$$
$$= |12m^4|$$
$$= 12m^4$$

47.
$$\sqrt{y^2 - 2y + 1} = \sqrt{(y-1)^2}$$
$$= |y - 1|$$

49.
$$\sqrt{a^4 + 6a^2 + 9} = \sqrt{(a^2 + 3)^2}$$
$$= a^2 + 3$$

51. $f(x) = -\sqrt{x}$

x	y
0	$-\sqrt{0} = 0$
1	$-\sqrt{1} = -1$
4	$-\sqrt{4} = -2$
9	$-\sqrt{9} = -3$
16	$-\sqrt{16} = -4$

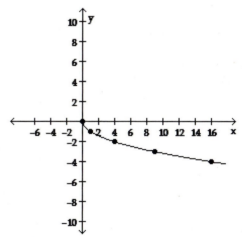

D: $[0, \infty)$ and R: $(-\infty, 0]$

53. $f(x) = \sqrt{x+4}$

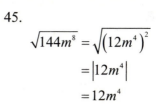

$D:[-4, \infty); \quad R:[0, \infty)$

55. $f(x) = \sqrt{x+6}$
$$x + 6 \geq 0$$
$$x \geq -6$$
$$D:[-6, \infty)$$

57. $g(x) = \sqrt{8 - 2x}$
$$8 - 2x \geq 0$$
$$-2x \geq -8$$
$$x \leq 4$$
$$D:(-\infty, 4]$$

59. $s(t) = \sqrt{9t - 4}$
$$9t - 4 \geq 0$$
$$9t \geq 4$$
$$x \geq \frac{4}{9}$$
$$D:\left[\frac{4}{9}, \infty\right)$$

61. $c(x) = \sqrt{0.5x - 20}$
$$0.5x - 20 \geq 0$$
$$0.5x \geq 20$$
$$x \geq 40$$
$$D:[40, \infty)$$

Section 9.1

63. $f(x) = \sqrt{3x+1}$

  a.
  $$f(8) = \sqrt{3(8)+1}$$
  $$= \sqrt{24+1}$$
  $$= \sqrt{25}$$
  $$= 5$$

  b.
  $$f(-2) = \sqrt{3(-2)+1}$$
  $$= \sqrt{-6+1}$$
  $$= \sqrt{-5}$$
  $$= \text{undefined}$$

65. $g(x) = \sqrt[3]{x-4}$

  a.
  $$g(12) = \sqrt[3]{12-4}$$
  $$= \sqrt[3]{8}$$
  $$= 2$$

  b.
  $$g(-23) = \sqrt[3]{-23-4}$$
  $$= \sqrt[3]{-27}$$
  $$= -3$$

67. $f(x) = \sqrt{x^2+1}$

  a.
  $$f(4) = \sqrt{4^2+1}$$
  $$= \sqrt{16+1}$$
  $$= \sqrt{17}$$
  $$= 4.1231$$

  b.
  $$f(2.35) = \sqrt{(2.35)^2+1}$$
  $$= \sqrt{5.5225+1}$$
  $$= \sqrt{6.5225}$$
  $$= 2.5539$$

69. $g(x) = \sqrt[3]{x^2+1}$

  a.
  $$g(6) = \sqrt[3]{6^2+1}$$
  $$= \sqrt[3]{36+1}$$
  $$= \sqrt[3]{37}$$
  $$= 3.3322$$

  b.
  $$g(21.57) = \sqrt[3]{(21.57)^2+1}$$
  $$= \sqrt[3]{465.2649+1}$$
  $$= \sqrt[3]{466.2649}$$
  $$= 7.7543$$

71.
$$\sqrt[3]{1} = 1$$

73.
$$\sqrt[3]{-125} = -5$$

75.
$$\sqrt[3]{\frac{8}{27}} = \frac{\sqrt[3]{8}}{\sqrt[3]{27}}$$
$$= \frac{2}{3}$$

77.
$$\sqrt[3]{64} = 4$$

79.
$$\sqrt[3]{-216a^3} = -6a$$

81.
$$\sqrt[3]{-1,000p^6q^3} = -10p^2q$$

83. $f(x) = \sqrt[3]{x} - 3$

x	y
-8	$\sqrt[3]{-8} - 3 = -2 - 3$ $= -5$
-1	$\sqrt[3]{-1} - 3 = -1 - 3$ $= -4$
0	$\sqrt[3]{0} - 3 = 0 - 3$ $= -3$
1	$\sqrt[3]{1} - 3 = 1 - 3$ $= -2$
8	$\sqrt[3]{8} - 3 = 2 - 3$ $= -1$

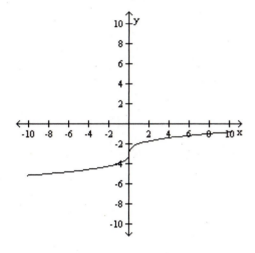

D: $(-\infty, \infty)$ and R: $(-\infty, \infty)$

85. $f(x) = \sqrt[3]{x - 3}$

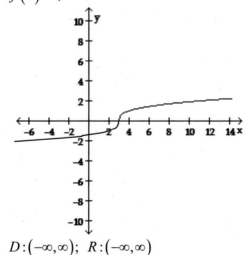

D: $(-\infty, \infty)$; R: $(-\infty, \infty)$

87.
$$\sqrt[4]{81} = \sqrt[4]{(3)^4}$$
$$= 3$$

89.
$$-\sqrt[5]{243} = -\sqrt[5]{(3)^5}$$
$$= -1(3)$$
$$= -3$$

91.
$\sqrt[4]{-256}$ is not a real number

93.
$$\sqrt[5]{-\frac{1}{32}} = \sqrt[5]{\frac{(1)^5}{(-2)^5}}$$
$$= -\frac{1}{2}$$

95.
$$\sqrt[5]{32a^5} = \sqrt[5]{(2a)^5}$$
$$= 2a$$

97.
$$\sqrt[4]{81a^4} = \sqrt[4]{(3a)^4}$$
$$= 3|a|$$

99.
$$\sqrt[6]{k^{12}} = \sqrt[6]{(k^2)^6}$$
$$= k^2$$

101.
$$\sqrt[4]{(m+4)^8} = \sqrt[4]{\left[(m+4)^2\right]^4}$$
$$= (m+4)^2$$

TRY IT YOURSELF

103.
$$\sqrt[3]{64s^9t^6} = \sqrt[3]{\left(4s^3t^2\right)^3}$$
$$= 4s^3t^2$$

105.
$$-\sqrt{49b^8} = -7b^4$$

- 585 -

107.

$$-\sqrt[5]{-\frac{1}{32}} = -\sqrt[5]{-\frac{(1)^5}{(2)^5}}$$
$$= -1\left(-\frac{1}{2}\right)$$
$$= \frac{1}{2}$$

109.

$$\sqrt[3]{-125m^6} = \sqrt[3]{\left(-5m^2\right)^3}$$
$$= -5m^2$$

111.

$$\sqrt{400m^{16}n^2} = \left|20m^8 n\right|$$
$$= 20m^8 \left|n\right|$$

113.

$$\sqrt[6]{64a^6b^6} = \sqrt[6]{\left(2ab\right)^6}$$
$$= \left|2ab\right|$$
$$= 2\left|ab\right|$$

115.

$\sqrt[4]{-81}$ is not a real number

117.

$$\sqrt{n^2 + 12n + 36} = \sqrt{\left(n+6\right)^2}$$
$$= \left|n+6\right|$$

119. a. $\sqrt{64} = 4$

b. $\sqrt[3]{64} = 4$

121. a. $\sqrt{81} = 9$

b. $\sqrt[4]{81} = 3$

123. **EMBROIDERY**

$$r = \sqrt{\frac{A}{\pi}}$$
$$= \sqrt{\frac{38.5}{\pi}}$$
$$\approx \sqrt{12.3}$$
$$\approx 3.5$$

$$d = 2r$$
$$= 2(3.5)$$
$$= 7.0 \text{ in.}$$

125. **SHOELACES**

$$S = 2\left[H + L + (p-1)\sqrt{H^2 + V^2}\right]$$
$$= 2\left[50 + 250 + (6-1)\sqrt{50^2 + 20^2}\right]$$
$$= 2\left[50 + 250 + (5)\sqrt{2{,}500 + 400}\right]$$
$$= 2\left[50 + 250 + (5)\sqrt{2{,}900}\right]$$
$$= 2\left[50 + 250 + (5)(53.851648)\right]$$
$$= 2\left[50 + 250 + 269.258\right]$$
$$= 2\left[569.23\right]$$
$$= 1{,}138.5 \text{ mm}$$

127. **PULSE RATES**
Change his height from feet to inches by multiplying by 12 since there are 12 inches in 1 foot.

$$8 \text{ ft } 5.5 \text{ in.} = 8(12) + 5.5 \text{ inches}$$
$$= 96 + 5.5 \text{ inches}$$
$$= 101.5 \text{ inches}$$

$$p(t) = \frac{590}{\sqrt{t}}$$
$$p(101.5) = \frac{590}{\sqrt{101.5}}$$
$$\approx 58.6 \text{ beats/min.}$$

129. BIOLOGY

The volume required for each rat is 125 ft^3, so to find the volume for 5 rats, multiply 125 times 5.

$$d(V) = \sqrt[3]{12\left(\frac{V}{\pi}\right)}$$

$$= \sqrt[3]{12\left(\frac{5 \cdot 125}{\pi}\right)}$$

$$= \sqrt[3]{12\left(\frac{625}{\pi}\right)}$$

$$= \sqrt[3]{12(198.9)}$$

$$= \sqrt[3]{2,386.8}$$

$$= 13.4 \text{ ft}$$

131. COLLECTIBLES

$$r = \sqrt[n]{\frac{A}{P}} - 1$$

$$= \sqrt[5]{\frac{950}{800}} - 1$$

$$= \sqrt[5]{1.1875} - 1$$

$$= 1.035 - 1$$

$$= 0.035$$

$$= 3.5\%$$

WRITING

133. Answers will vary.

135. Answers will vary.

REVIEW

137.

$$\frac{x^2 - 3xy - 4y^2}{x^2 + cx - 2yx - 2cy} \div \frac{x^2 - 2xy - 3y^2}{x^2 + cx - 4yx - 4cy}$$

$$= \frac{x^2 - 3xy - 4y^2}{x^2 + cx - 2yx - 2cy} \cdot \frac{x^2 + cx - 4yx - 4cy}{x^2 - 2xy - 3y^2}$$

$$= \frac{(x-4y)(x+y)}{x(x+c) - 2y(x+c)} \cdot \frac{x(x+c) - 4y(x+c)}{(x-3y)(x+y)}$$

$$= \frac{(x-4y)\cancel{(x+y)}}{(x-2y)\cancel{(x+c)}} \cdot \frac{(x-4y)\cancel{(x+c)}}{(x-3y)\cancel{(x+y)}}$$

$$= \frac{(x-4y)^2}{(x-2y)(x-3y)}$$

CHALLENGE PROBLEMS

139. $f(x) = -\sqrt{x-2} + 3$

D: $[2, \infty)$ and R: $(-\infty, 3]$

VOCABULARY

1. The expressions $4^{1/2}$ and $(-8)^{-2/3}$ have **rational (or fractional)** exponents.

3. We read $27^{-1/3}$ as "27 to the **negative** one-third power."

5. In the radical expression $\sqrt[4]{16x^8}$, 4 is the **index**, and $16x^8$ is the **radicand**.

CONCEPTS

7.

Radical form	Exponential form	Base	Exponent
$\sqrt[5]{25}$	$25^{1/5}$	25	$\dfrac{1}{5}$
$\left(\sqrt[3]{-27}\right)^2$	$(-27)^{2/3}$	-27	$\dfrac{2}{3}$
$\left(\sqrt[4]{16}\right)^{-3}$	$16^{-3/4}$	16	$-\dfrac{3}{4}$
$\left(\sqrt{81}\right)^3$	$81^{3/2}$	81	$\dfrac{3}{2}$
$-\sqrt{\dfrac{9}{64}}$	$-\left(\dfrac{9}{64}\right)^{1/2}$	$\dfrac{9}{64}$	$\dfrac{1}{2}$

9. Simplify each number.
$$8^{2/3} = 4$$
$$(-125)^{1/3} = -5$$
$$-16^{-1/4} = -\frac{1}{2}$$
$$4^{3/2} = 8$$
$$-\left(\frac{9}{100}\right)^{-1/2} = -\frac{10}{3}$$

11. $x^{1/n} = \boxed{\sqrt[n]{x}}$

13. $x^{-m/n} = \boxed{\dfrac{1}{x^{m/n}}}$

NOTATION

15.
$$\left(100a^4\right)^{3/2} = \left(\sqrt{\boxed{100a^4}}\right)^3$$
$$= \left(\boxed{10a^2}\right)^3$$
$$= 1{,}000a^6$$

GUIDED PRACTICE

17.
$$125^{1/3} = \sqrt[3]{125}$$
$$= 5$$

19.
$$81^{1/4} = \sqrt[4]{81}$$
$$= 3$$

21.
$$32^{1/5} = \sqrt[5]{32}$$
$$= 2$$

23.
$$(-216)^{1/3} = \sqrt[3]{-216}$$
$$= -6$$

25.
$$-16^{1/4} = -\sqrt[4]{16}$$
$$= -2$$

27.
$$\left(\frac{1}{4}\right)^{1/2} = \sqrt{\frac{1}{4}}$$
$$= \frac{\sqrt{1}}{\sqrt{4}}$$
$$= \frac{1}{2}$$

29.
$$\left(4x^4\right)^{1/2} = \sqrt{4x^4}$$
$$= 2x^2$$

31.
$$\left(x^2\right)^{1/2} = \sqrt{x^2}$$
$$= |x|$$

33.

$$\left(-64p^8\right)^{1/2} = \sqrt{-64p^8}$$

is not a real number

35.

$$\left(-27n^9\right)^{1/3} = \sqrt[3]{-27n^9}$$
$$= -3n^3$$

37.

$$\left(-64x^8\right)^{1/8} = \sqrt[8]{-64x^8}$$

is not a real number

39.

$$\left[(x+1)^6\right]^{1/6} = \sqrt[6]{(x+1)^6}$$
$$= |x+1|$$

41.

$$36^{3/2} = \left(\sqrt{36}\right)^3$$
$$= (6)^3$$
$$= 216$$

43.

$$16^{3/4} = \left(\sqrt[4]{16}\right)^3$$
$$= (2)^3$$
$$= 8$$

45.

$$\left(-\frac{1}{216}\right)^{2/3} = \left(\sqrt[3]{-\frac{1}{216}}\right)^2$$
$$= \left(-\frac{1}{6}\right)^2$$
$$= \frac{1}{36}$$

47.

$$-4^{5/2} = -\left(\sqrt{4}\right)^5$$
$$= -2^5$$
$$= -32$$

49.

$$\left(25x^4\right)^{3/2} = \left(\sqrt{25x^4}\right)^3$$
$$= \left(5x^2\right)^3$$
$$= 125x^6$$

51.

$$\left(-8x^6y^3\right)^{2/3} = \left(\sqrt[3]{-8x^6y^3}\right)^2$$
$$= \left(-2x^2y\right)^2$$
$$= 4x^4y^2$$

53.

$$\left(81x^4y^8\right)^{3/4} = \left(\sqrt[4]{81x^4y^8}\right)^3$$
$$= \left(3xy^2\right)^3$$
$$= 27x^3y^6$$

55.

$$-\left(\frac{x^5}{32}\right)^{4/5} = -\left(\sqrt[5]{\frac{x^5}{32}}\right)^4$$
$$= -\left(\frac{x}{2}\right)^4$$
$$= -\frac{x^4}{16}$$

57. $\sqrt[5]{8abc} = \left(8abc\right)^{1/5}$

59. $\sqrt[3]{a^2 - b^2} = \left(a^2 - b^2\right)^{1/3}$

61. $\left(6x^3y\right)^{1/4} = \sqrt[4]{6x^3y}$

63. $\left(2s^2 - r^2\right)^{1/2} = \sqrt{2s^2 - r^2}$

65.

$$4^{-1/2} = \frac{1}{4^{1/2}}$$
$$= \frac{1}{\sqrt{4}}$$
$$= \frac{1}{2}$$

Section 9.2

67.

$$125^{-1/3} = \frac{1}{125^{1/3}}$$

$$= \frac{1}{\sqrt[3]{125}}$$

$$= \frac{1}{5}$$

69.

$$-\left(1,000y^3\right)^{-2/3} = -\frac{1}{\left(1,000y^3\right)^{2/3}}$$

$$= -\frac{1}{\left(\sqrt[3]{1,000y^3}\right)^2}$$

$$= -\frac{1}{\left(10y\right)^2}$$

$$= -\frac{1}{100y^2}$$

71.

$$\left(-\frac{27}{8}\right)^{-4/3} = \left(-\frac{8}{27}\right)^{4/3}$$

$$= \left(\sqrt[3]{-\frac{8}{27}}\right)^4$$

$$= \left(-\frac{2}{3}\right)^4$$

$$= \frac{16}{81}$$

73.

$$\left(\frac{16}{81y^4}\right)^{-3/4} = \left(\frac{81y^4}{16}\right)^{3/4}$$

$$= \left(\sqrt[4]{\frac{81y^4}{16}}\right)^3$$

$$= \left(\frac{3y}{2}\right)^3$$

$$= \frac{27y^3}{8}$$

75.

$$\frac{1}{9^{-5/2}} = 9^{5/2}$$

$$= \left(\sqrt{9}\right)^5$$

$$= 3^5$$

$$= 243$$

77.

$$9^{3/7}9^{2/7} = 9^{3/7+2/7}$$

$$= 9^{5/7}$$

79.

$$6^{-2/3}6^{-4/3} = 6^{-6/3}$$

$$= 6^{-2}$$

$$= \frac{1}{6^2}$$

$$= \frac{1}{36}$$

81.

$$\left(m^{2/3}m^{1/3}\right)^6 = \left(m^{2/3+1/3}\right)^6$$

$$= \left(m^{3/3}\right)^6$$

$$= \left(m^1\right)^6$$

$$= m^6$$

83.

$$\left(a^{1/2}b^{1/3}\right)^{3/2} = a^{(1/2)(3/2)}b^{(1/3)(3/2)}$$

$$= a^{3/4}b^{1/2}$$

85.

$$\frac{3^{4/3}3^{1/3}}{3^{2/3}} = \frac{3^{4/3+1/3}}{3^{2/3}}$$

$$= \frac{3^{5/3}}{3^{2/3}}$$

$$= 3^{5/3-2/3}$$

$$= 3^{3/3}$$

$$= 3^1$$

$$= 3$$

87.

$$\frac{a^{3/4}a^{3/4}}{a^{1/2}} = \frac{a^{3/4+3/4}}{a^{1/2}}$$

$$= \frac{a^{6/4}}{a^{1/2}}$$

$$= \frac{a^{3/2}}{a^{1/2}}$$

$$= a^{3/2-1/2}$$

$$= a^{2/2}$$

$$= a$$

89.

$$y^{1/3}\left(y^{2/3} + y^{5/3}\right) = y^{1/3+2/3} + y^{1/3+5/3}$$

$$= y^{3/3} + y^{6/3}$$

$$= y + y^2$$

91.

$$x^{3/5}\left(x^{7/5} - x^{-3/5} + 1\right) = x^{3/5+7/5} - x^{3/5+(-3/5)} + x^{3/5}$$

$$= x^{10/5} - x^0 + x^{3/5}$$

$$= x^2 - 1 + x^{3/5}$$

93.

$$\sqrt[4]{5^2} = 5^{2/4}$$

$$= 5^{1/2}$$

$$= \sqrt{5}$$

95.

$$\sqrt[9]{11^3} = 11^{3/9}$$

$$= 11^{1/3}$$

$$= \sqrt[3]{11}$$

97.

$$\sqrt[6]{p^3} = p^{3/6}$$

$$= p^{1/2}$$

$$= \sqrt{p}$$

99.

$$\sqrt[10]{x^2 y^2} = \left(x^2 y^2\right)^{1/10}$$

$$= x^{2(1/10)} y^{2(1/10)}$$

$$= x^{2/10} y^{2/10}$$

$$= x^{1/5} y^{1/5}$$

$$= \sqrt[5]{xy}$$

101

$$\sqrt[9]{\sqrt{c}} = \left(c^{1/2}\right)^{1/9}$$

$$= c^{(1/2)(1/9)}$$

$$= c^{1/18}$$

$$= \sqrt[18]{c}$$

103.

$$\sqrt[5]{\sqrt[3]{7m}} = \left[\left(7m\right)^{1/3}\right]^{1/5}$$

$$= \left(7m\right)^{(1/3)(1/5)}$$

$$= \left(7m\right)^{1/15}$$

$$= \sqrt[15]{7m}$$

105.

$$15^{1/3} = \sqrt[3]{15}$$

$$= 2.47$$

107.

$$\left(1.045\right)^{2/5} = \left(\sqrt[5]{1.045}\right)^2$$

$$= \left(1.01\right)^2$$

$$= 1.02$$

TRY IT YOURSELF

109.

$$\left(25y^2\right)^{1/2} = \sqrt{25y^2}$$

$$= 5y$$

111.

$$-\left(\frac{a^4}{81}\right)^{3/4} = -\left(\sqrt[4]{\frac{a^4}{81}}\right)^3$$

$$= -\left(\frac{a}{3}\right)^3$$

$$= -\frac{a^3}{27}$$

Section 9.2

113.

$$16^{-3/2} = \frac{1}{16^{3/2}}$$

$$= \frac{1}{\left(\sqrt{16}\right)^3}$$

$$= \frac{1}{4^3}$$

$$= \frac{1}{64}$$

115.

$$\frac{p^{8/5}p^{7/5}}{p^2} = \frac{p^{8/5+7/5}}{p^2}$$

$$= \frac{p^{15/5}}{p^2}$$

$$= \frac{p^3}{p^2}$$

$$= p^{3-2}$$

$$= p$$

117.

$$\left(-27x^6\right)^{-1/3} =$$

$$= \left(-\frac{1}{27x^6}\right)^{1/3}$$

$$= \sqrt[3]{-\frac{1}{27x^6}}$$

$$= -\frac{1}{3x^2}$$

$$= -\frac{1}{3}x^2$$

119.

$$\frac{1}{32^{-1/5}} = 32^{1/5}$$

$$= \sqrt[5]{32}$$

$$= 2$$

121.

$$n^{1/5}\left(n^{2/5} - n^{-1/5}\right) = n^{1/5+2/5} - n^{1/5+(-1/5)}$$

$$= n^{3/5} - n^0$$

$$= n^{3/5} - 1$$

123.

$$\frac{1}{9^{-5/2}} = 9^{5/2}$$

$$= \left(\sqrt{9}\right)^5$$

$$= 3^5$$

$$= 243$$

125.

$$\left(m^4\right)^{1/2} = \sqrt{m^4}$$

$$= m^2$$

127.

$$\sqrt[4]{25b^2} = \left(5^2 b^2\right)^{1/4}$$

$$= 5^{2(1/4)} b^{2(1/4)}$$

$$= 5^{2/4} b^{2/4}$$

$$= 5^{1/2} b^{1/2}$$

$$= \sqrt{5b}$$

29

$$\left(16x^4\right)^{1/4} = \sqrt[4]{16x^4}$$

$$= |2x|$$

$$= 2|x|$$

131.

$$-\left(8a^3b^6\right)^{-2/3} = -\frac{1}{\left(8a^3b^6\right)^{2/3}}$$

$$= -\frac{1}{\left(\sqrt[3]{8a^3b^6}\right)^2}$$

$$= -\frac{1}{\left(2ab^2\right)^2}$$

$$= -\frac{1}{4a^2b^4}$$

133. **a.**

$$-125^{2/3} = -\left(\sqrt[3]{125}\right)^2$$

$$= -\left(5\right)^2$$

$$= -25$$

b.

$$\left(-125\right)^{2/3} = \left(\sqrt[3]{-125}\right)^2$$

$$= \left(-5\right)^2$$

$$= 25$$

c.

$$\left(-125\right)^{-2/3} = \frac{1}{\left(-125\right)^{2/3}}$$

$$= \frac{1}{\left(\sqrt[3]{-125}\right)^2}$$

$$= \frac{1}{\left(-5\right)^2}$$

$$= \frac{1}{25}$$

d.

$$\frac{1}{\left(-125\right)^{-2/3}} = \left(-125\right)^{2/3}$$

$$= \left(\sqrt[3]{-125}\right)^2$$

$$= \left(-5\right)^2$$

$$= 25$$

135. **a.**

$$\left(64a^4\right)^{1/2} = \sqrt{64a^4}$$

$$= 8a^2$$

b.

$$\left(64a^4\right)^{-1/2} = \frac{1}{\left(64a^4\right)^{1/2}}$$

$$= \frac{1}{\sqrt{64a^4}}$$

$$= \frac{1}{8a^2}$$

c.

$$-\left(64a^4\right)^{1/2} = -\sqrt{64a^4}$$

$$= -8a^2$$

d.

$$\frac{1}{\left(64a^4\right)^{1/2}} = \frac{1}{\sqrt{64a^4}}$$

$$= \frac{1}{8a^2}$$

APPLICATIONS

137. BALLISTIC PENDULUMS
Let $M = 6.0$, $m = 0.0625$, $g = 32$, and $h = 0.9$.

$$v = \frac{m+M}{m}\left(2gh\right)^{1/2}$$

$$= \frac{0.0625 + 6.0}{0.0625}\left(2 \cdot 32 \cdot 0.9\right)^{1/2}$$

$$= \frac{6.0625}{0.0625}\left(57.6\right)^{1/2}$$

$$= 97\left(\sqrt{57.6}\right)$$

$$= 97\left(7.589\right)$$

$$= 736 \text{ ft/sec}$$

139. RELATIVITY

Let $c = 186{,}000$, $v = 160{,}000$, and $m_0 = 1$.

$$m = m_0\left(1 - \frac{v^2}{c^2}\right)^{-1/2}$$

$$= 1\left(1 - \frac{160{,}000^2}{186{,}000^2}\right)^{-1/2}$$

$$= 1\left(1 - \frac{2.56 \times 10^{10}}{3.4596 \times 10^{10}}\right)^{-1/2}$$

$$= 1(1 - 0.74)^{-1/2}$$

$$= 1(0.26)^{-1/2}$$

$$= 1\left(\frac{1}{0.26}\right)^{1/2}$$

$$= 1\left(\sqrt{\frac{1}{0.26}}\right)$$

$$= 1(1.96)$$

$$= 1.96 \text{ units}$$

141. GENERAL CONTRACTOR

Let $a = 40$ and $b = 64$.

$$L = \left(a^{2/3} + b^{2/3}\right)^{3/2}$$

$$= \left(40^{2/3} + 64^{2/3}\right)^{3/2}$$

$$= \left[\left(\sqrt[3]{40}\right)^2 + \left(\sqrt[3]{64}\right)^2\right]^{3/2}$$

$$= \left[(3.42)^2 + (4)^2\right]^{3/2}$$

$$= [11.6964 + 16]^{3/2}$$

$$= (27.6964)^{3/2}$$

$$= \left(\sqrt{27.6964}\right)^3$$

$$= (5.26)^3$$

$$= 145.5 \text{ in.}$$

Divide by 12 since there are 12 inches in 1 foot.

$$L = \frac{145.5}{12}$$

$$= 12.1 \text{ ft}$$

WRITING

143. Answers will vary.

REVIEW

145. COMMUTING TIME

$$t = \frac{k}{r}$$

$$3 = \frac{k}{50}$$

$$50(3) = 50\left(\frac{k}{50}\right)$$

$$150 = k$$

$$t = \frac{150}{60}$$

$$t = 2.5 \text{ hours}$$

CHALLENGE PROBLEMS

147. Yes.

$$16^{2/4} = \left(\sqrt[4]{16}\right)^2 \qquad 16^{1/2} = \sqrt{16}$$

$$= 2^2 \qquad\qquad\quad = 4$$

$$= 4$$

SECTION 9.3

VOCABULARY

1. Radical expressions such as $\sqrt[3]{4}$ and $6\sqrt[3]{4}$ with the same index and the same radicand are called **like** radicals.

3. The largest perfect square **factor** of 27 is 9. The largest **perfect**-cube factor of 16 is 8.

CONCEPTS

5. The product rule for radicals:
$\sqrt[n]{ab} = \sqrt[n]{a}\sqrt[n]{b}$. In words, the nth root of the **product** of two numbers is equal to the product of their nth **roots**.

7. a. $\sqrt{4\cdot 5}$
   b. $\sqrt{4}\sqrt{5}$
   c. $\sqrt{4\cdot 5} = \sqrt{4}\sqrt{5}$

9. a. Answers will vary. Possible answers are $\sqrt{5}$ and $\sqrt[3]{5}$. No, the expressions cannot be added.
   b. Answers will vary. Possible answers are $\sqrt{5}$ and $\sqrt{6}$. No, the expressions cannot be added.

NOTATION

11.
$$\sqrt[3]{32k^4} = \sqrt[3]{\boxed{8k^3}\cdot 4k}$$
$$= \sqrt[3]{\boxed{8k^3}}\sqrt[3]{4k}$$
$$= 2k\sqrt[3]{\boxed{4k}}$$

GUIDED PRACTICE

13.
$$\sqrt{50} = \sqrt{25\cdot 2}$$
$$= \sqrt{25}\sqrt{2}$$
$$= 5\sqrt{2}$$

15.
$$8\sqrt{45} = 8\left(\sqrt{9\cdot 5}\right)$$
$$= 8\left(\sqrt{9}\sqrt{5}\right)$$
$$= 8\left(3\sqrt{5}\right)$$
$$= 24\sqrt{5}$$

17.
$$\sqrt[3]{32} = \sqrt[3]{8\cdot 4}$$
$$= \sqrt[3]{8}\sqrt[3]{4}$$
$$= 2\sqrt[3]{4}$$

19.
$$\sqrt[4]{48} = \sqrt[4]{16\cdot 3}$$
$$= \sqrt[4]{16}\sqrt[4]{3}$$
$$= 2\sqrt[4]{3}$$

21.
$$\sqrt{75a^2} = \sqrt{25a^2}\sqrt{3}$$
$$= 5a\sqrt{3}$$

23.
$$\sqrt{128a^3b^5} = \sqrt{64a^2b^4}\sqrt{2ab}$$
$$= 8ab^2\sqrt{2ab}$$

25.
$$2\sqrt[3]{-54x^6} = 2\sqrt[3]{-27x^6}\sqrt[3]{2}$$
$$= 2\left(-3x^2\sqrt[3]{2}\right)$$
$$= -6x^2\sqrt[3]{2}$$

27.
$$\sqrt[4]{32x^{12}y^4} = \sqrt[4]{16x^{12}y^4}\sqrt[4]{2}$$
$$= 2x^3y\sqrt[4]{2}$$

29.
$$\sqrt{242} = \sqrt{121}\sqrt{2}$$
$$= 11\sqrt{2}$$

31.
$$\sqrt{112a^3} = \sqrt{16a^2}\sqrt{7a}$$
$$= 4a\sqrt{7a}$$

33.

$$-\sqrt[5]{96a^4} = -\sqrt[5]{32}\sqrt[5]{3a^4}$$
$$= -2\sqrt[5]{3a^4}$$

35.

$$\sqrt[3]{405x^{12}y^4} = \sqrt[3]{27x^{12}y^3}\sqrt[3]{15y}$$
$$= 3x^4y\sqrt[3]{15y}$$

37.

$$\sqrt{\frac{11}{9}} = \frac{\sqrt{11}}{\sqrt{9}}$$
$$= \frac{\sqrt{11}}{3}$$

39.

$$\sqrt[4]{\frac{3}{625}} = \frac{\sqrt[4]{3}}{\sqrt[4]{625}}$$
$$= \frac{\sqrt[4]{3}}{5}$$

41

$$\sqrt[5]{\frac{3x^{10}}{32}} = \frac{\sqrt[5]{3x^{10}}}{\sqrt[5]{32}}$$
$$= \frac{x^2\sqrt[5]{3}}{2}$$

43.

$$\sqrt{\frac{z^2}{16x^2}} = \frac{\sqrt{z^2}}{\sqrt{16x^2}}$$
$$= \frac{z}{4x}$$

45.

$$\frac{\sqrt{500}}{\sqrt{5}} = \sqrt{\frac{500}{5}}$$
$$= \sqrt{100}$$
$$= 10$$

47.

$$\frac{\sqrt{98x^3}}{\sqrt{2x}} = \sqrt{\frac{98x^3}{2x}}$$
$$= \sqrt{49x^2}$$
$$= 7x$$

49.

$$\frac{\sqrt[3]{48x^7}}{\sqrt[3]{6x}} = \sqrt[3]{\frac{48x^7}{6x}}$$
$$= \sqrt[3]{8x^6}$$
$$= 2x^2$$

51.

$$\frac{\sqrt[3]{189a^5}}{\sqrt[3]{7a}} = \sqrt[3]{\frac{189a^5}{7a}}$$
$$= \sqrt[3]{27a^4}$$
$$= \sqrt[3]{27a^3}\sqrt[3]{a}$$
$$= 3a\sqrt[3]{a}$$

53.

$$5\sqrt{7} + 3\sqrt{7} = (5+3)\sqrt{7}$$
$$= 8\sqrt{7}$$

55.

$$20\sqrt[3]{4} - 15\sqrt[3]{4} = (20-15)\sqrt[3]{4}$$
$$= 5\sqrt[3]{4}$$

57.

$$4 + \sqrt{8} + \sqrt{2} + 8 = 4 + 2\sqrt{2} + \sqrt{2} + 8$$
$$= (4+8) + (2+1)\sqrt{2}$$
$$= 12 + 3\sqrt{2}$$

59.

$$\sqrt{98} - \sqrt{50} - \sqrt{72} = \sqrt{49}\sqrt{2} - \sqrt{25}\sqrt{2} - \sqrt{36}\sqrt{2}$$
$$= 7\sqrt{2} - 5\sqrt{2} - 6\sqrt{2}$$
$$= -4\sqrt{2}$$

61.

$$8 + \sqrt[3]{32} - \sqrt[3]{108} = 8 + \sqrt[3]{8}\sqrt[3]{4} - \sqrt[3]{27}\sqrt[3]{4}$$
$$= 8 + 2\sqrt[3]{4} - 3\sqrt[3]{4}$$
$$= 8 - \sqrt[3]{4}$$

63.

$$14\sqrt[4]{32} - 15\sqrt[4]{2} = 14\sqrt[4]{16}\sqrt[4]{2} - 15\sqrt[4]{2}$$
$$= 14\left(2\sqrt[4]{2}\right) - 15\left(\sqrt[4]{2}\right)$$
$$= 28\sqrt[4]{2} - 15\sqrt[4]{2}$$
$$= 13\sqrt[4]{2}$$

65.

$$4\sqrt{2x} + 6\sqrt{2x} = 10\sqrt{2x}$$

67.

$$\sqrt{18t} + \sqrt{300t} - \sqrt{243t}$$
$$= \sqrt{9}\sqrt{2t} + \sqrt{100}\sqrt{3t} - \sqrt{81}\sqrt{3t}$$
$$= 3\sqrt{2t} + 10\sqrt{3t} - 9\sqrt{3t}$$
$$= 3\sqrt{2t} + \sqrt{3t}$$

69.

$$2\sqrt[3]{16} - \sqrt[3]{54} - 3\sqrt[3]{128}$$
$$= 2\sqrt[3]{8}\sqrt[3]{2} - \sqrt[3]{27}\sqrt[3]{2} - 3\sqrt[3]{64}\sqrt[3]{2}$$
$$= 2\left(2\sqrt[3]{2}\right) - 3\sqrt[3]{2} - 3\left(4\sqrt[3]{2}\right)$$
$$= 4\sqrt[3]{2} - 3\sqrt[3]{2} - 12\sqrt[3]{2}$$
$$= -11\sqrt[3]{2}$$

71.

$$\sqrt[4]{64} + 5\sqrt[4]{4} - \sqrt[4]{324} = \sqrt[4]{16}\sqrt[4]{4} + 5\sqrt[4]{4} - \sqrt[4]{81}\sqrt[4]{4}$$
$$= 2\sqrt[4]{4} + 5\sqrt[4]{4} - 3\sqrt[4]{4}$$
$$= 4\sqrt[4]{4}$$

TRY IT YOURSELF

73.

$$\sqrt[6]{m^{11}} = \sqrt[6]{m^6}\sqrt[6]{m^5}$$
$$= m\sqrt[6]{m^5}$$

75.

$$2\sqrt[3]{64a} + 2\sqrt[3]{8a} = 2\sqrt[3]{64}\sqrt[3]{a} + 2\sqrt[3]{8}\sqrt[3]{a}$$
$$= 2\left(4\sqrt[3]{a}\right) + 2\left(2\sqrt[3]{a}\right)$$
$$= 8\sqrt[3]{a} + 4\sqrt[3]{a}$$
$$= 12\sqrt[3]{a}$$

77.

$$\sqrt{8y^7} + \sqrt{32y^7} - \sqrt{2y^7}$$
$$= \sqrt{4y^6}\sqrt{2y} + \sqrt{16y^6}\sqrt{2y} - \sqrt{y^6}\sqrt{2y}$$
$$= 2y^3\sqrt{2y} + 4y^3\sqrt{2y} - y^3\sqrt{2y}$$
$$= 5y^3\sqrt{2y}$$

79.

$$\sqrt{32b} = \sqrt{16}\sqrt{2b}$$
$$= 4\sqrt{2b}$$

81.

$$\sqrt{\frac{125n^5}{64n}} = \sqrt{\frac{125n^4}{64}}$$
$$= \frac{\sqrt{125n^4}}{\sqrt{64}}$$
$$= \frac{5n^2\sqrt{5}}{8}$$

83.

$$2\sqrt[3]{125} - 5\sqrt[3]{64} = 2(5) - 5(4)$$
$$= 10 - 20$$
$$= -10$$

85.

$$\sqrt{300xy} = \sqrt{100}\sqrt{3xy}$$
$$= 10\sqrt{3xy}$$

87.

$$\sqrt[4]{\frac{5x}{16z^4}} = \frac{\sqrt[4]{5x}}{\sqrt[4]{16z^4}}$$
$$= \frac{\sqrt[4]{5x}}{2z}$$

89.

$$8\sqrt[5]{7a^2} - 7\sqrt[5]{7a^2} = \sqrt[5]{7a^2}$$

91.

$$\sqrt[5]{x^6y^2} + \sqrt[5]{32x^6y^2} + \sqrt[5]{x^6y^2}$$
$$= \sqrt[5]{x^5}\sqrt[5]{xy^2} + \sqrt[5]{32x^5}\sqrt[5]{xy^2} + \sqrt[5]{x^5}\sqrt[5]{xy^2}$$
$$= x\sqrt[5]{xy^2} + 2x\sqrt[5]{xy^2} + x\sqrt[5]{xy^2}$$
$$= 4x\sqrt[5]{xy^2}$$

93.

$$\sqrt[4]{208m^4n} = \sqrt[4]{16m^4}\sqrt[4]{13n}$$
$$= 2m\sqrt[4]{13n}$$

Section 9.3

95.

$$\sqrt[3]{\frac{a^7}{64a}} = \sqrt[3]{\frac{a^6}{64}}$$

$$= \frac{\sqrt[3]{a^6}}{\sqrt[3]{64}}$$

$$= \frac{a^2}{4}$$

97.

$$\sqrt[3]{\frac{7}{64}} = \frac{\sqrt[3]{7}}{\sqrt[3]{64}}$$

$$= \frac{\sqrt[3]{7}}{4}$$

99.

$$\sqrt{80} + \sqrt{45} - \sqrt{27} = \sqrt{16}\sqrt{5} + \sqrt{9}\sqrt{5} - \sqrt{9}\sqrt{3}$$

$$= 4\sqrt{5} + 3\sqrt{5} - 3\sqrt{3}$$

$$= (4+3)\sqrt{5} - 3\sqrt{3}$$

$$= 7\sqrt{5} - 3\sqrt{3}$$

101.

$$\sqrt[5]{64t^{11}} = \sqrt[5]{32t^{10}}\,\sqrt[5]{2t}$$

$$= 2t^2\sqrt[5]{2t}$$

103.

$$\sqrt[3]{24x} + \sqrt[3]{3x} = \sqrt[3]{8}\sqrt[3]{3x} + \sqrt[3]{3x}$$

$$= 2\sqrt[3]{3x} + \sqrt[3]{3x}$$

$$= 3\sqrt[3]{3x}$$

105. a.

$$\sqrt{20} + \sqrt{20} = \sqrt{4}\sqrt{5} + \sqrt{4}\sqrt{5}$$

$$= 2\sqrt{5} + 2\sqrt{5}$$

$$= 4\sqrt{5}$$

b.

$$\sqrt{21} + \sqrt{21} = 2\sqrt{21}$$

107. a.

$$\sqrt{9x^2} - \sqrt{25x^2} + \sqrt{16x^2} = 3x - 5x + 4x$$

$$= 2x$$

b.

$$\sqrt{9x^3} - \sqrt{25x^3} + \sqrt{16x^3}$$

$$= 3x\sqrt{x} - 5x\sqrt{x} + 4x\sqrt{x}$$

$$= 2x\sqrt{x}$$

109. a.

$$3\sqrt{16} + \sqrt{54} = 3(4) + \sqrt{9}\sqrt{6}$$

$$= 12 + 3\sqrt{6}$$

b.

$$3\sqrt[3]{16} + \sqrt[3]{54} = 3\sqrt[3]{8}\sqrt[3]{2} + \sqrt[3]{27}\sqrt[3]{2}$$

$$= 6\sqrt[3]{2} + 3\sqrt[3]{2}$$

$$= 9\sqrt[3]{2}$$

111. a.

$$24\sqrt[5]{6x} + 16\sqrt[5]{6x} = 40\sqrt[5]{6x}$$

b.

$$24\sqrt[4]{6x} + 16\sqrt[4]{6x} = 40\sqrt[4]{6x}$$

APPLICATIONS

113.   GENERAL CONTRACTORS
Let $a = 20,\ b = 16,$ and $c = 14.$

$$L = \sqrt{\frac{b^2}{2} + \frac{c^2}{2} - \frac{a^2}{2}}$$

$$= \sqrt{\frac{16^2}{2} + \frac{14^2}{2} - \frac{20^2}{4}}$$

$$= \sqrt{\frac{256}{2} + \frac{196}{2} - \frac{400}{4}}$$

$$= \sqrt{128 + 98 - 100}$$

$$= \sqrt{126}$$

$$= \sqrt{9}\sqrt{14}$$

$$= 3\sqrt{14}\ \text{ft}$$

$$L \approx 11.2\ \text{ft}$$

115.   BLOW DRYERS
Let $P = 1{,}200$ and $R = 16.$

$$I = \sqrt{\frac{P}{R}}$$

$$= \sqrt{\frac{1{,}200}{16}}$$

$$= \sqrt{75}$$

$$= \sqrt{25}\sqrt{3}$$

$$= 5\sqrt{3}\ \text{amps}$$

$$I \approx 8.7\ \text{amps}$$

117. DUCTWORK
Add up all of the sides.

$$4\sqrt{80} + 2\sqrt{20} + 2\sqrt{45} + 2\sqrt{75}$$
$$= 4\sqrt{16}\sqrt{5} + 2\sqrt{4}\sqrt{5} + 2\sqrt{9}\sqrt{5} + 2\sqrt{25}\sqrt{3}$$
$$= 4\left(4\sqrt{5}\right) + 2\left(2\sqrt{5}\right) + 2\left(3\sqrt{5}\right) + 2\left(5\sqrt{3}\right)$$
$$= 16\sqrt{5} + 4\sqrt{5} + 6\sqrt{5} + 10\sqrt{3}$$
$$= \left(26\sqrt{5} + 10\sqrt{3}\right) \text{ in.}$$
$$\approx 75.5 \text{ in.}$$

WRITING

119. Answers will vary.

121. Answers will vary.

REVIEW

123.

$$3x^2 y^3\left(-5x^3 y^{-4}\right) = -15x^5 y^{-1}$$
$$= -\frac{15x^5}{y}$$

125.

$$
\begin{array}{r}
3p+4-\dfrac{5}{2p-5} \\
2p-5{\overline{\smash{\big)}\,6p^2-7p-25}} \\
\underline{6p^2-15p} \\
8p-25 \\
\underline{8p-20} \\
-5
\end{array}
$$

CHALLENGE PROBLEMS

127.

$$\frac{\sqrt{24}}{3} + \frac{\sqrt{6}}{5} = \frac{2\sqrt{6}}{3} + \frac{\sqrt{6}}{5}$$
$$= \frac{2\sqrt{6}}{3}\left(\frac{5}{5}\right) + \frac{\sqrt{6}}{5}\left(\frac{3}{3}\right)$$
$$= \frac{10\sqrt{6}}{15} + \frac{13\sqrt{6}}{15}$$
$$= \frac{23\sqrt{6}}{15}$$

129.

$$\frac{\sqrt[3]{3b}}{8} - 9\sqrt[3]{3b} = \frac{\sqrt[3]{3b}}{8} - \frac{72\sqrt[3]{3b}}{8}$$
$$= -\frac{71\sqrt[3]{3b}}{8}$$

131.

$$\sqrt{25x+25} - \sqrt{x+1} = \sqrt{25(x+1)} - \sqrt{x+1}$$
$$= \sqrt{25}\sqrt{x+1} - \sqrt{x+1}$$
$$= 5\sqrt{x+1} - \sqrt{x+1}$$
$$= 4\sqrt{x+1}$$

Section 9.3

SECTION 9.4

VOCABULARY

1. In this section, we used the **product** rule for radicals in reverse: $\sqrt[n]{a} \cdot \sqrt[n]{b} = \sqrt[n]{ab}$.

3. To **rationalize** the denominator of $\dfrac{4}{\sqrt{5}}$, we multiply the fraction by $\dfrac{\sqrt{5}}{\sqrt{5}}$.

5. To obtain a **perfect** cube radicand in the denominator of $\dfrac{\sqrt[3]{7}}{\sqrt[3]{5n}}$, we multiply the fraction by $\dfrac{\sqrt[3]{25n^2}}{\sqrt[3]{25n^2}}$.

CONCEPTS

7.

	Why isn't it in simplified form?	Simp. Form	Approx.
$\dfrac{3}{\sqrt{2}}$	A radical appears on the denominator.	$\dfrac{3\sqrt{2}}{2}$	2.121320344
$\dfrac{\sqrt{18}}{2}$	There is a perfect square factor in the radicand: 9.	$\dfrac{3\sqrt{2}}{2}$	2.121320344
$\sqrt{\dfrac{9}{2}}$	The radicand contains a fraction.	$\dfrac{3\sqrt{2}}{2}$	2.121320344

9.
$$\left(5 - \sqrt{x}\right)^2 = (5)^2 - 2(5)\left(\sqrt{x}\right) + \left(\sqrt{x}\right)^2$$
$$= 25 - 10\sqrt{x} + x$$

11. a. $4\sqrt{6} + 2\sqrt{6} = 6\sqrt{6}$
    b.
$$4\sqrt{6}\left(2\sqrt{6}\right) = 8\sqrt{36}$$
$$= 8(6)$$
$$= 48$$
    c. cannot be simplified
    d. $3\sqrt{2}\left(-2\sqrt{3}\right) = -6\sqrt{6}$

NOTATION

13.
$$5\sqrt{8} \cdot 7\sqrt{6} = 5(7)\sqrt{8}\boxed{\sqrt{6}}$$
$$= 35\sqrt{\boxed{48}}$$
$$= 35\sqrt{\boxed{16} \cdot 3}$$
$$= 35\left(\boxed{4}\right)\sqrt{3}$$
$$= 140\sqrt{3}$$

GUIDED PRACTICE

15.
$$\sqrt{3}\sqrt{15} = \sqrt{45}$$
$$= \sqrt{9}\sqrt{5}$$
$$= 3\sqrt{5}$$

17.
$$2\sqrt{3}\sqrt{6} = 2\sqrt{18}$$
$$= 2\sqrt{9}\sqrt{2}$$
$$= 2\left(3\sqrt{2}\right)$$
$$= 6\sqrt{2}$$

19.
$$\left(3\sqrt[3]{9}\right)\left(2\sqrt[3]{3}\right) = 3(2)\sqrt[3]{9}\sqrt[3]{3}$$
$$= 6\sqrt[3]{27}$$
$$= 6(3)$$
$$= 18$$

21.
$$\sqrt[3]{2} \cdot \sqrt[3]{12} = \sqrt[3]{24}$$
$$= \sqrt[3]{8}\sqrt[3]{3}$$
$$= 2\sqrt[3]{3}$$

23.
$$6\sqrt{ab^3}\left(8\sqrt{ab}\right) = 48\sqrt{a^2b^4}$$
$$= 48ab^2$$

25.
$$\sqrt[4]{5a^3}\sqrt[4]{125a^2} = \sqrt[4]{625a^5}$$
$$= \sqrt[4]{625a^4}\sqrt[4]{a}$$
$$= 5a\sqrt[4]{a}$$

27.

$$3\sqrt{5}\left(4-\sqrt{5}\right)=3\sqrt{5}\left(4\right)-3\sqrt{5}\left(\sqrt{5}\right)$$
$$=3(4)\sqrt{5}-3(5)$$
$$=12\sqrt{5}-15$$

29.

$$\sqrt{2}\left(4\sqrt{6}+2\sqrt{7}\right)=4\left(\sqrt{2}\sqrt{6}\right)+2\left(\sqrt{2}\sqrt{7}\right)$$
$$=4\sqrt{12}+2\sqrt{14}$$
$$=4\sqrt{4}\sqrt{3}+2\sqrt{14}$$
$$=4\left(2\sqrt{3}\right)+2\sqrt{14}$$
$$=8\sqrt{3}+2\sqrt{14}$$

31.

$$-2\sqrt{5x}\left(4\sqrt{2x}-3\sqrt{3}\right)$$
$$=-2(4)\sqrt{5x}\sqrt{2x}+2(3)\sqrt{5x}\sqrt{3}$$
$$=-8\sqrt{10x^2}+6\sqrt{15x}$$
$$=-8\sqrt{x^2}\sqrt{10}+6\sqrt{15x}$$
$$=-8x\sqrt{10}+6\sqrt{15x}$$

33.

$$\sqrt[3]{2}\left(4\sqrt[3]{4}+\sqrt[3]{12}\right)=\sqrt[3]{2}\left(4\sqrt[3]{4}\right)+\sqrt[3]{2}\left(\sqrt[3]{12}\right)$$
$$=4\sqrt[3]{2\cdot4}+\sqrt[3]{2\cdot12}$$
$$=4\sqrt[3]{8}+\sqrt[3]{24}$$
$$=4(2)+\sqrt[3]{8}\sqrt[3]{3}$$
$$=8+2\sqrt[3]{3}$$

35.

$$\left(\sqrt{2}+1\right)\left(\sqrt{2}-3\right)$$
$$=\sqrt{2}\sqrt{2}+\sqrt{2}\left(-3\right)+1\left(\sqrt{2}\right)+1\left(-3\right)$$
$$=\sqrt{4}-3\sqrt{2}+\sqrt{2}-3$$
$$=2-2\sqrt{2}-3$$
$$=-1-2\sqrt{2}$$

37.

$$\left(\sqrt{3x}-\sqrt{2y}\right)\left(\sqrt{3x}+\sqrt{2y}\right)$$
$$=\sqrt{3x}\sqrt{3x}+\sqrt{3x}\sqrt{2y}-\sqrt{2y}\sqrt{3x}-\sqrt{2y}\sqrt{2y}$$
$$=\sqrt{9x^2}+\sqrt{6xy}-\sqrt{6xy}-\sqrt{4y^2}$$
$$=3x-2y$$

39.

$$\left(2\sqrt[3]{4}-3\sqrt[3]{2}\right)\left(3\sqrt[3]{4}+2\sqrt[3]{10}\right)$$
$$=2\sqrt[3]{4}\left(3\sqrt[3]{4}\right)+2\sqrt[3]{4}\left(2\sqrt[3]{10}\right)-3\sqrt[3]{2}\left(3\sqrt[3]{4}\right)-3\sqrt[3]{2}\left(2\sqrt[3]{10}\right)$$
$$=6\sqrt[3]{16}+4\sqrt[3]{40}-9\sqrt[3]{8}-6\sqrt[3]{20}$$
$$=6\sqrt[3]{8}\sqrt[3]{2}+4\sqrt[3]{8}\sqrt[3]{5}-9\sqrt[3]{8}-6\sqrt[3]{20}$$
$$=6(2)\sqrt[3]{2}+4(2)\sqrt[3]{5}-9(2)-6\sqrt[3]{20}$$
$$=12\sqrt[3]{2}+8\sqrt[3]{5}-18-6\sqrt[3]{20}$$

41.

$$\left(\sqrt[3]{5z}+\sqrt[3]{3}\right)\left(\sqrt[3]{5z}+2\sqrt[3]{3}\right)$$
$$=\sqrt[3]{5z}\sqrt[3]{5z}+\sqrt[3]{5z}\left(2\sqrt[3]{3}\right)+\sqrt[3]{3}\left(\sqrt[3]{5z}\right)+\sqrt[3]{3}\left(2\sqrt[3]{3}\right)$$
$$=\sqrt[3]{25z^2}+2\sqrt[3]{15z}+\sqrt[3]{15z}+2\sqrt[3]{9}$$
$$=\sqrt[3]{25z^2}+3\sqrt[3]{15z}+2\sqrt[3]{9}$$

43.

$$\left(\sqrt{7}\right)^2=\sqrt{7}\sqrt{7}$$
$$=\sqrt{49}$$
$$=7$$

45.

$$\left(\sqrt[3]{12}\right)^3=\sqrt[3]{12}\sqrt[3]{12}\sqrt[3]{12}$$
$$=\sqrt[3]{1,728}$$
$$=12$$

47.

$$\left(3\sqrt{2}\right)^2=3\sqrt{2}\cdot3\sqrt{2}$$
$$=3(3)\sqrt{2}\sqrt{2}$$
$$=9\sqrt{4}$$
$$=9(2)$$
$$=18$$

49.

$$\left(-2\sqrt[3]{2x^2}\right)^3=\left(-2\sqrt[3]{2x^2}\right)\left(-2\sqrt[3]{2x^2}\right)\left(-2\sqrt[3]{2x^2}\right)$$
$$=-8\sqrt[3]{8x^6}$$
$$=-8\left(2x^2\right)$$
$$=-16x^2$$

Section 9.4

51.

$$\left(6-\sqrt{3}\right)^2 = \left(6\right)^2 - 2\left(6\right)\left(\sqrt{3}\right) + \left(\sqrt{3}\right)^2$$
$$= 36 - 12\sqrt{3} + 3$$
$$= 39 - 12\sqrt{3}$$

53.

$$\left(\sqrt{3x}+\sqrt{3}\right)^2$$
$$= \left(\sqrt{3x}+\sqrt{3}\right)\left(\sqrt{3x}+\sqrt{3}\right)$$
$$= \sqrt{3x}\sqrt{3x} + \sqrt{3x}\sqrt{3} + \sqrt{3}\sqrt{3x} + \sqrt{3}\sqrt{3}$$
$$= \sqrt{9x^2} + \sqrt{9x} + \sqrt{9x} + \sqrt{9}$$
$$= 3x + \sqrt{9}\sqrt{x} + \sqrt{9}\sqrt{x} + 3$$
$$= 3x + 3\sqrt{x} + 3\sqrt{x} + 3$$
$$= 3x + 6\sqrt{x} + 3$$

55.

$$\sqrt{\frac{2}{7}} = \frac{\sqrt{2}}{\sqrt{7}}$$
$$= \frac{\sqrt{2}}{\sqrt{7}} \cdot \frac{\sqrt{7}}{\sqrt{7}}$$
$$= \frac{\sqrt{14}}{\sqrt{49}}$$
$$= \frac{\sqrt{14}}{7}$$

57.

$$\sqrt{\frac{8}{3}} = \frac{\sqrt{8}}{\sqrt{3}}$$
$$= \frac{\sqrt{8}}{\sqrt{3}} \cdot \frac{\sqrt{3}}{\sqrt{3}}$$
$$= \frac{\sqrt{24}}{\sqrt{9}}$$
$$= \frac{\sqrt{4}\sqrt{6}}{3}$$
$$= \frac{2\sqrt{6}}{3}$$

59.

$$\frac{4}{\sqrt{6}} = \frac{4}{\sqrt{6}} \cdot \frac{\sqrt{6}}{\sqrt{6}}$$
$$= \frac{4\sqrt{6}}{\sqrt{36}}$$
$$= \frac{4\sqrt{6}}{6}$$
$$= \frac{2\sqrt{6}}{3}$$

61.

$$\frac{1}{\sqrt[3]{2}} = \frac{1}{\sqrt[3]{2}} \cdot \frac{\sqrt[3]{4}}{\sqrt[3]{4}}$$
$$= \frac{\sqrt[3]{4}}{\sqrt[3]{8}}$$
$$= \frac{\sqrt[3]{4}}{2}$$

63.

$$\frac{3}{\sqrt[3]{9}} = \frac{3}{\sqrt[3]{9}} \cdot \frac{\sqrt[3]{3}}{\sqrt[3]{3}}$$
$$= \frac{3\sqrt[3]{3}}{\sqrt[3]{27}}$$
$$= \frac{3\sqrt[3]{3}}{3}$$
$$= \sqrt[3]{3}$$

65.

$$\frac{1}{\sqrt[4]{4}} = \frac{1}{\sqrt[4]{4}} \cdot \frac{\sqrt[4]{4}}{\sqrt[4]{4}}$$
$$= \frac{\sqrt[4]{4}}{\sqrt[4]{16}}$$
$$= \frac{\sqrt[4]{4}}{2}$$

67.

$$\frac{\sqrt{10y^2}}{\sqrt{2y^3}} = \sqrt{\frac{10y^2}{2y^3}}$$

$$= \sqrt{\frac{5}{y}}$$

$$= \frac{\sqrt{5}}{\sqrt{y}}$$

$$= \frac{\sqrt{5}}{\sqrt{y}} \cdot \frac{\sqrt{y}}{\sqrt{y}}$$

$$= \frac{\sqrt{5y}}{y}$$

69.

$$\frac{\sqrt{48x^2}}{\sqrt{8x^2y}} = \sqrt{\frac{48x^2}{8x^2y}}$$

$$= \sqrt{\frac{6}{y}}$$

$$= \frac{\sqrt{6}}{\sqrt{y}} \cdot \frac{\sqrt{y}}{\sqrt{y}}$$

$$= \frac{\sqrt{6y}}{y}$$

71.

$$\frac{\sqrt[3]{12t^3}}{\sqrt[3]{54t^2}} = \sqrt[3]{\frac{12t^3}{54t^2}}$$

$$= \sqrt[3]{\frac{2t}{9}}$$

$$= \frac{\sqrt[3]{2t}}{\sqrt[3]{9}} \cdot \frac{\sqrt[3]{3}}{\sqrt[3]{3}}$$

$$= \frac{\sqrt[3]{6t}}{\sqrt[3]{27}}$$

$$= \frac{\sqrt[3]{6t}}{3}$$

73.

$$\frac{\sqrt[3]{4a^6}}{\sqrt[3]{2a^5b}} = \sqrt[3]{\frac{4a^6}{2a^5b}}$$

$$= \sqrt[3]{\frac{2a}{b}}$$

$$= \frac{\sqrt[3]{2a}}{\sqrt[3]{b}} \cdot \frac{\sqrt[3]{b^2}}{\sqrt[3]{b^2}}$$

$$= \frac{\sqrt[3]{2ab^2}}{\sqrt[3]{b^3}}$$

$$= \frac{\sqrt[3]{2ab^2}}{b}$$

75.

$$\frac{23}{\sqrt{50p^5}} = \frac{23}{\sqrt{25p^4}\sqrt{2p}}$$

$$= \frac{23}{5p^2\sqrt{2p}}$$

$$= \frac{23}{5p^2\sqrt{2p}} \cdot \frac{\sqrt{2p}}{\sqrt{2p}}$$

$$= \frac{23\sqrt{2p}}{5p^2(2p)}$$

$$= \frac{23\sqrt{2p}}{10p^3}$$

77.

$$\frac{7}{\sqrt{24b^3}} = \frac{7}{\sqrt{4b^2}\sqrt{6b}}$$

$$= \frac{7}{2b\sqrt{6b}} \cdot \frac{\sqrt{6b}}{\sqrt{6b}}$$

$$= \frac{7\sqrt{6b}}{2b(6b)}$$

$$= \frac{7\sqrt{6b}}{12b^2}$$

Section 9.4

79.

$$\sqrt[3]{\frac{5}{16}} = \frac{\sqrt[3]{5}}{\sqrt[3]{16}}$$

$$= \frac{\sqrt[3]{5}}{\sqrt[3]{8}\sqrt[3]{2}}$$

$$= \frac{\sqrt[3]{5}}{2\sqrt[3]{2}} \cdot \frac{\sqrt[3]{4}}{\sqrt[3]{4}}$$

$$= \frac{\sqrt[3]{20}}{2(2)}$$

$$= \frac{\sqrt[3]{20}}{4}$$

81.

$$\sqrt[3]{\frac{4}{81}} = \frac{\sqrt[3]{4}}{\sqrt[3]{81}}$$

$$= \frac{\sqrt[3]{4}}{\sqrt[3]{27}\sqrt[3]{3}}$$

$$= \frac{\sqrt[3]{4}}{3\sqrt[3]{3}} \cdot \frac{\sqrt[3]{9}}{\sqrt[3]{9}}$$

$$= \frac{\sqrt[3]{36}}{3\sqrt[3]{27}}$$

$$= \frac{\sqrt[3]{36}}{3(3)}$$

$$= \frac{\sqrt[3]{36}}{9}$$

83.

$$\frac{19}{\sqrt[3]{5c^2}} = \frac{19}{\sqrt[3]{5c^2}} \cdot \frac{\sqrt[3]{25c}}{\sqrt[3]{25c}}$$

$$= \frac{19\sqrt[3]{25c}}{\sqrt[3]{125c^3}}$$

$$= \frac{19\sqrt[3]{25c}}{5c}$$

85.

$$\frac{\sqrt[3]{3}}{\sqrt[3]{2r}} = \frac{\sqrt[3]{3}}{\sqrt[3]{2r}} \cdot \frac{\sqrt[3]{4r^2}}{\sqrt[3]{4r^2}}$$

$$= \frac{\sqrt[3]{12r^2}}{\sqrt[3]{8r^3}}$$

$$= \frac{\sqrt[3]{12r^2}}{2r}$$

87.

$$\frac{\sqrt[4]{2}}{\sqrt[4]{3t^2}} = \frac{\sqrt[4]{2}}{\sqrt[4]{3t^2}} \cdot \frac{\sqrt[4]{27t^2}}{\sqrt[4]{27t^2}}$$

$$= \frac{\sqrt[4]{54t^2}}{\sqrt[4]{81t^4}}$$

$$= \frac{\sqrt[4]{54t^2}}{3t}$$

89.

$$\frac{25}{\sqrt[4]{8a}} = \frac{25}{\sqrt[4]{8a}} \cdot \frac{\sqrt[4]{2a^3}}{\sqrt[4]{2a^3}}$$

$$= \frac{25\sqrt[4]{2a^3}}{\sqrt[4]{16a^4}}$$

$$= \frac{25\sqrt[4]{2a^3}}{2a}$$

91.

$$\frac{\sqrt{2}}{\sqrt{5}+3} = \frac{\sqrt{2}}{\sqrt{5}+3} \cdot \frac{\sqrt{5}-3}{\sqrt{5}-3}$$

$$= \frac{\sqrt{2}(\sqrt{5}-3)}{(\sqrt{5}+3)(\sqrt{5}-3)}$$

$$= \frac{\sqrt{10}-3\sqrt{2}}{\sqrt{25}-9}$$

$$= \frac{\sqrt{10}-3\sqrt{2}}{5-9}$$

$$= \frac{\sqrt{10}-3\sqrt{2}}{-4}$$

$$= \frac{-\sqrt{10}+3\sqrt{2}}{4}$$

$$= \frac{3\sqrt{2}-\sqrt{10}}{4}$$

93.

$$\frac{2}{\sqrt{x}+1} = \frac{2}{\sqrt{x}+1} \cdot \frac{\sqrt{x}-1}{\sqrt{x}-1}$$

$$= \frac{2(\sqrt{x}-1)}{(\sqrt{x}+1)\sqrt{x}-1}$$

$$= \frac{2\sqrt{x}-2}{\sqrt{x^2}-1}$$

$$= \frac{2(\sqrt{x}-1)}{x-1}$$

95.

$$\frac{\sqrt{7}-\sqrt{2}}{\sqrt{2}+\sqrt{7}} = \frac{\sqrt{7}-\sqrt{2}}{\sqrt{2}+\sqrt{7}} \cdot \frac{\sqrt{2}-\sqrt{7}}{\sqrt{2}-\sqrt{7}}$$

$$= \frac{\left(\sqrt{7}-\sqrt{2}\right)\left(\sqrt{2}-\sqrt{7}\right)}{\left(\sqrt{2}+\sqrt{7}\right)\left(\sqrt{2}-\sqrt{7}\right)}$$

$$= \frac{\sqrt{7}\sqrt{2}+\sqrt{7}\left(-\sqrt{7}\right)-\sqrt{2}\sqrt{2}-\sqrt{2}\left(-\sqrt{7}\right)}{\sqrt{2}\sqrt{2}-\sqrt{7}\sqrt{7}}$$

$$= \frac{\sqrt{14}-\sqrt{49}-\sqrt{4}+\sqrt{14}}{\sqrt{4}-\sqrt{49}}$$

$$= \frac{\sqrt{14}-7-2+\sqrt{14}}{2-7}$$

$$= \frac{-9+2\sqrt{14}}{-5}$$

$$= \frac{-\left(9-2\sqrt{14}\right)}{-5}$$

$$= \frac{9-2\sqrt{14}}{5}$$

97.

$$\frac{\sqrt{x}-\sqrt{y}}{\sqrt{x}+\sqrt{y}} = \frac{\sqrt{x}-\sqrt{y}}{\sqrt{x}+\sqrt{y}} \cdot \frac{\sqrt{x}-\sqrt{y}}{\sqrt{x}-\sqrt{y}}$$

$$= \frac{\left(\sqrt{x}-\sqrt{y}\right)\left(\sqrt{x}-\sqrt{y}\right)}{\left(\sqrt{x}+\sqrt{y}\right)\left(\sqrt{x}-\sqrt{y}\right)}$$

$$= \frac{\sqrt{x}\sqrt{x}-\sqrt{x}\sqrt{y}-\sqrt{y}\sqrt{x}+\sqrt{y}\sqrt{y}}{\sqrt{x}\sqrt{x}-\sqrt{y}\sqrt{y}}$$

$$= \frac{\sqrt{x^2}-\sqrt{xy}-\sqrt{xy}+\sqrt{y^2}}{\sqrt{x^2}-\sqrt{y^2}}$$

$$= \frac{x-2\sqrt{xy}+y}{x-y}$$

99.

$$\frac{\sqrt{x}+3}{x} = \frac{\sqrt{x}+3}{x} \cdot \frac{\sqrt{x}-3}{\sqrt{x}-3}$$

$$= \frac{\left(\sqrt{x}+3\right)\left(\sqrt{x}-3\right)}{x\left(\sqrt{x}-3\right)}$$

$$= \frac{\sqrt{x}\sqrt{x}-3(3)}{x\sqrt{x}-3x}$$

$$= \frac{\sqrt{x^2}-9}{x\sqrt{x}-3x}$$

$$= \frac{x-9}{x\sqrt{x}-3x}$$

$$= \frac{x-9}{x\left(\sqrt{x}-3\right)}$$

101.

$$\frac{\sqrt{x}+\sqrt{y}}{\sqrt{x}} = \frac{\sqrt{x}+\sqrt{y}}{\sqrt{x}} \cdot \frac{\sqrt{x}-\sqrt{y}}{\sqrt{x}-\sqrt{y}}$$

$$= \frac{\left(\sqrt{x}+\sqrt{y}\right)\left(\sqrt{x}-\sqrt{y}\right)}{\sqrt{x}\left(\sqrt{x}-\sqrt{y}\right)}$$

$$= \frac{\sqrt{x}\sqrt{x}-\sqrt{y}\sqrt{y}}{\sqrt{x}\left(\sqrt{x}-\sqrt{y}\right)}$$

$$= \frac{\sqrt{x^2}-\sqrt{y^2}}{\sqrt{x}\left(\sqrt{x}-\sqrt{y}\right)}$$

$$= \frac{x-y}{\sqrt{x}\left(\sqrt{x}-\sqrt{y}\right)}$$

TRY IT YOURSELF

103.

$$\sqrt{x}\left(\sqrt{14x}+\sqrt{2}\right) = \sqrt{14x^2}+\sqrt{2x}$$

$$= x\sqrt{14}+\sqrt{2x}$$

Section 9.4

105.

$$\frac{3\sqrt{2}-5\sqrt{3}}{2\sqrt{3}-3\sqrt{2}}$$

$$=\frac{3\sqrt{2}-5\sqrt{3}}{2\sqrt{3}-3\sqrt{2}}\cdot\frac{2\sqrt{3}+3\sqrt{2}}{2\sqrt{3}+3\sqrt{2}}$$

$$=\frac{\left(3\sqrt{2}-5\sqrt{3}\right)\left(2\sqrt{3}+3\sqrt{2}\right)}{\left(2\sqrt{3}-3\sqrt{2}\right)\left(2\sqrt{3}+3\sqrt{2}\right)}$$

$$=\frac{3\sqrt{2}\left(2\sqrt{3}\right)+3\sqrt{2}\left(3\sqrt{2}\right)-5\sqrt{3}\left(2\sqrt{3}\right)-5\sqrt{3}\left(3\sqrt{2}\right)}{2\sqrt{3}\left(2\sqrt{3}\right)-3\sqrt{2}\left(3\sqrt{2}\right)}$$

$$=\frac{6\sqrt{6}+9\sqrt{4}-10\sqrt{9}-15\sqrt{6}}{4\sqrt{9}-9\sqrt{4}}$$

$$=\frac{6\sqrt{6}+9(2)-10(3)-15\sqrt{6}}{4(3)-9(2)}$$

$$=\frac{6\sqrt{6}+18-30-15\sqrt{6}}{12-18}$$

$$=\frac{-12-9\sqrt{6}}{-6}$$

$$=\frac{-3\left(4+3\sqrt{6}\right)}{-3(2)}$$

$$=\frac{4+3\sqrt{6}}{2}$$

$$=\frac{3\sqrt{6}+4}{2}$$

107.

$$\left(10\sqrt[3]{2x}\right)^3=\left(10\sqrt[3]{2x}\right)\left(10\sqrt[3]{2x}\right)\left(10\sqrt[3]{2x}\right)$$

$$=1{,}000\sqrt[3]{8x^3}$$

$$=1{,}000\left(2x\right)$$

$$=2{,}000x$$

109.

$$-4\sqrt[3]{5r^2s}\left(5\sqrt[3]{2r}\right)=-20\sqrt[3]{10r^3s}$$

$$=-20\sqrt[3]{r^3}\sqrt[3]{10s}$$

$$=-20r\sqrt[3]{10s}$$

111.

$$\left(3p+\sqrt{5}\right)^2=\left(3p\right)^2+2\left(3p\right)\left(\sqrt{5}\right)+\left(\sqrt{5}\right)^2$$

$$=9p^2+6p\sqrt{5}+5$$

113.

$$\sqrt{\frac{72m^8}{25m^3}}=\sqrt{\frac{72m^5}{25}}$$

$$=\frac{\sqrt{72m^5}}{\sqrt{25}}$$

$$=\frac{\sqrt{36m^4}\sqrt{2m}}{5}$$

$$=\frac{6m^2\sqrt{2m}}{5}$$

115.

$$\sqrt[4]{3n^2}\sqrt[4]{27n^3}=\sqrt[4]{81n^5}$$

$$=\sqrt[4]{81n^4}\sqrt[4]{n}$$

$$=3n\sqrt[4]{n}$$

117.

$$\frac{\sqrt[3]{x}}{\sqrt[3]{9}}=\frac{\sqrt[3]{x}}{\sqrt[3]{9}}\cdot\frac{\sqrt[3]{3}}{\sqrt[3]{3}}$$

$$=\frac{\sqrt[3]{3x}}{\sqrt[3]{27}}$$

$$=\frac{\sqrt[3]{3x}}{3}$$

119.

$$\left(3\sqrt{2r}-2\right)^2$$

$$=\left(3\sqrt{2r}-2\right)\left(3\sqrt{2r}-2\right)$$

$$=3\sqrt{2r}\left(3\sqrt{2r}\right)+3\sqrt{2r}\left(-2\right)-2\left(3\sqrt{2r}\right)-2\left(-2\right)$$

$$=9\sqrt{4r^2}-6\sqrt{2r}-6\sqrt{2r}+4$$

$$=9\left(2r\right)-12\sqrt{2r}+4$$

$$=18r-12\sqrt{2r}+4$$

121.

$$\sqrt{x\left(x+3\right)}\sqrt{x^3\left(x+3\right)}=\sqrt{x^4\left(x+3\right)^2}$$

$$=x^2\left(x+3\right)$$

123.

$$\frac{2z-1}{\sqrt{2z}-1} = \frac{2z-1}{\sqrt{2z}-1} \cdot \frac{\sqrt{2z}+1}{\sqrt{2z}+1}$$

$$= \frac{(2z-1)(\sqrt{2z}+1)}{(\sqrt{2z}-1)(\sqrt{2z}+1)}$$

$$= \frac{(2z-1)(\sqrt{2z}+1)}{\sqrt{2z}\sqrt{2z}-1(1)}$$

$$= \frac{(2z-1)(\sqrt{2z}+1)}{\sqrt{4z^2}-1}$$

$$= \frac{(2z-1)(\sqrt{2z}+1)}{2z-1}$$

$$= \frac{(\sqrt{2z}+1)(2z-1)}{2z-1}$$

$$= \sqrt{2z}+1$$

125.a.

$$\left(3\sqrt{a}\right)^2 = \left(3\sqrt{a}\right)\left(3\sqrt{a}\right)$$

$$= 9\sqrt{a^2}$$

$$= 9a$$

b.

$$\left(3+\sqrt{a}\right)^2 = (3)^2 + 2(3)\left(\sqrt{a}\right) + \left(\sqrt{a}\right)^2$$

$$= 9 + 6\sqrt{a} + a$$

127.a.

$$\left(\sqrt{m-6}\right)^2 = \left(\sqrt{m-6}\right)\left(\sqrt{m-6}\right)$$

$$= m-6$$

b.

$$\left(\sqrt{m}-6\right)^2 = \left(\sqrt{m}\right)^2 - 2\left(\sqrt{m}\right)(6) + (-6)^2$$

$$= m - 12\sqrt{m} + 36$$

APPLICATIONS

129. STATISTICS

$$\frac{1}{\sigma\sqrt{2\pi}} = \frac{1}{\sigma\sqrt{2\pi}} \cdot \frac{\sqrt{2\pi}}{\sqrt{2\pi}}$$

$$= \frac{\sqrt{2\pi}}{\sigma\sqrt{4\pi^2}}$$

$$= \frac{\sqrt{2\pi}}{\sigma(2\pi)}$$

$$= \frac{\sqrt{2\pi}}{2\pi\sigma}$$

131. TRIGONOMETRY

$$\frac{\text{length of side } AC}{\text{length of side } AB} = \frac{1}{\sqrt{2}}$$

$$= \frac{1}{\sqrt{2}} \cdot \frac{\sqrt{2}}{\sqrt{2}}$$

$$= \frac{\sqrt{2}}{\sqrt{4}}$$

$$= \frac{\sqrt{2}}{2}$$

WRITING

133. Answers will vary.

135. Answers will vary.

137. Answers will vary.

REVIEW

Section 9.4

139.

$$\frac{8}{b-2} + \frac{3}{2-b} = -\frac{1}{b}$$

$$\frac{8}{b-2} - \frac{3}{b-2} = -\frac{1}{b}$$

$$b(b-2)\left(\frac{8}{b-2} - \frac{3}{b-2}\right) = b(b-2)\left(-\frac{1}{b}\right)$$

$$8b - 3b = -(b-2)$$

$$5b = -b + 2$$

$$6b = 2$$

$$b = \frac{2}{6}$$

$$b = \frac{1}{3}$$

CHALLENGE PROBLEMS

141.

$$\sqrt{2} \cdot \sqrt[3]{2} = 2^{1/2} \cdot 2^{1/3}$$

$$= 2^{(1/2)+(1/3)}$$

$$= 2^{3/6+2/6}$$

$$= 2^{5/6}$$

$$= \sqrt[6]{2^5}$$

$$= \sqrt[6]{32}$$

SECTION 9.5

VOCABULARY

1. Equations such as $\sqrt{x+4} - 4 = 5$ and $\sqrt[3]{x+1} = 12$ are called **radical** equations.

3. When we square both sides of a radical equation, we say we are **raising** both sides to the second power.

5. Proposed solutions of a radical equation that do not satisfy it are called **extraneous** solutions.

CONCEPTS

7. a. The power rule for solving radical equations states that if x, y, and n are real numbers and $x = y$, then $x^{\boxed{n}} = y^{\boxed{n}}$.

    b. $\sqrt[n]{a^n} = \boxed{a}$

9. a. square both sides
    b. subtract 3 from both sides
    c. add $\sqrt{2x+9}$ to both sides

11.
$$\left(\sqrt{x} - 3\right)^2 = \left(\sqrt{x}\right)^2 - 2(3)\left(\sqrt{x}\right) + (-3)^2$$
$$= x - 6\sqrt{x} + 9$$

NOTATION

13.
$$\sqrt{3x+3} - 1 = 5$$
$$\sqrt{3x+3} = \boxed{6}$$
$$\left(\sqrt{3x+3}\right)^{\boxed{2}} = (6)^{\boxed{2}}$$
$$\boxed{3x+3} = 36$$
$$3x = \boxed{33}$$
$$x = \boxed{11}$$
$$\boxed{\text{Yes}} \text{ it checks.}$$

GUIDED PRACTICE

15.
$$\sqrt{a-3} = 1$$
$$\left(\sqrt{a-3}\right)^2 = (1)^2$$
$$a - 3 = 1$$
$$a = 4$$

17.
$$\sqrt{4x+5} = 5$$
$$\left(\sqrt{4x+5}\right)^2 = 5^2$$
$$4x + 5 = 25$$
$$4x = 20$$
$$x = 5$$

19.
$$\sqrt{6x+13} = 7$$
$$\left(\sqrt{6x+13}\right)^2 = (7)^2$$
$$6x + 13 = 49$$
$$6x = 36$$
$$x = 6$$

21.
$$\sqrt{\frac{1}{3}x - 2} = 8$$
$$\left(\sqrt{\frac{1}{3}x - 2}\right)^2 = 8^2$$
$$\frac{1}{3}x - 2 = 64$$
$$\frac{1}{3}x = 66$$
$$3\left(\frac{1}{3}x\right) = 3(66)$$
$$x = 198$$

23.

$$\sqrt{2x+11} + 2 = x$$
$$\sqrt{2x+11} = x - 2$$
$$\left(\sqrt{2x+11}\right)^2 = (x-2)^2$$
$$2x + 11 = x^2 - 4x + 4$$
$$0 = x^2 - 6x - 7$$
$$0 = (x-7)(x+1)$$
$$x - 7 = 0 \quad \text{or} \quad x + 1 = 0$$
$$x = 7 \qquad x = \cancel{-1}$$

25.

$$\sqrt{2r-3} + 9 = r$$
$$\sqrt{2r-3} = r - 9$$
$$\left(\sqrt{2r-3}\right)^2 = (r-9)^2$$
$$2r - 3 = r^2 - 9r - 9r + 81$$
$$2r - 3 = r^2 - 18r + 81$$
$$0 = r^2 - 20r + 84$$
$$0 = (r-14)(r-6)$$
$$r - 14 = 0 \quad \text{or} \quad r - 6 = 0$$
$$r = 14 \qquad r = \cancel{6}$$

27.

$$\sqrt{3t+7} - t = 1$$
$$\sqrt{3t+7} = t + 1$$
$$\left(\sqrt{3t+7}\right)^2 = (t+1)^2$$
$$3t + 7 = t^2 + 2t + 1$$
$$0 = t^2 - t - 6$$
$$0 = (t-3)(t+2)$$
$$t - 3 = 0 \quad \text{or} \quad t + 2 = 0$$
$$t = 3 \qquad t = \cancel{-2}$$

29.

$$\sqrt{9-a} - a = 3$$
$$\sqrt{9-a} = a + 3$$
$$\left(\sqrt{9-a}\right)^2 = (a+3)^2$$
$$9 - a = a^2 + 6a + 9$$
$$0 = a^2 + 7a$$
$$0 = a(a+7)$$
$$a = 0 \quad \text{or} \quad a + 7 = 0$$
$$a = 0 \qquad a = \cancel{-7}$$

31.

$$\sqrt{5x} + 10 = 8$$
$$\sqrt{5x} = -2$$
$$\left(\sqrt{5x}\right) = (-2)^2$$
$$5x = 4$$
$$x = \cancel{\dfrac{4}{5}}$$

No Solution

33.

$$\sqrt{5-x} + 10 = 9$$
$$\sqrt{5-x} = -1$$
$$\left(\sqrt{5-x}\right)^2 = (-1)^2$$
$$5 - x = 1$$
$$-x = -4$$
$$x = \cancel{4}$$

No Solution

35.

$$\sqrt[3]{7n-1} = 3$$
$$\left(\sqrt[3]{7n-1}\right)^3 = (3)^3$$
$$7n - 1 = 27$$
$$7n = 28$$
$$n = 4$$

37.

$$\sqrt[3]{x^3 - 7} = x - 1$$

$$\left(\sqrt[3]{x^3 - 7}\right)^3 = (x-1)^3$$

$$x^3 - 7 = (x-1)(x-1)(x-1)$$

$$x^3 - 7 = (x-1)(x^2 - x - x + 1)$$

$$x^3 - 7 = (x-1)(x^2 - 2x + 1)$$

$$x^3 - 7 = x(x^2 - 2x + 1) - 1(x^2 - 2x + 1)$$

$$x^3 - 7 = x^3 - 2x^2 + x - x^2 + 2x - 1$$

$$x^3 - 7 = x^3 - 3x^2 + 3x - 1$$

$$x^3 - 7 - x^3 = x^3 - 3x^2 + 3x - 1 - x^3$$

$$-7 = -3x^2 + 3x - 1$$

$$3x^2 - 3x + 1 - 7 = 0$$

$$3x^2 - 3x - 6 = 0$$

$$\frac{3x^2}{3} - \frac{3x}{3} - \frac{6}{3} = \frac{0}{3}$$

$$x^3 - x - 2 = 0$$

$$(x - 2)(x + 1) = 0$$

$$x - 2 = 0 \quad \text{or} \quad x + 1 = 0$$

$$x = 2 \qquad\qquad x = -1$$

39.

$$\left(m^3 + 26\right)^{1/3} = m + 2$$

$$\sqrt[3]{m^3 + 26} = m + 2$$

$$\left(\sqrt[3]{m^3 + 26}\right)^3 = (m+2)^3$$

$$m^3 + 26 = (m+2)(m+2)(m+2)$$

$$m^3 + 26 = (m+2)(m^2 + 4m + 4)$$

$$m^3 + 26 = m^3 + 4m^2 + 4m + 2m^2 + 8m + 8$$

$$m^3 + 26 = m^3 + 6m^2 + 12m + 8$$

$$26 = 6m^2 + 12m + 8$$

$$0 = 6m^2 + 12m - 18$$

$$0 = 6(m^2 + 2m - 3)$$

$$0 = 6(m+3)(m-1)$$

$$m + 3 = 0 \quad \text{or} \quad m - 1 = 0$$

$$m = -3 \qquad\qquad m = 1$$

41.

$$(5r + 14)^{1/3} = 4$$

$$\sqrt[3]{5r + 14} = 4$$

$$\left(\sqrt[3]{5r + 14}\right)^3 = 4^3$$

$$5r + 14 = 64$$

$$5r = 50$$

$$r = 10$$

43. $f(x) = \sqrt[4]{3x + 1}$

$$\sqrt[4]{3x + 1} = 4$$

$$\left(\sqrt[4]{3x + 1}\right)^4 = 4^4$$

$$3x + 1 = 256$$

$$3x = 255$$

$$x = 85$$

45. $f(x) = \sqrt[3]{3x - 6}$

$$\sqrt[3]{3x - 6} = -3$$

$$\left(\sqrt[3]{3x - 6}\right)^3 = (-3)^3$$

$$3x - 6 = -27$$

$$3x = -21$$

$$x = -7$$

47.

$$\sqrt{3x + 12} = \sqrt{5x - 12}$$

$$\left(\sqrt{3x + 12}\right)^2 = \left(\sqrt{5x - 12}\right)^2$$

$$3x + 12 = 5x - 12$$

$$-2x + 12 = -12$$

$$-2x = -24$$

$$x = 12$$

49.

$$2\sqrt{4x + 1} = \sqrt{x + 4}$$

$$\left(2\sqrt{4x + 1}\right)^2 = \left(\sqrt{x + 4}\right)^2$$

$$4(4x + 1) = x + 4$$

$$16x + 4 = x + 4$$

$$15x + 4 = 4$$

$$15x = 0$$

$$x = 0$$

51.

$$\sqrt{6t+9} = 3\sqrt{t}$$
$$\left(\sqrt{6t+9}\right)^2 = \left(3\sqrt{t}\right)^2$$
$$6t+9 = 9t$$
$$9 = 3t$$
$$3 = t$$

53.

$$(34x+26)^{1/3} = 4(x-1)^{1/3}$$
$$\sqrt[3]{34x+26} = 4\sqrt[3]{x-1}$$
$$\left(\sqrt[3]{34x+26}\right)^3 = \left(4\sqrt[3]{x-1}\right)^3$$
$$34x+26 = 64(x-1)$$
$$34x+26 = 64x-64$$
$$-30x+26 = -64$$
$$-30x = -90$$
$$x = 3$$

55.

$$\sqrt{x-5} + \sqrt{x} = 5$$
$$\sqrt{x-5} = 5 - \sqrt{x}$$
$$\left(\sqrt{x-5}\right)^2 = \left(5 - \sqrt{x}\right)^2$$
$$x-5 = 25 - 5\sqrt{x} - 5\sqrt{x} + x$$
$$x-5 = 25 - 10\sqrt{x} + x$$
$$x-5-25-x = 25 - 10\sqrt{x} + x - 25 - x$$
$$-30 = -10\sqrt{x}$$
$$\frac{-30}{-10} = \frac{-10\sqrt{x}}{-10}$$
$$3 = \sqrt{x}$$
$$3^2 = \left(\sqrt{x}\right)^2$$
$$9 = x$$

57.

$$\sqrt{z+3} - \sqrt{z} = 1$$
$$\sqrt{z+3} = 1 + \sqrt{z}$$
$$\left(\sqrt{z+3}\right)^2 = \left(1 + \sqrt{z}\right)^2$$
$$z+3 = 1 + \sqrt{z} + \sqrt{z} + \sqrt{z^2}$$
$$z+3 = 1 + 2\sqrt{z} + z$$
$$z+3-z-1 = 1 + 2\sqrt{z} + z - z - 1$$
$$2 = 2\sqrt{z}$$
$$\frac{2}{2} = \frac{2\sqrt{z}}{2}$$
$$1 = \sqrt{z}$$
$$(1)^2 = \left(\sqrt{z}\right)^2$$
$$1 = z$$

59.

$$3 = \sqrt{y+4} - \sqrt{y+7}$$
$$3 + \sqrt{y+7} = \sqrt{y+4}$$
$$\left(3 + \sqrt{y+7}\right)^2 = \left(\sqrt{y+4}\right)^2$$
$$9 + 3\sqrt{y+7} + 3\sqrt{y+7} + y+7 = y+4$$
$$9 + 6\sqrt{y+7} + y+7 = y+4$$
$$16 + y + 6\sqrt{y+7} = y+4$$
$$6\sqrt{y+7} = -12$$
$$\frac{6\sqrt{y+7}}{6} = \frac{-12}{6}$$
$$\sqrt{y+7} = -2$$
$$\left(\sqrt{y+7}\right)^2 = (-2)^2$$
$$y+7 = 4$$
$$y = \cancel{-3}$$
No Solution

61.

$$2 = \sqrt{2u+7} - \sqrt{u}$$
$$2 + \sqrt{u} = \sqrt{2u+7}$$
$$\left(2 + \sqrt{u}\right)^2 = \left(\sqrt{2u+7}\right)^2$$
$$4 + 2\sqrt{u} + 2\sqrt{u} + \sqrt{u^2} = 2u+7$$
$$4 + 4\sqrt{u} + u = 2u+7$$
$$4 + 4\sqrt{u} + u - 4 - u = 2u+7-4-u$$
$$4\sqrt{u} = u+3$$
$$\left(4\sqrt{u}\right)^2 = (u+3)^2$$
$$16u = u^2 + 3u + 3u + 9$$
$$16u = u^2 + 6u + 9$$
$$16u - 16u = u^2 + 6u + 9 - 16u$$
$$0 = u^2 - 10u + 9$$
$$0 = (u-1)(u-9)$$
$$u-1=0 \quad \text{or} \quad u-9=0$$
$$u=1 \qquad\qquad u=9$$

63.

$$v = \sqrt{2gh}$$
$$v^2 = \left(\sqrt{2gh}\right)^2$$
$$v^2 = 2gh$$
$$\frac{v^2}{2g} = \frac{2gh}{2g}$$
$$\frac{v^2}{2g} = h$$

65.

$$T = 2\pi\sqrt{\frac{\ell}{32}}$$
$$(T)^2 = \left(2\pi\sqrt{\frac{\ell}{32}}\right)^2$$
$$T^2 = 4\pi^2 \cdot \frac{\ell}{32}$$
$$T^2 = \frac{4\pi^2 \ell}{32}$$
$$T^2 = \frac{\pi^2 \ell}{8}$$
$$8(T^2) = 8\left(\frac{\pi^2 \ell}{8}\right)$$
$$8T^2 = \pi^2 \ell$$
$$\frac{8T^2}{\pi^2} = \frac{\pi^2 \ell}{\pi^2}$$
$$\frac{8T^2}{\pi^2} = \ell$$

67.

$$r = \sqrt[3]{\frac{A}{P}} - 1$$
$$r + 1 = \sqrt[3]{\frac{A}{P}}$$
$$(r+1)^3 = \left(\sqrt[3]{\frac{A}{P}}\right)^3$$
$$(r+1)^3 = \frac{A}{P}$$
$$P(r+1)^3 = P\left(\frac{A}{P}\right)$$
$$P(r+1)^3 = A$$

69.

$$L_A = L_B \sqrt{1 - \frac{v^2}{c^2}}$$

$$\frac{L_A}{L_B} = \frac{L_B \sqrt{1 - \frac{v^2}{c^2}}}{L_B}$$

$$\frac{L_A}{L_B} = \sqrt{1 - \frac{v^2}{c^2}}$$

$$\left(\frac{L_A}{L_B}\right)^2 = \left(\sqrt{1 - \frac{v^2}{c^2}}\right)^2$$

$$\frac{L_A^{\,2}}{L_B^{\,2}} = 1 - \frac{v^2}{c^2}$$

$$\frac{L_A^{\,2}}{L_B^{\,2}} - 1 = -\frac{v^2}{c^2}$$

$$-c^2\left(\frac{L_A^{\,2}}{L_B^{\,2}} - 1\right) = -c^2\left(-\frac{v^2}{c^2}\right)$$

$$-c^2\left(\frac{L_A^{\,2}}{L_B^{\,2}} - 1\right) = v^2$$

$$c^2\left(-\frac{L_A^{\,2}}{L_B^{\,2}} + 1\right) = v^2$$

$$c^2\left(1 - \frac{L_A^{\,2}}{L_B^{\,2}}\right) = v^2$$

TRY IT YOURSELF

71.

$$2\sqrt{x} = \sqrt{5x - 16}$$

$$\left(2\sqrt{x}\right)^2 = \left(\sqrt{5x - 16}\right)^2$$

$$4x = 5x - 16$$

$$-x = -16$$

$$x = 16$$

73.

$$\sqrt{x+5} + \sqrt{x-3} = 4$$

$$\sqrt{x+5} = 4 - \sqrt{x-3}$$

$$\left(\sqrt{x+5}\right)^2 = \left(4 - \sqrt{x-3}\right)^2$$

$$x+5 = 16 - 4\sqrt{x-3} - 4\sqrt{x-3} + x - 3$$

$$x+5 = 13 - 8\sqrt{x-3} + x$$

$$x+5-x-13 = 13 - 8\sqrt{x-3} + x - x - 13$$

$$-8 = -8\sqrt{x-3}$$

$$\frac{-8}{-8} = \frac{-8\sqrt{x-3}}{-8}$$

$$1 = \sqrt{x-3}$$

$$1^2 = \left(\sqrt{x-3}\right)^2$$

$$1 = x - 3$$

$$4 = x$$

75.

$$n = \left(n^3 + n^2 - 1\right)^{1/3}$$

$$n = \sqrt[3]{n^3 + n^2 - 1}$$

$$(n)^3 = \left(\sqrt[3]{n^3 + n^2 - 1}\right)^3$$

$$n^3 = n^3 + n^2 - 1$$

$$n^3 - n^3 = n^3 + n^2 - 1 - n^3$$

$$0 = n^2 - 1$$

$$0 = (n+1)(n+1)$$

$$n+1 = 0 \quad \text{or} \quad n-1 = 0$$

$$n = -1 \qquad \qquad n = 1$$

77.

$$\sqrt{y+2} + y = 4$$

$$\sqrt{y+2} = 4 - y$$

$$\left(\sqrt{y+2}\right)^2 = (4-y)^2$$

$$y+2 = 16 - 4y - 4y + y^2$$

$$y+2 = 16 - 8y + y^2$$

$$y+2-y-2 = 16 - 8y + y^2 - y - 2$$

$$0 = y^2 - 9y + 14$$

$$0 = (y-2)(y-7)$$

$$y-2 = 0 \quad \text{or} \quad y-7 = 0$$

$$y = 2 \qquad \qquad y = \cancel{7}$$

79.

$$\sqrt[3]{x+8} = -2$$

$$\left(\sqrt[3]{x+8}\right)^3 = (-2)^3$$

$$x+8 = -8$$

$$x = -16$$

81.

$$2 = \sqrt{x+5} - \sqrt{x+1}$$

$$2 + \sqrt{x-1} = \sqrt{x+5} - \sqrt{x+1} + \sqrt{x-1}$$

$$1 + \sqrt{x} = \sqrt{x+5}$$

$$\left(1 + \sqrt{x}\right)^2 = \left(\sqrt{x+5}\right)^2$$

$$1 + \sqrt{x} + \sqrt{x} + x = x+5$$

$$1 + 2\sqrt{x} + x = x+5$$

$$1 + 2\sqrt{x} + x - x - 1 = x+5-x-1$$

$$2\sqrt{x} = 4$$

$$\frac{2\sqrt{x}}{2} = \frac{4}{2}$$

$$\sqrt{x} = 2$$

$$\left(\sqrt{x}\right)^2 = 2^2$$

$$x = 4$$

83.

$$x = \frac{\sqrt{12x-5}}{2}$$

$$2(x) = 2\left(\frac{\sqrt{12x-5}}{2}\right)$$

$$2x = \sqrt{12x-5}$$

$$\left(2x\right)^2 = \left(\sqrt{12x-5}\right)^2$$

$$4x^2 = 12x-5$$

$$4x^2 - 12x + 5 = 0$$

$$(2x-5)(2x-1) = 0$$

$$2x-5 = 0 \quad \text{or} \quad 2x-1 = 0$$

$$2x = 5 \qquad\qquad 2x = 1$$

$$x = \frac{5}{2} \qquad\qquad x = \frac{1}{2}$$

85.

$$\left(n^2 + 6n + 3\right)^{1/2} = \left(n^2 - 6n - 3\right)^{1/2}$$

$$\sqrt{n^2 + 6n + 3} = \sqrt{n^2 - 6n - 3}$$

$$\left(\sqrt{n^2 + 6n + 3}\right)^2 = \left(\sqrt{n^2 - 6n - 3}\right)^2$$

$$n^2 + 6n + 3 = n^2 - 6n - 3$$

$$n^2 + 6n + 3 - n^2 = n^2 - 6n - 3 - n^2$$

$$6n + 3 = -6n - 3$$

$$12n + 3 = -3$$

$$12n = -6$$

$$n = -\frac{6}{12}$$

$$n = -\frac{1}{2}$$

87.

$$\sqrt{x-5} - \sqrt{x+3} = 4$$

$$\sqrt{x-5} = 4 + \sqrt{x+3}$$

$$\left(\sqrt{x-5}\right)^2 = \left(4 + \sqrt{x+3}\right)^2$$

$$x-5 = 16 + 4\sqrt{x+3} + 4\sqrt{x+3} + x + 3$$

$$x-5 = 19 + 8\sqrt{x+3} + x$$

$$x-5-x-19 = 19 + 8\sqrt{x+3} + x - x - 19$$

$$-24 = 8\sqrt{x+3}$$

$$\frac{-24}{8} = \frac{8\sqrt{x+3}}{8}$$

$$-3 = \sqrt{x+3}$$

$$(-3)^2 = \left(\sqrt{x+3}\right)^2$$

$$9 = x+3$$

$$\cancel{6} = x$$

No Solution

89.

$$\sqrt[4]{10y+6} = 2\sqrt[4]{y}$$

$$\left(\sqrt[4]{10y+6}\right)^4 = \left(2\sqrt[4]{y}\right)^4$$

$$10y+6 = 16y$$

$$6 = 6y$$

$$1 = y$$

$$y = 1$$

Section 9.5

91.

$$\sqrt{-5x+24} = 6-x$$
$$\left(\sqrt{-5x+24}\right)^2 = \left(6-x\right)^2$$
$$-5x+24 = 36-6x-6x+x^2$$
$$-5x+24 = 36-12x+x^2$$
$$-5x+24+5x-24 = 36-12x+x^2+5x-24$$
$$0 = x^2-7x+12$$
$$0 = (x-3)(x-4)$$
$$x-3=0 \quad \text{or} \quad x-4=0$$
$$x=3 \qquad\qquad x=4$$

93.

$$\sqrt{2x}+5 = 1$$
$$\sqrt{2x} = -4$$
$$\left(\sqrt{2x}\right)^2 = (-4)^2$$
$$2x = 16$$
$$x = \cancel{8}$$
No Solution

95.

$$\sqrt{6x+2}-\sqrt{5x+3} = 0$$
$$\sqrt{6x+2} = \sqrt{5x+3}$$
$$\left(\sqrt{6x+2}\right)^2 = \left(\sqrt{5x+3}\right)^2$$
$$6x+2 = 5x+3$$
$$x+2 = 3$$
$$x = 1$$

97.

$$f(x) = g(x)$$
$$\sqrt{x+16} = 7-\sqrt{x+9}$$
$$\left(\sqrt{x+16}\right)^2 = \left(7-\sqrt{x+9}\right)^2$$
$$x+16 = 49-7\sqrt{x+9}-7\sqrt{x+9}+x+9$$
$$x+16 = 49-14\sqrt{x+9}+x+9$$
$$x+16 = 58+x-14\sqrt{x+9}$$
$$-42 = -14\sqrt{x+9}$$
$$\frac{-42}{-14} = \frac{-14\sqrt{x+9}}{-14}$$
$$3 = \sqrt{x+9}$$
$$3^2 = \left(\sqrt{x+9}\right)^2$$
$$9 = x+9$$
$$0 = x$$

99.

$$f(x) = 0$$
$$\sqrt[4]{x+8}-\sqrt[4]{2x} = 0$$
$$\sqrt[4]{x+8} = \sqrt[4]{2x}$$
$$\left(\sqrt[4]{x+8}\right)^4 = \left(\sqrt[4]{2x}\right)^4$$
$$x+8 = 2x$$
$$x+8-x = 2x-x$$
$$8 = x$$

101.a.

$$3\sqrt{5n+9} = \sqrt{5n}$$

$$\left(3\sqrt{5n+9}\right)^2 = \left(\sqrt{5n}\right)^2$$

$$9(5n+9) = 5n$$

$$45n+81 = 5n$$

$$81 = -40n$$

$$-\frac{81}{40} = n$$

b.

$$3+\sqrt{5n-9} = \sqrt{5n}$$

$$\left(3+\sqrt{5n-9}\right)^2 = \left(\sqrt{5n}\right)^2$$

$$(3)^2 + 2(3)\left(\sqrt{5n-9}\right) + \left(\sqrt{5n-9}\right)^2 = 5n$$

$$9 + 6\sqrt{5n-9} + 5n - 9 = 5n$$

$$6\sqrt{5n-9} + 5n = 5n$$

$$6\sqrt{5n-9} = 0$$

$$\frac{6\sqrt{5n-9}}{6} = \frac{0}{6}$$

$$\sqrt{5n-9} = 0$$

$$\left(\sqrt{5n-9}\right)^2 = (0)^2$$

$$5n-9 = 0$$

$$5n = 9$$

$$n = \frac{9}{5}$$

103.a.

$$\sqrt{2x} - 10 = 0$$

$$\sqrt{2x} = 10$$

$$\left(\sqrt{2x}\right)^2 = (10)^2$$

$$2x = 100$$

$$x = 50$$

b.

$$\sqrt{2x} + 10 = 0$$

$$\sqrt{2x} = -10$$

$$\left(\sqrt{2x}\right)^2 = (-10)^2$$

$$2x = 100$$

$$x = \cancel{50}; \text{ no solution}$$

105. HIGHWAY DESIGN

$$s = 3\sqrt{r}$$

$$40 = 3\sqrt{r}$$

$$(40)^2 = \left(3\sqrt{r}\right)^2$$

$$1,600 = 9r$$

$$\frac{1,600}{9} = \frac{9r}{9}$$

$$178 = r$$

$$r = 178 \text{ ft}$$

107. WIND POWER

$$v = \sqrt[3]{\frac{P}{0.02}}$$

$$29 = \sqrt[3]{\frac{P}{0.02}}$$

$$(29)^3 = \left(\sqrt[3]{\frac{P}{0.02}}\right)^3$$

$$24,389 = \frac{P}{0.02}$$

$$0.02(24,389) = 0.02\left(\frac{P}{0.02}\right)$$

$$488 = P$$

$$P = 488 \text{ watts}$$

109. GENERAL CONTRACTOR

$$\ell = \sqrt{f^2 + h^2}$$

$$10 = \sqrt{f^2 + 6^2}$$

$$10 = \sqrt{f^2 + 36}$$

$$(10)^2 = \left(\sqrt{f^2 + 36}\right)^2$$

$$100 = f^2 + 36$$

$$100 - 100 = f^2 + 36 - 100$$

$$0 = f^2 - 64$$

$$0 = (f+8)(f-8)$$

$$f+8 = 0 \quad \text{or} \quad f-8 = 0$$

$$f = \cancel{-8} \qquad f = 8 \text{ ft}$$

Section 9.5

111. SUPPLY AND DEMAND

$$\sqrt{5x} = \sqrt{100 - 3x^2}$$

$$\left(\sqrt{5x}\right)^2 = \left(\sqrt{100 - 3x^2}\right)^2$$

$$5x = 100 - 3x^2$$

$$3x^2 + 5x - 100 = 0$$

$$(3x + 20)(x - 5) = 0$$

$$3x + 20 = 0 \quad \text{or} \quad x - 5 = 0$$

$$3x = -20 \qquad x = 5$$

$$x = -\frac{20}{3}$$

The equilibrium price is $5.

WRITING

113.   Answers will vary.

115.   Answers will vary.

117.   Answers will vary.

119.   Answers will vary.

REVIEW

121.   Let I = intensity and d = distance from the bulb.

$$I = \frac{k}{d^2}$$

$$40 = \frac{k}{5^2}$$

$$40 = \frac{k}{25}$$

$$25(40) = 25\left(\frac{k}{25}\right)$$

$$1{,}000 = k$$

$$I = \frac{k}{d^2}$$

$$I = \frac{1{,}000}{20^2}$$

$$= \frac{1{,}000}{400}$$

$$= 2.5 \text{ foot-candles}$$

123.

$$\frac{12}{0.166044} = \frac{30}{x}$$

$$12x = 30(0.166044)$$

$$12x = 4.98132$$

$$x = 0.41511 \text{ in.}$$

CHALLENGE PROBLEMS

125.

$$\sqrt[4]{x} = \sqrt{\frac{x}{4}}$$

$$\left(\sqrt[4]{x}\right)^2 = \left(\sqrt{\frac{x}{4}}\right)^2$$

$$\sqrt[4]{x^2} = \frac{x}{4}$$

$$\left(\sqrt[4]{x^2}\right)^2 = \left(\frac{x}{4}\right)^2$$

$$\sqrt[4]{x^4} = \frac{x^2}{16}$$

$$x = \frac{x^2}{16}$$

$$16(x) = 16\left(\frac{x^2}{16}\right)$$

$$16x = x^2$$

$$0 = x^2 - 16x$$

$$0 = x(x - 16)$$

$$x = 0 \qquad x - 16 = 0$$

$$x = 0 \qquad\qquad x = 16$$

127.

$$\sqrt{x+2}+\sqrt{2x}=\sqrt{18-x}$$

$$\left(\sqrt{x+2}+\sqrt{2x}\right)^2=\left(\sqrt{18-x}\right)^2$$

$$x+2+\sqrt{x+2}\sqrt{2x}+\sqrt{x+2}\sqrt{2x}+2x=18-x$$

$$3x+2+2\sqrt{x+2}\sqrt{2x}=18-x$$

$$3x+2+2\sqrt{x+2}\sqrt{2x}-3x-2=18-x-3x-2$$

$$2\sqrt{x+2}\sqrt{2x}=-4x+16$$

$$\frac{2\sqrt{x+2}\sqrt{2x}}{2}=\frac{-4x}{2}+\frac{16}{2}$$

$$\sqrt{x+2}\sqrt{2x}=-2x+8$$

$$\left(\sqrt{x+2}\sqrt{2x}\right)^2=\left(-2x+8\right)^2$$

$$2x\left(x+2\right)=4x^2-16x-16x+64$$

$$2x^2+4x=4x^2-32x+64$$

$$2x^2+4x-2x^2-4x=4x^2-32x+64-2x^2-4x$$

$$0=2x^2-36x+64$$

$$\frac{0}{2}=\frac{2x^2}{2}-\frac{36x}{2}+\frac{64}{2}$$

$$0=x^2-18x+32$$

$$0=\left(x-2\right)\left(x-16\right)$$

$$x-2=0 \quad \text{or} \quad x-16=0$$

$$x=2 \qquad x=\cancel{16}$$

129.

$$\sqrt{2\sqrt{x+1}}=\sqrt{16-4x}$$

$$\left(\sqrt{2\sqrt{x+1}}\right)^2=\left(\sqrt{16-4x}\right)^2$$

$$2\sqrt{x+1}=16-4x$$

$$\frac{2\sqrt{x+1}}{2}=\frac{16-4x}{2}$$

$$\sqrt{x+1}=8-2x$$

$$\left(\sqrt{x+1}\right)^2=\left(8-2x\right)^2$$

$$x+1=64-32x+4x^2$$

$$0=4x^2-33x+63$$

$$0=\left(4x-21\right)\left(x-3\right)$$

$$4x-21=0 \quad x-3=0$$

$$4x=21 \qquad x=3$$

$$x=\cancel{\frac{21}{4}}$$

Section 9.5

VOCABULARY

1. In a right triangle, the side opposite the 90°
    angle is called the **hypotenuse**.

3. The **Pythagorean** Theorem states that in a
    right triangle, the sum of the squares of the
    lengths of the two legs is equal to the
    square of the hypotenuse.

CONCEPTS

5. If a and b are the lengths of the legs of a
    right triangle and c is the length of the
    hypotenuse, then $\boxed{a^2} + \boxed{b^2} = \boxed{c^2}$. This is
    called the Pythagorean **equation**.

7. In an isosceles right triangle, the length of
    the hypotenuse is $\sqrt{2}$ times the length of
    one leg.

9. The length of the longer leg of a 30°–60°–
    90° triangle is $\sqrt{3}$ times the length of the
    shorter leg.

11. The formula to find the distance between
    two points (x_1, y_1) and (x_2, y_2)
    is $d = \sqrt{(x_2 - x_1)^2 + (y_2 - y_1)^2}$.

NOTATION

13.
$$\sqrt{(-1-3)^2 + \left[2 - (-4)\right]^2} = \sqrt{(-4)^2 + \left[\boxed{6}\right]^2}$$
$$= \sqrt{\boxed{52}}$$
$$= \sqrt{\boxed{4} \cdot 13}$$
$$= \boxed{2}\sqrt{13}$$
$$\approx 7.21$$

GUIDED PRACTICE

15.
$$a^2 + b^2 = c^2$$
$$6^2 + 8^2 = c^2$$
$$36 + 64 = c^2$$
$$100 = c^2$$
$$\sqrt{100} = c$$
$$10 = c$$
$$c = 10 \text{ ft}$$

17.
$$a^2 + b^2 = c^2$$
$$8^2 + 15^2 = c^2$$
$$64 + 225 = c^2$$
$$289 = c^2$$
$$\sqrt{289} = c$$
$$17 = c$$
$$c = 17 \text{ ft}$$

19.
$$a^2 + b^2 = c^2$$
$$a^2 + 9^2 = 41^2$$
$$a^2 + 81 = 1{,}681$$
$$a^2 = 1{,}600$$
$$a = \sqrt{1{,}600}$$
$$a = 40 \text{ ft}$$

21.
$$a^2 + b^2 = c^2$$
$$10^2 + b^2 = 26^2$$
$$100 + b^2 = 676$$
$$b^2 = 576$$
$$b = \sqrt{576}$$
$$b = 24 \text{ cm}$$

23. In an isosceles right triangle, the length of the hypotenuse is the length of one leg times $\sqrt{2}$ and the lengths of the legs are equal.

$$x = 2$$

$$h = x\sqrt{2}$$
$$= 2\sqrt{2}$$
$$\approx 2.83$$

25. In an isosceles right triangle, the length of the hypotenuse is the length of one leg times $\sqrt{2}$ and the lengths of the legs are equal.

$$x = 3.2$$

$$h = x\sqrt{2}$$
$$= 3.2\sqrt{2}$$
$$\approx 4.53 \text{ ft}$$

27. In an isosceles right triangle, the length of the hypotenuse is the length of one leg times $\sqrt{2}$ and the lengths of the legs are equal.

$$3 = x\sqrt{2}$$
$$\frac{3}{\sqrt{2}} = \frac{x\sqrt{2}}{\sqrt{2}}$$
$$\frac{3}{\sqrt{2}} \cdot \frac{\sqrt{2}}{\sqrt{2}} = x$$
$$\frac{3\sqrt{2}}{\sqrt{4}} = x$$
$$\frac{3\sqrt{2}}{2} = x$$
$$2.12 \approx x$$

$$x = y$$
$$y = \frac{3\sqrt{2}}{2}$$
$$y \approx 2.12$$

29. The diagonal of the square is the hypotenuse of the right triangle.

$$10 = x\sqrt{2}$$
$$\frac{10}{\sqrt{2}} = \frac{x\sqrt{2}}{\sqrt{2}}$$
$$\frac{10}{\sqrt{2}} \cdot \frac{\sqrt{2}}{\sqrt{2}} = x$$
$$\frac{10\sqrt{2}}{\sqrt{4}} = x$$
$$\frac{10\sqrt{2}}{2} = x$$
$$5\sqrt{2} = x$$
$$7.07 \approx x$$

$$x = y$$
$$y = 5\sqrt{2}$$
$$y \approx 7.07 \text{ in.}$$

31. The shorter leg of a 30°–60°–90° triangle is half as long as the hypotenuse. The longer leg of a 30°–60°–90° triangle is the length of the shorter leg times $\sqrt{3}$.

$$x = 5\sqrt{3} \qquad h = 2(5)$$
$$x \approx 8.66 \qquad h = 10$$

33. The shorter leg of a 30°–60°–90° triangle is half as long as the hypotenuse. The longer leg of a 30°–60°–90° triangle is the length of the shorter leg times $\sqrt{3}$.

$$x = 75\sqrt{3} \qquad h = 2(75)$$
$$x \approx 129.90 \text{ cm} \qquad h = 150 \text{ cm}$$

35.

$$x\sqrt{3} = 40$$

$$\frac{x\sqrt{3}}{\sqrt{3}} = \frac{40}{\sqrt{3}}$$

$$x = \frac{40}{\sqrt{3}} \cdot \frac{\sqrt{3}}{\sqrt{3}}$$

$$x = \frac{40\sqrt{3}}{3}$$

$$\approx 23.09$$

$$h = 2x$$

$$= 2\left(\frac{40\sqrt{3}}{3}\right)$$

$$= \frac{80\sqrt{3}}{3}$$

$$\approx 46.19$$

37.

$$x\sqrt{3} = 55$$

$$\frac{x\sqrt{3}}{\sqrt{3}} = \frac{55}{\sqrt{3}}$$

$$x = \frac{55}{\sqrt{3}} \cdot \frac{\sqrt{3}}{\sqrt{3}}$$

$$x = \frac{55\sqrt{3}}{3}$$

$$\approx 31.75 \text{ mm}$$

$$h = 2x$$

$$= 2\left(\frac{55\sqrt{3}}{3}\right)$$

$$= \frac{110\sqrt{3}}{3}$$

$$\approx 63.51 \text{ mm}$$

39.

$$2x = 100$$

$$x = 50$$

$$y = x\sqrt{3}$$

$$= 50\sqrt{3}$$

$$\approx 86.60$$

41.

$$2x = 1.5$$

$$x = 0.75$$

$$y = x\sqrt{3}$$

$$= 0.75\sqrt{3}$$

$$\approx 1.30 \text{ ft}$$

43.

$$d = \sqrt{(x_2 - x_1)^2 + (y_2 - y_1)^2}$$

$$= \sqrt{(3-0)^2 + (-4-0)^2}$$

$$= \sqrt{(3)^2 + (-4)^2}$$

$$= \sqrt{9 + 16}$$

$$= \sqrt{25}$$

$$= 5$$

45.

$$d = \sqrt{(x_2 - x_1)^2 + (y_2 - y_1)^2}$$

$$= \sqrt{[3-(-2)]^2 + [4-(-8)]^2}$$

$$= \sqrt{(5)^2 + (12)^2}$$

$$= \sqrt{25 + 144}$$

$$= \sqrt{169}$$

$$= 13$$

47.

$$d = \sqrt{(x_2 - x_1)^2 + (y_2 - y_1)^2}$$

$$= \sqrt{(12-6)^2 + (16-8)^2}$$

$$= \sqrt{(6)^2 + (8)^2}$$

$$= \sqrt{36 + 64}$$

$$= \sqrt{100}$$

$$= 10$$

49.

$$d = \sqrt{(x_2 - x_1)^2 + (y_2 - y_1)^2}$$
$$= \sqrt{[3-(-2)]^2 + (4-1)^2}$$
$$= \sqrt{(5)^2 + (3)^2}$$
$$= \sqrt{25+9}$$
$$= \sqrt{34}$$

51.

$$d = \sqrt{(x_2 - x_1)^2 + (y_2 - y_1)^2}$$
$$= \sqrt{[3-(-1)]^2 + [-4-(-6)]^2}$$
$$= \sqrt{(4)^2 + (2)^2}$$
$$= \sqrt{16+4}$$
$$= \sqrt{20}$$
$$= \sqrt{4 \cdot 5}$$
$$= 2\sqrt{5}$$

53.

$$d = \sqrt{(x_2 - x_1)^2 + (y_2 - y_1)^2}$$
$$= \sqrt{[-5-(-2)]^2 + [8-(-1)]^2}$$
$$= \sqrt{(-3)^2 + (9)^2}$$
$$= \sqrt{9+81}$$
$$= \sqrt{90}$$
$$= \sqrt{9 \cdot 10}$$
$$= 3\sqrt{10}$$

55.

$$M = \left(\frac{x_1 + x_2}{2}, \frac{y_1 + y_2}{2}\right)$$
$$= \left(\frac{0+6}{2}, \frac{0+8}{2}\right)$$
$$= \left(\frac{6}{2}, \frac{8}{2}\right)$$
$$= (3,4)$$

57.

$$M = \left(\frac{x_1 + x_2}{2}, \frac{y_1 + y_2}{2}\right)$$
$$= \left(\frac{6+12}{2}, \frac{8+16}{2}\right)$$
$$= \left(\frac{18}{2}, \frac{24}{2}\right)$$
$$= (9,12)$$

59.

$$M = \left(\frac{x_1 + x_2}{2}, \frac{y_1 + y_2}{2}\right)$$
$$= \left(\frac{-2+3}{2}, \frac{-8+(-8)}{2}\right)$$
$$= \left(\frac{1}{2}, \frac{-16}{2}\right)$$
$$= \left(\frac{1}{2}, -8\right)$$

61.

$$M = \left(\frac{x_1 + x_2}{2}, \frac{y_1 + y_2}{2}\right)$$
$$= \left(\frac{7+(-10)}{2}, \frac{1+4}{2}\right)$$
$$= \left(\frac{-3}{2}, \frac{5}{2}\right)$$
$$= \left(-\frac{3}{2}, \frac{5}{2}\right)$$

63.

$$M = \left(\frac{x_1 + x_2}{2}, \frac{y_1 + y_2}{2}\right)$$
$$(-2,3) = \left(\frac{-8+x_2}{2}, \frac{5+y_2}{2}\right)$$
$$-2 = \frac{-8+x_2}{2} \quad \text{and} \quad 3 = \frac{5+y_2}{2}$$
$$2(-2) = 2\left(\frac{-8+x_2}{2}\right) \quad 2(3) = 2\left(\frac{5+y_2}{2}\right)$$
$$-4 = -8+x_2 \qquad\qquad 6 = 5+y_2$$
$$4 = x_2 \qquad\qquad\qquad 1 = y_2$$
$$Q = (4,1)$$

65.

$$M = \left(\frac{x_1 + x_2}{2}, \frac{y_1 + y_2}{2}\right)$$

$$(-7, -3) = \left(\frac{6 + x_2}{2}, \frac{-3 + y_2}{2}\right)$$

$$-7 = \frac{6 + x_2}{2} \quad \text{and} \quad -3 = \frac{-3 + y_2}{2}$$

$$2(-7) = 2\left(\frac{6 + x_2}{2}\right) \quad 2(-3) = 2\left(\frac{-3 + y_2}{2}\right)$$

$$-14 = 6 + x_2 \qquad\qquad -6 = -3 + y_2$$

$$-20 = x_2 \qquad\qquad -3 = y_2$$

$$P = (-20, -3)$$

APPLICATIONS

67. SOCCER

a. The diagonal would be the hypotenuse of the right triangle with equal sides of 64 and 100 m.

$$a^2 + b^2 = c^2$$

$$64^2 + 100^2 = c^2$$

$$4{,}096 + 10{,}000 = c^2$$

$$14{,}096 = c^2$$

$$\sqrt{14{,}096} = c$$

$$118.73 \text{ meters} \approx c$$

b. The diagonal would be the hypotenuse of the right triangle with equal sides of 75 and 110 m.

$$a^2 + b^2 = c^2$$

$$75^2 + 110^2 = c^2$$

$$5{,}625 + 12{,}100 = c^2$$

$$17{,}725 = c^2$$

$$\sqrt{17{,}725} = c$$

$$133.14 \text{ meters} \approx c$$

69. CUBES

Find the length of the blue diagonal. It would be the hypotenuse of a 45°–45°–90° triangle whose legs (x) are 7.

$$d = x\sqrt{2}$$

$$= 7\sqrt{2} \text{ cm}$$

The diagonal of the cube is the hypotenuse of the right triangle that has one leg (side of the cube) whose length is 7 cm and the other leg (blue diagonal) whose length is $7\sqrt{2}$ cm. Use the Pythagorean Theorem to find the length of the green diagonal.

$$(7)^2 + \left(7\sqrt{2}\right)^2 = c^2$$

$$49 + 49(2) = c^2$$

$$147 = c^2$$

$$c = \sqrt{147}$$

$$c = \sqrt{49 \cdot 3}$$

$$c = 7\sqrt{3} \text{ cm.}$$

71. WASHINGTON, D.C.

The x– and y–axes divide the square into 4 isosceles triangles (45°–45°–90°). If the area of the square is 100, then the length of the sides would be 10. Thus, the hypotenuse is 10. Find the length of the legs and that would be the distance of a corner from the origin.

$$x\sqrt{2} = 10$$

$$\frac{x\sqrt{2}}{\sqrt{2}} = \frac{10}{\sqrt{2}}$$

$$x = \frac{10}{\sqrt{2}} \cdot \frac{\sqrt{2}}{\sqrt{2}}$$

$$x = \frac{10\sqrt{2}}{2}$$

$$x = 5\sqrt{2}$$

The coordinates would all be $5\sqrt{2}$ units from the origin in all four directions:

$$\left(5\sqrt{2}, 0\right) = (7.07, 0)$$

$$\left(-5\sqrt{2}, 0\right) = (-7.07, 0)$$

$$\left(0, 5\sqrt{2}\right) = (0, 7.07)$$

$$\left(0, -5\sqrt{2}\right) = (0, -7.07)$$

73. HARDWARE

If the sides of the nut are 10 mm, then half of the side is 5 mm, which would be the length of the side opposite the 30° angle.

The height is one–half the side opposite the 60° angle.

To find the total height, multiply 2 times the length of the side opposite the 60° angle.

$$h = 2\left(x\sqrt{3}\right)$$
$$= 2\left(5\sqrt{3}\right)$$
$$= 10\sqrt{3} \text{ mm}$$
$$\approx 17.32 \text{ mm}$$

75. BASEBALL

Find the distance from 3rd base to 1st base.

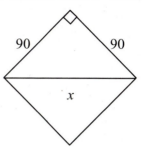

Use the Pythagorean Theorem to find x.
$$90^2 + 90^2 = x^2$$
$$8,100 + 8,100 = x^2$$
$$16,200 = x^2$$
$$\sqrt{16,200} = x$$
$$\sqrt{8,100 \cdot 2} = x$$
$$90\sqrt{2} = x$$
$$x = 90\sqrt{2} \text{ ft}$$
$$x \approx 127.3 \text{ ft}$$

If the ball lands 10 feet behind 3rd base, use the following triangle to find how far he must throw the ball.

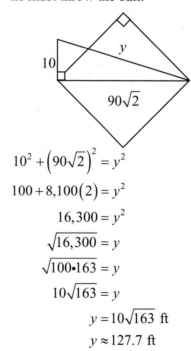

$$10^2 + \left(90\sqrt{2}\right)^2 = y^2$$
$$100 + 8,100(2) = y^2$$
$$16,300 = y^2$$
$$\sqrt{16,300} = y$$
$$\sqrt{100 \cdot 163} = y$$
$$10\sqrt{163} = y$$
$$y = 10\sqrt{163} \text{ ft}$$
$$y \approx 127.7 \text{ ft}$$

77. CLOTHESLINES

Find one–half of the amount stretched by using the Pythagorean Theorem. The blue line (amount stretched) is the hypotenuse (*c*). Let $a = 1$, and $b = \dfrac{15}{2} = 7.5$.

$$a^2 + b^2 = c^2$$
$$1^2 + (7.5)^2 = c^2$$
$$1 + 56.25 = c^2$$
$$57.25 = c^2$$
$$\sqrt{57.25} = c$$
$$c \approx 7.566$$
$$2c \approx 15.132$$

When stretched, the line is 15.132 ft. To find the amount the line is stretched, find the difference in the original length of the line and the length when stretched:

$$15.132 - 15 = 0.132$$
$$\approx 0.13 \text{ ft.}$$

79. ART HISTORY

a. Use the ordered pairs (5, 0) and (8, 21)

$$d = \sqrt{(x_2 - x_1)^2 + (y_2 - y_1)^2}$$
$$= \sqrt{(8 - 5)^2 + (21 - 0)^2}$$
$$= \sqrt{(3)^2 + (21)^2}$$
$$= \sqrt{9 + 441}$$
$$= \sqrt{450}$$
$$= 21.21 \text{ units}$$

b. Use the ordered pairs (10, 13) and (2, 11).

$$d = \sqrt{(x_2 - x_1)^2 + (y_2 - y_1)^2}$$
$$= \sqrt{(2 - 10)^2 + (11 - 13)^2}$$
$$= \sqrt{(-8)^2 + (-2)^2}$$
$$= \sqrt{64 + 4}$$
$$= \sqrt{68}$$
$$= 8.25 \text{ units}$$

c. Use the ordered pairs (7, 19) and (12, 7).

$$d = \sqrt{(x_2 - x_1)^2 + (y_2 - y_1)^2}$$
$$= \sqrt{(12 - 7)^2 + (7 - 19)^2}$$
$$= \sqrt{(5)^2 + (-12)^2}$$
$$= \sqrt{25 + 144}$$
$$= \sqrt{169}$$
$$= 13.00 \text{ units}$$

81. PACKAGING

Let $a = 24$, $b = 24$, and $c = 4$.

$$d = \sqrt{a^2 + b^2 + c^2}$$
$$= \sqrt{24^2 + 24^2 + 4^2}$$
$$= \sqrt{576 + 576 + 16}$$
$$= \sqrt{1{,}168}$$
$$\approx 34.2 \text{ in.}$$

Yes, the bone will fit in the box because the length of the diagonal (34.2 in.) is longer than the length of the bone (34 in.).

WRITING

83. Answers will vary.

85. Answers will vary.

REVIEW

87. DISCOUNT BUYING

Let c = cost of each unit and
let x = number of units purchased.
Unit cost $\cdot$ number = total cost

$$\text{unit cost } (c) = \frac{\text{total cost}}{\text{number of units}(x)}$$

	unit cost	number	total cost
first purchase	$c = \dfrac{224}{x}$	x	224
second purchase	$c - 4 = \dfrac{224}{x+1}$	$x + 1$	224

Solve the second unit cost for c.

$$c - 4 = \frac{224}{x+1}$$

$$c = \frac{224}{x+1} + 4$$

Set the costs equal and solve the equation for x.

$$\frac{224}{x} = \frac{224}{x+1} + 4$$

$$x(x+1)\left(\frac{224}{x}\right) = x(x+1)\left(\frac{224}{x+1} + 4\right)$$

$$224(x+1) = 224(x) + 4x(x+1)$$

$$224x + 224 = 224x + 4x^2 + 4x$$

$$224x + 224 - 224x = 224x + 4x^2 + 4x - 224x$$

$$224 = 4x^2 + 4x$$

$$224 - 224 = 4x^2 + 4x - 224$$

$$0 = 4x^2 + 4x - 224$$

$$\frac{0}{4} = \frac{4x^2}{4} + \frac{4x}{4} - \frac{224}{4}$$

$$0 = x^2 + x - 56$$

$$0 = (x - 7)(x + 8)$$

$$x - 7 = 0 \quad \text{or} \quad x + 8 = 0$$

$$x = 7 \qquad\qquad x = -8$$

Since the answer cannot be negative, he originally bought 7 motors.

CHALLENGE PROBLEMS

89. Using a 45°–45°–90° triangle on the face of the cube, you can draw a diagonal whose measure would be $a\sqrt{2}$ in. Now you have a triangle involving the following sides: base of the cubes whose measure is a, diagonal of a face whose length is $a\sqrt{2}$, and the diagonal of the cube (d). Use the Pythagorean Theorem to find the length of the diagonal (d).

$$(a)^2 + \left(a\sqrt{2}\right)^2 = d^2$$

$$a^2 + a^2(2) = d^2$$

$$a^2 + 2a^2 = d^2$$

$$3a^2 = d^2$$

$$d = \sqrt{3a^2}$$

$$d = a\sqrt{3} \text{ in.}$$

91.

$$d = \sqrt{(x_2 - x_1)^2 + (y_2 - y_1)^2}$$

$$= \sqrt{\left(\sqrt{12} - \sqrt{48}\right)^2 + \left(\sqrt{24} - \sqrt{150}\right)^2}$$

$$= \sqrt{\left(2\sqrt{3} - 4\sqrt{3}\right)^2 + \left(2\sqrt{6} - 5\sqrt{6}\right)^2}$$

$$= \sqrt{\left(-2\sqrt{3}\right)^2 + \left(-3\sqrt{6}\right)^2}$$

$$= \sqrt{12 + 54}$$

$$= \sqrt{66}$$

Section 9.6

VOCABULARY

1. The **imaginary** number i is defined as $i = \sqrt{-1}$. We call i^{25} a **power** of i.

3. For the complex number $2 + 5i$, we call 2 the **real** part and 5 the **imaginary** part.

CONCEPTS

5. a. $i = \boxed{\sqrt{-1}}$

   b. $i^2 = \boxed{-1}$

   c. $i^3 = \boxed{-i}$

   d. $i^4 = \boxed{1}$

   e. four

7. a. To add (or subtract) complex numbers, add (or subtract) their **real** parts and add (or subtract) their **imaginary** parts.

   b. To multiply two complex numbers, such as $(2 + 3i)(3 + 5i)$, we can use the **FOIL** method.

9. a. $2 + 3i$

   b. $2 - 0i$

   c. $0 + 3i$

11.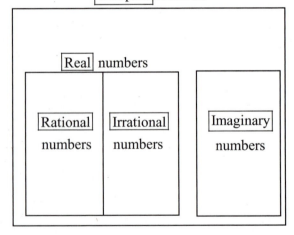

NOTATION

13.
$$(3 + 2i)(3 - i) = \boxed{9} - 3i + \boxed{6i} - 2i^2$$
$$= 9 + 3i + \boxed{2}$$
$$= \boxed{11} + 3i$$

15. a. true
    b. false
    c. false
    d. false

GUIDED PRACTICE

17.
$$\sqrt{-9} = \sqrt{-1 \cdot 9}$$
$$= 3i$$

19.
$$\sqrt{-7} = \sqrt{-1 \cdot 7}$$
$$= i\sqrt{7} \text{ or } \sqrt{7}i$$

21.
$$\sqrt{-24} = \sqrt{-1}\sqrt{4}\sqrt{6}$$
$$= 2i\sqrt{6}$$
$$= 2\sqrt{6}i$$

23.
$$-\sqrt{-72} = -\sqrt{-1}\sqrt{36}\sqrt{2}$$
$$= -6i\sqrt{2}$$
$$= -6\sqrt{2}i$$

25.
$$5\sqrt{-81} = 5\sqrt{-1}\sqrt{81}$$
$$= 5(9i)$$
$$= 45i$$

27.
$$\sqrt{-\frac{25}{9}} = \frac{\sqrt{-1}\sqrt{25}}{\sqrt{9}}$$
$$= \frac{5i}{3}$$
$$= \frac{5}{3}i$$

29. a. $5 + 0i$
    b. $0 + 7i$

31. a. $1 + 5i$
    b. $-3 + 2i\sqrt{2}$

33. a. $76 - 3i\sqrt{6}$
    b. $-7 + i\sqrt{19}$

35. a. $-6 - 3i$
    b. $3 + i\sqrt{6}$

37.
$$(3 + 4i) + (5 - 6i) = (3 + 5) + (4 - 6)i$$
$$= 8 - 2i$$

39.
$$(6 - i) + (9 + 3i) = (6 + 9) + (-1 + 3)i$$
$$= 15 + 2i$$

41.
$$(7 - 3i) - (4 + 2i) = (7 - 3i) + (-4 - 2i)$$
$$= (7 - 4) + (-3 - 2)i$$
$$= 3 - 5i$$

43.
$$\left(8 + \sqrt{-25}\right) - \left(7 + \sqrt{-4}\right) = (8 + 5i) - (7 + 2i)$$
$$= (8 - 7) + (5 - 2)i$$
$$= 1 + 3i$$

45.
$$\sqrt{-1}\sqrt{-36} = (i)(6i)$$
$$= 6i^2$$
$$= -6$$

47.
$$\sqrt{-2}\sqrt{-12} = \left(i\sqrt{2}\right)\left(i\sqrt{12}\right)$$
$$= i^2\sqrt{24}$$
$$= i^2\sqrt{4}\sqrt{6}$$
$$= 2i^2\sqrt{6}$$
$$= -2\sqrt{6}$$

49.
$$3(2 - 9i) = 6 - 27i$$

51.
$$7(5 - 4i) = 35 - 28i$$

53.
$$2i(7 - 3i) = 14i - 6i^2$$
$$= 14i - 6(-1)$$
$$= 14i + 6$$
$$= 6 + 14i$$

55.
$$-5i(5 - 5i) = -25i + 25i^2$$
$$= -25i - 25$$
$$= -25 - 25i$$

57.
$$(2 + i)(3 - i) = 6 - 2i + 3i - i^2$$
$$= 6 + i - (-1)$$
$$= 6 + i + 1$$
$$= 7 + i$$

59.
$$(3 - 2i)(2 + 3i) = 6 + 9i - 4i - 6i^2$$
$$= 6 + 5i - 6(-1)$$
$$= 6 + 5i + 6$$
$$= 12 + 5i$$

61.
$$(4 + i)(3 - i) = 12 - 4i + 3i - i^2$$
$$= 12 - i - (-1)$$
$$= 12 - i + 1$$
$$= 13 - i$$

63.
$$(2 + i)^2 = (2 + i)(2 + i)$$
$$= 4 + 2i + 2i + i^2$$
$$= 4 + 4i - 1$$
$$= 3 + 4i$$

65.
$$(2 + 6i)(2 - 6i) = 4 - 36i^2$$
$$= 4 - 36(-1)$$
$$= 4 + 36$$
$$= 40$$

Section 9.7

67.

$$(-4 - 7i)(-4 + 7i) = 16 - 49i^2$$
$$= 16 - 49(-1)$$
$$= 16 + 49$$
$$= 65$$

69.

$$\frac{9}{5+i} = \frac{9}{5+i} \cdot \frac{(5-i)}{(5-i)}$$
$$= \frac{9(5-i)}{(5+i)(5-i)}$$
$$= \frac{45 - 9i}{25 - i^2}$$
$$= \frac{45 - 9i}{25 - (-1)}$$
$$= \frac{45 - 9i}{26}$$
$$= \frac{45}{26} - \frac{9}{26}i$$

71.

$$\frac{11i}{4-7i} = \frac{11i}{4-7i} \cdot \frac{4+7i}{4+7i}$$
$$= \frac{11i(4+7i)}{(4-7i)(4+7i)}$$
$$= \frac{44i + 77i^2}{16 - 49i^2}$$
$$= \frac{44i + 77(-1)}{16 - 49(-1)}$$
$$= \frac{44i - 77}{16 + 49}$$
$$= \frac{-77 + 44i}{65}$$
$$= -\frac{77}{65} + \frac{44}{65}i$$

73.

$$\frac{3-2i}{4-i} = \frac{3-2i}{4-i} \cdot \frac{4+i}{4+i}$$
$$= \frac{(3-2i)(4+i)}{(4-i)(4+i)}$$
$$= \frac{12 + 3i - 8i - 2i^2}{16 - i^2}$$
$$= \frac{12 - 5i - 2(-1)}{16 - (-1)}$$
$$= \frac{12 - 5i + 2}{16 - (-1)}$$
$$= \frac{14 - 5i}{17}$$
$$= \frac{14}{17} - \frac{5}{17}i$$

75.

$$\frac{7+4i}{2-5i} = \frac{7+4i}{2-5i} \cdot \frac{2+5i}{2+5i}$$
$$= \frac{(7+4i)(2+5i)}{(2-5i)(2+5i)}$$
$$= \frac{14 + 35i + 8i + 20i^2}{4 - 25i^2}$$
$$= \frac{14 + 43i + 20(-1)}{4 - 25(-1)}$$
$$= \frac{14 + 43i - 20}{4 + 25}$$
$$= \frac{-6 + 43i}{29}$$
$$= -\frac{6}{29} + \frac{43}{29}i$$

77.

$$\frac{7+3i}{4-2i} = \frac{7+3i}{4-2i} \cdot \frac{(4+2i)}{(4+2i)}$$

$$= \frac{(7+3i)(4+2i)}{(4-2i)(4+2i)}$$

$$= \frac{28+14i+12i+6i^2}{16-4i^2}$$

$$= \frac{28+26i+6(-1)}{16-4(-1)}$$

$$= \frac{28+26i-6}{16+4}$$

$$= \frac{22+26i}{20}$$

$$= \frac{22}{20} + \frac{26}{20}i$$

$$= \frac{11}{10} + \frac{13}{10}i$$

79.

$$\frac{1-3i}{3+i} = \frac{1-3i}{3+i} \cdot \frac{3-i}{3-i}$$

$$= \frac{(1-3i)(3-i)}{(3+i)(3-i)}$$

$$= \frac{3-i-9i+3i^2}{9-i^2}$$

$$= \frac{3-10i+3(-1)}{9-(-1)}$$

$$= \frac{3-10i-3}{10}$$

$$= \frac{-10i}{10}$$

$$= -i$$

$$= 0-i$$

81.

$$\frac{8+\sqrt{-144}}{2+\sqrt{-9}} = \frac{8+12i}{2+3i}$$

$$= \frac{8+12i}{2+3i} \cdot \frac{2-3i}{2-3i}$$

$$= \frac{(8+12i)(2-3i)}{(2+3i)(2-3i)}$$

$$= \frac{16-24i+24i-36i^2}{4-9i^2}$$

$$= \frac{16-36(-1)}{4-9(-1)}$$

$$= \frac{16+36}{4+9}$$

$$= \frac{52}{13}$$

$$= 4+0i$$

83.

$$\frac{-4-\sqrt{-4}}{2+\sqrt{-1}} = \frac{-4-2i}{2+i}$$

$$= \frac{-4-2i}{2+i} \cdot \frac{2-i}{2-i}$$

$$= \frac{(-4-2i)(2-i)}{(2+i)(2-i)}$$

$$= \frac{-8+4i-4i+2i^2}{4-i^2}$$

$$= \frac{-8+2(-1)}{4-(-1)}$$

$$= \frac{-8-2}{4+1}$$

$$= \frac{-10}{5}$$

$$= -2+0i$$

85.

$$\frac{5}{3i} = \frac{5}{3i} \cdot \frac{i}{i}$$

$$= \frac{5i}{3i^2}$$

$$= \frac{5i}{3(-1)}$$

$$= -\frac{5i}{3}$$

$$= 0 - \frac{5}{3}i$$

Section 9.7

87.

$$-\frac{2}{7i} = -\frac{2}{7i} \cdot \frac{i}{i}$$

$$= -\frac{2i}{7i^2}$$

$$= -\frac{2i}{7(-1)}$$

$$= \frac{2i}{7}$$

$$= 0 + \frac{2}{7}i$$

89.

$$i^{21} = i^{4 \cdot 5 + 1}$$

$$= \left(i^4\right)^5 \cdot i^1$$

$$= (1)^5 \cdot i$$

$$= 1 \cdot i$$

$$= i$$

91.

$$i^{27} = i^{4 \cdot 6 + 3}$$

$$= \left(i^4\right)^6 \cdot i^3$$

$$= (1)^6 \cdot i^3$$

$$= 1 \cdot i^3$$

$$= i^3$$

$$= -i$$

93.

$$i^{100} = i^{4 \cdot 25}$$

$$= \left(i^4\right)^{25}$$

$$= (1)^{25}$$

$$= 1$$

95.

$$i^{42} = i^{4 \cdot 10 + 2}$$

$$= \left(i^4\right)^{10} \cdot i^2$$

$$= (1)^{10} \cdot i^2$$

$$= 1 \cdot i^2$$

$$= i^2$$

$$= -1$$

TRY IT YOURSELF

97.

$$(3-i)-(-1+10i) = 3 - i + 1 - 10i$$

$$= 4 - 11i$$

99.

$$\left(2-\sqrt{-16}\right)\left(3+\sqrt{-4}\right)$$

$$= (2-4i)(3+2i)$$

$$= 6 + 4i - 12i - 8i^2$$

$$= 6 - 8i - 8(-1)$$

$$= 6 - 8i + 8$$

$$= 14 - 8i$$

101.

$$(-6-9i)+(4+3i) = -6 - 9i + 4 + 3i$$

$$= -2 - 6i$$

103.

$$\frac{-2i}{3+2i} = \frac{-2i}{3+2i} \cdot \frac{3-2i}{3-2i}$$

$$= \frac{-2i(3-2i)}{(3+2i)(3-2i)}$$

$$= \frac{-6i + 4i^2}{9 - 4i^2}$$

$$= \frac{-6i + 4(-1)}{9 - 4(-1)}$$

$$= \frac{-6i - 4}{9 + 4}$$

$$= \frac{-4 - 6i}{13}$$

$$= -\frac{4}{13} - \frac{6}{13}i$$

105.

$$6i(2-3i) = 12i - 18i^2$$

$$= 12i - 18(-1)$$

$$= 12i + 18$$

$$= 18 + 12i$$

107.

$$\frac{4}{5i^{35}} = \frac{4}{5i^{35}} \cdot \frac{i}{i}$$

$$= \frac{4i}{5i^{36}}$$

$$= \frac{4i}{5(1)}$$

$$= \frac{4i}{5}$$

$$= 0 + \frac{4}{5}i$$

109.

$$\left(2 + i\sqrt{2}\right)\left(3 - i\sqrt{2}\right)$$

$$= 6 - 2i\sqrt{2} + 3i\sqrt{2} - 2i^2$$

$$= 6 + i\sqrt{2} - 2(-1)$$

$$= 6 + i\sqrt{2} + 2$$

$$= 8 + i\sqrt{2}$$

111.

$$\frac{5 + 9i}{1 - i} = \frac{5 + 9i}{1 - i} \cdot \frac{1 + i}{1 + i}$$

$$= \frac{(5 + 9i)(1 + i)}{(1 - i)(1 + i)}$$

$$= \frac{5 + 5i + 9i + 9i^2}{1 - i^2}$$

$$= \frac{5 + 14i + 9(-1)}{1 - (-1)}$$

$$= \frac{5 + 14i - 9}{2}$$

$$= \frac{-4 + 14i}{2}$$

$$= -\frac{4}{2} + \frac{14}{2}i$$

$$= -2 + 7i$$

113.

$$\left(4 - 8i\right)^2 = 4^2 - 2(4)(8i) + (-8i)^2$$

$$= 16 - 64i + 64i^2$$

$$= 16 - 64i + 64(-1)$$

$$= 16 - 64i - 64$$

$$= -48 - 64i$$

115.

$$\frac{\sqrt{5} - \sqrt{3}i}{\sqrt{5} + \sqrt{3}i} = \frac{\sqrt{5} - \sqrt{3}i}{\sqrt{5} + \sqrt{3}i} \cdot \frac{\sqrt{5} - \sqrt{3}i}{\sqrt{5} - \sqrt{3}i}$$

$$= \frac{\left(\sqrt{5} - \sqrt{3}i\right)\left(\sqrt{5} - \sqrt{3}i\right)}{\left(\sqrt{5} + \sqrt{3}i\right)\left(\sqrt{5} - \sqrt{3}i\right)}$$

$$= \frac{5 - \sqrt{15}i - \sqrt{15}i + 3i^2}{5 - 3i^2}$$

$$= \frac{5 - 2\sqrt{15}i + 3(-1)}{5 - 3(-1)}$$

$$= \frac{5 - 2\sqrt{15}i - 3}{5 + 3}$$

$$= \frac{2 - 2\sqrt{15}i}{8}$$

$$= \frac{2}{8} - \frac{2\sqrt{15}}{8}i$$

$$= \frac{1}{4} - \frac{\sqrt{15}}{4}i$$

117. a.

$$\sqrt{-8} = \sqrt{-1}\sqrt{4}\sqrt{2}$$

$$= 2i\sqrt{2}$$

b.

$$\sqrt[3]{-8} = \sqrt[3]{-1}\sqrt[3]{8}$$

$$= -2$$

119. a.

$$\left(2i\right)^2 = (2i)(2i)$$

$$= 4i^2$$

$$= 4(-1)$$

$$= -4$$

b.

$$\left(2 + i\right)^2 = (2)^2 + 2(2)(i) + (i)^2$$

$$= 4 + 4i - 1$$

$$= 3 + 4i$$

APPLICATIONS

121. **FRACTALS**

Step 1: $i^2 + i$

Step 2 $\quad (i^2 + i)^2 = (i^2 + i)(i^2 + i) + i$

$\qquad = i^4 + i^3 + i^3 + i^2 + i$

$\qquad = i^4 + 2i^3 + i^2 + i$

$\qquad = 1 + 2(-i) + (-1) + i$

$\qquad = 1 - 2i - 1 + i$

$\qquad = -i$

Step 3 $\quad (-i)^2 + i = i^2 + i$

$\qquad = -1 + i$

WRITING

123. Answers will vary.

125. Answers will vary.

REVIEW

127. **WIND SPEEDS**

Let $x =$ the rate of the wind.

	distance	rate	time
with a tail wind	330	$200 + x$	$t_1 = \dfrac{d}{r}$ $= \dfrac{330}{200+x}$
against the wind	330	$200 - x$	$t_1 = \dfrac{d}{r}$ $= \dfrac{330}{200-x}$

Since the total time for the trip is $3\dfrac{1}{3}$ hours, add the times and set the sum equal to $3\dfrac{1}{3}$.

$$t_1 + t_2 = 3\frac{1}{3}$$

$$\frac{330}{200+x} + \frac{330}{200-x} = \frac{10}{3}$$

$$3(200+x)(200-x)\left(\frac{330}{200+x} + \frac{330}{200-x}\right) = 3(200+x)(200-x)\left(\frac{10}{3}\right)$$

$$3\cdot 330(200-x) + 3\cdot 330(200+x) = 10(200+x)(200-x)$$

$$990(200-x) + 990(200+x) = 10(40,000 - x^2)$$

$$198,000 - 990x + 198,000 + 990x = 400,000 - 10x^2$$

$$396,000 = 400,000 - 10x^2$$

$$-4,000 = -10x^2$$

$$\frac{-4,000}{-10} = \frac{-10x^2}{-10}$$

$$400 = x^2$$

$$0 = x^2 - 400$$

$$0 = (x+20)(x-20)$$

$$x + 20 = 0 \quad \text{or} \quad x - 20 = 0$$

$$x = -20 \qquad x = 20$$

Since the rates cannot be negative, the rates of the wind is 20 mph.

CHALLENGE PROBLEMS

129.

$$\left(i^{349}\right)^{-i^{456}} = \left(i^{349}\right)^{-i^4} = \left(i^{349}\right)^{-1}$$

$$= i^{-349}$$

$$= \frac{1}{i^{349}}$$

$$= \frac{1}{i^1}$$

$$= \frac{1}{i}$$

$$= \frac{1}{i} \cdot \frac{i}{i}$$

$$= \frac{i}{i^2}$$

$$= \frac{i}{-1}$$

$$= -i$$

SECTION 9.1
Radical Expressions and Radical Functions

1. a. $f(1) = 2$
    b. $f(-3) = 0$
    c. $x = 6$
    d. D: $[-3, \infty)$; R: $[0, \infty)$

2. a. $\sqrt{100a^2} = 10|a|$
    b. $\sqrt{100a^2} = 10a$

3.
$$\sqrt{49} = 7$$

4.
$$-\sqrt{121} = -11$$

5.
$$\sqrt{\frac{225}{49}} = \frac{\sqrt{225}}{\sqrt{49}}$$
$$= \frac{15}{7}$$

6.
$$\sqrt{-4} \text{ is not real}$$

7.
$$\sqrt{100a^{12}} = 10a^6$$

8.
$$\sqrt{25x^2} = |5x|$$
$$= 5|x|$$

9.
$$\sqrt{x^8} = \sqrt{(x^4)^2}$$
$$= x^4$$

10.
$$\sqrt{x^2 + 4x + 4} = \sqrt{(x+2)^2}$$
$$= |x+2|$$

11.
$$\sqrt[3]{-27} = -3$$

12.
$$-\sqrt[3]{216} = -6$$

13.
$$\sqrt[3]{64x^6 y^3} = 4x^2 y$$

14.
$$\sqrt[3]{\frac{x^9}{125}} = \frac{\sqrt[3]{x^9}}{\sqrt[3]{125}}$$
$$= \frac{x^3}{5}$$

15.
$$\sqrt[6]{64} = 2$$

16.
$$\sqrt[5]{-32} = -2$$

17.
$$\sqrt[4]{256x^8 y^4} = |4x^2 y|$$
$$= 4x^2 |y|$$

18.
$$\sqrt[15]{(x+1)^{15}} = x+1$$

19.
$$-\sqrt[4]{\frac{1}{16}} = -\frac{\sqrt[4]{1}}{\sqrt[4]{16}}$$
$$= -\frac{1}{2}$$

20.
$$\sqrt[4]{-81} \text{ is not real}$$

21.
$$\sqrt[6]{-1} \text{ is not real}$$

22.
$$\sqrt[3]{0} = 0$$

23.
$$3x + 15 \geq 0$$
$$3x \geq -15$$
$$x \geq -5$$
$$D : [-5, \infty)$$

24. Find the length of each side.

$$s(A) = \sqrt{A}$$
$$s(169) = \sqrt{169}$$
$$= 13 \text{ ft}$$

25.

$$A(V) = 6\sqrt[3]{V^2}$$
$$A(8) = 6\sqrt[3]{8^2}$$
$$= 6\sqrt[3]{64}$$
$$= 6(4)$$
$$= 24 \text{ cm}^2$$

26.

$$g(-1.9) = \sqrt[3]{(-1.9)^2 + 9}$$
$$\approx 2.3276$$

27. $f(x) = \sqrt{x}$

$D: [0, \infty)$ $R: [0, \infty)$

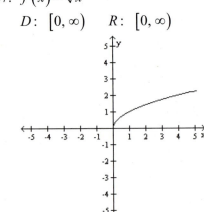

28. $f(x) = \sqrt[3]{x}$

$D: (-\infty, \infty)$ $R: (-\infty, \infty)$

29. $f(x) = \sqrt{x + 2}$

$D: [-2, \infty)$ $R: [0, \infty)$

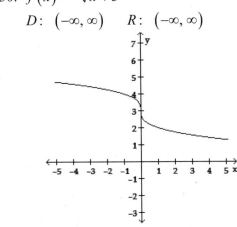

30. $f(x) = -\sqrt[3]{x} + 3$

$D: (-\infty, \infty)$ $R: (-\infty, \infty)$

SECTION 9.2
Rational Exponents

31.
$$t^{1/2} = \sqrt{t}$$

32.
$$\left(5xy^3\right)^{1/4} = \sqrt[4]{5xy^3}$$

33.
$$25^{1/2} = \sqrt{25}$$
$$= 5$$

34.
$$-36^{1/2} = -\sqrt{36}$$
$$= -6$$

35.
$$(-36)^{1/2} = \sqrt{-36}$$

is not a real number

36.

$$1^{1/5} = \sqrt[5]{1}$$
$$= 1$$

37.

$$\left(\frac{9}{x^2}\right)^{1/2} = \sqrt{\frac{9}{x^2}}$$
$$= \frac{\sqrt{9}}{\sqrt{x^2}}$$
$$= \frac{3}{x}$$

38.

$$(-8)^{1/3} = \sqrt[3]{-8}$$
$$= -2$$

39.

$$625^{1/4} = \sqrt[4]{625}$$
$$= 5$$

40.

$$\left(81c^4 d^4\right)^{1/4} = \sqrt[4]{81c^4 d^4}$$
$$= 3cd$$

41.

$$9^{3/2} = \left(\sqrt{9}\right)^3$$
$$= 3^3$$
$$= 27$$

42.

$$8^{-2/3} = \frac{1}{8^{2/3}}$$
$$= \frac{1}{\left(\sqrt[3]{8}\right)^2}$$
$$= \frac{1}{2^2}$$
$$= \frac{1}{4}$$

43.

$$-49^{5/2} = -\left(\sqrt{49}\right)^5$$
$$= -(7)^5$$
$$= -16,807$$

44.

$$\frac{1}{100^{-1/2}} = 100^{1/2}$$
$$= \sqrt{100}$$
$$= 10$$

45.

$$\left(\frac{4}{9}\right)^{-3/2} = \left(\frac{9}{4}\right)^{3/2}$$
$$= \left(\sqrt{\frac{9}{4}}\right)^3$$
$$= \left(\frac{3}{2}\right)^3$$
$$= \frac{27}{8}$$

46.

$$\frac{1}{25^{5/2}} = \frac{1}{\left(\sqrt{25}\right)^5}$$
$$= \frac{1}{5^5}$$
$$= \frac{1}{3,125}$$

47.

$$\left(25x^2 y^4\right)^{3/2} = \left(\sqrt{25x^2 y^4}\right)^3$$
$$= \left(5xy^2\right)^3$$
$$= 125x^3 y^6$$

48.

$$\left(8u^6 v^3\right)^{-2/3} = \frac{1}{\left(8u^6 v^3\right)^{2/3}}$$
$$= \frac{1}{\left(\sqrt[3]{8u^6 v^3}\right)^2}$$
$$= \frac{1}{\left(2u^2 v\right)^2}$$
$$= \frac{1}{4u^4 v^2}$$

49.

$$5^{1/4}5^{1/2} = 5^{1/4+1/2}$$
$$= 5^{1/4+2/4}$$
$$= 5^{3/4}$$

50.

$$a^{3/7}a^{-2/7} = a^{3/7+(-2/7)}$$
$$= a^{1/7}$$

51.

$$\left(k^{4/5}\right)^{10} = k^{(4/5)(10)}$$
$$= k^{8}$$

52.

$$\frac{3^{5/6}3^{1/3}}{3^{1/2}} = \frac{3^{5/6+1/3}}{3^{1/2}}$$
$$= \frac{3^{5/6+2/6}}{3^{1/2}}$$
$$= \frac{3^{7/6}}{3^{1/2}}$$
$$= 3^{7/6-1/2}$$
$$= 3^{7/6-3/6}$$
$$= 3^{4/6}$$
$$= 3^{2/3}$$

53.

$$u^{1/2}\left(u^{1/2} - u^{-1/2}\right) = u^{1/2}u^{1/2} - u^{1/2}u^{-1/2}$$
$$= u^{1/2+1/2} - u^{1/2+(-1/2)}$$
$$= u^{2/2} - u^{0}$$
$$= u^{1} - 1$$
$$= u - 1$$

54.

$$v^{2/3}\left(v^{1/3} + v^{4/3}\right) = v^{2/3}v^{1/3} + v^{2/3}v^{4/3}$$
$$= v^{2/3+1/3} + v^{2/3+4/3}$$
$$= v^{3/3} + v^{6/3}$$
$$= v^{1} + v^{2}$$
$$= v + v^{2}$$

55.

$$\sqrt[4]{a^2} = a^{2/4}$$
$$= a^{1/2}$$
$$= \sqrt{a}$$

56.

$$\sqrt[3]{\sqrt{c}} = \sqrt[3]{c^{1/2}}$$
$$= \left(c^{1/2}\right)^{1/3}$$
$$= c^{1/6}$$
$$= \sqrt[6]{c}$$

57.

$$d = 1.22a^{1/2}$$
$$= 1.22(22,500)^{1/2}$$
$$= 1.22\left(\sqrt{22,500}\right)$$
$$= 1.22(150)$$
$$= 183 \text{ miles}$$

58. Check (64, 64).

$$x^{2/3} + y^{2/3} = 32$$
$$(64)^{2/3} + (64)^{2/3} \overset{?}{=} 32$$
$$\left(\sqrt[3]{64}\right)^2 + \left(\sqrt[3]{64}\right)^2 \overset{?}{=} 32$$
$$4^2 + 4^2 \overset{?}{=} 32$$
$$16 + 16 \overset{?}{=} 32$$
$$32 = 32$$

Check (−64, 64).

$$x^{2/3} + y^{2/3} = 32$$
$$(-64)^{2/3} + (64)^{2/3} \overset{?}{=} 32$$
$$\left(\sqrt[3]{-64}\right)^2 + \left(\sqrt[3]{64}\right)^2 \overset{?}{=} 32$$
$$(-4)^2 + 4^2 \overset{?}{=} 32$$
$$16 + 16 \overset{?}{=} 32$$
$$32 = 32$$

SECTION 9.3
Simplifying and Combining Radical Expressions

59.

$$\sqrt{80} = \sqrt{16}\sqrt{5}$$
$$= 4\sqrt{5}$$

60.

$$\sqrt[3]{54} = \sqrt[3]{27}\sqrt[3]{2}$$
$$= 3\sqrt[3]{2}$$

61.

$$\sqrt[4]{160} = \sqrt[4]{16}\sqrt[4]{10}$$
$$= 2\sqrt[4]{10}$$

62.

$$\sqrt[5]{-96} = \sqrt[5]{-32}\sqrt[5]{3}$$
$$= -2\sqrt[5]{3}$$

63.

$$\sqrt{8x^5} = \sqrt{4x^4}\sqrt{2x}$$
$$= 2x^2\sqrt{2x}$$

64.

$$\sqrt[4]{r^{17}} = \sqrt[4]{r^{16}}\sqrt[4]{r}$$
$$= r^4\sqrt[4]{r}$$

65.

$$\sqrt[3]{-27j^7k} = \sqrt[3]{-27j^6}\sqrt[3]{jk}$$
$$= -3j^2\sqrt[3]{jk}$$

66.

$$\sqrt[3]{-16x^5y^4} = \sqrt[3]{-8x^3y^3}\sqrt[3]{2x^2y}$$
$$= -2xy\sqrt[3]{2x^2y}$$

67.

$$\sqrt{\frac{m}{144n^{12}}} = \frac{\sqrt{m}}{\sqrt{144n^{12}}}$$
$$= \frac{\sqrt{m}}{12n^6}$$

68.

$$\sqrt{\frac{17xy}{64a^4}} = \frac{\sqrt{17xy}}{\sqrt{64a^4}}$$
$$= \frac{\sqrt{17xy}}{8a^2}$$

69.

$$\frac{\sqrt[5]{64x^8}}{\sqrt[5]{2x^3}} = \sqrt[5]{\frac{64x^8}{2x^3}}$$
$$= \sqrt[5]{32x^5}$$
$$= 2x$$

70.

$$\frac{\sqrt[5]{243x^{16}}}{\sqrt[5]{x}} = \sqrt[5]{\frac{243x^{16}}{x}}$$
$$= \sqrt[5]{243x^{15}}$$
$$= 3x^3$$

71.

$$\sqrt{2} + 2\sqrt{2} = 3\sqrt{2}$$

72.

$$6\sqrt{20} - \sqrt{5} = 6\sqrt{4}\sqrt{5} - \sqrt{5}$$
$$= 6(2)\sqrt{5} - \sqrt{5}$$
$$= 12\sqrt{5} - \sqrt{5}$$
$$= 11\sqrt{5}$$

73.

$$2\sqrt[3]{3} - \sqrt[3]{24} = 2\sqrt[3]{3} - \sqrt[3]{8}\sqrt[3]{3}$$
$$= 2\sqrt[3]{3} - 2\sqrt[3]{3}$$
$$= 0$$

74.

$$-\sqrt[4]{32a^5} - 2\sqrt[4]{162a^5}$$
$$= -\sqrt[4]{16a^4}\sqrt[4]{2a} - 2\sqrt[4]{81a^4}\sqrt[4]{2a}$$
$$= -2a\sqrt[4]{2a} - 2(3a)\sqrt[4]{2a}$$
$$= -2a\sqrt[4]{2a} - 6a\sqrt[4]{2a}$$
$$= -8a\sqrt[4]{2a}$$

75.

$$2x\sqrt{8} + 2\sqrt{200x^2} + \sqrt{50x^2}$$
$$= 2x\sqrt{4}\sqrt{2} + 2\sqrt{100x^2}\sqrt{2} + \sqrt{25x^2}\sqrt{2}$$
$$= 2x(2)\sqrt{2} + 2(10x)\sqrt{2} + 5x\sqrt{2}$$
$$= 4x\sqrt{2} + 20x\sqrt{2} + 5x\sqrt{2}$$
$$= 29x\sqrt{2}$$

76.

$$\sqrt[3]{54x^3} - 3\sqrt[3]{16x^3} + 4\sqrt[3]{128x^3}$$
$$= \sqrt[3]{27x^3}\sqrt[3]{2} - 3\sqrt[3]{8x^3}\sqrt[3]{2} + 4\sqrt[3]{64x^3}\sqrt[3]{2}$$
$$= 3x\sqrt[3]{2} - 3(2x)\sqrt[3]{2} + 4(4x)\sqrt[3]{2}$$
$$= 3x\sqrt[3]{2} - 6x\sqrt[3]{2} + 16x\sqrt[3]{2}$$
$$= 13x\sqrt[3]{2}$$

Chapter 9 Review

77.

$$2\sqrt[4]{32t^3} - 8\sqrt[4]{6t^3} + 5\sqrt[4]{2t^3}$$

$$= 2\sqrt[4]{16}\sqrt[4]{2t^3} - 8\sqrt[4]{6t^3} + 5\sqrt[4]{2t^3}$$

$$= 2(2)\sqrt[4]{2t^3} - 8\sqrt[4]{6t^3} + 5\sqrt[4]{2t^3}$$

$$= 4\sqrt[4]{2t^3} - 8\sqrt[4]{6t^3} + 5\sqrt[4]{2t^3}$$

$$= 9\sqrt[4]{2t^3} - 8\sqrt[4]{6t^3}$$

78.

$$10\sqrt[4]{16x^9} - 8x^2\sqrt[4]{x} + 5\sqrt[4]{x^5}$$

$$= 10\sqrt[4]{16x^8}\sqrt[4]{x} - 8x^2\sqrt[4]{x} + 5\sqrt[4]{x^4}\sqrt[4]{x}$$

$$= 10\left(2x^2\right)\sqrt[4]{x} - 8x^2\sqrt[4]{x} + 5x\sqrt[4]{x}$$

$$= 20x^2\sqrt[4]{x} - 8x^2\sqrt[4]{x} + 5x\sqrt[4]{x}$$

$$= 12x^2\sqrt[4]{x} + 5x\sqrt[4]{x}$$

79.
  a. You do not add the radicands together.
  b. They are not like terms.
  c. It should be $2\sqrt[3]{y^2}$.
  d. You do not subtract the radicands.

80.

$$\sqrt{40} + \sqrt{32} + \sqrt{8}$$

$$= \sqrt{4}\sqrt{10} + \sqrt{16}\sqrt{2} + \sqrt{4}\sqrt{2}$$

$$= 2\sqrt{10} + 4\sqrt{2} + 2\sqrt{2}$$

$$= \left(2\sqrt{10} + 6\sqrt{2}\right) \text{ in.}$$

$$= 14.8 \text{ in.}$$

SECTION 9.4
Multiplying and Dividing Radical Expressions

81.

$$\sqrt{7}\sqrt{7} = \sqrt{49}$$

$$= 7$$

82.

$$\left(2\sqrt{5}\right)\left(3\sqrt{2}\right) = 6\sqrt{10}$$

83.

$$\left(-2\sqrt{8}\right)^2 = \left(-2\sqrt{8}\right)\left(-2\sqrt{8}\right)$$

$$= 4\sqrt{64}$$

$$= 4(8)$$

$$= 32$$

84.

$$2\sqrt{6}\sqrt{15} = 2\sqrt{90}$$

$$= 2\sqrt{9}\sqrt{10}$$

$$= 2(3)\sqrt{10}$$

$$= 6\sqrt{10}$$

85.

$$\sqrt{9x}\sqrt{x} = \sqrt{9x^2}$$

$$= 3x$$

86.

$$\left(\sqrt[3]{x+1}\right)^3 = x+1$$

87.

$$-\sqrt[3]{2x^2}\sqrt[3]{4x^8} = -\sqrt[3]{8x^{10}}$$

$$= -\sqrt[3]{8x^9}\sqrt[3]{x}$$

$$= -2x^3\sqrt[3]{x}$$

88.

$$\sqrt[5]{9}\cdot\sqrt[5]{27} = \sqrt[5]{243}$$

$$= 3$$

89.

$$3\sqrt{7t}\left(2\sqrt{7t} + 3\sqrt{3t^2}\right)$$

$$= 6\sqrt{49t^2} + 9\sqrt{21t^3}$$

$$= 6\sqrt{49t^2} + 9\sqrt{t^2}\sqrt{21t}$$

$$= 6(7t) + 9(t)\sqrt{21t}$$

$$= 42t + 9t\sqrt{21t}$$

90.

$$-\sqrt[4]{4x^5y^{11}}\sqrt[4]{8x^9y^3} = -\sqrt[4]{32x^{14}y^{14}}$$

$$= -\sqrt[4]{16x^{12}y^{12}}\sqrt[4]{2x^2y^2}$$

$$= -2x^3y^3\sqrt[4]{2x^2y^2}$$

91.

$$\left(\sqrt{3b}+\sqrt{3}\right)^2=\left(\sqrt{3b}+\sqrt{3}\right)\left(\sqrt{3b}+\sqrt{3}\right)$$
$$=3b+\sqrt{9b}+\sqrt{9b}+3$$
$$=3b+2\sqrt{9b}+3$$
$$=3b+2\sqrt{9}\sqrt{b}+3$$
$$=3b+2(3)\sqrt{b}+3$$
$$=3b+6\sqrt{b}+3$$

92.

$$\left(\sqrt[3]{3p}-2\sqrt[3]{2}\right)\left(\sqrt[3]{3p}+\sqrt[3]{2}\right)$$
$$=\sqrt[3]{9p^2}+\sqrt[3]{6p}-2\sqrt[3]{6p}-2\sqrt[3]{4}$$
$$=\sqrt[3]{9p^2}-\sqrt[3]{6p}-2\sqrt[3]{4}$$

93.

$$\frac{10}{\sqrt{3}}=\frac{10}{\sqrt{3}}\cdot\frac{\sqrt{3}}{\sqrt{3}}$$
$$=\frac{10\sqrt{3}}{\sqrt{9}}$$
$$=\frac{10\sqrt{3}}{3}$$

94.

$$\sqrt{\frac{3}{5xy}}=\frac{\sqrt{3}}{\sqrt{5xy}}$$
$$=\frac{\sqrt{3}}{\sqrt{5xy}}\cdot\frac{\sqrt{5xy}}{\sqrt{5xy}}$$
$$=\frac{\sqrt{15xy}}{\sqrt{25x^2y^2}}$$
$$=\frac{\sqrt{15xy}}{5xy}$$

95.

$$\frac{\sqrt[3]{6u}}{\sqrt[3]{u^5}}=\sqrt[3]{\frac{6u}{u^5}}$$
$$=\sqrt[3]{\frac{6}{u^4}}$$
$$=\sqrt[3]{\frac{6}{u^4}}\cdot\sqrt[3]{\frac{u^2}{u^2}}$$
$$=\sqrt[3]{\frac{6u^2}{u^6}}$$
$$=\frac{\sqrt[3]{6u^2}}{u^2}$$

96.

$$\frac{\sqrt[4]{a}}{\sqrt[4]{3b^2}}=\frac{\sqrt[4]{a}}{\sqrt[4]{3b^2}}\cdot\frac{\sqrt[4]{27b^2}}{\sqrt[4]{27b^2}}$$
$$=\frac{\sqrt[4]{27ab^2}}{\sqrt[4]{81b^4}}$$
$$=\frac{\sqrt[4]{27ab^2}}{3b}$$

97.

$$\frac{2}{\sqrt{2}-1}=\frac{2}{\sqrt{2}-1}\cdot\frac{\sqrt{2}+1}{\sqrt{2}+1}$$
$$=\frac{2\left(\sqrt{2}+1\right)}{\left(\sqrt{2}-1\right)\left(\sqrt{2}+1\right)}$$
$$=\frac{2\sqrt{2}+2}{2-1}$$
$$=\frac{2\sqrt{2}+2}{1}$$
$$=2\sqrt{2}+2$$
$$=2\left(\sqrt{2}+1\right)$$

98.

$$\frac{4\sqrt{x}-2\sqrt{z}}{\sqrt{z}+4\sqrt{x}}=\frac{4\sqrt{x}-2\sqrt{z}}{\sqrt{z}+4\sqrt{x}}\cdot\frac{\sqrt{z}-4\sqrt{x}}{\sqrt{z}-4\sqrt{x}}$$
$$=\frac{\left(4\sqrt{x}-2\sqrt{z}\right)\left(\sqrt{z}-4\sqrt{x}\right)}{\left(\sqrt{z}+4\sqrt{x}\right)\left(\sqrt{z}-4\sqrt{x}\right)}$$
$$=\frac{4\sqrt{xz}-16\sqrt{x^2}-2\sqrt{z^2}+8\sqrt{xz}}{\sqrt{z^2}-16\sqrt{x^2}}$$
$$=\frac{-16x-2z+12\sqrt{xz}}{z-16x}$$

Chapter 9 Review

99.

$$\frac{\sqrt{a}-\sqrt{b}}{\sqrt{a}} = \frac{\sqrt{a}-\sqrt{b}}{\sqrt{a}} \cdot \frac{\sqrt{a}+\sqrt{b}}{\sqrt{a}+\sqrt{b}}$$

$$= \frac{\left(\sqrt{a}-\sqrt{b}\right)\left(\sqrt{a}+\sqrt{b}\right)}{\sqrt{a}\left(\sqrt{a}+\sqrt{b}\right)}$$

$$= \frac{\sqrt{a^2}-\sqrt{b^2}}{\sqrt{a^2}+\sqrt{ab}}$$

$$= \frac{a-b}{a+\sqrt{ab}}$$

100.

$$r = \sqrt[3]{\frac{3V}{4\pi}}$$

$$= \frac{\sqrt[3]{3V}}{\sqrt[3]{4\pi}} \cdot \frac{\sqrt[3]{2\pi^2}}{\sqrt[3]{2\pi^2}}$$

$$= \frac{\sqrt[3]{6\pi^2 V}}{\sqrt[3]{8\pi^3}}$$

$$= \frac{\sqrt[3]{6\pi^2 V}}{2\pi}$$

SECTION 9.5
Solving Radical Equations

101.

$$\sqrt{7x-10}-1 = 11$$

$$\sqrt{7x-10} = 12$$

$$\left(\sqrt{7x-10}\right)^2 = 12^2$$

$$7x-10 = 144$$

$$7x = 154$$

$$x = 22$$

102.

$$u = \sqrt{25u-144}$$

$$u^2 = \left(\sqrt{25u-144}\right)^2$$

$$u^2 = 25u-144$$

$$u^2 - 25u+144 = 0$$

$$(u-9)(u-16) = 0$$

$$u-9=0 \quad \text{or} \quad u-16=0$$

$$u=9 \qquad\qquad u=16$$

103.

$$2\sqrt{y-3} = \sqrt{2y+1}$$

$$\left(2\sqrt{y-3}\right)^2 = \left(\sqrt{2y+1}\right)^2$$

$$4(y-3) = 2y+1$$

$$4y-12 = 2y+1$$

$$4y = 2y+13$$

$$2y = 13$$

$$y = \frac{13}{2}$$

104.

$$\sqrt{z+1}+\sqrt{z} = 2$$

$$\sqrt{z+1} = 2-\sqrt{z}$$

$$\left(\sqrt{z+1}\right)^2 = \left(2-\sqrt{z}\right)^2$$

$$z+1 = 4-2\sqrt{z}-2\sqrt{z}+z$$

$$z+1 = 4-4\sqrt{z}+z$$

$$z+1-z-4 = 4-4\sqrt{z}+z-z-4$$

$$-3 = -4\sqrt{z}$$

$$(-3)^2 = \left(-4\sqrt{z}\right)^2$$

$$9 = 16z$$

$$\frac{9}{16} = z$$

105.

$$\sqrt[3]{x^3 + 56} - 2 = x$$

$$\sqrt[3]{x^3 + 56} = x + 2$$

$$\left(\sqrt[3]{x^3 + 56}\right)^3 = (x + 2)^3$$

$$x^3 + 56 = (x + 2)(x + 2)^2$$

$$x^3 + 56 = (x + 2)(x^2 + 4x + 4)$$

$$x^3 + 56 = x^3 + 4x^2 + 4x + 2x^2 + 8x + 8$$

$$x^3 + 56 = x^3 + 6x^2 + 12x + 8$$

$$x^3 + 56 - x^3 = x^3 + 6x^2 + 12x + 8 - x^3$$

$$56 = 6x^2 + 12x + 8$$

$$56 - 56 = 6x^2 + 12x + 8 - 56$$

$$0 = 6x^2 + 12x - 48$$

$$\frac{0}{6} = \frac{6x^2}{6} + \frac{12x}{6} - \frac{48}{6}$$

$$0 = x^2 + 2x - 8$$

$$0 = (x + 4)(x - 2)$$

$$x + 4 = 0 \quad \text{or} \quad x - 2 = 0$$

$$x = -4 \qquad x = 2$$

106.

$$a = \sqrt{a^2 + 5a - 35}$$

$$a^2 = \left(\sqrt{a^2 + 5a - 35}\right)^2$$

$$a^2 = a^2 + 5a - 35$$

$$0 = 5a - 35$$

$$35 = 5a$$

$$7 = a$$

107.

$$(x + 2)^{1/2} - (4 - x)^{1/2} = 0$$

$$\sqrt{x + 2} - \sqrt{4 - x} = 0$$

$$\sqrt{x + 2} = \sqrt{4 - x}$$

$$\left(\sqrt{x + 2}\right)^2 = \left(\sqrt{4 - x}\right)^2$$

$$x + 2 = 4 - x$$

$$2x + 2 = 4$$

$$2x = 2$$

$$x = 1$$

108.

$$\sqrt{b^2 + b} = \sqrt{3 - b^2}$$

$$\left(\sqrt{b^2 + b}\right)^2 = \left(\sqrt{3 - b^2}\right)^2$$

$$b^2 + b = 3 - b^2$$

$$b^2 + b + b^2 = 3 - b^2 + b^2$$

$$2b^2 + b = 3$$

$$2b^2 + b - 3 = 0$$

$$(2b + 3)(b - 1) = 0$$

$$2b + 3 = 0 \quad \text{or} \quad b - 1 = 0$$

$$2b = -3 \qquad\qquad b = 1$$

$$b = -\frac{3}{2}$$

109.

$$\sqrt[4]{8x - 8} + 2 = 0$$

$$\sqrt[4]{8x - 8} = -2$$

$$\left(\sqrt[4]{8x - 8}\right)^4 = (-2)^4$$

$$8x - 8 = 16$$

$$8x = 24$$

$$x = \cancel{3}$$

No solution

110.

$$\sqrt{2m + 4} - \sqrt{m + 3} = 1$$

$$\sqrt{2m + 4} = \sqrt{m + 3} + 1$$

$$\left(\sqrt{2m + 4}\right)^2 = \left(\sqrt{m + 3} + 1\right)^2$$

$$2m + 4 = m + 3 + 2\sqrt{m + 3} + 1$$

$$2m + 4 = m + 4 + 2\sqrt{m + 3}$$

$$2m = m + 2\sqrt{m + 3}$$

$$m = 2\sqrt{m + 3}$$

$$m^2 = \left(2\sqrt{m + 3}\right)^2$$

$$m^2 = 4(m + 3)$$

$$m^2 = 4m + 12$$

$$m^2 - 4m - 12 = 0$$

$$(m - 6)(m + 2) = 0$$

$$m - 6 = 0 \quad \text{or} \quad m + 2 = 0$$

$$m = 6 \qquad\qquad m = \cancel{-2}$$

Chapter 9 Review

111.

$$f(x) = g(x)$$
$$\sqrt{5x+1} = x+1$$
$$\left(\sqrt{5x+1}\right)^2 = (x+1)^2$$
$$5x+1 = x^2 + 2x + 1$$
$$0 = x^2 + 2x + 1 - 5x - 1$$
$$0 = x^2 - 3x$$
$$0 = x(x-3)$$
$$x - 3 = 0 \quad \text{or} \quad x = 0$$
$$x = 3$$

112.

$$f(x) = \sqrt{2x^2 - 7x}$$
$$\sqrt{2x^2 - 7x} = 2$$
$$\left(\sqrt{2x^2 - 7x}\right)^2 = 2^2$$
$$2x^2 - 7x = 4$$
$$2x^2 - 7x - 4 = 0$$
$$(2x+1)(x-4) = 0$$
$$2x+1 = 0 \quad \text{or} \quad x - 4 = 0$$
$$2x = -1 \qquad\qquad x = 4$$
$$x = -\frac{1}{2}$$

113.

$$r = \sqrt{\frac{A}{P}} - 1$$
$$(r+1)^2 = \left(\sqrt{\frac{A}{P}}\right)^2$$
$$(r+1)^2 = \frac{A}{P}$$
$$P(r+1)^2 = P\left(\frac{A}{P}\right)$$
$$P(r+1)^2 = A$$
$$\frac{P(r+1)^2}{(r+1)^2} = \frac{A}{(r+1)^2}$$
$$P = \frac{A}{(r+1)^2}$$

114.

$$h = \sqrt[3]{\frac{12I}{b}}$$
$$(h)^3 = \left(\sqrt[3]{\frac{12I}{b}}\right)^3$$
$$h^3 = \frac{12I}{b}$$
$$b(h^3) = b\left(\frac{12I}{b}\right)$$
$$h^3 b = 12I$$
$$\frac{h^3 b}{12} = \frac{12I}{12}$$
$$\frac{h^3 b}{12} = I$$

SECTION 9.6
Geometric Applications of Radicals

115.   On the left right triangle, one leg is 8 and the other leg is $\frac{1}{2}(30) = 15$. The roof line, x, is the hypotenuse of the right triangle. Use the Pythagorean Theorem to find x.

$$a^2 + b^2 = c^2$$
$$8^2 + 15^2 = c^2$$
$$64 + 225 = c^2$$
$$289 = c^2$$
$$c = \sqrt{289}$$
$$c = 17$$

The roof line is 17 ft.

116. The hypotenuse is 125 and one leg is 117. Let x = the length of the other leg, which is one–half the d. Use the Pythagorean Theorem to find x.

$$a^2 + b^2 = c^2$$
$$x^2 + 117^2 = 125^2$$
$$x^2 + 13,689 = 15,625$$
$$x^2 = 1,936$$
$$x = \sqrt{1,936}$$
$$x = 44 \text{ yd}$$

$$d = 2x$$
$$= 2(44)$$
$$= 88 \text{ yd}$$

The distance the boat advances is 88 yards.

117. In an isosceles right triangle, the legs are equal. Let $a = b = 7$. Find c.

$$a^2 + b^2 = c^2$$
$$7^2 + 7^2 = c^2$$
$$49 + 49 = c^2$$
$$98 = c^2$$
$$c = \sqrt{98}$$
$$c = \sqrt{49}\sqrt{2}$$
$$c = 7\sqrt{2} \text{ meters}$$
$$c \approx 9.90 \text{ meters}$$

118. The hypotenuse is $\sqrt{2}$ times the length of the equal sides. Let x = the length of the equal sides.

$$x\sqrt{2} = 15$$
$$\frac{x\sqrt{2}}{\sqrt{2}} = \frac{15}{\sqrt{2}}$$
$$x = \frac{15}{\sqrt{2}} \cdot \frac{\sqrt{2}}{\sqrt{2}}$$
$$x = \frac{15\sqrt{2}}{2} \text{ yards}$$
$$x \approx 10.61 \text{ yards}$$

119. The hypotenuse is 12, which is twice the length of the shorter leg. The length of the longer leg is $\sqrt{3}$ times the length of the shorter leg. Let x = length of the shorter leg and y = length of the longer leg.

$$2x = 12$$
$$\frac{2x}{2} = \frac{12}{2}$$
$$x = 6 \text{ cm}$$

$$y = 6\left(\sqrt{3}\right)$$
$$= 6\sqrt{3}$$
$$\approx 10.39 \text{ cm}$$

120. The length of the longer leg is $\sqrt{3}$ times the length of the shorter leg. Let x = length of the shorter leg and y = length of the hypotenuse.

$$x\sqrt{3} = 60$$
$$\frac{x\sqrt{3}}{\sqrt{3}} = \frac{60}{\sqrt{3}}$$
$$x = \frac{60}{\sqrt{3}} \cdot \frac{\sqrt{3}}{\sqrt{3}}$$
$$x = \frac{60\sqrt{3}}{3}$$
$$x = 20\sqrt{3}$$
$$x \approx 34.64 \text{ feet}$$

$$y = 2x$$
$$y = 2\left(20\sqrt{3}\right)$$
$$y = 40\sqrt{3}$$
$$y \approx 69.28 \text{ feet}$$

121. The hypotenuse is $\sqrt{2}$ times the length of a leg.

$$x = 5\sqrt{2}$$
$$= 7.07$$
$$y = 5$$

Chapter 9 Review

122. The hypotenuse is 50, which is twice the length of the shorter leg. The length of the longer leg is $\sqrt{3}$ times the length of the shorter leg.

Let y = length of the shorter leg and x = length of the longer leg.

$$2y = 50$$
$$y = 25 \text{ cm}$$

$$x = 25\left(\sqrt{3}\right)$$
$$= 25\sqrt{3}$$
$$\approx 43.30 \text{ cm}$$

123.
$$d = \sqrt{\left(x_2 - x_1\right)^2 + \left(y_2 - y_1\right)^2}$$
$$= \sqrt{\left(6 - 1\right)^2 + \left(-9 - 3\right)^2}$$
$$= \sqrt{\left(5\right)^2 + \left(-12\right)^2}$$
$$= \sqrt{25 + 144}$$
$$= \sqrt{169}$$
$$= 13$$

124.
$$d = \sqrt{\left(x_2 - x_1\right)^2 + \left(y_2 - y_1\right)^2}$$
$$= \sqrt{\left[-2 - \left(-4\right)\right]^2 + \left(8 - 6\right)^2}$$
$$= \sqrt{\left(2\right)^2 + \left(2\right)^2}$$
$$= \sqrt{4 + 4}$$
$$= \sqrt{8}$$
$$= \sqrt{4}\sqrt{2}$$
$$= 2\sqrt{2}$$

125.
$$M = \left(\frac{x_1 + x_2}{2}, \frac{y_1 + y_2}{2}\right)$$
$$= \left(\frac{8 + 6}{2}, \frac{-2 + \left(-4\right)}{2}\right)$$
$$= \left(\frac{14}{2}, \frac{-6}{2}\right)$$
$$= \left(7, -3\right)$$

126.
$$M = \left(\frac{x_1 + x_2}{2}, \frac{y_1 + y_2}{2}\right)$$
$$\left(6, 1\right) = \left(\frac{10 + x_2}{2}, \frac{4 + y_2}{2}\right)$$
$$6 = \frac{10 + x_2}{2} \quad \text{and} \quad 1 = \frac{4 + y_2}{2}$$
$$2\left(6\right) = 2\left(\frac{10 + x_2}{2}\right) \quad 2\left(1\right) = 2\left(\frac{4 + y_2}{2}\right)$$
$$12 = 10 + x_2 \qquad\qquad 2 = 4 + y_2$$
$$2 = x_2 \qquad\qquad\qquad -2 = y_2$$
$$Q = \left(2, -2\right)$$

SECTION 9.7
Complex Numbers

127.
$$\sqrt{-25} = \sqrt{25}\sqrt{-1}$$
$$= 5i$$

128.
$$\sqrt{-18} = \sqrt{-1}\sqrt{9}\sqrt{2}$$
$$= 3i\sqrt{2}$$

129.
$$-\sqrt{-6} = -\sqrt{-1}\sqrt{6}$$
$$= -i\sqrt{6}$$

130.
$$\sqrt{-\frac{9}{64}} = \sqrt{-1}\sqrt{\frac{9}{64}}$$
$$= \frac{3}{8}i$$

131. Complex Numbers

Real Numbers	**Imaginary** Numbers

132. a. true
b. true
c. false
d. false

133. a. $3 - 6i$
b. $0 - 19i$

134.
a. $-1 + 7i$
b. $0 + i$

135.
$$(3 + 4i) + (5 - 6i) = (3 + 5) + (4 - 6)i$$
$$= 8 - 2i$$

136.
$$\left(7 - \sqrt{-9}\right) - \left(4 + \sqrt{-4}\right)$$
$$= (7 - 3i) - (4 + 2i)$$
$$= (7 - 3i) + (-4 - 2i)$$
$$= (7 - 4) + (-3 - 2)i$$
$$= 3 - 5i$$

137.
$$3i(2 - i) = 6i - 3i^2$$
$$= 6i - 3(-1)$$
$$= 6i + 3$$
$$= 3 + 6i$$

138.
$$(2 - 7i)(-3 + 4i)$$
$$= -6 + 8i + 21i - 28i^2$$
$$= -6 + 29i - 28(-1)$$
$$= -6 + 29i + 28$$
$$= 22 + 29i$$

139.
$$\sqrt{-3} \cdot \sqrt{-9} = i\sqrt{3} \cdot 3i$$
$$= 3i^2\sqrt{3}$$
$$= 3(-1)\sqrt{3}$$
$$= -3\sqrt{3}$$
$$= -3\sqrt{3} + 0i$$

140.
$$(9i)^2 = 81i^2$$
$$= 81(-1)$$
$$= -81$$
$$= -81 + 0i$$

141.
$$\frac{5 + 14i}{2 + 3i} = \frac{5 + 14i}{2 + 3i} \cdot \frac{2 - 3i}{2 - 3i}$$
$$= \frac{(5 + 14i)(2 - 3i)}{(2 + 3i)(2 - 3i)}$$
$$= \frac{10 - 15i + 28i - 42i^2}{4 - 9i^2}$$
$$= \frac{10 + 13i - 42(-1)}{4 - 9(-1)}$$
$$= \frac{10 + 13i + 42}{4 + 9}$$
$$= \frac{52 + 13i}{13}$$
$$= \frac{52}{13} + \frac{13}{13}i$$
$$= 4 + i$$

142.
$$\frac{3}{11i} = \frac{3}{11i} \cdot \frac{i}{i}$$
$$= \frac{3i}{11i^2}$$
$$= \frac{3i}{11(-1)}$$
$$= \frac{3i}{-11}$$
$$= 0 - \frac{3}{11}i$$

143.
$$i^{42} = i^{4 \cdot 10 + 2}$$
$$= \left(i^4\right)^{10} \cdot i^2$$
$$= (1)^{10} \cdot i^2$$
$$= 1 \cdot i^2$$
$$= i^2$$
$$= -1$$

144.
$$i^{97} = i^{4 \cdot 24 + 1}$$
$$= \left(i^4\right)^{24} \cdot i^1$$
$$= (1)^{24} \cdot i$$
$$= 1 \cdot i$$
$$= i$$

Chapter 9 Review

CHAPTER 9 TEST

1. a. The symbol $\sqrt{}$ is called a **radical** symbol.
    b. The **imaginary** number i is defined as $i = \sqrt{-1}$.
    c. Squaring both sides of an equation can introduce **extraneous** solutions.
    d. An **isosceles** right triangle is a right triangle with two legs of equal length.
    e. To **rationalize** the denominator of $\frac{4}{\sqrt{5}}$, we multiply the fraction by $\frac{\sqrt{5}}{\sqrt{5}}$.
    f. A **complex** number is any number that can be written in the form $a + bi$, where a and b are real numbers and $i = \sqrt{-1}$.

2. a. If $\sqrt[n]{a}$ and $\sqrt[n]{b}$ are real numbers, then $\sqrt[n]{ab} = \sqrt[n]{a}\,\sqrt[n]{b}$.
    b. If $\sqrt[n]{a}$ and $\sqrt[n]{b}$ are real numbers, then $\sqrt[n]{\dfrac{a}{b}} = \dfrac{\sqrt[n]{a}}{\sqrt[n]{b}}, \quad (b \ne 0)$.
    c. No real number raised to the fourth power is -16.

3. $f(x) = \sqrt{x-1}$
    D: $[1, \infty)$; R: $[0, \infty)$

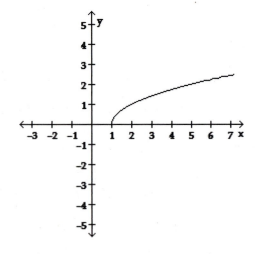

4.
$$v(d) = \sqrt{64.4d}$$
$$v(32.8) = \sqrt{64.4(32.8)}$$
$$= \sqrt{2{,}112.32}$$
$$= 46 \text{ mph}$$

5. a. $f(-1) = -1$
    b. $f(8) = 2$
    c. $x = 1$
    d. D: $(-\infty, \infty)$, R: $(-\infty, \infty)$

6. $10x + 50 \ge 0$
    $$10x \ge -50$$
    $$x \ge -5$$
    $$D : [-5, \infty)$$

7.
$$\left(49x^4\right)^{1/2} = \sqrt{49x^4}$$
$$= 7x^2$$

8.
$$-27^{2/3} = -\left(\sqrt[3]{27}\right)^2$$
$$= -(3)^2$$
$$= -9$$

9.
$$36^{-3/2} = \frac{1}{36^{3/2}}$$
$$= \frac{1}{\left(\sqrt{36}\right)^3}$$
$$= \frac{1}{6^3}$$
$$= \frac{1}{216}$$

10.
$$\left(-\frac{8}{125n^6}\right)^{-2/3} = \left(-\frac{125n^6}{8}\right)^{2/3}$$
$$= \left(-\sqrt[3]{\frac{125n^6}{8}}\right)^2$$
$$= \left(-\frac{5n^2}{2}\right)^2$$
$$= \frac{25n^4}{4}$$

11.

$$\frac{2^{5/3}2^{1/6}}{2^{1/2}} = \frac{2^{5/3+1/6}}{2^{1/2}}$$

$$= \frac{2^{10/6+1/6}}{2^{1/2}}$$

$$= \frac{2^{11/6}}{2^{1/2}}$$

$$= 2^{11/6-1/2}$$

$$= 2^{11/6-3/6}$$

$$= 2^{8/6}$$

$$= 2^{4/3}$$

12.

$$\left(a^{2/3}\right)^{1/6} = a^{(2/3)(1/6)}$$

$$= a^{2/18}$$

$$= a^{1/9}$$

13.

$$\sqrt{x^2} = |x|$$

14.

$$\sqrt{y^2 - 10y + 25} = \sqrt{(y-5)^2}$$

$$= |y-5|$$

15.

$$\sqrt[3]{-64x^3y^6} = -4xy^2$$

16.

$$\sqrt{\frac{4a^2}{9}} = \frac{2}{3}a$$

17.

$$\sqrt[5]{(t+8)^5} = t+8$$

18.

$$\sqrt{540x^3y^5} = \sqrt{36x^2y^4}\sqrt{15xy}$$

$$= 6xy^2\sqrt{15xy}$$

19.

$$\frac{\sqrt[3]{24x^{15}y^4}}{\sqrt[3]{y}} = \sqrt[3]{\frac{24x^{15}y^4}{y}}$$

$$= \sqrt[3]{24x^{15}y^3}$$

$$= \sqrt[3]{8x^{15}y^3}\sqrt[3]{3}$$

$$= 2x^5y\sqrt[3]{3}$$

20.

$$\sqrt[4]{32} = \sqrt[4]{16}\sqrt[4]{2}$$

$$= 2\sqrt[4]{2}$$

21.

$$2\sqrt{48y^5} - 3y\sqrt{12y^3}$$

$$= 2\sqrt{16y^4}\sqrt{3y} - 3y\sqrt{4y^2}\sqrt{3y}$$

$$= 2\left(4y^2\right)\sqrt{3y} - 3y(2y)\sqrt{3y}$$

$$= 8y^2\sqrt{3y} - 6y^2\sqrt{3y}$$

$$= 2y^2\sqrt{3y}$$

22.

$$2\sqrt[3]{40} - \sqrt[3]{5,000} + 4\sqrt[3]{625}$$

$$= 2\sqrt[3]{8}\sqrt[3]{5} - \sqrt[3]{1,000}\sqrt[3]{5} + 4\sqrt[3]{125}\sqrt[3]{5}$$

$$= 2(2)\sqrt[3]{5} - 10\sqrt[3]{5} + 4(5)\sqrt[3]{5}$$

$$= 4\sqrt[3]{5} - 10\sqrt[3]{5} + 20\sqrt[3]{5}$$

$$= 14\sqrt[3]{5}$$

23.

$$\sqrt[4]{243z^{13}} + z\sqrt[4]{48z^9} = \sqrt[4]{81z^{12}}\sqrt[4]{3z} + z\sqrt[4]{16z^8}\sqrt[4]{3z}$$

$$= 3z^3\sqrt[4]{3z} + z\left(2z^2\right)\sqrt[4]{3z}$$

$$= 3z^3\sqrt[4]{3z} + 2z^3\sqrt[4]{3z}$$

$$= 5z^3\sqrt[4]{3z}$$

24.

$$-2\sqrt{xy}\left(3\sqrt{x} + \sqrt{xy^3}\right) = -6\sqrt{x^2y} - 2\sqrt{x^2y^4}$$

$$= -6\sqrt{x^2}\sqrt{y} - 2\sqrt{x^2y^4}$$

$$-6x\sqrt{y} - 2xy^2$$

Chapter 9 Test

25.

$$\left(3\sqrt{2}+\sqrt{3}\right)\left(2\sqrt{2}-3\sqrt{3}\right)$$
$$=6\sqrt{4}-9\sqrt{6}+2\sqrt{6}-3\sqrt{9}$$
$$=6(2)-7\sqrt{6}-3(3)$$
$$=12-7\sqrt{6}-9$$
$$=3-7\sqrt{6}$$

26.

$$\left(\sqrt[3]{2a}+9\right)^2=\left(\sqrt[3]{2a}+9\right)\left(\sqrt[3]{2a}+9\right)$$
$$=\sqrt[3]{4a^2}+9\sqrt[3]{2a}+9\sqrt[3]{2a}+81$$
$$=\sqrt[3]{4a^2}+18\sqrt[3]{2a}+81$$

27.

$$\frac{8}{\sqrt{10}}=\frac{8}{\sqrt{10}}\cdot\frac{\sqrt{10}}{\sqrt{10}}$$
$$=\frac{8\sqrt{10}}{\sqrt{100}}$$
$$=\frac{8\sqrt{10}}{10}$$
$$=\frac{4\sqrt{10}}{5}$$

28.

$$\frac{\sqrt{x}+\sqrt{y}}{\sqrt{x}-\sqrt{y}}=\left(\frac{\sqrt{x}+\sqrt{y}}{\sqrt{x}-\sqrt{y}}\right)\left(\frac{\sqrt{x}+\sqrt{y}}{\sqrt{x}+\sqrt{y}}\right)$$
$$=\frac{\sqrt{x}\sqrt{x}+\sqrt{x}\sqrt{y}+\sqrt{x}\sqrt{y}+\sqrt{y}\sqrt{y}}{\sqrt{x}\sqrt{x}+\sqrt{x}\sqrt{y}-\sqrt{x}\sqrt{y}-\sqrt{y}\sqrt{y}}$$
$$=\frac{\sqrt{x^2}+2\sqrt{xy}+\sqrt{y^2}}{\sqrt{x^2}-\sqrt{y^2}}$$
$$=\frac{x+2\sqrt{xy}+y}{x-y}$$

29.

$$\sqrt[3]{\frac{9}{4a}}=\sqrt[3]{\frac{9}{4a}}\cdot\sqrt[3]{\frac{2a^2}{2a^2}}$$
$$=\frac{\sqrt[3]{18a^2}}{\sqrt[3]{8a^3}}$$
$$=\frac{\sqrt[3]{18a^2}}{2a}$$

30.

$$\frac{\sqrt{5}+3}{-4\sqrt{2}}=\frac{\sqrt{5}+3}{-4\sqrt{2}}\cdot\frac{\sqrt{5}-3}{\sqrt{5}-3}$$
$$=\frac{\left(\sqrt{5}+3\right)\left(\sqrt{5}-3\right)}{-4\sqrt{2}\left(\sqrt{5}-3\right)}$$
$$=\frac{5-9}{-4\sqrt{10}+12\sqrt{2}}$$
$$=\frac{-4}{-4\left(\sqrt{10}-3\sqrt{2}\right)}$$
$$=\frac{1}{\sqrt{10}-3\sqrt{2}}$$
$$=\frac{1}{\sqrt{2}\sqrt{5}-3\sqrt{2}}$$
$$=\frac{1}{\sqrt{2}\left(\sqrt{5}-3\right)}$$

31.

$$4\sqrt{x}=\sqrt{x+1}$$
$$\left(4\sqrt{x}\right)^2=\left(\sqrt{x+1}\right)^2$$
$$16x=x+1$$
$$15x=1$$
$$x=\frac{1}{15}$$

32.

$$\sqrt[3]{6n+4}-4=0$$
$$\sqrt[3]{6n+4}=4$$
$$\left(\sqrt[3]{6n+4}\right)^3=4^3$$
$$6n+4=64$$
$$6n=60$$
$$n=10$$

33.

$$1 = \sqrt{u-3} + \sqrt{u}$$
$$1 - \sqrt{u} = \sqrt{u-3}$$
$$\left(1 - \sqrt{u}\right)^2 = \left(\sqrt{u-3}\right)^2$$
$$1 - \sqrt{u} - \sqrt{u} + u = u - 3$$
$$1 - 2\sqrt{u} + u = u - 3$$
$$1 - 2\sqrt{u} + u - 1 - u = u - 3 - 1 - u$$
$$-2\sqrt{u} = -4$$
$$\frac{-2\sqrt{u}}{-2} = \frac{-4}{-2}$$
$$\sqrt{u} = 2$$
$$\left(\sqrt{u}\right)^2 = (2)^2$$
$$u = \cancel{4}$$

No solution

34.

$$\left(2m^2 - 9\right)^{1/2} = m$$
$$\sqrt{2m^2 - 9} = m$$
$$\left(\sqrt{2m^2 - 9}\right)^2 = (m)^2$$
$$2m^2 - 9 = m^2$$
$$2m^2 - 9 - m^2 = m^2 - m^2$$
$$m^2 - 9 = 0$$
$$(m+3)(m-3) = 0$$
$$m + 3 = 0 \quad \text{or} \quad m - 3 = 0$$
$$m = -\cancel{3} \qquad\qquad m = 3$$

35.

$$\sqrt{t-2} - t + 2 = 0$$
$$\sqrt{t-2} = t - 2$$
$$\sqrt{t-2} = t - 2$$
$$\left(\sqrt{t-2}\right)^2 = (t-2)^2$$
$$t - 2 = t^2 - 4t + 4$$
$$0 = t^2 - 5t + 6$$
$$0 = (t-2)(t-3)$$
$$t - 2 = 0 \quad \text{or} \quad t - 3 = 0$$
$$t = 2 \qquad\qquad t = 3$$

36.

$$\sqrt{x-8} + 10 = 0$$
$$\sqrt{x-8} = -10$$
$$\left(\sqrt{x-8}\right)^2 = (-10)^2$$
$$x - 8 = 100$$
$$x = \cancel{108}$$

no solution

37.

$$f(x) = g(x)$$
$$\sqrt[4]{15-x} = \sqrt[4]{13-2x}$$
$$\left(\sqrt[4]{15-x}\right)^4 = \left(\sqrt[4]{13-2x}\right)^4$$
$$15 - x = 13 - 2x$$
$$15 + x = 13$$
$$x = -2$$

38.

$$r = \sqrt[3]{\frac{GMt^2}{4\pi^2}}$$
$$(r)^3 = \left(\sqrt[3]{\frac{GMt^2}{4\pi^2}}\right)^3$$
$$r^3 = \frac{GMt^2}{4\pi^2}$$
$$4\pi^2 \left(r^3\right) = 4\pi^2 \left(\frac{GMt^2}{4\pi^2}\right)$$
$$4\pi^2 r^3 = GMt^2$$
$$\frac{4\pi^2 r^3}{Mt^2} = \frac{GMt^2}{Mt^2}$$
$$\frac{4\pi^2 r^3}{Mt^2} = G$$

Chapter 9 Test

39. The length of the longer side is $\sqrt{3}$ time the length of the shorter side. The length of the hypotenuse is twice the length of the shorter side. First find the length of the shorter side and use that to find h, the length of the hypotenuse.

Let x = the length of the shorter side.

$$x\sqrt{3} = 8$$

$$\frac{x\sqrt{3}}{\sqrt{3}} = \frac{8}{\sqrt{3}}$$

$$x = \frac{8}{\sqrt{3}}$$

$$x = \frac{8}{\sqrt{3}} \cdot \frac{\sqrt{3}}{\sqrt{3}}$$

$$x = \frac{8\sqrt{3}}{3}$$

$$x \approx 4.62 \text{ cm}$$

$$h = 2x$$

$$= 2(4.62)$$

$$\approx 9.24 \text{ cm}$$

40. The length of the hypotenuse is $\sqrt{2}$ times the length of a leg. Let x and y equal the length of a leg.

$$x\sqrt{2} = 12.26$$

$$\frac{x\sqrt{2}}{\sqrt{2}} = \frac{12.26}{\sqrt{2}}$$

$$x = \frac{12.26}{\sqrt{2}}$$

$$x = \frac{12.26}{\sqrt{2}} \cdot \frac{\sqrt{2}}{\sqrt{2}}$$

$$x = \frac{12.26\sqrt{2}}{2}$$

$$x = 6.13\sqrt{2}$$

$$x = 8.67 \text{ cm}$$

$$y = 6.13\sqrt{2}$$

$$y = 8.67 \text{ cm}$$

41.

$$d = \sqrt{(x_2 - x_1)^2 + (y_2 - y_1)^2}$$

$$= \sqrt{[22 - (-2)]^2 + (12 - 5)^2}$$

$$= \sqrt{(24)^2 + (7)^2}$$

$$= \sqrt{576 + 49}$$

$$= \sqrt{625}$$

$$= 25$$

42.

$$M = \left(\frac{x_1 + x_2}{2}, \frac{y_1 + y_2}{2}\right)$$

$$= \left(\frac{-2 + 7}{2}, \frac{-5 + (-11)}{2}\right)$$

$$= \left(\frac{5}{2}, \frac{-16}{2}\right)$$

$$= \left(\frac{5}{2}, -8\right)$$

43.

$$M = \left(\frac{x_1 + x_2}{2}, \frac{y_1 + y_2}{2}\right)$$

$$(3,0) = \left(\frac{-1 + x_2}{2}, \frac{-3 + y_2}{2}\right)$$

$$3 = \frac{-1 + x_2}{2} \quad \text{and} \quad 0 = \frac{-3 + y_2}{2}$$

$$2(3) = 2\left(\frac{-1 + x_2}{2}\right) \quad 2(0) = 2\left(\frac{-3 + y_2}{2}\right)$$

$$6 = -1 + x_2 \qquad 0 = -3 + y_2$$

$$7 = x_2 \qquad 3 = y_2$$

$$Q = (7,3)$$

44. Use the Pythagorean Theorem to find h.

$$a^2 + b^2 = c^2$$

$$(h)^2 + (45)^2 = (53)^2$$

$$h^2 + 2,025 = 2,809$$

$$h^2 = 784$$

$$h = \sqrt{784}$$

$$h = 28 \text{ in.}$$

45.

$$\sqrt{-45} = \sqrt{-9}\sqrt{5}$$
$$= 3i\sqrt{5}$$

46.

$$i^{106} = i^{4 \cdot 26 + 2}$$
$$= \left(i^4\right)^{26} \cdot i^2$$
$$= (1)^{26} \cdot i^2$$
$$= 1 \cdot i^2$$
$$= i^2$$
$$= -1$$

47.

$$(9 + 4i) + (-13 + 7i) = (9 - 13) + (4 + 7)i$$
$$= -4 + 11i$$

48.

$$\left(3 - \sqrt{-9}\right) - \left(-1 + \sqrt{-16}\right)$$
$$= (3 - 3i) - (-1 + 4i)$$
$$= (3 - 3i) + (1 - 4i)$$
$$= (3 + 1) + (-3 - 4)i$$
$$= 4 - 7i$$

49.

$$15i(3 - 5i) = 45i - 75i^2$$
$$= 45i - 75(-1)$$
$$= 45i + 75$$
$$= 75 + 45i$$

50.

$$(8 + 10i)(-7 - i) = -56 - 8i - 70i - 10i^2$$
$$= -56 - 78i - 10(-1)$$
$$= -56 - 78i + 10$$
$$= -46 - 78i$$

51.

$$\frac{1}{i\sqrt{2}} = \frac{1}{i\sqrt{2}} \cdot \frac{i\sqrt{2}}{i\sqrt{2}}$$
$$= \frac{i\sqrt{2}}{2i^2}$$
$$= \frac{i\sqrt{2}}{2(-1)}$$
$$= \frac{i\sqrt{2}}{-2}$$
$$= 0 - \frac{\sqrt{2}}{2}i$$

52.

$$\frac{2+i}{3-i} = \frac{2+i}{3-i} \cdot \frac{3+i}{3+i}$$
$$= \frac{(2+i)(3+i)}{(3-i)(3+i)}$$
$$= \frac{6 + 2i + 3i + i^2}{9 - i^2}$$
$$= \frac{6 + 5i + (-1)}{9 - (-1)}$$
$$= \frac{5 + 5i}{10}$$
$$= \frac{5}{10} + \frac{5}{10}i$$
$$= \frac{1}{2} + \frac{1}{2}i$$

Chapter 9 Test

SECTION 10.1

VOCABULARY

1. An equation of the form $ax^2 + bx + c = 0$, where $a \neq 0$, is called a **quadratic** equation.

3. When we add 16 to $x^2 + 8x$, we say that we have completed the **square** on $x^2 + 8x$.

CONCEPTS

5. For any nonnegative number c, if $x^2 = c$, then $\underline{x = \sqrt{c}}$ or $\underline{x = -\sqrt{c}}$.

7. a. $\left(\dfrac{12}{2}\right)^2 = 6^2 = 36$

   b. $\left(\dfrac{-5}{2}\right)^2 = \dfrac{25}{4}$

9. a. subtract 7 from both sides
   b. divide both sides by 4

11. Yes, it is a solution.
$$x^2 + 4x + 2 = 0$$
$$\left(-2+\sqrt{2}\right)^2 + 4\left(-2+\sqrt{2}\right) + 2 \stackrel{?}{=} 0$$
$$4 - 4\sqrt{2} + 2 - 8 + 4\sqrt{2} + 2 \stackrel{?}{=} 0$$
$$0 = 0$$

NOTATION

13. We read $8 \pm \sqrt{3}$ as "eight **plus or minus** the square root of 3."

GUIDED PRACTICE

15.
$$t^2 - 11 = 0$$
$$t^2 - 11 + 11 = 0 + 11$$
$$t^2 = 11$$
$$t = \sqrt{11} \quad \text{or} \quad t = -\sqrt{11}$$
$$t = \pm\sqrt{11}$$

17.
$$x^2 - 35 = 0$$
$$x^2 - 35 + 35 = 0 + 35$$
$$x^2 = 35$$
$$x = \sqrt{35} \quad \text{or} \quad x = -\sqrt{35}$$
$$x = \pm\sqrt{35}$$

19.
$$z^2 - 50 = 0$$
$$z^2 - 50 + 50 = 0 + 50$$
$$z^2 = 50$$
$$z = \pm\sqrt{50}$$
$$z = \pm\sqrt{25}\sqrt{2}$$
$$z = \pm 5\sqrt{2}$$
$$z = 5\sqrt{2} \quad \text{or} \quad z = -5\sqrt{2}$$

21.
$$3x^2 - 16 = 0$$
$$3x^2 = 16$$
$$x^2 = \dfrac{16}{3}$$
$$x = \sqrt{\dfrac{16}{3}} \quad \text{or} \quad x = -\sqrt{\dfrac{16}{3}}$$
$$x = \dfrac{4}{\sqrt{3}} \qquad\qquad x = -\dfrac{4}{\sqrt{3}}$$
$$x = \dfrac{4}{\sqrt{3}} \cdot \dfrac{\sqrt{3}}{\sqrt{3}} \qquad x = -\dfrac{4}{\sqrt{3}} \cdot \dfrac{\sqrt{3}}{\sqrt{3}}$$
$$x = \dfrac{4\sqrt{3}}{3} \qquad\qquad x = -\dfrac{4\sqrt{3}}{3}$$

23.
$$p^2 = -16$$
$$p = \sqrt{-16} \quad \text{or} \quad p = -\sqrt{-16}$$
$$p = 4i \qquad\qquad\quad p = -4i$$
$$p = \pm 4i$$

25.
$$a^2 + 8 = 0$$
$$a^2 + 8 - 8 = 0 - 8$$
$$a^2 = -8$$
$$a = \pm\sqrt{-8}$$
$$a = \pm\sqrt{-1}\sqrt{4}\sqrt{2}$$
$$a = \pm 2i\sqrt{2}$$

27.

$$4m^2 + 81 = 0$$

$$4m^2 = -81$$

$$m^2 = -\frac{81}{4}$$

$$m = \sqrt{-\frac{81}{4}} \quad \text{or} \quad m = -\sqrt{-\frac{81}{4}}$$

$$m = \frac{9}{2}i \qquad\qquad m = -\frac{9}{2}i$$

$$m = \pm\frac{9}{2}i$$

29.

$$6b^2 + 144 = 0$$

$$6b^2 = -144$$

$$b^2 = -\frac{144}{6}$$

$$b = \sqrt{-\frac{144}{6}} \quad \text{or} \quad b = -\sqrt{-\frac{144}{6}}$$

$$b = \frac{12}{\sqrt{6}}i \qquad\qquad b = -\frac{12}{\sqrt{6}}i$$

$$b = \frac{12}{\sqrt{6}}\left(\frac{\sqrt{6}}{\sqrt{6}}\right)i \qquad b = -\frac{12}{\sqrt{6}}\left(\frac{\sqrt{6}}{\sqrt{6}}\right)i$$

$$b = \frac{12\sqrt{6}}{6}i \qquad\qquad b = -\frac{12\sqrt{6}}{6}i$$

$$b = 2i\sqrt{6} \qquad\qquad b = -2i\sqrt{6}$$

$$b = \pm 2i\sqrt{6}$$

31.

$$(x+5)^2 = 9$$

$$\sqrt{(x+5)^2} = \sqrt{9}$$

$$x+5 = \pm 3$$

$$x+5 = 3 \quad \text{or} \quad x+5 = -3$$

$$x = -2 \qquad\qquad x = -8$$

33.

$$(t+4)^2 = 16$$

$$\sqrt{(t+4)^2} = \sqrt{16}$$

$$t+4 = \pm 4$$

$$t+4 = 4 \quad \text{or} \quad t+4 = -4$$

$$t = 0 \qquad\qquad t = -8$$

35.

$$(x+5)^2 = 3$$

$$\sqrt{(x+5)^2} = \sqrt{3}$$

$$x+5 = \sqrt{3} \quad \text{or} \quad x+5 = -\sqrt{3}$$

$$x = -5+\sqrt{3} \qquad x = -5-\sqrt{3}$$

$$x = -5 \pm \sqrt{3}$$

37.

$$(7a-2)^2 = 8$$

$$7a-2 = \sqrt{8} \quad \text{or} \quad 7a-2 = -\sqrt{8}$$

$$7a-2 = \sqrt{4}\sqrt{2} \qquad 7a-2 = -\sqrt{4}\sqrt{2}$$

$$7a-2 = 2\sqrt{2} \qquad 7a-2 = -2\sqrt{2}$$

$$7a = 2+2\sqrt{2} \qquad 7a = 2-2\sqrt{2}$$

$$a = \frac{2+2\sqrt{2}}{7} \qquad a = \frac{2-2\sqrt{2}}{7}$$

$$a = \frac{2 \pm 2\sqrt{2}}{7}$$

39.

$$\left(\frac{24}{2}\right)^2 = (12)^2 = 144$$

$$x^2 + 24x + 144 = (x+12)^2$$

41.

$$\left(\frac{-7}{2}\right)^2 = \frac{49}{4}$$

$$a^2 - 7a + \frac{49}{4} = \left(a - \frac{7}{2}\right)^2$$

43.

$$\frac{1}{2}\left(\frac{2}{3}\right) = \frac{1}{3}$$

$$\left(\frac{1}{3}\right)^2 = \frac{1}{9}$$

$$x^2 + \frac{2}{3}x + \frac{1}{9} = \left(x + \frac{1}{3}\right)^2$$

Section 10.1

45.

$$\frac{1}{2}\left(-\frac{5}{6}\right) = -\frac{5}{12}$$

$$\left(-\frac{5}{12}\right)^2 = \frac{25}{144}$$

$$m^2 - \frac{5}{6}m + \frac{25}{144} = \left(m - \frac{5}{12}\right)^2$$

47.

$$x^2 - 4x - 2 = 0$$

$$x^2 - 4x = 2$$

$$x^2 - 4x + \left(\frac{-4}{2}\right)^2 = 2 + \left(\frac{-4}{2}\right)^2$$

$$x^2 - 4x + (-2)^2 = 2 + (-2)^2$$

$$x^2 - 4x + 4 = 2 + 4$$

$$(x - 2)^2 = 6$$

$$x - 2 = \pm\sqrt{6}$$

$$x = 2 \pm \sqrt{6}$$

$$x \approx 4.45 \quad \text{and} \quad x \approx -0.45$$

49.

$$x^2 - 12x + 1 = 0$$

$$x^2 - 12x = -1$$

$$x^2 - 12x + \left(\frac{-12}{2}\right)^2 = -1 + \left(\frac{-12}{2}\right)^2$$

$$x^2 - 12x + (-6)^2 = -1 + (-6)^2$$

$$x^2 - 12x + 36 = -1 + 36$$

$$(x - 6)^2 = 35$$

$$x - 6 = \pm\sqrt{35}$$

$$x = 6 \pm \sqrt{35}$$

$$x \approx 11.92 \quad \text{and} \quad x \approx 0.08$$

51.

$$t^2 + 20t + 25 = 0$$

$$t^2 + 20t = -25$$

$$t^2 + 20t + \left(\frac{20}{2}\right)^2 = -25 + \left(\frac{20}{2}\right)^2$$

$$t^2 + 20t + (10)^2 = -25 + (10)^2$$

$$t^2 + 20t + 100 = -25 + 100$$

$$(t + 10)^2 = 75$$

$$t + 10 = \pm\sqrt{75}$$

$$t + 10 = \pm\sqrt{25}\sqrt{3}$$

$$t + 10 = \pm 5\sqrt{3}$$

$$t = -10 \pm 5\sqrt{3}$$

$$t \approx -1.34 \quad \text{and} \quad t \approx -18.66$$

53.

$$t^2 + 16t - 16 = 0$$

$$t^2 + 16t = 16$$

$$t^2 + 16t + \left(\frac{16}{2}\right)^2 = 16 + \left(\frac{16}{2}\right)^2$$

$$t^2 + 16t + (8)^2 = 16 + (8)^2$$

$$t^2 + 16t + 64 = 16 + 64$$

$$(t + 8)^2 = 80$$

$$t + 8 = \pm\sqrt{80}$$

$$t + 8 = \pm\sqrt{16}\sqrt{5}$$

$$t + 8 = \pm 4\sqrt{5}$$

$$t = -8 \pm 4\sqrt{5}$$

$$t \approx 0.94 \quad \text{and} \quad t \approx -16.94$$

55.

$$2x^2 - x - 1 = 0$$

$$2x^2 - x = 1$$

$$\frac{2x^2}{2} - \frac{x}{2} = \frac{1}{2}$$

$$x^2 - \frac{1}{2}x = \frac{1}{2}$$

$$x^2 - \frac{1}{2}x + \left(\frac{1}{2} \cdot \frac{-1}{2}\right)^2 = \frac{1}{2} + \left(\frac{1}{2} \cdot \frac{-1}{2}\right)^2$$

$$x^2 - \frac{1}{2}x + \left(\frac{-1}{4}\right)^2 = \frac{1}{2} + \left(\frac{-1}{4}\right)^2$$

$$x^2 - \frac{1}{2}x + \frac{1}{16} = \frac{1}{2} + \frac{1}{16}$$

$$\left(x - \frac{1}{4}\right)^2 = \frac{8}{16} + \frac{1}{16}$$

$$\left(x - \frac{1}{4}\right)^2 = \frac{9}{16}$$

$$x - \frac{1}{4} = \pm\sqrt{\frac{9}{16}}$$

$$x - \frac{1}{4} = \pm\frac{3}{4}$$

$$x = \frac{1}{4} \pm \frac{3}{4}$$

$$x = \frac{1}{4} + \frac{3}{4} \quad \text{or} \quad x = \frac{1}{4} - \frac{3}{4}$$

$$x = \frac{4}{4} \qquad\qquad x = -\frac{2}{4}$$

$$x = 1 \qquad\qquad x = -\frac{1}{2}$$

57.

$$12t^2 - 5t - 3 = 0$$

$$12t^2 - 5t = 3$$

$$\frac{12t^2}{12} - \frac{5t}{12} = \frac{3}{12}$$

$$t^2 - \frac{5}{12}t = \frac{1}{4}$$

$$t^2 - \frac{5}{12}t + \left(\frac{1}{2} \cdot \frac{-5}{12}\right)^2 = \frac{1}{4} + \left(\frac{1}{2} \cdot \frac{-5}{12}\right)^2$$

$$t^2 - \frac{5}{12}t + \left(\frac{-5}{24}\right)^2 = \frac{1}{4} + \left(\frac{-5}{24}\right)^2$$

$$t^2 - \frac{5}{12}t + \frac{25}{576} = \frac{1}{4} + \frac{25}{576}$$

$$\left(t - \frac{5}{24}\right)^2 = \frac{144}{576} + \frac{25}{576}$$

$$\left(t - \frac{5}{24}\right)^2 = \frac{169}{576}$$

$$t - \frac{5}{24} = \pm\sqrt{\frac{169}{576}}$$

$$t - \frac{5}{24} = \pm\frac{13}{24}$$

$$t = \frac{5}{24} \pm \frac{13}{24}$$

$$t = \frac{5}{24} + \frac{13}{24} \quad \text{or} \quad t = \frac{5}{24} - \frac{13}{24}$$

$$t = \frac{18}{24} \qquad\qquad t = -\frac{8}{24}$$

$$t = \frac{3}{4} \qquad\qquad t = -\frac{1}{3}$$

Section 10.1

59.

$$3x^2 - 12x + 1 = 0$$

$$\frac{3x^2}{3} - \frac{12x}{3} + \frac{1}{3} = \frac{0}{3}$$

$$x^2 - 4x + \frac{1}{3} = 0$$

$$x^2 - 4x = -\frac{1}{3}$$

$$x^2 - 4x + \left(\frac{1}{2}\cdot -4\right)^2 = -\frac{1}{3} + \left(\frac{1}{2}\cdot -4\right)^2$$

$$x^2 - 4x + (-2)^2 = -\frac{1}{3} + (-2)^2$$

$$x^2 - 4x + 4 = -\frac{1}{3} + 4$$

$$(x-2)^2 = -\frac{1}{3} + \frac{12}{3}$$

$$(x-2)^2 = \frac{11}{3}$$

$$x - 2 = \pm\sqrt{\frac{11}{3}}$$

$$x - 2 = \pm\frac{\sqrt{11}}{\sqrt{3}}\cdot\frac{\sqrt{3}}{\sqrt{3}}$$

$$x - 2 = \pm\frac{\sqrt{33}}{3}$$

$$x = 2 \pm \frac{\sqrt{33}}{3}$$

$$x = \frac{6}{3} \pm \frac{\sqrt{33}}{3}$$

$$x = \frac{6 \pm \sqrt{33}}{3}$$

$$x \approx 3.91 \quad \text{or} \quad x \approx 0.09$$

61.

$$2x^2 + 5x - 2 = 0$$

$$\frac{2x^2}{2} + \frac{5x}{2} - \frac{2}{2} = \frac{0}{2}$$

$$x^2 + \frac{5}{2}x - 1 = 0$$

$$x^2 + \frac{5}{2}x = 1$$

$$x^2 + \frac{5}{2}x + \left(\frac{1}{2}\cdot\frac{5}{2}\right)^2 = 1 + \left(\frac{1}{2}\cdot\frac{5}{2}\right)^2$$

$$x^2 + \frac{5}{2}x + \left(\frac{5}{4}\right)^2 = 1 + \left(\frac{5}{4}\right)^2$$

$$x^2 + \frac{5}{2}x + \frac{25}{16} = 1 + \frac{25}{16}$$

$$\left(x + \frac{5}{4}\right)^2 = \frac{16}{16} + \frac{25}{16}$$

$$\left(x + \frac{5}{4}\right)^2 = \frac{41}{16}$$

$$x + \frac{5}{4} = \pm\sqrt{\frac{41}{16}}$$

$$x + \frac{5}{4} = \pm\frac{\sqrt{41}}{4}$$

$$x = -\frac{5}{4} \pm \frac{\sqrt{41}}{4}$$

$$x = \frac{-5 \pm \sqrt{41}}{4}$$

$$x \approx 0.35 \quad \text{or} \quad x \approx -2.85$$

63.

$$p^2 + 2p + 2 = 0$$

$$p^2 + 2p = -2$$

$$p^2 + 2p + \left(\frac{2}{2}\right)^2 = -2 + \left(\frac{2}{2}\right)^2$$

$$p^2 + 2p + (1)^2 = -2 + (1)^2$$

$$p^2 + 2p + 1 = -2 + 1$$

$$(p+1)^2 = -1$$

$$p + 1 = \pm\sqrt{-1}$$

$$p + 1 = \pm i$$

$$p = -1 \pm i$$

65.

$$y^2 + 8y + 18 = 0$$
$$y^2 + 8y = -18$$
$$y^2 + 8y + \left(\frac{8}{2}\right)^2 = -18 + \left(\frac{8}{2}\right)^2$$
$$y^2 + 8y + (4)^2 = -18 + (4)^2$$
$$y^2 + 8y + 16 = -18 + 16$$
$$(y + 4)^2 = -2$$
$$y + 4 = \pm\sqrt{-2}$$
$$y + 4 = \pm i\sqrt{2}$$
$$y = -4 \pm i\sqrt{2}$$

67.

$$x^2 + \frac{2}{3}x + 7 = 0$$
$$x^2 + \frac{2}{3}x = -7$$
$$x^2 + \frac{2}{3}x + \left(\frac{1}{2} \cdot \frac{2}{3}\right)^2 = -7 + \left(\frac{1}{2} \cdot \frac{2}{3}\right)^2$$
$$x^2 + \frac{2}{3}x + \left(\frac{1}{3}\right)^2 = -7 + \left(\frac{1}{3}\right)^2$$
$$x^2 + \frac{2}{3}x + \frac{1}{9} = -7 + \frac{1}{9}$$
$$\left(x + \frac{1}{3}\right)^2 = -\frac{63}{9} + \frac{1}{9}$$
$$\left(x + \frac{1}{3}\right)^2 = -\frac{62}{9}$$
$$x + \frac{1}{3} = \pm\sqrt{-\frac{62}{9}}$$
$$x + \frac{1}{3} = \pm\frac{i\sqrt{62}}{3}$$
$$x = -\frac{1}{3} \pm \frac{i\sqrt{62}}{3}$$

69.

$$a^2 - \frac{1}{2}a + 1 = 0$$
$$a^2 - \frac{1}{2}a = -1$$
$$a^2 - \frac{1}{2}a + \left(\frac{-1}{2} \cdot \frac{-1}{2}\right)^2 = -1 + \left(\frac{-1}{2} \cdot \frac{-1}{2}\right)^2$$
$$a^2 - \frac{1}{2}a + \left(-\frac{1}{4}\right)^2 = -1 + \left(-\frac{1}{4}\right)^2$$
$$a^2 - \frac{1}{2}a + \frac{1}{16} = -1 + \frac{1}{16}$$
$$\left(a - \frac{1}{4}\right)^2 = -\frac{16}{16} + \frac{1}{16}$$
$$\left(a - \frac{1}{4}\right)^2 = -\frac{15}{16}$$
$$a - \frac{1}{4} = \pm\sqrt{-\frac{15}{16}}$$
$$a - \frac{1}{4} = \pm\frac{i\sqrt{15}}{4}$$
$$a = \frac{1}{4} \pm \frac{i\sqrt{15}}{4}$$

TRY IT YOURSELF

71.

$$(3x - 1)^2 = 25$$

$$3x - 1 = \sqrt{25} \quad \text{or} \quad 3x - 1 = -\sqrt{25}$$
$$3x - 1 = 5 \qquad\qquad 3x - 1 = -5$$
$$3x = 6 \qquad\qquad\quad 3x = -4$$
$$x = 2 \qquad\qquad\qquad x = -\frac{4}{3}$$

Section 10.1

73.

$$3x^2 - 6x = 1$$

$$\frac{3x^2}{3} - \frac{6x}{3} = \frac{1}{3}$$

$$x^2 - 2x = \frac{1}{3}$$

$$x^2 - 2x + \left(\frac{-2}{2}\right)^2 = \frac{1}{3} + \left(\frac{-2}{2}\right)^2$$

$$x^2 - 2x + 1 = \frac{1}{3} + 1$$

$$x^2 - 2x + 1 = \frac{1}{3} + \frac{3}{3}$$

$$(x-1)^2 = \frac{4}{3}$$

$$x - 1 = \pm\sqrt{\frac{4}{3}}$$

$$x - 1 = \pm\frac{2}{\sqrt{3}}$$

$$x - 1 = \pm\frac{2}{\sqrt{3}} \cdot \frac{\sqrt{3}}{\sqrt{3}}$$

$$x - 1 = \pm\frac{2\sqrt{3}}{3}$$

$$x = 1 \pm \frac{2\sqrt{3}}{3}$$

$$x = \frac{3}{3} \pm \frac{2\sqrt{3}}{3}$$

$$x = \frac{3 \pm 2\sqrt{3}}{3}$$

$$x \approx 2.15 \quad \text{and} \quad x \approx -0.15$$

75.

$$x^2 + 8x + 6 = 0$$

$$x^2 + 8x = -6$$

$$x^2 + 8x + \left(\frac{8}{2}\right)^2 = -6 + \left(\frac{8}{2}\right)^2$$

$$x^2 + 8x + (4)^2 = -6 + (4)^2$$

$$x^2 + 8x + 16 = -6 + 16$$

$$(x+4)^2 = 10$$

$$x + 4 = \pm\sqrt{10}$$

$$x = -4 \pm \sqrt{10}$$

$$x \approx -0.84 \quad \text{and} \quad x \approx -7.16$$

77.

$$6x^2 + 72 = 0$$

$$6x^2 = -72$$

$$\frac{6x^2}{6} = \frac{-72}{6}$$

$$x^2 = -12$$

$$x = \pm\sqrt{-12}$$

$$x = \pm i\sqrt{12}$$

$$x = \pm i\sqrt{4}\sqrt{3}$$

$$x = \pm 2i\sqrt{3}$$

79.

$$x^2 - 2x = 17$$

$$x^2 - 2x + \left(\frac{-2}{2}\right)^2 = 17 + \left(\frac{-2}{2}\right)^2$$

$$x^2 - 2x + (-1)^2 = 17 + (-1)^2$$

$$x^2 - 2x + 1 = 17 + 1$$

$$(x-1)^2 = 18$$

$$x - 1 = \pm\sqrt{18}$$

$$x - 1 = \pm\sqrt{9}\sqrt{2}$$

$$x - 1 = \pm 3\sqrt{2}$$

$$x = 1 \pm 3\sqrt{2}$$

$$x \approx 5.24 \quad \text{and} \quad x \approx -3.24$$

81.

$$m^2 - 7m + 3 = 0$$

$$m^2 - 7m = -3$$

$$m^2 - 7m + \left(\frac{-7}{2}\right)^2 = -3 + \left(\frac{-7}{2}\right)^2$$

$$m^2 - 7m + \frac{49}{4} = -3 + \frac{49}{4}$$

$$m^2 - 7m + \frac{49}{4} = -\frac{12}{4} + \frac{49}{4}$$

$$\left(m - \frac{7}{2}\right)^2 = \frac{37}{4}$$

$$m - \frac{7}{2} = \pm\sqrt{\frac{37}{4}}$$

$$m - \frac{7}{2} = \pm\frac{\sqrt{37}}{2}$$

$$m = \frac{7}{2} \pm \frac{\sqrt{37}}{2}$$

$$m = \frac{7 \pm \sqrt{37}}{2}$$

$$m \approx 6.54 \quad \text{and} \quad m \approx 0.46$$

83.

$$7h^2 = 35$$

$$\frac{7h^2}{7} = \frac{35}{7}$$

$$h^2 = 5$$

$$h = \pm\sqrt{5}$$

$$h \approx 2.24 \quad \text{and} \quad h \approx -2.24$$

85.

$$\frac{7x+1}{5} = -x^2$$

$$\frac{7x}{5} + \frac{1}{5} = -x^2$$

$$x^2 + \frac{7x}{5} = -\frac{1}{5}$$

$$x^2 + \frac{7}{5}x + \left(\frac{1}{2}\cdot\frac{7}{5}\right)^2 = -\frac{1}{5} + \left(\frac{1}{2}\cdot\frac{7}{5}\right)^2$$

$$x^2 + \frac{7}{5}x + \left(\frac{7}{10}\right)^2 = -\frac{1}{5} + \left(\frac{7}{10}\right)^2$$

$$x^2 + \frac{7}{5}x + \frac{49}{100} = -\frac{1}{5} + \frac{49}{100}$$

$$\left(x + \frac{7}{10}\right)^2 = -\frac{20}{100} + \frac{49}{100}$$

$$\left(x + \frac{7}{10}\right)^2 = \frac{29}{100}$$

$$x + \frac{7}{10} = \pm\sqrt{\frac{29}{100}}$$

$$x + \frac{7}{10} = \pm\frac{\sqrt{29}}{10}$$

$$x = -\frac{7}{10} \pm \frac{\sqrt{29}}{10}$$

$$x = \frac{-7 \pm \sqrt{29}}{10}$$

$$x \approx -0.16 \quad \text{and} \quad x \approx -1.24$$

87.

$$t^2 + t + 3 = 0$$

$$t^2 + t = -3$$

$$t^2 + t + \left(\frac{1}{2}\right)^2 = -3 + \left(\frac{1}{2}\right)^2$$

$$t^2 + t + \frac{1}{4} = -3 + \frac{1}{4}$$

$$t^2 + t + \frac{1}{4} = \frac{-12}{4} + \frac{1}{4}$$

$$\left(t + \frac{1}{2}\right)^2 = -\frac{11}{4}$$

$$t + \frac{1}{2} = \pm\sqrt{-\frac{11}{4}}$$

$$t + \frac{1}{2} = \pm\frac{\sqrt{-11}}{2}$$

$$t + \frac{1}{2} = \pm\frac{\sqrt{11}}{2}i$$

$$t = -\frac{1}{2} \pm \frac{\sqrt{11}}{2}i$$

89.

$$(8x + 5)^2 = 24$$

$$(8x + 5) = \pm\sqrt{24}$$

$$8x + 5 = \pm\sqrt{4}\sqrt{6}$$

$$8x + 5 = \pm 2\sqrt{6}$$

$$8x = -5 \pm 2\sqrt{6}$$

$$x = \frac{-5 \pm 2\sqrt{6}}{8}$$

$$x \approx -0.01 \quad \text{and} \quad x \approx -1.24$$

91.

$$r^2 - 6r - 27 = 0$$

$$r^2 - 6r = 27$$

$$r^2 - 6r + \left(\frac{-6}{2}\right)^2 = 27 + \left(\frac{-6}{2}\right)^2$$

$$r^2 - 6r + (-3)^2 = 27 + (-3)^2$$

$$r^2 - 6r + 9 = 27 + 9$$

$$(r - 3)^2 = 36$$

$$r - 3 = \pm\sqrt{36}$$

$$r - 3 = \pm 6$$

$$r = 3 \pm 6$$

$$r = 3 + 6 \quad \text{or} \quad r = 3 - 6$$

$$r = 9 \qquad \qquad r = -3$$

93.

$$4p^2 + 2p + 3 = 0$$

$$4p^2 + 2p = -3$$

$$\frac{4p^2}{4} + \frac{2p}{4} = \frac{-3}{4}$$

$$p^2 + \frac{1}{2}p = -\frac{3}{4}$$

$$p^2 + \frac{1}{2}p + \left(\frac{1}{2} \cdot \frac{1}{2}\right)^2 = -\frac{3}{4} + \left(\frac{1}{2} \cdot \frac{1}{2}\right)^2$$

$$p^2 + \frac{1}{2}p + \left(\frac{1}{4}\right)^2 = -\frac{3}{4} + \left(\frac{1}{4}\right)^2$$

$$p^2 + \frac{1}{2}p + \frac{1}{16} = -\frac{3}{4} + \frac{1}{16}$$

$$\left(p + \frac{1}{4}\right)^2 = -\frac{12}{16} + \frac{1}{16}$$

$$\left(p + \frac{1}{4}\right)^2 = -\frac{11}{16}$$

$$p + \frac{1}{4} = \pm\sqrt{-\frac{11}{16}}$$

$$p + \frac{1}{4} = \pm\frac{\sqrt{-11}}{4}$$

$$p + \frac{1}{4} = \pm\frac{\sqrt{11}}{4}i$$

$$p = -\frac{1}{4} \pm \frac{\sqrt{11}}{4}i$$

LOOK ALIKES...

95. a.

$$x^2 - 24 = 0$$

$$x^2 = 24$$

$$x = \pm\sqrt{24}$$

$$x = \pm\sqrt{4}\sqrt{6}$$

$$x = \pm 2\sqrt{6}$$

$$x \approx \pm 4.90$$

b.

$$x^2 + 24 = 0$$

$$x^2 = -24$$

$$x = \pm\sqrt{-24}$$

$$x = \pm\sqrt{-1}\sqrt{4}\sqrt{6}$$

$$x = \pm 2i\sqrt{6}$$

97. a.

$$2m^2 - 8m = 0$$

$$\frac{2m^2}{2} - \frac{8m}{2} = \frac{0}{2}$$

$$m^2 - 4m = 0$$

$$m^2 - 4m + \left(\frac{-4}{2}\right)^2 = 0 + \left(\frac{-4}{2}\right)^2$$

$$m^2 - 4m + 4 = 0 + 4$$

$$(m-2)^2 = 4$$

$$m - 2 = \pm 2$$

$$m - 2 = 2 \quad \text{and} \quad m - 2 = -2$$

$$m = 4 \quad \text{and} \quad m = 0$$

b.

$$2m^2 - 8m = 1$$

$$\frac{2m^2}{2} - \frac{8m}{2} = \frac{1}{2}$$

$$m^2 - 4m = \frac{1}{2}$$

$$m^2 - 4m + \left(\frac{-4}{2}\right)^2 = \frac{1}{2} + \left(\frac{-4}{2}\right)^2$$

$$m^2 - 4m + 4 = \frac{1}{2} + 4$$

$$(m-2)^2 = \frac{9}{2}$$

$$m - 2 = \pm\frac{3}{\sqrt{2}}$$

$$m - 2 = \pm\frac{3}{\sqrt{2}}\left(\frac{\sqrt{2}}{\sqrt{2}}\right)$$

$$m - 2 = \pm\frac{3\sqrt{2}}{2}$$

$$m = 2 \pm \frac{3\sqrt{2}}{2}$$

$$m = \frac{4}{2} \pm \frac{3\sqrt{2}}{2}$$

$$m = \frac{4 \pm 3\sqrt{2}}{2}$$

$$m \approx 4.12 \quad \text{and} \quad m \approx -0.12$$

99. a.

$$x^2 - 4x + 20 = 0$$

$$x^2 - 4x = -20$$

$$x^2 - 4x + \left(\frac{-4}{2}\right)^2 = -20 + \left(\frac{-4}{2}\right)^2$$

$$x^2 - 4x + 4 = -20 + 4$$

$$(x-2)^2 = -16$$

$$x - 2 = \pm\sqrt{-16}$$

$$x - 2 = \pm 4i$$

$$x = 2 \pm 4i$$

b.

$$x^2 - 4x - 20 = 0$$

$$x^2 - 4x = 20$$

$$x^2 - 4x + \left(\frac{-4}{2}\right)^2 = 20 + \left(\frac{-4}{2}\right)^2$$

$$x^2 - 4x + 4 = 20 + 4$$

$$(x-2)^2 = 24$$

$$x - 2 = \pm\sqrt{24}$$

$$x - 2 = \pm 2\sqrt{6}$$

$$x = 2 \pm 2\sqrt{6}$$

$$x \approx 6.90 \quad \text{and} \quad x \approx -2.90$$

101. a.

$$2r^2 - 4r + 3 = 0$$

$$2r^2 - 4r = -3$$

$$\frac{2r^2}{2} - \frac{4r}{2} = \frac{-3}{2}$$

$$r^2 - 2r = -\frac{3}{2}$$

$$r^2 - 2r + \left(\frac{-2}{2}\right)^2 = -\frac{3}{2} + \left(\frac{-2}{2}\right)^2$$

$$r^2 - 2r + 1 = -\frac{3}{2} + 1$$

$$(r-1)^2 = -\frac{1}{2}$$

$$r - 1 = \pm\sqrt{-\frac{1}{2}}$$

$$r - 1 = \pm i\frac{1}{\sqrt{2}}$$

$$r - 1 = \pm i\frac{1}{\sqrt{2}}\left(\frac{\sqrt{2}}{\sqrt{2}}\right)$$

$$r - 1 = \pm\frac{i\sqrt{2}}{2}$$

$$r = 1 \pm \frac{i\sqrt{2}}{2}$$

$$r = \frac{2}{2} \pm \frac{i\sqrt{2}}{2}$$

$$r = \frac{2 \pm i\sqrt{2}}{2}$$

b.

$$2r^2 - 4r - 3 = 0$$

$$2r^2 - 4r = 3$$

$$\frac{2r^2}{2} - \frac{4r}{2} = \frac{3}{2}$$

$$r^2 - 2r = \frac{3}{2}$$

$$r^2 - 2r + \left(\frac{-2}{2}\right)^2 = \frac{3}{2} + \left(\frac{-2}{2}\right)^2$$

$$r^2 - 2r + 1 = \frac{3}{2} + 1$$

$$(r-1)^2 = \frac{5}{2}$$

$$r - 1 = \pm\sqrt{\frac{5}{2}}$$

$$r - 1 = \pm\sqrt{\frac{5}{2}}\left(\frac{\sqrt{2}}{\sqrt{2}}\right)$$

$$r - 1 = \pm\frac{\sqrt{10}}{2}$$

$$r = 1 \pm \frac{\sqrt{10}}{2}$$

$$r = \frac{2}{2} \pm \frac{\sqrt{10}}{2}$$

$$r = \frac{2 \pm \sqrt{10}}{2}$$

$$r \approx 2.58 \quad \text{and} \quad r \approx -0.58$$

APPLICATIONS

103. MOVIE STUNTS

$$d = 16t^2$$

$$312 = 16t^2$$

$$\frac{312}{16} = \frac{16t^2}{16}$$

$$19.5 = t^2$$

$$t = \sqrt{19.5}$$

$$t = 4.4 \text{ sec}$$

105. **ACCIDENTS**

$$h = s - 16t^2$$
$$5 = 4(12) - 16t^2$$
$$5 = 48 - 16t^2$$
$$-43 = -16t^2$$
$$2.6875 = t^2$$
$$t = \sqrt{2.6875}$$
$$t = 1.6 \text{ sec}$$

107. **AUTOMOBILE ENGINES**

$$V = \pi r^2 h$$
$$47.75 = \pi r^2 (5.25)$$
$$\frac{47.75}{5.25\pi} = \frac{\pi r^2 (5.25)}{5.25\pi}$$
$$\frac{47.75}{16.5} = r^2$$
$$2.89 = r^2$$
$$r = \sqrt{2.89}$$
$$r = 1.70 \text{ in.}$$

109. **PHYSICS**

$$E = mc^2$$
$$\frac{E}{m} = \frac{mc^2}{m}$$
$$\frac{E}{m} = c^2$$
$$c^2 = \frac{E}{m}$$
$$c = \sqrt{\frac{E}{m}}$$
$$c = \frac{\sqrt{E}}{\sqrt{m}}\left(\frac{\sqrt{m}}{\sqrt{m}}\right)$$
$$c = \frac{\sqrt{Em}}{m}$$

WRITING

111. Answers will vary.

113. Answers will vary.

REVIEW

115.

$$\sqrt[3]{40a^3b^6} = \sqrt[3]{8a^3b^6}\sqrt[3]{5}$$
$$= 2ab^2\sqrt[3]{5}$$

117.

$$\sqrt[4]{\frac{16}{625}} = \sqrt[4]{\left(\frac{2}{5}\right)^4}$$
$$= \frac{2}{5}$$

CHALLENGE PROBLEMS

119. Take one-half of the coefficient of x and square it: $\left(\dfrac{\sqrt{3}}{2}\right)^2 = \dfrac{3}{4}$.

SECTION 10.2

VOCABULARY

1. The standard form of a **quadratic** equation is $ax^2 + bx + c = 0$.

CONCEPTS

3. a. $x^2 + 2x + 5 = 0$
   b. $3x^2 + 2x - 1 = 0$

5. a. true
   b. true
   c. false

7. a.
$$\frac{-2 \pm \sqrt{2^2 - 4(1)(-8)}}{2(1)}$$
$$= \frac{-2 \pm \sqrt{4 + 32}}{2}$$
$$= \frac{-2 \pm \sqrt{36}}{2}$$
$$= \frac{-2 \pm 6}{2}$$
$$x = \frac{-2+6}{2} \quad \text{or} \quad x = \frac{-2-6}{2}$$
$$x = \frac{4}{2} \qquad\qquad x = \frac{-8}{2}$$
$$x = 2 \qquad\qquad x = -4$$

b.
$$\frac{-(-1) \pm \sqrt{(-1)^2 - 4(2)(-4)}}{2(2)}$$
$$= \frac{1 \pm \sqrt{1 + 32}}{4}$$
$$= \frac{1 \pm \sqrt{33}}{4}$$

9. a.
$$\frac{3 \pm 6\sqrt{2}}{3} = \frac{\cancel{3}\left(1 \pm 2\sqrt{2}\right)}{\cancel{3}}$$
$$= \frac{1 \pm 2\sqrt{2}}{1}$$
$$= 1 \pm 2\sqrt{2}$$

b.
$$\frac{-12 \pm 4\sqrt{7}}{8} = \frac{\cancel{4}\left(-3 \pm \sqrt{7}\right)}{\cancel{4} \cdot 2}$$
$$= \frac{-3 \pm \sqrt{7}}{2}$$

NOTATION

11. a. The fraction bar wasn't drawn under both parts of the numerator.
    b. A $\pm$ sign wasn't written between b and the radical.

GUIDED PRACTICE

13. $x^2 - 3x + 2 = 0$
$a = 1, b = -3, \text{ and } c = 2$
$$x = \frac{-b \pm \sqrt{b^2 - 4ac}}{2a}$$
$$= \frac{-(-3) \pm \sqrt{(-3)^2 - 4(1)(2)}}{2(1)}$$
$$= \frac{3 \pm \sqrt{9 - 8}}{2}$$
$$= \frac{3 \pm \sqrt{1}}{2}$$
$$= \frac{3 \pm 1}{2}$$
$$x = \frac{3+1}{2} \quad \text{or} \quad x = \frac{3-1}{2}$$
$$x = \frac{4}{2} \qquad\qquad x = \frac{2}{2}$$
$$x = 2 \qquad\qquad x = 1$$

15. $x^2 + 12x = -36$
$x^2 + 12x + 36 = 0$
$a = 1, b = 12,$ and $c = 36$

$$x = \frac{-b \pm \sqrt{b^2 - 4ac}}{2a}$$

$$= \frac{-12 \pm \sqrt{(12)^2 - 4(1)(36)}}{2(1)}$$

$$= \frac{-12 \pm \sqrt{144 - 144}}{2}$$

$$= \frac{-12 \pm \sqrt{0}}{2}$$

$$= \frac{-12 \pm 0}{2}$$

$x = \frac{-12 + 0}{2}$ or $x = \frac{-12 - 0}{2}$

$x = \frac{-12}{2}$ $\qquad x = \frac{-12}{2}$

$x = -6$ $\qquad x = -6$

A repeated solution of –6

17. $2x^2 + x - 3 = 0$
$a = 2, b = 1,$ and $c = -3$

$$x = \frac{-b \pm \sqrt{b^2 - 4ac}}{2a}$$

$$= \frac{-1 \pm \sqrt{(1)^2 - 4(2)(-3)}}{2(2)}$$

$$= \frac{-1 \pm \sqrt{1 - (-24)}}{4}$$

$$= \frac{-1 \pm \sqrt{25}}{4}$$

$$= \frac{-1 \pm 5}{4}$$

$x = \frac{-1 + 5}{4}$ or $x = \frac{-1 - 5}{4}$

$x = \frac{4}{4}$ $\qquad x = \frac{-6}{4}$

$x = 1$ $\qquad x = -\frac{3}{2}$

19. $12t^2 - 5t - 2 = 0$
$a = 12, b = -5,$ and $c = -2$

$$t = \frac{-b \pm \sqrt{b^2 - 4ac}}{2a}$$

$$= \frac{-(-5) \pm \sqrt{(-5)^2 - 4(12)(-2)}}{2(12)}$$

$$= \frac{5 \pm \sqrt{25 + 96}}{24}$$

$$= \frac{5 \pm \sqrt{121}}{24}$$

$$= \frac{5 \pm 11}{24}$$

$t = \frac{5 + 11}{24}$ or $t = \frac{5 - 11}{24}$

$t = \frac{16}{24}$ $\qquad t = \frac{-6}{24}$

$t = \frac{2}{3}$ $\qquad t = -\frac{1}{4}$

21. $x^2 = x + 7$
$x^2 - x - 7 = 0$
$a = 1, b = -1,$ and $c = -7$

$$x = \frac{-b \pm \sqrt{b^2 - 4ac}}{2a}$$

$$= \frac{-(-1) \pm \sqrt{(-1)^2 - 4(1)(-7)}}{2(1)}$$

$$= \frac{1 \pm \sqrt{1 + 28}}{2}$$

$$= \frac{1 \pm \sqrt{29}}{2}$$

$x = \frac{1 + \sqrt{29}}{2}$ or $x = \frac{1 - \sqrt{29}}{2}$

$x \approx 3.19$ $\qquad x \approx -2.19$

Section 10.2

23. $5x^2 + 5x = -1$

    $5x^2 + 5x + 1 = 0$

    $a = 5$, $b = 5$, and $c = 1$

$$x = \frac{-b \pm \sqrt{b^2 - 4ac}}{2a}$$

$$= \frac{-5 \pm \sqrt{(5)^2 - 4(5)(1)}}{2(5)}$$

$$= \frac{-5 \pm \sqrt{25 - 20}}{10}$$

$$x = \frac{-5 \pm \sqrt{5}}{10}$$

$$x = \frac{-5 + \sqrt{5}}{10} \quad \text{or} \quad x = \frac{-5 - \sqrt{5}}{10}$$

$$x \approx -0.28 \qquad\qquad x \approx -0.72$$

25. $3y^2 + 1 = -6y$

    $3y^2 + 6y + 1 = 0$

    $a = 3$, $b = 6$, and $c = 1$

$$y = \frac{-b \pm \sqrt{b^2 - 4ac}}{2a}$$

$$= \frac{-6 \pm \sqrt{(6)^2 - 4(3)(1)}}{2(3)}$$

$$= \frac{-6 \pm \sqrt{36 - 12}}{6}$$

$$= \frac{-6 \pm \sqrt{24}}{6}$$

$$= \frac{-6 \pm \sqrt{4}\sqrt{6}}{6}$$

$$= \frac{-6 \pm 2\sqrt{6}}{6}$$

$$= \frac{\cancel{2}\left(-3 \pm \sqrt{6}\right)}{\cancel{2} \cdot 3}$$

$$y = \frac{-3 \pm \sqrt{6}}{3}$$

$$y = \frac{-3 + \sqrt{6}}{3} \quad \text{or} \quad y = \frac{-3 - \sqrt{6}}{3}$$

$$y \approx -0.18 \qquad\qquad y \approx -1.82$$

27. $4m^2 = 4m + 19$

    $4m^2 - 4m - 19 = 0$

    $a = 4$, $b = -4$, and $c = -19$

$$m = \frac{-b \pm \sqrt{b^2 - 4ac}}{2a}$$

$$= \frac{-(-4) \pm \sqrt{(-4)^2 - 4(4)(-19)}}{2(4)}$$

$$= \frac{4 \pm \sqrt{16 + 304}}{8}$$

$$= \frac{4 \pm \sqrt{320}}{8}$$

$$= \frac{4 \pm \sqrt{64}\sqrt{5}}{8}$$

$$= \frac{4 \pm 8\sqrt{5}}{8}$$

$$= \frac{\cancel{4}\left(1 \pm 2\sqrt{5}\right)}{\cancel{4} \cdot 2}$$

$$m = \frac{1 \pm 2\sqrt{5}}{2}$$

$$m = \frac{1 + 2\sqrt{5}}{2} \quad \text{or} \quad m = \frac{1 - 2\sqrt{5}}{2}$$

$$m \approx 2.74 \qquad\qquad m \approx -1.74$$

29. $2x^2 + x + 1 = 0$

    $a = 2$, $b = 1$, and $c = 1$

$$x = \frac{-b \pm \sqrt{b^2 - 4ac}}{2a}$$

$$= \frac{-1 \pm \sqrt{(1)^2 - 4(2)(1)}}{2(2)}$$

$$= \frac{-1 \pm \sqrt{1 - 8}}{4}$$

$$= \frac{-1 \pm \sqrt{-7}}{4}$$

$$= \frac{-1 \pm i\sqrt{7}}{4}$$

$$x = -\frac{1}{4} \pm \frac{\sqrt{7}}{4}i$$

31. $3x^2 - 2x + 1 = 0$

$a = 3$, $b = -2$, and $c = 1$

$$x = \frac{-b \pm \sqrt{b^2 - 4ac}}{2a}$$

$$= \frac{-(-2) \pm \sqrt{(-2)^2 - 4(3)(1)}}{2(3)}$$

$$= \frac{2 \pm \sqrt{4 - 12}}{6}$$

$$= \frac{2 \pm \sqrt{-8}}{6}$$

$$= \frac{2 \pm \sqrt{-4}\sqrt{2}}{6}$$

$$= \frac{2 \pm 2i\sqrt{2}}{6}$$

$$= \frac{2}{6} \pm \frac{2\sqrt{2}}{6}i$$

$$x = \frac{1}{3} \pm \frac{\sqrt{2}}{3}i$$

33. $x^2 - 2x + 2 = 0$

$a = 1$, $b = -2$, and $c = 2$

$$x = \frac{-b \pm \sqrt{b^2 - 4ac}}{2a}$$

$$= \frac{-(-2) \pm \sqrt{(-2)^2 - 4(1)(2)}}{2(1)}$$

$$= \frac{2 \pm \sqrt{4 - 8}}{2}$$

$$= \frac{2 \pm \sqrt{-4}}{2}$$

$$= \frac{2 \pm 2i}{2}$$

$$= \frac{2}{2} \pm \frac{2}{2}i$$

$$x = 1 \pm i$$

35. $4a^2 + 4a + 5 = 0$

$a = 4$, $b = 4$, and $c = 5$

$$a = \frac{-b \pm \sqrt{b^2 - 4ac}}{2a}$$

$$= \frac{-4 \pm \sqrt{(4)^2 - 4(4)(5)}}{2(4)}$$

$$= \frac{-4 \pm \sqrt{16 - 80}}{8}$$

$$= \frac{-4 \pm \sqrt{-64}}{8}$$

$$= \frac{-4 \pm 8i}{8}$$

$$= -\frac{4}{8} \pm \frac{8}{8}i$$

$$a = -\frac{1}{2} \pm i$$

37. a. $-5x^2 + 9x - 2 = 0$

divide each term by -1

$5x^2 - 9x + 2 = 0$

b. $1.6t^2 + 2.4t - 0.9 = 0$

multiply each term to 10

$16t^2 + 24t - 9 = 0$

39. a. $45x^2 + 30x - 15 = 0$

divide each term by 15

$3x^2 + 2x - 1 = 0$

b. $\frac{1}{3}m^2 - \frac{1}{2}m - \frac{1}{3} = 0$

multiply each term by 6

$2m^2 - 3m - 2 = 0$

TRY IT YOURSELF

41.

$$x^2 - \frac{14}{15}x = \frac{8}{15}$$

$$15\left(x^2 - \frac{14}{15}x\right) = 15\left(\frac{8}{15}\right)$$

$$15x^2 - 14x = 8$$

$$15x^2 - 14x - 8 = 0$$

$a = 15$, $b = -14$, and $c = -8$

$$x = \frac{-b \pm \sqrt{b^2 - 4ac}}{2a}$$

$$= \frac{-(-14) \pm \sqrt{(-14)^2 - 4(15)(-8)}}{2(15)}$$

$$= \frac{14 \pm \sqrt{196 - (-480)}}{30}$$

$$= \frac{14 \pm \sqrt{676}}{30}$$

$$= \frac{14 \pm 26}{30}$$

$$x = \frac{14 + 26}{30} \quad \text{or} \quad x = \frac{14 - 26}{30}$$

$$x = \frac{40}{30} \qquad\qquad x = -\frac{12}{30}$$

$$x = \frac{4}{3} \qquad\qquad x = -\frac{2}{5}$$

43.
$$3x^2 - 4x = -2$$
$$3x^2 - 4x + 2 = 0$$
$a = 3$, $b = -4$, and $c = 2$

$$x = \frac{-b \pm \sqrt{b^2 - 4ac}}{2a}$$

$$= \frac{-(-4) \pm \sqrt{(-4)^2 - 4(3)(2)}}{2(3)}$$

$$= \frac{4 \pm \sqrt{16 - 24}}{6}$$

$$= \frac{4 \pm \sqrt{-8}}{6}$$

$$= \frac{4 \pm \sqrt{-4}\sqrt{2}}{6}$$

$$= \frac{4 \pm 2i\sqrt{2}}{6}$$

$$= \frac{4}{6} \pm \frac{2\sqrt{2}}{6}i$$

$$x = \frac{2}{3} \pm \frac{\sqrt{2}}{3}i$$

45.
$$-16y^2 - 8y + 3 = 0$$
$$0 = 16y^2 + 8y - 3$$
$a = 16$, $b = 8$, $c = -3$

$$y = \frac{-b \pm \sqrt{b^2 - 4ac}}{2a}$$

$$= \frac{-8 \pm \sqrt{(8)^2 - 4(16)(-3)}}{2(16)}$$

$$= \frac{-8 \pm \sqrt{64 - (-192)}}{32}$$

$$= \frac{-8 \pm \sqrt{256}}{32}$$

$$= \frac{-8 \pm 16}{32}$$

$$y = \frac{-8 + 16}{32} \quad \text{or} \quad y = \frac{-8 - 16}{32}$$

$$y = \frac{8}{32} \qquad\qquad y = \frac{-24}{32}$$

$$y = \frac{1}{4} \qquad\qquad y = -\frac{3}{4}$$

47. $2x^2 - 3x - 1 = 0$

$a = 2$, $b = -3$, and $c = -1$

$$x = \frac{-b \pm \sqrt{b^2 - 4ac}}{2a}$$

$$= \frac{-(-3) \pm \sqrt{(-3)^2 - 4(2)(-1)}}{2(2)}$$

$$= \frac{3 \pm \sqrt{9 - (-8)}}{4}$$

$$x = \frac{3 \pm \sqrt{17}}{4}$$

$$x = \frac{3 + \sqrt{17}}{4} \quad \text{or} \quad x = \frac{3 - \sqrt{17}}{4}$$

$$x = 1.78 \qquad\qquad x = -0.28$$

49. $-x^2 + 10x = 18$

$0 = x^2 - 10x + 18$

$a = 1$, $b = -10$, and $c = 18$

$$x = \frac{-b \pm \sqrt{b^2 - 4ac}}{2a}$$

$$= \frac{-(-10) \pm \sqrt{(-10)^2 - 4(1)(18)}}{2(1)}$$

$$= \frac{10 \pm \sqrt{100 - 72}}{2}$$

$$= \frac{10 \pm \sqrt{28}}{2}$$

$$= \frac{10 \pm \sqrt{4}\sqrt{7}}{2}$$

$$= \frac{10 \pm 2\sqrt{7}}{2}$$

$$= \frac{\cancel{2}\left(5 \pm \sqrt{7}\right)}{\cancel{2}}$$

$$x = 5 \pm \sqrt{7}$$

$$x = 5 + \sqrt{7} \quad \text{or} \quad x = 5 - \sqrt{7}$$

$$x = 2.35 \qquad\qquad x = 7.65$$

51. $x(x - 6) = 391$

$x^2 - 6x = 391$

$x^2 - 6x - 391 = 0$

$a = 1$, $b = -6$, and $c = -391$

$$x = \frac{-b \pm \sqrt{b^2 - 4ac}}{2a}$$

$$= \frac{-(-6) \pm \sqrt{(-6)^2 - 4(1)(-391)}}{2(1)}$$

$$= \frac{6 \pm \sqrt{36 - (-1{,}564)}}{2}$$

$$= \frac{6 \pm \sqrt{1{,}600}}{2}$$

$$= \frac{6 \pm 40}{2}$$

$$x = \frac{6 + 40}{2} \quad \text{or} \quad x = \frac{6 - 40}{2}$$

$$x = \frac{46}{2} \qquad\qquad x = \frac{-34}{2}$$

$$x = 23 \qquad\qquad x = -17$$

53. $x^2 + 5x - 5 = 0$

$a = 1$, $b = 5$, and $c = -5$

$$x = \frac{-b \pm \sqrt{b^2 - 4ac}}{2a}$$

$$= \frac{-5 \pm \sqrt{(5)^2 - 4(1)(-5)}}{2(1)}$$

$$= \frac{-5 \pm \sqrt{25 + 20}}{2}$$

$$= \frac{-5 \pm \sqrt{45}}{2}$$

$$= \frac{-5 \pm \sqrt{9}\sqrt{5}}{2}$$

$$x = \frac{-5 \pm 3\sqrt{5}}{2}$$

$$x = \frac{-5 + 3\sqrt{5}}{2} \quad \text{or} \quad x = \frac{-5 - 3\sqrt{5}}{2}$$

$$x = 0.85 \qquad\qquad x = -5.85$$

Section 10.2

55. $9h^2 - 6h + 7 = 0$

$a = 9$, $b = -6$, and $c = 7$

$$h = \frac{-b \pm \sqrt{b^2 - 4ac}}{2a}$$

$$= \frac{-(-6) \pm \sqrt{(-6)^2 - 4(9)(7)}}{2(9)}$$

$$= \frac{6 \pm \sqrt{36 - 252}}{18}$$

$$= \frac{6 \pm \sqrt{-216}}{18}$$

$$= \frac{6 \pm i\sqrt{36}\sqrt{6}}{18}$$

$$h = \frac{6 \pm 6\sqrt{6}i}{18}$$

$$= \frac{6}{18} \pm \frac{6\sqrt{6}}{18}i$$

$$= \frac{1}{3} \pm \frac{\sqrt{6}}{3}i$$

57.

$$50x^2 + 30x - 10 = 0$$

$$\frac{50x^2}{10} + \frac{30x}{10} - \frac{10}{10} = 0$$

$$5x^2 + 3x - 1 = 0$$

$a = 5$, $b = 3$, and $c = -1$

$$x = \frac{-b \pm \sqrt{b^2 - 4ac}}{2a}$$

$$= \frac{-3 \pm \sqrt{(3)^2 - 4(5)(-1)}}{2(5)}$$

$$= \frac{-3 \pm \sqrt{9 - (-20)}}{10}$$

$$x = \frac{-3 \pm \sqrt{29}}{10}$$

$$x = \frac{-3 - \sqrt{29}}{10} \quad \text{or} \quad x = \frac{-3 - \sqrt{29}}{10}$$

$$x = 0.24 \qquad\qquad x = -0.84$$

59.

$$0.6x^2 + 0.03 - 0.4x = 0$$

$$0.6x^2 - 0.4x + 0.03 = 0$$

$$100(0.6x^2 - 0.4x + 0.03) = 100(0)$$

$$60x^2 - 40x + 3 = 0$$

$a = 60$, $b = -40$, and $c = 3$

$$x = \frac{-b \pm \sqrt{b^2 - 4ac}}{2a}$$

$$= \frac{-(-40) \pm \sqrt{(-40)^2 - 4(60)(3)}}{2(60)}$$

$$= \frac{40 \pm \sqrt{1,600 - 720}}{120}$$

$$= \frac{40 \pm \sqrt{880}}{120}$$

$$= \frac{40 \pm \sqrt{16}\sqrt{55}}{120}$$

$$= \frac{40 \pm 4\sqrt{55}}{120}$$

$$= \frac{\cancel{4}(10 \pm \sqrt{55})}{\cancel{4} \cdot 30}$$

$$x = \frac{10 \pm \sqrt{55}}{30}$$

$$x = \frac{10 + \sqrt{55}}{30} \quad \text{or} \quad x = \frac{10 - \sqrt{55}}{30}$$

$$x = 0.58 \qquad\qquad x = 0.09$$

61.

$$\frac{1}{8}x^2 - \frac{1}{2}x + 1 = 0$$

$$8\left(\frac{1}{8}x^2 - \frac{1}{2}x + 1\right) = 8(0)$$

$$x^2 - 4x + 8 = 0$$

$a = 1, b = -4,$ and $c = 8$

$$x = \frac{-b \pm \sqrt{b^2 - 4ac}}{2a}$$

$$= \frac{-(-4) \pm \sqrt{(-4)^2 - 4(1)(8)}}{2(1)}$$

$$= \frac{4 \pm \sqrt{16 - 32}}{2}$$

$$= \frac{4 \pm \sqrt{-16}}{2}$$

$$= \frac{4 \pm 4i}{2}$$

$$= \frac{4}{2} \pm \frac{4}{2}i$$

$$x = 2 \pm 2i$$

63.

$$\frac{a^2}{10} - \frac{3a}{5} + \frac{7}{5} = 0$$

$$10\left(\frac{a^2}{10} - \frac{3a}{5} + \frac{7}{5}\right) = 10(0)$$

$$a^2 - 6a + 14 = 0$$

$a = 1, b = -6,$ and $c = 14$

$$a = \frac{-b \pm \sqrt{b^2 - 4ac}}{2a}$$

$$= \frac{-(-6) \pm \sqrt{(-6)^2 - 4(1)(14)}}{2(1)}$$

$$= \frac{6 \pm \sqrt{36 - 56}}{2}$$

$$= \frac{6 \pm \sqrt{-20}}{2}$$

$$= \frac{6 \pm \sqrt{-4}\sqrt{5}}{2}$$

$$= \frac{6 \pm 2i\sqrt{5}}{2}$$

$$= \frac{6}{2} \pm \frac{2\sqrt{5}}{2}i$$

$$a = 3 \pm i\sqrt{5}$$

65.

$$\frac{x^2}{2} + \frac{5}{2}x = -1$$

$$2\left(\frac{x^2}{2} + \frac{5}{2}x\right) = 2(-1)$$

$$x^2 + 5x = -2$$

$$x^2 + 5x + 2 = 0$$

$a = 1, b = 5,$ and $c = 2$

$$x = \frac{-b \pm \sqrt{b^2 - 4ac}}{2a}$$

$$= \frac{-5 \pm \sqrt{(5)^2 - 4(1)(2)}}{2(1)}$$

$$= \frac{-5 \pm \sqrt{25 - 8}}{2}$$

$$x = \frac{-5 \pm \sqrt{17}}{2}$$

$$x = \frac{-5 + \sqrt{17}}{2} \quad \text{or} \quad x = \frac{-5 - \sqrt{17}}{2}$$

$$x = -0.44 \qquad\qquad x = -4.56$$

67.

$$900x^2 - 8{,}100x = 1{,}800$$

$$\frac{900x^2}{900} - \frac{8{,}100x}{900} = \frac{1{,}800}{900}$$

$$x^2 - 9x = 2$$

$$x^2 - 9x - 2 = 0$$

$a = 1, b = -9,$ and $c = -2$

$$x = \frac{-b \pm \sqrt{b^2 - 4ac}}{2a}$$

$$= \frac{-(-9) \pm \sqrt{(-9)^2 - 4(1)(-2)}}{2(1)}$$

$$= \frac{9 \pm \sqrt{81 - (-8)}}{2}$$

$$x = \frac{9 \pm \sqrt{89}}{2}$$

$$x = \frac{9 + \sqrt{89}}{2} \quad \text{or} \quad x = \frac{9 - \sqrt{89}}{2}$$

$$x = 9.22 \qquad\qquad x = -0.22$$

Section 10.2

69.

$$\frac{1}{4}x^2 - \frac{1}{6}x - \frac{1}{6} = 0$$

$$12\left(\frac{1}{4}x^2 - \frac{1}{6}x - \frac{1}{6}\right) = 12(0)$$

$$3x^2 - 2x - 2 = 0$$

$a = 3, b = -2,$ and $c = -2$

$$x = \frac{-b \pm \sqrt{b^2 - 4ac}}{2a}$$

$$= \frac{-(-2) \pm \sqrt{(-2)^2 - 4(3)(-2)}}{2(3)}$$

$$= \frac{2 \pm \sqrt{4 - (-24)}}{6}$$

$$= \frac{2 \pm \sqrt{28}}{6}$$

$$= \frac{2 \pm \sqrt{4}\sqrt{7}}{6}$$

$$= \frac{2 \pm 2\sqrt{7}}{6}$$

$$= \frac{1 \pm \sqrt{7}}{3}$$

$x = \dfrac{1 + \sqrt{7}}{3}$ or $x = \dfrac{1 - \sqrt{7}}{3}$

$x = 1.22$ $\qquad\qquad x = -0.55$

71.

$$f(x) = f(x)$$

$$0.7x^2 - 3.5x = 25$$

$$0.7x^2 - 3.5x - 25 = 0$$

$$10(0.7x^2 - 3.5x - 25) = 10(0)$$

$$7x^2 - 35x - 250 = 0$$

$a = 7, b = -35,$ and $c = -250$

$$x = \frac{-b \pm \sqrt{b^2 - 4ac}}{2a}$$

$$= \frac{-(-35) \pm \sqrt{(-35)^2 - 4(7)(-250)}}{2(7)}$$

$$= \frac{35 \pm \sqrt{1,225 - (-7,000)}}{14}$$

$$= \frac{35 \pm \sqrt{8,225}}{14}$$

$$= \frac{35 \pm \sqrt{25}\sqrt{329}}{14}$$

$$= \frac{35 \pm 5\sqrt{329}}{14}$$

$x = \dfrac{35 + 5\sqrt{329}}{14}$ or $x = \dfrac{35 - 5\sqrt{329}}{14}$

$x = 8.98$ $\qquad\qquad x = -3.98$

73. a. $a^2 + 4a - 7 = 0$
$a = 1,\ b = 4,\ \text{and}\ c = -7$

$$a = \frac{-b \pm \sqrt{b^2 - 4ac}}{2a}$$

$$= \frac{-4 \pm \sqrt{(4)^2 - 4(1)(-7)}}{2(1)}$$

$$= \frac{-4 \pm \sqrt{16 + 28}}{2}$$

$$= \frac{-4 \pm \sqrt{44}}{2}$$

$$= \frac{-4 \pm \sqrt{4}\sqrt{11}}{2}$$

$$= \frac{-4 \pm 2\sqrt{11}}{2}$$

$$= \frac{2\left(-2 \pm \sqrt{11}\right)}{2}$$

$a = -2 \pm \sqrt{11}$

$a = -2 + \sqrt{11}$ or $a = -2 - \sqrt{11}$

$a = 1.32$ $\qquad\qquad a = -5.32$

b. $a^2 - 4a - 7 = 0$
$a = 1,\ b = -4,\ \text{and}\ c = -7$

$$a = \frac{-b \pm \sqrt{b^2 - 4ac}}{2a}$$

$$= \frac{-(-4) \pm \sqrt{(-4)^2 - 4(1)(-7)}}{2(1)}$$

$$= \frac{4 \pm \sqrt{16 + 28}}{2}$$

$$= \frac{4 \pm \sqrt{44}}{2}$$

$$= \frac{4 \pm \sqrt{4}\sqrt{11}}{2}$$

$$= \frac{4 \pm 2\sqrt{11}}{2}$$

$$= \frac{2\left(2 \pm \sqrt{11}\right)}{2}$$

$a = 2 \pm \sqrt{11}$

$a = 2 + \sqrt{11}$ or $a = 2 - \sqrt{11}$

$a = 5.32$ $\qquad\qquad a = -1.32$

75. a. $(x + 2)(x - 4) = 16$
$x^2 - 4x + 2x - 8 = 16$
$x^2 - 2x - 8 = 16$
$x^2 - 2x - 24 = 0$
$a = 1,\ b = -2,\ \text{and}\ c = -24$

$$x = \frac{-b \pm \sqrt{b^2 - 4ac}}{2a}$$

$$= \frac{-(-2) \pm \sqrt{(-2)^2 - 4(1)(-24)}}{2(1)}$$

$$= \frac{2 \pm \sqrt{4 + 96}}{2}$$

$$= \frac{2 \pm \sqrt{100}}{2}$$

$$x = \frac{2 \pm 10}{2}$$

$x = \dfrac{2 + 10}{2}$ or $x = \dfrac{2 - 10}{2}$

$x = \dfrac{12}{2}$ $\qquad\qquad x = \dfrac{-8}{2}$

$x = 6$ $\qquad\qquad\quad x = -4$

b. $(x + 2)(x - 4) = -16$
$x^2 - 4x + 2x - 8 = -16$
$x^2 - 2x - 8 = -16$
$x^2 - 2x + 8 = 0$
$a = 1,\ b = -2,\ \text{and}\ c = 8$

$$x = \frac{-b \pm \sqrt{b^2 - 4ac}}{2a}$$

$$= \frac{-(-2) \pm \sqrt{(-2)^2 - 4(1)(8)}}{2(1)}$$

$$= \frac{2 \pm \sqrt{4 - 32}}{2}$$

$$= \frac{2 \pm \sqrt{-28}}{2}$$

$$= \frac{2 \pm i\sqrt{4}\sqrt{7}}{2}$$

$$= \frac{2 \pm 2i\sqrt{7}}{2}$$

$$= \frac{2\left(1 \pm i\sqrt{7}\right)}{2}$$

$x = 1 \pm i\sqrt{7}$

Section 10.2

77. a. $x^2 - 42x + 441 = 0$
$a = 1, b = -42,$ and $c = 441$

$$x = \frac{-b \pm \sqrt{b^2 - 4ac}}{2a}$$

$$= \frac{-(-42) \pm \sqrt{(-42)^2 - 4(1)(441)}}{2(1)}$$

$$= \frac{42 \pm \sqrt{1,764 - 1,764}}{2}$$

$$= \frac{42 \pm \sqrt{0}}{2}$$

$$= \frac{42 \pm 0}{2}$$

$x = \frac{42 + 0}{2}$ or $x = \frac{42 - 0}{2}$

$x = \frac{42}{2}$ $x = \frac{42}{2}$

$x = 21$ $x = 21$

A repeated solution of 21

b. $x^2 + 42x + 441 = 0$
$a = 1, b = 42,$ and $c = 441$

$$x = \frac{-b \pm \sqrt{b^2 - 4ac}}{2a}$$

$$= \frac{-42 \pm \sqrt{(42)^2 - 4(1)(441)}}{2(1)}$$

$$= \frac{-42 \pm \sqrt{1,764 - 1,764}}{2}$$

$$= \frac{-42 \pm \sqrt{0}}{2}$$

$$= \frac{-42 \pm 0}{2}$$

$x = \frac{-42 + 0}{2}$ or $x = \frac{-42 - 0}{2}$

$x = \frac{-42}{2}$ $x = \frac{-42}{2}$

$x = -21$ $x = -21$

A repeated solution of -21

APPLICATIONS

79. CROSSWALKS
Let x = length of First Avenue crosswalk, and $x + 7$ = length of the Main Street crosswalk. Use the Pythagorean Theorem to find x.

$$x^2 + (x + 7)^2 = 97^2$$

$$x^2 + x^2 + 14x + 49 = 97^2$$

$$2x^2 + 14x + 49 = 97^2$$

$$2x^2 + 14x - 9,360 = 0$$

$$\frac{2x^2}{2} + \frac{14x}{2} - \frac{9,360}{2} = 0$$

$$x^2 + 7x - 4,680 = 0$$

$a = 1, b = 7,$ and $c = -4,680$

$$x = \frac{-b \pm \sqrt{b^2 - 4ac}}{2a}$$

$$= \frac{-7 \pm \sqrt{(7)^2 - 4(1)(-4,680)}}{2(1)}$$

$$= \frac{-7 \pm \sqrt{49 + 18,720}}{2}$$

$$= \frac{-7 \pm \sqrt{18,769}}{2}$$

$$= \frac{-7 \pm 137}{2}$$

$x = \frac{-7 + 137}{2}$ or $x = \frac{-7 - 137}{2}$

$= \frac{130}{2}$ $= \frac{-144}{2}$

$= 65$ $= -72$

Since the length cannot be negative, the length of First Avenue crosswalk must be 65 feet and the length of Main Street crosswalk is 65 + 7 = 72 feet. The total distance walked this way is 65 + 72 = 137 feet. The distance the shopper saves would be 137 – 97 = 40 feet.

81. RIGHT TRIANGLES

The hypotenuse is 2.5.
Let the shorter leg $= x$.
The longer leg $= x + 1.7$.
Use the Pythagorean Theorem.

$$a^2 + b^2 = c^2$$
$$(x)^2 + (x+1.7)^2 = (2.5)^2$$
$$x^2 + x^2 + 1.7x + 1.7x + 2.89 = 6.25$$
$$2x^2 + 3.4x + 2.89 = 6.25$$
$$2x^2 + 3.4x - 3.36 = 0$$
$$100(2x^2 + 3.4x - 3.36) = 100(0)$$
$$200x^2 + 340x - 336 = 0$$

$a = 200$, $b = 340$, and $c = -336$

$$x = \frac{-b \pm \sqrt{b^2 - 4ac}}{2a}$$
$$= \frac{-340 \pm \sqrt{(340)^2 - 4(200)(-336)}}{2(200)}$$
$$= \frac{-340 \pm \sqrt{115,600 - (-268,800)}}{400}$$
$$= \frac{-340 \pm \sqrt{384,400}}{400}$$
$$= \frac{-340 \pm 620}{400}$$

$$x = \frac{-340 + 620}{400} \quad \text{or} \quad x = \frac{-340 - 620}{400}$$
$$x = \frac{280}{400} \qquad\qquad x = \frac{-960}{400}$$
$$x = 0.7 \qquad\qquad x = -2.4$$
$$x + 1.7 = 2.4$$

The shorter leg is 0.7 units, the longer leg is 2.4 units, and the hypotenuse is 2.5 units.

83. IMAX SCREENS

Let the width $= x$ and length $= x + 20$.
The area is 11,349.

$$(\text{width})(\text{length}) = \text{area}$$
$$x(x + 20) = 11,349$$
$$x^2 + 20x - 11,349 = 0$$

$a = 1$, $b = 20$, and $c = -11,349$

$$x = \frac{-b \pm \sqrt{b^2 - 4ac}}{2a}$$
$$= \frac{-20 \pm \sqrt{(20)^2 - 4(1)(-11,349)}}{2(1)}$$
$$= \frac{-20 \pm \sqrt{400 - (-45,396)}}{2}$$
$$= \frac{-20 \pm \sqrt{45,796}}{2}$$
$$= \frac{-20 \pm 214}{2}$$

$$x = \frac{-20 + 214}{2} \quad \text{or} \quad x = \frac{-20 - 214}{2}$$
$$x = \frac{194}{2} \qquad\qquad x = \frac{-234}{2}$$
$$x = 97 \qquad\qquad x = -117$$
$$x + 20 = 117$$

The dimensions are 97 ft by 117 ft.

85. PARKS

Let the width $= x$ and length $= 5x$.

Perimeter $= 2x + 2(5x) = 2x + 10x = 12x$.

Area $= x(5x) = 5x^2$

Perimeter $=$ Area $+ 4.75$

$$12x = 5x^2 + 4.75$$
$$0 = 5x^2 - 12x + 4.75$$

$a = 5$, $b = -12$, and $c = 4.75$

$$x = \frac{-b \pm \sqrt{b^2 - 4ac}}{2a}$$

$$= \frac{-(-12) \pm \sqrt{(-12)^2 - 4(5)(4.75)}}{2(5)}$$

$$= \frac{12 \pm \sqrt{144 - 95}}{10}$$

$$= \frac{12 \pm \sqrt{49}}{10}$$

$$= \frac{12 \pm 7}{10}$$

$x = \dfrac{12 + 7}{10}$ or $x = \dfrac{12 - 7}{10}$

$x = \dfrac{19}{10}$ $x = \dfrac{5}{10}$

$x = 1.9$ $x = 0.5$

$$5x = 5(0.5)$$
$$= 2.5$$

$x = 1.9$ is not possible because it states that the width is less than 1 mile. The width must be 0.5 miles and the length must be 2.5 miles.

87. POLYGONS

$$275 = \frac{n(n-3)}{2}$$

$$2 \cdot 275 = 2 \cdot \frac{n(n-3)}{2}$$

$$550 = n(n-3)$$

$$550 = n^2 - 3n$$

$$0 = n^2 - 3n - 550$$

$a = 1$, $b = -3$, $c = -550$

$$x = \frac{-b \pm \sqrt{b^2 - 4ac}}{2a}$$

$$= \frac{-(-3) \pm \sqrt{(-3)^2 - 4(1)(-550)}}{2(1)}$$

$$= \frac{3 \pm \sqrt{9 - (-2,200)}}{2}$$

$$= \frac{3 \pm \sqrt{2,209}}{2}$$

$$= \frac{3 \pm 47}{2}$$

$x = \dfrac{3 + 47}{2}$ or $x = \dfrac{3 - 47}{2}$

$x = \dfrac{50}{2}$ $x = \dfrac{-44}{2}$

$x = 25$ $x = -22$

The polygon has 25 sides.

89. DANCES

Let x = number of increases in ticket price.
New price: $4 + 0.10x$
Number tickets sold: $300 - 5x$
New price · # tickets sold = new receipts

$$(4 + 0.10x)(300 - 5x) = 1,248$$
$$1,200 - 20x + 30x - 0.5x^2 = 1,248$$
$$1,200 + 10x - 0.5x^2 = 1,248$$
$$0 = 0.5x^2 - 10x + 48$$
$$10(0) = 10(0.5x^2 - 10x + 48)$$
$$0 = 5x^2 - 100x + 480$$

$a = 5$, $b = -100$, $c = 480$

$$x = \frac{-b \pm \sqrt{b^2 - 4ac}}{2a}$$
$$= \frac{-(-100) \pm \sqrt{(-100)^2 - 4(5)(480)}}{2(5)}$$
$$= \frac{100 \pm \sqrt{10,000 - 9,600}}{10}$$
$$= \frac{100 \pm \sqrt{400}}{10}$$
$$= \frac{100 \pm 20}{10}$$

$x = \dfrac{100 + 20}{10}$ or $x = \dfrac{100 - 20}{10}$

$x = \dfrac{120}{10}$ $\qquad$ $x = \dfrac{80}{10}$

$x = 12$ $\qquad\qquad$ $x = 8$

$4 + 0.10x = 4 + 0.10(12)$ $\quad$ $4 + 0.10x = 4 + 0.10(8)$
$\qquad\qquad = 4 + 1.20$ $\qquad\qquad\qquad = 4 + 0.80$
$\qquad\qquad = \$5.20$ $\qquad\qquad\qquad\quad = \$4.80$

When the ticket prices are $5.20 or $4.80, the receipts will be $1,248.

91. MAGAZINE SALES

Let x = number of new subscribers
New price: $20 + 0.01x$
Number subscribers: $3,000 + x$
New price · # subscribers = total profit

$$(20 + 0.01x)(3,000 + x) = 120,000$$
$$60,000 + 20x + 30x + 0.01x^2 = 120,000$$
$$60,000 + 50x + 0.01x^2 = 120,000$$
$$0.01x^2 + 50x - 60,000 = 0$$
$$100(0.01x^2 + 50x - 60,000) = 100(0)$$
$$x^2 + 5,000x - 6,000,000 = 0$$

$a = 1$, $b = 5,000$, $c = -6,000,000$

$$x = \frac{-b \pm \sqrt{b^2 - 4ac}}{2a}$$
$$= \frac{-5,000 \pm \sqrt{(5,000)^2 - 4(1)(-6,000,000)}}{2(1)}$$
$$= \frac{-5,000 \pm \sqrt{25,000,000 - (-24,000,000)}}{2}$$
$$= \frac{-5,000 \pm \sqrt{49,000,000}}{2}$$
$$= \frac{-5,000 \pm 7,000}{2}$$

$x = \dfrac{-5,000 + 7,000}{2}$ or $x = \dfrac{-5,000 - 7,000}{2}$

$x = \dfrac{2,000}{2}$ $\qquad$ $x = \dfrac{-12,000}{2}$

$x = 1,000$ $\qquad\quad$ $x = \cancel{-6,000}$

$3,000 + x = 3,000 + 1,000$ $\quad$ There cannot be a
$\qquad\qquad = 4,000$ $\qquad\qquad$ number of subscribers.

4,000 subscribers will bring a profit of $120,000.

Section 10.2

93. PATROL OFFICER

$$f(t) = 75,000$$

$$-24t^2 + 1,534t + 72,065 = 75,000$$

$$-24t^2 + 1,534t - 2,935 = 0$$

$$24t^2 - 1,534t + 2,935 = 0$$

$$a = 24, \ b = -1,534, \ c = 2,935$$

$$t = \frac{-b \pm \sqrt{b^2 - 4ac}}{2a}$$

$$= \frac{-(-1,534) \pm \sqrt{(-1,534)^2 - 4(24)(2,935)}}{2(24)}$$

$$= \frac{1,534 \pm \sqrt{2,353,156 - 281,760}}{48}$$

$$= \frac{1,534 \pm \sqrt{2,071,396}}{48}$$

$$\approx \frac{1,534 \pm 1,439}{48}$$

$$t \approx \frac{1,534 + 1,439}{48} \quad \text{or} \quad t \approx \frac{1,534 - 1,439}{48}$$

$$t \approx 62 \qquad\qquad\qquad t \approx 2$$

$t = 62$ would represent the year 2062 (2000 + 62 = 2062) which hasn't occurred yet.

$t = 2$ would represent the year 2002 (2002 + 2 = 2002). Thus, the model indicates that there were 75,000 female officers in 2002.

95. PICTURE FRAMING

Let x = the width of the matting. Then the total length of the picture plus the mat would be $(2x + 5)$ in. since there is the same width of matting on both sides of the picture, and the total width of the picture plus the mat would be $(2x + 4)$ in.

The area of the mat would be the total area minus the area of the picture, which is 4(5) or 20 in.^{2}

$$\text{area of mat} = \text{area of picture}$$

$$\text{total area - area of picture} = \text{area of picture}$$

$$(2x+5)(2x+4) - 20 = 20$$

$$4x^2 + 8x + 10x + 20 - 20 = 20$$

$$4x^2 + 18x = 20$$

$$4x^2 + 18x - 20 = 0$$

$$a = 4, \ b = 18, \ c = -20$$

$$x = \frac{-b \pm \sqrt{b^2 - 4ac}}{2a}$$

$$= \frac{-18 \pm \sqrt{(18)^2 - 4(4)(-20)}}{2(4)}$$

$$= \frac{-18 \pm \sqrt{324 + 320}}{8}$$

$$= \frac{-18 \pm \sqrt{644}}{8}$$

$$= \frac{-18 \pm 25.38}{8}$$

$$x = \frac{-18 + 25.38}{8} \quad \text{or} \quad x = \frac{-18 - 25.38}{8}$$

$$x \approx 0.92 \qquad\qquad x \approx \cancel{-5.42}$$

The mat is approximately 0.92 inches wide.

97. DIMENSIONS OF A RECTANGLE
Let the width = x ft. Then the length would be $(x + 4)$ ft..

$$(\text{width})(\text{length}) = \text{Area}$$
$$x(x+4) = 20$$
$$x^2 + 4x = 20$$
$$x^2 + 4x - 20 = 0$$

$a = 1, b = 4, c = -20$

$$x = \frac{-b \pm \sqrt{b^2 - 4ac}}{2a}$$

$$= \frac{-4 \pm \sqrt{(4)^2 - 4(1)(-20)}}{2(1)}$$

$$= \frac{-4 \pm \sqrt{16 + 80}}{2}$$

$$= \frac{-4 \pm \sqrt{96}}{2}$$

$$= \frac{-4 \pm 9.8}{2}$$

$$x = \frac{-4 + 9.8}{2} \quad \text{or} \quad x = \frac{-4 - 9.8}{2}$$

$$x \approx 2.9 \qquad\qquad x \approx \cancel{-6.9}$$

The width is 2.9 ft.
The length is 2.9 + 4 or 6.9 ft.

WRITING

99. Answers will vary.

REVIEW

101. $\sqrt{n} = n^{1/2}$

103. $\sqrt[4]{3b} = (3b)^{1/4}$

105. $t^{1/3} = \sqrt[3]{t}$

107. $(3t)^{1/4} = \sqrt[4]{3t}$

CHALLENGE PROBLEMS

109. $\quad x^2 + 2\sqrt{2}x - 6 = 0$

$a = 1, b = 2\sqrt{2}, c = -6$

$$x = \frac{-b \pm \sqrt{b^2 - 4ac}}{2a}$$

$$= \frac{-2\sqrt{2} \pm \sqrt{(2\sqrt{2})^2 - 4(1)(-6)}}{2(1)}$$

$$= \frac{-2\sqrt{2} \pm \sqrt{8 - (-24)}}{2}$$

$$= \frac{-2\sqrt{2} \pm \sqrt{32}}{2}$$

$$= \frac{-2\sqrt{2} \pm \sqrt{16}\sqrt{2}}{2}$$

$$= \frac{-2\sqrt{2} \pm 4\sqrt{2}}{2}$$

$$x = \frac{-2\sqrt{2} + 4\sqrt{2}}{2} \quad \text{or} \quad x = \frac{-2\sqrt{2} - 4\sqrt{2}}{2}$$

$$x = \frac{2\sqrt{2}}{2} \qquad\qquad x = \frac{-6\sqrt{2}}{2}$$

$$x = \sqrt{2} \qquad\qquad x = -3\sqrt{2}$$

111. $\quad x^2 - 3ix - 2 = 0$

$a = 1, b = -3i, c = -2$

$$x = \frac{-b \pm \sqrt{b^2 - 4ac}}{2a}$$

$$= \frac{-(-3i) \pm \sqrt{(-3i)^2 - 4(1)(-2)}}{2(1)}$$

$$= \frac{3i \pm \sqrt{9i^2 - (-8)}}{2}$$

$$= \frac{3i \pm \sqrt{-9 + 8}}{2}$$

$$= \frac{3i \pm \sqrt{-1}}{2}$$

$$= \frac{3i \pm i}{2}$$

$$x = \frac{3i + i}{2} \quad \text{or} \quad x = \frac{3i - i}{2}$$

$$x = \frac{4i}{2} \qquad\qquad x = \frac{2i}{2}$$

$$x = 2i \qquad\qquad x = i$$

SECTION 10.3

VOCABULARY

1. For the quadratic equation $ax^2 + bx + c = 0$, the **discriminant** is $b^2 - 4ac$.

CONCEPTS

3. If $b^2 - 4ac < 0$, the solutions of the equation are two different imaginary numbers that are complex **conjugates**.

5. If $b^2 - 4ac$ is a perfect square, the solutions of the equation are two different **rational** numbers.

7. a. $y = x^2$

   b. $y = \sqrt{x}$

   c. $y = x^{1/3}$

   d. $y = \dfrac{1}{x}$

   e. $y = x + 1$

NOTATION

9.
$$b^2 - \boxed{4ac} = \boxed{5}^2 - 4(1)\left(\boxed{6}\right)$$
$$= 25 - \boxed{24}$$
$$= 1$$

Since a, b, and c are rational numbers and the value of the discriminant is a perfect square, the solutions are two different **rational** numbers.

GUIDED PRACTICE

11.
$$4x^2 - 4x + 1 = 0$$
$$b^2 - 4ac = (-4)^2 - 4(4)(1)$$
$$= 16 - 16$$
$$= 0$$
one repeated rational-number solution

13.
$$5x^2 + x + 2 = 0$$
$$b^2 - 4ac = (1)^2 - 4(5)(2)$$
$$= 1 - 40$$
$$= -39$$
two imaginary -number solutions (complex conjugates)

15.
$$2x^2 = 4x - 1$$
$$2x^2 - 4x + 1 = 0$$
$$b^2 - 4ac = (-4)^2 - 4(2)(1)$$
$$= 16 - 8$$
$$= 8$$
two different irrational-number solutions

17.
$$x(2x - 3) = 20$$
$$2x^2 - 3x = 20$$
$$2x^2 - 3x - 20 = 0$$
$$b^2 - 4ac = (-3)^2 - 4(2)(-20)$$
$$= 9 - (-160)$$
$$= 169 \text{ (perfect square)}$$
two different rational-number solutions

19.
$$3x^2 - 10 = 0$$
$$b^2 - 4ac = 0^2 - 4(3)(-10)$$
$$= 0 + 120$$
$$= 120$$
two different irrational-number solutions

21.
$$x^2 - \frac{14}{15}x = \frac{8}{15}$$
$$x^2 - \frac{14}{15}x - \frac{8}{15} = 0$$
$$b^2 - 4ac = \left(-\frac{14}{15}\right)^2 - 4(1)\left(-\frac{8}{15}\right)$$
$$= \frac{196}{225} + \frac{32}{15}$$
$$= \frac{676}{225}$$
two different rational-number solutions

23. Let $y = x^2$.

$$x^4 - 17x^2 + 16 = 0$$
$$(x^2)^2 - 17x^2 + 16 = 0$$
$$y^2 - 17y + 16 = 0$$
$$(y - 16)(y - 1) = 0$$
$$y - 16 = 0 \quad \text{or} \quad y - 1 = 0$$
$$y = 16 \qquad y = 1$$
$$x^2 = 16 \qquad x^2 = 1$$
$$x = \pm\sqrt{16} \qquad x = \pm\sqrt{1}$$
$$x = \pm 4 \qquad x = \pm 1$$
$$x = 4, \ -4, \ 1, \ -1$$

25. Let $y = x^2$.

$$x^4 + 5x^2 - 36 = 0$$
$$(x^2)^2 + 5x^2 - 36 = 0$$
$$y^2 + 5y - 36 = 0$$
$$(y + 9)(y - 4) = 0$$
$$y + 9 = 0 \quad \text{or} \quad y - 4 = 0$$
$$y = -9 \qquad y = 4$$
$$x^2 = -9 \qquad x^2 = 4$$
$$x = \pm\sqrt{-9} \qquad x = \pm\sqrt{2}$$
$$x = \pm 3i \qquad x = \pm 2$$
$$x = 3i, \ -3i, \ 2, \ -2$$

27. Let $y = \sqrt{x}$.

$$x - 13\sqrt{x} + 40 = 0$$
$$(\sqrt{x})^2 - 13\sqrt{x} + 40 = 0$$
$$y^2 - 13y + 40 = 0$$
$$(y - 8)(y - 5) = 0$$
$$y - 8 = 0 \quad \text{or} \quad y - 5 = 0$$
$$y = 8 \qquad y = 5$$
$$\sqrt{x} = 8 \qquad \sqrt{x} = 5$$
$$(\sqrt{x})^2 = (8)^2 \quad (\sqrt{x})^2 = (5)^2$$
$$x = 64 \qquad x = 25$$
$$x = 64, \ 25$$

29. Let $y = \sqrt{x}$.

$$2x + \sqrt{x} - 3 = 0$$
$$2(\sqrt{x})^2 + \sqrt{x} - 3 = 0$$
$$2y^2 + y - 3 = 0$$
$$(2y + 3)(y - 1) = 0$$
$$2y + 3 = 0 \quad \text{or} \quad y - 1 = 0$$
$$y = -\frac{3}{2} \qquad y = 1$$
$$\sqrt{x} = -\frac{3}{2} \qquad \sqrt{x} = 1$$
$$(\sqrt{x})^2 = \left(-\frac{3}{2}\right)^2 \ (\sqrt{x})^2 = (1)^2$$
$$x = \frac{9}{4} \qquad x = 1$$

$\dfrac{9}{4}$ is extraneous;

$$x = 1$$

31. Let $y = a^{1/3}$.

$$a^{2/3} - 2a^{1/3} = 3$$
$$a^{2/3} - 2a^{1/3} - 3 = 0$$
$$(a^{1/3})^2 - 2a^{1/3} - 3 = 0$$
$$y^2 - 2y - 3 = 0$$
$$(y + 1)(y - 3) = 0$$
$$y + 1 = 0 \quad \text{or} \quad y - 3 = 0$$
$$y = -1 \qquad y = 3$$
$$a^{1/3} = -1 \qquad a^{1/3} = 3$$
$$(a^{1/3})^3 = (-1)^3 \quad (a^{1/3})^3 = (3)^3$$
$$a = -1 \qquad a = 27$$

33. Let $y = x^{1/3}$.

$$x^{2/3} + 2x^{1/3} - 8 = 0$$
$$(x^{1/3})^2 + 2x^{1/3} - 8 = 0$$
$$y^2 + 2y - 8 = 0$$
$$(y - 2)(y + 4) = 0$$
$$y - 2 = 0 \quad \text{or} \quad y + 4 = 0$$
$$y = 2 \qquad y = -4$$
$$x^{1/3} = 2 \qquad x^{1/3} = -4$$
$$(x^{1/3})^3 = (2)^3 \quad (x^{1/3})^3 = (-4)^3$$
$$x = 8 \qquad x = -64$$
$$x = 8, \ -64$$

- 683 -

35. Let $y = c + 1$.

$$(c+1)^2 - 4(c+1) + 3 = 0$$

$$y^2 - 4y + 3 = 0$$

$$(y-3)(y-1) = 0$$

$y - 3 = 0$	$y - 1 = 0$
$y = 3$	$y = 1$
$c + 1 = 3$	$c + 1 = 1$
$c + 1 - 1 = 3 - 1$	$c + 1 - 1 = 1 - 1$
$c = 2$	$c = 0$

37. Let $y = 2x + 1$.

$$2(2x+1)^2 - 7(2x+1) + 6 = 0$$

$$2y^2 - 7y + 6 = 0$$

$$(2y-3)(y-2) = 0$$

$$2y - 3 = 0 \quad \text{or} \quad y - 2 = 0$$

$y = \dfrac{3}{2}$	$y = 2$
$2x + 1 = \dfrac{3}{2}$	$2x + 1 = 2$
$2x = \dfrac{1}{2}$	$2x = 1$
$x = \dfrac{1}{4}$	$x = \dfrac{1}{2}$

39. Let $y = \dfrac{1}{m}$.

$$m^{-2} + m^{-1} - 6 = 0$$

$$\left(\frac{1}{m}\right)^2 + \left(\frac{1}{m}\right) - 6 = 0$$

$$y^2 + y - 6 = 0$$

$$(y+3)(y-2) = 0$$

$$y + 3 = 0 \quad \text{or} \quad y - 2 = 0$$

$y = -3$	$y = 2$
$\dfrac{1}{m} = -3$	$\dfrac{1}{m} = 2$
$m\left(\dfrac{1}{m}\right) = m(-3)$	$m\left(\dfrac{1}{m}\right) = m(2)$
$1 = -3m$	$1 = 2m$
$-\dfrac{1}{3} = m$	$\dfrac{1}{2} = m$

$$m = -\frac{1}{3}, \ \frac{1}{2}$$

41. Let $y = \dfrac{1}{x}$.

$$8x^{-2} - 10x^{-1} - 3 = 0$$

$$8\left(\frac{1}{x}\right)^2 - 10\left(\frac{1}{x}\right) - 3 = 0$$

$$8y^2 - 10y - 3 = 0$$

$$(2y-3)(4y+1) = 0$$

$$2y - 3 = 0 \quad \text{or} \quad 4y + 1 = 0$$

$y = \dfrac{3}{2}$	$y = -\dfrac{1}{4}$
$\dfrac{1}{x} = \dfrac{3}{2}$	$\dfrac{1}{x} = -\dfrac{1}{4}$
$2 = 3x$	$4 = -x$
$\dfrac{2}{3} = x$	$-4 = x$

$$x = \frac{2}{3}, \ -4$$

43.

$$1 - \frac{5}{x} = \frac{10}{x^2}$$

$$x^2\left(1 - \frac{5}{x}\right) = x^2\left(\frac{10}{x^2}\right)$$

$$x^2 - 5x = 10$$

$$x^2 - 5x - 10 = 0$$

$$x = \frac{-b \pm \sqrt{b^2 - 4ac}}{2a}$$

$$= \frac{-(-5) \pm \sqrt{(-5)^2 - 4(1)(-10)}}{2(1)}$$

$$= \frac{5 \pm \sqrt{25 + 40}}{2}$$

$$x = \frac{5 \pm \sqrt{65}}{2}$$

45.

$$\frac{1}{2}+\frac{1}{b}=\frac{1}{b-7}$$

$$2b(b-7)\left(\frac{1}{2}+\frac{1}{b}\right)=2b(b-7)\left(\frac{1}{b-7}\right)$$

$$b(b-7)+2(b-7)=2b$$

$$b^2-7b+2b-14=2b$$

$$b^2-5b-14=2b$$

$$b^2-7b-14=0$$

$$b=\frac{-b\pm\sqrt{b^2-4ac}}{2a}$$

$$=\frac{-(-7)\pm\sqrt{(-7)^2-4(1)(-14)}}{2(1)}$$

$$=\frac{7\pm\sqrt{49+56}}{2}$$

$$b=\frac{7\pm\sqrt{105}}{2}$$

TRY IT YOURSELF

47. Let $y=\sqrt{x}$.

$$2x-\sqrt{x}=3$$

$$2x-\sqrt{x}-3=0$$

$$2\left(\sqrt{x}\right)^2-\sqrt{x}-3=0$$

$$2y^2-y-3=0$$

$$(2y-3)(y+1)=0$$

$$2y-3=0 \quad \text{or} \quad y+1=0$$

$$y=\frac{3}{2} \qquad\qquad y=-1$$

$$\sqrt{x}=\frac{3}{2} \qquad\qquad \sqrt{x}=-1$$

$$\left(\sqrt{x}\right)^2=\left(\frac{3}{2}\right)^2 \quad \left(\sqrt{x}\right)^2=(-1)^2$$

$$x=\frac{9}{4} \qquad\qquad x=\not{1}$$

1 is extraneous

$$x=\frac{9}{4}$$

49. Let $y=\dfrac{1}{x}$.

$$x^{-2}+2x^{-1}-3=0$$

$$\left(\frac{1}{x}\right)^2+2\left(\frac{1}{x}\right)-3=0$$

$$y^2+2y-3=0$$

$$(y+3)(y-1)=0$$

$$y+3=0 \quad \text{or} \quad y-1=0$$

$$y=-3 \qquad\qquad y=1$$

$$\frac{1}{x}=-3 \qquad\qquad \frac{1}{x}=1$$

$$1=-3x \qquad\qquad 1=x$$

$$-\frac{1}{3}=x$$

$$x=-\frac{1}{3},\ 1$$

51. Let $y=x^2$.

$$x^4+19x^2+18=0$$

$$\left(x^2\right)^2+19x^2+18=0$$

$$y^2+19y+18=0$$

$$(y+18)(y+1)=0$$

$$y+18=0 \quad \text{or} \quad y+1=0$$

$$y=-18 \qquad\qquad y=-1$$

$$x^2=-18 \qquad\qquad x^2=-1$$

$$x=\pm\sqrt{-18} \qquad x=\pm\sqrt{-1}$$

$$x=\pm3i\sqrt{2} \qquad\quad x=\pm i$$

$$x=3i\sqrt{2},\ -3i\sqrt{2},\ i,\ -i$$

53. Let $y = k - 7$.

$$(k-7)^2 + 6(k-7) + 10 = 0$$
$$y^2 + 6y + 10 = 0$$
$$y = \frac{-b \pm \sqrt{b^2 - 4ac}}{2a}$$
$$y = \frac{-6 \pm \sqrt{(6)^2 - 4(1)(10)}}{2(1)}$$
$$y = \frac{-6 \pm \sqrt{36 - 40}}{2}$$
$$y = \frac{-6 \pm \sqrt{-4}}{2}$$
$$y = \frac{-6 \pm 2i}{2}$$
$$y = \frac{2(-3 \pm i)}{2}$$
$$y = -3 \pm i$$
$$k - 7 = -3 \pm i$$
$$k - 7 + 7 = -3 \pm i + 7$$
$$k = 4 \pm i$$

55.

$$\frac{2}{x-1} + \frac{1}{x+1} = 3$$
$$(x-1)(x+1)\left(\frac{2}{x-1} + \frac{1}{x+1}\right) = 3(x-1)(x+1)$$
$$2(x+1) + (x-1) = 3(x^2-1)$$
$$2x + 2 + x - 1 = 3x^2 - 3$$
$$3x + 1 = 3x^2 - 3$$
$$0 = 3x^2 - 3x - 4$$
$$x = \frac{-b \pm \sqrt{b^2 - 4ac}}{2a}$$
$$x = \frac{-(-3) \pm \sqrt{(-3)^2 - 4(3)(-4)}}{2(3)}$$
$$x = \frac{3 \pm \sqrt{9 - (-48)}}{6}$$
$$x = \frac{3 \pm \sqrt{57}}{6}$$

57. Let $y = x^{1/2}$.

$$x - 6x^{1/2} = -8$$
$$x - 6x^{1/2} + 8 = 0$$
$$\left(x^{1/2}\right)^2 - 6x^{1/2} + 8 = 0$$
$$y^2 - 6y + 8 = 0$$
$$(y-2)(y-4) = 0$$
$$y - 2 = 0 \quad \text{or} \quad y - 4 = 0$$
$$y = 2 \qquad\qquad y = 4$$
$$x^{1/2} = 2 \qquad\qquad x^{1/2} = 4$$
$$\left(x^{1/2}\right)^2 = (2)^2 \quad \left(x^{1/2}\right)^2 = (4)^2$$
$$x = 4 \qquad\qquad x = 16$$

59. Let $x = y^2 - 9$.

$$\left(y^2-9\right)^2 + 2\left(y^2-9\right) - 99 = 0$$
$$x^2 + 2x - 99 = 0$$
$$(x+11)(x-9) = 0$$
$$x + 11 = 0 \quad \text{or} \quad x - 9 = 0$$
$$x = -11 \qquad\qquad x = 9$$
$$y^2 - 9 = -11 \quad y^2 - 9 = 9$$
$$y^2 = -2 \qquad\qquad y^2 = 18$$
$$y = \pm\sqrt{-2} \qquad y = \pm\sqrt{18}$$
$$y = \pm i\sqrt{2} \qquad y = \pm 3\sqrt{2}$$
$$y = i\sqrt{2}, \ -i\sqrt{2}, \ 3\sqrt{2}, \ -3\sqrt{2}$$

61. Let $y = \dfrac{1}{x^2}$.

$$x^{-4} - 2x^{-2} + 1 = 0$$
$$\frac{1}{x^4} - 2\left(\frac{1}{x^2}\right) + 1 = 0$$
$$\left(\frac{1}{x^2}\right)^2 - 2\left(\frac{1}{x^2}\right) + 1 = 0$$
$$y^2 - 2y + 1 = 0$$
$$(y-1)(y-1) = 0$$
$$y - 1 = 0 \quad \text{or} \quad y - 1 = 0$$
$$y = 1 \qquad\qquad y = 1$$
$$\frac{1}{x^2} = \frac{1}{1} \qquad\qquad \frac{1}{x^2} = \frac{1}{1}$$
$$1 = x^2 \qquad\qquad 1 = x^2$$
$$x = \pm\sqrt{1} \qquad x = \pm\sqrt{1}$$
$$x = \pm 1 \qquad\qquad x = \pm 1$$
$$x = 1, \ -1, \ 1, \ -1$$

63. Let $y = t^2$.

$$t^4 + 3t^2 = 28$$
$$t^4 + 3t^2 - 28 = 0$$
$$\left(t^2\right)^2 + 3t^2 - 28 = 0$$
$$y^2 + 3y - 28 = 0$$
$$(y-4)(y+7) = 0$$

$y - 4 = 0 \quad$ or $\quad y + 7 = 0$

$\begin{array}{ll} y = 4 & y = -7 \\ t^2 = 4 & t^2 = -7 \\ t = \pm\sqrt{4} & t = \pm\sqrt{-7} \\ t = \pm 2 & t = \pm i\sqrt{7} \end{array}$

$$t = 2, \ -2, \ i\sqrt{7}, \ -i\sqrt{7}$$

65. Let $y = x^{1/5}$.

$$2x^{2/5} - 5x^{1/5} = -3$$
$$2x^{2/5} - 5x^{1/5} + 3 = 0$$
$$2\left(x^{1/5}\right)^2 - 5x^{1/5} + 3 = 0$$
$$2y^2 - 5y + 3 = 0$$
$$(2y - 3)(y - 1) = 0$$

$2y - 3 = 0 \quad$ or $\quad y - 1 = 0$

$\begin{array}{ll} y = \dfrac{3}{2} & y = 1 \\[2mm] x^{1/5} = \dfrac{3}{2} & x^{1/5} = 1 \\[2mm] \left(x^{1/5}\right)^5 = \left(\dfrac{3}{2}\right)^5 & \left(x^{1/5}\right)^5 = (1)^5 \\[2mm] x = \dfrac{243}{32} & x = 1 \end{array}$

67. Let $y = \dfrac{3m+2}{m}$.

$$9\left(\frac{3m+2}{m}\right)^2 - 30\left(\frac{3m+2}{m}\right) + 25 = 0$$
$$9y^2 - 30y + 25 = 0$$
$$(3y - 5)(3y - 5) = 0$$

$3y - 5 = 0 \quad$ or $\quad 3y - 5 = 0$

$\begin{array}{ll} y = \dfrac{5}{3} & y = \dfrac{5}{3} \\[2mm] \dfrac{3m+2}{m} = \dfrac{5}{3} & \dfrac{3m+2}{m} = \dfrac{5}{3} \\[2mm] 3(3m+2) = 5m & 3(3m+2) = 5m \\ 9m + 6 = 5m & 9m + 6 = 5m \\ 6 = -4m & 6 = -4m \\ -\dfrac{3}{2} = m & -\dfrac{3}{2} = m \end{array}$

$$m = -\frac{3}{2}, \ -\frac{3}{2}$$

69.

$$\frac{3}{a-1} = 1 - \frac{2}{a}$$
$$a(a-1)\left(\frac{3}{a-1}\right) = a(a-1)\left(1 - \frac{2}{a}\right)$$
$$3a = a(a-1) - 2(a-1)$$
$$3a = a^2 - a - 2a + 2$$
$$3a = a^2 - 3a + 2$$
$$0 = a^2 - 6a + 2$$
$$a = \frac{-b \pm \sqrt{b^2 - 4ac}}{2a}$$
$$a = \frac{-(-6) \pm \sqrt{(-6)^2 - 4(1)(2)}}{2(1)}$$
$$= \frac{6 \pm \sqrt{36 - 8}}{2}$$
$$= \frac{6 \pm \sqrt{28}}{2}$$
$$= \frac{6 \pm \sqrt{4}\sqrt{7}}{2}$$
$$= \frac{6 \pm 2\sqrt{7}}{2}$$
$$= \frac{2\left(3 \pm \sqrt{7}\right)}{2}$$
$$a = 3 \pm \sqrt{7}$$

Section 10.3

71. Let $y = 8 - \sqrt{a}$.

$$\left(8-\sqrt{a}\right)^2 + 6\left(8-\sqrt{a}\right) - 7 = 0$$
$$y^2 + 6y - 7 = 0$$
$$(y+7)(y-1) = 0$$
$$y + 7 = 0 \quad \text{or} \quad y - 1 = 0$$
$$y = -7 \qquad y = 1$$
$$8 - \sqrt{a} = -7 \quad 8 - \sqrt{a} = 1$$
$$-\sqrt{a} = -15 \quad -\sqrt{a} = -7$$
$$\sqrt{a} = 15 \qquad \sqrt{a} = 7$$
$$\left(\sqrt{a}\right)^2 = (15)^2 \quad \left(\sqrt{a}\right)^2 = (7)^2$$
$$a = 225 \qquad a = 49$$
$$a = 225, \ 49$$

73.

$$x + \frac{2}{x-2} = 0$$
$$(x-2)\left(x + \frac{2}{x-2}\right) = (x-2)0$$
$$x(x-2) + 2 = 0$$
$$x^2 - 2x + 2 = 0$$
$$x = \frac{-b \pm \sqrt{b^2 - 4ac}}{2a}$$
$$x = \frac{-(-2) \pm \sqrt{(-2)^2 - 4(1)(2)}}{2(1)}$$
$$x = \frac{2 \pm \sqrt{4-8}}{2}$$
$$x = \frac{2 \pm \sqrt{-4}}{2}$$
$$x = \frac{2 \pm 2i}{2}$$
$$x = \frac{2}{2} \pm \frac{2}{2}i$$
$$x = 1 \pm i$$

75. Let $y = \sqrt{x}$.

$$3x + 5\sqrt{x} + 2 = 0$$
$$3\left(\sqrt{x}\right)^2 + 5\sqrt{x} + 2 = 0$$
$$3y^2 + 5y + 2 = 0$$
$$(3y+2)(y+1) = 0$$
$$3y + 2 = 0 \quad \text{or} \quad y + 1 = 0$$
$$y = -\frac{2}{3} \qquad y = -1$$
$$\sqrt{x} = -\frac{2}{3} \qquad \sqrt{x} = -1$$
$$\left(\sqrt{x}\right)^2 = \left(-\frac{2}{3}\right)^2 \quad \sqrt{x} = (-1)^2$$
$$x = \cancel{\frac{4}{9}} \qquad x = \cancel{1}$$

$\dfrac{4}{9}$ and 1 are extraneous

No Solution

77. Let $y = x^2$.

$$x^4 - 6x^2 + 5 = 0$$
$$\left(x^2\right)^2 - 6x^2 + 5 = 0$$
$$y^2 - 6y + 5 = 0$$
$$(y-5)(y-1) = 0$$
$$y - 5 = 0 \quad \text{or} \quad y - 1 = 0$$
$$y = 5 \qquad y = 1$$
$$x^2 = 5 \qquad x^2 = 1$$
$$x = \pm\sqrt{5} \qquad x = \pm\sqrt{1}$$
$$x = \pm\sqrt{5} \qquad x = \pm 1$$
$$x = \sqrt{5}, \ -\sqrt{5}, \ 1, \ -1$$

79. Let $y = \dfrac{1}{t+1}$.

$$8(t+1)^{-2} - 30(t+1)^{-1} + 7 = 0$$

$$8\left(\frac{1}{t+1}\right)^2 - 30\left(\frac{1}{t+1}\right) + 7 = 0$$

$$8y^2 - 30y + 7 = 0$$

$$(4y-1)(2y-7) = 0$$

$$4y-1=0 \quad \text{or} \quad 2y-7=0$$

$$y = \frac{1}{4} \qquad\qquad y = \frac{7}{2}$$

$$\frac{1}{t+1} = \frac{1}{4} \qquad \frac{1}{t+1} = \frac{7}{2}$$

$$4 = t+1 \qquad 7(t+1) = 2$$

$$3 = t \qquad\qquad 7t+7 = 2$$

$$t = 3 \qquad\qquad t = -\frac{5}{7}$$

$$t = 3,\ -\frac{5}{7}$$

81.

$$\frac{1}{x+2} + \frac{24}{x+3} = 13$$

$$(x+2)(x+3)\left(\frac{1}{x+2} + \frac{24}{x+3}\right) = (x+2)(x+3)(13)$$

$$x+3 + 24(x+2) = 13(x+2)(x+3)$$

$$x+3 + 24x + 48 = 13(x^2 + 5x + 6)$$

$$25x + 51 = 13x^2 + 65x + 78$$

$$0 = 13x^2 + 40x + 27$$

$$0 = (13x+27)(x+1)$$

$$13x + 27 = 0 \quad \text{or} \quad x+1 = 0$$

$$x = -\frac{27}{13} \qquad x = -1$$

83. a. Let $x = y^{1/3}$.

$$y^{2/3} + y^{1/3} - 20 = 0$$

$$\left(y^{1/3}\right)^2 + y^{1/3} - 20 = 0$$

$$x^2 + x - 20 = 0$$

$$(x-4)(x+5) = 0$$

$$x-4=0 \quad \text{or} \quad x+5=0$$

$$x = 4 \qquad\qquad x = -5$$

$$y^{1/3} = 4 \qquad\qquad y^{1/3} = -5$$

$$\left(y^{1/3}\right)^3 = (4)^3 \qquad \left(y^{1/3}\right)^3 = (-5)^3$$

$$y = 64 \qquad\qquad y = -125$$

b. Let $x = y^{-1}$.

$$y^{-2} + y^{-1} - 20 = 0$$

$$\left(y^{-1}\right)^2 + y^{-1} - 20 = 0$$

$$x^2 + x - 20 = 0$$

$$(x-4)(x+5) = 0$$

$$x-4=0 \quad \text{or} \quad x+5=0$$

$$x = 4 \qquad\qquad x = -5$$

$$y^{-1} = 4 \qquad\qquad y^{-1} = -5$$

$$\left(y^{-1}\right)^{-1} = (4)^{-1} \qquad \left(y^{-1}\right)^{-1} = (-5)^{-1}$$

$$y = \frac{1}{4} \qquad\qquad y = -\frac{1}{5}$$

85. a.

$$\frac{1}{x} = \frac{2x}{x+1}$$

$$x(x+1)\left(\frac{1}{x}\right) = x(x+1)\left(\frac{2x}{x+1}\right)$$

$$(x+1)(1) = x(2x)$$

$$x+1 = 2x^2$$

$$0 = 2x^2 - x - 1$$

$$0 = (2x+1)(x-1)$$

$$2x+1 = 0 \quad \text{or} \quad x-1 = 0$$

$$2x = -1 \qquad\qquad x = 1$$

$$x = -\frac{1}{2}$$

b.

$$\frac{1}{x} + \frac{2x}{x+1} = 1$$

$$x(x+1)\left(\frac{1}{x} + \frac{2x}{x+1}\right) = x(x+1)(1)$$

$$(x+1)(1) + x(2x) = x(x+1)$$

$$x+1 + 2x^2 = x^2 + x$$

$$x^2 + 1 = 0$$

$$x^2 = -1$$

$$x = \pm\sqrt{-1}$$

$$x = \pm i$$

APPLICATIONS

87. CROWD CONTROL

Let x = the number of minutes it takes to empty the parking lot.

West exit's work in 1 minute = $\dfrac{1}{x}$

East exit's work in 1 minute = $\dfrac{1}{x+40}$

Amount of work together in 1 minute = $\dfrac{1}{60}$

$$\frac{1}{x}+\frac{1}{x+40}=\frac{1}{60}$$

$$60x(x+40)\left(\frac{1}{x}+\frac{1}{x+40}\right)=60x(x+40)\left(\frac{1}{60}\right)$$

$$60(x+40)+60x=x(x+40)$$

$$60x+2{,}400+60x=x^2+40x$$

$$120x+2{,}400=x^2+40x$$

$$0=x^2-80x-2{,}400$$

$$x=\frac{-b\pm\sqrt{b^2-4ac}}{2a}$$

$$x=\frac{-(-80)\pm\sqrt{(-80)^2-4(1)(-2400)}}{2(1)}$$

$$x=\frac{80\pm\sqrt{6{,}400-(-9{,}600)}}{2}$$

$$x=\frac{80\pm\sqrt{16{,}000}}{2}$$

$$x=\frac{80\pm40\sqrt{10}}{2}$$

$$x=40\pm20\sqrt{10}$$

$$x=40+20\sqrt{10}\ \text{or}\ x=40-20\sqrt{10}$$

$$x=103\qquad\qquad x=-23$$

It takes 103 minutes to clear the parking lot using only the west exit.

89. ASSEMBLY LINES

Let r = rate.

distance = (rate)(time)

Original Time:
$$300=rt$$
$$\frac{300}{r}=t$$

Faster Time:
$$300=(r+5)(t-3)$$
$$\frac{300}{r+5}=t-3$$
$$\frac{300}{r+5}+3=t$$

Substitute $\dfrac{300}{r}=t$ into the last equation and solve for r.

$$\frac{300}{r+5}+3=\frac{300}{r}$$

$$r(r+5)\left(\frac{300}{r+5}+3\right)=r(r+5)\left(\frac{300}{r}\right)$$

$$300r+3r(r+5)=300(r+5)$$

$$300r+3r^2+15r=300r+1{,}500$$

$$3r^2+15r-1{,}500=0$$

$$\frac{3r^2}{3}+\frac{15r}{3}-\frac{1{,}500}{3}=\frac{0}{3}$$

$$r^2+5r-500=0$$

$$(r+25)(r-20)=0$$

$$r+25=0\quad\text{or}\quad r-20=0$$
$$r=-25\qquad\qquad r=20$$

Since the rate cannot be negative, r = 20 feet per second.

91. ARCHITECTURE

$$\frac{\ell}{w} = \frac{w}{\ell - w}$$

$$\frac{\ell}{20} = \frac{20}{\ell - 20}$$

$$\ell(\ell - 20) = 20(20)$$

$$\ell^2 - 20\ell = 400$$

$$\ell^2 - 20\ell - 400 = 0$$

$$\ell = \frac{-b \pm \sqrt{b^2 - 4ac}}{2a}$$

$$\ell = \frac{-(-20) \pm \sqrt{(-20)^2 - 4(1)(-400)}}{2(1)}$$

$$\ell = \frac{20 \pm \sqrt{400 - (-1600)}}{2}$$

$$\ell = \frac{20 \pm \sqrt{2,000}}{2}$$

$$\ell = \frac{20 \pm 20\sqrt{5}}{2}$$

$$\ell = \frac{20}{2} \pm \frac{20}{2}\sqrt{5}$$

$$\ell = 10 \pm 10\sqrt{5}$$

$$\ell = 10 + 10\sqrt{5} \quad \text{or} \quad \ell = 10 - 10\sqrt{5}$$

$$\ell = 32.4 \text{ ft.} \qquad \ell = -12.4$$

The length should be 32.4 ft.

WRITING

93. Answers will vary.

REVIEW

95. $x = 3$

97. $m = \dfrac{2}{3}$ and $b = (0, 0)$

$$y = mx + b$$

$$y = \frac{2}{3}x + 0$$

$$y = \frac{2}{3}x$$

CHALLENGE PROBLEMS

99. Let $y = x^2$.

$$x^4 - 3x^2 - 2 = 0$$

$$(x^2)^2 - 3x^2 - 2 = 0$$

$$y^2 - 3y - 2 = 0$$

$$a = 1, b = -3, c = -2$$

$$y = \frac{-b \pm \sqrt{b^2 - 4ac}}{2a}$$

$$y = \frac{-(-3) \pm \sqrt{(-3)^2 - 4(1)(-2)}}{2(1)}$$

$$y = \frac{3 \pm \sqrt{9 - (-8)}}{2}$$

$$y = \frac{3 \pm \sqrt{17}}{2}$$

$$y = \frac{3 + \sqrt{17}}{2} \quad \text{or} \quad y = \frac{3 - \sqrt{17}}{2}$$

$$x^2 = \frac{3 + \sqrt{17}}{2} \qquad\qquad x^2 = \frac{3 - \sqrt{17}}{2}$$

$$x = \pm\sqrt{\frac{3 + \sqrt{17}}{2}} \qquad x = \pm\sqrt{\frac{3 - \sqrt{17}}{2}}$$

$$x = \pm\frac{\sqrt{3 + \sqrt{17}}}{\sqrt{2}} \cdot \frac{\sqrt{2}}{\sqrt{2}} \qquad x = \pm\frac{\sqrt{3 - \sqrt{17}}}{\sqrt{2}} \cdot \frac{\sqrt{2}}{\sqrt{2}}$$

$$x = \pm\frac{\sqrt{6 + 2\sqrt{17}}}{2} \qquad x = \pm\frac{\sqrt{6 - 2\sqrt{17}}}{2}$$

Since $x = \pm\dfrac{\sqrt{6 - 2\sqrt{17}}}{2}$ is not real, the only

solutions are $x = \pm\dfrac{\sqrt{6 + 2\sqrt{17}}}{2}$.

101.

$$x^3 - x^2 + 16x - 16 = 0$$

$$x^2(x - 1) + 16(x - 1) = 0$$

$$(x^2 + 16)(x - 1) = 0$$

$$x^2 + 16 = 0 \quad \text{or} \quad x - 1 = 0$$

$$x^2 = -16 \qquad\qquad x = 1$$

$$\sqrt{x^2} = \pm\sqrt{-16}$$

$$x = \pm 4i$$

Section 10.3

VOCABULARY

1. $f(x) = 2x^2 - 4x + 1$ is called a **quadratic** function. Its graph is a cup–shaped figure called a **parabola.**

3. The vertical line $x = 1$ divides the parabola into two halves. This line is called the **axis of symmetry**.

CONCEPTS

5. a. (1, 0) and (3, 0)
    b. (0, –3)
    c. (2, 1)
    d. $x = 2$
    e. D: $(-\infty, \infty)$; R: $(-\infty, 1]$

7.

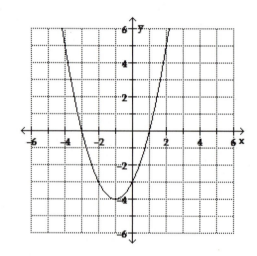

9. a. $f(x) = \boxed{2}\left(x^2 + 6x\right) + 11$

    b. $f(x) = \boxed{2}\left(x^2 + 6x + \boxed{9}\right) + 11 - \boxed{18}$

11. The x–intercepts are (–3, 0) and (5, 0), so the solutions of the function are $x = -3$ and $x = 5$.

NOTATION

13. $h = -1$

$$f(x) = 2\left[x - (-1)\right]^2 + 6$$

GUIDED PRACTICE

15.

x	$f(x) = x^2$
–3	$(-3)^2 = 9$
–2	$(-2)^2 = 4$
–1	$(-1)^2 = 1$
0	$(0)^2 = 0$
1	$(1)^2 = 1$
2	$(2)^2 = 4$
3	$(3)^2 = 9$

x	$g(x) = 2x^2$
–3	$2(-3)^2 = 18$
–2	$2(-2)^2 = 8$
–1	$2(-1)^2 = 2$
0	$2(0)^2 = 0$
1	$2(1)^2 = 2$
2	$2(2)^2 = 8$
3	$2(3)^2 = 18$

x	$s(x) = \dfrac{1}{2}x^2$
–3	$\dfrac{1}{2}(-3)^2 = \dfrac{9}{2}$
–2	$\dfrac{1}{2}(-2)^2 = 2$
–1	$\dfrac{1}{2}(-1)^2 = \dfrac{1}{2}$
0	$\dfrac{1}{2}(0)^2 = 0$
1	$\dfrac{1}{2}(1)^2 = \dfrac{1}{2}$
2	$\dfrac{1}{2}(2)^2 = 2$
3	$\dfrac{1}{2}(3)^2 = \dfrac{9}{2}$

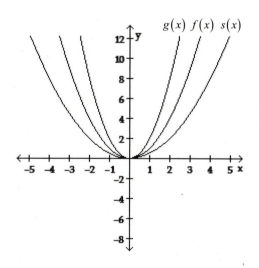

17.

x	$f(x) = 2x^2$
–3	$2(-3)^2 = 18$
–2	$2(-2)^2 = 8$
–1	$2(-1)^2 = 2$
0	$2(0)^2 = 0$
1	$2(1)^2 = 2$
2	$2(2)^2 = 8$
3	$2(3)^2 = 18$

x	$g(x) = -2x^2$
–3	$-2(-3)^2 = -18$
–2	$-2(-2)^2 = -8$
–1	$-2(-1)^2 = -2$
0	$-2(0)^2 = 0$
1	$-2(1)^2 = -2$
2	$-2(2)^2 = -8$
3	$-2(3)^2 = -18$

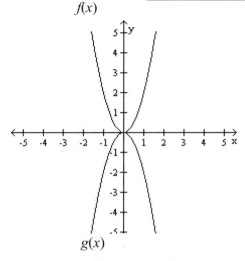

19.

x	$f(x) = 4x^2$
–3	$4(-3)^2 = 36$
–2	$4(-2)^2 = 16$
–1	$4(-1)^2 = 4$
0	$4(0)^2 = 0$
1	$4(1)^2 = 4$
2	$4(2)^2 = 16$
3	$4(3)^2 = 36$

The graph of $g(x) = 4x^2 + 3$ is shifted 3 units up, and the graph of $s(x) = 4x^2 - 2$ is shifted 2 units down.

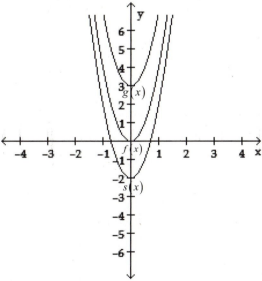

21.

x	$f(x) = 3x^2$
–3	$3(-3)^2 = 27$
–2	$3(-2)^2 = 12$
–1	$3(-1)^2 = 3$
0	$3(0)^2 = 0$
1	$3(1)^2 = 3$
2	$3(2)^2 = 12$
3	$3(3)^2 = 27$

The graph of $g(x) = 3(x+2)^2$ would be shifted 2 units left.

The graph of $s(x) = 3(x-3)^2$ would be shifted 3 units right.

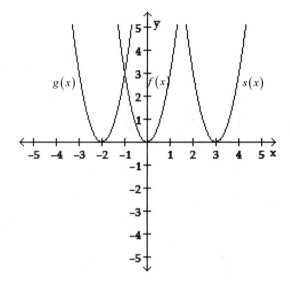

23. $f(x) = (x-1)^2 + 2$
The vertex is (1, 2).
The axis of symmetry is $x = 1$.
The parabola opens upward.

25. $f(x) = -2(x+3)^2 - 4$
The vertex is (–3, –4).
The axis of symmetry is $x = -3$.
The parabola opens downward.

27. $f(x) = -0.5(x-7.5)^2 + 8.5$
The vertex is (7.5, 8.5).
The axis of symmetry is $x = 7.5$.
The parabola opens downward.

29. $f(x) = 2x^2 - 4$

$\quad f(x) = 2(x-0)^2 - 4$

The vertex is (0, –4).
The axis of symmetry is $x = 0$.
The parabola opens upward.

31. $f(x) = (x-3)^2 + 2$

The vertex is (3, 2).
The axis of symmetry is $x = 3$.
The parabola opens upward.

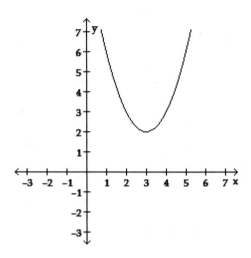

33. $f(x) = -(x-2)^2$

$\quad f(x) = -(x-2)^2 + 0$

The vertex is (2, 0).
The axis of symmetry is $x = 2$.
The parabola opens downward.

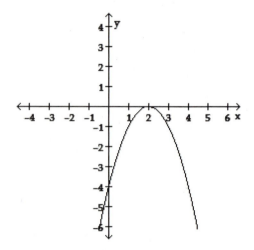

35. $f(x) = -2(x+3)^2 + 4$

The vertex is (–3, 4).
The axis of symmetry is $x = -3$.
The parabola opens downward.

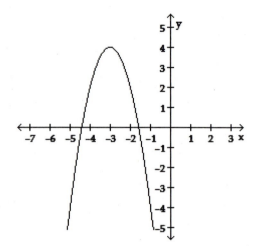

37. $f(x) = \dfrac{1}{2}(x+1)^2 - 3$

The vertex is (–1, –3).
The axis of symmetry is $x = -1$.
The parabola opens upward.

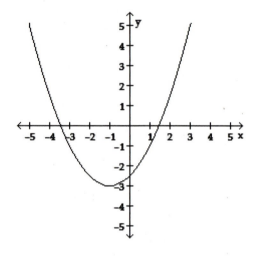

39. $f(x) = x^2 + 4x + 5$

$\quad f(x) = \left(x^2 + 4x + \left(\dfrac{4}{2}\right)^2 \right) + 5 - \left(\dfrac{4}{2}\right)^2$

$\quad f(x) = \left(x^2 + 4x + 4 \right) + 5 - 4$

$\quad f(x) = (x+2)^2 + 1$

The vertex is (–2, 1).
The axis of symmetry is $x = -2$.
The parabola opens upward.

41. $f(x) = -x^2 + 6x - 15$

$f(x) = -\left(x^2 - 6x + \left(\dfrac{-6}{2}\right)^2\right) - 15 + \left(\dfrac{-6}{2}\right)^2$

$f(x) = -\left(x^2 - 6x + 9\right) - 15 + 9$

$f(x) = -(x-3)^2 - 6$

The vertex is (3, –6).
The axis of symmetry is $x = 3$.
The parabola opens downward.

43. $f(x) = x^2 + 2x - 3$

$f(x) = \left(x^2 + 2x\right) - 3$

$f(x) = \left(x^2 + 2x + 1\right) - 3 - 1$

$f(x) = (x+1)^2 - 4$

The vertex is (–1, –4).
The axis of symmetry is $x = -1$.
The parabola opens upward.

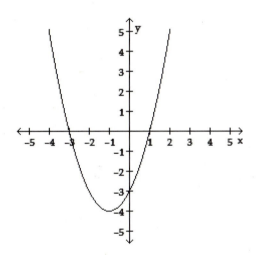

45. $f(x) = 4x^2 + 24x + 37$

$f(x) = 4\left(x^2 + 6x\right) + 37$

$f(x) = 4\left(x^2 + 6x + 9\right) + 37 - 4(9)$

$f(x) = 4(x+3)^2 + 37 - 36$

$f(x) = 4(x+3)^2 + 1$

The vertex is (–3, 1).
The axis of symmetry is $x = -3$.
The parabola opens upward.

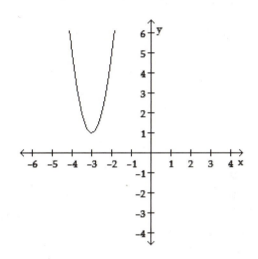

47. $f(x) = x^2 + x - 6$

$f(x) = \left(x^2 + x\right) - 6$

$f(x) = \left(x^2 + x + \dfrac{1}{4}\right) - 6 - \dfrac{1}{4}$

$f(x) = \left(x + \dfrac{1}{2}\right)^2 - \dfrac{25}{4}$

The vertex is $\left(-\dfrac{1}{2}, -\dfrac{25}{4}\right)$.

The axis of symmetry is $x = -\dfrac{1}{2}$.

The parabola opens upward.

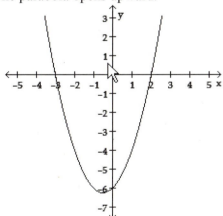

- 695 -

49. $f(x) = -4x^2 + 16x - 10$

    $f(x) = -4(x^2 - 4x) - 10$

    $f(x) = -4(x^2 - 4x + 4) - 10 - (-4)(4)$

    $f(x) = -4(x - 2)^2 - 10 + 16$

    $f(x) = -4(x - 2)^2 + 6$

The vertex is (2, 6).
The axis of symmetry is $x = 2$.
The parabola opens downward.

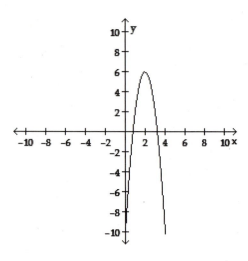

53. $f(x) = -x^2 - 8x - 17$

    $f(x) = -(x^2 + 8x) - 17$

    $f(x) = -(x^2 + 8x + 16) - 17 - (-1)(16)$

    $f(x) = -(x + 4)^2 - 17 + 16$

    $f(x) = -(x + 4)^2 - 1$

The vertex is (–4, –1).
The axis of symmetry is $x = -4$.
The parabola opens downward.

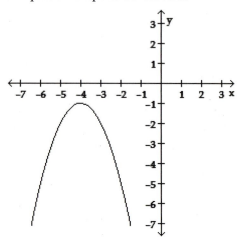

51. $f(x) = 2x^2 + 8x + 6$

    $f(x) = 2(x^2 + 4x) + 6$

    $f(x) = 2(x^2 + 4x + 4) + 6 - 2(4)$

    $f(x) = 2(x + 2)^2 + 6 - 8$

    $f(x) = 2(x + 2)^2 - 2$

The vertex is (–2, –2).
The axis of symmetry is $x = -2$.
The parabola opens upward.

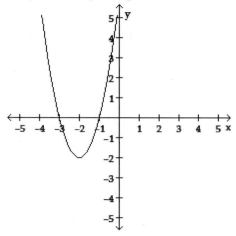

55. $f(x) = x^2 + 2x - 5$

$$x = \frac{-b}{2a}$$

$$= \frac{-2}{2(1)}$$

$$= \frac{-2}{2}$$

$$= -1$$

$$f(-1) = (-1)^2 + 2(-1) - 5$$

$$= 1 - 2 - 5$$

$$= -6$$

The vertex is (–1, –6).

57. $f(x) = 2x^2 - 3x + 4$

$$x = \frac{-b}{2a}$$

$$= \frac{-(-3)}{2(2)}$$

$$= \frac{3}{4}$$

$$f\left(\frac{3}{4}\right) = 2\left(\frac{3}{4}\right)^2 - 3\left(\frac{3}{4}\right) + 4$$

$$= 2\left(\frac{9}{16}\right) - \frac{9}{4} + 4$$

$$= \frac{18}{16} - \frac{9}{4} + 4$$

$$= \frac{18}{16} - \frac{36}{16} + \frac{64}{16}$$

$$= \frac{46}{16}$$

$$= \frac{23}{8}$$

The vertex is $\left(\frac{3}{4}, \frac{23}{8}\right)$.

59. x–intercept: Let $f(x) = 0$.

$$0 = x^2 - 2x - 35$$

$$0 = (x - 7)(x + 5)$$

$$x - 7 = 0 \quad \text{or} \quad x + 5 = 0$$

$$x = 7 \qquad\qquad x = -5$$

x–intercepts are $(7, 0)$ and $(-5, 0)$

y–intercept $= (0, c)$

$c = -35$

y–intercept is $(0, -35)$

61. x–intercept: Let $f(x) = 0$.

$$0 = -2x^2 + 4x$$

$$0 = -2x(x - 2)$$

$$-2x = 0 \quad \text{or} \quad x - 2 = 0$$

$$x = 0 \qquad\qquad x = 2$$

x–intercepts are $(0, 0)$ and $(2, 0)$

y–intercept $= (0, c)$

$c = 0$

y–intercept is $(0, 0)$

63. $f(x) = x^2 + 4x + 4$

Step 1: Since $a = 1 > 0$, the parabola opens upward.

Step 2: Find the vertex and axis of symmetry.

$$x = \frac{-b}{2a} = \frac{-4}{2(1)} = \frac{-4}{2} = -2$$

$$y = (-2)^2 + 4(-2) + 4 = 4 - 8 + 4 = 0$$

vertex: $(-2, 0)$

The axis of symmetry is $x = -2$.

Step 3: Find the x– and y–intercepts.

Since $c = 4$, the y–intercept is $(0, 4)$.

To find the x–intercepts, let $y = 0$ and solve the equation for x.

$$0 = x^2 + 4x + 4$$

$$0 = (x + 2)(x + 2)$$

$$x + 2 = 0 \quad \text{or} \quad x + 2 = 0$$

$$x = -2 \qquad\qquad x = -2$$

x–intercept: $(-2, 0)$

y–intercept: $(0, 4)$

Step 4: Find another point(s).

Let $x = -1$.

$$y = (-1)^2 + 4(-1) + 4 = 1 - 4 + 4 = 1$$

$(-1, 1)$

Use symmetry to find the points $(-3, 1)$ and $(-4, 4)$.

Step 5: Plot the points and draw the parabola.

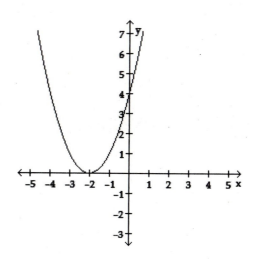

65. $f(x) = -x^2 + 2x - 1$

Step 1: Since $a = -1 < 0$, the parabola opens downward.

Step 2: Find the vertex and axis of symmetry.
$$x = \frac{-b}{2a} = \frac{-2}{2(-1)} = \frac{-2}{-2} = 1$$
$$y = -(1)^2 + 2(1) - 1 = -1 + 2 - 1 = 0$$
vertex: $(1, 0)$
The axis of symmetry is $x = 1$.

Step 3: Find the x– and y–intercepts.
Since $c = -1$, the y–intercept is $(0, -1)$.
To find the x–intercepts, let $y = 0$ and solve the equation for x.
$$0 = -x^2 + 2x - 1$$
$$x^2 - 2x + 1 = 0$$
$$(x-1)(x-1) = 0$$
$$x - 1 = 0 \quad \text{or} \quad x - 1 = 0$$
$$x = 1 \qquad\qquad x = 1$$
x–intercept: $(1, 0)$
y–intercept: $(0, -1)$

Step 4: Find another point(s).
Let $x = -1$.
$$y = -(-1)^2 + 2(-1) - 1 = -1 - 2 - 1 = -4$$
$(-1, -4)$
Use symmetry to find the points $(2, -1)$ and $(3, -4)$.

Step 5: Plot the points and draw the parabola.

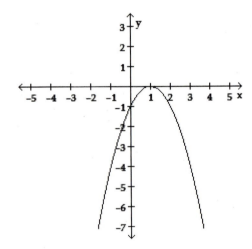

67. $f(x) = x^2 - 2x$

Step 1: Since $a = 1 > 0$, the parabola opens upward.

Step 2: Find the vertex and axis of symmetry.
$$x = \frac{-b}{2a} = \frac{-(-2)}{2(1)} = \frac{2}{2} = 1$$
$$y = (1)^2 - 2(1) = 1 - 2 = -1$$
vertex: $(1, -1)$
The axis of symmetry is $x = 1$.

Step 3: Find the x– and y–intercepts.
Since $c = 0$, the y–intercept is $(0, 0)$.
To find the x–intercepts, let $y = 0$ and solve the equation for x.
$$0 = x^2 - 2x$$
$$0 = x(x - 2)$$
$$x = 0 \quad \text{or} \quad x - 2 = 0$$
$$x = 0 \qquad\qquad x = 2$$
x–intercepts: $(0, 0)$ and $(2, 0)$
y–intercept: $(0, 0)$

Step 4: Find another point(s).
Let $x = -1$.
$$y = (-1)^2 - 2(-1) = 1 + 2 = 3$$
$(-1, 3)$
Use symmetry to find the point $(3, 3)$.

Step 5: Plot the points and draw the parabola.

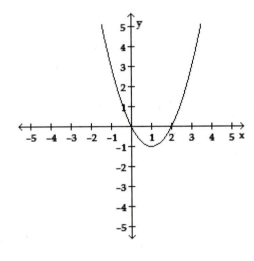

69. $f(x) = 2x^2 - 8x + 6$

   Step 1: Since $a = 2 > 0$, the parabola opens upward.

   Step 2: Find the vertex and axis of symmetry.

   $$x = \frac{-b}{2a} = \frac{-(-8)}{2(2)} = \frac{8}{4} = 2$$

   $$y = 2(2)^2 - 8(2) + 6 = 8 - 16 + 6 = -2$$

   vertex: $(2, -2)$

   The axis of symmetry is $x = 2$.

   Step 3: Find the x– and y–intercepts.
   Since $c = 6$, the y–intercept is $(0, 6)$.
   To find the x–intercepts, let $y = 0$ and solve the equation for x.

   $$0 = 2x^2 - 8x + 6$$

   $$0 = 2(x^2 - 4x + 3)$$

   $$0 = (x - 1)(x - 3)$$

   $$x - 1 = 0 \quad \text{or} \quad x - 3 = 0$$

   $$x = 1 \qquad\qquad x = 3$$

   x–intercepts: $(1, 0)$ and $(3, 0)$
   y–intercept: $(0, 6)$

   Step 4: Use symmetry to find the point $(4, 6)$

   Step 5: Plot the points and draw the parabola.

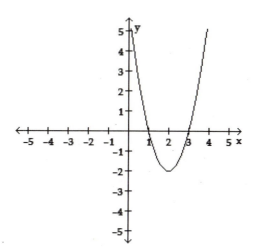

71. $f(x) = -6x^2 - 12x - 8$

   Step 1: Since $a = -6 < 0$, the parabola opens downward.

   Step 2: Find the vertex and axis of symmetry.

   $$x = \frac{-b}{2a} = \frac{-(-12)}{2(-6)} = \frac{12}{-12} = -1$$

   $$y = -6(-1)^2 - 12(-1) - 8$$

   $$= -6 + 12 - 8$$

   $$= -2$$

   vertex: $(-1, -2)$

   The axis of symmetry is $x = -1$.

   Step 3: Find the x– and y–intercepts.
   Since $c = -8$, the y–intercept is $(0, -8)$.
   Since the vertex is located below the x–axis and the parabola opens downward, there are no x–intercepts.

   x–intercepts: none
   y–intercept: $(0, -8)$

   Step 4: Find another point(s).
   Let $x = 1$.

   $$y = -6(1)^2 - 12(1) - 8$$

   $$= -6 - 12 - 8$$

   $$= -26$$

   $(1, -26)$

   Use symmetry to find the points $(-2, -8)$ and $(-3, -26)$.

   Step 5: Plot the points and draw the parabola.

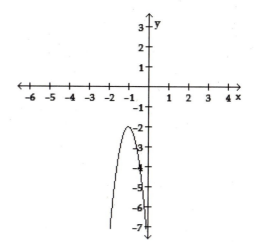

73. $f(x) = 4x^2 - 12x + 9$

Step 1: Since $a = 4 > 0$, the parabola opens upward.

Step 2: Find the vertex and axis of symmetry.

$$x = \frac{-b}{2a} = \frac{-(-12)}{2(4)} = \frac{12}{8} = \frac{3}{2}$$

$$y = 4\left(\frac{3}{2}\right)^2 - 12\left(\frac{3}{2}\right) + 9$$

$$= 4\left(\frac{9}{4}\right) - 12\left(\frac{3}{2}\right) + 9$$

$$= 9 - 18 + 9$$

$$= 0$$

vertex: $\left(\frac{3}{2}, 0\right)$

The axis of symmetry is $x = \frac{3}{2}$.

Step 3: Find the x– and y–intercepts. Since $c = 9$, the y–intercept is $(0, 9)$. To find the x–intercepts, let $y = 0$ and solve the equation for x.

$$0 = 4x^2 - 12x + 9$$

$$0 = (2x - 3)(2x - 3)$$

$2x - 3 = 0$ or $2x - 3 = 0$

$\quad 2x = 3 \qquad\qquad 2x = 3$

$\quad x = \frac{3}{2} \qquad\qquad x = \frac{3}{2}$

x–intercept: $\left(\frac{3}{2}, 0\right)$

y–intercept: $(0, 9)$

Step 4: Find another point(s).

Let $x = 1$.

$$y = 4(1)^2 - 12(1) + 9 = 4 - 12 + 9 = 1$$

$(1, 1)$

Use symmetry to find the points $(2, 1)$ and $(3, 9)$.

Step 5: Plot the points and draw the parabola.

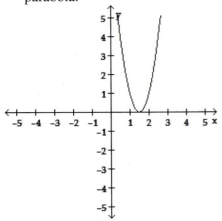

75. $f(x) = 2x^2 - x + 1$

vertex: $(0.25, 0.88)$

77. $f(x) = -x^2 + x + 7$

vertex: $(0.50, 7.25)$

79. $x^2 + x - 6 = 0$

The x–intercepts are $(-3, 0)$ and $(2, 0)$, so the solutions of the equation are $x = -3$ and $x = 2$.

81. $0.5x^2 - 0.7x - 3 = 0$

The x–intercepts are $(-1.85, 0)$ and $(3.25, 0)$, so the solutions of the equation are $x = -1.85$ and $x = 3.25$.

APPLICATIONS

83. CROSSWORD PUZZLES

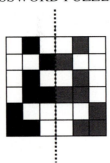

85. OPERATING COSTS

The vertex of the parabola represented by the function would be its minimum cost.

$$C(n) = 2.2n^2 - 66n + 655$$

$$n = \frac{-b}{2a}$$

$$= \frac{-(-66)}{2(2.2)}$$

$$= \frac{66}{4.4}$$

$$= 15 \text{ minutes}$$

$$C(15) = 2.2(15)^2 - 66(15) + 655$$

$$= 2.2(225) - 66(15) + 655$$

$$= 495 - 990 + 655$$

$$= \$160$$

The cost of running the machine is a minimum at 15 minutes. The minimum cost is $160.

87. FIREWORKS

The vertex of the parabola would be its maximum height and time of explosion.

$$s = 120t - 16t^2$$
$$s = -16t^2 + 120t$$
$$t = \frac{-b}{2a}$$
$$= \frac{-120}{2(-16)}$$
$$= \frac{-120}{-32}$$
$$= 3.75$$

$$s(3.75) = 120(3.75) - 16(3.75)^2$$
$$= 450 - 16(14.0625)$$
$$= 450 - 225$$
$$= 225$$

The shell reaches its maximum height of 225 ft in 3.75 seconds.

89. POLICE PATROL OFFICER

Let the width = x and the length = y.
The sum of the four sides is 300 ft.

$$x + y + x + y = 300$$
$$2x + 2y = 300$$
$$2y = -2x + 300$$
$$y = -x + 150$$

To find a function for the area, multiply length times width.

$$\text{Area} = \text{width} \cdot \text{length}$$
$$= x(y)$$
$$= x(-x + 150)$$
$$A(x) = -x^2 + 150x$$

Use the vertex formula to find the width (x) and substitution to find the length (y).

$$A(x) = -x^2 + 150x$$
$$x = \frac{-b}{2a}$$
$$= \frac{-150}{2(-1)}$$
$$= \frac{-150}{-2}$$
$$= 75 \text{ ft}$$
$$y = -x + 150$$
$$= -75 + 150$$
$$= 75$$

The dimensions are 75 ft by 75 ft.

To find the maximum area, find $A(75)$.

$$A(75) = -(75)^2 + 150(75)$$
$$= -5,625 + 11,250$$
$$= 5,625 \text{ ft}^2$$

The maximum area is 5,625 square ft.

Section 10.4

91. MILITARY HISTORY

The vertex of the parabola represented by the function would be its maximum strength.

$$N(x) = -0.0534x^2 + 0.337x + 0.97$$

$$x = \frac{-b}{2a}$$

$$= \frac{-0.337}{2(-0.0534)}$$

$$= \frac{-0.337}{-0.1068}$$

$$= 3.16$$

$$N(3.16) = -0.0534(3.16)^2 + 0.337(3.16) + 0.97$$

$$= -0.0534(9.9856) + 0.337(3.16) + 0.97$$

$$= -0.533 + 1.065 + 0.97$$

$$= 1.502$$

It maximum strength would be in the 3rd year, which represents 1968. During that year, the army's personnel strength level was 1.5 million. At this time in history, the U.S. involvement in the Vietnam War was at its peak.

93. MAXIMIZING REVENUE

The vertex of the parabola represented by the function would be its maximum revenue.

$$R(x) = -\frac{x^2}{5} + 80x - 1{,}000$$

$$R(x) = -\frac{1}{5}x^2 + 80x - 1{,}000$$

$$R(x) = -0.2x^2 + 80x - 1{,}000$$

$$x = \frac{-b}{2a}$$

$$= \frac{-80}{2(-0.2)}$$

$$= \frac{-80}{-0.4}$$

$$= 200$$

$$R(200) = -\frac{(200)^2}{5} + 80(200) - 1{,}000$$

$$= -\frac{40{,}000}{5} + 80(200) - 1{,}000$$

$$= -8{,}000 + 16{,}000 - 1{,}000$$

$$= 7{,}000$$

To obtain the maximum revenue, they must sell 200 stereos. The maximum revenue is $7,000.

WRITING

95. Answers will vary.

97. Answers will vary.

99. Answers will vary.

REVIEW

101.

$$\frac{\sqrt{3}}{\sqrt{50}} = \frac{\sqrt{3}}{\sqrt{25}\sqrt{2}}$$

$$= \frac{\sqrt{3}}{5\sqrt{2}}$$

$$= \frac{\sqrt{3}}{5\sqrt{2}} \cdot \frac{\sqrt{2}}{\sqrt{2}}$$

$$= \frac{\sqrt{6}}{5\sqrt{4}}$$

$$= \frac{\sqrt{6}}{5(2)}$$

$$= \frac{\sqrt{6}}{10}$$

103.

$$3\left(\sqrt{5b} - \sqrt{3}\right)^2 = 3\left(\sqrt{5b} - \sqrt{3}\right)\left(\sqrt{5b} - \sqrt{3}\right)$$

$$= 3\left(\sqrt{25b^2} - \sqrt{15b} - \sqrt{15b} + \sqrt{9}\right)$$

$$= 3\left(5b - 2\sqrt{15b} + 3\right)$$

$$= 15b - 6\sqrt{15b} + 9$$

CHALLENGE PROBLEMS

105.

$$f(x) = x - x^2$$

$$f(x) = -x^2 + x$$

$$\frac{-b}{2a} = \frac{-1}{2(-1)}$$

$$= \frac{1}{2}$$

SECTION 10.5

VOCABUALRY

1. $x^2 + 3x - 18 < 0$ is an example of a **quadratic** inequality in one variable.

3. $y \le x^2 - 4x + 3$ is an example of a nonlinear inequality in **two** variables.

CONCEPTS

5. $(-\infty, -1), (1, 4), (4, \infty)$

7. a. yes
   b. no
   c. yes
   d. no

9. a. $(-3, 2)$
   b. $(-\infty, -1] \cup [1, \infty)$

11. a. solid
    b. yes
    $$y \le x^2 + 2x + 1$$
    $$0 \overset{?}{\le} 0^2 + 2(0) + 1$$
    $$0 \overset{?}{\le} 0 + 0 + 1$$
    $$0 \le 1$$

NOTATION

13. $x^2 - 6x - 7 \ge 0$

GUIDED PRACTICE

15.
$$x^2 - 5x + 4 < 0$$
$$x^2 - 5x + 4 = 0$$
$$(x-1)(x-4) = 0$$
$$x - 1 = 0 \quad \text{or} \quad x - 4 = 0$$
$$x = 1 \qquad\qquad x = 4$$

$(1, 4)$

17.
$$x^2 - 8x + 15 > 0$$
$$x^2 - 8x + 15 = 0$$
$$(x-5)(x-3) = 0$$
$$x - 5 = 0 \quad \text{or} \quad x - 3 = 0$$
$$x = 5 \qquad\qquad x = 3$$

$(-\infty, 3) \cup (5, \infty)$

19.
$$x^2 - x \ge 42$$
$$x^2 - x - 42 \ge 0$$
$$x^2 - x - 42 = 0$$
$$(x+6)(x-7) = 0$$
$$x + 6 = 0 \quad \text{or} \quad x - 7 = 0$$
$$x = -6 \qquad\qquad x = 7$$

$(-\infty, -6] \cup [7, \infty)$

21.
$$x^2 + x \le 12$$
$$x^2 + x - 12 \le 0$$
$$x^2 + x - 12 = 0$$
$$(x-3)(x+4) = 0$$
$$x - 3 = 0 \quad \text{or} \quad x + 4 = 0$$
$$x = 3 \qquad\qquad x = -4$$

$[-4, 3]$

23.

$$\frac{1}{x} < 2$$

$$\frac{1}{x} = 2$$

$$\frac{1}{x} - 2 = 0$$

$$\frac{1}{x} - \frac{2x}{x} = 0$$

$$\frac{1 - 2x}{x} = 0$$

$$x\left(\frac{1 - 2x}{x}\right) = x(0)$$

$$1 - 2x = 0$$

$$-2x = -1$$

$$\frac{1}{2} = x$$

Set the denominator $= 0$.

$$x = 0$$

Critical numbers $= 0$ and $\dfrac{1}{2}$

$$(-\infty, 0) \cup \left(\frac{1}{2}, \infty\right)$$

25.

$$\frac{5}{x} \geq -3$$

$$\frac{5}{x} = -3$$

$$\frac{5}{x} + 3 = 0$$

$$\frac{5}{x} + \frac{3x}{x} = 0$$

$$\frac{5 + 3x}{x} = 0$$

$$x\left(\frac{5 + 3x}{x}\right) = x(0)$$

$$5 + 3x = 0$$

$$3x = -5$$

$$x = -\frac{5}{3}$$

Set the denominator $= 0$.

$$x = 0$$

Critical numbers $= 0$ and $-\dfrac{5}{3}$

$$\left(-\infty, -\frac{5}{3}\right] \cup (0, \infty)$$

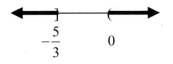

27.

$$\frac{x^2 - x - 12}{x - 1} < 0$$

$$\frac{x^2 - x - 12}{x - 1} = 0$$

$$(x-1)\left(\frac{x^2 - x - 12}{x - 1}\right) = (x-1)(0)$$

$$x^2 - x - 12 = 0$$

$$(x-4)(x+3) = 0$$

$$x - 4 = 0 \quad \text{or} \quad x + 3 = 0$$

$$x = 4 \qquad\qquad x = -3$$

Set the denominator $= 0$.

$$x - 1 = 0$$

$$x = 1$$

Critical numbers $= -3,\ 1,\ \text{and } 4$

$$(-\infty, -3) \cup (1, 4)$$

29.

$$\frac{6x^2 - 5x + 1}{2x + 1} > 0$$

$$\frac{6x^2 - 5x + 1}{2x + 1} = 0$$

$$(2x+1)\left(\frac{6x^2 - 5x + 1}{2x + 1}\right) = (2x+1)(0)$$

$$6x^2 - 5x + 1 = 0$$

$$(2x-1)(3x-1) = 0$$

$$2x - 1 = 0 \quad \text{or} \quad 3x - 1 = 0$$

$$x = \frac{1}{2} \qquad\qquad x = \frac{1}{3}$$

Set the denominator $= 0$.

$$2x + 1 = 0$$

$$x = -\frac{1}{2}$$

Critical numbers $= -\frac{1}{2},\ \frac{1}{3},\ \text{and } \frac{1}{2}$

$$\left(-\frac{1}{2}, \frac{1}{3}\right) \cup \left(\frac{1}{2}, \infty\right)$$

31.

$$\frac{3}{x - 2} < \frac{4}{x}$$

$$\frac{3}{x - 2} - \frac{4}{x} < 0$$

$$\frac{3}{x - 2} - \frac{4}{x} = 0$$

$$\frac{3}{x - 2} \cdot \frac{x}{x} - \frac{4}{x} \cdot \frac{x - 2}{x - 2} = 0$$

$$\frac{3x - 4(x - 2)}{x(x - 2)} = 0$$

$$\frac{3x - 4x + 8}{x(x - 2)} = 0$$

$$\frac{-x + 8}{x(x - 2)} = 0$$

$$x(x-2)\left(\frac{-x + 8}{x(x - 2)}\right) = x(x-2)(0)$$

$$-x + 8 = 0$$

$$8 = x$$

Set the denominator $= 0$.

$$x(x - 2) = 0$$

$$x = 0 \quad \text{or} \quad x - 2 = 0$$

$$x = 0 \qquad\qquad x = 2$$

Critical numbers $= 0,\ 2,\ \text{and } 8$

$$(0, 2) \cup (8, \infty)$$

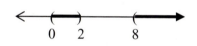

Section 10.5

33.

$$\frac{7}{x-3} \geq \frac{2}{x+4}$$

$$\frac{7}{x-3} - \frac{2}{x+4} \geq 0$$

$$\frac{7}{x-3} - \frac{2}{x+4} = 0$$

$$\frac{7}{x-3} \cdot \frac{x+4}{x+4} - \frac{2}{x+4} \cdot \frac{x-3}{x-3} = 0$$

$$\frac{7(x+4) - 2(x-3)}{(x-3)(x+4)} = 0$$

$$\frac{7x+28 - 2x+6}{(x-3)(x+4)} = 0$$

$$\frac{5x+34}{(x-3)(x+4)} = 0$$

$$(x-3)(x+4)\left(\frac{5x+34}{(x-3)(x+4)}\right) = (x-3)(x+4)(0)$$

$$5x+34 = 0$$

$$5x = -34$$

$$x = -\frac{34}{5}$$

Set the denominator $= 0$.

$$(x-3)(x+4) = 0$$

$$x-3 = 0 \quad \text{or} \quad x+4 = 0$$

$$x = 3 \qquad\qquad x = -4$$

Critical numbers $= -\dfrac{34}{5}, -4,$ and 3

$$\left[-\frac{34}{5}, -4\right) \cup (3, \infty)$$

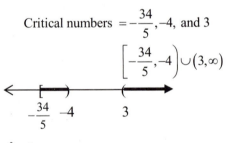

35. $y < x^2 + 1$

Complete a table of values and use $(0, 0)$ as a check point.

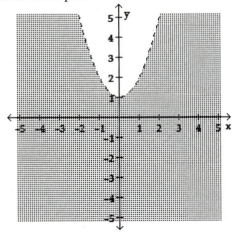

37. $y \leq x^2 + 5x + 6$

Complete a table of values and use $(0, 0)$ as a check point.

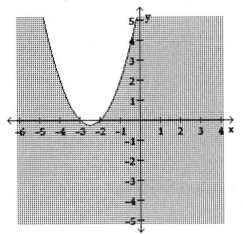

39. $y < |x+4|$

Complete a table of values and use $(0, 0)$ as a check point.

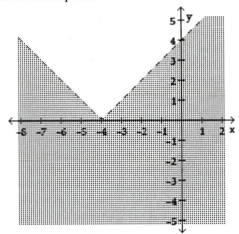

41. $y \leq -|x| + 2$

Complete a table of values and use $(0, 0)$ as a check point.

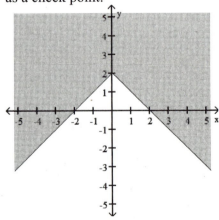

43. $x^2 - 2x - 3 < 0$

$(-1, 3)$

45. $\dfrac{x+3}{x-2} > 0$

$(-\infty, -3) \cup (2, \infty)$

TRY IT YOURSELF

47.

$$\frac{x}{x+4} \le \frac{1}{x+1}$$

$$\frac{x}{x+4} - \frac{1}{x+1} \le 0$$

$$\frac{x}{x+4} - \frac{1}{x+1} = 0$$

$$\frac{x}{x+4} \cdot \frac{x+1}{x+1} - \frac{1}{x+1} \cdot \frac{x+4}{x+4} = 0$$

$$\frac{x(x+1) - (x+4)}{(x+1)(x+4)} = 0$$

$$\frac{x^2 + x - x - 4}{(x+1)(x+4)} = 0$$

$$\frac{x^2 - 4}{(x+1)(x+4)} = 0$$

$$(x+1)(x+4)\left(\frac{x^2-4}{(x+1)(x+4)}\right) = (x+1)(x+4)(0)$$

$$x^2 - 4 = 0$$

$$(x+2)(x-2) = 0$$

$$x+2 = 0 \quad \text{or} \quad x-2 = 0$$

$$x = -2 \qquad x = 2$$

Set the denominator $= 0$.

$$(x+1)(x+4) = 0$$

$$x+1 = 0 \quad \text{or} \quad x+4 = 0$$

$$x = -1 \qquad x = -4$$

Critical numbers $= -4, -2, -1,$ and 2

$$(-4, -2] \cup (-1, 2]$$

49.

$$x^2 \ge 9$$

$$x^2 - 9 \ge 0$$

$$x^2 - 9 = 0$$

$$(x+3)(x-3) = 0$$

$$x+3 = 0 \quad \text{or} \quad x-3 = 0$$

$$x = -3 \qquad x = 3$$

$$(-\infty, -3] \cup [3, \infty)$$

51.

$$x^2 + 6x \ge -9$$

$$x^2 + 6x + 9 \ge 0$$

$$x^2 + 6x + 9 = 0$$

$$(x+3)(x+3) = 0$$

$$x+3 = 0 \quad \text{or} \quad x+3 = 0$$

$$x = -3 \qquad x = -3$$

$$(-\infty, \infty)$$

53.

$$\frac{x^2 + x - 2}{x - 3} > 0$$

$$\frac{x^2 + x - 2}{x - 3} = 0$$

$$(x-3)\left(\frac{x^2+x-2}{x-3}\right) = (x-3)(0)$$

$$x^2 + x - 2 = 0$$

$$(x-1)(x+2) = 0$$

$$x-1 = 0 \quad \text{or} \quad x+2 = 0$$

$$x = 1 \qquad x = -2$$

Set the denominator $= 0$.

$$x - 3 = 0$$

$$x = 3$$

Critical numbers $= -2, 1,$ and 3

$$(-2, 1) \cup (3, \infty)$$

Section 10.5

55.

$$2x^2 - 50 < 0$$
$$2(x^2 - 25) = 0$$
$$2(x+5)(x-5) = 0$$
$$x + 5 = 0 \quad \text{or} \quad x - 5 = 0$$
$$x = -5 \qquad\qquad x = 5$$

$$(-5, 5)$$

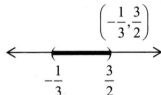

57.

$$\frac{2x-3}{3x+1} < 0$$
$$\frac{2x-3}{3x+1} = 0$$
$$(3x+1)\left(\frac{2x-3}{3x+1}\right) = (3x+1)0$$
$$2x - 3 = 0$$
$$2x = 3$$
$$x = \frac{3}{2}$$

Set the denominator $= 0$.
$$3x + 1 = 0$$
$$3x = -1$$
$$x = -\frac{1}{3}$$

Critical numbers $= -\frac{1}{3}$ and $\frac{3}{2}$

$$\left(-\frac{1}{3}, \frac{3}{2}\right)$$

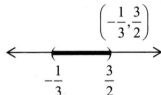

59.

$$x^2 - 6x + 9 < 0$$
$$x^2 - 6x + 9 = 0$$
$$(x-3)(x-3) = 0$$
$$x - 3 = 0 \quad \text{or} \quad x - 3 = 0$$
$$x = 3 \qquad\qquad x = 3$$
$$\varnothing$$

No Solution

61.

$$\frac{5}{x+1} > \frac{3}{x-4}$$
$$\frac{5}{x+1} - \frac{3}{x-4} > 0$$
$$\frac{5}{x+1} - \frac{3}{x-4} = 0$$
$$\frac{x-4}{x-4} \cdot \frac{5}{x+1} - \frac{3}{x-4} \cdot \frac{x+1}{x+1} = 0$$
$$\frac{5(x-4) - 3(x+1)}{(x+1)(x-4)} = 0$$
$$\frac{5x - 20 - 3x - 3}{(x+1)(x-4)} = 0$$
$$\frac{2x - 23}{(x+1)(x-4)} = 0$$
$$(x+1)(x-4)\left(\frac{2x-23}{(x+1)(x-4)}\right) = (x+1)(x-4)(0)$$
$$2x - 23 = 0$$
$$2x = 23$$
$$x = \frac{23}{2}$$

Set the denominator $= 0$.
$$(x+1)(x-4) = 0$$
$$x + 1 = 0 \quad \text{or} \quad x - 4 = 0$$
$$x = -1 \qquad\qquad x = 4$$

Critical numbers $= -1, 4$, and $\frac{23}{2}$

$$(-1, 4) \cup \left(\frac{23}{2}, \infty\right)$$

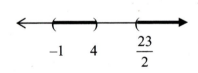

APPLICATIONS

63. BRIDGES

$$L = \frac{1}{9,000}x^2 + 5$$

$$\frac{1}{9,000}x^2 + 5 > 95$$

$$\frac{1}{9,000}x^2 + 5 = 95$$

$$\frac{1}{9,000}x^2 - 90 = 0$$

$$\frac{x^2}{9,000} - \frac{90}{1} \cdot \frac{9,000}{9,000} = 0$$

$$\frac{x^2 - 810,000}{9,000} = 0$$

$$9,000\left(\frac{x^2 - 810,000}{9,000}\right) = 9,000(0)$$

$$x^2 - 810,000 = 0$$

$$(x - 900)(x + 900) = 0$$

$$x - 900 = 0 \quad \text{or} \quad x + 900 = 0$$

$$x = 900 \qquad x = -900$$

$$(-2,100, -900) \cup (900, 2100)$$

CHALLENGE PROBLEMS

73. a.

$$x^2 - x - 12 > 0$$

$$x^2 - x - 12 = 0$$

$$(x - 4)(x + 3) = 0$$

$$x - 4 = 0 \quad \text{or} \quad x + 3 = 0$$

$$x = 4 \qquad x = -3$$

$$(-\infty, -3) \cup (4, \infty)$$

b. Answers will vary.

$$\frac{x - 4}{x + 3} > 0$$

WRITING

65. Answers will vary.

67. Answers will vary.

REVIEW

69. $x = ky$

71. $t = kxy$

Section 10.5

SECTION 10.1
The Square Root Property and Completing the Square

1.
$$x^2 + 9x + 20 = 0$$
$$(x+4)(x+5) = 0$$
$$x + 4 = 0 \quad \text{or} \quad x + 5 = 0$$
$$x = -4 \qquad x = -5$$

2.
$$6x^2 + 17x + 5 = 0$$
$$(2x+5)(3x+1) = 0$$
$$2x + 5 = 0 \quad \text{or} \quad 3x + 1 = 0$$
$$2x = -5 \qquad 3x = -1$$
$$x = -\frac{5}{2} \qquad x = -\frac{1}{3}$$

3.
$$x^2 = 28$$
$$x = \pm\sqrt{28}$$
$$x = \pm\sqrt{4}\sqrt{7}$$
$$x = \pm 2\sqrt{7}$$

4.
$$(t+2)^2 = 36$$
$$t + 2 = \pm\sqrt{36}$$
$$t + 2 = \pm 6$$
$$t + 2 = 6 \quad \text{or} \quad t + 2 = -6$$
$$t = 4 \qquad t = -8$$

5.
$$a^2 + 25 = 0$$
$$a^2 = -25$$
$$a = \pm\sqrt{-25}$$
$$a = \pm 5i$$

6.
$$5x^2 - 49 = 0$$
$$5x^2 = 49$$
$$x^2 = \frac{49}{5}$$
$$x = \pm\sqrt{\frac{49}{5}}$$
$$x = \pm\frac{7}{\sqrt{5}}$$
$$x = \pm\frac{7}{\sqrt{5}} \cdot \frac{\sqrt{5}}{\sqrt{5}}$$
$$x = \pm\frac{7\sqrt{5}}{5}$$

7.
$$A = \pi r^2$$
$$\frac{A}{\pi} = \frac{\pi r^2}{\pi}$$
$$\frac{A}{\pi} = r^2$$
$$r^2 = \frac{A}{\pi}$$
$$r = \sqrt{\frac{A}{\pi}}$$
$$r = \frac{\sqrt{A}}{\sqrt{\pi}}$$
$$r = \frac{\sqrt{A}}{\sqrt{\pi}} \cdot \frac{\sqrt{\pi}}{\sqrt{\pi}}$$
$$r = \frac{\sqrt{A\pi}}{\pi}$$

8. $\frac{1}{4}$ must be added to both sides because
$$\left(\frac{-1}{2}\right)^2 = \frac{1}{4}$$
$$x^2 - x + \frac{1}{4} = \left(x - \frac{1}{2}\right)^2$$

9.

$$x^2 + 6x + 8 = 0$$
$$x^2 + 6x = -8$$
$$x^2 + 6x + \left(\frac{6}{2}\right)^2 = -8 + \left(\frac{6}{2}\right)^2$$
$$x^2 + 6x + (3)^2 = -8 + (3)^2$$
$$x^2 + 6x + 9 = -8 + 9$$
$$(x+3)^2 = 1$$
$$x + 3 = \pm\sqrt{1}$$
$$x + 3 = \pm 1$$
$$x + 3 = 1 \quad \text{or} \quad x + 3 = -1$$
$$x = -2 \qquad\qquad x = -4$$

10.

$$2x^2 - 6x + 3 = 0$$
$$2x^2 - 6x = -3$$
$$\frac{2x^2}{2} - \frac{6x}{2} = \frac{-3}{2}$$
$$x^2 - 3x = -\frac{3}{2}$$
$$x^2 - 3x + \left(\frac{-3}{2}\right)^2 = -\frac{3}{2} + \left(\frac{-3}{2}\right)^2$$
$$x^2 - 3x + \frac{9}{4} = -\frac{3}{2} + \frac{9}{4}$$
$$\left(x - \frac{3}{2}\right)^2 = -\frac{6}{4} + \frac{9}{4}$$
$$\left(x - \frac{3}{2}\right)^2 = \frac{3}{4}$$
$$x - \frac{3}{2} = \pm\sqrt{\frac{3}{4}}$$
$$x - \frac{3}{2} = \pm\frac{\sqrt{3}}{\sqrt{4}}$$
$$x - \frac{3}{2} = \pm\frac{\sqrt{3}}{2}$$
$$x = \frac{3}{2} \pm \frac{\sqrt{3}}{2}$$
$$x = \frac{3 \pm \sqrt{3}}{2}$$

11.

$$6a^2 - 12a = -1$$
$$\frac{6a^2}{6} - \frac{12a}{6} = \frac{-1}{6}$$
$$a^2 - 2a = -\frac{1}{6}$$
$$a^2 - 2a + \left(\frac{-2}{2}\right)^2 = -\frac{1}{6} + \left(\frac{-2}{2}\right)^2$$
$$a^2 - 2a + 1 = -\frac{1}{6} + 1$$
$$(a-1)^2 = \frac{5}{6}$$
$$a - 1 = \pm\sqrt{\frac{5}{6}}$$
$$a - 1 = \pm\sqrt{\frac{5 \cdot 6}{6 \cdot 6}}$$
$$a - 1 = \pm\sqrt{\frac{30}{36}}$$
$$a - 1 = \pm\frac{\sqrt{30}}{6}$$
$$a = 1 \pm \frac{\sqrt{30}}{6}$$
$$a = \frac{6 \pm \sqrt{30}}{6}$$

12.

$$x^2 - 2x = -13$$
$$x^2 - 2x + \left(\frac{-2}{2}\right)^2 = -13 + \left(\frac{-2}{2}\right)^2$$
$$x^2 - 2x + (-1)^2 = -13 + (-1)^2$$
$$x^2 - 2x + 1 = -13 + 1$$
$$(x-1)^2 = -12$$
$$x - 1 = \pm\sqrt{-12}$$
$$x - 1 = \pm\sqrt{-4}\sqrt{3}$$
$$x - 1 = \pm 2i\sqrt{3}$$
$$x = 1 \pm 2i\sqrt{3}$$

Chapter 10 Review

13.

$$x^2 = 32$$
$$x = \pm\sqrt{32}$$
$$x = \pm\sqrt{16}\sqrt{2}$$
$$x = \pm 4\sqrt{2}$$

14.

$$(7x + 51)^2 = 11$$
$$\sqrt{(7x + 51)^2} = \pm\sqrt{11}$$
$$7x + 51 = \pm\sqrt{11}$$
$$7x = -51 \pm \sqrt{11}$$
$$x = \frac{-51 \pm \sqrt{11}}{7}$$

15. Because 7 is an odd number and not divisible by 2, the computations involved in completed the square on $x^2 + 7x$ involve fractions. The computations involved in completing the square on $x^2 + 6x$ do not.

16.

$$d = 16t^2$$
$$605 = 16t^2$$
$$\frac{605}{16} = \frac{16t^2}{16}$$
$$37.8125 = t^2$$
$$t = \pm\sqrt{37.8125}$$
$$t = \pm 6$$

Since the time cannot be negative, the ball should be dropped 6 seconds before midnight.

17.

$$2x^2 + 13x = 7$$
$$2x^2 + 13x - 7 = 0$$
$$x = \frac{-b \pm \sqrt{b^2 - 4ac}}{2a}$$
$$= \frac{-13 \pm \sqrt{(13)^2 - 4(2)(-7)}}{2(2)}$$
$$= \frac{-13 \pm \sqrt{169 - (-56)}}{4}$$
$$= \frac{-13 \pm \sqrt{225}}{4}$$
$$= \frac{-13 \pm 15}{4}$$

$$x = \frac{-13 + 15}{4} \quad \text{or} \quad x = \frac{-13 - 15}{4}$$
$$x = \frac{2}{4} \qquad\qquad x = \frac{-28}{4}$$
$$x = \frac{1}{2} \qquad\qquad x = -7$$

18.

$$-x^2 + 10x - 18 = 0$$
$$0 = x^2 - 10x + 18$$
$$x = \frac{-b \pm \sqrt{b^2 - 4ac}}{2a}$$
$$= \frac{-(-10) \pm \sqrt{(-10)^2 - 4(1)(18)}}{2(1)}$$
$$= \frac{10 \pm \sqrt{100 - 72}}{2}$$
$$= \frac{10 \pm \sqrt{28}}{2}$$
$$= \frac{10 \pm \sqrt{4}\sqrt{7}}{2}$$
$$= \frac{10 \pm 2\sqrt{7}}{2}$$
$$= \frac{10}{2} \pm \frac{2\sqrt{7}}{2}$$
$$x = 5 \pm \sqrt{7}$$
$$x = 5 + \sqrt{7} \quad \text{or} \quad x = 5 - \sqrt{7}$$
$$x \approx 7.65 \qquad\qquad x \approx 2.35$$

19.

$$x^2 - 10x = 0$$

$$x = \frac{-b \pm \sqrt{b^2 - 4ac}}{2a}$$

$$= \frac{-(-10) \pm \sqrt{(-10)^2 - 4(1)(0)}}{2(1)}$$

$$= \frac{10 \pm \sqrt{100 - 0}}{2}$$

$$= \frac{10 \pm \sqrt{100}}{2}$$

$$= \frac{10 \pm 10}{2}$$

$$x = \frac{10 + 10}{2} \qquad \text{or} \qquad x = \frac{10 - 10}{2}$$

$$x = \frac{20}{2} \qquad\qquad\qquad x = \frac{0}{2}$$

$$x = 10 \qquad\qquad\qquad x = 0$$

20.

$$3y^2 = 26y - 2$$

$$3y^2 - 26y + 2 = 0$$

$$y = \frac{-b \pm \sqrt{b^2 - 4ac}}{2a}$$

$$= \frac{-(-26) \pm \sqrt{(-26)^2 - 4(3)(2)}}{2(3)}$$

$$= \frac{26 \pm \sqrt{676 - 24}}{6}$$

$$= \frac{26 \pm \sqrt{652}}{6}$$

$$= \frac{26 \pm \sqrt{4}\sqrt{163}}{6}$$

$$= \frac{26 \pm 2\sqrt{163}}{6}$$

$$= \frac{\cancel{2}\left(13 \pm \sqrt{163}\right)}{\cancel{2} \cdot 3}$$

$$y = \frac{13 \pm \sqrt{163}}{3}$$

$$y = \frac{13 + \sqrt{163}}{3} \qquad \text{or} \quad y = \frac{13 - \sqrt{163}}{3}$$

$$y \approx 8.59 \qquad\qquad\qquad y \approx 0.08$$

21.

$$\frac{1}{3}p^2 + \frac{1}{2}p + \frac{1}{2} = 0$$

$$6\left(\frac{1}{3}p^2 + \frac{1}{2}p + \frac{1}{2}\right) = 6(0)$$

$$2p^2 + 3p + 3 = 0$$

$$p = \frac{-b \pm \sqrt{b^2 - 4ac}}{2a}$$

$$= \frac{-3 \pm \sqrt{(3)^2 - 4(2)(3)}}{2(2)}$$

$$= \frac{-3 \pm \sqrt{9 - 24}}{4}$$

$$= \frac{-3 \pm \sqrt{-15}}{4}$$

$$= \frac{-3 \pm i\sqrt{15}}{4}$$

$$p = -\frac{3}{4} \pm \frac{\sqrt{15}}{4}i$$

22.

$$3,000t^2 - 4,000t = -2,000$$

$$\frac{3,000t^2}{1,000} - \frac{4,000t}{1,000} = \frac{-2,000}{1,000}$$

$$3t^2 - 4t = -2$$

$$3t^2 - 4t + 2 = 0$$

$$t = \frac{-b \pm \sqrt{b^2 - 4ac}}{2a}$$

$$= \frac{-(-4) \pm \sqrt{(-4)^2 - 4(3)(2)}}{2(3)}$$

$$= \frac{4 \pm \sqrt{16 - 24}}{6}$$

$$= \frac{4 \pm \sqrt{-8}}{6}$$

$$= \frac{4 \pm \sqrt{-4}\sqrt{2}}{6}$$

$$= \frac{4 \pm 2i\sqrt{2}}{6}$$

$$= \frac{4}{6} \pm \frac{2\sqrt{2}}{6}i$$

$$t = \frac{2}{3} \pm \frac{\sqrt{2}}{3}i$$

Chapter 10 Review

23.

$$0.5x^2 + 0.3x - 0.1 = 0$$
$$10(0.5x^2 + 0.3x - 0.1) = 10(0)$$
$$5x^2 + 3x - 1 = 0$$
$$x = \frac{-b \pm \sqrt{b^2 - 4ac}}{2a}$$
$$= \frac{-3 \pm \sqrt{(3)^2 - 4(5)(-1)}}{2(5)}$$
$$= \frac{-3 \pm \sqrt{9 - (-20)}}{10}$$
$$= \frac{-3 \pm \sqrt{29}}{10}$$
$$x = \frac{-3 + \sqrt{29}}{10} \quad \text{or} \quad x = \frac{-3 - \sqrt{29}}{10}$$
$$x \approx 0.24 \qquad\qquad x \approx -0.84$$

24.

$$x^2 - 3x - 27 = 0$$
$$x = \frac{-b \pm \sqrt{b^2 - 4ac}}{2a}$$
$$= \frac{-(-3) \pm \sqrt{(-3)^2 - 4(1)(-27)}}{2(1)}$$
$$= \frac{3 \pm \sqrt{9 + 108}}{2}$$
$$= \frac{3 \pm \sqrt{117}}{2}$$
$$= \frac{3 \pm 3\sqrt{13}}{2}$$
$$x = \frac{3 + 3\sqrt{13}}{2} \quad \text{or} \quad x = \frac{3 - 3\sqrt{13}}{2}$$
$$x \approx 6.91 \qquad\qquad x \approx -3.91$$

25.

$$x^2 + 3x - 7 = 1$$
$$x^2 + 3x - 8 = 0$$
$$x = \frac{-b \pm \sqrt{b^2 - 4ac}}{2a}$$
$$= \frac{-3 \pm \sqrt{(3)^2 - 4(1)(-8)}}{2(1)}$$
$$= \frac{-3 \pm \sqrt{9 + 32}}{2}$$
$$= \frac{-3 \pm \sqrt{41}}{2}$$

26.

$$-4x^2 - 2x = -3$$
$$0 = 4x^2 + 2x - 3$$
$$x = \frac{-b \pm \sqrt{b^2 - 4ac}}{2a}$$
$$= \frac{-2 \pm \sqrt{(2)^2 - 4(4)(-3)}}{2(4)}$$
$$= \frac{-2 \pm \sqrt{4 + 48}}{8}$$
$$= \frac{-2 \pm \sqrt{52}}{8}$$
$$= \frac{-2 \pm 2\sqrt{13}}{8}$$
$$= \frac{-1 \pm \sqrt{13}}{4}$$

27. –2 is not a factor of the numerator—it is a term. Only common factors of the numerator and denominator can be removed.

28. a. $(x + x + 2)$ ft $= (2x + 2)$ ft
    b. $(x + x + 6)$ ft $= (2x + 6)$ ft

29. Let the border on top/bottom be x. Then the borders on the side would be ½x. The length of the picture is $35 - 2(x)$ and the width of the picture is $23 - 2($½$x)$.

$$(\text{length})(\text{width}) = \text{Area}$$

$$\left(35 - 2x\right)\left(23 - 2 \cdot \frac{1}{2}x\right) = 615$$

$$\left(35 - 2x\right)\left(23 - x\right) = 615$$

$$805 - 35x - 46x + 2x^2 = 615$$

$$2x^2 - 81x + 805 = 615$$

$$2x^2 - 81x + 190 = 0$$

$$x = \frac{-b \pm \sqrt{b^2 - 4ac}}{2a}$$

$$= \frac{-(-81) \pm \sqrt{(-81)^2 - 4(2)(190)}}{2(2)}$$

$$= \frac{81 \pm \sqrt{6,561 - 1,520}}{4}$$

$$= \frac{81 \pm \sqrt{5,041}}{4}$$

$$= \frac{81 \pm 71}{4}$$

$$x = \frac{81 + 71}{4} \quad \text{or} \quad x = \frac{81 - 71}{4}$$

$$x = \frac{152}{4} \qquad\qquad x = \frac{10}{4}$$

$$x = 38 \qquad\qquad x = 2.5$$

Since the total length is only 35 inches, the border cannot be 38 inches. So, the border on top and bottom is 2.5 inches and the border on the sides is ½ of 2.5, which is 1.25 inches.

30. The # of students would be $300 - 5x$. The price per student would be $20 + 0.50x$. (# of students)(price per student) = revenue

$$\left(300 - 5x\right)\left(20 + 0.5x\right) = 6,240$$

$$6,000 + 150x - 100x - 2.5x^2 = 6,240$$

$$6,000 + 50x - 2.5x^2 = 6,240$$

$$0 = 2.5x^2 - 50x + 240$$

$$\frac{0}{2.5} = \frac{2.5x^2}{2.5} - \frac{50x}{2.5} + \frac{240}{2.5}$$

$$0 = x^2 - 20x + 96$$

$$x = \frac{-b \pm \sqrt{b^2 - 4ac}}{2a}$$

$$= \frac{-(-20) \pm \sqrt{(-20)^2 - 4(1)(96)}}{2(1)}$$

$$= \frac{20 \pm \sqrt{400 - 384}}{2}$$

$$= \frac{20 \pm \sqrt{16}}{2}$$

$$= \frac{20 \pm 4}{2}$$

$$x = \frac{20 + 4}{2} \quad \text{or} \quad x = \frac{20 - 4}{2}$$

$$x = \frac{24}{2} \qquad\qquad x = \frac{16}{2}$$

$$x = 12 \qquad\qquad x = 8$$

$$\begin{aligned}
\text{price} &= 20 + 0.5x & \text{price} &= 20 + 0.5x \\
&= 20 + 0.5(12) & &= 20 + 0.5(8) \\
&= 20 + 6 & &= 20 + 4 \\
&= \$26 & &= \$24
\end{aligned}$$

31.

$$d = -16t^2 + 40t + 5$$

$$25 = -16t^2 + 40t + 5$$

$$16t^2 - 40t + 20 = 0$$

$$t = \frac{-b \pm \sqrt{b^2 - 4ac}}{2a}$$

$$= \frac{-(-40) \pm \sqrt{(-40)^2 - 4(16)(20)}}{2(16)}$$

$$= \frac{40 \pm \sqrt{1,600 - 1,280}}{32}$$

$$= \frac{40 \pm \sqrt{320}}{32}$$

$$= \frac{40 \pm \sqrt{64}\sqrt{5}}{32}$$

$$= \frac{40 \pm 8\sqrt{5}}{32}$$

$$= \frac{8(5 \pm \sqrt{5})}{32}$$

$$= \frac{5 \pm \sqrt{5}}{4}$$

$$t = \frac{5 + \sqrt{5}}{4} \quad \text{or} \quad t = \frac{5 - \sqrt{5}}{4}$$

$$t = 1.8 \qquad\qquad t = 0.7$$

He will be able to grab it in 0.7 seconds and 1.8 seconds.

32. Let x = the length of the shorter leg. Then $x + 23$ is the length of the longer leg.

$$(\text{short leg})^2 + (\text{long leg})^2 = (\text{hypotenuse})^2$$

$$x^2 + (x + 23)^2 = 65^2$$

$$x^2 + x^2 + 46x + 529 = 4,225$$

$$2x^2 + 46x - 3,696 = 0$$

$$\frac{2x^2}{2} + \frac{46x}{2} - \frac{3,696}{2} = \frac{0}{2}$$

$$x^2 + 23x - 1,848 = 0$$

$$x = \frac{-b \pm \sqrt{b^2 - 4ac}}{2a}$$

$$x = \frac{-23 \pm \sqrt{23^2 - 4(1)(-1,848)}}{2(1)}$$

$$= \frac{-23 \pm \sqrt{529 + 7,392}}{2}$$

$$= \frac{-23 \pm \sqrt{7,921}}{2}$$

$$= \frac{-23 \pm 89}{2}$$

$$x = \frac{-23 + 89}{2} \quad \text{or} \quad x = \frac{-23 - 89}{2}$$

$$= \frac{66}{2} \qquad\qquad = \frac{-112}{2}$$

$$= 33 \qquad\qquad\quad = -56$$

Since the length cannot be negative, then the length of the shorter leg is 33 in.
Short leg = x = 33 in.
Long leg = $x + 23$ = 33 + 23 = 56 in.

SECTION 10.3
The Discriminant and Equations That Can be Written in Quadratic Form

33. $3x^2 + 4x - 3 = 0$

$$b^2 - 4ac = (4)^2 - 4(3)(-3)$$

$$= 16 - (-36)$$

$$= 52$$

two different irrational-number solutions

34. $4x^2 - 5x + 7 = 0$

$$b^2 - 4ac = (-5)^2 - 4(4)(7)$$

$$= 25 - 112$$

$$= -87$$

two imaginary-number solutions that are complex conjugates

35.
$$3x^2 - 4x + \frac{4}{3} = 0$$
$$3\left(3x^2 - 4x + \frac{4}{3}\right) = 3(0)$$
$$9x^2 - 12x + 4 = 0$$
$$b^2 - 4ac = (-12)^2 - 4(9)(4)$$
$$= 144 - 144$$
$$= 0$$
one repeated solution, a rational number

36.
$$m(2m - 3) = 20$$
$$2m^2 - 3m = 20$$
$$2m^2 - 3m - 20 = 0$$
$$b^2 - 4ac = (-3)^2 - 4(2)(-20)$$
$$= 9 - (-160)$$
$$= 169$$
two different rational-number solutions

37. Let $y = \sqrt{x}$.
$$x - 13\sqrt{x} + 12 = 0$$
$$\left(\sqrt{x}\right)^2 - 13\sqrt{x} + 12 = 0$$
$$y^2 - 13y + 12 = 0$$
$$(y - 12)(y - 1) = 0$$
$$y - 12 = 0 \quad \text{and} \quad y - 1 = 0$$
$$y = 12 \qquad\qquad y = 1$$
$$\sqrt{x} = 12 \qquad\qquad \sqrt{x} = 1$$
$$\left(\sqrt{x}\right)^2 = (12)^2 \quad \left(\sqrt{x}\right)^2 = (1)^2$$
$$x = 144 \qquad\qquad x = 1$$

38. Let $y = a^{1/3}$.
$$a^{2/3} + a^{1/3} - 6 = 0$$
$$\left(a^{1/3}\right)^2 + a^{1/3} - 6 = 0$$
$$y^2 + y - 6 = 0$$
$$(y + 3)(y - 2) = 0$$
$$y + 3 = 0 \quad \text{or} \quad y - 2 = 0$$
$$y = -3 \qquad\qquad y = 2$$
$$a^{1/3} = -3 \qquad\qquad a^{1/3} = 2$$
$$\left(a^{1/3}\right)^3 = (-3)^3 \quad \left(a^{1/3}\right)^3 = (2)^3$$
$$a = -27 \qquad\qquad a = 8$$

39. Let $y = x^2$.
$$3x^4 + x^2 - 2 = 0$$
$$3\left(x^2\right)^2 + x^2 - 2 = 0$$
$$3y^2 + y - 2 = 0$$
$$(3y - 2)(y + 1) = 0$$
$$3y - 2 = 0 \quad \text{or} \quad y + 1 = 0$$
$$3y = 2$$
$$y = \frac{2}{3} \qquad\qquad y = -1$$
$$x^2 = \frac{2}{3} \qquad\qquad x^2 = -1$$
$$x = \pm\sqrt{\frac{2}{3}} \qquad x = \pm\sqrt{-1}$$
$$x = \pm\sqrt{\frac{2}{3} \cdot \frac{3}{3}} \quad x = \pm i$$
$$x = \pm\frac{\sqrt{6}}{3}$$
$$x = \frac{\sqrt{6}}{3}, \ -\frac{\sqrt{6}}{3}, \ i, \ -i$$

40.
$$\frac{6}{x + 2} + \frac{6}{x + 1} = 5$$
$$(x + 2)(x + 1)\left(\frac{6}{x + 2} + \frac{6}{x + 1}\right) = 5(x + 2)(x + 1)$$
$$6(x + 1) + 6(x + 2) = 5\left(x^2 + 2x + x + 2\right)$$
$$6x + 6 + 6x + 12 = 5\left(x^2 + 3x + 2\right)$$
$$12x + 18 = 5x^2 + 15x + 10$$
$$0 = 5x^2 + 3x - 8$$
$$0 = (5x + 8)(x - 1)$$
$$5x + 8 = 0 \quad \text{or} \quad x - 1 = 0$$
$$5x = -8 \qquad\qquad x = 1$$
$$x = -\frac{8}{5}$$

Chapter 10 Review

41. Let $y = (x - 7)$.

$(x-7)^2 + 6(x-7) + 10 = 0$

$y^2 + 6y + 10 = 0$

$$y = \frac{-b \pm \sqrt{b^2 - 4ac}}{2a}$$

$$y = \frac{-6 \pm \sqrt{(6)^2 - 4(1)(10)}}{2(1)}$$

$$y = \frac{-6 \pm \sqrt{36 - 40}}{2}$$

$$y = \frac{-6 \pm \sqrt{-4}}{2}$$

$$y = \frac{-6 \pm 2i}{2}$$

$$y = \frac{-6}{2} \pm \frac{2}{2}i$$

$$y = -3 \pm i$$

$$x - 7 = -3 \pm i$$

$$x = -3 \pm i + 7$$

$$x = 4 \pm i$$

42. Let $y = \dfrac{1}{m^2}$.

$m^{-4} - 2m^{-2} + 1 = 0$

$\dfrac{1}{m^4} - 2\left(\dfrac{1}{m^2}\right) + 1 = 0$

$\left(\dfrac{1}{m^2}\right)^2 - 2\left(\dfrac{1}{m^2}\right) + 1 = 0$

$y^2 - 2y + 1 = 0$

$(y - 1)(y - 1) = 0$

$y - 1 = 0 \quad$ or $\quad y - 1 = 0$

$\quad y = 1 \qquad\qquad y = 1$

$\quad \dfrac{1}{m^2} = \dfrac{1}{1} \qquad \dfrac{1}{m^2} = \dfrac{1}{1}$

$\quad m^2 = 1 \qquad\quad m^2 = 1$

$\quad m = \pm\sqrt{1} \qquad m = \pm\sqrt{1}$

$\quad m = \pm 1 \qquad\quad m = \pm 1$

$\quad m = 1, \ -1, \ 1, \ -1$

43. Let $y = \dfrac{x+1}{x}$.

$4\left(\dfrac{x+1}{x}\right)^2 + 12\left(\dfrac{x+1}{x}\right) + 9 = 0$

$4y^2 + 12y + 9 = 0$

$(2y + 3)(2y + 3) = 0$

$2y + 3 = 0 \quad$ or $\quad 2y + 3 = 0$

$\quad 2y = -3 \qquad\qquad 2y = -3$

$\quad y = -\dfrac{3}{2} \qquad\qquad y = -\dfrac{3}{2}$

$\quad \dfrac{x+1}{x} = -\dfrac{3}{2} \qquad \dfrac{x+1}{x} = -\dfrac{3}{2}$

$\quad 2(x+1) = -3x \qquad 2(x+1) = -3x$

$\quad 2x + 2 = -3x \qquad 2x + 2 = -3x$

$\quad 2 = -5x \qquad\qquad 2 = -5x$

$\quad -\dfrac{2}{5} = x \qquad\qquad -\dfrac{2}{5} = x$

44. Let $y = m^{1/5}$.

$2m^{2/5} - 5m^{1/5} + 2 = 0$

$2\left(m^{1/5}\right)^2 - 5m^{1/5} + 2 = 0$

$2y^2 - 5y + 2 = 0$

$(2y - 1)(y - 2) = 0$

$2y - 1 = 0 \quad$ or $\quad y - 2 = 0$

$\quad 2y = 1 \qquad\qquad y = 2$

$\quad y = \dfrac{1}{2}$

$\quad m^{1/5} = \dfrac{1}{2} \qquad\qquad m^{1/5} = 2$

$\quad \left(m^{1/5}\right)^5 = \left(\dfrac{1}{2}\right)^5 \qquad \left(m^{1/5}\right)^5 = (2)^5$

$\quad m = \dfrac{1}{32} \qquad\qquad m = 32$

45. Let x = the number of minutes it takes the younger girl to do the yard work.

Younger girl's work in 1 minute = $\dfrac{1}{x}$

Older girl's work in 1 minute = $\dfrac{1}{x-20}$

Work together in 1 minute = $\dfrac{1}{45}$

Younger girl + older girl = work together

$$\frac{1}{x}+\frac{1}{x-20}=\frac{1}{45}$$

$$45x(x-20)\left(\frac{1}{x}+\frac{1}{x-20}\right)=45x(x-20)\left(\frac{1}{45}\right)$$

$$45(x-20)+45x=x(x-20)$$

$$45x-900+45x=x^2-20x$$

$$90x-900=x^2-20x$$

$$0=x^2-110x+900$$

$$x=\frac{-b\pm\sqrt{b^2-4ac}}{2a}$$

$$x=\frac{-(-110)\pm\sqrt{(-110)^2-4(1)(900)}}{2(1)}$$

$$x=\frac{110\pm\sqrt{12,100-3,600}}{2}$$

$$x=\frac{110\pm\sqrt{8,500}}{2}$$

$$x=\frac{110\pm92.2}{2}$$

$$x=\frac{110+92.2}{2}\quad\text{or}\quad x=\frac{110-92.2}{2}$$

$$x=\frac{202}{2}\qquad\qquad x=\frac{17}{2}$$

$$x=101\qquad\qquad x=8$$

$$x-20=101-20\quad x-20=8-20$$
$$=81\qquad\qquad =-12$$

Since the older sister's time cannot be negative, her time must be about 81 min.

46. Recall $t=\dfrac{d}{r}$.

	distance	rate	time
original rate	150	r	$t=\dfrac{150}{r}$
increased rate	150	$r+20$	$t-2=\dfrac{150}{r+20}$

Solve the second time for t.

$$t-2=\frac{150}{r+20}$$

$$t=\frac{150}{r+20}+2$$

Set the times equal and solve for r.

$$\frac{150}{r}=\frac{150}{r+20}+2$$

$$r(r+20)\left(\frac{150}{r}\right)=r(r+20)\left(\frac{150}{r+20}+2\right)$$

$$150(r+20)=150(r)+2r(r+20)$$

$$150r+3,000=150r+2r^2+40r$$

$$150r+3,000-150r=150r+2r^2+40r-150r$$

$$3,000=2r^2+40r$$

$$0=2r^2+40r-3,000$$

$$0=\frac{2r^2}{2}+\frac{40r}{2}-\frac{3,000}{2}$$

$$0=r^2+20r-1,500$$

$$r=\frac{-b\pm\sqrt{b^2-4ac}}{2a}$$

$$=\frac{-20\pm\sqrt{20^2-4(1)(-1,500)}}{2(1)}$$

$$=\frac{-20\pm\sqrt{400+6,000}}{2}$$

$$=\frac{-20\pm\sqrt{6,400}}{2}$$

$$=\frac{-20\pm80}{2}$$

$$r=\frac{-20+80}{2}\quad\text{or}\quad r=\frac{-20-80}{2}$$

$$r=\frac{60}{2}\qquad\qquad r=\frac{-100}{2}$$

$$r=30\qquad\qquad r=-50$$

Since the rate cannot be negative, her original rate is 30 mph.

47. Since $2005 - 1980 = 25$, let $x = 25$.
$$A(x) = 0.03x^2 - 0.88x + 37.3$$
$$A(25) = 0.03x^2 - 0.88x + 37.3$$
$$= 0.03(25)^2 - 0.88(25) + 37.3$$
$$= 0.03(625) - 0.88(25) + 37.3$$
$$= 18.75 - 22 + 37.3$$
$$= 34.05$$
$$\approx 34.1 \text{ million}$$

48. The graph of the quadratic function $f(x) = a(x-h)^2 + k$ is a parabola with vertex at (h, k). The axis of symmetry is the line $x = h$. The parabola opens upward when $a > 0$ and downward when $a < 0$.

49.

x	$f(x) = 2x^2$
-2	$2(-2)^2 = 8$
-1	$2(-1)^2 = 2$
0	$2(0)^2 = 0$
1	$2(1)^2 = 2$
2	$2(2)^2 = 8$

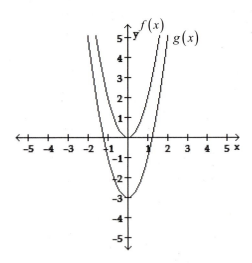

The graph of $g(x)$ is shifted down 3 units.

50.

x	$f(x) = -\dfrac{1}{4}x^2$
-2	$-\dfrac{1}{4}(-2)^2 = -1$
-1	$-\dfrac{1}{4}(-1)^2 = -\dfrac{1}{4}$
0	$-\dfrac{1}{4}(0)^2 = 0$
1	$-\dfrac{1}{4}(1)^2 = -\dfrac{1}{4}$
2	$-\dfrac{1}{4}(2)^2 = -1$

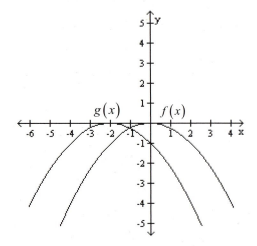

The graph of $g(x)$ is shifted left 2 units.

51. $f(x) = -2(x-1)^2 + 4$
Vertex: $(1, 4)$
Axis of symmetry: $x = 1$

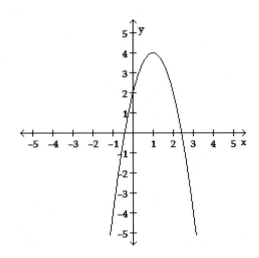

52. $f(x) = 4x^2 + 16x + 9$

$f(x) = 4(x^2 + 4x) + 9$

$f(x) = 4(x^2 + 4x + 4) + 9 - 4(4)$

$f(x) = 4(x + 2)^2 + 9 - 16$

$f(x) = 4(x + 2)^2 - 7$

Vertex: $(-2, -7)$

Axis of symmetry: $x = -2$

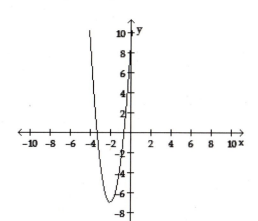

53. $f(x) = -2x^2 + 4x - 8$

$x = \dfrac{-b}{2a}$

$= \dfrac{-4}{2(-2)}$

$= \dfrac{-4}{-4}$

$= 1$

$f(1) = -2(1)^2 + 4(1) - 8$

$= -2(1) + 4(1) - 8$

$= -2 + 4 - 8$

$= -6$

The vertex is $(1, -6)$.

54. $f(x) = x^2 + x - 2$

Step 1: Since $a = 1 > 0$, the parabola opens upward.

Step 2: Find the vertex and axis of symmetry.

$x = \dfrac{-b}{2a} = \dfrac{-1}{2(1)} = -\dfrac{1}{2}$

$y = \left(-\dfrac{1}{2}\right)^2 + \left(-\dfrac{1}{2}\right) - 2$

$= \dfrac{1}{4} - \dfrac{1}{2} - 2$

$= -\dfrac{9}{4}$

vertex: $\left(-\dfrac{1}{2}, -\dfrac{9}{4}\right)$

The axis of symmetry is $x = -\dfrac{1}{2}$.

Step 3: Find the x– and y–intercepts.
Since $c = -2$, the y–intercept is $(0, -2)$.
To find the x–intercepts, let $y = 0$ and solve the equation for x.

$0 = x^2 + x - 2$

$0 = (x + 2)(x - 1)$

$x + 2 = 0 \quad$ or $\quad x - 1 = 0$

$x = -2 \qquad\qquad x = 1$

x–intercepts: $(-2, 0)$ and $(1, 0)$

y–intercept: $(0, -2)$

Step 4: Use symmetry to find the point $(-1, -2)$.

Step 5: Plots the points and draw the parabola.

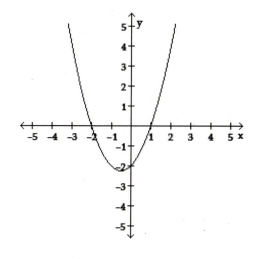

Chapter 10 Review

55. The vertex would represent the maximum of the function. Use the vertex formula to find the value of x and use substitution to find the number of farms.

$$x = \frac{-b}{2a}$$
$$= \frac{-155,652}{2(-1,526)}$$
$$= \frac{-155,652}{-3,052}$$
$$= 51$$
$$1870 + 51 = 1921$$

$$N(51) = -1,526(51)^2 + 155,652(51) + 2,500,200$$
$$= -1,526(2,601) + 155,652(51) + 2,500,200$$
$$= -3,969,126 + 7,938,252 + 2,500,200$$
$$= 6,469,326 \text{ farms}$$

56. The x–intercepts are $(-2, 0)$ and $\left(\frac{1}{3}, 0\right)$, so the solutions to the equation are $x = -2$ and $x = \frac{1}{3}$.

SECTION 10.5
Quadratic and Other Nonlinear Inequalities

57.
$$x^2 + 2x - 35 > 0$$
$$x^2 + 2x - 35 = 0$$
$$(x + 7)(x - 5) = 0$$
$$x + 7 = 0 \quad \text{or} \quad x - 5 = 0$$
$$x = -7 \qquad x = 5$$
Critical numbers $= -7$ and 5.
$$(-\infty, -7) \cup (5, \infty)$$

58.
$$x^2 \leq 81$$
$$x^2 - 81 \leq 0$$
$$x^2 - 81 = 0$$
$$(x + 9)(x - 9) = 0$$
$$x + 9 = 0 \quad \text{or} \quad x - 9 = 0$$
$$x = -9 \qquad x = 9$$
Critical numbers $= -9$ and 9
$$[-9, 9]$$

59.
$$\frac{3}{x} \leq 5$$
$$\frac{3}{x} = 5$$
$$\frac{3}{x} - 5 = 0$$
$$\frac{3}{x} - \frac{5x}{x} = 0$$
$$\frac{3 - 5x}{x} = 0$$
$$x\left(\frac{3 - 5x}{x}\right) = x(0)$$
$$3 - 5x = 0$$
$$-5x = -3$$
$$x = \frac{3}{5}$$
Set the denominator $= 0$.
$$x = 0$$
Critical numbers $= 0$ and $\frac{3}{5}$
$$(-\infty, 0) \cup \left[\frac{3}{5}, \infty\right)$$

60.

$$\frac{2x^2 - x - 28}{x - 1} > 0$$

$$\frac{2x^2 - x - 28}{x - 1} = 0$$

$$(x-1)\left(\frac{2x^2 - x - 28}{x - 1}\right) = (x-1)(0)$$

$$2x^2 - x - 28 = 0$$

$$(x-4)(2x+7) = 0$$

$$x - 4 = 0 \quad \text{or} \quad 2x + 7 = 0$$

$$x = 4 \qquad x = -\frac{7}{2}$$

Set the denominator $= 0$.

$$x - 1 = 0$$

$$x = 1$$

Critical numbers $= -\dfrac{7}{2}$, 1, and 4

$$\left(-\frac{7}{2}, 1\right) \cup (4, \infty)$$

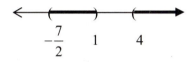

64. $y \geq -|x|$

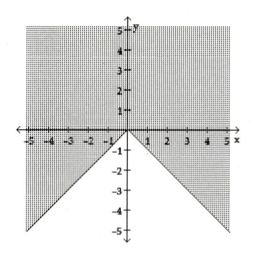

61. $\left[-4, \dfrac{2}{3}\right]$

62. $(-\infty, 0) \cup (1, \infty)$

63. $y < \dfrac{1}{2}x^2 - 1$

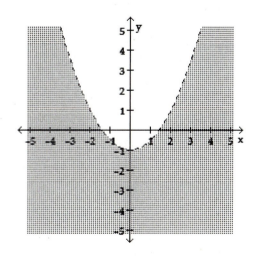

Chapter 10 Review

CHAPTER 10 TEST

1. a. An equation of the form
$ax^2 + bx + c = 0$, where $a \neq 0$, is called a **quadratic** equation.

   b. When we add 81 to $x^2 + 18x$, we say that we have **completed** the **square** on $x^2 + 18x$.

   c. The lowest point on a parabola that opens upward, or the highest point on a parabola that opens downward, is called the **vertex** of the parabola.

   d. $\dfrac{x-5}{x^2-x-56} > 0$ is an example of a _ **rational** inequality in one variable.

   e. $y \leq x^2 - 4x + 3$ is an example of a **nonlinear** inequality in two variables.

2.
$$x^2 - 63 = 0$$
$$x^2 = 63$$
$$x = \pm\sqrt{63}$$
$$x = \pm\sqrt{9}\sqrt{7}$$
$$x = \pm 3\sqrt{7}$$
$$x \approx \pm 7.94$$

3.
$$(a+7)^2 = 50$$
$$a + 7 = \pm\sqrt{50}$$
$$a + 7 = \pm\sqrt{25}\sqrt{2}$$
$$a + 7 = \pm 5\sqrt{2}$$
$$a = -7 \pm 5\sqrt{2}$$

4.
$$m^2 + 4 = 0$$
$$m^2 = -4$$
$$m = \pm\sqrt{-4}$$
$$m = \pm 2i$$

5.
$$\left(\frac{11}{2}\right)^2 = \frac{121}{4}$$
$$x^2 + 11x + \frac{121}{4} = \left(x + \frac{11}{2}\right)^2$$

6.
$$x^2 + 3x - 2 = 0$$
$$x^2 + 3x = 2$$
$$x^2 + 3x + \left(\frac{3}{2}\right)^2 = 2 + \left(\frac{3}{2}\right)^2$$
$$x^2 + 3x + \frac{9}{4} = 2 + \frac{9}{4}$$
$$\left(x + \frac{3}{2}\right)^2 = \frac{17}{4}$$
$$\sqrt{\left(x + \frac{3}{2}\right)^2} = \pm\sqrt{\frac{17}{4}}$$
$$x + \frac{3}{2} = \pm\frac{\sqrt{17}}{2}$$
$$x = -\frac{3}{2} \pm \frac{\sqrt{17}}{2}$$
$$x = -\frac{3}{2} + \frac{\sqrt{17}}{2} \quad \text{or} \quad x = -\frac{3}{2} - \frac{\sqrt{17}}{2}$$
$$x \approx 0.56 \qquad\qquad x \approx -3.56$$

7.
$$2x^2 + 8x + 12 = 0$$
$$\frac{2x^2}{2} + \frac{8x}{2} + \frac{12}{2} = \frac{0}{2}$$
$$x^2 + 4x + 6 = 0$$
$$x^2 + 4x = -6$$
$$x^2 + 4x + \left(\frac{4}{2}\right)^2 = -6 + \left(\frac{4}{2}\right)^2$$
$$x^2 + 4x + 4 = -6 + 4$$
$$(x + 2)^2 = -2$$
$$\sqrt{(x+2)^2} = \pm\sqrt{-2}$$
$$x + 2 = \pm i\sqrt{2}$$
$$x = -2 \pm i\sqrt{2}$$

8.

$$(3x-2)^2 = 18$$

$$\sqrt{(3x-2)^2} = \sqrt{18}$$

$$3x-2 = \pm\sqrt{9}\sqrt{2}$$

$$3x-2 = \pm 3\sqrt{2}$$

$$3x = 2 \pm 3\sqrt{2}$$

$$x = \frac{2 \pm 3\sqrt{2}}{3}$$

9.

$$4x^2 + 4x - 1 = 0$$

$$x = \frac{-b \pm \sqrt{b^2 - 4ac}}{2a}$$

$$= \frac{-4 \pm \sqrt{(4)^2 - 4(4)(-1)}}{2(4)}$$

$$= \frac{-4 \pm \sqrt{16 + 16}}{8}$$

$$= \frac{-4 \pm \sqrt{32}}{8}$$

$$= \frac{-4 \pm \sqrt{16}\sqrt{2}}{8}$$

$$= \frac{-4 \pm 4\sqrt{2}}{8}$$

$$= \frac{4\left(-1 \pm \sqrt{2}\right)}{4 \cdot 2}$$

$$= \frac{-1 \pm \sqrt{2}}{2}$$

10.

$$\frac{t^2}{8} - \frac{t}{4} = \frac{1}{2}$$

$$8\left(\frac{t^2}{8} - \frac{t}{4}\right) = 8\left(\frac{1}{2}\right)$$

$$t^2 - 2t = 4$$

$$t^2 - 2t - 4 = 0$$

$$t = \frac{-b \pm \sqrt{b^2 - 4ac}}{2a}$$

$$= \frac{-(-2) \pm \sqrt{(-2)^2 - 4(1)(-4)}}{2(1)}$$

$$= \frac{2 \pm \sqrt{4 - (-16)}}{2}$$

$$= \frac{2 \pm \sqrt{20}}{2}$$

$$= \frac{2 \pm \sqrt{4}\sqrt{5}}{2}$$

$$= \frac{2 \pm 2\sqrt{5}}{2}$$

$$= \frac{\cancel{2}\left(1 \pm \sqrt{5}\right)}{\cancel{2}}$$

$$t = 1 \pm \sqrt{5}$$

11.

$$-t^2 + 4t - 13 = 0$$

$$0 = t^2 - 4t + 13$$

$$t = \frac{-b \pm \sqrt{b^2 - 4ac}}{2a}$$

$$= \frac{-(-4) \pm \sqrt{(-4)^2 - 4(1)(13)}}{2(1)}$$

$$= \frac{4 \pm \sqrt{16 - 52}}{2}$$

$$= \frac{4 \pm \sqrt{-36}}{2}$$

$$= \frac{4 \pm 6i}{2}$$

$$= \frac{4}{2} \pm \frac{6}{2}i$$

$$t = 2 \pm 3i$$

Chapter 10 Test

12.

$$0.01x^2 = -0.08x - 0.15$$

$$100\left(0.01x^2\right) = 100\left(-0.08x - 0.15\right)$$

$$x^2 = -8x - 15$$

$$x^2 + 8x + 15 = 0$$

$$x = \frac{-b \pm \sqrt{b^2 - 4ac}}{2a}$$

$$= \frac{-8 \pm \sqrt{(8)^2 - 4(1)(15)}}{2(1)}$$

$$= \frac{-8 \pm \sqrt{64 - 60}}{2}$$

$$= \frac{-8 \pm \sqrt{4}}{2}$$

$$= \frac{-8 \pm 2}{2}$$

$$x = \frac{-8 + 2}{2} \quad \text{or} \quad x = \frac{-8 - 2}{2}$$

$$x = \frac{-6}{2} \qquad\qquad x = \frac{-10}{2}$$

$$x = -3 \qquad\qquad x = -5$$

13.

$$m^2 - 94m = -2,209$$

$$m^2 - 94m + 2,209 = 0$$

$$m = \frac{-b \pm \sqrt{b^2 - 4ac}}{2a}$$

$$= \frac{-(-94) \pm \sqrt{(-94)^2 - 4(1)(2,209)}}{2(1)}$$

$$= \frac{94 \pm \sqrt{8,836 - 8,836}}{2}$$

$$= \frac{94 \pm \sqrt{0}}{2}$$

$$= \frac{94 \pm 0}{2}$$

$$x = \frac{94 + 0}{2} \quad \text{or} \quad x = \frac{94 - 0}{2}$$

$$x = \frac{94}{2} \qquad\qquad x = \frac{94}{2}$$

$$x = 47 \qquad\qquad x = 47$$

A repeated solution of 47

14.

$$3x^2 - 20 = 10$$

$$3x^2 = 30$$

$$\frac{3x^2}{3} = \frac{30}{3}$$

$$x^2 = 10$$

$$\sqrt{x^2} = \sqrt{10}$$

$$x = \pm\sqrt{10}$$

15. Let $x = \sqrt{y}$

$$2y - 3\sqrt{y} + 1 = 0$$

$$2\left(\sqrt{y}\right)^2 - 3\sqrt{y} + 1 = 0$$

$$2x^2 - 3x + 1 = 0$$

$$(2x - 1)(x - 1) = 0$$

$$2x - 1 = 0 \quad \text{or} \quad x - 1 = 0$$

$$2x = 1 \qquad\qquad x - 1 + 1 = 0 + 1$$

$$x = \frac{1}{2} \qquad\qquad x = 1$$

$$\sqrt{y} = \frac{1}{2} \qquad\qquad \sqrt{y} = 1$$

$$\left(\sqrt{y}\right)^2 = \left(\frac{1}{2}\right)^2 \quad \left(\sqrt{y}\right)^2 = (1)^2$$

$$y = \frac{1}{4} \qquad\qquad y = 1$$

16. Let $y = \dfrac{1}{m}$.

$$3 = m^{-2} - 2m^{-1}$$
$$0 = m^{-2} - 2m^{-1} - 3$$
$$m^{-2} - 2m^{-1} - 3 = 0$$
$$\left(\dfrac{1}{m}\right)^2 - 2\left(\dfrac{1}{m}\right) - 3 = 0$$
$$y^2 - 2y - 3 = 0$$
$$(y-3)(y+1) = 0$$
$$y - 3 = 0 \quad \text{or} \quad y + 1 = 0$$
$$y = 3 \qquad\qquad y = -1$$

$$\dfrac{1}{m} = 3 \qquad\qquad \dfrac{1}{m} = -1$$
$$m\left(\dfrac{1}{m}\right) = m(3) \quad m\left(\dfrac{1}{m}\right) = m(-1)$$
$$1 = 3m \qquad\qquad 1 = -m$$
$$\dfrac{1}{3} = m \qquad\qquad -1 = m$$

17. Let $y = x^2$
$$x^4 - x^2 - 12 = 0$$
$$\left(x^2\right)^2 - x^2 - 12 = 0$$
$$y^2 - y - 12 = 0$$
$$(y-4)(y+3) = 0$$
$$y - 4 = 0 \quad \text{or} \quad y + 3 = 0$$
$$y = 4 \qquad\qquad y = -3$$
$$x^2 = 4 \qquad\qquad x^2 = -3$$
$$x = \pm\sqrt{4} \qquad\quad x = \pm\sqrt{-3}$$
$$x = \pm 2 \qquad\qquad x = \pm i\sqrt{3}$$
$$x = 2,\ -2,\ i\sqrt{3},\ -i\sqrt{3}$$

18. Let $y = \dfrac{x+2}{3x}$.

$$4\left(\dfrac{x+2}{3x}\right)^2 - 4\left(\dfrac{x+2}{3x}\right) - 3 = 0$$
$$4y^2 - 4y - 3 = 0$$
$$(2y+1)(2y-3) = 0$$
$$2y + 1 = 0 \quad \text{or} \quad 2y - 3 = 0$$
$$2y = -1 \qquad\qquad 2y = 3$$
$$y = -\dfrac{1}{2} \qquad\qquad y = \dfrac{3}{2}$$
$$\dfrac{x+2}{3x} = -\dfrac{1}{2} \qquad \dfrac{x+2}{3x} = \dfrac{3}{2}$$
$$2(x+2) = -1(3x) \quad 2(x+2) = 3(3x)$$
$$2x + 4 = -3x \qquad 2x + 4 = 9x$$
$$4 = -5x \qquad\qquad 4 = 7x$$
$$-\dfrac{4}{5} = x \qquad\qquad \dfrac{4}{7} = x$$

19.
$$\dfrac{1}{n+2} = \dfrac{1}{3} - \dfrac{1}{n}$$
$$3n(n+2)\left(\dfrac{1}{n+2}\right) = 3n(n+2)\left(\dfrac{1}{3} - \dfrac{1}{n}\right)$$
$$3n = n(n+2) - 3(n+2)$$
$$3n = n^2 + 2n - 3n - 6$$
$$3n = n^2 - n - 6$$
$$0 = n^2 - 4n - 6$$
$$n = \dfrac{-b \pm \sqrt{b^2 - 4ac}}{2a}$$
$$= \dfrac{-(-4) \pm \sqrt{(-4)^2 - 4(1)(-6)}}{2(1)}$$
$$= \dfrac{4 \pm \sqrt{16 + 24}}{2}$$
$$= \dfrac{4 \pm \sqrt{40}}{2}$$
$$= \dfrac{4 \pm \sqrt{4}\sqrt{10}}{2}$$
$$= \dfrac{4 \pm 2\sqrt{10}}{2}$$
$$n = 2 \pm \sqrt{10}$$

20. Let $y = a^{1/3}$.

$$5a^{2/3} + 11a^{1/3} = -2$$
$$5a^{2/3} + 11a^{1/3} + 2 = 0$$
$$5\left(a^{1/3}\right)^2 + 11a^{1/3} + 2 = 0$$
$$5y^2 + 11y + 2 = 0$$
$$(5y + 1)(y + 2) = 0$$

$$5y + 1 = 0 \quad \text{or} \quad y + 2 = 0$$
$$5y = -1 \qquad\qquad y = -2$$
$$y = -\frac{1}{5}$$

$$a^{1/3} = -\frac{1}{5} \qquad a^{1/3} = -2$$

$$\left(a^{1/3}\right)^3 = \left(-\frac{1}{5}\right)^3 \qquad \left(a^{1/3}\right)^3 = (-2)^3$$

$$a = -\frac{1}{125} \qquad\qquad a = -8$$

21.

$$10x(x + 1) = -3$$
$$10x^2 + 10x + 3 = 0$$

$$x = \frac{-b \pm \sqrt{b^2 - 4ac}}{2a}$$
$$= \frac{-10 \pm \sqrt{(10)^2 - 4(10)(3)}}{2(10)}$$
$$= \frac{-10 \pm \sqrt{100 - 120}}{20}$$
$$= \frac{-10 \pm \sqrt{-20}}{20}$$
$$= \frac{-10 \pm 2i\sqrt{5}}{20}$$
$$= \frac{-5 \pm i\sqrt{5}}{10}$$

22.

$$a^3 - 3a^2 + 8a - 24 = 0$$
$$\left(a^3 - 3a^2\right) + (8a - 24) = 0$$
$$a^2(a - 3) + 8(a - 3) = 0$$
$$\left(a^2 + 8\right)(a - 3) = 0$$

$$a^2 + 8 = 0 \quad \text{or} \quad a - 3 = 0$$
$$a^2 = -8 \qquad\qquad a = 3$$
$$a = \pm\sqrt{-8}$$
$$a = \pm 2i\sqrt{2}$$

23.

$$E = mc^2$$
$$\frac{E}{m} = \frac{mc^2}{m}$$
$$\frac{E}{m} = c^2$$
$$c = \sqrt{\frac{E}{m}}$$
$$c = \frac{\sqrt{E}}{\sqrt{m}} \cdot \frac{\sqrt{m}}{\sqrt{m}}$$
$$c = \frac{\sqrt{Em}}{m}$$

24. a. $\quad 3x^2 + 5x + 17 = 0$
$$b^2 - 4ac = (5)^2 - 4(3)(17)$$
$$= 25 - 204$$
$$= -179$$

two different imaginary-number solutions that are complex conjugates

b. $\quad 9m^2 - 12m = -4$
$$9m^2 - 12m + 4 = 0$$
$$b^2 - 4ac = (-12)^2 - 4(9)(4)$$
$$= 144 - 144$$
$$= 0$$

one repeated solution, a rational number

25. Let the width $= x$.
    Then length $= 332x + 8$.

$$(\text{width})(\text{length}) = \text{Area}$$
$$x(332x + 8) = 6,759$$
$$332x^2 + 8x - 6,759 = 0$$

$$x = \frac{-b \pm \sqrt{b^2 - 4ac}}{2a}$$

$$= \frac{-8 \pm \sqrt{(8)^2 - 4(332)(-6,759)}}{2(332)}$$

$$= \frac{-8 \pm \sqrt{64 - (-8,975,952)}}{664}$$

$$= \frac{-8 \pm \sqrt{8,976,016}}{664}$$

$$= \frac{-8 \pm 2,996}{664}$$

$$x = \frac{-8 + 2,996}{664} \quad \text{or} \quad x = \frac{-8 - 2,996}{664}$$

$$x = \frac{2,988}{664} \qquad\qquad x = \frac{-3,004}{664}$$

$$x = 4.5 \text{ ft} \qquad\qquad x = \cancel{-4.52}$$

$$332x + 8 = 332(4.5) + 8$$
$$= 1,502 \text{ ft}$$

The dimensions are 4.5 ft by 1,502 ft.

26. Let x = the number of minutes it takes the assistant to make the pastry dessert.

Assistant's work in 1 minute $= \dfrac{1}{x}$

Chef's work in 1 minute $= \dfrac{1}{x - 8}$

Work together in 1 minute $= \dfrac{1}{25}$

Assistant + chef = work together

$$\frac{1}{x} + \frac{1}{x - 8} = \frac{1}{25}$$

$$25x(x-8)\left(\frac{1}{x} + \frac{1}{x-8}\right) = 25x(x-8)\left(\frac{1}{25}\right)$$

$$25(x-8) + 25x = x(x-8)$$
$$25x - 200 + 25x = x^2 - 8x$$
$$50x - 200 = x^2 - 8x$$
$$0 = x^2 - 58x + 200$$

$$x = \frac{-b \pm \sqrt{b^2 - 4ac}}{2a}$$

$$x = \frac{-(-58) \pm \sqrt{(-58)^2 - 4(1)(200)}}{2(1)}$$

$$x = \frac{58 \pm \sqrt{3,364 - 800}}{2}$$

$$x = \frac{58 \pm \sqrt{2,564}}{2}$$

$$x = \frac{58 \pm 50.6}{2}$$

$$x = \frac{58 + 50.6}{2} \quad \text{or} \quad x = \frac{58 - 50.6}{2}$$

$$x = \frac{108.6}{2} \qquad\qquad x = \frac{7.4}{2}$$

$$x = 54.3 \qquad\qquad x = 3.7$$

$$x - 8 = 54.3 - 8 \qquad x - 8 = 3.7 - 8$$
$$= 46.3 \qquad\qquad = \cancel{4.3}$$

It takes the assistant about 54 minutes and the chef about 46 minutes to make the pastry dessert.

Chapter 10 Test

27. Let x = shorter side of a triangle.
Then $x + 14$ = longer leg of a triangle.
Use Pythagorean Theorem to find x.

$$x^2 + (x+14)^2 = 26^2$$
$$x^2 + x^2 + 28x + 196 = 676$$
$$2x^2 + 28x - 480 = 0$$
$$\frac{2x^2}{2} + \frac{28x}{2} - \frac{480}{2} = \frac{0}{2}$$
$$x^2 + 14x - 240 = 0$$
$$(x-10)(x+24) = 0$$
$$x - 10 = 0 \quad \text{or} \quad x + 24 = 0$$
$$x = 10 \qquad\qquad x = \cancel{-24}$$

The segment extending from ground to the top of the building is the sum of 2 shorter legs. Since the length of one shorter leg is 10 in., the length of the segment is 20 in.

28. Let x = width of border. Then the new length of the mirror with border is $2x + 18$ and the new width of the mirror with border is $2x + 24$.

Total area - area of picture = area of border
$$(2x+18)(2x+24) - (18)(24) = (18)(24)$$
$$4x^2 + 48x + 36x + 432 - 432 = 432$$
$$4x^2 + 84x - 432 = 0$$
$$x^2 + 21x - 108 = 0$$
$$x = \frac{-b \pm \sqrt{b^2 - 4ac}}{2a}$$
$$= \frac{-21 \pm \sqrt{(21)^2 - 4(1)(-108)}}{2(1)}$$
$$= \frac{-21 \pm \sqrt{441 + 432}}{2}$$
$$= \frac{-21 \pm \sqrt{873}}{2}$$
$$x = \frac{-21 + \sqrt{873}}{2} \quad \text{or} \quad x = \frac{-21 - \sqrt{873}}{2}$$
$$x \approx 4.3 \qquad\qquad x \approx \cancel{-25.3}$$

The border should be approximately 4.3 in. wide.

29. Find when $E(t) = 5,000$.
$$2.29t^2 - 75.72t + 5,206.95 = 5,000$$
$$2.29t^2 - 75.72t + 206.95 = 0$$
$$t = \frac{-b \pm \sqrt{b^2 - 4ac}}{2a}$$
$$t = \frac{-(-75.72) \pm \sqrt{(-75.72)^2 - 4(2.29)(206.95)}}{2(2.29)}$$
$$t = \frac{75.72 \pm \sqrt{3837.8564}}{4.58}$$
$$t = \frac{75.72 + \sqrt{3837.8564}}{4.58} \quad \text{or} \quad t = \frac{75.72 - \sqrt{3837.8564}}{4.58}$$
$$t \approx 30 \qquad\qquad t \approx 3$$

Since 30 years from 1990 has occurred yet, then $t = 3$ which would be 1993.

30. The vertex is $(0, 6)$ and the parabola is going down. The mathematical model that has those characteristics is **iii**.

31. $f(x) = -3(x-1)^2 - 2$
Vertex: $(1, -2)$
Axis of Symmetry: $x = 1$

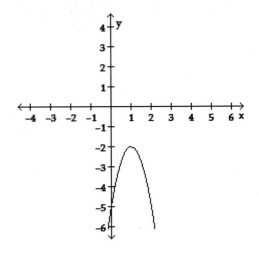

32. $f(x) = 5x^2 + 10x - 1$

$f(x) = 5(x^2 + 2x) - 1$

$f(x) = 5(x^2 + 2x + 1) - 1 - 5(1)$

$f(x) = 5(x+1)^2 - 1 - 5$

$f(x) = 5(x+1)^2 - 6$

Vertex: $(-1, -6)$

Axis of Symmetry: $x = -1$

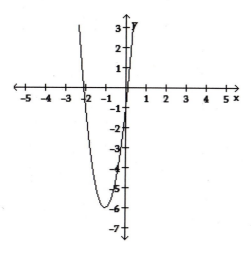

33. $f(x) = 2x^2 + x - 1$

Step 1: Since $a = 2 > 0$, the parabola opens upward.

Step 2: Find the vertex and axis of symmetry.

$$x = \frac{-b}{2a} = \frac{-1}{2(2)} = -\frac{1}{4}$$

$$y = 2\left(-\frac{1}{4}\right)^2 + \left(-\frac{1}{4}\right) - 1$$

$$= 2\left(\frac{1}{16}\right) - \frac{1}{4} - 1$$

$$= \frac{1}{8} - \frac{2}{8} - \frac{8}{8}$$

$$= -\frac{9}{8}$$

vertex: $\left(-\frac{1}{4}, -\frac{9}{8}\right)$

The axis of symmetry is $x = -\frac{1}{4}$.

Step 3: Find the x– and y–intercepts. Since $c = -1$, the y–intercept is $(0, -1)$. To find the x–intercepts, let $y = 0$ and solve the equation for x.

$$0 = 2x^2 + x - 1$$

$$0 = (2x - 1)(x + 1)$$

$$2x - 1 = 0 \quad \text{or} \quad x + 1 = 0$$

$$x = \frac{1}{2} \qquad x = -1$$

x–intercepts: $\left(\frac{1}{2}, 0\right)$ and $(-1, 0)$

y–intercept: $(0, -1)$

Step 4: Let $x = 1$ and $x = -2$ to find two more points..

$$y = 2(1)^2 + (1) - 1 = 2 + 1 - 1 = 2$$

$$y = 2(-2)^2 + (-2) - 1 = 8 - 2 - 1 = 5$$

$(1, 2)$ and $(-2, 5)$

Step 5: Plot the points and draw the parabola.

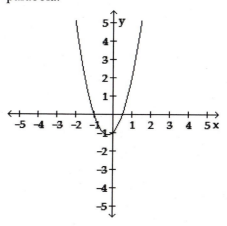

Chapter 10 Test

34. The vertex of the parabola represented by the equation would be the highest point. Use the vertex formula to find the vertex.

$$h = -16t^2 + 112t + 15$$

$$x = \frac{-b}{2a} = \frac{-112}{2(-16)} = \frac{-112}{-32} = 3.5$$

$$y = -16(3.5)^2 + 112(3.5) + 15$$

$$= -16(12.25) + 112(3.5) + 15$$

$$= -196 + 392 + 15$$

$$= 211 \text{ ft}$$

The flare explodes at 211 ft.

35.

$$x^2 - 2x > 8$$

$$x^2 - 2x - 8 > 0$$

$$x^2 - 2x - 8 = 0$$

$$(x - 4)(x + 2) = 0$$

$$x - 4 = 0 \quad \text{or} \quad x + 2 = 0$$

$$x = 4 \qquad x = -2$$

critical numbers = 4 and -2

$$(-\infty, -2) \cup (4, \infty)$$

36.

$$\frac{x - 2}{x + 3} \leq 0$$

$$\frac{x - 2}{x + 3} = 0$$

$$(x + 3)\left(\frac{x - 2}{x + 3}\right) = (x + 3)(0)$$

$$x - 2 = 0$$

$$x = 2$$

Set the denominator $= 0$

$$x + 3 = 0$$

$$x = -3$$

critical numbers = -3 and 2

$$(-3, 2]$$

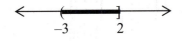

37. Since July is the 7th month, find $T(7)$.

$$T(m) = -1.1m^2 + 15.3m + 29.5$$

$$T(7) = -1.1(7)^2 + 15.3(7) + 29.5$$

$$= -1.1(49) + 15.3(7) + 29.5$$

$$= -53.9 + 107.1 + 29.5$$

$$= 82.7° \text{ F}$$

38. $y \leq -x^2 + 3$

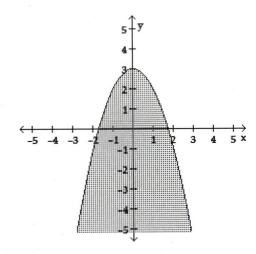

39. The x–intercepts are $(3, 0)$ and $(-2, 0)$, so the solutions to the equation are $x = 3$ and $x = -2$.

40. $[-2, 3]$

1. a. True
    b. False
    c. True

2.

$$\frac{-3(3+2)^2-(-5)}{17-|-22|}=\frac{-3(5)^2-(-5)}{17-|-22|}$$

$$=\frac{-3(25)-(-5)}{17-22}$$

$$=\frac{-75+5}{-5}$$

$$=\frac{-70}{-5}$$

$$=14$$

3.

$$-9(3a-9)-7(2a-7)=-27a+81-14a+49$$

$$=-41a+130$$

4. Yes.

$$\frac{3(-11+2)}{2}\overset{?}{=}\frac{4(-11)-10}{4}$$

$$\frac{3(-9)}{2}\overset{?}{=}\frac{-44-10}{4}$$

$$\frac{-27}{2}\overset{?}{=}\frac{-54}{4}$$

$$-13.5=-13.5$$

5.

$$2-(4x+7)=3+2(x+2)$$

$$2-4x-7=3+2x+4$$

$$-5-4x=7+2x$$

$$-5-4x-2x=7+2x-2x$$

$$-5-6x=7$$

$$-5-6x+5=7+5$$

$$-6x=12$$

$$x=-2$$

6.

$$2(6n+5)=4(3n+2)+2$$

$$12n+10=12n+8+2$$

$$12n+10=12n+10$$

$$10=10$$

all real numbers

7. 20% of 85 lbs is what?

$$0.20\cdot85=x$$

$$17=x$$

17 lbs

8.

$$A=2lw+2wh+2lh$$

$$202=2(9)(5)+2(5)h+2(9)h$$

$$202=90+10h+18h$$

$$202=90+28h$$

$$112=28h$$

$$4=h$$

9. In an isosceles triangle, there are two equal base angles and one vertex angles, and the sum of all three is 180°. Let $x=$ the measure of a base angle.

$$x+x+53=180$$

$$2x+53=180$$

$$2x=127$$

$$x=63.5°$$

10. Let x be the amount of time. Since d = rt, the team going north would go a distance of $2x$ and the team going south would go a distance of $4x$. The sum of their distances is 21 miles.

$$2x+4x=21$$

$$6x=21$$

$$x=3.5\text{ hr}$$

11. Let $x=$ liters of 1% glucose solution.

$$0.01x+0.05(2)=0.02(x+2)$$

$$0.01x+0.1=0.02x+0.04$$

$$100(0.01x+0.1)=100(0.02x+0.04)$$

$$x+10=2x+4$$

$$x+10-4=2x+4-4$$

$$x+6=2x$$

$$x+6-x=2x-x$$

$$6=x$$

6 Liters of 1% glucose solution are needed

12. Let x = pounds of regular coffee needed. The value of the regular coffee would be $8x$. The value of the gourmet coffee would be $14(40) = \$560$. The value of the mixture would be $\$10(x + 40)$.

$$8x + 560 = 10(x + 40)$$
$$8x + 560 = 10x + 400$$
$$560 = 2x + 400$$
$$160 = 2x$$
$$80 = x$$

80 pounds of regular coffee is needed.

16. $3x - 4y = 12$

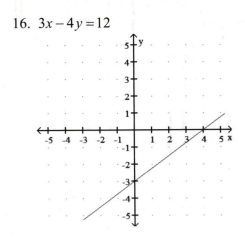

13.
$$3 - 3x \geq 6 + x$$
$$3 - 4x \geq 6$$
$$-4x \geq 3$$
$$x \leq -\frac{3}{4}$$
$$\left(-\infty, -\frac{3}{4}\right]$$

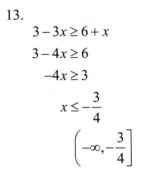

17. $x = 5$

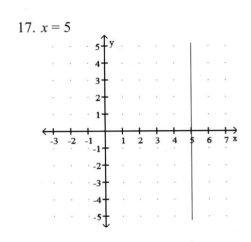

14.
$$4x - 3y = -4$$
$$4(-6) - 3(-7) \overset{?}{=} -4$$
$$-24 + 21 \overset{?}{=} -4$$
$$-3 \neq -4$$

No, it is not a solution.

18. $y = 2x^2 - 3$

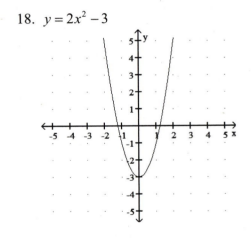

15. $y = \frac{1}{2}x$

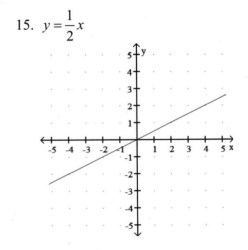

19. x-intercept:

Let $y = 0$.

$$5x - 3y = 6$$
$$5x - 3(0) = 6$$
$$5x = 6$$
$$x = \frac{6}{5}$$
$$\left(\frac{6}{5}, 0\right)$$

y-intercept:

Let $x = 0$.

$$5x - 3y = 6$$
$$5(0) - 3y = 6$$
$$-3y = 6$$
$$y = -2$$
$$(0, -2)$$

20. Use the ordered pairs (2010, 559) and (2007, 217) to find the slope.

$$m = \frac{y_2 - y_1}{x_2 - x_1}$$
$$= \frac{559 - 217}{2010 - 2007}$$
$$= \frac{342}{3}$$
$$= 114$$

An increase of 114 million subscribers per year

21. a. $y = 3x - 7$

$m = 3$

b. $5x - 6y = 13$

$$-6y = -5x + 13$$
$$y = \frac{5}{6}x - \frac{13}{6}$$
$$m = \frac{5}{6}$$

22. $y = -3$ is a horizontal line so its slope is 0.

23. Use point–slope formula.

$$y - y_1 = m(x - x_1)$$
$$y - (-4) = 3\left[x - (-2)\right]$$
$$y + 4 = 3(x + 2)$$
$$y + 4 = 3x + 6$$
$$y = 3x + 2$$

24. Parallel lines have the same slope, so find the slope of the given equation by solving the equation for y.

$$2x + 3y = 6$$
$$3y = -2x + 6$$
$$y = -\frac{2}{3}x + 2$$
$$m = -\frac{2}{3}$$

Use the point–slope formula.

$$y - y_1 = m(x - x_1)$$
$$y - (-2) = -\frac{2}{3}(x - 0)$$
$$y + 2 = -\frac{2}{3}x - 0$$
$$y = -\frac{2}{3}x - 2$$

25. $f(x) = 2x^2 - 3x + 1$

$$f(-3) = 2(-3)^2 - 3(-3) + 1$$
$$= 2(9) - 3(-3) + 1$$
$$= 18 + 9 + 1$$
$$= 28$$

26. Yes. It passes the vertical line test.

27. $\begin{cases} x - 3y = 1 \\ -2x + 6 = -6y \end{cases}$

$x - 3y = 1$ $\qquad$ $-2x + 6 = -6y$

$2 - 3\left(\dfrac{1}{3}\right) \overset{?}{=} 1$ $\qquad$ $-2(2) + 6 \overset{?}{=} -6\left(\dfrac{1}{3}\right)$

$2 - 1 \overset{?}{=} 1$ $\qquad\qquad$ $-4 + 6 \overset{?}{=} -2$

$1 = 1$ $\qquad\qquad\qquad$ $2 \neq -2$

No, it is not a solution.

28. $\begin{cases} x + y = 4 \\ y = x + 6 \end{cases}$

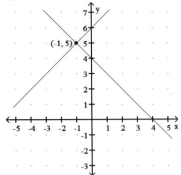

The solution is $(-1, 5)$.

29. $\begin{cases} x = y + 4 \\ 2x + y = 5 \end{cases}$

Substitute the first equation into the second and solve for y.

$$2x + y = 5$$
$$2(y + 4) + y = 5$$
$$2y + 8 + y = 5$$
$$3y + 8 = 5$$
$$3y = -3$$
$$y = -1$$

Substitute $y = -1$ into the first equation and solve for x.

$$x = y + 4$$
$$x = -1 + 4$$
$$x = 3$$

The solution is $(3, -1)$.

30. $\begin{cases} 3s + 4t = 5 \\ 2s - 3t = -8 \end{cases}$

Multiply the first equation by 3 and the second equation by 4. Add the two equations together to solve for s.

$$9s + 12t = 15$$
$$\underline{8s - 12t = -32}$$
$$17s = -17$$
$$s = -1$$

Substitute $s = -1$ into the first equation and solve for t.

$$3(-1) + 4t = 5$$
$$-3 + 4t = 5$$
$$4t = 8$$
$$t = 2$$

The solution is $(-1, 2)$.

31. Let x = amount invested at 6% and y = amount invested 12%.

$$\begin{cases} x + y = 6,000 \\ 0.06x + 0.12y = 540 \end{cases}$$

Multiply the first equation by -0.06 and add the equations to eliminate x.

$$-0.06x - 0.06y = -360$$
$$\underline{0.06x + 0.12y = 540}$$
$$0.06y = 180$$
$$y = \$3,000$$

Substitute $y = \$3,000$ into the first equation and solve for x.

$$x + 3,000 = 6,000$$
$$x = 3,000$$

$3,000 was invested in both accounts.

32. $\begin{cases} 3x + 2y \geq 6 \\ x + 3y \leq 6 \end{cases}$

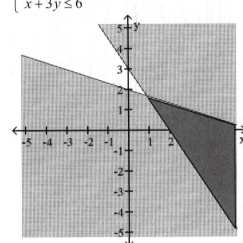

33.
$$\left(x^5\right)^2 \left(x^7\right)^3 = \left(x^{10}\right)\left(x^{21}\right)$$
$$= x^{10+21}$$
$$= x^{31}$$

34.
$$\left(\frac{a^3 b}{c^4}\right)^5 = \frac{\left(a^3\right)^5 \left(b\right)^5}{\left(c^4\right)^5}$$
$$= \frac{a^{15} b^5}{c^{20}}$$

35.
$$4^{-3} \cdot 4^{-2} \cdot 4^5 = 4^{-3+(-2)+5}$$
$$= 4^0$$
$$= 1$$

36.

$$\left(\frac{2a^2b^3c^{-4}}{5a^{-2}b^{-1}c^3}\right)^{-3} = \left(\frac{5a^{-2}b^{-1}c^3}{2a^2b^3c^{-4}}\right)^3$$

$$= \left(\frac{5}{2}a^{-2-2}b^{-1-3}c^{3-(-4)}\right)^3$$

$$= \left(\frac{5}{2}a^{-4}b^{-4}c^7\right)^3$$

$$= \left(\frac{5c^7}{2a^4b^4}\right)^3$$

$$= \frac{5^3c^{7\cdot3}}{2^3a^{4\cdot3}b^{4\cdot3}}$$

$$= \frac{125c^{21}}{8a^{12}b^{12}}$$

37.

$$5n^0 = 5(1)$$
$$= 5$$

38.

$$\frac{4b^{-4}}{\left(a^{-2}\right)^{-5}} = \frac{4b^{-4}}{a^{10}}$$

$$= \frac{4}{a^{10}b^4}$$

39. $\dfrac{(1,280,000,000)(2,700,000)}{(240,000)}$

$$= \frac{\left(1.28\times10^9\right)\left(2.7\times10^6\right)}{2.4\times10^5}$$

$$= \frac{(1.28\times2.7)\left(10^9\times10^6\right)}{2.4\times10^5}$$

$$= \frac{3.456\times10^{15}}{2.4\times10^5}$$

$$= \frac{3.456}{2.4}\times10^{15-5}$$

$$= 1.44\times10^{10}$$

$$= 14,400,000,000$$

40.

$$\frac{3}{5}s^2 - \frac{2}{5}t^2 - \frac{1}{2}s^2 - \frac{7}{10}st - \frac{3}{10}st = \frac{1}{10}s^2 - \frac{2}{5}t^2 - \frac{10}{10}st$$

$$= \frac{1}{10}s^2 - \frac{2}{5}t^2 - st$$

41.

$$\left(-8.9t^3 - 2.4t\right) - \left(2.1t^3 + 0.8t^2 - t\right)$$
$$= -8.9t^3 - 2.4t - 2.1t^3 - 0.8t^2 + t$$
$$= -11t^3 - 0.8t^2 - 1.4t$$

42.

$$(2a - b)\left(4a^2 + 2ab + b^2\right)$$
$$= 2a\left(4a^2 + 2ab + b^2\right) - b\left(4a^2 + 2ab + b^2\right)$$
$$= 8a^3 + 4a^2b + 2ab^2 - 4a^2b - 2ab^2 - b^3$$
$$= 8a^3 - b^3$$

43.

$$(-3t + 2s)(2t - 3s) = -6t^2 + 9st + 4st - 6s^2$$
$$= -6t^2 + 13st - 6s^2$$

44.

$$(4b - 8)^2 = (4b)^2 - 2(4b)(8) + (8)^2$$
$$= 16b^2 - 64b + 64$$

45.

$$\left(6b + \frac{1}{2}\right)\left(6b - \frac{1}{2}\right) = (6b)^2 - \left(\frac{1}{2}\right)^2$$
$$= 36b^2 - \frac{1}{4}$$

46.

$$\begin{array}{r} 2x - 1 \\ x+2\overline{)2x^2 + 3x - 2} \\ \underline{2x^2 + 4x} \\ -x - 2 \\ \underline{-x - 2} \\ 0 \end{array}$$

47.

$$12uvw^3 - 18uv^2w^2 = 6uvw^2(2w - 3v)$$

48.

$$x^2 + 4y - xy - 4x = x^2 - xy - 4x + 4y$$
$$= x(x - y) - 4(x - y)$$
$$= (x - y)(x - 4)$$

49.

$$x^2 + 7x + 10 = (x + 2)(x + 5)$$

Chapter 10 Cumulative Review

50.

$6 + 3x^2 + x = 3x^2 + x + 6$ is prime

51.

$6a^2 - 7a - 20 = (3a + 4)(2a - 5)$

52.

$$30a^4 - 4a^3 - 16a^2 = 2a^2(15a^2 - 2a - 8)$$
$$= 2a^2(3a + 2)(5a - 4)$$

53.

$49s^6 - 84s^3n^2 + 36n^4 = (7s^3 - 6n^2)^2$

54.

$$x^4 - 16y^4 = (x^2 + 4y^2)(x^2 - 4y^2)$$
$$= (x^2 + 4y^2)(x + 2y)(x - 2y)$$

55.

$$x^3 - 64 = (x)^3 - (4)^3$$
$$= (x - 4)\left((x)^2 + 4(x) + (4)^2\right)$$
$$= (x - 4)(x + 4x + 16)$$

56.

$$8x^6 + 125y^3$$
$$= (2x^2 + 5y)\left[(2x^2)^2 - (2x^2)(5y) + (5y)^2\right]$$
$$= (2x^2 + 5y)(4x^4 - 10x^2y + 25y^2)$$

57.

$$x^2 + 3x + 2 = 0$$
$$(x + 1)(x + 2) = 0$$
$$x + 1 = 0 \quad \text{or} \quad x + 2 = 0$$
$$x = -1 \qquad x = -2$$

58.

$$5x^2 = 10x$$
$$5x^2 - 10x = 0$$
$$5x(x - 10) = 0$$
$$5x = 0 \quad \text{or} \quad x - 10 = 0$$
$$x = 0 \qquad\qquad x = 10$$

59.

$$6x^2 - x = 2$$
$$6x^2 - x - 2 = 0$$
$$(3x - 2)(2x + 1) = 0$$
$$3x - 2 = 0 \quad \text{or} \quad 2x + 1 = 0$$
$$3x = 2 \qquad\qquad 2x = -1$$
$$x = \frac{2}{3} \qquad\qquad x = -\frac{1}{2}$$

60.

$$a^2 - 25 = 0$$
$$(a - 5)(a + 5) = 0$$
$$a - 5 = 0 \quad \text{or} \quad a + 5 = 0$$
$$a = 5 \qquad\qquad a = -5$$

61.

$$(m + 4)(2m + 3) - 22 = 10m$$
$$2m^2 + 3m + 8m + 12 - 22 = 10m$$
$$2m^2 + 11m - 10 = 10m$$
$$2m^2 + m - 10 = 0$$
$$(2m + 5)(m - 2) = 0$$
$$2m + 5 = 0 \quad \text{or} \quad m - 2 = 0$$
$$m = -\frac{5}{2} \qquad\qquad m = 2$$

62.

$$6a^3 - 2a = a^2$$
$$6a^3 - a^2 - 2a = 0$$
$$a(6a^2 - a - 2) = 0$$
$$a(2a + 1)(3a - 2) = 0$$
$$a = 0 \quad 2a + 1 = 0 \quad \text{or} \quad 3a - 2 = 0$$
$$a = 0 \qquad a = -\frac{1}{2} \qquad a = \frac{2}{3}$$

63. Let x = length, then the width is $x - 1$.

$$x(x-1) = 20$$
$$x^2 - x = 20$$
$$x^2 - x - 20 = 0$$
$$(x-5)(x+4) = 0$$
$$x - 5 = 0 \quad \text{or} \quad x + 4 = 0$$
$$x = 5 \qquad\qquad x = -4$$

Since the length cannot be negative, the length must be 5 cm.

64.

$$\frac{x^2 + 2x + 1}{x^2 - 1} = \frac{(x+1)(x+1)}{(x+1)(x-1)}$$
$$= \frac{x+1}{x-1}$$

65.

$$\frac{p^2 - p - 6}{3p - 9} \div \frac{p^2 + 6p + 9}{p^2 - 9}$$
$$= \frac{p^2 - p - 6}{3p - 9} \cdot \frac{p^2 - 9}{p^2 + 6p + 9}$$
$$= \frac{(p-3)(p+2)}{3(p-3)} \cdot \frac{(p-3)(p+3)}{(p+3)(p+3)}$$
$$= \frac{(p-3)(p+2)}{3(p+3)}$$

66.

$$\frac{12x^2}{7-x} \cdot \frac{x-7}{20x^3} = \frac{3(4)(x)(x)}{7-x} \cdot \frac{-1(7-x)}{5(4)(x)(x)(x)}$$
$$= -\frac{3}{5x}$$

67.

$$\frac{13}{15a} - \frac{8}{15a} = \frac{5}{15a}$$
$$= \frac{1}{3a}$$

68.

$$\frac{x+2}{x+5} - \frac{x-3}{x+7} = \frac{x+2}{x+5}\left(\frac{x+7}{x+7}\right) - \frac{x-3}{x+7}\left(\frac{x+5}{x+5}\right)$$
$$= \frac{x^2 + 9x + 14}{(x+5)(x+7)} - \frac{x^2 + 2x - 15}{(x+5)(x+7)}$$
$$= \frac{x^2 + 9x + 14 - x^2 - 2x + 15}{(x+5)(x+7)}$$
$$= \frac{7x + 29}{(x+5)(x+7)}$$

69.

$$\frac{1}{6b^4} - \frac{8}{9b^2} = \frac{1}{6b^4}\left(\frac{3}{3}\right) - \frac{8}{9b^2}\left(\frac{2b^2}{2b^2}\right)$$
$$= \frac{3}{18b^4} - \frac{16b^2}{18b^4}$$
$$= \frac{3 - 16b^2}{18b^4}$$

70.

$$\frac{\dfrac{1}{x} + \dfrac{1}{y}}{\dfrac{1}{x} - \dfrac{1}{y}} = \frac{xy\left(\dfrac{1}{x} + \dfrac{1}{y}\right)}{xy\left(\dfrac{1}{x} - \dfrac{1}{y}\right)}$$
$$= \frac{y + x}{y - x[}$$

71.

$$\frac{7}{a^2 - a - 2} + \frac{1}{a+1} = \frac{3}{a-2}$$
$$\frac{7}{(a-2)(a+1)} + \frac{1}{a+1} = \frac{3}{a-2}$$
$$(a-2)(a+1)\left(\frac{7}{(a-2)(a+1)} + \frac{1}{a+1}\right) = (a-2)(a+1)\left(\frac{3}{a-2}\right)$$
$$7 + (a-2) = 3(a+1)$$
$$7 + a - 2 = 3a + 3$$
$$a + 5 = 3a + 3$$
$$-2a + 5 = 3$$
$$-2a = -2$$
$$a = 1$$

Chapter 10 Cumulative Review

72.

$$\frac{x-4}{x-3}+\frac{x-2}{x-3}=x-3$$

$$x-3\left(\frac{x-4}{x-3}+\frac{x-2}{x-3}\right)=(x-3)(x-3)$$

$$x-4+x-2=x^2-3x-3x+9$$

$$2x-6=x^2-6x+9$$

$$0=x^2-8x+15$$

$$0=(x-3)(x-5)$$

$$x-3=0 \quad \text{or} \quad x-5=0$$

$$x=\cancel{3} \qquad\qquad x=5$$

3 is extraneous

73.

$$\frac{1}{R}=\frac{1}{R_1}+\frac{1}{R_2}+\frac{1}{R_3}$$

$$RR_1R_2R_3\left(\frac{1}{R}\right)=RR_1R_2R_3\left(\frac{1}{R_1}+\frac{1}{R_2}+\frac{1}{R_3}\right)$$

$$R_1R_2R_3=RR_2R_3+RR_1R_3+RR_1R_2$$

$$R_1R_2R_3=R(R_2R_3+R_1R_3+R_1R_2)$$

$$\frac{R_1R_2R_3}{R_2R_3+R_1R_3+R_1R_2}=\frac{R(R_2R_3+R_1R_3+R_1R_2)}{R_2R_3+R_1R_3+R_1R_2}$$

$$\frac{R_1R_2R_3}{R_2R_3+R_1R_3+R_1R_2}=R$$

74. Let $x =$ time it takes to fill the pool if the pipes are working together.

$$\frac{1}{5}+\frac{1}{4}=\frac{1}{x}$$

$$20x\left(\frac{1}{5}+\frac{1}{4}\right)=20x\left(\frac{1}{x}\right)$$

$$4x+5x=20$$

$$9x=20$$

$$x=\frac{20}{9}$$

$$x=2\frac{2}{9}\text{ hr}$$

75. Let $x =$ # of sales transactions.

$$\frac{9}{500}=\frac{x}{360,000}$$

$$3,240,000=500x$$

$$6,480=x$$

76

$$\frac{x}{8}=\frac{25}{10}$$

$$10x=200$$

$$x=20$$

77.

$$\frac{1}{6}(a+12)+1=\frac{a}{3}$$

$$\frac{1}{6}a+2+1=\frac{a}{3}$$

$$\frac{1}{6}a+3=\frac{a}{3}$$

$$6\left(\frac{1}{6}a+3\right)=6\left(\frac{a}{3}\right)$$

$$a+18=2a$$

$$18=a$$

78.

$$2x+1=5(0.2x+1)-(4-x)$$

$$2x+1=x+5-4+x$$

$$2x+1=2x+1$$

$$1=1$$

all real numbers

$\Re$; identity

79.

$$\frac{75+83+91+68+x}{5}\geq 80$$

$$\frac{317+x}{5}\geq 80$$

$$5\left(\frac{317+x}{5}\right)\geq 5(80)$$

$$317+x\geq 400$$

$$x\geq 83$$

80. Given a relation in x and y, if to each value of x in the domain there corresponds exactly one value of y in the range, then y is said to be a **function** of x.

81. a. 0

b. 2

82. D: The set of real numbers

R: The set of real numbers

83. D: $\{-6, 0, 1, 5\}$
    R: $\{-12, 4, 7, 8$:
    No, it is not a function.

84. No; (4, 2) and (4, –2)

85.
$$h(x) = 0$$
$$-\frac{1}{5}x - 12 = 0$$
$$5\left(-\frac{1}{5}x - 12\right) = 5(0)$$
$$-x - 60 = 0$$
$$-x = 60$$
$$x = -60$$

86. a. $g(x) = -2x - 1$
$$g(-2) = -2(-2) - 1$$
$$= 4 - 1$$
$$= 3$$
   b. $f(x) = 3x^2 + 2$
$$f(-r) = 3(-r)^2 + 2$$
$$= 3r^2 + 2$$

87. Set the denominator equal to 0 and solve for x to see what cannot be a member of the domain.
$$x + 1 = 0$$
$$x = -1$$
   D: The set of all real numbers except –1.

88. Use $y = mx + b$ for $f(x) = 6x + 15$ to find that the slope is 6 and the y-intercept is (0, 15),

89. $y = -\dfrac{7}{8}$ so the function would be
$$f(x) = -\frac{7}{8}.$$

90. a. Use the ordered pairs (10, 5.76) and (25, 8.31) to find the slope.
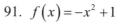
$$m = \frac{y_2 - y_1}{x_2 - x_1}$$
$$= \frac{8.31 - 5.76}{25 - 10}$$
$$= \frac{2.55}{15}$$
$$= 0.17$$
   Use point–slope formula.
$$y - y_1 = m(x - x_1)$$
$$y - 8.31 = 0.17(x - 25)$$
$$y - 8.31 = 0.17x - 4.25$$
$$y = 0.17x + 4.06$$
$$M(t) = 0.17t + 4.06$$
   b. Let $t = 40$.
$$M(40) = 0.17(40) + 4.06$$
$$= 10.86$$

91. $f(x) = -x^2 + 1$

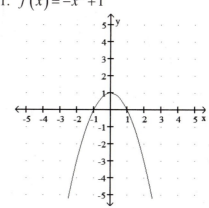

   D: the set of real numbers
   R: the set of real numbers less than or equal to 1

92. $g(x) = |x - 3| - 4$

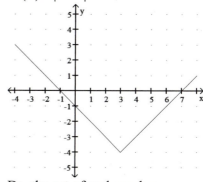

   D: the set of real numbers
   R: the set of real numbers greater than or equal to –4

Chapter 10 Cumulative Review

93.

$$2|4x-3|+1=19$$
$$2|4x-3|=18$$
$$|4x-3|=9$$
$$4x-3=9 \quad \text{or} \quad 4x-3=-9$$
$$4x=12 \qquad\qquad 4x=-6$$
$$x=3 \qquad\qquad x=-\frac{3}{2}$$

94.

$$|2x-1|=|3x+4|$$
$$2x-1=3x+4 \quad \text{and} \quad 2x-1=-(3x+4)$$
$$-x-1=4 \qquad\qquad 2x-1=-3x-4$$
$$-x-1=4 \qquad\qquad 5x-1=-4$$
$$-x=5 \qquad\qquad 5x=-3$$
$$x=-5 \qquad\qquad x=-\frac{3}{5}$$

95.

$$5(-2x+2)>20-x$$
$$-10x+10>20-x$$
$$-9x+10>20$$
$$-9x>10$$
$$x<-\frac{10}{9}$$
$$\left(-\infty,-\frac{10}{9}\right)$$

96.

$$3x+4<-2 \quad \text{or} \quad 3x+4>10$$
$$3x<-6 \qquad\qquad 3x>6$$
$$x<-2 \qquad\qquad x>2$$
$$(-\infty,-2)\cup(2,\infty)$$

97.

$$5x-3\geq 2 \qquad \text{and} \qquad 6\geq 4x-3$$
$$5x-3+3\geq 2+3 \qquad 6+3\geq 4x-3+3$$
$$5x\geq 5 \qquad\qquad 9\geq 4x$$
$$\frac{5x}{5}\geq\frac{5}{5} \qquad\qquad \frac{9}{4}\geq\frac{4x}{4}$$
$$x\geq 1 \qquad\qquad \frac{9}{4}\geq x$$
$$x\leq\frac{9}{4}$$
$$\left[1,\frac{9}{4}\right]$$

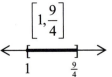

98.

$$|2x-5|\geq 25$$
$$2x-5\geq 25 \quad \text{or} \quad 2x-5\leq -25$$
$$2x\geq 30 \qquad\qquad 2x\leq -20$$
$$x\geq 15 \qquad\qquad x\leq -10$$
$$(-\infty,-10]\cup[15,\infty)$$

99.

$$|3x-2|\leq 4$$
$$-4\leq 3x-2\leq 4$$
$$-2\leq 3x\leq 6$$
$$-\frac{2}{3}\leq x\leq 2$$
$$\left[-\frac{2}{3},2\right]$$

100.

$$5|4-x|+6<1$$
$$5|4-x|<-4$$
$$|4-x|<-\frac{4}{5}$$

Since the absolute value is never less than a negative number, there is NO SOLUTION to this inequality.

101. Let $m = (x + y)$.
$$(x+y)^2 + 7(x+y) + 12 = m^2 + 7m + 12$$
$$= (m+3)(m+4)$$
$$= (x+y+3)(x+y+4)$$

102.
$$d_1 d_2 - fd_2 = fd_1$$
$$d_1 d_2 - fd_2 + fd_2 = fd_1 + fd_2$$
$$d_1 d_2 = fd_1 + fd_2$$
$$d_1 d_2 = f(d_1 + d_2)$$
$$\frac{d_1 d_2}{d_1 + d_2} = \frac{f(d_1 + d_2)}{d_1 + d_2}$$
$$\frac{d_1 d_2}{d_1 + d_2} = f$$

103.
$$x^6 - 1 = (x^3 - 1)(x^3 + 1)$$
$$= (x-1)(x^2 + x + 1)(x+1)(x^2 - x + 1)$$

104. Set the denominator equal to 0 and solve for x.
$$x^2 - 2x = 0$$
$$x(x-2) = 0$$
$$x = 0 \quad \text{or} \quad x - 2 = 0$$
$$x = 0 \qquad\qquad x = 2$$
The domain is all real numbers except for 0 and 2: $(-\infty, 0) \cup (0, 2) \cup (2, \infty)$.

105.
   a. iii
   b. i
   c. iv
   d. ii

106. Let c = cost, n = number of trucks, and h = number of hours used.
$$c = nhk$$
$$3,600 = 8(12)k$$
$$3,600 = 96k$$
$$37.5 = k$$
$$c = nkh$$
$$c = 20(12)(37.5)$$
$$c = \$9,000$$

107. $f(x) = \sqrt{x-2}$
$$D:[2,\infty), R:[0,\infty)$$

108.
$$\sqrt[3]{-27x^3} = -3x$$

109.
$$\sqrt{48t^3} = \sqrt{16t^2}\sqrt{3t}$$
$$= 4t\sqrt{3t}$$

110.
$$64^{-2/3} = \frac{1}{64^{2/3}}$$
$$= \frac{1}{\left(\sqrt[3]{64}\right)^2}$$
$$= \frac{1}{(4)^2}$$
$$= \frac{1}{16}$$

111.
$$\frac{x^{5/3}x^{1/2}}{x^{3/4}} = \frac{x^{5/3+1/2}}{x^{3/4}}$$
$$= \frac{x^{10/6+3/6}}{x^{3/4}}$$
$$= \frac{x^{13/6}}{x^{3/4}}$$
$$= x^{13/6-3/4}$$
$$= x^{26/12-9/12}$$
$$= x^{17/12}$$

Chapter 10 Cumulative Review

112.

$$-3\sqrt[4]{32} - 2\sqrt[4]{162} + 5\sqrt[4]{48}$$
$$= -3\sqrt[4]{16}\sqrt[4]{2} - 2\sqrt[4]{81}\sqrt[4]{2} + 5\sqrt[4]{16}\sqrt[4]{3}$$
$$= -3(2)\sqrt[4]{2} - 2(3)\sqrt[4]{2} + 5(2)\sqrt[4]{3}$$
$$= -6\sqrt[4]{2} - 6\sqrt[4]{2} + 10\sqrt[4]{3}$$
$$= -12\sqrt[4]{2} + 10\sqrt[4]{3}$$

113.

$$3\sqrt{2}\left(2\sqrt{3} - 4\sqrt{12}\right) = 6\sqrt{6} - 12\sqrt{24}$$
$$= 6\sqrt{6} - 12\sqrt{4}\sqrt{6}$$
$$= 6\sqrt{6} - 12(2)\sqrt{6}$$
$$= 6\sqrt{6} - 24\sqrt{6}$$
$$= -18\sqrt{6}$$

114.

$$\frac{\sqrt{x}+2}{\sqrt{x}-1} = \frac{\sqrt{x}+2}{\sqrt{x}-1} \cdot \frac{\sqrt{x}+1}{\sqrt{x}+1}$$
$$= \frac{\left(\sqrt{x}+2\right)\left(\sqrt{x}+1\right)}{\left(\sqrt{x}-1\right)\left(\sqrt{x}+1\right)}$$
$$= \frac{\sqrt{x^2}+\sqrt{x}+2\sqrt{x}+2}{\sqrt{x^2}-1}$$
$$= \frac{x+3\sqrt{x}+2}{x-1}$$

115.

$$\frac{5}{\sqrt[3]{x}} = \frac{5}{\sqrt[3]{x}} \cdot \frac{\sqrt[3]{x^2}}{\sqrt[3]{x^2}}$$
$$= \frac{5\sqrt[3]{x^2}}{\sqrt[3]{x^3}}$$
$$= \frac{5\sqrt[3]{x^2}}{x}$$

116.

$$5\sqrt{x+2} = x+8$$
$$\left(5\sqrt{x+2}\right)^2 = (x+8)^2$$
$$25(x+2) = x^2 + 8x + 8x + 64$$
$$25x + 50 = x^2 + 16x + 64$$
$$0 = x^2 - 9x + 14$$
$$0 = (x-2)(x-7)$$
$$x - 2 = 0 \quad \text{or} \quad x - 7 = 0$$
$$x = 2 \qquad \qquad x = 7$$

117.

$$f(x) = g(x)$$
$$\sqrt[3]{x^2 + 2x} = 2\sqrt[3]{x-1}$$
$$\left(\sqrt[3]{x^2 + 2x}\right)^3 = \left(2\sqrt[3]{x-1}\right)^3$$
$$x^2 + 2x = 8(x-1)$$
$$x^2 + 2x = 8x - 8$$
$$x^2 - 6x + 8 = 0$$
$$(x-2)(x-4) = 0$$
$$x - 2 = 0 \quad \text{and} \quad x - 4 = 0$$
$$x = 2 \qquad \qquad x = 4$$

118.

$$\sqrt{x} + \sqrt{x+2} = 2$$
$$\sqrt{x+2} = 2 - \sqrt{x}$$
$$\left(\sqrt{x+2}\right)^2 = \left(2 - \sqrt{x}\right)^2$$
$$x + 2 = 4 - 2\sqrt{x} - 2\sqrt{x} + \sqrt{x^2}$$
$$x + 2 = 4 - 4\sqrt{x} + x$$
$$x + 2 - 4 - x = 4 - 4\sqrt{x} + x - 4 - x$$
$$-2 = -4\sqrt{x}$$
$$(-2)^2 = \left(-4\sqrt{x}\right)^2$$
$$4 = 16x$$
$$\frac{4}{16} = x$$
$$\frac{1}{4} = x$$

119. a. The length of the hypotenuse in an isosceles right triangles is the length of the leg times $\sqrt{2}$. The length of the hypotenuse is $3\sqrt{2}$ in.

b. The length of the longer leg is $\sqrt{3}$ times the length of the shorter leg. The length of the hypotenuse is 2 times the length of the shorter side. Let x = length of the shorter leg.

longer leg:

$$x\sqrt{3} = 3$$

$$\frac{x\sqrt{3}}{\sqrt{3}} = \frac{3}{\sqrt{3}}$$

$$x = \frac{3}{\sqrt{3}} \cdot \frac{\sqrt{3}}{\sqrt{3}}$$

$$x = \frac{3\sqrt{3}}{3}$$

$$x = \sqrt{3} \text{ in.}$$

hypotenuse:

$$2x = 2\sqrt{3} \text{ in.}$$

120. a.
$$d = \sqrt{(x_1 - x_2)^2 + (y_1 - y_2)^2}$$

$$= \sqrt{[4 - (-2)]^2 + (14 - 6)^2}$$

$$= \sqrt{(6)^2 + (8)^2}$$

$$= \sqrt{36 + 64}$$

$$= \sqrt{100}$$

$$= 10$$

b.
$$M = \left(\frac{x_1 + x_2}{2}, \frac{y_1 + y_2}{2} \right)$$

$$= \left(\frac{7 + (-10)}{2}, \frac{1 + 4}{2} \right)$$

$$= \left(\frac{-3}{2}, \frac{5}{2} \right)$$

121.
$$i^{43} = i^{40} i^2 i$$

$$= 1(-1)i$$

$$= -i$$

122.
$$\left(-7 + \sqrt{-81} \right) - \left(-2 - \sqrt{-64} \right)$$

$$= (-7 + 9i) - (-2 - 8i)$$

$$= -7 + 9i + 2 + 8i$$

$$= -5 + 17i$$

123.
$$\frac{5}{3-i} = \frac{5}{3-i} \cdot \frac{3+i}{3+i}$$

$$= \frac{5(3+i)}{(3-i)(3+i)}$$

$$= \frac{15 + 5i}{9 - i^2}$$

$$= \frac{15 + 5i}{9 - (-1)}$$

$$= \frac{15 + 5i}{10}$$

$$= \frac{15}{10} + \frac{5}{10}i$$

$$= \frac{3}{2} + \frac{1}{2}i$$

124.
$$(2+i)^2 = (2+i)(2+i)$$

$$= 4 + 2i + 2i + i^2$$

$$= 4 + 4i + (-1)$$

$$= 3 + 4i$$

125.
$$\frac{-4}{6i^7} = \frac{-4}{6i^7} \cdot \frac{i}{i}$$

$$= \frac{-4i}{6i^8}$$

$$= \frac{-4i}{6(1)}$$

$$= -\frac{4}{6}i$$

$$= -\frac{2}{3}i$$

$$= 0 - \frac{2}{3}i$$

126.

$$x^2 = 28$$

$$x = \pm\sqrt{28}$$

$$x = \pm\sqrt{4}\sqrt{7}$$

$$x = \pm 2\sqrt{7}$$

127.

$$(x-19)^2 = -5$$

$$x - 19 = \sqrt{-5} \quad \text{or} \quad x - 19 = -\sqrt{-5}$$

$$x - 19 = i\sqrt{5} \qquad x - 19 = -i\sqrt{5}$$

$$x = 19 + i\sqrt{5} \qquad x = 19 - i\sqrt{5}$$

$$x = 19 \pm i\sqrt{5}$$

128.

$$2x^2 - 6x + 3 = 0$$

$$2x^2 - 6x = -3$$

$$\frac{2x^2}{2} - \frac{6x}{2} = -\frac{3}{2}$$

$$x^2 - 3x = -\frac{3}{2}$$

$$x^2 - 3x + \left(\frac{-3}{2}\right)^2 = -\frac{3}{2} + \left(\frac{-3}{2}\right)^2$$

$$x^2 - 3x + \frac{9}{4} = -\frac{3}{2} + \frac{9}{4}$$

$$\left(x - \frac{3}{2}\right)^2 = \frac{3}{4}$$

$$x - \frac{3}{2} = \pm\sqrt{\frac{3}{4}}$$

$$x - \frac{3}{2} = \pm\frac{\sqrt{3}}{2}$$

$$x = \frac{3}{2} \pm \frac{\sqrt{3}}{2}$$

$$x = \frac{3 \pm \sqrt{3}}{2}$$

129.

$$a^2 - \frac{2}{5}a = -\frac{1}{5}$$

$$5\left(a^2 - \frac{2}{5}a\right) = 5\left(-\frac{1}{5}\right)$$

$$5a^2 - 2a = -1$$

$$5a^2 - 2a + 1 = 0$$

$$a = \frac{-b \pm \sqrt{b^2 - 4ac}}{2a}$$

$$= \frac{-(-2) \pm \sqrt{(-2)^2 - 4(5)(1)}}{2(5)}$$

$$= \frac{2 \pm \sqrt{4 - 20}}{10}$$

$$= \frac{2 \pm \sqrt{-16}}{10}$$

$$= \frac{2 \pm 4i}{10}$$

$$= \frac{2}{10} \pm \frac{4i}{10}$$

$$= \frac{1}{5} \pm \frac{2}{5}i$$

130. Let x = width of the uniform sidewalk. To find the length or width, you must take the total length or width and subtract $2x$ since there is a sidewalk of width x on both ends of the garden. The length is $(24 - 2x)$ and the width is $(16 - 2x)$. Solve the inequality to find x.

$$(\text{length})(\text{width}) \leq 180$$

$$(24 - 2x)(16 - 2x) \leq 180$$

$$384 - 48x - 32x + 4x^2 \leq 180$$

$$4x^2 - 80x + 384 \leq 180$$

$$4x^2 - 80x + 204 \leq 0$$

$$4x^2 - 80x + 204 = 0$$

$$\frac{4x^2}{4} - \frac{80x}{4} + \frac{204}{4} = \frac{0}{4}$$

$$x^2 - 20x + 51 = 0$$

$$(x - 3)(x - 17) = 0$$

$$x - 3 = 0 \quad \text{or} \quad x - 17 = 0$$

$$x = 3 \qquad\qquad x = 17$$

17 is not possible: it would make the length and width negative.

Length:
$$24 - 2(3) = 24 - 6$$
$$= 18 \text{ ft}$$

Width:
$$16 - 2(3) = 16 - 6$$
$$= 10 \text{ ft}$$

The dimension of the largest possible garden is 10 ft by 18 ft.

131. Let x = length of shorter leg. Then $170 - x$ = length of longer leg. Use Pythagorean Theorem to find x.

$$x^2 + (170 - x)^2 = 130^2$$

$$x^2 + 28,900 - 170x - 170x + x^2 = 16,900$$

$$2x^2 - 340x + 28,900 = 16,900$$

$$2x^2 - 340x + 12,000 = 0$$

$$\frac{2x^2}{2} - \frac{340x}{2} + \frac{12,000}{2} = 0$$

$$x^2 - 170x + 6,000 = 0$$

$$x = \frac{-(-170) \pm \sqrt{(-170)^2 - 4(1)(6,000)}}{2(1)}$$

$$x = \frac{170 \pm \sqrt{28,900 - 24,000}}{2}$$

$$x = \frac{170 \pm \sqrt{4,900}}{2}$$

$$x = \frac{170 \pm 70}{2}$$

$$x = \frac{170 + 70}{2} \quad \text{or} \quad x = \frac{170 - 70}{2}$$

$$x = \frac{240}{2} \qquad\qquad x = \frac{100}{2}$$

$$x = 120 \qquad\qquad x = 50$$

The two segments are 50 m and 120 m.

132. Let $y = x^{1/3}$.

$$t^{2/3} - t^{1/3} = 6$$

$$t^{2/3} - t^{1/3} - 6 = 0$$

$$\left(t^{1/3}\right)^2 - t^{1/3} - 6 = 0$$

$$y^2 - y - 6 = 0$$

$$(y - 3)(y + 2) = 0$$

$$y - 3 = 0 \quad \text{or} \quad y + 2 = 0$$

$$y = 3 \qquad\qquad y = -2$$

$$x^{1/3} = 3 \qquad\qquad x^{1/3} = -2$$

$$\left(x^{1/3}\right)^3 = (3)^3 \qquad \left(x^{1/3}\right)^3 = (-2)^3$$

$$x = 27 \qquad\qquad x = -8$$

$$x = 27, \ -8$$

Chapter 10 Cumulative Review

133. Let $y = \dfrac{1}{x^2}$.

$$x^{-4} - 2x^{-2} + 1 = 0$$

$$\frac{1}{x^4} - 2\left(\frac{1}{x^2}\right) + 1 = 0$$

$$\left(\frac{1}{x^2}\right)^2 - 2\left(\frac{1}{x^2}\right) + 1 = 0$$

$$y^2 - 2y + 1 = 0$$

$$(y-1)(y-1) = 0$$

$$y - 1 = 0 \quad \text{or} \quad y - 1 = 0$$

$$y = 1 \qquad\qquad y = 1$$

$$\frac{1}{x^2} = 1 \qquad\qquad \frac{1}{x^2} = 1$$

$$x^2 = 1 \qquad\qquad x^2 = 1$$

$$x = \pm\sqrt{1} \qquad x = \pm\sqrt{1}$$

$$x = \pm 1 \qquad\qquad x = \pm 1$$

$$x = 1, \ -1, \ 1, \ -1$$

134. $f(x) = -x^2 - 4x$

Step 1: Since $a = -1 < 0$, the parabola opens downward.

Step 2: Find the vertex and axis of symmetry.
$$x = \frac{-b}{2a} = \frac{-(-4)}{2(-1)} = \frac{4}{-2} = -2$$
$$y = -(-2)^2 - 4(-2) = -4 + 8 = 4$$
vertex: (–2, 4)
The axis of symmetry is $x = -2$.

Step 3: Find the x– and y–intercepts.
Since $c = 0$, the y–intercept is (0, 0).
To find the x–intercept, let $y = 0$.
$$0 = -x^2 - 4x$$
$$0 = -x(x + 4)$$
$$-x = 0 \quad \text{or} \quad x + 4 = 0$$
$$x = 0 \qquad\qquad x = -4$$
x–intercepts: (0, 0) and (–4, 0)
y–intercepts: (0, 0)

Step 4: Find another point(s).
Let $x = -1$.
$$y = -(-1)^2 - 4(-1) = -1 + 4 = 3$$
(–1, 3)
Use symmetry to find the point (–3, 3)

Step 5: Plots the points and draw the parabola.

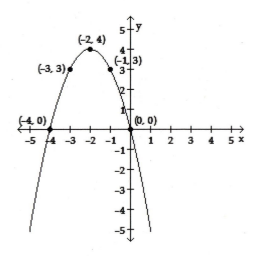

135.

$$x^2 - 81 < 0$$
$$x^2 - 81 = 0$$
$$(x-9)(x+9) = 0$$
$$x - 9 = 0 \quad \text{or} \quad x + 9 = 0$$
$$x = 9 \qquad\qquad x = -9$$

$$(-9, 9)$$

137. **a.** The x–intercept is $\left(-\dfrac{3}{4}, 0\right)$, so the solution to the equation is $x = -\dfrac{3}{4}$.

b. Use $x = -\dfrac{3}{4}$ as the critical number. The answer would be **no solution**.

136.

$$\frac{1}{x+1} \geq \frac{x}{x+4}$$

$$\frac{1}{x+1} - \frac{x}{x+4} \geq 0$$

$$\frac{1}{x+1} - \frac{x}{x+4} = 0$$

$$\frac{1}{x+1} \cdot \frac{x+4}{x+4} - \frac{x}{x+4} \cdot \frac{x+1}{x+1} = 0$$

$$\frac{(x+4) - x(x+1)}{(x+1)(x+4)} = 0$$

$$\frac{x + 4 - x^2 - x}{(x+1)(x+4)} = 0$$

$$\frac{4 - x^2}{(x+1)(x+4)} = 0$$

$$(x+1)(x+4)\left(\frac{4 - x^2}{(x+1)(x+4)}\right) = (x+1)(x+4)(0)$$

$$4 - x^2 = 0$$

$$(2+x)(2-x) = 0$$

$$2 + x = 0 \quad \text{or} \quad 2 - x = 0$$

$$x = -2 \qquad\qquad 2 = x$$

Set the denominator $= 0$.

$$(x+1)(x+4) = 0$$

$$x + 1 = 0 \quad \text{or} \quad x + 4 = 0$$

$$x = -1 \qquad\qquad x = -4$$

Critical numbers $= -4, -2, -1,$ and 2

$$(-4, -2] \cup (-1, 2]$$

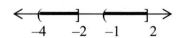

Chapter 10 Cumulative Review

SECTION 11.1

VOCABULARY

1. The **sum** of f and g, denoted as $f + g$, is defined by $(f+g)(x) = \boxed{f(x)+g(x)}$

   and the **difference** of f and g, denoted as $f - g$, is defined by

   $(f-g)(x) = \boxed{f(x)-g(x)}$.

3. The **domain** of the function $f + g$ is the set of real numbers x that are in the domain of both f and g.

5. When we write $(f \circ g)(x)$ as $f(g(x))$, we have changed from $\circ$ notation to **nested** parentheses notation.

CONCEPTS

7. a. $(f \circ g)(3) = f\left(\boxed{g(3)}\right)$

   b. To find $f(g(3))$, we first find $\boxed{g(3)}$ and then substitute that value for x in $f(x)$.

9.
$$f(-2) = 2 - 3(-2)^2$$
$$= 2 - 3(4)$$
$$= 2 - 12$$
$$= -10$$
$$g(-10) = -10 + 10$$
$$= 0$$

NOTATION

11.
$$(f \cdot g)(x) = f(x) \cdot \boxed{g(x)}$$
$$= \boxed{(3x-1)}(2x+3)$$
$$= 6x^2 + \boxed{9x} - \boxed{2x} - 3$$
$$(f \cdot g)(x) = 6x^2 + 7x - 3$$

GUIDED PRACTICE

13. $f + g$
$$(f+g)(x) = f(x) + g(x)$$
$$= 2x + 1 + x - 3$$
$$= 3x - 2$$
$$D:(-\infty, \infty)$$

15. $g - f$
$$(g-f)(x) = g(x) - f(x)$$
$$= (x-3) - (2x+1)$$
$$= x - 3 - 2x - 1$$
$$= -x - 4$$
$$D:(-\infty, \infty)$$

17. $f \cdot g$
$$(f \cdot g)(x) = f(x) \cdot g(x)$$
$$= (2x+1)(x-3)$$
$$= 2x^2 - 6x + x - 3$$
$$= 2x^2 - 5x - 3$$
$$D:(-\infty, \infty)$$

19. g / f
$$(g/f)(x) = \frac{g(x)}{f(x)}$$
$$= \frac{x-3}{2x+1}$$
$$D:\left(-\infty, -\frac{1}{2}\right) \cup \left(-\frac{1}{2}, \infty\right)$$

21. $f + g$
$$(f+g)(x) = f(x) + g(x)$$
$$= 3x + 4x$$
$$= 7x$$
$$D:(-\infty, \infty)$$

23. $g - f$
$$(g-f)(x) = g(x) - f(x)$$
$$= 4x - 3x$$
$$= x$$
$$D:(-\infty, \infty)$$

25. $f \cdot g$

$$\begin{aligned}(f \bullet g)(x) &= f(x) \bullet g(x)\\ &= (3x)(4x)\\ &= 12x^2\\ D&:(-\infty, \infty)\end{aligned}$$

27. g / f

$$\begin{aligned}(g/f)(x) &= \frac{g(x)}{f(x)}\\ &= \frac{4x}{3x}\\ &= \frac{4}{3}\\ D&:(-\infty, 0) \cup (0, \infty)\end{aligned}$$

29. $(f + g)(8)$

$$\begin{aligned}f(8) + g(8) &= (2(8) - 5) + (8 + 1)\\ &= (16 - 5) + (9)\\ &= 11 + 9\\ &= 20\end{aligned}$$

31. $(f \cdot g)(0)$

$$\begin{aligned}f(0)g(0) &= (2(0) - 5)(0 + 1)\\ &= (0 - 5)(1)\\ &= (-5)(1)\\ &= -5\end{aligned}$$

33. $(s \cdot t)(-2)$

$$\begin{aligned}s(-2)t(-2) &= (3 - (-2))((-2)^2 - (-2) - 6)\\ &= (3 + 2)(4 + 2 - 6)\\ &= (5)(0)\\ &= 0\end{aligned}$$

35. $(s/t)(1)$

$$\begin{aligned}s(1)/t(1) &= \frac{3 - (1)}{(1)^2 - (1) - 6}\\ &= \frac{2}{1 - 1 - 6}\\ &= \frac{2}{-6}\\ &= -\frac{1}{3}\end{aligned}$$

37. $(f \circ g)(2)$

$$\begin{aligned}(f \circ g)(2) &= f(g(2))\\ &= f(2^2 - 1)\\ &= f(4 - 1)\\ &= f(3)\\ &= 2(3) + 1\\ &= 6 + 1\\ &= 7\end{aligned}$$

39. $(g \circ f)(-3)$

$$\begin{aligned}(g \circ f)(-3) &= g(f(-3))\\ &= g(2(-3) + 1)\\ &= g(-6 + 1)\\ &= g(-5)\\ &= (-5)^2 - 1\\ &= 25 - 1\\ &= 24\end{aligned}$$

41. $(f \circ g)\left(\frac{1}{2}\right)$

$$\begin{aligned}(f \circ g)\left(\frac{1}{2}\right) &= f\left(g\left(\frac{1}{2}\right)\right)\\ &= f\left(\left(\frac{1}{2}\right)^2 - 1\right)\\ &= f\left(\frac{1}{4} - 1\right)\\ &= f\left(-\frac{3}{4}\right)\\ &= 2\left(-\frac{3}{4}\right) + 1\\ &= \frac{-3}{2} + 1\\ &= -\frac{1}{2}\end{aligned}$$

Section 11.1

43. $(g \circ f)(2x)$

$(g \circ f)(2x) = g(f(2x))$

$\qquad = g(2(2x) + 1)$

$\qquad = g(4x + 1)$

$\qquad = (4x + 1)^2 - 1$

$\qquad = 16x^2 + 8x + 1 - 1$

$\qquad = 16x^2 + 8x$

45. a. $(f + g)(-5) = f(-5) + g(-5)$

$\qquad = 3 + (-3)$

$\qquad = 0$

b. $(f - g)(3) = f(3) - g(3)$

$\qquad = 3 - (-3)$

$\qquad = 6$

c. $(f \cdot g)(-3) = f(-3)g(-3)$

$\qquad = 1(-3)$

$\qquad = -3$

47. a. $(g + f)(2) = g(2) + f(2)$

$\qquad = -3 + 2$

$\qquad = -1$

b. $(g - f)(-5) = g(-5) - f(-5)$

$\qquad = 3 - (-3)$

$\qquad = 6$

c. $(g \cdot f)(1) = g(1)f(1)$

$\qquad = (-3)(2)$

$\qquad = -6$

49. Let $f(x) = x^2$ and $g(x) = x + 15$.

$h(x) = (f \circ g)(x)$

$\qquad = f(g(x))$

$\qquad = f(x + 15)$

$\qquad = (x + 15)^2$

51. Let $f(x) = x + 9$ and $g(x) = x^5$.

$h(x) = (f \circ g)(x)$

$\qquad = f(g(x))$

$\qquad = f(x^5)$

$\qquad = x^5 + 9$

53. Let $f(x) = \sqrt{x}$ and $g(x) = 16x - 1$

$h(x) = (f \circ g)(x)$

$\qquad = f(g(x))$

$\qquad = f(16x - 1)$

$\qquad = \sqrt{16x - 1}$

55. Let $f(x) = \dfrac{1}{x}$ and $g(x) = x - 4$

$h(x) = (f \circ g)(x)$

$\qquad = f(g(x))$

$\qquad = f(x - 4)$

$\qquad = \dfrac{1}{x - 4}$

TRY IT YOURSELF

57. $(f \circ g)(4)$

$(f \circ g)(4) = f(g(4))$

$\qquad = f((4)^2 + 4)$

$\qquad = f(16 + 4)$

$\qquad = f(20)$

$\qquad = 3(20) - 2$

$\qquad = 60 - 2$

$\qquad = 58$

59. $(g \circ f)(-3)$

$(g \circ f)(-3) = g(f(-3))$

$\qquad = g(3(-3) - 2)$

$\qquad = g(-9 - 2)$

$\qquad = g(-11)$

$\qquad = (-11)^2 - 11$

$\qquad = 121 - 11$

$\qquad = 110$

61. $(g \circ f)(0)$

$(g \circ f)(0) = g(f(0))$

$= g(3(0) - 2)$

$= g(0 - 2)$

$= g(-2)$

$= (-2)^2 - 2$

$= 4 - 2$

$= 2$

63. $(g \circ f)(x)$

$(g \circ f)(x) = g(f(x))$

$= g(3x - 2)$

$= (3x - 2)^2 + 3x - 2$

$= 9x^2 - 12x + 4 + 3x - 2$

$= 9x^2 - 9x + 2$

65. $f - g$

$(f - g)(x) = f(x) - g(x)$

$= (3x - 2) - (2x^2 + 1)$

$= 3x - 2 - 2x^2 - 1$

$= -2x^2 + 3x - 3$

$D : (-\infty, \infty)$

67. f / g

$(f / g)(x) = \dfrac{f(x)}{g(x)}$

$= \dfrac{3x - 2}{2x^2 + 1}$

$D : (-\infty, \infty)$

69. $(g \circ f)\left(\dfrac{1}{3}\right)$

$(g \circ f)\left(\dfrac{1}{3}\right) = g\left(f\left(\dfrac{1}{3}\right)\right)$

$= g\left(\dfrac{1}{\frac{1}{3}}\right)$

$= g\left(1 \div \dfrac{1}{3}\right)$

$= g\left(1 \cdot \dfrac{3}{1}\right)$

$= g(3)$

$= \dfrac{1}{(3)^2}$

$= \dfrac{1}{9}$

71. $(g \circ f)(8x)$

$(g \circ f)(8x) = g(f(8x))$

$= g\left(\dfrac{1}{8x}\right)$

$= \dfrac{1}{\left(\dfrac{1}{8x}\right)^2}$

$= \dfrac{1}{\dfrac{1}{64x^2}}$

$= 1 \div \dfrac{1}{64x^2}$

$= 1 \cdot \dfrac{64x^2}{1}$

$= 64x^2$

73. $f - g$

$(f - g)(x) = f(x) - g(x)$

$= (x^2 - 1) - (x^2 - 4)$

$= x^2 - 1 - x^2 + 4$

$= 3$

$D : (-\infty, \infty)$

Section 11.1

75. g/f

$$(g/f)(x) = \frac{g(x)}{f(x)}$$

$$= \frac{x^2 - 4}{x^2 - 1}$$

$$D: (-\infty, -1) \cup (-1, 1) \cup (1, \infty)$$

77. $(h \circ k)(18)$

$$(h \circ k)(18) = h(k(18))$$

$$= h(18 - 5)$$

$$= h(13)$$

$$= \sqrt{13 + 3}$$

$$= \sqrt{16}$$

$$= 4$$

79. $(k \circ h)(22)$

$$(k \circ h)(22) = k(h(22))$$

$$= k(\sqrt{22 + 3})$$

$$= k(\sqrt{25})$$

$$= k(5)$$

$$= 5 - 5$$

$$= 0$$

81. a.

$$(f + g)(1) = f(1) + g(1)$$

$$= 3 + 4$$

$$= 7$$

b.

$$(f - g)(5) = f(5) - g(5)$$

$$= 8 - 0$$

$$= 8$$

c.

$$(f \cdot g)(1) = f(1)g(1)$$

$$= 3 \cdot 4$$

$$= 12$$

d.

$$(g/f)(5) = \frac{g(5)}{f(5)}$$

$$= \frac{0}{8}$$

$$= 0$$

83.

$$(f \circ g)(x) = f(g(x))$$

$$= f(2x - 5)$$

$$= (2x - 5) + 1$$

$$= 2x - 4$$

$$(g \circ f)(x) = g(f(x))$$

$$= g(x + 1)$$

$$= 2(x + 1) - 5$$

$$= 2x + 2 - 5$$

$$= 2x - 3$$

So, $(f \circ g)(x) \neq (g \circ f)(x)$.

APPLICATIONS

85. SAT SCORES
   a. In 2004, the average combined math and reading score was 1,026.

$$(m + r)(4) = m(4) + r(4)$$

$$= 518 + 508$$

$$= 1,026$$

   b. In 2004, the average difference in the math and reading scores was 10.

$$(m - r)(4) = m(4) - r(4)$$

$$= 518 - 508$$

$$= 10$$

   c.

$$(m + r)(9) = m(9) + r(9)$$

$$= 515 + 501$$

$$= 1,016$$

   d.

$$(m - r)(9) = m(9) - r(9)$$

$$= 515 - 501$$

$$= 14$$

87. METALLURGY

$$F(t) = -200t + 2,700$$

$$C(F) = \frac{5}{9}(F - 32)$$

$$(C \circ F)(t) = C(F(t))$$

$$= C(-200t + 2,700)$$

$$= \frac{5}{9}\left[(-200t + 2,700) - 32\right]$$

$$= \frac{5}{9}(-200t + 2,668)$$

$$C(t) = \frac{5}{9}(2,668 - 200t)$$

89. VACATION MILEAGE COSTS

a. On the first graph, 500 miles corresponds with about 25 gallons consumed. On the second graph, 25 gallons consumed corresponds with about $75.

b. Find $(C \circ G)(m)$.

$$(C \circ G)(m) = C(G(m))$$

$$= C\left(\frac{m}{20}\right)$$

$$= 3\left(\frac{m}{20}\right)$$

$$C(m) = \frac{3m}{20}$$

$$C(m) = 0.15m$$

WRITING

91. Answers will vary.

93. Answers will vary.

95.

$$\frac{\dfrac{ac - ad - c + d}{a^3 - 1}}{\dfrac{c^2 - 2cd + d^2}{a^2 + a + 1}} = \frac{ac - ad - c + d}{a^3 - 1} \div \frac{c^2 - 2cd + d^2}{a^2 + a + 1}$$

$$= \frac{ac - ad - c + d}{a^3 - 1} \cdot \frac{a^2 + a + 1}{c^2 - 2cd + d^2}$$

$$= \frac{a(c - d) - 1(c - d)}{(a - 1)(a^2 + a + 1)} \cdot \frac{a^2 + a + 1}{(c - d)(c - d)}$$

$$= \frac{\cancel{(c - d)}\,\cancel{(a - 1)}}{\cancel{(a - 1)}\,\cancel{(a^2 + a + 1)}} \cdot \frac{\cancel{a^2 + a + 1}}{\cancel{(c - d)}(c - d)}$$

$$= \frac{1}{c - d}$$

CHALLENGE PROBLEMS

97. If $f(x) = x^2$ and $g(x) = \boxed{2x + 5}$, then

$$(f \circ g)(x) = 4x^2 + 20x + 25.$$

99.

$$f(x) = \frac{1}{12}x + \frac{7}{2};\ \ g(x) = 2$$

$$f(x) + g(x) = \frac{1}{12}x + \frac{7}{2} + 2$$

$$= \frac{1}{12}x + \frac{11}{2}$$

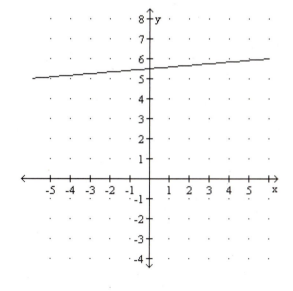

SECTION 11.2

VOCABULARY

1. A function is called a **one–to–one** function if different inputs determine different outputs.

3. The functions f and f^{-1} are **inverses**.

CONCEPTS

5. If any horizontal line that intersects the graph of a function does so more than once, the function is not **one–to–one**.

7. If f is a one–to–one function, the domain of f is the **range** of f^{-1}, and the range of f is the **domain** of f^{-1}.

9. If f is a one-to-one function, and if $f(1)=6$, then $f^{-1}(6)=\boxed{1}$.

11. a. no
    b. no

13.

x	$f^{-1}(x)$
–2	**–4**
0	**0**
4	**8**

NOTATION

15.

$$\boxed{y}=2x-3$$
$$x=\boxed{2y}-3$$
$$x+\boxed{3}=2y$$
$$\frac{x+3}{2}=\boxed{y}$$

The inverse of $f(x)=2x-3$ is $\boxed{f^{-1}}(x)=\dfrac{x+3}{2}$.

17. The symbol f^{-1} is read as "the **inverse of** f" or "f **inverse**."

GUIDED PRACTICE

19. Yes, each output corresponds to exactly one input, so the function is one–to–one.

21. No. $2^4=(-2)^4=16$ so the output 16 corresponds to two different inputs, 2 and –2.

23. No. $-(3)^2+3(3)=-(0)+3(0)=0$ so the output 0 corresponds to two different inputs, 0 and –3.

25. No. The output 1 corresponds to more than one input, 1, 2, 3 and 4.

27. one–to–one

29. not one–to–one

31. not one–to–one

33. one-to-one

35.
$$f(x)=2x+4$$
$$y=2x+4$$
$$x=2y+4$$
$$x-4=2y$$
$$\frac{x-4}{2}=y$$
$$y=\frac{x-4}{2} \quad \text{or} \quad y=\frac{1}{2}x-2$$
$$f^{-1}(x)=\frac{x-4}{2} \quad \text{or} \quad f^{-1}(x)=\frac{1}{2}x-2$$

37.
$$f(x)=\frac{x}{5}+\frac{4}{5}$$
$$y=\frac{x}{5}+\frac{4}{5}$$
$$x=\frac{y}{5}+\frac{4}{5}$$
$$5(x)=5\left(\frac{y}{5}+\frac{4}{5}\right)$$
$$5x=y+4$$
$$5x-4=y$$
$$y=5x-4$$
$$f^{-1}(x)=5x-4$$

39.

$$f(x) = \frac{x-4}{5}$$

$$y = \frac{x-4}{5}$$

$$x = \frac{y-4}{5}$$

$$5(x) = 5\left(\frac{y-4}{5}\right)$$

$$5x = y - 4$$

$$5x + 4 = y$$

$$y = 5x + 4$$

$$f^{-1}(x) = 5x + 4$$

41.

$$f(x) = \frac{2}{x-3}$$

$$y = \frac{2}{x-3}$$

$$x = \frac{2}{y-3}$$

$$(y-3)(x) = (y-3)\left(\frac{2}{y-3}\right)$$

$$xy - 3x = 2$$

$$xy = 2 + 3x$$

$$y = \frac{2+3x}{x}$$

$$y = \frac{2}{x} + \frac{3x}{x}$$

$$y = \frac{2}{x} + 3$$

$$f^{-1}(x) = \frac{2}{x} + 3$$

43.

$$f(x) = \frac{4}{x}$$

$$y = \frac{4}{x}$$

$$x = \frac{4}{y}$$

$$y(x) = y\left(\frac{4}{y}\right)$$

$$xy = 4$$

$$y = \frac{4}{x}$$

$$f^{-1}(x) = \frac{4}{x}$$

45.

$$f(x) = x^3 + 8$$

$$y = x^3 + 8$$

$$x = y^3 + 8$$

$$x - 8 = y^3$$

$$y^3 = x - 8$$

$$\sqrt[3]{y^3} = \sqrt[3]{x-8}$$

$$y = \sqrt[3]{x-8}$$

$$f^{-1}(x) = \sqrt[3]{x-8}$$

47.

$$f(x) = \sqrt[3]{x}$$

$$y = \sqrt[3]{x}$$

$$x = \sqrt[3]{y}$$

$$(x)^3 = \left(\sqrt[3]{y}\right)^3$$

$$x^3 = y$$

$$y = x^3$$

$$f^{-1}(x) = x^3$$

Section 11.2

49.

$$f(x) = (x+10)^3$$
$$y = (x+10)^3$$
$$x = (y+10)^3$$
$$\sqrt[3]{x} = \sqrt[3]{(y+10)^3}$$
$$\sqrt[3]{x} = y+10$$
$$\sqrt[3]{x} - 10 = y$$
$$y = \sqrt[3]{x} - 10$$
$$f^{-1}(x) = \sqrt[3]{x} - 10$$

51.

$$f(x) = 2x^3 - 3$$
$$y = 2x^3 - 3$$
$$x = 2y^3 - 3$$
$$x + 3 = 2y^3$$
$$\frac{x+3}{2} = y^3$$
$$\sqrt[3]{\frac{x+3}{2}} = \sqrt[3]{y^3}$$
$$\sqrt[3]{\frac{x+3}{2}} = y$$
$$y = \sqrt[3]{\frac{x+3}{2}}$$
$$f^{-1}(x) = \sqrt[3]{\frac{x+3}{2}}$$

53.

$$f(x) = \frac{x^7}{2}$$
$$y = \frac{x^7}{2}$$
$$x = \frac{y^7}{2}$$
$$2(x) = 2\left(\frac{y^7}{2}\right)$$
$$2x = y^7$$
$$\sqrt[7]{2x} = \sqrt[7]{y^7}$$
$$\sqrt[7]{2x} = y$$
$$y = \sqrt[7]{2x}$$
$$f^{-1}(x) = \sqrt[7]{2x}$$

55.

$$(f \circ f^{-1})(x) = f\left(f^{-1}(x)\right)$$
$$= f\left(\frac{x-9}{2}\right)$$
$$= 2\left(\frac{x-9}{2}\right) + 9$$
$$= x - 9 + 9$$
$$= x$$

$$(f^{-1} \circ f)(x) = f^{-1}\left(f(x)\right)$$
$$= f^{-1}(2x+9)$$
$$= \frac{2x+9-9}{2}$$
$$= \frac{2x}{2}$$
$$= x$$

57.

$$(f \circ f^{-1})(x) = f\left(f^{-1}(x)\right)$$
$$= f\left(\frac{2}{x} + 3\right)$$
$$= \frac{2}{\left(\frac{2}{x} + 3\right) - 3}$$
$$= \frac{2}{\frac{2}{x}}$$
$$= \frac{2}{2} \cdot \frac{x}{x}$$
$$= \frac{2x}{2}$$
$$= x$$

$$(f^{-1} \circ f)(x) = f^{-1}\left(f(x)\right)$$
$$= f^{-1}\left(\frac{2}{x-3}\right)$$
$$= \frac{2}{\frac{2}{x-3}} + 3$$
$$= \frac{2}{1} \cdot \frac{x-3}{2} + 3$$
$$= \frac{2(x-3)}{2} + 3$$
$$= x - 3 + 3$$
$$= x$$

59.

$$f(x) = 2x$$
$$y = 2x$$
$$x = 2y$$
$$\frac{x}{2} = y$$
$$y = \frac{x}{2} \text{ or } y = \frac{1}{2}x$$
$$f^{-1}(x) = \frac{x}{2} \text{ or } f^{-1}(x) = \frac{1}{2}x$$

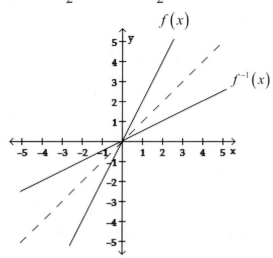

61.

$$f(x) = 4x + 3$$
$$y = 4x + 3$$
$$x = 4y + 3$$
$$x - 3 = 4y$$
$$\frac{x - 3}{4} = y$$
$$y = \frac{x - 3}{4}$$
$$f^{-1}(x) = \frac{x - 3}{4}$$

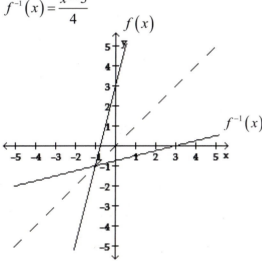

63.

$$f(x) = -\frac{2}{3}x + 3$$
$$y = -\frac{2}{3}x + 3$$
$$x = -\frac{2}{3}y + 3$$
$$3(x) = 3\left(-\frac{2}{3}y + 3\right)$$
$$3x = -2y + 9$$
$$3x - 9 = -2y$$
$$\frac{3x - 9}{-2} = y$$
$$-\frac{3}{2}x + \frac{9}{2} = y$$
$$y = -\frac{3}{2}x + \frac{9}{2}$$
$$f^{-1}(x) = -\frac{3}{2}x + \frac{9}{2}$$

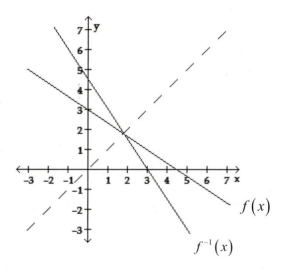

Section 11.2

65.

$$f(x)=x^3$$
$$y=x^3$$
$$x=y^3$$
$$\sqrt[3]{x}=\sqrt[3]{y^3}$$
$$\sqrt[3]{x}=y$$
$$y=\sqrt[3]{x}$$
$$f^{-1}(x)=\sqrt[3]{x}$$

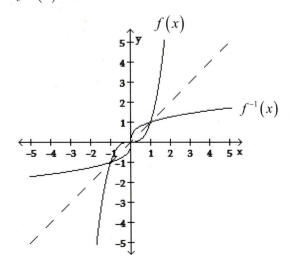

67.

$$f(x)=x^2-1$$
$$y=x^2-1$$
$$x=y^2-1$$
$$x+1=y^2$$
$$\sqrt{x+1}=\sqrt{y^2}$$
$$\sqrt{x+1}=y$$
$$y=\sqrt{x+1}$$
$$f^{-1}(x)=\sqrt{x+1}$$

69. INTERPERSONAL RELATIONSHIPS
   a. The graph is a function, but its inverse is not.
   b. No. Twice during this period, the person's anxiety level was at the maximum threshold value.

WRITING

71. Answers will vary.

73. Answers will vary.

75. Answers will vary.

REVIEW

77.

$$3-\sqrt{-64}=3-8i$$

79.

$$(3+4i)(2-3i)=6-9i+8i-12i^2$$
$$=6-i-12(-1)$$
$$=6-i+12$$
$$=18-i$$

81.

$$(6-8i)^2=(6-8i)(6-8i)$$
$$=36-48i-48i+64i^2$$
$$=36-96i+64(-1)$$
$$=36-96i-64$$
$$=-28-96i$$

CHALLENGE PROBLEMS

83.

$$f(x) = \frac{x+1}{x-1}$$

$$y = \frac{x+1}{x-1}$$

$$x = \frac{y+1}{y-1}$$

$$(y-1)(x) = (y-1)\left(\frac{y+1}{y-1}\right)$$

$$xy - x = y + 1$$

$$xy = y + 1 + x$$

$$xy - y = x + 1$$

$$y(x-1) = x + 1$$

$$y = \frac{x+1}{x-1}$$

$$f^{-1}(x) = \frac{x+1}{x-1}$$

85.

$$f^{-1}(f(4)) = f^{-1}(10)$$
$$= 4$$

$$f(f^{-1}(2)) = f(0)$$
$$= 2$$

VOCABULARY

1. $f(x)=2^x$ and $f(x)=\left(\dfrac{1}{4}\right)^x$ are examples
    of **exponential** functions.

3. The graph of $f(x)=3^x$ approaches, but
    never touches, the negative portion of the
    x-axis. Thus, the x-axis is an **asymptote** of
    the graph.

5. a. exponential
    b. $(-\infty, \infty)$
    c. $(0, \infty)$
    d. $(0, 1)$; none
    e. yes
    f. the x–axis $(y = 0)$
    g. increasing
    h. $y = 3$

CONCEPTS

7. a. $3^{-2}=\dfrac{1}{3^2}=\dfrac{1}{9}$

    b. $\left(\dfrac{1}{2}\right)^4=\dfrac{1^4}{2^4}=\dfrac{1}{16}$

    c. $\left(\dfrac{1}{5}\right)^{-2}=(5)^2=25$

9. a. iii
    b. iv
    c. ii
    d. i

11. a. D: $(-\infty,\infty)$, R: $(-3,\infty)$
    b. D: $(-\infty,\infty)$, R: $(1,\infty)$

13. $f(x)=5^x$

x	$f(x)$
-3	$5^{-3}=\dfrac{1}{5^3}$ $=\dfrac{1}{125}$
-2	$5^{-2}=\dfrac{1}{5^2}$ $=\dfrac{1}{25}$
-1	$5^{-1}=\dfrac{1}{5}$
0	$5^0=1$
1	$5^1=5$
2	$5^2=25$
3	$5^3=125$

15. a. ii.
    b. i.
    c. ii.
    d. ii.
    e. i.

NOTATION

17. $b > 0$ and $b \neq 1$

GUIDED PRACTICE

19. $f(x)=3^x$

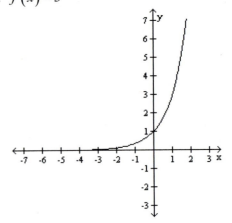

21. $f(x) = 5^x$

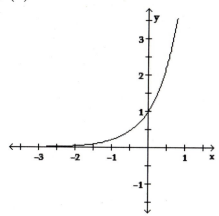

27. $g(x) = 3^x - 2$

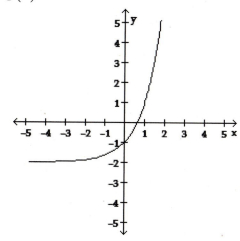

23. $f(x) = \left(\dfrac{1}{4}\right)^x$

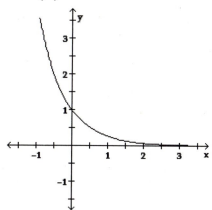

29. $g(x) = 2^{x+1}$

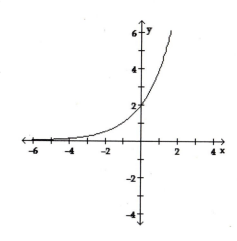

25. $f(x) = \left(\dfrac{1}{6}\right)^x$

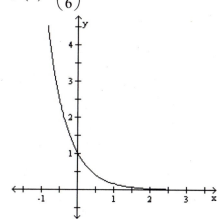

31. $g(x) = 4^{x-1} + 2$

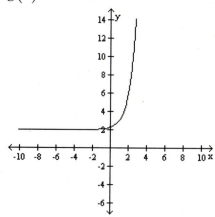

Section 11.3

33. $g(x) = -2^x$

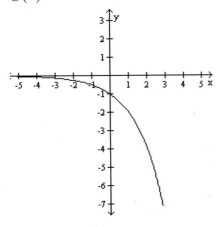

35. $f(x) = \dfrac{1}{2}\left(3^{x/2}\right)$

The function is increasing.

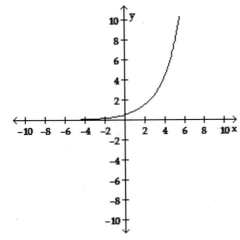

37. $y = 2\left(3^{-x/2}\right)$

The function is decreasing.

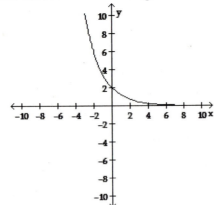

APPLICATIONS

39. CO_2 CONCENTRATION
   a. 280 ppm, 295 ppm, 370 ppm
   b. about 1970

41. VALUE OF A CAR
   a. at the end of the 2nd year
   b. at the end of the 4th year
   c. during the 7th year

43. COMPUTER VIRUSES
   a. $c(t) = 5(1.034)^t$

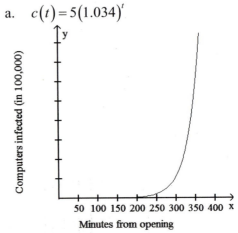

   b. Let $t = 480$.
   $$c(480) = 5(1.034)^{480}$$
   $$= 5(9,329,505.043)$$
   $$= 46,647,545.22$$

45. GUITARS
   $$f(n) = 650(0.94)^n$$
   $$f(7) = 650(0.94)^7$$
   $$= 650(0.648477594)$$
   $$\approx 422 \text{ mm}$$

47. RADIOACTIVE DECAY
   Let $t = 5$.
   $$A = 500\left(\frac{2}{3}\right)^t$$
   $$= 500\left(\frac{2}{3}\right)^{10}$$
   $$= 500\left(\frac{1,024}{59,049}\right)$$
   $$= 8.7 \text{ gm}$$

49. SOCIAL WORKER

$$P(t) = 35.8(1.06)^t$$

For 2006, let $t = 0$.

$$P(0) = 35.8(1.06)^0$$
$$= 35.8(1)$$
$$= 35.8$$

For 2007, let $t = 1$.

$$P(1) = 35.8(1.06)^1$$
$$= 35.8(1.06)$$
$$= 37.9$$

For 2008, let $t = 2$.

$$P(2) = 35.8(1.06)^2$$
$$= 35.8(1.1236)$$
$$= 40.2$$

For 2009, let $t = 3$.

$$P(3) = 35.8(1.06)^3$$
$$= 35.8(1.191016)$$
$$= 42.6$$

51. POPULATION GROWTH

Let $t = 6\dfrac{9}{12} = 6\dfrac{3}{4} = 6.75$.

$$P = 3,745(0.93)^t$$
$$= 3,745(0.93)^{6.75}$$
$$= 3,745(0.612717)$$
$$\approx 2,295$$

53. COMPOUND INTEREST

$P = 10,000,\ r = 0.08,\ k = 4,\ t = 10$

$$A = P\left(1 + \frac{r}{k}\right)^{kt}$$
$$= 10,000\left(1 + \frac{0.08}{4}\right)^{4(10)}$$
$$= 10,000(1 + 0.02)^{40}$$
$$= 10,000(1.02)^{40}$$
$$= 10,000(2.20804)$$
$$= \$22,080.40$$

55. COMPARING INTEREST RATES

Find the amount earned at $5\dfrac{1}{2}\%$.

$P = 1,000,\ r = 0.055,\ k = 4,\ t = 5$

$$A = P\left(1 + \frac{r}{k}\right)^{kt}$$
$$= 1,000\left(1 + \frac{0.055}{4}\right)^{4(5)}$$
$$= 1,000(1.01375)^{20}$$
$$= 1,000(1.314066)$$
$$= \$1,314.07$$

Find the amount earned at 5%.

$P = 1,000,\ r = 0.05,\ k = 4,\ t = 5$

$$A = P\left(1 + \frac{r}{k}\right)^{kt}$$
$$= 1,000\left(1 + \frac{0.05}{4}\right)^{4(5)}$$
$$= 1,000(1.0125)^{20}$$
$$= 1,000(1.282037)$$
$$= \$1,282.04$$

Find the difference in the two amounts.

$$\$1,314.07 - 1,282.04 = \$32.03$$

57. COMPOUND INTEREST

$P = 1,\ r = 0.05,\ k = 1,\ t = 300$

$$A = P\left(1 + \frac{r}{k}\right)^{kt}$$
$$= 1\left(1 + \frac{0.05}{1}\right)^{1(300)}$$
$$= 1(1 + 0.05)^{300}$$
$$= 1(1.05)^{300}$$
$$= 1(2,273,996.129)$$
$$= \$2,273,996.13$$

WRITING

59. Answers will vary.

61. Answers will vary.

63. Answers will vary.

65. Answers will vary.

67. Answers will vary.

Section 11.3

69. The sum of same–side interior angles is 180°.

$$3x + (2x - 20) = 180$$
$$5x - 20 = 180$$
$$5x = 200$$
$$x = 40°$$

71. Angle 2 is equal to $3x$ because they are alternate interior angles.

$$3x = 3(40)$$
$$= 120°$$

CHALLENGE PROBLEMS

73. $f(x) = 3^x$

$f(1.5) \approx 5.2$

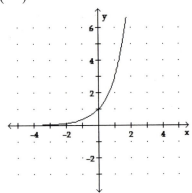

75. For $b = \dfrac{1}{50}$, the graph would be correct.

77. $f(x) = 2^{|x|}$

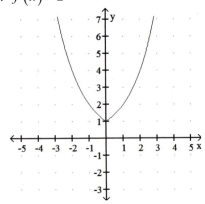

VOCABULARY

1. $f(x) = \log_2 x$ and $g(x) = \log x$ are examples of **logarithmic** functions.

3. The graph of $f(x) = \log_2 x$ approaches, but never touches, the negative portion of the y-axis. Thus the y-axis is an **asymptote** of the graph.

CONCEPTS

5. a. logarithmic
   b. D: $(0, \infty)$; R: $(-\infty, \infty)$
   c. none, $(1, 0)$
   d. yes
   e. the y–axis $(x = 0)$
   f. increasing
   g. 1

7. a. ii.
   b. iii.
   c. i.
   d. iv.

9. $\log_6 36 = 2$ means $6^2 = 36$.

11. $\log_b x$ is the **exponent** to which b is raised to get x.

13. The inverse of an exponential function is called a **logarithmic** function.

15. $f(x) = \log x$

x	$f(x)$
100	2
$\dfrac{1}{100}$	-2

17. $f(x) = \log_6 x$

input	output
6	1
-6	undefined
0	Undefined

19. a. $f(x) = \log x$

x	$f(x)$
0.5	-0.30
1	0
2	0.30
4	0.60
6	0.78
8	0.90
10	1

b.

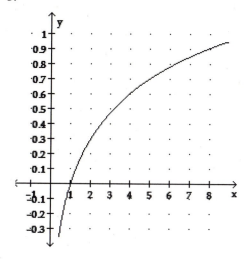

NOTATION

21. a. $\log x = \log_{\boxed{10}} x$

   b. $\log_{10} 10^x = \boxed{x}$

GUIDED PRACTICE

23.
$$\log_3 81 = 4$$
$$3^4 = 81$$

25.
$$\log_{10} 10 = 1$$
$$10^1 = 10$$

27.
$$\log_4 \frac{1}{64} = -3$$
$$4^{-3} = \frac{1}{64}$$

Section 11.4

29.

$$\log_5 \sqrt{5} = \frac{1}{2}$$
$$5^{1/2} = \sqrt{5}$$

31.

$$\log 0.1 = -1$$
$$10^{-1} = 0.1$$

33.

$$x = \log_8 64$$
$$8^x = 64$$

35.

$$t = \log_b T_1$$
$$b^t = T_1$$

37.

$$\log_n C = -42$$
$$n^{-42} = C$$

39.

$$8^2 = 64$$
$$\log_8 64 = 2$$

41.

$$4^{-2} = \frac{1}{16}$$
$$\log_4 \frac{1}{16} = -2$$

43.

$$\left(\frac{1}{2}\right)^{-5} = 32$$
$$\log_{1/2} 32 = -5$$

45.

$$x^y = z$$
$$\log_x z = y$$

47.

$$y^t = 8.6$$
$$\log_y 8.6 = t$$

49.

$$7^{4.3} = B + 1$$
$$\log_7(B+1) = 4.3$$

51.

$$\log_x 81 = 2$$
$$x^2 = 81$$
$$x^2 = 9^2$$
$$x = 9$$

53.

$$\log_8 x = 2$$
$$8^2 = x$$
$$64 = x$$

55.

$$\log_5 125 = x$$
$$5^x = 125$$
$$5^x = 5^3$$
$$x = 3$$

57.

$$\log_5 x = -2$$
$$5^{-2} = x$$
$$\frac{1}{5^2} = x$$
$$\frac{1}{25} = x$$

59.

$$\log_{36} x = -\frac{1}{2}$$
$$36^{-1/2} = x$$
$$\frac{1}{36^{1/2}} = x$$
$$\frac{1}{\sqrt{36}} = x$$
$$\frac{1}{6} = x$$

61.

$$\log_x 0.01 = -2$$
$$x^{-2} = 0.01$$
$$\left(x^{-2}\right)^{-1/2} = (0.01)^{-1/2}$$
$$x = \left(\frac{1}{100}\right)^{-1/2}$$
$$x = (100)^{1/2}$$
$$x = \sqrt{100}$$
$$x = 10$$

63.

$$\log_{27} 9 = x$$
$$27^x = 9$$
$$3^{3x} = 3^2$$
$$3x = 2$$
$$x = \frac{2}{3}$$

65.

$$\log_x 5^3 = 3$$
$$x^3 = 5^3$$
$$x^3 = 125$$
$$\sqrt[3]{x^3} = \sqrt[3]{125}$$
$$x = 5$$

67.

$$\log_{100} x = \frac{3}{2}$$
$$100^{3/2} = x$$
$$\left(\sqrt{100}\right)^3 = x$$
$$(10)^3 = x$$
$$1,000 = x$$

69.

$$\log_x \frac{1}{64} = -3$$
$$x^{-3} = \frac{1}{64}$$
$$\left(x^{-3}\right)^{-1/3} = \left(\frac{1}{64}\right)^{-1/3}$$
$$x = (64)^{1/3}$$
$$x = \sqrt[3]{64}$$
$$x = 4$$

71.

$$\log_8 x = 0$$
$$8^0 = x$$
$$1 = x$$

73.

$$\log_x \frac{\sqrt{3}}{3} = \frac{1}{2}$$
$$x^{1/2} = \frac{\sqrt{3}}{3}$$
$$\left(x^{1/2}\right)^2 = \left(\frac{\sqrt{3}}{3}\right)^2$$
$$x = \frac{\sqrt{9}}{9}$$
$$x = \frac{3}{9}$$
$$x = \frac{1}{3}$$

75.

$$\log_2 8 = x$$
$$2^x = 8$$
$$2^x = 2^3$$
$$x = 3$$

77.

$$\log_4 16 = x$$
$$4^x = 16$$
$$4^x = 4^2$$
$$x = 2$$

79.

$$\log 1,000,000 = x$$
$$10^x = 1,000,000$$
$$10^x = 10^6$$
$$x = 6$$

81.

$$\log \frac{1}{10} = x$$
$$10^x = \frac{1}{10}$$
$$10^x = 10^{-1}$$
$$x = -1$$

Section 11.4

83.
$$\log_{1/2} \frac{1}{32} = x$$
$$\left(\frac{1}{2}\right)^x = \frac{1}{32}$$
$$2^{-x} = 2^{-5}$$
$$-x = -5$$
$$x = 5$$

85.
$$\log_9 3 = x$$
$$9^x = 3$$
$$3^{2x} = 3^1$$
$$2x = 1$$
$$x = \frac{1}{2}$$

87.
$$\log 3.25 = 0.5119$$

89.
$$\log 0.00467 = -2.3307$$

91.
$$\log x = 3.7813$$
$$x = 6{,}043.6597$$

93.
$$\log x = -0.7630$$
$$x = 0.1726$$

95.
$$\log x = -0.5$$
$$x = 0.3162$$

97.
$$\log x = -1.71$$
$$x = 0.0195$$

99. $f(x) = \log_3 x$
increasing

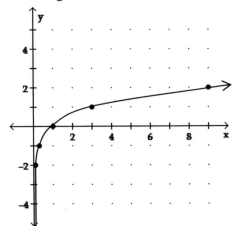

101. $y = \log_{1/2} x$
decreasing

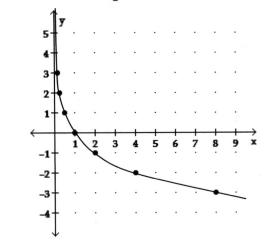

103. $f(x) = 3 + \log_3 x$

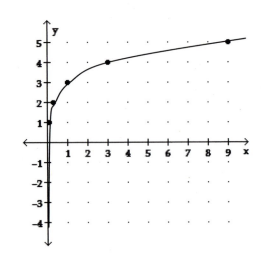

105. $y = \log_{1/2}(x - 2)$

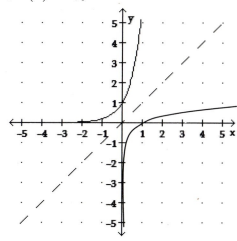

107.

$f(x) = 6^x$

$f^{-1}(x) = \log_6 x$

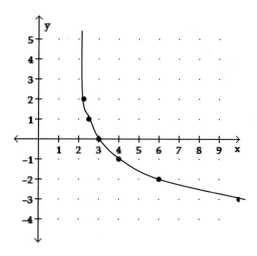

109. $f(x) = 5^x$ and $f^{-1}(x) = \log_5 x$

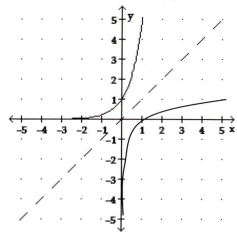

APPLICATIONS

111. db GAIN

$E_O = 30$ and $E_I = 0.1$

$$\text{db gain} = 20\left(\log \frac{E_O}{E_I}\right)$$

$$= 20\left(\log \frac{30}{0.1}\right)$$

$$= 20(\log 300)$$

$$= 20(2.477)$$

$$= 49.5 \text{ db}$$

113. EARTHQUAKES

$A = 5,000$ and $P = 0.2$

$$R = \log \frac{A}{P}$$

$$= \log \frac{5,000}{0.2}$$

$$= \log 25,000$$

$$\approx 4.4$$

115. SOCIAL WORKER

Let $m = 6$.

$$c(m) = 500\log(m + 1)$$

$$c(6) = 500\log(6 + 1)$$

$$= 500\log 7$$

$$\approx 422 \text{ cases}$$

117. STOCKING LAKES

Let $t = 2.5$.

$$f(t) = 75 + 45\log(t + 1)$$

$$f(2.5) = 75 + 45\log(2.5 + 1)$$

$$= 75 + 45\log 3.5$$

$$\approx 99 \text{ sunfish}$$

119. CHILDREN'S HEIGHT

$$h(A) = 29 + 48.8\log(A + 1)$$

$$h(9) = 29 + 48.8\log(9 + 1)$$

$$= 29 + 48.8\log(10)$$

$$= 29 + 48.8$$

$$= 77.8$$

77.8% of his adult height.

Section 11.4

121. INVESTING

$P = 1{,}000$, $A = 20{,}000$, and $r = 0.12$

$$n = \frac{\log\left(\dfrac{Ar}{P}+1\right)}{\log(1+r)}$$

$$= \frac{\log\left(\dfrac{20{,}000 \cdot 0.12}{1{,}000}+1\right)}{\log(1+0.12)}$$

$$= \frac{\log(2.4+1)}{\log(1+0.12)}$$

$$= \frac{\log 3.4}{\log 1.12}$$

$$\approx \frac{0.5315}{0.0492}$$

$$\approx 10.8 \text{ years}$$

WRITING

123. Answers will vary.

125. Answers will vary.

REVIEW

127.

$$\sqrt[3]{6x+4} = 4$$

$$\left(\sqrt[3]{6x+4}\right)^3 = (4)^3$$

$$6x+4 = 64$$

$$6x = 60$$

$$x = 10$$

129.

$$\sqrt{a+1}-1 = 3a$$

$$\sqrt{a+1} = 3a+1$$

$$\left(\sqrt{a+1}\right)^2 = (3a+1)^2$$

$$a+1 = 9a^2+6a+1$$

$$0 = 9a^2+5a$$

$$0 = a(9a+5)$$

$$a = 0 \quad \text{or} \quad 9a+5 = 0$$

$$a = 0 \qquad\qquad a = -\frac{5}{9}$$

CHALLENGE PROBLEMS

131. The domain would be

$$(-\infty,-1)\cup(1,\infty)$$

133.

$$R = \log\frac{A}{P}$$

$$10^R = \frac{A}{P}$$

1985:

$$10^{8.1} = \frac{A}{P}$$

1989:

$$10^{7.1} = \frac{A}{P}$$

If the period remains constant, the amplitude of an earthquake must change by a factor of 10 to increase its severity by 1 point on the Richter scale.

SECTION 11.5

VOCABULARY

1. $f(x) = e^x$ is called the natural **exponential** function. The base is **e**.

3. If a bank pays interest infinitely many times a year, we say that the interest is compounded **continuously**.

CONCEPTS

5. a. the natural exponential function
    b. D: $(-\infty, \infty)$; R: $(0, \infty)$
    c. $(0, 1)$; none
    d. yes
    e. the x–axis $(y = 0)$
    f. increasing
    g. e

7. $e = 2.718281828459\ldots$

9. If n gets larger and larger, the value of $\left(1 + \dfrac{1}{n}\right)^n$ approaches the value of **e**.

11. To find $\ln e^2$, we ask, "To what power must we raise **e** to get e^2?" Since the answer is the 2nd power, $\ln e^2 = 2$.

13. $\sqrt{2} \approx 1.41$; $\ e \approx 2.718$; $\ \pi \approx 3.14$

15. a. The y–coordinate of the point on the graph having an x–coordinate of 1 is $2.7182818\ldots$ The name of that number is e.
    b. As x decreases, the values of $f(x)$ decreases. The value of $f(x)$ will never be 0 or negative.

17. $f^{-1}(x) = e^x$

NOTATION

19. $P = 1,000$, $r = 0.09$, and $t = 10$

$$A = \boxed{1{,}000}\,e^{(0.09)\left(\boxed{10}\right)}$$
$$= 1{,}000\,e^{\boxed{0.9}}$$
$$\approx \boxed{2{,}459.6}$$

21. We read $\ln x$ letter-by-letter as "$\boxed{\ell}\ldots\boxed{n}\ldots$ of x."

23. To evaluate a base-10 logarithm with a calculator, use the **LOG** key. To evaluate the base-e logarithm, use the **LN** key.

GUIDED PRACTICE

25. $f(x) = e^x$

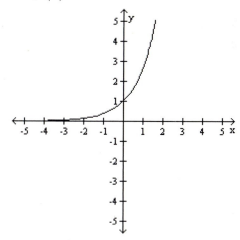

27. $f(x) = e^x + 1$

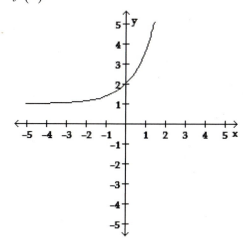

29. $y = e^{x+3}$

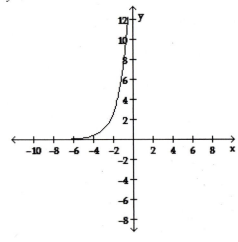

31. $f(x) = 2e^x$

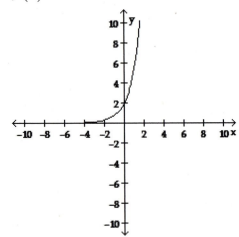

33. $A = Pe^{rt}$

$A = (5,000)e^{0.08(20)}$

$= (5,000)e^{1.6}$

$= 5,000(4.953)$

$= 24,765.16$

35. $A = Pe^{rt}$

$A = 20,000e^{0.105(50)}$

$= 20,000e^{5.25}$

$= 20,000(190.566)$

$= 3,811,325.37$

37. $A = Pe^{rt}$

$A = 15,895e^{-0.02(16)}$

$= 15,895e^{-0.32}$

$= 15,895(0.726)$

$= 11,542.14$

39. $A = Pe^{rt}$

$A = 565e^{-0.005(8)}$

$= 565e^{-0.04}$

$= 565(0.961)$

$= 542.85$

41. $\ln e^5 = 5$

43. $\ln e^6 = 6$

45. $\ln \dfrac{1}{e} = \ln e^{-1} = -1$

47. $\ln \sqrt[4]{e} = \ln e^{1/4} = \dfrac{1}{4}$

49. $\ln \sqrt[3]{e^2} = \ln e^{2/3} = \dfrac{2}{3}$

51. $\ln e^{-7} = -7$

53. $\ln 35.15 = 3.5596$

55. $\ln 0.00465 = -5.3709$

57. $\ln 1.72 = 0.5423$

59. $\ln (-0.1) = \text{undefined}$

61.

$\ln x = 1.4023$

$e^{1.4023} = x$

$4.0645 = x$

63.

$\ln x = 4.24$

$e^{4.24} = x$

$69.4079 = x$

65.
$$\ln x = -3.71$$
$$e^{-3.71} = x$$
$$0.0245 = x$$

67.
$$1.001 = \ln x$$
$$x = e^{1.001}$$
$$x = 2.7210$$

69. $f(x) = \ln\left(\dfrac{1}{2}x\right)$

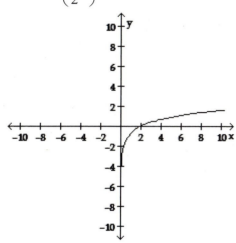

71. $f(x) = \ln(-x)$

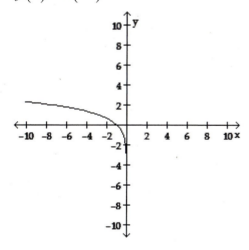

73. CONTINUOUS COMPOUND INTEREST
$P = 5,000$, $r = 0.082$, and $t = 12$
$$A = Pe^{rt}$$
$$= 5,000e^{(0.082)(12)}$$
$$= 5,000e^{0.984}$$
$$\approx 5,000(2.675135411)$$
$$\approx \$13,375.68$$

75. COMPARISON OF COMPOUNDING METHODS
Annual Compounding:
$PV = 5,000$, $i = 0.085$, and $n = 5$
$$FV = PV(1+i)^n$$
$$= 5,000(1+0.085)^5$$
$$= 5,000(1.085)^5$$
$$= 5,000(1.5037)$$
$$= \$7,518.28$$

Continuous Compounding:
$P = 5,000$, $r = 0.085$, and $t = 5$
$$A = Pe^{rt}$$
$$= 5,000e^{(0.085)(5)}$$
$$= 5,000e^{0.425}$$
$$\approx 5,000(1.52959042)$$
$$\approx \$7,647.95$$

77. DETERMINING INITIAL DEPOSIT
$A = 11,180$, $r = 0.07$, and $t = 7$
$$A = Pe^{rt}$$
$$11,180 = Pe^{(0.07)(7)}$$
$$11,180 = Pe^{0.49}$$
$$\frac{11,180}{e^{0.49}} = \frac{Pe^{0.49}}{e^{0.49}}$$
$$\frac{11,180}{e^{0.49}} = P$$
$$P \approx \frac{11,180}{1.63231622}$$
$$P \approx \$6,849.16$$

Section 11.5

79. The 20th CENTURY

$$A(t) = 123e^{0.0117t}$$

$$1937 - A(7) = 123e^{0.0117(7)}$$
$$= 133 \text{ million}$$

$$1941 - A(11) = 123e^{0.0117(11)}$$
$$= 140 \text{ million}$$

$$1955 - A(25) = 123e^{0.0117(25)}$$
$$= 165 \text{ million}$$

$$1969 - A(39) = 123e^{0.0117(39)}$$
$$= 194 \text{ million}$$

$$1974 - A(44) = 123e^{0.0117(44)}$$
$$= 206 \text{ million}$$

$$1986 - A(56) = 123e^{0.0117(56)}$$
$$= 237 \text{ million}$$

$$1997 - A(67) = 123e^{0.0117(67)}$$
$$= 269 \text{ million}$$

81. HIGHS AND LOWS

Kuwait:

$P = 2,789,132$, $r = 0.03501$, and $t = 15$

$$A = Pe^{rt}$$
$$= 2,789,132e^{(0.03501)(15)}$$
$$= 2,789,132e^{0.52515}$$
$$\approx 4,715,620$$

Bulgaria:

$P = 7,148,785$, $r = -0.00768$, and $t = 15$

$$A = Pe^{rt}$$
$$= 7,148,785e^{(-0.00768)(15)}$$
$$= 7,148,785e^{-0.1152}$$
$$\approx 6,370,911$$

83. EPIDEMICS

$$P(t) = 2e^{0.27t}$$

$$P(12) = 2e^{0.27(12)}$$
$$= 2e^{3.24}$$
$$\approx 2(25.53372175)$$
$$\approx 51$$

85. ANTS

Let $t = 40$.

$$a(t) = 1.36\left(\frac{e}{2.5}\right)^t$$

$$a(40) = 1.36(1.087)^{40}$$
$$= 1.36(28.4563)$$
$$\approx 39$$

87. SOCIAL WORKERS

The trainee had to assemble 14 chairs before meeting company standards.

89. DISINFECTANTS

$$A(t) = 2,000,000e^{-0.588t}$$

$$A(5) = 2,000,000e^{-0.588(5)}$$
$$= 2,000,000e^{-2.94}$$
$$= 105,731$$

91. SKYDIVING

$$f(t) = 50(1 - e^{-0.2t})$$

$$f(20) = 50(1 - e^{-0.2(20)})$$
$$= 50(1 - e^{-4})$$
$$\approx 50(1 - 0.018315639)$$
$$\approx 50(0.981684361)$$
$$\approx 49 \text{ mps}$$

93. THE TARHEEL STATE

Let $r = 0.043$.

$$t = \frac{\ln 2}{r}$$
$$= \frac{\ln 2}{0.043}$$
$$\approx \frac{0.6931}{0.043}$$
$$\approx 16 \text{ years}$$

95. THE EQUALITY STATE

Let $r = 0.0213$.

$$t = \frac{\ln 2}{r}$$
$$= \frac{\ln 2}{0.0213}$$
$$\approx 33 \text{ years}$$

33 years from 2009 is the year 2042.

97. POPULATION GROWTH

Let $r = 0.12$.

$$t = \frac{\ln 3}{r}$$

$$= \frac{\ln 3}{0.12}$$

$$\approx \frac{1.0986}{0.12}$$

$$\approx 9.2 \text{ years}$$

99. FORENSIC MEDICINE

Let $T_s = 70$.

$$t = \frac{1}{0.25} \ln \frac{98.6 - T_s}{82 - T_s}$$

$$= \frac{1}{0.25} \ln \frac{98.6 - 70}{82 - 70}$$

$$= \frac{1}{0.25} \ln \frac{28.6}{12}$$

$$= \frac{1}{0.25} \ln 2.383$$

$$\approx \frac{1}{0.25}(0.8684)$$

$$\approx 3.5 \text{ hours}$$

101. CROSS COUNTRY SKIING

Let $s = 7.5$.

$$H(s) = -47.73 + 107.38 \ln s$$

$$H(7.5) = -47.73 + 107.38 \ln(7.5)$$

$$= -47.73 + 216.36$$

$$\approx 169 \text{ beats per minute}$$

103. THE PACE OF LIFE

Use $s(p) = 0.05 + 0.37 \ln p$.

For New York City:

$$s(8{,}392{,}000) = 0.05 + 0.37 \ln 8{,}392{,}000$$

$$= 0.05 + 5.8988$$

$$\approx 5.9 \text{ ft.sec}$$

For Atlanta:

$$s(541{,}000) = 0.05 + 0.37 \ln 541{,}000$$

$$= 0.05 + 488$$

$$\approx 4.9 \text{ ft.sec}$$

The average pedestrian in NYC walks about 1 foot per second faster than the average pedestrian in Atlanta.

WRITING

105. Answers will vary.

107. Answers will vary.

109. Answers will vary.

111. Answers will vary.

113. Answers will vary.

REVIEW

115.
$$\sqrt{240x^5} = \sqrt{16x^4}\sqrt{15x}$$
$$= 4x^2\sqrt{15x}$$

117.
$$4\sqrt{48y^3} - 3y\sqrt{12y} = 4\sqrt{16y^2}\sqrt{3y} - 3y\sqrt{4}\sqrt{3y}$$
$$= 4(4y)\sqrt{3y} - 3y(2)\sqrt{3y}$$
$$= 16y\sqrt{3y} - 6y\sqrt{3y}$$
$$= 10y\sqrt{3y}$$

CHALLENGE PROBLEMS

119. False.

$$e^e \overset{?}{>} e^3$$

$$e^{2.71829} \overset{?}{>} e^3$$

$$15.15 \not> 20.09$$

121.
$$e^{t+5} = ke^t$$

$$\frac{e^{t+5}}{e^t} = \frac{ke^t}{e^t}$$

$$e^{t+5-t} = k$$

$$e^5 = k$$

$$k = e^5$$

123.
$$P = P_0 e^{rt}$$

$$P = P_0 e^{r\left(\frac{\ln 2}{r}\right)}$$

$$P = P_0 e^{\ln 2}$$

$$P = P_0(2)$$

$$P = 2P_0$$

Section 11.5

125. $y = f(x) = \dfrac{1}{1 + e^{-2x}}$

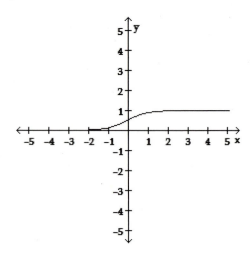

SECTION 11.6

VOCABULARY

1. The logarithm of a **product**, such as $\log_3 4x$, equals the sum of the logarithms of the factors.

3. The logarithm of a **power**, such as $\log_4 5^3$, equals the power times the logarithm of the number.

CONCEPTS

5. a. $\log_b 1 = \boxed{0}$

   b. $\log_b b = \boxed{1}$

   c. $\log_b b^x = \boxed{x}$

   d. $b^{\log_b x} = \boxed{x}$

7.
$$\log\left[(2.5)(3.7)\right] = \log 2.5 + \log 3.7$$
$$\log(9.25) = 0.3979 + 0.5682$$
$$0.9661 = 0.9661$$

9.
$$\ln\frac{11.3}{6.1} = \ln 11.3 - \ln 6.1$$
$$\ln 1.8525 = 2.4248 - 1.8082$$
$$0.617 = 0.617$$

11. c. apply the product rule

13. b. apply the power rule

NOTATION

15.
$$\log_8 8a^3 = \log_8 \boxed{8} + \log_8 \boxed{a^3}$$
$$= \log_8 8 + \boxed{3}\log_8 a$$
$$= \boxed{1} + 3\log_8 a$$

17. True

GUIDED PRACTICE

19. $\log_6 1 = 0$

21. $\log_4 4^7 = 7$

23. $5^{\log_5 10} = 10$

25. $\log_5 5^2 = 2$

27. $\ln e = 1$

29. $\log_3 3^7 = 7$

31.
$$\log_2(4\cdot 5) = \log_2 4 + \log_2 5$$
$$= \log_2 2^2 + \log_2 5$$
$$= 2 + \log_2 5$$

33.
$$\log 25y = \log 25 + \log y$$

35.
$$\log 100 pq = \log 100 + \log p + \log q$$
$$= \log 10^2 + \log p + \log q$$
$$= 2 + \log p + \log q$$

37.
$$\log 5xyz = \log 5 + \log x + \log y + \log z$$

39.
$$\log\frac{100}{9} = \log 100 - \log 9$$
$$= \log 10^2 - \log 9$$
$$= 2 - \log 9$$

41.
$$\log_6\frac{x}{36} = \log_6 x - \log_6 36$$
$$= \log_6 x - \log_6 6^2$$
$$= \log_6 x - 2$$

43.
$$\log\frac{7c}{2} = \log 7c - \log 2$$
$$= \log 7 + \log c - \log 2$$

45.
$$\log\frac{10x}{y} = \log 10x - \log y$$
$$= \log 10 + \log x - \log y$$
$$= 1 + \log x - \log y$$

47.

$$\ln \frac{exy}{z} = \ln exy - \ln z$$
$$= \ln e + \ln x + \ln y - \ln z$$
$$= 1 + \ln x + \ln y - \ln z$$

49.

$$\log_8 \frac{1}{8m} = \log_8 1 - \log_8 8m$$
$$= \log_8 1 - \log_8 8 - \log_8 m$$
$$= \log_8 1 - 1 - \log_8 m$$
$$= 0 - 1 - \log_8 m$$
$$= -1 - \log_8 m$$

51.

$$\ln y^7 = 7\ln y$$

53.

$$\log \sqrt{5} = \log 5^{1/2}$$
$$= \frac{1}{2}\log 5$$

55.

$$\log e^{-3} = -3\log e$$

57.

$$\log_7 \left(\sqrt[5]{100}\right)^3 = \log_7 (100)^{3/5}$$
$$= \frac{3}{5}\log_7 100$$

59.

$$\log xyz^2 = \log x + \log y + \log z^2$$
$$= \log x + \log y + 2\log z$$

61.

$$\log_2 \frac{2\sqrt[3]{x}}{y} = \log_2 \frac{2x^{1/3}}{y}$$
$$= \log_2 \left(2x^{1/3}\right) - \log_2 y$$
$$= \log_2 2 + \log_2 x^{1/3} - \log_2 y$$
$$= 1 + \frac{1}{3}\log_2 x - \log_2 y$$

63.

$$\log x^3 y^2 = \log x^3 + \log y^2$$
$$= 3\log x + 2\log y$$

65.

$$\log_b \sqrt{xy} = \log_b (xy)^{1/2}$$
$$= \frac{1}{2}\log_b (xy)$$
$$= \frac{1}{2}\left(\log_b x + \log_b y\right)$$

67.

$$\log_a \frac{\sqrt[3]{x}}{\sqrt[4]{yz}} = \log_a \frac{x^{1/3}}{(yz)^{1/4}}$$
$$= \log_a x^{1/3} - \log_a (yz)^{1/4}$$
$$= \frac{1}{3}\log_a x - \frac{1}{4}\log_a (yz)$$
$$= \frac{1}{3}\log_a x - \frac{1}{4}\left(\log_a y + \log_a z\right)$$
$$= \frac{1}{3}\log_a x - \frac{1}{4}\log_a y - \frac{1}{4}\log_a z$$

69.

$$\ln x^{20}\sqrt{z} = \ln x^{20} + \ln \sqrt{z}$$
$$= \ln x^{20} + \ln z^{1/2}$$
$$= 20\ln x + \frac{1}{2}\ln z$$

71.

$$\log_5 \left(\frac{1}{t^3}\right)^d = \log_5 \left(t^{-3}\right)^d$$
$$= \log_5 t^{-3d}$$
$$= -3d\log_5 t$$

73.

$$\ln \sqrt{ex} = \ln (ex)^{1/2}$$
$$= \frac{1}{2}\ln (ex)$$
$$= \frac{1}{2}\ln e + \frac{1}{2}\ln x$$
$$= \frac{1}{2}(1) + \frac{1}{2}\ln x$$
$$= \frac{1}{2} + \frac{1}{2}\ln x$$

75.

$$\log_2 (x+1) + 9\log_2 x = \log_2 (x+1) + \log_2 x^9$$
$$= \log_2 x^9 (x+1)$$

77.

$$\log_3 x + \log_3 (x+2) - \log_3 8$$
$$= \log_3 x(x+2) - \log_3 8$$
$$= \log_3 \frac{x(x+2)}{8}$$

79.

$$-3\log_b x - 2\log_b y + \frac{1}{2}\log_b z$$
$$= \frac{1}{2}\log_b z - 3\log_b x - 2\log_b y$$
$$= \log_b z^{1/2} - \log_b x^3 - \log_b y^2$$
$$= \log_b z^{1/2} - \left(\log_b x^3 + \log_b y^2\right)$$
$$= \log_b z^{1/2} - \left(\log_b x^3 y^2\right)$$
$$= \log_b \frac{z^{1/2}}{x^3 y^2}$$
$$= \log_b \frac{\sqrt{z}}{x^3 y^2}$$

81.

$$\frac{1}{3}\left[\log_b (M^2-9) - \log_b (M+3)\right]$$
$$= \log_b (M^2-9)^{1/3} - \log_b (M+3)^{1/3}$$
$$= \log_b \frac{(M^2-9)^{1/3}}{(M+3)^{1/3}}$$
$$= \log_b \left(\frac{M^2-9}{M+3}\right)^{1/3}$$
$$= \log_b \left(\frac{(M+3)(M-3)}{M+3}\right)^{1/3}$$
$$= \log_b (M-3)^{1/3}$$
$$= \log_b \sqrt[3]{M-3}$$

83.

$$\ln\left(\frac{x}{z}+x\right) - \ln\left(\frac{y}{z}+y\right) = \ln\left(\frac{\frac{x}{z}+x}{\frac{y}{z}+y}\right)$$
$$= \ln\left(\frac{\frac{x}{z}+x}{\frac{y}{z}+y} \cdot \frac{z}{z}\right)$$
$$= \ln\left(\frac{x+xz}{y+yz}\right)$$
$$= \ln\left(\frac{x(1+z)}{y(1+z)}\right)$$
$$= \ln\frac{x}{y}$$

85.

$$\frac{1}{2}\log_6 (x^2+1) - \log_6 (x^2+2)$$
$$= \log_6 (x^2+1)^{1/2} - \log_6 (x^2+2)$$
$$= \log_6 \frac{(x^2+1)^{1/2}}{x^2+2}$$
$$= \log_6 \frac{\sqrt{x^2+1}}{x^2+2}$$

87.

$$\log_b 28 = \log_b (4 \cdot 7)$$
$$= \log_b 4 + \log_b 7$$
$$= 0.6021 + 0.8451$$
$$= 1.4472$$

89.

$$\log_b \frac{4}{63} = \log_b 4 - \log_b 63$$
$$= \log_b 4 - \log_b (7 \cdot 9)$$
$$= \log_b 4 - \left(\log_b 7 + \log_b 9\right)$$
$$= \log_b 4 - \log_b 7 - \log_b 9$$
$$= 0.6021 - 0.8451 - 0.9542$$
$$= -1.1972$$

Section 11.6

91.

$$\log_b \frac{63}{4} = \log_b 63 - \log_b 4$$
$$= \log_b (7 \cdot 9) - \log_b 4$$
$$= \log_b 7 + \log_b 9 - \log_b 4$$
$$= 0.8451 + 0.9542 - 0.6021$$
$$= 1.1972$$

93.

$$\log_b 64 = \log_b 4^3$$
$$= 3\log_b 4$$
$$= 3(0.6021)$$
$$= 1.8063$$

95.

$$\log_3 7 = \frac{\log 7}{\log 3}$$
$$\approx \frac{0.84509}{0.47712}$$
$$\approx 1.7712$$

97.

$$\log_{1/3} 3 = \frac{\log 3}{\log \frac{1}{3}}$$
$$\approx \frac{0.47712}{-0.47712}$$
$$\approx -1.0000$$

99.

$$\log_3 8 = \frac{\log 8}{\log 3}$$
$$\approx \frac{0.90309}{0.47712}$$
$$\approx 1.8928$$

101.

$$\log_{\sqrt{2}} \sqrt{5} = \frac{\log \sqrt{5}}{\log \sqrt{2}}$$
$$\approx \frac{0.349485}{0.150514}$$
$$\approx 2.3219$$

103.   a. and c. are equivalent
$$\log(9 \cdot 3) = \log 9 + \log 3$$

105.   a. and b. are equivalent
$$\log_2 11^4 = 4\log_2 11$$

APPLICATIONS

107.   pH OF A SOLUTION
$$pH = -\log[H^+]$$
$$= -\log[1.7 \times 10^{-5}]$$
$$= -(\log 1.7 + \log 10^{-5})$$
$$= -(\log 1.7 - 5\log 10)$$
$$= -\log 1.7 + 5\log 10$$
$$= -0.2304 + 5 \cdot 1$$
$$= -0.2304 + 5$$
$$= 4.8$$

109.   FORMULAS
a.
$$B = 10(\log I - \log I_0)$$
$$= 10\left(\log \frac{I}{I_0}\right)$$
$$= \log\left(\frac{I}{I_0}\right)^{10}$$

b.
$$T = \frac{1}{k}(\ln C_2 - \ln C_1)$$
$$= \frac{1}{k}\left(\ln \frac{C_2}{C_1}\right)$$
$$= \ln \sqrt[k]{\frac{C_2}{C_1}}$$

WRITING

111.   Answers will vary.

113.   Answers will vary.

115.   Answers will vary.

117.   Answers will vary.

119.

$$m = \frac{y_2 - y_1}{x_2 - x_1}$$

$$= \frac{-4 - 3}{4 - (-2)}$$

$$= \frac{-7}{6}$$

$$= -\frac{7}{6}$$

121.

$$\left(\frac{x_1 + x_2}{2}, \frac{y_1 + y_2}{2} \right) = \left(\frac{-2 + 4}{2}, \frac{3 + (-4)}{2} \right)$$

$$= \left(\frac{2}{2}, \frac{-1}{2} \right)$$

$$= \left(1, -\frac{1}{2} \right)$$

CHALLENGE PROBLEMS

123.   Answers will vary.

125.

$$\log_{b^2} x = \frac{1}{2} \log_b x$$

$$\log_{b^2} x = \log_b x^{1/2}$$

$$\log_{b^2} x = \log_{b^2} x$$

VOCABULARY

1. An equation with a variable in its exponent, such as $3^{2x} = 8$, is called an **exponential** equation.

CONCEPTS

3. a. If two exponential expressions with the same base are equal, their exponents are **equal**.

   $b^x = b^y$ is equivalent to $\boxed{x = y}$

   b. If the logarithms base–b of two numbers are equal, the numbers are **equal**.

   $\log_b x = \log_b y$ is equivalent to $\boxed{x = y}$

5. If $6^{4x} = 6^{-2}$, then $4x = \boxed{-2}$.

7. To solve $5^x = 2$, we can take the **logarithm** of both sides of the equation to get $\log 5^x = \log 2$.

9. If $e^{x+2} = 4$, then $\ln e^{x+2} = \boxed{\ln 4}$.

11. No, it is not a solution.

   $$\log_5 (x+1) = 2$$

   $$\log_5 (4+1) \overset{?}{=} 2$$

   $$\log_5 (5) \overset{?}{=} 2$$

   $$5^2 \overset{?}{=} 5$$

   $$25 \neq 5$$

13. a. Divide both sides by $\ln 3$.

    b. $\dfrac{\ln 5}{\ln 3}$

    c. 1.4650

15. a. $\dfrac{\log 8}{\log 5} \approx 1.2920$

    b. $\dfrac{3 \ln 12}{\ln 4 - \ln 2} \approx 10.7549$

17. a. $\mathrm{pH} = -\log \left[H^+ \right]$

    b. $A = \boxed{A_0 2^{-t/h}}$

    c. $P = \boxed{P_0 e^{kt}}$

NOTATION

19.

$$2^x = 7$$

$$\boxed{\log} 2^x = \log 7$$

$$x \boxed{\log 2} = \log 7$$

$$x = \dfrac{\log 7}{\log 2}$$

$$x \approx \boxed{2.8074}$$

GUIDED PRACTICE

21.

$$6^{x-2} = 36$$

$$6^{x-2} = 6^2$$

$$x - 2 = 2$$

$$x = 4$$

23.

$$5^{4x} = \dfrac{1}{125}$$

$$5^{4x} = \dfrac{1}{5^3}$$

$$5^{4x} = 5^{-3}$$

$$4x = -3$$

$$x = -\dfrac{3}{4}$$

$$x = -0.75$$

25.

$$2^{x^2 - 2x} = 8$$

$$2^{x^2 - 2x} = 2^3$$

$$x^2 - 2x = 3$$

$$x^2 - 2x - 3 = 0$$

$$(x-3)(x+1) = 0$$

$$x - 3 = 0 \quad \text{or} \quad x + 1 = 0$$

$$x = 3 \qquad\qquad x = -1$$

27.

$$3^{x^2+4x} = \frac{1}{81}$$

$$3^{x^2+4x} = \frac{1}{3^4}$$

$$3^{x^2+4x} = 3^{-4}$$

$$x^2 + 4x = -4$$

$$x^2 + 4x + 4 = 0$$

$$(x+2)(x+2) = 0$$

$$x + 2 = 0 \quad \text{or} \quad x + 2 = 0$$

$$x = -2 \qquad\qquad x = -2$$

29.

$$4^x = 5$$

$$\log 4^x = \log 5$$

$$x \log 4 = \log 5$$

$$\frac{x \log 4}{\log 4} = \frac{\log 5}{\log 4}$$

$$x \approx 1.1610$$

31.

$$13^{x-1} = 2$$

$$\log 13^{x-1} = \log 2$$

$$(x-1)\log 13 = \log 2$$

$$x \log 13 - \log 13 = \log 2$$

$$x \log 13 = \log 2 + \log 13$$

$$\frac{x \log 13}{\log 13} = \frac{\log 2 + \log 13}{\log 13}$$

$$x \approx 1.2702$$

33.

$$2^{x+1} = 3^x$$

$$\log 2^{x+1} = \log 3^x$$

$$(x+1)\log 2 = x \log 3$$

$$x \log 2 + \log 2 = x \log 3$$

$$\log 2 = x \log 3 - x \log 2$$

$$\log 2 = x(\log 3 - \log 2)$$

$$\frac{\log 2}{\log 3 - \log 2} = \frac{x(\log 3 - \log 2)}{\log 3 - \log 2}$$

$$x = \frac{\log 2}{\log 3 - \log 2}$$

$$x \approx 1.7095$$

35.

$$5^{x-3} = 3^{2x}$$

$$\log 5^{x-3} = \log 3^{2x}$$

$$(x-3)\log 5 = 2x \log 3$$

$$x \log 5 - 3 \log 5 = 2x \log 3$$

$$-3 \log 5 = 2x \log 3 - x \log 5$$

$$-3 \log 5 = x(2 \log 3 - \log 5)$$

$$\frac{-3 \log 5}{2 \log 3 - \log 5} = \frac{x(2 \log 3 - \log 5)}{2 \log 3 - \log 5}$$

$$x = \frac{-3 \log 5}{2 \log 3 - \log 5}$$

$$x \approx -8.2144$$

37.

$$e^{2.9x} = 4.5$$

$$\ln e^{2.9x} = \ln 4.5$$

$$2.9x \ln e = \ln 4.5$$

$$2.9x \cdot 1 = \ln 4.5$$

$$2.9x = \ln 4.5$$

$$x = \frac{\ln 4.5}{2.9}$$

$$x \approx 0.5186$$

39.

$$e^{-0.2t} = 14.2$$

$$\ln e^{-0.2t} = \ln 14.2$$

$$-0.2t \ln e = \ln 14.2$$

$$-0.2t \cdot 1 = \ln 14.2$$

$$-0.2t = \ln 14.2$$

$$t = \frac{\ln 14.2}{-0.2}$$

$$t \approx -13.2662$$

41.

$$\log 2x = 4$$

$$10^4 = 2x$$

$$10,000 = 2x$$

$$5,000 = x$$

43.

$$\log_3 (x-3) = 2$$

$$3^2 = x - 3$$

$$9 = x - 3$$

$$12 = x$$

Section 11.7

45.
$$\log(7-x)=2$$
$$10^2=7-x$$
$$100=7-x$$
$$93=-x$$
$$-93=x$$

47.
$$\log\frac{1}{8}x=-2$$
$$10^{-2}=\frac{1}{8}x$$
$$\frac{1}{100}=\frac{1}{8}x$$
$$800\left(\frac{1}{100}\right)=800\left(\frac{1}{8}x\right)$$
$$8=100x$$
$$\frac{8}{100}=x$$
$$0.08=x$$

49.
$$\log(3-2x)=\log(x+24)$$
$$3-2x=x+24$$
$$3-3x=24$$
$$-3x=21$$
$$x=-7$$

51.
$$\ln(3x+1)=\ln(x+7)$$
$$3x+1=x+7$$
$$2x+1=7$$
$$2x=6$$
$$x=3$$

53.
$$\log x+\log(x-48)=2$$
$$\log_{10}\left[x(x-48)\right]=2$$
$$10^2=x(x-48)$$
$$100=x^2-48x$$
$$0=x^2-48x-100$$
$$0=(x-50)(x+2)$$
$$x-50=0 \quad\text{or}\quad x+2=0$$
$$x=50 \qquad x=\cancel{-2}$$

A negative number does not have a logarithm, so the only answer is 50.

55.
$$\log_5(4x-1)+\log_5 x=1$$
$$\log_5\left[x(4x-1)\right]=1$$
$$5^1=x(4x-1)$$
$$5=4x^2-x$$
$$0=4x^2-x-5$$
$$0=(4x-5)(x+1)$$
$$4x-5=0 \quad\text{or}\quad x+1=0$$
$$4x=5 \qquad\qquad x=\cancel{-1}$$
$$x=\frac{5}{4}$$

A negative number does not have a logarithm, so the only answer is $\frac{5}{4}$.

57.
$$\log 5-\log x=1$$
$$\log\frac{5}{x}=1$$
$$10^1=\frac{5}{x}$$
$$10=\frac{5}{x}$$
$$10x=5$$
$$x=0.5$$

59.
$$\log_3 4x-\log_3 7=2$$
$$\log_3\frac{4x}{7}=2$$
$$3^2=\frac{4x}{7}$$
$$9=\frac{4x}{7}$$
$$63=4x$$
$$15.75=x$$

61.
$$\log 2x=\log 4$$
$$2x=4$$
$$x=2$$

63.

$$\ln x = 1$$
$$e^1 = x$$
$$x = e$$
$$x \approx 2.7183$$

65.

$$7^{x^2} = 10$$
$$\log 7^{x^2} = \log 10$$
$$x^2 \log 7 = \log 10$$
$$x^2 = \frac{\log 10}{\log 7}$$
$$x = \pm\sqrt{\frac{\log 10}{\log 7}}$$
$$x \approx \pm 1.0878$$

67.

$$\log(x+90) + \log x = 3$$
$$\log_{10}\left[x(x+90)\right] = 3$$
$$10^3 = x(x+90)$$
$$1,000 = x^2 + 90x$$
$$0 = x^2 + 90x - 1,000$$
$$0 = (x+100)(x-10)$$
$$x+100 = 0 \quad \text{or} \quad x-10 = 0$$
$$x = \cancel{-100} \qquad x = 10$$

Since A negative number does not have a logarithm, the only possible answer is 10.

69.

$$3^{x-6} = 81$$
$$3^{x-6} = 3^4$$
$$x - 6 = 4$$
$$x = 10$$

71.

$$\log\frac{4x+1}{2x+9} = 0$$
$$10^0 = \frac{4x+1}{2x+9}$$
$$1 = \frac{4x+1}{2x+9}$$
$$2x+9 = 4x+1$$
$$9 = 2x+1$$
$$8 = 2x$$
$$4 = x$$

73.

$$15 = 9^{x+2}$$
$$\log 15 = \log 9^{x+2}$$
$$\log 15 = (x+2)\log 9$$
$$\log 15 = x\log 9 + 2\log 9$$
$$\log 15 - 2\log 9 = x\log 9$$
$$\frac{\log 15 - 2\log 9}{\log 9} = x$$
$$x \approx -0.7675$$

75.

$$\log x^2 = 2$$
$$10^2 = x^2$$
$$100 = x^2$$
$$x^2 = 100$$
$$x = \pm\sqrt{100}$$
$$x = \pm 10$$
$$x = 10, -10$$

77.

$$\log(x-6) - \log(x-2) = \log\frac{5}{x}$$
$$\log\left(\frac{x-6}{x-2}\right) = \log\frac{5}{x}$$
$$\frac{x-6}{x-2} = \frac{5}{x}$$
$$x(x-2)\left(\frac{x-6}{x-2}\right) = x(x-2)\left(\frac{5}{x}\right)$$
$$x(x-6) = 5(x-2)$$
$$x^2 - 6x = 5x - 10$$
$$x^2 - 6x - 5x + 10 = 0$$
$$x^2 - 11x + 10 = 0$$
$$(x-1)(x-10) = 0$$
$$x-1 = 0 \quad \text{or} \quad x-10 = 0$$
$$x = \cancel{1} \qquad\qquad x = 10$$

When $x = 1$, the two numbers on the left side would be negative, which is not possible. The only possible answer for x is 10.

Section 11.7

79.

$$\log_3 x = \log_3\left(\frac{1}{x}\right) + 4$$

$$\log_3 x - \log_3\left(\frac{1}{x}\right) = 4$$

$$\log_3 \frac{x}{\frac{1}{x}} = 4$$

$$\log_3 x^2 = 4$$

$$3^4 = x^2$$

$$81 = x^2$$

$$x^2 = 81$$

$$x = \pm\sqrt{81}$$

$$x = \pm 9$$

$$x = 9, \cancel{-9}$$

A negative number does not have a logarithm, so the only answer is $x = 9$.

81.

$$2\log_2 x = 3 + \log_2(x-2)$$

$$2\log_2 x - \log_2(x-2) = 3$$

$$\log_2 x^2 - \log_2(x-2) = 3$$

$$\log_2 \frac{x^2}{x-2} = 3$$

$$2^3 = \frac{x^2}{x-2}$$

$$8 = \frac{x^2}{x-2}$$

$$(x-2)(8) = (x-2)\left(\frac{x^2}{x-2}\right)$$

$$8x - 16 = x^2$$

$$0 = x^2 - 8x + 16$$

$$0 = (x-4)(x-4)$$

$$x - 4 = 0 \quad \text{or} \quad x - 4 = 0$$

$$x = 4 \qquad\qquad x = 4$$

83.

$$\log(7y+1) = 2\log(y+3) - \log 2$$

$$\log(7y+1) = \log(y+3)^2 - \log 2$$

$$\log(7y+1) = \log\frac{(y+3)^2}{2}$$

$$7y+1 = \frac{(y+3)^2}{2}$$

$$2(7y+1) = 2\left(\frac{(y+3)^2}{2}\right)$$

$$14y + 2 = (y+3)^2$$

$$14y + 2 = y^2 + 6y + 9$$

$$0 = y^2 + 6y + 9 - 14y - 2$$

$$0 = y^2 - 8y + 7$$

$$0 = (y-1)(y-7)$$

$$y - 1 = 0 \quad \text{or} \quad y - 7 = 0$$

$$y = 1 \qquad\qquad y = 7$$

85.

$$e^{3x} = 9$$

$$\ln e^{3x} = \ln 9$$

$$3x \ln e = \ln 9$$

$$3x \cdot 1 = \ln 9$$

$$3x = \ln 9$$

$$x = \frac{\ln 9}{3}$$

$$x \approx 0.7324$$

87.

$$\frac{\log(5x+6)}{2} = \log x$$

$$2\left(\frac{\log(5x+6)}{2}\right) = 2(\log x)$$

$$\log(5x+6) = \log x^2$$

$$5x + 6 = x^2$$

$$0 = x^2 - 5x - 6$$

$$0 = (x-6)(x+1)$$

$$x - 6 = 0 \quad \text{or} \quad x + 1 = 0$$

$$x = 6 \qquad\qquad x = \cancel{-1}$$

A negative number does not have a logarithm, so the only answer is $x = 6$.

89. a.

$$\log 5x = 1.7$$
$$10^{1.7} = 5x$$
$$x = \frac{10^{1.7}}{5}$$
$$x \approx 10.0237$$

b.

$$\ln 5x = 1.7$$
$$e^{1.7} = 5x$$
$$x = \frac{e^{1.7}}{5}$$
$$x \approx 1.0948$$

91. a.

$$4^{3x-5} = 90$$
$$\log 4^{3x-5} = \log 90$$
$$(3x - 5)\log 4 = \log 90$$
$$3x\log 4 - 5\log 4 = \log 90$$
$$3x\log 4 = \log 90 + 5\log 4$$
$$\frac{3x\log 4}{3\log 4} = \frac{\log 90 + 5\log 4}{3\log 4}$$
$$x = \frac{\log 90 + 5\log 4}{3\log 4}$$
$$x \approx 2.7486$$

b.

$$e^{3x-5} = 90$$
$$\ln e^{3x-5} = \ln 90$$
$$(3x - 5)\ln e = \ln 90$$
$$3x - 5 = \ln 90$$
$$3x = \ln 90 + 5$$
$$x = \frac{\ln 90 + 5}{3}$$
$$x \approx 3.1666$$

93. a.

$$\log_2 (x + 5) - \log_2 4x = \log_2 x$$
$$\log_2 \frac{x + 5}{4x} = \log_2 x$$
$$\frac{x + 5}{4x} = x$$
$$4x\left(\frac{x + 5}{4x}\right) = 4x(x)$$
$$x + 5 = 4x^2$$
$$0 = 4x^2 - x - 5$$
$$0 = (4x - 5)(x + 1)$$
$$4x - 5 = 0 \quad \text{and} \quad x + 1 = 0$$
$$4x = 5 \qquad x = \cancel{-1}$$
$$x = \frac{5}{4}$$

b.

$$\ln(x + 5) - \ln 4x = \ln x$$
$$\ln \frac{x + 5}{4x} = \ln x$$
$$\frac{x + 5}{4x} = x$$
$$4x\left(\frac{x + 5}{4x}\right) = 4x(x)$$
$$x + 5 = 4x^2$$
$$0 = 4x^2 - x - 5$$
$$0 = (4x - 5)(x + 1)$$
$$4x - 5 = 0 \quad \text{and} \quad x + 1 = 0$$
$$4x = 5 \qquad x = \cancel{-1}$$
$$x = \frac{5}{4}$$

Section 11.7

95. a.

$$\left(\frac{2}{3}\right)^{6-x} = \frac{8}{27}$$

$$\left(\frac{2}{3}\right)^{6-x} = \left(\frac{2}{3}\right)^3$$

$$6 - x = 3$$

$$-x = -3$$

$$x = 3$$

b.

$$\left(\frac{2}{3}\right)^{6-x} = \frac{16}{81}$$

$$\left(\frac{2}{3}\right)^{6-x} = \left(\frac{2}{3}\right)^4$$

$$6 - x = 4$$

$$-x = -2$$

$$x = 2$$

97.

$$2^{x+1} = 7$$

$$2^{x+1} - 7 = 0$$

$$y = 2^{x+1} - 7$$

The x–intercept is (1.8, 0) so the solution is $x \approx 1.8$.

99.

$$\log x + \log(x - 15) = 2$$

$$\log_{10}\left[x(x-15)\right] = 2$$

$$10^2 = x(x - 15)$$

$$100 = x^2 - 15x$$

$$0 = x^2 - 15x - 100$$

$$0 = (x - 20)(x + 5)$$

$$x - 20 = 0 \quad \text{or} \quad x + 5 = 0$$

$$x = 20 \qquad x = \cancel{-5}$$

A negative number does not have a logarithm, so the only possible answer is 20.

APPLICATIONS

101. HYDROGEN ION CONCENTRATION

$$pH = -\log\left[H^+\right]$$

$$13.2 = -\log\left[H^+\right]$$

$$-13.2 = \log\left[H^+\right]$$

$$\left[H^+\right] = 10^{-13.2}$$

$$\left[H^+\right] \approx 6.3 \times 10^{-14} \text{ grams-ions per liter}$$

103. TRITIUM DECAY

If 25% of the tritium has decayed, then 75%, or 0.75, of the tritium remains. (100% − 25% = 75%)
Let $h = 12.4$ and $A = 0.75A_0$.

$$A = A_0 2^{-t/h}$$

$$0.75A_0 = A_0 2^{-t/12.4}$$

$$\frac{0.75A_0}{A_0} = \frac{A_0 2^{-t/12.4}}{A_0}$$

$$0.75 = 2^{-t/12.4}$$

$$\log 0.75 = \log 2^{-t/12.4}$$

$$\log 0.75 = -\frac{t}{12.4} \log 2$$

$$-12.4(\log 0.75) = -12.4\left(-\frac{t}{12.4} \log 2\right)$$

$$-12.4(\log 0.75) = t \log 2$$

$$\frac{-12.4(\log 0.75)}{\log 2} = \frac{t \log 2}{\log 2}$$

$$\frac{-12.4(\log 0.75)}{\log 2} = t$$

$$t = \frac{-12.4(\log 0.75)}{\log 2}$$

$$t \approx 5.1 \text{ years}$$

105. **THORIUM DECAY**
If 80% of the thorium has decayed,
then 20%, or 0.20, of the thorium
remains.
$(100\% - 80\% = 20\%)$

Let $h = 18.4$ and $A = 0.20A_0$.

$$A = A_0 2^{-t/h}$$
$$0.20A_0 = A_0 2^{-t/18.4}$$
$$\frac{0.20A_0}{A_0} = \frac{A_0 2^{-t/18.4}}{A_0}$$
$$0.20 = 2^{-t/18.4}$$
$$\log 0.20 = \log 2^{-t/18.4}$$
$$\log 0.20 = -\frac{t}{18.4}\log 2$$
$$-18.4(\log 0.20) = -18.4\left(-\frac{t}{18.4}\log 2\right)$$
$$-18.4(\log 0.20) = t\log 2$$
$$\frac{-18.4(\log 0.20)}{\log 2} = \frac{t\log 2}{\log 2}$$
$$\frac{-18.4(\log 0.20)}{\log 2} = t$$
$$t = \frac{-18.4(\log 0.20)}{\log 2}$$
$$t \approx 42.7 \text{ days}$$

107. **CARBON–14 DATING**
60% of the carbon–14 remains and the
half–life is 5,700 years (from Example
10).

Let $h = 5,700$ and $A = 0.60A_0$.

$$A = A_0 2^{-t/h}$$
$$0.60A_0 = A_0 2^{-t/5,700}$$
$$\frac{0.60A_0}{A_0} = \frac{A_0 2^{-t/5,700}}{A_0}$$
$$0.60 = 2^{-t/5,700}$$
$$\log 0.60 = \log 2^{-t/5,700}$$
$$\log 0.60 = -\frac{t}{5,700}\log 2$$
$$-5,700(\log 0.60) = -5,700\left(-\frac{t}{5,700}\log 2\right)$$
$$-5,700(\log 0.60) = t\log 2$$
$$\frac{-5,700(\log 0.60)}{\log 2} = \frac{t\log 2}{\log 2}$$
$$\frac{-5,700(\log 0.60)}{\log 2} = t$$
$$t = \frac{-5,700(\log 0.60)}{\log 2}$$
$$t \approx 4,200 \text{ years}$$

109. COMPOUND INTEREST
Let $P = 800$, $P_0 = 500$, $r = 0.085$, and $k = 2$.

$$P = P_0\left(1 + \frac{r}{k}\right)^{kt}$$

$$800 = 500\left(1 + \frac{0.085}{2}\right)^{2t}$$

$$800 = 500(1 + 0.0425)^{2t}$$

$$800 = 500(1.0425)^{2t}$$

$$\frac{800}{500} = \frac{500(1.0425)^{2t}}{500}$$

$$1.6 = (1.0425)^{2t}$$

$$\log 1.6 = \log 1.0425^{2t}$$

$$\log 1.6 = 2t \log 1.0425$$

$$\frac{\log 1.6}{\log 1.0425} = 2t$$

$$2t = \frac{\log 1.6}{\log 1.0425}$$

$$t = \frac{\log 1.6}{2\log 1.0425}$$

$$t \approx 5.6 \text{ years}$$

111. COMPOUND INTEREST
Let $P = 2{,}100$, $P_0 = 1{,}300$, $r = 0.09$, and $k = 4$.

$$P = P_0\left(1 + \frac{r}{k}\right)^{kt}$$

$$2{,}100 = 1{,}300\left(1 + \frac{0.09}{4}\right)^{4t}$$

$$2{,}100 = 1{,}300(1 + 0.0225)^{4t}$$

$$2{,}100 = 1{,}300(1.0225)^{4t}$$

$$\frac{2{,}100}{1{,}300} = \frac{1{,}300(1.0225)^{4t}}{1{,}300}$$

$$1.6154 = (1.0225)^{4t}$$

$$\log 1.6154 = \log(1.0225)^{4t}$$

$$\log 1.6154 = 4t \log 1.0225$$

$$\frac{\log 1.6154}{\log 1.0225} = 4t$$

$$4t = \frac{\log 1.6154}{\log 1.0225}$$

$$t = \frac{\log 1.6154}{4\log 1.0225}$$

$$t \approx 5.4 \text{ years}$$

113. RULE OF SEVENTY
This formula works because $\ln 2 \approx 0.7$.

115. RODENT CONTROL
Let $P_0 = 30{,}000$, $P = 2(30{,}000) = 60{,}000$, and $t = 5$.

$$P = P_0 e^{kt}$$
$$60{,}000 = 30{,}000 e^{k5}$$
$$60{,}000 = 30{,}000 e^{5k}$$
$$\frac{60{,}000}{30{,}000} = \frac{30{,}000 e^{5k}}{30{,}000}$$
$$2 = e^{5k}$$
$$5k = \ln 2$$
$$k = \frac{\ln 2}{5}$$

Let $P_0 = 30{,}000$, $P = 1{,}000{,}000$, and $k = \dfrac{\ln 2}{5}$.

$$P = P_0 e^{kt}$$
$$1{,}000{,}000 = 30{,}000 e^{\left(\frac{\ln 2}{5}\right)t}$$
$$\frac{1{,}000{,}000}{30{,}000} = \frac{30{,}000 e^{\left(\frac{\ln 2}{5}\right)t}}{30{,}000}$$
$$\frac{100}{3} = e^{\left(\frac{\ln 2}{5}\right)t}$$
$$\left(\frac{\ln 2}{5}\right)t = \ln \frac{100}{3}$$
$$\left(\frac{5}{\ln 2}\right)\left(\frac{\ln 2}{5}\right)t = \left(\frac{5}{\ln 2}\right)\left(\ln \frac{100}{3}\right)$$
$$t \approx 25.3 \text{ years}$$

117. BACTERIA CULTURE
Let $P_0 = 1$, $P = 2(1) = 2$, and $t = 24$.

$$P = P_0 e^{kt}$$
$$2 = 1 e^{k24}$$
$$2 = e^{24k}$$
$$\ln 2 = \ln e^{24k}$$
$$\ln 2 = 24k$$
$$24k = \ln 2$$
$$k = \frac{\ln 2}{24}$$

Let $P_0 = 1$, $t = 36$, and $k = \dfrac{\ln 2}{24}$.

$$P = P_0 e^{kt}$$
$$P = 1 e^{\left(\frac{\ln 2}{24}\right)(36)}$$
$$P = e^{\frac{3\ln 2}{2}}$$
$$P \approx 2.828 \text{ times larger}$$

119. NEWTON'S LAW OF COOLING
Let $t = 3$ and $T = 90$.

$$T = 60 + 40 e^{kt}$$
$$90 = 60 + 40 e^{3k}$$
$$90 - 60 = 60 + 40 e^{3k} - 60$$
$$30 = 40 e^{3k}$$
$$\frac{30}{40} = \frac{40 e^{3k}}{40}$$
$$0.75 = e^{3k}$$
$$\ln 0.75 = \ln e^{3k}$$
$$\ln 0.75 = 3k$$
$$\frac{\ln 0.75}{3} = \frac{3k}{3}$$
$$\frac{\ln 0.75}{3} = k$$
$$k = \frac{\ln 0.75}{3}$$
$$k = \frac{1}{3}\ln 0.75$$
$$k \approx -0.0959$$

WRITING

121. Answers will vary.

123. Answers will vary.

Section 11.7

125. Use the Pythagorean Theorem.

$$a^2 + b^2 = c^2$$

$$\left(\sqrt{7}\right)^2 + b^2 = (12)^2$$

$$7 + b^2 = 144$$

$$b^2 = 137$$

$$b = \sqrt{137} \text{ in.}$$

CHALLENGE PROBLEMS

127.

$$\log_3 x + \log_3 (x+2) = 2$$

$$\log_3 x(x+2) = 2$$

$$\log_3 \left(x^2 + 2x\right) = 2$$

$$3^2 = x^2 + 2x$$

$$9 = x^2 + 2x$$

$$0 = x^2 + 2x - 9$$

$$x = \frac{-b \pm \sqrt{b^2 - 4ac}}{2a}$$

$$x = \frac{-2 \pm \sqrt{2^2 - 4(1)(-9)}}{2(1)}$$

$$x = \frac{-2 \pm \sqrt{4 + 36}}{2}$$

$$x = \frac{-2 \pm \sqrt{40}}{2}$$

$$x = \frac{-2 \pm 2\sqrt{10}}{2}$$

$$x = -1 + \sqrt{10} \quad \text{or} \quad x = \cancel{-1 - \sqrt{10}}$$

$$x \approx 2.1623$$

129.

$$\frac{\log_2 (6x-8)}{\log_2 x} = 2$$

$$\left(\log_2 x\right)\left(\frac{\log_2 (6x-8)}{\log_2 x}\right) = \left(\log_2 x\right)(2)$$

$$\log_2 (6x-8) = 2\log_2 x$$

$$\log_2 (6x-8) = \log_2 x^2$$

$$6x - 8 = x^2$$

$$0 = x^2 - 6x + 8$$

$$0 = (x-2)(x-4)$$

$$x - 2 = 0 \quad \text{or} \quad x - 4 = 0$$

$$x = 2 \qquad\qquad x = 4$$

SECTION 11.1
Algebra and Composition of Functions

1. $f+g$
$$(f+g)(x) = f(x)+g(x)$$
$$= (2x)+(x+1)$$
$$= 3x+1$$
$$D:(-\infty,\infty)$$

2. $f-g$
$$(f-g)(x) = f(x)-g(x)$$
$$= (2x)-(x+1)$$
$$= 2x-x-1$$
$$= x-1$$
$$D:(-\infty,\infty)$$

3. $f\cdot g$
$$(f\cdot g)(x) = f(x)\cdot g(x)$$
$$= (2x)(x+1)$$
$$= 2x^2+2x$$
$$D:(-\infty,\infty)$$

4. f/g
$$(f/g)(x) = \frac{f(x)}{g(x)}$$
$$= \frac{2x}{x+1}$$
$$D:(-\infty,-1)\cup(-1,\infty)$$

5.
$$(f\circ g)(-1) = f(g(-1))$$
$$= f(2(-1)+1)$$
$$= f(-2+1)$$
$$= f(-1)$$
$$= (-1)^2+2$$
$$= 1+2$$
$$= 3$$

6.
$$(g\circ f)(0) = g(f(0))$$
$$= g(0^2+2)$$
$$= g(0+2)$$
$$= g(2)$$
$$= 2(2)+1$$
$$= 4+1$$
$$= 5$$

7.
$$(f\circ g)(x) = f(g(x))$$
$$= f(2x+1)$$
$$= (2x+1)^2+2$$
$$= 4x^2+2x+2x+1+2$$
$$= 4x^2+4x+3$$

8.
$$(g\circ f)(x) = g(f(x))$$
$$= g(x^2+2)$$
$$= 2(x^2+2)+1$$
$$= 2x^2+4+1$$
$$= 2x^2+5$$

9.
a.
$$(f+g)(2) = f(2)+g(2)$$
$$= 2+(-2)$$
$$= 0$$

b.
$$(f\cdot g)(-4) = f(-4)\cdot g(-4)$$
$$= 4(-2)$$
$$= -8$$

c.
$$(f\circ g)(4) = f(g(4))$$
$$= f(4)$$
$$= 0$$

d.
$$(g\circ f)(6) = g(f(6))$$
$$= g(-2)$$
$$= -1$$

10.

$$(C \circ f)(m) = C(f(m))$$

$$= C\left(\frac{m}{8}\right)$$

$$= 2.85\left(\frac{m}{8}\right)$$

$$C(m) = \frac{2.85m}{8}$$

SECTION 11.2
Inverse Functions

11. No. $f(1) = 4$ and $f(-1) = 4$

12. Yes.

13. Yes.

14. No. The output -5 corresponds with more than one input, 0 and 4.

15. Yes.

16. No. It does not pass the horizontal line test.

17.

x	$f^{-1}(x)$
-6	-6
-3	-1
12	7
3	20

18.

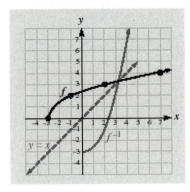

19.

$$f(x) = 6x - 3$$

$$y = 6x - 3$$

$$x = 6y - 3$$

$$x + 3 = 6y$$

$$\frac{x+3}{6} = y$$

$$y = \frac{x+3}{6}$$

$$f^{-1}(x) = \frac{x+3}{6}$$

20.

$$f(x) = \frac{4}{x-1}$$

$$y = \frac{4}{x-1}$$

$$x = \frac{4}{y-1}$$

$$(y-1)(x) = (y-1)\left(\frac{4}{y-1}\right)$$

$$xy - x = 4$$

$$xy = 4 + x$$

$$y = \frac{4+x}{x}$$

$$y = \frac{4}{x} + \frac{x}{x}$$

$$y = \frac{4}{x} + 1$$

$$f^{-1}(x) = \frac{4}{x} + 1$$

21.

$$f(x) = (x+2)^3$$

$$y = (x+2)^3$$

$$x = (y+2)^3$$

$$\sqrt[3]{x} = \sqrt[3]{(y+2)^3}$$

$$\sqrt[3]{x} = y + 2$$

$$\sqrt[3]{x} - 2 = y$$

$$y = \sqrt[3]{x} - 2$$

$$f^{-1}(x) = \sqrt[3]{x} - 2$$

22.

$$f(x) = \frac{x}{6} - \frac{1}{6}$$

$$y = \frac{x}{6} - \frac{1}{6}$$

$$x = \frac{y}{6} - \frac{1}{6}$$

$$x + \frac{1}{6} = \frac{y}{6}$$

$$6\left(x + \frac{1}{6}\right) = 6\left(\frac{y}{6}\right)$$

$$6x + 1 = y$$

$$y = 6x + 1$$

$$f^{-1}(x) = 6x + 1$$

23.

$$f(x) = \sqrt[3]{x - 1}$$

$$y = \sqrt[3]{x - 1}$$

$$x = \sqrt[3]{y - 1}$$

$$(x)^3 = \left(\sqrt[3]{y - 1}\right)^3$$

$$x^3 = y - 1$$

$$x^3 + 1 = y$$

$$y = x^3 + 1$$

$$f^{-1}(x) = x^3 + 1$$

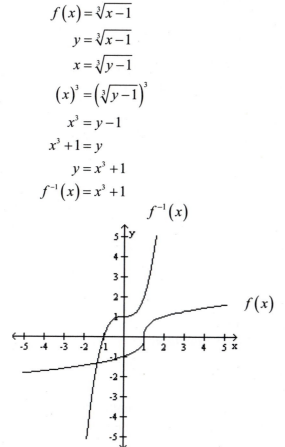

24.

$$(f \circ f^{-1})(x) = f\left(f^{-1}(x)\right)$$

$$= f\left(-\frac{x - 5}{4}\right)$$

$$= 5 - 4\left(-\frac{x - 5}{4}\right)$$

$$= 5 + x - 5$$

$$= x$$

$$(f^{-1} \circ f)(x) = f^{-1}\left(f(x)\right)$$

$$= f^{-1}(5 - 4x)$$

$$= -\frac{(5 - 4x) - 5}{4}$$

$$= -\frac{5 - 4x - 5}{4}$$

$$= -\frac{-4x}{4}$$

$$= x$$

SECTION 11.3
Exponential Functions

25. a. $n(x) = 2^x$; $s(t) = 1.08^t$

b. $0.9(1.42)^{14} \approx 121.9774$

26. a. exponential decay

b. exponential growth

27. $f(x) = 3^x$

D: $(-\infty, \infty)$, R: $(0, \infty)$

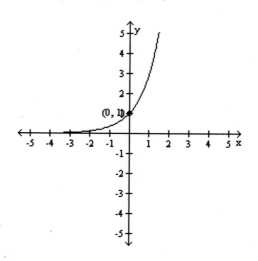

28. $f(x) = \left(\dfrac{1}{3}\right)^x$

    D: $(-\infty, \infty)$, R: $(0, \infty)$

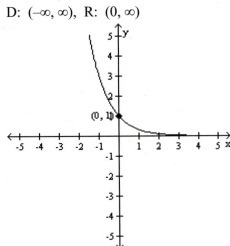

29. $f(x) = \left(\dfrac{1}{2}\right)^x - 2$

    D: $(-\infty, \infty)$, R: $(-2, \infty)$

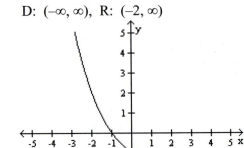

30. $f(x) = 3^{x-1}$

    D: $(-\infty, \infty)$, R: $(0, \infty)$

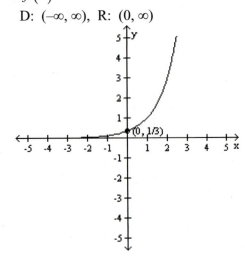

31. the x–axis ($y = 0$)

32. a. Let $t = 0$.
$$c(t) = 128{,}000(1.08)^t$$
$$c(0) = 128{,}000(1.08)^0$$
$$= 128{,}000(1)$$
$$= 128{,}000 \text{ tons}$$

    b. Let $t = 100$.
$$c(t) = 128{,}000(1.08)^t$$
$$c(0) = 128{,}000(1.08)^{100}$$
$$\approx 128{,}000(2199.76)$$
$$\approx 281{,}569{,}441 \text{ tons}$$

33. Let $P = 10{,}500$, $r = 0.09$, $k = 4$, and $t = 60$
$$A = 10{,}500\left(1 + \frac{0.09}{4}\right)^{4(60)}$$
$$= 10{,}500(1 + 0.0225)^{240}$$
$$= 10{,}500(1.0225)^{240}$$
$$= 10{,}500(208.543186)$$
$$= \$2{,}189{,}703.45$$

34. Let $t = 5$.
$$V(t) = 12{,}000\left(10^{-0.155t}\right)$$
$$V(5) = 12{,}000\left(10^{-0.155(5)}\right)$$
$$= 12{,}000\left(10^{-0.775}\right)$$
$$= 12{,}000(0.16788)$$
$$\approx \$2{,}015$$

SECTION 11.4
Logarithmic Functions

35. D: $(0, \infty)$; R: $(-\infty, \infty)$

36. Since there is no real number such that
    $10^? = 0$, log 0 is undefined.

37. $4^3 = 64$

38. $\log_7 \dfrac{1}{7} = -1$

39.

$$\log_3 9 = x$$
$$3^x = 9$$
$$3^x = 3^2$$
$$x = 2$$

40.

$$\log_9 \frac{1}{81} = x$$
$$9^x = \frac{1}{81}$$
$$9^x = \frac{1}{9^2}$$
$$9^x = 9^{-2}$$
$$x = -2$$

41.

$$\log_{1/2} 1 = x$$
$$\left(\frac{1}{2}\right)^x = 1$$
$$x = 0$$

42.

$$\log_5 (-25) = x$$
$$5^x = -25$$
not possible

43.

$$\log_6 \sqrt{6} = x$$
$$6^x = \sqrt{6}$$
$$6^x = 6^{1/2}$$
$$x = \frac{1}{2}$$

44.

$$\log 1,000 = x$$
$$\log_{10} 1,000 = x$$
$$10^x = 1,000$$
$$10^x = 10^3$$
$$x = 3$$

45.

$$\log_2 x = 5$$
$$2^5 = x$$
$$32 = x$$

46.

$$\log_3 x = -4$$
$$3^{-4} = x$$
$$\frac{1}{3^4} = x$$
$$\frac{1}{81} = x$$

47.

$$\log_x 16 = 2$$
$$x^2 = 16$$
$$\sqrt{x^2} = \sqrt{16}$$
$$x = 4$$

48.

$$\log_x \frac{1}{100} = -2$$
$$x^{-2} = \frac{1}{100}$$
$$\left(x^{-2}\right)^{-1/2} = \left(\frac{1}{100}\right)^{-1/2}$$
$$x = (100)^{1/2}$$
$$x = \sqrt{100}$$
$$x = 10$$

49.

$$\log_9 3 = x$$
$$9^x = 3$$
$$\left(3^2\right)^x = 3^1$$
$$3^{2x} = 3^1$$
$$2x = 1$$
$$x = \frac{1}{2}$$

50.

$$\log_{27} 3 = x$$
$$27^x = 3$$
$$\left(3^3\right)^x = 3$$
$$(3)^{3x} = 3^1$$
$$3x = 1$$
$$x = \frac{1}{3}$$

Chapter 11 Review

51.

$$\log 4.51 = x$$
$$0.6542 = x$$

52.

$$\log x = 1.43$$
$$10^{1.43} = x$$
$$x = 26.9153$$

53. $f(x) = \log_4 x$ and $g(x) = 4^x$

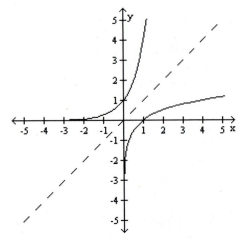

54. $f(x) = \log_{1/3} x$ and $g(x) = \left(\dfrac{1}{3}\right)^x$

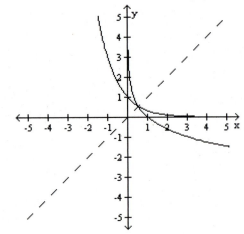

55. $f(x) = \log(x-2)$

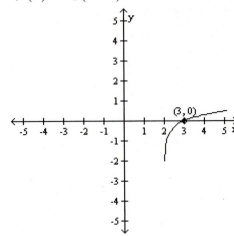

56. $f(x) = 3 + \log x$

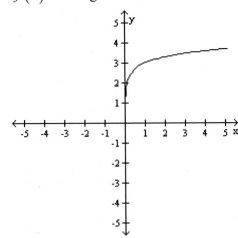

57. Let $E_O = 18$ and $E_I = 0.04$.

$$\text{db gain} = 20\log\frac{E_O}{E_I}$$
$$= 20\log\frac{18}{0.04}$$
$$= 20\log 450$$
$$\approx 20(2.65321)$$
$$\approx 53$$

58. Let $P = 0.3$ and $A = 7,500$.

$$R = \log\frac{A}{P}$$
$$= \log\frac{7,500}{0.3}$$
$$= \log 25,000$$
$$\approx 4.4$$

59.a. Let $n = 1$.
$$h(1) = 52 + 25\log 1$$
$$= 52 + 25(0)$$
$$= 52 + 0$$
$$= 52 \text{ cm}$$

b. Let $n = 8$.
$$h(8) = 52 + 25\log 8$$
$$= 52 + 25(0.9031)$$
$$\approx 74.6 \text{ cm}$$

60. Let $A = 13$.
$$P(13) = 61.8 + 34.9\log(13 - 4)$$
$$= 61.8 + 34.9\log 9$$
$$\approx 95\%$$

SECTION 11.5
Base-e Exponential and Logarithmic Functions

61. a. $e \approx 2.72$

b. $\ln 15 \approx 2.7081$ means $e^{\boxed{2.7081}} \approx \boxed{15}$

62. 16.6704

63. $f(x) = e^x + 1$
D: $(-\infty, \infty)$, R: $(1, \infty)$

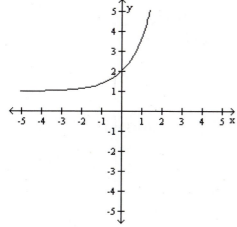

64. $f(x) = e^{x-3}$
D: $(-\infty, \infty)$, R: $(0, \infty)$

65. Let $P_0 = 10{,}500$, $k = 0.09$, and $t = 60$.
$$P = P_0 e^{kt}$$
$$= 10{,}500 e^{0.09(60)}$$
$$= 10{,}500 e^{5.4}$$
$$= 2{,}324{,}767.37$$

66. Let $P_0 = 142.5$ billion, $k = 0.066$, and $t = 6$.
$$P = P_0 e^{kt}$$
$$= 142.5 e^{0.066(6)}$$
$$= 142.5 e^{0.396}$$
$$\approx \$211.7 \text{ billion}$$

67. For 1980, let $t = 0$.
$$r(t) = 13.9 e^{-0.035t}$$
$$r(0) = 13.9 e^{-0.035(0)}$$
$$= 13.9 e^0$$
$$= 13.9\%$$
For 1985, let $t = 5$.
$$r(t) = 13.9 e^{-0.035t}$$
$$r(5) = 13.9 e^{-0.035(5)}$$
$$= 13.9 e^{-0.175}$$
$$= 11.67\%$$
For 1990, let $t = 10$.
$$r(t) = 13.9 e^{-0.035t}$$
$$r(10) = 13.9 e^{-0.035(10)}$$
$$= 13.9 e^{-0.35}$$
$$= 9.8\%$$

Chapter 11 Review

68. The exponent on the base e is negative.

69.
$$\ln e = 1$$

70.
$$\ln e^2 = 2$$

71.
$$\ln \frac{1}{e^5} = \ln e^{-5}$$
$$= -5$$

72.
$$\ln \sqrt{e} = \ln e^{1/2}$$
$$= \frac{1}{2}$$

73.
$$\ln (-e) = \text{undefined}$$

74.
$$\ln 0 = \text{undefined}$$

75.
$$\ln 1 = 0$$

76.
$$\ln e^{-7} = -7$$

77.
$$\ln 452 = 6.1137$$

78.
$$\ln 0.85 = -0.1625$$

79.
$$\ln x = 2.336$$
$$e^{\ln x} = e^{2.336}$$
$$x = 10.3398$$

80.
$$\ln x = -8.8$$
$$e^{\ln x} = e^{-8.8}$$
$$x = 0.0002$$

81. They have different bases:
$$\log x = \log_{10} x \quad \text{and} \quad \ln x = \log_e x$$

82. $f^{-1}(x) = e^x$

83. $f(x) = 1 + \ln x$

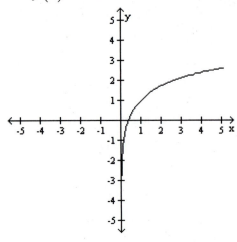

84. $f(x) = \ln(x+1)$

85. Let $r = 0.01118$.
$$t = \frac{\ln 2}{0.01118}$$
$$\approx 62$$

86. Let $a = 19$.
$$H(a) = 13 + 20.03 \ln a$$
$$H(19) = 13 + 20.03 \ln 19$$
$$\approx 13 + 20.03(2.94444)$$
$$\approx 72 \text{ in. or } 6 \text{ ft.}$$

SECTION 11.6
Properties of Logarithms

87.
$$\log_2 1 = 0$$

88.
$$\log_9 9 = 1$$

89.

$$\log 10^3 = 3$$

90.

$$7^{\log_7 4} = 4$$

91.

$$\begin{aligned}\log_3 27x &= \log_3 27 + \log_3 x\\ &= 3 + \log_3 x\end{aligned}$$

92.

$$\begin{aligned}\log \frac{100}{x} &= \log 100 - \log x\\ &= 2 - \log x\end{aligned}$$

93.

$$\begin{aligned}\log_5 \sqrt{27} &= \log_5 27^{1/2}\\ &= \frac{1}{2}\log_5 27\end{aligned}$$

94.

$$\begin{aligned}\log_b 10ab &= \log_b 10 + \log_b a + \log_b b\\ &= \log_b 10 + \log_b a + 1\end{aligned}$$

95.

$$\begin{aligned}\log_b \frac{x^2 y^3}{z} &= \log_b x^2 + \log_b y^3 - \log_b z\\ &= 2\log_b x + 3\log_b y - \log_b z\end{aligned}$$

96.

$$\begin{aligned}\ln \sqrt{\frac{x}{yz^2}} &= \ln\left(\frac{x}{yz^2}\right)^{1/2}\\ &= \frac{1}{2}\ln \frac{x}{yz^2}\\ &= \frac{1}{2}\left(\ln x - \ln y - \ln z^2\right)\\ &= \frac{1}{2}\left(\ln x - \ln y - 2\ln z\right)\end{aligned}$$

97.

$$\begin{aligned}3\log_2 x - 5\log_2 y + 7\log_2 z &\\ = \log_2 x^3 - \log_2 y^5 + \log_2 z^7 &\\ = \log_2 \frac{x^3 z^7}{y^5} &\end{aligned}$$

98.

$$\begin{aligned}-3\log_b y - 7\log_b z + \frac{1}{2}\log_b(x+2) &\\ = -\log_b y^3 - \log_b z^7 + \log_b (x+2)^{1/2} &\\ = -\log_b y^3 - \log_b z^7 + \log_b \sqrt{x+2} &\\ = \log \frac{\sqrt{x+2}}{y^3 z^7} &\end{aligned}$$

99.

$$\begin{aligned}\log_b(a^2 - 25) - \log_b(a+5) &\\ = \log_b \frac{a^2 - 25}{a+5} &\\ = \log_b \frac{(a+5)(a-5)}{a+5} &\\ = \log_b(a-5) &\end{aligned}$$

100.

$$\begin{aligned}3\log_8 x + 4\log_8 x &= \log_8 x^3 + \log_8 x^4\\ &= \log_8\left(x^3 x^4\right)\\ &= \log_8 x^7\end{aligned}$$

101.

$$\begin{aligned}\log_b 40 &= \log_b(5\cdot 8)\\ &= \log_b 5 + \log_b 8\\ &= 1.1609 + 1.5000\\ &= 2.6609\end{aligned}$$

102.

$$\begin{aligned}\log_b 64 &= \log_b(8\cdot 8)\\ &= \log_b 8 + \log_b 8\\ &= 1.5000 + 1.5000\\ &= 3.0000\end{aligned}$$

103.

$$\begin{aligned}\log_5 17 &= \frac{\log 17}{\log 5}\\ &\approx \frac{1.23045}{0.69897}\\ &\approx 1.7604\end{aligned}$$

Chapter 11 Review

104.

$$pH = -\log\left[H^+\right]$$
$$pH = -\log\left[7.9 \times 10^{-4}\right]$$
$$pH = -(-3.1)$$
$$pH = 3.1$$

SECTION 11.7
Exponential and Logarithmic Equations

105.

$$5^{x+6} = 25$$
$$5^{x+6} = 5^2$$
$$x + 6 = 2$$
$$x = -4$$

106.

$$2^{x^2+4x} = \frac{1}{8}$$
$$2^{x^2+4x} = 2^{-3}$$
$$x^2 + 4x = -3$$
$$x^2 + 4x + 3 = 0$$
$$(x+1)(x+3) = 0$$
$$x + 1 = 0 \quad \text{or} \quad x + 3 = 0$$
$$x = -1 \qquad x = -3$$

107.

$$3^x = 7$$
$$\log 3^x = \log 7$$
$$x \log 3 = \log 7$$
$$x = \frac{\log 7}{\log 3}$$
$$x \approx 1.7712$$

108.

$$2^x = 3^{x-4}$$
$$\log 2^x = \log 3^{x-4}$$
$$x \log 2 = (x-4) \log 3$$
$$x \log 2 = x \log 3 - 4 \log 3$$
$$x \log 2 - x \log 3 = -4 \log 3$$
$$x(\log 2 - \log 3) = -4 \log 3$$
$$x = \frac{-4 \log 3}{\log 2 - \log 3}$$
$$x \approx 10.8380$$

109.

$$e^x = 7$$
$$\ln e^x = \ln 7$$
$$x = \ln 7$$
$$x = 1.9459$$

110.

$$e^{-0.4t} = 25$$
$$\ln e^{-0.4t} = \ln 25$$
$$-0.4t = \ln 25$$
$$t = \frac{\ln 25}{-0.4}$$
$$t = -8.0472$$

111.

$$\left(\frac{2}{5}\right)^{3x-4} = \frac{8}{125}$$
$$\left(\frac{2}{5}\right)^{3x-4} = \left(\frac{2}{5}\right)^3$$
$$3x - 4 = 3$$
$$3x = 7$$
$$x = \frac{7}{3}$$

112.

$$9^{x^2} = 33$$
$$\log 9^{x^2} = \log 33$$
$$x^2 \log 9 = \log 33$$
$$x^2 = \frac{\log 33}{\log 9}$$
$$x = \sqrt{\frac{\log 33}{\log 9}}$$
$$x \approx 1.2615$$

113.

$$\log(x-4) = 2$$
$$10^2 = x - 4$$
$$100 = x - 4$$
$$104 = x$$

114.

$$\ln(2x-3)=\ln 15$$
$$e^{\ln(2x-3)}=e^{\ln 15}$$
$$2x-3=15$$
$$2x=18$$
$$x=9$$

115.

$$\log x+\log(29-x)=2$$
$$\log[x(29-x)]=2$$
$$10^2=x(29-x)$$
$$100=29x-x^2$$
$$x^2-29x+100=0$$
$$(x-25)(x-4)=0$$
$$x-25=0 \quad \text{or} \quad x-4=0$$
$$x=25 \qquad x=4$$

116.

$$\log_2 x+\log_2(x-2)=3$$
$$\log_2 x(x-2)=3$$
$$2^3=x(x-2)$$
$$8=x^2-2x$$
$$0=x^2-2x-8$$
$$0=(x-4)(x+2)$$
$$x-4=0 \quad \text{or} \quad x+2=0$$
$$x=4 \qquad x=\cancel{-2}$$

117.

$$\frac{\log(7x-12)}{\log x}=2$$
$$\log x\left(\frac{\log(7x-12)}{\log x}\right)=2\log x$$
$$\log(7x-12)=\log x^2$$
$$7x-12=x^2$$
$$0=x^2-7x+12$$
$$0=(x-3)(x-4)$$
$$x-3=0 \quad \text{or} \quad x-4=0$$
$$x=3 \qquad x=4$$

118.

$$\log_2(x+2)+\log_2(x-1)=2$$
$$\log_2[(x+2)(x-1)]=2$$
$$2^2=(x+2)(x-1)$$
$$4=x^2-x+2x-2$$
$$4=x^2+x-2$$
$$0=x^2+x-6$$
$$0=(x+3)(x-2)$$
$$x+3=0 \quad \text{or} \quad x-2=0$$
$$x=\cancel{-3} \qquad x=2$$

119.

$$\log x+\log(x-5)=\log 6$$
$$\log x(x-5)=\log 6$$
$$x(x-5)=6$$
$$x^2-5x=6$$
$$x^2-5x-6=0$$
$$(x-6)(x+1)=0$$
$$x-6=0 \quad \text{or} \quad x+1=0$$
$$x=6 \qquad x=\cancel{-1}$$

120.

$$\log 3-\log(x-1)=-1$$
$$\log\frac{3}{x-1}=-1$$
$$10^{-1}=\frac{3}{x-1}$$
$$\frac{1}{10}=\frac{3}{x-1}$$
$$10(x-1)\left(\frac{1}{10}\right)=10(x-1)\left(\frac{3}{x-1}\right)$$
$$x-1=30$$
$$x=31$$

121.

$$\frac{\log 8}{\log 15}\neq\log 8-\log 15$$
$$0.7679\neq-0.2730$$

122. Let $h = 5{,}700$ and $A = \dfrac{2}{3}A_0$.

$$A = A_0 2^{-t/h}$$

$$\frac{2}{3}A_0 = A_0 2^{-t/5{,}700}$$

$$\frac{2A_0}{3A_0} = \frac{A_0 2^{-t/5{,}700}}{A_0}$$

$$\frac{2}{3} = 2^{-t/5{,}700}$$

$$\log\frac{2}{3} = \log 2^{-t/5{,}700}$$

$$\log\frac{2}{3} = -\frac{t}{5{,}700}\log 2$$

$$-5{,}700\left(\log\frac{2}{3}\right) = -5{,}700\left(-\frac{t}{5{,}700}\log 2\right)$$

$$-5{,}700\left(\log\frac{2}{3}\right) = t\log 2$$

$$\frac{-5{,}700\left(\log\frac{2}{3}\right)}{\log 2} = \frac{t\log 2}{\log 2}$$

$$\frac{-5{,}700\left(\log\frac{2}{3}\right)}{\log 2} = t$$

$$t = \frac{-5{,}700\left(\log\frac{2}{3}\right)}{\log 2}$$

$$t \approx 3{,}300 \text{ years}$$

123. Let $P_0 = 800$, $P = 3(800) = 2{,}400$, and $t = 14$.

$$P = P_0 e^{kt}$$

$$2{,}400 = 800e^{k(14)}$$

$$3 = e^{14k}$$

$$\ln 3 = \ln e^{14k}$$

$$\ln 3 = 14k$$

$$14k = \ln 3$$

$$k = \frac{\ln 3}{14}$$

Let $P_0 = 800$, $P = 1{,}000{,}000$, and $k = \dfrac{\ln 3}{14}$.

$$P = P_0 e^{kt}$$

$$1{,}000{,}000 = 800e^{\left(\frac{\ln 3}{14}\right)t}$$

$$1{,}250 = e^{\left(\frac{\ln 3}{14}\right)t}$$

$$\ln 1{,}250 = \ln e^{\left(\frac{\ln 3}{14}\right)t}$$

$$\ln 1{,}250 = \left(\frac{\ln 3}{14}\right)t$$

$$\left(\frac{14}{\ln 3}\right)(\ln 1{,}250) = \left(\frac{14}{\ln 3}\right)\left(\frac{\ln 3}{14}\right)t$$

$$\frac{14\ln 1{,}250}{\ln 3} = t$$

$$t = \frac{14\ln 1{,}250}{\ln 3}$$

$$t \approx 91 \text{ days}$$

124. $x = 2$ and $x = 5$

For $x = 2$:

$$\log x = 1 - \log(7 - x)$$

$$\log 2 \overset{?}{=} 1 - \log(7 - 2)$$

$$\log 2 \overset{?}{=} 1 - \log 5$$

$$0.3010 = 0.3010$$

For $x = 5$:

$$\log x = 1 - \log(7 - x)$$

$$\log 5 \overset{?}{=} 1 - \log(7 - 5)$$

$$\log 5 \overset{?}{=} 1 - \log 2$$

$$0.6990 = 0.6990$$

1. a. A **composite** function is denoted by $f \circ g$.

   b. $f(x) = e^x$ is the **natural** exponential function.

   c. In **continuous** compound interest, the number of compoundings is infinitely large.

   d. The functions $f(x) = \log_{10} x$ and $f(x) = 10^x$ are **inverse** functions.

   e. $f(x) = \log_4 x$ is **logarithmic** a function.

2. a. f composed with g of x
   b. g of f of eight
   c. f inverse of x

3. $f + g$
$$(f + g)(x) = f(x) + g(x)$$
$$= x + 9 + 4x^2 - 3x + 2$$
$$= 4x^2 - 2x + 11$$
$$D : (-\infty, \infty)$$

4. g / f
$$(g / f)(x) = \frac{g(x)}{f(x)}$$
$$= \frac{4x^2 - 3x + 2}{x + 9}$$
$$D : (-\infty, -9) \cup (-9, \infty)$$

5.
$$(g \circ f)(-3) = g(f(-3))$$
$$= g\left(2(-3)^2 + 3\right)$$
$$= g(2(9) + 3)$$
$$= g(18 + 3)$$
$$= g(21)$$
$$= 4(21) - 8$$
$$= 84 - 8$$
$$= 76$$

6.
$$(f \circ g)(x) = f(g(x))$$
$$= f(4x - 8)$$
$$= 2(4x - 8)^2 + 3$$
$$= 2(16x^2 - 32x - 32x + 64) + 3$$
$$= 2(16x^2 - 64x + 64) + 3$$
$$= 32x^2 - 128x + 128 + 3$$
$$= 32x^2 - 128x + 131$$

7. a.
$$(f \cdot g)(9) = f(9) \cdot g(9)$$
$$= -1 \cdot 16$$
$$= -16$$

   b.
$$(f \circ g)(-3) = f(g(-3))$$
$$= f(10)$$
$$= 17$$

8.
   a. $(g / f)(-4) = \dfrac{g(-4)}{f(-4)}$
$$= \frac{0}{4}$$
$$= 0$$

   b. $(f \circ g)(1) = f(g(1))$
$$= f(-2)$$
$$= 3$$

   c. $(f + g)(2) = f(2) + g(2)$
$$= 4 + (-3)$$
$$= 1$$

   d. $(f \cdot g)(0) = (f)(0) \cdot (g)(0)$
$$= 2(-1)$$
$$= -2$$

   e. $(g - f)(1) = (g)(1) - (f)(1)$
$$= -2 - 3$$
$$= -5$$

9. a. No
   b. Yes
   c. Yes
   d. No

10.

$$f(x) = -\frac{1}{3}x$$

$$y = -\frac{1}{3}x$$

$$x = -\frac{1}{3}y$$

$$3(x) = 3\left(-\frac{1}{3}y\right)$$

$$3x = -y$$

$$-3x = y$$

$$y = -3x$$

$$f^{-1}(x) = -3x$$

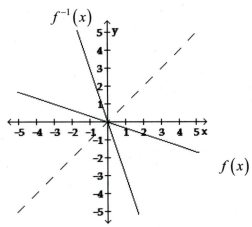

11. one-to-one function

$$f(x) = \frac{1}{3}x + 2$$

$$y = \frac{1}{3}x + 2$$

$$x = \frac{1}{3}y + 2$$

$$3(x) = 3\left(\frac{1}{3}y + 2\right)$$

$$3x = y + 6$$

$$3x - 6 = y$$

$$y = 3x - 6$$

$$f^{-1}(x) = 3x - 6$$

12.

$$f(x) = (x - 15)^3$$

$$y = (x - 15)^3$$

$$x = (y - 15)^3$$

$$\sqrt[3]{x} = \sqrt[3]{(y - 15)^3}$$

$$\sqrt[3]{x} = y - 15$$

$$\sqrt[3]{x} + 15 = y$$

$$y = \sqrt[3]{x} + 15$$

$$f^{-1}(x) = \sqrt[3]{x} + 15$$

13.

$$(f \circ f^{-1})(x) = f\left(f^{-1}(x)\right)$$

$$= f\left(\frac{x - 4}{4}\right)$$

$$= 4\left(\frac{x - 4}{4}\right) + 4$$

$$= x - 4 + 4$$

$$= x$$

$$(f^{-1} \circ f)(x) = f^{-1}\left(f(x)\right)$$

$$= f^{-1}(4x + 4)$$

$$= \frac{4x + 4 - 4}{4}$$

$$= \frac{4x}{4}$$

$$= x$$

14. a. Yes.
    b. Yes.
    c. $f^{-1}(260) = 80$; when the temperature of the tire tread is 260°, the vehicle is traveling 80 mph.

15. $f(x) = 2^x + 1$
     D: $(-\infty, \infty)$, R: $(1, \infty)$

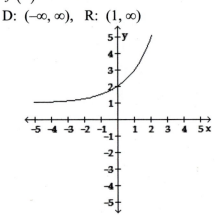

16. $f(x) = 3^{-x}$

D: $(-\infty, \infty)$, R: $(0, \infty)$

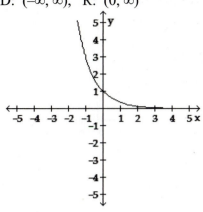

17. Let $t = 6$ and $A_0 = 3$.

$A = A_0(2)^{-t}$

$ = 3(2)^{-6}$

$ = 3(0.015625)$

$ = 0.046875$ g

18. Let $P = 1{,}000$, $r = 0.06$, $k = 2$, and $t = 1$.

$A = P\left(1 + \dfrac{r}{k}\right)^{kt}$

$ = 1{,}000\left(1 + \dfrac{0.06}{2}\right)^{2(1)}$

$ = 1{,}000(1 + 0.03)^2$

$ = 1{,}000(1.03)^2$

$ = 1{,}000(1.0609)$

$ = \$1{,}060.90$

19. a. $f(x) = e^x$

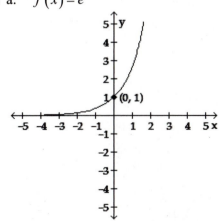

(0, 1)

b. D: $(-\infty, \infty)$; R: $(0, \infty)$

c. $f^{-1}(x) = \ln x$

20. Let $P_0 = 1{,}173{,}108{,}018$, $r = 0.01376$, and $t = 10$.

$P = P_0 e^{kt}$

$ = (1{,}173{,}108{,}018)e^{0.01376(10)}$

$ = (1{,}173{,}108{,}018)e^{0.1376}$

$ \approx 1{,}346{,}161{,}000$

21. $H(a) = 1{,}425(1.052)^{-a}$

$H(55) = 1{,}425(1.052)^{-55}$

$ \approx 88$ micrograms per day

22.

$\log_6 \dfrac{1}{36} = -2$

$6^{-2} = \dfrac{1}{36}$

23 a. D: $(0, \infty)$, R: $(-\infty, \infty)$

b. $f^{-1}(x) = 10^x$

24. logarithmic growth

25.

$\log_5 25 = x$

$5^x = 25$

$5^x = 5^2$

$x = 2$

26.

$\log_9 \dfrac{1}{81}$

$9^x = \dfrac{1}{81}$

$9^x = 9^{-2}$

$x = -2$

27. $\log(-100)$ is undefined since $10^? \neq -100$

28.

$\ln \dfrac{1}{e^6} = \ln e^{-6}$

$\phantom{\ln \dfrac{1}{e^6}} = -6$

Chapter 11 Test

29.

$$\log_4 2 = x$$

$$4^x = 2$$

$$2^{2x} = 2^1$$

$$2x = 1$$

$$x = \frac{1}{2}$$

30.

$$\log_{1/3} 1 = x$$

$$\frac{1}{3}^x = 1$$

$$3^{-x} = 1$$

$$x = 0$$

31.

$$\log_x 32 = 5$$

$$x^5 = 32$$

$$\sqrt[5]{x^5} = \sqrt[5]{32}$$

$$x = 2$$

32.

$$\log_8 x = \frac{4}{3}$$

$$8^{4/3} = x$$

$$\left(\sqrt[3]{8}\right)^4 = x$$

$$(2)^4 = x$$

$$16 = x$$

33.

$$\log_3 x = -3$$

$$3^{-3} = x$$

$$\frac{1}{3^3} = x$$

$$\frac{1}{27} = x$$

34.

$$\ln x = 1$$

$$e^{\ln x} = e^1$$

$$x = e$$

35. $f(x) = -\log_3 x$

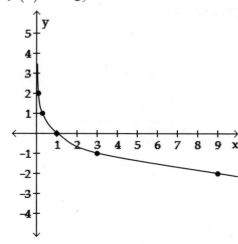

36. $f(x) = \ln x$

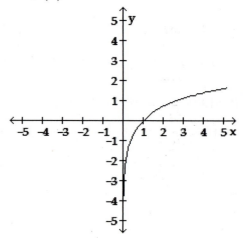

37. Let $H^+ = 3.7 \times 10^{-7}$.

$$pH = -\log\left[H^+\right]$$

$$= -\log\left(3.7 \times 10^{-7}\right)$$

$$\approx -(-6.4)$$

$$\approx 6.4$$

38. Let $E_O = 60$ and $E_I = 0.3$.

$$\text{db gain} = 20\log\left(\frac{E_O}{E_I}\right)$$

$$= 20\log\left(\frac{60}{0.3}\right)$$

$$= 20\log 200$$

$$\approx 20(2.3)$$

$$\approx 46$$

39.

$$\log x = -1.06$$
$$10^{-1.06} = x$$
$$x = 10^{-1.06}$$
$$x \approx 0.0871$$

40.

$$\log_7 3 = \frac{\log 3}{\log 7}$$
$$\approx 0.5646$$

41.

$$\log_b a^2 bc^3 = \log_b a^2 + \log_b b + \log_b c^3$$
$$= 2\log_b a + 1 + 3\log_b c$$

42.

$$\frac{1}{2}\ln(a+2) + \ln b - 3\ln c$$
$$= \ln(a+2)^{1/2} + \ln b - \ln c^3$$
$$= \ln\sqrt{a+2} + \ln b - \ln c^3$$
$$= \ln b\sqrt{a+2} - \ln c^3$$
$$= \ln\frac{b\sqrt{a+2}}{c^3}$$

43.

$$5^x = 3$$
$$\log 5^x = \log 3$$
$$x\log 5 = \log 3$$
$$x = \frac{\log 3}{\log 5}$$
$$x \approx 0.6826$$

44.

$$3^{x-1} = 27$$
$$3^{x-1} = 3^3$$
$$x-1 = 3$$
$$x = 4$$

45.

$$\left(\frac{3}{2}\right)^{6x+2} = \frac{27}{8}$$
$$\left(\frac{3}{2}\right)^{6x+2} = \left(\frac{3}{2}\right)^3$$
$$6x+2 = 3$$
$$6x = 1$$
$$x = \frac{1}{6}$$

46.

$$e^{0.08t} = 4$$
$$\ln e^{0.08t} = \ln 4$$
$$0.08t = \ln 4$$
$$t = \frac{\ln 4}{0.08}$$
$$t \approx 17.3287$$

47.

$$2\log x = \log 25$$
$$\log x^2 = \log 25$$
$$x^2 = 25$$
$$x = \pm\sqrt{25}$$
$$x = 5 \quad \text{or} \quad x = -5$$

48.

$$\log_2(x+2) - \log_2(x-5) = 3$$
$$\log_2\frac{x+2}{x-5} = 3$$
$$\frac{x+2}{x-5} = 2^3$$
$$\frac{x+2}{x-5} = 8$$
$$(x-5)\left(\frac{x+2}{x-5}\right) = (x-5)(8)$$
$$x+2 = 8x - 40$$
$$-7x+2 = -40$$
$$-7x = -42$$
$$x = 6$$

Chapter 11 Test

49.

$$\ln(5x+2)=\ln(2x+5)$$
$$5x+2=2x+5$$
$$3x+2=5$$
$$3x=3$$
$$x=1$$

50.

$$\log x + \log(x-9)=1$$
$$\log x(x-9)=1$$
$$10^1 = x(x-9)$$
$$10 = x^2 - 9x$$
$$0 = x^2 - 9x - 10$$
$$0 = (x-10)(x+1)$$
$$x-10=0 \quad \text{or} \quad x+1=0$$
$$x=10 \qquad\qquad x=\cancel{-1}$$

The log cannot be negative, so the only solution is 10.

51.

$$x \approx 5$$
$$\frac{1}{2}\ln(x-1)=\ln 2$$
$$\frac{1}{2}\ln(5-1)\overset{?}{=}\ln 2$$
$$\frac{1}{2}\ln 4 \overset{?}{=}\ln 2$$
$$\frac{1}{2}(1.3863)\overset{?}{=}0.6931$$
$$0.6931 = 0.6931$$

52. Let $P_0 = 5$, $P = 4(5) = 20$, and $t = 6$.

$$P = P_0 e^{kt}$$
$$20 = 5e^{k(6)}$$
$$4 = e^{6k}$$
$$\ln 4 = \ln e^{6k}$$
$$\ln 4 = 6k$$
$$6k = \ln 4$$
$$k = \frac{\ln 4}{6}$$

Let $P_0 = 5$, $P = 500$, and $k = \dfrac{\ln 4}{6}$.

$$P = P_0 e^{kt}$$
$$500 = 5e^{\left(\frac{\ln 4}{6}\right)t}$$
$$100 = e^{\left(\frac{\ln 4}{6}\right)t}$$
$$\ln 100 = \ln e^{\left(\frac{\ln 4}{6}\right)t}$$
$$\ln 100 = \left(\frac{\ln 4}{6}\right)t$$
$$\left(\frac{6}{\ln 4}\right)(\ln 100) = \left(\frac{6}{\ln 4}\right)\left(\frac{\ln 4}{6}\right)t$$
$$\frac{6\ln 100}{\ln 4} = t$$
$$t = \frac{6\ln 100}{\ln 4}$$
$$t \approx 20 \text{ minutes}$$

VOCABULARY

1. The pair of equations $\begin{cases} x - y = -1 \\ 2x - y = 1 \end{cases}$ is called

    a **system** of equations.

3. When the graphs of the equations of a system are identical lines, the equations are called dependent and the system has **infinitely** many solutions.

CONCEPTS

5. a. true
    b. false
    c. true
    d. true

7. a. 3, –4 (answers will vary)
    b. 2; –3 (answers will vary)

NOTATION

9. Solve $\begin{cases} y = 3x - 7 \\ x + y = 5 \end{cases}$.

    $x + \left(\boxed{3x - 7} \right) = 5$

    $x + 3x - 7 = \boxed{5}$

    $\boxed{4}x - 7 = 5$

    $4x = \boxed{12}$

    $x = 3$

    $y = 3x - 7$

    $y = 3\left(\boxed{3} \right) - 7$

    $y = \boxed{2}$

    The solution is (3, 2).

11. Yes.

$$4x - y = -19 \qquad\qquad 3x + 2y = -6$$
$$4(-4) - 3 \stackrel{?}{=} -19 \qquad 3(-4) + 2(3) \stackrel{?}{=} -6$$
$$-16 - 3 \stackrel{?}{=} -19 \qquad\qquad -12 + 6 \stackrel{?}{=} -6$$
$$-19 = -19 \qquad\qquad\qquad -6 = -6$$

13. No

$$y + 2 = \frac{1}{2}x \qquad\qquad 3x + 2y = 0$$
$$-3 + 2 \stackrel{?}{=} \frac{1}{2}(2) \qquad 3(2) + 2(-3) \stackrel{?}{=} 0$$
$$-1 \neq 1 \qquad\qquad\qquad 6 - 6 \stackrel{?}{=} 0$$
$$\qquad\qquad\qquad\qquad 0 = 0$$

15. No

$$2x + 3y = 2 \qquad\qquad 4x - 9y = 1$$
$$2\left(\frac{1}{2}\right) + 3\left(\frac{1}{3}\right) \stackrel{?}{=} 2 \qquad 4\left(\frac{1}{2}\right) - 9\left(\frac{1}{3}\right) \stackrel{?}{=} 1$$
$$1 + 1 \stackrel{?}{=} 2 \qquad\qquad 2 - 3 \stackrel{?}{=} 1$$
$$2 = 2 \qquad\qquad\qquad -1 \neq 1$$

17. Yes

$$2x + 5y = 2.1 \qquad\qquad 5x + y = -0.5$$
$$2(-0.2) + 5(0.5) \stackrel{?}{=} 2.1 \qquad 5(-0.2) + 0.5 \stackrel{?}{=} -0.5$$
$$-0.4 + 2.5 \stackrel{?}{=} 2.1 \qquad -1 + 0.5 \stackrel{?}{=} -0.5$$
$$2.1 = 2.1 \qquad\qquad\qquad -0.5 = -0.5$$

19. $\begin{cases} x + y = 6 \\ x - y = 2 \end{cases}$ (4, 2)

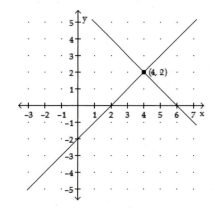

21. $\begin{cases} y = -2x + 1 \\ x - 2y = -7 \end{cases}$ $(-1, 3)$

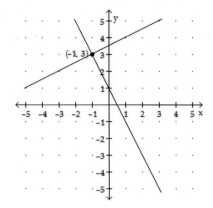

23. $\begin{cases} 3x - 3y = 4 \\ x - y = 4 \end{cases}$

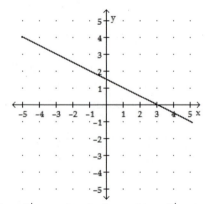

No solution, $\varnothing$, inconsistent system

25. $\begin{cases} x = 3 - 2y \\ 2x + 4y = 6 \end{cases}$

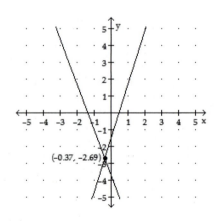

$\{(x, y)| \ x = 3 - 2y\}$ or $\{(x, y)| \ x + 2y = 3\}$

infinitely many solutions;
dependent equations

27. $\begin{cases} \dfrac{1}{6}x = \dfrac{1}{3}y + \dfrac{1}{2} \\ y = x \end{cases}$ $(-3, -3)$

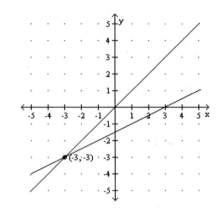

29. $\begin{cases} \dfrac{1}{3}x - \dfrac{7}{6}y = \dfrac{1}{2} \\ \dfrac{1}{5}y = \dfrac{1}{3}x + \dfrac{7}{15} \end{cases}$ $(-2, -1)$

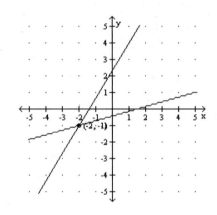

31. $\begin{cases} y = 3.2x - 1.5 \\ y = -2.7x - 3.7 \end{cases}$ $(-0.37, -2.69)$

33. $\begin{cases} 1.7x + 2.3y = 3.2 \\ y = 0.25x + 8.95 \end{cases}$ $(-7.64, 7.04)$

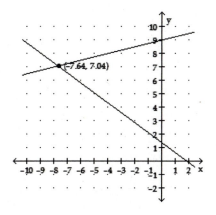

35.
$\begin{cases} y = 3x \\ x + y = 8 \end{cases}$

$x + y = 8$ The 2nd equation

$x + (3x) = 8$

$4x = 8$

$\boxed{x = 2}$

$y = 3x$ The 1st equation

$y = 3(2)$

$\boxed{y = 6}$

The solution is $(2, 6)$.

37.
$\begin{cases} x = 2 + y \\ 2x + y = 13 \end{cases}$

$2x + y = 13$ The 2nd equation

$2(2 + y) + y = 13$

$4 + 2y + y = 13$

$4 + 3y = 13$

$3y = 9$

$\boxed{y = 3}$

$x = 2 + y$ The 1st equation

$x = 2 + 3$

$\boxed{x = 5}$

The solution is $(5, 3)$.

39.
$\begin{cases} x + 2y = 6 \\ 3x - y = -10 \end{cases}$

Solve the first equation for x.

$\begin{cases} x = -2y + 6 \\ 3x - y = -10 \end{cases}$

$3x - y = -10$ The 2nd equation

$3(-2y + 6) - y = -10$

$-6y + 18 - y = -10$

$-7y + 18 = -10$

$-7y = -28$

$\boxed{y = 4}$

$x = -2y + 6$ The 1st equation

$x = -2(4) + 6$

$x = -8 + 6$

$\boxed{x = -2}$

The solution is $(-2, 4)$.

41.
$\begin{cases} 5x + 3y = -26 \\ 3x + y = -14 \end{cases}$

Solve the second equation for y.

$\begin{cases} 5x + 3y = -26 \\ y = -3x - 14 \end{cases}$

$5x + 3y = -26$ 1st equation

$5x + 3(-3x - 14) = -26$

$5x - 9x - 42 = -26$

$-4x - 42 = -26$

$-4x = 16$

$x = -4$

$y = -3x - 14$ 2nd equation

$y = -3(-4) - 14$

$y = 12 - 14$

$y = -2$

The solution is $(-4, -2)$.

43.

$$\begin{cases} x - y = 7 \\ x + y = 11 \end{cases}$$

eliminate the y

$$x - y = 7$$
$$\underline{x + y = 11}$$
$$2x = 18$$

$$\boxed{x = 9}$$

$$x - y = 7 \quad 1^{\text{st}} \text{ equation}$$
$$9 - y = 7$$
$$9 - y - 9 = 7 - 9$$
$$-y = -2$$
$$\frac{-y}{-1} = \frac{-2}{-1}$$

$$\boxed{y = 2}$$

The solution is (9, 2).

45.

$$\begin{cases} 2s + 3t = -8 \\ 2s - 3t = -8 \end{cases}$$

eliminate the t

$$2s + 3t = -8$$
$$\underline{2s - 3t = -8}$$
$$4s = -16$$

$$\boxed{s = -4}$$

$$2s + 3t = -8 \quad 1^{\text{st}} \text{ equation}$$
$$2(-4) + 3t = -8$$
$$-8 + 3t = -8$$
$$-8 + 3t + 8 = -8 + 8$$
$$3t = 0$$

$$\boxed{t = 0}$$

The solution is $(-4, 0)$.

47.

$$\begin{cases} 3x + 4y = -24 \\ 5x + 12y = -72 \end{cases}$$

multiply 1st equation by -3

$$\begin{cases} -9x - 12y = 72 \\ 5x + 12y = -72 \end{cases}$$

eliminate the y

$$-9x - 12y = 72$$
$$\underline{5x + 12y = -72}$$
$$-4x = 0$$

$$\boxed{x = 0}$$

$$3x + 4y = -24 \quad 1^{\text{st}} \text{ equation}$$
$$3(0) + 4y = -24$$
$$0 + 4y = -24$$
$$4y = -24$$

$$\boxed{y = -6}$$

The solution is $(0, -6)$.

49.

$$\begin{cases} \dfrac{5}{6}x + \dfrac{1}{2}y = 12 \\ 0.3x + 0.5y = 5.6 \end{cases}$$

multiply 1st equation by 6

multiply 2nd equation by 10

$$\begin{cases} 5x + 3y = 72 \\ 3x + 5y = 56 \end{cases}$$

multiply 1st equation by 3

multiply 2nd equation by -5

$$\begin{cases} 15x + 9y = 216 \\ -15x - 25y = -280 \end{cases}$$

eliminate the x

$$15x + 9y = 216$$
$$\underline{-15x - 25y = -280}$$
$$-16y = -64$$

$$\boxed{y = 4}$$

$$5x + 3y = 72 \quad 1^{\text{st}} \text{ equation}$$
$$5x + 3(4) = 72$$
$$5x + 12 = 72$$
$$5x + 12 - 12 = 72 - 12$$
$$5x = 60$$

$$\boxed{x = 12}$$

The solution is (12, 4).

51.

$$\begin{cases} 6x + 3y = 18 \\ y = -2x + 5 \end{cases}$$

Substitute the second equation into the first equation and solve for x.

$$6x + 3y = 18 \quad 1^{st} \text{ equation}$$
$$6x + 3(-2x + 5) = 18$$
$$6x - 6x + 15 = 18$$
$$15 \neq 18$$

No solution; $\varnothing$

Inconsistent system

53.

$$\begin{cases} 3x - y = 5 \\ 21x = 7(y + 5) \end{cases}$$

Solve the first equation for y.
Simplify the second equation.

$$\begin{cases} y = -5 + 3x \\ 21x = 7y + 35 \end{cases}$$
$$21x = 7y + 35 \quad \text{The } 2^{nd} \text{ equation}$$
$$21x = 7(-5 + 3x) + 35$$
$$21x = -35 + 21x + 35$$
$$21x = 21x$$
$$21x - 21x = 21x - 21x$$
$$0 = 0$$
$$\text{true}$$

Infinitely many solutions

$$\{(x, y) \mid 3x - y = 5\}$$

Dependent Equations

55.

$$\begin{cases} 4x - 3y = 5 \\ y = -2x \end{cases}$$

Substitute the second equation for y in the first equation.

$$4x - 3y = 5$$
$$4x - 3(-2x) = 5$$
$$4x + 6x = 5$$
$$10x = 5$$
$$x = \frac{1}{2}$$

Substitute $x = \frac{1}{2}$ into the second equation and solve for y.

$$y = -2\left(\frac{1}{2}\right)$$
$$= -1$$

The solution is $\left(\frac{1}{2}, -1\right)$.

57.

$$\begin{cases} y = -\dfrac{5}{2}x + \dfrac{1}{2} \\ 2x - \dfrac{3}{2}y = 5 \end{cases}$$

Multiply the equations by 2 and write in standard form.

$$\begin{cases} 5x + 2y = 1 \\ 4x - 3y = 10 \end{cases}$$

Multiply the first equation by 3 and the second equation by 2.

$$15x + 6y = 3$$
$$\underline{8x - 6y = 20}$$
$$23x = 23$$
$$x = 1$$

Substitute $x = 1$ into the first equation.

$$5x + 2y = 1$$
$$5(1) + 2y = 1$$
$$5 + 2y = 1$$
$$2y = -4$$
$$y = -2$$

The solution is $(1, -2)$.

Section 12.1

59.

$$\begin{cases} x = \dfrac{11-2y}{3} \\ y = \dfrac{11-6x}{4} \end{cases}$$

Multiply the first equation by 3 and the second equation by 4 and write in standard form.

$$\begin{cases} 3x + 2y = 11 \\ 6x + 4y = 11 \end{cases}$$

Multiply the first equation by –2.

$$-6x - 4y = -22$$
$$\underline{6x + 4y = 11}$$
$$0 \neq -11$$

No solution; $\varnothing$
Inconsistent system

61.

$$\begin{cases} x = 13 - 4y \\ 3x = 4 + 2y \end{cases}$$

Substitute $x = 13 - 4y$ into the second equation.

$$3x = 4 + 2y$$
$$3(13 - 4y) = 4 + 2y$$
$$39 - 12y = 4 + 2y$$
$$39 - 14y = 4$$
$$-14y = -35$$
$$y = \frac{5}{2}$$

Substitute $y = \dfrac{5}{2}$ into the first equation and solve for x.

$$x = 13 - 4y$$
$$x = 13 - 4\left(\frac{5}{2}\right)$$
$$x = 13 - 10$$
$$x = 3$$

The solution is $\left(3, \dfrac{5}{2}\right)$.

63.

$$\begin{cases} x = 2 \\ y = -\dfrac{1}{2}x + 2 \end{cases}$$

Substitute the first equation into the second equation.

$$y = -\frac{1}{2}x + 2$$
$$y = -\frac{1}{2}(2) + 2$$
$$y = -1 + 2$$
$$y = 1$$

The solution is (2, 1).

65.

$$\begin{cases} x + 3y = 6 \\ y = -\dfrac{1}{3}x + 2 \end{cases}$$

Substitute the second equation into the first equation.

$$x + 3y = 6$$
$$x + 3\left(-\frac{1}{3}x + 2\right) = 6$$
$$x - x + 6 = 6$$
$$6 = 6$$

Infinitely many solutions;
$$\{(x, y) | x + 3y = 6\}$$
dependent equations

67.

$$\begin{cases} 2x + 3y = 8 \\ 3x - 2y = -1 \end{cases}$$

Multiply the 1st equation by 2 and multiply the 2nd equation by 3.

$$\begin{cases} 4x + 6y = 16 \\ 9x - 6y = -3 \end{cases}$$

Add the equations to eliminate y.

$$\begin{array}{r} 4x + 6y = 16 \\ \underline{9x - 6y = -3} \\ 13x = 13 \\ x = 1 \end{array}$$

Substitute $x = 1$ into the first equation and solve for y.

$$2x + 3y = 8 \quad \text{1st equation}$$
$$2(1) + 3y = 8$$
$$2 + 3y = 8$$
$$2 + 3y - 2 = 8 - 2$$
$$3y = 6$$
$$y = 2$$

The solution is (1, 2).

69.

$$\begin{cases} 4(x - 2) = -9y \\ 2(x - 3y) = -3 \end{cases}$$

$$\begin{cases} 4x - 8 = -9y \\ 2x - 6y = -3 \end{cases}$$

Write the 1st equation in standard form.

$$\begin{cases} 4x + 9y = 8 \\ 2x - 6y = -3 \end{cases}$$

Multiply the 2nd equation by -2.

$$\begin{cases} 4x + 9y = 8 \\ -4x + 12y = 6 \end{cases}$$

Add the equations to eliminate x.

$$\begin{array}{r} 4x + 9y = 8 \\ \underline{-4x + 12y = 6} \\ 21y = 14 \\ y = \dfrac{14}{21} \\ y = \dfrac{2}{3} \end{array}$$

Substitute $y = \dfrac{2}{3}$ into the first equation and solve for x.

$$4x + 9y = 8 \quad \text{1st equation}$$
$$4x + 9\left(\dfrac{2}{3}\right) = 8$$
$$4x + 6 = 8$$
$$4x + 6 - 6 = 8 - 6$$
$$4x = 2$$
$$x = \dfrac{2}{4}$$
$$x = \dfrac{1}{2}$$

The solution is $\left(\dfrac{1}{2}, \dfrac{2}{3}\right)$.

Section 12.1

71.

$$\begin{cases} 0.3a + 0.1b = 0.5 \\ \dfrac{4}{3}a + \dfrac{1}{3}b = 3 \end{cases}$$

Multiply the 1st equation by 10 to eliminate decimals and multiply the 2nd equation by 3 to eliminate fractions.

$$\begin{cases} 3a + b = 5 \\ 4a + b = 9 \end{cases}$$

Solve the 1st equation for b by subtracting $-3a$ from both sides.

$$\begin{cases} b = 5 - 3a \\ 4a + b = 9 \end{cases}$$

$$4a + b = 9 \quad \text{The 2}^{\text{nd}} \text{ equation}$$

$$4a + (5 - 3a) = 9$$

$$4a + 5 - 3a = 9$$

$$a + 5 = 9$$

$$\boxed{a = 4}$$

$$b = 5 - 3a \quad \text{The 1}^{\text{st}} \text{ equation}$$

$$b = 5 - 3(4)$$

$$b = 5 - 12$$

$$\boxed{b = -7}$$

The solution is $(4, -7)$.

73.

$$\begin{cases} \dfrac{x}{2} + \dfrac{y}{2} = 6 \\ \dfrac{x}{3} + \dfrac{y}{3} = 4 \end{cases}$$

Multiply the equations by 2 and 3 to eliminate fractions.

$$\begin{cases} x + y = 12 \\ x + y = 12 \end{cases}$$

Multiply the first equation by -1 and add the equations to eliminate y.

$$-x - y = -12$$

$$\underline{x + y = 12}$$

$$0 = 0$$

Infinitely many solutions

$$\{(x, y) \mid x + y = 12\}$$

dependent equations

75.

$$\begin{cases} x = \dfrac{2}{3}y \\ y = 4x + 5 \end{cases}$$

Write each equation in standard form.

$$\begin{cases} 3x - 2y = 0 \\ -4x + y = 5 \end{cases}$$

Multiply the second equation by 2 and add to eliminate y.

$$3x - 2y = 0$$

$$\underline{-8x + 2y = 10}$$

$$-5x = 10$$

$$x = -2$$

Substitute $x = -2$ into the 2nd equation and solve for y.

$$y = 4x + 5$$

$$y = 4(-2) + 5$$

$$y = -8 + 5$$

$$y = -3$$

The solution is $(-2, -3)$.

77.

$$\begin{cases} 3x - 4y = 9 \\ x + 2y = 8 \end{cases}$$

Solve the second equation for x.

$$\begin{cases} 3x - 4y = 9 \\ x = -2y + 8 \end{cases}$$

Substitute $x = -2y + 8$ into the first equation and solve for y.

$$3x - 4y = 9 \qquad \text{the 1}^{\text{st}} \text{ equation}$$
$$3(-2y + 8) - 4y = 9$$
$$-6y + 24 - 4y = 9$$
$$-10y + 24 = 9$$
$$-10y = -15$$
$$y = \frac{15}{10}$$
$$y = \frac{3}{2}$$

Substitute $y = \frac{3}{2}$ into the second equation and solve for x.

$$x + 2y = 8 \qquad \text{the 2}^{\text{nd}} \text{ equation}$$
$$x + 2\left(\frac{3}{2}\right) = 8$$
$$x + 3 = 8$$
$$x = 5$$

The solution is $\left(5, \dfrac{3}{2}\right)$.

79.

$$\begin{cases} x - \dfrac{4y}{5} = 4 \\ \dfrac{y}{3} = \dfrac{x}{2} - \dfrac{5}{2} \end{cases}$$

clear both equations of fractions

$$\begin{cases} 5(x) - 5\left(\dfrac{4y}{5}\right) = 5(4) \\ 6\left(\dfrac{y}{3}\right) = 6\left(\dfrac{x}{2}\right) - 6\left(\dfrac{5}{2}\right) \end{cases}$$

$$\begin{cases} 5x - 4y = 20 \\ 2y = 3x - 15 \end{cases}$$

$$\begin{cases} 5x - 4y = 20 \\ -3x + 2y = -15 \end{cases}$$

Multiply the 2nd equation by 2.

$$5x - 4y = 20$$
$$\underline{-6x + 4y = -30}$$
$$-x = -10$$
$$\boxed{x = 10}$$

Substitute $x = 10$ into the 1st equation.

$$5x - 4y = 20$$
$$5(10) - 4y = 20$$
$$50 - 4y = 20$$
$$-4y = -30$$
$$\boxed{y = \frac{15}{2}}$$

The solution is $\left(10, \dfrac{15}{2}\right)$.

81.

$$\begin{cases} \dfrac{2}{3}x - \dfrac{1}{4}y = -8 \\ 0.5x - 0.375y = -9 \end{cases}$$

Multiply the 1st equation by 12 to eliminate fractions and the 2nd equation by 1,000 to eliminate decimals.

$$\begin{cases} 8x - 3y = -96 \\ 500x - 375y = -9,000 \end{cases}$$

Multiply the 1st equation by -125 and add the equations to eliminate y.

$$-1,000x + 375y = 12,000$$
$$\underline{500x - 375y = -9,000}$$
$$-500x = 3,000$$
$$x = -6$$

Substitute $x = -6$ into the 1st equation and solve for y.

$$8x - 3y = -96$$
$$8(-6) - 3y = -96$$
$$-48 - 3y = -96$$
$$-3y = -48$$
$$y = 16$$

The solution is $(-6, 16)$.

83.

$$\begin{cases} \dfrac{3}{2}p + \dfrac{1}{3}q = 2 \\ \dfrac{2}{3}p + \dfrac{1}{9}q = 1 \end{cases}$$

Multiply the 1st equation by 6 and the 2nd equation by 9 to eliminate fractions.

$$\begin{cases} 9p + 2q = 12 \\ 6p + q = 9 \end{cases}$$

Multiply the 2nd equation by -2 and add the equations to eliminate q.

$$9p + 2q = 12$$
$$\underline{-12p - 2q = -18}$$
$$-3p = -6$$
$$p = 2$$

Substitute $p = 2$ into the 2nd equation and solve for q.

$$6p + q = 9$$
$$6(2) + q = 9$$
$$12 + q = 9$$
$$q = -3$$

The solution is $(2, -3)$.

85.

$$\begin{cases} \dfrac{m-n}{5} + \dfrac{m+n}{2} = 6 \\ \dfrac{m-n}{2} - \dfrac{m+n}{4} = 3 \end{cases}$$

Multiply the 1st equation by 10 and the 2nd equation by 4 to eliminate fractions.

$$\begin{cases} 2m - 2n + 5m + 5n = 60 \\ 2m - 2n - m - n = 12 \end{cases}$$

Combine like terms and write in general form.

$$\begin{cases} 7m + 3n = 60 \\ m - 3n = 12 \end{cases}$$

Add the equations to eliminate n.

$$7m + 3n = 60$$
$$\underline{m - 3n = 12}$$
$$8m = 72$$
$$m = 9$$

Substitute $m = 9$ into the 2nd equation and solve for n.

$$m - 3n = 12$$
$$9 - 3n = 12$$
$$-3n = 3$$
$$n = -1$$

The solution is $(9, -1)$.

87.

$$\begin{cases} \dfrac{1}{x} + \dfrac{1}{y} = \dfrac{5}{6} \\ \dfrac{1}{x} - \dfrac{1}{y} = \dfrac{1}{6} \end{cases}$$

Let $a = \dfrac{1}{x}$ and $b = \dfrac{1}{y}$.

$$\begin{cases} a + b = \dfrac{5}{6} \\ a - b = \dfrac{1}{6} \end{cases}$$

Add equations to eliminate b.

$$a + b = \dfrac{5}{6}$$
$$\underline{a - b = \dfrac{1}{6}}$$
$$2a = \dfrac{6}{6}$$
$$2a = 1$$
$$a = \dfrac{1}{2}$$

Substitute $a = \dfrac{1}{2}$ in the 1st equation and solve for b.

$$a + b = \dfrac{5}{6}$$
$$\dfrac{1}{2} + b = \dfrac{5}{6}$$
$$\dfrac{1}{2} + b - \dfrac{1}{2} = \dfrac{5}{6} - \dfrac{1}{2}$$
$$b = \dfrac{1}{3}$$

$$a = \dfrac{1}{2} = \dfrac{1}{x} \text{ so } x = 2$$
$$b = \dfrac{1}{3} = \dfrac{1}{y} \text{ so } y = 3$$

The solution is $(2, 3)$.

Section 12.1

89.

$$\begin{cases} \dfrac{1}{x} + \dfrac{2}{y} = -1 \\[2mm] \dfrac{2}{x} - \dfrac{1}{y} = -7 \end{cases}$$

Let $a = \dfrac{1}{x}$ and $b = \dfrac{1}{y}$.

$$\begin{cases} a + 2b = -1 \\ 2a - b = -7 \end{cases}$$

Multiply the 2nd equation by 2 and add equations to eliminate b.

$$a + 2b = -1$$
$$\underline{4a - 2b = -14}$$
$$5a = -15$$
$$a = -3$$

Substitute $a = -3$ in the 1st equation and solve for b.

$$a + 2b = -1$$
$$(-3) + 2b = -1$$
$$2b = 2$$
$$b = 1$$

$$a = \dfrac{-3}{1} = \dfrac{1}{x} \text{ so } x = -\dfrac{1}{3}$$
$$b = \dfrac{1}{1} = \dfrac{1}{y} \text{ so } y = 1$$

The solution is $\left(-\dfrac{1}{3}, 1\right)$.

APPLICATIONS

91. HEARING TESTS
    a. (1978, 50%)
    b. In 1978, the percent share of the U.S. footwear market for shoes produced in the United States and imports was the same: 50%.

93. SUPPLY AND DEMAND
    a.

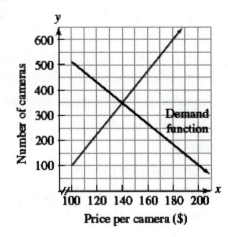

    b. \$140 – the x–value at the point of intersection
    c. As supply increases, demand decreases.

95. BUSINESS
The breakeven point is (250, 2,500). If the company makes 250 widgets, the cost to make them and the revenue obtained from their sale will be equal: \$2,500.

97. TICKET SALES
Let x = the number of adult tickets and y = the number of children tickets.

$$\begin{cases} x + y = 390 \\ 5x + 3y = 1,470 \end{cases}$$

Multiply the first equation by –5.
$$-5x - 5y = -1,950$$
$$\underline{5x + 3y = 1,470}$$
$$-2y = -480$$
$$y = 240$$

Substitute $y = 240$ into the first equation and solve for x.
$$x + y = 390$$
$$x + 240 = 390$$
$$x = 150$$

They sold 150 adult tickets and 240 children's tickets.

99. FASHION DESIGNER

Let x = the number of Gap stores and y = the number of Aéropostale stores.

$$\begin{cases} x + y = 4{,}046 \\ x = 3\frac{1}{4}y \end{cases}$$

Substitute $x = 3\frac{1}{4}y$ into the first equation and solve for y.

$$x + y = 4{,}046$$
$$3\frac{1}{4}y + y = 4{,}046$$
$$\frac{13}{4}y + y = 4{,}046$$
$$\frac{17}{4}y = 4{,}046$$
$$4\left(\frac{17}{4}y\right) = 4(4{,}046)$$
$$17y = 16{,}184$$
$$y = 952$$

Substitute $y = 952$ into the second equation to solve for x.

$$x = 3\frac{1}{4}(952)$$
$$x = 3{,}094$$

101. GEOMETRY

The sum of the two acute angles in a triangle is 90°.

$$\begin{cases} x + y = 90 \\ y = 2x + 15 \end{cases}$$

Use substitution.

$$x + y = 90$$
$$x + (2x + 15) = 90$$
$$3x + 15 = 90$$
$$3x = 75$$
$$x = 25$$

Substitute $x = 25$ in the 2nd equation and solve for y.

$$y = 2(25) + 15$$
$$y = 50 + 15$$
$$y = 65$$

The two angles are 25° and 65°.

103. INVESTMENT CLUBS

Let x = amount at 10% and y = amount at 12%.

$$\begin{cases} x + y = 8{,}000 \\ 0.10x + 0.12y = 900 \end{cases}$$

Multiply the 1st equation by -10 and multiply the 2nd equation by 100.

$$-10x - 10y = -80{,}000$$
$$\underline{10x + 12y = 90{,}000}$$
$$2y = 10{,}000$$
$$y = 5{,}000$$

Substitute $y = 5{,}000$ in the 1st equation and solve for x.

$$x + y = 8{,}000$$
$$x + 5{,}000 = 8{,}000$$
$$x = 3{,}000$$

$3{,}000 was invested at 10% and $5{,}000 was invested at 12%.

105. SNOWMOBILING

Let x = time snowmobiling and y = time skiing. Then $20x$ would be the distance snowmobiling and $4y$ would be the distance skiing.

$$\begin{cases} x + y = 6 \\ 20x + 4y = 48 \end{cases}$$

Multiply the first equation by -20 and add the equations to eliminate x.

$$-20x - 20y = -120$$
$$\underline{20x + 4y = 48}$$
$$-16y = -72$$
$$y = 4.5$$
$$x + y = 6$$
$$x + 4.5 = 6$$
$$x = 1.5$$

He rode the snowmobile for 1.5 hours and skied for 4.5 hours.

Section 12.1

107. PRODUCTION PLANNING

Let x = number of racing bikes and y = number of mountain bikes.

	Racing	Mountain	Combined
Materials	$110x$	$140y$	26,150
Labor	$120x$	$180y$	31,800

$$\begin{cases} 110x + 140y = 26{,}150 \\ 120x + 180y = 31{,}800 \end{cases}$$

Multiply the 1st equation by -120 and the 2nd equation by 110. Add equations to eliminate x.

$$-13{,}200x - 16{,}800y = -3{,}138{,}000$$
$$\underline{13{,}200x + 19{,}800y = 3{,}498{,}000}$$
$$3{,}000y = 360{,}000$$
$$y = 120$$

Substitute $y = 120$ in the 1st equation and solve for x.

$$110x + 140y = 26{,}150$$
$$110x + 140(120) = 26{,}150$$
$$110x + 16{,}800 = 26{,}150$$
$$110x = 9{,}350$$
$$x = 85$$

85 racing bikes and 120 mountain bikes

109. COSMETOLOGY

Let x = number of permanents.

$$\begin{cases} C_1 = 23.60x + 2{,}101.20 \\ C_2 = 44x \end{cases}$$

Breakeven point:

$$C_1 = C_2$$
$$23.60x + 2{,}101.20 = 44x$$
$$2{,}101.20 = 20.4x$$
$$103 = x$$

The breakeven point is 103 permanents.

111. DERMATOLOGY

Let x = # of grams of 0.2% cream and y = # of grams of 0.7% cream

	0.2% solution	0.7% solution	0.3% mixture
# of grams	x	y	185
% Triclosan	$0.002x$	$0.007y$	$0.003(185)$

$$\begin{cases} x + y = 185 \\ 0.002x + 0.007y = 0.003(185) \end{cases}$$

Multiply the 2nd equation by 1,000 to eliminate decimals.

$$\begin{cases} x + y = 185 \\ 2x + 7y = 555 \end{cases}$$

Multiply the 1st equation by -2. Add equations to eliminate x.

$$-2x - 2y = -370$$
$$\underline{2x + 7y = 555}$$
$$5y = 185$$
$$y = 37$$

Substitute $y = 37$ in the 1st equation and solve for x.

$$x + y = 185$$
$$x + 37 = 185$$
$$x = 148$$

148 grams of 0.2% and 37 grams of 0.7%

WRITING

113. Answers will vary.

115. Answers will vary.

117.

$$\frac{V_2}{V_1} = \frac{P_1}{P_2}$$

$$P_2 V_1 \left(\frac{V_2}{V_1}\right) = P_2 V_1 \left(\frac{P_1}{P_2}\right)$$

$$P_2 V_2 = P_1 V_1$$

$$\frac{P_2 V_2}{V_1} = \frac{P_1 V_1}{V_1}$$

$$\frac{P_2 V_2}{V_1} = P_1$$

$$P_1 = \frac{P_2 V_2}{V_1}$$

119.

$$S = \frac{a - lr}{1 - r}$$

$$(1-r)(S) = (1-r)\left(\frac{a - lr}{1 - r}\right)$$

$$S - Sr = a - lr$$

$$S - a = Sr - lr$$

$$\frac{S - a}{S - l} = \frac{r(S - l)}{S - l}$$

$$\frac{S - a}{S - l} = r$$

121.

$$\begin{cases} Ax + By = -2 \\ Bx - Ay = -26 \end{cases}$$

Let $x = -3$ and $y = 5$.

$$\begin{cases} A(-3) + B(5) = -2 \\ B(-3) - A(5) = -26 \end{cases}$$

$$\begin{cases} -3A + 5B = -2 \\ -5A - 3B = -26 \end{cases}$$

Multiply the first equation by 3 and the second by 5.

$$-9A + 15B = -6$$
$$\underline{-25A - 15B = -130}$$
$$-34A = -136$$
$$A = 4$$

Substitute $A = 4$ into the first equation and solve for B.

$$-3A + 5B = -2$$
$$-3(4) + 5B = -2$$
$$-12 + 5B = -2$$
$$5B = 10$$
$$B = 2$$

$A = 4$ and $B = 2$.

VOCABULARY

1. $\begin{cases} 2x+y-3z=0 \\ 3x-y+4z=5 \\ 4x+2y-6z=0 \end{cases}$ is called a **system** of

   three linear equations. Each equation is written in **standard** $Ax + By + Cz = D$ form.

3. Solutions of a system of three equations in three variables, x, y, and z are written in the form (x, y, z) and are called ordered **triples**.

5. When three planes coincide, the equations of the system are **dependent**, and there are infinitely many solutions.

CONCEPTS

7. a. no solution
   b. no solution

NOTATION

9.
   $\begin{cases} x+y+4z=3 \\ 7x-2y+8z=15 \\ 3x+2y-z=4 \end{cases}$

GUIDED PRACTICE

11. Yes.

$x-y+z=2$

$2-1+1\overset{?}{=}2$

$2=2$

$2x+y-z=4$

$2(2)+1-1\overset{?}{=}4$

$4+1-1\overset{?}{=}4$

$4=4$

$2x-3y+z=2$

$2(2)-3(1)+1\overset{?}{=}2$

$4-3+1\overset{?}{=}2$

$2=2$

13. No.

$3x-2y-z=37$

$3(6)-2(-7)-(-5)\overset{?}{=}37$

$18+14+5\overset{?}{=}37$

$37=37$

$x-3y=27$

$(6)-3(-7)\overset{?}{=}27$

$6+21\overset{?}{=}27$

$27=27$

$2x+7y+2z=-48$

$2(6)+7(-7)+2(-5)\overset{?}{=}-48$

$12-49-10\overset{?}{=}-48$

$-47 \neq -48$

15.

$$\begin{cases} x+y+z=4 & (1) \\ 2x+y-z=1 & (2) \\ 2x-3y+z=1 & (3) \end{cases}$$

Add Equations 1 and 2 to eliminate z.

$$x+y+z=4 \quad (1)$$
$$\underline{2x+y-z=1 \quad (2)}$$
$$3x+2y=5 \quad (4)$$

Add Equations 3 and 2 to eliminate z.

$$2x-3y+z=1 \quad (3)$$
$$\underline{2x+y-z=1 \quad (2)}$$
$$4x-2y=2 \quad (5)$$

Add Equations 4 and 5 to eliminate y.

$$3x+2y=5 \quad (4)$$
$$\underline{4x-2y=2 \quad (5)}$$
$$7x=7$$
$$x=1$$

Substitute $x=1$ into Equation 4 to find y.

$$3x+2y=5 \quad (4)$$
$$3(1)+2y=5$$
$$3+2y=5$$
$$2y=2$$
$$y=1$$

Substitute $x=1$ and $y=1$ into Equation 1 to find z.

$$x+y+z=4 \quad (1)$$
$$1+1+z=4$$
$$2+z=4$$
$$z=2$$

The solution is $(1, 1, 2)$.

17.

$$\begin{cases} 3x+2y-5z=3 & (1) \\ 4x-2y-3z=-10 & (2) \\ 5x-2y-2z=-11 & (3) \end{cases}$$

Add Equations 1 and 2 to eliminate y.

$$3x+2y-5z=3 \quad (1)$$
$$\underline{4x-2y-3z=-10 \quad (2)}$$
$$7x-8z=-7 \quad (4)$$

Add Equations 1 and 3 to eliminate y.

$$3x+2y-5z=3 \quad (1)$$
$$\underline{5x-2y-2z=-11 \quad (3)}$$
$$8x-7z=-8 \quad (5)$$

Multiply Equation 4 by 8 and Equation 5 by -7. Add to eliminate x.

$$56x-64z=-56$$
$$\underline{-56x+49z=56}$$
$$-15z=0$$
$$z=0$$

Substitute $z=0$ into Equation 4 and solve for x.

$$7x-8z=-7 \quad (4)$$
$$7x-8(0)=-7$$
$$7x-0=-7$$
$$7x=-7$$
$$x=-1$$

Substitute $x=-1$ and $z=0$ into Equation 1 and solve for y.

$$3x+2y-5z=3 \quad (1)$$
$$3(-1)+2y-5(0)=3$$
$$-3+2y-0=3$$
$$-3+2y=3$$
$$2y=6$$
$$y=3$$

The solution is $(-1, 3, 0)$.

Section 12.2

19.

$$\begin{cases} 2x + 6y + 3z = 9 & (1) \\ 5x - 3y - 5z = 3 & (2) \\ 4x + 3y + 2z = 15 & (3) \end{cases}$$

Multiply Equation 2 by 2 and add Equations 1 and 2 to eliminate y.

$$2x + 6y + 3z = 9 \quad (1)$$
$$\underline{10x - 6y - 10z = 6} \quad (2)$$
$$12x - 7z = 15 \quad (4)$$

Multiply Equation 3 by -2 and add Equations 1 and 3 to eliminate y.

$$2x + 6y + 3z = 9 \quad (1)$$
$$\underline{-8x - 6y - 4z = -30} \quad (3)$$
$$-6x - z = -21 \quad (5)$$

Multiply Equation 5 by -7. Add Equations 4 and 5 to eliminate z.

$$12x - 7z = 15$$
$$\underline{42x + 7z = 147}$$
$$54x = 162$$
$$x = 3$$

Substitute $x = 3$ into Equation 4 and solve for z.

$$12x - 7z = 15 \quad (4)$$
$$12(3) - 7z = 15$$
$$36 - 7z = 15$$
$$-7z = -21$$
$$z = 3$$

Substitute $x = 3$ and $z = 3$ into Equation 1 and solve for y.

$$2x + 6y + 3z = 9 \quad (1)$$
$$2(3) + 6y + 3(3) = 9$$
$$6 + 6y + 9 = 9$$
$$6y + 15 = 9$$
$$6y = -6$$
$$y = -1$$

The solution is $(3, -1, 3)$.

21.

$$\begin{cases} 4x - 5y - 8z = -52 & (1) \\ 2x - 3y - 4z = -26 & (2) \\ 3x + 7y + 8z = 31 & (3) \end{cases}$$

Multiply Equation 2 by -2 and add Equations 1 and 2 to eliminate x and z.

$$4x - 5y - 8z = -52 \quad (1)$$
$$\underline{-4x + 6y + 8z = 52} \quad (2)$$
$$y = 0$$

Add Equations 1 and 3 to eliminate z.

$$4x - 5y - 8z = -52 \quad (1)$$
$$\underline{3x + 7y + 8z = 31} \quad (3)$$
$$7x + 2y = -21 \quad (4)$$

Substitute $y = 0$ into Equation 4 and solve for x.

$$7x + 2y = -21 \quad (4)$$
$$7x + 2(0) = -21$$
$$7x + 0 = -21$$
$$7x = -21$$
$$x = -3$$

Substitute $x = -3$ and $y = 0$ into Equation 1 and solve for z.

$$4x - 5y - 8z = -52 \quad (1)$$
$$4(-3) - 5(0) - 8z = -52$$
$$-12 - 0 - 8z = -52$$
$$-12 - 8z = -52$$
$$-8z = -40$$
$$z = 5$$

The solution is $(-3, 0, 5)$.

23.

$$\begin{cases} 3x + 3z = 6 - 4y \\ 7x - 5z = 46 + 2y \\ 4x = 31 - z \end{cases}$$

Write each equation in standard form.

$$\begin{cases} 3x + 4y + 3z = 6 & (1) \\ 7x - 2y - 5z = 46 & (2) \\ 4x + z = 31 & (3) \end{cases}$$

Multiply Equation 2 by 2 and add Equations 1 and 2 to eliminate y.

$$3x + 4y + 3z = 6$$
$$\underline{14x - 4y - 10z = 92}$$
$$17x - 7z = 98 \quad (4)$$

Multiply Equation 3 by 7 and add Equations 3 and 4 to eliminate z.

$$28x + 7z = 217$$
$$\underline{17x - 7z = 98} \quad (4)$$
$$45x = 315$$
$$x = 7$$

Substitute $x = 7$ into Equation 4 and solve for z.

$$17x - 7z = 98 \quad (4)$$
$$17(7) - 7z = 98$$
$$119 - 7z = 98$$
$$-7z = -21$$
$$z = 3$$

Substitute $x = 7$ and $z = 3$ into Equation 1 and solve for y.

$$3x + 4y + 3z = 6 \quad (1)$$
$$3(7) + 4y + 3(3) = 6$$
$$21 + 4y + 9 = 6$$
$$4y + 30 = 6$$
$$4y = -24$$
$$y = -6$$

The solution is $(7, -6, 3)$.

25.

$$\begin{cases} 2x + z = -2 + y \\ 8x - 3y = -2 \\ 6x - 2y + 3z = -4 \end{cases}$$

Write Equation 1 in standard form.

$$\begin{cases} 2x - y + z = -2 & (1) \\ 8x - 3y = -2 & (2) \\ 6x - 2y + 3z = -4 & (3) \end{cases}$$

Multiply Equation 1 by -3 and add Equations 1 and 3 to eliminate x and z.

$$-6x + 3y - 3z = 6$$
$$\underline{6x - 2y + 3z = -4}$$
$$y = 2$$

Substitute $y = 2$ into Equation 2 and solve for x.

$$8x - 3y = -2$$
$$8x - 3(2) = -2$$
$$8x - 6 = -2$$
$$8x = 4$$
$$x = \frac{1}{2}$$

Substitute $x = \frac{1}{2}$ and $y = 2$ into Equation 1 and solve for z.

$$2x - y + z = -2$$
$$2\left(\frac{1}{2}\right) - (2) + z = -2$$
$$1 - 2 + z = -2$$
$$-1 + z = -2$$
$$z = -1$$

The solution is $\left(\frac{1}{2}, 2, -1\right)$.

27.

$$\begin{cases} x + y + 3z = 35 & (1) \\ -x - 3y = 20 & (2) \\ 2y + z = -35 & (3) \end{cases}$$

Add Equations 1 and 2 to eliminate x.

$$x + y + 3z = 35 \quad (1)$$
$$\underline{-x - 3y = 20 \quad (2)}$$
$$-2y + 3z = 55 \quad (4)$$

Add Equation 3 and 4 to eliminate y.

$$2y + z = -35$$
$$\underline{-2y + 3z = 55}$$
$$4z = 20$$
$$z = 5$$

Substitute $z = 5$ into Equation 3 and solve for y.

$$2y + z = -35$$
$$2y + 5 = -35$$
$$2y = -40$$
$$y = -20$$

Substitute $y = -20$ into Equation 2 and solve for x.

$$-x - 3y = 20$$
$$-x - 3(-20) = 20$$
$$-x + 60 = 20$$
$$-x = -40$$
$$x = 40$$

The solution is $(40, -20, 5)$.

29.

$$\begin{cases} 3x + 2y - z = 7 & (1) \\ 6x - 3y = -2 & (2) \\ 3y - 2z = 8 & (3) \end{cases}$$

Multiply Equation 1 by -2 and add Equations 1 and 2 to eliminate x.

$$-6x - 4y + 2z = -14$$
$$\underline{6x - 3y \qquad = -2}$$
$$-7y + 2z = -16 \quad (4)$$

Add Equations 3 and 4 to eliminate z.

$$3y - 2z = 8 \quad (3)$$
$$\underline{-7y + 2z = -16 \quad (4)}$$
$$-4y = -8$$
$$y = 2$$

Substitute $y = 2$ into Equation 3 and solve for z.

$$3y - 2z = 8 \quad (3)$$
$$3(2) - 2z = 8$$
$$6 - 2z = 8$$
$$-2z = 2$$
$$z = -1$$

Substitute $y = 2$ into Equation 2 and solve for x.

$$6x - 3y = -2 \quad (2)$$
$$6x - 3(2) = -2$$
$$6x - 6 = -2$$
$$6x = 4$$
$$x = \frac{4}{6}$$
$$x = \frac{2}{3}$$

The solution is $\left(\frac{2}{3}, 2, -1 \right)$.

31.

$$\begin{cases} r + s - 3t = 21 & (1) \\ r + 4s = 9 & (2) \\ 5s + t = -4 & (3) \end{cases}$$

Multiply Equation 2 by -1 and add Equations 1 and 2 to eliminate r.

$$r + \;\; s - 3t = 21$$
$$\underline{-r - 4s \qquad\;\; = -9}$$
$$-3s - 3t = 12 \quad (4)$$

Multiply Equation 3 by 3 and add Equations 3 and 4 to eliminate t.

$$15s + 3t = -12$$
$$\underline{-3s - 3t = 12}$$
$$12s = 0$$
$$s = 0$$

Substitute $s = 0$ into Equation 3 and solve for t.

$$5s + t = -4 \quad (3)$$
$$5(0) + t = -4$$
$$0 + t = -4$$
$$t = -4$$

Substitute $s = 0$ into Equation 2 and solve for r.

$$r + 4s = 9 \quad (2)$$
$$r + 4(0) = 9$$
$$r + 0 = 9$$
$$r = 9$$

The solution is $(9, 0, -4)$.

33.

$$\begin{cases} x - 8z = -30 & (1) \\ 3x + y - 4z = 5 & (2) \\ y + 7z = 30 & (3) \end{cases}$$

Multiply Equation 3 by -1 and add Equations 2 and 3 to eliminate y.

$$3x + y - 4z = 5 \quad (2)$$
$$\underline{-y - 7z = -30}$$
$$3x - 11z = -25 \quad (4)$$

Multiply Equation 1 by -3 and add Equations 1 and 4 to eliminate x.

$$-3x + 24z = 90$$
$$\underline{3x - 11z = -25} \quad (4)$$
$$13z = 65$$
$$z = 5$$

Substitute $z = 5$ into Equation 1 and solve for x.

$$x - 8z = -30 \quad (1)$$
$$x - 8(5) = -30$$
$$x - 40 = -30$$
$$x = 10$$

Substitute $z = 5$ into Equation 3 and solve for y.

$$y + 7z = 30 \quad (3)$$
$$y + 7(5) = 30$$
$$y + 35 = 30$$
$$y = -5$$

The solution is $(10, -5, 5)$.

35.

$$\begin{cases} 7a + 9b - 2c = -5 & (1) \\ 5a + 14b - c = -11 & (2) \\ 2a - 5b - c = 3 & (3) \end{cases}$$

Multiply Equation 2 by –1 and add Equations 2 and 3 to eliminate c.

$$-5a - 14b + c = 11 \quad (2)$$
$$\underline{2a - 5b - c = 3} \quad (3)$$
$$-3a - 19b = 14 \quad (4)$$

Multiply Equation 2 by –2 and add Equations 1 and 2 to eliminate c.

$$7a + 9b - 2c = -5 \quad (1)$$
$$\underline{-10a - 28b + 2c = 22}$$
$$-3a - 19b = 17 \quad (5)$$

Multiply Equation 4 by –1 and add Equations 4 and 5 to eliminate a.

$$3a + 19b = -14 \quad (4)$$
$$\underline{-3a - 19b = 17} \quad (5)$$
$$0 \neq 3$$

No solution; $\varnothing$

Inconsistent system

37.

$$\begin{cases} 7x - y - z = 10 & (1) \\ x - 3y + z = 2 & (2) \\ x + 2y - z = 1 & (3) \end{cases}$$

Add Equations 1 and 2 to eliminate z.

$$7x - y - z = 10 \quad (1)$$
$$\underline{x - 3y + z = 2} \quad (2)$$
$$8x - 4y = 12 \quad (4)$$

Add Equations 2 and 3 to eliminate z.

$$x - 3y + z = 2 \quad (2)$$
$$\underline{x + 2y - z = 1} \quad (3)$$
$$2x - y = 3 \quad (5)$$

Multiply Equation 5 by –4 and add equations 4 and 5 to eliminate x.

$$8x - 4y = 12 \quad (4)$$
$$\underline{-8x + 4y = -12} \quad (5)$$
$$0 = 0$$

Infinitely many solutions; dependent equations

39.

$$\begin{cases} 2a + 3b - 2c = 18 & (1) \\ 5a - 6b + c = 21 & (2) \\ 4b - 2c = 6 & (3) \end{cases}$$

Multiply Equation 3 by –1 and add Equations 1 and 3 to eliminate c.

$$2a + 3b - 2c = 18 \quad (1)$$
$$\underline{-4b + 2c = -6} \quad (3)$$
$$2a - b = 12 \quad (4)$$

Multiply Equation 2 by 2 and add Equations 1 and 2 to eliminate c.

$$2a + 3b - 2c = 18 \quad (1)$$
$$\underline{10a - 12b + 2c = 42} \quad (2)$$
$$12a - 9b = 60 \quad (5)$$

Multiply Equation 4 by –6 and add Equations 4 and 5 to eliminate a.

$$-12a + 6b = -72 \quad (4)$$
$$\underline{12a - 9b = 60} \quad (5)$$
$$-3b = -12$$
$$b = 4$$

Substitute $b = 4$ into Equation 3 and solve for c.

$$4b - 2c = 6 \quad (3)$$
$$4(4) - 2c = 6$$
$$16 - 2c = 6$$
$$-2c = -10$$
$$c = 5$$

Substitute $b = 4$ and $c = 5$ into Equation 1 and solve for a.

$$2a + 3b - 2c = 18 \quad (1)$$
$$2a + 3(4) - 2(5) = 18$$
$$2a + 12 - 10 = 18$$
$$2a + 2 = 18$$
$$2a = 16$$
$$a = 8$$

The solution is (8, 4, 5).

41.

$$\begin{cases} 2x+2y-z=2 & (1) \\ x+3z-24=0 & (2) \\ y=7-4z & (3) \end{cases}$$

Solve Equation 2 for x.

$$\begin{cases} 2x+2y-z=2 & (1) \\ x=-3z+24 & (2) \\ y=7-4z & (3) \end{cases}$$

Use substitution. Substitute $x=-3z+24$ and $y=7-4z$ into Equation 1 and solve for z.

$$2x+2y-z=2$$
$$2(-3z+24)+2(7-4z)-z=2$$
$$-6z+48+14-8z-z=2$$
$$-15z+62=2$$
$$-15z=-60$$
$$z=4$$

Substitute $z=4$ into Equation 3 and solve for y.

$$y=7-4z \qquad (3)$$
$$y=7-4(4)$$
$$y=7-16$$
$$y=-9$$

Substitute $z=4$ into Equation 2 and solve for x.

$$x+3z-24=0 \quad (2)$$
$$x+3(4)-24=0$$
$$x+12-24=0$$
$$x-12=0$$
$$x=12$$

The solution is $(12, -9, 4)$.

43.

$$\begin{cases} b+2c=7-a \\ a+c=2(4-b) \\ 2a+b+c=9 \end{cases}$$

Write all equations in general form.

$$\begin{cases} a+b+2c=7 & (1) \\ a+2b+c=8 & (2) \\ 2a+b+c=9 & (3) \end{cases}$$

Multiply Equation 2 by -2 and add Equations 1 and 2 to eliminate c.

$$\begin{array}{rl} a+\;\;b+2c=7 & (1) \\ \underline{-2a-4b-2c=-16} & (2) \\ -a-3b\quad\;\;=-9 & (4) \end{array}$$

Multiply Equation 2 by -1 and add Equations 2 and 3 to eliminate c.

$$\begin{array}{rl} -a-2b-c=-8 & (2) \\ \underline{2a+\;b+\;c=9} & (3) \\ a-\;b\quad\;\;=1 & (5) \end{array}$$

Add Equations 4 and 5 to eliminate a.

$$\begin{array}{rl} -a-3b=-9 & (4) \\ \underline{a-\;\;b=1} & (5) \\ -4b=-8 \\ b=2 \end{array}$$

Substitute $b=2$ into Equation 5 and solve for a.

$$a-b=1 \quad (5)$$
$$a-2=1$$
$$a=3$$

Substitute $a=3$ and $b=2$ into Equation 1 and solve for c.

$$a+b+2c=7 \quad (1)$$
$$3+2+2c=7$$
$$5+2c=7$$
$$2c=2$$
$$c=1$$

The solution is $(3, 2, 1)$.

45.

$$\begin{cases} 2x + y - z = 1 & (1) \\ x + 2y + 2z = 2 & (2) \\ 4x + 5y + 3z = 3 & (3) \end{cases}$$

Multiply Equation 1 by 2 and
add Equations 1 and 2 to eliminate z.

$$\begin{array}{ll} 4x + 2y - 2z = 2 & (1) \\ \underline{x + 2y + 2z = 2} & (2) \\ 5x + 4y = 4 & (4) \end{array}$$

Multiply Equation 1 by 3 and
add Equations 1 and 3 to eliminate z.

$$\begin{array}{ll} 6x + 3y - 3z = 3 & (1) \\ \underline{4x + 5y + 3z = 3} & (3) \\ 10x + 8y \quad = 6 & (5) \end{array}$$

Multiply Equaiton 4 by -2 and
add Equations 4 and 5 to eliminate x.

$$\begin{array}{l} -10x - 8y = -8 \\ \underline{10x + 8y = 6} \\ 0 \neq -2 \end{array}$$

Since $0 \neq -2$, the system is inconsistent and
has no solution.

47.

$$\begin{cases} 0.4x + 0.3z = 0.4 & (1) \\ 2y - 6z = -1 & (2) \\ 4(2x + y) = 9 - 3z & (3) \end{cases}$$

Multiply Equation 1 by 10 to clear decimals
and write Equation 3 in standard form.

$$\begin{cases} 4x + 3z = 4 & (1) \\ 2y - 6z = -1 & (2) \\ 8x + 4y + 3z = 9 & (3) \end{cases}$$

Multiply Equation 1 by -2 and
add Equations 1 and 3 to eliminate x.

$$\begin{array}{ll} -8x \quad\quad - 6z = -8 & (1) \\ \underline{8x + 4y + 3z = 9} & (3) \\ 4y - 3z = 1 & (4) \end{array}$$

Multiply Equation 4 by -2 and
add Equations 4 and 2 to eliminate z.

$$\begin{array}{ll} -8y + 6z = -2 & (4) \\ \underline{2y - 6z = -1} & (2) \\ -6y = -3 \\ y = \dfrac{1}{2} \end{array}$$

Substitute $y = \dfrac{1}{2}$ into Equation 2 and
solve for z.

$$2y - 6z = -1 \quad (2)$$

$$2\left(\dfrac{1}{2}\right) - 6z = -1$$

$$1 - 6z = -1$$

$$-6z = -2$$

$$z = \dfrac{1}{3}$$

Substitute $z = \dfrac{1}{3}$ into Equation 1 and
solve for x.

$$4x + 3z = 4 \quad (1)$$

$$4x + 3\left(\dfrac{1}{3}\right) = 4$$

$$4x + 1 = 4$$

$$4x = 3$$

$$x = \dfrac{3}{4}$$

The solution is $\left(\dfrac{3}{4}, \dfrac{1}{2}, \dfrac{1}{3}\right)$.

49.

$$\begin{cases} r + s + 4t = 3 & (1) \\ 3r + 7t = 0 & (2) \\ 3s + 5t = 0 & (3) \end{cases}$$

Multiply Equation 1 by -3 and add
Equations 1 and 2 to eliminate r.

$$\begin{array}{ll} -3r - 3s - 12t = -9 & (1) \\ \underline{3r + 7t = 0} & (2) \\ -3s - 5t = -9 & (4) \end{array}$$

Add Equations 3 and 4 to eliminate s.

$$\begin{array}{ll} 3s + 5t = 0 & (3) \\ \underline{-3s - 5t = -9} & (4) \\ 0 \neq -9 \end{array}$$

No solution; $\varnothing$
Inconsistent System

51.

$$\begin{cases} 0.5a + 0.3b = 2.2 & (1) \\ 1.2c - 8.5b = -24.4 & (2) \\ 3.3c + 1.3a = 29 & (3) \end{cases}$$

Multiply each equation by 10 to eliminate
decimials and write in standard form.

$$\begin{cases} 5a + 3b = 22 & (1) \\ -85b + 12c = -244 & (2) \\ 13a + 33c = 290 & (3) \end{cases}$$

Multiply Equation 1 by 85 and
Equation 2 by 3. Add the equations.

$$\begin{array}{ll} 425a + 255b = 1{,}870 & (1) \\ \underline{ -255b + 36c = -732} & (2) \\ 425a + 36c = 1{,}138 & (4) \end{array}$$

Multiply Equation 3 by -36 and
Equation 4 by 33. Add the equations.

$$\begin{array}{ll} -468a - 1{,}188c = -10{,}440 & (3) \\ \underline{14{,}025a + 1{,}188c = 37{,}554} & (4) \\ 13{,}557a \phantom{+ 1{,}188c} = 27{,}114 \\ \phantom{13{,}557}a = 2 \end{array}$$

Substitute $a = 2$ into Equation 1 and
solve for b.

$$\begin{array}{ll} 5a + 3b = 22 & (1) \\ 5(2) + 3b = 22 \\ 10 + 3b = 22 \\ 3b = 12 \\ b = 4 \end{array}$$

Substitute $a = 2$ into Equation 3 and
solve for c.

$$\begin{array}{ll} 13a + 33c = 290 & (3) \\ 13(2) + 33c = 290 \\ 26 + 33c = 290 \\ 33c = 264 \\ c = 8 \end{array}$$

The solution is $(2, 4, 8)$.

Section 12.2

53.

$$\begin{cases} 2x + 3y = 6 - 4z & (1) \\ 2x = 3y + 4z - 4 & (2) \\ 4x + 6y + 8z = 12 & (3) \end{cases}$$

Write each equation in standard form.

$$\begin{cases} 2x + 3y + 4z = 6 & (1) \\ 2x - 3y - 4z = -4 & (2) \\ 4x + 6y + 8z = 12 & (3) \end{cases}$$

Multiply Equation 1 by –2 and add Equations 1 and 3 to eliminate x.

$$\begin{aligned} -4x - 6y - 8z &= -12 & (1) \\ \underline{4x + 6y + 8z} &= \underline{12} & (3) \\ 0 &= 0 \end{aligned}$$

Infinitely many solutions
dependent equations

55.

$$\begin{cases} a + b = 2 + c \\ a = 3 + b - c \\ -a + b + c - 4 = 0 \end{cases}$$

Write each equation in standard form.

$$\begin{cases} a + b - c = 2 & (1) \\ a - b + c = 3 & (2) \\ a - b - c = -4 & (3) \end{cases}$$

Add Equations 1 and 2 to eliminate b and c.

$$\begin{aligned} a + b - c &= 2 & (1) \\ \underline{a - b + c} &= \underline{3} & (2) \\ 2a &= 5 \\ a &= 2.5 \end{aligned}$$

Multiply Equation 2 by –1 and add Equations 2 and 3 to eliminate a and b.

$$\begin{aligned} -a + b - c &= -3 & (2) \\ \underline{a - b - c} &= \underline{-4} & (3) \\ -2c &= -7 \\ c &= 3.5 \end{aligned}$$

Substitute $a = 2.5$ and $c = 3.5$ into Equation 1 and solve for b.

$$\begin{aligned} a + b - c &= 2 & (1) \\ 2.5 + b - 3.5 &= 2 \\ -1 + b &= 2 \\ b &= 3 \end{aligned}$$

The solution is $(2.5, 3, 3.5)$.

57.

$$\begin{cases} x + \dfrac{1}{3}y + z = 13 \\ \dfrac{1}{2}x - y + \dfrac{1}{3}z = -2 \\ x + \dfrac{1}{2}y - \dfrac{1}{3}z = 2 \end{cases}$$

Write each equation in standard form by muliplying by a LCD to eliminate fractions.

$$\begin{cases} 3x + y + 3z = 39 & (1) \\ 3x - 6y + 2z = -12 & (2) \\ 6x + 3y - 2z = 12 & (3) \end{cases}$$

Multiply Equation 2 by -1 and add Equations 1 and 2.

$$\begin{array}{ll} -3x + 6y - 2z = 12 & (2) \\ \underline{3x + \ y + 3z = 39} & (1) \\ \qquad\quad 7y + z = 51 & (4) \end{array}$$

Multiply Equation 1 by -2 and add Equations 1 and 3.

$$\begin{array}{ll} -6x - 2y - 6z = -78 & (1) \\ \underline{6x + 3y - 2z = 12} & (3) \\ \qquad\quad y - 8z = -66 & (5) \end{array}$$

Multiply Equation 4 by 8 and add Equations 4 and 5.

$$\begin{array}{ll} 56y + 8z = 408 & (4) \\ \underline{y - 8z = -66} & (5) \\ 57y = 342 \\ y = 6 \end{array}$$

Substitute $y = 6$ into Equation 4.

$$\begin{array}{ll} 7y + z = 51 & (4) \\ 7(6) + z = 51 \\ 42 + z = 51 \\ z = 9 \end{array}$$

Substitute $y = 6$ and $z = 9$ into Equation 1.

$$\begin{array}{ll} 3x + y + 3z = 39 & (1) \\ 3x + 6 + 3(9) = 39 \\ 3x + 33 = 39 \\ 3x = 6 \\ x = 2 \end{array}$$

The solution is $(2, \ 6, \ 9)$.

APPLICATIONS

59. a. infinitely many solutions, all lying on the line running down the binding
   b. 3 parallel planes (shelves); no solution
   c. each pair of planes (cards) intersects; no solution
   d. 3 planes (faces of die) intersect at a corner; 1 solution

61. NBA RECORDS

$$\begin{cases} x + y + z = 259 & (1) \\ x - y = 19 & (2) \\ x - z = 22 & (3) \end{cases}$$

Add Equations 1 and 2 to eliminate y.

$$\begin{array}{ll} x + y + z = 259 & (1) \\ \underline{x - y \qquad\;\; = \;\; 19} & (2) \\ 2x \quad + z = 278 & (4) \end{array}$$

Add Equations 3 and 4 to eliminate z.

$$\begin{array}{ll} x - z = 22 & (3) \\ \underline{2x + z = 278} & (4) \\ 3x = 300 \\ x = 100 \end{array}$$

Substitute $x = 100$ into Equation 2 and solve for y.

$$\begin{array}{ll} x - y = 19 & (2) \\ 100 - y = 19 \\ -y = -81 \\ y = 81 \end{array}$$

Substitute $x = 100$ into Equation 3 and solve for z.

$$\begin{array}{ll} x - z = 22 & (3) \\ 100 - z = 22 \\ -z = -78 \\ z = 78 \end{array}$$

The solutions are 100, 81, and 78.

WRITING

63. Answers will vary.

65. Answers will vary.

Section 12.2

REVIEW

67. $f(x) = |x|$

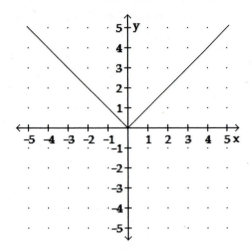

69. $h(x) = x^3$

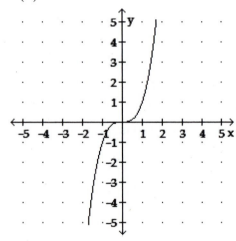

71.

$$\begin{cases} w + x + y + z = 3 & (1) \\ w - x + y + z = 1 & (2) \\ w + x - y + z = 1 & (3) \\ w + x + y - z = 3 & (4) \end{cases}$$

Add Equations 3 and 4.

$$w + x - y + z = 1 \quad (3)$$
$$\underline{w + x + y - z = 3} \quad (4)$$
$$2w + 2x = 4 \quad (5)$$

Multiply Equation 1 by -1 and add Equations 1 and 2.

$$-w - x - y - z = -3 \quad (1)$$
$$\underline{w - x + y + z = 1} \quad (2)$$
$$-2x = -2$$
$$x = 1$$

Substitute $x = 1$ into Equation 5.

$$2w + 2x = 4 \quad (5)$$
$$2w + 2(1) = 4$$
$$2w + 2 = 4$$
$$2w = 2$$
$$w = 1$$

Add Equations 2 and 3.

$$w - x + y + z = 1 \quad (2)$$
$$\underline{w + x - y + z = 1} \quad (3)$$
$$2w + 2z = 2 \quad (6)$$

Substitute $w = 1$ into Equation 6.

$$2w + 2z = 2 \quad (6)$$
$$2(1) + 2z = 2$$
$$2 + 2z = 2$$
$$2z = 0$$
$$z = 0$$

Substitute $x = 1$, $w = 1$, and $z = 0$ into Equation 1.

$$w + x + y + z = 3 \quad (1)$$
$$1 + 1 + y + 0 = 3$$
$$y + 2 = 3$$
$$y = 1$$

The solution is $(1, \ 1, \ 1, \ 0)$.

73.

$$\begin{cases} 4a + b + 2c - 3d = -16 & \text{(Eq.1)} \\ 3a - 3b + c - 4d = -20 & \text{(Eq.2)} \\ a - 2b - 5c - d = 4 & \text{(Eq.3)} \\ 5a + 4b + 3c - d = -10 & \text{(Eq.4)} \end{cases}$$

Multiply Equation 3 by -3 and add Equation 1 and 3 to get Equation 5.

$$\begin{aligned} 4a + b + 2c - 3d &= -16 \\ \underline{-3a + 6b + 15c + 3d} &= \underline{-12} \\ a + 7b + 17c &= -28 \quad \text{(Eq.5)} \end{aligned}$$

Multiply Equation 3 by -4 and add Equation 2 and 3 to get Equation 6.

$$\begin{aligned} 3a - 3b + c - 4d &= -20 \\ \underline{-4a + 8b + 20c + 4d} &= \underline{-16} \\ -a + 5b + 21c &= -36 \quad \text{(Eq.6)} \end{aligned}$$

Multiply Equation 3 by -1 and add Equation 4 and 3 to get Equation 7.

$$\begin{aligned} 5a + 4b + 3c - d &= -10 \\ \underline{-a + 2b + 5c + d} &= \underline{-4} \\ 4a + 6b + 8c &= -14 \quad \text{(Eq.7)} \end{aligned}$$

Add Equations 5 and 6 to get Equation 8.

$$\begin{aligned} a + 7b + 17c &= -28 \\ \underline{-a + 5b + 21c} &= \underline{-36} \\ 12b + 38c &= -64 \quad \text{(Eq.8)} \end{aligned}$$

Multiply Equation 6 by 4 and add Equation 6 and 7 to get Equation 9.

$$\begin{aligned} -4a + 20b + 84c &= -144 \\ \underline{4a + 6b + 8c} &= \underline{-14} \\ 26b + 92c &= -158 \quad \text{(Eq.9)} \end{aligned}$$

Multiply Equation 8 by 13 and multiply Equation 9 by -6. Add to solve for c.

$$\begin{aligned} 156b + 494c &= -832 \\ \underline{-156b - 552c} &= \underline{948} \\ -58c &= 116 \\ c &= -2 \end{aligned}$$

Substitute $c = -2$ into Equation 9 and solve for b.

$$\begin{aligned} 26b + 92c &= -158 \\ 26b + 92(-2) &= -158 \\ 26b - 184 &= -158 \\ 26b &= 26 \\ b &= 1 \end{aligned}$$

Substitute $c = -2$ and $b = 1$ into Equation 5 and solve for a.

$$\begin{aligned} a + 7b + 17c &= -28 \\ a + 7(1) + 17(-2) &= -28 \\ a + 7 - 34 &= -28 \\ a - 27 &= -28 \\ a &= -1 \end{aligned}$$

Substitute $c = -2$, $b = 1$ and $a = -1$ into Equation 3 and solve for d.

$$\begin{aligned} a - 2b - 5c - d &= 4 \\ (-1) - 2(1) - 5(-2) - d &= 4 \\ -1 - 2 + 10 - d &= 4 \\ 7 - d &= 4 \\ -d &= -3 \\ d &= 3 \end{aligned}$$

The solution is $(-1, 1, -2, 3)$.

Section 12.2

VOCABULARY

1. If a point lies on the graph of an equation, it is a solution of the equation and the coordinates of the point **satisfy** the equation.

CONCEPTS

3.
$$\begin{cases} x + y + z = 50 \\ 5x + 6y + 7z = 295 \\ 2x + 3y + 4z = 145 \end{cases}$$

5.
$$y = ax^2 + bx + c$$
$$-3 = a(2)^2 + b(2) + c$$
$$-3 = 4a + 2b + c$$

APPLICATIONS

7. MAKING STATUES
    Let x = large, y = medium, and z = small type.
    $$\begin{cases} x + y + z = 180 & (1) \\ 5x + 4y + 3z = 650 & (2) \\ 20x + 12y + 9z = 2{,}100 & (3) \end{cases}$$

    Multiply Equation 1 by -5 and add Equations 1 and 2.
    $$\begin{aligned} -5x - 5y - 5z &= -900 & (1) \\ 5x + 4y + 3z &= 650 & (2) \\ \hline -y - 2z &= -250 & (4) \end{aligned}$$

    Multiply Equation 1 by -20 and add Equations 1 and 3.
    $$\begin{aligned} -20x - 20y - 20z &= -3{,}600 & (1) \\ 20x + 12y + 9z &= 2{,}100 & (3) \\ \hline -8y - 11z &= -1{,}500 & (5) \end{aligned}$$

    Multiply Equation 4 by -8 and add Equations 4 and 5.
    $$\begin{aligned} 8y + 16z &= 2{,}000 & (4) \\ -8y - 11z &= -1{,}500 & (5) \\ \hline 5z &= 500 \\ z &= 100 \end{aligned}$$

    Substitute $z = 100$ into Equation 4.
    $$\begin{aligned} -y - 2z &= -250 & (4) \\ -y - 2(100) &= -250 \\ -y - 200 &= -250 \\ -y &= -50 \\ y &= 50 \end{aligned}$$

    Substitute $z = 100$ and $y = 50$ into Equation 4.
    $$\begin{aligned} x + y + z &= 180 & (1) \\ x + 50 + 100 &= 180 \\ x + 150 &= 180 \\ x &= 30 \end{aligned}$$

    He must sell 30 large types, 50 medium types, and 100 small types.

9. NUTRITION

a.

Name of food	# of oz used	Oz. of fat	Oz. of carbs	Oz. of protein
A	a	$2a$	$3a$	$2a$
B	b	$3b$	$2b$	b
C	c	c	c	$2c$
Total		14	13	9

b. Write the 3 equations.

$$\begin{cases} 2a + 3b + c = 14 & (1) \\ 3a + 2b + c = 13 & (2) \\ 2a + b + 2c = 9 & (3) \end{cases}$$

Multiply Equation 1 by –2 and add Equations 1 and 3.

$$-4a - 6b - 2c = -28 \quad (1)$$
$$\underline{2a + b + 2c = 9} \quad (3)$$
$$-2a - 5b = -19 \quad (4)$$

Multiply Equation 2 by –2 and add Equations 2 and 3.

$$-6a - 4b - 2c = -26 \quad (2)$$
$$\underline{2a + b + 2c = 9} \quad (3)$$
$$-4a - 3b = -17 \quad (5)$$

Multiply Equation 4 by –2 and add Equations 4 and 5.

$$4a + 10b = 38 \quad (4)$$
$$\underline{-4a - 3b = -17} \quad (5)$$
$$7b = 21$$
$$b = 3$$

Substitute $b = 3$ into Equation 4 and solve for a.

$$-2a - 5(3) = -19$$
$$-2a - 15 = -19$$
$$-2a = -4$$
$$a = 2$$

Substitute $a = 2$ and $b = 3$ into Equation 1 and solve for c.

$$2(2) + 3(3) + c = 14$$
$$4 + 9 + c = 14$$
$$13 + c = 14$$
$$c = 1$$

She needs 2 oz of Food A, 3 oz of Food B, and 1 oz of Food C.

11. FASHION DESIGNER

Let x = # of coats, y = number of shirts, and z = # of slacks.

Change the time available from hours to minutes by multiplying by 60.

$$\begin{cases} 20x + 15y + 10z = 6,900 & (1) \\ 60x + 30y + 24z = 16,800 & (2) \\ 5x + 12y + 6z = 3,900 & (3) \end{cases}$$

Divide Equation 1 by 5 and Equation 2 by 6.

$$\begin{cases} 4x + 3y + 2z = 1,380 & (1) \\ 10x + 5y + 4z = 2,800 & (2) \\ 5x + 12y + 6z = 3,900 & (3) \end{cases}$$

Multiply Equation 1 by –2 and add Equations 1 and 2.

$$-8x - 6y - 4z = -2,760 \quad (1)$$
$$\underline{10x + 5y + 4z = 2,800} \quad (2)$$
$$2x - y = 40 \quad (4)$$

Multiply Equation 1 by –3 and add Equations 1 and 3.

$$-12x - 9y - 6z = -4,140 \quad (1)$$
$$\underline{5x + 12y + 6z = 3,900} \quad (3)$$
$$-7x + 3y = -240 \quad (5)$$

Multiply Equation 4 by 3 and add Equations 4 and 5.

$$6x - 3y = 120 \quad (4)$$
$$\underline{-7x + 3y = -240} \quad (5)$$
$$-x = -120$$
$$x = 120$$

Substitute $x = 120$ into Equation 4.

$$2x - y = 40 \quad (4)$$
$$2(120) - y = 40$$
$$240 - y = 40$$
$$-y = -200$$
$$y = 200$$

Substitute $x = 120$ and $y = 200$ into Equation 1.

$$4x + 3y + 2z = 1,380 \quad (1)$$
$$4(120) + 3(200) + 2z = 1,380$$
$$1,080 + 2z = 1,380$$
$$2z = 300$$
$$z = 150$$

120 coats, 200 shirts, and 150 slacks should be made.

Section 12.3

13. NFL RECORDS

Let x = # of passes from Young, y = # of passes from Montana, and z = # of passes from Gannon.

$$\begin{cases} x = y + 30 \\ y = z + 39 \\ x + y + z = 156 \end{cases}$$

Solve the 2nd equation for z.

$$\begin{cases} x = y + 30 & (1) \\ z = y - 39 & (2) \\ x + y + z = 156 & (3) \end{cases}$$

Substitute $x = y + 30$ and $z = y - 39$ into Equation 3 and solve for y.

$$x + y + z = 156 \quad (3)$$
$$(y + 30) + (y) + (y - 39) = 156$$
$$3y - 9 = 156$$
$$3y = 165$$
$$y = 55$$

Substitute $y = 55$ into Equation 1.

$$x = y + 30 \quad (1)$$
$$x = 55 + 30$$
$$x = 85$$

Substitute $y = 55$ into Equation 2.

$$z = y - 39 \quad (2)$$
$$z = 55 - 39$$
$$z = 16$$

Rice caught 85 passes from Young, 55 passes from Montana, and 16 from Gannon.

15. EARTH'S ATMOSPHERE

Let x = % nitrogen, y = % oxygen, and z = % other gases.

$$\begin{cases} x = 12 + 3(y + z) & (1) \\ z = y - 20 & (2) \\ x + y + z = 100 & (3) \end{cases}$$

Write each equation in standard form.

$$\begin{cases} x - 3y - 3z = 12 & (1) \\ y - z = 20 & (2) \\ x + y + z = 100 & (3) \end{cases}$$

Multiply Equation 1 by -1 and add Equations 1 and 3.

$$\begin{aligned} -x + 3y + 3z &= -12 \quad (1) \\ \underline{x + y + z} &= \underline{100} \quad (2) \\ 4y + 4z &= 88 \quad (4) \end{aligned}$$

Multiply Equation 2 by 4 and add Equations 2 and 4.

$$\begin{aligned} 4y + 4z &= 88 \quad (4) \\ \underline{4y - 4z} &= \underline{80} \quad (2) \\ 8y &= 168 \\ y &= 21 \end{aligned}$$

Substitute $y = 21$ into Equation 2.

$$y - z = 20 \quad (2)$$
$$21 - z = 20$$
$$-z = -1$$
$$z = 1$$

Substitute $z = 1$ and $y = 21$ into Equation 3.

$$x + y + z = 100 \quad (3)$$
$$x + 21 + 1 = 100$$
$$x + 22 = 100$$
$$x = 78$$

The Earth's atmosphere is 78% nitrogen, 21 % oxygen, and 1% other gases.

17. TRIANGLES

$$\begin{cases} A+B+C=180 \\ A=(B+C)-100 \\ C=2B-40 \end{cases}$$

Simplify.

$$\begin{cases} A+B+C=180 & (1) \\ A-B-C=-100 & (2) \\ -2B+C=-40 & (3) \end{cases}$$

Add Equations 1 and 2.

$$A+B+C=180 \quad (1)$$
$$\underline{A-B-C=-100} \quad (2)$$
$$2A=80$$
$$A=40$$

Substitute $A=40$ into Equation 2.

$$A-B-C=-100 \quad (2)$$
$$40-B-C=-100$$
$$-B-C=-140 \quad (4)$$

Add Equations 3 and 4.

$$-2B+C=-40 \quad (3)$$
$$\underline{-B-C=-140} \quad (4)$$
$$-3B=-180$$
$$B=60$$

Substitute $B=60$ into Equation 3.

$$-2B+C=-40 \quad (3)$$
$$-2(60)+C=-40$$
$$-120+C=-40$$
$$C=80$$

The measure of the angles are
$\angle A=40°, \angle B=60°,$ and $\angle C=80°.$

19. TV HISTORY

Let $x=$ number of episodes of *X-Files*, $y=$ number of episodes of *Will & Grace*, and $z=$ number of episodes of *Seinfeld*.

$$\begin{cases} x+y+z=575 & (1) \\ x=21+z & (2) \\ y-z=14 & (3) \end{cases}$$

Solve Equation 3 for y.

$$\begin{cases} x+y+z=575 & (1) \\ x=21+z & (2) \\ y=14+z & (3) \end{cases}$$

Substitute Equations 2 and 3 into Equation 1 and solve for z.

$$x+y+z=575 \quad (1)$$
$$(21+z)+(14+z)+z=575$$
$$3z+35=575$$
$$3z=540$$
$$z=180$$

Substitute $z=180$ into Equation 2.

$$x=21+z \quad (2)$$
$$x=21+180$$
$$x=201$$

Substitute $z=180$ into Equation 3.

$$y=14+z \quad (3)$$
$$y=14+180$$
$$y=194$$

There were 201 episodes of *X-Files*, 194 episodes of *Will & Grace*, and 180 episodes of *Seinfeld*.

Section 12.3

21. ICE SKATING

Let x = radius of left circle, y = radius of middle circle, and z = radius of right circle.

$$\begin{cases} x + y = 10 & (1) \\ y + z = 14 & (2) \\ x + z = 18 & (3) \end{cases}$$

Multiply Equation 1 by –1 and add to Equation 2.

$$\begin{aligned} -x - y &= -10 \quad (1) \\ y + z &= 14 \quad (2) \\ \hline -x + z &= 4 \quad (4) \end{aligned}$$

Add Equations 3 and 4 to eliminate x.

$$\begin{aligned} x + z &= 18 \quad (3) \\ -x + z &= 4 \quad (4) \\ \hline 2z &= 22 \\ z &= 11 \end{aligned}$$

Substitute $z = 11$ into Equation 2 to solve for y.

$$\begin{aligned} y + z &= 14 \quad (2) \\ y + 11 &= 14 \\ y &= 3 \end{aligned}$$

Substitute $z = 11$ into Equation 3 to solve for x.

$$\begin{aligned} x + z &= 18 \quad (3) \\ x + 11 &= 18 \\ x &= 7 \end{aligned}$$

The radius of the circle on the left is 7 yards, the radius of the circle in the middle is 3 yards, and the radius of the circle on the right is 11 yards.

23. POTPOURRI

Let x = rose petals, y = lavender, and z = buck–wheat hulls.

$$\begin{cases} x + y + z = 10 & (1) \\ 6x + 5y + 4z = 5.5(10) & (2) \\ x = 2y & (3) \end{cases}$$

Multiply Equation 1 by -4 and add Equations 1 and 2.

$$\begin{aligned} -4x - 4y - 4z &= -40 \quad (1) \\ 6x + 5y + 4z &= 55 \quad (2) \\ \hline 2x + y &= 15 \quad (4) \end{aligned}$$

Substitute $x = 2y$ into Equation 4 and solve for y.

$$\begin{aligned} 2x + y &= 15 \quad (4) \\ 2(2y) + y &= 15 \\ 4y + y &= 15 \\ 5y &= 15 \\ y &= 3 \end{aligned}$$

Substitute $y = 3$ into Equation 3 and solve for x.

$$\begin{aligned} x &= 2y \quad (3) \\ x &= 2(3) \\ x &= 6 \end{aligned}$$

Substitute $y = 3$ and $x = 6$ into Equation 1 and solve for x.

$$\begin{aligned} x + y + z &= 10 \quad (1) \\ 6 + 3 + z &= 10 \\ 9 + z &= 10 \\ z &= 1 \end{aligned}$$

She should use 6 pounds of rose petals, 3 pounds of lavender, and 1 pound of buck–wheat hulls.

25. PIGGY BANKS

Let x = number of nickels, y = number of dimes and z = number of quarters.

$$\begin{cases} x + y + z = 64 & (1) \\ 0.05x + 0.10y + 0.25z = 6 & (2) \\ 0.10x + 0.05y + 0.25z = 5 & (3) \end{cases}$$

Multiply Equations 2 and 3 by 100 to clear decimals.

$$\begin{cases} x + y + z = 64 & (1) \\ 5x + 10y + 25z = 600 & (2) \\ 10x + 5y + 25z = 500 & (3) \end{cases}$$

Multiply Equation 1 by -5 and add Equations 1 and 2.

$$\begin{aligned} -5x - 5y - 5z &= -320 \quad (1) \\ \underline{5x + 10y + 25z} &= \underline{600} \quad (2) \\ 5y + 20z &= 280 \quad (4) \end{aligned}$$

Multiply Equation 1 by -10 and add Equations 1 and 3.

$$\begin{aligned} -10x - 10y - 10z &= -640 \quad (1) \\ \underline{10x + 5y + 25z} &= \underline{500} \quad (3) \\ -5y + 15z &= -140 \quad (5) \end{aligned}$$

Add equations 4 and 5.

$$\begin{aligned} 5y + 20z &= 280 \quad (4) \\ \underline{-5y + 15z} &= \underline{-140} \quad (5) \\ 35z &= 140 \\ z &= 4 \end{aligned}$$

Substitute $z = 4$ into Equation 5.

$$\begin{aligned} -5y + 15z &= -140 \quad (5) \\ -5y + 15(4) &= -140 \\ -5y + 60 &= -140 \\ -5y &= -200 \\ y &= 40 \end{aligned}$$

Substitute $z = 4$ and $y = 40$ into Equation 1.

$$\begin{aligned} x + y + z &= 64 \quad (1) \\ x + 40 + 4 &= 64 \\ x + 44 &= 64 \\ x &= 20 \end{aligned}$$

She had 20 nickels, 40 dimes, and 4 quarters in her piggy bank.

27. ASTRONOMY

Substitute the coordinates of $(-2, 5)$, $(2, -3)$, and $(4, -1)$ into $y = ax^2 + bx + c$.

$$\begin{cases} a(-2)^2 + b(-2) + c = 5 \\ a(2)^2 + b(2) + c = -3 \\ a(4)^2 + b(4) + c = -1 \end{cases}$$

Simplify.

$$\begin{cases} 4a - 2b + c = 5 & (1) \\ 4a + 2b + c = -3 & (2) \\ 16a + 4b + c = -1 & (3) \end{cases}$$

Add Equations 1 and 2.

$$\begin{aligned} 4a - 2b + c &= 5 \quad (1) \\ \underline{4a + 2b + c} &= \underline{-3} \quad (2) \\ 8a + 2c &= 2 \quad (4) \end{aligned}$$

Multiply Equation 1 by 2 and add Equations 1 and 3.

$$\begin{aligned} 8a - 4b + 2c &= 10 \quad (1) \\ \underline{16a + 4b + c} &= \underline{-1} \quad (2) \\ 24a + 3c &= 9 \quad (5) \end{aligned}$$

Multiply Equation 4 by -3 and add Equations 4 and 5.

$$\begin{aligned} -24a - 6c &= -6 \quad (4) \\ \underline{24a + 3c} &= \underline{9} \quad (5) \\ -3c &= 3 \\ c &= -1 \end{aligned}$$

Substitute $c = -1$ into Equation 4.

$$\begin{aligned} 8a + 2c &= 2 \quad (4) \\ 8a + 2(-1) &= 2 \\ 8a - 2 &= 2 \\ 8a &= 4 \\ a &= \frac{1}{2} \end{aligned}$$

Substitute $c = -1$ and $a = \frac{1}{2}$ into Equation 2.

$$\begin{aligned} 4a + 2b + c &= -3 \quad (2) \\ 4\left(\frac{1}{2}\right) + 2b + (-1) &= -3 \\ 2 + 2b - 1 &= -3 \\ 2b + 1 &= -3 \\ 2b &= -4 \\ b &= -2 \end{aligned}$$

The equation is $y = \frac{1}{2}x^2 - 2x - 1$.

Section 12.3

29. WALKWAYS

Substitute the coordinates of $(1, 3)$, $(3, 1)$, and $(1, -1)$ into $x^2 + y^2 + Cx + Dy + E = 0$.

$$\begin{cases} (1)^2 + (3)^2 + C(1) + D(3) + E = 0 \\ (3)^2 + (1)^2 + C(3) + D(1) + E = 0 \\ (1)^2 + (-1)^2 + C(1) + D(-1) + E = 0 \end{cases}$$

Simplify.

$$\begin{cases} 1 + 9 + C + 3D + E = 0 & (1) \\ 9 + 1 + 3C + D + E = 0 & (2) \\ 1 + 1 + C - D + E = 0 & (3) \end{cases}$$

$$\begin{cases} C + 3D + E = -10 & (1) \\ 3C + D + E = -10 & (2) \\ C - D + E = -2 & (3) \end{cases}$$

Multiply Equation 2 by -1 and add Equations 1 and 2.

$$\begin{array}{ll} C + 3D + E = -10 & (1) \\ \underline{-3C - D - E = 10} & (2) \\ -2C + 2D = 0 & (4) \end{array}$$

Multiply Equation 2 by -1 and add Equations 3 and 2.

$$\begin{array}{ll} C - D + E = -2 & (3) \\ \underline{-3C - D - E = 10} & (2) \\ -2C - 2D = 8 & (5) \end{array}$$

Add Equations 4 and 5.

$$\begin{array}{ll} -2C + 2D = 0 & (4) \\ \underline{-2C - 2D = 8} & (5) \\ -4C = 8 \\ C = -2 \end{array}$$

Substitute $C = -2$ into Equation 4.

$$\begin{array}{ll} -2C + 2D = 0 & (4) \\ -2(-2) + 2D = 0 \\ 4 + 2D = 0 \\ 2D = -4 \\ D = -2 \end{array}$$

Substitute $C = -2$ and $D = -2$ into Equation 3.

$$\begin{array}{ll} C - D + E = -2 & (3) \\ -2 - (-2) + E = -2 \\ -2 + 2 + E = -2 \\ E = -2 \end{array}$$

The equation is $x^2 + y^2 - 2x - 2y - 2 = 0$.

WRITING

31. Answers will vary.

REVIEW

33. yes

35. yes

37. no; $(1, 2)$ and $(1, -2)$

39. no; $(4, 2)$ and $(4, -2)$

CHALLENGE PROBLEMS

41. DIGITS PROBLEM

Let x = hundreds digit, y = tens digit, and z = ones digit.

$$\begin{cases} x + y + z = 8 \\ 2x + y = z \\ 100z + 10y + x = 82 + 2(100x + 10y + z) \end{cases}$$

Write the equations in standard form.

$$\begin{cases} x + y + z = 8 & (1) \\ 2x + y - z = 0 & (2) \\ 199x + 10y - 98z = -82 & (3) \end{cases}$$

Add Equations 1 and 2.

$$\begin{array}{ll} x + y + z = 8 & (1) \\ \underline{2x + y - z = 0} & (2) \\ 3x + 2y = 8 & (4) \end{array}$$

Multiply Equation 1 by 98 and add Equations 1 and 3.

$$\begin{array}{ll} 98x + 98y + 98z = 784 & (1) \\ \underline{199x + 10y - 98z = -82} & (3) \\ 297x + 108y = 702 & (5) \end{array}$$

Multiply Equation 4 by –99 and add Equations 4 and 5.

$$\begin{array}{ll} -297x - 198y = -792 & (4) \\ \underline{297x + 108y = 702} & (5) \\ -90y = -90 & \\ y = 1 & \end{array}$$

Substitute $y = 1$ into Equation 4.

$$3x + 2y = 8 \quad (4)$$
$$3x + 2(1) = 8$$
$$3x + 2 = 8$$
$$3x = 6$$
$$x = 2$$

Substitute $x = 2$ and $y = 1$ into Equation 1.

$$x + y + z = 8 \quad (1)$$
$$2 + 1 + z = 8$$
$$3 + z = 8$$
$$z = 5$$

The number would be 215.

SECTION 12.4

VOCABULARY

1. A **matrix** is a rectangular array of numbers written within brackets.

3. Of the order of a matrix is 3x4, it had 3 **rows** and 4 **columns**. We read 3x4 as "3 **by** 4."

5. Elementary **row** operations are used on the augmented matrix to produce a simpler matrix that gives to the solution of a system. This process is called **Gauss-Jordan**.

CONCEPTS

7. a. 2×3
   b. 3×4

9. a. $\begin{cases} x = -10 \\ y = 6 \end{cases}$; The solution of the system

   is $\left(\boxed{10}, \boxed{6} \right)$.

   b. $\begin{cases} x = -16 \\ y = 8 \\ z = 4 \end{cases}$; The solution of the system

   is $\left(\boxed{-16}, \boxed{8}, \boxed{4} \right)$.

NOTATION

11. a. interchange rows 1 and 2

    b. multiply row 1 by $\dfrac{1}{2}$

    c. add row 3 to 6 times row 2

GUIDED PRACTICE

13. $\left[\begin{array}{cc|c} 1 & 2 & 6 \\ 3 & -1 & -10 \end{array} \right]$

15. $\begin{cases} x + 6y = 7 \\ \qquad y = 4 \end{cases}$

17. $\left[\begin{array}{cc|c} 1 & -4 & 4 \\ -3 & 1 & -6 \end{array} \right]$

19. $\left[\begin{array}{cc|c} 1 & -\dfrac{1}{3} & 2 \\ 1 & -4 & 4 \end{array} \right]$

21. $\left[\begin{array}{ccc|c} 3 & 6 & -9 & 0 \\ -2 & 2 & -2 & 5 \\ 1 & 5 & -2 & 1 \end{array} \right]$

23. $\left[\begin{array}{ccc|c} 1 & 2 & -3 & 0 \\ 0 & 3 & 1 & 1 \\ -2 & 2 & -2 & 5 \end{array} \right]$

25.

$\left[\begin{array}{cc|c} 1 & 1 & 2 \\ 1 & -1 & 0 \end{array} \right]$

$-R_1 + R_2$

$\left[\begin{array}{cc|c} 1 & 1 & 2 \\ 0 & -2 & -2 \end{array} \right]$

$-\dfrac{1}{2}R_2$

$\left[\begin{array}{cc|c} 1 & 1 & 2 \\ 0 & 1 & 1 \end{array} \right]$

$R_1 - R_2$

$\left[\begin{array}{cc|c} 1 & 0 & 1 \\ 0 & 1 & 1 \end{array} \right]$

This matrix represents the system

$\begin{cases} x = 1 \\ y = 1 \end{cases}$

The solution is $(1, 1)$.

27.

$$\begin{bmatrix} 2 & 1 & | & 1 \\ 1 & 2 & | & -4 \end{bmatrix}$$

$R_1 \leftrightarrow R_2$

$$\begin{bmatrix} 1 & 2 & | & -4 \\ 2 & 1 & | & 1 \end{bmatrix}$$

$-2R_1 + R_2$

$$\begin{bmatrix} 1 & 2 & | & -4 \\ 0 & -3 & | & 9 \end{bmatrix}$$

$-\dfrac{1}{3} R_2$

$$\begin{bmatrix} 1 & 2 & | & -4 \\ 0 & 1 & | & -3 \end{bmatrix}$$

$-2R_2 + R_1$

$$\begin{bmatrix} 1 & 0 & | & 2 \\ 0 & 1 & | & -3 \end{bmatrix}$$

This matrix represents the system

$$\begin{cases} x = -2 \\ y = -3 \end{cases}$$

The solution is $(2, -3)$.

29.

$$\begin{bmatrix} 1 & 1 & 1 & | & 6 \\ 1 & 2 & 1 & | & 8 \\ 1 & 1 & 2 & | & 7 \end{bmatrix}$$

$-1R_1 + R_2$

$$\begin{bmatrix} 1 & 1 & 1 & | & 6 \\ 0 & 1 & 0 & | & 2 \\ 1 & 1 & 2 & | & 7 \end{bmatrix}$$

$-1R_1 + R_3$

$$\begin{bmatrix} 1 & 1 & 1 & | & 6 \\ 0 & 1 & 0 & | & 2 \\ 0 & 0 & 1 & | & 1 \end{bmatrix}$$

$-1R_2 + R_1$

$$\begin{bmatrix} 1 & 0 & 1 & | & 4 \\ 0 & 1 & 0 & | & 2 \\ 0 & 0 & 1 & | & 1 \end{bmatrix}$$

$-1R_3 + R_1$

$$\begin{bmatrix} 1 & 0 & 0 & | & 3 \\ 0 & 1 & 0 & | & 2 \\ 0 & 0 & 1 & | & 1 \end{bmatrix}$$

This matrix represents the system

$$\begin{cases} x = 3 \\ y = 2 \\ z = 1 \end{cases}$$

The solution is $(3, 2, 1)$.

Section 12.4

31.

$$\begin{bmatrix} 3 & 1 & -3 & \vdots & 5 \\ 1 & -2 & 4 & \vdots & 10 \\ 1 & 1 & 1 & \vdots & 13 \end{bmatrix}$$

$R_1 \leftrightarrow R_3$

$$\begin{bmatrix} 1 & 1 & 1 & \vdots & 13 \\ 1 & -2 & 4 & \vdots & 10 \\ 3 & 1 & -3 & \vdots & 5 \end{bmatrix}$$

$-1R_1 + R_2$

$$\begin{bmatrix} 1 & 1 & 1 & \vdots & 13 \\ 0 & -3 & 3 & \vdots & -3 \\ 3 & 1 & -3 & \vdots & 5 \end{bmatrix}$$

$-\dfrac{1}{3}R_2$

$$\begin{bmatrix} 1 & 1 & 1 & \vdots & 13 \\ 0 & 1 & -1 & \vdots & 1 \\ 3 & 1 & -3 & \vdots & 5 \end{bmatrix}$$

$-3R_1 + R_3$

$$\begin{bmatrix} 1 & 1 & 1 & \vdots & 13 \\ 0 & 1 & -1 & \vdots & 1 \\ 0 & -2 & -6 & \vdots & -34 \end{bmatrix}$$

$2R_2 + R_3$

$$\begin{bmatrix} 1 & 1 & 1 & \vdots & 13 \\ 0 & 1 & -1 & \vdots & 1 \\ 0 & 0 & -8 & \vdots & -32 \end{bmatrix}$$

$-\dfrac{1}{8}R_3$

$$\begin{bmatrix} 1 & 1 & 1 & \vdots & 13 \\ 0 & 1 & -1 & \vdots & 1 \\ 0 & 0 & 1 & \vdots & 4 \end{bmatrix}$$

$-R_2 + R_1$

$$\begin{bmatrix} 1 & 0 & 2 & \vdots & 12 \\ 0 & 1 & -1 & \vdots & 1 \\ 0 & 0 & 1 & \vdots & 4 \end{bmatrix}$$

$-2R_3 + R_1$

$$\begin{bmatrix} 1 & 0 & 0 & \vdots & 4 \\ 0 & 1 & -1 & \vdots & 1 \\ 0 & 0 & 1 & \vdots & 4 \end{bmatrix}$$

$R_3 + R_2$

$$\begin{bmatrix} 1 & 0 & 0 & \vdots & 4 \\ 0 & 1 & 0 & \vdots & 5 \\ 0 & 0 & 1 & \vdots & 4 \end{bmatrix}$$

The solution is $(4, 5, 4)$.

33.

$$\begin{bmatrix} 1 & -3 & \vdots & 9 \\ -2 & 6 & \vdots & 18 \end{bmatrix}$$

$2R_1 + R_2$

$$\begin{bmatrix} 1 & -3 & \vdots & 9 \\ 0 & 0 & \vdots & 36 \end{bmatrix}$$

This matrix represents the system

$$\begin{cases} x - 3y = 9 \\ 0 + 0 = 36 \end{cases}$$

No solution. The system is inconsistent.

35.

$$\begin{bmatrix} -4 & -4 & \vdots & -12 \\ 1 & 1 & \vdots & 3 \end{bmatrix}$$

$\dfrac{1}{4}R_1$

$$\begin{bmatrix} -1 & -1 & \vdots & -3 \\ 1 & 1 & \vdots & 3 \end{bmatrix}$$

$R_1 \leftrightarrow R_2$

$$\begin{bmatrix} 1 & 1 & \vdots & 3 \\ -1 & -1 & \vdots & -3 \end{bmatrix}$$

$R_1 + R_2$

$$\begin{bmatrix} 1 & 1 & \vdots & 3 \\ 0 & 0 & \vdots & 0 \end{bmatrix}$$

This matrix represents the system

$$\begin{cases} x + y = 3 \\ 0 + 0 = 0 \end{cases}$$

$$\{(x, y) \mid x + y = 3\}$$

Infinitely many solutions.

The equations are dependent.

37.

$$\begin{bmatrix} 2 & 3 & -1 & \vdots & -8 \\ 1 & -1 & -1 & \vdots & -2 \\ -4 & 3 & 1 & \vdots & 6 \end{bmatrix}$$

$R_1 \leftrightarrow R_2$

$$\begin{bmatrix} 1 & -1 & -1 & \vdots & -2 \\ 2 & 3 & -1 & \vdots & -8 \\ -4 & 3 & 1 & \vdots & 6 \end{bmatrix}$$

$-2R_1 + R_2$

$$\begin{bmatrix} 1 & -1 & -1 & \vdots & -2 \\ 0 & 5 & 1 & \vdots & -4 \\ -4 & 3 & 1 & \vdots & 6 \end{bmatrix}$$

$4R_1 + R_3$

$$\begin{bmatrix} 1 & -1 & -1 & \vdots & -2 \\ 0 & 5 & 1 & \vdots & -4 \\ 0 & -1 & -3 & \vdots & -2 \end{bmatrix}$$

$-R_3 \leftrightarrow R_2$

$$\begin{bmatrix} 1 & -1 & -1 & \vdots & -2 \\ 0 & 1 & 3 & \vdots & 2 \\ 0 & 5 & 1 & \vdots & -4 \end{bmatrix}$$

$-5R_2 + R_3$

$$\begin{bmatrix} 1 & -1 & -1 & \vdots & -2 \\ 0 & 1 & 3 & \vdots & 2 \\ 0 & 0 & -14 & \vdots & -14 \end{bmatrix}$$

$-\dfrac{1}{14}R_3$

$$\begin{bmatrix} 1 & -1 & -1 & \vdots & -2 \\ 0 & 1 & 3 & \vdots & 2 \\ 0 & 0 & 1 & \vdots & 1 \end{bmatrix}$$

$R_2 + R_1$

$$\begin{bmatrix} 1 & 0 & 2 & \vdots & 0 \\ 0 & 1 & 3 & \vdots & 2 \\ 0 & 0 & 1 & \vdots & 1 \end{bmatrix}$$

$-2R_3 + R_1$

$$\begin{bmatrix} 1 & 0 & 0 & \vdots & -2 \\ 0 & 1 & 3 & \vdots & 2 \\ 0 & 0 & 1 & \vdots & 1 \end{bmatrix}$$

$-3R_3 + R_2$

$$\begin{bmatrix} 1 & 0 & 0 & \vdots & -2 \\ 0 & 1 & 0 & \vdots & -1 \\ 0 & 0 & 1 & \vdots & 1 \end{bmatrix}$$

The solution is $\left(-2, -1,\ 1\right)$.

39.

$$\begin{bmatrix} 2 & -1 & \vdots & -1 \\ 1 & -2 & \vdots & 1 \end{bmatrix}$$

$R_1 \leftrightarrow R_2$

$$\begin{bmatrix} 1 & -2 & \vdots & 1 \\ 2 & -1 & \vdots & -1 \end{bmatrix}$$

$-2R_1 + R_2$

$$\begin{bmatrix} 1 & -2 & \vdots & 1 \\ 0 & 3 & \vdots & -3 \end{bmatrix}$$

$\dfrac{1}{3}R_2$

$$\begin{bmatrix} 1 & -2 & \vdots & 1 \\ 0 & 1 & \vdots & -1 \end{bmatrix}$$

$2R_2 + R_1$

$$\begin{bmatrix} 1 & 0 & \vdots & -1 \\ 0 & 1 & \vdots & -1 \end{bmatrix}$$

The solution is $\left(-1,\ -1\right)$.

41.

$$\begin{bmatrix} 3 & 4 & \vdots & -12 \\ 9 & -2 & \vdots & 6 \end{bmatrix}$$

$\dfrac{1}{3}R_1$

$$\begin{bmatrix} 1 & \frac{4}{3} & \vdots & -4 \\ 9 & -2 & \vdots & 6 \end{bmatrix}$$

$-9R_1 + R_2$

$$\begin{bmatrix} 1 & \frac{4}{3} & \vdots & -4 \\ 0 & -14 & \vdots & 42 \end{bmatrix}$$

$-\dfrac{1}{14}R_2$

$$\begin{bmatrix} 1 & \frac{4}{3} & \vdots & -4 \\ 0 & 1 & \vdots & -3 \end{bmatrix}$$

$-\dfrac{4}{3}R_2 + R_1$

$$\begin{bmatrix} 1 & 0 & \vdots & 0 \\ 0 & 1 & \vdots & -3 \end{bmatrix}$$

The solution is $\left(0,\ -3\right)$.

43.

$$\begin{bmatrix} 2 & 1 & -1 & \vdots & 1 \\ 1 & 2 & 2 & \vdots & 2 \\ 4 & 5 & 3 & \vdots & 3 \end{bmatrix}$$

$R_1 \leftrightarrow R_2$

$$\begin{bmatrix} 1 & 2 & 2 & \vdots & 2 \\ 2 & 1 & -1 & \vdots & 1 \\ 4 & 5 & 3 & \vdots & 3 \end{bmatrix}$$

$-2R_1 + R_2$

$$\begin{bmatrix} 1 & 2 & 2 & \vdots & 2 \\ 0 & -3 & -5 & \vdots & -3 \\ 4 & 5 & 3 & \vdots & 3 \end{bmatrix}$$

$-\dfrac{1}{3}R_2$

$$\begin{bmatrix} 1 & 2 & 2 & \vdots & 2 \\ 0 & 1 & \frac{5}{3} & \vdots & 1 \\ 4 & 5 & 3 & \vdots & 3 \end{bmatrix}$$

$-4R_1 + R_3$

$$\begin{bmatrix} 1 & 2 & 2 & \vdots & 2 \\ 0 & 1 & \frac{5}{3} & \vdots & 1 \\ 0 & -3 & -5 & \vdots & -5 \end{bmatrix}$$

$3R_2 + R_3$

$$\begin{bmatrix} 1 & 2 & 2 & \vdots & 2 \\ 0 & 1 & \frac{5}{3} & \vdots & 1 \\ 0 & 0 & 0 & \vdots & -2 \end{bmatrix}$$

This system is inconsistent.

No solution.

45.

$$\begin{bmatrix} 8 & -2 & \vdots & 4 \\ 4 & -1 & \vdots & 2 \end{bmatrix}$$

$\dfrac{1}{8}R_1$

$$\begin{bmatrix} 1 & -\frac{1}{4} & \vdots & \frac{1}{2} \\ 4 & -1 & \vdots & 2 \end{bmatrix}$$

$-4R_1 + R_2$

$$\begin{bmatrix} 1 & -\frac{1}{4} & \vdots & \frac{1}{2} \\ 0 & 0 & \vdots & 0 \end{bmatrix}$$

Infinitely many solutions.

The equations are dependent.

47.

$$\begin{bmatrix} 2 & 1 & -2 & \vdots & 6 \\ 4 & -1 & 1 & \vdots & -1 \\ 6 & -2 & 3 & \vdots & -5 \end{bmatrix}$$

$\dfrac{1}{2}R_1$

$$\begin{bmatrix} 1 & \frac{1}{2} & -1 & \vdots & 3 \\ 4 & -1 & 1 & \vdots & -1 \\ 6 & -2 & 3 & \vdots & -5 \end{bmatrix}$$

$-4R_1 + R_2$

$$\begin{bmatrix} 1 & \frac{1}{2} & -1 & \vdots & 3 \\ 0 & -3 & 5 & \vdots & -13 \\ 6 & -2 & 3 & \vdots & -5 \end{bmatrix}$$

$-6R_1 + R_3$

$$\begin{bmatrix} 1 & \frac{1}{2} & -1 & \vdots & 3 \\ 0 & -3 & 5 & \vdots & -13 \\ 0 & -5 & 9 & \vdots & -23 \end{bmatrix}$$

$-\dfrac{1}{3}R_2$

$$\begin{bmatrix} 1 & \frac{1}{2} & -1 & \vdots & 3 \\ 0 & 1 & -\frac{5}{3} & \vdots & \frac{13}{3} \\ 0 & -5 & 9 & \vdots & -23 \end{bmatrix}$$

$5R_2 + R_3$

$$\begin{bmatrix} 1 & \frac{1}{2} & -1 & \vdots & 3 \\ 0 & 1 & -\frac{5}{3} & \vdots & \frac{13}{3} \\ 0 & 0 & \frac{2}{3} & \vdots & -\frac{4}{3} \end{bmatrix}$$

$\dfrac{3}{2}R_3$

$$\begin{bmatrix} 1 & \frac{1}{2} & -1 & \vdots & 3 \\ 0 & 1 & -\frac{5}{3} & \vdots & \frac{13}{3} \\ 0 & 0 & 1 & \vdots & -2 \end{bmatrix}$$

$-\dfrac{1}{2}R_2 + R_1$

$$\begin{bmatrix} 1 & 0 & -\frac{1}{6} & \vdots & \frac{5}{6} \\ 0 & 1 & -\frac{5}{3} & \vdots & \frac{13}{3} \\ 0 & 0 & 1 & \vdots & -2 \end{bmatrix}$$

$\dfrac{1}{6}R_3 + R_1$

$$\begin{bmatrix} 1 & 0 & 0 & \vdots & \frac{1}{2} \\ 0 & 1 & -\frac{5}{3} & \vdots & \frac{13}{3} \\ 0 & 0 & 1 & \vdots & -2 \end{bmatrix}$$

$\dfrac{5}{3}R_3 + R_2$

$$\begin{bmatrix} 1 & 0 & 0 & \vdots & \frac{1}{2} \\ 0 & 1 & 0 & \vdots & 1 \\ 0 & 0 & 1 & \vdots & -2 \end{bmatrix}$$

The solution is $\left(\dfrac{1}{2}, 1, -2 \right)$.

49.

$$\begin{bmatrix} 6 & 1 & -1 & \vdots & -2 \\ 1 & 2 & 1 & \vdots & 5 \\ 0 & 5 & -1 & \vdots & 2 \end{bmatrix}$$

$R_1 \leftrightarrow R_2$

$$\begin{bmatrix} 1 & 2 & 1 & \vdots & 5 \\ 6 & 1 & -1 & \vdots & -2 \\ 0 & 5 & -1 & \vdots & 2 \end{bmatrix}$$

$-6R_1 + R_2$

$$\begin{bmatrix} 1 & 2 & 1 & \vdots & 5 \\ 0 & -11 & -7 & \vdots & -32 \\ 0 & 5 & -1 & \vdots & 2 \end{bmatrix}$$

$-\dfrac{1}{11}R_2$

$$\begin{bmatrix} 1 & 2 & 1 & \vdots & 5 \\ 0 & 1 & \frac{7}{11} & \vdots & \frac{32}{11} \\ 0 & 5 & -1 & \vdots & 2 \end{bmatrix}$$

$-5R_2 + R_3$

$$\begin{bmatrix} 1 & 2 & 1 & \vdots & 5 \\ 0 & 1 & \frac{7}{11} & \vdots & \frac{32}{11} \\ 0 & 0 & -\frac{46}{11} & \vdots & -\frac{138}{11} \end{bmatrix}$$

$-\dfrac{11}{46}R_3$

$$\begin{bmatrix} 1 & 2 & 1 & \vdots & 5 \\ 0 & 1 & \frac{7}{11} & \vdots & \frac{32}{11} \\ 0 & 0 & 1 & \vdots & 3 \end{bmatrix}$$

$-\dfrac{7}{11}R_3 + R_2$

$$\begin{bmatrix} 1 & 2 & 1 & \vdots & 5 \\ 0 & 1 & 0 & \vdots & 1 \\ 0 & 0 & 1 & \vdots & 3 \end{bmatrix}$$

$-2R_2 + R_1$

$$\begin{bmatrix} 1 & 0 & 1 & \vdots & 3 \\ 0 & 1 & 0 & \vdots & 1 \\ 0 & 0 & 1 & \vdots & 3 \end{bmatrix}$$

$-R_3 + R_1$

$$\begin{bmatrix} 1 & 0 & 0 & \vdots & 0 \\ 0 & 1 & 0 & \vdots & 1 \\ 0 & 0 & 1 & \vdots & 3 \end{bmatrix}$$

The solution is $(0, 1, 3)$.

51.

$$\begin{bmatrix} 5 & 3 & 0 & \vdots & 4 \\ 0 & 3 & -4 & \vdots & 4 \\ 1 & 0 & 1 & \vdots & 1 \end{bmatrix}$$

$\dfrac{1}{5}R_1$

$$\begin{bmatrix} 1 & \frac{3}{5} & 0 & \vdots & \frac{4}{5} \\ 0 & 3 & -4 & \vdots & 4 \\ 1 & 0 & 1 & \vdots & 1 \end{bmatrix}$$

$-R_1 + R_3$

$$\begin{bmatrix} 1 & \frac{3}{5} & 0 & \vdots & \frac{4}{5} \\ 0 & 3 & -4 & \vdots & 4 \\ 0 & -\frac{3}{5} & 1 & \vdots & \frac{1}{5} \end{bmatrix}$$

$\dfrac{1}{3}R_2$

$$\begin{bmatrix} 1 & \frac{3}{5} & 0 & \vdots & \frac{4}{5} \\ 0 & 1 & -\frac{4}{3} & \vdots & \frac{4}{3} \\ 0 & -\frac{3}{5} & 1 & \vdots & \frac{1}{5} \end{bmatrix}$$

$\dfrac{3}{5}R_2 + R_3$

$$\begin{bmatrix} 1 & \frac{3}{5} & 0 & \vdots & \frac{4}{5} \\ 0 & 1 & -\frac{4}{3} & \vdots & \frac{4}{3} \\ 0 & 0 & \frac{1}{5} & \vdots & 1 \end{bmatrix}$$

$5R_3$

$$\begin{bmatrix} 1 & \frac{3}{5} & 0 & \vdots & \frac{4}{5} \\ 0 & 1 & -\frac{4}{3} & \vdots & \frac{4}{3} \\ 0 & 0 & 1 & \vdots & 5 \end{bmatrix}$$

$\dfrac{4}{3}R_3 + R_2$

$$\begin{bmatrix} 1 & \frac{3}{5} & 0 & \vdots & \frac{4}{5} \\ 0 & 1 & 0 & \vdots & 8 \\ 0 & 0 & 1 & \vdots & 5 \end{bmatrix}$$

$-\dfrac{3}{5}R_2 + R_1$

$$\begin{bmatrix} 1 & 0 & 0 & \vdots & -4 \\ 0 & 1 & 0 & \vdots & 8 \\ 0 & 0 & 1 & \vdots & 5 \end{bmatrix}$$

The solution is $(-4, 8, 5)$.

APPLICATIONS

53. DIGITAL PHOTOGRAPHY

$512(512) = 262{,}144$

55. COMPLEMENTARY ANGLES

$$\begin{cases} x + y = 90 \\ y = x + 46 \end{cases}$$

Write the equations in standard form.

$$\begin{cases} x + y = 90 \\ -x + y = 46 \end{cases}$$

$$\begin{bmatrix} 1 & 1 & \vdots & 90 \\ -1 & 1 & \vdots & 46 \end{bmatrix}$$

$R_1 + R_2 = R_2$

$$\begin{bmatrix} 1 & 1 & \vdots & 90 \\ 0 & 2 & \vdots & 136 \end{bmatrix}$$

$\dfrac{1}{2} R_2$

$$\begin{bmatrix} 1 & 1 & \vdots & 90 \\ 0 & 1 & \vdots & 68 \end{bmatrix}$$

$-R_2 + R_1$

$$\begin{bmatrix} 1 & 0 & \vdots & 22 \\ 0 & 1 & \vdots & 68 \end{bmatrix}$$

The measures of the angles are 22° and 68°.

57. TRIANGLES

$$\begin{cases} A + B + C = 180 \\ B = 25 + A \\ C = 2A - 5 \end{cases}$$

Write the equations in standard form.

$$\begin{cases} A + B + C = 180 \\ A - B = -25 \\ 2A - C = 5 \end{cases}$$

$$\begin{bmatrix} 1 & 1 & 1 & \vdots & 180 \\ 1 & -1 & 0 & \vdots & -25 \\ 2 & 0 & -1 & \vdots & 5 \end{bmatrix}$$

$-R_1 + R_2$

$$\begin{bmatrix} 1 & 1 & 1 & \vdots & 180 \\ 0 & -2 & -1 & \vdots & -205 \\ 2 & 0 & -1 & \vdots & 5 \end{bmatrix}$$

$-\dfrac{1}{2} R_2$

$$\begin{bmatrix} 1 & 1 & 1 & \vdots & 180 \\ 0 & 1 & \frac{1}{2} & \vdots & 102.5 \\ 2 & 0 & -1 & \vdots & 5 \end{bmatrix}$$

$-2R_1 + R_3$

$$\begin{bmatrix} 1 & 1 & 1 & \vdots & 180 \\ 0 & 1 & \frac{1}{2} & \vdots & 102.5 \\ 0 & -2 & -3 & \vdots & -355 \end{bmatrix}$$

$2R_2 + R_3$

$$\begin{bmatrix} 1 & 1 & 1 & \vdots & 180 \\ 0 & 1 & \frac{1}{2} & \vdots & 102.5 \\ 0 & 0 & -2 & \vdots & -150 \end{bmatrix}$$

$-\dfrac{1}{2} R_3$

$$\begin{bmatrix} 1 & 1 & 1 & \vdots & 180 \\ 0 & 1 & \frac{1}{2} & \vdots & 102.5 \\ 0 & 0 & 1 & \vdots & 75 \end{bmatrix}$$

$-\dfrac{1}{2} R_3 + R_2$

$$\begin{bmatrix} 1 & 1 & 1 & \vdots & 180 \\ 0 & 1 & 0 & \vdots & 65 \\ 0 & 0 & 1 & \vdots & 75 \end{bmatrix}$$

$-R_2 + R_1$

$$\begin{bmatrix} 1 & 0 & 1 & \vdots & 115 \\ 0 & 1 & 0 & \vdots & 65 \\ 0 & 0 & 1 & \vdots & 75 \end{bmatrix}$$

$-R_3 + R_1$

$$\begin{bmatrix} 1 & 0 & 0 & \vdots & 40 \\ 0 & 1 & 0 & \vdots & 65 \\ 0 & 0 & 1 & \vdots & 75 \end{bmatrix}$$

The measures of the angles are 40°, 65°, and 75°.

Section 12.4

WRITING

59. Answers will vary.

61. Answers will vary.

REVIEW

63. $m = \dfrac{y_2 - y_1}{x_2 - x_1}\left(x_2 \neq x_1\right)$

65. $y - y_1 = m\left(x - x_1\right)$

CHALLENGE PROBLEMS

67. Let $a = x^2$, $b = y^2$, and $c = z^2$.

$$\begin{bmatrix} 1 & 1 & 1 & | & 14 \\ 2 & 3 & -2 & | & -7 \\ 1 & -5 & 1 & | & 8 \end{bmatrix}$$

$-2R_1 + R_2$

$$\begin{bmatrix} 1 & 1 & 1 & | & 14 \\ 0 & 1 & -4 & | & -35 \\ 1 & -5 & 1 & | & 8 \end{bmatrix}$$

$-R_1 + R_3$

$$\begin{bmatrix} 1 & 1 & 1 & | & 14 \\ 0 & 1 & -4 & | & -35 \\ 0 & -6 & 0 & | & -6 \end{bmatrix}$$

$-\dfrac{1}{6}R_3$

$$\begin{bmatrix} 1 & 1 & 1 & | & 14 \\ 0 & 1 & -4 & | & -35 \\ 0 & 1 & 0 & | & 1 \end{bmatrix}$$

$R_2 \leftrightarrow R_3$

$$\begin{bmatrix} 1 & 1 & 1 & | & 14 \\ 0 & 1 & 0 & | & 1 \\ 0 & 1 & -4 & | & -35 \end{bmatrix}$$

$-R_2 + R_3$

$$\begin{bmatrix} 1 & 1 & 1 & | & 14 \\ 0 & 1 & 0 & | & 1 \\ 0 & 0 & -4 & | & -36 \end{bmatrix}$$

$-\dfrac{1}{4}R_3$

$$\begin{bmatrix} 1 & 1 & 1 & | & 14 \\ 0 & 1 & 0 & | & 1 \\ 0 & 0 & 1 & | & 9 \end{bmatrix}$$

$-R_2 + R_1$

$$\begin{bmatrix} 1 & 0 & 1 & | & 13 \\ 0 & 1 & 0 & | & 1 \\ 0 & 0 & 1 & | & 9 \end{bmatrix}$$

$-R_3 + R_1$

$$\begin{bmatrix} 1 & 0 & 0 & | & 4 \\ 0 & 1 & 0 & | & 1 \\ 0 & 0 & 1 & | & 9 \end{bmatrix}$$

Use substitution to solve for x, y, and z.

$x^2 = a$

$x^2 = 4$

$x^2 = \pm 2$

$y^2 = b$

$y^2 = 1$

$y = \pm 1$

$z^2 = c$

$z^2 = 9$

$z = \pm 3$

The solution is $\left(\pm 2, \pm 1, \pm 3\right)$.

VOCABULARY

1. $\begin{vmatrix} 4 & 9 \\ -6 & 1 \end{vmatrix}$ is a 2×2 **determinant**. The numbers 4 and 1 lie along its main **diagonal**.

3. The **minor** of b_1 in $\begin{vmatrix} a_1 & b_1 & c_1 \\ a_2 & b_2 & c_2 \\ a_3 & b_3 & c_3 \end{vmatrix}$ is $\begin{vmatrix} a_2 & c_2 \\ a_3 & c_3 \end{vmatrix}$.

CONCEPTS

5. $\begin{vmatrix} a & b \\ c & d \end{vmatrix} = \boxed{ad} - \boxed{bc}$

7. It was expanded about the third column

9. $\begin{vmatrix} 1 & 2 & 0 \\ 3 & 1 & -1 \\ 8 & 4 & -1 \end{vmatrix}$

11. $\left(\dfrac{-28}{14}, \dfrac{-14}{14}, \dfrac{14}{14} \right) = (-2, -1, 1)$

NOTATION

13.
$$\begin{vmatrix} 5 & -2 \\ -2 & 6 \end{vmatrix} = 5\left(\boxed{6}\right) - (-2)(-2)$$
$$= \boxed{30} - 4$$
$$= 26$$

PRACTICE

15.
$$\begin{vmatrix} 2 & 3 \\ 2 & 5 \end{vmatrix} = 2(5) - 3(2)$$
$$= 10 - 6$$
$$= 4$$

17.
$$\begin{vmatrix} -9 & 7 \\ 4 & -2 \end{vmatrix} = -9(-2) - 7(4)$$
$$= 18 - 28$$
$$= -10$$

19.
$$\begin{vmatrix} 5 & 20 \\ 10 & 6 \end{vmatrix} = 5(6) - 20(10)$$
$$= 30 - 200$$
$$= -170$$

21.
$$\begin{vmatrix} -6 & -2 \\ 15 & 4 \end{vmatrix} = -6(4) - (-2)(15)$$
$$= -24 - (-30)$$
$$= 6$$

23.
$$\begin{vmatrix} -9 & -1 \\ -10 & -5 \end{vmatrix} = -9(-5) - (-1)(-10)$$
$$= 45 - 10$$
$$= 35$$

25.
$$\begin{vmatrix} 8 & 8 \\ -9 & -9 \end{vmatrix} = 8(-9) - 8(-9)$$
$$= -72 + 72$$
$$= 0$$

27.
$$\begin{vmatrix} 3 & 2 & 1 \\ 4 & 1 & 2 \\ 5 & 3 & 1 \end{vmatrix} = 3\begin{vmatrix} 1 & 2 \\ 3 & 1 \end{vmatrix} - 2\begin{vmatrix} 4 & 2 \\ 5 & 1 \end{vmatrix} + 1\begin{vmatrix} 4 & 1 \\ 5 & 3 \end{vmatrix}$$
$$= 3(1 - 6) - 2(4 - 10) + 1(12 - 5)$$
$$= 3(-5) - 2(-6) + 1(7)$$
$$= -15 + 12 + 7$$
$$= 4$$

29.

$$\begin{vmatrix} 1 & -2 & 3 \\ -2 & 1 & 1 \\ -3 & -2 & 1 \end{vmatrix} = 1\begin{vmatrix} 1 & 1 \\ -2 & 1 \end{vmatrix} - (-2)\begin{vmatrix} -2 & 1 \\ -3 & 1 \end{vmatrix} + 3\begin{vmatrix} -2 & 1 \\ -3 & -2 \end{vmatrix}$$

$$= 1(1+2) + 2(-2+3) + 3(4+3)$$
$$= 1(3) + 2(1) + 3(7)$$
$$= 3 + 2 + 21$$
$$= 26$$

31.

$$\begin{vmatrix} -2 & 5 & 1 \\ 0 & 3 & 4 \\ -1 & 2 & 6 \end{vmatrix} = -2\begin{vmatrix} 3 & 4 \\ 2 & 6 \end{vmatrix} - 5\begin{vmatrix} 0 & 4 \\ -1 & 6 \end{vmatrix} + 1\begin{vmatrix} 0 & 3 \\ -1 & 2 \end{vmatrix}$$

$$= -2(18-8) - 5(0+4) + 1(0+3)$$
$$= -2(10) - 5(4) + 1(3)$$
$$= -20 - 20 + 3$$
$$= -37$$

33.

$$\begin{vmatrix} 1 & -4 & 1 \\ 3 & 0 & -2 \\ 3 & 1 & -2 \end{vmatrix} = 1\begin{vmatrix} 0 & -2 \\ 1 & -2 \end{vmatrix} - (-4)\begin{vmatrix} 3 & -2 \\ 3 & -2 \end{vmatrix} + 1\begin{vmatrix} 3 & 0 \\ 3 & 1 \end{vmatrix}$$

$$= 1(0+2) + 4(-6+6) + 1(3-0)$$
$$= 1(2) + 4(0) + 1(3)$$
$$= 2 + 0 + 3$$
$$= 5$$

35.

$$\begin{vmatrix} 1 & 2 & 1 \\ -3 & 7 & 3 \\ -4 & 3 & -5 \end{vmatrix} = 1\begin{vmatrix} 7 & 3 \\ 3 & -5 \end{vmatrix} - 2\begin{vmatrix} -3 & 3 \\ -4 & -5 \end{vmatrix} + 1\begin{vmatrix} -3 & 7 \\ -4 & 3 \end{vmatrix}$$

$$= 1(-35-9) - 2(15+12) + 1(-9+28)$$
$$= 1(-44) - 2(27) + 1(19)$$
$$= -44 - 54 + 19$$
$$= -79$$

37.

$$\begin{vmatrix} 1 & 2 & 0 \\ 0 & 1 & 2 \\ 0 & 0 & 1 \end{vmatrix} = 1\begin{vmatrix} 1 & 2 \\ 0 & 1 \end{vmatrix} - 2\begin{vmatrix} 0 & 2 \\ 0 & 1 \end{vmatrix} + 0\begin{vmatrix} 0 & 1 \\ 0 & 0 \end{vmatrix}$$

$$= 1(1-0) - 2(0-0) + 0(0-0)$$
$$= 1(1) - 2(0) + 0(0)$$
$$= 1 - 0 + 0$$
$$= 1$$

39.

$$x = \frac{D_x}{D} \qquad\qquad y = \frac{D_y}{D}$$

$$= \frac{\begin{vmatrix} 6 & 1 \\ 2 & -1 \end{vmatrix}}{\begin{vmatrix} 1 & 1 \\ 1 & -1 \end{vmatrix}} \qquad\qquad = \frac{\begin{vmatrix} 1 & 6 \\ 1 & 2 \end{vmatrix}}{\begin{vmatrix} 1 & 1 \\ 1 & -1 \end{vmatrix}}$$

$$= \frac{6(-1)-1(2)}{1(-1)-1(1)} \qquad = \frac{1(2)-6(1)}{1(-1)-1(1)}$$

$$= \frac{-6-2}{-1-1} \qquad\qquad = \frac{2-6}{-1-1}$$

$$= \frac{-8}{-2} \qquad\qquad\quad = \frac{-4}{-2}$$

$$= 4 \qquad\qquad\qquad = 2$$

The solution is (4, 2).

41.

$$x = \frac{D_x}{D} \qquad\qquad y = \frac{D_y}{D}$$

$$= \frac{\begin{vmatrix} -21 & 2 \\ 11 & -2 \end{vmatrix}}{\begin{vmatrix} 1 & 2 \\ 1 & -2 \end{vmatrix}} \qquad = \frac{\begin{vmatrix} 1 & -21 \\ 1 & 11 \end{vmatrix}}{\begin{vmatrix} 1 & 2 \\ 1 & -2 \end{vmatrix}}$$

$$= \frac{-21(-2)-2(11)}{1(-2)-2(1)} \qquad = \frac{1(11)-(-21)(1)}{1(-2)-2(1)}$$

$$= \frac{42-22}{-2-2} \qquad\qquad = \frac{11+21}{-2-2}$$

$$= \frac{20}{-4} \qquad\qquad\quad = \frac{32}{-4}$$

$$= -5 \qquad\qquad\qquad = -8$$

The solution is (−5, −8).

43.

$$x = \frac{D_x}{D}$$

$$= \frac{\begin{vmatrix} 9 & -4 \\ 8 & 2 \end{vmatrix}}{\begin{vmatrix} 3 & -4 \\ 1 & 2 \end{vmatrix}}$$

$$= \frac{9(2) - (-4)(8)}{3(2) - (-4)(1)}$$

$$= \frac{18 + 32}{6 + 4}$$

$$= \frac{50}{10}$$

$$= 5$$

$$y = \frac{D_y}{D}$$

$$= \frac{\begin{vmatrix} 3 & 9 \\ 1 & 8 \end{vmatrix}}{\begin{vmatrix} 3 & -4 \\ 1 & 2 \end{vmatrix}}$$

$$= \frac{3(8) - 9(1)}{3(2) - (-4)(1)}$$

$$= \frac{24 - 9}{6 + 4}$$

$$= \frac{15}{10}$$

$$= \frac{3}{2}$$

The solution is $\left(5, \dfrac{3}{2}\right)$.

45.

$$x = \frac{D_x}{D}$$

$$= \frac{\begin{vmatrix} 31 & 3 \\ 39 & 2 \end{vmatrix}}{\begin{vmatrix} 2 & 3 \\ 3 & 2 \end{vmatrix}}$$

$$= \frac{31(2) - 3(39)}{2(2) - 3(3)}$$

$$= \frac{62 - 117}{4 - 9}$$

$$= \frac{-55}{-5}$$

$$= 11$$

$$y = \frac{D_y}{D}$$

$$= \frac{\begin{vmatrix} 2 & 31 \\ 3 & 39 \end{vmatrix}}{\begin{vmatrix} 2 & 3 \\ 3 & 2 \end{vmatrix}}$$

$$= \frac{2(39) - 31(3)}{2(2) - 3(3)}$$

$$= \frac{78 - 93}{4 - 9}$$

$$= \frac{-15}{-5}$$

$$= 3$$

The solution is (11, 3).

47.

$$x = \frac{D_x}{D}$$

$$= \frac{\begin{vmatrix} 11 & 2 \\ 11 & 4 \end{vmatrix}}{\begin{vmatrix} 3 & 2 \\ 6 & 4 \end{vmatrix}}$$

$$= \frac{11(4) - 2(11)}{3(4) - 2(6)}$$

$$= \frac{44 - 22}{12 - 12}$$

$$= \frac{22}{0}$$

undefined

$$y = \frac{D_y}{D}$$

$$= \frac{\begin{vmatrix} 3 & 11 \\ 6 & 11 \end{vmatrix}}{\begin{vmatrix} 3 & 2 \\ 6 & 4 \end{vmatrix}}$$

$$= \frac{3(11) - 11(6)}{3(4) - 2(6)}$$

$$= \frac{33 - 66}{12 - 12}$$

$$= \frac{-33}{0}$$

undefined

The system has no solution and is an inconsistent system.

49. Write the equations in standard form.

$$\begin{cases} 5x + 6y = 12 \\ 10x + 12y = 24 \end{cases}$$

$$x = \frac{D_x}{D}$$

$$= \frac{\begin{vmatrix} 12 & 6 \\ 24 & 12 \end{vmatrix}}{\begin{vmatrix} 5 & 6 \\ 10 & 12 \end{vmatrix}}$$

$$= \frac{12(12) - 6(24)}{5(12) - 6(10)}$$

$$= \frac{144 - 144}{60 - 60}$$

$$= \frac{0}{0}$$

$$y = \frac{D_y}{D}$$

$$= \frac{\begin{vmatrix} 5 & 12 \\ 10 & 24 \end{vmatrix}}{\begin{vmatrix} 5 & 6 \\ 10 & 12 \end{vmatrix}}$$

$$= \frac{5(24) - 12(10)}{5(12) - 6(10)}$$

$$= \frac{120 - 120}{60 - 60}$$

$$= \frac{0}{0}$$

$$\{(x, y) \mid 5x + 6y = 12\}$$

This system has infinitely many solutions and has dependent equations.

51.

$$x = \frac{D_x}{D}$$

$$= \frac{\begin{vmatrix} 4 & 1 & 1 \\ 0 & 1 & -1 \\ 2 & -1 & 1 \end{vmatrix}}{\begin{vmatrix} 1 & 1 & 1 \\ 1 & 1 & -1 \\ 1 & -1 & 1 \end{vmatrix}}$$

$$= \frac{4 \begin{vmatrix} 1 & -1 \\ -1 & 1 \end{vmatrix} - 1 \begin{vmatrix} 0 & -1 \\ 2 & 1 \end{vmatrix} + 1 \begin{vmatrix} 0 & 1 \\ 2 & -1 \end{vmatrix}}{1 \begin{vmatrix} 1 & -1 \\ -1 & 1 \end{vmatrix} - 1 \begin{vmatrix} 1 & -1 \\ 1 & 1 \end{vmatrix} + 1 \begin{vmatrix} 1 & 1 \\ 1 & -1 \end{vmatrix}}$$

$$= \frac{4(1-1) - 1(0+2) + 1(0-2)}{1(1-1) - 1(1+1) + 1(-1-1)}$$

$$= \frac{4(0) - 1(2) + 1(-2)}{1(0) - 1(2) + 1(-2)}$$

$$= \frac{0 - 2 - 2}{0 - 2 - 2}$$

$$= \frac{-4}{-4}$$

$$= 1$$

$$y = \frac{D_y}{D}$$

$$= \frac{\begin{vmatrix} 1 & 4 & 1 \\ 1 & 0 & -1 \\ 1 & 2 & 1 \end{vmatrix}}{\begin{vmatrix} 1 & 1 & 1 \\ 1 & 1 & -1 \\ 1 & -1 & 1 \end{vmatrix}}$$

$$= \frac{1 \begin{vmatrix} 0 & -1 \\ 2 & 1 \end{vmatrix} - 4 \begin{vmatrix} 1 & -1 \\ 1 & 1 \end{vmatrix} + 1 \begin{vmatrix} 1 & 0 \\ 1 & 2 \end{vmatrix}}{1 \begin{vmatrix} 1 & -1 \\ -1 & 1 \end{vmatrix} - 1 \begin{vmatrix} 1 & -1 \\ 1 & 1 \end{vmatrix} + 1 \begin{vmatrix} 1 & 1 \\ 1 & -1 \end{vmatrix}}$$

$$= \frac{1(0+2) - 4(1+1) + 1(2-0)}{1(1-1) - 1(1+1) + 1(-1-1)}$$

$$= \frac{1(2) - 4(2) + 1(2)}{1(0) - 1(2) + 1(-2)}$$

$$= \frac{2 - 8 + 2}{0 - 2 - 2}$$

$$= \frac{-4}{-4}$$

$$= 1$$

$$z = \frac{D_z}{D}$$

$$= \frac{\begin{vmatrix} 1 & 1 & 4 \\ 1 & 1 & 0 \\ 1 & -1 & 2 \end{vmatrix}}{\begin{vmatrix} 1 & 1 & 1 \\ 1 & 1 & -1 \\ 1 & -1 & 1 \end{vmatrix}}$$

$$= \frac{1 \begin{vmatrix} 1 & 0 \\ -1 & 2 \end{vmatrix} - 1 \begin{vmatrix} 1 & 0 \\ 1 & 2 \end{vmatrix} + 4 \begin{vmatrix} 1 & 1 \\ 1 & -1 \end{vmatrix}}{1 \begin{vmatrix} 1 & -1 \\ -1 & 1 \end{vmatrix} - 1 \begin{vmatrix} 1 & -1 \\ 1 & 1 \end{vmatrix} + 1 \begin{vmatrix} 1 & 1 \\ 1 & -1 \end{vmatrix}}$$

$$= \frac{1(2-0) - 1(2-0) + 4(-1-1)}{1(1-1) - 1(1+1) + 1(-1-1)}$$

$$= \frac{1(2) - 1(2) + 4(-2)}{1(0) - 1(2) + 1(-2)}$$

$$= \frac{2 - 2 - 8}{0 - 2 - 2}$$

$$= \frac{-8}{-4}$$

$$= 2$$

The solution is (1, 1, 2).

53.

$$x = \frac{D_x}{D}$$

$$= \frac{\begin{vmatrix} -8 & 2 & -1 \\ 10 & -1 & 7 \\ -10 & 2 & -3 \end{vmatrix}}{\begin{vmatrix} 3 & 2 & -1 \\ 2 & -1 & 7 \\ 2 & 2 & -3 \end{vmatrix}}$$

$$= \frac{-8\begin{vmatrix} -1 & 7 \\ 2 & -3 \end{vmatrix} - 2\begin{vmatrix} 10 & 7 \\ -10 & -3 \end{vmatrix} - 1\begin{vmatrix} 10 & -1 \\ -10 & 2 \end{vmatrix}}{3\begin{vmatrix} -1 & 7 \\ 2 & -3 \end{vmatrix} - 2\begin{vmatrix} 2 & 7 \\ 2 & -3 \end{vmatrix} - 1\begin{vmatrix} 2 & -1 \\ 2 & 2 \end{vmatrix}}$$

$$= \frac{-8(3-14) - 2(-30+70) - 1(20-10)}{3(3-14) - 2(-6-14) - 1(4+2)}$$

$$= \frac{-8(-11) - 2(40) - 1(10)}{3(-11) - 2(-20) - 1(6)}$$

$$= \frac{88 - 80 - 10}{-33 + 40 - 6}$$

$$= \frac{-2}{1}$$

$$= -2$$

$$y = \frac{D_y}{D}$$

$$= \frac{\begin{vmatrix} 3 & -8 & -1 \\ 2 & 10 & 7 \\ 2 & -10 & -3 \end{vmatrix}}{\begin{vmatrix} 3 & 2 & -1 \\ 2 & -1 & 7 \\ 2 & 2 & -3 \end{vmatrix}}$$

$$= \frac{3\begin{vmatrix} 10 & 7 \\ -10 & -3 \end{vmatrix} - (-8)\begin{vmatrix} 2 & 7 \\ 2 & -3 \end{vmatrix} - 1\begin{vmatrix} 2 & 10 \\ 2 & -10 \end{vmatrix}}{3\begin{vmatrix} -1 & 7 \\ 2 & -3 \end{vmatrix} - 2\begin{vmatrix} 2 & 7 \\ 2 & -3 \end{vmatrix} - 1\begin{vmatrix} 2 & -1 \\ 2 & 2 \end{vmatrix}}$$

$$= \frac{3(-30+70) + 8(-6-14) - 1(-20-20)}{3(3-14) - 2(-6-14) - 1(4+2)}$$

$$= \frac{3(40) + 8(-20) - 1(-40)}{3(-11) - 2(-20) - 1(6)}$$

$$= \frac{120 - 160 + 40}{-33 + 40 - 6}$$

$$= \frac{0}{1}$$

$$= 0$$

$$z = \frac{D_z}{D}$$

$$= \frac{\begin{vmatrix} 3 & 2 & -8 \\ 2 & -1 & 10 \\ 2 & 2 & -10 \end{vmatrix}}{\begin{vmatrix} 3 & 2 & -1 \\ 2 & -1 & 7 \\ 2 & 2 & -3 \end{vmatrix}}$$

$$= \frac{3\begin{vmatrix} -1 & 10 \\ 2 & -10 \end{vmatrix} - 2\begin{vmatrix} 2 & 10 \\ 2 & -10 \end{vmatrix} - 8\begin{vmatrix} 2 & -1 \\ 2 & 2 \end{vmatrix}}{3\begin{vmatrix} -1 & 7 \\ 2 & -3 \end{vmatrix} - 2\begin{vmatrix} 2 & 7 \\ 2 & -3 \end{vmatrix} - 1\begin{vmatrix} 2 & -1 \\ 2 & 2 \end{vmatrix}}$$

$$= \frac{3(10-20) - 2(-20-20) - 8(4+2)}{3(3-14) - 2(-6-14) - 1(4+2)}$$

$$= \frac{3(-10) - 2(-40) - 8(6)}{3(-11) - 2(-20) - 1(6)}$$

$$= \frac{-30 + 80 - 48}{-33 + 40 - 6}$$

$$= \frac{2}{1}$$

$$= 2$$

The solution is (–2, 0, 2).

55.

$$x = \frac{D_x}{D}$$

$$= \frac{\begin{vmatrix} 5 & 1 & 1 \\ 10 & -2 & 3 \\ -3 & 1 & -4 \end{vmatrix}}{\begin{vmatrix} 2 & 1 & 1 \\ 1 & -2 & 3 \\ 1 & 1 & -4 \end{vmatrix}}$$

$$= \frac{5\begin{vmatrix} -2 & 3 \\ 1 & -4 \end{vmatrix} - 1\begin{vmatrix} 10 & 3 \\ -3 & -4 \end{vmatrix} + 1\begin{vmatrix} 10 & -2 \\ -3 & 1 \end{vmatrix}}{2\begin{vmatrix} -2 & 3 \\ 1 & -4 \end{vmatrix} - 1\begin{vmatrix} 1 & 3 \\ 1 & -4 \end{vmatrix} + 1\begin{vmatrix} 1 & -2 \\ 1 & 1 \end{vmatrix}}$$

$$= \frac{5(8-3) - 1(-40+9) + 1(10-6)}{2(8-3) - 1(-4-3) + 1(1+2)}$$

$$= \frac{5(5) - 1(-31) + 1(4)}{2(5) - 1(-7) + 1(3)}$$

$$= \frac{25+31+4}{10+7+3}$$

$$= \frac{60}{20}$$

$$= 3$$

$$y = \frac{D_y}{D}$$

$$= \frac{\begin{vmatrix} 2 & 5 & 1 \\ 1 & 10 & 3 \\ 1 & -3 & -4 \end{vmatrix}}{\begin{vmatrix} 2 & 1 & 1 \\ 1 & -2 & 3 \\ 1 & 1 & -4 \end{vmatrix}}$$

$$= \frac{2\begin{vmatrix} 10 & 3 \\ -3 & -4 \end{vmatrix} - 5\begin{vmatrix} 1 & 3 \\ 1 & -4 \end{vmatrix} + 1\begin{vmatrix} 1 & 10 \\ 1 & -3 \end{vmatrix}}{2\begin{vmatrix} -2 & 3 \\ 1 & -4 \end{vmatrix} - 1\begin{vmatrix} 1 & 3 \\ 1 & -4 \end{vmatrix} + 1\begin{vmatrix} 1 & -2 \\ 1 & 1 \end{vmatrix}}$$

$$= \frac{2(-40+9) - 5(-4-3) + 1(-3-10)}{2(8-3) - 1(-4-3) + 1(1+2)}$$

$$= \frac{2(-31) - 5(-7) + 1(-13)}{2(5) - 1(-7) + 1(3)}$$

$$= \frac{-62+35-13}{10+7+3}$$

$$= \frac{-40}{20}$$

$$= -2$$

$$z = \frac{D_z}{D}$$

$$= \frac{\begin{vmatrix} 2 & 1 & 5 \\ 1 & -2 & 10 \\ 1 & 1 & -3 \end{vmatrix}}{\begin{vmatrix} 2 & 1 & 1 \\ 1 & -2 & 3 \\ 1 & 1 & -4 \end{vmatrix}}$$

$$= \frac{2\begin{vmatrix} -2 & 10 \\ 1 & -3 \end{vmatrix} - 1\begin{vmatrix} 1 & 10 \\ 1 & -3 \end{vmatrix} + 5\begin{vmatrix} 1 & -2 \\ 1 & 1 \end{vmatrix}}{2\begin{vmatrix} -2 & 3 \\ 1 & -4 \end{vmatrix} - 1\begin{vmatrix} 1 & 3 \\ 1 & -4 \end{vmatrix} + 1\begin{vmatrix} 1 & -2 \\ 1 & 1 \end{vmatrix}}$$

$$= \frac{2(6-10) - 1(-3-10) + 5(1+2)}{2(8-3) - 1(-4-3) + 1(1+2)}$$

$$= \frac{2(-4) - 1(-13) + 5(3)}{2(5) - 1(-7) + 1(3)}$$

$$= \frac{-8+13+15}{10+7+3}$$

$$= \frac{20}{20}$$

$$= 1$$

The solution is (3, –2, 1).

57. Write the first equation in standard form.

$$y = \frac{-2x+1}{3}$$

$$3y = -2x+1$$

$$2x+3y = 1$$

$$\begin{cases} 2x+3y = 1 \\ 3x-2y = 8 \end{cases}$$

$$x = \frac{D_x}{D} \qquad\qquad y = \frac{D_y}{D}$$

$$= \frac{\begin{vmatrix} 1 & 3 \\ 8 & -2 \end{vmatrix}}{\begin{vmatrix} 2 & 3 \\ 3 & -2 \end{vmatrix}} \qquad = \frac{\begin{vmatrix} 2 & 1 \\ 3 & 8 \end{vmatrix}}{\begin{vmatrix} 2 & 3 \\ 3 & -2 \end{vmatrix}}$$

$$= \frac{1(-2)-3(8)}{2(-2)-3(3)} \qquad = \frac{2(8)-1(3)}{2(-2)-3(3)}$$

$$= \frac{-2-24}{-4-9} \qquad\qquad = \frac{16-3}{-4-9}$$

$$= \frac{-26}{-13} \qquad\qquad = \frac{13}{-13}$$

$$= 2 \qquad\qquad\qquad = -1$$

The solution is $(2, -1)$.

59.

$$x = \frac{D_x}{D}$$

$$= \frac{\begin{vmatrix} 1 & -3 & 0 \\ 1 & 0 & -8 \\ 0 & 2 & -4 \end{vmatrix}}{\begin{vmatrix} 4 & -3 & 0 \\ 6 & 0 & -8 \\ 0 & 2 & -4 \end{vmatrix}}$$

$$= \frac{1\begin{vmatrix} 0 & -8 \\ 2 & -4 \end{vmatrix} - (-3)\begin{vmatrix} 1 & -8 \\ 0 & -4 \end{vmatrix} + 0\begin{vmatrix} 1 & 0 \\ 0 & 2 \end{vmatrix}}{4\begin{vmatrix} 0 & -8 \\ 2 & -4 \end{vmatrix} - (-3)\begin{vmatrix} 6 & -8 \\ 0 & -4 \end{vmatrix} + 0\begin{vmatrix} 6 & 0 \\ 0 & 2 \end{vmatrix}}$$

$$= \frac{1(0+16)+3(-4-0)+0(2-0)}{4(0+16)+3(-24-0)+0(12-0)}$$

$$= \frac{1(16)+3(-4)+0(2)}{4(16)+3(-24)+0(12)}$$

$$= \frac{16-12+0}{64-72+0}$$

$$= \frac{4}{-8}$$

$$= -\frac{1}{2}$$

$$y = \frac{D_y}{D}$$

$$= \frac{\begin{vmatrix} 4 & 1 & 0 \\ 6 & 1 & -8 \\ 0 & 0 & -4 \end{vmatrix}}{\begin{vmatrix} 4 & -3 & 0 \\ 6 & 0 & -8 \\ 0 & 2 & -4 \end{vmatrix}}$$

$$= \frac{4\begin{vmatrix} 1 & -8 \\ 0 & -4 \end{vmatrix} - 1\begin{vmatrix} 6 & -8 \\ 0 & -4 \end{vmatrix} + 0\begin{vmatrix} 6 & 1 \\ 0 & 0 \end{vmatrix}}{4\begin{vmatrix} 0 & -8 \\ 2 & -4 \end{vmatrix} - (-3)\begin{vmatrix} 6 & -8 \\ 0 & -4 \end{vmatrix} + 0\begin{vmatrix} 6 & 0 \\ 0 & 2 \end{vmatrix}}$$

$$= \frac{4(-4-0)-1(-24-0)+0(0-0)}{4(0+16)+3(-24-0)+0(12-0)}$$

$$= \frac{4(-4)-1(-24)+0(0)}{4(16)+3(-24)+0(12)}$$

$$= \frac{-16+24+0}{64-72+0}$$

$$= \frac{8}{-8}$$

$$= -1$$

$$z = \frac{D_z}{D}$$

$$= \frac{\begin{vmatrix} 4 & -3 & 1 \\ 6 & 0 & 1 \\ 0 & 2 & 0 \end{vmatrix}}{\begin{vmatrix} 4 & -3 & 0 \\ 6 & 0 & -8 \\ 0 & 2 & -4 \end{vmatrix}}$$

$$= \frac{4\begin{vmatrix} 0 & 1 \\ 2 & 0 \end{vmatrix} - (-3)\begin{vmatrix} 6 & 1 \\ 0 & 0 \end{vmatrix} + 1\begin{vmatrix} 6 & 0 \\ 0 & 2 \end{vmatrix}}{4\begin{vmatrix} 0 & -8 \\ 2 & -4 \end{vmatrix} - (-3)\begin{vmatrix} 6 & -8 \\ 0 & -4 \end{vmatrix} + 0\begin{vmatrix} 6 & 0 \\ 0 & 2 \end{vmatrix}}$$

$$= \frac{4(0-2)+3(0-0)+1(12-0)}{4(0+16)+3(-24-0)+0(12-0)}$$

$$= \frac{4(-2)+3(0)+1(12)}{4(16)+3(-24)+0(12)}$$

$$= \frac{-8+0+12}{64-72+0}$$

$$= \frac{4}{-8}$$

$$= -\frac{1}{2}$$

The solution is $\left(-\frac{1}{2}, -1, -\frac{1}{2}\right)$.

Section 12.5

61. Write the equations in standard form.

$$\begin{cases} 2x + y - z = 1 \\ x + 2y + 2z = 2 \\ 4x + 5y + 3z = 3 \end{cases}$$

$$x = \frac{D_x}{D}$$

$$= \frac{\begin{vmatrix} 1 & 1 & -1 \\ 2 & 2 & 2 \\ 3 & 5 & 3 \end{vmatrix}}{\begin{vmatrix} 2 & 1 & -1 \\ 1 & 2 & 2 \\ 4 & 5 & 3 \end{vmatrix}}$$

$$= \frac{1\begin{vmatrix} 2 & 2 \\ 5 & 3 \end{vmatrix} - 1\begin{vmatrix} 2 & 2 \\ 3 & 3 \end{vmatrix} - 1\begin{vmatrix} 2 & 2 \\ 3 & 5 \end{vmatrix}}{2\begin{vmatrix} 2 & 2 \\ 5 & 3 \end{vmatrix} - 1\begin{vmatrix} 1 & 2 \\ 4 & 3 \end{vmatrix} - 1\begin{vmatrix} 1 & 2 \\ 4 & 5 \end{vmatrix}}$$

$$= \frac{1(6-10) - 1(6-6) - 1(10-6)}{2(6-10) - 1(3-8) - 1(5-8)}$$

$$= \frac{1(-4) - 1(0) - 1(4)}{2(-4) - 1(-5) - 1(-3)}$$

$$= \frac{-4 - 0 - 4}{-8 + 5 + 3}$$

$$= \frac{-8}{0} \text{ is undefined}$$

Since the denominator determinant D is 0 and the numerator determinant D_x is not 0, the system is inconsistent and has no solutions.

63. Write the equations in standard form.

$$\begin{cases} 3x - 5y = 16 \\ -3x + 5y = 33 \end{cases}$$

$$x = \frac{D_x}{D} \qquad\qquad y = \frac{D_y}{D}$$

$$= \frac{\begin{vmatrix} 16 & -5 \\ 33 & 5 \end{vmatrix}}{\begin{vmatrix} 3 & -5 \\ -3 & 5 \end{vmatrix}} \qquad = \frac{\begin{vmatrix} 3 & 16 \\ -3 & 33 \end{vmatrix}}{\begin{vmatrix} 3 & -5 \\ -3 & 5 \end{vmatrix}}$$

$$= \frac{16(5) - (-5)(33)}{3(5) - (-5)(-3)} \qquad = \frac{3(33) - 16(-3)}{3(5) - (-5)(-3)}$$

$$= \frac{80 + 165}{15 - 15} \qquad\qquad = \frac{99 + 48}{15 - 15}$$

$$= \frac{245}{0} \qquad\qquad = \frac{147}{0}$$

undefined $\qquad\qquad$ undefined

Since the denominator determinant D is 0 and the numerator determinant D_x is not 0, the system is inconsistent and has no solutions.

65. Write the equations in standard form. Multiply the second equation by 2 to eliminate fractions.

$$\begin{cases} x+y=1 \\ y+2z=5 \\ x-z=-3 \end{cases}$$

$$x=\frac{D_x}{D}$$

$$=\frac{\begin{vmatrix} 1 & 1 & 0 \\ 5 & 1 & 2 \\ -3 & 0 & -1 \end{vmatrix}}{\begin{vmatrix} 1 & 1 & 0 \\ 0 & 1 & 2 \\ 1 & 0 & -1 \end{vmatrix}}$$

$$=\frac{1\begin{vmatrix} 1 & 2 \\ 0 & -1 \end{vmatrix}-1\begin{vmatrix} 5 & 2 \\ -3 & -1 \end{vmatrix}+0\begin{vmatrix} 5 & 1 \\ -3 & 0 \end{vmatrix}}{1\begin{vmatrix} 1 & 2 \\ 0 & -1 \end{vmatrix}-1\begin{vmatrix} 0 & 2 \\ 1 & -1 \end{vmatrix}+0\begin{vmatrix} 0 & 1 \\ 1 & 0 \end{vmatrix}}$$

$$=\frac{1(-1-0)-1(-5+6)+0(0+3)}{1(-1-0)-1(0-2)+0(0-1)}$$

$$=\frac{1(-1)-1(1)+0(3)}{1(-1)-1(-2)+0(-1)}$$

$$=\frac{-1-1+0}{-1+2+0}$$

$$=\frac{-2}{1}$$

$$=-2$$

$$y=\frac{D_y}{D}$$

$$=\frac{\begin{vmatrix} 1 & 1 & 0 \\ 0 & 5 & 2 \\ 1 & -3 & -1 \end{vmatrix}}{\begin{vmatrix} 1 & 1 & 0 \\ 0 & 1 & 2 \\ 1 & 0 & -1 \end{vmatrix}}$$

$$=\frac{1\begin{vmatrix} 5 & 2 \\ -3 & -1 \end{vmatrix}-1\begin{vmatrix} 0 & 2 \\ 1 & -1 \end{vmatrix}+0\begin{vmatrix} 0 & 5 \\ 1 & -3 \end{vmatrix}}{1\begin{vmatrix} 1 & 2 \\ 0 & -1 \end{vmatrix}-1\begin{vmatrix} 0 & 2 \\ 1 & -1 \end{vmatrix}+0\begin{vmatrix} 0 & 1 \\ 1 & 0 \end{vmatrix}}$$

$$=\frac{1(-5+6)-1(0-2)+0(0-5)}{1(-1-0)-1(0-2)+0(0-1)}$$

$$=\frac{1(1)-1(-2)+0(-5)}{1(-1)-1(-2)+0(-1)}$$

$$=\frac{1+2+0}{-1+2+0}$$

$$=\frac{3}{1}$$

$$=3$$

$$z=\frac{D_z}{D}$$

$$=\frac{\begin{vmatrix} 1 & 1 & 1 \\ 0 & 1 & 5 \\ 1 & 0 & -3 \end{vmatrix}}{\begin{vmatrix} 1 & 1 & 0 \\ 0 & 1 & 2 \\ 1 & 0 & -1 \end{vmatrix}}$$

$$=\frac{1\begin{vmatrix} 1 & 5 \\ 0 & -3 \end{vmatrix}-1\begin{vmatrix} 0 & 5 \\ 1 & -3 \end{vmatrix}+1\begin{vmatrix} 0 & 1 \\ 1 & 0 \end{vmatrix}}{1\begin{vmatrix} 1 & 2 \\ 0 & -1 \end{vmatrix}-1\begin{vmatrix} 0 & 2 \\ 1 & -1 \end{vmatrix}+0\begin{vmatrix} 0 & 1 \\ 1 & 0 \end{vmatrix}}$$

$$=\frac{1(-3-0)-1(0-5)+1(0-1)}{1(-1-0)-1(0-2)+0(0-1)}$$

$$=\frac{1(-3)-1(-5)+1(-1)}{1(-1)-1(-2)+0(-1)}$$

$$=\frac{-3+5-1}{-1+2+0}$$

$$=\frac{1}{1}$$

$$=1$$

The solution is (–2, 3, 1).

Section 12.5

67.

$$x = \frac{D_x}{D} \qquad\qquad y = \frac{D_y}{D}$$

$$= \frac{\begin{vmatrix} 0 & 3 \\ -4 & -6 \end{vmatrix}}{\begin{vmatrix} 2 & 3 \\ 4 & -6 \end{vmatrix}} \qquad\qquad = \frac{\begin{vmatrix} 2 & 0 \\ 4 & -4 \end{vmatrix}}{\begin{vmatrix} 2 & 3 \\ 4 & -6 \end{vmatrix}}$$

$$= \frac{0(-6)-3(-4)}{2(-6)-3(4)} \qquad = \frac{2(-4)-0(4)}{2(-6)-3(4)}$$

$$= \frac{0+12}{-12-12} \qquad\qquad = \frac{-8-0}{-12-12}$$

$$= \frac{12}{-24} \qquad\qquad = \frac{-8}{-24}$$

$$= -\frac{1}{2} \qquad\qquad = \frac{1}{3}$$

The solution is $\left(-\dfrac{1}{2}, \dfrac{1}{3}\right)$.

69.

$$x = \frac{D_x}{D}$$

$$= \frac{\begin{vmatrix} 6 & 3 & 4 \\ -4 & -3 & -4 \\ 12 & 6 & 8 \end{vmatrix}}{\begin{vmatrix} 2 & 3 & 4 \\ 2 & -3 & -4 \\ 4 & 6 & 8 \end{vmatrix}}$$

$$= \frac{6\begin{vmatrix} -3 & -4 \\ 6 & 8 \end{vmatrix} - 3\begin{vmatrix} -4 & -4 \\ 12 & 8 \end{vmatrix} + 4\begin{vmatrix} -4 & -3 \\ 12 & 6 \end{vmatrix}}{2\begin{vmatrix} -3 & -4 \\ 6 & 8 \end{vmatrix} - 3\begin{vmatrix} 2 & -4 \\ 4 & 8 \end{vmatrix} + 4\begin{vmatrix} 2 & -3 \\ 4 & 6 \end{vmatrix}}$$

$$= \frac{6(-24+24)-3(-32+48)+4(-24+36)}{2(-24+24)-3(16+16)+4(12+12)}$$

$$= \frac{6(0)-3(16)+4(12)}{2(0)-3(32)+4(24)}$$

$$= \frac{0-48+48}{0-96+96}$$

$$= \frac{0}{0}$$

Since the denominator determinant D is 0 and the numerator determinant D_x is 0, the system is consistent, but the equations are dependent. There are infinitely many solutions.

71. –46,811

73 –60,527,941

APPLICATIONS

75.

$$\begin{cases} 2x+y=180 \\ y=30+x \end{cases}$$

$$\begin{cases} 2x+y=180 \\ -x+y=30 \end{cases}$$

$$x = \frac{D_x}{D}$$

$$= \frac{\begin{vmatrix} 180 & 1 \\ 30 & 1 \end{vmatrix}}{\begin{vmatrix} 2 & 1 \\ -1 & 1 \end{vmatrix}}$$

$$= \frac{180(1)-1(30)}{2(1)-1(-1)}$$

$$= \frac{180-30}{2+1}$$

$$= \frac{150}{3}$$

$$= 50$$

$$y = \frac{D_y}{D}$$

$$= \frac{\begin{vmatrix} 2 & 180 \\ -1 & 30 \end{vmatrix}}{\begin{vmatrix} 2 & 1 \\ -1 & 1 \end{vmatrix}}$$

$$= \frac{2(30)-180(-1)}{2(1)-1(-1)}$$

$$= \frac{60+180}{2+1}$$

$$= \frac{240}{3}$$

$$= 80$$

$x = 50°$ and $y = 80°$.

77. Answers will vary.

79. Answers will vary.

81. Answers will vary.

REVIEW

83. Find the slope of each using $y = mx + b$.

$y = 2x - 7$ $x - 2y = 7$
$m = 2$ $-2y = -x + 7$

$$y = \frac{1}{2}x - \frac{7}{2}$$

$$m = \frac{1}{2}$$

No. Since the slopes are not opposite reciprocals, the lines are NOT perpendicular.

85. The graph of g is 2 units below the graph of f. $g(x)$ is shifted down 2 units.

87. The y–intercept since the x–coordinate is 0.

89. The independent variable is x and the dependent variable is y.

91.

$$\begin{vmatrix} x & y & 1 \\ -2 & 3 & 1 \\ 3 & 5 & 1 \end{vmatrix} = 0$$

$$x\begin{vmatrix} 3 & 1 \\ 5 & 1 \end{vmatrix} - y\begin{vmatrix} -2 & 1 \\ 3 & 1 \end{vmatrix} + 1\begin{vmatrix} -2 & 3 \\ 3 & 5 \end{vmatrix} = 0$$

$$x(3-5) - y(-2-3) + 1(-10-9) = 0$$

$$x(-2) - y(-5) + 1(-19) = 0$$

$$-2x + 5y - 19 = 0$$

$$2x - 5y + 19 = 0$$

$$2x - 5y = -19$$

For $(-2, 3)$:

$$2(-2) - 5(3) \overset{?}{=} -19$$

$$-4 - 15 \overset{?}{=} -19$$

$$-19 = -19$$

For $(3, 5)$:

$$2(3) - 5(5) \overset{?}{=} -19$$

$$6 - 25 \overset{?}{=} -19$$

$$-19 = -19$$

SECTION 12.1
Solving Systems of Equations in Two Variable

1. Yes.

$$x + 2y = 0 \qquad\qquad x + 4y = 1$$

$$-1 + 2\left(\frac{1}{2}\right) \overset{?}{=} 0 \qquad -1 + 4\left(\frac{1}{2}\right) \overset{?}{=} 1$$

$$-1 + 1 \overset{?}{=} 0 \qquad\qquad -1 + 2 \overset{?}{=} 1$$

$$0 = 0 \qquad\qquad\qquad 1 = 1$$

2. No.

$$3a - 2b + 7 = 0 \qquad -2a + b = -4$$

$$3(13) - 2(23) + 7 \overset{?}{=} 0 \qquad -2(13) + 23 \overset{?}{=} -4$$

$$39 - 46 + 7 \overset{?}{=} 0 \qquad -26 + 23 \overset{?}{=} -4$$

$$0 = 0 \qquad\qquad\qquad -3 \neq -4$$

3. a. Answers may vary.
      (1, 3), (2, 1), and (4, –3)
   b. Answers may vary.
      (0, –4), (2, –2), and (4, 0)
   c. The lines intersect at the point (3, –1).

4. The point of intersection is (2019, 15). In 2019, the weekly amount of time spent viewing live broadcast television, approximately 15 hours, will be the same as that spent viewing Internet video.

5. $\begin{cases} 2x + y = 11 \\ -x + 2y = 7 \end{cases}$

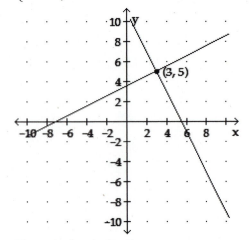

The solution is (3, 5).

6. $\begin{cases} y = -\dfrac{3}{2}x \\ 2x - 3y + 13 = 0 \end{cases}$

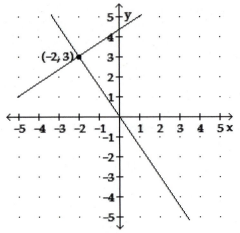

The solution is (–2, 3).

7. $\begin{cases} \dfrac{1}{2}x + \dfrac{1}{3}y = 2 \\ y = 6 - \dfrac{3}{2}x \end{cases}$

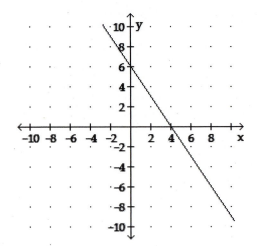

There are infinitely many solutions; the equations are dependent.

8. $\begin{cases} \dfrac{x}{3} - \dfrac{y}{2} = 1 \\ 6x - 9y = 3 \end{cases}$

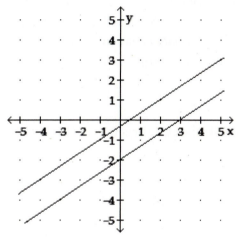

There are no solutions; the system is inconsistent.

9. $\begin{cases} x = y - 4 \\ 2x + 3y = 7 \end{cases}$

Substitute $x = y - 4$ into the second equation and solve for y.

$$2x + 3y = 7$$
$$2(y - 4) + 3y = 7$$
$$2y - 8 + 3y = 7$$
$$5y - 8 = 7$$
$$5y = 15$$
$$y = 3$$

Subsitute $y = 3$ into the first equation and solve the x.

$$x = y - 4$$
$$x = 3 - 4$$
$$x = -1$$

The solution is $(-1, 3)$.

10. $\begin{cases} y = 2x + 5 \\ 3x - 5y = -4 \end{cases}$

Substitute $y = 2x + 5$ into the second equation and solve for y.

$$3x - 5y = -4$$
$$3x - 5(2x + 5) = -4$$
$$3x - 10x - 25 = -4$$
$$-7x - 25 = -4$$
$$-7x = 21$$
$$x = -3$$

Subsitute $x = -3$ into the first equation and solve for y.

$$y = 2x + 5$$
$$y = 2(-3) + 5$$
$$y = -6 + 5$$
$$y = -1$$

The solution is $(-3, -1)$.

11. $\begin{cases} x - 2y = 11 \quad (1) \\ x + 2y = -21 \quad (2) \end{cases}$

Add Equations 1 and 2.

$$x - 2y = 11 \quad (1)$$
$$\underline{x + 2y = -21} \quad (2)$$
$$2x = -10$$
$$x = -5$$

Substitute $x = -5$ into Equation 1.

$$x - 2y = 11$$
$$-5 - 2y = 11$$
$$-2y = 16$$
$$y = -8$$

The solution is $(-5, -8)$.

12. $\begin{cases} 4a + 5b = -9 \\ 6a = 3b - 3 \end{cases}$

Write the equations in standard form.

$\begin{cases} 4a + 5b = -9 & (1) \\ 6a - 3b = -3 & (2) \end{cases}$

Multiply Equation 1 by 3 and Equation 2 by 5.

$12a + 15b = -27$

$\underline{30a - 15b = -15}$

$42a = -42$

$a = -1$

Substitute $a = -1$ into Equation 1.

$4a + 5b = -9$

$4(-1) + 5b = -9$

$-4 + 5b = -9$

$5b = -5$

$b = -1$

The solution is $(-1, -1)$.

13. $\begin{cases} \dfrac{1}{2}a - \dfrac{3}{8}b = -9 \\ \dfrac{2}{3}a - \dfrac{1}{4}b = -8 \end{cases}$

Multiply the first equation by 8 and the second equation by 12 to eliminate fractions.

$\begin{cases} 4a - 3b = -72 & (1) \\ 8a - 3b = -96 & (2) \end{cases}$

Multiply Equation 1 by -1.

$-4a + 3b = 72 \quad (1)$

$\underline{8a - 3b = -96 \quad (2)}$

$4a = -24$

$a = -6$

Substitute $a = -6$ into Equation 1.

$4a - 3b = -72 \quad (1)$

$4(-6) - 3b = -72$

$-24 - 3b = -72$

$-3b = -48$

$b = 16$

The solution is $(-6, 16)$.

14. $\begin{cases} y = \dfrac{x-3}{2} \\ x = \dfrac{2y+7}{2} \end{cases}$

Multiply both equations by 2 to eliminate fractions

$\begin{cases} 2y = x - 3 \\ 2x = 2y + 7 \end{cases}$

Write the equations in standard form.

$\begin{cases} -x + 2y = -3 \\ 2x - 2y = 7 \end{cases}$

Add the equations to eliminate y.

$-x + 2y = -3$

$\underline{2x - 2y = 7}$

$x = 4$

Substitute $x = 4$ into the first equation and solve for y.

$2y = x - 3$

$2y = 4 - 3$

$2y = 1$

$y = \dfrac{1}{2}$

The solution is $\left(4, \dfrac{1}{2}\right)$.

15. $\begin{cases} 4x = 8y + 5 \\ 8x = 1 - 2y \end{cases}$

Write the equations in standard form.

$\begin{cases} 4x - 8y = 5 \quad (1) \\ 8x + 2y = 1 \quad (2) \end{cases}$

Multiply Equation 2 by 4.

$4x - 8y = 5 \quad (1)$

$\underline{32x + 8y = 4 \quad (2)}$

$36x = 9$

$x = \dfrac{1}{4}$

Substitute $x = \dfrac{1}{4}$ into Equation 1.

$4x - 8y = 5$

$4\left(\dfrac{1}{4}\right) - 8y = 5$

$1 - 8y = 5$

$-8y = 4$

$y = -\dfrac{1}{2}$

The solution is $\left(\dfrac{1}{4}, -\dfrac{1}{2}\right)$.

16. $\begin{cases} x + 3y = -2 \\ -2(x + 3y) = 4 \end{cases}$

Distribute -2 on Equation 2.

$\begin{cases} x + 3y = -2 \quad (1) \\ -2x - 6y = 4 \quad (2) \end{cases}$

Multiply Equation 1 by 2.

$2x + 6y = -4 \quad (1)$

$\underline{-2x - 6y = 4 \quad (2)}$

$0 = 0$

$\{(x, y) \mid x + 3y = -2\}$

Infinitely many solutions
Dependent equations

17. $\begin{cases} 0.07x = 0.05 + 0.09y \\ 7x - 9y = 8 \end{cases}$

Multiply the first equation by 100 to eliminate decimals.

$\begin{cases} 7x = 5 + 9y \quad (1) \\ 7x - 9y = 8 \quad (2) \end{cases}$

Write the equations in standard form.

$7x - 9y = 5 \quad (1)$

$7x - 9y = 8 \quad (2)$

Multiply Equation 1 by -1.

$-7x + 9y = -5 \quad (1)$

$\underline{7x - 9y = 8 \quad (2)}$

$0 \neq 3$

No solution; $\varnothing$
Inconsistent system

18.

$\begin{cases} 0.1x + 0.2y = 1.1 \\ 2x - y = 2 \end{cases}$

Multiply the first equation by 10 to eliminate decimals.

$\begin{cases} x + 2y = 11 \\ 2x - y = 2 \end{cases}$

Solve the first equation for x.

$\begin{cases} x = 11 - 2y \\ 2x - y = 2 \end{cases}$

Substitute $x = 11 - 2y$ into the second equation and solve for y.

$2x - y = 2$

$2(11 - 2y) - y = 2$

$22 - 4y - y = 2$

$22 - 5y = 2$

$-5y = -20$

$y = 4$

Substitute $y = 4$ into the first equation and solve for x.

$x = 11 - 2y$

$x = 11 - 2(4)$

$x = 11 - 8$

$x = 3$

The solution is $(3, 4)$.

19.

$$\begin{cases} y = -\dfrac{2}{3}x \\ 2x - 3y = -4 \end{cases}$$

Substitute $y = -\dfrac{2}{3}x$ into the second

equation and solve for y.

$$2x - 3y = -4$$

$$2x - 3\left(-\dfrac{2}{3}x\right) = -4$$

$$2x + 2x = -4$$

$$4x = -4$$

$$x = -1$$

Subsitute $x = -1$ into the first equation

and solve for y.

$$y = -\dfrac{2}{3}(-1)$$

$$y = \dfrac{2}{3}$$

The solution is $\left(-1, \dfrac{2}{3}\right)$.

20. Answers will vary.

21. Let x = # of residents and y = # of
nonresidents.

$$\begin{cases} x + y = 11 \\ 3x + 7y = 41 \end{cases}$$

Solve the first equation for x and substitute
into the second equation to solve for y.

$$x = 11 - y$$

$$3x + 7y = 41$$

$$3(11 - y) + 7y = 41$$

$$33 - 3y + 7y = 41$$

$$33 + 4y = 41$$

$$4y = 8$$

$$y = 2$$

Substitute $y = 2$ into the first equation and
solve for x.

$$x + y = 11$$

$$x + 2 = 11$$

$$x = 9$$

9 residents and 2 nonresidents

22. Let x = distance between Austin and
Houston and y = distance Austin and San
Antonio.

$$\begin{cases} x = 2y - 4 \\ x + y + 197 = 442 \end{cases}$$

Write the equations in standard form.

$$\begin{cases} x - 2y = -4 \\ x + y = 245 \end{cases}$$

Multiply the first equation by -1 and
add the equations to eliminate x.

$$-x + 2y = 4$$

$$\underline{x + y = 245}$$

$$3y = 249$$

$$y = 83$$

Substitute $y = 83$ into the first equation
and solve for x.

$$x = 2y - 4$$

$$x = 2(83) - 4$$

$$x = 166 - 4$$

$$x = 162$$

It is 162 miles from Austin to Houston and
83 miles from Austin to San Antonio.

23. Let b = boat's rate and c = current's rate.

	Rate · time = distance		
downstream	$b + c$	3	30
upstream	$b - c$	5	30

$$\begin{cases} 3(b+c) = 30 \\ 5(b-c) = 30 \end{cases}$$

Write equations in standard form.

$$\begin{cases} 3b + 3c = 30 \\ 5b - 5c = 30 \end{cases}$$

Multiply the first equation by -5 and the second equation by 3 to eliminate x.

$$\begin{array}{r} -15b - 15c = -150 \\ \underline{15b - 15c = 90} \\ -30c = -60 \\ c = 2 \end{array}$$

Substitute $c = 2$ into the first equation and solve for b.

$$3b + 3c = 30$$
$$3b + 3(2) = 30$$
$$3b + 6 = 30$$
$$3b = 24$$
$$b = 8$$

The rate of the boat is 8 mph and the rate of the current is 2 mph.

24. Let x = oz of 6% solution and y = oz of 18% solution.

$$\begin{cases} x + y = 750 \\ 0.06x + 0.18y = 0.10(750) \end{cases}$$

Multiply the second equation by 100 to eliminate decimals.

$$\begin{cases} x + y = 750 & (1) \\ 6x + 18y = 7,500 & (2) \end{cases}$$

Multiply Equation 1 by –6.

$$\begin{array}{r} -6x - 6y = -4,500 \quad (1) \\ \underline{6x + 18y = 7,500 \quad (2)} \\ 12y = 3,000 \\ y = 250 \end{array}$$

Substitute $y = 250$ into Equation 1.

$$x + y = 750 \quad (1)$$
$$x + 250 = 750$$
$$x = 500$$

500 oz of the 6% solution and 250 oz of the 18% solution are needed.

25. Let x = amount invested at 6% and y = amount invested at 12%

$$\begin{cases} x + y = 10,000 \\ 0.06x + 0.12y = 960 \end{cases}$$

Multiply the second equation by 100 to eliminate decimals.

$$\begin{cases} x + y = 10,000 & (1) \\ 6x + 12y = 96,000 & (2) \end{cases}$$

Multiply Equation 1 by –6.

$$\begin{array}{r} -6x - 6y = -60,000 \quad (1) \\ \underline{6x + 12y = 96,000 \quad (2)} \\ 6y = 36,000 \\ y = 6,000 \end{array}$$

Substitute $y = 6,000$ into Equation 1.

$$x + y = 10,000 \quad (1)$$
$$x + 6,000 = 10,000$$
$$x = 4,000$$

They invested \$4,000 at 6% and \$6,000 at 12%.

26. Let x = milliliters in one teaspoon and y = milliliters in one tablespoon.

$$\begin{cases} 2x + 5y = 85 & (1) \\ 5x + 2y = 55 & (2) \end{cases}$$

Multiply Equation 1 by –5 and Equation 2 by 2.

$$\begin{array}{r} -10x - 25y = -425 \quad (1) \\ \underline{10x + 4y = 110 \quad (2)} \\ -21y = -315 \\ y = 15 \end{array}$$

Substitute $y = 15$ into Equation 1.

$$2x + 5y = 85 \quad (1)$$
$$2x + 5(15) = 85$$
$$2x + 75 = 85$$
$$2x = 10$$
$$x = 5$$

There are 5 mL in one teaspoon and 15 mL in one tablespoon.

27. No.

$$x - y + z = 4 \qquad\qquad x + 2y - z = -1$$

$$2 - (-1) + 1 \overset{?}{=} 4 \qquad 2 + 2(-1) - 1 \overset{?}{=} -1$$

$$2 + 1 + 1 \overset{?}{=} 4 \qquad\quad 2 - 2 - 1 \overset{?}{=} -1$$

$$4 = 4 \qquad\qquad\qquad -1 = -1$$

$$x + y - 3z = -1$$

$$2 + (-1) - 3(1) \overset{?}{=} -1$$

$$2 - 1 - 3 \overset{?}{=} -1$$

$$-2 \neq -1$$

28. Yes; one solution

29.

$$\begin{cases} x - 2y + 3z = -7 & (1) \\ -x + 3y + 2z = -8 & (2) \\ 2x - y - z = 7 & (3) \end{cases}$$

Add Equations 1 and 2.

$$x - 2y + 3z = -7 \qquad (1)$$
$$\underline{-x + 3y + 2z = -8} \qquad (2)$$
$$y + 5z = -15 \qquad (4)$$

Multiply Equation 2 by 2 and add Equation 3.

$$-2x + 6y + 4z = -16 \quad (2)$$
$$\underline{2x - y - z = 7} \qquad (3)$$
$$5y + 3z = -9 \quad (5)$$

Multiply Equation 4 by –5 and add Equations 4 and 5.

$$-5y - 25z = 75 \quad (4)$$
$$\underline{5y + 3z = -9} \quad (5)$$
$$-22z = 66$$
$$z = -3$$

Substitute $z = -3$ into Equation 4.

$$y + 5z = -15 \quad (4)$$
$$y + 5(-3) = -15$$
$$y - 15 = -15$$
$$y = 0$$

Substitute $y = 0$ and $z = -3$ into Equation 1.

$$x - 2y + 3z = -7$$
$$x - 2(0) + 3(-3) = -7$$
$$x - 0 - 9 = -7$$
$$x - 9 = -7$$
$$x = 2$$

The solution is (2, 0, –3).

30. $$\begin{cases} x + y + z = 4 & (1) \\ x - 2y - z = 1 & (2) \\ 2x - y - 2z = -1 & (3) \end{cases}$$

Add Equations 1 and 2.

$$x + y + z = 4 \qquad (1)$$
$$\underline{x - 2y - z = 1} \qquad (2)$$
$$2x - y = 5 \qquad (4)$$

Multiply Equation 1 by 2 and add Equations 1 and 3.

$$2x + 2y + 2z = 8 \quad (1)$$
$$\underline{2x - y - 2z = -1} \quad (3)$$
$$4x + y = 7 \quad (5)$$

Add Equations 4 and 5.

$$2x - y = 5 \quad (4)$$
$$\underline{4x + y = 7} \quad (5)$$
$$6x = 12$$
$$x = 2$$

Substitute $x = 2$ into Equation 4.

$$2x - y = 5 \quad (4)$$
$$2(2) - y = 5$$
$$4 - y = 5$$
$$-y = 1$$
$$y = -1$$

Substitute $x = 2$ and $y = -1$ into Equation 1.

$$x + y + z = 4$$
$$2 + (-1) + z = 4$$
$$1 + z = 4$$
$$z = 3$$

The solution is (2, –1, 3).

31. $\begin{cases} x+y-z=-3 & (1) \\ x+z=2 & (2) \\ 2x-y+2z=3 & (3) \end{cases}$

Add Equations 1 and 3 to eliminate y.

$$\begin{array}{r} x+y-z=-3 \quad (1) \\ \underline{2x-y+2z=3} \quad (3) \\ 3x+z=0 \quad (4) \end{array}$$

Multiply Equation 2 by -1 and add Equations 2 and 4 to eliminate z.

$$\begin{array}{r} -x-z=-2 \quad (2) \\ \underline{3x+z=0} \quad (4) \\ 2x=-2 \\ x=-1 \end{array}$$

Substitute $x=-1$ into Equation 4 and solve for z.

$$\begin{array}{r} 3x+z=0 \quad (4) \\ 3(-1)+z=0 \\ -3+z=0 \\ z=3 \end{array}$$

Substitute $z=3$ and $x=-1$ into Equation 1 and solve for y.

$$\begin{array}{r} x+y-z=-3 \quad (1) \\ -1+y-3=-3 \\ y-4=-3 \\ y=1 \end{array}$$

The solution is $(-1,\,1,\,3)$.

32. $\begin{cases} b-4c=2 \\ a-b+2c=1 \\ 2a-2b=-2-5c \end{cases}$

Write Equation 3 in standard form.

$\begin{cases} b-4c=2 & (1) \\ a-b+2c=1 & (2) \\ 2a-2b+5c=-2 & (3) \end{cases}$

Multiply Equation 2 by -2 and add Equations 2 and 3.

$$\begin{array}{r} -2a+2b-4c=-2 \quad (2) \\ \underline{2a-2b+5c=-2} \quad (3) \\ c=-4 \end{array}$$

Substitute $c=-4$ into Equation 1.

$$\begin{array}{r} b-4c=2 \quad (1) \\ b-4(-4)=2 \\ b+16=2 \\ b=-14 \end{array}$$

Substitute $b=-14$ and $c=-4$ into Equation 2.

$$\begin{array}{r} a-b+2c=1 \quad (2) \\ a-(-14)+2(-4)=1 \\ a+14-8=1 \\ a+6=1 \\ a=-5 \end{array}$$

The solution is $(-5,\,-14,\,-4)$.

Chapter 12 Review

33. $\begin{cases} x + 2z = 10 & (1) \\ 3x + 2y - 3z = 8 & (2) \\ y + 4z = 6 & (3) \end{cases}$

Multiply Equation 1 by –3 and add
Equations 1 and 2.

$$-3x - 6z = -30 \quad (1)$$
$$\underline{3x + 2y - 3z = 8} \quad (2)$$
$$2y - 9z = -22 \quad (4)$$

Multiply Equation 3 by –2 and add
Equations 3 and 4.

$$-2y - 8z = -12 \quad (3)$$
$$\underline{2y - 9z = -22} \quad (4)$$
$$-17z = -34$$
$$z = 2$$

Substitute $z = 2$ into Equation 1.

$$x + 2z = 10 \quad (1)$$
$$x + 2(2) = 10$$
$$x + 4 = 10$$
$$x = 6$$

Substitute $z = 2$ into Equation 3.

$$y + 4z = 6 \quad (3)$$
$$y + 4(2) = 6$$
$$y + 8 = 6$$
$$y = -2$$

The solution is (6, –2, 2).

34. $\begin{cases} x + 3y + z = 14 & (1) \\ x - 5y = -19 & (2) \\ 3y + z = 13 & (3) \end{cases}$

Multiply Equation 2 by –1 and add
Equations 1 and 2.

$$x + 3y + z = 14 \quad (1)$$
$$\underline{-x + 5y = 19} \quad (2)$$
$$8y + z = 33 \quad (4)$$

Multiply Equation 4 by –1 and add
Equations 3 and 4.

$$3y + z = 13 \quad (3)$$
$$\underline{-8y - z = -33} \quad (4)$$
$$-5y = -20$$
$$y = 4$$

Substitute $y = 4$ into Equation 3.

$$3y + z = 13 \quad (3)$$
$$3(4) + z = 13$$
$$12 + z = 13$$
$$z = 1$$

Substitute $y = 4$ into Equation 2.

$$x - 5y = -19 \quad (2)$$
$$x - 5(4) = -19$$
$$x - 20 = -19$$
$$x = 1$$

The solution is (1, 4, 1).

35.
$$\begin{cases} 2x+3y+z=-5 & (1) \\ -x+2y-z=-6 & (2) \\ 3x+y+2z=4 & (3) \end{cases}$$

Multiply Equation 2 by 2 and add Equations 1 and 2 to eliminate x.

$$2x+3y+z=-5 \quad (1)$$
$$\underline{-2x+4y-2z=-12 \quad (2)}$$
$$7y-z=-17 \quad (4)$$

Multiply Equation 2 by 3 and add Equations 2 and 3 to eliminate x.

$$-3x+6y-3z=-18 \quad (2)$$
$$\underline{3x+y+2z=4 \quad (3)}$$
$$7y-z=-14 \quad (5)$$

Multiply Equation 4 by -1 and add Equations 4 and 5 to eliminate y and z.

$$-7y+z=17 \quad (4)$$
$$\underline{7y-z=-14 \quad (5)}$$
$$0 \neq 3$$

No solution; Inconsistent system

36.
$$\begin{cases} 3x+3y+6z=-6 & (1) \\ -x-y-2z=2 & (2) \\ 2x+2y+4z=-4 & (3) \end{cases}$$

Multiply Equation 2 by 3 and add Equations 1 and 2 to eliminate x.

$$3x+3y+6z=-6 \quad (1)$$
$$\underline{-3x-3y-6z=6 \quad (2)}$$
$$0=0$$

Infinitely many solutions; Dependent equations

SECTION 12.3
Problem Solving Using Systems of Three Equations

37. Let x = the small bears, y = medium bears, and z = large bears.

$$\begin{cases} 3x+5y+10z=850 & (1) \\ 2x+3y+5z=480 & (2) \\ 6x+8y+12z=1{,}260 & (3) \end{cases}$$

Multiply Equation 1 by −2 and add Equations 1 and 3.

$$-6x-10y-20z=-1{,}700 \quad (1)$$
$$\underline{6x+8y+12z=1{,}260 \quad (3)}$$
$$-2y-8z=-440 \quad (4)$$

Multiply Equation 2 by −3 and add Equations 2 and 3.

$$-6x-9y-15z=-1{,}440 \quad (2)$$
$$\underline{6x+8y+12z=1{,}260 \quad (3)}$$
$$-y-3z=-180 \quad (5)$$

Multiply Equation 5 by −2 and add Equations 4 and 5.

$$-2y-8z=-440 \quad (4)$$
$$\underline{2y+6z=360 \quad (5)}$$
$$-2z=-80$$
$$z=40$$

Substitute $z = 40$ into Equation 4.

$$-2y-8z=-440 \quad (4)$$
$$-2y-8(40)=-440$$
$$-2y-320=-440$$
$$-2y=-120$$
$$y=60$$

Substitute $y = 60$ and $z = 40$ into Equation 1.

$$3x+5y+10z=850 \quad (1)$$
$$3x+5(60)+10(40)=850$$
$$3x+300+400=850$$
$$3x+700=850$$
$$3x=150$$
$$x=50$$

The toy company produces 50 small bears, 60 medium bears, and 40 large bears.

38. Let x = amount invested at 5%, y = amount at 6%, and z = amount invested at 7%.

$$\begin{cases} x + y + z = 22,000 \\ y = x + 2,000 \\ 0.05x + 0.06y + 0.07z = 1,370 \end{cases}$$

Multiply Equation 3 by 100 to clear decimals.

$$\begin{cases} x + y + z = 22,000 & (1) \\ y = x + 2,000 & (2) \\ 5x + 6y + 7z = 137,000 & (3) \end{cases}$$

Multiply Equation 1 by –7 and add Equation 1 and 3.

$$-7x - 7y - 7z = -154,000 \quad (1)$$
$$\underline{5x + 6y + 7z = 137,000 \quad (3)}$$
$$-2x - y = -17,000 \quad (4)$$

Substitute Equation 2 into Equation 4.

$$-2x - y = -17,000 \quad (4)$$
$$-2x - (x + 2,000) = -17,000$$
$$-2x - x - 2,000 = -17,000$$
$$-3x = -15,000$$
$$x = 5,000$$

Substitute $x = 5,000$ into Equation 2.

$$y = x + 2,000 \quad (2)$$
$$y = 5,000 + 2,000$$
$$y = 7,000$$

Substitute $x = 5,000$ and $y = 7,000$ into Equation 1.

$$x + y + z = 22,000 \quad (1)$$
$$5,000 + 7,000 + z = 22,000$$
$$12,000 + z = 22,000$$
$$z = 10,000$$

She invested $5,000 at 5%, $7,000 at 6%, and $10,000 at 7%.

39. Substitute the coordinates of $(0, 0)$, $(8, 12)$, and $(12, 15)$ into $y = ax^2 + bx + c$.

$$\begin{cases} a(0)^2 + b(0) + c = 0 \\ a(8)^2 + b(8) + c = 12 \\ a(12)^2 + b(12) + c = 15 \end{cases}$$

Simplify.

$$\begin{cases} c = 0 & (1) \\ 64a + 8b + c = 12 & (2) \\ 144a + 12b + c = 15 & (3) \end{cases}$$

Multiply Equation 2 by 3 and Equation 4 by -2. Add Equations 2 and 3.

$$192a + 24b + 3c = 36 \quad (1)$$
$$\underline{-288a - 24b - 2c = -30 \quad (2)}$$
$$-96a + c = 6 \quad (4)$$

Substitute $c = 0$ into Equation 4.

$$-96a + c = 6 \quad (4)$$
$$-96a + 0 = 6$$
$$-96a = 6$$
$$a = -\frac{1}{16}$$

Substitute $c = 0$ and $a = -\frac{1}{16}$ into Equation 2.

$$64a + 8b + c = 12 \quad (2)$$
$$64\left(-\frac{1}{16}\right) + 8b + 0 = 12$$
$$-4 + 8b = 12$$
$$8b = 16$$
$$b = 2$$

$$\left(-\frac{1}{16}, 2, 0\right)$$

40. Let x = number of cups from Mix A,
y = cups from Mix B, and z = cups from
Mix C.

$$\begin{cases} 5x + 6y + 8z = 24 \quad (1) \\ 2x + 3y + 3z = 10 \quad (2) \\ x + 2y + z = 5 \qquad (3) \end{cases}$$

Multiply Equation 3 by –5 and add
Equations 1 and 3.

$$\begin{aligned} 5x + 6y + 8z &= 24 \quad &(1) \\ \underline{-5x - 10y - 5z} &= -25 \quad &(3) \\ -4y + 3z &= -1 \quad &(4) \end{aligned}$$

Multiply Equation 3 by –2 and add
Equations 2 and 3.

$$\begin{aligned} 2x + 3y + 3z &= 10 \quad &(2) \\ \underline{-2x - 4y - 2z} &= -10 \quad &(3) \\ -y + z &= 0 \quad &(5) \end{aligned}$$

Multiply Equation 5 by –4 and add
Equations 4 and 5.

$$\begin{aligned} -4y + 3z &= -1 \quad &(4) \\ \underline{4y - 4z} &= 0 \quad &(5) \\ -z &= -1 \\ z &= 1 \end{aligned}$$

Substitute $z = 1$ into Equation 5.

$$\begin{aligned} -y + z &= 0 \quad (5) \\ -y + 1 &= 0 \\ -y &= -1 \\ y &= 1 \end{aligned}$$

Substitute $y = 1$ and $z = 1$ into
Equation 3.

$$\begin{aligned} x + 2y + z &= 5 \quad (3) \\ x + 2(1) + 1 &= 5 \\ x + 2 + 1 &= 5 \\ x + 3 &= 5 \\ x &= 2 \end{aligned}$$

You need 2 cups of Mix A, 1 cup of Mix
B, and 1 cup of Mix C.

SECTION 12.4
Solving Systems of Equations Using Matrices

41. $\begin{bmatrix} 5 & 4 & \vdots & 3 \\ 1 & -1 & \vdots & -3 \end{bmatrix}$

42. $\begin{bmatrix} 1 & 2 & 3 & \vdots & 6 \\ 1 & -3 & -1 & \vdots & 4 \\ 6 & 1 & -2 & \vdots & -1 \end{bmatrix}$

43. a. $\begin{bmatrix} 1 & 3 & -2 \\ 6 & 12 & -6 \end{bmatrix}$

b. $\begin{bmatrix} 1 & 2 & -1 \\ 1 & 3 & -2 \end{bmatrix}$

c. $\begin{bmatrix} 0 & -6 & 6 \\ 1 & 3 & -2 \end{bmatrix}$

44. a. $\begin{bmatrix} 1 & 1 & 0 & -1 \\ 2 & -1 & 1 & 3 \\ 3 & -1 & -2 & 7 \end{bmatrix}$

b. $\begin{bmatrix} 2 & -1 & 1 & 3 \\ 3 & 3 & 0 & -3 \\ 3 & -1 & -2 & 7 \end{bmatrix}$

c. $\begin{bmatrix} 0 & -3 & 1 & 5 \\ 1 & 1 & 0 & -1 \\ 3 & -1 & -2 & 7 \end{bmatrix}$

45.

$\begin{bmatrix} 1 & -1 & 4 \\ 3 & 7 & -18 \end{bmatrix}$

$-3R_1 + R_2$

$\begin{bmatrix} 1 & -1 & 4 \\ 0 & 10 & -30 \end{bmatrix}$

$\dfrac{1}{10} R_2$

$\begin{bmatrix} 1 & -1 & 4 \\ 0 & 1 & -3 \end{bmatrix}$

This matrix represents the system

$$\begin{cases} x - y = 4 \\ y = -3 \end{cases}$$

$$\begin{aligned} x - y &= 4 \\ x - (-3) &= 4 \\ x + 3 &= 4 \\ x &= 1 \end{aligned}$$

The solution is $(1, -3)$.

46.

$$\begin{bmatrix} 1 & 2 & -3 & | & 5 \\ 1 & 1 & 1 & | & 0 \\ 3 & 4 & 2 & | & -1 \end{bmatrix}$$

$R_1 - R_2$

$$\begin{bmatrix} 1 & 2 & -3 & | & 5 \\ 0 & 1 & -4 & | & 5 \\ 3 & 4 & 2 & | & -1 \end{bmatrix}$$

$-3R_1 + R_3$

$$\begin{bmatrix} 1 & 2 & -3 & | & 5 \\ 0 & 1 & -4 & | & 5 \\ 0 & -2 & 11 & | & -16 \end{bmatrix}$$

$2R_2 + R_3$

$$\begin{bmatrix} 1 & 2 & -3 & | & 5 \\ 0 & 1 & -4 & | & 5 \\ 0 & 0 & 3 & | & -6 \end{bmatrix}$$

$\dfrac{1}{3}R_3$

$$\begin{bmatrix} 1 & 2 & -3 & | & 5 \\ 0 & 1 & -4 & | & 5 \\ 0 & 0 & 1 & | & -2 \end{bmatrix}$$

$$\begin{cases} x + 2y - 3z = 5 \\ y - 4z = 5 \\ z = -2 \end{cases}$$

$$y - 4(-2) = 5$$
$$y + 8 = 5$$
$$y = -3$$
$$x + 2(-3) - 3(-2) = 5$$
$$x - 6 + 6 = 5$$
$$x = 5$$

The solution is $(5, -3, -2)$.

47.

$$\begin{bmatrix} 16 & -8 & | & 32 \\ -2 & 1 & | & -4 \end{bmatrix}$$

$\dfrac{1}{16}R_1$

$$\begin{bmatrix} 1 & -\frac{1}{2} & | & 2 \\ -2 & 1 & | & -4 \end{bmatrix}$$

$2R_1 + R_2$

$$\begin{bmatrix} 1 & -\frac{1}{2} & | & 2 \\ 0 & 0 & | & 0 \end{bmatrix}$$

This matrix represents the system

$$\begin{cases} x - \dfrac{1}{2}y = 2 \\ 0 + 0 = 0 \end{cases}$$

Infinitely many solutions;
Dependent equations

48.

$$\begin{bmatrix} 1 & 2 & -1 & | & 4 \\ 1 & 3 & 4 & | & 1 \\ 2 & 4 & -2 & | & 3 \end{bmatrix}$$

$-R_1 + R_2$

$$\begin{bmatrix} 1 & 2 & -1 & | & 4 \\ 0 & 1 & 5 & | & -3 \\ 2 & 4 & -2 & | & 3 \end{bmatrix}$$

$-2R_1 + R_3$

$$\begin{bmatrix} 1 & 2 & -1 & | & 4 \\ 0 & 1 & 5 & | & -3 \\ 0 & 0 & 0 & | & -5 \end{bmatrix}$$

This matrix represents the system

$$\begin{cases} x + 2y - z = 4 \\ y + 5z = -3 \\ 0 + 0 + 0 = -5 \end{cases}$$

No solution; $\varnothing$;
Inconsistent system

49.
$$2(3) - 3(-4) = 6 + 12$$
$$= 18$$

50.
$$-3(-6) - (-4)(5) = 18 + 20$$
$$= 38$$

51.
$$-1\begin{vmatrix} -1 & 3 \\ -2 & 2 \end{vmatrix} - 2\begin{vmatrix} 2 & 3 \\ 1 & 2 \end{vmatrix} - 1\begin{vmatrix} 2 & -1 \\ 1 & -2 \end{vmatrix}$$
$$= -1(-2 + 6) - 2(4 - 3) - 1(-4 + 1)$$
$$= -1(4) - 2(1) - 1(-3)$$
$$= -4 - 2 + 3$$
$$= -3$$

52.
$$3\begin{vmatrix} -2 & -2 \\ 1 & -1 \end{vmatrix} + 2\begin{vmatrix} 1 & -2 \\ 2 & -1 \end{vmatrix} + 2\begin{vmatrix} 1 & -2 \\ 2 & 1 \end{vmatrix}$$
$$= 3(2 + 2) + 2(-1 + 4) + 2(1 + 4)$$
$$= 3(4) + 2(3) + 2(5)$$
$$= 12 + 6 + 10$$
$$= 28$$

53.
$$x = \frac{D_x}{D} \qquad\qquad y = \frac{D_y}{D}$$

$$= \frac{\begin{vmatrix} 10 & 4 \\ 1 & -3 \end{vmatrix}}{\begin{vmatrix} 3 & 4 \\ 2 & -3 \end{vmatrix}} \qquad = \frac{\begin{vmatrix} 3 & 10 \\ 2 & 1 \end{vmatrix}}{\begin{vmatrix} 3 & 4 \\ 2 & -3 \end{vmatrix}}$$

$$= \frac{10(-3) - 4(1)}{3(-3) - 4(2)} \qquad = \frac{3(1) - 10(2)}{3(-3) - 4(2)}$$

$$= \frac{-30 - 4}{-9 - 8} \qquad = \frac{3 - 20}{-9 - 8}$$

$$= \frac{-34}{-17} \qquad\qquad = \frac{-17}{-17}$$

$$= 2 \qquad\qquad\qquad = 1$$

The solution is (2, 1).

54.
$$x = \frac{D_x}{D}$$

$$= \frac{\begin{vmatrix} -6 & -4 \\ 5 & 2 \end{vmatrix}}{\begin{vmatrix} -6 & -4 \\ 3 & 2 \end{vmatrix}}$$

$$= \frac{-6(2) - (-4)(5)}{-6(2) - (-4)(3)}$$

$$= \frac{-12 + 20}{-12 + 12}$$

$$= \frac{8}{0}$$

No solution
inconsistent system

55.

$$x = \frac{D_x}{D}$$

$$= \frac{\begin{vmatrix} 0 & 2 & 1 \\ 3 & 1 & 1 \\ 5 & 1 & 2 \end{vmatrix}}{\begin{vmatrix} 1 & 2 & 1 \\ 2 & 1 & 1 \\ 1 & 1 & 2 \end{vmatrix}}$$

$$= \frac{0\begin{vmatrix} 1 & 1 \\ 1 & 2 \end{vmatrix} - 2\begin{vmatrix} 3 & 1 \\ 5 & 2 \end{vmatrix} + 1\begin{vmatrix} 3 & 1 \\ 5 & 1 \end{vmatrix}}{1\begin{vmatrix} 1 & 1 \\ 1 & 2 \end{vmatrix} - 2\begin{vmatrix} 2 & 1 \\ 1 & 2 \end{vmatrix} + 1\begin{vmatrix} 2 & 1 \\ 1 & 1 \end{vmatrix}}$$

$$= \frac{0(2-1) - 2(6-5) + 1(3-5)}{1(2-1) - 2(4-1) + 1(2-1)}$$

$$= \frac{0(1) - 2(1) + 1(-2)}{1(1) - 2(3) + 1(1)}$$

$$= \frac{0 - 2 - 2}{1 - 6 + 1}$$

$$= \frac{-4}{-4}$$

$$= 1$$

$$y = \frac{D_y}{D}$$

$$= \frac{\begin{vmatrix} 1 & 0 & 1 \\ 2 & 3 & 1 \\ 1 & 5 & 2 \end{vmatrix}}{\begin{vmatrix} 1 & 2 & 1 \\ 2 & 1 & 1 \\ 1 & 1 & 2 \end{vmatrix}}$$

$$= \frac{1\begin{vmatrix} 3 & 1 \\ 5 & 2 \end{vmatrix} - 0\begin{vmatrix} 2 & 1 \\ 1 & 2 \end{vmatrix} + 1\begin{vmatrix} 2 & 3 \\ 1 & 5 \end{vmatrix}}{1\begin{vmatrix} 1 & 1 \\ 1 & 2 \end{vmatrix} - 2\begin{vmatrix} 2 & 1 \\ 1 & 2 \end{vmatrix} + 1\begin{vmatrix} 2 & 1 \\ 1 & 1 \end{vmatrix}}$$

$$= \frac{1(6-5) - 0(4-1) + 1(10-3)}{1(2-1) - 2(4-1) + 1(2-1)}$$

$$= \frac{1(1) - 0(3) + 1(7)}{1(1) - 2(3) + 1(1)}$$

$$= \frac{1 - 0 + 7}{1 - 6 + 1}$$

$$= \frac{8}{-4}$$

$$= -2$$

$$z = \frac{D_z}{D}$$

$$= \frac{\begin{vmatrix} 1 & 2 & 0 \\ 2 & 1 & 3 \\ 1 & 1 & 5 \end{vmatrix}}{\begin{vmatrix} 1 & 2 & 1 \\ 2 & 1 & 1 \\ 1 & 1 & 2 \end{vmatrix}}$$

$$= \frac{1\begin{vmatrix} 1 & 3 \\ 1 & 5 \end{vmatrix} - 2\begin{vmatrix} 2 & 3 \\ 1 & 5 \end{vmatrix} + 0\begin{vmatrix} 2 & 1 \\ 1 & 1 \end{vmatrix}}{1\begin{vmatrix} 1 & 1 \\ 1 & 2 \end{vmatrix} - 2\begin{vmatrix} 2 & 1 \\ 1 & 2 \end{vmatrix} + 1\begin{vmatrix} 2 & 1 \\ 1 & 1 \end{vmatrix}}$$

$$= \frac{1(5-3) - 2(10-3) + 0(2-1)}{1(2-1) - 2(4-1) + 1(2-1)}$$

$$= \frac{1(2) - 2(7) + 0(1)}{1(1) - 2(3) + 1(1)}$$

$$= \frac{2 - 14 + 0}{1 - 6 + 1}$$

$$= \frac{-12}{-4}$$

$$= 3$$

The solution is $(1, -2, 3)$.

56.

$$x = \frac{D_x}{D}$$

$$= \frac{\begin{vmatrix} 2 & 3 & 1 \\ 7 & 3 & 2 \\ -7 & -1 & -1 \end{vmatrix}}{\begin{vmatrix} 2 & 3 & 1 \\ 1 & 3 & 2 \\ 1 & -1 & -1 \end{vmatrix}}$$

$$= \frac{2\begin{vmatrix} 3 & 2 \\ -1 & -1 \end{vmatrix} - 3\begin{vmatrix} 7 & 2 \\ -7 & -1 \end{vmatrix} + 1\begin{vmatrix} 7 & 3 \\ -7 & -1 \end{vmatrix}}{2\begin{vmatrix} 3 & 2 \\ -1 & -1 \end{vmatrix} - 3\begin{vmatrix} 1 & 2 \\ 1 & -1 \end{vmatrix} + 1\begin{vmatrix} 1 & 3 \\ 1 & -1 \end{vmatrix}}$$

$$= \frac{2(-3+2) - 3(-7+14) + 1(-7+21)}{2(-3+2) - 3(-1-2) + 1(-1-3)}$$

$$= \frac{2(-1) - 3(7) + 1(14)}{2(-1) - 3(-3) + 1(-4)}$$

$$= \frac{-2-21+14}{-2+9-4}$$

$$= \frac{-9}{3}$$

$$= -3$$

$$y = \frac{D_y}{D}$$

$$= \frac{\begin{vmatrix} 2 & 2 & 1 \\ 1 & 7 & 2 \\ 1 & -7 & -1 \end{vmatrix}}{\begin{vmatrix} 2 & 3 & 1 \\ 1 & 3 & 2 \\ 1 & -1 & -1 \end{vmatrix}}$$

$$= \frac{2\begin{vmatrix} 7 & 2 \\ -7 & -1 \end{vmatrix} - 2\begin{vmatrix} 1 & 2 \\ 1 & -1 \end{vmatrix} + 1\begin{vmatrix} 1 & 7 \\ 1 & -7 \end{vmatrix}}{2\begin{vmatrix} 3 & 2 \\ -1 & -1 \end{vmatrix} - 3\begin{vmatrix} 1 & 2 \\ 1 & -1 \end{vmatrix} + 1\begin{vmatrix} 1 & 3 \\ 1 & -1 \end{vmatrix}}$$

$$= \frac{2(-7+14) - 2(-1-2) + 1(-7-7)}{2(-3+2) - 3(-1-2) + 1(-1-3)}$$

$$= \frac{2(7) - 2(-3) + 1(-14)}{2(-1) - 3(-3) + 1(-4)}$$

$$= \frac{14+6-14}{-2+9-4}$$

$$= \frac{6}{3}$$

$$= 2$$

$$z = \frac{D_z}{D}$$

$$= \frac{\begin{vmatrix} 2 & 3 & 2 \\ 1 & 3 & 7 \\ 1 & -1 & -7 \end{vmatrix}}{\begin{vmatrix} 2 & 3 & 1 \\ 1 & 3 & 2 \\ 1 & -1 & -1 \end{vmatrix}}$$

$$= \frac{2\begin{vmatrix} 3 & 7 \\ -1 & -7 \end{vmatrix} - 3\begin{vmatrix} 1 & 7 \\ 1 & -7 \end{vmatrix} + 2\begin{vmatrix} 1 & 3 \\ 1 & -1 \end{vmatrix}}{2\begin{vmatrix} 3 & 2 \\ -1 & -1 \end{vmatrix} - 3\begin{vmatrix} 1 & 2 \\ 1 & -1 \end{vmatrix} + 1\begin{vmatrix} 1 & 3 \\ 1 & -1 \end{vmatrix}}$$

$$= \frac{2(-21+7) - 3(-7-7) + 2(-1-3)}{2(-3+2) - 3(-1-2) + 1(-1-3)}$$

$$= \frac{2(-14) - 3(-14) + 2(-4)}{2(-1) - 3(-3) + 1(-4)}$$

$$= \frac{-28+42-8}{-2+9-4}$$

$$= \frac{6}{3}$$

$$= 2$$

The solution is (–3, 2, 2).

CHAPTER 12 TEST

1. a. $\begin{cases} 2x - 7y = 1 \\ 4x - y = -8 \end{cases}$ is called a **system** of linear equations.

   b. The matrix $\begin{bmatrix} -10 & 3 \\ 4 & 9 \end{bmatrix}$ has 2 **rows** and 2 **columns**.

   c. Solutions of a system of three equations in three variables x, y, and z, are written in the form (x, y, z) and are called ordered **triples**.

   d. The graph of the equation $2x + 3y + 4z = 5$ is a flat surface called a **plane**.

   e. A **matrix** is a rectangular array of numbers written with brackets.

2. $\begin{cases} 2x + y = 5 \\ y = 2x - 3 \end{cases}$

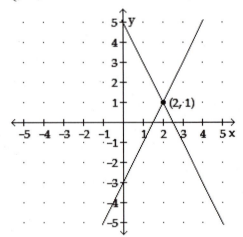

(2, 1)

3. Yes, it is a solution.

$$10x - 12y = 3 \qquad\qquad 18x - 15y = 1$$

$$10\left(-\frac{1}{2}\right) - 12\left(\frac{2}{3}\right) \overset{?}{=} 3 \qquad 18\left(-\frac{1}{2}\right) - 15\left(-\frac{2}{3}\right) \overset{?}{=} 1$$

$$-5 + 8 \overset{?}{=} 3 \qquad\qquad -9 + 10 \overset{?}{=} 1$$

$$3 = 3 \qquad\qquad\qquad 1 = 1$$

4. The point of intersection is (June 2006, 44%). That means that Governor Schwarzenegger's job approval and disapproval ratings were the same in June of 2006; approximately 44%.

5. a. inconsistent system; no solution, $\varnothing$

   b. dependent equations; infinitely many solutions

6. The graphs intersect at $x = 3$.

7. $\begin{cases} 2x - 4y = 14 \\ x + 2y = 7 \end{cases}$

   Solve the second equation for x.

   $\begin{cases} 2x - 4y = 14 \\ x = -2y + 7 \end{cases}$

   Substitute $x = -2y + 7$ into the first equation and solve for y.

   $$2x + 4y = 14$$
   $$2(-2y + 7) - 4y = 14$$
   $$-4y + 14 - 4y = 14$$
   $$-8y + 14 = 14$$
   $$-8y = 0$$
   $$y = 0$$

   Substitute $y = 0$ into the second equation and solve for x.

   $$x = -2y + 7$$
   $$x = -2(0) + 7$$
   $$x = 0 + 7$$
   $$x = 7$$

   The solution is $(7, 0)$.

8. $\begin{cases} 2c + 3d = -5 \\ 3c - 2d = 12 \end{cases}$

   Multiply the first equation by -3 and the second equation by 2 to eliminate c.

   $$-6c - 9d = 15$$
   $$\underline{6c - 4d = 24}$$
   $$-13d = 39$$
   $$d = -3$$

   Substitute $d = -3$ into the first equation and solve for c.

   $$2c + 3(-3) = -5$$
   $$2c - 9 = -5$$
   $$2c = 4$$
   $$c = 2$$

   The solution is $(2, -3)$.

9.

$$\begin{cases} 3(x+y)=x-3 \\ -y=\dfrac{2x+3}{3} \end{cases}$$

Write the equations in standard form.

$$\begin{cases} 3x+3y=x-3 \\ -3y=2x+3 \end{cases}$$

$$\begin{cases} 2x+3y=-3 \\ 2x+3y=-3 \end{cases}$$

$$\{(x,y)\mid 2x+3y=-3\}$$

The equations are the same so they are dependent and there are infinitely many solutions.

10.

$$\begin{cases} 0.6x+0.5y=1.2 \\ x-\dfrac{4}{9}y+\dfrac{5}{9}=0 \end{cases}$$

Multiply the first equation by 10 to clear decimals and the second equation by 9 to eliminate fractions and write in standard form.

$$\begin{cases} 6x+5y=12 \quad (1) \\ 9x-4y=-5 \quad (2) \end{cases}$$

Multiply Equation 1 by 4 and Equation 2 by 5.

$$24x+20y=48 \quad (1)$$
$$\underline{45x-20y=-25} \quad (2)$$
$$69x=23$$
$$x=\dfrac{23}{69}$$
$$x=\dfrac{1}{3}$$

Substitute $x=\dfrac{1}{3}$ into Equation 1 and

solve for y.
$$6x+5y=12 \quad (1)$$
$$6\left(\dfrac{1}{3}\right)+5y=12$$
$$2+5y=12$$
$$5y=10$$
$$y=2$$

The solution is $\left(\dfrac{1}{3},2\right)$.

11. The sum of the angles of a triangle is 180°.

$$\begin{cases} 2x+y=180 \\ y=x+15 \end{cases}$$

Substitute $y=x+15$ into the first equation and solve for x.

$$2x+(x+15)=180$$
$$3x+15=180$$
$$3x=165$$
$$x=55$$

Substitute $x=55$ into the second equation and solve for y.

$$y=55+15$$
$$y=70$$

The measure of the angles are $55°$ and $70°$.

12. Let x = gallons of 40% solution and y = gallons of 80% solution.

$$\begin{cases} x+y=20 \\ 0.40x+0.80y=0.50(20) \end{cases}$$

Multiply the second equation by 100 and write in standard form.

$$\begin{cases} x+y=20 \\ 40x+80y=50(20) \end{cases}$$

$$\begin{cases} x+y=20 \\ 40x+80y=1{,}000 \end{cases}$$

Multiply the first equation by -40 and add the equations to eliminate x.

$$-40x-40y=-800$$
$$\underline{40x+80y=1{,}000}$$
$$40y=200$$
$$y=5$$

Substitute $y=5$ into the first equation and solve for x.

$$x+5=20$$
$$x=15$$

15 gallons of 40% solution and 5 gallons of 80% solution are needed.

13. Let $x = $ # of avalanches in US and $y = $ # in Canada.
$$\begin{cases} x + y = 48 \\ x = 3y \end{cases}$$
Substitute $x = 3y$ into the first equation and solve for y.
$$x + y = 48$$
$$3y + y = 48$$
$$4y = 48$$
$$y = 12$$
Substitute $y = 12$ into the second equation and solve for x.
$$x = 3y$$
$$x = 3(12)$$
$$x = 36$$
36 avalanches in US and 12 in Canada

14. Let $x = $ speed walking and $y = $ speed of the walkway.
$$\begin{cases} 40(x + y) = 320 \\ 80(x - y) = 320 \end{cases}$$
Distribute.
$$\begin{cases} 40x + 40y = 320 \\ 80x - 80y = 320 \end{cases}$$
Multiply the first equation by 2 and add both equations to eliminate y.
$$80x + 80y = 640$$
$$\underline{80x - 80y = 320}$$
$$160x = 960$$
$$x = 6$$
Substitute $x = 6$ into the first equation and solve for y.
$$40(6 + y) = 320$$
$$240 + 40y = 320$$
$$40y = 80$$
$$y = 2$$
His walking pace is 6 ft per sec and the walkway moves at 2 ft per sec.

15. Let $x = $ amount of 2.5% loan and $y = $ amount of the 4% loan.
$$\begin{cases} x + y = 8,500 & (1) \\ 0.025x + 0.04y = 265 & (2) \end{cases}$$
Multiply the Equation 2 by 1000 to eliminate decimals.
$$\begin{cases} x + y = 8,500 & (1) \\ 25x + 40y = 265,000 & (2) \end{cases}$$
Multiply the Equation 1 by -25 and add Equations 1 and 2 to solve for y.
$$-25x - 25y = -212,500$$
$$\underline{25x + 40y = \ \ 265,000}$$
$$15y = 52,500$$
$$y = 3,500$$
Substitute $y = 3,500$ into Equation 1 and solve for x.
$$x + y = 8,500$$
$$x + 3,500 = 8,500$$
$$x = 5,000$$
The amount of the 2.5% loan was \$5,000 and the amount of the 4% loan was \$3,500.

16. Let $x = $ the cost of one pear and $y = $ the cost of one apple.
$$\begin{cases} 6x + 4y = 25.50 & (1) \\ 4x + 10y = 31.30 & (2) \end{cases}$$
Multiply Equation 1 by 4 and Equation 2 by -6. Add the equations to solve for y.
$$24x + 16y = \ \ \ 102$$
$$\underline{-24x - 60y = -187.8}$$
$$-44y = -85.8$$
$$y = 1.95$$
Substitute $y = 1.95$ into Equation 1 and solve for x.
$$6x + 4y = 22.5$$
$$6x + 4(1.95) = 25.5$$
$$6x + 7.8 = 25.5$$
$$6x = 17.7$$
$$x = 2.95$$
The price of one pear is \$2.95 and the price of one apple is \$1.95.

17. No.

Check Equation 1.
$$x - 2y + z = 5$$
$$-1 - 2\left(-\frac{1}{2}\right) + 5 \overset{?}{=} 5$$
$$-1 + 1 + 5 \overset{?}{=} 5$$
$$5 = 5$$

Check Equation 2.
$$2x + 4y = -4$$
$$2(-1) + 4\left(-\frac{1}{2}\right) \overset{?}{=} -4$$
$$-2 - 2 \overset{?}{=} -4$$
$$-4 = -4$$

Check Equation 3.
$$-6y + 4z = 22$$
$$-6\left(-\frac{1}{2}\right) + 4(5) \overset{?}{=} 22$$
$$3 + 20 \overset{?}{=} 23$$
$$23 \ne 22$$

18. It has no solutions since the three lines do not all intersect at the same point.

19.
$$\begin{cases} x + y + z = 4 & (1) \\ x + y - z = 6 & (2) \\ 2x - 3y + z = -1 & (3) \end{cases}$$

Multiply Equation 1 by -1 and add Equations 1 and 2.

$$\begin{aligned} -x - y - z &= -4 \quad (1) \\ x + y - z &= 6 \quad (2) \\ \hline -2z &= 2 \\ z &= -1 \end{aligned}$$

Multiply Equation 1 by -2 and add Equations 1 and 3 to eliminate x.

$$\begin{aligned} -2x - 2y - 2z &= -8 \quad (1) \\ 2x - 3y + z &= -1 \quad (3) \\ \hline -5y - z &= -9 \quad (4) \end{aligned}$$

Substitute $z = -1$ into Equation 4 and solve for y.

$$\begin{aligned} -5y - z &= -9 \quad (4) \\ -5y - (-1) &= -9 \\ -5y + 1 &= -9 \\ -5y &= -10 \\ y &= 2 \end{aligned}$$

Substitute $z = -1$ and $y = 2$ into Equation 1 and solve for x.

$$\begin{aligned} x + y + z &= 4 \quad (1) \\ x + 2 - 1 &= 4 \\ x + 1 &= 4 \\ x &= 3 \end{aligned}$$

The solution is $(3, 2, -1)$.

20. Write the equations in standard form.

$$\begin{cases} -2y + z = 1 & (1) \\ x + y + z = 1 & (2) \\ x + 5y = 4 & (3) \end{cases}$$

Multiply Equation 1 by -1 and add Equations 1 and 2.

$$2y - z = -1 \quad (1)$$
$$\underline{x + y + z = 1 \quad (2)}$$
$$x + 3y = 0 \quad (4)$$

Multiply Equation 3 by -1 and add Equations 3 and 4.

$$-x - 5y = -4 \quad (3)$$
$$\underline{x + 3y = 0 \quad (4)}$$
$$-2y = -4$$
$$y = 2$$

Substitute $y = 2$ into Equation 1.

$$-2y + z = 1 \quad (1)$$
$$-2(2) + z = 1$$
$$-4 + z = 1$$
$$z = 5$$

Substitute $y = 2$ into Equation 3.

$$x + 5y = 4 \quad (3)$$
$$x + 5(2) = 4$$
$$x + 10 = 4$$
$$x = -6$$

The solution is $(-6, 2, 5)$.

21. Let x = children's tickets sold, y = general admission tickets, and z = seniors tickets.

$$\begin{cases} x + y + z = 100 \\ 3x + 6y + 5z = 410 \\ x = 2y \end{cases}$$

Write the third equation in standard form.

$$\begin{cases} x + y + z = 100 & (1) \\ 3x + 6y + 5z = 410 & (2) \\ x - 2y = 0 & (3) \end{cases}$$

Multiply Equation 1 by -5 to eliminate z and add Equations 1 and 2.

$$-5x - 5y - 5z = -500$$
$$\underline{3x + 6y + 5z = 410}$$
$$-2x + y = -90 \quad (4)$$

Multiply Equation 3 by 2 and add Equations 3 and 4 to eliminate x.

$$2x - 4y = 0 \quad (3)$$
$$\underline{-2x + y = -90 \quad (4)}$$
$$-3y = -90$$
$$y = 30$$

Substitute $y = 30$ into Equation 3 and solve for x.

$$x - 2y = 0 \quad (3)$$
$$x - 2(30) = 0$$
$$x - 60 = 0$$
$$x = 60$$

Substitute $y = 30$ and $x = 60$ into Equation 1 and solve for z.

$$x + y + z = 100 \quad (1)$$
$$60 + 30 + z = 100$$
$$90 + z = 100$$
$$z = 10$$

They sold 60 children's ticket, 30 general admission tickets, and 10 seniors' tickets.

22. Let x = the weight of the bar, y = the weight of the small plate, z = the weight of the large plate.

$$\begin{cases} x+2y+2z=155 & (1) \\ x+2y+4z=245 & (2) \\ x+6y+6z=375 & (3) \end{cases}$$

Multiply Equation 1 by -1 and add Equations 1 and 2.

$$\begin{aligned} -x-2y-2z &= -155 \\ \underline{x+2y+4z =\;\; 245} \\ 2z &= -90 \end{aligned}$$

$$z = 45$$

Multiply Equation 3 by -1 and add Equations 1 and 3.

$$\begin{aligned} -x-6y-6z &= -375 \\ \underline{x+2y+4z =\;\; 245} \\ -4y-2z &= -130 \quad (4) \end{aligned}$$

Substitute $z = 45$ into Equation 4 and solve for y.

$$-4y-2z = -130 \quad (4)$$
$$-4y-2(45) = -130$$
$$-4y-90 = -130$$
$$-4y-90+90 = -130+90$$
$$-4y = -40$$
$$y = 10$$

Substitute $y = 10$ and $z = 45$ into Equation 1 and solve for x.

$$x+2y+2z = 155$$
$$x+2(10)+2(45) = 155$$
$$x+20+90 = 155$$
$$x+110 = 155$$
$$x+110-110 = 155-110$$
$$x = 45$$

The bar weighs 45 lbs, the small plates weighs 10 lbs, and the large plate weights 45 lbs.

23. $\begin{bmatrix} 1 & 7 & -3 \\ 0 & -22 & 22 \end{bmatrix}$

24.

$\begin{bmatrix} 1 & 1 & 4 \\ 2 & -1 & 2 \end{bmatrix}$

$-2R_1 + R_2$

$\begin{bmatrix} 1 & 1 & 4 \\ 0 & -3 & -6 \end{bmatrix}$

$-\dfrac{1}{3}R_2$

$\begin{bmatrix} 1 & 1 & 4 \\ 0 & 1 & 2 \end{bmatrix}$

This matrix represents the system

$$\begin{cases} x+y=4 \\ y=2 \end{cases}$$
$$x+2=4$$
$$x=2$$

The solution is $(2, 2)$.

25.

$\begin{bmatrix} 1 & -3 & 2 & 1 \\ 1 & -2 & 3 & 5 \\ 2 & -6 & 4 & 3 \end{bmatrix}$

$-1R_1 + R_2$

$\begin{bmatrix} 1 & -3 & 2 & 1 \\ 0 & 1 & 1 & 4 \\ 2 & -6 & 4 & 3 \end{bmatrix}$

$-2R_1 + R_3$

$\begin{bmatrix} 1 & -3 & 2 & 1 \\ 0 & 1 & 1 & 4 \\ 0 & 0 & 0 & 1 \end{bmatrix}$

This matrix represents the system

$$\begin{cases} x-3y+2z=1 \\ y+z=4 \\ 0=1 \end{cases}$$

There is No Solution, $\varnothing$.

Inconsistent system

26. $\begin{cases} a+2b+2c=10 & (1) \\ 2a+3b-c=6 & (2) \\ 3a+b+5c=8 & (3) \end{cases}$

Multiply the first equation by –2 and add the first two equations.

$-2a-4b-4c=-20$

$\underline{2a+3b-c=6}$

$-b-5c=-14 \quad (4)$

Multiply the first equation by –3 and add the first equation and the third equation.

$-3a-6b-6c=-30$

$\underline{3a+b+5c=8}$

$-5b-c=-22 \quad (5)$

Multiply Equation 4 by –5 and add Equations 4 and 5.

$5b+25c=70$

$\underline{-5b-c=-22}$

$24c=48$

$c=2$

Substitute $c=2$ into Equation 4 and solve for b.

$-b-5(2)=-14$

$-b-10=-14$

$-b=-4$

$b=4$

Substitute $b=4$ and $c=2$ into Equation 1 and solve for a.

$a+2b+2c=10$

$a+2(4)+2(2)=10$

$a+8+4=10$

$a+12=10$

$a=-2$

The solution is (–2, 4, 2).

27. $\begin{vmatrix} 2 & -3 \\ -4 & 5 \end{vmatrix} = 2(5)-(-3)(-4)$

$=10-12$

$=-2$

28. $1\begin{vmatrix} 0 & 3 \\ -2 & 2 \end{vmatrix} - 2\begin{vmatrix} 2 & 3 \\ 1 & 2 \end{vmatrix} + 0\begin{vmatrix} 2 & 0 \\ 1 & -2 \end{vmatrix}$

$=1(0+6)-2(4-3)+0(-4-0)$

$=1(6)-2(1)+0(-4)$

$=6-2+0$

$=4$

29. The solution is (–3, 3).

$x=\dfrac{D_x}{D}$ $\qquad\qquad$ $y=\dfrac{D_y}{D}$

$=\dfrac{\begin{vmatrix} -6 & -1 \\ -6 & 1 \end{vmatrix}}{\begin{vmatrix} 1 & -1 \\ 3 & 1 \end{vmatrix}}$ $\qquad$ $=\dfrac{\begin{vmatrix} 1 & -6 \\ 3 & -6 \end{vmatrix}}{\begin{vmatrix} 1 & -1 \\ 3 & 1 \end{vmatrix}}$

$=\dfrac{-6(1)-(-1)(-6)}{1(1)-(-1)(3)}$ $\qquad$ $=\dfrac{1(-6)-(-6)(3)}{1(1)-(-1)(3)}$

$=\dfrac{-6-6}{1+3}$ $\qquad\qquad$ $=\dfrac{-6+18}{1+3}$

$=\dfrac{-12}{4}$ $\qquad\qquad$ $=\dfrac{12}{4}$

$=-3$ $\qquad\qquad\qquad$ $=3$

30. $z=\dfrac{D_z}{D}$

$=\dfrac{\begin{vmatrix} 1 & 1 & 4 \\ 1 & 1 & 6 \\ 2 & -3 & -1 \end{vmatrix}}{\begin{vmatrix} 1 & 1 & 1 \\ 1 & 1 & -1 \\ 2 & -3 & 1 \end{vmatrix}}$

$=\dfrac{1\begin{vmatrix} 1 & 6 \\ -3 & -1 \end{vmatrix} - 1\begin{vmatrix} 1 & 6 \\ 2 & -1 \end{vmatrix} + 4\begin{vmatrix} 1 & 1 \\ 2 & -3 \end{vmatrix}}{1\begin{vmatrix} 1 & -1 \\ -3 & 1 \end{vmatrix} - 1\begin{vmatrix} 1 & -1 \\ 2 & 1 \end{vmatrix} + 1\begin{vmatrix} 1 & 1 \\ 2 & -3 \end{vmatrix}}$

$=\dfrac{1(-1+18)-1(-1-12)+4(-3-2)}{1(1-3)-1(1+2)+1(-3-2)}$

$=\dfrac{1(17)-1(-13)+4(-5)}{1(-2)-1(3)+1(-5)}$

$=\dfrac{17+13-20}{-2-3-5}$

$=\dfrac{10}{-10}$

$=-1$

1. a. true
    b. true
    c. true

2.

$$252 + 51 - 24 + 127 + 109 + 32 + 14 - 48 - 149 - 36$$
$$= 328$$

There was a net inflow of $328 billion.

3.

$$-4 + 2\left[-7 - 3(-9)\right] = -4 + 2\left[-7 + 27\right]$$
$$= -4 + 2\left[20\right]$$
$$= -4 + 40$$
$$= 36$$

4.

$$\left|\frac{4}{5} \cdot 10 - 12\right| = \left|8 - 12\right|$$
$$= \left|-4\right|$$
$$= 4$$

5.

$$(x-a)^2 + (y-b)^2 = (-2-5)^2 + (1-(-3))^2$$
$$= (-7)^2 + (4)^2$$
$$= 49 + 16$$
$$= 65$$

6.

$$3p - 6(p-9) + p = 3p - 6p + 54 + p$$
$$= -2p + 54$$

7.

$$\frac{5}{6}k = 10$$
$$\frac{6}{5}\left(\frac{5}{6}k\right) = \frac{6}{5}(10)$$
$$k = 12$$

8.

$$-(3a+1) + a = 2$$
$$-3a - 1 + a = 2$$
$$-2a - 1 = 2$$
$$-2a = 3$$
$$a = -\frac{3}{2}$$

9. $60 - 54.12 = $5.88, so he was charged $5.88 to use the machine.

5.88 is what percent of 60?

$$5.88 = x(60)$$
$$\frac{5.88}{60} = x$$
$$0.098 = x$$
$$9.8\% = x$$

10.

$$T = 2r + 2t$$
$$T - 2t = 2r + 2t - 2t$$
$$T - 2t = 2r$$
$$\frac{T - 2t}{2} = \frac{2r}{2}$$
$$\frac{T - 2t}{2} = r$$

11. Let x = listing price.
    List price – commission = profit
$$x - 0.04x = 330,000$$
$$0.96x = 330,000$$
$$x = 343,750$$
    It should be listed at $343,750.

12. Let x = amount lent at 7%.
    Then $(28,000 - x)$ is amount lent at 10%.
$$0.07x + 0.10(28,000 - x) = 2,560$$
$$0.07x + 2,800 - 0.10x = 2,560$$
$$2,800 - 0.03x = 2,560$$
$$-0.03x = -240$$
$$x = 8,000$$
$$28,000 - 8,000 = 20,000$$
    $8,000 was borrowed at 7% and $20,000 at 10%.

13.

$$5x + 7 < 2x + 1$$
$$3x + 7 < 1$$
$$3x < -6$$
$$x < -2$$
$$(-\infty, -2)$$

Chapter 12 Cumulative Review

14. $2x - 3y = -1;\quad (-5, -3)$

$$2(-5) - 3(-3) \overset{?}{=} -1$$

$$-10 + 9 \overset{?}{=} -1$$

$$-1 = -1$$

Yes, it is a solution.

15. $y = -x + 2$

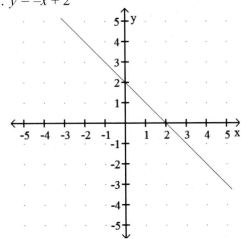

16. $2y - 2x = 6$

$$y = x + 3$$

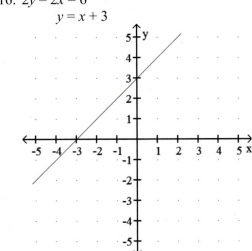

17. $y = -3$

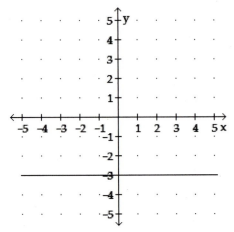

18. $y < 3x$

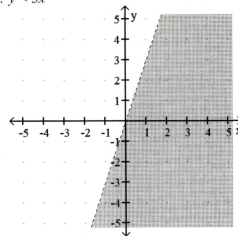

19.

$$m = \frac{y_2 - y_1}{x_2 - x_1}$$

$$= \frac{-8 - (-2)}{-12 - (-2)}$$

$$= \frac{-6}{-10}$$

$$= \frac{3}{5}$$

20. Find the rate of change (slope) using the ordered pairs (2009, 22,900,000) and (1995, 35,500,000).

$$m = \frac{y_2 - y_1}{x_2 - x_1}$$

$$= \frac{35,500,000 - 22,900,000}{1995 - 2009}$$

$$= \frac{12,600,000}{-14}$$

$$= -900,000$$

A decrease of 900,000 viewers per year

21. Write the equation is slope-intercept form.
$$4x + 5y = 6$$

$$5y = -4x + 6$$

$$y = \frac{-4}{5}x + \frac{6}{5}$$

$$m = -\frac{4}{5}$$

22. $m = -2$ and $b = 1$

$$y = mx + b$$

$$y = -2x + 1$$

23. Compare the slopes.

$y = 4x + 9$ $x + 4y = -10$

$m = 4$ $4y = -x - 10$

$$y = -\frac{1}{4}x - \frac{5}{2}$$

$$m = -\frac{1}{4}$$

The slopes are opposite reciprocals so the lines are perpendicular.

24. $m = \dfrac{1}{4}$, $(8, 1)$

$$y - y_1 = m(x - x_1)$$

$$y - 1 = \frac{1}{4}(x - 8)$$

$$y - 1 = \frac{1}{4}x - 2$$

$$y = \frac{1}{4}x - 1$$

25. $(-2, -1)$ and $m = \dfrac{4}{3}$

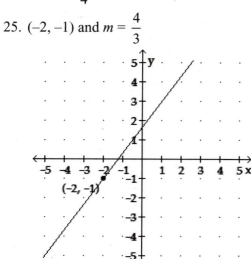

26. $f(x) = 3x^2 + 3x - 8$

$f(-1) = 3(-1)^2 + 3(-1) - 8$

$\qquad = 3(1) + 3(-1) - 8$

$\qquad = 3 - 3 - 8$

$\qquad = -8$

27. D: $\{-4, 1, 4, 5\}$
   R: $\{-3, 2, 8\}$

28. a. Yes. It passes the vertical line test.
   b. No. $(0, 10)$ and $(0, 12)$

29. $\begin{cases} x + y = 1 \\ y = x + 5 \end{cases}$

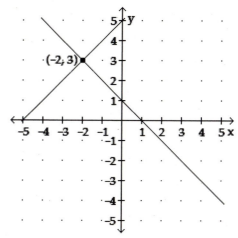

The solution is $(-2, 3)$.

30. $\begin{cases} y = 2x + 5 \\ x + 2y = -5 \end{cases}$

Substitute $y = 2x + 5$ into the second equation and solve for x.

$$x + 2y = -5$$

$$x + 2(2x + 5) = -5$$

$$x + 4x + 10 = -5$$

$$5x + 10 = -5$$

$$5x = -15$$

$$x = -3$$

Substitute $x = -3$ into the first equation and solve for y.

$$y = 2x + 5$$

$$y = 2(-3) + 5$$

$$y = -6 + 5$$

$$y = -1$$

The solution is $(-3, -1)$.

Chapter 12 Cumulative Review

31. $\begin{cases} \dfrac{3}{5}s + \dfrac{4}{5}t = 1 \\ -\dfrac{1}{4}s + \dfrac{3}{8}t = 1 \end{cases}$

Multiply each equation by the LCD to eliminate fractions.

$\begin{cases} 5\left(\dfrac{3}{5}s + \dfrac{4}{5}t\right) = 5(1) \\ 8\left(-\dfrac{1}{4}s + \dfrac{3}{8}t\right) = 8(1) \end{cases}$

$\begin{cases} 3s + 4t = 5 \\ -2s + 3t = 8 \end{cases}$

Multiply the first equation by 2 and the second equation by 3 to create opposites, and then add the two equations.

$\begin{array}{r} 6s + 8t = 10 \\ -6s + 9t = 24 \\ \hline 17t = 34 \\ t = 2 \end{array}$

Substitute $t = 2$ into the first equation and solve for s.

$$3s + 4t = 5$$
$$3s + 4(2) = 5$$
$$3s + 8 = 5$$
$$3s = -3$$
$$s = -1$$

The solution is $(-1, 2)$.

32. Let s = the speed of the plane in still air and w = the speed of the wind. Then the speed of the plane flying with the wind is $s + w$ and the speed of the plane flying against the wind is $s - w$.

$\begin{cases} 5(s + w) = 3{,}000 \\ 6(s - w) = 3{,}000 \end{cases}$

Distribute.

$\begin{cases} 5s + 5w = 3{,}000 \\ 6s - 6w = 3{,}000 \end{cases}$

Multiply the 1st equation by 6 and the 2nd equation by 5 and add equations to eliminate the w.

$\begin{array}{r} 30s + 30w = 18{,}000 \\ 30s - 30w = 15{,}000 \\ \hline 60s = 33{,}000 \\ s = 550 \end{array}$

The speed of the plane is 550 mph.

33.

	Hard Candy	Soft Candy	Mixture
# of lbs	x	y	48
Value	$4x$	$6y$	$4.50(48) =$ 216

$\begin{cases} x + y = 48 \\ 4x + 6y = 216 \end{cases}$

Multiply the first equation by -4, and add the equations to eliminate x.

$\begin{array}{r} -4x - 4y = -192 \\ 4x + 6y = 216 \\ \hline 2y = 24 \\ y = 12 \end{array}$

Substitute $y = 12$ into the first equation and solve for x.

$$x + y = 48$$
$$x + 12 = 48$$
$$x = 36$$

36 pounds of hard candy and 12 pounds of soft candy are needed.

34. $\begin{cases} 3x + 4y \geq -7 \\ 2x - 3y \geq 1 \end{cases}$

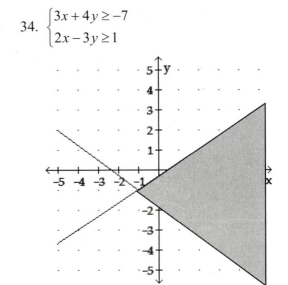

35.
$$y^3\left(y^2 y^4\right) = y^3\left(y^{2+4}\right)$$
$$= y^3\left(y^6\right)$$
$$= y^{3+6}$$
$$= y^9$$

36.

$$\left(\frac{b^2}{3a}\right)^3 = \frac{\left(b^2\right)^3}{(3a)^3}$$

$$= \frac{b^{2(3)}}{3^3 a^3}$$

$$= \frac{b^6}{27a^3}$$

37.

$$\frac{10a^4 a^{-2}}{5a^2 a^0} = \frac{10a^{4-2}}{5a^{2+0}}$$

$$= \frac{10a^2}{5a^2}$$

$$= 2a^{2-2}$$

$$= 2$$

38.

$$\left(\frac{21x^{-2}y^2 z^{-2}}{7x^3 y^{-1}}\right)^{-2} = \left(\frac{7x^3 y^{-1}}{21x^{-2}y^2 z^{-2}}\right)^2$$

$$= \left(\frac{7}{21}x^{3-(-2)}y^{-1-2}z^{0-(-2)}\right)^2$$

$$= \left(\frac{1}{3}x^5 y^{-3} z^2\right)^2$$

$$= \left(\frac{x^5 z^2}{3y^3}\right)^2$$

$$= \frac{x^{5\cdot 2}z^{2\cdot 2}}{3^2 y^{3\cdot 2}}$$

$$= \frac{x^{10}z^4}{9y^6}$$

39. $2.6 \times 10^6 = 2{,}600{,}000$

40. $0.00073 = 7.3 \times 10^{-4}$

41. $y = x^3 - 2$

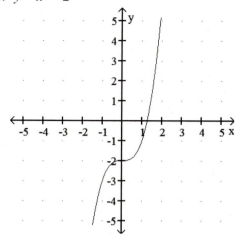

42.

$$P = 2L + 2W$$

$$= 2\left(x^3 + 3x\right) + 2\left(2x^3 - x\right)$$

$$= 2x^3 + 6x + 4x^3 - 2x$$

$$= 6x^3 + 4x$$

43.

$$4\left(4x^3 + 2x^2 - 3x - 8\right) - 5\left(2x^3 - 3x + 8\right)$$

$$= 16x^3 + 8x^2 - 12x - 32 - 10x^3 + 15x - 40$$

$$= 6x^3 + 8x^2 + 3x - 72$$

44.

$$\left(-2a^3\right)\left(3a^2\right) = -6a^{3+2}$$

$$= -6a^5$$

45.

$$(2b - 1)(3b + 4) = 6b^2 + 8b - 3b - 4$$

$$= 6b^2 + 5b - 4$$

46.

$$(3x + y)\left(2x^2 - 3xy + y^2\right)$$

$$= 3x\left(2x^2 - 3xy + y^2\right) + y\left(2x^2 - 3xy + y^2\right)$$

$$= 6x^3 - 9x^2 y + 3xy^2 + 2x^2 y - 3xy^2 + y^3$$

$$= 6x^3 - 7x^2 y + y^3$$

47.

$$(2x + 5y)^2 = (2x + 5y)(2x + 5y)$$

$$= 4x^2 + 10xy + 10xy + 25y^2$$

$$= 4x^2 + 20xy + 25y^2$$

Chapter 12 Cumulative Review

48.
$$(9m^2 - 1)(9m^2 + 1) = 81m^4 - 1$$

49.
$$\frac{12a^3b - 9a^2b^2 + 3ab}{6a^2b} = \frac{12a^3b}{6a^2b} - \frac{9a^2b^2}{6a^2b} + \frac{3ab}{6a^2b}$$
$$= 2a - \frac{3}{2}b + \frac{1}{2a}$$

50. Write the problem in descending order before dividing.

$$\begin{array}{r} 2x+1 \\ x-3 \overline{\smash{\big)}\ 2x^2 - 5x - 3} \\ \underline{2x^2 - 6x} \\ x - 3 \\ \underline{x - 3} \\ 0 \end{array}$$

The answer is $2x + 1$.

51.
$$6a^2 - 12a^3b + 36ab = 6a\left(a - 2a^2b + 6b\right)$$

52.
$$2x + 2y + ax + ay = (2x + 2y) + (ax + ay)$$
$$= 2(x + y) + a(x + y)$$
$$= (x + y)(2 + a)$$

53.
$$x^2 - 6x - 16 = (x - 8)(x + 2)$$

54.
$$30y^5 + 63y^4 - 30y^3 = 3y^3\left(10y^2 + 21y - 10\right)$$
$$= 3y^3(5y - 2)(2y + 5)$$

55.
$$t^4 - 16 = \left(t^2 + 4\right)\left(t^2 - 4\right)$$
$$= \left(t^2 + 4\right)(t + 2)(t - 2)$$

56.
$$b^3 + 125 = (b)^3 + (5)^3$$
$$= (b + 5)\left(b^2 - 5b + 25\right)$$

57.
$$3x^2 + 8x = 0$$
$$x(3x + 8) = 0$$
$$x = 0 \quad \text{or} \quad 3x + 8 = 0$$
$$3x = -8$$
$$x = -\frac{8}{3}$$

58.
$$15x^2 - 2 = 7x$$
$$15x^2 - 7x - 2 = 0$$
$$(3x - 2)(5x + 1) = 0$$
$$3x - 2 = 0 \quad \text{or} \quad 5x + 1 = 0$$
$$3x = 2 \qquad\qquad 5x = -1$$
$$x = \frac{2}{3} \qquad\qquad x = -\frac{1}{5}$$

59.
$$A = \frac{1}{2}bh$$
$$22.5 = \frac{1}{2}(x + 4)(x)$$
$$2(22.5) = 2\left(\frac{1}{2}\right)(x + 4)(x)$$
$$45 = x(x + 4)$$
$$45 = x^2 + 4x$$
$$0 = x^2 + 4x - 45$$
$$0 = (x + 9)(x - 5)$$
$$x + 9 = 0 \quad \text{or} \quad x - 5 = 0$$
$$x = \cancel{-9} \qquad\qquad x = 5$$

Since the height cannot be negative, the height of the triangle is 5 inches.

60.
$$x + 8 = 0$$
$$x = -8$$

61.
$$\frac{3x^2 - 27}{x^2 + 3x - 18} = \frac{3\left(x^2 - 9\right)}{x^2 + 3x - 18}$$
$$= \frac{3(x + 3)\cancel{(x - 3)}}{(x + 6)\cancel{(x - 3)}}$$
$$= \frac{3(x + 3)}{x + 6}$$

62.

$$\frac{a-15}{15-a} = \frac{a-15}{-(a-15)}$$
$$= -1$$

63.

$$\frac{x^2 - x - 6}{2x^2 + 9x + 10} \div \frac{x^2 - 25}{2x^2 + 15x + 25}$$

$$= \frac{x^2 - x - 6}{2x^2 + 9x + 10} \cdot \frac{2x^2 + 15x + 25}{x^2 - 25}$$

$$= \frac{(x-3)\cancel{(x+2)}}{\cancel{(2x+5)}\cancel{(x+2)}} \cdot \frac{\cancel{(2x+5)}\cancel{(x+5)}}{\cancel{(x+5)}(x-5)}$$

$$= \frac{x-3}{x-5}$$

64.

$$\frac{1}{s^2 - 4s - 5} + \frac{s}{s^2 - 4s - 5} = \frac{s+1}{s^2 - 4s - 5}$$

$$= \frac{s+1}{(s+1)(s-5)}$$

$$= \frac{1}{s-5}$$

65.

$$\frac{x+5}{xy} - \frac{x-1}{x^2 y} = \frac{x+5}{xy}\left(\frac{x}{x}\right) - \frac{x-1}{x^2 y}$$

$$= \frac{x^2 + 5x}{x^2 y} - \frac{x-1}{x^2 y}$$

$$= \frac{x^2 + 5x - (x-1)}{x^2 y}$$

$$= \frac{x^2 + 5x - x + 1}{x^2 y}$$

$$= \frac{x^2 + 4x + 1}{x^2 y}$$

66.

$$\frac{x}{x-2} + \frac{3x}{x^2 - 4} = \frac{x}{x-2} + \frac{3x}{(x-2)(x+2)}$$

$$= \frac{x}{x-2}\left(\frac{x+2}{x+2}\right) + \frac{3x}{(x-2)(x+2)}$$

$$= \frac{x^2 + 2x}{(x-2)(x+2)} + \frac{3x}{(x-2)(x+2)}$$

$$= \frac{x^2 + 5x}{(x-2)(x+2)}$$

67.

$$\frac{\dfrac{9m-27}{m^6}}{\dfrac{2m-6}{m^8}} = \frac{\dfrac{9m-27}{m^6}}{\dfrac{2m-6}{m^8}} \cdot \frac{m^8}{m^8}$$

$$= \frac{m^2(9m-27)}{2m-6}$$

$$= \frac{9m^2\cancel{(m-3)}}{2\cancel{(m-3)}}$$

$$= \frac{9m^2}{2}$$

68.

$$\frac{\dfrac{5}{y} + \dfrac{4}{y+1}}{\dfrac{4}{y} - \dfrac{5}{y+1}} = \frac{\dfrac{5}{y} + \dfrac{4}{y+1}}{\dfrac{4}{y} - \dfrac{5}{y+1}} \cdot \frac{y(y+1)}{y(y+1)}$$

$$= \frac{\dfrac{5y(y+1)}{y} + \dfrac{4y(y+1)}{y+1}}{\dfrac{4y(y+1)}{y} - \dfrac{5y(y+1)}{y+1}}$$

$$= \frac{5(y+1) + 4y}{4(y+1) - 5y}$$

$$= \frac{5y + 5 + 4y}{4y + 4 - 5y}$$

$$= \frac{9y + 5}{-y + 4}$$

$$= \frac{9y + 5}{4 - y}$$

69.

$$\frac{2p}{3} - \frac{1}{p} = \frac{2p-1}{3}$$

$$3p\left(\frac{2p}{3} - \frac{1}{p}\right) = 3p\left(\frac{2p-1}{3}\right)$$

$$2p^2 - 3 = 2p^2 - p$$

$$2p^2 - 3 - 2p^2 = 2p^2 - p - 2p^2$$

$$-3 = -p$$

$$3 = p$$

Chapter 12 Cumulative Review

70.

$$\frac{7}{q^2 - q - 2} + \frac{1}{q+1} = \frac{3}{q-2}$$

$$\frac{7}{(q+1)(q-2)} + \frac{1}{q+1} = \frac{3}{q-2}$$

$$(q+1)(q-2)\left(\frac{7}{(q+1)(q-2)} + \frac{1}{q+1}\right) = (q+1)(q-2)\left(\frac{3}{q-2}\right)$$

$$7 + (q-2) = 3(q+1)$$

$$5 + q = 3q + 3$$

$$5 - 2q = 3$$

$$-2q = -2$$

$$q = 1$$

71.

$$\frac{1}{a} + \frac{1}{b} = 1$$

$$ab\left(\frac{1}{a} + \frac{1}{b}\right) = ab(1)$$

$$b + a = ab$$

$$b = ab - a$$

$$b = a(b-1)$$

$$\frac{b}{b-1} = a$$

72. Let x = # days it will take them to roof the house together.

Homeowner's work in 1 day = $\frac{1}{7}$

Roofer's work in 1 day = $\frac{1}{4}$

Work together in 1 day = $\frac{1}{x}$

Owner's work + roofer's work = work together

$$\frac{1}{7} + \frac{1}{4} = \frac{1}{x}$$

$$28x\left(\frac{1}{7} + \frac{1}{4}\right) = 28x\left(\frac{1}{x}\right)$$

$$4x + 7x = 28$$

$$11x = 28$$

$$x = \frac{28}{11}$$

$$x = 2\frac{6}{11}$$

It will take $2\frac{6}{11}$ days to roof the house working together.

73. Let x = days to lose 25 pounds.

$$\frac{10 \text{ pounds}}{350 \text{ days}} = \frac{25 \text{ pounds}}{x \text{ days}}$$

$$10(x) = 25(350)$$

$$10x = 8,750$$

$$x = 875$$

74.

$$\frac{6}{4} = \frac{x}{26}$$

$$6(26) = x(4)$$

$$156 = 4x$$

$$39 = x$$

75.

$$\frac{8}{5} - \frac{x+2}{5} = \frac{x+2}{5} - 4x$$

$$5\left(\frac{8}{5} - \frac{x+2}{5}\right) = 5\left(\frac{x+2}{5} - 4x\right)$$

$$8 - (x+2) = x + 2 - 20x$$

$$8 - x - 2 = x + 2 - 20x$$

$$-x + 6 = -19x + 2$$

$$18x + 6 = 2$$

$$18x = -4$$

$$x = -\frac{4}{18}$$

$$x = -\frac{2}{9}$$

76.

$$-6.2(-a-1) - 4 = 4.2a - (-2a)$$

$$6.2a + 6.2 - 4 = 4.2a + 2a$$

$$6.2a + 2.2 = 6.2a$$

$$2.2 \neq 0$$

No Solution

77.
$$A = \frac{1}{2}h(b_1 + b_2)$$
$$2 \cdot A = 2 \cdot \frac{1}{2}h(b_1 + b_2)$$
$$2A = h(b_1 + b_2)$$
$$2A = hb_1 + hb_2$$
$$2A - hb_2 = hb_1$$
$$\frac{2A - hb_2}{h} = b_1$$

78. Let x = # of students.
$$1,810 - 15.50x = 1,500$$
$$-15.50x = -310$$
$$x = 20$$

79.
$$2(5x - 6) > 4x - 15 + 6x$$
$$10x - 12 > 10x - 15$$
$$-12 > -15$$
$$(-\infty, \infty); \ \Re$$

0

80.
$$-100,000t + 3,600,000 < 2,400,000$$
$$-100,000t < -1,200,000$$
$$t > 12$$
$$1990 + 12 = 2002$$
Years after 2002

81. $f(x) = -x^2 - \dfrac{x}{2}$

a.
$$f(10) = -(10)^2 - \frac{10}{2}$$
$$= -100 - 5$$
$$= -105$$

b.
$$f(r) = -r^2 - \frac{r}{2}$$

82. a. Find the slope using (5, 4,500) and (15, 2,800).
$$m = \frac{y_2 - y_1}{x_2 - x_1}$$
$$= \frac{4,500 - 2,800}{5 - 15}$$
$$= \frac{1,700}{-10}$$
$$= -170$$
Use $m = -170$ and (5, 4,500).
$$y - y_1 = m(x - x_1)$$
$$y - 4,500 = -170(x - 5)$$
$$y - 4,500 = -170x + 850$$
$$y = -170x + 5,350$$
$$A(t) = -170t + 5,350$$

b. Let $t = 20$.
$$A(25) = -170(25) + 5,350$$
$$= 1,100 \text{ accidents}$$

83. a. 4
b. 5
c. −2, 1
d. D: $(-\infty, \infty)$; R: $(-\infty, \infty)$

84. $h(t) = -16t^2 + 144t$
$$h(3) = -16(3)^2 + 144(3)$$
$$= -16(9) + 432$$
$$= -144 + 432$$
$$= 288 \text{ ft}$$

85. a. Yes.
b. No. It does not pass the vertical line test.

Chapter 12 Cumulative Review

86. a.   D: the set of real numbers
         R: the set of real numbers greater than
            or equal to 0

$$f(x) = x^2 \quad \text{and} \quad g(x) = (x+4)^2$$

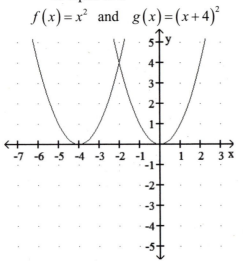

   b. $g(x) = |x| + 3$

87.

$$\left|\frac{x-2}{3}\right| - 4 \le 0$$

$$\left|\frac{x-2}{3}\right| \le 4$$

$$-4 \le \frac{x-2}{3} \le 4$$

$$3(-4) \le 3\left(\frac{x-2}{3}\right) \le 3(4)$$

$$-12 \le x - 2 \le 12$$

$$-12 + 2 \le x - 2 + 2 \le 12 + 2$$

$$-10 \le x \le 14$$

$$[-10, 14]$$

```
  ←————[————————]————→
     -10          14
```

88.

$$3x + 2 < 8 \quad \text{or} \quad 2x - 3 > 11$$

$$3x < 6 \qquad\qquad 2x > 14$$

$$x < 2 \qquad\qquad x > 7$$

$$(-\infty, 2) \cup (7, \infty)$$

```
  ←————————)———(————————→
           2   7
```

89.

$$x^2 + 4x + 4 - y^2 = (x+2)^2 - y^2$$

$$= (x+2-y)(x+2+y)$$

90.

$$b^4 - 17b^2 + 16 = (b^2 - 1)(b^2 - 16)$$

$$= (b+1)(b-1)(b+4)(b-4)$$

91. $$f(x) = \frac{5x-10}{x^3 - 3x}$$

$$x^3 - 3x \ne 0$$

$$x(x^2 - 9) \ne 0$$

$$x(x-3)(x+3) \ne 0$$

$$x \ne 0 \quad x - 3 \ne 0 \quad x + 3 \ne 0$$

$$x \ne 0 \qquad x \ne 3 \qquad x \ne -3$$

$$(-\infty, -3) \cup (-3, 0) \cup (0, 3) \cup (3, \infty)$$

92.

$$(10n - n^2) \cdot \frac{n^6}{n^4 - 10n^3 - 2n^2 + 20n}$$

$$= n(10 - n) \cdot \frac{n^6}{n(n-10)(n^2 - 2)}$$

$$= -\frac{n^6}{n^2 - 2}$$

93.

$$\frac{2x^2y + xy - 6y}{3x^2y + 5xy - 2y} = \frac{\cancel{y}(2x-3)\cancel{(x+2)}}{\cancel{y}(3x-1)\cancel{(x+2)}}$$

$$= \frac{2x-3}{3x-1}$$

94. $f(x) = \dfrac{1}{x}$

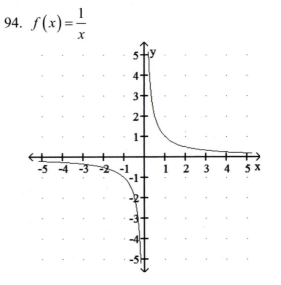

95. Find k.

$w = kx$

$1.2 = k(4)$

$\dfrac{1.2}{4} = k$

$0.3 = k$

Find w.

$w = kx$

$w = 0.3(30)$

$w = 9$

96. Let s = speed of the gears and t = # of teeth.

Find k when $s = 3$ and $t = 10$

$s = \dfrac{k}{t}$

$3 = \dfrac{k}{10}$

$10(3) = 10\left(\dfrac{k}{10}\right)$

$30 = k$

Find s when $k = 30$ and $t = 25$.

$s = \dfrac{k}{t}$

$s = \dfrac{30}{25}$

$s = 1.2$

97. D: $[2, \infty)$; R: $[0, \infty)$

98. $C(L) = 1.25\sqrt{L}$

$C(6.5) = 1.25\sqrt{6.5}$

≈ 3.2 meters per second

99. $\sqrt{4x^2} = 2|x|$

100.

$(-8)^{-4/3} = \left(-\dfrac{1}{8}\right)^{4/3}$

$= \left(\sqrt[3]{-\dfrac{1}{8}}\right)^4$

$= \left(-\dfrac{1}{2}\right)^4$

$= \dfrac{1}{16}$

101.

$\sqrt{100a^6 b^4} = 10a^3 b^2$

102.

$\sqrt[4]{16x^7 y^4} = \sqrt[4]{16x^4 y^4}\,\sqrt[4]{x^3}$

$= 2xy\sqrt[4]{x^3}$

103.

$3\sqrt{24} + \sqrt{54} = 3\sqrt{4}\sqrt{6} + \sqrt{9}\sqrt{6}$

$= 3(2)\sqrt{6} + 3\sqrt{6}$

$= 6\sqrt{6} + 3\sqrt{6}$

$= 9\sqrt{6}$

104.

$-3\sqrt[4]{32} - 2\sqrt[4]{162} + 5\sqrt[4]{48}$

$= -3\sqrt[4]{16}\sqrt[4]{2} - 2\sqrt[4]{81}\sqrt[4]{2} + 5\sqrt[4]{16}\sqrt[4]{3}$

$= -3(2)\sqrt[4]{2} - 2(3)\sqrt[4]{2} + 5(2)\sqrt[4]{3}$

$= -6\sqrt[4]{2} - 6\sqrt[4]{2} + 10\sqrt[4]{3}$

$= -12\sqrt[4]{2} + 10\sqrt[4]{3}$

105.

$\sqrt{\dfrac{72x^3}{y^2}} = \dfrac{\sqrt{72x^3}}{\sqrt{y^2}}$

$= \dfrac{\sqrt{36x^2}\sqrt{2x}}{\sqrt{y^2}}$

$= \dfrac{6x\sqrt{2x}}{y}$

Chapter 12 Cumulative Review

106.

$$\sqrt[3]{\frac{27m^3}{8n^6}} = \frac{\sqrt[3]{27m^3}}{\sqrt[3]{8n^6}}$$

$$= \frac{3m}{2n^2}$$

107.

$$\frac{2}{\sqrt[3]{a}} = \frac{2}{\sqrt[3]{a}} \cdot \frac{\sqrt[3]{a^2}}{\sqrt[3]{a^2}}$$

$$= \frac{2\sqrt[3]{a^2}}{\sqrt[3]{a^3}}$$

$$= \frac{2\sqrt[3]{a^2}}{a}$$

108.

$$\frac{\sqrt{x}-\sqrt{y}}{\sqrt{x}+\sqrt{y}} = \frac{\sqrt{x}-\sqrt{y}}{\sqrt{x}+\sqrt{y}}\left(\frac{\sqrt{x}-\sqrt{y}}{\sqrt{x}-\sqrt{y}}\right)$$

$$= \frac{\sqrt{x^2}-\sqrt{xy}-\sqrt{xy}+\sqrt{y^2}}{\sqrt{x^2}-\sqrt{y^2}}$$

$$= \frac{x-2\sqrt{xy}+y}{x-y}$$

109.

$$2+\sqrt{u} = \sqrt{2u+7}$$

$$\left(2+\sqrt{u}\right)^2 = \left(\sqrt{2u+7}\right)^2$$

$$4+2\sqrt{u}+2\sqrt{u}+\sqrt{u^2} = 2u+7$$

$$4+4\sqrt{u}+u = 2u+7$$

$$4+4\sqrt{u}+u-4-u = 2u+7-4-u$$

$$4\sqrt{u} = u+3$$

$$\left(4\sqrt{u}\right)^2 = \left(u+3\right)^2$$

$$16u = u^2+6u+9$$

$$0 = u^2-10u+9$$

$$0 = \left(u-1\right)\left(u-9\right)$$

$$u-1=0 \quad \text{or} \quad u-9=0$$

$$u=1 \qquad\qquad u=9$$

110. Use the Pythagorean Theorem to find x.

$$x^2+x^2 = 15^2$$

$$2x^2 = 225$$

$$x^2 = 112.5$$

$$x = \sqrt{112.5}$$

$$x \approx 10.6$$

The height of the entire storage arrangement is $x + x = 2x$.

$$2x \approx 2\left(10.6\right)$$

$$\approx 21.2$$

The height of the entire storage arrangement is about 21.2 inches.

111.

$$\sqrt{-49} = 7i$$

112.

$$\sqrt{-54} = \sqrt{-9}\sqrt{6}$$

$$= 3i\sqrt{6}$$

113. a.

$$\left(2+3i\right)-\left(1-2i\right) = 2+3i-1+2i$$

$$= 1+5i$$

b.

$$\left(7-4i\right)+\left(9+2i\right) = 16-2i$$

114. a.

$$\left(3-2i\right)\left(4-3i\right) = 12-9i-8i+6i^2$$

$$= 12-17i+6\left(-1\right)$$

$$= 12-17i-6$$

$$= 6-17i$$

b.

$$\frac{3-i}{2+i} = \frac{3-i}{2+i}\cdot\frac{2-i}{2-i}$$

$$= \frac{6-3i-2i+i^2}{4-i^2}$$

$$= \frac{6-5i+\left(-1\right)}{4-\left(-1\right)}$$

$$= \frac{5-5i}{5}$$

$$= \frac{5}{5}-\frac{5}{5}i$$

$$= 1-i$$

115.

$$x^2 + 8x + 12 = 0$$
$$x^2 + 8x = -12$$
$$x^2 + 8x + 16 = -12 + 16$$
$$(x+4)^2 = 4$$
$$\sqrt{(x+4)^2} = \pm\sqrt{4}$$
$$x + 4 = 2 \quad \text{or} \quad x + 4 = -2$$
$$x = -2 \qquad x = -6$$

116. $\quad 4x^2 - x - 2 = 0$
$$a = 4, b = -1, c = -2$$
$$x = \frac{-b \pm \sqrt{b^2 - 4ac}}{2a}$$
$$= \frac{-(-1) \pm \sqrt{(-1)^2 - 4(4)(-2)}}{2(4)}$$
$$= \frac{1 \pm \sqrt{1 + 32}}{8}$$
$$= \frac{1 \pm \sqrt{33}}{8}$$
$$x = \frac{1 + \sqrt{33}}{8} \quad \text{or} \quad x = \frac{1 - \sqrt{33}}{8}$$
$$x \approx 0.84 \qquad x \approx -0.59$$

117.

$$x^2 + 16 = 0$$
$$x^2 = -16$$
$$x = \pm\sqrt{-16}$$
$$x = \pm 4i$$
$$x = 0 \pm 4i$$

118.

$$x^2 - 4x = -5$$
$$x^2 - 4x + 5 = 0$$
$$a = 1, b = -4, c = 5$$
$$x = \frac{-b \pm \sqrt{b^2 - 4ac}}{2a}$$
$$= \frac{-(-4) \pm \sqrt{(-4)^2 - 4(1)(5)}}{2(1)}$$
$$= \frac{4 \pm \sqrt{16 - 20}}{2}$$
$$= \frac{4 \pm \sqrt{-4}}{2}$$
$$= \frac{4 \pm 2i}{2}$$
$$= \frac{4}{2} \pm \frac{2}{2}i$$
$$x = 2 \pm i$$

119. $\quad$ Let $x = a^{1/3}$.

$$a^{2/3} + a^{1/3} - 6 = 0$$
$$\left(a^{1/3}\right)^2 + a^{1/3} - 6 = 0$$
$$x^2 + x - 6 = 0$$
$$(x+3)(x-2) = 0$$
$$x + 3 = 0 \quad \text{or} \quad x - 2 = 0$$
$$x = -3 \qquad x = 2$$
$$a^{1/3} = -3 \qquad a^{1/3} = 2$$
$$\left(a^{1/3}\right)^3 = (-3)^3 \quad \left(a^{1/3}\right)^3 = (2)^3$$
$$a = -27 \qquad a = 8$$

Chapter 12 Cumulative Review

120. $y = 2x^2 + 8x + 6$

Step 1: Since $a = 2 > 0$, the parabola opens upward.

Step 2: Find the vertex and axis of symmetry.

$$x = \frac{-b}{2a} = \frac{-8}{2(2)} = -\frac{8}{4} = -2$$

$$y = 2(-2)^2 + 8(-2) + 6$$
$$= 8 - 16 + 6$$
$$= -2$$

vertex: $(-2, \; -2)$

The axis of symmetry is $x = -2$.

Step 3: Find the x– and y–intercepts.
Since $c = 6$, the y–intercept is $(0, 6)$.
To find the x–intercepts, let $y = 0$ and solve the equation for x.

$$0 = 2x^2 + 8x + 6$$
$$0 = 2(x^2 + 4x + 3)$$
$$0 = (x + 3)(x + 1)$$
$$x + 3 = 0 \quad \text{or} \quad x + 1 = 0$$
$$x = -3 \qquad \qquad x = -1$$

x–intercepts: $(-3, 0)$ and $(-1, 0)$
y–intercept: $(0, 6)$

Step 4: Use symmetry to find the point $(-4, 6)$.

Step 5: Plot the points and draw the parabola.

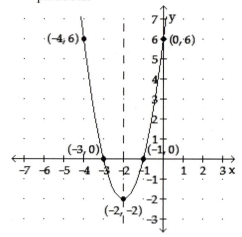

121.
$$(f \circ g)(-3) = f(g(-3))$$
$$= f\left((-3)^2 + (-3)\right)$$
$$= f(9 - 3)$$
$$= f(6)$$
$$= 3(6) - 2$$
$$= 18 - 2$$
$$= 16$$

122.
$$f(x) = -\frac{3}{2}x + 3$$
$$y = -\frac{3}{2}x + 3$$
$$x = -\frac{3}{2}y + 3$$
$$x - 3 = -\frac{3}{2}y$$
$$-\frac{2}{3}(x - 3) = -\frac{2}{3}\left(-\frac{3}{2}y\right)$$
$$-\frac{2}{3}x + 2 = y$$
$$y = -\frac{2}{3}x + 2$$
$$f^{-1}(x) = -\frac{2}{3}x + 2$$

123. $\quad f(x) = 5^x$

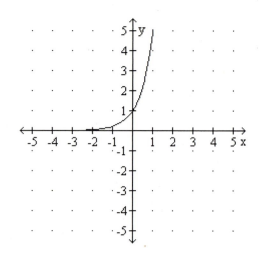

$$D : (-\infty, \; \infty); \; R : (0, \; \infty)$$

124. $f(x) = \ln x$

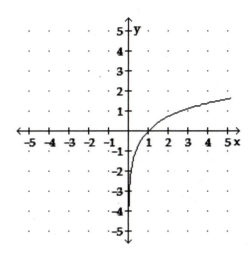

$$D:(0,\,\infty);\ R:(-\infty,\,\infty)$$

125. $e \approx 2.7$

126. WORLD POPULATION GROWTH
$P = 6.9$, $r = 0.01092$, and $t = 20$
$$A = Pe^{rt}$$
$$= 6.9e^{(0.01092)(20)}$$
$$= 6.9e^{0.2184}$$
$$\approx 8.6 \text{ billion}$$

127.
$$\log_x 5 = 1$$
$$x^1 = 5$$
$$x = 5$$

128.
$$\log_8 x = 2$$
$$8^2 = x$$
$$64 = x$$

129.
$$\log_9 \frac{1}{81} = x$$
$$9^x = \frac{1}{81}$$
$$9^x = 9^{-2}$$
$$x = -2$$

130. $\ln e = 1$

131.
$$\ln \frac{y^3 \sqrt{x}}{z} = \ln \frac{y^3 x^{1/2}}{z}$$
$$= \ln\left(y^3 x^{1/2}\right) - \ln z$$
$$= \ln y^3 + \ln x^{1/2} - \ln z$$
$$= 3\ln y + \frac{1}{2}\ln x - \ln z$$

132.
$$2\log x - 3\log y + \log z = \log x^2 - \log y^3 + \log z$$
$$= \log \frac{x^2 z}{y^3}$$

133.
$$5^{x-3} = 3^{2x}$$
$$\log 5^{x-3} = \log 3^{2x}$$
$$(x-3)\log 5 = (2x)\log 3$$
$$x\log 5 - 3\log 5 = 2x\log 3$$
$$-3\log 5 = 2x\log 3 - x\log 5$$
$$-3\log 5 = x(2\log 3 - \log 5)$$
$$\frac{-3\log 5}{2\log 3 - \log 5} = \frac{x(2\log 3 - \log 5)}{2\log 3 - \log 5}$$
$$\frac{-3\log 5}{2\log 3 - \log 5} = x$$
$$x = \frac{-3\log 5}{2\log 3 - \log 5}$$
$$x \approx -8.2144$$

134.
$$\log(x + 90) = 3 - \log x$$
$$\log(x + 90) + \log x = 3$$
$$\log x(x + 90) = 3$$
$$\log_{10}\left(x^2 + 90x\right) = 3$$
$$10^3 = x^2 + 90x$$
$$1,000 = x^2 + 90x$$
$$0 = x^2 + 90x - 1,000$$
$$0 = (x + 100)(x - 10)$$
$$x + 100 = 0 \quad \text{or} \quad x - 10 = 0$$
$$x = \cancel{-100} \qquad x = 10$$
Since the log of a negative number is undefined, $x \ne -100$. So, $x = 10$.

135.

$$\begin{cases} x - y + z = 4 & (1) \\ x + 2y - z = -1 & (2) \\ x + y - 3z = -2 & (3) \end{cases}$$

Add Equation 1 and 2 to eliminate z.

$$\begin{array}{rl} x - y + z = & 4 \quad (1) \\ x + 2y - z = & -1 \quad (2) \\ \hline 2x + y \quad = & 3 \quad (4) \end{array}$$

Multiply Equation 2 by -3 and
add Equations 2 and 3 to eliminate z.

$$\begin{array}{rl} -3x - 6y + 3z = 3 & (2) \\ x + \ y - 3z = -2 & (3) \\ \hline -2x - 5y \quad = 1 & (5) \end{array}$$

Add Equations 4 and 5 to eliminate x.

$$\begin{array}{rl} 2x + \ y = 3 & (4) \\ -2x - 5y = 1 & (5) \\ \hline -4y = 4 & (5) \\ y = -1 \end{array}$$

Substitute $y = -1$ into Equation 4 to find x.

$$\begin{array}{rl} 2x + y = 3 & (4) \\ 2x - 1 = 3 \\ 2x = 4 \\ x = 2 \end{array}$$

Substitute $x = 2$ and $y = -1$ into
Equation 1 to find z.

$$\begin{array}{rl} x - y + z = 4 & (1) \\ 2 - (-1) + z = 4 \\ 3 + z = 4 \\ z = 1 \end{array}$$

The solution is $(2, -1, 1)$.

136.

$$\begin{cases} 2x + y = 1 \\ x + 2y = -4 \end{cases}$$

$$\begin{bmatrix} 2 & 1 & \vdots & 1 \\ 1 & 2 & \vdots & -4 \end{bmatrix}$$

$R_1 \leftrightarrow R_2$

$$\begin{bmatrix} 1 & 2 & \vdots & -4 \\ 2 & 1 & \vdots & 1 \end{bmatrix}$$

$-2R_1 + R_2$

$$\begin{bmatrix} 1 & 2 & \vdots & -4 \\ 0 & -3 & \vdots & 9 \end{bmatrix}$$

$-\dfrac{1}{3}R_2$

$$\begin{bmatrix} 1 & 2 & \vdots & -4 \\ 0 & 1 & \vdots & -3 \end{bmatrix}$$

This matrix represents the system

$$\begin{cases} x + 2y = -4 \\ y = -3 \end{cases}$$

$$x - 6 = -4$$
$$x = 2$$

The solution is $(2, -3)$.

137.

$$\begin{cases} 3x - 4y = 9 \\ x + 2y = 8 \end{cases}$$

$$x = \frac{D_x}{D} \qquad\qquad y = \frac{D_y}{D}$$

$$= \frac{\begin{vmatrix} 9 & -4 \\ 8 & 2 \end{vmatrix}}{\begin{vmatrix} 3 & -4 \\ 1 & 2 \end{vmatrix}} \qquad\qquad = \frac{\begin{vmatrix} 3 & 9 \\ 1 & 8 \end{vmatrix}}{\begin{vmatrix} 3 & -4 \\ 1 & 2 \end{vmatrix}}$$

$$= \frac{9(2) - (-4)(8)}{3(2) - (-4)(1)} \qquad = \frac{3(8) - 9(1)}{3(2) - (-4)(1)}$$

$$= \frac{18 + 32}{6 + 4} \qquad\qquad = \frac{24 - 9}{6 + 4}$$

$$= \frac{50}{10} \qquad\qquad\qquad = \frac{15}{10}$$

$$= 5 \qquad\qquad\qquad = \frac{3}{2}$$

The solution is $\left(5, \dfrac{3}{2}\right)$.

138.   TRIANGLES

$$\begin{cases} D+E+F=180 \\ D=(E+F)-100 \\ F=2E-40 \end{cases}$$

Simplify.

$$\begin{cases} D+E+F=180 & (1) \\ D-E-F=-100 & (2) \\ -2E+F=-40 & (3) \end{cases}$$

Add Equations 1 and 2.

$$D+E+F=180 \quad (1)$$
$$\underline{D-E-F=-100} \quad (2)$$
$$2D=80$$
$$D=40$$

Substitute $D=40$ into Equation 2.

$$D-E-F=-100 \quad (2)$$
$$40-E-F=-100$$
$$-E-F=-140 \quad (4)$$

Add Equations 3 and 4.

$$-2E+F=-40 \quad (3)$$
$$\underline{-E-F=-140} \quad (4)$$
$$-3E=-180$$
$$E=60$$

Substitute $E=60$ into Equation 3.

$$-2E+F=-40 \quad (3)$$
$$-2(60)+F=-40$$
$$-120+F=-40$$
$$F=80$$

The measure of the angles are
$\angle D=40°, \angle E=60°,$ and $\angle F=80°.$

SECTION 13.1

VOCABULARY

1. The curves formed by the intersection of a plane with an infinite right–circular cone are called **conic sections**.

3. A **circle** is the set of all points in a plane that are a fixed distance from a fixed point called its center. The fixed distance is called the **radius**.

CONCEPTS

5. a. $(x-h)^2+(y-k)^2=r^2$
   b. $x^2+y^2=r^2$

7. a. $(2, -1);\ r=4$
   b. $(x-2)^2+(y-(-1))^2=4^2$
   $(x-2)^2+(y+1)^2=16$

9. a. $y=a(x-h)^2+k$
   b. $x=a(y-k)^2+h$

11. a. circle
    b. parabola
    c. parabola
    d. circle

NOTATION

13. $h=6,\ k=-2,$ and $r=3$

GUIDED PRACTICE

15. $x^2+y^2=9$
    center (0, 0) and $r=3$

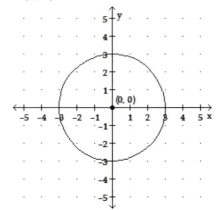

17. $x^2+(y+3)^2=1$
    center (0, –3) and $r=1$

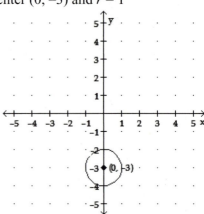

19. $(x+3)^2+(y-1)^2=16$
    center (–3, 1) and $r=4$

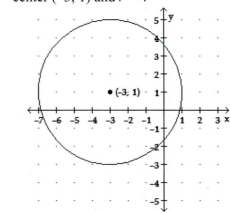

21. $x^2+y^2=6$
    center (0, 0) and $r=\sqrt{6}\approx2.4$

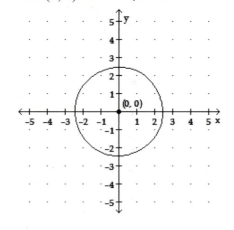

23. $h=0,\ k=0,\ r=1$
    $(x-h)^2+(y-k)^2=r^2$
    $(x-0)^2+(y-0)^2=1^2$
    $x^2+y^2=1$

25. $h = 6, k = 8, r = 5$

$$\left(x-h\right)^2 + \left(y-k\right)^2 = r^2$$
$$\left(x-6\right)^2 + \left(y-8\right)^2 = 5^2$$
$$\left(x-6\right)^2 + \left(y-8\right)^2 = 25$$

27. $h = -2, k = 6, r = 12$

$$\left(x-h\right)^2 + \left(y-k\right)^2 = r^2$$
$$\left(x-\left(-2\right)\right)^2 + \left(y-6\right)^2 = 12^2$$
$$\left(x+2\right)^2 + \left(y-6\right)^2 = 144$$

29. $h = 0, k = 0, r = \dfrac{1}{4}$

$$\left(x-h\right)^2 + \left(y-k\right)^2 = r^2$$
$$\left(x-0\right)^2 + \left(y-0\right)^2 = \left(\dfrac{1}{4}\right)^2.$$
$$x^2 + y^2 = \dfrac{1}{16}$$

31. $h = \dfrac{2}{3}, k = -\dfrac{7}{8}, r = \sqrt{2}$

$$\left(x-h\right)^2 + \left(y-k\right)^2 = r^2$$
$$\left(x-\dfrac{2}{3}\right)^2 + \left(y-\left(-\dfrac{7}{8}\right)\right)^2 = \left(\sqrt{2}\right)^2.$$
$$\left(x-\dfrac{2}{3}\right)^2 + \left(y+\dfrac{7}{8}\right)^2 = 2$$

33. $h = 0, \ k = 0, \ d = 4\sqrt{2}$ so $r = \dfrac{4\sqrt{2}}{2} = 2\sqrt{2}$

$$\left(x-h\right)^2 + \left(y-k\right)^2 = r^2$$
$$\left(x-0\right)^2 + \left(y-0\right)^2 = \left(2\sqrt{2}\right)^2$$
$$x^2 + y^2 = 4\left(2\right)$$
$$x^2 + y^2 = 8$$

35.

$$x^2 + y^2 - 2x + 4y = -1$$
$$\left(x^2 - 2x\right) + \left(y^2 + 4y\right) = -1$$
$$\left(x^2 - 2x + 1\right) + \left(y^2 + 4y + 4\right) = -1 + 1 + 4$$
$$\left(x-1\right)^2 + \left(y+2\right)^2 = 4$$

Center at $\left(1, \ -2\right)$; radius $= 2$

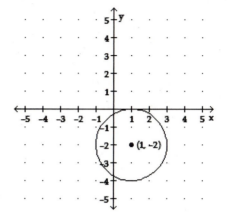

37.

$$x^2 + y^2 + 4x + 2y = 4$$
$$\left(x^2 + 4x\right) + \left(y^2 + 2y\right) = 4$$
$$\left(x^2 + 4x + 4\right) + \left(y^2 + 2y + 1\right) = 4 + 4 + 1$$
$$\left(x+2\right)^2 + \left(y+1\right)^2 = 9$$

Center at $\left(-2, \ -1\right)$; radius $= 3$

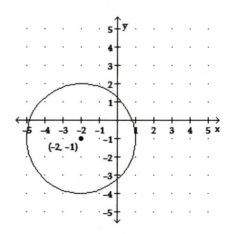

Section 13.1

39.

$$y = 2x^2 - 4x + 5$$

$$y = 2(x^2 - 2x) + 5$$

$$y = 2(x^2 - 2x + 1) + 5 - 2(1)$$

$$y = 2(x - 1)^2 + 5 - 2$$

$$y = 2(x - 1)^2 + 3$$

$$a = 2,\ h = 1,\ k = 3$$

$$\text{vertex at}\ (1, 3)$$

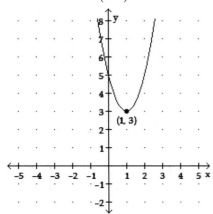

41.

$$y = -x^2 - 2x + 3$$

$$y = -(x^2 + 2x) + 3$$

$$y = -(x^2 + 2x + 1) + 3 - (-1)(1)$$

$$y = -(x + 1)^2 + 3 + 1$$

$$y = -(x + 1)^2 + 4$$

$$a = -1,\ h = -1,\ k = 4$$

$$\text{vertex at}\ (-1, 4)$$

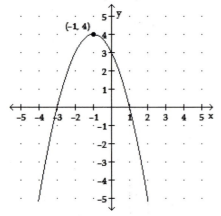

43.

$$x = y^2$$

$$a = 1,\ h = 0,\ k = 0$$

$$\text{vertex at}\ (0, 0)$$

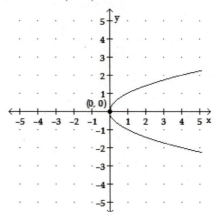

45.

$$x = 2(y + 1)^2 + 3$$

$$a = 2,\ h = 3,\ k = -1$$

$$\text{vertex at}\ (3,\ -1)$$

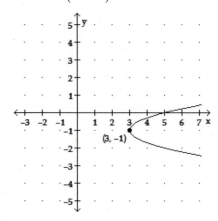

47.

$x = y^2 - 2y + 5$

$x = y^2 - 2y + 1 + 5 - 1$

$x = (y-1)^2 + 4$

$a = 1, h = 4, k = 1$

vertex at $(4, 1)$

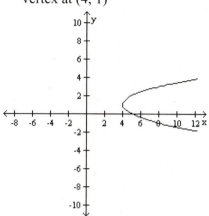

51. $x^2 + y^2 = 7$

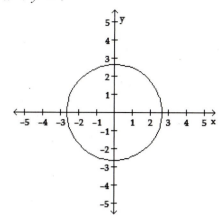

53. $(x+1)^2 + y^2 = 16$

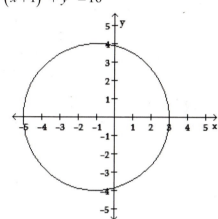

49.

$x = -3y^2 + 18y - 25$

$x = -3(y^2 - 6y) - 25$

$x = -3(y^2 - 6y + 9) - 25 - (-3)(9)$

$x = -3(y-3)^2 - 25 + 27$

$x = -3(y-3)^2 + 2$

$a = -3, h = 2, k = 3$

vertex at $(2, 3)$

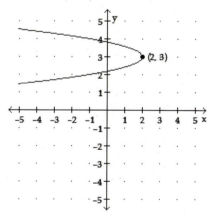

55. $x = 2y^2$

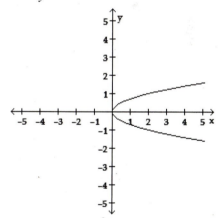

57. $x^2 - 2x + y = 6$

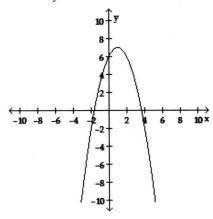

TRY IT YOURSELF

59.

$x = y^2 - 6y + 4$

$x = y^2 - 6y + 9 + 4 - 9$

$x = (y-3)^2 - 5$

$a = 1,\ h = -5,\ k = 3$

vertex at $(-5, 3)$

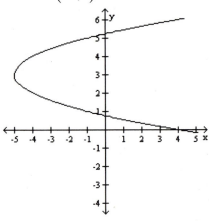

61. $(x-2)^2 + y^2 = 25$

center $(2, 0)$ and $r = 5$

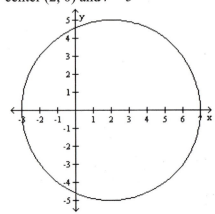

63.

$x^2 + y^2 - 6x + 8y + 18 = 0$

$(x^2 - 6x) + (y^2 + 8y) = -18$

$(x^2 - 6x + 9) + (y^2 + 8y + 16) = -18 + 9 + 16$

$(x-3)^2 + (y+4)^2 = 7$

Center at $(3,\ -4)$; radius $= \sqrt{7}$

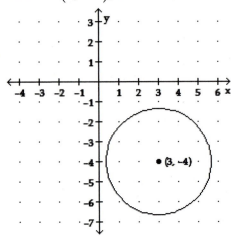

65.

$y = 4x^2 - 16x + 17$

$y = 4(x^2 - 4x) + 17$

$y = 4(x^2 - 4x + 4) + 17 - 16$

$y = 4(x-2)^2 + 1$

vertex at $(2,\ 1)$

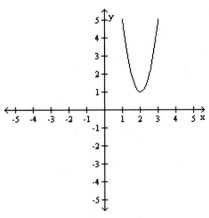

67. $(x-1)^2 + (y-3)^2 = 15$

    center $(1, 3)$ and $r = \sqrt{15} \approx 3.9$

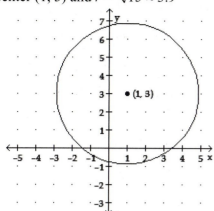

73.

$$x = -\frac{1}{4}y^2$$

$$a = -\frac{1}{4},\ h = 0,\ k = 0$$

    vertex at $(0, 0)$

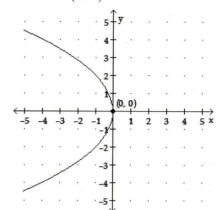

69.

$$x = -y^2 + 1$$

$$a = -1,\ h = 1,\ k = 0$$

    vertex at $(1, 0)$

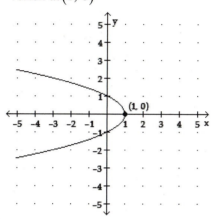

75.

$$x = -6(y-1)^2 + 3$$

$$a = -6,\ h = 3,\ k = 1$$

    vertex at $(3, 1)$

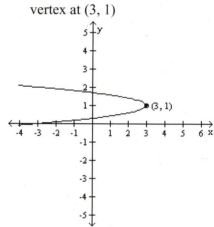

71. $(x-2)^2 + (y-4)^2 = 36$

    center $(2, 4)$ and $r = 6$

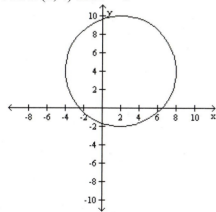

Section 13.1

77.

$$x^2 + y^2 + 2x - 8 = 0$$
$$\left(x^2 + 2x\right) + y^2 = 0 + 8$$
$$\left(x^2 + 2x + 1\right) + y^2 = 8 + 1$$
$$\left(x+1\right)^2 + y^2 = 9$$

Center at $\left(-1, 0\right)$; radius $= 3$

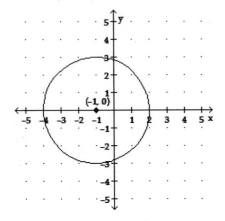

79.

$$x = \frac{1}{2}y^2 + 2y$$
$$x = \frac{1}{2}\left(y^2 + 4y\right)$$
$$x = \frac{1}{2}\left(y^2 + 4y + 4\right) - \frac{1}{2}(4)$$
$$x = \frac{1}{2}\left(y + 2\right)^2 - 2$$
$$a = \frac{1}{2},\ h = -2,\ k = -2$$

vertex at $\left(-2, -2\right)$

81.

$$y = -4(x+5)^2 + 5$$
$$a = -4,\ h = -5,\ k = 5$$

vertex at $(-5, 5)$

83. a. $\left(x-4\right)^2 + \left(y+7\right)^2 = 28$

center: $(4, -7)$, r $= 2\sqrt{7}$

b. $\left(x+4\right)^2 + \left(y-7\right)^2 = 28$

center: $(-4, 7)$, r $= 2\sqrt{7}$

85. a. $y = 8\left(x-3\right)^2 + 6$

V(3, 6); opens upward

b. $x = 8\left(y-3\right)^2 + 6$

V(6, 3); opens to the right

APPLICATIONS

87. **BROADCAST RANGES**
   Find the standard equation for each coverage and graph to see if they overlap.

   $$x^2 + y^2 - 8x - 20y + 16 = 0$$
   $$\left(x^2 - 8x\right) + \left(y^2 - 20y\right) = -16$$
   $$\left(x^2 - 8x + 16\right) + \left(y^2 - 20y + 100\right) = -16 + 16 + 100$$
   $$\left(x - 4\right)^2 + \left(y - 10\right)^2 = 100$$

   center at $(4, 10)$; radius $= 10$

   $$x^2 + y^2 + 2x + 4y - 11 = 0$$
   $$\left(x^2 + 2x\right) + \left(y^2 + 4y\right) = 11$$
   $$\left(x^2 + 2x + 1\right) + \left(y^2 + 4y + 4\right) = 11 + 1 + 4$$
   $$\left(x + 1\right)^2 + \left(y + 2\right)^2 = 16$$

   center at $(-1, -2)$; radius $= 4$

   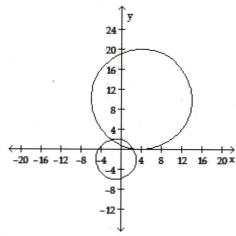

   Since the coverage areas overlap, they cannot be licensed for the same frequency.

89. **CIVIL ENGINEER**
   Find the center and radius of the circle.

   $$x^2 + y^2 - 10x - 12y + 52 = 0$$
   $$\left(x^2 - 10x\right) + \left(y^2 - 12y\right) = -52$$
   $$\left(x^2 - 10x + 25\right) + \left(y^2 - 12y + 36\right) = -52 + 25 + 36$$
   $$\left(x - 5\right)^2 + \left(y - 6\right)^2 = 9$$

   center at $(5, 6)$; radius $= 3$

   To find the locations of the intersections, add the radius of 3 to the x- and y-coordinates of the center.

   a. The intersection with State (x-value) is $5 + 3 = 8$ miles.
   b. The intersection with Highway 60 (y-value) is $6 + 3 = 9$ miles.

91. **PROJECTILES**
   Its landing position is an x–intercept, so let $y = 0$, and solve the equation for x.

   $$y = 30x - x^2$$
   $$0 = 30x - x^2$$
   $$0 = x\left(30 - x\right)$$
   $$x = 0 \quad \text{or} \quad 30 - x = 0$$
   $$x = 0 \qquad\qquad 30 = x$$

   If it lands 30 feet away from its starting point, it is $35 - 30 = 5$ ft from the castle.

93. **COMETS**
   Find the vertex of the comet's orbit.

   $$2y^2 - 9x = 18$$
   $$-9x = -2y^2 + 18$$
   $$x = \frac{2}{9}y^2 - 2$$
   $$k = 0 \ \text{ and } h = -2$$
   $$\text{vertex} = \left(-2, 0\right)$$

   Find the distance between the sun $(0, 0)$ and the comet $(-2, 0)$ using the distance formula.

   $$d = \sqrt{\left(x_2 - x_1\right)^2 + \left(y_2 - y_1\right)^2}$$
   $$= \sqrt{\left(-2 - 0\right)^2 + \left(0 - 0\right)^2}$$
   $$= \sqrt{4}$$
   $$= 2 \text{ AU}$$

WRITING

95. Answers will vary.

97. Answers will vary.

REVIEW

99.
   $$|3x - 4| = 11$$
   $$3x - 4 = 11 \ \text{ or } \ 3x - 4 = -11$$
   $$3x = 15 \qquad\qquad 3x = -7$$
   $$x = 5 \qquad\qquad x = -\frac{7}{3}$$

101.

$$|3x+4|=|5x-2|$$

$$3x+4=5x-2 \quad \text{or} \quad 3x+4=-(5x-2)$$

$$3x+4=5x-2 \qquad\qquad 3x+4=-5x+2$$

$$-2x+4=-2 \qquad\qquad\quad 8x+4=2$$

$$-2x=-6 \qquad\qquad\qquad 8x=-2$$

$$x=3 \qquad\qquad\qquad x=-\frac{1}{4}$$

CHALLENGE PROBLEMS

103.   Yes. Answers will vary.

105.   To find the center of the circle, find the midpoint of the points (–2, –6) and (8, 10).

$$\left(\frac{x_1+x_2}{2},\frac{y_1+y_2}{2}\right)=\left(\frac{-2+8}{2},\frac{-6+10}{2}\right)$$

$$=\left(\frac{6}{2},\frac{4}{2}\right)$$

$$=(3,2)$$

To find the radius, use the distance formula to find the distance from (8, 10) and the center (3, 2).

$$d=\sqrt{(x_2-x_1)^2+(y_2-y_1)^2}$$

$$=\sqrt{(3-8)^2+(2-10)^2}$$

$$=\sqrt{(-5)^2+(-8)^2}$$

$$=\sqrt{25+64}$$

$$=\sqrt{89}$$

Write the equation of the circle with center (3, 2) and radius $\sqrt{89}$.

$$(x-3)^2+(y-2)^2=\left(\sqrt{89}\right)^2$$

$$(x-3)^2+(y-2)^2=89$$

SECTION 13.2

VOCABULARY

1. The curve graphed below is an **ellipse**.

3. In the graph above, F_1 and F_2 are the **foci** of the ellipse. Each one is called a **focus** of the ellipse.

5. The line segment joining the vertices of an ellipse is called the **major** axis of the ellipse.

CONCEPTS

7. $\dfrac{x^2}{a^2} + \dfrac{y^2}{b^2} = 1$

9. x–intercepts: $(a, 0)$ and $(-a, 0)$
   y–intercepts: $(0, b)$ and $(0, -b)$

11. a. $(-2, 1)$; $a = 2$ and $b = 5$
    b. vertical
    c. $\dfrac{(x+2)^2}{4} + \dfrac{(y-1)^2}{25} = 1$

13.

$$4(x-1)^2 + 64(y+5)^2 = 64$$

$$\frac{4(x-1)^2}{64} + \frac{64(y+5)^2}{64} = \frac{64}{64}$$

$$\frac{(x-1)^2}{16} + \frac{(y+5)^2}{1} = 1$$

NOTATION

15. $h = -8$, $k = 6$, $a = \sqrt{100} = 10$,
    $b = \sqrt{144} = 12$

GUIDED PRACTICE

17. $\dfrac{x^2}{25} + \dfrac{y^2}{4} = 1$

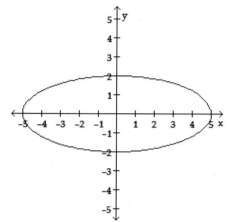

19. $\dfrac{x^2}{4} + \dfrac{y^2}{9} = 1$

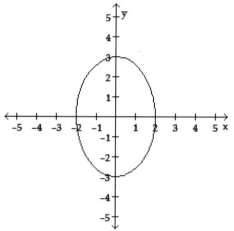

21.

$$x^2 + 9y^2 = 9$$

$$\frac{x^2}{9} + \frac{9y^2}{9} = \frac{9}{9}$$

$$\frac{x^2}{9} + \frac{y^2}{1} = 1$$

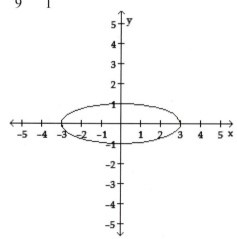

23.

$$16x^2 + 4y^2 = 64$$

$$\frac{16x^2}{64} + \frac{4y^2}{64} = \frac{64}{64}$$

$$\frac{x^2}{4} + \frac{y^2}{16} = 1$$

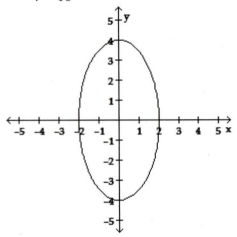

27. $\dfrac{(x+2)^2}{64} + \dfrac{(y-2)^2}{100} = 1$

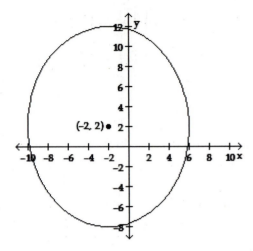

29.

$$(x+1)^2 + 4(y+2)^2 = 4$$

$$\frac{(x+1)^2}{4} + \frac{4(y+2)^2}{4} = \frac{4}{4}$$

$$\frac{(x+1)^2}{4} + \frac{(y+2)^2}{1} = 1$$

25. $\dfrac{(x-2)^2}{9} + \dfrac{(y-1)^2}{4} = 1$

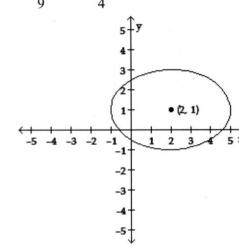

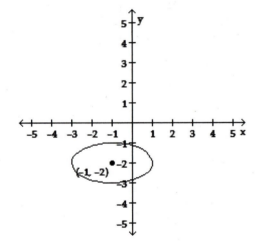

31.

$$16(x-2)^2 + 4(y+4)^2 = 256$$

$$\frac{16(x-2)^2}{256} + \frac{4(y+4)^2}{256} = \frac{256}{256}$$

$$\frac{(x-2)^2}{16} + \frac{(y+4)^2}{64} = 1$$

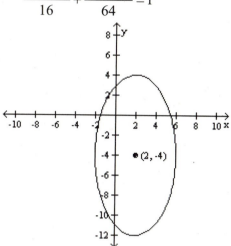

37. $(x+1)^2 + (y-2)^2 = 16$

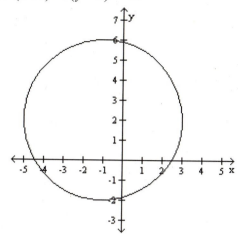

33.

$$\frac{x^2}{9} + \frac{y^2}{4} = 1$$

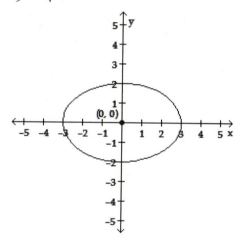

39. $\dfrac{x^2}{16} + \dfrac{y^2}{1} = 1$

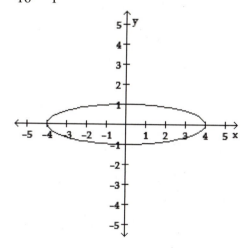

35.

$$\frac{x^2}{4} + \frac{(y-1)^2}{9} = 1$$

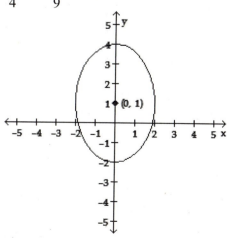

41. $x = \dfrac{1}{2}(y-1)^2 - 2$

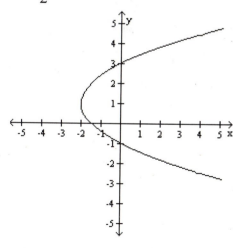

Section 13.2

43. $x^2 + y^2 - 25 = 0$

    $x^2 + y^2 = 25$

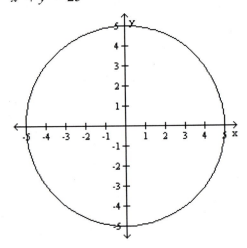

47.

$y = -3x^2 - 24x - 43$

$y = -3\left(x^2 + 8x\right) - 43$

$y = -3\left(x^2 + 8x + 16\right) - 43 + 48$

$y = -3\left(x + 4\right)^2 + 5$

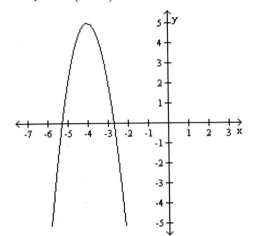

45.

$x^2 = 100 - 4y^2$

$x^2 + 4y^2 = 100$

$\dfrac{x^2}{100} + \dfrac{4y^2}{100} = \dfrac{100}{100}$

$\dfrac{x^2}{100} + \dfrac{y^2}{25} = 1$

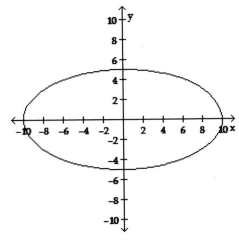

49.

$x^2 + y^2 - 2x + 4y - 4 = 0$

$\left(x^2 - 2x\right) + \left(y^2 + 4y\right) = 4$

$\left(x^2 - 2x + 1\right) + \left(y^2 + 4y + 4\right) = 4 + 1 + 4$

$\left(x - 1\right)^2 + \left(y + 2\right)^2 = 9$

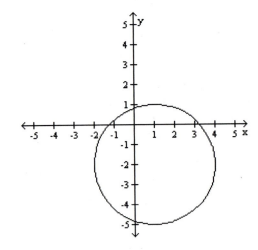

51.

$$9(x-1)^2 + 4(y+2)^2 = 36$$

$$9(x-1)^2 + 4(y+2)^2 = 36$$

$$\frac{9(x-1)^2}{36} + \frac{4(y+2)^2}{36} = \frac{36}{36}$$

$$\frac{(x-1)^2}{4} + \frac{(y+2)^2}{9} = 1$$

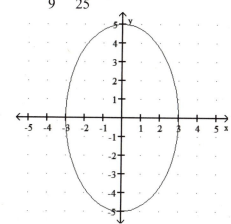

53. a. $\dfrac{x^2}{9} + \dfrac{y^2}{25} = 1$

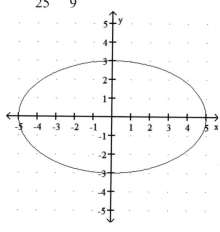

b. $\dfrac{x^2}{25} + \dfrac{y^2}{9} = 1$

55. a. $\dfrac{(x-3)^2}{100} + \dfrac{(y-2)^2}{36} = 1$

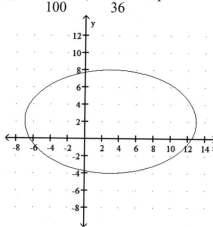

b. $\dfrac{(x+3)^2}{100} + \dfrac{(y+2)^2}{36} = 1$

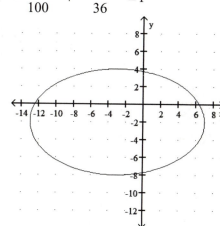

APPLICATIONS

57. CIVIL ENGINEER

    a. $a = \pm 20, \ b \pm 10$

$$\frac{x^2}{400} + \frac{y^2}{100} = 1$$

    b. Let $x = 10$.

$$\frac{10^2}{400} + \frac{y^2}{100} = 1$$

$$\frac{100}{400} + \frac{y^2}{100} = 1$$

$$0.25 + \frac{y^2}{100} = 1$$

$$\frac{y^2}{100} = 0.75$$

$$y^2 = 75$$

$$y = 5\sqrt{3}$$

$$y \approx 8.7 \text{ ft}$$

Section 13.2

59. KOI PONDS

$a = 110/2 - 15 = 40$
$b = 100/2 - 15 = 35$
$a = \pm 40, \; b \pm 35$

$$\frac{x^2}{1,600} + \frac{y^2}{1,125} = 1$$

61. AREA OF AN ELLIPSE

$$9x^2 + 16y^2 = 144$$
$$\frac{9x^2}{144} + \frac{16y^2}{144} = \frac{144}{144}$$
$$\frac{x^2}{16} + \frac{y^2}{9} = 1$$
$$a = 4 \text{ and } b = 3$$
$$A = \pi ab$$
$$= \pi(4)(3)$$
$$= 12\pi \text{ sq. units}$$
$$\approx 37.7 \text{ sq. units}$$

WRITING

63. Answers will vary.

65. Answers will vary.

REVIEW

67.

$$3x^{-2}y^2\left(4x^2 + 3y^{-2}\right) = 12x^0y^2 + 9x^{-2}y^0$$
$$= 12(1)y^2 + 9x^{-2}(1)$$
$$= 12y^2 + \frac{9}{x^2}$$

69.

$$\frac{x^{-2} + y^{-2}}{x^{-2} - y^{-2}} = \frac{\dfrac{1}{x^2} + \dfrac{1}{y^2}}{\dfrac{1}{x^2} - \dfrac{1}{y^2}}$$

$$= \frac{\dfrac{1}{x^2} + \dfrac{1}{y^2}}{\dfrac{1}{x^2} - \dfrac{1}{y^2}} \cdot \frac{x^2y^2}{x^2y^2}$$

$$= \frac{\dfrac{x^2y^2}{x^2} + \dfrac{x^2y^2}{y^2}}{\dfrac{x^2y^2}{x^2} - \dfrac{x^2y^2}{y^2}}$$

$$= \frac{y^2 + x^2}{y^2 - x^2}$$

$$= \frac{y^2 + x^2}{(y+x)(y-x)}$$

CHALLENGE PROBLEMS

71. It forms a circle.

73.

$$9x^2 + 4y^2 - 18x + 16y = 11$$
$$\left(9x^2 - 18x\right) + \left(4y^2 + 16y\right) = 11$$
$$9\left(x^2 - 2x\right) + 4\left(y^2 + 4y\right) = 11$$
$$9\left(x^2 - 2x + 1\right) + 4\left(y^2 + 4y + 4\right) = 11 + 9(1) + 4(4)$$
$$9\left(x-1\right)^2 + 4\left(y+2\right)^2 = 36$$
$$\frac{9\left(x-1\right)^2}{36} + \frac{4\left(y+2\right)^2}{36} = \frac{36}{36}$$
$$\frac{\left(x-1\right)^2}{4} + \frac{\left(y+2\right)^2}{9} = 1$$

SECTION 13.3

VOCABULARY

1. The two–branch curve graphed below is a **hyperbola**.

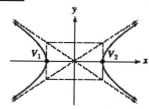

3. In the graph above, V_1 and V_2 are called the **vertices** of the hyperbola.

5. The extended **diagonals** of the central rectangle are asymptotes of the hyperbola.

CONCEPTS

7. $\dfrac{x^2}{a^2} - \dfrac{y^2}{b^2} = 1$

9. $\dfrac{(x-h)^2}{a^2} - \dfrac{(y-k)^2}{b^2} = 1$

11. a. $(-1, -2)$; $a = 3$, $b = 1$

　　b. $\dfrac{(y+2)^2}{9} - \dfrac{(x+1)^2}{1} = 1$

13.
$$100(x+1)^2 - 25(y-5)^2 = 100$$
$$\frac{100(x+1)^2}{100} - \frac{25(y-5)^2}{100} = \frac{100}{100}$$
$$\frac{(x+1)^2}{1} - \frac{(y-5)^2}{4} = 1$$

NOTATION

15. $h = 5$, $k = -11$, $a = 5$, $b = 6$

GUIDED PRACTICE

17. $\dfrac{x^2}{9} - \dfrac{y^2}{4} = 1$

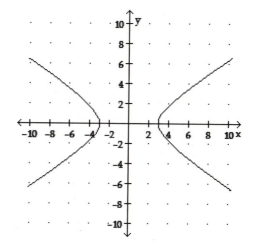

19. $\dfrac{y^2}{4} - \dfrac{x^2}{9} = 1$

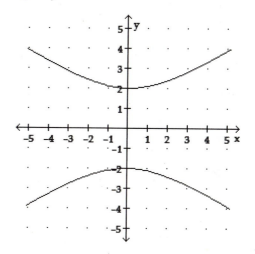

Section 13.3

21.

$$y^2 - 4x^2 = 16$$

$$\frac{y^2}{16} - \frac{4x^2}{16} = \frac{16}{16}$$

$$\frac{y^2}{16} - \frac{x^2}{4} = 1$$

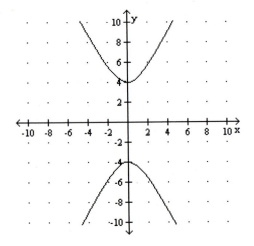

23.

$$25x^2 - y^2 = 25$$

$$\frac{25x^2}{25} - \frac{y^2}{25} = \frac{25}{25}$$

$$\frac{x^2}{1} - \frac{y^2}{25} = 1$$

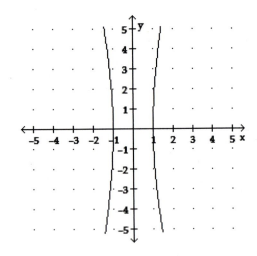

25. $\dfrac{(x-2)^2}{9} - \dfrac{y^2}{16} = 1$

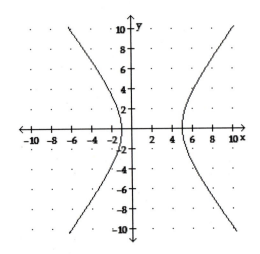

27. $\dfrac{(y+1)^2}{1} - \dfrac{(x-2)^2}{4} = 1$

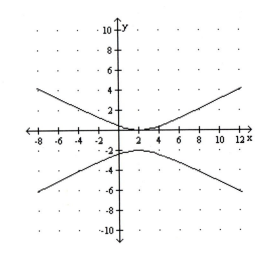

29. $\dfrac{(x+1)^2}{9} - \dfrac{(y+1)^2}{9} = 1$

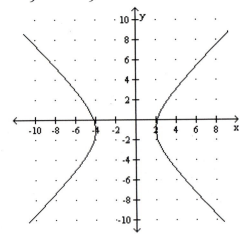

31. $\dfrac{(y-3)^2}{25} - \dfrac{x^2}{25} = 1$

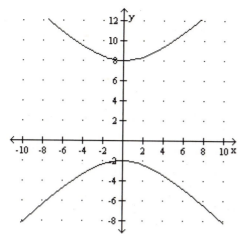

33.

$xy = 8$

$y = \dfrac{8}{x}$

x	$y = \dfrac{8}{x}$
1	$\dfrac{8}{1} = 8$
2	$\dfrac{8}{2} = 4$
4	$\dfrac{8}{4} = 2$
-4	$\dfrac{8}{-4} = -2$
-2	$\dfrac{8}{-2} = -4$
-1	$\dfrac{8}{-1} = -8$

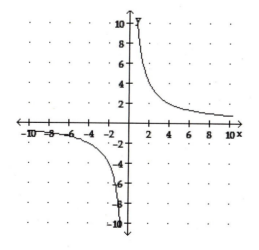

35.

$xy = -10$

$y = \dfrac{-10}{x}$

x	$y = \dfrac{-10}{x}$
1	$\dfrac{-10}{1} = -10$
2	$\dfrac{-10}{2} = -5$
5	$\dfrac{-10}{5} = -2$
-5	$\dfrac{-10}{-5} = 2$
-2	$\dfrac{-10}{-2} = 5$
-1	$\dfrac{-10}{-1} = 10$

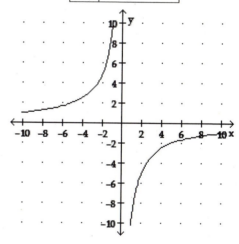

37. $\dfrac{x^2}{9} - \dfrac{y^2}{4} = 1$

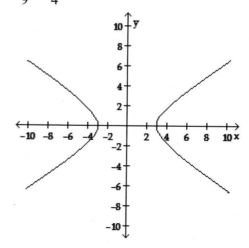

Section 13.3

39. $\dfrac{x^2}{4} - \dfrac{(y-1)^2}{9} = 1$

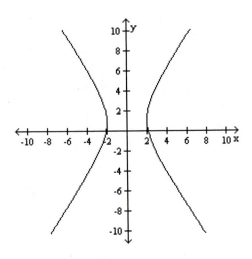

43.

$$9x^2 - 49y^2 = 441$$

$$\dfrac{9x^2}{441} - \dfrac{49y^2}{441} = \dfrac{441}{441}$$

$$\dfrac{x^2}{49} - \dfrac{y^2}{9} = 1$$

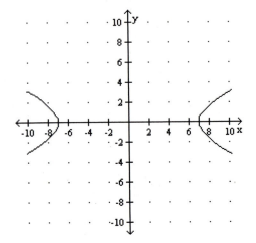

TRY IT YOURSELF

41. $(x+1)^2 + (y-2)^2 = 16$

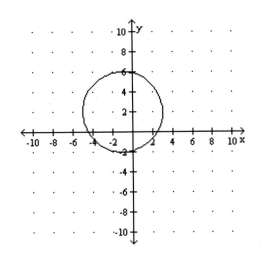

45.

$$4(x+1)^2 + 9(y+1)^2 = 36$$

$$\dfrac{4(x+1)^2}{36} + \dfrac{9(y+1)^2}{36} = \dfrac{36}{36}$$

$$\dfrac{(x+1)^2}{9} + \dfrac{(y+1)^2}{4} = 1$$

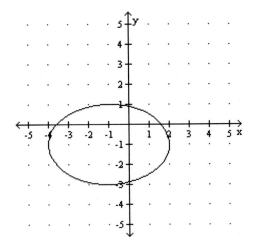

47.

$$4(x+3)^2 - (y-1)^2 = 4$$

$$\frac{4(x+3)^2}{4} - \frac{(y-1)^2}{4} = \frac{4}{4}$$

$$\frac{(x+3)^2}{1} - \frac{(y-1)^2}{4} = 1$$

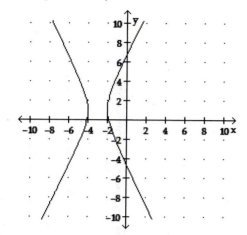

49.

$$xy = -6$$

$$y = \frac{-6}{x}$$

x	$y = \dfrac{-6}{x}$
1	$\dfrac{-6}{1} = -6$
2	$\dfrac{-6}{2} = -3$
3	$\dfrac{-6}{3} = -2$
-3	$\dfrac{-6}{-3} = 2$
-2	$\dfrac{-6}{-2} = 3$
-1	$\dfrac{-6}{-1} = 6$

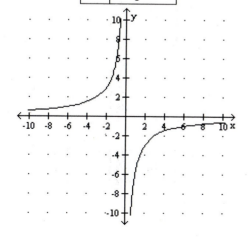

51. $x = \dfrac{1}{2}(y-1)^2 - 2$

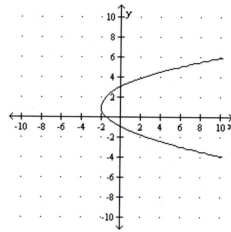

53. $\dfrac{y^2}{25} - \dfrac{(x-2)^2}{4} = 1$

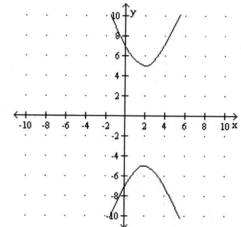

55.

$$y = -x^2 + 6x - 4$$

$$y = -(x^2 - 6x) - 4$$

$$y = -(x^2 - 6x + 9) - 4 + 9$$

$$y = -(x-3)^2 + 5$$

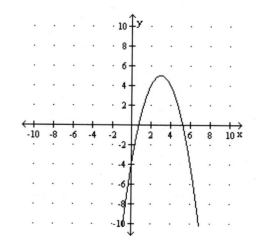

- 929 -

Section 13.3

57. $\dfrac{x^2}{1} + \dfrac{y^2}{36} = 1$

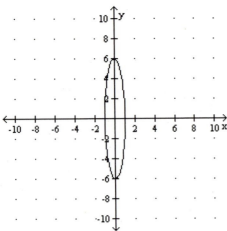

59.

$$x^2 + y^2 + 4x - 6y - 23 = 0$$

$$\left(x^2 + 4x\right) + \left(y^2 - 6y\right) = 23$$

$$\left(x^2 + 4x + 4\right) + \left(y^2 - 6y + 9\right) = 23 + 4 + 9$$

$$\left(x + 2\right)^2 + \left(y - 3\right)^2 = 36$$

APPLICATIONS

61. ALPHA PARTICLES
Find the coordinates of the vertex.

$$9y^2 - x^2 = 81$$

$$\dfrac{9y^2}{81} - \dfrac{x^2}{81} = \dfrac{81}{81}$$

$$\dfrac{y^2}{9} - \dfrac{x^2}{81} = 1$$

$$a = 3 \text{ and } b = 9$$

The vertex is (3, 0), so the particle comes within 3 units of the nucleus.

63. CIVIL ENGINEER
Let $x = 5$ and solve for y to find half of the width of the hyperbola.

$$y^2 - x^2 = 25$$

$$y^2 - 5^2 = 25$$

$$y^2 - 25 = 25$$

$$y^2 = 50$$

$$y = \sqrt{50}$$

$$y = 5\sqrt{2}$$

The width of the hyperbola would be $2y$.

$$2y = 2\left(5\sqrt{2}\right)$$

$$= 10\sqrt{2} \text{ miles}$$

$$\approx 14.1 \text{ miles}$$

65. NUCLEAR POWER
The tower is in the shape of a hyperbola.

WRITING

67. Answers will vary.

69. Answers will vary.

REVIEW

71.

$$\log_8 x = 2$$

$$8^2 = x$$

$$64 = x$$

73.

$$\log_{1/2} \dfrac{1}{8} = x$$

$$\left(\dfrac{1}{2}\right)^x = \dfrac{1}{8}$$

$$2^{-x} = 2^{-3}$$

$$-x = -3$$

$$x = 3$$

75.

$$\log_x \frac{9}{4} = 2$$

$$x^2 = \frac{9}{4}$$

$$x = \sqrt{\frac{9}{4}}$$

$$x = \frac{3}{2}$$

77.

$$\log_x 1,000 = 3$$

$$x^3 = 1,000$$

$$x^3 = 10^3$$

$$x = 10$$

CHALLENGE PROBLEMS

79.

$$x^2 - y^2 - 2x + 4y = 12$$

$$x^2 - 2x - y^2 + 4y = 12$$

$$(x^2 - 2x) - (y^2 - 4y) = 12$$

$$(x^2 - 2x + 1) - (y^2 - 4y + 4) = 12 + 1 - 4$$

$$(x-1)^2 - (y-2)^2 = 9$$

$$\frac{(x-1)^2}{9} - \frac{(y-2)^2}{9} = 1$$

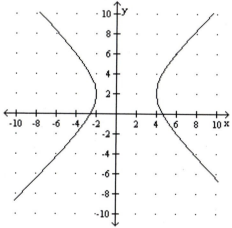

81.

$$36x^2 - 25y^2 - 72x - 100y = 964$$

$$(36x^2 - 72x) - (25y^2 + 100y) = 964$$

$$36(x^2 - 2x) - 25(y^2 + 4y) = 964$$

$$36(x^2 - 2x + 1) - 25(y^2 + 4y + 4) = 964 + 36(1) - 25(4)$$

$$36(x-1)^2 - 25(y+2)^2 = 900$$

$$\frac{36(x-1)^2}{900} - \frac{25(y+2)^2}{900} = \frac{900}{900}$$

$$\frac{(x-1)^2}{25} - \frac{(y+2)^2}{36} = 1$$

83.

$$16x^2 - 25y^2 = 1$$

$$\frac{x^2}{\frac{1}{16}} - \frac{y^2}{\frac{1}{25}} = 1$$

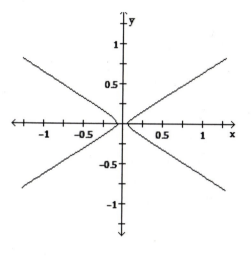

VOCABULARY

1. $\begin{cases} 4x^2 + 6y^2 = 24 \\ 9x^2 - y^2 = 9 \end{cases}$ is a **system** of two nonlinear equations.

3. When solving a system by graphing, it is often difficult to determine the coordinates of the points of **intersection** of the graphs.

5. A **secant** is a line that intersects a circle at two points.

CONCEPTS

7. a. A line can intersect an ellipse in at most **two** points.
   b. An ellipse can intersect a parabola in at most **four** points.
   c. An ellipse can intersect a circle in at most **four** points.
   d. A hyperbola can intersect a circle in at most **four** points.

9. They intersect at $(-3, 2)$, $(3, 2)$, $(-3, -2)$, and $(3, -2)$.

11. a. -4
    b. -2

NOTATION

13.
$$x^2 + y^2 = 5$$
$$x^2 + \left(\boxed{2x}\right)^2 = 5$$
$$x^2 + 4x^2 = \boxed{5}$$
$$\boxed{5}x^2 = 5$$
$$x^2 = \boxed{1}$$
$$x = 1 \quad \text{or} \quad x = -1$$

If $x = 1$, then
$$y = 2\left(\boxed{1}\right) = 2$$

If $x = -1$, then
$$y = 2\left(\boxed{-1}\right) = -2$$

The solutions are $(1, 2)$ and $\left(-1, \boxed{-2}\right)$.

GUIDED PRACTICE

15. $\begin{cases} x^2 + y^2 = 9 \\ y - x = 3 \end{cases}$

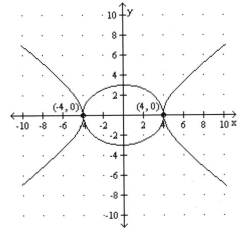

The solutions are $(0, 3)$ and $(-3, 0)$.

17. $\begin{cases} 9x^2 + 16y^2 = 144 \\ 9x^2 - 16y^2 = 144 \end{cases} \Rightarrow \begin{cases} \dfrac{x^2}{16} + \dfrac{y^2}{9} = 1 \\ \dfrac{x^2}{16} - \dfrac{y^2}{9} = 1 \end{cases}$

The solutions are $(-4, 0)$ and $(4, 0)$.

19. $\begin{cases} y = x^2 - 4x \\ x^2 + y = 0 \end{cases} \Rightarrow \begin{cases} y = x^2 - 4x \\ y = -x^2 \end{cases}$

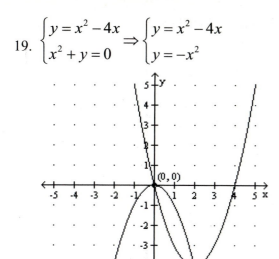

The solutions are $(0, 0)$ and $(2, -4)$.

21. $\begin{cases} x^2 + 4y^2 = 4 \\ x = 2y^2 - 2 \end{cases} \Rightarrow \begin{cases} \dfrac{x^2}{4} + \dfrac{y^2}{1} = 1 \\ x = 2y^2 - 2 \end{cases}$

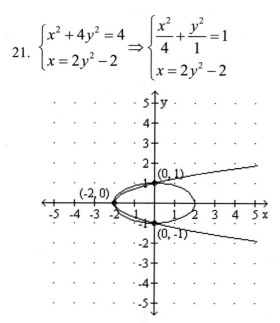

The solutions are $(-2, 0)$, $(0, -1)$, and $(0, 1)$.

23.

$\begin{cases} x^2 + y^2 = 5 \\ x + y = 3 \end{cases}$

Solve the second equation for y.

$x + y = 3$

$\quad y = -x + 3$

Substitute $y = -x + 3$ into the first equation and solve for x.

$$x^2 + y^2 = 5$$
$$x^2 + (-x + 3)^2 = 5$$
$$x^2 + (x^2 - 3x - 3x + 9) = 5$$
$$x^2 + (x^2 - 6x + 9) = 5$$
$$2x^2 - 6x + 9 = 5$$
$$2x^2 - 6x + 4 = 0$$
$$2(x^2 - 3x + 2) = 0$$
$$2(x - 1)(x - 2) = 0$$
$$x - 1 = 0 \quad \text{or} \quad x - 2 = 0$$
$$x = 1 \qquad\qquad x = 2$$

Substitute $x = 1$ into $x + y = 2$ and solve for y.

$(1) + y = 3 \qquad (2) + y = 3$

$\quad y = 2 \qquad\qquad y = 1$

The solutions are $(1, 2)$ and $(2, 1)$.

Section 13.4

25.

$$\begin{cases} y = x^2 + 6x + 7 \\ 2x + y = -5 \end{cases}$$

Substitute $y = x^2 + 6x + 7$ into the second equation and solve for x.

$$2x + y = -5$$
$$2x + \left(x^2 + 6x + 7\right) = -5$$
$$x^2 + 8x + 7 = -5$$
$$x^2 + 8x + 12 = 0$$
$$(x + 2)(x + 6) = 0$$
$$x + 2 = 0 \quad \text{and} \quad x + 6 = 0$$
$$x = -2 \qquad x = -6$$

Substitute $x = -2$ and $x = -6$ into the second equation and solve for y.

$$2x + y = -5 \qquad 2x + y = -5$$
$$2(-2) + y = -5 \qquad 2(-6) + y = -5$$
$$-4 + y = -5 \qquad -12 + y = -5$$
$$y = -1 \qquad y = 7$$

The solutions are $(-2, -1)$ and $(-6, 7)$.

27.

$$\begin{cases} x^2 + y^2 = 13 \\ y = x^2 - 1 \end{cases}$$

Substitute $y = x^2 - 1$ into the first equation and solve for x.

$$x^2 + y^2 = 13$$
$$x^2 + \left(x^2 - 1\right)^2 = 13$$
$$x^2 + \left(x^4 - x^2 - x^2 + 1\right) = 13$$
$$x^2 + \left(x^4 - 2x^2 + 1\right) = 13$$
$$x^4 - x^2 + 1 = 13$$
$$x^4 - x^2 - 12 = 0$$
$$\left(x^2 - 4\right)\left(x^2 + 3\right) = 0$$
$$x^2 - 4 = 0 \quad \text{or} \quad x^2 + 3 = 0$$
$$x^2 = 4 \qquad x^2 = -3$$
$$x = \sqrt{4} \qquad x = \sqrt{-3}$$
$$x = \pm 2 \qquad \text{not real}$$

Substitute $x = 2$ and $x = -2$ into $y = x^2 - 1$ and solve for y.

$$y = (2)^2 - 1 \qquad y = (-2)^2 - 1$$
$$y = 4 - 1 \qquad y = 4 - 1$$
$$y = 3 \qquad y = 3$$

The solutions are $(2, 3)$ and $(-2, 3)$.

29.

$$\begin{cases} x^2 + y^2 = 30 \\ y = x^2 \end{cases}$$

Substitute $y = x^2$ into the first equation and solve for x.

$$x^2 + y^2 = 30$$
$$x^2 + \left(x^2\right)^2 = 30$$
$$x^2 + x^4 = 30$$
$$x^4 + x^2 - 30 = 0$$
$$\left(x^2 + 6\right)\left(x^2 - 5\right) = 0$$
$$x^2 + 6 = 0 \quad \text{and} \quad x^2 - 5 = 0$$
$$x^2 = -6 \qquad x^2 = 5$$
$$x = \pm\sqrt{-6} \qquad x = \pm\sqrt{5}$$
$$\text{not real}$$

Substitute $x = \pm\sqrt{5}$ into the second equation and solve for y.

$$y = x^2 \qquad y = x^2$$
$$y = \left(\sqrt{5}\right)^2 \qquad y = \left(-\sqrt{5}\right)^2$$
$$y = 5 \qquad y = 5$$

The solutions are $\left(\sqrt{5}, 5\right)$ and $\left(-\sqrt{5}, 5\right)$.

31.

$$\begin{cases} x^2 + y^2 = 20 \\ x^2 - y^2 = -12 \end{cases}$$

Add the equations and use elimination to solve for x.

$$x^2 + y^2 = 20$$
$$\underline{x^2 - y^2 = -12}$$
$$2x^2 = 8$$
$$x^2 = 4$$
$$x = \sqrt{4}$$
$$x = \pm 2$$

Substitute $x = 2$ and $x = -2$ into $x^2 + y^2 = 20$ and solve for y.

$$x^2 + y^2 = 20 \qquad x^2 + y^2 = 20$$
$$(2)^2 + y^2 = 20 \qquad (-2)^2 + y^2 = 20$$
$$4 + y^2 = 20 \qquad 4 + y^2 = 20$$
$$y^2 = 16 \qquad y^2 = 16$$
$$y = \sqrt{16} \qquad y = \sqrt{16}$$
$$y = \pm 4 \qquad y = \pm 4$$

The solutions are $(2, 4)$, $(2, -4)$, $(-2, 4)$ and $(-2, -4)$.

33.

$$\begin{cases} 9x^2 - 7y^2 = 81 \\ x^2 + y^2 = 9 \end{cases}$$

Multiply the second equation by -9
use elimination to solve for y.

$$9x^2 - 7y^2 = 81$$
$$\underline{-9x^2 - 9y^2 = -81}$$
$$-16y^2 = 0$$
$$y^2 = 0$$
$$y = \sqrt{0}$$
$$y = 0$$

Substitute $y = 0$ into $x^2 + y^2 = 9$
and solve for x.

$$x^2 + y^2 = 9$$
$$x^2 + 0^2 = 9$$
$$x^2 = 9$$
$$x = \sqrt{9}$$
$$x = \pm 3$$

The solutions are $(3,0)$ and $(-3,0)$.

35.

$$\begin{cases} 2x^2 + y^2 = 6 \\ x^2 - y^2 = 3 \end{cases}$$

Add the two equations and use
elimination to solve for x.

$$2x^2 + y^2 = 6$$
$$\underline{x^2 - y^2 = 3}$$
$$3x^2 = 9$$
$$x^2 = 3$$
$$x = \pm\sqrt{3}$$

Substitute $x = \sqrt{3}$ and $x = -\sqrt{3}$ into
$x^2 - y^2 = 3$ and solve for y.

$$x^2 - y^2 = 3 \qquad\qquad x^2 - y^2 = 3$$
$$\left(\sqrt{3}\right)^2 - y^2 = 3 \qquad \left(-\sqrt{3}\right)^2 - y^2 = 3$$
$$3 - y^2 = 3 \qquad\qquad 3 - y^2 = 3$$
$$-y^2 = 0 \qquad\qquad -y^2 = 0$$
$$y^2 = 0 \qquad\qquad y^2 = 0$$
$$y = \sqrt{0} \qquad\qquad y = \sqrt{0}$$
$$y = 0 \qquad\qquad y = 0$$

The solutions are $\left(\sqrt{3},0\right)$ and $\left(-\sqrt{3},0\right)$.

37.

$$\begin{cases} x^2 - y^2 = -5 \\ 3x^2 + 2y^2 = 30 \end{cases}$$

Multiply the first equation by 2 and use
elimination to solve for x.

$$2x^2 - 2y^2 = -10$$
$$\underline{3x^2 + 2y^2 = 30}$$
$$5x^2 = 20$$
$$x^2 = 4$$
$$x = \sqrt{4}$$
$$x = \pm 2$$

Substitute $x = 2$ and $x = -2$ into
$x^2 - y^2 = -5$ and solve for y.

$$x^2 - y^2 = -5 \qquad\qquad x^2 - y^2 = -5$$
$$(2)^2 - y^2 = -5 \qquad (-2)^2 - y^2 = -5$$
$$4 - y^2 = -5 \qquad\qquad 4 - y^2 = -5$$
$$-y^2 = -9 \qquad\qquad -y^2 = -9$$
$$y^2 = 9 \qquad\qquad y^2 = 9$$
$$y = \pm\sqrt{9} \qquad\qquad y = \pm\sqrt{9}$$
$$y = \pm 3 \qquad\qquad y = \pm 3$$

The solutions are $(2, 3)$, $(2,-3)$, $(-2, 3)$,
and $(-2,-3)$.

39.

$$\begin{cases} x^2 - 6x - y = -5 \\ x^2 - 6x + y = -5 \end{cases}$$

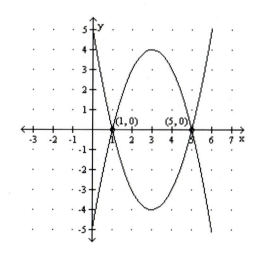

The solutions are $(1, 0)$ and $(5, 0)$.

TRY IT YOURSELF

41. $\begin{cases} 2x^2 - 3y^2 = 5 \\ 3x^2 + 4y^2 = 16 \end{cases}$

Multiply the first equation by 4 and the second equation by 3. Add the equations and use elimination to solve for x.

$$8x^2 - 12y^2 = 20$$
$$\underline{9x^2 + 12y^2 = 48}$$
$$17x^2 = 68$$
$$x^2 = 4$$
$$x = \pm 2$$

Substitute $x = 2$ and $x = -2$ into the first equation and solve for y.

$2x^2 - 3y^2 = 5$	$2x^2 - 3y^2 = 5$
$2(2)^2 - 3y^2 = 5$	$2(-2)^2 - 3y^2 = 5$
$2(4) - 3y^2 = 5$	$2(4) - 3y^2 = 5$
$8 - 3y^2 = 5$	$8 - 3y^2 = 5$
$-3y^2 = -3$	$-3y^2 = -3$
$y^2 = 1$	$y^2 = 1$
$y = \pm 1$	$y = \pm 1$

The solutions are $(2, 1)$, $(2, -1)$, $(-2, 1)$, and $(-2, -1)$.

43. $\begin{cases} y = x^2 - 4 \\ x^2 - y^2 = -16 \end{cases}$

Write the first equation in standard form and use elimination.

$$-x^2 + y = -4$$
$$\underline{x^2 - y^2 = -16}$$
$$-y^2 + y = -20$$
$$y^2 - y - 20 = 0$$
$$(y + 4)(y - 5) = 0$$
$$y + 4 = 0 \quad \text{or} \quad y - 5 = 0$$
$$y = -4 \qquad\qquad y = 5$$

Substitute $y = -4$ and $y = 5$ into $y = x^2 - 4$ and solve for x.

$-4 = x^2 - 4$	$5 = x^2 - 4$
$x^2 - 4 = -4$	$x^2 - 4 = 5$
$x^2 = 0$	$x^2 = 9$
$x = \sqrt{0}$	$x = \sqrt{9}$
$x = 0$	$x = \pm 3$

The solutions are $(0, -4), (3, 5),$ and $(-3, 5)$.

45. $\begin{cases} 3y^2 = xy \\ 2x^2 + xy - 84 = 0 \end{cases}$

Solve the first equation for x.

$$3y^2 = xy$$
$$\frac{3y^2}{y} = x$$
$$3y = x$$

Substitute $3y = x$ into the second equation and solve for y.

$$2(3y)^2 + (3y)y - 84 = 0$$
$$2(9y^2) + 3y^2 - 84 = 0$$
$$18y^2 + 3y^2 - 84 = 0$$
$$21y^2 - 84 = 0$$
$$21y^2 = 84$$
$$y^2 = 4$$
$$y = \sqrt{4}$$
$$y = \pm 2$$

Substitute $y = 2$, and $y = -2$ into the first equation to solve for x.

$3y^2 = xy$	$3y^2 = xy$
$3(2)^2 = x(2)$	$3(-2)^2 = x(-2)$
$12 = 2x$	$12 = -2x$
$6 = x$	$-6 = x$

Set the denominator of $\dfrac{3y^2}{y} = 0$ and you also get $y = 0$. Substitute $y = 0$ into the second equation and solve for x.

$$2x^2 + xy - 84 = 0$$
$$2x^2 + x(0) - 84 = 0$$
$$2x^2 - 84 = 0$$
$$2(x^2 - 42) = 0$$
$$x^2 = 42$$
$$x = \pm\sqrt{42}$$

The solutions are $(6, 2), (-6, -2), (\sqrt{42}, 0)$ and $(-\sqrt{42}, 0)$.

Section 13.4   - 936 -

47.

$$\begin{cases} y^2 = 40 - x^2 \\ y = x^2 - 10 \end{cases}$$

Write both equations in standard form and use elimination.

$$x^2 + y^2 = 40$$
$$\underline{-x^2 + y = -10}$$
$$y^2 + y = 30$$

$$y^2 + y - 30 = 0$$
$$(y+6)(y-5) = 0$$
$$y + 6 = 0 \quad \text{or} \quad y - 5 = 0$$
$$y = -6 \qquad\qquad y = 5$$

Substitute $y = -6$ and $y = 5$ into $y = x^2 - 10$ and solve for x.

$$-6 = x^2 - 10 \qquad\qquad 5 = x^2 - 10$$
$$x^2 - 10 = -6 \qquad\qquad x^2 - 10 = 5$$
$$x^2 = 4 \qquad\qquad x^2 = 15$$
$$x = \sqrt{4} \qquad\qquad x = \sqrt{15}$$
$$x = \pm 2 \qquad\qquad x = \pm\sqrt{15}$$

The solutions are $(2, -6), (-2, -6), (\sqrt{15}, 5)$, and $(-\sqrt{15}, 5)$.

49.

$$\begin{cases} 3x - y = -3 \\ 25y^2 - 9x^2 = 225 \end{cases}$$

Solve the first equation for y.

$$\begin{cases} y = 3x + 3 \\ 25y^2 - 9x^2 = 225 \end{cases}$$

Substitute $y = 3x + 3$ into the second equation and solve for x.

$$25y^2 - 9x^2 = 225$$
$$25(3x+3)^2 - 9x^2 = 225$$
$$25(9x^2 + 18x + 9) - 9x^2 = 225$$
$$225x^2 + 450x + 225 - 9x^2 = 225$$
$$216x^2 + 450x = 0$$
$$18x(12x + 25) = 0$$
$$18x = 0 \quad \text{and} \quad 12x + 25 = 0$$
$$x = 0 \qquad\qquad x = -\frac{25}{12}$$

Substitute $x = 0$ and $x = -\dfrac{25}{12}$ into the first equation and solve for y.

$$y = 3x + 3 \qquad\qquad y = 3x + 3$$
$$y = 3(0) + 3 \qquad\quad y = 3\left(-\frac{25}{12}\right) + 3$$
$$y = 0 + 3 \qquad\qquad$$
$$y = 3 \qquad\qquad y = -\frac{25}{4} + 3$$
$$\qquad\qquad\qquad\qquad y = -\frac{13}{4}$$

The solutions are $(0, 3)$ and $\left(-\dfrac{25}{12}, -\dfrac{13}{4}\right)$.

Section 13.4

51.

$$\begin{cases} x^2 - y = 0 \\ x^2 - 4x + y = 0 \end{cases}$$

Add the two equations and solve for x.

$$x^2 \quad - y = 0$$
$$\underline{x^2 - 4x + y = 0}$$
$$2x^2 - 4x = 0$$
$$2x(x - 2) = 0$$
$$2x = 0 \quad \text{and} \quad x - 2 = 0$$
$$x = 0 \qquad\qquad x = 2$$

Substitute $x = 0$ and $x = 2$ into the first equation and solve for y.

$$\begin{array}{ll} x^2 - y = 0 & x^2 - y = 0 \\ 0^2 - y = 0 & 2^2 - y = 0 \\ -y = 0 & 4 - y = 0 \\ y = 0 & 4 = y \end{array}$$

The solutions are (0, 0) and (2, 4).

53.

$$\begin{cases} x^2 - 2y^2 = 6 \\ x^2 + 2y^2 = 2 \end{cases}$$

Add the equations and solve for x.

$$x^2 - 2y^2 = 6$$
$$\underline{x^2 + 2y^2 = 2}$$
$$2x^2 = 8$$
$$x^2 = 4$$
$$x = \pm 2$$

Substitute $x = 2$ and $x = -2$ into the first equation and solve for y.

$$\begin{array}{ll} x^2 - 2y^2 = 6 & x^2 - 2y^2 = 6 \\ (2)^2 - 2y^2 = 6 & (-2)^2 - 2y^2 = 6 \\ 4 - 2y^2 = 6 & 4 - 2y^2 = 6 \\ -2y^2 = 2 & -2y^2 = 2 \\ y^2 = -1 & y^2 = -1 \\ y = \sqrt{-1} & y = \sqrt{-1} \\ y = i & y = i \\ \text{not real} & \text{not real} \end{array}$$

No solution; $\varnothing$

55.

$$\begin{cases} y = x^2 - 4 \\ 6x - y = 13 \end{cases}$$

Substitute $y = x^2 - 4$ into the second equation and solve for x.

$$6x - y = 13$$
$$6x - (x^2 - 4) = 13$$
$$6x - x^2 + 4 = 13$$
$$0 = x^2 - 6x + 9$$
$$0 = (x - 3)(x - 3)$$
$$x - 3 = 0 \quad \text{and} \quad x - 3 = 0$$
$$x = 3 \qquad\qquad x = 3$$

Substitute $x = 3$ into the second equation and solve for y.

$$6x - y = 13$$
$$6(3) - y = 13$$
$$18 - y = 13$$
$$-y = -5$$
$$y = 5$$

The solution is (3, 5).

57. $\begin{cases} x^2 + y^2 = 4 \\ 9x^2 + y^2 = 9 \end{cases}$

Multiply the first equation by -1 and add the equations to solve for x.

$$-x^2 - y^2 = -4$$
$$\underline{9x^2 + y^2 = 9}$$
$$8x^2 = 5$$
$$x^2 = \frac{5}{8}$$
$$x = \pm\sqrt{\frac{5}{8}}$$
$$x = \pm\sqrt{\frac{5 \cdot 2}{8 \cdot 2}}$$
$$x = \pm\frac{\sqrt{10}}{4}$$

Substitute $x = \pm\dfrac{\sqrt{10}}{4}$ into the first equation and solve for y.

$$x^2 + y^2 = 4$$
$$\left(\frac{\sqrt{10}}{4}\right)^2 + y^2 = 4$$
$$\frac{10}{16} + y^2 = 4$$
$$\frac{5}{8} + y^2 = 4$$
$$y^2 = 4 - \frac{5}{8}$$
$$y^2 = \frac{27}{8}$$
$$y = \pm\sqrt{\frac{27}{8}}$$
$$y = \pm\sqrt{\frac{27 \cdot 2}{8 \cdot 2}}$$
$$y = \pm\frac{\sqrt{54}}{\sqrt{16}}$$
$$y = \pm\frac{3\sqrt{6}}{4}$$

The solutions are $\left(\dfrac{\sqrt{10}}{4}, \dfrac{3\sqrt{6}}{4}\right)$,

$\left(-\dfrac{\sqrt{10}}{4}, \dfrac{3\sqrt{6}}{4}\right)$, $\left(\dfrac{\sqrt{10}}{4}, -\dfrac{3\sqrt{6}}{4}\right)$, and

$\left(-\dfrac{\sqrt{10}}{4}, -\dfrac{3\sqrt{6}}{4}\right)$.

59.
$$\begin{cases} xy = \dfrac{1}{6} \\ y + x = 5xy \end{cases}$$

Solve the first equation for y.

$$xy = \frac{1}{6}$$
$$\frac{1}{x}(xy) = \frac{1}{x}\left(\frac{1}{6}\right)$$
$$y = \frac{1}{6x}$$

Substitute $y = \dfrac{1}{6x}$ into the second equation and solve for x.

$$y + x = 5xy$$
$$\frac{1}{6x} + x = 5x\left(\frac{1}{6x}\right)$$
$$\frac{1}{6x} + x = \frac{5}{6}$$
$$6x\left(\frac{1}{6x} + x\right) = 6x\left(\frac{5}{6}\right)$$
$$1 + 6x^2 = 5x$$
$$6x^2 - 5x + 1 = 0$$
$$(2x - 1)(3x - 1) = 0$$
$$2x - 1 = 0 \quad \text{or} \quad 3x - 1 = 0$$
$$x = \frac{1}{2} \qquad\qquad x = \frac{1}{3}$$

Substitute $x = \dfrac{1}{2}$ and $x = \dfrac{1}{3}$ into

$xy = \dfrac{1}{6}$ and solve for y.

$$xy = \frac{1}{6} \qquad\qquad xy = \frac{1}{6}$$
$$\frac{1}{2}y = \frac{1}{6} \qquad\qquad \frac{1}{3}y = \frac{1}{6}$$
$$6\left(\frac{1}{2}y\right) = 6\left(\frac{1}{6}\right) \quad 6\left(\frac{1}{3}y\right) = 6\left(\frac{1}{6}\right)$$
$$3y = 1 \qquad\qquad 2y = 1$$
$$y = \frac{1}{3} \qquad\qquad y = \frac{1}{2}$$

The solutions are $\left(\dfrac{1}{2}, \dfrac{1}{3}\right)$ and $\left(\dfrac{1}{3}, \dfrac{1}{2}\right)$.

Section 13.4

61. $\begin{cases} x^2 = 4 - y \\ y = x^2 + 2 \end{cases}$

Substitute $y = x^2 + 2$ into the first equation and solve for x.

$$x^2 = 4 - y$$
$$x^2 = 4 - \left(x^2 + 2\right)$$
$$x^2 = 4 - x^2 - 2$$
$$2x^2 = 2$$
$$x^2 = 1$$
$$x = \pm 1$$

Substitute $x = 1$ and $x = -1$ into the second equation and solve for y.

$$y = x^2 + 2 \qquad\qquad y = x^2 + 2$$
$$y = (1)^2 + 2 \qquad\quad y = (-1)^2 + 2$$
$$y = 1 + 2 \qquad\qquad y = 1 + 2$$
$$y = 3 \qquad\qquad\quad\; y = 3$$

The solutions are $(1, 3)$ and $(-1, 3)$.

63. $\begin{cases} x^2 - y^2 = 4 \\ x + y = 4 \end{cases}$

Solve the second equation for x.

$$\begin{cases} x^2 - y^2 = 4 \\ x = -y + 4 \end{cases}$$

Substitute $x = -y + 4$ into the first equation and solve for y.

$$x^2 - y^2 = 4$$
$$\left(-y + 4\right)^2 - y^2 = 4$$
$$y^2 - 8y + 16 - y^2 = 4$$
$$-8y + 16 = 4$$
$$-8y = -12$$
$$y = \frac{3}{2}$$

Substitute $y = \frac{3}{2}$ into the second equation and solve for x.

$$x + y = 4$$
$$x + \frac{3}{2} = 4$$
$$x = \frac{5}{2}$$

The solution is $\left(\frac{5}{2}, \frac{3}{2}\right)$.

APPLICATIONS

65. INTEGER PROBLEM

Let the integers be x and y.

$$\begin{cases} xy = 32 \\ x + y = 12 \end{cases}$$

Solve the first equation for y.

$$xy = 32$$
$$y = \frac{32}{x}$$

Substitute $y = \frac{32}{x}$ into $x + y = 12$ and solve for x.

$$x + y = 12$$
$$x + \frac{32}{x} = 12$$
$$x\left(x + \frac{32}{x}\right) = x(12)$$
$$x^2 + 32 = 12x$$
$$x^2 - 12x + 32 = 0$$
$$(x - 4)(x - 8) = 0$$
$$x - 4 = 0 \quad \text{or} \quad x - 8 = 0$$
$$x = 4 \qquad\qquad\quad x = 8$$

Substitute $x = 4$ and $x = 8$ into the first equation to solve for y.

$$xy = 32 \qquad\qquad xy = 32$$
$$4y = 32 \qquad\qquad 8y = 32$$
$$y = 8 \qquad\qquad\; y = 4$$

The two integers are 4 and 8.

67. ARCHERY

$$\begin{cases} y = -\dfrac{1}{6}x^2 + 2x \\ y = \dfrac{1}{3}x \end{cases}$$

Substitute $y = \dfrac{1}{3}x$ into the first equation

and solve for x.

$$y = -\frac{1}{6}x^2 + 2x$$

$$\frac{1}{3}x = -\frac{1}{6}x^2 + 2x$$

$$6\left(\frac{1}{3}x\right) = 6\left(-\frac{1}{6}x^2 + 2x\right)$$

$$2x = -x^2 + 12x$$

$$x^2 - 10x = 0$$

$$x(x - 10) = 0$$

$$x = 0 \quad \text{and} \quad x - 10 = 0$$

$$x = 0 \qquad\qquad x = 10$$

Substitute $x = 10$ into the second equation.

$$y = \frac{1}{3}x$$

$$y = \frac{1}{3}(10)$$

$$y = \frac{10}{3}$$

The coordinates of the point of impact are

$\left(10, \dfrac{10}{3}\right)$.

To find the distance of the gun from the point of impact, you must use the Pythagorean Theorem. 10 is one leg (the distance on the x–axis), $\dfrac{10}{3}$ is the other leg (the height from the x–axis of the hill at the point of impact), and the hypotenuse is the distance on the hill from the archer to the point of impact.

$$a^2 + b^2 = c^2$$

$$10^2 + \left(\frac{10}{3}\right)^2 = c^2$$

$$100 + \frac{100}{9} = c^2$$

$$\frac{1,000}{9} = c^2$$

$$c = \sqrt{\frac{1,000}{9}}$$

$$c = \frac{10\sqrt{10}}{3} \text{ miles}$$

69. FENCING PASTURES

Let x = length and y = width.

Area = length · width

The perimeter is 2 widths + 1 length since the other length is riverbank.

$$\begin{cases} xy = 8,000 \\ x + 2y = 260 \end{cases}$$

Solve the first equation for y.

$$xy = 63$$

$$y = \frac{8,000}{x}$$

Substitute $y = \dfrac{8,000}{x}$ into the second

equation and solve for x.

$$x + 2y = 260$$

$$x + 2\left(\frac{8,000}{x}\right) = 260$$

$$x + \frac{16,000}{x} = 260$$

$$x\left(x + \frac{16,000}{x}\right) = x(260)$$

$$x^2 + 16,000 = 260x$$

$$x^2 - 260x + 16,000 = 0$$

$$(x - 100)(x - 160) = 0$$

$$x - 100 = 0 \quad \text{or} \quad x - 160 = 0$$

$$x = 100 \qquad\qquad x = 160$$

Substitute $x = 100$ and $x = 160$ into the first equation and solve for y.

$$\begin{array}{ll} xy = 8,000 & xy = 8,000 \\ 100y = 8,000 & 160y = 8,000 \\ y = 80 & y = 50 \end{array}$$

The dimensions are 80 ft by 100 ft or 50 ft by 160 ft.

71. INVESTING

Let x = the amount invested and y = rate of investment.

Grant's investment:
$$xy = 225$$

Jeff's investment:
$$(x + 500)(y - 0.01) = 240$$

$$\begin{cases} xy = 225 \\ (x+500)(y-0.01) = 240 \end{cases}$$

Solve the first equation for y and simplify the second equation

$$\begin{cases} y = \dfrac{225}{x} \\ xy - 0.01x + 500y - 5 = 240 \end{cases}$$

$$\begin{cases} y = \dfrac{225}{x} \\ xy - 0.01x + 500y = 245 \end{cases}$$

$$xy - 0.01x + 500y = 245$$

$$x\left(\frac{225}{x}\right) - 0.01x + 500\left(\frac{225}{x}\right) = 245$$

$$225 - 0.01x + \frac{112{,}500}{x} = 245$$

$$-0.01x + \frac{112{,}500}{x} = 20$$

$$x\left(-0.01x + \frac{112{,}500}{x}\right) = x(20)$$

$$-0.01x^2 + 112{,}500 = 20x$$

$$-0.01x^2 - 20x + 112{,}500 = 0$$

$$-100(-0.01x^2 - 20x + 112{,}500) = -100(0)$$

$$x^2 + 2{,}000x - 11{,}250{,}000 = 0$$

$$(x + 4{,}500)(x - 2{,}500) = 0$$

$$x + 4{,}500 = 0 \quad \text{or} \quad x - 2{,}500 = 0$$

$$x = \cancel{-4{,}500} \qquad x = 2{,}500$$

$$y = \frac{225}{x}$$

$$y = \frac{225}{2{,}50}$$

$$y = 0.09$$

$$y = 9\%$$

Since the amount invested cannot be negative, Grant invested $2,500 at 9%.

WRITING

73. Answers will vary.

REVIEW

75.
$$\log 5x = 4$$
$$10^4 = 5x$$
$$10{,}000 = 5x$$
$$2{,}000 = x$$

77.
$$\frac{\log(8x - 7)}{\log x} = 2$$
$$\log(8x - 7) = 2\log x$$
$$\log(8x - 7) = \log x^2$$
$$8x - 7 = x^2$$
$$0 = x^2 - 8x + 7$$
$$0 = (x - 7)(x - 1)$$
$$x - 7 = 0 \quad \text{or} \quad x - 1 = 0$$
$$x = 7 \qquad\qquad x = \cancel{1}$$

$x \neq 1$ because it make the denominator undefined

CHALLENGE PROBLEMS

79. a. The system may have 0, 1, 2, 3, or 4 solutions.

b. The system may have 0, 1, 2, 3, or 4 solutions.

81.

$$\begin{cases} \dfrac{1}{x} + \dfrac{3}{y} = 4 \\ \dfrac{2}{x} - \dfrac{1}{y} = 7 \end{cases}$$

Multiply the second equation by 3 and add the two equations to eliminate y.

$$\dfrac{1}{x} + \dfrac{3}{y} = 4$$

$$\dfrac{6}{x} - \dfrac{3}{y} = 21$$

$$\dfrac{7}{x} = 25$$

$$x\left(\dfrac{7}{x}\right) = x(25)$$

$$7 = 25x$$

$$\dfrac{7}{25} = x$$

Substitute $x = \dfrac{7}{25}$ into the first equation and solve for y.

$$\dfrac{1}{\dfrac{7}{25}} + \dfrac{3}{y} = 4$$

$$\dfrac{25}{7} + \dfrac{3}{y} = 4$$

$$7y\left(\dfrac{25}{7} + \dfrac{3}{y}\right) = 7y(4)$$

$$25y + 21 = 28y$$

$$21 = 3y$$

$$7 = y$$

The solution is $\left(\dfrac{7}{25}, 7\right)$.

SECTION 13.1
The Circle and the Parabola

1. $x^2 + y^2 = 16$

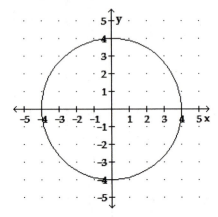

2. $(x-4)^2 + (y+3)^2 = 4$

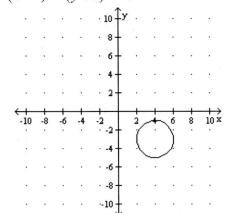

3.

$$x^2 + y^2 + 4x - 2y = 4$$
$$\left(x^2 + 4x\right) + \left(y^2 - 2y\right) = 4$$
$$\left(x^2 + 4x + 4\right) + \left(y^2 - 2y + 1\right) = 4 + 4 + 1$$
$$(x+2)^2 + (y-1)^2 = 9$$

4.

$$(x-9)^2 + (y-9)^2 = (9)^2$$
$$(x-9)^2 + (y-9)^2 = 81$$

5. center: $(-6, 0)$
   radius: $\sqrt{24} = 2\sqrt{6}$

6. A circle is the set of all points in a plane that are a fixed distance from a point called its **center**. The fixed distance is called the **radius** of the circle.

7. $x = y^2$
   vertex: $(0, 0)$

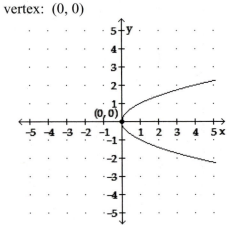

8. $x = 2(y+1)^2 - 2$
   vertex: $(-2, -1)$

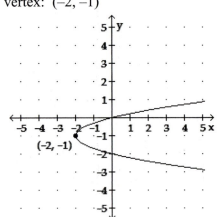

9.

$$x = -3y^2 + 12y - 7$$

$$x = -3(y^2 - 4y + 4) - 7 + 12$$

$$x = -3(y - 2)^2 + 5$$

vertex: $(5,\ 2)$

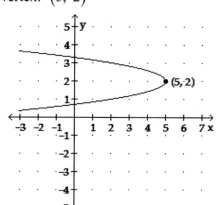

10.

$$y = x^2 + 8x + 11$$

$$y = (x^2 + 8x + 16) + 11 - 16$$

$$y = (x + 4)^2 - 5$$

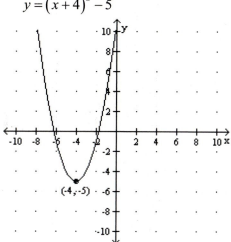

11. $(2, -2)$ and $(8, -3)$

12. When $x = 22$, $y = 0$.

$$y = -\frac{5}{121}(x - 11)^2 + 5$$

$$0 \overset{?}{=} -\frac{5}{121}(22 - 11)^2 + 5$$

$$0 \overset{?}{=} -\frac{5}{121}(11)^2 + 5$$

$$0 \overset{?}{=} -\frac{5}{121}(121) + 5$$

$$0 \overset{?}{=} -5 + 5$$

$$0 = 0$$

SECTION 13.2
The Ellipse

13.

$$\frac{x^2}{16} + \frac{y^2}{9} = 1$$

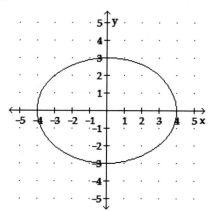

14.

$$\frac{(x - 2)^2}{4} + \frac{(y - 1)^2}{25} = 1$$

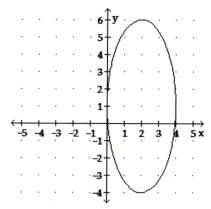

Chapter 13 Review

15.

$$4(x+1)^2 + 9(y-1)^2 = 36$$

$$\frac{4(x+1)^2}{36} + \frac{9(y-1)^2}{36} = \frac{36}{36}$$

$$\frac{(x+1)^2}{9} + \frac{(y-1)^2}{4} = 1$$

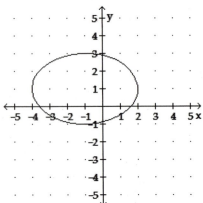

SECTION 13.3
The Hyperbola

21.

$$\frac{y^2}{9} - \frac{x^2}{1} = 1$$

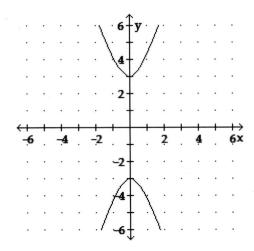

16.

$$\frac{x^2}{144} + \frac{y^2}{1} = 1$$

$$\frac{x^2}{12^2} + \frac{y^2}{1^2} = 1$$

22.

$$9(x-1)^2 - 4(y+1)^2 = 36$$

$$\frac{9(x-1)^2}{36} - \frac{4(y+1)^2}{36} = \frac{36}{36}$$

$$\frac{(x-1)^2}{4} - \frac{(y+1)^2}{9} = 1$$

17. a. circle
  b. ellipse
  c. parabola
  d. ellipse

18.

$$\frac{x^2}{5^2} + \frac{y^2}{3^2} = 1$$

$$\frac{x^2}{25} + \frac{y^2}{9} = 1$$

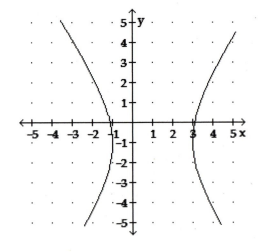

19. An **ellipse** is the set of all points in a plane for which the sum of the distances from two fixed points is a constant. Each of the fixed points is called a **focus**.

20. Answers may vary.

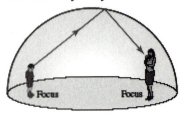

Chapter 13 Review   **- 946 -**

23. $\dfrac{(y-2)^2}{25}-\dfrac{(x+1)^2}{25}=1$

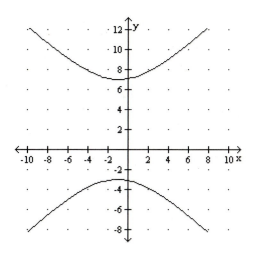

24. $xy=9$

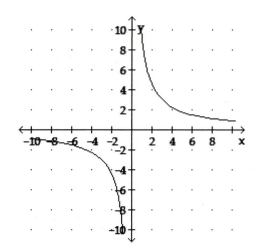

25.

$$x^2-4y^2=4$$

$$\dfrac{x^2}{4}-\dfrac{4y^2}{4}=\dfrac{4}{4}$$

$$\dfrac{x^2}{4}-\dfrac{y^2}{1}=1$$

$a=2$, so they were $2^2=4$ units apart.

26. a. ellipse
    b. hyperbola
    c. parabola
    d. circle

SECTION 13.4
Solving Systems of Nonlinear Equations

27. Yes, it is a solution.

$$x^2+y^2=20 \qquad\qquad x^2-y^2=2$$

$$\left(-\sqrt{11}\right)^2+\left(-3\right)^2\overset{?}{=}20 \quad \left(-\sqrt{11}\right)^2-\left(-3\right)^2\overset{?}{=}2$$

$$11+9\overset{?}{=}20 \qquad\qquad 11-9\overset{?}{=}2$$

$$20=20 \qquad\qquad\qquad 2=2$$

28. The graphs intersect at $(0, 3)$ and $(0, -3)$.

29. $\begin{cases} xy=4 \\ y=2x-2 \end{cases}$

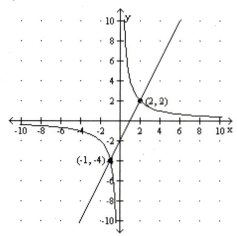

The solutions are $(2, 2)$ and $(-1, -4)$.

30. a. 2
    b. 4
    c. 4
    d. 4

31. Let $x=0$ in both equations to find the y–coordinate.

$$x^2+y^2=1 \qquad 4y^2-x^2=4$$

$$0^2+y^2=1 \qquad 4y^2-0^2=4$$

$$y^2=1 \qquad\qquad 4y^2=4$$

$$y=\sqrt{1} \qquad\qquad y^2=1$$

$$y=\pm1 \qquad\qquad y=\sqrt{1}$$

$$\qquad\qquad\qquad y=\pm1$$

The solutions are $(0, 1)$ and $(0, -1)$.

32. Solve the second equation for y: $y=3x-1$

Chapter 13 Review

33. $\begin{cases} y^2 - x^2 = 16 \\ y + 4 = x^2 \end{cases}$

Substitute $y + 4 = x^2$ into the first equation and solve for y.

$$y^2 - x^2 = 16$$
$$y^2 - (y + 4) = 16$$
$$y^2 - y - 4 = 16$$
$$y^2 - y - 20 = 0$$
$$(y - 5)(y + 4) = 0$$
$$y - 5 = 0 \quad \text{or} \quad y + 4 = 0$$
$$y = 5 \qquad\qquad y = -4$$

Substitute $y = 5$ and $y = -4$ into the second equation and solve for x.

$$y + 4 = x^2 \qquad\qquad y + 4 = x^2$$
$$5 + 4 = x^2 \qquad\qquad -4 + 4 = x^2$$
$$9 = x^2 \qquad\qquad\qquad 0 = x^2$$
$$x = \sqrt{9} \qquad\qquad\qquad x = \sqrt{0}$$
$$x = \pm 3 \qquad\qquad\qquad x = 0$$

The solutions are $(3, 5), (-3, 5)$, and $(0, -4)$.

34. $\begin{cases} y = -x^2 + 2 \\ x^2 - y - 2 = 0 \end{cases}$

Substitute $y = -x^2 + 2$ into the second equation and solve for x.

$$x^2 - y - 2 = 0$$
$$x^2 - (-x^2 + 2) - 2 = 0$$
$$x^2 + x^2 - 2 - 2 = 0$$
$$2x^2 - 4 = 0$$
$$2(x^2 - 2) = 0$$
$$x^2 - 2 = 0$$
$$x^2 = 2$$
$$x = \pm\sqrt{2}$$

Substitute $x = \sqrt{2}$ and $x = -\sqrt{2}$ into the first equation and solve for y.

$$y = -x^2 + 2 \qquad\qquad y = -x^2 + 2$$
$$y = -\left(\sqrt{2}\right)^2 + 2 \qquad y = -\left(-\sqrt{2}\right)^2 + 2$$
$$y = -2 + 2 \qquad\qquad y = -2 + 2$$
$$y = 0 \qquad\qquad\qquad y = 0$$

The solutions are $\left(\sqrt{2}, 0\right)$ and $\left(-\sqrt{2}, 0\right)$.

35. $\begin{cases} x^2 + 2y^2 = 12 \\ 2x - y = 2 \end{cases}$

Solve the second equation for y.

$$2x - y = 2$$
$$-y = -2x + 2$$
$$y = 2x - 2$$

Substitute $y = 2x - 2$ into the first equation and solve for x.

$$x^2 + 2y^2 = 12$$
$$x^2 + 2(2x - 2)^2 = 12$$
$$x^2 + 2(4x^2 - 4x - 4x + 4) = 12$$
$$x^2 + 2(4x^2 - 8x + 4) = 12$$
$$x^2 + 8x^2 - 16x + 8 = 12$$
$$9x^2 - 16x - 4 = 0$$
$$(9x + 2)(x - 2) = 12$$
$$9x + 2 = 0 \quad \text{or} \quad x - 2 = 0$$
$$9x = -2 \qquad\qquad x = 2$$
$$x = -\frac{2}{9}$$

Substitute $x = -\frac{2}{9}$ and $x = 2$ into the second equation and solve for y.

$$y = 2x - 2 \qquad\qquad y = 2x - 2$$
$$y = 2\left(-\frac{2}{9}\right) - 2 \qquad y = 2(2) - 2$$
$$y = -\frac{4}{9} - 2 \qquad\qquad y = 4 - 2$$
$$y = -\frac{22}{9} \qquad\qquad\qquad y = 2$$

The solutions are $\left(-\frac{2}{9}, -\frac{22}{9}\right)$ and $(2, 2)$.

36.

$$\begin{cases} 3x^2 + y^2 = 52 \\ x^2 - y^2 = 12 \end{cases}$$

Add the equations and solve for x.

$$3x^2 + y^2 = 52$$
$$\underline{x^2 - y^2 = 12}$$
$$4x^2 = 64$$
$$x^2 = 16$$
$$x = \sqrt{16}$$
$$x = \pm 4$$

Substitute $x = 4$ and $x = -4$ into the second equation and solve for y.

$$x^2 - y^2 = 12 \qquad\qquad x^2 - y^2 = 12$$
$$(4)^2 - y^2 = 12 \qquad (-4)^2 - y^2 = 12$$
$$16 - y^2 = 12 \qquad\quad 16 - y^2 = 12$$
$$-y^2 = -4 \qquad\qquad -y^2 = -4$$
$$y^2 = 4 \qquad\qquad\quad y^2 = 4$$
$$y = \sqrt{4} \qquad\qquad\quad y = \sqrt{4}$$
$$y = \pm 2 \qquad\qquad\quad y = \pm 2$$

The solutions are $(4, 2), (4, -2),$
$(-4, 2),$ and $(-4, -2)$

37.

$$\begin{cases} \dfrac{x^2}{16} + \dfrac{y^2}{12} = 1 \\ \dfrac{x^2}{1} - \dfrac{y^2}{3} = 1 \end{cases}$$

Multiply the first equation by 48 and the second equation to 3 to eliminate fractions.

$$\begin{cases} 3x^2 + 4y^2 = 48 \\ 3x^2 - y^2 = 3 \end{cases}$$

Multiply the second equation by 4.
Add the equations and solve for x.

$$3x^2 + 4y^2 = 48$$
$$\underline{12x^2 - 4y^2 = 12}$$
$$15x^2 = 60$$
$$x^2 = 4$$
$$x = \sqrt{4}$$
$$x = \pm 2$$

Substitute $x = 2$ and $x = -2$ into the second equation and solve for y.

$$3x^2 - y^2 = 3 \qquad\qquad 3x^2 - y^2 = 3$$
$$3(2)^2 - y^2 = 3 \qquad 3(-2)^2 - y^2 = 3$$
$$3(4) - y^2 = 3 \qquad\quad 3(4) - y^2 = 3$$
$$12 - y^2 = 3 \qquad\qquad 12 - y^2 = 3$$
$$-y^2 = -9 \qquad\qquad\quad -y^2 = -9$$
$$y^2 = 9 \qquad\qquad\qquad y^2 = 9$$
$$y = \sqrt{9} \qquad\qquad\qquad y = \sqrt{9}$$
$$y = \pm 3 \qquad\qquad\qquad y = \pm 3$$

The solutions are $(2, 3), (2, -3),$
$(-2, 3),$ and $(-2, -3)$

38.

$$\begin{cases} xy = 4 \\ \dfrac{x^2}{1} + \dfrac{y^2}{2} = 9 \end{cases}$$

Solve the first equation for y and multiply the second equation by 2 to eliminate fractions.

$$\begin{cases} y = \dfrac{4}{x} \\ 2x^2 + y^2 = 18 \end{cases}$$

Substitute $y = \dfrac{4}{x}$ into the second equation and solve for x.

$$2x^2 + y^2 = 18$$

$$2x^2 + \left(\dfrac{4}{x}\right)^2 = 18$$

$$2x^2 + \dfrac{16}{x^2} = 18$$

$$x^2\left(2x^2 + \dfrac{16}{x^2}\right) = x^2(18)$$

$$2x^4 + 16 = 18x^2$$

$$2x^4 - 18x^2 + 16 = 0$$

$$2\left(x^4 - 9x^2 + 8\right) = 0$$

$$2\left(x^2 - 1\right)\left(x^2 - 8\right) = 0$$

$$x^2 - 1 = 0 \quad \text{or} \quad x^2 - 8 = 0$$
$$x^2 = 1 \qquad\qquad x^2 = 8$$
$$x = \sqrt{1} \qquad\qquad x = \sqrt{8}$$
$$x = \pm 1 \qquad\qquad x = \pm 2\sqrt{2}$$

Substitute all four values into the first equation and solve for y.

$$\begin{array}{ll} xy = 4 & xy = 4 \\ 1y = 4 & -1y = 4 \\ y = 4 & y = -4 \end{array}$$

$$\begin{array}{ll} xy = 4 & xy = 4 \\ 2\sqrt{2}\cdot y = 4 & -2\sqrt{2}\cdot y = 4 \\ y = \dfrac{4}{2\sqrt{2}} & y = \dfrac{4}{-2\sqrt{2}} \\ y = \dfrac{2}{\sqrt{2}} & y = -\dfrac{2}{\sqrt{2}} \\ y = \dfrac{2\sqrt{2}}{2} & y = -\dfrac{2\sqrt{2}}{2} \\ y = \sqrt{2} & y = -\sqrt{2} \end{array}$$

The solutions are $(1, 4), (-1, -4),$

$\left(2\sqrt{2}, \sqrt{2}\right),$ and $\left(-2\sqrt{2}, -\sqrt{2}\right)$

39.

$$\begin{cases} y = -x^2 + 1 \\ x + y = 5 \end{cases}$$

Substitute $y = -x^2 + 1$ into the second equation and solve for x.

$$x + y = 5$$

$$x + \left(-x^2 + 1\right) = 5$$

$$-x^2 + x + 1 = 5$$

$$0 = x^2 - x + 4$$

Since the quadratic equation $x^2 - x + 4$ cannot be factored using real numbers, there is no solution to this system.

40.

$$\begin{cases} x = y^2 - 3 \\ x = y^2 - 3y \end{cases}$$

Substitute $x = y^2 - 3$ into the second equation for x.

$$x = y^2 - 3y$$

$$y^2 - 3 = y^2 - 3y$$

$$-3 = -3y$$

$$1 = y$$

Substitute $y = 1$ into the first equation and solve for x.

$$x = y^2 - 3$$

$$x = 1^2 - 3$$

$$x = 1 - 3$$

$$x = -2$$

The solution is $(-2, 1)$.

CHAPTER 13 TEST

1. a. The curves formed by the intersection of a plane with an infinite right-circular cone are called **conic** sections.

    b. A circle is the set of all points in a plane that are a fixed distance from a point called its **center**. The fixed distance is called the **radius** of the circle.

    c. The standard form for the equation of a **hyperbola** centered at the origin that opens left and right is
$$\frac{x^2}{a^2} - \frac{y^2}{b^2} = 1.$$

    d. $\begin{cases} y = x^2 + x - 4 \\ x^2 + y^2 = 36 \end{cases}$ is a(n) **nonlinear** system of equations

    e. The standard form for the equation of an **ellipse** centered at the origin is
$$\frac{x^2}{a^2} + \frac{y^2}{b^2} = 1.$$

2. center: (0, 0); radius: $\sqrt{100} = 10$

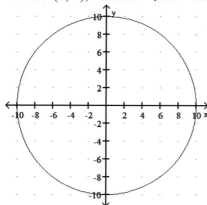

3.
$$x^2 + y^2 + 4x - 6y = 5$$
$$\left(x^2 + 4x\right) + \left(y^2 - 6y\right) = 5$$
$$\left(x^2 + 4x + 4\right) + \left(y^2 - 6y + 9\right) = 5 + 4 + 9$$
$$\left(x + 2\right)^2 + \left(y - 3\right)^2 = 18$$
center: $\left(-2, 3\right)$
radius: $\sqrt{18} = 3\sqrt{2}$

4. The center is (4, 3) and the radius is 3. The equation of the circle is
$$\left(x - 4\right)^2 + \left(y - 3\right)^2 = 3^2$$
$$\left(x - 4\right)^2 + \left(y - 3\right)^2 = 9$$

5.
$$r^2 = \frac{441}{16}$$
$$r = \sqrt{\frac{441}{16}}$$
$$r = \frac{21}{4}$$
diameter $= 2r$
$$= 2\left(\frac{21}{4}\right)$$
$$= \frac{21}{2} \text{ in.}$$
$$= 10.5 \text{ in.}$$

6.
$$x = y^2 + 8y + 10$$
$$x = \left(y^2 + 8y + \boxed{16}\right) + 10 - \boxed{16}$$
$$x = \left(y + \boxed{4}\right)^2 - 6$$

7. $\left(x + 2\right)^2 + \left(y - 1\right)^2 = 9$

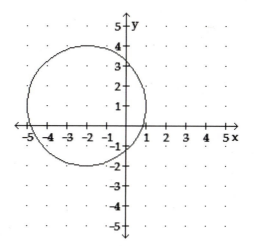

8.

$$x = y^2 - 4y + 3$$
$$x = \left(y^2 - 4y\right) + 3$$
$$x = \left(y^2 - 4y + 4\right) + 3 - 4$$
$$x = (y - 2)^2 - 1$$

vertex: $(-1, 2)$
axis of symmetry: $y = 2$

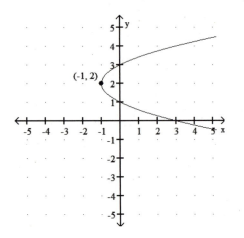

10.

$$xy = 4$$
$$y = \frac{4}{x}$$

x	$y = \dfrac{4}{x}$
1	$\dfrac{4}{1} = 4$
2	$\dfrac{4}{2} = 2$
4	$\dfrac{4}{4} = 1$
-4	$\dfrac{4}{-4} = -1$
-2	$\dfrac{4}{-2} = -2$
-1	$\dfrac{4}{-1} = -4$

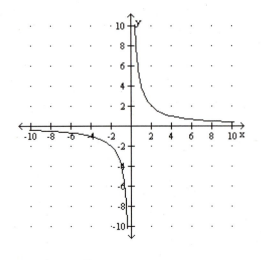

9.

$$y = -2x^2 - 4x + 5$$
$$y = -2\left(x^2 + 2x\right) + 5$$
$$y = -2\left(x^2 + 2x + 1\right) + 5 - (-2)(1)$$
$$y = -2(x + 1)^2 + 5 + 2$$
$$y = -2(x + 1)^2 + 7$$

vertex: $(-1, 7)$

axis of symmetry: $x = -1$

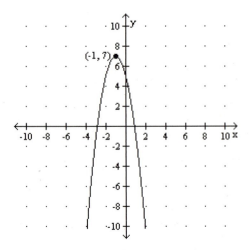

11.

$$9x^2 + 4y^2 = 36$$

$$\frac{9x^2}{36} + \frac{4y^2}{36} = \frac{36}{36}$$

$$\frac{x^2}{4} + \frac{y^2}{9} = 1$$

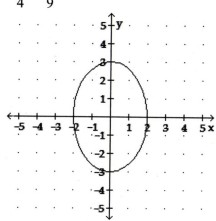

12. $\dfrac{(x-2)^2}{9} - \dfrac{y^2}{1} = 1$

13. $\dfrac{(x-3)^2}{49} + \dfrac{(y+2)^2}{16} = 1$

14.

$$x^2 + y^2 = 7$$

$$x^2 + y^2 = \left(\sqrt{7}\right)^2$$

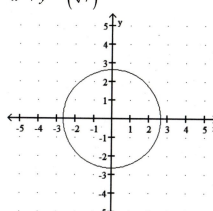

15. $x = -\dfrac{1}{2}y^2$

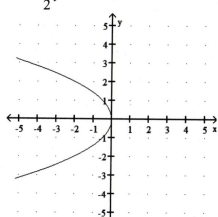

16. $\dfrac{y^2}{25} - \dfrac{x^2}{9} = 1$

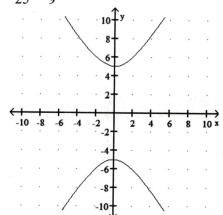

17. The center is $(1, -2)$, $a = 4$, and $b = 3$.
The equation of the ellipse is

$$\frac{(x-1)^2}{16} + \frac{(y+2)^2}{9} = 1.$$

Chapter 13 Test

18. If the width is 10, then $y = 5$.

$$x = \frac{1}{10}y^2$$

$$x = \frac{1}{10}(5)^2$$

$$x = \frac{1}{10}(25)$$

$$x = 2.5$$

19. The ellipse is centered at the origin and extends on the x-axis to -28 and 28 (half of $60 - 4$ for the 2 inches on each side) and extends on the y-axis to 16 and -16 (half of $36 - 4$ for the 2 inches on each side).

$$\frac{x^2}{28^2} + \frac{y^2}{16^2} = 1$$

$$\frac{x^2}{784} + \frac{y^2}{256} = 1$$

20.

$$(x+1)^2 - (y-1)^2 = 4$$

center: $(-1, 1)$

radius: 2

horizonal dimensions: 4 units

vertical dimensions: 4 units

21. The center is $(0, 0)$. The y-value is 4 and the x-value is 6. It opens up.

$$\frac{y^2}{9} - \frac{x^2}{36} = 1$$

22. a. ellipse
    b. hyperbola
    c. circle
    d. parabola

23. $\begin{cases} x^2 + y^2 = 25 \\ y - x = 1 \end{cases}$

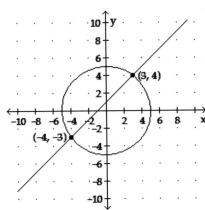

24.

$$\begin{cases} 2x - y = -2 \\ x^2 + y^2 = 16 + 4y \end{cases}$$

Solve the first equation for y.

$$\begin{cases} y = 2x + 2 \\ x^2 + y^2 = 16 + 4y \end{cases}$$

Substitute $y = 2x + 2$ into the second equation and solve for x.

$$x^2 + y^2 = 16 + 4y$$

$$x^2 + (2x+2)^2 = 16 + 4(2x+2)$$

$$x^2 + (4x^2 + 4x + 4x + 4) = 16 + 8x + 8$$

$$5x^2 + 8x + 4 = 8x + 24$$

$$5x^2 - 20 = 0$$

$$5(x^2 - 4) = 0$$

$$5(x-2)(x+2) = 0$$

$$x - 2 = 0 \quad \text{or} \quad x + 2 = 0$$

$$x = 2 \qquad\qquad x = -2$$

Substitute $x = 2$ and $x = -2$ into the first equation and solve for y.

$$\begin{array}{ll} y = 2x + 2 & y = 2x + 2 \\ y = 2(2) + 2 & y = 2(-2) + 2 \\ y = 4 + 2 & y = -4 + 2 \\ y = 6 & y = -2 \end{array}$$

The solutions are $(2, 6)$ and $(-2, -2)$.

25.

$$\begin{cases} 5x^2 - y^2 - 3 = 0 \\ x^2 + 2y^2 = 5 \end{cases}$$

Write the first equation in standard form

$$\begin{cases} 5x^2 - y^2 = 3 \\ x^2 + 2y^2 = 5 \end{cases}$$

Multiply the second equation by -5 and add the equations to solve for y.

$$\begin{array}{r} 5x^2 - y^2 = 3 \\ -5x^2 - 10y^2 = -25 \\ \hline -11y^2 = -22 \\ y^2 = 2 \\ y = \pm\sqrt{2} \end{array}$$

Substitute $y = \sqrt{2}$ and $y = -\sqrt{2}$ into the second equation and solve for x.

$$\begin{array}{ll} x^2 + 2y^2 = 5 & x^2 + 2y^2 = 5 \\ x^2 + 2\left(\sqrt{2}\right)^2 = 5 & x^2 + 2\left(-\sqrt{2}\right)^2 = 5 \\ x^2 + 2(2) = 5 & x^2 + 2(2) = 5 \\ x^2 + 4 = 5 & x^2 + 4 = 5 \\ x^2 = 1 & x^2 = 1 \\ x = \sqrt{1} & x = \sqrt{1} \\ x = \pm 1 & x = \pm 1 \end{array}$$

The solutions are $\left(1, \sqrt{2}\right), \left(-1, \sqrt{2}\right),$ $\left(1, -\sqrt{2}\right),$ and $\left(-1, -\sqrt{2}\right).$

26.

$$\begin{cases} xy = -\dfrac{9}{2} \\ 3x + 2y = 6 \end{cases}$$

Solve the first equation for y.

$$\begin{cases} y = -\dfrac{9}{2x} \\ 3x + 2y = 6 \end{cases}$$

Substitute $y = -\dfrac{9}{2x}$ into the second equation and solve for x.

$$3x + 2y = 6$$
$$3x + 2\left(-\dfrac{9}{2x}\right) = 6$$
$$3x - \dfrac{9}{x} = 6$$
$$x\left(3x - \dfrac{9}{x}\right) = x(6)$$
$$3x^2 - 9 = 6x$$
$$3x^2 - 6x - 9 = 0$$
$$3\left(x^2 - 2x - 3\right) = 0$$
$$3(x - 3)(x + 1) = 0$$
$$x - 3 = 0 \quad \text{or} \quad x + 1 = 0$$
$$x = 3 \qquad\qquad x = -1$$

Substitute $x = 3$ and $x = -1$ into the first equation and solve for y.

$$\begin{array}{ll} xy = -\dfrac{9}{2} & xy = -\dfrac{9}{2} \\ 3y = -\dfrac{9}{2} & -1y = -\dfrac{9}{2} \\ \dfrac{1}{3}(3y) = \dfrac{1}{3}\left(-\dfrac{9}{2}\right) & -1(-1y) = -1\left(-\dfrac{9}{2}\right) \\ y = -\dfrac{3}{2} & y = \dfrac{9}{2} \end{array}$$

The solutions are $\left(3, -\dfrac{3}{2}\right)$ and $\left(-1, \dfrac{9}{2}\right).$

27. $\begin{cases} y = x+1 \\ x^2 - y^2 = 1 \end{cases}$

Substitute $y = x + 1$ into the second equation and solve for x.

$$x^2 - y^2 = 1$$
$$x^2 - (x+1)^2 = 1$$
$$x^2 - (x^2 + 2x + 1)^2 = 1$$
$$x^2 - x^2 - 2x - 1 = 1$$
$$-2x = 2$$
$$x = -1$$

Substitute $x = -1$ into the first equation and solve for y.

$$y = x + 1$$
$$y = -1 + 1$$
$$y = 0$$

The solution is $(-1, 0)$.

28. $\begin{cases} x^2 + 3y^2 = 6 \\ x^2 + y = 8 \end{cases}$

Solve the second equation for x^2

$$\begin{cases} x^2 + 3y^2 = 6 \\ x^2 = -y + 8 \end{cases}$$

Substitute $x^2 = -y + 8$ into the first equation and solve for y.

$$-y + 8 + 3y^2 = 6$$
$$3y^2 - y + 2 = 0$$

No Solution; $\varnothing$

The quadratic equation is not factorable so the solutions are not real

SECTION 14.1

VOCABULARY

1. The two-term polynomial expression $a + b$ is called a **binomial**.

3. We can use the **binomial** theorem to raise binomials to positive-integer powers without doing the actual multiplication.

5. $n!$ (read as "n **factorial**") is the product of consecutive **decreasing** natural numbers from n to 1.

CONCEPTS

7. The binomial expansion of $(m+n)^6$ has **one** more term than the power of the binomial.

9. The first term of the expansion of $(r+s)^{20}$ is $\boxed{r^{20}}$ and the last term is $\boxed{s^{20}}$.

11. The coefficients of the terms of the expansion of $(c+d)^{20}$ begin with $\boxed{1}$, increase through some values, and then decrease through those same values, back to $\boxed{1}$.

13. $n \cdot \boxed{(n-1)!} = n!$

15. $0! = \boxed{1}$

17. The coefficient of the fourth term of the expansion of $(a+b)^9$ is 9! Divided by $\underline{3!(9-3)!}$.

19. The exponent on the a in the fifth term of the expansion of $(a+b)^6$ is $\boxed{2}$ and the exponent on b is $\boxed{4}$.

21. $(x+y)^3$

$$= x^{\boxed{3}} + \frac{\boxed{3!}}{1!(3-1)!} x^2 \boxed{y} + \frac{\boxed{3!}}{\boxed{2}!(3-2)!} xy^{\boxed{2}} + y^{\boxed{3}}$$

NOTATION

23. $n! = n \cdot \left(\boxed{n-1}\right)(n-2)\ldots 3 \cdot 2 \cdot 1$

GUIDED PRACTICE

25.
$$(a+b)^3 = a^3 + 3a^2b + 3ab^2 + b^3$$

27.
$$(m-p)^5 = m^5 - 5m^4 p + 10m^3 p^2 - 10m^2 p^3 + 5mp^4 - p^5$$

29.
$$3! = 3 \cdot 2 \cdot 1$$
$$= 6$$

31.
$$5! = 5 \cdot 4 \cdot 3 \cdot 2 \cdot 1$$
$$= 120$$

33.
$$3! + 4! = 3 \cdot 2 \cdot 1 + 4 \cdot 3 \cdot 2 \cdot 1$$
$$= 6 + 24$$
$$= 30$$

35.
$$3!(4!) = (3 \cdot 2 \cdot 1)(4 \cdot 3 \cdot 2 \cdot 1)$$
$$= (6)(24)$$
$$= 144$$

37.
$$8(7!) = 8(7 \cdot 6 \cdot 5 \cdot 4 \cdot 3 \cdot 2 \cdot 1)$$
$$= 8(5,040)$$
$$= 40,320$$

39.
$$\frac{49!}{47!} = \frac{49 \cdot 48 \cdot \cancel{47!}}{\cancel{47!}}$$
$$= \frac{49 \cdot 48}{1}$$
$$= 2,352$$

41.

$$\frac{9!}{11!} = \frac{\cancel{9}\cdot\cancel{8}\cdot\cancel{7}\cdot\cancel{6}\cdot\cancel{5}\cdot\cancel{4}\cdot\cancel{3}\cdot\cancel{2}\cdot\cancel{1}}{11\cdot10\cdot\cancel{9}\cdot\cancel{8}\cdot\cancel{7}\cdot\cancel{6}\cdot\cancel{5}\cdot\cancel{4}\cdot\cancel{3}\cdot\cancel{2}\cdot\cancel{1}}$$

$$= \frac{1}{11\cdot10}$$

$$= \frac{1}{110}$$

43.

$$\frac{9!}{7!0!} = \frac{9\cdot8\cdot\cancel{7!}}{\cancel{7!}0!}$$

$$= \frac{9\cdot8}{1}$$

$$= 72$$

45.

$$\frac{5!}{1!(5-1)!} = \frac{5!}{1!4!}$$

$$= \frac{5\cdot\cancel{4!}}{1\cdot\cancel{4!}}$$

$$= 5$$

47.

$$\frac{5!}{3!(5-3)!} = \frac{5!}{3!2!}$$

$$= \frac{5\cdot4\cdot\cancel{3!}}{\cancel{3!}(2\cdot1)}$$

$$= \frac{20}{2}$$

$$= 10$$

49.

$$\frac{5!(8-5)!}{4!7!} = \frac{5!\,\cancel{3!}}{(4\cdot\cancel{3!})(7\cdot6\cdot\cancel{5!})}$$

$$= \frac{1}{4\cdot7\cdot6}$$

$$= \frac{1}{168}$$

51.

$$\frac{7!}{5!(7-5)!} = \frac{7!}{5!2!}$$

$$= \frac{7\cdot6\cdot\cancel{5!}}{\cancel{5!}(2\cdot1)}$$

$$= \frac{42}{2}$$

$$= 21$$

53.

$$11! = 39{,}916{,}800$$

55.

$$20! = 2.432902008 \times 10^{18}$$

57.

$$(m+n)^4$$

$$= m^4 + \frac{4!}{1!(4-1)!}m^3n + \frac{4!}{2!(4-2)!}m^2n^2$$

$$\quad + \frac{4!}{3!(4-3)!}mn^3 + n^4$$

$$= m^4 + \frac{4\cdot\cancel{3!}}{1!\,\cancel{3!}}m^3n + \frac{4\cdot3\cdot\cancel{2!}}{2!\,\cancel{2!}}m^2n^2$$

$$\quad + \frac{4\cdot\cancel{3!}}{\cancel{3!}1!}mn^3 + n^4$$

$$= m^4 + \frac{4}{1}m^3n + \frac{12}{2}m^2n^2 + \frac{4}{1}mn^3 + n^4$$

$$= m^4 + 4m^3n + 6m^2n^2 + 4mn^3 + n^4$$

59.

$$(c-d)^5$$

$$= c^5 + \frac{5!}{1!(5-1)!}c^4(-d) + \frac{5!}{2!(5-2)!}c^3(-d)^2$$

$$\quad + \frac{5!}{3!(5-3)!}c^2(-d)^3 + \frac{5!}{4!(5-4)!}c(-d)^4 + (-d)^5$$

$$= c^5 + \frac{5\cdot\cancel{4!}}{1!\,\cancel{4!}}c^4(-d) + \frac{5\cdot4\cdot\cancel{3!}}{2!\,\cancel{3!}}c^3(-d)^2$$

$$\quad + \frac{5\cdot4\cdot\cancel{3!}}{\cancel{3!}2!}c^2(-d)^3 + \frac{5\cdot\cancel{4!}}{\cancel{4!}1!}c(-d)^4 + (-d)^5$$

$$= c^5 - \frac{5}{1}c^4d + \frac{20}{2}c^3d^2 - \frac{20}{2}c^2d^3 + \frac{5}{1}cd^4 - d^5$$

$$= c^5 - 5c^4d + 10c^3d^2 - 10c^2d^3 + 5cd^4 - d^5$$

61.

$$(a-b)^9 = \left(a+(-b)\right)^9$$

$$= a^9 + \frac{9!}{1!(9-1)!}a^8(-b) + \frac{9!}{2!(9-2)!}a^7(-b)^2$$

$$+ \frac{9!}{3!(9-3)!}a^6(-b)^3 + \frac{9!}{4!(9-4)!}a^5(-b)^4$$

$$+ \frac{9!}{5!(9-5)!}a^4(-b)^5 + \frac{9!}{6!(9-6)!}a^3(-b)^6$$

$$+ \frac{9!}{7!(9-7)!}a^2(-b)^7 + \frac{9!}{8!(9-8)!}a(-b)^8 + (-b)^9$$

$$= a^9 + \frac{9!}{1!8!}a^8(-b) + \frac{9!}{2!7!}a^7(-b)^2 + \frac{9!}{3!6!}a^6(-b)^3$$

$$+ \frac{9!}{4!5!}a^5(-b)^4 + \frac{9!}{5!4!}a^4(-b)^5 + \frac{9!}{6!3!}a^3(-b)^6$$

$$+ \frac{9!}{7!2!}a^2(-b)^7 + \frac{9!}{8!1!}a(-b)^8 + (-b)^9$$

$$= a^9 - \frac{9}{1}a^8b + \frac{72}{2}a^7b^2 - \frac{504}{6}a^6b^3 + \frac{3{,}024}{24}a^5b^4$$

$$- \frac{3{,}024}{24}a^4b^5 + \frac{504}{6}a^3b^6 - \frac{72}{2}a^2b^7 + \frac{9}{1}ab^8 - b^9$$

$$= a^9 - 9a^8b + 36a^7b^2 - 84a^6b^3 + 126a^5b^4 - 126a^4b^5$$

$$+ 84a^3b^6 - 36a^2b^7 + 9ab^8 - b^9$$

63.

$$(s+t)^6$$

$$= s^6 + \frac{6!}{1!(6-1)!}s^5t + \frac{6!}{2!(6-2)!}s^4t^2$$

$$+ \frac{6!}{3!(6-3)!}s^3t^3 + \frac{6!}{4!(6-4)!}s^2t^4$$

$$+ \frac{6!}{5!(6-5)!}st^5 + t^6$$

$$= s^6 + \frac{6!}{1!5!}s^5t + \frac{6!}{2!4!}s^4t^2$$

$$+ \frac{6!}{3!3!}s^3t^3 + \frac{6!}{4!2!}s^2t^4$$

$$+ \frac{6!}{5!1!}st^5 + t^6$$

$$= s^6 + \frac{6}{1}s^5t + \frac{30}{2}s^4t^2 + \frac{120}{6}s^3t^3 + \frac{30}{2}s^2t^4$$

$$+ \frac{6}{1}st^5 + t^6$$

$$= s^6 + 6s^5t + 15s^4t^2 + 20s^3t^3 + 15s^2t^4 + 6st^5 + t^6$$

65.

$$(2x+y)^3$$

$$= (2x)^3 + \frac{3!}{1!(3-1)!}(2x)^2y + \frac{3!}{2!(3-2)!}(2x)y^2 + y^3$$

$$= 8x^3 + \frac{3!}{1!2!}(4x^2)y + \frac{3!}{2!1!}(2x)y^2 + y^3$$

$$= 8x^3 + \frac{3}{1}(4x^2)y + \frac{3}{1}(2x)y^2 + y^3$$

$$= 8x^3 + 12x^2y + 6xy^2 + y^3$$

67.

$$(2t-3)^5$$

$$= (2t)^5 + \frac{5!}{1!(5-1)!}(2t)^4(-3) + \frac{5!}{2!(5-2)!}(2t)^3(-3)^2$$

$$+ \frac{5!}{3!(5-3)!}(2t)^2(-3)^3 + \frac{5!}{4!(5-4)!}(2t)(-3)^4 + (-3)^5$$

$$= (2t)^5 + \frac{5\cdot4!}{1!\,4!}(2t)^4(-3) + \frac{5\cdot4\cdot3!}{2!\,3!}(2t)^3(-3)^2$$

$$+ \frac{5\cdot4\cdot3!}{3!\,2!}(2t)^2(-3)^3 + \frac{5\cdot4!}{4!\,1!}(2t)(-3)^4 + (-3)^5$$

$$= 32t^5 + \frac{5}{1}(16t^4)(-3) + \frac{20}{2}(8t^3)(9) + \frac{20}{2}(4t^2)(-27)$$

$$+ \frac{5}{1}(2t)(81) - 243$$

$$= 32t^5 - 240t^4 + 720t^3 - 1{,}080t^2 + 810t - 243$$

69.

$$(5m-2n)^4$$

$$=(5m)^4+\frac{4!}{1!(4-1)!}(5m)^3(-2n)$$

$$+\frac{4!}{2!(4-2)!}(5m)^2(-2n)^2+\frac{4!}{3!(4-3)!}(5m)(-2n)^3$$

$$+(-2n)^4$$

$$=625m^4+\frac{4\bullet 3\!\!\!/!}{1!\,3\!\!\!/!}(125m^3)(-2n)$$

$$+\frac{4\bullet 3\bullet 2\!\!\!/!}{2!\,2\!\!\!/!}(25m^2)(4n^2)+\frac{4\bullet 3\!\!\!/!}{3\!\!\!/!1!}(5m)(-8n^3)+16n^4$$

$$=625m^4+\frac{4}{1}(-250m^3n)+\frac{12}{2}(100m^2n^2)$$

$$+\frac{4}{1}(-40mn^3)+16n^4$$

$$=625m^4-1{,}000m^3n+600m^2n^2-160mn^3+16n^4$$

71.

$$\left(\frac{x}{3}+\frac{y}{2}\right)^3$$

$$=\left(\frac{x}{3}\right)^3+\frac{3!}{1!(3-1)!}\left(\frac{x}{3}\right)^2\left(\frac{y}{2}\right)+\frac{3!}{2!(3-2)!}\left(\frac{x}{3}\right)\left(\frac{y}{2}\right)^2$$

$$+\left(\frac{y}{2}\right)^3$$

$$=\frac{x^3}{27}+\frac{3!}{1!2!}\left(\frac{x^2}{9}\right)\left(\frac{y}{2}\right)+\frac{3!}{2!1!}\left(\frac{x}{3}\right)\left(\frac{y^2}{4}\right)+\frac{y^3}{8}$$

$$=\frac{x^3}{27}+\frac{3}{1}\left(\frac{x^2y}{18}\right)+\frac{3}{1}\left(\frac{xy^2}{12}\right)+\frac{y^3}{8}$$

$$=\frac{x^3}{27}+\frac{x^2y}{6}+\frac{xy^2}{4}+\frac{y^3}{8}$$

73.

$$\left(\frac{x}{3}-\frac{y}{2}\right)^4$$

$$=\left(\frac{x}{3}\right)^4+\frac{4!}{1!(4-1)!}\left(\frac{x}{3}\right)^3\left(-\frac{y}{2}\right)$$

$$+\frac{4!}{2!(4-2)!}\left(\frac{x}{3}\right)^2\left(-\frac{y}{2}\right)^2+\frac{4!}{3!(4-3)!}\left(\frac{x}{3}\right)\left(-\frac{y}{2}\right)^3$$

$$+\left(-\frac{y}{2}\right)^4$$

$$=\frac{x^4}{81}+\frac{4\bullet 3\!\!\!/!}{1!\,3\!\!\!/!}\left(\frac{x^3}{27}\right)\left(-\frac{y}{2}\right)+\frac{4\bullet 3\bullet 2\!\!\!/!}{2!\,2\!\!\!/!}\left(\frac{x^2}{9}\right)\left(\frac{y^2}{4}\right)$$

$$+\frac{4\bullet 3\!\!\!/!}{3\!\!\!/!1!}\left(\frac{x}{3}\right)\left(-\frac{y^3}{8}\right)+\frac{y^4}{16}$$

$$=\frac{x^4}{81}+\frac{4}{1}\left(-\frac{x^3y}{54}\right)+\frac{12}{2}\left(\frac{x^2y^2}{36}\right)$$

$$+\frac{4}{1}\left(-\frac{xy^3}{24}\right)+\frac{y^4}{16}$$

$$=\frac{x^4}{81}-\frac{2x^3y}{27}+\frac{x^2y^2}{6}-\frac{xy^3}{6}+\frac{y^4}{16}$$

75.

$$\left(c^2-d^2\right)^5$$

$$=\left(c^2\right)^5+\frac{5!}{1!(5-1)!}\left(c^2\right)^4\left(-d^2\right)$$

$$+\frac{5!}{2!(5-2)!}\left(c^2\right)^3\left(-d^2\right)^2+\frac{5!}{3!(5-3)!}\left(c^2\right)^2\left(-d^2\right)^3$$

$$+\frac{5!}{4!(5-4)!}\left(c^2\right)\left(-d^2\right)^4+\left(-d^2\right)^5$$

$$=c^{10}+\frac{5\bullet4!}{1!\,4!}\left(c^8\right)\left(-d^2\right)+\frac{5\bullet4\bullet3!}{2!\,3!}\left(c^6\right)\left(d^4\right)$$

$$+\frac{5\bullet4\bullet3!}{3!\,2!}\left(c^4\right)\left(-d^6\right)+\frac{5\bullet4!}{4!\,1!}\left(c^2\right)\left(d^8\right)-d^{10}$$

$$=c^{10}+\frac{5}{1}\left(-c^8d^2\right)+\frac{20}{2}\left(c^6d^4\right)+\frac{20}{2}\left(-c^4d^6\right)$$

$$+\frac{5}{1}\left(c^2d^8\right)-d^{10}$$

$$=c^{10}-5c^8d^2+10c^6d^4-10c^4d^6+5c^2d^8-d^{10}$$

77. The 3rd term of $(x+y)^8$

$$\frac{8!}{2!(8-2)!}x^6y^2=\frac{8\bullet7\bullet6!}{2!\,6!}x^6y^2$$

$$=\frac{56}{2}x^6y^2$$

$$=28x^6y^2$$

79. The 5th term of $(r+s)^6$

$$\frac{6!}{4!(6-4)!}r^2s^4=\frac{6\bullet5\bullet4!}{4!\,2!}r^2s^4$$

$$=\frac{30}{2}r^2s^4$$

$$=15r^2s^4$$

81. The 3rd term of $(x-1)^{13}$

$$\frac{13!}{2!(13-2)!}x^{11}\left(-1\right)^2=\frac{13\bullet12\bullet11!}{2!\,11!}x^{11}(1)$$

$$=\frac{156}{2}x^{11}$$

$$=78x^{11}$$

83. The 2nd term of $(x-3y)^4$

$$\frac{4!}{1!(4-1)!}x^3\left(-3y\right)=\frac{4\bullet3!}{1!\,3!}\left(-3x^3y\right)$$

$$=\frac{4}{1}\left(-3x^3y\right)$$

$$=-12x^3y$$

85. The 5th term of $(2x-3y)^5$

$$\frac{5!}{4!(5-4)!}\left(2x\right)\left(-3y\right)^4=\frac{5\bullet4!}{4!\,1!}\left(2x\right)\left(81y^4\right)$$

$$=\frac{5}{1}\left(162xy^4\right)$$

$$=810xy^4$$

87. The 2nd term of $\left(\dfrac{c}{2}-\dfrac{d}{3}\right)^4$

$$\frac{4!}{1!(4-1)!}\left(\frac{c}{2}\right)^3\left(-\frac{d}{3}\right)=\frac{4\bullet3!}{1!\,3!}\left(\frac{c^3}{8}\right)\left(-\frac{d}{3}\right)$$

$$=\frac{4}{1}\left(-\frac{c^3d}{24}\right)$$

$$=-\frac{c^3d}{6}$$

$$=-\frac{1}{6}c^3d$$

89. The 4th term of $(2t-5)^7$

$$\frac{7!}{3!(7-3)!}\left(2t\right)^4\left(-5\right)^3=\frac{7\bullet6\bullet5\bullet4!}{3!\,4!}\left(16t^4\right)\left(-125\right)$$

$$=\frac{210}{6}\left(-2,000t^4\right)$$

$$=35\left(-2,000t^4\right)$$

$$=-70,000t^4$$

91. The 2nd term of $\left(a^2+b^2\right)^6$

$$\frac{6!}{1!(6-1)!}\left(a^2\right)^5\left(b^2\right)=\frac{6\bullet5!}{1!\,5!}\left(a^{10}\right)\left(b^2\right)$$

$$=\frac{6}{1}\left(a^{10}b^2\right)$$

$$=6a^{10}b^2$$

WRITING

93. Answers will vary.

95. Answers will vary.

REVIEW

97.

$$2\log x + \frac{1}{2}\log y = \log x^2 + \log y^{1/2}$$
$$= \log x^2 y^{1/2}$$

99.

$$\ln\left(xy + y^2\right) - \ln\left(xz + yz\right) + \ln z$$
$$= \ln z\left(xy + y^2\right) - \ln\left(xz + yz\right)$$
$$= \ln\frac{z\left(xy + y^2\right)}{\left(xz + yz\right)}$$
$$= \ln\frac{zy\left(x + y\right)}{z\left(x + y\right)}$$
$$= \ln y$$

CHALLENGE PROBLEMS

101.   The constant term is the coefficient of $x^{10}x^{-10} = x^0 = 1$. It is the 10th term in the expansion.

$$\frac{10!}{5!(10-5)!}(x)^5\left(\frac{1}{x}\right)^5 = \frac{10\cdot9\cdot8\cdot7\cdot6\cdot5!}{5!5!}(x)^5\left(x^{-1}\right)^5$$
$$= \frac{30,240}{120}(a^5)(a^{-5})$$
$$= 252x^{5+(-5)}$$
$$= 252x^0$$
$$= 252(1)$$
$$= 252$$

103.   a.
$$\frac{n!}{0!(n-0)!} = \frac{n!}{(1)\,n!}$$
$$= 1$$

b.
$$\frac{n!}{n!(n-n)!} = \frac{n!}{n!\,0!}$$
$$= \frac{1}{1}$$
$$= 1$$

SECTION 14.2

VOCABULARY

1. A **sequence** is a function whose domain is the set of natural numbers.

3. Each term of an **arithmetic** sequence is found by adding the same number to the previous term.

5. If a single number is inserted between a and b to form an arithmetic sequence, the number is called the arithmetic **mean** between a and b.

CONCEPTS

7. 1, 7, 13

9. a. $a_n = a_1 + (n-1)d$
    b. `

NOTATION

11. The notation a_n represents the **nth** term of a sequence.

13. The symbol Σ is the Greek letter **sigma**.

15. We read $\sum_{k=1}^{10} 3k$ as "the **summation** of $3k$ as k **runs** from 1 to 10."

PRACTICE

17. $a_n = 4n - 1$
$$a_1 = 4(1) - 1 = 3$$
$$a_2 = 4(2) - 1 = 7$$
$$a_3 = 4(3) - 1 = 11$$
$$a_4 = 4(4) - 1 = 15$$
$$a_5 = 4(5) - 1 = 19$$
$$a_{40} = 4(40) - 1 = 159$$

19. $a_n = -3n + 1$
$$a_1 = -3(1) + 1 = -2$$
$$a_2 = -3(2) + 1 = -5$$
$$a_3 = -3(3) + 1 = -8$$
$$a_4 = -3(4) + 1 = -11$$
$$a_5 = -3(5) + 1 = -14$$
$$a_{30} = -3(30) + 1 = -89$$

21. $a_n = -n^2$
$$a_1 = -1^2 = -1$$
$$a_2 = -2^2 = -4$$
$$a_3 = -3^2 = -9$$
$$a_4 = -4^2 = -16$$
$$a_5 = -5^2 = -25$$
$$a_{20} = -20^2 = -400$$

23. $a_n = \dfrac{n-1}{n}$
$$a_1 = \frac{1-1}{1} = \frac{0}{1} = 0$$
$$a_2 = \frac{2-1}{2} = \frac{1}{2}$$
$$a_3 = \frac{3-1}{3} = \frac{2}{3}$$
$$a_4 = \frac{4-1}{4} = \frac{3}{4}$$
$$a_5 = \frac{5-1}{5} = \frac{4}{5}$$
$$a_{12} = \frac{12-1}{12} = \frac{11}{12}$$

25. $a_n = \dfrac{(-1)^n}{3^n}$
$$a_1 = \frac{(-1)^1}{3^1} = \frac{-1}{3} = -\frac{1}{3}$$
$$a_2 = \frac{(-1)^2}{3^2} = \frac{1}{9}$$
$$a_3 = \frac{(-1)^3}{3^3} = \frac{-1}{27} = -\frac{1}{27}$$
$$a_4 = \frac{(-1)^4}{3^4} = \frac{1}{81}$$

27. $a_n = (-1)^n (n+6)$

$a_1 = (-1)^1 (1+6) = -1(7) = -7$

$a_2 = (-1)^2 (2+6) = 1(8) = 8$

$a_3 = (-1)^3 (3+6) = -1(9) = -9$

$a_4 = (-1)^4 (4+6) = 1(10) = 10$

29. $a_n = a_1 + (n-1)d$

$a_n = 3 + 2(n-1)$

$a_1 = 3 + 2(1-1) = 3 + 0 = 3$

$a_2 = 3 + 2(2-1) = 3 + 2 = 5$

$a_3 = 3 + 2(3-1) = 3 + 4 = 7$

$a_4 = 3 + 2(4-1) = 3 + 6 = 9$

$a_5 = 3 + 2(5-1) = 3 + 8 = 11$

$a_{10} = 3 + 2(10-1) = 3 + 18 = 21$

31. $a_n = a_1 + (n-1)d$

$a_n = -5 - 3(n-1)$

$a_1 = -5 - 3(1-1) = -5 + 0 = -5$

$a_2 = -5 - 3(2-1) = -5 - 3 = -8$

$a_3 = -5 - 3(3-1) = -5 - 6 = -11$

$a_4 = -5 - 3(4-1) = -5 - 9 = -14$

$a_5 = -5 - 3(5-1) = -5 - 12 = -17$

$a_{15} = -5 - 3(15-1) = -5 - 42 = -47$

33. $a_n = a_1 + (n-1)d$

$a_n = 7 + 12(n-1)$

$a_1 = 7 + 12(1-1) = 7 + 0 = 7$

$a_2 = 7 + 12(2-1) = 7 + 12 = 19$

$a_3 = 7 + 12(3-1) = 7 + 24 = 31$

$a_4 = 7 + 12(4-1) = 7 + 36 = 43$

$a_5 = 7 + 12(5-1) = 7 + 48 = 55$

$a_{30} = 7 + 12(30-1) = 7 + 348 = 355$

35. $a_n = a_1 + (n-1)d$

$a_n = -7 - 2(n-1)$

$a_1 = -7 - 2(1-1) = -7 - 0 = -7$

$a_2 = -7 - 2(2-1) = -7 - 2 = -9$

$a_3 = -7 - 2(3-1) = -7 - 4 = -11$

$a_4 = -7 - 2(4-1) = -7 - 6 = -13$

$a_5 = -7 - 2(5-1) = -7 - 8 = -15$

$a_{15} = -7 - 2(15-1) = -7 - 28 = -35$

37.

$a_1 = 1, \quad d = 4 - 1 = 3, \quad n = 30$

$a_n = 1 + 3(30-1)$

$\quad = 1 + 3(29)$

$\quad = 88$

39.

$a_1 = -5, \quad d = -1 - (-5) = 4, \quad n = 17$

$a_n = -5 + 4(17-1)$

$\quad = -5 + 4(16)$

$\quad = 59$

41. $a_n = a_1 + (n-1)d$

Find d.

$29 = 5 + d(5-1)$

$29 = 5 + d(4)$

$24 = 4d$

$6 = d$

$a_1 = 5 + 6(1-1) = 5 + 0 = 5$

$a_2 = 5 + 6(2-1) = 5 + 6 = 11$

$a_3 = 5 + 6(3-1) = 5 + 12 = 17$

$a_4 = 5 + 6(4-1) = 5 + 18 = 23$

$a_5 = 5 + 6(5-1) = 5 + 24 = 29$

The first five terms are 5, 11, 17, 23, and 29.

43. $a_n = a_1 + (n-1)d$

Find d.

$-39 = -4 + d(6-1)$

$-39 = -4 + d(5)$

$-35 = 5d$

$-7 = d$

$a_1 = -4 - 7(1-1) = -4 - 0 = -4$

$a_2 = -4 - 7(2-1) = -4 - 7 = -11$

$a_3 = -4 - 7(3-1) = -4 - 14 = -18$

$a_4 = -4 - 7(4-1) = -4 - 21 = -25$

$a_5 = -4 - 7(5-1) = -4 - 28 = -32$

The first five terms are $-4, -11, -18, -25,$ and -32.

45. Let $a_1 = 2$ and $a_4 = 11$

$a_2 = 2 + d, a_3 = 2 + 2d$

$a_4 = a_1 + (4-1)d$

$11 = 2 + 3d$

$9 = 3d$

$3 = d$

$a_2 = 2 + 3(2-1) = 2 + 3 = 5$

$a_3 = 2 + 3(3-1) = 2 + 6 = 8$

47. Let $a_1 = 10$ and $a_6 = 20$

$a_2 = 10 + d, a_3 = 10 + 2d, a_4 = 10 + 3d, a_5 = 10 + 4d$

$a_6 = a_1 + (6-1)d$

$20 = 10 + 5d$

$10 = 5d$

$2 = d$

$a_2 = 10 + d = 10 + 2 = 12$

$a_3 = 10 + 2d = 10 + 2(2) = 14$

$a_4 = 10 + 3d = 10 + 3(2) = 16$

$a_5 = 10 + 4d = 10 + 4(2) = 18$

49. Let $a_1 = 20$ and $a_5 = 30$

$a_2 = 20 + d, a_3 = 20 + 2d, a_4 = 20 + 3d$

$a_5 = a_1 + (5-1)d$

$30 = 20 + 4d$

$10 = 4d$

$\dfrac{5}{2} = d$

$a_2 = 20 + d = 20 + \dfrac{5}{2} = \dfrac{45}{2}$

$a_3 = 20 + 2d = 20 + 2\left(\dfrac{5}{2}\right) = 25$

$a_4 = 20 + 3d = 20 + 3\left(\dfrac{5}{2}\right) = \dfrac{55}{2}$

51. Let $a_1 = -4.5$ and $a_3 = 7$

$a_2 = -4.5 + d$

$a_3 = a_1 + (3-1)d$

$7 = -4.5 + 2d$

$11.5 = 2d$

$5.75 = d$

$a_2 = -4.5 + d = -4.5 + 5.75 = 1.25 = \dfrac{5}{4}$

53. Find a_{35}.

$a_1 = 5, d = 4, n = 35$

$a_n = a_1 + (n-1)d$

$a_{35} = 5 + 4(35-1) = 5 + 4(34) = 141$

$S_n = \dfrac{n(a_1 + a_n)}{2}$

$S_{35} = \dfrac{35(5 + 141)}{2}$

$= \dfrac{35(146)}{2}$

$= 2,555$

Section 14.2

55. Find a_{40}.

$a_1 = -5, d = 4, n = 40$

$a_n = a_1 + (n-1)d$

$a_{40} = -5 + 4(40-1) = -5 + 4(39) = 151$

$S_n = \dfrac{n(a_1 + a_n)}{2}$

$S_{40} = \dfrac{40(-5+151)}{2}$

$= 2,920$

57.

$\displaystyle\sum_{k=1}^{4}(3k) = 3(1) + 3(2) + 3(3) + 3(4)$

$= 3 + 6 + 9 + 12$

59.

$\displaystyle\sum_{k=2}^{4}k^2 = 2^2 + 3^2 + 4^2$

$= 4 + 9 + 16$

61.

$\displaystyle\sum_{k=1}^{4}(6k) = 6(1) + 6(2) + 6(3) + 6(4)$

$= 6 + 12 + 18 + 24$

$= 60$

63.

$\displaystyle\sum_{k=3}^{4}k^3 = 3^3 + 4^3$

$= 27 + 64$

$= 91$

65.

$\displaystyle\sum_{k=3}^{4}(k^2 + 3) = (3^2 + 3) + (4^2 + 3)$

$= (9 + 3) + (16 + 3)$

$= (12) + (19)$

$= 31$

67.

$\displaystyle\sum_{k=4}^{4}(2k + 4) = 2(4) + 4$

$= 8 + 4$

$= 12$

69.

$\displaystyle\sum_{k=2}^{5}(5k) = 5(2) + 5(3) + 5(4) + 5(5)$

$= 10 + 15 + 20 + 25$

$= 70$

71.

$\displaystyle\sum_{k=4}^{6}(4k - 1) = (4 \cdot 4 - 1) + (4 \cdot 5 - 1) + (4 \cdot 6 - 1)$

$= 15 + 19 + 23$

$= 57$

TRY IT YOURSELF

73. Let $n = 44$ and $a_n = 556$, $a_1 = 40$

$556 = 40 + d(44 - 1)$

$556 = 40 + d(43)$

$516 = 43d$

$12 = d$

75.

$a_2 = 7, \quad a_3 = 12, \quad d = 12 - 7 = 5, \quad n = 12$

$7 = a_1 + 5(2 - 1)$

$7 = a_1 + 5(1)$

$7 = a_1 + 5$

$2 = a_1$

$a_{12} = 2 + 5(12 - 1)$

$= 2 + 5(11)$

$= 57$

$S_n = \dfrac{n(a_1 + a_n)}{2}$

$S_{17} = \dfrac{12(2 + 57)}{2}$

$= \dfrac{12(59)}{2}$

$= 354$

77. $a_n = a_1 + (n-1)d$

Find a_1.

$-83 = a_1 + 7(6-1)$

$-83 = a_1 + 7(5)$

$-83 = a_1 + 35$

$-118 = a_1$

$a_1 = -118 + 7(1-1) = -118 + 0 = -118$

$a_2 = -118 + 7(2-1) = -118 + 7 = -111$

$a_3 = -118 + 7(3-1) = -118 + 14 = -104$

$a_4 = -118 + 7(4-1) = -118 + 21 = -97$

$a_5 = -118 + 7(5-1) = -118 + 28 = -90$

The first five terms are $-118, -111, -104,$ $-97,$ and -90.

79. $a_n = a_1 + (n-1)d$

Find a_1.

$10 = a_1 - 3(9-1)$

$10 = a_1 - 3(8)$

$10 = a_1 - 24$

$34 = a_1$

$a_1 = 34 - 3(1-1) = 34 - 0 = 34$

$a_2 = 34 - 3(2-1) = 34 - 3 = 31$

$a_3 = 34 - 3(3-1) = 34 - 6 = 28$

$a_4 = 34 - 3(4-1) = 34 - 9 = 25$

$a_5 = 34 - 3(5-1) = 34 - 12 = 22$

The first five terms are $34, 31, 28, 25,$ and 22.

81. Find a_{200}.

$a_1 = 5, d = 7, n = 200$

$a_n = a_1 + (n-1)d$

$a_{200} = 5 + 7(200-1) = 5 + 7(199) = 1,398$

83.

$a_1 = 1, \ a_n = 50, \ n = 50$

$S_n = \dfrac{n(a_1 + a_n)}{2}$

$= \dfrac{50(1+50)}{2}$

$= \dfrac{50(51)}{2}$

$= 1,275$

85. Find d

$-9 - (-4),$ so $d = -5$.

Find a_1 if $a_n = -4$ and $n = 2$.

$-4 = a_1 - 5(2-1)$

$-4 = a_1 - 5(1)$

$-4 = a_1 - 5$

$1 = a_1$

Find a_{37}

$a_{37} = 1 - 5(37-1)$

$= 1 - 5(36)$

$= 1 - 180$

$= -179$

87. Let $d = 11, n = 27$ and $a_n = 263$.

$263 = a_1 + 11(27-1)$

$263 = a_1 + 11(26)$

$263 = a_1 + 286$

$-23 = a_1$

89. Find a_{15}.

$a_1 = \dfrac{1}{2}, d = -\dfrac{1}{4}, n = 15$

$a_n = a_1 + (n-1)d$

$a_{15} = \dfrac{1}{2} - \dfrac{1}{4}(15-1) = \dfrac{1}{2} - \dfrac{1}{4}(14) = -3$

91.

$a_1 = 1, \ a_n = 99, \ n = 50$

$S_n = \dfrac{n(a_1 + a_n)}{2}$

$= \dfrac{50(1+99)}{2}$

$= \dfrac{50(100)}{2}$

$= 2,500$

APPLICATIONS

93. SAVING MONEY

Let $a_1 = 60$ and $d = 50$.

Let $n =$ the month. Find a_n to find the amount in the account each month with $a_1 = 60$ and $d = 50$.

$a_n = a_1 + d(n-1)$

$a_1 = 60$

$a_2 = 60 + 50(2-1) = 60 + 50(1) = 110$

$a_3 = 60 + 50(3-1) = 60 + 50(2) = 160$

$a_4 = 60 + 50(4-1) = 60 + 50(3) = 210$

$a_5 = 60 + 50(5-1) = 60 + 50(4) = 260$

$a_6 = 60 + 50(6-1) = 60 + 50(5) = 310$

To find the amount after 10 years, or $10(12) = 120$ months, plus the first deposit, let $n = 120 + 1 = 121$ and find a_{121}.

$a_{121} = 60 + 50(121-1) = 60 + 50(120) = 6,060$

She will have a savings of $6,060 after 10 years.

95. DESIGNING PATIOS

$a_1 = 1, \quad a_n = 150, \quad n = 150$

$S_n = \dfrac{n(a_1 + a_n)}{2}$

$= \dfrac{150(1+150)}{2}$

$= \dfrac{150(151)}{2}$

$= 11,325$

11,325 bricks are needed.

97. HOLIDAY SONGS

$S_n = \dfrac{n(a_1 + a_n)}{2}$

$= \dfrac{12(1+12)}{2}$

$= \dfrac{12(13)}{2}$

$= \dfrac{156}{2}$

$= 78$

WRITING

99. Answers will vary.

101. Answers will vary.

REVIEW

103.

$\log_2 \dfrac{2x}{y} = \log_2 2 + \log_2 x - \log_2 y$

$= 1 + \log_2 x - \log_2 y$

105.

$\log x^3 y^2 = \log x^3 + \log y^2$

$= 3\log x + 2\log y$

CHALLENGE PROBLEMS

107.

$1 + 4 + 9 + 16 + 25 = 1^2 + 2^2 + 3^2 + 4^2 + 5^2$

$= \displaystyle\sum_{k=1}^{5} k^2$

109. Let $x = 9$. Then,

$x - 2 = 9 - 2 = 7$

$2x + 4 = 2(9) + 4 = 22$

$5x - 8 = 5(9) - 8 = 37$

It forms an arithmetic sequence with a difference of 15.

VOCABULARY

1. Each term of a **geometric** sequence is found by multiplying the previous term by the same number.

3. If a single number is inserted between a and b to form a geometric sequence, the number is called the geometric **mean** between a and b.

CONCEPTS

5.
$$a_1 = 16$$
$$a_2 = 16\left(\frac{1}{4}\right) = 4$$
$$a_3 = 16\left(\frac{1}{4}\right)^2 = 1$$
16, 4, 1

7. $a_n = a_1 r^{n-1}$

9. a. yes
   b. no, $3 \not< 1$
   c. no, $6 \not< 1$
   d. yes

NOTATION

11. An infinite geometric sequence is of the form $a_1, a_1 r, \boxed{a_1 r^2}, a_1 r^3, \boxed{a_1 r^4}, \dots$

13. To find the common ratio of a geometric sequence, we use the formula $r = \dfrac{a_{\boxed{n+1}}}{a_{\boxed{n}}}$.

15. $a_1 = 3, r = 2$
$$a_1 = 3$$
$$a_2 = a_1 r = 3(2) = 6$$
$$a_3 = a_1 r^2 = 3(2)^2 = 12$$
$$a_4 = a_1 r^3 = 3(2)^3 = 24$$
$$a_5 = a_1 r^4 = 3(2)^4 = 48$$
$$a_9 = a_1 r^8 = 3(2)^8 = 768$$

17. $a_1 = -5, r = \dfrac{1}{5}$
$$a_1 = -5$$
$$a_2 = a_1 r = -5\left(\frac{1}{5}\right) = -1$$
$$a_3 = a_1 r^2 = -5\left(\frac{1}{5}\right)^2 = -\frac{1}{5}$$
$$a_4 = a_1 r^3 = -5\left(\frac{1}{5}\right)^3 = -\frac{1}{25}$$
$$a_5 = a_1 r^4 = -5\left(\frac{1}{5}\right)^4 = -\frac{1}{125}$$
$$a_8 = a_1 r^7 = -5\left(\frac{1}{5}\right)^7 = -\frac{1}{15,625}$$

19. $a_1 = 2, r = 3, n = 7$
$$a_n = a_1 r^{n-1}$$
$$a_7 = 2(3)^{7-1}$$
$$= 2(3)^6$$
$$= 2(729)$$
$$= 1,458$$

21.

$$a_1 = \frac{1}{2}, a_2 = -\frac{5}{2}, n = 6$$

$$r = \frac{a_2}{a_1} = \frac{-\frac{5}{2}}{\frac{1}{2}} = -5$$

$$a_n = a_1 r^{n-1}$$

$$a_6 = \frac{1}{2}(-5)^{6-1}$$

$$= \frac{1}{2}(-5)^5$$

$$= \frac{1}{2}(-3,125)$$

$$= -\frac{3,125}{2}$$

23. $a_1 = 2$, $r > 0$, third term is 18

Use $a_n = ar^{n-1}$ to find r.

$$a = 2, \ a_3 = 18 \text{ and } n = 3$$

$$18 = 2r^{3-1}$$

$$18 = 2r^2$$

$$9 = r^2$$

$$3 = r \text{ since } r > 0$$

$$a_1 = 2$$

$$a_2 = a_1 r = 2(3) = 6$$

$$a_3 = a_1 r^2 = 2(3)^2 = 18$$

$$a_4 = a_1 r^3 = 2(3)^3 = 54$$

$$a_5 = a_1 r^4 = 2(3)^4 = 162$$

$$2, \ 6, 18, \ 54, \ 162$$

25. $a_1 = 3$, fourth term is 24

Use $a_n = ar^{n-1}$ to find r.

$$a = 3, \ a_4 = 24 \text{ and } n = 4$$

$$24 = 3r^{4-1}$$

$$24 = 3r^3$$

$$8 = r^3$$

$$2 = r$$

$$a_1 = 3$$

$$a_2 = a_1 r = 3(2) = 6$$

$$a_3 = a_1 r^2 = 3(2)^2 = 12$$

$$a_4 = a_1 r^3 = 3(2)^3 = 24$$

$$a_5 = a_1 r^4 = 3(2)^4 = 48$$

$$3, 6, 12, 24, 48$$

27. $a_1 = 2$, $a_5 = 162$ for three geometric means between 2 and 162

Use $a_n = ar^{n-1}$ to find r.

$$a_1 = 2, \ a_5 = 162 \text{ and } n = 5$$

$$162 = 2(r)^{5-1}$$

$$162 = 2r^4$$

$$81 = r^4$$

$$3 = r$$

$$a_1 = 2$$

$$a_2 = 2(3) = 6$$

$$a_3 = 2(3)^2 = 18$$

$$a_4 = 2(3)^3 = 54$$

$$a_5 = 2(3)^4 = 162$$

$$6, 18, 54$$

29. $a_1 = -4$, $a_6 = -12{,}500$ for four geometric
means between -4 and $-12{,}500$

Use $a_n = ar^{n-1}$ to find r.

$a_1 = -4$, $a_6 = -12{,}500$ and $n = 6$

$-12{,}500 = -4(r)^{6-1}$

$-12{,}500 = -4r^5$

$3{,}125 = r^5$

$5 = r$

$a_1 = -4$

$a_2 = -4(5) = -20$

$a_3 = -4(5)^2 = -100$

$a_4 = -4(5)^3 = -500$

$a_5 = -4(5)^4 = -2{,}500$

$a_6 = -4(5)^5 = -12{,}500$

$-20, -100, -500, -2{,}500$

31. $a_1 = 2$, $a_3 = 128$ for the geometric mean
between 2 and 128

Use $a_n = ar^{n-1}$ to find r.

$a_1 = 2$, $a_3 = 128$ and $n = 3$

$128 = 2(r)^{3-1}$

$128 = 2r^2$

$64 = r^2$

$\pm 8 = r$

$a_1 = 2$

$a_2 = 2(\pm 8) = \pm 16$

The geometric mean is 16 or -16.

33. $a_1 = 10$, $a_3 = 20$ for the geometric mean
between 10 and 20

Use $a_n = ar^{n-1}$ to find r.

$a_1 = 10$, $a_3 = 20$ and $n = 3$

$20 = 10(r)^{3-1}$

$20 = 10r^2$

$2 = r^2$

$\pm\sqrt{2} = r$

$a_1 = 10$

$a_2 = 10\left(\pm\sqrt{2}\right) = \pm 10\sqrt{2}$

The geometric mean is $\pm 10\sqrt{2}$.

35. Find r.

$$r = \frac{a_2}{a_1} = \frac{6}{2} = 3$$

Find the sum with $a_1 = 2$, $r = 3$, and $n = 6$.

$$S_n = \frac{a_1(1 - r^n)}{1 - r}$$

$$= \frac{2(1 - 3^6)}{1 - 3}$$

$$= \frac{2(1 - 729)}{-2}$$

$$= \frac{2(-728)}{-2}$$

$$= 728$$

37. Find r.

$$r = \frac{a_2}{a_1} = \frac{-6}{2} = -3$$

Find the sum with $a_1 = 2$, $r = -3$, and $n = 5$.

$$S_n = \frac{a_1(1 - r^n)}{1 - r}$$

$$= \frac{2(1 - (-3)^5)}{1 - (-3)}$$

$$= \frac{2(1 + 243)}{4}$$

$$= \frac{2(244)}{4}$$

$$= 122$$

39. Find r.

$$r = \frac{a_2}{a_1} = \frac{-6}{3} = -2$$

Find the sum with $a_1 = 3$, $r = -2$, and $n = 8$.

$$S_n = \frac{a_1(1 - r^n)}{1 - r}$$

$$= \frac{3(1 - (-2)^8)}{1 - (-2)}$$

$$= \frac{3(1 - 256)}{3}$$

$$= \frac{3(-255)}{3}$$

$$= -255$$

41. Find r.

$$r = \frac{a_2}{a_1} = \frac{6}{3} = 2$$

Find the sum with $a_1 = 3$, $r = 2$, and $n = 7$.

$$S_n = \frac{a_1(1-r^n)}{1-r}$$

$$= \frac{3(1-2^7)}{1-2}$$

$$= \frac{3(1-128)}{-1}$$

$$= \frac{3(-127)}{-1}$$

$$= 381$$

43. Find r.

$$r = \frac{a_2}{a_1} = \frac{4}{8} = \frac{1}{2}$$

$$|r| = \left|\frac{1}{2}\right| = \frac{1}{2} < 1 \text{ so the sum does exist}$$

Find the sum with $a_1 = 8$ and $r = \frac{1}{2}$.

$$S = \frac{a_1}{1-r}$$

$$= \frac{8}{1-\frac{1}{2}}$$

$$= \frac{8}{\frac{1}{2}}$$

$$= 8 \cdot \frac{2}{1}$$

$$= 16$$

45. Find r.

$$r = \frac{a_2}{a_1} = \frac{18}{54} = \frac{1}{3}$$

$$|r| = \left|\frac{1}{3}\right| = \frac{1}{3} < 1 \text{ so the sum does exist}$$

Find the sum with $a_1 = 54$ and $r = \frac{1}{3}$.

$$S = \frac{a_1}{1-r}$$

$$= \frac{54}{1-\frac{1}{3}}$$

$$= \frac{54}{\frac{2}{3}}$$

$$= 54 \cdot \frac{3}{2}$$

$$= 81$$

47. Find r.

$$r = \frac{a_2}{a_1} = \frac{-9}{-\frac{27}{2}} = \frac{2}{3}$$

$$|r| = \left|\frac{2}{3}\right| = \frac{2}{3} < 1 \text{ so the sum does exist}$$

Find the sum with $a_1 = -\frac{27}{2}$ and $r = \frac{2}{3}$.

$$S = \frac{a_1}{1-r}$$

$$= \frac{-\frac{27}{2}}{1-\frac{2}{3}}$$

$$= \frac{-\frac{27}{2}}{\frac{1}{3}}$$

$$= -\frac{27}{2} \cdot 3$$

$$= -\frac{81}{2}$$

49. Find r.

$$r = \frac{a_2}{a_1} = \frac{6}{\frac{9}{2}} = \frac{4}{3}$$

Since $|r| = \left|\frac{4}{3}\right| = \frac{4}{3} \geq 1$, the sum of the terms

of the sequence, S, does not exist

51. Find r.

$$r = \frac{a_2}{a_1} = \frac{-6}{12} = -\frac{1}{2}$$

$$|r| = \left|-\frac{1}{2}\right| = \frac{1}{2} < 1 \text{ so the sum does exist}$$

Find the sum with $a_1 = 12$ and $r = -\frac{1}{2}$.

$$S = \frac{a_1}{1-r}$$

$$= \frac{12}{1-\left(-\frac{1}{2}\right)}$$

$$= \frac{12}{\frac{3}{2}}$$

$$= 12 \cdot \frac{2}{3}$$

$$= 8$$

53. Find r.

$$r = \frac{a_2}{a_1} = \frac{15}{-45} = -\frac{1}{3}$$

$$|r| = \left|-\frac{1}{3}\right| = \frac{1}{3} < 1 \text{ so the sum does exist}$$

Find the sum with $a_1 = -45$ and $r = -\frac{1}{3}$.

$$S = \frac{a_1}{1-r}$$

$$= \frac{-45}{1-\left(-\frac{1}{3}\right)}$$

$$= \frac{-45}{\frac{4}{3}}$$

$$= -45 \cdot \frac{3}{4}$$

$$= -\frac{135}{4}$$

55.

$$0.\bar{1} = 0.111... = \frac{1}{10} + \frac{1}{100} + \frac{1}{1,000} + ...$$

where $a_1 = \frac{1}{10}$ and $r = \frac{1}{10}$.

$$|r| = \left|\frac{1}{10}\right| = \frac{1}{10} < 1 \text{ so the sum does exist.}$$

Find the sum with $a_1 = \frac{1}{10}$ and $r = \frac{1}{10}$.

$$S = \frac{a_1}{1-r}$$

$$= \frac{\frac{1}{10}}{1-\frac{1}{10}}$$

$$= \frac{\frac{1}{10}}{\frac{9}{10}}$$

$$= \frac{1}{10} \cdot \frac{10}{9}$$

$$= \frac{1}{9}$$

$$0.\bar{1} = \frac{1}{9}$$

Section 14.3

57.

$$0.\overline{3} = 0.333... = \frac{3}{10} + \frac{3}{100} + \frac{3}{1,000} + ...$$

where $a_1 = \dfrac{3}{10}$ and $r = \dfrac{1}{10}$.

$|r| = \left|\dfrac{1}{10}\right| = \dfrac{1}{10} < 1$ so the sum does exist.

Find the sum with $a_1 = \dfrac{3}{10}$ and $r = \dfrac{1}{10}$.

$$S = \frac{a_1}{1-r}$$

$$= \frac{\dfrac{3}{10}}{1 - \dfrac{1}{10}}$$

$$= \frac{\dfrac{3}{10}}{\dfrac{9}{10}}$$

$$= \frac{3}{10} \cdot \frac{10}{9}$$

$$= \frac{1}{3}$$

$$0.\overline{3} = \frac{1}{3}$$

59.

$$0.\overline{12} = 0.121212... = \frac{12}{100} + \frac{12}{10,000} + \frac{12}{1,000,000} + ...$$

where $a_1 = \dfrac{12}{100}$ and $r = \dfrac{1}{100}$.

$|r| = \left|\dfrac{1}{100}\right| = \dfrac{1}{100} < 1$ so the sum does exist.

Find the sum with $a_1 = \dfrac{12}{100}$ and $r = \dfrac{1}{100}$.

$$S = \frac{a_1}{1-r}$$

$$= \frac{\dfrac{12}{100}}{1 - \dfrac{1}{100}}$$

$$= \frac{\dfrac{12}{100}}{\dfrac{99}{100}}$$

$$= \frac{12}{100} \cdot \frac{100}{99}$$

$$= \frac{12}{99}$$

$$= \frac{4}{33}$$

$$= \frac{4}{33}$$

61.

$$0.\overline{75} = 0.757575\ldots = \frac{75}{100} + \frac{75}{10{,}000} + \frac{75}{1{,}000{,}000} + \ldots$$

where $a_1 = \dfrac{75}{100}$ and $r = \dfrac{1}{100}$.

$|r| = \left|\dfrac{1}{100}\right| = \dfrac{1}{100} < 1$ so the sum does exist.

Find the sum with $a_1 = \dfrac{75}{100}$ and $r = \dfrac{1}{100}$.

$$S = \frac{a_1}{1-r}$$

$$= \frac{\dfrac{75}{100}}{1 - \dfrac{1}{100}}$$

$$= \frac{\dfrac{75}{100}}{\dfrac{99}{100}}$$

$$= \frac{75}{100} \cdot \frac{100}{99}$$

$$= \frac{75}{99}$$

$$= \frac{25}{33}$$

$$0.\overline{75} = \frac{25}{33}$$

63. $a_1 = -8$, $a_6 = -1{,}944$

Use $a_n = ar^{n-1}$ to find a_1.

$a_1 = -8$, $a_6 = -1{,}944$ and $n = 6$

$$-1{,}944 = -8(r)^{6-1}$$

$$-1{,}944 = -8r^5$$

$$243 = r^5$$

$$3 = r$$

65. $a_1 = -64$, $r < 0$, fifth term is -4

Use $a_n = ar^{n-1}$ to find r.

$a = -64$, $a_5 = -4$ and $n = 5$

$$-4 = -64r^{5-1}$$

$$-4 = -64r^4$$

$$\frac{1}{16} = r^4$$

$$-\frac{1}{2} = r \text{ since } r < 0$$

$$a_1 = -64$$

$$a_2 = a_1 r = -64\left(-\frac{1}{2}\right) = 32$$

$$a_3 = a_1 r^2 = -64\left(-\frac{1}{2}\right)^2 = -16$$

$$a_4 = a_1 r^3 = -64\left(-\frac{1}{2}\right)^3 = 8$$

$$a_5 = a_1 r^4 = -64\left(-\frac{1}{2}\right)^4 = -4$$

$$-64, 32, -16, 8, -4$$

Section 14.3

67. $a_1 = -50$, $a_3 = 10$ for the geometric mean between -50 and 10

Use $a_n = ar^{n-1}$ to find r.

$a_1 = -50$, $a_3 = 10$ and $n = 3$

$10 = -50(r)^{3-1}$

$10 = -50r^2$

$-\dfrac{1}{5} = r^2$

There is no geometric mean because no real number squared is $-\dfrac{1}{5}$. When you square a real number, it is always positive.

69. $a_1 = 7, r = 2$

Use $a_n = ar^{n-1}$ to find a_{10}.

$r = 2$, $a_1 = 7$ and $n = 10$

$a_{10} = 7(2)^{10-1}$

$a_{10} = 7(2)^9$

$a_{10} = 7(512)$

$a_{10} = 3,584$

71. $a_1 = -64$, sixth term is -2

Use $a_n = ar^{n-1}$ to find r.

$a_1 = -64$, $a_6 = -2$ and $n = 6$

$-2 = -64r^{6-1}$

$-2 = -64r^5$

$\dfrac{1}{32} = r^5$

$\dfrac{1}{2} = r$

$a_1 = -64$

$a_2 = a_1 r = -64\left(\dfrac{1}{2}\right) = -32$

$a_3 = a_1 r^2 = -64\left(\dfrac{1}{2}\right)^2 = -16$

$a_4 = a_1 r^3 = -64\left(\dfrac{1}{2}\right)^3 = -8$

$-64, -32, -16, -8$

73. Find r.

$r = \dfrac{a_2}{a_1} = \dfrac{\frac{3}{4}}{3} = \dfrac{1}{4}$

$|r| = \left|\dfrac{1}{4}\right| = \dfrac{1}{4} < 1$ so the sum does exist

Find the sum with $a_1 = 3$ and $r = \dfrac{1}{4}$.

$S = \dfrac{a_1}{1-r}$

$= \dfrac{3}{1 - \frac{1}{4}}$

$= \dfrac{3}{\frac{3}{4}}$

$= 3 \cdot \dfrac{4}{3}$

$= 4$

75. $r = -3$, $a_8 = -81$

Use $a_n = ar^{n-1}$ to find a_1.

$r = -3$, $a_8 = -81$ and $n = 8$

$-81 = a_1 (-3)^{8-1}$

$-81 = a_1 (-3)^7$

$-81 = a_1 (-2,187)$

$\dfrac{-81}{-2,187} = a_1$

$\dfrac{1}{27} = a_1$

77. Find the sum with $a_1 = 3$, $r = 2$, and $n = 5$.

$S_n = \dfrac{a_1(1-r^n)}{1-r}$

$= \dfrac{3(1-2^5)}{1-2}$

$= \dfrac{3(1-32)}{-1}$

$= \dfrac{3(-31)}{-1}$

$= 93$

APPLICATIONS

79. DECLINING SAVINGS
If he spends 12% of the funds each year, then 88% of the funds remain. The amount of money in the savings box after 15 years is represented by the 16th term of a geometric series, where $a_1 = 10,000$, $n = 16$, and $r = 0.88$.

$$a_n = a_1 r^{n-1}$$
$$a_{16} = (10,000)(0.88)^{16-1}$$
$$a_{16} = (10,000)(0.88)^{15}$$
$$a_{16} = (10,000)(0.1469738)$$
$$a_{16} = \$1,469.74$$

He will have $1,469.74 after 15 years.

81. REAL ESTATE SALES AGENT
If the house appreciates 8% each year, then it will be worth 108% of the preceding years' value (100% + 8% interest). The value of the house after 10 years is represented by the 11th term of a geometric series, where $a_1 = 250,000$, $n = 11$, and $r = 1.08$.

$$a_n = a_1 r^{n-1}$$
$$a_{11} = (250,000)(1.08)^{11-1}$$
$$a_{11} = (250,000)(1.08)^{10}$$
$$a_{11} = (250,000)(2.158924997)$$
$$a_{11} = \$539,731.25$$

The house will be worth approximately $539,731 in 10 years.

83. INSCRIBED SQUARES
The area of each new is ½ the area of the previous square. The area of the 12th square is represented by the 12th term of a geometric series, where $a_1 = 1$, $r = ½$, and $n = 12$.

$$a_n = a_1 r^{n-1}$$
$$a_{12} = (1)\left(\frac{1}{2}\right)^{12-1}$$
$$a_{12} = (1)\left(\frac{1}{2}\right)^{11}$$
$$a_{12} = (1)(0.00048828)$$
$$a_{12} \approx 0.0005$$

85. BOUNCING BALLS
The total distance the ball travels is the sum of two motions, falling and rebounding. The distance the ball falls is given by the sum $10 + \frac{1}{2}\cdot 10 + \frac{1}{2}\left(\frac{1}{2}\cdot 10\right) + ...$

or $10 + 5 + \frac{5}{2} + ...$ The distance the ball rebounds begins with the first bounce up which is 5 m, and then is one–half the distance afterwards or $5 + \frac{5}{2} + \frac{5}{4} + ...$

Since these are infinite geometric series, use the formula $S = \frac{a_1}{1-r}$ to find the sum.

Falling:

$a_1 = 10$ and $r = \frac{1}{2}$

$$S = \frac{a_1}{1-r}$$
$$= \frac{10}{1-\frac{1}{2}}$$
$$= \frac{10}{\frac{1}{2}}$$
$$= 10\cdot\frac{2}{1}$$
$$= 20$$

Rebounding:

$a_1 = 5$ and $r = \frac{1}{2}$

$$S = \frac{a_1}{1-r}$$
$$= \frac{5}{1-\frac{1}{2}}$$
$$= \frac{5}{\frac{1}{2}}$$
$$= 5\cdot\frac{2}{1}$$
$$= 10$$

The distance the ball travels is the sum of the distances falling and rebounding:

$20 + 10 = 30$ m.
The ball travels a total of 30 m.

87. PEST CONTROL
To find the sum of the given infinite geometric series use the formula $S = \frac{a_1}{1-r}$ with $a_1 = 1,000$ and $r = 0.8$.

$$S = \frac{a_1}{1-r}$$
$$= \frac{1,000}{1-0.8}$$
$$= \frac{1,000}{0.2}$$
$$= 5,000$$

The long–term population is 5,000.

89. Answers will vary.

91. Answers will vary.

REVIEW

93.

$$x^2 - 5x - 6 \le 0$$

$$(x-6)(x+1) = 0$$

$$x - 6 = 0 \quad \text{or} \quad x + 1 = 0$$

$$x = 6 \qquad x = -1$$

The critical numbers are 6 and –1. Pick numbers in all three parts and substitute to determine which areas to shade.

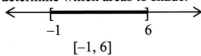

$$[-1, 6]$$

95.

$$\frac{x-4}{x+3} > 0$$

$$\frac{x-4}{x+3} = 0$$

$$(x+3)\left(\frac{x-4}{x+3}\right) = (x+3)(0)$$

$$x - 4 = 0$$

$$x = 4$$

Set the denominator $= 0$.

$$x + 3 = 0$$

$$x = -3$$

Critical numbers $= -3$ and 4

$$(-\infty, -3) \cup (4, \infty)$$

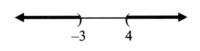

97.

$$f(x) = 1 + x + x^2 + x^3 + x^4 + \dots$$

$$f\left(\frac{1}{2}\right) = 1 + \frac{1}{2} + \left(\frac{1}{2}\right)^2 + \left(\frac{1}{2}\right)^3 + \left(\frac{1}{2}\right)^4 + \dots$$

$$f\left(\frac{1}{2}\right) = 1 + \text{infinite geometric series} \left(a = \frac{1}{2}, r = \frac{1}{2}\right)$$

$$f\left(\frac{1}{2}\right) = 1 + \frac{\frac{1}{2}}{1 - \frac{1}{2}}$$

$$f\left(\frac{1}{2}\right) = 1 + \frac{\frac{1}{2}}{\frac{1}{2}}$$

$$f\left(\frac{1}{2}\right) = 1 + 1$$

$$f\left(\frac{1}{2}\right) = 2$$

$$f\left(-\frac{1}{2}\right) = 1 + \left(-\frac{1}{2}\right) + \left(-\frac{1}{2}\right)^2 + \left(-\frac{1}{2}\right)^3 + \left(-\frac{1}{2}\right)^4 + \dots$$

$$f\left(-\frac{1}{2}\right) = 1 + \text{infinite geometric series} \left(a = -\frac{1}{2}, r = -\frac{1}{2}\right)$$

$$f\left(-\frac{1}{2}\right) = 1 + \frac{-\frac{1}{2}}{1 - \left(-\frac{1}{2}\right)}$$

$$f\left(-\frac{1}{2}\right) = 1 + \frac{-\frac{1}{2}}{\frac{3}{2}}$$

$$f\left(-\frac{1}{2}\right) = 1 - \frac{1}{3}$$

$$f\left(-\frac{1}{2}\right) = \frac{2}{3}$$

99. the arithmetic mean

CHAPTER 14 REVIEW

SECTION 14.1
The Binomial Theorem

1. The coefficients for the expansion of $(a + b)^5$ are 1, 5, 10, 10, 5, 1.

$$
\begin{array}{ccccccccccccc}
 & & & & & & 1 & & & & & & \\
 & & & & & \boxed{1} & & 1 & & & & & \\
 & & & & 1 & & 2 & & 1 & & & & \\
 & & & 1 & & \boxed{3} & & 3 & & \boxed{1} & & & \\
 & & \boxed{1} & & 4 & & 6 & & 4 & & 1 & & \\
 & 1 & & 5 & & \boxed{10} & & 10 & & 5 & & 1 & \\
1 & & 6 & & 15 & & 20 & & \boxed{15} & & 6 & & 1
\end{array}
$$

2. a. 13 terms
    b. 12
    c. The first term is a^{12} and the last term is b^{12}.
    d. The exponents of a decrease and the exponents of b increase.

3.
$$
\begin{aligned}
(4!)(3!) &= (4 \cdot 3 \cdot 2 \cdot 1)(3 \cdot 2 \cdot 1) \\
&= (24)(6) \\
&= 144
\end{aligned}
$$

4.
$$
\begin{aligned}
\frac{5!}{3!} &= \frac{5 \cdot 4 \cdot \cancel{3!}}{\cancel{3!}} \\
&= 20
\end{aligned}
$$

5.
$$
\begin{aligned}
\frac{6!}{2!(6-2)!} &= \frac{6!}{2!4!} \\
&= \frac{6 \cdot 5 \cdot \cancel{4!}}{2 \cdot 1 \cdot \cancel{4!}} \\
&= \frac{30}{2} \\
&= 15
\end{aligned}
$$

6.
$$
\begin{aligned}
\frac{12!}{3!(12-3)!} &= \frac{12!}{3!9!} \\
&= \frac{12 \cdot 11 \cdot 10 \cdot \cancel{9!}}{3 \cdot 2 \cdot 1 \cdot \cancel{9!}} \\
&= \frac{1,320}{6} \\
&= 220
\end{aligned}
$$

7.
$$
\begin{aligned}
(n-n)! &= 0! \\
&= 1
\end{aligned}
$$

8.
$$
\begin{aligned}
\frac{8!}{7!} &= \frac{8 \cdot \cancel{7!}}{\cancel{7!}} \\
&= 8
\end{aligned}
$$

9.
$$
\begin{aligned}
(x&+y)^5 \\
&= x^5 + \frac{5!}{1!(5-1)!}x^4 y + \frac{5!}{2!(5-2)!}x^3 y^2 \\
&\quad + \frac{5!}{3!(5-3)!}x^2 y^3 + \frac{5!}{4!(5-4)!}xy^4 + y^5 \\
&= x^5 + \frac{5 \cdot \cancel{4!}}{1! \cancel{4!}}x^4 y + \frac{5 \cdot 4 \cdot \cancel{3!}}{2! \cancel{3!}}x^3 y^2 \\
&\quad + \frac{5 \cdot 4 \cdot \cancel{3!}}{\cancel{3!}2!}x^2 y^3 + \frac{5 \cdot \cancel{4!}}{\cancel{4!}1!}xy^4 + y^5 \\
&= x^5 + \frac{5}{1}x^4 y + \frac{20}{2}x^3 y^2 + \frac{20}{2}x^2 y^3 + \frac{5}{1}xy^4 + y^5 \\
&= x^5 + 5x^4 y + 10x^3 y^2 + 10x^2 y^3 + 5xy^4 + y^5
\end{aligned}
$$

10.

$$(x-y)^9 = (x+(-y))^9$$

$$= x^9 + \frac{9!}{1!(9-1)!}x^8(-y) + \frac{9!}{2!(9-2)!}x^7(-y)^2$$

$$+ \frac{9!}{3!(9-3)!}x^6(-y)^3 + \frac{9!}{4!(9-4)!}x^5(-y)^4$$

$$+ \frac{9!}{5!(9-5)!}x^4(-y)^5 + \frac{9!}{6!(9-6)!}x^3(-y)^6$$

$$+ \frac{9!}{7!(9-7)!}x^2(-y)^7 + \frac{9!}{8!(9-8)!}x(-y)^8 + (-y)^9$$

$$= x^9 + \frac{9!}{1!8!}x^8(-y) + \frac{9!}{2!7!}x^7(-y)^2 + \frac{9!}{3!6!}x^6(-y)^3$$

$$+ \frac{9!}{4!5!}x^5(-y)^4 + \frac{9!}{5!4!}x^4(-y)^5 + \frac{9!}{6!3!}x^3(-y)^6$$

$$+ \frac{9!}{7!2!}x^2(-y)^7 + \frac{9!}{8!1!}x(-y)^8 + (-y)^9$$

$$= x^9 - \frac{9}{1}x^8 y + \frac{72}{2}x^7 y^2 - \frac{504}{6}x^6 y^3 + \frac{3,024}{24}x^5 y^4$$

$$- \frac{3,024}{24}x^4 y^5 + \frac{504}{6}x^3 y^6 - \frac{72}{2}x^2 y^7 + \frac{9}{1}xy^8 - y^9$$

$$= x^9 - 9x^8 y + 36x^7 y^2 - 84x^6 y^3 + 126x^5 y^4 - 126x^4 y^5$$

$$+ 84x^3 y^6 - 36x^2 y^7 + 9xy^8 - y^9$$

11.

$$(4x-y)^3$$

$$= (4x)^3 + \frac{3!}{1!(3-1)!}(4x)^2(-y) + \frac{3!}{2!(3-2)!}(4x)(-y)^2 + (-y)^3$$

$$= 64x^3 + \frac{3!}{1!2!}(16x^2)(-y) + \frac{3!}{2!1!}(4x)(y^2) - y^3$$

$$= 64x^3 - \frac{3}{1}(16x^2)y + \frac{3}{1}(4x)y^2 - y^3$$

$$= 64x^3 - 48x^2 y + 12xy^2 - y^3$$

12.

$$\left(\frac{c}{2}+\frac{d}{3}\right)^4$$

$$= \left(\frac{c}{2}\right)^4 + \frac{4!}{1!(4-1)!}\left(\frac{c}{2}\right)^3\left(\frac{d}{3}\right)$$

$$+ \frac{4!}{2!(4-2)!}\left(\frac{c}{2}\right)^2\left(\frac{d}{3}\right)^2 + \frac{4!}{3!(4-3)!}\left(\frac{c}{2}\right)\left(\frac{d}{3}\right)^3$$

$$+ \left(\frac{d}{3}\right)^4$$

$$= \frac{c^4}{16} + \frac{4 \cdot 3!}{1! 3!}\left(\frac{c^3}{8}\right)\left(\frac{d}{3}\right) + \frac{4 \cdot 3 \cdot 2!}{2! 2!}\left(\frac{c^2}{4}\right)\left(\frac{d^2}{9}\right)$$

$$+ \frac{4 \cdot 3!}{3! 1!}\left(\frac{c}{2}\right)\left(\frac{d^3}{27}\right) + \frac{d^4}{81}$$

$$= \frac{c^4}{16} + \frac{4}{1}\left(\frac{c^3 d}{24}\right) + \frac{12}{2}\left(\frac{c^2 d^2}{36}\right)$$

$$+ \frac{4}{1}\left(\frac{cd^3}{54}\right) + \frac{d^4}{81}$$

$$= \frac{c^4}{16} + \frac{c^3 d}{6} + \frac{c^2 d^2}{6} + \frac{2cd^3}{27} + \frac{d^4}{81}$$

13. The 3rd term of $(x+y)^4$

$$\frac{4!}{2!(4-2)!}x^2 y^2 = \frac{4 \cdot 3 \cdot 2!}{2! 2!}x^2 y^2$$

$$= \frac{12}{2}x^2 y^2$$

$$= 6x^2 y^2$$

14. The 4th term of $(x-y)^6$

$$\frac{6!}{3!(6-3)!}x^3(-y)^3 = \frac{6 \cdot 5 \cdot 4 \cdot 3!}{3! 3 \cdot 2 \cdot 1}x^3(-y)^3$$

$$= -\frac{120}{6}x^3 y^3$$

$$= -20x^3 y^3$$

15. The 2nd term of $(3x-4y)^3$

$$\frac{3!}{1!(3-1)!}(3x)^2(-4y) = \frac{3 \cdot 2!}{1! 2!}(9x^2)(-4y)$$

$$= \frac{3}{1}(-36x^2 y)$$

$$= -108x^2 y$$

16. The 5ᵗʰ term of $\left(u^2 - v^2\right)^5$

$$\frac{5!}{4!(5-4)!}\left(u^2\right)\left(-v^3\right)^4 = \frac{5 \cdot \cancel{4!}}{\cancel{4!}1!}\left(u^2\right)\left(v^{12}\right)$$

$$= \frac{5}{1}\left(u^2 v^{12}\right)$$

$$= 5u^2 v^{12}$$

SECTION 14.2
Arithmetic Sequences and Series

17. $a_n = 2n - 4$

$a_1 = 2(1) - 4 = -2$

$a_2 = 2(2) - 4 = 0$

$a_3 = 2(3) - 4 = 2$

$a_4 = 2(4) - 4 = 4$

The first four terms are –2, 0, 2, and 4.

18. $a_n = \dfrac{(-1)^n}{n+1}$

$a_1 = \dfrac{(-1)^1}{1+1} = -\dfrac{1}{2}$

$a_2 = \dfrac{(-1)^2}{2+1} = \dfrac{1}{3}$

$a_3 = \dfrac{(-1)^3}{3+1} = -\dfrac{1}{4}$

$a_4 = \dfrac{(-1)^4}{4+1} = \dfrac{1}{5}$

$a_5 = \dfrac{(-1)^5}{5+1} = -\dfrac{1}{6}$

The first five terms are $-\dfrac{1}{2}, \dfrac{1}{3}, -\dfrac{1}{4}, \dfrac{1}{5}, -\dfrac{1}{6}$.

19. $a_n = 100 - \dfrac{n}{2}$

$a_{50} = 100 - \dfrac{50}{2}$

$= 100 - 25$

$= 75$

20. Find a_8 if $a_1 = 7$ and $d = 5$.

$a_8 = a_1 + (n-1)(d)$

$a_8 = 7 + (8-1)(5)$

$a_8 = 7 + (7)(5)$

$a_8 = 7 + 35$

$a_8 = 42$

21. Find d

$242 - 212 = 30$, so $2d = 30$ and $d = 15$.

Find a_1 if $a_n = 212$ and $n = 7$.

$212 = a_1 + 15(7-1)$

$212 = a_1 + 15(6)$

$212 = a_1 + 90$

$122 = a_1$

$a_1 = 122 + 15(1-1) = 122 + 0 = 122$

$a_2 = 122 + 15(2-1) = 122 + 15 = 137$

$a_3 = 122 + 15(3-1) = 122 + 30 = 152$

$a_4 = 122 + 15(4-1) = 122 + 45 = 167$

$a_5 = 122 + 15(5-1) = 122 + 16 = 182$

22. Find d

$-6 - 6 = -12$, so $d = -12$.

Find a_{101} if $a_1 = 6$, $d = -12$ and $n = 101$.

$a_n = a_1 + d(n-1)$

$a_{101} = 6 - 12(101-1)$

$= 6 - 12(100)$

$= 6 - 1,200$

$= -1,194$

23. Let $n = 23$ and $a_n = -625$, $a_1 = -515$

$-625 = -515 + d(23-1)$

$-625 = -515 + d(22)$

$-110 = 22d$

$-5 = d$

24. Let $a_1 = 8$ and $a_4 = 25$.

$a_2 = 2 + d, a_3 = 2 + 2d, a_4 = 2 + 3d$

$a_4 = a_1 + (4 - 1)d$

$25 = 8 + 3d$

$17 = 3d$

$\dfrac{17}{3} = d$

$a_2 = 8 + d = 8 + \dfrac{17}{3} = \dfrac{41}{3}$

$a_3 = 8 + 2d = 8 + 2\left(\dfrac{17}{3}\right) = \dfrac{58}{3}$

25.

$a_1 = 9, \quad d = 6\dfrac{1}{2} - 9 = -2\dfrac{1}{2} = -\dfrac{5}{2}, \quad n = 10$

$a_{10} = 9 + -\dfrac{5}{2}(10 - 1)$

$\quad = 9 + -\dfrac{5}{2}(9)$

$\quad = -\dfrac{27}{2}$

$\quad = -13.5$

$S_n = \dfrac{n(a_1 + a_n)}{2}$

$S_{10} = \dfrac{10(9 - 13.5)}{2}$

$\quad = \dfrac{10(-4.5)}{2}$

$\quad = -22.5$

$\quad = -\dfrac{45}{2}$

26. Find d

$22 - 6 = 16$, so $4d = 16$ and $d = 4$.

Find a_1 if $a_n = 6$ and $n = 2$.

$6 = a_1 + 4(2 - 1)$

$6 = a_1 + 4(1)$

$6 = a_1 + 4$

$2 = a_1$

Find a_{28}.

$a_1 = 2, \quad d = 4, \quad n = 28$

$a_{28} = 2 + 4(28 - 1)$

$\quad = 2 + 4(27)$

$\quad = 110$

$S_n = \dfrac{n(a_1 + a_n)}{2}$

$S_{28} = \dfrac{28(2 + 110)}{2}$

$\quad = \dfrac{28(112)}{2}$

$\quad = 1,568$

27.

$\displaystyle\sum_{k=4}^{6} \dfrac{1}{2}k = \dfrac{1}{2}(4) + \dfrac{1}{2}(5) + \dfrac{1}{2}(6)$

$\quad = 2 + \dfrac{5}{2} + 3$

$\quad = \dfrac{15}{2}$

28.

$\displaystyle\sum_{k=2}^{5} 7k^2 = 7(2^2) + 7(3^2) + 7(4^2) + 7(5^2)$

$\quad = 7(4) + 7(9) + 7(16) + 7(25)$

$\quad = 28 + 63 + 112 + 175$

$\quad = 378$

29.

$\displaystyle\sum_{k=1}^{4} (3k - 4) = (3 \cdot 1 - 4) + (3 \cdot 2 - 4) + (3 \cdot 3 - 4) + (3 \cdot 4 - 4)$

$\quad = -1 + 2 + 5 + 8$

$\quad = 14$

30.

$$\sum_{k=10}^{10} 36k = 36(10)$$

$$= 360$$

31.

$$a_1 = 1,\ a_n = 200,\ n = 200,$$

$$S_n = \frac{n(a_1 + a_n)}{2}$$

$$S_{100} = \frac{200(1 + 200)}{2}$$

$$= \frac{200(201)}{2}$$

$$= 20,100$$

32. $a_1 = 10,\ a_2 = 12$, so $d = 2$.

Let $n = 30$ and find a_{30} to find the number of seats on the last row.

$$a_{30} = 10 + 2(30 - 1)$$

$$= 10 + 2(29)$$

$$= 68$$

Find the sum of all of the seats.

$$S_n = \frac{n(a_1 + a_n)}{2}$$

$$S_{30} = \frac{30(10 + 68)}{2}$$

$$= \frac{30(78)}{2}$$

$$= 1,170$$

SECTION 14.3
Geometric Sequences and Series

33. $a_1 = \dfrac{1}{8},\ r = 2$

$$a_n = a_1 r^{n-1}$$

$$a_6 = a_1 r^{6-1}$$

$$= \frac{1}{8}(2)^{6-1}$$

$$= \frac{1}{8}(2)^5$$

$$= \frac{1}{8}(32)$$

$$= 4$$

34. 4th term is 3, 5th term is $\dfrac{3}{2}$

Use $r = \dfrac{a_{n+1}}{a_n}$ to find r.

$$a_4 = 3 \text{ and } a_5 = \frac{3}{2}$$

$$r = \frac{\dfrac{3}{2}}{3}$$

$$r = \frac{3}{2} \cdot \frac{1}{3}$$

$$r = \frac{1}{2}$$

Use $a_n = a_1 r^{n-1}$ to find a_1.

$$r = \frac{1}{2},\ a_4 = 3 \text{ and } n = 4$$

$$3 = a_1 \left(\frac{1}{2}\right)^{4-1}$$

$$3 = a_1 \left(\frac{1}{2}\right)^3$$

$$3 = \frac{1}{8}a_1$$

$$24 = a_1$$

$$a_1 = 24$$

$$a_2 = a_1 r = 24\left(\frac{1}{2}\right) = 12$$

$$a_3 = a_1 r^2 = 24\left(\frac{1}{2}\right)^2 = 6$$

$$a_4 = a_1 r^3 = 24\left(\frac{1}{2}\right)^3 = 3$$

$$a_5 = a_1 r^4 = 24\left(\frac{1}{2}\right)^4 = \frac{3}{2}$$

$$24,\ 12,\ 6,\ 3,\ \frac{3}{2}$$

35. $r = -3, a_9 = 243$

Use $a_n = ar^{n-1}$ to find a_1.

$r = -3,\ a_9 = 243$ and $n = 9$

$243 = a_1(-3)^{9-1}$

$243 = a_1(-3)^8$

$243 = a_1(6,561)$

$\dfrac{243}{6,561} = a_1$

$\dfrac{1}{27} = a_1$

36. $a_1 = -6, a_4 = 384$ for two geometric means between -6 and 384

Use $a_n = ar^{n-1}$ to find r.

$a_1 = -6,\ a_4 = 384$ and $n = 4$

$384 = -6(r)^{4-1}$

$384 = -6r^3$

$-64 = r^3$

$-4 = r$

$a_1 = -6$

$a_2 = -6(-4) = 24$

$a_3 = -6(-4)^2 = -96$

$a_4 = -6(-4)^3 = 384$

The geometric means are 24 and -96.

37. Find r.

$r = \dfrac{a_2}{a_1} = \dfrac{54}{162} = \dfrac{1}{3}$

Find the sum of $a_1 = 162$, $r = \dfrac{1}{3}$, and $n = 7$.

$S_n = \dfrac{a_1(1 - r^n)}{1 - r}$

$= \dfrac{162\left(1 - \left(\dfrac{1}{3}\right)^7\right)}{1 - \dfrac{1}{3}}$

$= \dfrac{162\left(1 - \dfrac{1}{2,187}\right)}{\dfrac{2}{3}}$

$= \dfrac{162\left(\dfrac{2,186}{2,187}\right)}{\dfrac{2}{3}}$

$= \dfrac{\dfrac{4,372}{27}}{\dfrac{2}{3}}$

$= \dfrac{4,372}{27} \cdot \dfrac{3}{2}$

$= \dfrac{2,186}{9}$

38. Find r.

$$r = \frac{a_2}{a_1} = \frac{-\dfrac{1}{4}}{\dfrac{1}{8}} = -2$$

Find the sum of $a_1 = \dfrac{1}{8}$, $r = -2$, and $n = 8$.

$$S_n = \frac{a_1\left(1-r^n\right)}{1-r}$$

$$= \frac{\dfrac{1}{8}\left(1-(-2)^8\right)}{1-(-2)}$$

$$= \frac{\dfrac{1}{8}(1-256)}{3}$$

$$= \frac{\dfrac{1}{8}(-255)}{3}$$

$$= \frac{-255}{8} \cdot \frac{1}{3}$$

$$= \frac{-255}{24}$$

$$= -\frac{85}{8}$$

39. If he uses 25% of the birdseed each month, then 75% of the birdseed remains. The amount of in the bag after 12 months is represented by the 13th term of a geometric series, where $a_1 = 50$, $n = 13$, and $r = 0.75$.

$$a_n = a_1 r^{n-1}$$

$$a_{13} = (50)(0.75)^{13-1}$$

$$a_{13} = (50)(0.75)^{12}$$

$$a_{13} = (50)(0.031676352)$$

$$a_{13} \approx 1.6 \text{ lb}$$

He will have about 1.6 lbs of birdseed left.

40. Find r.

$$r = \frac{a_2}{a_1} = \frac{20}{25} = \frac{4}{5}$$

Find the sum of the infinite geometric sequence with $a_1 = 25$ and $r = \dfrac{4}{5}$.

$$S = \frac{a_1}{1-r}$$

$$= \frac{25}{1-\dfrac{4}{5}}$$

$$= \frac{25}{\dfrac{1}{5}}$$

$$= 25 \cdot \frac{5}{1}$$

$$= 125$$

41.

$$0.\overline{05} = 0.050505\ldots = \frac{5}{100} + \frac{5}{10,000} + \frac{5}{1,000,000} + \ldots$$

where $a_1 = \dfrac{5}{100}$ and $r = \dfrac{1}{100}$.

Find the sum with $a_1 = \dfrac{5}{100}$ and $r = \dfrac{1}{100}$.

$$S = \frac{a_1}{1-r}$$

$$= \frac{\dfrac{5}{100}}{1-\dfrac{1}{100}}$$

$$= \frac{\dfrac{5}{100}}{\dfrac{99}{100}}$$

$$= \frac{5}{100} \cdot \frac{100}{99}$$

$$= \frac{5}{99}$$

$$0.\overline{05} = \frac{5}{99}$$

42. The total distance the ball travels is the sum of two motions, falling and rebounding. The distance the ball falls is given by the sum

$$10 + \frac{9}{10} \cdot 10 + \frac{9}{10}\left(\frac{9}{10} \cdot 10\right) + \ldots \text{ or}$$

$$10 + 9 + \frac{81}{10} + \ldots.$$ The distance the ball rebounds begins with the first bounce up which is 9 ft, and then is nine-tenths the distance afterwards or $9 + \frac{81}{10} + \frac{729}{100} + \ldots.$

Since these are infinite geometric series, use the formula $S = \frac{a_1}{1-r}$ to find the sum.

Falling:

$a_1 = 10$ and $r = \frac{9}{10}$

$$S = \frac{a_1}{1-r}$$

$$= \frac{10}{1 - \frac{9}{10}}$$

$$= \frac{10}{\frac{1}{10}}$$

$$= 10 \cdot \frac{10}{1}$$

$$= 100$$

Rebounding:

$a_1 = 9$ and $r = \frac{9}{10}$

$$S = \frac{a_1}{1-r}$$

$$= \frac{9}{1 - \frac{9}{10}}$$

$$= \frac{9}{\frac{1}{10}}$$

$$= 9 \cdot \frac{10}{1}$$

$$= 90$$

The distance the ball travels is the sum of the distances falling and rebounding:

$$100 + 90 = 190 \text{ ft.}$$

The ball travels a total of 190 ft.

1. a. The array of numbers that gives the coefficients of the terms of a binomial expansion is called **Pascal's** triangle.
    b. In the expansion $a^3 - 3a^2b + 3ab^2 - b^3$, the signs **alternate** between + and –.
    c. Each term of an **arithmetic** sequence is found by adding the same number to the previous term.
    d. The sum of the terms of an arithmetic sequence is called an arithmetic **series**.
    e. Each term of a **geometric** sequence is found by multiplying the previous term by the same number.

2. $a_n = -6n + 8$
    $a_1 = -6(1) + 8 = 2$
    $a_2 = -6(2) + 8 = -4$
    $a_3 = -6(3) + 8 = -10$
    $a_4 = -6(4) + 8 = -16$
    The first four terms are 2, –4, –10, and –16.

3. $a_n = \dfrac{(-1)^n}{n^3}$
    $a_1 = \dfrac{(-1)^1}{1^3} = \dfrac{-1}{1} = -1$
    $a_2 = \dfrac{(-1)^2}{2^3} = \dfrac{1}{8}$
    $a_3 = \dfrac{(-1)^3}{3^3} = -\dfrac{1}{27}$
    $a_4 = \dfrac{(-1)^4}{4^3} = \dfrac{1}{64}$
    $a_5 = \dfrac{(-1)^5}{5^3} = -\dfrac{1}{125}$

4.
$$\frac{10!}{6!(10-6)!} = \frac{10 \cdot 9 \cdot 8 \cdot 7 \cdot \cancel{6!}}{\cancel{6!}\,4!}$$
$$= \frac{5{,}040}{24}$$
$$= 210$$

5.
$$(a-b)^6 = \left(a + (-b)\right)^6$$
$$= a^6 + \frac{6!}{1!(6-1)!}a^5(-b) + \frac{6!}{2!(6-2)!}a^4(-b)^2$$
$$+ \frac{6!}{3!(6-3)!}a^3(-b)^3 + \frac{6!}{4!(6-4)!}a^2(-b)^4$$
$$+ \frac{6!}{5!(6-5)!}a(-b)^5 + (-b)^6$$
$$= a^6 + \frac{6!}{1!5!}a^5(-b) + \frac{6!}{2!4!}a^4(-b)^2$$
$$+ \frac{6!}{3!3!}a^3(-b)^3 + \frac{6!}{4!2!}a^2(-b)^4$$
$$+ \frac{6!}{5!1!}a(-b)^5 + (-b)^6$$
$$= a^6 - \frac{6}{1}a^5b + \frac{30}{2}a^4b^2 - \frac{120}{6}a^3b^3 + \frac{30}{2}a^2b^4$$
$$- \frac{6}{1}ab^5 + b^6$$
$$= a^6 - 6a^5b + 15a^4b^2 - 20a^3b^3 + 15a^2b^4 - 6ab^5 + b^6$$

6. The 3rd term of $(x^2 + 2y)^4$
$$\frac{4!}{2!(4-2)!}(x^2)^2(2y)^2 = \frac{4 \cdot 3 \cdot \cancel{2!}}{2!\,\cancel{2!}}(x^4)(4y^2)$$
$$= \frac{12}{2}(4x^4y^2)$$
$$= 24x^4y^2$$

7. Find d.
    $10 - 3 = 7$, so $d = 7$.
    Find a_{10} if $a_1 = 3$, $d = 7$ and $n = 10$.
    $$a_n = a_1 + d(n-1)$$
    $$a_{10} = 3 + 7(10 - 1)$$
    $$= 3 + 7(9)$$
    $$= 3 + 63$$
    $$= 66$$

8.

$$a_1 = -2, \ d = 3 - (-2) = 5, \ n = 12$$
$$a_{12} = -2 + 5(12 - 1)$$
$$= -2 + 5(11)$$
$$= -2 + 55$$
$$= 53$$

$$S_n = \frac{n(a_1 + a_n)}{2}$$
$$S_{12} = \frac{12(-2 + 53)}{2}$$
$$= \frac{12(51)}{2}$$
$$= 306$$

9. Let $a_1 = 10$ and $a_3 = 19$
$$a_2 = 10 + d$$
Let $a_1 = 2$ and $a_4 = 98$.
$$a_4 = a_1 + (4 - 1)d$$
$$98 = 2 + 3d$$
$$96 = 3d$$
$$32 = d$$

$$a_n = a_1 + (n - 1)d$$
$$a_2 = 2 + (2 - 1)(32) = 2 + 1(32) = 34$$
$$a_3 = 2 + (3 - 1)(32) = 2 + 2(32) = 66$$
The arithmetic means are 34 and 66.

10.
$$\left(5 - \frac{5}{4}\right) = (17 - 2)d$$
$$\frac{15}{4} = 15d$$
$$\frac{1}{15} \cdot \frac{15}{4} = \frac{1}{15} \cdot 15d$$
$$\frac{1}{4} = d$$

11. Find d
$$(-75 - (-11)) = (20 - 4)d$$
$$-64 = 16d$$
$$-4 = d$$

Find a_1 if $a_n = -11, d = -4$ and $n = 4$.
$$-11 = a_1 - 4(4 - 1)$$
$$-11 = a_1 - 4(3)$$
$$-11 = a_1 - 12$$
$$1 = a_1$$

Find a_{27}.
$$a_1 = 1, \ d = 4, \ n = 27$$
$$a_{27} = 1 - 4(27 - 1)$$
$$= 1 - 4(26)$$
$$= -103$$

$$S_n = \frac{n(a_1 + a_n)}{2}$$
$$S_{27} = \frac{27(1 - 103)}{2}$$
$$= \frac{27(-102)}{2}$$
$$= -1,377$$

12. The sum of all 25 rows minus the sum of the first 15 rows equals the number of pipe left in the stack after removing the first 15 rows. For S_{25}, let $a_1 = 1$, $a_{25} = 25$, and $n = 25$. For S_{15}, let $a_1 = 1$, $a_{15} = 15$, and $n = 15$.

$$S_n = \frac{n(a_1 + a_n)}{2}$$
$$S_{25} - S_{15} = \frac{25(1 + 25)}{2} - \frac{15(1 + 15)}{2}$$
$$= \frac{25(26)}{2} - \frac{15(16)}{2}$$
$$= 325 - 120$$
$$= 205$$
There are 205 pipes remaining in the stack.

13. Find d if $a_1 = 16$, $a_2 = 48$, and $n = 2$.

$$a_n = a_1 + d(n-1)$$
$$48 = 16 + d(2-1)$$
$$48 = 16 + d$$
$$32 = d$$

Find a_{10} to find out how far the object fell during the 10th second.

$$a_1 = 16, \quad d = 32, \quad n = 10$$
$$a_{10} = 16 + 32(10-1)$$
$$= 16 + 32(9)$$
$$= 304$$

Find the sum of the distances fell during the first 10 seconds.

$$S_n = \frac{n(a_1 + a_n)}{2}$$
$$S_{10} = \frac{10(16 + 304)}{2}$$
$$= \frac{10(320)}{2}$$
$$= 1,600$$

14.

$$\sum_{k=1}^{3}(2k-3) = (2\cdot 1 - 3) + (2\cdot 2 - 3) + (2\cdot 3 - 3)$$
$$= -1 + 1 + 3$$
$$= 3$$

15.

Use $r = \dfrac{a_{n+1}}{a_n}$ to find r.

$$a_2 = -\frac{1}{3} \text{ and } a_1 = -\frac{1}{9}$$

$$r = \frac{-\dfrac{1}{3}}{-\dfrac{1}{9}}$$

$$r = -\frac{1}{3}\cdot -\frac{9}{1}$$

$$r = 3$$

Use $a_n = a_1 r^{n-1}$ to find a_7.

$$r = 3, \ a_1 = -\frac{1}{9} \text{ and } n = 7$$

$$a_7 = -\frac{1}{9}(3)^{7-1}$$
$$a_7 = -\frac{1}{9}(3)^6$$
$$a_7 = -\frac{1}{9}(729)$$
$$a_7 = -81$$

16. Find r.

$$r = \frac{a_2}{a_1} = \frac{\dfrac{1}{9}}{\dfrac{1}{27}} = \frac{1}{9}\cdot\frac{27}{1} = 3$$

Find the sum of $a_1 = \dfrac{1}{27}$, $r = 3$, and $n = 6$.

$$S_n = \frac{a_1(1-r^n)}{1-r}$$

$$= \frac{\dfrac{1}{27}(1-3^6)}{1-3}$$

$$= \frac{\dfrac{1}{27}(1-729)}{-2}$$

$$= \frac{\dfrac{1}{27}(-728)}{-2}$$

$$= \frac{-728}{27}\cdot\frac{1}{-2}$$

$$= \frac{364}{27}$$

Chapter 14 Test

17. Let $r = -\dfrac{2}{3}$ and $a_4 = -\dfrac{16}{9}$

$$a_n = a_1 r^{n-1}$$
$$a_4 = a_1 r^{4-1}$$
$$a_4 = a_1 r^3$$
$$-\dfrac{16}{9} = a_1\left(-\dfrac{2}{3}\right)^3$$
$$-\dfrac{16}{9} = -\dfrac{8}{27}a_1$$
$$-\dfrac{27}{8} \cdot -\dfrac{16}{9} = -\dfrac{27}{8} \cdot -\dfrac{8}{27}a_1$$
$$6 = a_1$$

18. $a_1 = 3$, $a_4 = 648$ for two geometric means between 3 and 648

Use $a_n = ar^{n-1}$ to find r.

$$a_1 = 3,\ a_4 = 648 \text{ and } n = 4$$
$$648 = 3(r)^{4-1}$$
$$648 = 3r^3$$
$$216 = r^3$$
$$\sqrt[3]{216} = \sqrt[3]{r^3}$$
$$6 = r$$
$$a_1 = 3$$
$$a_2 = 3(6) = 18$$
$$a_3 = 3(6)^2 = 108$$
$$a_4 = 3(6)^3 = 648$$

The geometric means are 18 and 108.

19. Find r.

$$r = \dfrac{a_2}{a_1} = \dfrac{3}{9} = \dfrac{1}{3}$$

Find the sum of the infinite geometric series with $a_1 = 9$ and $r = \dfrac{1}{3}$.

$$S = \dfrac{a_1}{1-r}$$
$$= \dfrac{9}{1-\dfrac{1}{3}}$$
$$= \dfrac{9}{\dfrac{2}{3}}$$
$$= 9 \cdot \dfrac{3}{2}$$
$$= \dfrac{27}{2}$$

20. DEPRECIATION

If the yacht depreciates 8% of its value each year, then 92% of its value remains at the end of the first year. The value of the boat in 10 years is represented by the 11th term of a geometric series, where $a_1 = 1{,}500{,}000$, $n = 11$, and $r = 0.92$.

$$a_n = a_1 r^{n-1}$$
$$a_{11} = (1{,}500{,}000)(0.92)^{11-1}$$
$$a_{11} = (1{,}500{,}000)(0.92)^{10}$$
$$a_{11} \approx (1{,}500{,}000)(0.43)$$
$$a_{11} \approx \$651{,}583$$

The boat is worth \$651,583 after 10 years.

21. PENDULUMS

Let $a_1 = 60$, and $r = \dfrac{97}{100}$.

$$S = \dfrac{a_1}{1-r}$$
$$= \dfrac{60}{1-\dfrac{97}{100}}$$
$$= \dfrac{60}{\dfrac{3}{100}}$$
$$= 2{,}000 \text{ in.}$$

22.

$$0.\overline{7} = 0.777... = \frac{7}{10} + \frac{7}{100} + \frac{7}{1,000} + ...$$

where $a_1 = \frac{7}{10}$ and $r = \frac{1}{10}$.

$|r| = \left|\frac{1}{10}\right| = \frac{1}{10} < 1$ so the sum does exist.

Find the sum with $a_1 = \frac{7}{10}$ and $r = \frac{1}{10}$.

$$S = \frac{a_1}{1-r}$$

$$= \frac{\frac{7}{10}}{1 - \frac{1}{10}}$$

$$= \frac{\frac{7}{10}}{\frac{9}{10}}$$

$$= \frac{7}{10} \cdot \frac{10}{9}$$

$$= \frac{7}{9}$$

$$0.\overline{7} = \frac{7}{9}$$

Chapter 14 Test

1. a. 0

    b. $-\dfrac{4}{3}, 5.6, 0, -23$

    c. $\pi, \sqrt{2}, e$

    d. $-\dfrac{4}{3}, \pi, 5.6, \sqrt{2}, 0, -23, e$

2.
$$6\big[x - (2 - x)\big] = -4(8x + 3)$$
$$6\big[x - 2 + x\big] = -32x - 12$$
$$6\big[2x - 2\big] = -32x - 12$$
$$12x - 12 = -32x - 12$$
$$12x - 12 + 32x = -32x - 12 + 32x$$
$$44x - 12 = -12$$
$$44x - 12 + 12 = -12 + 12$$
$$44x = 0$$
$$\frac{44x}{44} = \frac{0}{44}$$
$$x = 0$$

3.
$$A = \frac{1}{2}h(b_1 + b_2)$$
$$2 \cdot A = 2 \cdot \frac{1}{2}h(b_1 + b_2)$$
$$2A = h(b_1 + b_2)$$
$$2A = hb_1 + hb_2$$
$$2A - hb_1 = hb_1 + hb_2 - hb_1$$
$$2A - hb_1 = hb_2$$
$$\frac{2A - hb_1}{h} = \frac{hb_2}{h}$$
$$\frac{2A - hb_1}{h} = b_2$$

4.
Let x = measure of angle C.
Then $\angle B = 5 + 5x$ and
$\angle A = 5 + (5 + 5x) = 10 + 5x$. The sum of
the angles of the triangle is 180°.
$$\angle A + \angle B + \angle C = 180$$
$$(10 + 5x) + (5 + 5x) + (x) = 180$$
$$11x + 15 = 180$$
$$11x = 165$$
$$x = 15$$
$$\angle C = 15°$$
$$\angle B = 5 + 5(15) = 80°$$
$$\angle A = 10 + 5(15) = 85°$$

5. Let x = the amount of money she has to
    invest. Then $x + 3,000$ = the needed
    amount of money to earn the greater
    interest.

	principal	rate	Interest
original	x	0.075	0.075x
larger amount	$x + 3,000$	0.11	0.11($x + 3,000$)

11% investment = 2(7.5% investment)
$$0.11(x + 3,000) = 2(0.075x)$$
$$0.11x + 330 = 0.15x$$
$$330 = 0.04x$$
$$8,250 = x$$

She had $8,250 to invest originally.

6.
$$-5x + 7 \le 12$$
$$-5x \le 5$$
$$x \ge -1$$
$$[-1, \infty)$$

7.
$$A = \frac{1}{2}bh$$
$$\frac{1}{2}(16)(h) \le 100$$
$$8h \le 100$$
$$h \le 12.5$$

8. Use the points (1,600, 68) and (2,800, 78).

rate of change $= \dfrac{y_2 - y_1}{x_2 - x_1}$

$= \dfrac{78 - 68}{2,800 - 1,600}$

$= \dfrac{10}{1,200}$

$= \dfrac{1}{120}$ db/rpm

9. a. Find the slope of each line.

$3x - 4y = 12$

$-4y = -3x + 12$

$y = \dfrac{3}{4}x - 3$

$m = \dfrac{3}{4}$

$y = \dfrac{3}{4}x - 5$

$m = \dfrac{3}{4}$

The slopes are the same, so parallel.

b. Find the slope of each line.

$y = 3x + 4$

$m = 3$

$x = -3y + 4$

$x - 4 = -3y$

$y = -\dfrac{1}{3}x + \dfrac{4}{3}$

$m = -\dfrac{1}{3}$

The slopes are opposite reciprocals, so the lines are perpendicular.

10. SALVAGE VALUE

Write ordered pairs in the form (age, value). Use (0, 28,000) for the purchase value and (6, 7,600) for the predicted value.
Find the rate of change (m).

$m = \dfrac{y_2 - y_1}{x_2 - x_1}$

$= \dfrac{7,600 - 28,000}{6 - 0}$

$= \dfrac{-20,400}{6}$

$= -3,400$

Use the point (0, 28,000) and $m = -3,400$.

$y - y_1 = m(x - x_1)$

$y - 28,000 = -3,400(x - 0)$

$y - 28,000 = -3,400x + 0$

$y = -3,400x + 28,000$

11. $y = mx + b;\ m = -2$ and $b = 5$

$y = -2x + 5$

12. Find m.

$m = \dfrac{y_2 - y_1}{x_2 - x_1}$

$= \dfrac{4 - (-5)}{-5 - 8}$

$= \dfrac{9}{-13}$

$= -\dfrac{9}{13}$

Use $m = -\dfrac{9}{13}$ and $(8, -5)$.

$y - y_1 = m(x - x_1)$

$y - (-5) = -\dfrac{9}{13}(x - 8)$

$y + 5 = -\dfrac{9}{13}x + \dfrac{72}{13}$

$y = -\dfrac{9}{13}x + \dfrac{7}{13}$

13. It doesn't pass the vertical line test. The graph passes through (0, 2) and (0, –2).

14. $f(x) = 3x^5 - 2x^2 + 1$

$f(-1) = 3(-1)^5 - 2(-1)^2 + 1$

$= 3(-1) - 2(1) + 1$

$= -3 - 2 + 1$

$= -4$

$f(a) = 3a^5 - 2a^2 + 1$

Chapter 14 Cumulative Review

15.

$$\begin{cases} 2x - y = -21 \\ 4x + 5y = 7 \end{cases}$$

Solve the first equation for y.

$$2x - y = -21$$
$$-y = -2x - 21$$
$$y = 2x + 21$$

Substitute $y = 2x + 2y$ into the second equation and solve for x.

$$4x + 5y = 7$$
$$4x + 5(2x + 21) = 7$$
$$4x + 10x + 105 = 7$$
$$14x = -98$$
$$x = -7$$
$$y = 2x + 21$$
$$y = 2(-7) + 21$$
$$y = -14 + 21$$
$$y = 7$$

The solution is $(-7, 7)$.

16.

$$\begin{cases} 4y + 5x - 7 = 0 \\ \dfrac{10}{7}x - \dfrac{4}{9}y = \dfrac{17}{21} \end{cases}$$

Write the equations in standard form. Multiply the second equation by 63 to eliminate fractions.

$$\begin{cases} 5x + 4y = 7 \\ 90x - 28y = 51 \end{cases}$$

Multiply the first equation by 7 and add the equations to solve for x.

$$35x + 28y = 49$$
$$\underline{90x - 28y = 51}$$
$$125x = 100$$
$$x = \frac{4}{5}$$

Substitute $x = \dfrac{4}{5}$ into the first equation and solve for y.

$$5x + 4y = 7$$
$$5\left(\frac{4}{5}\right) + 4y = 7$$
$$4 + 4y = 7$$
$$4y = 3$$
$$y = \frac{3}{4}$$

The solution is $\left(\dfrac{4}{5}, \dfrac{3}{4}\right)$.

17. MIXING COFFEE

Let x = # of lb of regular coffee and y = # of lb of Brazilian coffee.

	Regular coffee	Brazilian coffee	Mixture
# of pounds	x	y	40
Value	$4x$	$11.50y$	$6(40)$

$$\begin{cases} x + y = 40 \\ 4x + 11.50y = 6(40) \end{cases}$$

$$\begin{cases} x + y = 40 \\ 4x + 11.50y = 240 \end{cases}$$

Multiply the 1st equation by -4.

Add equations to eliminate x.

$$-4x - 4y = -160$$
$$\underline{4x + 11.5y = 240}$$
$$7.5y = 80$$
$$y = 10\frac{2}{3}$$
$$x + y = 40$$
$$x + 10\frac{2}{3} = 40$$
$$x = 29\frac{1}{3}$$

$29\frac{1}{3}$ pounds of regular coffee and $10\frac{2}{3}$ pounds of Brazilian coffee are needed.

18.

$$\begin{cases} 3x - 2y \leq 6 \\ y < -x + 2 \end{cases}$$

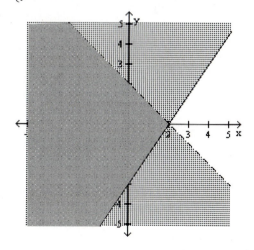

19.

$$(x^2)^5 y^7 y^3 x^{-2} y^0 = x^{2\cdot5} y^7 y^3 x^{-2} y^0$$
$$= x^{10} y^7 y^3 x^{-2} y^0$$
$$= x^{10+(-2)} y^{7+3+0}$$
$$= x^8 y^{10}$$

20.

$$\left(\frac{3x^5 y^2}{6x^5 y^{-2}}\right)^{-4} = \left(\frac{6x^5 y^{-2}}{3x^5 y^2}\right)^4$$
$$= \left(2x^{5-5} y^{-2-2}\right)^4$$
$$= \left(2x^0 y^{-4}\right)^4$$
$$= \left(2\cdot1y^{-4}\right)^4$$
$$= 2^4 y^{-4\cdot4}$$
$$= 16y^{-16}$$
$$= \frac{16}{y^{16}}$$

21. 1.73×10^{14}; 4.6×10^{-8}

22.

$$\frac{(0.00024)(96,000,000)}{(640,000,000)(0.025)} = \frac{(2.4 \times 10^{-4})(9.6 \times 10^7)}{(6.4 \times 10^8)(2.5 \times 10^{-2})}$$
$$= \frac{(2.4 \times 9.6)(10^{-4+7})}{(6.4 \times 2.5)(10^{8+(-2)})}$$
$$= \frac{23.04 \times 10^3}{16.0 \times 10^6}$$
$$= \frac{23.04}{16.0} \times 10^{3-6}$$
$$= 1.44 \times 10^{-3}$$
$$= 0.00144$$

23. $\dfrac{9}{4}rt^2 - \dfrac{5}{3}rt - \dfrac{1}{2}rt^2 + \dfrac{5}{6}rt$
$$= \frac{7}{4}rt^2 - \frac{5}{6}rt$$

Chapter 14 Cumulative Review

24.

$$\left(3x^2+4x-7\right)+\left[\left(-2x^2-7x+1\right)-\left(-4x^2+8x-1\right)\right]$$
$$=\left(3x^2+4x-7\right)+\left[-2x^2-7x+1+4x^2-8x+1\right]$$
$$=\left(3x^2+4x-7\right)+\left[2x^2-15x+2\right]$$
$$=3x^2+4x-7+2x^2-15x+2$$
$$=5x^2-11x-5$$

25.

$$\left(-2x^2y^3+6xy+5y^2\right)-\left(-4x^2y^3-7xy+2y^2\right)$$
$$=-2x^2y^3+6xy+5y^2+4x^2y^3+7xy-2y^2$$
$$=2x^2y^3+13xy+3y^2$$

26.

$$\left(x-3y\right)\left(x^2+3xy+9y^2\right)$$
$$=x\left(x^2+3xy+9y^2\right)-3y\left(x^2+3xy+9y^2\right)$$
$$=x^3+3x^2y+9xy^2-3x^2y-9xy^2-27y^3$$
$$=x^3-27y^3$$

27.

$$\left(2m^5-7\right)\left(3m^5-1\right)=6m^{10}-2m^5-21m^5+7$$
$$=6m^{10}-23m^5+7$$

28.

$$\left(9ab^2-4\right)^2=\left(9ab^2-4\right)\left(9ab^2-4\right)$$
$$=81a^2b^4-36ab^2-36ab^2+16$$
$$=81a^2b^4-72ab^2+16$$

29.

$$b^3-4b^2-3b+12=b^2\left(b-4\right)-3\left(b-4\right)$$
$$=\left(b-4\right)\left(b^2-3\right)$$

30.

$$12y^2+23y+10=\left(3y+2\right)\left(4y+5\right)$$

31.

$$144x^2-y^2=\left(12x-y\right)\left(12x+y\right)$$

32.

$$27t^3+u^3=\left(3t+u\right)\left(\left(3t\right)^2+3tu+u^2\right)$$
$$=\left(3t+u\right)\left(9t^2+3tu+u^2\right)$$

33.

$$a^4b^2-20a^2b^2+64b^2$$
$$=b^2\left(a^4-20a^2+64\right)$$
$$=b^2\left(a^2-4\right)\left(a^2-16\right)$$
$$=b^2\left(a+2\right)\left(a-2\right)\left(a+4\right)\left(a-4\right)$$

34.

$$3x^2+5x-2=0$$
$$\left(3x-1\right)\left(x+2\right)=0$$
$$3x-1=0 \quad \text{and} \quad x+2=0$$
$$3x=1 \qquad\qquad x=-2$$
$$x=\frac{1}{3}$$

35. Let $u=(x+7)$.

$$\left(x+7\right)^2=-2\left(x+7\right)-1$$
$$\left(x+7\right)^2+2\left(x+7\right)+1=0$$
$$u^2+2u+1=0$$
$$\left(u+1\right)\left(u+1\right)=0$$
$$u+1=0 \quad \text{or} \quad u+1=0$$
$$u=-1 \qquad\qquad u=-1$$
$$x+7=-1 \qquad x+7=-1$$
$$x=-8 \qquad\qquad x=-8$$

36.

$$x^3+8x^2=9x$$
$$x^3+8x^2-9x=0$$
$$x\left(x^2+8x-9\right)=0$$
$$x\left(x-1\right)\left(x+9\right)=0$$
$$x=0, \ x-1=0, \ \text{or} \ x+9=0$$
$$x=1 \qquad\qquad x=-9$$

37. PAINTING

Let the width $= w$.

Then the length $= 5w + 1$.

$$(\text{width})(\text{length}) = \text{Area}$$

$$w(5w + 1) = 84$$

$$5w^2 + w = 84$$

$$5w^2 + w - 84 = 0$$

$$(5w + 21)(w - 4) = 0$$

$$5w + 21 = 0 \quad \text{or} \quad w - 4 = 0$$

$$w = -\frac{21}{5} \qquad w = 4$$

Since length and width cannot be negative, the width is 4 inches and the length is $5(4) + 1 = 21$ inches.

38.

$$h(t) = -16t^2 + 64t + 80$$

$$0 = -16t^2 + 64t + 80$$

$$0 = -16(t^2 - 4t - 5)$$

$$0 = -16(t - 5)(t + 1)$$

$$t - 5 = 0 \quad \text{or} \quad t + 1 = 0$$

$$t = 5 \qquad \quad t = \cancel{-1}$$

It will take 5 seconds.

39. Use the Pythagorean Theorem if the shorter leg is x, the longer leg is $x + 2$, and the hypotenuse is $x + 4$.

$$x^2 + (x + 2)^2 = (x + 4)^2$$

$$x^2 + x^2 + 4x + 4 = x^2 + 8x + 16$$

$$2x^2 + 4x + 4 = x^2 + 8x + 16$$

$$x^2 - 4x - 12 = 0$$

$$(x - 6)(x + 2) = 0$$

$$x - 6 = 0 \quad \text{and} \quad x + 2 = 0$$

$$x = 6 \qquad \qquad x = \cancel{-2}$$

The lengths are 6, 8, and 10.

40.

$$\frac{x^2 - 10x + 21}{6x^2 - 41x - 7} = \frac{(x - 3)\cancel{(x - 7)}}{(6x + 1)\cancel{(x - 7)}}$$

$$= \frac{x - 3}{6x + 1}$$

41.

$$\frac{p^3 - q^3}{q^2 - p^2} \cdot \frac{q^2 + pq}{p^3 + p^2 q + pq^2}$$

$$= \frac{\cancel{(p - q)}(p^2 + pq + q^2)}{-\cancel{(p - q)}(q + p)} \cdot \frac{q(q + p)}{p(p^2 + pq + q^2)}$$

$$= -\frac{q}{p}$$

42.

$$\frac{2x + 1}{x^4 - 81} + \frac{2 - x}{x^4 - 81} = \frac{2x + 1 + 2 - x}{x^4 - 81}$$

$$= \frac{x + 3}{x^4 - 81}$$

$$= \frac{x + 3}{(x^2 + 9)(x^2 - 9)}$$

$$= \frac{\cancel{x + 3}}{(x^2 + 9)(x - 3)\cancel{(x + 3)}}$$

$$= \frac{1}{(x^2 + 9)(x - 3)}$$

43.

$$\frac{2}{a - 2} + \frac{3}{a + 2} - \frac{a - 1}{a^2 - 4}$$

$$= \frac{2}{a - 2} + \frac{3}{a + 2} - \frac{a - 1}{(a + 2)(a - 2)}$$

$$= \frac{2}{a - 2}\left(\frac{a + 2}{a + 2}\right) + \frac{3}{a + 2}\left(\frac{a - 2}{a - 2}\right) - \frac{a - 1}{(a + 2)(a - 2)}$$

$$= \frac{2a + 4}{(a + 2)(a - 2)} + \frac{3a - 6}{(a + 2)(a - 2)} - \frac{a - 1}{(a + 2)(a - 2)}$$

$$= \frac{2a + 4 + 3a - 6 - (a - 1)}{(a + 2)(a - 2)}$$

$$= \frac{2a + 4 + 3a - 6 - a + 1}{(a + 2)(a - 2)}$$

$$= \frac{4a - 1}{(a + 2)(a - 2)}$$

44.

$$\dfrac{\dfrac{y}{x}-\dfrac{x}{y}}{\dfrac{1}{x}+\dfrac{1}{y}} = \dfrac{xy\left(\dfrac{y}{x}-\dfrac{x}{y}\right)}{xy\left(\dfrac{1}{x}+\dfrac{1}{y}\right)}$$

$$= \dfrac{y^2 - x^2}{y + x}$$

$$= \dfrac{\cancel{(y+x)}(y-x)}{\cancel{y+x}}$$

$$= y - x$$

45.

$$\dfrac{3}{4n}+\dfrac{2}{n}=1$$

$$4n\left(\dfrac{3}{4n}+\dfrac{2}{n}\right)=4n(1)$$

$$3 + 8 = 4n$$

$$11 = 4n$$

$$\dfrac{11}{4}=n$$

46.

$$\dfrac{1}{R}=\dfrac{1}{R_1}+\dfrac{1}{R_2}+\dfrac{1}{R_3}$$

$$(RR_1R_2R_3)\left(\dfrac{1}{R}\right)=(RR_1R_2R_3)\left(\dfrac{1}{R_1}+\dfrac{1}{R_2}+\dfrac{1}{R_3}\right)$$

$$R_1R_2R_3 = RR_2R_3 + RR_1R_3 + RR_1R_2$$

$$R_1R_2R_3 = R(R_2R_3 + R_1R_3 + R_1R_2)$$

$$\dfrac{R_1R_2R_3}{R_2R_3 + R_1R_3 + R_1R_2} = \dfrac{R(R_2R_3 + R_1R_3 + R_1R_2)}{R_2R_3 + R_1R_3 + R_1R_2}$$

$$\dfrac{R_1R_2R_3}{R_2R_3 + R_1R_3 + R_1R_2} = R$$

$$R = \dfrac{R_1R_2R_3}{R_2R_3 + R_1R_3 + R_1R_2}$$

47. Let x = the amount of time it takes when the machines work together.

$$\dfrac{1}{6}+\dfrac{1}{4}=\dfrac{1}{x}$$

$$12x\left(\dfrac{1}{6}+\dfrac{1}{4}\right)=12x\left(\dfrac{1}{x}\right)$$

$$2x + 3x = 12$$

$$5x = 12$$

$$x = \dfrac{12}{5}$$

$$x = 2.4 \text{ hours}$$

48. Use a ratio.

$$\dfrac{24}{x}=\dfrac{8}{31}$$

$$24(31) = 8(x)$$

$$744 = 8x$$

$$93 = x$$

49.

$$3(x-4)+6 = -2(x+4)+5x$$

$$3x - 12 + 6 = -2x - 8 + 5x$$

$$3x - 6 = 3x - 8$$

$$-6 \neq -8$$

No Solution, $\varnothing$

Contradiction

50.

$$\dfrac{3}{2}(y+4)=\dfrac{20-y}{2}$$

$$\dfrac{3}{2}(-4+4)=\dfrac{20-(-4)}{2}$$

$$\dfrac{3}{2}(0)=\dfrac{20+4}{2}$$

$$0 \neq 12$$

No, it is not a solution.

51. WORK SCHEDULES

job	pay per hour	number of hr worked	Amount earned
bookstore	$12	x	$12x$
Airport	$22.5	$25 - x$	$22.5(25 - x)$

$$12x + 22.5(25 - x) \geq 450$$
$$12x + 562.5 - 22.5x \geq 450$$
$$-10.5x + 562.5 \geq 450$$
$$-10.5x + 562.5 - 562.5 \geq 450 - 562.5$$
$$-10.5x \geq -112.5$$
$$\frac{-10.5x}{-10.5} \leq \frac{-112.5}{-10.5}$$
$$x \leq 10.7$$

She can work **10 hours** at the bookstore.

52. No. $(2, 2)$ and $(2, -2)$

53. a. 4
 b. 3
 c. 0, 2
 d. 1

54. a. Find the slope using $(5, 10.3)$ and $(10, 9)$.

$$m = \frac{y_2 - y_1}{x_2 - x_1}$$
$$= \frac{10.3 - 9}{5 - 10}$$
$$= \frac{1.3}{-5}$$
$$= -0.26$$

Use $m = -0.26$ and $(10, 9)$.

$$y - y_1 = m(x - x_1)$$
$$y - 9 = -0.26(x - 10)$$
$$y - 9 = -0.26x + 2.6$$
$$y = -0.26x + 11.6$$
$$P(t) = -0.26t + 11.6$$

 b. Let $t = 40$.
$$P(40) = -0.26(40) + 11.6$$
$$= 1.2\%$$

55. The square base has 6 canteloupes on each side so let $c = 6$.

$$n(c) = \frac{1}{3}c^3 + \frac{1}{2}c^2 + \frac{1}{6}c$$
$$n(6) = \frac{1}{3}(6)^3 + \frac{1}{2}(6)^2 + \frac{1}{6}(6)$$
$$= \frac{1}{3}(216) + \frac{1}{2}(36) + \frac{1}{6}(6)$$
$$= 72 + 18 + 1$$
$$= 91 \text{ cantaloupes}$$

56. $g(x) = (x - 6)^2$

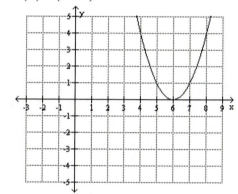

D: the set of real numbers $(-\infty, \infty)$
R: the set of nonnegative real numbers $[0, \infty)$

57.
$$4.5x - 1 < -10 \quad \text{or} \quad 6 - 2x \geq 12$$
$$4.5x < -9 \qquad\qquad -2x \geq 6$$
$$x < -2 \qquad\qquad\quad x \leq -3$$
$$(-\infty, -2)$$

58.
$$|x + 5| \geq 7$$
$$x + 5 \geq 7 \quad \text{or} \quad x + 5 \leq -7$$
$$x \geq 2 \qquad\qquad x \leq -12$$
$$(-\infty, -12] \cup [2, \infty)$$

Chapter 14 Cumulative Review

59.

$5(x+1) \le 4(x+3)$ and $x+12 < -3$

$5x+5 \le 4x+12$ $x < -15$

$x+5 \le 12$

$x \le 7$

$(-\infty, -15)$

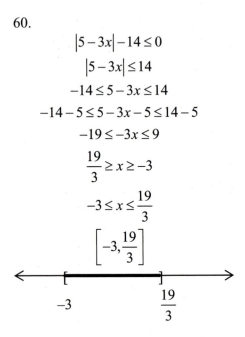

-15

60.

$|5-3x|-14 \le 0$

$|5-3x| \le 14$

$-14 \le 5-3x \le 14$

$-14-5 \le 5-3x-5 \le 14-5$

$-19 \le -3x \le 9$

$\dfrac{19}{3} \ge x \ge -3$

$-3 \le x \le \dfrac{19}{3}$

$\left[-3, \dfrac{19}{3}\right]$

-3 $\dfrac{19}{3}$

61.

$|2x+1| = |3(x+1)|$

$2x+1 = 3(x+1)$ or $2x+1 = -3(x+1)$

$2x+1 = 3x+3$ $2x+1 = -3x-3$

$-x+1 = 3$ $5x+1 = -3$

$-x = 2$ $5x = -4$

$x = -2$ $x = -\dfrac{4}{5}$

62. Let $m = (x-t)$.

$14m^2 - 17m - 6 = (7m+2)(2m-3)$

$= (7(x-t)+2)(2(x-t)-3)$

$= (7x-7t+2)(2x-2t-3)$

63.

$\dfrac{A\lambda}{2} + 1 = 2d + 3\lambda$

$\dfrac{A\lambda}{2} - 3\lambda = 2d - 1$

$2\left(\dfrac{A\lambda}{2} - 3\lambda\right) = 2(2d-1)$

$A\lambda - 6\lambda = 4d - 2$

$\lambda(A-6) = 4d - 2$

$\dfrac{\lambda(A-6)}{A-6} = \dfrac{4d-2}{A-6}$

$\lambda = \dfrac{4d-2}{A-6}$

64.

$x^2 + 10x + 25 - 16z^2 = (x+5)^2 - 16z^2$

$= (x+5+4z)(x+5-4z)$

65.

$64a^6 - 27b^3$

$= \left(4a^2\right)^3 - (3b)^3$

$= \left(4a^2 - 3b\right)\left(\left(4a^2\right)^2 + \left(4a^2\right)(3b) + (3b)^2\right)$

$= \left(4a^2 - 3b\right)\left(16a^4 + 12a^2b + 9b^2\right)$

66.

$\dfrac{x^3+64}{x^3+4x^2+3x+12} = \dfrac{\cancel{(x+4)}\left(x^2-4x+16\right)}{\left(x^2+3\right)\cancel{(x+4)}}$

$= \dfrac{x^2-4x+16}{x^2+3}$

67.

$$\frac{1}{a+5} = \frac{1}{3a+6} - \frac{a+2}{a^2+7a+10}$$

$$\frac{1}{a+5} = \frac{1}{3(a+2)} - \frac{a+2}{(a+5)(a+2)}$$

$$3(a+5)(a+2)\left(\frac{1}{a+5}\right) = 3(a+5)(a+2)\left(\frac{1}{3(a+2)} - \frac{a+2}{(a+5)(a+2)}\right)$$

$$3(a+2) = a+5-3(a+2)$$

$$3a+6 = a+5-3a-6$$

$$3a+6 = -2a-1$$

$$3a+6+2a = -2a-1+2a$$

$$5a+6 = -1$$

$$5a+6-6 = -1-6$$

$$5a = -7$$

$$a = -\frac{7}{5}$$

68. Set the denominator equal to zero.

$$0 \neq x^2 - 25$$

$$0 \neq (x-5)(x+5)$$

$$x-5 \neq 0 \quad \text{and} \quad x+5 \neq 0$$

$$x \neq 5 \qquad\qquad x \neq -5$$

The domain is all real numbers except for and −5.

69. Let I = intensity of the light and
d = distance from the source.

$$I = \frac{k}{d^2}$$

$$18 = \frac{k}{4^2}$$

$$18 = \frac{k}{16}$$

$$16(18) = 16\left(\frac{k}{16}\right)$$

$$288 = k$$

Find I if $d = 12$ and $k = 288$.

$$I = \frac{k}{d^2}$$

$$I = \frac{288}{12^2}$$

$$I = \frac{288}{144}$$

$$I = 2 \text{ lumens}$$

70. $f(x) = \sqrt{x+3}$

$$\sqrt{x+3} \geq 0$$

$$\left(\sqrt{x+3}\right)^2 \geq (0)^2$$

$$x+3 \geq 0$$

$$x \geq -3$$

D: $[-3, \infty)$

71.

$$L(g) = \sqrt[3]{\frac{g}{7.5}}$$

$$L(480) = \sqrt[3]{\frac{480}{7.5}}$$

$$= \sqrt[3]{64}$$

$$= 4 \text{ ft}$$

72. $f(x) = \sqrt{x} + 2$

$D: [0, \infty); \quad R: [2, \infty)$

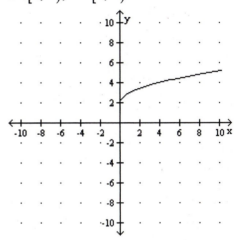

73.

$$\left(\frac{25}{49}\right)^{-3/2} = \left(\frac{49}{25}\right)^{3/2}$$

$$= \left(\sqrt{\frac{49}{25}}\right)^3$$

$$= \left(\frac{7}{5}\right)^3$$

$$= \frac{343}{125}$$

74.

$$\sqrt{112a^3b^5} = \sqrt{16a^2b^4}\sqrt{7ab}$$

$$= 4ab^2\sqrt{7ab}$$

75.

$$\sqrt{98} + \sqrt{8} - \sqrt{32} = \sqrt{49}\sqrt{2} + \sqrt{4}\sqrt{2} - \sqrt{16}\sqrt{2}$$
$$= 7\sqrt{2} + 2\sqrt{2} - 4\sqrt{2}$$
$$= 5\sqrt{2}$$

76.

$$12\sqrt[3]{648x^4} + 3\sqrt[3]{81x^4} = 12\sqrt[3]{216x^3}\sqrt[3]{3x} + 3\sqrt[3]{27x^3}\sqrt[3]{3x}$$
$$= 12(6x)\sqrt[3]{3x} + 3(3x)\sqrt[3]{3x}$$
$$= 72x\sqrt[3]{3x} + 9x\sqrt[3]{3x}$$
$$= 81x\sqrt[3]{3x}$$

77.

$$\left(2\sqrt{7} + 1\right)\left(\sqrt{7} - 1\right) = 2\sqrt{49} - 2\sqrt{7} + \sqrt{7} - 1$$
$$= 14 - \sqrt{7} - 1$$
$$= 13 - \sqrt{7}$$

78.

$$3\left(\sqrt{5x} - \sqrt{3}\right)^2 = 3\left(\sqrt{5x} - \sqrt{3}\right)\left(\sqrt{5x} - \sqrt{3}\right)$$
$$= 3\left(\sqrt{25x^2} - \sqrt{15x} - \sqrt{15x} + \sqrt{9}\right)$$
$$= 3\left(5x - 2\sqrt{15x} + 3\right)$$
$$= 15x - 6\sqrt{15x} + 9$$

79.

$$\frac{\sqrt[3]{4}}{\sqrt[3]{b}} = \sqrt[3]{\frac{4}{b}}$$
$$= \sqrt[3]{\frac{4}{b} \cdot \frac{b^2}{b^2}}$$
$$= \sqrt[3]{\frac{4b^2}{b^3}}$$
$$= \frac{\sqrt[3]{4b^2}}{b}$$

80.

$$\frac{3t - 1}{\sqrt{3t} + 1} = \frac{3t - 1}{\sqrt{3t} + 1} \cdot \frac{\sqrt{3t} - 1}{\sqrt{3t} - 1}$$
$$= \frac{3t\sqrt{3t} - 3t - \sqrt{3t} + 1}{\sqrt{9t^2} - 1}$$
$$= \frac{3t\left(\sqrt{3t} - 1\right) - \left(\sqrt{3t} - 1\right)}{3t - 1}$$
$$= \frac{\left(\sqrt{3t} - 1\right)\left(3t - 1\right)}{3t - 1}$$
$$= \sqrt{3t} - 1$$

81.

$$2x = \sqrt{16x - 12}$$
$$(2x)^2 = \left(\sqrt{16x - 12}\right)^2$$
$$4x^2 = 16x - 12$$
$$4x^2 - 16x + 12 = 0$$
$$4\left(x^2 - 4x + 3\right) = 0$$
$$4(x - 1)(x - 3) = 0$$
$$x - 1 = 0 \quad \text{or} \quad x - 3 = 0$$
$$x = 1 \qquad\qquad x = 3$$

82.

$$\sqrt[3]{12m + 4} = 4$$
$$\left(\sqrt[3]{12m + 4}\right)^3 = 4^3$$
$$12m + 4 = 64$$
$$12m + 4 - 4 = 64 - 4$$
$$12m = 60$$
$$m = 5$$

83.

$$f(x) = g(x)$$
$$\sqrt{x + 3} - \sqrt{3} = \sqrt{x}$$
$$\left(\sqrt{x + 3} - \sqrt{3}\right)^2 = \left(\sqrt{x}\right)^2$$
$$x + 3 - 2\sqrt{3(x + 3)} + 3 = x$$
$$x + 6 - 2\sqrt{3x + 9} = x$$
$$-2\sqrt{3x + 9} = -6$$
$$\frac{-2\sqrt{3x + 9}}{-2} = \frac{-6}{-2}$$
$$\sqrt{3x + 9} = 3$$
$$\left(\sqrt{3x + 9}\right)^2 = 3^2$$
$$3x + 9 = 9$$
$$3x = 0$$
$$x = 0$$

84. Use the Pythagorean Theorem. The legs of the triangle formed are both 16 (sides of the square) and the waist size is the hypotenuse, c.

$$a^2 + b^2 = c^2$$
$$16^2 + 16^2 = c^2$$
$$256 + 256 = c^2$$
$$512 = c^2$$
$$\sqrt{512} = \sqrt{c^2}$$
$$22.6 = c$$

Subtract 1 inch from 22.6 to allow for the pin, and the largest waist size the diaper can wrap around is about 21.6 inches.

85.
$$\sqrt{-25} = \sqrt{25}\sqrt{-1}$$
$$= 5i$$

86.
$$i^{42} = i^{40} \cdot i^2$$
$$= \left(i^4\right)^{10} \cdot i^2$$
$$= (1)^{10} \cdot (-1)$$
$$= 1 \cdot (-1)$$
$$= -1$$

87.
$$(-7 + 9i) - (-2 - 8i) = -7 + 9i + 2 + 8i$$
$$= -5 + 17i$$

88.
$$\frac{2 - 5i}{2 + 5i} = \frac{2 - 5i}{2 + 5i} \cdot \frac{2 - 5i}{2 - 5i}$$
$$= \frac{4 - 10i - 10i + 25i^2}{4 - 25i^2}$$
$$= \frac{4 - 20i + 25(-1)}{4 - 25(-1)}$$
$$= \frac{4 - 20i - 25}{4 + 25}$$
$$= \frac{-21 - 20i}{29}$$
$$= -\frac{21}{29} - \frac{20}{29}i$$

89.
$$t^2 = 24$$
$$\sqrt{t^2} = \pm\sqrt{24}$$
$$t = \pm\sqrt{4}\sqrt{6}$$
$$t = \pm 2\sqrt{6}$$

90.
$$m^2 + 10m - 7 = 0$$
$$m^2 + 10m = 7$$
$$m^2 + 10m + 25 = 7 + 25$$
$$(m + 5)^2 = 32$$
$$\sqrt{(m + 5)^2} = \pm\sqrt{32}$$
$$m + 5 = \pm\sqrt{16}\sqrt{2}$$
$$m + 5 = \pm 4\sqrt{2}$$
$$m = -5 \pm 4\sqrt{2}$$

91.
$$4w^2 + 6w + 1 = 0$$
$$w = \frac{-b \pm \sqrt{b^2 - 4ac}}{2a}$$
$$w = \frac{-6 \pm \sqrt{6^2 - 4(4)(1)}}{2(4)}$$
$$w = \frac{-6 \pm \sqrt{36 - 16}}{8}$$
$$w = \frac{-6 \pm \sqrt{20}}{8}$$
$$w = \frac{-6 \pm 2\sqrt{5}}{8}$$
$$w = -\frac{6}{8} \pm \frac{2\sqrt{5}}{8}$$
$$w = -\frac{3}{4} \pm \frac{\sqrt{5}}{4}$$
$$w = \frac{-3 \pm \sqrt{5}}{4}$$

92.

$$3x^2 - 4x = -2$$
$$3x^2 - 4x + 2 = 0$$

$$x = \frac{-b \pm \sqrt{b^2 - 4ac}}{2a}$$

$$x = \frac{-(-4) \pm \sqrt{(-4)^2 - 4(3)(2)}}{2(3)}$$

$$x = \frac{4 \pm \sqrt{16 - 24}}{6}$$

$$x = \frac{4 \pm \sqrt{-8}}{6}$$

$$x = \frac{4 \pm 2i\sqrt{2}}{6}$$

$$x = \frac{4}{6} \pm \frac{2\sqrt{2}}{6}i$$

$$x = \frac{2}{3} \pm \frac{\sqrt{2}}{3}i$$

93. Let $u = 2x + 1$.

$$2(2x+1)^2 - 7(2x+1) + 6 = 0$$
$$2u^2 - 7u + 6 = 0$$
$$(2u - 3)(u - 2) = 0$$
$$2u - 3 = 0 \quad \text{or} \quad u - 2 = 0$$
$$u = \frac{3}{2} \qquad\qquad u = 2$$
$$2x + 1 = \frac{3}{2} \qquad 2x + 1 = 2$$
$$2x = \frac{1}{2} \qquad\qquad 2x = 1$$
$$x = \frac{1}{4} \qquad\qquad x = \frac{1}{2}$$

94.

$$x^4 + 19x^2 + 18 = 0$$
$$(x^2 + 1)(x^2 + 18) = 0$$
$$x^2 + 1 = 0 \quad \text{or} \quad x^2 + 18 = 0$$
$$x^2 = -1 \qquad\qquad x^2 = -18$$
$$x = \sqrt{-1} \qquad\qquad x = \sqrt{-18}$$
$$x = \pm i \qquad\qquad x = \pm 3i\sqrt{2}$$
$$x = -i, i, 3i\sqrt{2}, -3i\sqrt{2}$$

95. a. a quadratic function

b. at about 85% and 120% of the suggested inflation

96. $f(x) = -6x^2 - 12x - 8$

Step 1: Since $a = -6 < 0$, the parabola opens downward.

Step 2: Find the vertex and axis of symmetry.

$$x = \frac{-b}{2a} = \frac{-(-12)}{2(-6)} = \frac{12}{-12} = -1$$

$$f(-1) = -6(-1)^2 - 12(-1) - 8$$
$$= -6 + 12 - 8$$
$$= -2$$

vertex: $(-1, -2)$

The axis of symmetry is $x = -1$.

Step 3: Find the x- and y-intercepts. Since $c = -8$, the y-intercept is $(0, -8)$. The vertex is located below the x-axis and it opens downward, so there are no x-intercepts.

Step 4: Use symmetry to find the point $(-4, -8)$.

Step 5: Plots the points and draw the parabola.

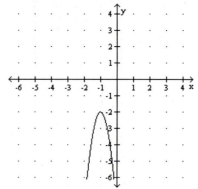

97.

$$x^2 - 8x \le -15$$
$$x^2 - 8x + 15 \le 0$$
$$(x - 3)(x - 5) = 0$$
$$x - 3 = 0 \quad \text{or} \quad x - 5 = 0$$
$$x = 3 \qquad\qquad x = 5$$

$$[3, 5]$$

98. a.
$$(f+g)(x) = f(x) + g(x)$$
$$= x^2 - 2 + 2x + 1$$
$$= x^2 + 2x - 1$$

b.
$$(f \cdot g)(x) = f(x)g(x)$$
$$= (x^2 - 2)(2x + 1)$$
$$= 2x^3 + x^2 - 4x - 2$$

c.
$$(f \circ g)(x) = f(g(x))$$
$$= f(2x + 1)$$
$$= (2x + 1)^2 - 2$$
$$= 4x^2 + 4x + 1 - 2$$
$$= 4x^2 + 4x - 1$$

99.
$$f(x) = 2x^3 - 1$$
$$y = 2x^3 - 1$$
$$x = 2y^3 - 1$$
$$x + 1 = 2y^3$$
$$\frac{x+1}{2} = y^3$$
$$\sqrt[3]{\frac{x+1}{2}} = y$$
$$y = \sqrt[3]{\frac{x+1}{2}}$$
$$f^{-1}(x) = \sqrt[3]{\frac{x+1}{2}}$$

100. $f(x) = \left(\dfrac{1}{2}\right)^x$

$D:(-\infty, \infty); \quad R:(0, \infty)$

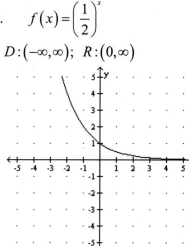

101. $f(x) = e^x$

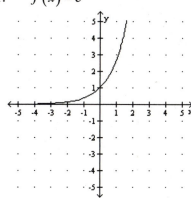

102. POPULATION GROWTH
$P = 114$ million, $r = 0.01102$, and
$t = 25$
$$A = Pe^{rt}$$
$$= 114e^{(0.01102)(25)}$$
$$= 114e^{0.2755}$$
$$\approx 150 \text{ million}$$

103.
$$\log 1,000 = x$$
$$\log_{10} 1,000 = x$$
$$10^x = 1,000$$
$$10^x = 10^3$$
$$x = 3$$

104.
$$\log_8 64 = x$$
$$8^x = 64$$
$$8^x = 8^2$$
$$x = 2$$

105.
$$\log_3 x = -3$$
$$3^{-3} = x$$
$$\frac{1}{3^3} = x$$
$$\frac{1}{27} = x$$

106.
$$\log_x 25 = 2$$
$$x^2 = 25$$
$$x^2 = 5^2$$
$$x = 5$$

Chapter 14 Cumulative Review

107.

$$\ln e = x$$
$$\log_e e = x$$
$$e^x = e^1$$
$$x = 1$$

108.

$$\ln \frac{1}{e} = x$$
$$\log_e \frac{1}{e} = x$$
$$e^x = \frac{1}{e}$$
$$e^x = e^{-1}$$
$$x = -1$$

109. $\log 0$ is undefined

110.

$$\log_6 \frac{36}{x^3} = \log_6 36 - \log_6 x^3$$
$$= \log_6 6^2 - \log_6 x^3$$
$$= 2\log_6 6 - 3\log_6 x$$
$$= 2(1) - 3\log_6 x$$
$$= 2 - 3\log_6 x$$

111.

$$\frac{1}{2}\ln x + \ln y - \ln z = \ln x^{1/2} + \ln y - \ln z$$
$$= \ln \frac{x^{1/2}y}{z}$$
$$= \ln \frac{\sqrt{x} \cdot y}{z}$$
$$= \ln \frac{y\sqrt{x}}{z}$$

112. BACTERIA GROWTH

Let $P_0 = 200$, $P = 600$, and $t = 4$.

$$P = P_0 e^{kt}$$
$$600 = 200e^{k \cdot 4}$$
$$600 = 200e^{4k}$$
$$3 = e^{4k}$$
$$\ln 3 = \ln e^{4k}$$
$$\ln 3 = 4k \ln e$$
$$\ln 3 = 4k$$
$$k = \frac{\ln 3}{4}$$

Let $P_0 = 200$, $P = 8{,}000$, and $k = \frac{\ln 3}{4}$.

$$P = P_0 e^{kt}$$
$$8{,}000 = 200e^{\left(\frac{\ln 3}{4}\right)t}$$
$$40 = e^{\frac{\ln 3}{4}t}$$
$$\ln 40 = \ln e^{\frac{\ln 3}{4}t}$$
$$\ln 40 = \frac{\ln 3}{4}t$$
$$\left(\frac{4}{\ln 3}\right)(\ln 40) = \left(\frac{4}{\ln 3}\right)\left(\frac{\ln 3}{4}t\right)$$
$$\frac{4\ln 40}{\ln 3} = t$$
$$t = 13.4 \text{ hours}$$

113.

$$5^{4x} = \frac{1}{125}$$
$$5^{4x} = 5^{-3}$$
$$4x = -3$$
$$x = -\frac{3}{4}$$

114.

$$2^{x+2} = 3^x$$
$$\log 2^{x+2} = \log 3^x$$
$$(x+2)\log 2 = x\log 3$$
$$x\log 2 + 2\log 2 = x\log 3$$
$$2\log 2 = x\log 3 - x\log 2$$
$$2\log 2 = x(\log 3 - \log 2)$$
$$\frac{2\log 2}{\log 3 - \log 2} = x$$
$$x = \frac{2\log 2}{\log 3 - \log 2}$$
$$x \approx 3.4190$$

115.

$$\log x + \log(x+9) = 1$$
$$\log(x^2 + 9x) = 1$$
$$10^1 = x^2 + 9x$$
$$0 = x^2 + 9x - 10$$
$$0 = (x+10)(x-1)$$
$$x + 10 = 0 \quad \text{or} \quad x - 1 = 0$$
$$x = \cancel{-10} \qquad x = 1$$

116.

$$\log_3 x = \log_3\left(\frac{1}{x}\right) + 4$$
$$\log_3 x - \log_3\left(\frac{1}{x}\right) = 4$$
$$\log_3 \frac{x}{\frac{1}{x}} = 4$$
$$\log_3 x^2 = 4$$
$$3^4 = x^2$$
$$81 = x^2$$
$$9 = x$$

117.
a. ii.
b. iv.
c. i.
d. iii.

118.

$$\begin{cases} 2x + y = 1 \\ x + 2y = -4 \end{cases}$$

$$\begin{bmatrix} 2 & 1 & \vdots & 1 \\ 1 & 2 & \vdots & -4 \end{bmatrix}$$
$$R_1 \leftrightarrow R_2$$
$$\begin{bmatrix} 1 & 2 & \vdots & -4 \\ 2 & 1 & \vdots & 1 \end{bmatrix}$$
$$-2R_1 + R_2$$
$$\begin{bmatrix} 1 & 2 & \vdots & -4 \\ 0 & -3 & \vdots & 9 \end{bmatrix}$$
$$-\frac{1}{3}R_2$$
$$\begin{bmatrix} 1 & 2 & \vdots & -4 \\ 0 & 1 & \vdots & -3 \end{bmatrix}$$

This matrix represents the system

$$\begin{cases} x + 2y = -4 \\ y = -3 \end{cases}$$
$$x - 6 = -4$$
$$x = 2$$

The solution is $(2, -3)$.

119.

Write the first equation in standard form.

$$\begin{cases} 2x + 2y = -1 \\ 3x + 4y = 0 \end{cases}$$

$$x = \frac{D_x}{D} \qquad\qquad y = \frac{D_y}{D}$$

$$= \frac{\begin{vmatrix} -1 & 2 \\ 0 & 4 \end{vmatrix}}{\begin{vmatrix} 2 & 2 \\ 3 & 4 \end{vmatrix}} \qquad = \frac{\begin{vmatrix} 2 & -1 \\ 3 & 0 \end{vmatrix}}{\begin{vmatrix} 2 & 2 \\ 3 & 4 \end{vmatrix}}$$

$$= \frac{-1(4)-2(0)}{2(4)-2(3)} \qquad = \frac{2(0)-(-1)(3)}{2(4)-2(3)}$$

$$= \frac{-4-0}{8-6} \qquad\qquad = \frac{0+3}{8-6}$$

$$= \frac{-4}{2} \qquad\qquad\quad = \frac{3}{2}$$

$$= -2$$

The solution is $\left(-2, \frac{3}{2}\right)$.

Chapter 14 Cumulative Review

120.

$$\begin{cases} b + 2c = 7 - a \\ a + c = 8 - 2b \\ 2a + b + c = 9 \end{cases}$$

Write the equations in standard form.

$$\begin{cases} a + b + 2c = 7 & (1) \\ a + 2b + c = 8 & (2) \\ 2a + b + c = 9 & (3) \end{cases}$$

Multiply Equation 2 by -2 and add Equations 1 and 2.

$$\begin{array}{r} a + b + 2c = 7 \quad (1) \\ -2a - 4b - 2c = -16 \quad (2) \\ \hline -a - 3b = -9 \quad (4) \end{array}$$

Multiply Equation 2 by -1 and add Equations 2 and 3.

$$\begin{array}{r} -a - 2b - c = -8 \quad (2) \\ 2a + b + c = 9 \quad (3) \\ \hline a - b = 1 \quad (5) \end{array}$$

Add Equations 4 and 5.

$$\begin{array}{r} -a - 3b = -9 \quad (4) \\ a - b = 1 \quad (5) \\ \hline -4b = -8 \\ b = 2 \end{array}$$

Substitute $b = 2$ into Equation 5.

$$a - b = 1 \quad (5)$$
$$a - 2 = 1$$
$$a = 3$$

Substitute $b = 2$ and $a = 3$ into Equation 1.

$$a + b + 2c = 7 \quad (1)$$
$$3 + 2 + 2c = 7$$
$$5 + 2c = 7$$
$$2c = 2$$
$$c = 1$$

The solution is $(3, 2, 1)$.

121. SPORTS SOCKS

Let x = number of ankle socks, y = number of low cut socks, and z = number of crew socks.

$$\begin{cases} x + y + z = 500 \\ 2x + 3y + 4z = 1,650 \\ 3x + 5y + 6z = 2,550 \end{cases}$$

Multiply the first equation by -2 and add Equations 1 and 2.

$$\begin{array}{r} -2x - 2y - 2z = -1,000 \\ 2x + 3y + 4z = 1,650 \\ \hline y + 2z = 650 \end{array}$$

Multiply the first equation by -3 and add Equations 1 and 3.

$$\begin{array}{r} -3x - 3y - 3z = -1,500 \\ 3x + 5y + 6z = 2,550 \\ \hline 2y + 3z = 1,050 \end{array}$$

Multiply the first new equation by -2 and add it to the other new equation..

$$\begin{array}{r} -2y - 4z = -1,300 \\ 2y + 3z = 1,050 \\ \hline -z = -250 \\ z = 250 \end{array}$$

Substitute $z = 250$ into the first new equation to solve for y.

$$y + 2z = 650$$
$$y + 2(250) = 650$$
$$y + 500 = 650$$
$$y = 150$$

Substitute $y = 150$ and $z = 250$ into the frist equation and solve for x.

$$x + y + z = 500$$
$$x + 150 + 250 = 500$$
$$x + 400 = 500$$
$$x = 100$$

They make 100 ankle socks, 150 low cut, and 250 crew socks each day.

122. center $(1, -3)$, radius $= 2$

$(x-1)^2 + (y+3)^2 = 2^2$

$(x-1)^2 + (y+3)^2 = 4$

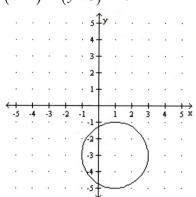

123.

$y^2 + 4x - 6y = -1$

$4x = -y^2 + 6y - 1$

$x = -\dfrac{1}{4}y^2 + \dfrac{3}{2}y - \dfrac{1}{4}$

$x = -\dfrac{1}{4}(y^2 - 6y) - \dfrac{1}{4}$

$x = -\dfrac{1}{4}(y^2 - 6y + 9) - \dfrac{1}{4} + \dfrac{1}{4}(9)$

$x = -\dfrac{1}{4}(y-3)^2 + 2$

vertex: $(2, 3)$

axis of symmetry: $y = 3$

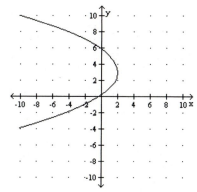

124. $\dfrac{(x+1)^2}{4} + \dfrac{(y-3)^2}{16} = 1$

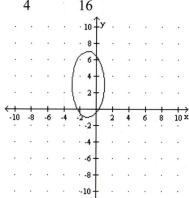

125.

$(x-2)^2 - 9y^2 = 9$

$\dfrac{(x-2)^2}{9} - \dfrac{9y^2}{9} = \dfrac{9}{9}$

$\dfrac{(x-2)^2}{9} - \dfrac{y^2}{1} = 1$

126.

$(3a - b)^4$

$= (3a)^4 + \dfrac{4!}{1!(4-1)!}(3a)^3(-b)$

$\quad + \dfrac{4!}{2!(4-2)!}(3a)^2(-b)^2 + \dfrac{4!}{3!(4-3)!}(3a)(-b)^3$

$\quad + (-b)^4$

$= 81a^4 + \dfrac{4 \cdot 3!}{1! \, 3!}(27a^3)(-b) + \dfrac{4 \cdot 3 \cdot 2!}{2! \, 2!}(9a^2)(b^2)$

$\quad + \dfrac{4 \cdot 3!}{3! 1!}(3a)(-b^3) + b^4$

$= 81a^4 - 108a^3b + 54a^2b^2 - 12ab^3 + b^4$

Chapter 14 Cumulative Review

127. The 7th term of $(2x-y)^8$

$$\frac{8!}{6!(8-6)!}(2x)^2(-y)^6 = \frac{8 \cdot 7 \cdot \cancel{6!}}{\cancel{6!}2!}(4x^2)(y^6)$$

$$= \frac{56}{2}(4x^2y^6)$$

$$= 28(4x^2y^6)$$

$$= 112x^2y^6$$

128.

$$\frac{12!}{10!(12-10)!} = \frac{12 \cdot 11 \cdot \cancel{10!}}{\cancel{10!}2!}$$

$$= \frac{132}{2}$$

$$= 66$$

129. Let $n=20$, $a_1=-11$, and $d=6$.

$$a_n = a_1 + d(n-1)$$

$$a_{20} = -11 + 6(20-1)$$

$$a_{20} = -11 + 6(19)$$

$$a_{20} = -11 + 114$$

$$a_{20} = 103$$

130. Let $n=20$, $a_1=6$, and $d=3$.

$$a_n = a_1 + d(n-1)$$

$$a_{20} = 6 + 3(20-1)$$

$$a_{20} = 6 + 3(19)$$

$$a_{20} = 6 + 57$$

$$a_{20} = 63$$

Find S if $a_{20}=63$ and $a_1=6$.

$$S_n = \frac{n(a_1 + a_n)}{2}$$

$$S_{20} = \frac{20(6+63)}{2}$$

$$= \frac{20(69)}{2}$$

$$= 690$$

131.

$$\sum_{k=3}^{5}(2k+1) = (2\cdot3+1)+(2\cdot4+1)+(2\cdot5+1)$$

$$= 7+9+11$$

$$= 27$$

132. If the boat depreciates 12% of its value each year, then 88% of its value remains. The value of the boat in 9 years is represented by the 10th term of a geometric series, where $a_1 = 9,000$, $n = 10$, and $r = 0.88$.

$$a_n = a_1 r^{n-1}$$

$$a_{10} = (9,000)(0.88)^{10-1}$$

$$a_{10} = (9,000)(0.88)^9$$

$$a_{10} = (9,000)(0.316478)$$

$$a_{10} = \$2,848.31$$

The boat is worth \$2,848.31 after 9 years.

133. Find r.

$$r = \frac{a_2}{a_1} = \frac{\dfrac{1}{32}}{\dfrac{1}{64}} = \frac{1}{32} \cdot \frac{64}{1} = 2$$

Find the sum with $a_1 = \dfrac{1}{64}$ $n = 10$, and $r = 2$.

$$S_n = \frac{a_1(1-r^n)}{1-r}$$

$$= \frac{\dfrac{1}{64}(1-2^{10})}{1-2}$$

$$= \frac{\dfrac{1}{64}(1-1,024)}{-1}$$

$$= \frac{\dfrac{1}{64}(-1,023)}{-1}$$

$$= \frac{-\dfrac{1,023}{64}}{-1}$$

$$= \frac{1,023}{64}$$

134. Find r.

$$r = \frac{a_2}{a_1} = \frac{3}{9} = \frac{1}{3}$$

Find the sum with $a_1 = 9$ and $r = \frac{1}{3}$.

$$S = \frac{a_1}{1-r}$$

$$= \frac{9}{1 - \frac{1}{3}}$$

$$= \frac{9}{\frac{2}{3}}$$

$$= \frac{9}{1} \cdot \frac{3}{2}$$

$$= \frac{27}{2}$$

Chapter 14 Cumulative Review